# COLD SPRING HARBOR SYMPOSIA ON QUANTITATIVE BIOLOGY

# VOLUME LXVIII

# COLD SPRING HARBOR SYMPOSIA ON QUANTITATIVE BIOLOGY

# VOLUME LXVIII

# The Genome of *Homo sapiens*

www.cshl-symposium.org

Meeting Organized by Bruce Stillman and David Stewart

COLD SPRING HARBOR LABORATORY PRESS
2003

# COLD SPRING HARBOR SYMPOSIA ON QUANTITATIVE BIOLOGY VOLUME LXVII


Cold Spring Harbor, New York
International Standard Book Number 0-87969-709-1 (cloth)
International Standard Book Number 0-87969-710-5 (paper)
International Standard Serial Number 0091-7451
Library of Congress Catalog Card Number 34-8174

*Printed in China*


*COLD SPRING HARBOR SYMPOSIA ON QUANTITATIVE BIOLOGY*
*Founded in 1933 by*
REGINALD G. HARRIS
*Director of the Biological Laboratory 1924 to 1936*
*Previous Symposia Volumes*

I (1933) Surface Phenomena
II (1934) Aspects of Growth
III (1935) Photochemical Reactions
IV (1936) Excitation Phenomena
V (1937) Internal Secretions
VI (1938) Protein Chemistry
VII (1939) Biological Oxidations
VIII (1940) Permeability and the Nature of Cell Membranes
IX (1941) Genes and Chromosomes: Structure and Organization
X (1942) The Relation of Hormones to Development
XI (1946) Heredity and Variation in Microorganisms
XII (1947) Nucleic Acids and Nucleoproteins
XIII (1948) Biological Applications of Tracer Elements
XIV (1949) Amino Acids and Proteins
XV (1950) Origin and Evolution of Man
XVI (1951) Genes and Mutations
XVII (1952) The Neuron
XVIII (1953) Viruses
XIX (1954) The Mammalian Fetus: Physiological Aspects of Development
XX (1955) Population Genetics: The Nature and Causes of Genetic Variability in Population
XXI (1956) Genetic Mechanisms: Structure and Function
XXII (1957) Population Studies: Animal Ecology and Demography
XXIII (1958) Exchange of Genetic Material: Mechanism and Consequences
XXIV (1959) Genetics and Twentieth Century Darwinism
XXV (1960) Biological Clocks
XXVI (1961) Cellular Regulatory Mechanisms
XXVII (1962) Basic Mechanisms in Animal Virus Biology
XXVIII (1963) Synthesis and Structure of Macromolecules
XXIX (1964) Human Genetics
XXX (1965) Sensory Receptors
XXXI (1966) The Genetic Code
XXXII (1967) Antibodies
XXXIII (1968) Replication of DNA in Microorganisms
XXXIV (1969) The Mechanism of Protein Synthesis
XXXV (1970) Transcription of Genetic Material
XXXVI (1971) Structure and Function of Proteins at the Three-dimensional Level
XXXVII (1972) The Mechanism of Muscle Contraction
XXXVIII (1973) Chromosome Structure and Function
XXXIX (1974) Tumor Viruses
XL (1975) The Synapse
XLI (1976) Origins of Lymphocyte Diversity
XLII (1977) Chromatin
XLIII (1978) DNA: Replication and Recombination
XLIV (1979) Viral Oncogenes
XLV (1980) Movable Genetic Elements
XLVI (1981) Organization of the Cytoplasm
XLVII (1982) Structures of DNA
XLVIII (1983) Molecular Neurobiology
XLIX (1984) Recombination at the DNA Level
L (1985) Molecular Biology of Development
LI (1986) Molecular Biology of *Homo sapiens*
LII (1987) Evolution of Catalytic Function
LIII (1988) Molecular Biology of Signal Transduction
LIV (1989) Immunological Recognition
LV (1990) The Brain
LVI (1991) The Cell Cycle
LVII (1992) The Cell Surface
LVIII (1993) DNA and Chromosomes
LVIX (1994) The Molecular Genetics of Cancer
LX (1995) Protein Kinesis: The Dynamics of Protein Trafficking and Stability
LXI (1996) Function & Dysfunction in the Nervous System
LXII (1997) Pattern Formation during Development
LXIII (1998) Mechanisms of Transcription
LXIV (1999) Signaling and Gene Expression in the Immune System
LXV (2000) Biological Responses to DNA Damage
LXVI (2001) The Ribosome
LXVII (2002) The Cardiovascular System

***Front Cover*** (*Paperback*): "Spirals Time—Time Spirals" by Charles A. Jencks. Sculpture is located north of Airslie House. (Photo by Miriam Chua.)



All Cold Spring Harbor Laboratory Press publications may be ordered directly from Cold Spring Harbor Laboratory Press, 500 Sunnyside Boulevard, Woodbury, NY 11797-2924. Phone: 1-800-843-4388 in Continental U.S. and Canada. All other locations: (516) 422-4100. FAX: (516) 422-4097. E-mail: cshpress@cshl.edu. For a complete catalog of all Cold Spring Harbor Laboratory Press publications, visit our World Wide Web Site http://www.cshlpress.com/

# Symposium Participants

Aburatani, Hiroyuki, Div. of Genome Science, Research Center for Advanced Science and Technology, University of Tokyo, Tokyo, Japan

Adams, Mark, Celera Genomics, Rockville, Maryland

Afonina, Irina, Epoch Biosciences, Inc., Bothell, Washington

Agarwal, Pankaj, Dept. of Bioinformatics, GlaxoSmithKline, King of Prussia, Pennsylvania

Aitman, Timothy, Dept. of Physiological Genomics and Medicine, Medical Research Council, London, United Kingdom

Altshuler, David, Dept. of Molecular Biology, Harvard Medical School, Massachusetts General Hospital, Boston, Massachusetts

Amero, Sally, Genome Study Section, Center for Scientific Review, National Institutes of Health, Bethesda, Maryland

Andrés, Aida, Unitat de Biologia Evolutiva, Ciències Experimentals i de la Salut, Universitat Pompeu Fabra, Barcelona, Spain

Antonarakis, Stylianos, Div. of Medical Genetics, University of Geneva, Geneva, Switzerland

Arai, Yoshio, Dept. of Radiation Oncology, University of Pittsburgh, Pittsburgh, Pennsylvania

Ardlie, Kristin, Genomics Collaborative, Inc., Cambridge, Massachusetts

Arkin, Miriam

Armengol, Lluís, Genes and Disease Program, Center for Genomic Regulation, Universitat Pompeu Fabra, Barcelona, Spain

Ashburner, Michael, Dept. of Genetics, European Bioinformatics Institute, Cambridge University, Cambridge, United Kingdom

Ashurst, Jennifer, Dept. of Informatics, Genome Campus, The Wellcome Trust Sanger Institute, Hinxton, Cambridge, United Kingdom

Austin, Finley, Hoffmann-La Roche, Nutley, New Jersey

Baertsch, Robert, Dept. of Computer Science, University of California, Santa Cruz

Bai, Yuchen, Dept. of Genomics, Wyeth Pharmaceuticals, Collegeville, Pennsylvania

Bailey, David, Dept. of Bioinformatics, Genpath Pharmaceuticals, Cambridge, Massachusetts

Bailey, Jeff, Dept. of Human Genetics, Case Western Reserve University, Cleveland, Ohio

Baillie, David, Dept. of Molecular Biology and Biochemistry, Simon Fraser University, Burnaby, British Columbia, Canada

Bair, Thomas, Center for Bioinformatics and Computational Biology, University of Iowa, Iowa City

Balija, Vivekenand, Genome Center, Cold Spring Harbor Laboratory, Woodbury, New York

Baroukh, Nadine, Dept. of Genome Sciences, Lawrence Berkeley National Laboratory, Berkeley, California

Barve, Ruteja, Dept. of Computational Biology, Pfizer, Inc., St. Louis, Missouri

Batzer, Mark, Dept. of Biological Sciences, Louisiana State University, Baton Rouge, Louisiana

Bayne, Marvin, Discovery Technologies, Schering-Plough Research Institute, Kenilworth, New Jersey

Beaudet, Lucille, Perkin Elmer BioSignal, Inc., Montréal, Québec, Canada

Bejerano, Gill, School of Engineering, Center for Biomolecular Science and Engineering, University of California, Santa Cruz

Beleza, Sandra, Dept. of Population Genetics, Instituto de Patologia e Immunologia Molecular da Universidade do Porto, Porto, Portugal

Bell, George, Biocomputing Group, Whitehead Institute for Biomedical Research, Massachusetts Institute of Technology, Cambridge, Massachusetts

Bentley, David, Dept. of Human Genetics, Genome Campus, The Wellcome Trust Sanger Institute, Hinxton, Cambridge, United Kingdom

Bernardi, Giorgio, Stazione Zoologica Anton Dohrn, Naples, Italy

Bertranpetit, Jaume, Ciències Experimentals i de la Salut, Unitat de Biologia Evolutiva, Universitat Pompeu Fabra, Barcelona, Spain

Bibikova, Marina, Dept. of Molecular Biology, Illumina, Inc., San Diego, California

Bina, Minou, Dept. of Chemistry, Purdue University, West Lafayette, Indiana

Bingham, Paul, Dept. of Biochemistry and Cell Biology, State University of New York, Stony Brook

Bini, Alessandra, Dept. of Extramural Affairs, Scientific Review Office, National Institute on Aging, National Institutes of Health, Bethesda, Maryland

Birney, Ewan, EMBL-European Bioinformatics Institute, Wellcome Trust Genome Campus, Hinxton, Cambridge, United Kingdom

Biro, Jan, Homulus Informatics, San Francisco, California

Birren, Bruce, Center for Genome Research, Whitehead Institute for Biomedical Research, Massachusetts Institute of Technology, Cambridge, Massachusetts

Blanchette, Mathieu, School of Engineering, Center for Biomolecular Science and Engineering, University of California, Santa Cruz

BLOOM, THEODORA, *Journal of Biology*, BioMed Central, London, United Kingdom
BOGUSKI, MARK, Vulcan, Inc., Seattle, Washington
BONALDO, MARIA DE FATIMA, Dept. of Pediatrics, College of Medicine, University of Iowa, Iowa City
BOTCHERBY, MARC, HGMP- Resource Center, Genome Campus, Hinxton, Cambridge, United Kingdom
BOTSTEIN, DAVID, Dept. of Genetics, Stanford University School of Medicine, Stanford, California
BOYLSTON, ARTHUR, Dept. of Molecular Medicine, St. James's Hospital, University of Leeds, Leeds, United Kingdom
BRÄNDSTRÖM, HELENA, Génome Québec Innovation Centre, Mc Gill University, Montréal, Québec, Canada
BRANSCOMB, ELBERT, Department of Energy Joint Genome Institute, Walnut Creek, California
BRENT, MICHAEL, Dept. of Computer Science, Washington University, St. Louis, Missouri
BROOKS, LISA, National Human Genome Research Institute, National Institutes of Health, Bethesda, Maryland
BUCK, GREGORY, Dept. of Microbiology and Immunology, Virginia Commonwealth University, Richmond, Virginia
BURKE, ADRIENNE, *Genome Technology*, GenomeWeb, New York, New York
BURT, DAVE, Dept. of Genomics and Bioinformatics, Roslin Institute, Edinburgh, Scotland, United Kingdom
BURTON, JOHN, Dept. of Large Genome Sequencing Production, Genome Campus, The Wellcome Trust Sanger Institute, Hinxton, Cambridge, United Kingdom
BUTLAND, STEFANIE, Bioinformatics Centre, University of British Columbia, Vancouver, British Columbia, Canada
BUXBAUM, JOEL, Dept. of Molecular and Experimental Medicine, The Scripps Research Institute, La Jolla, California
CAI, WEIWEN, Dept. of Molecular and Human Genetics, Baylor College of Medicine, Houston, Texas
CALLINAN, PAULINE, Dept. of Biological Sciences, Louisiana State University, Baton Rouge, Louisiana
CAPORALE, LYNN, www.DarwinGenome.info, New York, New York
CARGILL, MICHELE, Discovery Research, Celera Diagnostics, Alameda, California
CARLETON, STEPHEN, Dept. of Anatomy and Cell Biology, Health Science Center, State University of New York, Brooklyn
CASTELLANO, SERGI, Dept. of Bioinformatics, Institut Municipal d'Investigacio Medica, Universitat Pompeu Fabra, Barcelona, Spain
CHAKRAVARTI, ARAVINDA, McKusick-Nathans Institute of Genetic Medicine, Johns Hopkins University School of Medicine, Baltimore, Maryland
CHAN, SHIRLEY, Dolan DNA Learning Center, Cold Spring Harbor Laboratory, Cold Spring Harbor, New York
CHARLIER, CAROLE, Dept. of Genetics, University of Liège, Liège, Belgium
CHEE, MARK, Science Research and Development, Illumina, Inc., San Diego, California
CHEN, FENG, Dept. of Genome Biochemistry, Exelixis, Inc., South San Francisco, California
CHEN, GENGXIN, Cold Spring Harbor Laboratory, Cold Spring Harbor, New York
CHEN, JACK, Cold Spring Harbor Laboratory, Cold Spring Harbor, New York
CHEN, ZUGEN, Dept. of Human Genetics, University of California, Los Angeles
CHEUNG, JOSEPH, Dept. of Genetics and Genomic Biology, The Hospital for Sick Children, Toronto, Ontario, Canada
CHEUNG, VIVIAN, Dept. of Pediatrics, University of Pennsylvania, Philadelphia
CHIBA-FALEK, ORNIT, Genetic Disease Research Branch, National Human Genome Research Institute, National Institutes of Health, Bethesda, Maryland
CHISSOE, STEPHANIE, Discovery Genetics, GlaxoSmithKline, Research Triangle Park, North Carolina
CHONG, WAI-PO, Dept. of Pedriatrics and Adolescent Medicine, University of Hong Kong, Hong Kong, China
CLARK, ANDREW, Dept. of Molecular Biology and Genetics, Cornell University, Ithaca, New York
CLARK, HILLARY, Dept. of Bioinformatics, Genentech, Inc., South San Francisco, California
CLIFTON, SANDRA, Dept. of Genetics, Genome Sequencing Center, School of Medicine, Washington University, St. Louis, Missouri
COLLINS, FRANCIS, National Human Genome Research Institute, National Institutes of Health, Bethesda, Maryland
CORSARO, CHERYL, Genome Study Section, Center for Scientific Review, National Institutes of Health, Bethesda, Maryland
COURONNE, OLIVIER, Dept. of Life Science, Lawrence Berkeley National Laboratory, Berkeley, California
COX, DAVID, Perlegen Sciences, Inc., Mountain View, California
CUI, XIAOLIN, Dept. of Critical Care Medicine, National Institutes of Health, Bethesda, Maryland
CUNNINGHAM, FIONA, Cold Spring Harbor Laboratory, Cold Spring Harbor, New York
CUTLER, DAVID, McKusick-Nathans Institute of Genetic Medicine, Johns Hopkins University School of Medicine, Baltimore, Maryland
DAHARY, DVIR, Dept. of Human Genetics, Tel-Aviv University, Tel-Aviv, Israel
DALY, MARK, Whitehead Institute for Biomedical Research, Massachusetts Institute of Technology, Cambridge, Massachusetts
DAY, ALLEN, Dept. of Human Genetics, University of California, Los Angeles
DEAR, PAUL, PNAC Biotech, MRC Laboratory of Molecular Biology, Cambridge, United Kingdom
DEIFALLA, ABDEL, Dept. of Biological Sciences, Louisiana State University, Baton Rogue, Louisiana
DE JONG, PIETER, Dept. of BACPAC Resources, Children's Hospital and Oakland Research Institute, Oakland, California
DE KONING, JASON, Dept. of Biological Sciences, State University of New York, Albany

De La Vega, Francisco, Applied Biosystems, Foster City, California
Delehaunty, Kim, Dept. of Genetics, Genome Sequencing Center, Washington University School of Medicine, St. Louis, Missouri
Deloukas, Panos, Dept. of Human Genetics, Genome Campus, The Wellcome Trust Sanger Institute, Hinxton, Cambridge, United Kingdom
Demiya, Sven, Human Genome Research Group, RIKEN Genomic Sciences Center, Yokohama, Japan
Dennehey, Briana, Dept. of Molecular, Cellular, and Developmental Biology, University of Colorado, Boulder
Dermitzakis, Emmanouil, Div. of Medical Genetics, University of Geneva, Geneva, Switzerland
Deschamps, Stephane, Dept. of Chemistry and Biochemistry, University of Oklahoma, Norman
Devine, Scott, Dept. of Biochemistry, Emory University School of Medicine, Atlanta, Georgia
Dickinson, Todd, Marketing, Illumina, Inc., San Diego, California
Di Rienzo, Anna, Dept. of Human Genetics, University of Chicago, Chicago, Illinois
Doggett, Norman, Center for Human Genome Studies, Los Alamos National Laboratory, Los Alamos, New Mexico
Donnelly, Peter, Dept. of Statistics, University of Oxford, Oxford, United Kingdom
Drayna, Dennis, National Institute on Deafness and Other Communication Disorders, National Institutes of Health, Rockville, Maryland
Du, Lei, Dept. of Bioinformatics, 454 Life Sciences, Branford, Connecticut
Dugaiczyk, Achilles, Dept. of Biochemistry, University of California, Riverside
Dunham, Andrew, Dept. of Human Genetics, Genome Campus, The Wellcome Trust Sanger Institute, Hinxton, Cambridge, United Kingdom
Edwards, Yvonne, HGMP- Resource Center, Genome Campus, Hinxton, Cambridge, United Kingdom
Eichler, Evan, Dept. of Genetics, Case Western Reserve University, Cleveland, Ohio
Elnitski, Laura, Dept. of Computer Science and Engineering, Pennsylvania State University, University Park, Pennsylvania
Elsik, Christine, Dept. of Animal Science, Texas A&M University, College Station, Texas
Emes, Richard, Dept. of Human Anatomy and Genetics, MRC Functional Genetics Unit, Oxford University, Oxford, United Kingdom
Estivill, Xavier, Genes and Disease Program, Center for Genomic Regulation, Universitat Pompeu Fabra, Barcelona, Spain
Feingold, Elise, National Human Genome Research Institute, National Institutes of Health, Bethesda, Maryland
Felsenfeld, Adam, National Human Genome Research Institute, National Institutes of Health, Bethesda, Maryland
Ferretti, Vincent, Génome Québec Innovation Centre, McGill University, Montréal, Québec, Canada
Fewell, Ginger, Dept. of Genetics, Genome Sequencing Center, Washington University School of Medicine, St. Louis, Missouri
Fossella, John, Dept. of Psychiatry, Sackler Institute, Weill Medical College of Cornell University, New York, New York
Foxall, Roger, Dept. of Research, Genome British Columbia, Vancouver, British Columbia, Canada
Fraser, Claire, The Institute for Genomic Research, Rockville, Maryland
Frazer, Kelly, Dept. of Functional Genomics, Perlegen Sciences, Inc., Mountain View, California
Freudenberg, Jan, Institute of Human Genetics, University of Bonn, Bonn, Germany
Friddle, Carl, Expression Analysis, Lexicon Genetics, Inc., The Woodlands, Texas
Frisse, Linda, Dept. of Human Genetics, University of Chicago, Chicago, Illinois
Fujiyama, Asao, Dept. of Bioinformatics, National Institute of Informatics, Tokyo, Japan
Fulton, Lucinda, Dept. of Genetics, Genome Sequencing Center, Washington University School of Medicine, St. Louis, Missouri
Furey, Terrence, Dept. of Computer Science, Baskin Engineering, Howard Hughes Medical Institute, University of California, Santa Cruz
Galibert, Francis, Lab. of Genetics and Development, Unité Mixte de Recherche, Centre National de la Recherche Scientifique, Rennes, France
Gan, Y. Thomas, Center for Bioinformatics, University of Pennsylvania, Philadelphia
Garcia, Roman, Dept. of Biological Sciences, Florida International University, Miami, Florida
Gardiner, Kathleen, Eleanor Roosevelt Institute, Denver, Colorado
Ge, Wei, Dept. of Informatics, Rosetta Inpharmatics, Merck and Co., Kirkland, Washington
Georges, Michel, Dept. of Genetics, Faculty of Veterinary Medicine, University of Liège, Liège, Belgium
Gibbons, Gary, Cardiovascular Research Institute, Morehouse School of Medicine, Atlanta, Georgia
Gibbons, Rosaleen, Dept. of Biochemistry, University of California, Riverside
Gibbs, Richard, Dept. of Molecular and Human Genetics, Baylor College of Medicine, Houston, Texas
Gish, Warren, Dept. of Genetics, Genome Sequencing Center, Washington University School of Medicine, St. Louis, Missouri
Glöckner, Gernot, Dept. of Genome Analysis, Institute of Molecular Biotechnology, Jena, Germany
Gnirke, Andreas, Center for Genome Research, Whitehead Institute for Biomedical Research, Massachusetts Institute of Technology, Cambridge, Massachusetts
Godson, G. Nigel, Dept. of Biochemistry, New York University Medical School, New York, New York
Gold, Bert, Lab. of Genomic Diversity, Human Genetics Section, National Cancer Institute, Frederick, Maryland
Goldstein, David, Dept of Biology, University College London, London, United Kingdom
Good, Peter, National Human Genome Research Institute, National Institutes of Health, Bethesda, Maryland
Goodman, Laurie, *Genome Research*, Cold Spring Harbor Laboratory Press, Woodbury, New York

GOODSTADT, LEO, Dept. of Human Anatomy and Genetics, MRC Functional Genetics Unit, Oxford University, Oxford, United Kingdom

GOPALAKRISHNAN, VANATHI, Dept. of Medicine, University of Pittsburgh, Pittsburgh, Pennsylvania

GOPINATHRAO, GOPAL, Bioinformatics Research Center, Medical College of Wisconsin, Milwaukee, Wisconsin

GRAHAM, BETTIE, National Human Genome Research Institute, National Institutes of Health, Bethesda, Maryland

GRAVES, TINA, Dept. of Genetics, Genome Sequencing Center, Washington University School of Medicine, St. Louis, Missouri

GREEN, ERIC, Genome Technology Branch, National Human Genome Research Institute, National Institutes of Health, Bethesda, Maryland

GREEN, PHILIP, Dept. of Molecular Biotechnology, University of Washington, Seattle

GRIMWOOD, JANE, Stanford University School of Medicine, Stanford Human Genome Center, Palo Alto, California

GUHATHAKURTA, DEBRAJ, Dept. Research Genetics, Rosetta Inpharmatics, Merck and Co., Kirkland, Washington

GUIGO, RODERIC, Research Group in Biomedical Informatics, Institut Municipal d'Investigacio Medica, Universitat Pompeu Fabra, Barcelona, Spain

GUNDERSON, KEVIN, Dept. of Genomics, Illumina, Inc., San Diego, California

GUNTER, CHRIS, Nature, Nature Publishing Group, Washington, D.C.

GUYER, MARK, National Human Genome Research Institute, National Institutes of Health, Bethesda, Maryland

HACKER, KEVIN, Electrophoresis Research and Development, Applied Biosystems, Foster City, California

HAN, CLIFF, Bioscience Division, Los Alamos National Laboratory, Los Alamos, New Mexico

HANNON, GREGORY, Cold Spring Harbor Laboratory, Cold Spring Harbor, New York

HARDISON, ROSS, Dept. of Biochemistry and Molecular Biology, Pennsylvania State University, University Park, Pennsylvania

HARMON, CYRUS, Dept. of Molecular and Cell Biology, University of California, Berkeley

HATTORI, MASAHIRA, Human Genome Research Group, RIKEN Genomic Sciences Center, Yokohama, Japan

HATZIGEORGIOU, ARTEMIS, Dept. of Genetics, University of Pennsylvania, Philadelphia

HAUSSLER, DAVID, Center for Biomolecular Science and Engineering, Howard Hughes Medical Institute, University of California, Santa Cruz

HAYASHI, KENSHI, Medical Institute of Bioregulation, Kyushu University, Fukuoka, Japan

HAYASHIZAKI, YOSHIHIDE, Genome Exploration Research Group, RIKEN Genomic Sciences Center, Yokohama, Japan

HEIN, JOTUN, Dept. of Statistics, Oxford University, Oxford, United Kingdom

HEINER, CHERYL, Sequencing Research and Development, Applied Biosystems, Foster City, California

HELLER, JAMES, James G. Heller Consulting, Inc., Toronto, Ontario, Canada

HENKE, JÜRGEN, Dept. of Genetics, Institut für Blutgruppenforschung, Köln, Germany

HENKE, LOTTE, Dept. of Genetics, Institut für Blutgruppenforschung, Köln, Germany

HIETER, PHILIP, Lab. of Biotechnology, University of British Columbia, Vancouver, British Columbia, Canada

HILL, JEFFREY, Dept. of Genomic and Proteomic Sciences, GlaxoSmithKline, Harlow, United Kingdom

HIMMELBAUER, HEINZ, Dept. of Vertebrate Genomics, Max-Planck-Institute for Molecular Genetics, Berlin-Dahlem, Germany

HOFKER, MARTEN, Dept. of Molecular Genetics, Maastricht University, Maastricht, The Netherlands

HOLDEN, ARTHUR, The SNP Consortium, Deerfield, Illinois

HOOD, LEROY, Institute for Systems Biology, Seattle, Washington

HOWELL, GARETH, The Jackson Laboratory, Bar Harbor, Maine

HUA, AXIN, Dept. of Chemistry and Biochemistry, University of Oklahoma, Norman

HUBNER, NORBERT, Dept. of Cardiovascular Genetics, Max-Delbruck-Center for Molecular Medicine, Berlin, Germany

HUDSON, THOMAS, Génome Québec Innovation Centre, McGill University, Montréal, Québec, Canada

HUMPHRAY, SEAN, Dept. of Human Genetics, Genome Campus, The Wellcome Trust Sanger Institute, Hinxton, Cambridge, United Kingdom

HUNKAPILLER, TIM, Discovery Biosciences Corp., Seattle, Washington

HURWITZ, BONNIE, Dept. of Medical Informatics, Third Wave Technologies, Madison, Wisconsin

HWANG, BYUNG JOON, Div. of Biology, Howard Hughes Medical Institute, California Institute of Technology, Pasadena, California

JASNY, BARBARA, *Science*, American Association for the Advancement of Science, Washington, D.C.

JENKINS, JOHN, *Oncogene*, Nature Pulishing Group, Edenbridge, Kent, United Kingdom

JIANG, TAO, Dept. of Human and Molecular Genetics, Baylor College of Medicine, Houston, Texas

JIN, PEI, Incyte Pharmaceuticals, Inc., Palo Alto, California

JOHN, MARKUS, Arthritis and Bone Metabolism, Novartis Institutes of Biomedical Research, Basel, Switzerland,

JORDAN, BARBARA, Genomics Collaborative, Inc., Cambridge, Massachusetts

KAMHOLZ, SANDRA, National Human Genome Research Institute, National Institutes of Health, Bethesda, Maryland

KASIF, SIMON, Center for Advanced Genomic Technology, Boston University, Boston, Massachusetts

KATOH, MASARU, Genetics and Cell Biology Section, National Cancer Center Research Institute, Tokyo, Japan

KAUL, RAJINDER, Dept. of Medicine, Genome Center, University of Washington, Seattle

KENEALY, SHANNON, Program in Human Genetics, Vanderbilt University, Nashville, Tennessee

KENMOCHI, NAOYA, Central Research Laboratories, Miyazaki Medical College, Miyazaki, Japan

KENNEDY, KAREN, Dept. of Ventures and Initiatives, The Wellcome Trust, London, United Kingdom

KENT, JAMES, Center for Biomolecular Science and Engineering, School of Engineering, University of California, Santa Cruz

KHAJA, RAZI, Dept. of Genetics and Genomic Biology, The Hospital for Sick Children, Toronto, Ontario, Canada

KIM, JUNG-HYUN, Laboratory of Biosystems and Cancer, National Cancer Institute, National Institutes of Health, Bethesda, Maryland

KIM, SOO JUNG, Digital Bio Laboratory, Samsung, Yongin, South Korea

KIM, UN-KYUNG, Lab. of Molecular Genetics, National Institute on Deafness and Other Communication Disorders, National Institutes of Health, Rockville, Maryland

KIMURA, KOUICHI, Dept. of Biosystems Research, Central Research Laboratory, Hitachi Ltd., Tokyo, Japan

KLANNEMARK, MIA, British Columbia Cancer Research Center, British Columbia Cancer Agency, Vancouver, British Columbia, Canada

KOENIG, BARBARA, Dept. of Neurology and Neurological Sciences, Stanford Humanities Center, Stanford, California

KOLBE, DIANA, Depts. of Computer Science and Engineering and Biochemistry and Molecular Biology, Pennsylvania State University, University Park, Pennsylvania

KONG, HUIMIN, Dept. of Research, New England Biolabs, Beverly, Massachusetts

KOONIN, EUGENE, National Center for Biotechnology Information, National Library of Medicine, National Institutes of Health, Bethesda, Maryland

KOUPRINA, NATALYA, Lab. of Biosystems and Cancer, National Cancer Institute, National Institutes of Health, Bethesda, Maryland

KOZYAVKIN, SERGEI, Fidelity Systems, Inc., Gaithersburg, Maryland

KUROKI, YOKO, Human Genome Research Group, RIKEN Genomic Sciences Center, Yokohama, Japan

KWOK, PUI-YAN, Dept. of Dermatology, Cardiovascular Research Institute, University of California, San Francisco

KWON, TAEJOON, Bioinformatics Team, Samsung Advanced Institute of Technology, Suwon, South Korea

LAI, ERIC, Discovery Genetics, GlaxoSmithKline, Research Triangle Park, North Carolina

LANDER, ERIC, Center for Genome Research, Whitehead Institute for Biomedical Research, Massachusetts Institute of Technology, Cambridge, Massachusetts

LARIONOV, VLADIMIR, Lab. of Biosystems and Cancer, National Cancer Institute, National Institutes of Health, Bethesda, Maryland

LARSEN, JENNIE, Lab. of Molecular Genetics, National Institute on Deafness and Other Communication Disorders, National Institutes of Health, Rockville, Maryland

LEAKE, JONATHAN, *The Sunday Times*, London, United Kingdom

LEE, DAN, The Institute for Genomic Research, Rockville, Maryland

LEE, SEONG-GENE, Dept. of Genomics, Asan Institute for Life Sciences, Seoul, South Korea

LE FANU, JAMES, *The Daily Telegraph*, London, United Kingdom

LEEM, SUN-HEE, Lab. of Biosystems and Cancer, National Cancer Institute, National Institutes of Health, Bethesda, Maryland

LEVIS, BAILEY, Lab. of Molecular Genetics, National Institute on Deafness and Other Communication Disorders, National Institutes of Health, Rockville, Maryland

LEVY, SAMUEL, Dept. of Bioinformatics, Center for the Advancement of Genomics, Rockville, Maryland

LEWIS, SUZANNA, Dept. of Molecular and Cell Biology, University of California, Berkeley

LI, JIANGZHEN, Dept. of Molecular and Human Genetics, Baylor College of Medicine, Houston, Texas

LIMBORSKA, SVETLANA, Dept. of Molecular Bases of Human Genetics, Institute of Molecular Genetics-RAS, Moscow, Russia

LIND, DENISE, Cardiovascular Research Institute, University of California, San Francisco

LINDBLAD-TOH, KERSTIN, Center for Genome Research, Whitehead Institute for Biomedical Research, Massachusetts Institute of Technology, Cambridge, Massachusetts

LING, VINCENT, Dept. of Musculoskeletal Sciences, Wyeth Pharmaceuticals, Cambridge, Massachusetts

LIPOVICH, LEONARD, Dept. of Genome Sciences, University of Washington, Seattle

LIRA, MARUJA, Dept. of Genomics and Proteomics Sciences, Pfizer Global Research and Development, Groton, Connecticut

LITTLE, PETER, Dept. of Botany, School of Biotechnology and Biomolecular Sciences, University of New South Wales, Sydney, New South Wales, Australia

LIU, GUOYING, Dept. of Bioinformatics, Affymetrix, Inc., Emeryville, California

LIU, HONG, Aventis Pharmaceuticals, Bridgewater, New Jersey

LIU, IRENE (YUEYI), Dept. of Biomedical Informatics, Stanford University, Stanford, California

LIU, X. SHIRLEY, Dept. of Biostatistical Science, Harvard School of Public Health, Dana-Farber Cancer Institute, Boston, Massachusetts

LOCKE, DEVIN, Dept. of Genetics, Case Western Reserve University, Cleveland, Ohio

LOOTS, GABRIELA, Div. of Genome Biology, Lawrence Livermore National Laboratory, Livermore, California

LOPES, ALEXANDRA, Dept. of Population Genetics, Instituto de Patologia e Immunologia Molecular da Universidade do Porto, Porto, Portugal

LOPEZ-BIGAS, NURIA, Computational Genomics Group, EMBL-European Bioinformatics Institute, Wellcome Trust Genome Campus, Hinxton, Cambridge, United Kingdom

LUCAS, SUSAN, Production Sequencing, Department of Energy Joint Genome Institute, Walnut Creek, California

LUPSKI, JAMES, Dept. of Molecular and Human Genetics, Baylor College of Medicine, Houston, Texas

LYONS, LESLIE, School of Veterinary Medicine-Population Health and Reproduction, University of California, Davis

MACDONALD, JEFF, Dept. of Genetics and Genomic Biology, The Hospital for Sick Children, Toronto, Ontario, Canada

MALEK, JOEL, Dept. of Genomics and Proteomics, Agencourt Bioscience Corp., Beverly, Massachusetts

MARCHINI, JONATHAN, Dept. of Statistics and Mathematical Genetics, Oxford University, Oxford, United Kingdom

MARDIS, ELAINE, Genome Sequencing Center, Washington University School of Medicine, St. Louis, Missouri

MARIÑO-RAMIREZ, Leonardo, National Center for Biotechnology Information, National Library of Medicine, National Institutes of Health, Bethesda, Maryland

MARX, STEPHEN, Metabolic Diseases Branch, National Institute of Diabetes and Digestive and Kidney Diseases, National Institutes of Health, Bethesda, Maryland

MCCAULEY, JACOB, Dept. of Molecular Physiology and Biophysics, Vanderbilt University, Nashville, Tennessee

MCKERNAN, KEVIN, Agencourt Bioscience Corp., Beverly, Massachusetts

MCPHERSON, JOHN, Dept. of Genetics, Genome Sequencing Center, Washington University School of Medicine, St. Louis, Missouri

MCVEAN, GILEAN, Department of Statistics, University of Oxford, Oxford, United Kingdom

MERIKANGAS, KATHLEEN, Mood and Anxiety Disorders Program, National Institute of Mental Health, National Institutes of Health, Bethesda, Maryland

MERRIMAN, BARRY, Dept. of Human Genetics, University of California, Los Angeles

MESIROV, JILL, Center for Genome Research, Whitehead Institute for Biomedical Research, Massachusetts Institute of Technology, Cambridge, Massachusetts

METSPALU, ANDRES, Dept. of Biotechnology, Estonian Biocentre, University of Tartu, Tartu, Estonia

MICKLE, JOHN, Dept of Pediatrics, Institute of Genetic Medicine, Johns Hopkins University School of Medicine, Baltimore, Maryland

MILLER, MARY, Center for Media and Communication, Exploratorium, San Francisco, California

MILLER, RAYMOND, Dept. of Dermatology, Washington University School of Medicine, St. Louis, Missouri

MILOSAVLJEVIC, ALEKSANDAR, Dept. of Molecular and Human Genetics, Baylor College of Medicine, Houston, Texas

MINER, TRACIE, Dept. of Genetics, Genome Sequencing Center, Washington University School of Medicine, St. Louis, Missouri

MINOSHIMA, SHINSEI, Dept. of Molecular Biology, Keio University School of Medicine, Tokyo, Japan

MIZUNO, MASAHIKO, Computational Biology Research Center, National Institute of Advanced Industrial Science and Technology, Tokyo, Japan

MONTPETIT, ALEXANDRE, Dept. of Human Genetics, McGill University, Montréal, Québec, Canada

MORRIS, DAVID, Dept. of Medicine, University of California, San Francisco

MORRIS, DAVID W., Sagres Discovery, Davis, California

MOYZIS, ROBERT, Dept. of Biological Chemistry, College of Medicine, University of California, Irvine

MULLIKIN, JAMES, Genome Technology Branch, National Human Genome Research Institute, National Institutes of Health, Bethesda, Maryland

MUNGALL, ANDREW, Dept. of Human Genetics, Genome Campus, The Wellcome Trust Sanger Institute, Hinxton, Cambridge, United Kingdom

MURAL, RICHARD, Scientific Content and Analysis, Celera Genomics, Rockville, Maryland

MURPHY, ELLEN, Dept. of Genomics, Wyeth Pharmaceuticals, Pearl River, New York

MYERS, EUGENE, Dept. of Computer Science, University of California, Berkeley

MYERS, JEREMY, Dept. of Biological Sciences, Louisiana State University, Baton Rouge, Louisiana

MYERS, RICHARD, Dept. of Genetics, Stanford University School of Medicine, Stanford, California

NAKAI, KENTA, Human Genome Center, Institute of Medical Science, University of Tokyo, Tokyo, Japan

NAKAMURA, KENJI, National Human Genome Research Institute, National Institutes of Health, Bethesda, Maryland

NG, PAULINE, Marketing, Illumina, Inc., San Diego, California

NOBREGA, MARCELO, Dept. of Genome Sciences, Lawrence Berkeley National Laboratory, Berkeley, California

NOGUCHI, HIDEKI, Human Genome Research Group, RIKEN Genomic Sciences Center, Yokohama, Japan

NUSBAUM, CHAD, Center for Genome Research, Whitehead Institute for Biomedical Research, Massachusetts Institute of Technology, Cambridge, Massachusetts

OLSON, MAYNARD, Dept. of Medicine, University of Washington, Seattle

OSOEGAWA, KAZUTOYO, Children's Hospital and Oakland Research Institute, Oakland, California

OSTRANDER, ELAINE, Div. of Clinical and Human Biology, Fred Hutchinson Cancer Research Center, Seattle, Washington

OTILLAR, ROBERT, Dept. of Genetics, Stanford University School of Medicine, Stanford, California

PÄÄBO, SVANTE, Max-Planck-Institute for Evolutionary Anthropology, Leipzig, Germany

PADDOCK, MARCIA, Dept. of Medicine, Genome Center, University of Washington, Seattle

PAGE, DAVID, Howard Hughes Medical Institute, Center for Genome Research, Whitehead Institute for Biomedical Research, Massachusetts Institute of Technology, Cambridge, Massachusetts

PARKER, HEIDI, Div. of Human Biology, Fred Hutchinson Cancer Research Center, Seattle, Washington

PARKHILL, JULIAN, Genome Campus, The Wellcome Trust Sanger Institute, Hinxton, Cambridge, United Kingdom

PARMENTIER, LAURENT, Dept. of Medical Research, GlaxoSmithKline, Marly le Roi, France

PARRA, GENIS, Dept. of Bioinformatics, Center for Genomic Regulation, Institut Municipal d'Investigacio Medica, Universitat Pompeu Fabra, Barcelona, Spain

PASKO, DEAN, Bioinformatics Research Center, Medical College of Wisconsin, Milwaukee, Wisconsin
PASTINEN, TOMI, Génome Québec Innovation Centre, McGill University, Montréal, Québec, Canada
PATIL, NILA, Dept. of Genetics, Perlegen Sciences, Inc., Mountain View, California
PATIL, SANDEEP, Dept. of Neuroscience, Eli Lilly and Co., Indianapolis, Indiana
PEARSON, WILLIAM, Dept. of Biochemistry and Molecular Genetics, University of Virginia, Charlottesville
PECK, ALLISON, National Human Genome Research Institute, National Institutes of Health, Bethesda, Maryland
PENNISI, ELIZABETH, *Science*, American Association for the Advancement of Science, Washington, D.C.
PEPKE, SHIRLEY, Dept. of Computational Biology, Berlex Biosciences, Richmond, California
PETERSON, JANE, National Human Genome Research Institute, National Institutes of Health, Bethesda, Maryland
PETRYSHEN, TRACEY, Center for Genome Research, Whitehead Institute for Biomedical Research, Massachusetts Institute of Technology, Cambridge, Massachusetts
PETUKHOVA, LYNN, Genotyping Center, Rockefeller University, New York, New York
PLATT, DARREN, Computational Target Discovery Informatics, Exelixis, Inc., South San Francisco, California
PLATZER, MATTHIAS, Dept. of Genome Analysis, Institute of Molecular Biotechnology, Jena, Germany
PLENGE, ROBERT, Dept. of Medical and Population Genetics, Center for Genome Research, Whitehead Institute for Biomedical Research, Massachusetts Institute of Technology, Cambridge, Massachusetts
PONTING, CHRISTOPHER, MRC Functional Genetics Unit, University of Oxford, Oxford, United Kingdom
PORCEL, BETINA, Dept. of Bioinformatics, Genoscope-Centre National de Séquençage, Evry, France
PORTEOUS, DAVID, Medical Genetics Section, Molecular Medicine Centre, Western General Hospital, Edinburgh, Scotland, United Kingdom
PORTNOY, MATTHEW, Genome Technology Branch, National Human Genome Research Institute, National Institutes of Health, Bethesda, Maryland
POWLEDGE, TABITHA, *The Scientist*, Philadelphia, Pennsylvania
POZZATTI, RUDY, Scientific Review Branch, National Human Genome Research Institute, National Institutes of Health, Bethesda, Maryland
PRAVENEC, MICHAL, Institute of Physiology, Czech Academy of Sciences, Prague, Czech Republic
QIAN, HUI-RONG, Dept. of Global Statistical Sciences, Eli Lilly and Co., Indianapolis, Indiana
QIN, XIAOLI, Dept. of Microbiology and Immunology, Stanford University, Stanford, California
RAY, DAVID, Dept. of Biological Sciences, Louisiana State University, Baton Rouge, Louisiana
REBATCHOUK, DMITRI, Dept. of Bioinformatics, Aventis Pharmaceuticals, Bridgewater, New Jersey
REBSCHER, HANS, Div. of Genotyping, Febit AG, Mannheim, Germany
REESE, MARTIN, Omicia, Inc., Oakland, California
REITER, LAWRENCE, Dept. of Biology, University of California at San Diego, La Jolla
RIAZ, NAVEEDA, National Institute on Deafness and Other Communication Disorders, National Institutes of Health, Rockville, Maryland
RICHARDSON, PAUL, Dept. of Functional Genomics, Department of Energy Joint Genome Institute, Walnut Creek, California
RIETHMAN, HAROLD, Dept. of Molecular and Cellular Oncogenesis, The Wistar Institute, Philadelphia, Pennsylvania
RISCH, NEIL, Dept. of Genetics, Stanford University School of Medicine, Stanford, California
ROBERTS, JERRY, National Human Genome Research Institute, National Institutes of Health, Bethesda, Maryland
ROE, BRUCE, Dept. of Chemistry and Biochemistry, University of Oklahoma, Norman
ROGATCHEVA, MARGARITA, Lab. of Comparative Genomics, University of Illinois, Urbana
ROGERS, JANE, Dept. of DNA Sequencing, Genome Campus, The Wellcome Trust Sanger Institute, Hinxton, Cambridge, United Kingdom
ROGOZIN, IGOR, National Center for Biotechnology Information, National Institutes of Health, Bethesda, Maryland
ROKHSAR, DANIEL, Dept. of Computational Genomics, Department of Energy Joint Genome Institute, Walnut Creek, California
ROSE, ANN, Dept. of Medical Genetics, University of British Columbia, Vancouver, British Columbia, Canada
ROSS, MARK, Genome Campus, The Wellcome Trust Sanger Institute, Hinxton, Cambridge, United Kingdom
ROWEN, LEE, Multimegabase Sequencing Center, Institute for Systems Biology, Seattle, Washington
RUBIN, EDWARD, Dept. of Genome Sciences, Lawrence Berkeley National Laboratory, Berkeley, California
RUBIN, GERALD, Janelia Farm Research Campus, Howard Hughes Medical Institute, Chevy Chase, Maryland
RUBINSTEIN, JOANNA, Div. of Health Sciences, Columbia University, New York, New York
RUBINSTEIN, ELLIS, New York Academy of Sciences, New York, New York
SAAVEDRA, RAUL, Scientific Review Branch, National Institute of Neurological Disorders and Stroke, National Institutes of Health, Bethesda, Maryland
SABOL, STEVEN, Molecular Disease Branch, National Heart, Lung, and Blood Institute, National Institutes of Health, Bethesda, Maryland
SAKAKI, YOSHIYUKI, Human Genome Research Group, RIKEN Genomic Sciences Center, Yokohama, Japan
SCHEETZ, TODD, Center for Bioinformatics and Computational Biology, University of Iowa, Iowa City
SCHERER, STEPHEN, Dept. of Genetics, Hospital for Sick Children, Toronto, Ontario, Canada
SCHERER, STEVEN, Dept. of Molecular and Human Genetics, Human Genome Sequencing Center, Baylor College of Medicine, Houston, Texas
SCHMUTZ, JEREMY, Dept. of Genetics, Stanford Human Genome Center, Palo Alto, California
SCHUELER, MARY, Genome Technology Branch, National Human Genome Research Institute, National Institutes of Health, Bethesda, Maryland

SCHWAB, SIBYLLE, Dept. of Psychiatry, University of Bonn, Bonn, Germany

SEIELSTAD, MARK, Dept. of Population Genetics, Genome Institute of Singapore, Singapore, China

SHAIKH, SANOBER, Dept. of Biostatistics and Epidemiology, Genomics Research Center, Genset, Evry, France

SHEARMAN, AMANDA, Center for Cancer Research, Massachusetts Institute of Technology, Cambridge, Massachusetts

SHEPHERD, NANCY, Genetics and Discovery Alliances, GlaxoSmithKline, Research Triangle Park, North Carolina

SHIA, MICHAEL, Dept. of Molecular Biology and Biochemistry, U.S. Genomics, Inc., Woburn, Massachusetts

SHIMIZU, ATSUSHI, Dept. of Molecular Biology, Keio University School of Medicine, Kanagawa, Japan

SHIMIZU, NOBUYOSHI, Dept. of Molecular Biology, Keio University School of Medicine, Tokyo, Japan

SHIZUYA, HIROAKI, Div. of Biology, California Institute of Technology, Pasadena, California

SIEPEL, ADAM, Center for Biomolecular Science and Engineering, University of California, Santa Cruz

SIMON, JASON, Discovery Technologies, Schering-Plough Research Institute, Kenilworth, New Jersey

SINGER, JONATHAN, Center for Genome Research, Whitehead Institute for Biomedical Research, Massachusetts Institute of Technology, Cambridge, Massachusetts

SKLAR, PAMELA, Dept. of Psychiatry, Harvard Medical School, Charlestown, Massachusetts

SLAUGENHAUPT, SUSAN, Harvard Institute of Human Genetics, Massachusetts General Hospital, Boston, Massachusetts

SLESAREV, ALEXEI, Fidelity Systems, Inc., Gaithersburg, Maryland

SMITH, DESMOND, Dept. of Molecular and Medical Pharmacology, School of Medicine, University of California, Los Angeles

SMITH, DOUGLAS, Agencourt Biosciences Corp., Beverly, Massachusetts

SMITH, RANDALL, Bioinformatics Discovery, GlaxoSmithKline, King of Prussia, Pennsylvania

SNYDER, MICHAEL, Dept. of Molecular, Cellular, and Developmental Biology, Yale University, New Haven, Connecticut

SOLOVYEV, VICTOR, Dept. of Computational Genomics, Department of Energy Joint Genome Institute, Walnut Creek, California

SONG, KYUYOUNG, Dept. of Biochemistry and Molecular Biology, University of Ulsan College of Medicine, Seoul, South Korea

SOREK, ROTEM, Dept. of Human Genetics and Molecular Medicine, Tel Aviv University and Compugen, Ltd., Tel Aviv, Israel

SPIELMAN, RICHARD, Dept. of Genetics, University of Pennsylvania, Philadelphia

SPURR, NIGEL, Discovery Genetics, GlaxoSmithKline, Research Triangle Park, North Carolina

STAMBOLIAN, DWIGHT, Dept. of Opthalmology, University of Pennsylvania, Philadelphia

STAUB, EIKE, Div. of Colon Cancer Research, metaGen Pharmaceuticals GmbH, Berlin, Germany

STEIN, ARNOLD, Dept. of Biological Sciences, Purdue University, West Lafayette, Indiana

STEIN, LINCOLN, Cold Spring Harbor Laboratory, Cold Spring Harbor, New York

STEWART, CARO-BETH, Dept. of Biological Sciences, State University of New York, Albany

STEWART, DAVID, Meetings and Courses Programs, Cold Spring Harbor Laboratory, Cold Spring Harbor, New York

STILLMAN, BRUCE, Cold Spring Harbor Laboratory, Cold Spring Harbor, New York

STUBBS, LISA, Genome Biology Division, Lawrence Livermore National Laboratory, Livermore, California

STUVE, LAURA, Genome Biology, Incyte Pharmaceuticals, Inc., Palo Alto, California

SUGANO, SUMIO, Dept. of Genome Structure Analysis, Human Genome Center, Institute of Medical Science, Unversity of Tokyo, Tokyo, Japan

SUGAWARA, HIDEAKI, Center for Information Biology and DNA Data Bank of Japan, National Institute of Genetics, Mishima, Shizuoka, Japan

SUGNET, CHARLES, Dept. of Computational Biology, University of California, Santa Cruz

SUNG, WEN-CHING, Dept. of Anthropology, Harvard University, Cambridge, Massachusetts

SUTTER, NATHAN, Dept. of Human Biology, Fred Hutchinson Cancer Research Center, Seattle, Washington

SWERDLOW, HAROLD, Solexa, Ltd., Little Chesterford, Essex, United Kingdom

TAILLON-MILLER, PATRICIA, Dept. of Dermatology, Washington University Medical School, St. Louis, Missouri

TAKACS, LASZLO, Dept. of Genomics and Bioinformatics, Global Research and Development, Pfizer, Inc., Fresnes, France

TAYLOR, TODD, Human Genome Research Group, RIKEN Genomic Sciences Center, Yokohama, Japan

TELLO-RUIZ, MARCELA, Cold Spring Harbor Laboratory, Cold Spring Harbor, New York

TERRILL, BRONWYN, Dolan DNA Learning Center, Cold Spring Harbor Laboratory, Cold Spring Harbor, New York

TEWARI, MUNEESH, Dept. of Cancer Biology, Dana-Farber Cancer Institute, Boston, Massachusetts

THOMAS, ELIZABETH, Watson School of Biological Sciences, Cold Spring Harbor Laboratory, Cold Spring Harbor, New York

THOMAS, PAUL, Celera Diagnostics, Alameda, California

THOMAS, SANDY, Nuffield Council on Bioethics, London, United Kingdom

THOMPSON, JOHN, Dept. of Cardiovascular and Metabolic Diseases, Pfizer Global Research and Development, Groton, Connecticut

THOMPSON, LARRY, Communications and Public Liaison Branch, National Human Genome Research Institute, National Institutes of Health, Bethesda, Maryland

THORISSON, GUDMUNDUR, Cold Spring Harbor Laboratory, Cold Spring Harbor, New York

TOTOKI, YASUSHI, Human Genome Research Group, RIKEN Genomic Sciences Center, Yokohama, Japan
TSO, HOI WAN, Dept. of Pediatrics and Adolescent Medicine, University of Hong Kong, Hong Kong, China
TYERS, MICHAEL, Mount Sinai Hospital, Samuel Lunenfeld Research Institute, Toronto, Ontario, Canada
TYLER-SMITH, CHRIS, Dept. of Biochemistry, University of Oxford, Oxford, United Kingdom
UBERBACHER, EDWARD, Div. of Life Sciences, Oak Ridge National Laboratory, Oak Ridge, Tennessee
ULANOVSKY, LEVY, Div. of Bioscience, Argonne National Laboratory, Argonne, New Mexico
UNDERHILL, PETER, Dept. of Genetics, Stanford UniversityStanford, California
VAN ES, HELMUTH, Dept. of Cellular and Molecular Biology, Galapagos Genomics BV, Leiden, The Netherlands
VELAZQUEZ, JOSE, Dept. of Genetics and Proteomics, National Institute on Alcohol Abuse and Alcoholism, National Institutes of Health, Bethesda, Maryland
VENTER, J. CRAIG, Center for the Advancement of Genomics, Rockville, Maryland
VESTAL, MARVIN, Discovery Proteomics and Small Molecules, Applied Biosystems, Framingham, Massachusetts
VINCENT, BETHANEY, Dept. of Biological Sciences, Louisiana State University, New Orleans, Louisiana
VITT, URSULA, Product Science, Incyte Pharmaceuticals, Inc., Palo Alto, California
VLIETINCK, ROBERT, Dept. of Human Genetics, University of Leuven, Leuven, Belgium
VYSOTSKAIA, VALENTINA, Dept. of Genome Biochemistry, Exelixis, Inc., South San Francisco, California
WADE, NICHOLAS, *The New York Times*, New York, New York
WAINSTOCK, DANIEL, *Cell*, Cell Press, Cambridge, Massachusetts
WALLACE, DOUGLAS, Center for Molecular and Mitochondrial Medicine and Genetics, University of California, Irvine
WANG, JINHUA, Cold Spring Harbor Laboratory, Cold Spring Harbor, New York
WANG, WEI, Dept. of Genetics, Stanford University, Stanford, California
WANG, YONGHONG, Dept. of Mammalian Genomics, The Institute for Genomic Research, Rockville, Maryland
WARBURTON, PETER, Dept. of Human Genetics, Mount Sinai School of Medicine, New York, New York
WATANABE, HIDEMI, Dept. of Bioinformatics and Genomics, Nara Institute of Science and Technology, Ikoma, Nara, Japan
WATERMAN, MICHAEL, Dept. of Biological Sciences, University of Southern California, Los Angeles, California
WATERSTON, ROBERT, Dept. of Genome Sciences, School of Medicine, University of Washington, Seattle
WATSON, JAMES, Cold Spring Harbor Laboratory, Cold Spring Harbor, New York
WEINER, MICHAEL, Guilford, Connecticut
WEISSENBACH, JEAN, Genoscope-Centre National de Séquençage, Evry, France
WETTERSTRAND, KRIS, National Human Genome Research Institute, National Institutes of Health, Bethesda, Maryland
WHITE, CHARLES, Omicia, Inc., Oakland, California
WHITFIELD, MICHAEL, Dept. of Genetics, Stanford University, Stanford, California
WIGLER, MICHAEL, Cold Spring Harbor Laboratory, Cold Spring Harbor, New York
WIJMENGA, CISCA, Dept. of Biomedical Genetics, Medical Center,University of Utrecht, Utrecht, The Netherlands
WILDENAUER, DIETER, Dept. of Psychiatry, University of Bonn, Bonn, Germany
WILLARD, HUNTINGTON, Institute for Genome Sciences and Policy, Duke University, Durham, North Carolina
WILLIAMSON, ALAN, Beaconsfield, United Kingdom
WILSON, RICHARD, Genome Sequencing Center, Washington University School of Medicine, St. Louis, Missouri
WINDEMUTH, ANDREAS, Genaissance Pharmaceuticals, Inc., New Haven, Connecticut
WINTER, EITAN, Dept. of Human Anatomy and Genetics, Oxford University, Oxford, United Kingdom
WITKOWSKI, JAN, Banbury Center, Cold Spring Harbor Laboratory, Cold Spring Harbor, New York
WU, KUNSHENG, Dept. of Breeding Technology, Monsanto, St. Louis, Missouri
Wu, Shujian, Dept. of Applied Genomics, Pharmaceutical Research Institute, Bristol-Myers Squibb, Princeton, New Jersey
XIAO, MING, Cardiovascular Research Institute, University of California, San Francisco
XIAO, YONGHONG, Target Discovery, Lion Bioscience Research, Inc., Cambridge, Massachusetts
XIE, HEHUANG, Dept. of Pediatrics, College of Medicine, University of Iowa, Iowa City
XU, CHUN-FANG, Medicines Research Center, GlaxoSmithKline, Stevenage, United Kingdom
XU, PING, Dept. of Microbiology and Immunology, Virginia Commonwealth University, Richmond, Virginia
XUAN, ZEHNYU, Cold Spring Harbor Laboratory, Cold Spring Harbor, New York
YAKUB, IMTIAZ, Human Genome Sequencing Center, Baylor College of Medicine, Houston, Texas
YAMASHITA, RIU, Human Genome Center, Institute of Medical Science, University of Tokyo, Tokyo, Japan
YAMAZAKI, SATORU, Dept. of Molecular Biology, Keio University School of Medicine, Kanagawa, Japan
YAN, CHUNHUA, Dept. of Bioinformatics, Celera Genomics, Rockville, Maryland
YANCOPOULOS, SOPHIA, *The Scientist* (Free Lance), Philadelphia, Pennsylvania
YANG, LEI, Div. of Molecular and Cellular Pathology, University of Alabama, Birmingham
YANG, ALICIA, Applied Biosystems, Foster City, California
YANG, CANZHU, Genome Research Center, Cold Spring Harbor Laboratory, Woodbury, New York
YANG, SHAN, Dept. of Biochemistry and Molecular Biology, Pennsylvania State University, State College, Pennsylvania

Young, Janet, Dept. of Human Biology, Fred Hutchinson Cancer Research Center, Seattle, Washington

Young, Richard, Dept. of Biology, Whitehead Institute for Biomedical Research, Massachusetts Institute of Technology, Cambridge, Massachusetts

Yu, Fuli, Human Genome Sequencing Center, Baylor College of Medicine, Houston, Texas

Yuan, Bingbing, Biocomputing Group, Whitehead Institute for Biomedical Research, Massachusetts Institute of Technology, Cambridge, Massachusetts

Zeng, Changqing, Beijing Genomics Institute, Beijing, China

Zhang, Jinghui, Lab. of Population Genetics, National Cancer Institute, National Institutes of Health, Bethesda, Maryland

Zhang, Michael, Cold Spring Harbor Laboratory, Cold Spring Harbor, New York

Zhang, Theresa, Dept. of Bioinformatics, Merck Research Laboratories, Rahway, New Jersey

Zhang, Xinmin, Dept. of Biological Sciences, Columbia University, New York, New York

Zhang, Zemin, Dept. of Bioinformatics, Genentech, Inc., South San Francisco, California

Zhao, Fang, Cold Spring Harbor Laboratory, Cold Spring Harbor, New York

Zhao, Shaying, Dept. of Mammalian Genomics, The Institute for Genomic Research, Rockville, Maryland

Zhao, Sheng, Dept. of Biomathematics, M.D. Anderson Cancer Center, University of Texas, Houston

Zhao, Wenqing, Dept. of Bioinformatics, Merck Research Laboratories, Rahway, New Jersey

Zhou, Yang, Dept. of Medicine, Genome Center, University of Washington, Seattle, Washington

Ziaugra, Liuda, Center for Genome Research, Whitehead Institute for Biomedical Research, Massachusetts Institute of Technology, Cambridge, Massachusetts

Zidanic, Michael, Dept. of Experimental Therapeutics, Walter Reed Army Institute of Research, Silver Spring, Maryland

*First row:* S. Thomas, M. Ashburner, N. Wade; M. Adams, C. Fraser, M. Waterman
*Second row:* L. Stein, S. Lewis, C. Harmon; W.R. McCombie signing the "genome" guitar
*Third row:* P. Hieter, D. Drayna, M. Olson; P. Underhill, P. Hardenbol, M. Faham
*Fourth row:* B. Vincent, P. Callinan, J. Myers; Y. Kuroki, B. Porcel

*First row:* A. Stein, M. Blanchette; J. Cheung, J. Ashurst, J. MacDonald
*Second row:* D. Bentley, G. Howell; R. Foxall, S. Humphray, A. Mungall
*Third row:* G. Rubin, P. Hieter, P. Good; D. Platt, J. Mullikin, P. Green
*Fourth row:* S. Limborska, D. Wallace; L. Du, M. Weiner

*First row:* S. Pääbo, A. Di Rienzo; M. Mullokandov, I. Rogozin, E. Koonin; E. Myers
*Second row:* B. Cooke, L. Hood; D. Steward, M. Olson; E. Green
*Third row:* A. Day, B. Gold; D. Page, E. Eichler
*Fourth row:* K. Ardlie, D. Cox; A. Andrés, J. Bertranpetit

*First row:* P. Deloukas, M. Chee; D. Ray, J. Myers, A. Delfalla
*Second row:* E. Lander, F. Collins, J.D. Watson, J. Witkowski; E. Mardis, W.R. McCombie
*Third row:* H. Riethman, R. Kaul; S. Chissoe, N. Spurr
*Fourth row:* L. Hood; L. Reiter, J. de Koning; C. Wijmenga, M. Hofker

*First row:* S. Deschamps, R. Gibbons; J. Kent, M. Ashburner
*Second row:* E. Rubin, R. Gibbs; C. Venter, C. Prizzi; K. Janssen, M. Flavell
*Third row:* L. Lyons, S. Slaugenhaupt; C. Tyler-Smith, M. Ross; Symposium cocktails

# Foreword

In 2001, as we considered topics for future Symposia, it rapidly became clear what the focus of attention should be in 2003. It had escaped no-one's notice that there was a momentous event to be celebrated that year—the 50th anniversary of the proposal by James Watson and Francis Crick of a structure for DNA now famously known as the double helix. And of great significance at Cold Spring Harbor was the 35th anniversary in 2003 of Jim Watson's appointment as Laboratory Director and the beginning of his unique influence on all aspects of this institution. These anniversaries would clearly require special attention and were indeed recognized in February at a remarkable meeting at the Laboratory devoted to DNA and at a spectacular gala in New York.

For the Symposium topic, however, we turned to another achievement that, even two years in advance, could be seen as the most significant milestone in biological science since the discovery of the double helix: the completion of the sequence of the human genome. Begun formally in 1990, with Jim Watson as its first director, the federally funded effort to map and sequence the entire human DNA molecule had resulted in the publication of a draft sequence in 2001. The predicted availability of the complete sequence in the first half of 2003 provided the ideal backdrop to a Symposium that focused not just on the details of the sequence, but on the power of information it contains to transform scientific investigations into fundamental biological processes and the causes of human disease.

In planning the Symposium, I was fortunate to have as co-organizers Jane Rogers, Edward Rubin, and my colleague David Stewart, whose advice and help in choosing speakers and themes were invaluable. The opening night of the meeting provided a glimpse of both the history of genomics and its extraordinary potential for the future as we listened to incisive lectures from Jane Rogers, David Cox, David Page, and David Botstein. In total, there were 81 oral presentations on the program and 180 poster presentations, all of striking quality. The lecture named for the Symposium series founder, Reginald G. Harris, was given in lively style by Claire Fraser. The annual Dorcas Cummings Lecture for our neighbors was a special treat, another demonstration of the masterful ability of Francis Collins, Director of the Human Genome Research Institute at the National Insitutes of Health, to convey the excitement of this kind of science to the general public—on this occasion not just with words, but with songs and his own guitar accompaniment. I am also particularly grateful to Maynard Olson for summarizing the meeting in his typically lucid and thoughtful way despite an electrical blackout that took place throughout the entire Summary presentation.

More than 525 scientists from 30 countries attended the meeting. Managing such an event is a complex challenge, but the staff of the Meetings and Courses Division, particularly Jenna Williams and Mary Smith, rose to it in their customary professional style. The audiovisual staff, Ed Campodonico, William Dickerson, Jonathan Parsons, and Gerald McCloskey, delivered flawless service throughout the meeting.

I thank the staff of Cold Spring Harbor Laboratory Press, particularly Joan Ebert, Patricia Barker, Melissa Frey, and Susan Schaefer, for their diligence in preparing the papers in this volume for publication. For the first time in its 68-year history, the proceedings of the Symposium are being published online as well as in print. As a result of this initiative by Executive Director John Inglis and colleagues Kathryn Fitzpatrick, Alex Gann, Jan Argentine, and Denise Weiss, institutional libraries that purchase this year's addition to the long line of famous red-covered volumes will also be able to make the information contained in the book available electronically throughout their campuses, along with reports created during the meeting itself, and archival material including the contents of the past five Symposia. The Web site for the Symposium information is www.cshl-symposium.org.

Finally, I am pleased to acknowledge the funding from companies, foundations, and the federal government, noted on the following page, that made this year's Symposium possible.

Bruce Stillman
*March 2004*

This meeting was funded in part by the **National Human Genome Research Institute** and the **National Cancer Institute**, branches of the **National Institutes of Health**; and the **U.S. Department of Energy.***

Contributions from the following companies provide core support for the Cold Spring Harbor Laboratory meetings program.

**Corporate Benefactors**

Amgen, Inc.
Aventis Pharma AG
Bristol-Myers Squibb Company
Eli Lilly and Company
GlaxoSmithKline
Novartis Pharma AG
Pfizer Inc.

**Corporate Sponsors**

Applied Biosystems
AstraZeneca
BioVentures, Inc.
Cogene BioTech Ventures, Ltd.
Diagnostic Products Corporation
Forest Laboratories, Inc.
Genentech, Inc.
Hoffmann-La Roche, Inc.
Johnson & Johnson Pharmaceutical Research & Development, L.L.C.
Kyowa Hakko Kogyo Co., Ltd.
Lexicon Genetics, Inc.
Merck Research Laboratories
New England BioLabs, Inc.
OSI Pharmaceuticals, Inc.
Pall Corporation
Schering-Plough Research Institute
Wyeth Genetics Institute

**Plant Corporate Associates**

MeadWestvaco Corporation
Monsanto Company
Pioneer Hi-Bred International, Inc.

**Corporate Associates**

Affymetrix, Inc.

**Corporate Contributors**

Biogen, Inc.
ImmunoRx
KeyGene

**Foundations**

Albert B. Sabin Vaccine Institute, Inc.

*This conference was supported by the Office of Science (BER), U.S. Department of Energy, Grant No. DE-FG02-03ER63519.

# Contents

# The Finished Genome Sequence of *Homo sapiens*

J. Rogers
*The Wellcome Trust Sanger Institute, Hinxton, Cambridge CB10 1SA, United Kingdom*

In April, 2003, just 50 years after the publication of the double helical structure of DNA by James Watson and Francis Crick, the International Human Genome Sequencing Consortium (IHGSC) announced that the genome sequence of *Homo sapiens* was essentially complete. This marked the culmination of the Human Genome Project (HGP) as conceived originally in 1985. For the thousands of people throughout the world who have played a part in the project over the past 15–20 years, generating the maps and the sequence of this large (3,000,000,000 base) genome, the success of the project has been a tribute to international cooperation. For everyone, this achievement represents a major turning point in our quest to learn how all the components of the human genome interact and contribute to biological processes and physiological complexity. In April, 1953, the journalist writing in the *Cambridge News Chronicle* concluded his piece on the discovery of the double helix with the comment: "...discovering how these ... chemical cards are shuffled will keep the scientists busy for the next 50 years." It has taken the past 50 years to spell out the human code. We can speculate on whether we shall need a further 50 years to understand its full complexity.

## OVERVIEW OF THE HUMAN GENOME PROJECT

The foundations for the HGP were laid in the late 1970s and early 1980s with the demonstration that small fragments of DNA could be sequenced and assembled into complete genomes (Sanger et al. 1977, 1982; Fiers et al. 1978; Anderson et al. 1981). For larger genomes, genome-wide genetic maps (Botstein et al. 1980) and physical clone maps (Coulson et al. 1986; Olson et al. 1986) could be built, and these provided the foundations for obtaining the complete DNA sequence of living organisms. The idea of sequencing the entire human genome arose from meetings organized between 1984 and1986 by the U.S. Department of Energy (Palca 1986; Sinsheimer 1989). In 1988, a committee appointed by the U.S. National Research Council endorsed the concept (National Research Council 1988), but recommended a broader program that included the generation of genetic, physical, and sequence maps of the human, and parallel efforts in key model organisms such as yeast, worms, flies, and mice. Technology development was to be a key component of the program, and research into the ethical, legal, and social issues raised by human genome research was also promoted. The program was launched as a joint effort of the Department of Energy and the National Institutes of Health in the US, at the same time as initiatives were launched in France, the UK, and Japan. With the establishment of the Human Genome Organisation, HUGO, in spring 1988, additional countries became involved with the program, notably Germany and China, who later contributed to the sequencing effort.

In April, 1988, the first Genome Mapping and Sequencing Meeting was held at Cold Spring Harbor. Since that time, the meeting has served as a primary focus for the HGP, and over the course of the past 15 years it has witnessed the passing of each successive milestone (see Table 1). These include the construction of genetic maps for the human (Weissenbach et al. 1992; Dib et al.1996) and mouse genomes (Dietrich et al. 1994), the first global genome maps. They have provided key tools for the localization of disease genes, and a framework for construction of gene maps using radiation hybrids (Schuler et al. 1996; Deloukas et al. 1998), which led in turn to the clone-based physical maps (Hudson et al. 1995; Nusbaum et al. 1999). The genetic and gene-based markers have served subsequently as important anchor points for the organization of genomic sequence. Also critical for the sequencing of the human genome was the experience gained from projects to sequence the genomes of model organisms, including yeast (Oliver et al. 1992; Yeast Genome Directory 1997), worm (Wilson et al. 1994; Consortium 1998), and the fruit fly (Adams et al. 2000). In 1998, the Cold Spring Harbor meeting witnessed the events that catalyzed the acceleration of the public sequencing effort alongside the announcement of plans to produce a privately owned human genome sequence by Celera Genomics. The subsequent scale-up in sequencing efforts around the world had a significant impact on the field. It accelerated the production of sequences of the human, mouse, and rat genomes, and also provided the capacity to discover millions of single nucleotide polymorphisms (SNPs) in the human genome sequence. These efforts have culminated in 2003 with the celebration of the finished human genome sequence and the beginnings of its translation into genome biology.

## INTERNATIONAL HGP COORDINATION

From the outset, the HGP attracted international involvement. Coordination of the systematic sequencing phase of the HGP was initiated by the Wellcome Trust, the National Human Genome Research Institute, and the U.S. Department of Energy in Bermuda in 1996. In three annual meetings, groups from six countries (US, UK, France, Germany, Japan, and, later, China) agreed on a

**Table 1.** Major Mapping and Sequencing Milestones of the Human Genome Project

| Date | Genome mapping | Genome sequences | Publication |
|---|---|---|---|
| 1987 | First-generation human genetic map | | Donis-Keller et al. (1987) |
| 1988 | Yeast artificial chromosome (YAC) cloning developed | | Burke et al. (1987) |
| 1989 | Sequence-tagged site (STS) mapping concept established | | Olson et al. (1989) |
| 1992 | Second-generation human genetic map | | Weissenbach et al. (1992) |
| 1994 | High-resolution human genetic map: 5,264 microsatellites | | Dib et al. (1996) |
| 1995 | STS-based physical map: human genome | | Hudson et al. (1995) |
| | | first bacterial genome: *Haemophilus influenzae* | Fleishmann et al. (1995) |
| 1996 | First human gene map | | Schuler et al. (1996) |
| | | first yeast genome: *Saccharomyces cerevisiae* | Yeast Genome Directory (1997) |
| | Mouse genetic map | | Dietrich et al. (1994) |
| 1997 | | *Escherichia coli* genome | Blattner et al. (1997) |
| 1998 | Second human gene map: 30,000 genes | | Deloukas et al. (1998) |
| | | first metazoan genome: *Caenorhabditis elegans* | *C. elegans* Sequencing Consortium (1998) |
| 1999 | YAC-based physical map: mouse genome | | Nusbaum et al. (1999) |
| | | scale-up of human genome sequencing | |
| | | first human chromosome: Chromosome 22 | Dunham et al. (1999) |
| 2000 | Bacterial clone map: human genome | | McPherson et al. (2001) |
| | | draft sequence of human genome | Lander et al. (2001); Venter et al. (2001) |
| | | first plant genome (*Arabidopsis thaliana*) | *Arabidopsis* Genome Initiative (2000) |
| | | *Drosophila melanogaster* genome | Adams et al. (2000) |
| 2001 | First single-nucleotide polymorphism (SNP) map of the human genome | | Sachidanandam et al. (2001) |
| 2002 | Bacterial clone map: mouse genome | | Gregory et al. (2002) |
| | | draft sequence of the mouse genome | Waterston et al. (2002) |
| | | draft version of rat genome | |
| 2003 | | finished human genome sequence | |

strategy for generating the sequence of the human genome and on the quality of sequence that should be achieved. They also agreed that to stimulate research and development and maximize the benefit to society of this ambitious and costly project, sequence data should be released immediately in the public domain and no patents should be filed on the sequence. As a result, sequence assemblies of 1–2 kb or greater, generated from mapped bacterial clones, were deposited every 24 hours in high-throughput genome sequence (HTGS) divisions of the public databases (EMBL, GenBank, or DDBJ). To improve the utility of the sequence, sequence assemblies were graded by quality: Phase 1 sequence represented assembled reads without ordering information; phase 2 sequence consisted of read assemblies (contigs) that were ordered and oriented; and phase 3 sequence was finished with no gaps and an error rate of less than 1:10,000 bases. The ultimate goal of the project was phase 3 sequence covering at least 95% of euchromatin. It was not the aim of the HGP to sequence the heterochromatin, such as the centromeres, telomeres, and tandemly repeated sequences on short arms of acrocentric chromosomes, although this is a possible goal for the future.

## HUMAN GENOME SEQUENCING STRATEGY

The IHGSC adopted a hierarchical shotgun strategy to sequence the human genome as follows (see Fig. 1). Maps were developed at progressively higher resolution up to the genomic sequence itself, using several independent methods. Over 30,000 landmarks from the genetic (Dib et al. 1996) and radiation hybrid maps of the genome (Deloukas et al. 1998; http://www.ncbi.nlm.nih. gov/genemap99) were used to anchor physical maps constructed from large-insert bacterial clones. These were mostly P1-derived or bacterial artificial chromosome (PACs [Ioannou et al. 1994] or BACs [Shizuya et al. 1992], respectively), clones in whole-genome libraries made with DNA from anonymous donors (Osoegawa et al. 2001). Overlaps between clones were identified primarily on the basis of shared restriction fingerprints (Bentley et al. 2001; Bruls et al. 2001; McPherson et al. 2001; Montgomery et al. 2001; Tilford et al. 2001). For each chromosome, a set of minimally overlapping bacterial clones (also referred to as the "tilepath") was selected from the map. Each clone was sequenced individually by the random shotgun approach. In this hierarchical strategy, in-

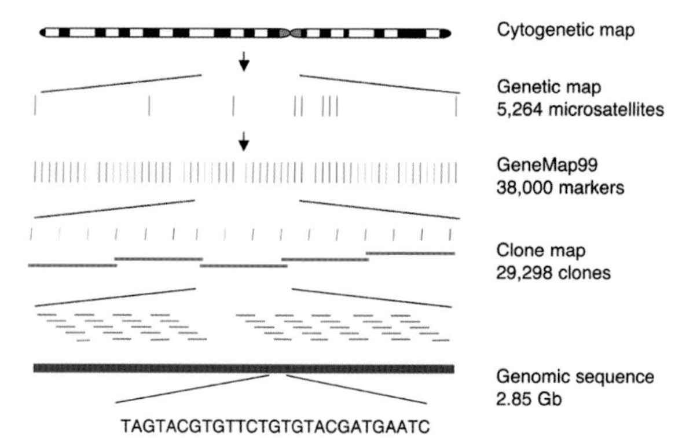


**Figure 1.** Schematic representation of the hierarchical mapping and sequencing strategy used for the Human Genome Project.

formation from the different maps was integrated by markers that were shared between them, thus providing independent confirmation of each level. Because most markers had sequence information attached to them (i.e., they were sequence-tagged sites, or STSs, in the genome), they were also integrated into the genomic sequence. This approach made it possible to select clones for sequencing that ensured maximal coverage of the genome, while minimizing sequencing redundancy. In addition, obtaining all the sequence on a clone-by-clone basis ensured that any regions which proved difficult to sequence could be resolved locally within the 40- to 200-kb segment of the large-insert bacterial clone.

Clones identified for sequencing were shotgun subcloned into single-stranded M13 bacteriophage or double-stranded plasmid vectors (e.g., pUC18). Sequence reads from one or both ends were generated after propagation of the subclones in *Escherichia coli* and extraction of the DNA. For efficient generation of high-quality finished sequence, each BAC or PAC clone was typically sequenced to give a minimum of six- or eightfold coverage in random shotgun reads, which were assembled using the program PHRAP (http://www.phrap.org/) to generate on average between six and ten contigs. This assembled unfinished sequence was then manually assessed in order to determine the best strategy for closing gaps and resolving all ambiguities. Additional sequence was generated from subclones or PCR products that spanned gaps. A range of directed sequencing strategies, such as the use of small-insert (McMurray et al. 1998) or transposon-tagged libraries (Devine et al. 1997), were developed to obtain sequence in particularly difficult regions. At the end of the finishing process, the sequence of each bacterial clone was at least 99.99% accurate and contained no gaps, achieving the standard agreed upon by the international consortium.

Through the course of the HGP, considerable improvements in sequencing chemistry and hardware were implemented to improve the overall quality of the product. Examples of chemistry improvements include the replacement of fluorescent dye primer sequencing by robust, fluorescent dye terminator sequencing, and the development of sequencing strategies for tracts of DNA that can adopt secondary structures inhibitory to polymerase progression. Early sequencing hardware improvements included increasing the number of lanes on each polyacrylamide slab gel, and extending read lengths. Substantial further improvement was provided by the later introduction of capillary sequencers, which had the advantages of very accurate lane tracking, higher throughput (8 runs/day compared with 3 runs/day on slab gel sequencers), longer read lengths, and no requirement for manual preparation of polyacrylamide gels. Most of the large sequencing laboratories also invested considerable effort in automating many of the steps in the sequencing process, including plaque/colony picking, DNA extraction, and sequencing. As a result of these improvements, the total capacity of sequencing centers engaged in the project had risen to over 100 million reads per year by August, 2000.

In 1999 and 2000, the increased efficiency in shotgun sequence generation was used to accelerate the production of sequence for the whole genome by generating draft sequence (assemblies of at least fourfold sequence coverage in high-quality sequence) for every clone in the tilepath (see Fig. 2). By October 2000, a "working draft" of the genome was assembled comprising sequence from 29,298 large bacterial clones (Lander et al. 2001). Of the clones, 8,277, representing approximately 30% of the genome, were already finished (see above). This included the sequences of Chromosomes 22 and 21, published in 1999 and 2000, respectively (Dunham et al. 1999; Hattori et al. 2000).

With the completion of the working draft, the IHGSC continued the process of generating a finished genome sequence. At this stage, completion of the clone tilepaths for each chromosome was assigned to chromosome coordinators distributed among the sequencing centers (see Table 2). To finish the map, gaps were closed wherever possible by exhaustive screening of multiple libraries (BAC, PAC, cosmid, or YAC) to identify additional clones to be sequenced. In the late stages of this process, the sizes of the remaining gaps were estimated by fluorescent in situ hybridization of extended DNA fibers, interphase nuclei, or metaphase chromosomes. In addition, when the mouse and rat draft genome sequences became available, it was possible to align the human sequence in the vicinity of the gap to the syntenic position in the rodent sequence and obtain an estimate of gap size. To finish the sequence of each BAC or PAC clone, at least twofold additional shotgun sequence coverage of each clone was generated to bring the overall coverage to at least sixfold. The iterative steps of manual review and directed sequencing (described above) were then taken to complete the sequence to the standards agreed upon by the IHGSC.

Further work on chromosome closure continues to be pursued at the genome centers, taking advantage of new resources and new techniques for manipulating DNA to close gaps that have been refractory to the methods used previously. For example, using sequence reads generated from flow-sorted Chromosome 20 for SNP discovery, it has already been possible to add a further 120 kb of sequence and to close three of the gaps that were not represented in large-insert clone libraries (P. Deloukas, pers. comm.).

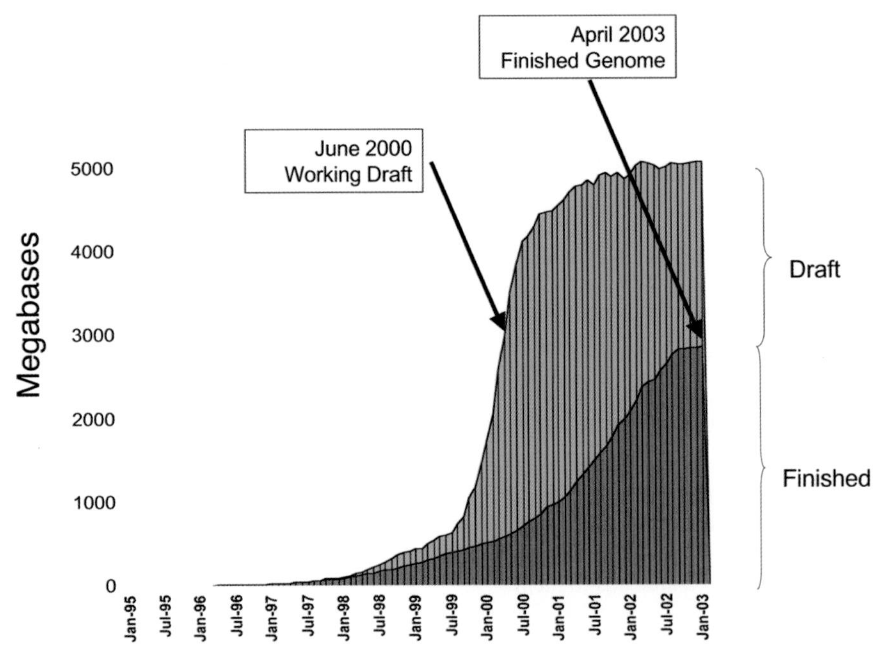


**Figure 2.** Accumulation of human genome sequence in the High Throughput Genome Sequence Division of GenBank over the course of the HGP. Draft sequence is shown in green and finished sequence in brown.

**Table 2.** Sequencing Centers Responsible for Coordination of Chromosome Finishing

| Sequencing center | Chromosome coordination |
|---|---|
| Sanger Institute | 1, 6, 9, 10, 13, 20, 22, X |
| Washington University Genome Sequencing Center, St. Louis | 2, 4, 7, Y |
| Whitehead Institute for Biomedical Research | 8, 15, 17, 18 |
| Baylor College of Medicine Genome Center | 3, 12 |
| Joint Genome Institute | 5, 16, 19 |
| RIKEN | 11, 21 |
| Genoscope | 14 |

## SEQUENCE QUALITY

The goal agreed upon by all participants in the public domain HGP was that the finished reference sequence of the human genome should have an accuracy of 99.99% or better. To assess the quality of each base in an individual sequence read, the PHRED quality base-call score (Q) was introduced: (Q = –10 x log10(p), where p is an estimated error probability calculated from the sequence trace) (Ewing and Green 1998; Ewing et al. 1998). This score has become a widely accepted metric for the assessment of sequence accuracy. To assess the quality of each base of the consensus of an assembled sequence, PHRAP scores were used in a similar manner. These scores are calculated as the product of error probabilities for bases in independent sequence reads at a given position. For the HGP, it was agreed that only base-calls of PHRED Q20 or better should be used to calculate sequence coverage. (A Q20 base-call has a probability of 1/100 of being incorrect; a Q30 base-call has a probability of 1/1000, or 0.1%, etc.). To meet the criteria for finished sequence, assembled sequences must have a PHRAP score of 40 or higher. Calculations undertaken for the draft human genome assembly showed an average base-call accuracy of greater than 99.5% for draft sequence and well above 99.99% for finished sequence. To ensure that this high sequence quality was achieved uniformly across the genome, a series of checking exercises on individual clones sequenced by different centers has been carried out (Felsenfeld et al. 1999). The most recent quality assessment exercise was reported at this meeting (Schmutz, this volume). These exercises showed that there was some variation between sequencing centers, but that the overall standard was lower than 1 error in 100,000 bases.

In addition to meeting criteria for base-calling accuracy, finished sequence is subjected to procedures to check the assemblies of individual clones and the compilation of clones into contiguous stretches of chromosomal sequence. An example of one method used to assess the assembly of each clone sequence is illustrated in Figure 3. Here the CONFIRM program is used to compare an in silico restriction digest of a clone (i.e., the pattern of fragments generated computationally from the assembled sequence) with the experimentally derived restriction digest of the clone (visualized after separation of the fragments by agarose gel electrophoresis). In the example shown, the in silico digestion of the initial assembly of clone RP-11 298A8 is missing a fragment present in the *Eco*RI digest of the clone, due to merging two exact copies of a 12-kb sequence. The correct assembly of the clone is confirmed in the lower trace.

The compilation of individual BAC or PAC sequences to generate contiguous sequence along chromosomes is

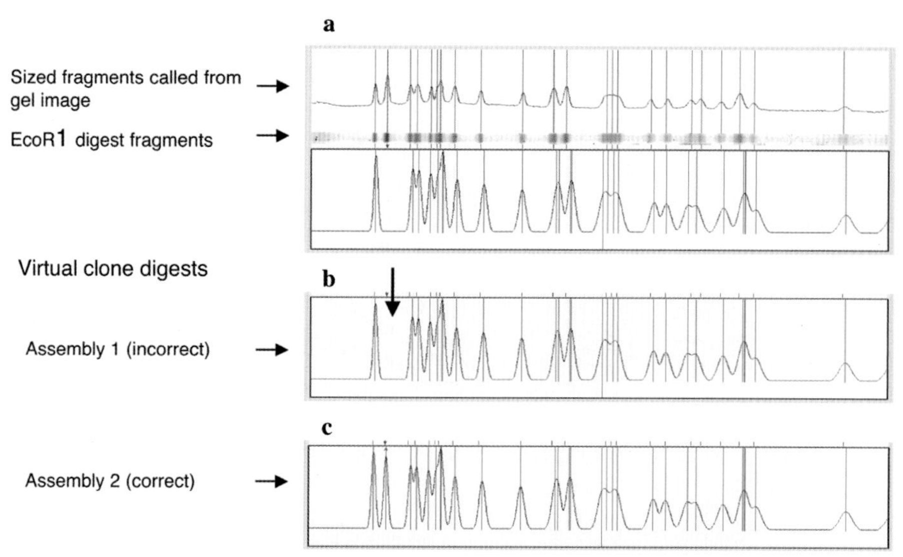


**Figure 3.** Restriction digest checking of the assembled sequence of large insert bacterial clones. Comparison of experimentally determined *Eco*RI digest of clone RP-11 298A8 (*a*) with the in silico restriction pattern derived from the original assembly of the clone sequence (*b*) shows that a fragment is missing from the assembly (*arrow*). The later assembly, in which two exact copies of a 12- kb repeat sequence have been resolved, shows exact correspondence of the new in silico pattern (*c*) with the experimental data.

checked in different ways. First, the overlaps of clones with their left and right neighbors are checked for sequence fidelity. The exact overlap positions are annotated in the EMBL/GenBank/DDBJ database entries for each clone. Because each BAC or PAC clone is isolated independently, two neighboring clones may be derived from unrelated individuals. If so, they will harbor different variant alleles at polymorphic sites. Sequence overlaps with a mismatch of more than 0.4% are judged to contain more differences than expected for average levels of polymorphism, and these are investigated further. The most common reasons for these apparent discrepancies are insertions or deletions of mono- or dinucleotide runs in repeat motifs, regions that are known to be polymorphic in the human population. High levels of single-base discrepancies can indicate that clones do not have authentic overlaps and that they should be positioned elsewhere in the genome. In some cases, the high levels of polymorphism have been confirmed to exist in the population by sequencing PCR products generated from a Human Diversity Panel set of 24 DNAs. In regions of the genome that have undergone duplication, such as the pericentromeric regions of Chromosomes 9 and 1, or segments of Chromosome 7 and the Y chromosome, assessment of clone overlaps at the sequence level has proved to be almost the only way to build up contiguous sequences over long repetitive regions (Skaletsky et al. 2003; Waterston et al., this volume).

A second method that has been used to provide confidence in the long-range compilation of chromosome sequences is the alignment of paired-end sequences from a whole-genome fosmid library (generated by the Whitehead Institute) to the human genome sequence assembly. The packaging process used to construct fosmid libraries imposes strict constraints on the permitted size of the cloned insert (of 30–50 kb). Therefore, if a pair of end sequence reads from a single fosmid clone do not align to the human sequence within 30–50 kb of each other, this points to a region of possible mis-assembly, or missing sequence (unless it is a genuine length polymorphism, where different length alleles are present in the fosmid and the assembled sequence).

A third method to assess the completeness and assembly of the genome has been developed by alignment of the assembled sequence to matching complementary DNA (cDNA), expressed-sequence tag (EST), and curated RefSeq gene sequences and markers positioned on genetic and radiation-hybrid maps (T. Furey, unpubl.). For example, any well-characterized cDNA that is missing indicated a possible gap and simultaneously provided a new probe to screen for a genomic clone. Misalignments between a cDNA and the genomic sequence highlight areas for checking both the genomic and the cDNA sequence, to resolve the conflict.

Using these tools to assess and validate the human genome sequence assembly, the IHGSC concluded in April 2003 that 2.85 Gb of sequence representing 99% of euchromatin has been finished. The contributions of the participating sequencing centers are illustrated in Figure 4. Comparison with the draft genome assembly showed that the finishing process had reduced the number of gaps in the human genome sequence assembly from around 148,000 to 393 regions that are apparently unclonable by methods used to date. (These comprise 281 gaps in the euchromatic regions and 106 in non-euchromatic regions including the centromeres—build 34, July, 2003 (http://genome.ucsc.edu/goldenPath/stats.html). The finishing process has resulted in an increase in the N50 length (defined as the length L such that at least 50% of all nucleotides are contained in contigs of size at least L) from 81 kb to 29 Mb. Thus, the average length of contiguous sequence is now approximately 1000 times the average gene length.

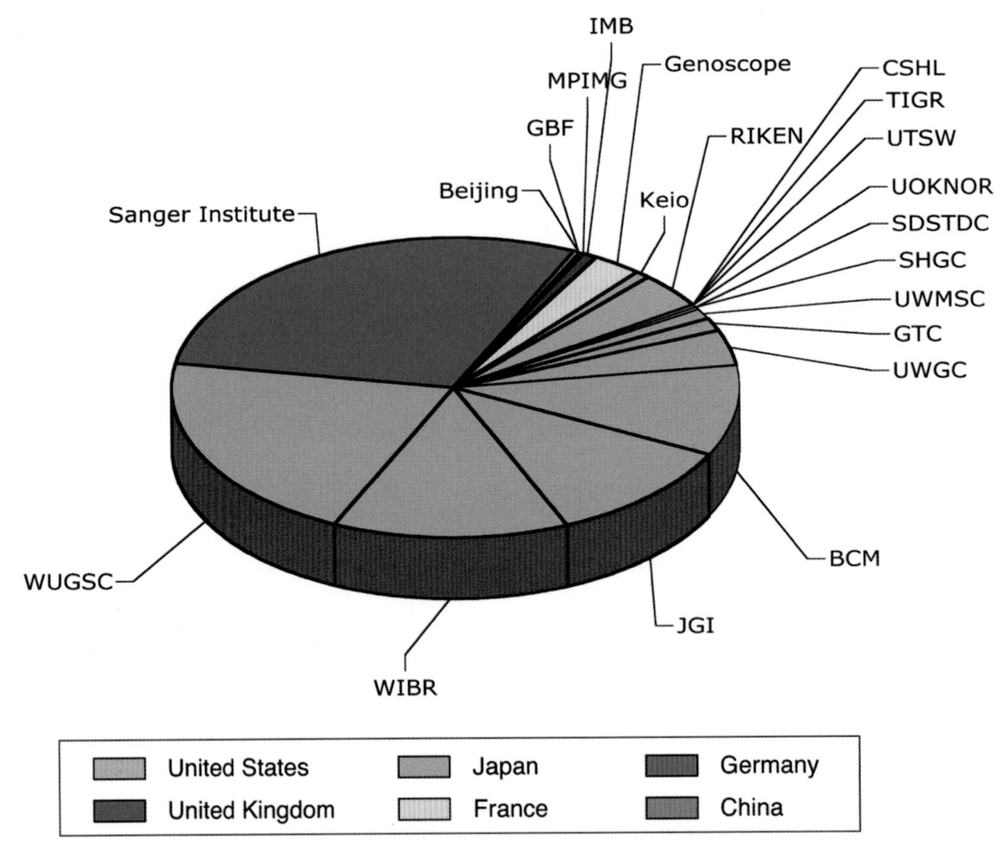


**Figure 4.** Sequencing center contributions to the finished human genome sequence. Abbreviations: (BCM) Baylor College of Medicine; (Beijing) Human Genome Center, Institute of Genetics, Chinese Academy of Sciences; (CSHL) Cold Spring Harbor Laboratory; (GBF) Gesellschaft fur Biotechnologische Forschung,mbH; (GTC) Genome Therapeutics Corporation; (IMB) Institute for Molecular Biology, Jena; (JGI) Joint Genome Institute, U.S. Department of Energy; (Keio) Keio University; (MPIMG) Max Planck Institute for Molecular Genetics; (RIKEN) RIKEN Genomic Sciences Center; (Sanger Institute) Wellcome Trust Sanger Institute; (SDSTDC) Stanford DNA Sequencing and Technology Center; (SHGC) Stanford Human Genome Center; (TIGR) The Institute for Genome Research; (UOKNOR) University of Oklahoma; (UTSW) University of Texas, Southwestern Medical Center; (UWGC) University of Washington Genome Center; University of Washington Multimegabase Sequencing Center; (WIBR) Whitehead Institute for Biomedical Research, MIT; (WUGSC) Washington University Genome Sequencing Center.

## SEQUENCE ANNOTATION

With the finished sequence in hand, the next key step in understanding the biology of the genome is accurate annotation of the human gene set. Ultimately, we require all genome features, including genes, alternative transcripts, sequence variations, promoters, enhancers, and other regulatory motifs to be accurately defined and displayed along the single metric of the genome sequence.

The assembly of the working draft sequence provided the basis for the first completely automated and nonredundant annotation of the human genome. Key features of the sequence analysis method were that it was based on using unfinished sequence and that the annotation could be updated rapidly as the sequence evolved. Ensembl (a joint project between the European Bioinformatics Institute and the Sanger Institute) (Hubbard and Birney 2000; Hubbard et al. 2002) annotates known genes and predicts novel genes, with functional annotation from the InterPro (Apweiler et al. 2001) protein family databases and from disease expression and gene family databases (Enright et al. 1999; Antonarakis and McKusick 2000; Wheeler et al. 2002). The Ensembl gene build system uses a three-step process, incorporating information on exon structure and placement from alignment of genes predicted from human proteins in SPTREMBL (Bairoch and Apweiler 2000); the alignment of paralogous human proteins and proteins from other organisms; and ab initio gene prediction using genscan (Burge and Karlin 1997). Exons that are supported by more than one prediction are clustered to form genes. "Ensembl genes" are regarded as being accurate predicted gene structures with a low false-positive rate, since they are all supported by experimental evidence from protein and nucleotide databases.

Genome annotation can be viewed in genome browsers developed by Ensembl (http://www.ensembl.org), the University of California at Santa Cruz (http://genome.ucsc.edu), and the National Center for Biotechnology

Information (NCBI) (http://www.ncbi.nlm.nih./gov/genome/guide/human/). The browsers give users access to the vast amount of human sequence and map data and enable them to scan the genome for features that include gene structures, repeat families, STS markers positioned on genetic and physical maps, ESTs, coding and noncoding RNAs, C+G content, single-nucleotide polymorphisms (SNPs), and sequence similarities with other organisms. Since the first assembly of the draft sequence in October, 2000, the browsers have continued to provide regularly updated views of the human genome sequence, taking account of new genomic sequence information and new supporting data from ongoing programs such as full-length cDNA sequencing (http://genome.rtc.riken.go.jp/, http://www.ncbi.nlm.nih.gov/MGC/), which help to improve the gene builds. Another recent improvement has been the integration of EST data into gene building, which aids in the prediction of noncoding exons, especially those located within the 3′UTR, and the prediction of pseudogenes (see Birney et al., this volume). The browsers also regularly add new features that provide improved functionality for biologists using genomic information.

Although the genome browsers provide consistent views of annotation across the most recent assembly of the human genome sequence, it is clear that further work is needed to achieve the highest levels of accuracy. Comparison of Ensembl genes with regions of the genome that have been manually annotated and experimentally investigated, such as Chromosomes 20 and 22, shows currently that approximately 70% of gene loci are identified by the automated system. Features that are underrepresented include single-exon genes, pseudogenes, splice variants, and sites of splicing and polyadenylation (Ashurst and Collins 2003). To provide the most comprehensive set of human genes, systematic manual annotation has been taken on so far by the sequencing centers coordinating the finishing process of each finished human chromosome. The Sanger Institute carried out the initial annotation of Chromosome 22 (Dunham et al. 1999), updated in 2003 (Collins et al. 2003), and this has been followed by systematic annotation of Chromosomes 20 (Deloukas et al. 2001), 6 (Mungall et al. 2003), 9, 10, 13, X, and 1. Manual annotation of Chromosomes 21 (Hattori et al. 2000), 14 (Heilig et al. 2003), 7 (Hillier et al. 2003), and Y (Skaletsky et al. 2003) has been published by centers that have produced those sequences. Annotation for the remaining chromosomes will be generated by participating centers over the coming months.

The products of the manual annotation process are available in the Vertebrate and Genome Analysis (VEGA) database, http://vega.sanger.ac.uk, and can also be viewed in Ensembl. A view of VEGA gene predictions and Ensembl genes in a region of Chromosome 20 is shown in Figure 5. The VEGA browser takes advantage of the Ensembl database framework and displays experimental evidence associated with each prediction. In contrast to gene structures generated by automated systems, local amendments can be made to gene predictions rapidly by annotators in response to user feedback, and the database is intended to provide an accurate, continually updated resource for the research community.

To achieve consistency in primary manual annotation across the genome, guidelines have been established in a series of Human Annotation Workshop (HAWK) meetings. The model provided by the VEGA database and HAWK guidelines gives a possible mechanism for future long-term re-annotation and curation of the human genome sequence. The experience on Chromosome 22, re-annotated three years after the initial annotation was produced, indicates the value to the research community of establishing systematic ongoing curation on a genome-wide basis: Using data from EST databases, comparative genome analysis, and experimental verification to fuse previously fragmented genes and identify novel genes resulted in a 74% increase in total annotated exon sequence length on this chromosome. This included one new gene, 100 pseudogenes, and 31 non-protein-coding transcripts, of which 16 are likely antisense RNAs (Collins et al. 2003).

## FUTURE PROSPECTS

Annotating the genes is only the beginning, and already it is clear that a truly comprehensive picture of the complexity of the human gene set, including all the biologically important alternative transcripts and promoters, will take some years to emerge. What about other functionally important sequences? An initial comparison between the human and mouse genome sequences suggested that approximately 5% of the human genome is under purifying selection and may therefore comprise functionally important conserved sequences (Waterston et al. 2002). One-third of this contains nearly all the protein-coding exons, and the other two-thirds have non-protein-coding sequence. The use of comparative sequence analysis was taken further in a multispecies comparison of a 1.8-Mb region of human Chromosome 7, which led to the identification of multiple-species conserved sites (MCSs) that are strong candidates for these sequences (Margulies et al. 2003; Thomas et al. 2003; Margulies et al., this volume). Other studies have also illustrated the value of comparing genomes of different vertebrate species to discover new genes or regulatory elements ("phylogenetic footprinting") (Gottgens et al. 2000; Pennacchio et al. 2001; Pennacchio and Rubin 2001). The same principle has been extended to comparisons between multiple primate sequences ("phylogenetic shadowing") (Boffelli et al. 2003). This work has formed the basis for a plan to evaluate in detail different methods to extract as much biological information as possible in selected regions of the genome comprising a total of 30 Mb (1%), the ENCyclopedia of DNA Elements (ENCODE) project (http://www.genome.gov/10005107).

The finished human genome sequence provides resources for further experimental investigation. For example, the large-insert genomic clones that formed the map (BACs and PACs) have been used to generate genomic microarrays for the investigation of chromosomal aberrations that accompany disease (Solinas-Toldo et al. 1997;

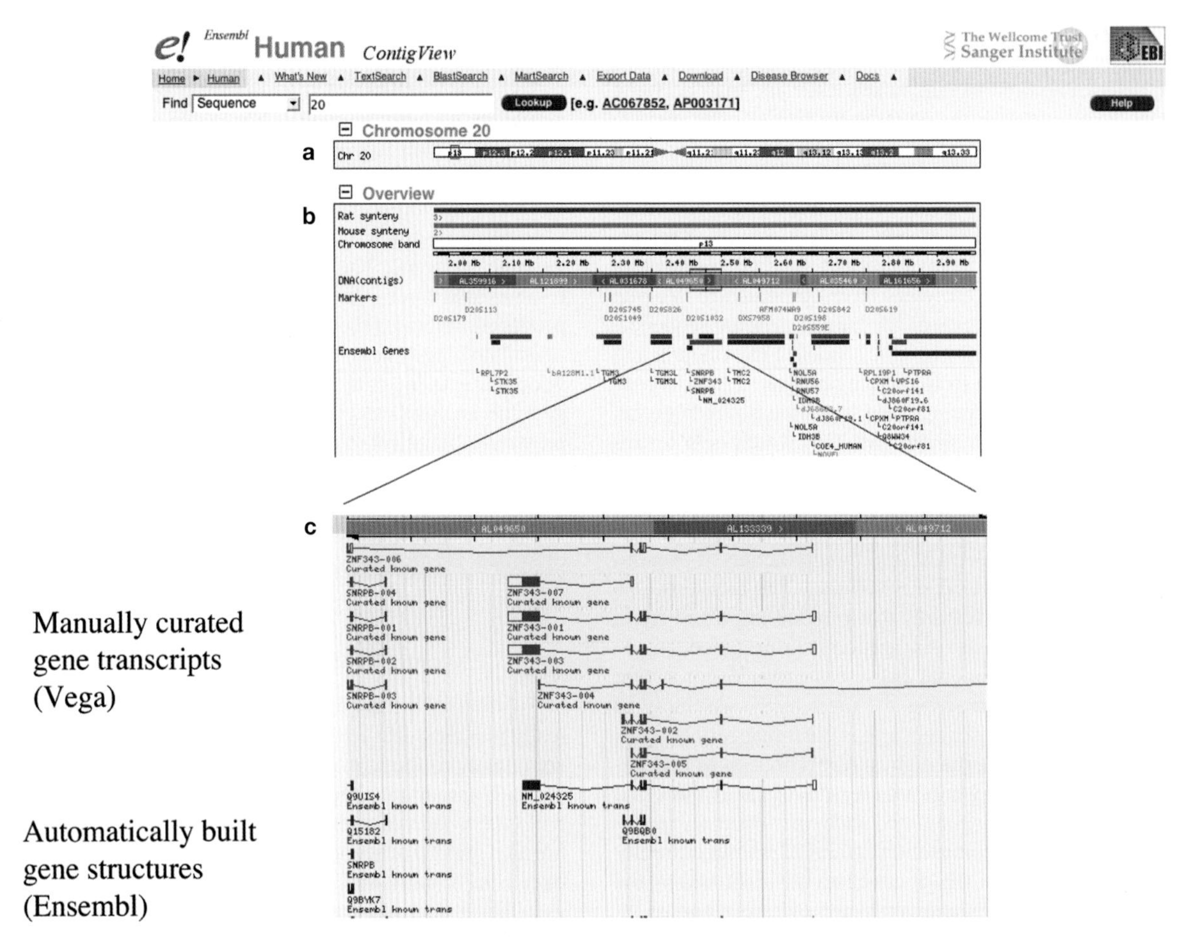


**Figure 5.** Ensembl view of a region of Chromosome 20 showing a comparison of gene structures produced by the automatic annotation process (*red*) and structures resolved after manual annotation (*blue*) (see text for details). Note the resolution of multiple alternative transcripts by manual annotation in the detailed view. In the Chromosome 20 ideogram (*a*), the region highlighted in the red box is expanded and shown in the overview (*b*). Similarly, the central region of view *b* (boxed in *red*) is expanded and shown in view *c*.

Fiegler et al. 2003). The same arrays provide the means to carry out an initial survey of replication timing in human cells (Woodfine et al. 2004). Using information from the sequence to make higher-resolution arrays (e.g., with 1-kb genomic segments) is proving valuable to investigate sites of chromatin modification that may accompany regulation. For example, histone acetylation or transcription factor binding may be detected by immunoprecipitation of chromatin and recovery of the underlying chromosomal DNA as a small fragment, which can then be labeled and hybridized to the array to determine its localization in the genome sequence. DNA methylation is also a new area for study that may be of great importance in understanding the function of the human genome. DNA methylation is involved in imprinting, regulation, chromatin structure, X inactivation, and disease, and the human genome sequence provides a basis for the study of epigenetics on a genome-wide basis (for review, see Novik et al. 2002).

The human genome sequence provides an archival reference for studying sequence variation in the human population. Sequence variation underlies the genetic predisposition to health and disease and the variable response of individuals to environmental factors including toxins and drugs, and will provide important information on human evolutionary history. Already, the draft sequence, and now the finished sequence, have been used to identify over 5 million SNPs and a range of other sequence variants in the genome (Sachidanandam et al. 2001); http://ncbi.nlm.nih.gov/SNP/snp_summary.cgi). Two directions are made possible from this research. First, a much deeper search for additional sequence variants will reveal the true nature of variation in the human species; and second, correlating the variants with genes and other features of the genome annotation will enable detection of the functional variants that are important in health and disease (for further review, see Bentley 2004).

## ACKNOWLEDGMENTS

This paper discusses work that has been carried out by the enormous number of people who have contributed to the Human Genome Project. The author is grateful for the opportunity to review it. Special thanks go to my colleagues who have provided valuable input in discussion and material: Jennifer Ashurst, Stephan Beck, David

Bentley, Christine Bird, Ewan Birney, Nigel Carter, Francis Collins, Panos Deloukas, Ian Dunham, Heike Fiegler, Darren Grafham, Tim Hubbard, Greg Schuler, Bob Waterston, and Kathryn Woodfine.

## REFERENCES

Adams M.D., Celniker S.E., Holt R.A., Evans C.A., Gocayne J.D., Amanatides P.G., Scherer S.E., Li P.W., Hoskins R.A., Galle R.F., George R.A., Lewis S.E., Richards S., Ashburner M., Henderson S.N., Sutton G.G., Wortman J.R., Yandell M.D., Zhang Q., Chen L.X., Brandon R.C., Rogers Y.H., Blazej R.G., Champe M., and Pfeiffer B.D., et al. 2000. The genome sequence of *Drosophila melanogaster*. *Science* **287:** 2185.

Anderson S., Bankier A.T., Barrell B.G., de Bruijn M.H., Coulson A.R., Drouin J., Eperon I.C., Nierlich D.P., Roe B.A., Sanger F., Schreier P.H., Smith A.J., Staden R., and Young I.G. 1981. Sequence and organization of the human mitochondrial genome. *Nature* **290:** 457.

Antonarakis S.E. and McKusick V.A. 2000. OMIM passes the 1,000-disease-gene mark. *Nat. Genet.* **25:** 11.

Apweiler R., Attwood T.K., Bairoch A., Bateman A., Birney E., Biswas M., Bucher P., Cerutti L., Corpet F., Croning M.D., Durbin R., Falquet L., Fleischmann W., Gouzy J., Hermjakob H., Hulo N., Jonassen I., Kahn D., Kanapin A., Karavidopoulou Y., Lopez R., Marx B., Mulder N.J., Oinn T.M., and Pagni M., et al. 2001. The InterPro database, an integrated documentation resource for protein families, domains and functional sites. *Nucleic Acids Res.* **29:** 37.

*Arabidopsis* Genome Initiative. 2000. Analysis of the genome sequence of the flowering plant *Arabidopsis thaliana*. *Nature* **408:** 796.

Ashurst J.L. and Collins J.E. 2003. Gene annotation: Prediction and testing. *Annu. Rev. Genomics Hum. Genet.* **4:** 69.

Bairoch A. and Apweiler R. 2000. The SWISS-PROT protein sequence database and its supplement TrEMBL in 2000. *Nucleic Acids Res.* **28:** 45.

Bentley D.R. 2004. Genome vision. *Nature* (in press).

Bentley D.R., Deloukas P., Dunham A., French L., Gregory S.G., Humphray S.J., Mungall A.J., Ross M.T., Carter N.P., Dunham I., Scott C.E., Ashcroft K.J., Atkinson A.L., Aubin K., Beare D.M., Bethel G., Brady N., Brook J.C., Burford D.C., Burrill W.D., Burrows C., Butler A.P., Carder C., Catanese J.J., and Clee C.M., et al. 2001. The physical maps for sequencing human chromosomes 1, 6, 9, 10, 13, 20 and X. *Nature* **409:** 942.

Blattner F.R., Plunkett G., III, Bloch C.A., Perna N.T., Burland V., Riley M., Collado-Vides J., Glasner J.D., Rode C.K., Mayhew G.F., Gregor J., Davis N.W., Kirkpatrick H.A., Goeden M.A., Rose D.J., Mau B., and Shao Y. 1997. The complete genome sequence of *Escherichia coli* K-12. *Science* **277:** 1453.

Boffelli D., McAuliffe J., Ovcharenko D., Lewis K.D., Ovcharenko I., Pachter L., and Rubin E.M. 2003. Phylogenetic shadowing of primate sequences to find functional regions of the human genome. *Science* **299:** 1391.

Botstein D., White R.L., Skolnick M., and Davis R.W. 1980. Construction of a genetic linkage map in man using restriction fragment length polymorphisms. *Am. J. Hum. Genet.* **32:** 314.

Bruls T., Gyapay G., Petit J.L., Artiguenave F., Vico V., Qin S., Tin-Wollam A.M., Da Silva C., Muselet D., Mavel D., Pelletier E., Levy M., Fujiyama A., Matsuda F., Wilson R., Rowen L., Hood L., Weissenbach J., Saurin W., and Heilig R. 2001. A physical map of human chromosome 14. *Nature* **409:** 947.

Burge C. and Karlin S. 1997. Prediction of complete gene structures in human genomic DNA. *J. Mol. Biol.* **268:** 78.

Burke D.T., Carle G.F., and Olson M.V. 1987. Cloning of large segments of exogenous DNA into yeast by means of artificial chromosome vectors. *Science* **236:** 806.

Collins J.E., Goward M.E., Cole C.G., Smink L.J., Huckle E.J., Knowles S., Bye J.M., Beare D.M., and Dunham I. 2003. Reevaluating human gene annotation: A second-generation analysis of chromosome 22. *Genome Res.* **13:** 27.

The *C. elegans* Sequencing Consortium. 1998. Genome sequence of the nematode *C. elegans:* A platform for investigating biology. *Science* **282:** 2012.

Coulson A., Sulston J., Brenner S., and Karn J. 1986. Toward a physical map of the genome of the nematode *Caenorhabditis elegans*. *Proc. Natl. Acad. Sci.* **83:** 7821.

Deloukas P., Matthews L.H., Ashurst J., Burton J., Gilbert J.G., Jones M., Stavrides G., Almeida J.P., Babbage A.K., Bagguley C.L., Bailey J., Barlow K.F., Bates K.N., Beard L.M., Beare D.M., Beasley O.P., Bird C.P., Blakey S.E., Bridgeman A.M., Brown A.J., Buck D., Burrill W., Butler A.P., Carder C., and Carter N.P., et al. 2001. The DNA sequence and comparative analysis of human chromosome 20. *Nature* **414:** 865.

Deloukas P., Schuler G.D., Gyapay G., Beasley E.M., Soderlund C., Rodriguez-Tome P., Hui L., Matise T.C., McKusick K.B., Beckmann J.S., Bentolila S., Bihoreau M., Birren B.B., Browne J., Butler A., Castle A.B., Chiannilkulchai N., Clee C., Day P.J., Dehejia A., Dibling T., Drouot N., Duprat S., Fizames C., and Bentley D.R., et al. 1998. A physical map of 30,000 human genes. *Science* **282:** 744.

Devine S.E., Chissoe S.L., Eby Y., Wilson R.K., and Boeke J.D. 1997. A transposon-based strategy for sequencing repetitive DNA in eukaryotic genomes. *Genome Res.* **7:** 551.

Dib C., Faure S., Fizames C., Samson D., Drouot N., Vignal A., Millasseau P., Marc S., Hazan J., Seboun E., Lathrop M., Gyapay G., Morissette J., and Weissenbach J. 1996. A comprehensive genetic map of the human genome based on 5,264 microsatellites. *Nature* **380:** 152.

Dietrich W.F., Miller J.C., Steen R.G., Merchant M., Damron D., Nahf R., Gross A., Joyce D.C., Wessel M., and Dredge R.D., et al. 1994. A genetic map of the mouse with 4,006 simple sequence length polymorphisms. *Nat. Genet.* **7:** 220.

Donis-Keller H., Green P., Helms C., Cartinhour S., Weiffenbach B., Stephens K., Keith T.P., Bowden D.W., Smith D.R., and Lander E.S., et al. 1987. A genetic linkage map of the human genome. *Cell* **51:** 319.

Dunham I., Shimizu N., Roe B.A., Chissoe S., Hunt A.R., Collins J.E., Bruskiewich R., Beare D.M., Clamp M., Smink L.J., Ainscough R., Almeida J.P., Babbage A., Bagguley C., Bailey J., Barlow K., Bates K.N., Beasley O., Bird C.P., Blakey S., Bridgeman A.M., Buck D., Burgess J., Burrill W.D., and O'Brien K.P., et al. 1999. The DNA sequence of human chromosome 22. *Nature* **402:** 489.

Enright A.J., Iliopoulos I., Kyrpides N.C., and Ouzounis C.A. 1999. Protein interaction maps for complete genomes based on gene fusion events. *Nature* **402:** 86.

Ewing B. and Green P. 1998. Base-calling of automated sequencer traces using phred. II. Error probabilities. *Genome Res.* **8:** 186.

Ewing B., Hillier L., Wendl M.C., and Green P. 1998. Base-calling of automated sequencer traces using phred. I. Accuracy assessment. *Genome Res.* **8:** 175.

Felsenfeld A., Peterson J., Schloss J., and Guyer M. 1999. Assessing the quality of the DNA sequence from the Human Genome Project. *Genome Res.* **9:** 1.

Fiegler H., Gribble S.M., Burford D.C., Carr P., Prigmore E., Porter K.M., Clegg S., Crolla J.A., Dennis N.R., Jacobs P., and Carter N.P. 2003. Array painting: A method for the rapid analysis of aberrant chromosomes using DNA microarrays. *J. Med. Genet.* **40:** 664.

Fiers W., Contreras R., Haegemann G., Rogiers R., Van de Voorde A., Van Heuverswyn H., Van Herreweghe J., Volckaert G., and Ysebaert M. 1978. Complete nucleotide sequence of SV40 DNA. *Nature* **273:** 113.

Fleischmann R.D., Adams M.D., White O., Clayton R.A., Kirkness E.F., Kerlavage A.R., Bult C.J., Tomb J.F., Dougherty B.A., Merrick J.M., et al. 1995. Whole-genome random sequencing and assembly of *Haemophilus influenzae* Rd. *Science* **269:** 496.

Gottgens B., Barton L.M., Gilbert J.G., Bench A.J., Sanchez M.J., Bahn S., Mistry S., Grafham D., McMurray A., Vaudin M., Amaya E., Bentley D.R., Green A.R., and Sinclair A.M.

2000. Analysis of vertebrate SCL loci identifies conserved enhancers. *Nat. Biotechnol.* **18:** 181.

Gregory S.G., Sekhon M., Schein J., Zhao S., Osoegawa K., Scott C.E., Evans R.S., Burridge P.W., Cox T.V., Fox C.A., Hutton R.D., Mullenger I.R., Phillips K.J., Smith J., Stalker J., Threadgold G.J., Birney E., Wylie K., Chinwalla A., Wallis J., Hillier L., Carter J., Gaige T., Jaeger S., and Kremitzki C., et al. 2002. A physical map of the mouse genome. *Nature* **418:** 743.

Hattori M., Fujiyama A., Taylor T.D., Watanabe H., Yada T., Park H.S., Toyoda A., Ishii K., Totoki Y., Choi D.K., Soeda E., Ohki M., Takagi T., Sakaki Y., Taudien S., Blechschmidt K., Polley A., Menzel U., Delabar J., Kumpf K., Lehmann R., Patterson D., Reichwald K., Rump A., and Schillhabel M., et al. 2000. The DNA sequence of human chromosome 21. The chromosome 21 mapping and sequencing consortium. *Nature* **405:** 311.

Heilig R., Eckenberg R., Petit J.L., Fonknechten N., Da Silva C., Cattolico L., Levy M., Barbe V., de Berardinis V., Ureta-Vidal A., Pelletier E., Vico V., Anthouard V., Rowen L., Madan A., Qin S., Sun H., Du H., Pepin K., Artiguenave F., Robert C., Cruaud C., Bruls T., Jaillon O., and Friedlander L., et al. 2003. The DNA sequence and analysis of human chromosome 14. *Nature* **421:** 601.

Hillier L.W., Fulton R.S., Fulton L.A., Graves T.A., Pepin K.H., Wagner-McPherson C., Layman D., Maas J., Jaeger S., Walker R., Wylie K., Sekhon M., Becker M.C., O'Laughlin M.D., Schaller M.E., Fewell G.A., Delehaunty K.D., Miner T.L., Nash W.E., Cordes M., Du H., Sun H., Edwards J., Bradshaw-Cordum H., and Ali J., et al. 2003. The DNA sequence of human chromosome 7. *Nature* **424:** 157.

Hubbard T. and Birney E. 2000. Open annotation offers a democratic solution to genome sequencing. *Nature* **403:** 825.

Hubbard T., Barker D., Birney E., Cameron G., Chen Y., Clark L., Cox T., Cuff J., Curwen V., Down T., Durbin R., Eyras E., Gilbert J., Hammond M., Huminiecki L., Kasprzyk A., Lehvaslaiho H., Lijnzaad P., Melsopp C., Mongin E., Pettett R., Pocock M., Potter S., Rust A., and Schmidt E., et al. 2002. The Ensembl genome database project. *Nucleic Acids Res.* **30:** 38.

Hudson T.J., Stein L.D., Gerety S.S., Ma J., Castle A.B., Silva J., Slonim D.K., Baptista R., Kruglyak L., and Xu S.H., et al. 1995. An STS-based map of the human genome. *Science* **270:** 1945.

Ioannou P.A., Amemiya C.T., Garnes J., Kroisel P.M., Shizuya H., Chen C., Batzer M.A., and de Jong P.J. 1994. A new bacteriophage P1-derived vector for the propagation of large human DNA fragments. *Nat. Genet.* **6:** 84.

Lander E.S., Linton L.M., Birren B., Nusbaum C., Zody M.C., Baldwin J., Devon K., Dewar K., Doyle M., FitzHugh W., Funke R., Gage D., Harris K., Heaford A., Howland J., Kann L., Lehoczky J., LeVine R., McEwan P., McKernan K., Meldrim J., Mesirov J.P., Miranda C., Morris W., and Naylor J., et al. (International Human Genome Sequencing Consortium). 2001. Initial sequencing and analysis of the human genome. *Nature* **409:** 860.

Margulies E.H., Blanchette M., Haussler D., and Green E.D. 2003. Identification and characterization of multi-species conserved sequences. *Genome Res.* **13:** 2507.

McMurray A.A., Sulston J.E., and Quail M.A. 1998. Short-insert libraries as a method of problem solving in genome sequencing. *Genome Res.* **8:** 562.

McPherson J.D., Marra M., Hillier L., Waterston R.H., Chinwalla A., Wallis J., Sekhon M., Wylie K., Mardis E.R., Wilson R.K., Fulton R., Kucaba T.A., Wagner-McPherson C., Barbazuk W.B., Gregory S.G., Humphray S.J., French L., Evans R.S., Bethel G., Whittaker A., Holden J.L., McCann O.T., Dunham A., Soderlund C., and Scott C.E., et al. (International Human Genome Mapping Consortium). 2001. A physical map of the human genome. *Nature* **409:** 934.

Montgomery K.T., Lee E., Miller A., Lau S., Shim C., Decker J., Chiu D., Emerling S., Sekhon M., Kim R., Lenz J., Han J., Ioshikhes I., Renault B., Marondel I., Yoon S.J., Song K., Murty V.V., Scherer S., Yonescu R., Kirsch I.R., Ried T., McPherson J., Gibbs R., and Kucherlapati R. 2001. A high-resolution map of human chromosome 12. *Nature* **409:** 945.

Mungall A.J., Palmer S.A., Sims S.K., Edwards C.A., Ashurst K.L., Wilming L., Jones M.C., Horton R., Hunt S.E., Scott C.E., Gilbert J.G., Clamp M.E., Bethel G., Milne S., Ainscough R., Almeida J.P., Ambrose K.D., Andrews T.D., Ashwell R.I., Babbage A.K., Bagguley C.L., Bailey J., Banerjee R., Barker D.J., and Barlow K.F., et al. 2003. The DNA sequence and analysis of human chromosome 6. *Nature* **425:** 805.

National Research Council. 1988. Mapping and sequencing the human genome (Committee on Mapping and Sequencing the Human Genome). *National Academies Press, Washington D.C.*

Novik K.L., Nimmrich I., Genc B., Maier S., Piepenbrock C., Olek A., and Beck S. 2002. Epigenomics: Genome-wide study of methylation phenomena. *Curr. Issues Mol. Biol.* **4:** 111.

Nusbaum C., Slonim D.K., Harris K.L., Birren B.W., Steen R.G., Stein L.D., Miller J., Dietrich W.F., Nahf R., Wang V., Merport O., Castle A.B., Husain Z., Farino G., Gray D., Anderson M.O., Devine R., Horton L.T., Jr., Ye W., Wu X., Kouyoumjian V., Zemsteva I.S., Wu Y., Collymore A.J., and Courtney D.F., et al. 1999. A YAC-based physical map of the mouse genome. *Nat. Genet.* **22:** 388.

Oliver S.G., van der Aart Q.J., Agostoni-Carbone M.L., Aigle M., Alberghina L., Alexandraki D., Antoine G., Anwar R., Ballesta J. P., and Benit P., et al. 1992. The complete DNA sequence of yeast chromosome III. *Nature* **357:** 38.

Olson M., Hood L., Cantor C., and Botstein D. 1989. A common language for physical mapping of the human genome. *Science* **245:** 1434.

Olson M.V., Dutchik J.E., Graham M.Y., Brodeur G.M., Helms C., Frank M., MacCollin M., Scheinman R., and Frank T. 1986. Random-clone strategy for genomic restriction mapping in yeast. *Proc. Natl. Acad. Sci.* **83:** 7826.

Osoegawa K., Mammoser A.G., Wu C., Frengen E., Zeng C., Catanese J.J., and de Jong P.J. 2001. A bacterial artificial chromosome library for sequencing the complete human genome. *Genome Res.* **11:** 483.

Palca J. 1986. Human genome. Department of Energy on the map. *Nature* **321:** 371.

Pennacchio L.A. and Rubin E.M. 2001. Genomic strategies to identify mammalian regulatory sequences. *Nat. Rev. Genet.* **2:** 100.

Pennacchio L.A., Olivier M., Hubacek J.A., Cohen J.C., Cox D.R., Fruchart J.C., Krauss R.M., and Rubin E.M. 2001. An apolipoprotein influencing triglycerides in humans and mice revealed by comparative sequencing. *Science* **294:** 169.

Sachidanandam R., Weissman D., Schmidt S.C., Kakol J.M., Stein L.D., Marth G., Sherry S., Mullikin J.C., Mortimore B.J., Willey D.L., Hunt S.E., Cole C.G., Coggill P.C., Rice C.M., Ning Z., Rogers J., Bentley D.R., Kwok P.Y., Mardis E.R., Yeh R.T., Schultz B., Cook L., Davenport R., Dante M., and Fulton L., et al. 2001. A map of human genome sequence variation containing 1.42 million single nucleotide polymorphisms. *Nature* **409:** 928.

Sanger F., Coulson A.R., Hong G.F., Hill D.F., and Petersen G.B. 1982. Nucleotide sequence of bacteriophage lambda DNA. *J. Mol. Biol.* **162:** 729.

Sanger F., Air G.M., Barrell B.G., Brown N.L., Coulson A.R., Fiddes C.A., Hutchison C.A., Slocombe P.M., and Smith M. 1977. Nucleotide sequence of bacteriophage phi X174 DNA. *Nature* **265:** 687.

Schuler G.D., Boguski M.S., Stewart E.A., Stein L.D., Gyapay G., Rice K., White R.E., Rodriguez-Tome P., Aggarwal A., Bajorek E., Bentolila S., Birren B.B., Butler A., Castle A.B., Chiannilkulchai N., Chu A., Clee C., Cowles S., Day P.J., Dibling T., Drouot N., Dunham I., Duprat S., East C., and Hudson T.J., et al. 1996. A gene map of the human genome. *Science* **274:** 540.

Shizuya H., Birren B., Kim U.J., Mancino V., Slepak T., Tachiiri Y., and Simon M. 1992. Cloning and stable maintenance of 300-kilobase-pair fragments of human DNA in *Escherichia coli* using an F-factor-based vector. *Proc. Natl. Acad. Sci.* **89:** 8794.

Sinsheimer R.L. 1989. The Santa Cruz Workshop, May 1985. *Genomics* **5:** 954.

Skaletsky H., Kuroda-Kawaguchi T., Minx P.J., Cordum H.S., Hillier L., Brown L.G., Repping S., Pyntikova T., Ali J., Bieri T., Chinwalla A., Delehaunty A., Delehaunty K., Du H., Fewell G., Fulton L., Fulton R., Graves T., Hou S.F., Latrielle P., Leonard S., Mardis E., Maupin R., McPherson J., and Miner T., et al. 2003. The male-specific region of the human Y chromosome is a mosaic of discrete sequence classes. *Nature* **423:** 825.

Solinas-Toldo S., Lampel S., Stilgenbauer S., Nickolenko J., Benner A., Dohner H., Cremer T., and Lichter P. 1997. Matrix-based comparative genomic hybridization: Biochips to screen for genomic imbalances. *Genes Chromosomes Cancer* **20:** 399.

Thomas J.W., Touchman J.W., Blakesley R.W., Bouffard G.G., Beckstrom-Sternberg S.M., Margulies E.H., Blanchette M., Siepel A.C., Thomas P.J., McDowell J.C., Maskeri B., Hansen N.F., Schwartz M.S., Weber R.J., Kent W.J., Karolchik D., Bruen T.C., Bevan R., Cutler D.J., Schwartz S., Elnitski L., Idol J.R., Prasad A.B., Lee-Lin S.Q., and Maduro V.V., et al. 2003. Comparative analyses of multi-species sequences from targeted genomic regions. *Nature* **424:** 788.

Tilford C.A., Kuroda-Kawaguchi T., Skaletsky H., Rozen S., Brown L.G., Rosenberg M., McPherson J.D., Wylie K., Sekhon M., Kucaba T.A., Waterston R.H., and Page D.C. 2001. A physical map of the human Y chromosome. *Nature* **409:** 943.

Venter J.C., Adams M.D., Myers E.W., Li P.W., Mural R.J., Sutton G.G., Smith H.O., Yandell M., Evans C.A., Holt R.A., Gocayne J.D., Amanatides P., Ballew R.M., Huson D.H., Wortman J.R., Zhang Q., Kodira C.D., Zheng X.H., Chen L., Skupski M., Subramanian G., Thomas P.D., Zhang J., Gabor Miklos G.L., and Nelson C., et al. 2001. The sequence of the human genome. *Science* **291:** 1304.

Waterston R.H., Lindblad-Toh K., Birney E., Rogers J., Abril J.F., Agarwal P., Agarwala R., Ainscough R., Alexandersson M., An P., Antonarakis S.E., Attwood J., Baertsch R., Bailey J., Barlow K., Beck S., Berry E., Birren B., Bloom T., Bork P., Botcherby M., Bray N., Brent M.R., Brown D.G., and Brown S.D., et al. (Mouse Genome Sequencing Consortium). 2002. Initial sequencing and comparative analysis of the mouse genome. *Nature* **420:** 520.

Weissenbach J., Gyapay G., Dib C., Vignal A., Morissette J., Millasseau P., Vaysseix G., and Lathrop M. 1992. A second-generation linkage map of the human genome. *Nature* **359:** 794.

Wheeler D.L., Church D.M., Lash A.E., Leipe D.D., Madden T.L., Pontius J.U., Schuler G.D., Schriml L.M., Tatusova T.A., Wagner L., and Rapp B.A. 2002. Database resources of the National Center for Biotechnology Information: 2002 update. *Nucleic Acids Res.* **30:** 13.

Wilson R., Ainscough R., Anderson K., Baynes C., Berks M., Bonfield J., Burton J., Connell M., Copsey T., and Cooper J., et al. 1994. 2.2 Mb of contiguous nucleotide sequence from chromosome III of *C. elegans*. *Nature* **368:** 32.

Woodfine K., Fiegler H., Beare D.M., Collins J.E., McCann O.T., Young B.D., Debernardi S., Mott R., Dunham I., and Carter N.P. 2004. Replication timing of the human genome. *Hum. Mol. Genet.* **13:** 191.

Yeast Genome Directory. 1997. The yeast genome directory. *Nature* (suppl.) **387:** 5.

# The Human Genome: Genes, Pseudogenes, and Variation on Chromosome 7

R.H. WATERSTON,* L.W. HILLIER,† L.A. FULTON,† R.S. FULTON,† T.A. GRAVES,† K.H. PEPIN,† P. BORK,‡ M. SUYAMA,‡ D. TORRENTS,‡ A.T. CHINWALLA,† E.R. MARDIS,† J.D. MCPHERSON,†¶ AND R.K. WILSON†

*Department of Genome Sciences, University of Washington School of Medicine, Seattle, Washington 98195; †Genome Sequencing Center, Department of Genetics, Washington University, St. Louis, Missouri 63108; ‡EMBL, Heidelberg 69117, Germany

When the idea of sequencing the entire human genome was initially put forward (Dulbecco 1986; DeLisi 1988; Sinsheimer 1989), DNA sequencing was an expensive, laborious process. The largest genome to have been sequenced at that time was that of the Epstein-Barr virus, which at 172,282 bases, or about 1/20,000 the size of the human genome, had taken a group of about a dozen people several years to complete (Baer et al. 1984). The only systematic analyses of larger, more complex genomes had sought to produce ordered clone sets (clone-based physical maps) of the *Caenorhabditis elegans* and *Saccharomyces cerevisiae* genomes (Coulson et al. 1986; Olson et al. 1986). Today, less than 20 years later and just 2 1/2 years after publications describing draft sequences (Lander et al. 2001; Venter et al. 2001), we have in hand the essentially complete sequence of the human genome (Rogers, this volume). Groups from across the globe contributed to the sequence (Table 1) in a coordinated effort called the International Human Genome Project (IHGP).

The groups exploited advances contributed by many laboratories for every step of the sequence process (Table 2), but none was more important than the steady advance in fluorescence-based DNA sequencing instruments (Smith et al. 1985; Ansorge et al. 1987; Prober et al. 1987; Brumbaugh et al. 1988). Initially these machines required sophisticated users and large amounts of carefully quantified, high-quality template DNA to produce sequence of limited accuracy over 300–400 bases, and capacity was limited—for example, 16 samples daily on the ABI373 machine. With steady improvements not only in the instruments, but also in each step of the process before and after sequence collection, a typical ABI3730 in a large genome center today produces 1300 samples daily with highly accurate reads of 750–900 bases with minimal attendance. Automation with sophisticated LIMS systems and sample tracking allow these machines to be fed continuously 24 hours a day, 7 days a week, 52 weeks a year, with little down time.

The clone-based hierarchical shotgun strategy employed by the IHGP has resulted in a sequence that contains more than 99% of the euchromatic sequence in highly accurate form. The details of the sequence are presented elsewhere, but it is clear that the sequence includes some large, complicated, repeated sequences that would

**Table 1.** Contributions to the Finished Sequence by Center

| Center | Bases |
|---|---|
| Wellcome Trust Sanger Institute | 824,879,622 |
| Washington University Genome Sequencing Center | 584,507,988 |
| U.S. DOE Joint Genome Institute | 310,982,691 |
| Whitehead Institute Center for Genome Research | 374,404,412 |
| Baylor College of Medicine Human Genome Sequencing Center | 278,229,771 |
| RIKEN Genomic Sciences Center | 110,409,096 |
| University of Washington Genome Center | 108,221,396 |
| Genoscope and CNRS UMR-8030 | 77,479,268 |
| Keio University | 21,285,289 |
| GTC Sequencing Center | 36,061,549 |
| The Institute for Systems Biology | 26,871,969 |
| Max Planck Institute for Molecular Genetics | 20,095,113 |
| Beijing Genomics Institute/Human Genome Center | 17,141,643 |
| University of Oklahoma's Advanced Center for Genome Technology | 5,444,166 |
| Stanford Human Genome Center | 8,374,342 |
| Max Planck Institute for Molecular Genetics | 5,024,546 |
| GBF German Research Center for Biotechnology | 5,547,223 |
| Others | 17,049,046 |

**Table 2.** Advances in Sequencing

| Advance | Reference |
|---|---|
| Libraries | Shizuya et al. (1992) |
| DNA template prep | Hawkins et al. (1994) |
| Cycle sequencing | Axelrod and Majors (1989); Craxton (1993) |
| Fluorescent dyes | Ju et al. (1995) |
| Taq polymerase variants | Tabor and Richardson (1995) |
| Capillary sequencing | Lu et al. (1994) |
| Base calling | Ewing et al. (1998) |
| Assembly | P. Green and L. Hillier (unpubl.) |
| Databases | R. Durbin and J. Thierry-Mieg (unpubl.); FlyBase Consortium (1994) |

The principal steps in the DNA sequence process are listed along with examples of early contributions to improvements to the process. The continual improvement of each step has been critical to the success of the Human Genome Project (Axelrod and Majors 1989; Shizuya et al. 1992; Craxton 1993; Hawkins et al. 1994; Lu et al. 1994; Ju et al. 1995; Tabor and Richardson 1995; Ewing et al. 1998).

¶Present address: Department of Molecular and Human Genetics and Human Genome Sequencing Center, Baylor College of Medicine, One Baylor Plaza, N1519, Houston, Texas 77030.

have been difficult to obtain with the alternative whole-genome shotgun strategy. The single-base miscall rate is estimated at about 1/100,000 bases, and small deletion/insertion differences arising from errors in propagation and assembly occur on the order of 1 per 5 Mb (Schmutz et al., this volume), both rates much lower than observed polymorphism rates for these types of differences. Indeed, many of the small insertion/deletion artifacts occur in short, tandemly repeated sequences, which are often themselves polymorphic in the population.

Some regions of the genome are not accounted for in the current sequence. The short arms of the acrocentric chromosomes are not represented at all, and the large heterochromatic regions of Chromosomes 1, 9, and 16 and the centromeric α-satellite sequences are only sparsely represented. There are also infrequent gaps in the euchromatic regions, whose size generally is estimated from FISH or comparison with orthologous mouse/rat genome sequence. These gaps in sequence coverage reflect the absence of clones in available libraries and are particularly prevalent near the centromeres and telomeres and in heterochromatic regions (Dunham et al. 1999; Deloukas et al. 2001; Heilig et al. 2003; Hillier et al. 2003; Mungall et al. 2003; Skaletsky et al. 2003). These gaps are often flanked by locally highly repetitive sequence, by segmentally duplicated sequence, or by regions with exceptionally high GC content. Efforts to close these few remaining gaps and to correct detected errors continue at many of the centers.

With a highly accurate, essentially complete sequence of the genome now in hand, efforts are shifting to the understanding of its contents. A full understanding will take decades, but two immediate tasks face us now: defining those sequences that function in the specification of humans—the "parts list'"—and defining human variation. We comment in this paper on aspects of these tasks that we faced in annotating the finished Chromosome 7 sequence (Hillier et al. 2003). One aspect deals with the impact of human variation on the assembly of the human sequence and the extremes of variation observed in clone overlaps. These more highly variant regions complicated assembly of the genome, since such overlaps actually might have represented distinct regions of the genome that had resulted from segmental duplication, rather than the same region of the genome. With appropriate tests, they have been clearly shown to derive from a single region of the genome and thus represent curiously highly variant regions of the genome. Another aspect deals with efforts to improve the set of protein-coding gene predictions, the first and most critical element of the parts list. Gene-finding in mammalian genomes requires a combined approach, employing gene prediction programs, experimental evidence, and comparative sequence analysis. The best sources of experimental evidence currently are the RefSeq (Pruitt and Maglott 2001) and MGC (Strausberg et al. 1999) full-length cDNA sequence collections. However, in carefully aligning these sequences against the genome, we have noted discrepancies between the cDNA and genomic sequences that alter the reading frame. In investigating the source of these differences, we have discovered instances of sequence variants in the population that must alter the protein product of the gene. We have also used the mouse sequence for gene prediction in the human genome as a means of greatly reducing the problem of pseudogene contamination and other false-positive predictions in the predicted set.

## ASSEMBLY AND VARIATION

Assembling the human genome sequence from individual BAC clones presented challenges not faced in assembling simpler genomes like those of yeast (Johnston et al. 1997; Mewes et al. 1997) and *C. elegans* (Consortium 1998). The extensive amount of repeated sequence, both from interspersed repeats (44% of the genome) and from segmental duplications (another 5% of the genome) (Lander et al. 2001), means that different BAC clones may contain similar sequence, but derive from entirely different regions of the genome. On the other hand, clones may derive from the same region of the genome, but because they derive from different genomes, they may differ in sequence. About 1 in 1300 bases is expected to differ between any two copies of the genome (Sachidanandam et al. 2001), and the sequenced clones came from multiple, different diploid individuals (with 70% coming from a single individual). Even within clones from a single library, only half the overlaps are expected to derive from the same chromosomal copy or haplotype. Thus, in judging whether two clones overlap on the basis of sequence, there is inevitably a balance to be found between accepting clones that truly should overlap and rejecting clones that derive from two similar but distinct regions of the genome.

Fortunately, the physical map largely mitigates the problem (McPherson et al. 2001). Here, the large size of the BAC clones (average insert length = 150–175 kb) and their extensive overlaps (>90% on average) allow all but the largest, most recent duplications to be sorted out. Even nearly identical repeats of greater than the BAC insert length may be recognized by the overabundance of clone coverage in a region and targeted for special attention. As a further check, each overlap can be examined at the sequence level. For Chromosomes 2, 4, and 7, we required that the overlap extend at least 2 kb through the ends of the clones with at least 99.8% sequence identity, with similar criteria applied for other chromosomes. Overlaps not meeting these criteria were subjected to additional scrutiny.

In extreme cases, such as the Y chromosome (Tilford et al. 2001; Skaletsky et al. 2003) and the Williams-Beuren Syndrome region (Hillier et al. 2003), the region could not be sorted out by the physical map alone, and its resolution required sequence information from clones from the same haplotype with extensive overlap. Using these procedures on Chromosome 7 allowed us to detect 8.2% of the sequence as segmentally duplicated, 7.0% within the chromosome and 2.2% between 7 and other chromosomes (Hillier et al. 2003).

In the course of the clone assembly, we encountered examples where the sequence variation between two clones was unusually high, and yet the physical map supported the join. To determine whether these clones derived from the same region or instead arose from unrecognized large, highly similar duplications, we used PCR to recover a segment of the variant region from both the parent clones and the genome from 24 individuals, a subset of the DNA Polymorphism Discovery Resource (Collins et al. 1998). We reasoned that if the sequences derived from the same region, the two variants would be allelic and would segregate in the population, with heterozygotes and homozygotes of each allele in proportion to its frequency in the population. If, however, the sequences derived from distinct copies of a repeated sequence, both sequences would be present in all individuals. In rare instances, the same site might be variant within both copies; even here we would find unusual ratios of the two sequences in the population samples.

In toto for Chromosome 7 we analyzed 39 overlaps with a difference rate of at least 3 events per kilobase. These overlaps totaled 1911 kb, with an average variation of 4.5 differences per kilobase. In all cases, the variant bases were found to be polymorphic in the population sample, with homozygotes evident for at least one form. The proportion of heterozygotes to homozygotes was generally in accord with Hardy-Weinberg equilibrium. Furthermore, when several variants were present in the region assayed, they often behaved as a haplotype block; that is, the presence of a particular base in one variant position was predictive of the other variant bases assayed. Thus, these overlaps all appeared to sample highly variant regions of the genome.

To investigate more broadly the frequency and extent of such highly variant regions, we examined the pattern of variation within 2,718 overlaps between clones on Chromosomes 2, 4, and 7. For each overlap of at least 5 kb, we looked at the variation within 5-kb nonoverlapping windows. On average, just fewer than 4 differences in each window would be expected if the clones derived from different haplotypes. Although many segments conformed to expectation, we saw 302 intervals with more than 18 differences (2 standard deviations from the mean). Often we found multiple successive 5-kb segments with high variation (Fig. 1), as might be expected from a series of segments exhibiting linkage disequilibrium (Daly et al. 2001). To determine whether the set of variations arise from two distinct copies of a single haplotype block will require more experimental data across the region.

These investigations support two conclusions. First, the finding that none of the tested regions of high variability represented segmental duplications, but rather highly variable segments of the genome, suggests that

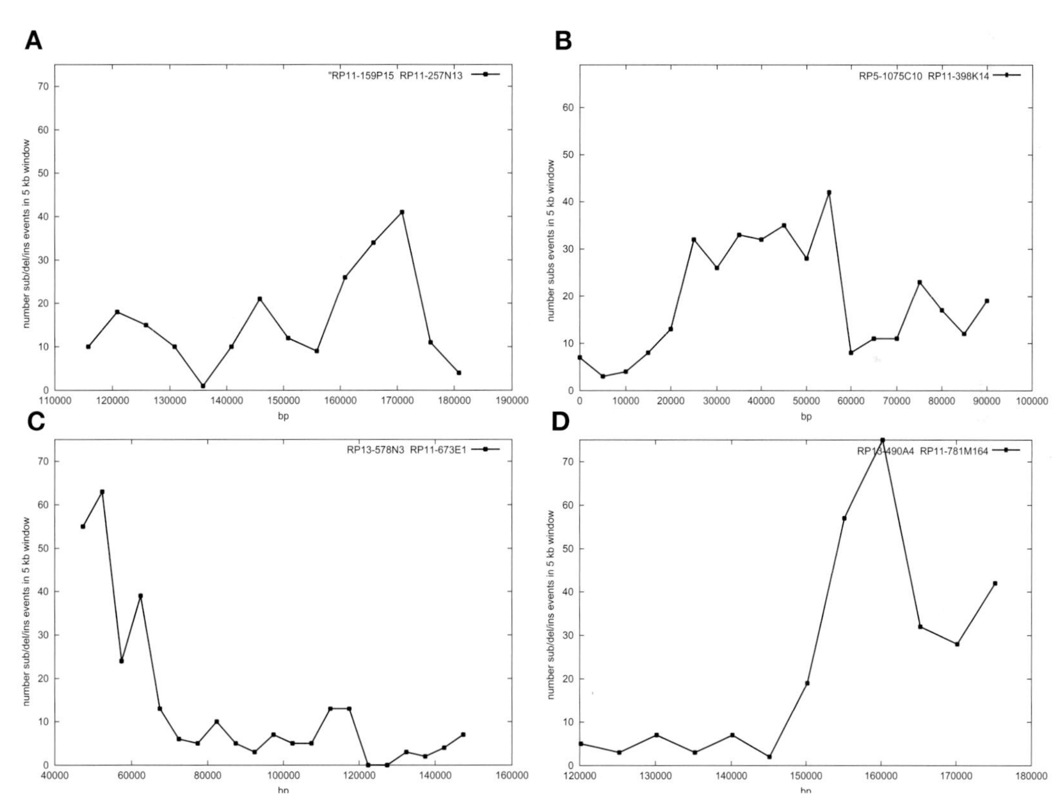


**Figure 1.** Regions of high variation in the overlap between adjacent clones. Four different overlaps are depicted. Each point represents the number of substitution events in 5-kb nonoverlapping windows, including insertion/deletion events for 3 of them. The full extent of the overlap is shown. The expected number of events per 5 kb between two clones of different origin would be about 4, and the mean number of events plus 2 standard deviations for a 5-kb window is 18 for this set (15 for substitutions alone). For much of the overlap the variation is within the normal limits, but in each case several successive 5-kb segments are present, with one case extending over about 40 kb. The region of high variation in *A* lies within a large intron of the LRP1B gene (low-density lipoprotein receptor-related protein 1B) and has several regions conserved in mouse and rat. The region of high variation in *B* shows no genes but does contain several regions conserved with mouse or rat. The region in *C* corresponds to the 5′ half of the GYPE gene (a surface glycophorin on red cells related to glycophorin A and B genes that specify the MN and Ss blood group antigens). The region in *D* contains no feature and exhibits no regions conserved with mouse and rat.

there are unlikely to be many large, recently duplicated segments of the genome yet to be identified in regions where the above criteria have been used to judge overlaps. Second, there are multiple regions of the genome where there is locally high variation, sometimes extending over tens of kilobases. Whether these are present because of balanced selection or because they represent by chance the tail end of the distribution from the last coalescent is uncertain (Charlesworth et al. 1997).

We have examined the regions for features that might suggest a basis for such balancing selection. Fifteen were in the region of some gene, including the known genes CRYPTIC, TSSC1, LRP1B, and GYPE. However, others contain no known features. Regardless of the basis, if variation has been accumulating at neutral rates, these regions have been maintained independently in the population since well before the founding of the species and, in some cases, almost equal to the time of the chimp–human divergence.

## GENES

Finding genes in the draft human genome sequence was challenging, and the results were often inconsistent (Hogenesch et al. 2001). Even with finished, highly accurate genomic sequence, the extensive amount of noncoding sequence and the abundant pseudogenes present a significant challenge to gene prediction. The concerted efforts to obtain full-length sequences for cDNAs have been invaluable in annotation (Strausberg et al. 2002), providing direct evidence for an increasing number of genes. However, as yet, the collections for human remain incomplete and contain errors. Furthermore, the assignment of the sequence to its source in the genome may be complicated by the presence of close paralogs, recently formed pseudogenes, and polymorphisms. Another potentially powerful complementary approach comes from comparative sequence analysis, where conserved features such as genes stand out against neutrally evolving sequence. The recent release of a draft mouse sequence (Waterston et al. 2002) provides an opportunity to apply this approach genome-wide to human. As the first of many other mammalian sequences that will become available in the next few years, it provides a taste of the power to come from comparative analysis.

In annotating the finished sequence of Chromosome 7, we combined cDNAs, gene models from gene prediction programs, and comparison with the mouse sequence to derive a gene set that accurately reflects available experimental evidence and offers improved discrimination between genes and pseudogenes. In the course of these studies, we uncovered genes with apparently active and inactive forms present in the population. Such variants might be usefully exploited to learn more about the role of the gene in human biology.

## KNOWN GENES: cDNAs

We began our efforts to define the genes on Chromosome 7 with the extensive RefSeq (Pruitt and Maglott 2001) and the MGC (Strausberg et al. 2002) cDNA collections. We aligned all 14,769 RefSeq and 10,047 MGC sequences against the entire genome sequence, allowing each cDNA to confirm only a single genomic locus. Spliced forms were favored over single or minimally spliced forms to avoid processed pseudogenes, and only the single match with the highest percent identity was kept to avoid confusion with recent paralogs.

In aligning these cDNAs against the genome, we noticed certain discrepancies that complicated interpretation. Alignment of the cDNAs to the genome, although often straightforward, was erroneous in many cases using any of the several programs. In our experience, Spidey (Wheelan et al. 2001) gave the most reliable results, aligning about 65% of the cDNAs and 79% of their exons accurately against the genomic sequence. But for any given gene, BLAT (Kent 2002) or EST_GENOME (Mott 1997) may give a better alignment (where the quality of the alignment is judged by minimizing base substitutions between the cDNA and genomic sequence and the avoidance of noncanonical splice sites). Every case was manually reviewed, and in some cases, manual review uncovered better alignments than did any of the available programs.

Even after exhaustive efforts to obtain an optimal alignment, some problems remained. For example, 23 mRNAs had no similarity to any mouse gene, and the translation product had no similarity to any known protein. Although these could be true genes, it seems more likely that they represent untranslated segments of bona fide genes. Nonetheless, these were kept in the current gene set. Eight others contained only a single, very short open reading frame (<20 amino acids), where again the translation product had no similarity to any known protein. Inspection of the genomic region generally showed that these instances were immediately downstream of another gene. We suspect these represent incomplete cDNAs from alternative 3′ exons, and we excluded these genes from the Chromosome 7 gene list.

Other genes contained differences between the cDNA and the aligned genomic sequence as expected, since the cDNA and genomic sequences derived from different individuals. Whereas most of these left the gene intact, at most changing a codon or two, some disrupted the reading frame of either the cDNA or the genomic sequence, either by changing the frame of the translation or by directly introducing a stop codon. For Chromosome 7, this amounted to some 60 genes (10% of the total genomic sites to which cDNAs aligned). To investigate which of the two sequences (the cDNA and the genomic) was correct, or if indeed both versions of the sequence are commonly represented in the population, we attempted to resequence the region in question from the original clone and related clones and also in a panel of 24 diverse individual DNAs from the DNA Polymorphism Discovery Resource (Collins et al. 1998). We also compared the sequence to the orthologous region from the mouse where available.

Because of repeated sequence and other technical issues, a few regions failed to give results. We found 35 of the cDNA/genomic pairs had a likely error in the cDNA sequence; that is, sequence from all tested individual

DNAs and the clones agreed with the original genomic sequence. Of these, 16 had multiple base insertion/deletion differences such that the reading frame between the two sequences was eventually restored. In such cases, comparison with the mouse sequence revealed that in each instance the reading frame of the genomic sequence yielded the amino acid sequence better conserved in the mouse translation. The other 19 had simple insertion/deletion or missense differences that shifted or truncated the reading frame in the cDNA relative to the genomic sequence. Again, translation of the orthologous mouse sequence supported the genomic reading frame.

We found eight instances where the genomic sequence was likely in error. For five of these, reexamination and/or resequencing revealed an error in the original BAC sequence, either in sequencing or during propagation of the clone. For an additional three, resequencing of the BAC agreed with the original genomic sequence, but all of the individual DNAs agreed with the cDNA sequence. No other clones from the same library were available in these cases to determine whether the difference might be an individual polymorphism.

Surprisingly, in three instances, we found the panel of 24 to contain both versions of the sequence; that is, the human population is polymorphic at the site. These cases involved different kinds of changes. In one, a single base deletion in the genomic sequence produced a frameshift in exon 31 of more than 40 exons in the gene encoding zonadhesin, a protein found on the surface of the sperm head (Fig. 2) (Hardy and Garbers 1995). In a second case, a premature stop codon (codon 60) was found in the genomic sequence where there is a glutamine in the cDNA for transmembrane protein induced by tumor necrosis factor (TMPIT, NM_031925) (Table 3). The consequence, if any, of these disrupted or altered translations on human biology will require further investigation.

The accurate alignment of 1,073 RefSeq and MGC cDNA sequences onto Chromosome 7 has provided a clear evidence for 605 genes. The alignment of the cDNAs against the finished genomic sequence also provides a valuable means of improving the cDNA resources. A few seem unlikely to represent independent or functional genes. Probable frameshift errors or other chain-terminating changes in genes have been recognized and investigated. The compensated frameshift errors found here have led to incorrect translation products and masked otherwise conserved regions of the protein. Finally, the careful comparison of cDNA and genomic sequence has led to the discovery of some variants that disrupt the reading frame. These genes may be in transition to becoming pseudogenes, or perhaps both forms of the protein may be of selective benefit in certain circumstances, leading to their longer-term persistence.

## PREDICTED GENES

With a solid set of known genes established for Chromosome 7, we set out to find additional likely protein-coding genes. We wanted the set to be comprehensive, yet as free as possible of false predictions, particularly pseudogenes, which can be significant contaminants of predicted gene sets. For example, in initial analysis of the mouse genome, almost 20% of the identified genes were estimated to be pseudogenes. In one dramatic case, the mouse genome contains just one functional copy of the GAPDH gene and more than 400 related sequences. Of these, 118 were found contaminating the predicted gene set.

Pseudogenes are of two principal types, arising by distinct mechanisms. Processed pseudogenes result from the reverse transcription of mRNA and the subsequent insertion of the copy into the genome. Unprocessed pseudogenes arise from the duplication of segments of the genome that include functional genes or from the degradation of genes that are no longer subject to selection. Pseudogenes of either type may be complete or partial and, over time, will accumulate mutations that make it obvious they are inactive. Deletions, insertions, and other rearrangements can also obscure their origin.

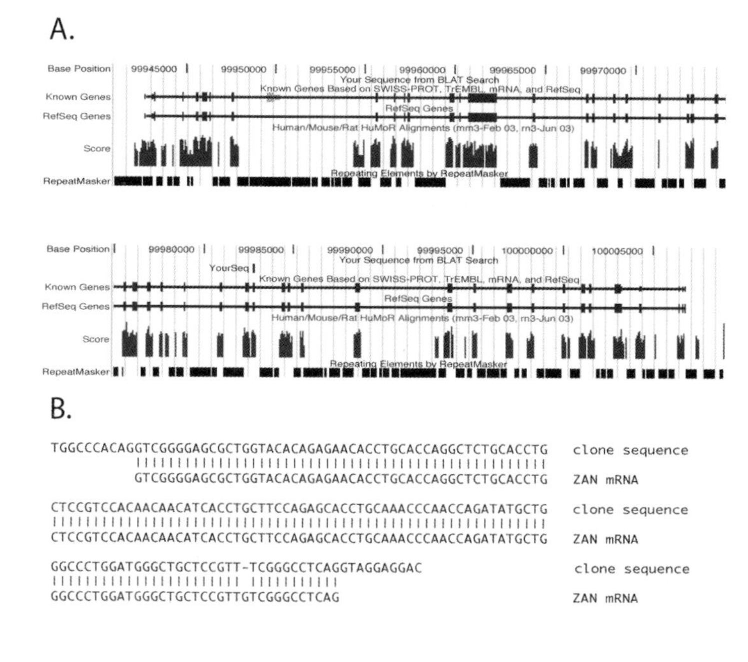


**Figure 2.** The gene encoding zonadhesin contains a variable base that alters the reading frame. (*A*) The gene structure as displayed on the UCSC browser (labeled "Known Genes" and "RefSeq Genes"), along with tracks showing the location along Chromosome 7, the position of the genomic region containing the frameshifting variation (labeled "Your sequence"), the regions conserved with mouse and rat, and the repeat content of the sequence. (*B*) Alignment of the genomic sequence and exon 31 of the zonadhesin cDNA, demonstrating the single-base deletion in the genomic sequence relative to the cDNA.

**Table 3.** Genes with Disruptive Variants

| Gene ID<br>RefSeq ID | Gene name | cDNA<br>sequence | Genomic<br>sequence | Codon | BACS<br>gs/cdna | DNA PDR<br>gs/het/cDNA |
|---|---|---|---|---|---|---|
| gi13994299<br>NM_031925 | TMPIT | cttCAGaac<br>L Q N | CttTAGacc<br>L * N | 42 | 3:3 | 0:0:24 |
| gi15706488<br>BC012777 | hypothetical | gaatgCGAgcc<br>M R A | gaatgTGAgcc<br>M * A | 2 | nd | 2:11:9 |
| gi16554448<br>NM_003386 | zonadhesin | ccgtTGTcgg<br>R C R | CcgtT*Ttcgg<br>R F G | 1,922 | nd | 9:7:7 |

To avoid pseudogenes and other false positives in the Chromosome 7 predicted gene set, we developed a protocol that exploited newly available draft genome sequence of the mouse. We reasoned that recently derived pseudogenes are problematic because they most closely resemble functional genes, whereas older pseudogenes have degraded through neutral drift and are less likely to be confused with genes. With 75 million years separating mouse and human from their last common ancestor, only pseudogenes that arose after the split would present problems. Since processed pseudogenes in general do not insert near their source gene, any processed pseudogene that arose in one organism thus would not have a counterpart in the corresponding portion of the other genome. (For convenience, we call regions in the mouse and human genomes that descended from a common ancestor orthologous, similar to the usage that has been adopted for genes.) Thus, for regions where the orthologous relationships have been established between the two genomes (and this covers 90% of the human genome), genes are likely to have counterparts in both organisms, whereas processed pseudogenes will not. Furthermore, having arisen since the divergence of the two species, the closest relative of the pseudogene will lie within the same genome, rather than in the second genome.

The situation is less clear with pseudogenes arising from duplication. Duplications may be at a distance, in which case orthology may be a useful discriminator, but often they are near and even tandemly arranged with respect to the source gene. In some cases, both copies may be functional members of a gene family, with one or both copies adopting new functions. Nonetheless, those copies that become pseudogenes may be recognized, because the copy may be incomplete or may have drifted significantly from the original, often with frameshift mutations or other chain-termination mutations disrupting the reading frame. The functional gene will remain most similar to the orthologous gene in the second organism.

With this screening procedure in mind, we set about to obtain a comprehensive set of putative genes, reasoning that false positives could be recognized and removed by careful examination of the orthologous regions between mouse and human. We used three different gene prediction programs, FGENESH2 (Solovyev 2001), TwinScan (Korf et al. 2001), and GeneWise (Birney and Durbin 2000), to recognize as many elements as possible. The first two use comparative sequence information (without using orthology) to modify ab initio predictions and improve accuracy, whereas GeneWise uses protein homologies to seed predictions. We used the mouse sequence as the informant sequence for TwinScan and FGENESH2 and all available protein predictions for GeneWise. TwinScan produced 1,350 gene models, FGENESH2 3,793 models, and GeneWise 22,326 models. (GeneWise predicts multiple, alternatively spliced models in any given region.) The combined output included 90% of all known exons and at least one exon from 98% of the known genes.

Although the combined output from the three programs is thus reasonably sensitive, there are undoubtedly a large number of false-positive predictions, including many pseudogenes. To eliminate most of these unwanted predictions, we used each prediction to search the orthologous region of the mouse genome for a matching gene. In turn, the matching mouse genes were used to search for matches in the orthologous region of the human genome. The matches in the orthologous regions had to be the best or close to the best in the entire genome. Furthermore, single-exon genes were removed if they had matches to multi-exon genes in either genome. Generally, at any one site only the best reciprocally matching pair was kept (near-best for local duplications). This left us with 728 FGENESH2, 284 TwinScan, and 400 GeneWise predictions.

To eliminate redundancy between the various gene models, for each region we accepted only one prediction, taking in order known genes, FGENESH2, TwinScan, and GeneWise (the order was based on the performance of the three programs against known genes) and giving priority to models with the best reciprocal matches. Models that showed signs of nonfunctionality (early truncation compared to closely related genes, absence of introns in gene models with closely related genes with exons) and high homology to L1/reverse transcriptase were also removed. This yielded an additional 545 predicted genes.

We examined this gene set in several ways to determine how complete and accurate it might be. On average, the predicted genes, when compared to the known genes, have fewer exons (6.7 vs. 9.5), have fewer coding bases (1,231 vs. 1,457), and span less of the genome (28.5 kb vs. 61.4 kb), suggesting that the predicted genes either lack terminal exons or include some fragmented genes that reduce the average. A high fraction of both known genes (92%) and predicted genes (95%) have reciprocal best matches in the mouse genome at the orthologous position, whereas the others have near-best matches. This is expected, given the methods used to build the set, but nonetheless their conservation strongly supports their validity.

Since we had not used ESTs in building the gene set, we used them as an independent measure of the representation of the set. We found 41,399 spliced ESTs with their best match to Chromosome 7. Of these, 93% at least partially overlapped an exon of the gene set, and an additional 1% lay near or within existing genes, suggesting that they might represent alternative splice forms or exons missing from the predicted genes. The remainder lacked significant open reading frames, and none satisfied the reciprocal match criteria used in making the gene predictions. Only 5% of the remainder had any match to mouse sequence. These unmatched spliced ESTs may nonetheless represent missed genes, although at present there is little corroborating evidence that they derive from protein-coding genes.

## PSEUDOGENES

We attempted to identify pseudogenes directly, adapting a method used previously (Waterston et al. 2002; Zdobnov et al. 2002). We inspected all the intervals between known and predicted genes for sequence that yielded translation products with similarity to known proteins. Altogether we identified 941 such regions. More than one pseudogene may lie in any one interval, and old, largely degraded pseudogenes would be missed using the thresholds used here. As a result, we probably have undercounted pseudogenes.

We then evaluated the validity of our classifications to determine how often likely pseudogenes were included in the gene set and how many excluded genes were later found in the pseudogene set. We reasoned that genes should largely be under purifying selection, and pseudogenes should be subject to neutral drift. These differences in evolutionary pressures would produce differences in the ratio of synonymous vs. nonsynonymous substitutions ($K_a/K_s$ ratio) in the coding portion of the genes or pseudogenes (Ohta and Ina 1995). Positive selection acting on genes will increase the $K_a/K_s$ ratio, but generally the positive selection is limited to specific domains. Only rarely will positive selection act so broadly across a gene as to elevate the $K_a/K_s$ ratio to near or above that of neutrally evolving sequence.

Of the 941 regions identified as containing likely pseudogenes, nearly all (97% ± 3%) had $K_a/K_s$ ratios consistent with neutrally evolving sequence, supporting our classification. As with the mouse genome analysis, a significant fraction of the predicted pseudogenes (33%) had as yet no disruption to the reading frame. Virtually all the predicted pseudogenes could be aligned to another region of the human genome with higher sequence identity than to any region of the mouse genome, consistent with an origin after the mouse–human divergence.

We also attempted to classify the pseudogenes by origin, by using the orthologous mouse region for related sequence. For 88% (573/654) of the identified pseudogenes, no related sequence in the orthologous mouse region was found; these are likely to represent processed pseudogenes and are broadly distributed through the chromosome. Another 12% (81/654) did have related sequence in the orthologous mouse region, suggesting they were derived by segmental duplication. Indeed, these lie predominantly in segmentally duplicated regions of the chromosome.

We carried out the same analysis on the 1,152 members of the gene set. Only 5% ± 3% had a ratio consistent with neutral selection, suggesting the set is relatively free of pseudogenes. The total of 1,152 genes is a relatively modest number. Extrapolating to the genome, this would suggest that the human genome contains some 25,000 genes. The total number of genes is only slightly more than the number of pseudogenes found and is about 40% less than the number of genes predicted in another analysis of Chromosome 7 sequence (Scherer et al. 2003). Our approach has been deliberately conservative, but several points of our analysis suggest that our count is fairly accurate. By $K_a/K_s$ analysis, only a few (0–60) of the pseudogenes are likely to be functional. The gene set covers the vast majority of the ESTs that have their best match to Chromosome 7, and those few that fall outside the gene set do not seem likely to be protein-coding. Perhaps much of the difference between our estimate and that of others lies in our treatment of pseudogenes.

## CONCLUSION

Our initial analysis of the content of human Chromosome 7 illustrates some of the challenges that lie ahead, even with an accurate, complete sequence. It also suggests some avenues available for understanding the genome.

An immediate goal must be defining the parts list, that is, all the functional elements of the genome. At present, obtaining even the protein-coding gene set remains a difficult and complex task. The available experimentally determined cDNA sequences remain incomplete, and both these and, to a lesser extent, the genome sequence contain errors. Alignment of the cDNA sequences to the genome is not always straightforward and is complicated by polymorphism. Gene prediction programs give only partial answers and are often confounded by the abundant pseudogenes in the human genome. Comparative sequence analysis, using the mouse as the informative sequence, improves the accuracy of exon prediction in new programs such as TwinScan and FGENESH2. Further processing of the results exploiting the conserved synteny between mouse and human to establish orthologous relationships helps substantially in distinguishing genes from pseudogenes. As the sequences of additional mammalian genomes become available over the next few years, the description of the gene set will become increasingly accurate and complete. These additional sequences will also facilitate the identification of other functional sequences, such as those for noncoding RNAs and regulation of gene expression. This pathway, combined with ongoing experimental testing and validation, holds the promise of a relatively complete parts list in just a few years' time.

Understanding how those parts function and how they contribute to human disease when they fail to function normally will take much longer. Studies of homologous elements in experimentally tractable animals will be critical in this analysis, along with studies of human genes in tissue culture and in vitro. Ultimately, we will need to understand these elements in the context of the human. Natural variation in the human population provides a powerful tool for achieving this understanding. Thus, a key challenge in the coming years will be to define the variation in the human population, focusing first on common variations and then, as methods improve, to extend this to less common variations. In turn the impact of these variations, if any, on the human phenotype and in particular on health and disease must be established, leading to an understanding of the role of each element in the whole.

The complete sequence of the human genome is thus only a beginning. Our efforts on Chromosome 7 were, of course, focused on obtaining this reference sequence, but variation was encountered in the course of the analysis, and our results suggest two avenues of exploration.

In the comparison of the cDNAs to genomic sequence, we encountered examples of variations with predictable effects on gene function; that is, alterations to the reading frame that are expected to result in loss of function. This strong prediction is testable, and examination of the impact of such loss-of-function variations on the human phenotype would undoubtedly be highly informative about the normal role of the gene, just as loss-of-function mutations have been important in the genetic dissection of function in experimental animals.

The number of genes with such variations was small, but came at almost no extra cost and with the comparison of only two different sequences. As additional human genome sequences are described, other examples will come to light. Furthermore, for many genes, missense substitutions can have an equally predictable effect on protein function, expanding the range of changes that can be used. Comparative sequence analysis will reveal those residues under purifying selection, adding still more candidates. As resequencing of human genomes becomes routine, "acquisition by genotype," as this approach might be called, will become more widespread, complementing the more traditional approach of "acquisition by phenotype."

The regions with unusually high density of variation may point to functionally important regions of the genome. Differences in SNP density have been noted previously, with low density noted in some regions (Miller and Kwok 2001; Miller et al. 2001) and high density in HLA region (Mungall et al. 2003) and around the gene underlying the ABO blood group antigen (Yip 2002). The SNP "deserts" may reflect the relatively recent fixation of a single variant in the population, either through selective mechanisms or through founder effects and chance. The regions of high SNP density around HLA and the ABO gene are thought to represent instances of balanced selection, where multiple versions of the region have been maintained in the population over millions of years for HLA and hundreds of thousands of years for the ABO locus.

The regions described here have a lower density of variation than observed in the HLA region and are more similar in density to that observed in the ABO locus. Their size (tens of kilobases) is similar to that of haplotype blocks described elsewhere in the genome (Gabriel et al. 2002), but whether their boundaries correlate with the boundaries of haplotype blocks remains to be determined. Of course, such regions might simply represent the chance persistence of two different haplotype blocks over extensive evolutionary time. However, if balancing selection is responsible for the maintenance of these regions for the hundreds of thousands to millions of years required to accumulate this density of variation, understanding their role in the generation of the variable human phenotype will be important.

Both the enumeration of a parts list and a description of human variation require the high-quality, essentially complete reference human genome sequence that is now available to all without restriction. Distinguishing genes from pseudogenes was an almost impossible task with the draft human sequence and without a high-quality mouse sequence for comparison. The variants that alter the reading of genes are sufficiently unusual that errors in sequence would have obscured them in earlier versions, and even with a high-quality human genome, errors in the other data sets predominated. Recognition of areas of high SNP density requires accurate assembly, so that segmentally duplicated regions are not mistaken for such regions.

The completion of the human genome sequence marks a major milestone in science and comes just 50 years after the discovery of the structure of DNA. For the first time, we as a species have before us the genetic instruction set that molds us. Knowledge of the sequence presents the scientific community with the challenge to understand its content. The human genome sequence also presents society with enormous challenges, not the least of which is to use the knowledge for the betterment of humankind. Undoubtedly, the tasks will take longer than some have forecast, but the potential is clear and the implications profound.

## ACKNOWLEDGMENTS

We thank the dedicated individuals at the Washington University Genome Sequencing Center and other centers across the world for their contributions throughout this project. The work was supported through grants from the National Human Genome Research Institute.

## REFERENCES

Ansorge W., Sproat B., Stegemann J., Schwager C., and Zenke M. 1987. Automated DNA sequencing: Ultrasensitive detection of fluorescent bands during electrophoresis. *Nucleic Acids Res.* **15:** 4593.

Axelrod J.D. and Majors J. 1989. An improved method for photofootprinting yeast genes in vivo using Taq polymerase. *Nucleic Acids Res.* **17:** 171.

Baer R., Bankier A.T., Biggin M.D., Deininger P.L., Farrell P.J., Gibson T.J., Hatfull G., Hudson G.S., Satchwell S.C., and Seguin C., et al. 1984. DNA sequence and expression of the B95-8 Epstein-Barr virus genome. *Nature* **310:** 207.

Birney E. and R. Durbin R. 2000. Using GeneWise in the

*Drosophila* annotation experiment. *Genome Res.* **10:** 547.
Brumbaugh J.A., Middendorf L.R., Grone D.L., and Ruth J.L. 1988. Continuous, on-line DNA sequencing using oligodeoxynucleotide primers with multiple fluorophores. *Proc. Natl. Acad. Sci.* **85:** 5610.
Charlesworth B., Nordborg M., and Charlesworth D. 1997. The effects of local selection, balanced polymorphism and background selection on equilibrium patterns of genetic diversity in subdivided populations. *Genet. Res.* **70:** 155.
Collins F.S., Brooks L.D., and Chakravarti A. 1998. A DNA polymorphism discovery resource for research on human genetic variation. *Genome Res.* **8:** 1229.
Consortium (The *C. elegans* Sequencing Consortium). 1998. Genome sequence of the nematode *C. elegans:* A platform for investigating biology. *Science* **282:** 2012.
Coulson A., Sulston J., Brenner S., and Karn J. 1986. Towards a physical map of the genome of the nematode *Caenorhabditis elegans. Proc. Natl. Acad. Sci.* **83:** 7831.
Craxton M. 1993. Cosmid sequencing. *Methods Mol. Biol.* **23:** 149.
Daly M.J., Rioux J.D., Schaffner S.F., Hudson T.J., and Lander E.S. 2001. High-resolution haplotype structure in the human genome. *Nat. Genet.* **29:** 229.
DeLisi C. 1988. The Human Genome Project. *Am. Sci.* **76:** 488.
Deloukas P., Matthews L.H., Ashurst J., Burton J., Gilbert J.G., Jones M., Stavrides G., Almeida J.P., Babbage A.K., Bagguley C.L., Bailey J., Barlow K.F., Bates K.N., Beard L.M., Beare D.M., Beasley O.P., Bird C.P., Blakey S.E., Bridgeman A.M., Brown A.J., Buck D., Burrill W., Butler A.P., Carder C., and Carter N.P., et al. 2001. The DNA sequence and comparative analysis of human chromosome 20. *Nature* **414:** 865.
Dulbecco R. 1986. A turning point in cancer research: Sequencing the human genome. *Science* **231:** 1055.
Dunham I., Shimizu N., Roe B.A., Chissoe S., Hunt A.R., Collins J.E., Bruskiewich R., Beare D.M., Clamp M., Smink L.J., Ainscough R., Almeida J.P., Babbage A., Bagguley C., Bailey J., Barlow K., Bates K.N., Beasley O., Bird C.P., Blakey S., Bridgeman A.M., Buck D., Burgess J., Burrill W.D., and O'Brien K.P., et al. 1999. The DNA sequence of human chromosome 22. *Nature* **402:** 489.
Ewing B., Hillier L., Wendl M.C., and Green P. 1998. Base-calling of automated sequencer traces using phred. I. Accuracy assessment. *Genome Res.* **8:** 175.
FlyBase Consortium. 1994. FlyBase—The *Drosophila* database (The FlyBase Consortium). *Nucleic Acids Res.* **22:** 3456.
Gabriel S.B., Schaffner S.F., Nguyen H., Moore J.M., Roy J., Blumenstiel B., Higgins J., DeFelice M., Lochner A., Faggart M., Liu-Cordero S.N., Rotimi C., Adeyemo A., Cooper R., Ward R., Lander E.S., Daly M.J., and Altshuler D. 2002. The structure of haplotype blocks in the human genome. *Science* **296:** 2225.
Hardy D.M. and Garbers D.L. 1995. A sperm membrane protein that binds in a species-specific manner to the egg extracellular matrix is homologous to von Willebrand factor. *J. Biol. Chem.* **270:** 26025.
Hawkins T.L., O'Connor-Morin T., Roy A., and Santillan C. 1994. DNA purification and isolation using a solid-phase. *Nucleic Acids Res.* **22:** 4543.
Heilig R., Eckenberg R., Petit J.L., Fonknechten N., Da Silva C., Cattolico L., Levy M., Barbe V., de Berardinis V., Ureta-Vidal A., Pelletier E., Vico V., Anthouard V., Rowen L., Madan A., Qin S., Sun H., Du H., Pepin K., Artiguenave F., Robert C., Cruaud C., Bruls T., Jaillon O., and Friedlander L., et al. 2003. The DNA sequence and analysis of human chromosome 14. *Nature* **421:** 601.
Hillier L.W., Fulton R.S., Fulton L.A., Graves T.A., Pepin K.H., Wagner-McPherson C., Layman D., Maas J., Jaeger S., Walker R., Wylie K., Sekhon M., Becker M.C., O'Laughlin M.D., Schaller M.E., Fewell G.A., Delehaunty K.D., Miner T.L., Nash W.E., Cordes M., Du H., Sun H., Edwards J., Bradshaw-Cordum H., and Ali J., et al. 2003. The DNA sequence of human chromosome 7. *Nature* **424:** 157.
Hogenesch J.B., Ching K.A., Batalov S., Su A.I., Walker J.R., Zhou Y., Kay S.A., Schultz P.G., and Cooke M.P. 2001. A comparison of the Celera and Ensembl predicted gene sets reveals little overlap in novel genes. *Cell* **106:** 413.
Johnston M., Hillier L., Riles L., Albermann K., Andre B., Ansorge W., Benes V., Bruckner M., Delius H., Dubois E., Dusterhoft A., Entian K.D., Floeth M., Goffeau A., Hebling U., Heumann K., Heuss-Neitzel D., Hilbert H., Hilger F., Kleine K., Kotter P., Louis E.J., Messenguy F., Mewes H.W., and Hoheisel J.D., et al. 1997. The nucleotide sequence of *Saccharomyces cerevisiae* chromosome XII. *Nature* **387:** 87.
Ju J., Ruan C., Fuller C.W., Glazer A.N., and Mathies R.A. 1995. Fluorescence energy transfer dye-labeled primers for DNA sequencing and analysis. *Proc. Natl. Acad. Sci.* **92:** 4347.
Kent W.J. 2002. BLAT—The BLAST-like alignment tool. *Genome Res.* **12:** 656.
Korf I., Flicek P., Duan D, and Brent M.R. 2001. Integrating genomic homology into gene structure prediction. *Bioinformatics* (suppl. 1) **17:** S140.
Lander E.S., Linton L.M., Birren B., Nusbaum C., Zody M.C., Baldwin J., Devon K., Dewar K., Doyle M., FitzHugh W., Funke R., Gage D., Harris K., Heaford A., Howland J., Kann L., Lehoczky J., LeVine R., McEwan P., McKernan K., Meldrim J., Mesirov J.P., Miranda C., Morris W., and Naylor J., et al. (International Human Genome Sequencing Consortium). 2001. Initial sequencing and analysis of the human genome. *Nature* **409:** 860.
Lu H., Arriaga E., Chen D.Y., and Dovichi N.J. 1994. High-speed and high-accuracy DNA sequencing by capillary gel electrophoresis in a simple, low cost instrument. Two-color peak-height encoded sequencing at 40 degrees C. *J. Chromatogr. A* **680:** 497.
McPherson J.D., Marra M., Hillier L., Waterston R.H., Chinwalla A., Wallis J., Sekhon M., Wylie K., Mardis E.R., Wilson R.K., Fulton R., Kucaba T.A., Wagner-McPherson C., Barbazuk W.B., Gregory S.G., Humphray S.J., French L., Evans R.S., Bethel G., Whittaker A., Holden J.L., McCann O.T., Dunham A., Soderlund C., and Scott C.E., et al. (International Human Genome Mapping Consortium). 2001. A physical map of the human genome. *Nature* **409:** 934.
Mewes H.W., Albermann K., Bahr M., Frishman D., Gleissner A., Hani J., Heumann K., Kleine K., Maierl A., Oliver S.G., Pfeiffer F., and Zollner A. 1997. Overview of the yeast genome. *Nature* **387:** 7.
Miller R.D. and Kwok P.Y. 2001. The birth and death of human single-nucleotide polymorphisms: New experimental evidence and implications for human history and medicine. *Hum. Mol. Genet.* **10:** 2195.
Miller R.D., Taillon-Miller P., and Kwok P.Y. 2001. Regions of low single-nucleotide polymorphism incidence in human and orangutan xq: Deserts and recent coalescences. *Genomics* **71:** 78.
Mott R. 1997. EST_GENOME: A program to align spliced DNA sequences to unspliced genomic DNA. *Comput. Appl. Biosci.* **13:** 477.
Mungall A.J., Palmer S.A., Sims S.K., Edwards C.A., Ashurst K.L., Wilming L., Jones M.C., Horton R., Hunt S.E., Scott C.E., Gilbert J.G., Clamp M.E., Bethel G., Milne S., Ainscough R., Almeida J.P., Ambrose K.D., Andrews T.D., Ashwell R.I., Babbage A.K., Bagguley C.L., Bailey J., Banerjee R., Barker D.J., and Barlow K.F., et al. 2003. The DNA sequence and analysis of human chromosome 6. *Nature* **425:** 805.
Ohta T. and Ina Y. 1995. Variation in synonymous substitution rates among mammalian genes and the correlation between synonymous and nonsynonymous divergences. *J. Mol. Evol.* **41:** 717.
Olson M.V., Dutchik J.E., Graham M.Y., Brodeur G.M., Helms C., Frank M., MacCollin M., Scheinman R., and Frank T. 1986. Random-clone strategy for genomic restriction mapping in yeast. *Proc. Natl. Acad. Sci.* **83:** 7826.
Prober J.M., Trainor G.L., Dam R.J., Hobbs F.W., Robertson C.W., Zagursky R.J., Cocuzza A.J., Jensen M.A., and Baumeister K. 1987. A system for rapid DNA sequencing with fluorescent chain-terminating dideoxynucleotides. *Science* **238:** 336.

Pruitt K.D. and Maglott D.R. 2001. RefSeq and LocusLink: NCBI gene-centered resources. *Nucleic Acids Res.* **29:** 137.

Sachidanandam R., Weissman D., Schmidt S.C., Kakol J.M., Stein L.D., Marth G., Sherry S., Mullikin J.C., Mortimore B.J., Willey D.L., Hunt S.E., Cole C.G., Coggill P.C., Rice C.M., Ning Z., Rogers J., Bentley D.R., Kwok P.Y., Mardis E.R., Yeh R.T., Schultz B., Cook L., Davenport R., Dante M., and Fulton L., et al. 2001. A map of human genome sequence variation containing 1.42 million single nucleotide polymorphisms. *Nature* **409:** 928.

Scherer S.W., Cheung J., MacDonald J.R., Osborne L.R., Nakabayashi K., Herbrick J.A., Carson A.R., Parker-Katiraee L., Skaug J., Khaja R., Zhang J., Hudek A.K., Li M., Haddad M., Duggan G.E., Fernandez B.A., Kanematsu E., Gentles S., Christopoulos C.C., Choufani S., Kwasnicka D., Zheng X.H., Lai Z., Nusskern D., and Zhang Q., et al. 2003. Human chromosome 7: DNA sequence and biology. *Science* **300:** 767.

Shizuya H., Birren B., Kim U.J., Mancino V., Slepak T., Y. Tachiiri Y., and Simon M. 1992. Cloning and stable maintenance of 300-kilobase-pair fragments of human DNA in *Escherichia coli* using an F-factor-based vector. *Proc. Natl. Acad. Sci.* **89:** 8794.

Sinsheimer, R.L. 1989. The Santa Cruz Workshop, May 1985. *Genomics* **5:** 954.

Skaletsky H., Kuroda-Kawaguchi T., Minx P.J., Cordum H.S., Hillier L., Brown L.G., Repping S., Pyntikova T., Ali J., Bieri T., Chinwalla A., Delehaunty A., Delehaunty K., Du H., Fewell G., Fulton L., Fulton R., Graves T., Hou S.F., Latreille P., Leonard S., Mardis E., Maupin R., McPherson J., and Miner T., et al. 2003. The male-specific region of the human Y chromosome is a mosaic of discrete sequence classes. *Nature* **423:** 825.

Smith L.M., Fung S., Hunkapiller M.W., Hunkapiller T.J., and Hood L.E. 1985. The synthesis of oligonucleotides containing an aliphatic amino group at the 5′ terminus: Synthesis of fluorescent DNA primers for use in DNA sequence analysis. *Nucleic Acids Res.* **13:** 2399.

Solovyev V.V. 2001. Statistical approaches in eukaryotic gene prediction. In *Handbook of statistical genetics* (ed. D.J. Balding et al.), p. 83. Wiley, New York.

Strausberg, R.L., E.A. Feingold, R.D. Klausner, and F.S. Collins. 1999. The mammalian gene collection. *Science* **286:** 455-7.

Strausberg R.L., Feingold E.A., Grouse L.H., Derge J.G., Klausner R.D., Collins F.S., Wagner L., Shenmen C.M., Schuler G.D., Altschul S.F., Zeeberg B., Buetow K.H., Schaefer C.F., Bhat N.K., Hopkins R.F., Jordan H., Moore T., Max S.I., Wang J., Hsieh F., Diatchenko L., Marusina K., Farmer A.A., Rubin G.M., and Hong L., et al. 2002. Generation and initial analysis of more than 15,000 full-length human and mouse cDNA sequences. *Proc. Natl. Acad. Sci.* **99:** 16899.

Tabor S. and Richardson C.C. 1995. A single residue in DNA polymerases of the *Escherichia coli* DNA polymerase I family is critical for distinguishing between deoxy- and dideoxyribonucleotides. *Proc. Natl. Acad. Sci.* **92:** 6339.

Tilford C.A., Kuroda-Kawaguchi T., Skaletsky H., Rozen S., Brown L.G., Rosenberg M., McPherson J.D., Wylie K., Sekhon M., Kucaba T.A., Waterston R.H., and Page D.C. 2001. A physical map of the human Y chromosome. *Nature* **409:** 943.

Venter J.C., Adams M.D., Myers E.W., Li P.W., Mural R.J., Sutton G.G., Smith H.O., Yandell M., Evans C.A., Holt R.A., Gocayne J.D., Amanatides P., Ballew R.M., Huson D.H., Wortman J.R., Zhang Q., Kodira C.D., Zheng X.H., Chen L., Skupski M., Subramanian G., Thomas P.D., Zhang J., Gabor Miklos G.L., and Nelson C., et al. 2001. The sequence of the human genome. *Science* **291:** 1304.

Waterston R.H., Lindblad-Toh K., Birney E., Rogers J., Abril J.F., Agarwal P., Agarwala R., Ainscough R., Alexandersson M., An P., Antonarakis S.E., Attwood J., Baertsch R., Bailey J., Barlow K., Beck S., Berry E., Birren B., Bloom T., Bork P., Botcherby M., Bray N., Brent M.R., Brown D.G., and Brown S.D., et al. (Mouse Genome Sequencing Consortium). 2002. Initial sequencing and comparative analysis of the mouse genome. *Nature* **420:** 520.

Wheelan S.J., Church D.M., and Ostell J.M. 2001. Spidey: A tool for mRNA-to-genomic alignments. *Genome Res.* **11:** 1952.

Yip S.P. 2002. Sequence variation at the human ABO locus. *Ann. Hum. Genet.* **66:** 1.

Zdobnov E.M., von Mering C., Letunic I., Torrents D., Suyama M., Copley R.R., Christophides G.K., Thomasova D., Holt R.A., Subramanian G.M., Mueller H.M., Dimopoulos G., Law J.H., Wells M.A., Birney E., Charlab R., Halpern A.L., Kokoza E., Kraft C.L., Lai Z., Lewis S., Louis C., Barillas-Mury C., Nusskern D., and G.M. Rubin, et al. 2002. Comparative genome and proteome analysis of *Anopheles gambiae* and *Drosophila melanogaster*. *Science* **298:** 149.

# Mutational Profiling in the Human Genome

R.K. Wilson, T.J. Ley, F.S. Cole, J.D. Milbrandt, S. Clifton, L. Fulton, G. Fewell, P. Minx, H. Sun, M. McLellan, C. Pohl, and E.R. Mardis
*Washington University School of Medicine, St. Louis, Missouri 63108*

Prior to its completion as an initial draft in 2000 (Lander et al. 2001), and a finished entity in 2003 (Dunham et al. 1999; Hattori et al. 2000; Deloukas et al. 2001; Heilig et al. 2003; Hillier et al. 2003; Mungall et al. 2003; Skaletsky et al. 2003), the human genome sequence offered many promises for the advancement of biological research and the improvement of human health care. Current efforts are largely aimed at comprehensive analysis and annotation of genes and other functional elements contained within the sequence. These efforts include algorithmic approaches to gene discovery and characterization, as well as comparative genome sequencing that will utilize conserved elements in other mammalian and vertebrate genomes to highlight human sequences of functional importance. Every new genome sequenced and every optimization of gene-finding software will incrementally improve our view of the human genome landscape.

As these discoveries are realized, one crucial activity for the association of genes and their regulatory elements to the initiation and progression of disease is the profiling of sequence mutations and variations for a specific subset of genome elements within the human population. For those diseases involving mutation of a single gene (e.g., cystic fibrosis), this activity will be relatively straightforward once the associated gene has been identified. Here, re-sequencing of all or part of the gene in a large number of affected and unaffected individuals will reveal the common mutation(s) responsible for the disease phenotype (Fig. 1a). Likewise, once the components of any multigenic disease have been identified, systematic sequence-based profiling of affected and unaffected individuals will result in a better understanding of the underlying molecular mechanism of disease initiation and ultimately to various strategies for better diagnosis and therapy (Fig. 1b).

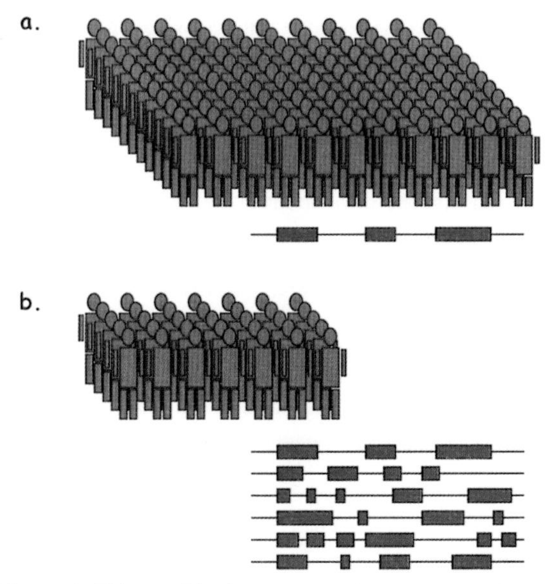


**Figure 1.** This graphic demonstrates the two types of high-throughput mutational profiling studies: (*a*) a "focused" study involving a single gene candidate with multiple patient samples to be sequenced for the gene of interest and (*b*) a "multifocal" study involving multiple candidate or suspect genes and multiple patient samples.

Unfortunately, for most diseases, we haven't the luxury of waiting for multigenic components to be identified so that these types of *mutational profiling* studies can be initiated. In this paradigm, high-throughput mutational profiling is performed across a range of candidate or "suspect" genes in large numbers of samples, either comparing cases to controls, or normal to tumor tissue from each patient. The ensuing data analysis seeks to develop testable hypotheses of causality by correlating the frequency of specific mutations to the occurrence, severity, or outcome of the disease. The appropriate follow-on studies to establish or refute causality can then be designed and carried out.

## MATERIALS AND METHODS

Our current approach to mutational profiling centers around the techniques of PCR amplification with exon-specific primer pairs for each gene, and DNA sequencing of the resulting products, followed by analysis of the sequence data (Fig. 2). Our previous experience with establishing a high-throughput DNA sequencing pipeline has now been brought to bear on this experimental approach, enabling us to use the aforementioned techniques, along with liquid-handling robots and high-throughput DNA sequencing instruments, to begin the optimization of the associated work flow. Specifically, some of the challenges of establishing a high-throughput mutational profiling pipeline center around (1) aspects of sample acquisition and preparation of genomic template DNA, including compliance with the new federal regulations aimed at the protection of patient privacy (HIPAA); (2) the need for technology-enhanced throughput at a number of rate-limiting steps; (3) the development of appropriate

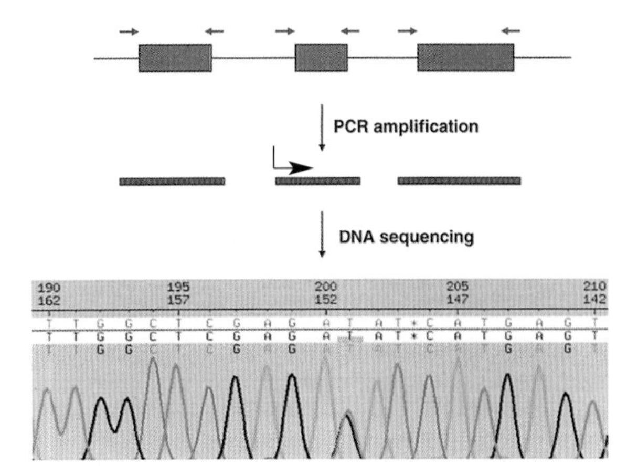


**Figure 2.** A generalized overview of the mutational profiling process, including the design and use of PCR amplification primer sets to amplify individual exons from genomic samples, the purification of PCR amplicons, and their use in sequencing reactions to generate data from each patient.

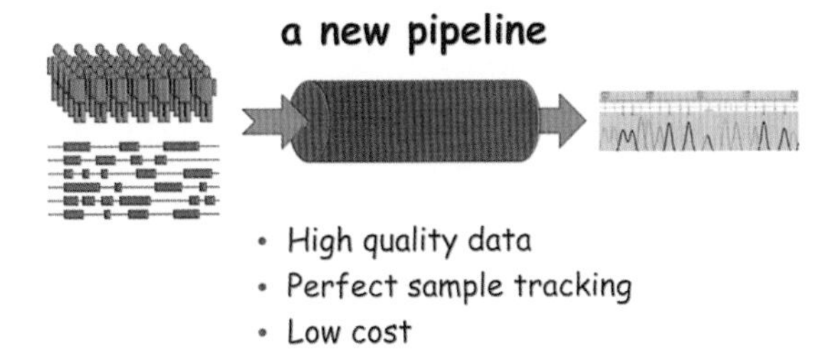


**Figure 3.** Our prototype of a high-throughput mutation profiling pipeline where multiple patient samples are sequenced to provide the highest quality data, with unequivocal sample tracking and low cost per sample.

software and data-management tools that enable data visualization, data organization, and production of meaningful statistics; and (4) the ability to assemble and correlate the acquired sequence data and other patient-specific data such as phenotype, symptoms, and outcomes. It is the latter challenge that poses perhaps the most significant barrier to the effective application of mutational profiling to the study of disease (Fig. 3). Specific methodological aspects of our current mutational profiling pipeline are described below.

***Primer design aspects.*** The availability of a reference human genome sequence with annotated genes is an invaluable resource in the bioinformatic exercise to design primers for PCR amplification of exonic sequences. Furthermore, the selection of primer sets requires a robust algorithm to render this activity as automated as possible. In some instances, a secondary attempt at manual primer design is required to enhance the robustness of PCR in difficult areas, such as those with high GC content, to provide needed specificity to amplify unique members of gene families, etc. In addition to unique sequence being selected, a uniform amplicon size range of 400–700 bases is attempted, in order to enhance PCR robustness and yield uniformity. Such smaller PCR products will require multiple primer pairs across large exons, each with a small amount of sequence overlap with adjoining amplicons to ensure proper assembly. However, the benefits of shorter amplicons outweigh the expense of more primer sets, due to the increased reliability of obtaining these products in high yield from a given genomic DNA sample. Another aspect of primer design that is important for work flow aspects downstream of PCR is the inclusion of universal primer "tails" onto the gene-specific portions of the PCR primers. This approach, described previously (Roy and Steffens 1997), provides an overall lower background in sequencing data and thus enhances the automated detection of true polymorphisms in downstream analysis steps.

Once primer selection is accomplished, however, considerable effort goes into validating the primer sets prior to their use in the mutational profiling pipeline. Given the number of genes in the human genome and the numbers of genes that will be required for certain mutational profiling studies, the combination of primer selection and validation as an iterative process represents a substantial effort. It is perhaps the most time-consuming and expensive aspect, yet one that must be done effectively for the greatest efficiency and data quality to result from the mutational profiling pipeline.

***PCR optimization aspects.*** The process of developing validated PCR amplification primer sets is intertwined with the process of determining a robust PCR amplification method. Ideally, the method selected should be applicable to most sequence content and should be developed in anticipation that each sequence to be amplified will pose significant challenges. Although this is ostensibly an expensive approach in terms of the cost per reaction, the resulting data quality from a wide range of template/product combinations will justify the cost increase. For high-throughput purposes, we also favor the use of 384-well plates and low-volume reactions (~10 μl) for performing amplification and sequencing reactions. The overall goal is to produce data of the highest quality possible, so that the downstream conclusions reached are solidly based.

***DNA sequencing aspects.*** Our overall goal in developing DNA sequencing methods for mutational profiling has been to achieve the lowest background possible (to reduce/eliminate false polymorphism calls), with read length as a secondary concern because most PCR amplicons average 500 bp in length and can be fully resolved by sequence across their length on a standard sequencing separation. As mentioned earlier, the use of PCR amplicons with "forward" and "reverse" universal primer tails has demonstrated the highest read quality/lowest background overall, in our experience. An alternative to this approach is to utilize nested sequencing primers that anneal internal to the PCR primer sequence. However, we favor the universal primer tailing approach to the nested primer approach because of the overall facility it provides to streamlining the sequencing pipeline without adding cost for additional amplicon-specific primers.

Another aspect of DNA sequencing is focused on obtaining the highest fidelity of sequence assignment, or base-calling, to the resulting data. Specifically, base-calling algorithms must be tuned to detect polymorphisms and flag them for downstream analysis. This presents somewhat of a challenge, since polymorphic sites typically exhibit two unique characteristics that are not recognized well by most DNA sequence base-calling algorithms. Namely, a polymorphic site exhibits a significant drop in peak height relative to neighboring (non-polymorphic) peaks, and contains a smaller underlying peak produced by the second allele. Although the number of commercial software packages aimed at polymorphism detection is increasing, we have used the publicly available *polyphred* package of Nickerson (Nickerson et al. 1997) to identify and tag polymorphic sites in mutational profiling data. This algorithm has its functions fully integrated within the *phred/phrap/consed* software packages (Ewing and Green 1998; Ewing et al. 1998; Gordon et al. 1998) that we have used in the production, assembly, and viewing of de novo genome sequence data. As such, we have the ability to tag polymorphic sites and assemble (via *phrap*) the component reads that constitute the exonic (or full gene) sequences for a given patient sample. Large-scale comparisons of cases versus controls or normal versus tumor samples can then be examined using the *consed* sequence alignment viewer, in which the *polyphred* tags can be readily found and the underlying trace data examined to validate the *polyphred* assignment. Ultimately, these data can be combined into a *visual genotypes* output (Nickerson et al. 1998; Rieder et al. 1998) that enables large-scale examination and clustering of polymorphic sites for multiple genes in multiple patient samples.

Underlying all of the data collection aspects of a mutational profiling project is the crucial aspect of sample tracking, which must be perfect and unequivocal in order for results and conclusions to be accurate and meaningful. In our laboratory, we rely on an Oracle database and accompanying schema to enable barcode-based sample tracking of patient samples and associated data throughout the mutational profiling pipeline.

## MUTATIONAL PROFILING PROJECTS AT WASHINGTON UNIVERSITY

### Pulmonary Surfactant Protein B Deficiency

Newborn respiratory distress syndrome (N-RDS) is the most frequent respiratory cause of death and morbidity in infants under one year of age in the United States (Guyer et al. 1998). N-RDS has most commonly been attributed in premature infants to pulmonary surfactant deficiency due to developmental immaturity of surfactant production (Avery and Mead 1959; Whitsett and Stahlman 1998), since type II pneumocytes do not appear prior to 32–34 weeks of gestation and lack the ability to produce functional surfactant. Surfactant replacement therapy has been associated with a decline in N-RDS-associated mortality, but not in long-term respiratory morbidity, hence revealing the contribution of genetic causes of N-RDS in infancy. Genetic disruption of surfactant protein B gene expression has provided unambiguous human and murine respiratory phenotypes in newborn infants and pups. Surfactant protein B deficiency was the first identified genetic cause of lethal N-RDS in infants and was attributed to a mutation resulting from a 1-bp deletion and a 3-bp insertion at codon 121 of the gene (121ins2) (Nogee et al. 1993). This mutation creates a new *Sfu*I restriction site, allowing rapid detection of the mutation, and leads to a translation stop signal at codon 214. A truncated, unstable surfactant protein B transcript results, but no protein is synthesized (Nogee et al. 1994; Beers et al. 2000). The clinical phenotype of infants homozygous for the 121ins2 mutation is consistent: They are born at full-term, develop RDS within the first 12–24 hours after birth, have no associated extrapulmonary organ dysfunction, and without lung transplantation will expire within the first 1–6 months of life (Hamvas et al. 1997; Nogee 1997). Studies of compound heterozygotes of 121ins2 indicate that a minimum net production of surfactant protein B between 10% and 50% is essential for survival (Ballard et al. 1995; Yusen et al. 1999).

The DNA sequence of the surfactant protein B gene and its regulatory regions span 11 kb of the human genome and are contained in 11 exons, with exons 6 and 7 encoding the mature protein and exon 11 providing a 3′-untranslated region (Whitsett et al. 1995). The mutational spectrum of this gene in affecteds has been profiled by previous studies, with the 121ins2 mutation found on ~60% of chromosomes (Nogee et al. 2000). In addition to this predominant mutation, 16 exonic mutations have been characterized. Studies of murine knockout or transgenic lineages with disrupted or altered surfactant protein B gene expression have mimicked many of the histopathologic features of the disease in humans, suggesting that mutations resulting in altered processing, folding, intracellular itinerary, or phospholipid interaction may be clinically significant. Therefore, we have initiated a project to perform mutational profiling across this gene in multiple patient samples, in an effort to better characterize those mutations in and around the surfactant protein B gene that appear to predispose infants to RDS.

Because of the relatively compact nature of the SP-B gene, and our desire to understand both coding and noncoding mutational profiles for affecteds in this study, we took an approach that focused on the entirety of the gene sequence, including regions just 5′ of the first exon and the 3′ UTR, represented as ~500-bp PCR amplicons having a small amount of sequence overlap with adjoining products. Most of the patient DNAs used in these studies are obtained from Chelex-based extraction of dried blood spots taken at birth. The patient samples designated for this study represent large numbers of patients (~10,000 each) from a diverse geographical collection, including the state of Missouri; Cape Town, South Africa; Seoul, South Korea; and Oslo, Norway. Comparisons of the mutational profiles for these distinct geographic and ethnic groups, when correlated to phenotype, will enable an estimation of ethnic frequency of the predominant mutation subtypes, including 121ins2.

### Prostate Cancer

Prostate cancer is responsible for 3% of all deaths in men over 55 years of age and is second only to lung cancer as a cause of cancer death in the western world (Chakravarti and Zhai 2003). Here, prostate cancer is the most commonly diagnosed cancer, predominately through the use of the serum marker prostate-specific antigen (PSA) assays. Numerous epidemiological and molecular biological studies have accumulated evidence that favors a significant but heterogeneous hereditary component in prostate cancer susceptibility (Stanford and Ostrander 2001). There have been seven susceptibility loci mapped to date, with three loci having been characterized for specific genes and their constituent mutations. The role of these loci in hereditary and sporadic disease is still debatable, however, and it could be very modest (Verhage and Kiemeney 2003). However, besides age and race, family history is the only well-established risk factor, where epidemiological studies show that first-degree relatives of prostate cancer patients have a two- to threefold increased risk of prostate cancer (Steinberg et al. 1990). The mode of heritability is still subject to debate and, indeed, may be different for early-onset versus late-onset disease.

Recently, multiple genes have been identified with putative relevance to prostate carcinogenesis, with a model for prostate cancer progression that includes the potential contribution of inflammation to the development of preneoplastic or neoplastic lesions. The different stages of prostate carcinogenesis are characterized more by the accumulation of multiple somatic genome alterations than by any one genetic lesion, although specific changes may increase the likelihood of further neoplastic transformation. Overall, an understanding of the contributions of somatic gene defects to prostate carcinogenesis may facilitate the development of novel targeted therapeutic approaches for clinical management or more specific diagnosis (Gonzalgo and Isaacs 2003). To this end, we have elected to profile the mutational spectrum of two broad groups of the human "kinome" (Manning et al. 2002); namely, the genes that encode tyrosine kinase and tyrosine-kinase-like proteins. The role of kinases as biological control points, the causal role of protein kinase mutations and dysregulation in human disease, and their tractability as drug targets make these genes logical choices for our study of prostate carcinogenesis.

Our approach to this analysis has focused on mRNA transcripts purified from tumor tissue and reverse-transcribed into cDNA templates. This strategy was chosen in order to minimize the number of primers required for amplification and sequencing, thereby maximizing the number of patient samples that can be evaluated.

### Acute Myeloid Leukemia

Acute myeloid leukemia (AML) is the most common form of leukemia and the most common cause of death from leukemia. Although conventional chemotherapy can cure 25–45% of patients, most either die of relapse or from complications associated with treatment (Stirewalt et al. 2003). AML is a heterogeneous disease, characterized by a myriad of genetic defects that have been found to include translocations involving oncogenes and transcription factors, activation of signal transduction pathways, and alterations to growth factor receptors. It appears that these multiple affected pathways interact in the onset of leukemogenesis, such that multiple genetic abnormalities are necessary for the development of overt leukemia (Kelly et al. 2002). Consequently, it is thought that a single target of therapy may be widely useful in treating the spectrum of AML subtypes.

Mutations in several receptor tyrosine kinases (RTKs) have been frequently described in AML, including FMS, KIT, and FLT3. FLT3 is the most frequently mutated RTK in AML, found in ~15–40% of AML patients (Nakao et al. 1996; Kiyoi et al. 1997, 1988; Yokota et al. 1997; Iwai et al. 1999; Xu et al. 1999; Abu-Duhier et al. 2000, 2001; Meshinchi et al. 2001; Stirewalt et al. 2001; Yamamoto et al. 2001; Schnittger et al. 2002; Thiede et al. 2002). Two common mutations have been found in FLT3, both of which constitutively activate the receptor. One is an internal tandem duplication (ITD) within exons 11 and 12 whose length varies from 3 to >400 bp (Schnittger et al. 2002). The prevalence of FLT3 ITDs varies with age, from 10–15% in pediatrics to 25–35% in older adults. The other is a point mutation in exon 17, typically in codon 835, which is found in 5–10% of AML patients (Abu-Duhier et al. 2001; Yamamoto et al. 2001).

In addition to the FLT3 mutations, mutations in the RAS gene superfamily occur in ~15–25% of AML cases (Bos et al. 1987; Farr et al. 1988; Bartram et al. 1989; Radich et al. 1990; Vogelstein et al. 1990; Neubauer et al. 1994). These are predominantly point mutations in codons 12, 13, and 61 and act to prevent the conversion of RAS from an active to an inactive form, thus constitutively activating downstream pathways (Byrne and Marshall 1998). In general, our hypothesis is that AML cells contain a variety of acquired mutations, such as those described above, but that only some of them are relevant for disease pathogenesis. To characterize these mutations and to begin associating their presence with disease initiation and progression, we are producing mutational profiling data for ~450 candidate genes, initially using samples derived from a cohort of 47 AML patients for which banked tumor (bone marrow) and germ-line (skin) samples are in hand. The resulting mutational profiles will be correlated back to the AML subtypes and clinical outcomes for these patients, as a means to discover and characterize the mutations (and other polymorphisms) that can be putatively associated with AML pathogenesis. This set of activities constitutes a "discovery phase" of the AML project, and will culminate in the selection of a smaller number of genes to be the subjects of further mutational profiling work on a second set of AML patients with matched tumor and germ-line samples. Other types of AML studies, funded by this project, also will be considered in the selection of genes to be examined in this second, or "validation phase" of the project. These studies include functional genomics-based approaches using several mouse models of AML, as well as microarray-based comparative genome hybridization studies of

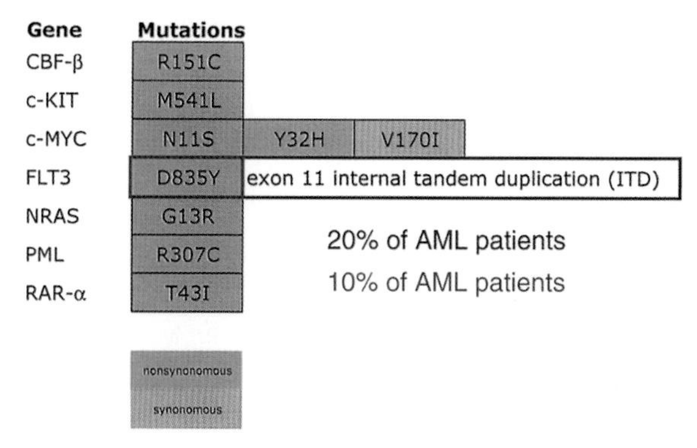


**Figure 4.** Results from our pilot study of AML suspect genes in 42 patient samples, indicating the gene sequences (*left* column), and the synonomous and nonsynonomous amino acid-altering mutations found. In the case of the FLT3 gene, we found the exon 11 ITD in 20% of patients as well as the exon 17 (D835Y) mutation in 10% of the patients studied.

mouse model and patient samples. Overall, the validation phase will serve as a means of increasing the size of our AML data set of mutational profiles centered on a smaller number of candidate genes. Following on, a second round of correlation of these profiles back to AML subtypes and clinical outcomes should serve to further focus our attention on a handful of genes now highly suspect for their involvement in AML pathogenesis.

Our preliminary findings from this study, using 13 genes, have been described elsewhere (Ley et al. 2003) and are summarized in Figure 4. Briefly, for the 13 genes studied in 46 patient samples (tumor and somatic/control DNA samples) representing different AML subtypes (M0/M1, M2, M3/APL, and M4), we found that previously described mutations in CBF-β, FLT3, c-KIT, c-MYC, N-RAS, PML, and RARα were also found in our patient samples, as indicated. Most notably, in FLT3 we detected the previously described internal tandem duplication (ITD) in exon 11 but also found another FLT3 mutation causing a nonsynonomous amino acid change (D835Y) in 10% of the patient samples sequenced.

## FUTURE DIRECTIONS

Sequence-based mutational profiling represents an immediate application of the human genome sequence, and the technology developed to produce it, toward the study of human health and disease. In the near term, we can utilize mutational profiling to better understand the molecular basis of many human diseases for which there are candidate genes, or at least a reasonable number of suspected candidates. For genes such as SPB that play a role in both moderate and severe forms of a disease, such an understanding will be critical to improved diagnosis, treatment, and management of patients who carry these sequence changes. Likewise, as we begin to discover and study the interplay of sequence changes and mutations in the bevy of genes that underlie various human cancers, we will create new opportunities for earlier diagnosis that will substantially save lives and decrease health care costs even with current state-of-the-art cancer treatments. Ultimately, a better understanding of the genes involved in cancer and other diseases will allow the development of targeted therapeutics (such as Gleevec™) that are aimed directly at the molecular flaw.

Although the technology and its associated expense currently require that we focus our efforts and sequencing pipelines nearly exclusively on the exons of candidate genes, it is clear that we must in the future develop mutational profiling strategies and methods that will allow us to cast a broader net, aimed at discovering causal mutations and sequence changes that lie outside of coding sequence and candidate genes. Ideally, we would prefer to sequence a patient's complete genome (both "germ line" and from the affected tissue), cross-reference all sequence variations and discovered mutations, and characterize any epigenomic changes as well. Such a comprehensive analysis should not only be relatively inexpensive, it should also be performed in a sufficiently narrow time frame so as to allow a physician to quickly respond to acute symptoms. Of course, we are currently far from being able to offer patients such a comprehensive genome-based diagnostic analysis, but the promise of genomic medicine is vast, and we believe that exon-based mutational profiling such as we describe here represents an early step on the road to such promise.

## REFERENCES

Abu-Duhier F.M., Goodeve A.C., Wilson G.A., Care R.S., Peake I.R., and Reilly J.T. 2001. Identification of novel FLT-3 Asp835 mutations in adult acute myeloid leukaemia. *Br. J. Haematol.* **113:** 983.

Abu-Duhier F.M., Goodeve A.C., Wilson G.A., Gari M.A., Peake I.R., Rees D.C., Vandenberghe E.A., Winship P.R., and Reilly J.T. 2000. FLT3 internal tandem duplication mutations in adult acute myeloid leukaemia define a high-risk group. *Br. J. Haematol.* **111:** 190.

Avery M.E. and Mead J. 1959. Surface properties in relation to atelectasis and hyaline membrane disease. *AMA J. Dis. Child.* **97:** 517.

Ballard P.L., Nogee L.M., Beers M.F., Ballard R.A., Planer B.C., Polk L., deMello D.E., Moxley M.A., and Longmore W.J. 1995. Partial deficiency of surfactant protein B in an infant with chronic lung disease. *Pediatrics* **96:** 1046.

Bartram C.R., Ludwig W.D., Hiddemann W., Lyons J., Buschle M., Ritter J., Harbott J., Frohlich A., and Janssen J.W. 1989. Acute myeloid leukemia: Analysis of ras gene mutations and clonality defined by polymorphic X-linked loci. *Leukemia* **3:** 247.

Beers M.F., Hamvas A., Moxley M.A., Gonzales L.W., Guttentag S.H., Solarin K.O., Longmore W.J., Nogee L.M., and Ballard P.L. 2000. Pulmonary surfactant metabolism in infants lacking surfactant protein B. *Am. J. Respir. Cell Mol. Biol.* **22:** 380.

Bos J.L., Verlaan-de Vries M., van der Eb A.J., Janssen J.W., Delwel R., Lowenberg B., and Colly L.P. 1987. Mutations in N-ras predominate in acute myeloid leukemia. *Blood* **69:** 1237.

Byrne J.L. and Marshall C.J. 1998. The molecular pathophysiology of myeloid leukaemias: Ras revisited. *Br. J. Haematol.* **100:** 256.

Chakravarti A. and Zhai G.G. 2003. Molecular and genetic prognostic factors of prostate cancer. *World J. Urol.* **21:** 265.

Deloukas P., Matthews L.H., Ashurst J., Burton J., Gilbert J.G., Jones M., Stavrides G., Almeida J.P., Babbage A.K., Bagguley C.L., Bailey J., Barlow K.F., Bates K.N., Beard L.M., Beare D.M., Beasley O.P., Bird C.P., Blakey S.E., Bridgeman A.M., Brown A.J., Buck D., Burrill W., Butler A.P., Carder

C., and Carter N.P., et al. 2001. The DNA sequence and comparative analysis of human chromosome 20. *Nature* **414:** 865.

Dunham I., Shimizu N., Roe .A., Chissoe S., Hunt A.R., Collins J.E., Bruskiewich R., Beare D.M., Clamp M., Smink L.J., Ainscough R., Almeida J.P., Babbage A., Bagguley C., Bailey J., Barlow K., Bates K.N., Beasley O., Bird C.P., Blakey S., Bridgeman A.M., Buck D., Burgess J., Burrill W.D., and O'Brien K.P., et al. 1999. The DNA sequence of human chromosome 22 (erratum in *Nature* [2000] **404:** 904). *Nature* **402:** 489.

Ewing B. and Green P. 1998. Base-calling of automated sequencer traces using phred. II. Error probabilities. *Genome Res.* **8:** 186.

Ewing B., Hillier L., Wendl. M.C., and Green P. 1998. Base-calling of automated sequencer traces using phred. I. Accuracy assessment. *Genome Res.* **8:** 175.

Farr C.J., Saiki R.K., Erlich H.A., McCormick F., and Marshall C.J. 1988. Analysis of RAS gene mutations in acute myeloid leukemia by polymerase chain reaction and oligonucleotide probes. *Proc. Natl. Acad. Sci.* **85:** 1629.

Gonzalgo M.L. and Isaacs W.B. 2003. Molecular pathways to prostate cancer. *J. Urol.* **170:** 2444.

Gordon D., Abajian C., and Green P. 1998. Consed: A graphical tool for sequence finishing. *Genome Res.* **8:** 195.

Guyer B., MacDorman M.F., Martin J.A., Peters K.D., and Strobino D.M. 1998. Annual summary of vital statistics - 1997. *Pediatrics* **102:** 1333.

Hamvas A., Nogee L.M., Mallory G.B., Jr., Spray T.L., Huddleston C.B., August A., Dehner L.P., deMello D.E., Moxley M., Nelson R., Cole F.S., and Colten H.R. 1997. Lung transplantation for treatment of infants with surfactant protein B deficiency. *J. Pediatr.* **130:** 231.

Hattori M., Fujiyama A., Taylor T.D., Watanabe H., Yada T., Park H.S., Toyoda A., Ishii K., Totoki Y., Choi D.K., Soeda E., Ohki M., Takagi T., Sakaki Y., Taudien S., Blechschmidt K., Polley A., Menzel U., Delabar J., Kumpf K., Lehmann R., Patterson D., Reichwald K., Rump A., and Schillhabel M., et al. 2000. The DNA sequence of human chromosome 21. The chromosome 21 mapping and sequencing consortium. *Nature* **405:** 311.

Heilig R., Eckenberg R., Petit J.L., Fonknechten N., Da Silva C., Cattolico L., Levy M., Barbe V., de Berardinis V., Ureta-Vidal A., Pelletier E., Vico V., Anthouard V., Rowen L., Madan A., Qin S., Sun H., Du H., Pepin K., Artiguenave F., Robert C., Cruaud C., Bruls T., Jaillon O., and Friedlander L., et al. 2003. The DNA sequence and analysis of human chromosome 14. *Nature* **421:** 601.

Hillier L.W., Fulton R.S., Fulton L.A., Graves T.A., Pepin K.H., Wagner-McPherson C., Layman D., Maas J., Jaeger S., Walker R., Wylie K., Sekhon M., Becker M.C., O'Laughlin M.D., Schaller M.E., Fewell G.A., Delehaunty K.D., Miner T.L., Nash W.E., Cordes M., Du H., Sun H., Edwards J., Bradshaw-Cordum H., and Ali J., et al. 2003. The DNA sequence of human chromosome 7. *Nature* **424:** 157.

Iwai T., S. Yokota S., Nakao M., Okamoto T., Taniwaki M., Onodera N., Watanabe A., Kikuta A., Tanaka A., Asami K., Sekine I., Mugishima H., Nishimura Y., Koizumi S., Horikoshi Y., Mimaya J., Ohta S., Nishikawa K., Iwai A., Shimokawa T., Nakayama M., Kawakami K., Gushiken T., Hyakuna N., and Fujimoto T., et al. 1999. Internal tandem duplication of the FLT3 gene and clinical evaluation in childhood acute myeloid leukemia. The Children's Cancer and Leukemia Study Group, Japan. *Leukemia* **13:** 38.

Kelly L.M., Yu J.C., Boulton C.L., Apatira M., Li J., Sullivan C.M., Williams I., Amaral S.M., Curley D.P., Duclos N., Neuberg D., Scarborough R.M., Pandey A., Hollenbach S., Abeb K., Lokker N.A., Gilliland D.G., and Giese N.A. 2002. CT53518, a novel selective FLT3 antagonist for the treatment of acute myelogenous leukemia (AML). *Cancer Cell* **1:** 421.

Kiyoi H., Towatari M., Yokota S., Hamaguchi M., Ohno R., Saito H., and Naoe T. 1998. Internal tandem duplication of the FLT3 gene is a novel modality of elongation mutation which causes constitutive activation of the product. *Leukemia* **12:** 1333.

Kiyoi H., Naoe T., Yokota S., Nakao M., Minami S., Kuriyama K., Takeshita A., Saito K., Hasegawa S., Shimodaira S., Tamura J., Shimazaki C., Matsue K., Kobayashi H., Arima N., Suzuki R., Morishita H., Saito H., Ueda R., and Ohno R. 1997. Internal tandem duplication of FLT3 associated with leukocytosis in acute promyelocytic leukemia. Leukemia Study Group of the Ministry of Health and Welfare (Kohseisho). *Leukemia* **11:** 1447.

Lander E.S., Linton L.M., Birren B., Nusbaum C., Zody M.C., Baldwin J., Devon K., Dewar K., Doyle M., FitzHugh W., Funke R., Gage D., Harris K., Heaford A., Howland J., Kann L., Lehoczky J., LeVine R., McEwan P., McKernan K., Meldrim J., Mesirov J.P., Miranda C., Morris W., and Naylor J., et al. (International Human Genome Sequencing Consortium). 2001. Initial sequencing and analysis of the human genome. *Nature* **409:** 860.

Ley T.J., Minx P.J., Walter M.J., Ries R.E., Sun H., McLellan M., DiPersio J.F., Link D.C., Tomasson M.H., Graubert T.A., McLeod H., Khoury H., Watson M., Shannon W., Trinkaus K., Heath S., Vardiman J.W., Caligiuri M.A., Bloomfield C.D., Milbrandt J.D., Mardis E.R., and Wilson R.K. 2003. A pilot study of high-throughput, sequence-based mutational profiling of primary human acute myeloid leukemia cell genomes. *Proc. Natl. Acad. Sci.* **100:** 14275.

Manning G., Whyte D.B., Martinez R., Hunter T., and Sudarsanam S. 2002. The protein kinase complement of the human genome. *Science* **298:** 1912.

Meshinchi S., Woods W.G., Stirewalt D.L., Sweetser D.A., Buckley J.D., Tjoa T.K., Bernstein I.D., and Radich J.P. 2001. Prevalence and prognostic significance of Flt3 internal tandem duplication in pediatric acute myeloid leukemia. *Blood* **97:** 89.

Mungall A.J., Palmer S.A., Sims S.K., Edwards C.A., Ashurst K.L., Wilming L., Jones M.C., Horton R., Hunt S.E., Scott C.E., Gilbert J.G., Clamp M.E., Bethel G., Milne S., Ainscough R., Almeida J.P., Ambrose K.D., Andrews T.D., Ashwell R.I., Babbage A.K., Bagguley C.L., Bailey J., Banerjee R., Barker D.J., and Barlow K.F., et al. 2003. The DNA sequence and analysis of human chromosome 6. *Nature* **425:** 805.

Nakao M., Yokota S., Iwai T., Kaneko H., Horiike S., Kashima K., Sonoda Y., Fujimoto T., and Misawa S. 1996. Internal tandem duplication of the flt3 gene found in acute myeloid leukemia. *Leukemia* **10:** 1911.

Neubauer A., Dodge R.K., George S.L., Davey F.R., Silver R.T., Schiffer C.A., Mayer R.J., Ball E.D., Wurster-Hill D., and Bloomfield C.D., et al. 1994. Prognostic importance of mutations in the ras proto-oncogenes in de novo acute myeloid leukemia. *Blood* **83:** 1603.

Nickerson D.A., Tobe V.O., and Taylor S.L. 1997. PolyPhred: Automating the detection and genotyping of single nucleotide substitutions using fluorescence-based resequencing. *Nucleic Acids Res.* **25:** 2745.

Nickerson D.A., Taylor S.L., Weiss K.M., Clark A.G., Hutchinson R.G., Stengard J., Salomaa V., Vartiainen E., Boerwinkle E., and Sing C.F. 1998. DNA sequence diversity in a 9.7-kb region of the human lipoprotein lipase gene (see comments). *Nat. Genet.* **19:** 233.

Nogee L.M. 1997. Surfactant protein-B deficiency. *Chest* (suppl. 6) **111:** 129S.

Nogee L.M., de Mello D.E., Dehner L.P., and Colten H.R. 1993. Brief report: Deficiency of pulmonary surfactant protein B in congenital alveolar proteinosis. *N. Engl. J. Med.* **328:** 406.

Nogee L.M., Wert S.E., Proffit S.A., Hull W.M., and Whitsett J.A. 2000. Allelic heterogeneity in hereditary surfactant protein B (SP-B) deficiency. *Am. J. Respir. Crit. Care Med.* **161:** 973.

Nogee L.M., Garnier G., Dietz H.C., Singer L., Murphy A.M., deMello D.E., and Colten H.R. 1994. A mutation in the surfactant protein B gene responsible for fatal neonatal respiratory disease in multiple kindreds. *J. Clin. Invest.* **93:** 1860.

Radich J.P., Kopecky K.J., Willman C.L., Weick J., Head D., Appelaum F., and Collins S.J. 1990. N-ras mutations in adult de novo acute myelogenous leukemia: Prevalence and clinical significance. *Blood* **76:** 801.

Rieder M.J., Taylor S.L., Toe V.O., and Nickerson D.A. 1998. Automating the identification of DNA variations using quality-based fluorescence re-sequencing: Analysis of the human mitochondrial genome. *Nucleic Acids Res.* **26:** 967.

Roy R. and Steffens D.L. 1997. Infrared fluorescent detection of PCR amplified gender identifying alleles. *J. Forensic Sci.* **42:** 452.

Schnittger S., Schoch C., Dugas M., Kern W., Staib P. Wuchter C., Loffler H., Sauerland C.M., Serve H., Buchner T., Haferlach T., and Hiddemann W. 2002. Analysis of FLT3 length mutations in 1003 patients with acute myeloid leukemia: Correlation to cytogenetics, FAB subtype, and prognosis in the AMLCG study and usefulness as a marker for the detection of minimal residual disease. *Blood* **100:** 59.

Skaletsky H., Kuroda-Kawaguchi T., Minx P.J., Cordum H.S., Hillier L., Brown L.G., Repping S., Pyntikova T., Ali J., Bieri T., Chinwalla A., Delehaunty A., Delehaunty K., Du H., Fewell G., Fulton L., Fulton R., Graves T., Hou S.F., Latrielle P., Leonard S., Mardis E., Maupin R., McPherson J., and Miner T., et al. 2003. The male-specific region of the human Y chromosome is a mosaic of discrete sequence classes. *Nature* **423:** 825.

Stanford J.L. and Ostrander E.A. 2001. Familial prostate cancer. *Epidemiol. Rev.* **23:** 19.

Steinberg G.D., Carter B.S., Beaty T.H., Childs B., and Walsh P.C. 1990. Family history and the risk of prostate cancer. *Prostate* **17:** 337.

Stirewalt D.L., Meshinchi S., and Radich J.P. 2003. Molecular targets in acute myelogenous leukemia. *Blood Rev.* **17:** 15.

Stirewalt D.L., Kopecky K.J., Meshinchi S., Appelbaum F.R., Slovak M.L., Willman C.L., and Radich J.P. 2001. FLT3, RAS, and TP53 mutations in elderly patients with acute myeloid leukemia. *Blood* **97:** 3589.

Thiede C., Steudel C., Mohr B., Schaich M., Schakel U., Platzbecker U., Wermke M., Bornhauser M., Ritter M., Neubauer A., Ehninger G., and Illmer T. 2002. Analysis of FLT3-activating mutations in 979 patients with acute myelogenous leukemia: Association with FAB subtypes and identification of subgroups with poor prognosis. *Blood* **99:** 4326.

Verhage B.A. and Kiemeney L.A. 2003. Genetic susceptibility to prostate cancer: A review. *Fam. Cancer* **2:** 57.

Vogelstein B., Civin C.I., Preisinger A.C., Krischer J.P., Steuber P., Ravindranath Y., Weinstein H., Elfferich P., and Bos J. 1990. RAS gene mutations in childhood acute myeloid leukemia: A Pediatric Oncology Group study. *Genes Chromosomes Cancer* **2:** 159.

Whitsett J.A. and Stahlman M.T. 1998. Impact of advances in physiology, biochemistry, and molecular biology on pulmonary disease in neonates. *Am. J. Respir. Crit. Care Med.* **157:** S67.

Whitsett J.A., Nogee L.M., Weaver T.E., and Horowitz A.D. 1995. Human surfactant protein B: Structure, function, regulation, and genetic disease. *Physiol. Rev.* **75:** 749.

Xu F., Taki T., Yang H.W., Hanada R., Hongo T., Ohnishi H., Kobayashi M., Bessho F., Yanagisawa M., and Hayashi Y. 1999. Tandem duplication of the FLT3 gene is found in acute lymphoblastic leukaemia as well as acute myeloid leukaemia but not in myelodysplastic syndrome or juvenile chronic myelogenous leukaemia in children. *Br. J. Haematol.* **105:** 155.

Yamamoto Y., Kiyoi H., Nakano Y., Suzuki R., Kodera Y., Miyawaki S., Asou N., Kuriyama K., Yagasaki F., Shimazaki C., Akiyama H., Saito K., Nishimura M., Motoji T., Shinagawa K., Takeshita A., Saito H., Ueda R., Ohno R., and Naoe T. 2001. Activating mutation of D835 within the activation loop of FLT3 in human hematologic malignancies. *Blood* **97:** 2434.

Yokota S., Kiyoi H., Nakao M., Iwai T., Misawa S., Okuda T., Sonoda Y., Abe T., Kahsima K., Matsuo Y., and Naoe T. 1997. Internal tandem duplication of the FLT3 gene is preferentially seen in acute myeloid leukemia and myelodysplastic syndrome among various hematological malignancies. A study on a large series of patients and cell lines. *Leukemia* **11:** 1605.

Yusen R.D., Cohen A.H., and Hamvas A. 1999. Normal lung function in subjects heterozygous for surfactant protein-B deficiency. *Am. J. Respir. Crit. Care Med.* **159:** 411.

# Assessing the Quality of Finished Genomic Sequence

J. SCHMUTZ, J. WHEELER, J. GRIMWOOD, M. DICKSON, AND R.M. MYERS
*Stanford Human Genome Center, Stanford University School of Medicine, Palo Alto, California 94304*

This April, the Human Genome Project (HGP) announced the essential completion of the human genome sequence. In just a few years, from 2001 to 2003, the percentage of finished *Homo sapiens* sequence jumped from 25% to 99%. This represented a dramatic increase in the production finishing capacity of genome centers worldwide and a shift from a primary focus on the production of draft shotgun sequence (a streamlined production pipeline) to the generation of complete and accurate finished genomic sequence (a difficult process involving decision-making and consecutive rounds of experiments). By 2001, the large genome centers had proven that they could reduce the cost of the sequencing read through increased automation, conservation of reagents, and 24/7 production level processes, but could they do the same thing for producing finished sequence? Although it is a significant challenge to maintain a production level of millions of shotgun sequencing reads per month, it is arguably more difficult to maintain a steady output of finished sequence that meets a defined accuracy standard. Perhaps surprising ourselves, we did it, overcoming the complexities of the finishing process and the allelic variation in the human genome to produce an essentially complete human genome sequence.

Now that 2.82 billion base pairs of finished human sequence have been generated, how can we be assured that the production of finished genomic sequence merited the enormous investment? Because of the cost and immensity of the project, the finished human reference sequence is not likely to be reproduced at any time in the near future. Therefore, to assess the general quality of the finished product, we must examine small portions of it and extrapolate to the rest of the completed human genome. During the project, the Stanford Human Genome Center was given a mandate by the National Human Genome Research Institute (NHGRI) to perform such an examination. In the process, we learned much about the problem of assessing the quality of the product generated by such a complex scientific process as the HGP. In this paper, we summarize the results of our quality assessment of the finished human genome sequence and offer suggestions as to how to apply these lessons to the problem of evaluating the quality of future genome sequencing projects.

## HISTORICAL MEASUREMENTS OF SEQUENCE ACCURACY IN THE HGP

In 1997, world standards for sequence accuracy were established at a meeting of HGP participants in Bermuda (now known as the "Bermuda Standards"). At this meeting, it was decided that any clone from the human genome sequence submitted as finished should have less than one error per 10,000 bases and that the sequence should be contiguous with no gaps (http://www.gene.ucl.ac.uk/hugo/bermuda2.htm). At the time, very few centers were submitting clone sequences that had no gaps, and laboratory and data analysis mechanisms for finishing all clones with no gaps were not in common practice among the data producers, and a sufficient protocol for measuring the sequence accuracy component was also unknown. Due to the prohibitively high cost of producing finished sequence, it was impractical to independently resequence and refinish many clones to establish a firm error rate for the finished sequence that was being produced.

Since this time, two principal methods for estimating sequence accuracy have been employed by the HGP genome centers: the use of quality scores from Phred processed by Phrap (Ewing and Green 1998; Ewing et al. 1998) and examintion of potential overlapping sequence from different clones for errors. To better understand the limitations, it is helpful to understand how the accuracy is generally estimated with these methods.

In the first of these methods, Phred assigns error probabilities to each base pair in every sequencing trace, based on large training data sets. Traces base-called by Phred are subsequently assembled by the assembly algorithm Phrap, which propagates these single-base-pair error scores to the consensus sequence constructed from many overlapping sequence reads. Although Phred quality scores for measuring the accuracy of single sequencing traces have been extensively validated (Ewing et al. 1998; Richterich 1998) and are used to monitor the quality of production sequencing, the cumulative Phrap score for finished bases appearing in a consensus sequence has not been similarly examined. The Phrap score provides an indication of the value of the underlying base, but because of the complexity of the data assembly process, one cannot simply add up the Phrap estimated errors for all of the bases in a finished clone to determine the error rate. In our experience, the Phrap error rates are tenfold lower than the actual errors in the clone sequences. The simple reason for this is that Phrap error rates are assigned only to bases appearing in the consensus, and potential problem base pairs tend not to be included in the Phrap-derived consensus for the very reason that they are poor-quality bases. These problem bases include compressed or stretched peaks, and errors created by the Phrap assembly algorithm itself, such as badly aligned base pairs. In addition, the cumulative error rate estimation generated by Phrap is applicable only to the closed data set of the assembled sequence contig, and as such, has no way of estimating the incidence of potential errors caused by deleted sequence or erroneous assembly. We believe that the Phrap quality scores should be used solely as a guide toward addressing the problems in constructing a consensus sequence (Gordon et al. 1998) from many short stretches of sequence, and that cumulative Phrap quality scores fall short of giving an accurate picture of the underlying sequence accuracy to the biological template from which it was derived.

The second method of determining base-pair accuracy is the examination of overlapping clones, which relies on the availability of independently finished redundant clone sequences. The discrepancies between clones covering the same genomic region are counted and then determined to be due to either finishing errors or polymorphic variation. Errors found in this manner are then used to calculate the accuracy of the finished sequence. Although this method gives a reasonable estimate of the base-pair error rate, it relies on the principle of independently finished redundant data, and it is often unusual for the HGP to find redundantly finished clones that were not finished using information from both clones. The problem is compounded by the difficulty of finding adjoining clones from the same chromosome due to the mosaic of individual sources used to construct the final HGP consensus. To reduce the cost of finishing, efforts were made to minimize redundancy, and for much of the HGP, only 100 bp of overlapping data between clones were submitted to the sequence databases, leaving few independently sequenced data sets available to compare sequences for the purposes of quality analysis.

Because of the difficulties and inherent bias with each center measuring its own sequence quality, the NHGRI conducted three rounds of "round-robin" quality exchanges, with the results of the first two rounds having been described (Felsenfeld et al. 1999). Centers exchanged sequence trace data from 2–4 finished clones, and then assembled and verified the finished consensus. They also attempted to fix low-quality bases and any gaps left by the original finishing center. In the second round, which was performed in 1998, 1.7 Mb were surveyed and 22 of 36 clones were found to meet the Bermuda Standards. With only 61% of the clones meeting the quality standard and 2.6 billion base pairs remaining to be finished, the genome centers had a significant amount of work to do to raise the accuracy of their finished sequence and reduce the number of final gaps in the genomic sequence.

## SUMMARY OF OUR QUALITY ASSESSMENT

The NHGRI released an RFA in 1999 soliciting researchers to establish a central quality assessment center, and did not fund such a center from that search. In 2001 the NHGRI commissioned the Stanford Human Genome Center (SHGC) to perform a large-scale quality assessment of the finished sequence produced by the NHGRI-funded sequencing centers. The SHGC was in the unique position of being the only U.S. large-scale finishing center not funded by NHGRI, with funding from the Department of Energy (DOE) to finish chromosomes 19, 5, and 16 (in collaboration with the DOE's production sequencing center, the Joint Genome Institute). Because we were in the midst of contributing 11% of the finished human sequence ourselves and had participated in the previous quality exercises organized by NHGRI, we had the experience necessary to evaluate the finishing processes and finished sequence of the other centers and the ability to do so independently from funding concerns.

We designed a procedure to perform quality assessment of finished sequence to accurately measure the number of errors in finished genomic sequences, irrespective of the specific techniques employed to produce the original data sets. We aimed not only to calculate an accurate error estimate for the center's finished sequence, but also to identify any specific error-contributing laboratory or analysis techniques employed by each genome center. We requested a glycerol stock of the large-insert clone, created sized plasmid libraries for these clones, and sequenced both ends of these subclones to 3–4x coverage in high-quality bases (Phred 20 bases). These additional sequence data were then combined with the center's original trace data to create a "gold standard" assembly. We finished these clones to a high degree of accuracy with direct primer walks on the large-insert clones and compared our consensus sequence to the original submitted by the center. Using this robust approach, we surveyed 34 Mb of finished sequence from seven major contributors to the human genome and subsequently analyzed the resulting errors.

The results of this analysis have been published elsewhere (in press), and only a summary of the results is presented here. We identified 466 errors in this finished sequence, or about 1 error per 73,000 bp. Adjusting this error rate based on the amount of sequence contributed by a given center to the final human reference sequence gives an estimate of an average of 1 error per 143,000 bp in the finished human genome. In total, 184 of 197 large-insert clones substantially exceeded the standard base-pair accuracy rate of 1 error in 10,000 bp. This is a dramatic improvement over the results reported from the round-robin study cited above, with 93% of the clones exceeding the accuracy requirement (up from 61% 4 years earlier). Nevertheless, 12 of the 13 clones that did not meet the accuracy standard were in error because they contained a misassembly or deleted a portion of the true consensus. The remaining clone did not meet the Bermuda accuracy standards due to erroneous base-pair content.

## ASSESSING A LARGE COMPLEX FINISHED SEQUENCE

Sequencing and complete finishing of a large-insert clone is a difficult process, akin to correctly assembling a 3000+ piece jigsaw puzzle with many near-identical pieces and no available picture on the box cover. Much of the HGP was devoted to developing techniques to accurately reconstruct large pieces of genomic sequence from small, noisy overlapping reads of subclone sequence. The experimental and computational methods employed to finish genomic sequence clones have changed dramatically over the course of the project. Read length and read quality have improved with advances in sequence chemistry and detection platforms, and the cost of producing shotgun reads has decreased exponentially. This marked reduction in cost and effort per lane of sequence has resulted in a corresponding shift in sequencing strategy. When sequencing costs were a major constraint, it was desirable to minimize the draft redundancy of samplings per base pair, and to complete the sequence using directed finishing sequence reads and extensive hand-curation of the sequence. Now that sequencing costs are dramatically reduced, it is customary to generate 10x or more shotgun coverage sampling and to minimize the use of expensive directed reads and human review of the sequence. The preferred cloning system for high-throughput sequencing has also changed from M13 bacteriophage clones to bacterial plasmid clones, from which two paired reads can be generated per DNA isolation. Finishing strategies have evolved such that gaps can now be closed by sequencing directly from the BAC clones, and difficult regions can be closed by using chemically enhanced sequencing reactions. These changes have increased the overall quality and throughput of the genome centers and have, in general, increased the accuracy of the finished sequence. Although the majority of finished human genomic sequence was generated using these modern production-scale techniques, the entire human sequence is not finished to the same quality standards because these improvements were incremental over the course of the HGP.

There are two possible experimental procedures for performing a quality assessment of a large vertebrate genome sequence, each with its own set of difficulties and practical concerns: time-based sampling of sequence production over the duration of the project, and "geographic" sampling of the completed genome sequence. A thorough time-interval-based quality analysis of a completed genome would require sampling finished sequence data at regular time intervals throughout the duration of the entire project. The advantage to this approach is that quality analysis can proceed concurrently with the generation of finished sequence and provides a convenient metric for evaluating the efficiency of production finishing processes. The disadvantage is that it is difficult to ensure sufficient geographic coverage of the genome for global quality assessments. This difficulty can be overcome theoretically by sampling a sufficient number of clones from each time period to ensure that some of the more geographically difficult regions of the genome will be covered by the analysis. A geographic approach would sample the finished genome at regular spatial intervals along the chromosomes. The advantage to this approach is that geographic coverage of the genome is assured, and global statements about the quality of the genome as a whole can be readily drawn. The disadvantage is that the quality analysis can only occur after the entire genome has been completed.

Our quality analysis used the time-based approach, spanning two distinct time periods, or rounds. Figure 1 depicts the total nonredundant finished human sequence submitted to GenBank over time, and the amount of that sequence that was available to us for analysis in the given time periods. During each round, we sampled a percentage of clones finished monthly by each center for the time period indicated. As mentioned previously, there have been substantial and incremental improvements to the finishing process during the HGP. Thus, our sampling is merely a snapshot of the tools and procedures in place during the time period analyzed, and it is possible that had we sampled sequence generated prior to 2001, our results might have reflected a higher error rate. For some of the

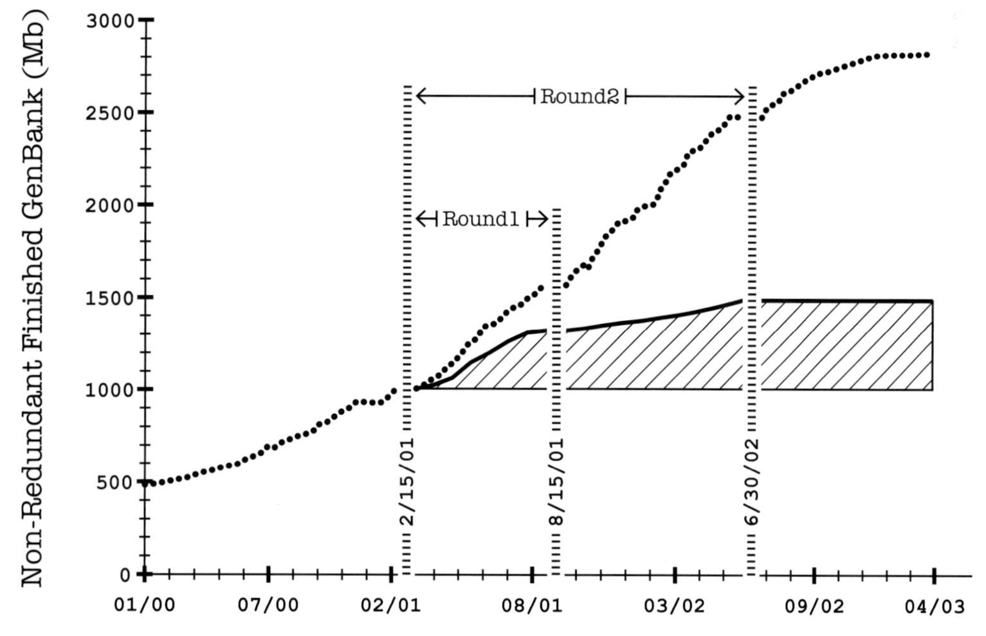


**Figure 1.** Time-interval-based sampling of finished human sequence.

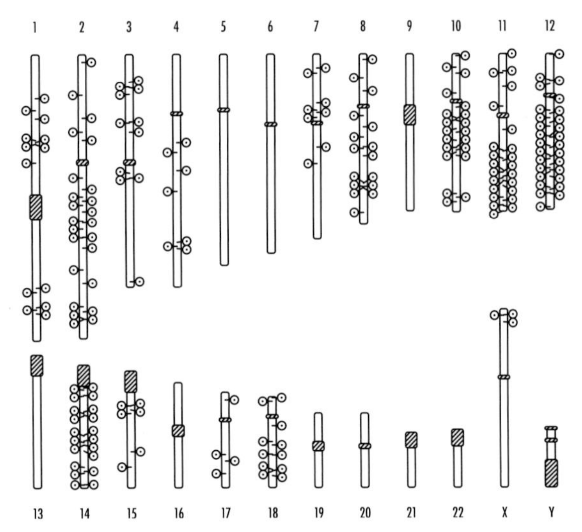


**Figure 2.** Genomic distribution of QA sampled clones.

smaller centers in our QA study we surveyed a substantial amount of their total contribution to the finished genome, and indeed, error rates in their finished clones were found to decrease over time.

For comparison of a geographic sampling to that of our QA, see Figure 2, which shows the location of large-insert clones from our random time-interval-based assay. The distribution of surveyed clones is far from even across the genome, and reflects the limitations of a time-based survey that does not span the entire duration of the genome sequencing project. We encountered few clones near the telomeres, which have large, difficult-to-assemble repeat structures that were in some cases misassembled. Much of the genomic sequence varies in self-similarity and the corresponding assembly difficulty. A geographic sampling would need to take this variation into account, and it may be difficult to select a set of clones that are truly representative of the genome without re-sequencing a rather large portion of it. Additionally, our QA discovered that the variation in sequence quality was most significantly correlated to the sequencing center that generated the data, and a geographic sample would need to include sufficient clones from each contributing center to accurately estimate the quality of the genome as a whole.

## AUGMENTING THE FINISHING STANDARDS

Building on the Bermuda Standards, the Finishing Working Group, chaired by Dr. Rick Wilson of Washington University, progressively revised and expanded the criteria by which a sequence is considered to be a finished product for the human genome. The current version of "best practice standards" is available from http://www.genome.wustl.edu/Overview/g16stand.php. It states in detail the definition of a finished clone and describes how to annotate difficult issues in the finished sequence. These finishing rules address the specifications of the final product but do not provide direct guidelines toward achieving this product quality. As they were revised along with the technical abilities of the contributing centers, it was assumed that centers utilizing these guidelines of the HGP understood the steps necessary to meet the specifications. However, these guidelines are now rapidly being adopted by other large-scale sequencing projects and by many new sequencing centers and, as shown by our QA, not all centers contributing to the human genome equally applied these finishing guidelines. We believe that it is important in future large-scale sequencing projects to include some revisions in the finishing standards which close loopholes in the current standards in order to improve and ensure the quality of finished sequence currently being deposited in the public databases.

We would add the concept of completeness to the finishing standards, mindful of the notion that, although a singular high-quality consensus sequence may have been constructed "end to end" of the large-insert clone, there may be sequence missing due to false joins. A thorough accounting for all paired-plasmid end links in a sequenced subclone library is usually a sufficient measure of completeness. Such a criterion is now possible because all of the large-scale sequencing centers have switched their production to plasmid-based sequencing from sized libraries. The two paired-plasmid reads provide two sequences spanning an expected distance in the assembly, and a thorough accounting for these reads, including their relative order and orientation, is usually sufficient to catch deleted regions of sequence. Thorough accounting would be mindful of the fact that the entire assembly should be covered by multiple plasmid subclones, all falling within this defined library size range, and all contigs containing reads from plasmid subclones that are linked to the main sequence contig must be assembled into the finished consensus. To achieve completeness in a finished assembly, there should be no breaks in the paired-link coverage over the entire clone, and all spurious links should be addressed. Figure 3 gives an example of a false join from a QA project where the small sequence contig has not been included in the main consensus due to a false join. Using the information provided by the paired end links, it is not difficult to determine that the small piece should be inserted into the larger consensus.

Although the finishing rules address the concept of a single read from a single subclone, we suggest that the rules for single subclones be further expanded. The original intention of allowing a single high-quality plasmid read of less than 100 bp to appear in the finished clone was to allow large-insert clones to be deposited that had, through random distribution, some bases not covered by multiple reads. This rule, however, has been taken to mean that no regions of the consensus generated by reads from single subclones less than 100 bp need to be addressed during the finishing process. Figure 4 shows a common occurrence found in the QA clones that led in some cases to significant errors in the consensus sequence. The sequencing reads from both directions have hard stops due to simple sequence runs or "hairpin" structures, and an attempt was made to close this sequence gap in the finishing process. After many repeated attempts, a sequencing reaction was performed on a single subclone that reads through the sequence stop and agrees with the

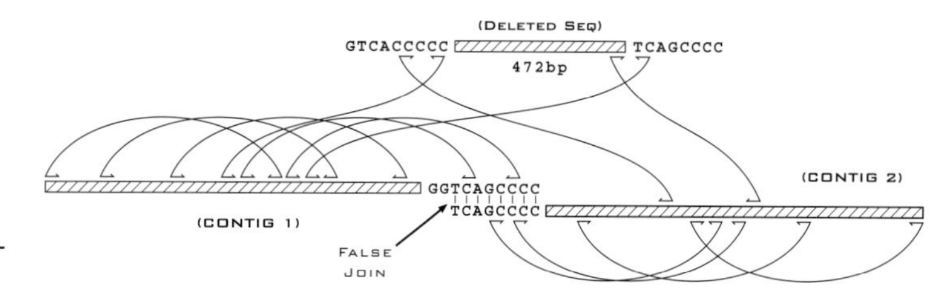


**Figure 3.** False join, resolvable through paired-end examination.

consensus on the other side of the gap. Although there is only a single high-quality read that bridges the gap, the clone meets the existing finishing criteria because the extent of the single subclone read is less than 100 bp. On many occasions, we found that the single subclone read "closed" the gap merely due to the fact that the sequencing reaction skipped the hairpin, deleting portions of sequence to bridge the gap, or was a reaction performed on a deleted template. This is not identifiable through paired end-link examination, as usually the sequence missing is not enough to skew the size distribution of the paired ends. In these cases, another subclone should always be sampled to confirm that the join across the difficult region is accurate. As a general rule, single subclone regions of <100 bp are acceptable in regions of low sequence coverage, but not in regions for which attempts to close a gap have been made.

## WHERE ARE WE GOING WITH FINISHED SEQUENCE?

We found it difficult to assess the finished sequence quality of a complex vertebrate genome like that of human, yet we are moving toward producing data sets that will be much harder to assess than the clone-by-clone-based approach used to finish the human genome. Although the debate over whether the whole genome shotgun (WGS) approach or the publicly funded clone-by-clone approach produced a better draft human sequence continues (Green 2002; Mardis et al. 2002; Myers et al. 2002; Olson 2002; Waterston et al. 2002a, 2003), the trend among large-scale sequencing centers is toward producing WGS data sets. Doing so minimizes the need to even create a minimal map before sequencing. The current approach, being used for the important model organisms like mouse (Waterston et al. 2002b), zebrafish, and chicken, is to produce some level of WGS data and follow up with directed clone-by-clone finishing. This has many of the benefits of WGS, in that sequence is produced rapidly and very cheaply and the reads can be used in subsequent clone assemblies, and still provides a level of assurance and correction of the assembly with long stretches of finished clones. The first large genome, *Drosophila*, has been produced in this manner (Adams et al. 2000) and resulted in a high-quality finished genome. The mouse genome, of similar size and complexity to human, is also rapidly being finished with this hybrid approach. These hybrid genomes can be evaluated similarly to our human QA by assessing the individual finished clones using either the temporal or geographical strategies outlined above.

In some cases, however, genomes have been sequenced using only WGS reads, and the assembly of these reads forms the end product. In the case of the two *Ciona* sequences (Dehal et al. 2002), the extreme allelic variation makes producing a unique reference sequence neither possible nor desirable. For *Fugu*, the mosquito, and a certain poodle, the WGS is the finished product (Aparicio et al. 2002; Holt et al. 2002; Kirkness et al. 2003). As more of the biologically important genomes are completed, there will be less need to finish a sequence to the human standard, and WGS-produced data sets will dominate the available public sequence. How can we assess the sequence accuracy of these shotgun-only genomes? It is difficult to definitively answer this question, as there are many parts of the procedure for generating WGS genomes that will need to be standardized first.

The largest of these issues is the assembly process itself. Details of the many recently developed genome assembly programs have been described (Huson et al. 2001; Batzoglou et al. 2002; Jaffe et al. 2003; Mullikin and Ning 2003), but little or no benchmarking has been done for whole-genome assemblies. The most comparisons between any two algorithms were performed with Arachne and Phusian during the development and production of the mouse WGS data set; the others have not yet been

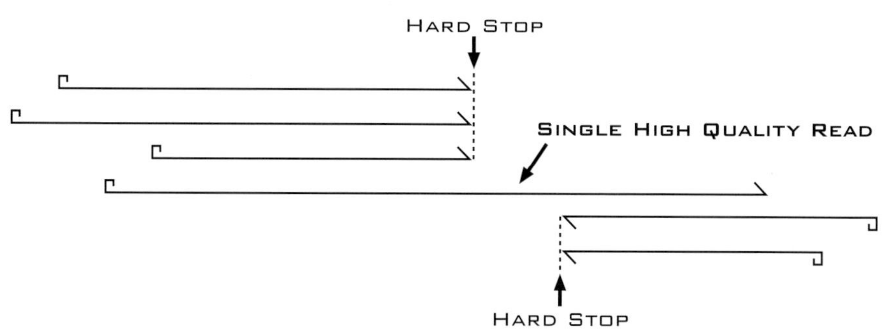


**Figure 4.** Potential false join on a single high-quality sequence read.

compared. Some of these algorithms are designed to address a single allelic copy WGS and do not work on the polymorphic data sets. In addition, the algorithms vary in their ability to resolve repeats based on paired end links. The resulting assemblies of vertebrates are large, and the metrics for comparing them have yet to be established. We need a standard polymorphic and a standard inbred WGS data set, along with finished genomes for both, to be able to compare the effectiveness and accuracy of the WGS assembly. As we venture farther away from human sequence in evolutionary terms, we will uncover new and difficult problems in other species, which will add to the problem of defining the genome quality. The sequence coverage levels for published assemblies for nonhuman WGS data sets vary from 1.5x for the poodle to 10.2x for mosquito, with *Fugu* falling in between at 5.6x. The coverage level is likely to vary for most WGS organisms due to funding pressures and the desirability of completeness for the genome. The variation in sequence coverage, together with cloning bias and repeated sequences, combine to cause significant issues with evaluating a final WGS product. Furthermore, the finishing rules in place for human cannot be applied to the resulting assemblies. This includes our notion of completeness, as the WGS assemblers must through necessity discard many sequence traces that cannot be placed due to repeats or that are contradictory to already assembled contigs. In addition, because contamination is common in the large data sets, many reads will not be placed in the finished product.

One approach for quality control and assessment for the WGS data sets is to finish a randomly selected set of large-insert clones to a high degree and compare the resulting consensus sequence to the WGS assembly. The finished clones will both provide a training set of sequence for the assembly process and serve to verify a portion of the WGS consensus. One percent or more of the total organism should be finished in a clone-based approach to ensure the quality and accuracy of the draft assembly.

## CONCLUSIONS

With our evaluation of the finished human genome sequence, we determined that the majority of the human genome sequence exceeded the accuracy standards; however, there are improvements that can be made to the finished genome and to the current finishing standards. The finished genome of *Drosophila* has had its third finished release (Celniker et al. 2002), and the human genome has already had one new assembly since the April finished date. We expect that improvements to the final human genome will continue as new techniques are developed which can address the serious remaining issues of segmental duplications and sequences that are unclonable in existing vector systems. The data producers face the difficult challenge of maintaining the finished sequence quality standards used for the evaluation of the human genome sequence as many more genomic sequences, each having their own unique experimental problems to overcome, are produced and finished. As it becomes even faster and easier to produce draft genomes, we must, as a community, address the concerns of sequence accuracy of the WGS assemblies and provide benchmarks of acceptable quality for these genomic sequences, given the intended scientific uses of these data sets. Regardless, once finishing standards are defined and agreed upon, it would be wise to integrate a quality assessment strategy into the production process, so that global statements of genome quality can be made for each future genome project.

## ACKNOWLEDGMENT

We thank James Retterer for the illustrations included in this manuscript.

## REFERENCES

Adams M.D., Celniker S.E., Holt R.A., Evans C.A., Gocayne J.D., Amanatides P.G., Scherer S.E., Li P.W., Hoskins R.A., Galle R.F., George R.A., Lewis S.E., Richards S., Ashburner M., Henderson S.N., Sutton G.G., Wortman J.R., Yandell M.D., Zhang Q., Chen L.X., Brandon R.C., Rogers Y.H., Blazej R.G., Champe M., and Pfeiffer B.D., et al. 2000. The genome sequence of *Drosophila melanogaster. Science* **287:** 2185.

Aparicio S., Chapman J., Stupka E., Putnam N., Chia J.M., Dehal P., Christoffels A., Rash S., Hoon S., Smit A., Gelpke M.D., Roach J., Oh T., Ho I.Y., Wong M., Detter C., Verhoef F., Predki P., Tay A., Lucas S., Richardson P., Smith S.F., Clark M.S., Edwards Y.J., and Doggett N., et al. 2002. Whole-genome shotgun assembly and analysis of the genome of *Fugu rubripes. Science* **297:** 1301.

Batzoglou S., Jaffe D.B., Stanley K., Butler J., Gnerre S., Mauceli E., Berger B., Mesirov J.P., and Lander E.S. 2002. ARACHNE: A whole-genome shotgun assembler. *Genome Res.* **12:** 177.

Celniker S.E., Wheeler D.A., Kronmiller B., Carlson J.W., Halpern A., Patel S., Adams M., Champe M., Dugan S.P., Frise E., Hodgson A., George R.A., Hoskins R.A., Laverty T., Muzny D.M., Nelson C.R., Pacleb J.M., Park S., Pfeiffer B.D., Richards S., Sodergren E.J., Svirskas R., Tabor P.E., Wan K., and Stapleton M., et al. 2002. Finishing a whole-genome shotgun: Release 3 of the *Drosophila melanogaster* euchromatic genome sequence. *Genome Biol.* **3:** RESEARCH0079.

Dehal P., Satou Y., Campbell R.K., Chapman J., Degnan B., De Tomaso A., Davidson B., Di Gregorio A., Gelpke M., Goodstein D.M., Harafuji N., Hastings K.E., Ho I., Hotta K., Huang W., Kawashima T., Lemaire P., Martinez D., Meinertzhagen I.A., Necula S., Nonaka M., Putnam N., Rash S., Saiga H., and Satake M., et al. 2002. The Draft genome of *Ciona intestinalis:* Insights into chordate and vertebrate origins. *Science* **298:** 2157.

Ewing B. and Green P. 1998. Base-calling of automated sequencing traces using *Phred.* II. Error probabilities. *Genome Res.* **8:** 186.

Ewing B., Hiller L., Wendl M., and Green P. 1998. Base-calling of automated sequence traces using *Phred.* I. Accuracy assessment. *Genome Res.* **8:** 175.

Felsenfeld A., Peterson J., Schloss J., and Guyer M. 1999. Assessing the quality of the DNA sequence from the human genome project. *Genome Res.* **9:** 1.

Gordon D., Abajian C., and Green P. 1998. Consed: A graphical tool for sequence finishing. *Genome Res.* **8:** 195.

Green P. 2002. Whole-genome disassembly. *Proc. Natl. Acad. Sci.* **99:** 4143.

Holt R.A., Subramanian G.M., Halpern A., Sutton G.G., Charlab R., Nusskern D.R., Wincker P., Clark A.G., Ribeiro J.M., Wides R., Salzberg S.L., Loftus B., Yandell M., Majoros W.H., Rusch D.B., Lai Z., Kraft C.L., Abril J.F., Anthouard V., Arensburger P., Atkinson P.W., Baden H., de Berardinis V., Baldwin D., and Benes V., et al. 2002. The genome sequence of the malaria mosquito *Anopheles gambiae*. *Science* **298:** 129.

Huson D.H., Reinert K., Kravitz S.A., Remington K.A., Delcher A.L., Dew I.M., Flanigan M., Halpern A.L., Lai Z., Mobarry C.M., Sutton G.G., and Myers E.W. 2001. Design of a compartmentalized shotgun assembler for the human genome. *Bioinformatics* (suppl. 1) **17:** S132.

Jaffe D.B., Butler J., Gnerre S., Mauceli E., Lindblad-Toh K., Mesirov J.P., Zody M.C., and Lander E.S. 2003. Whole-genome sequence assembly for mammalian genomes: Arachne 2. *Genome Res.* **13:** 91.

Kirkness E.F., Bafna V., Halpern A.L., Levy S., Remington K., Rusch D.B., Delcher A.L., Pop M., Wang W., Fraser C.M., and Venter J.C. 2003. The dog genome: Survey sequencing and comparative analysis. *Science* **301:** 1898.

Mardis E., McPherson J., Martienssen R., Wilson R.K., and McCombie W.R. 2002. What is finished, and why does it matter. *Genome Res.* **12:** 669.

Mullikin J.C. and Ning Z. 2003. The phusion assembler. *Genome Res.* **13:** 81.

Myers E.W., Sutton G.G., Smith H.O., Adams M.D., and Venter J.C. 2002. On the sequencing and assembly of the human genome. *Proc. Natl. Acad. Sci.* **99:** 4145.

Olson M.V. 2002. The human genome project: A player's perspective. *J. Mol. Biol.* **319:** 931.

Richterich P. 1998. Estimation of errors in "raw" DNA sequences: A validation study. *Genome Res.* **8:** 251.

Waterston R.H., Lander E.S., and Sulston J.E. 2002a. On the sequencing of the human genome. *Proc. Natl. Acad. Sci.* **99:** 3712.

———. 2003. More on the sequencing of the human genome. *Proc. Natl. Acad. Sci.* **100:** 3022.

Waterston R.H., Lindblad-Toh K., Birney E., Rogers J., Abril J.F., Agarwal P., Agarwala R., Ainscough R., Alexandersson M., An P., Antonarakis S.E., Attwood J., Baertsch R., Bailey J., Barlow K., Beck S., Berry E., Birren B., Bloom T., Bork P., Botcherby M., Bray N., Brent M.R., Brown D.G., and Brown S.D., et al. (Mouse Genome Sequencing Consortium). 2002b. Initial sequencing and comparative analysis of the mouse genome. *Nature* **420:** 520.

# Human Subtelomeric DNA

H. RIETHMAN, A. AMBROSINI, C. CASTANEDA,* J.M. FINKLESTEIN,†
X.-L. HU, S. PAUL, AND J. WEI
*The Wistar Institute, Philadelphia, Pennsylvania 19104*

Telomeres are dynamic and complex chromosomal structures. They are essential for genome stability and faithful chromosome replication, and they mediate key biological activities including cell cycle regulation, cellular aging and immortalization, movements and localization of chromosomes within the nucleus, and transcriptional regulation of subtelomeric genes (Blasco et al. 1999; Feuerbach et al. 2002).

The DNA at each human chromosome terminus is a simple repeat sequence tract (TTAGGG)n, typically 5 to 15 kb in length in somatic cells (Moyzis et al. 1988), that ends with a single-stranded extension of the G-strand of DNA (Griffith et al. 1999). The lengths of the terminal repeat tracts are dynamically modulated in a tissue-specific and individual-specific manner. Adjacent to this "terminal repeat" is a subtelomeric repeat region comprising a mosaic patchwork of segmentally duplicated DNA. This class of low-copy repeat DNA is characterized by very high sequence similarity (90% to >99.5%) between duplicated tracts, and variably sized, but often very large, duplicated segment lengths (1 kb to >200 kb). Some of the segmental duplications are unique to subtelomeric repeat regions, some are shared with a subset of pericentromeric repeat regions, and some are shared with one or several interstitial chromosomal loci (Riethman et al. 2001). The aggregate size of a subtelomeric repeat region varies according to the specific telomere; the shortest subtelomeric repeat region is 2 kb in length and the longest is greater than 500 kb. At many individual telomeres, allelic differences in the sizes of subtelomeric repeat regions can be quite large, on the order of hundreds of kilobases in length.

The lengths of (TTAGGG)n tracts vary from telomere to telomere within the same cell (Lansdorp et al. 1996; Zijlmans et al. 1997) and, strikingly, between alleles at the same telomere (Baird et al. 2003). In humans, these differences are not apparently inherited in Mendelian fashion but rather are preset in the zygote in a stochastic fashion and attrited from the zygotic set points with progressive cell divisions thereafter (Baird et al. 2003). Importantly, the effects of (TTAGGG)n tract loss on cell viability and chromosome stability are attributable to the shortest telomere in a cell, rather than to average telomere length (Hemann et al. 2001). The implications of this recent work are that measurement of bulk telomere length is insufficient for detecting biologically critical telomere loss, that the global complement of variable telomere (TTAGGG)n tract lengths in a genome is likely to be individual-specific, and that stochastically preset zygotic telomere lengths might influence the rates at which cells age and become susceptible to genome instability. Molecular tools to measure specific telomere tract lengths are essential to further characterize this fascinating aspect of human telomere dynamics and its relationship to cellular and organismal aging. Subtelomeric DNA sequences immediately adjacent to human terminal (TTAGGG)n tracts are required, along with information on telomere-specific and allele-specific sequence differences in this DNA, to develop these tools.

The unusual sequence organization of human telomere regions has led to complications with respect to mapping and sequencing, due both to underrepresentation of subtelomeric clones in genomic libraries and to localization and assembly problems caused by the large size and close similarity of duplicated subtelomeric DNA segments. Whole-genome shotgun sequence assembly of these regions is expected to be especially unreliable. In addition, segmental duplications can predispose associated chromosome segments to genetic instability and have been connected with genetic diseases (Bailey et al. 2002). Finally, evolutionarily recent duplicative transposition of these large DNA tracts has led to the generation of new gene families and to the formation of fusion transcripts with potentially new functions (Bailey et al. 2002).

In this work, half-YAC cloning and physical mapping data were combined with public draft and finished sequences to derive subtelomeric sequence assemblies for each of the 41 genetically distinct human telomere regions (Fig. 1). Sequence gaps that remain on the reference telomeres are generally small, well-defined, and, for the most part, restricted to regions directly adjacent to the terminal (TTAGGG)n tract. Of the 18.87 Mb of subtelomeric DNA analyzed, 2.35 Mb are subtelomeric repeat sequences (Srpt) and an additional 1.57 Mb are segmental duplications. The subtelomeric sequence assemblies are highly enriched in short, internal (TTAGGG)n-like sequences relative to the rest of the genome. A total of 99 (TTAGGG)n-like islands were found in subtelomeric DNA; 45 within Srpt regions, 38 within 1-copy regions, 7 at 1-copy/Srpt or Srpt/segmental duplication boundaries, and 9 at the telomeric ends of assemblies. Known and potential protein-encoding transcripts were annotated in each assembly, noting their mapping coordinates relative

Present addresses: *Program in Molecular Biophysics, Johns Hopkins University, Baltimore, Maryland 21218; †Cell and Molecular Biology Program, University of Pennsylvania, Philadelphia, Pennsylvania 19104.

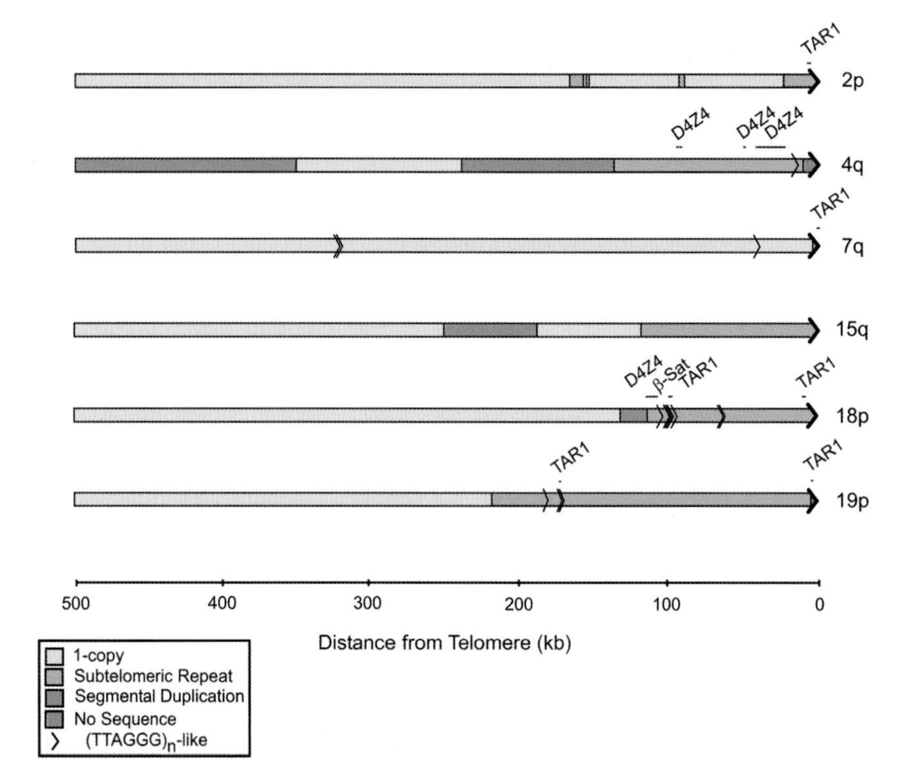


**Figure 1.** Examples of sequence organization in human subtelomeric DNA regions. The terminal (TTAGGG)n repeat tract consists of 2–15 kb of the simple repeat sequence (TTAGGG)n and is indicated by the black arrow at the 0 coordinate. The subtelomeric repeat (Srpt) region (*blue*) comprises a mosaic patchwork of segmentally duplicated DNA tracts that occur in two or more subtelomere regions. The Srpt region can be shorter than 10 kb or longer than 300 kb, depending on the specific telomere. Centromeric to the Srpt region is chromosome-specific genomic DNA, typically with a high GC content and high gene density. Stretches of segmentally duplicated DNA (sequence >90% identical, >1 kb in size) that are not subtelomere-specific (*green*) are interspersed in a telomere-specific fashion with 1-copy subtelomeric regions (*yellow*). Short (50–250 bp) and often degenerate (TTAGGG) tracts are interspersed within the Srpt region and, occasionally, within the adjacent 1-copy subtelomeric DNA (*internal black arrows*).

to their respective telomere and whether they originate in duplicated DNA or single-copy DNA. Seven hundred thirty-two transcripts were found in 14.95 Mb of 1-copy DNA, 82 transcripts in 1.57 Mb of segmentally duplicated DNA, and 198 transcripts in 2.35 Mb of Srpt sequence. This overall transcript density is about 20% greater than that which is found genome-wide. Zinc finger-containing genes and olfactory receptor genes are duplicated within and between multiple telomere regions. Detailed maps, subtelomeric assemblies (FASTA format), and transcript annotations are available at our laboratory Web site (http://www.wistar.upenn.edu/Riethman/).

## CLONING AND SEQUENCING

Our basic approach to resolving the subtelomeric mapping and sequencing problem has been to isolate each telomere region using a specialized yeast artificial chromosome (YAC) system that permits propagation of large telomere-terminal human DNA fragments as linear plasmids in yeast (Riethman et al. 1989). The result of a successful cloning event is a yeast strain that carries a large "half-YAC" (Riethman et al. 2001; Riethman 2003); the terminal repeat tract, the entire subtelomeric repeat region, and the adjacent single-copy DNA region are physically linked on a single large DNA segment that has been purified from the rest of the human genome. From this starting material, the most distal single-copy segments of each chromosome arm can be identified and analyzed, and the particular subtelomeric repeat organization of each cloned fragment can be deciphered without interference from duplicons derived from elsewhere in the genome.

A major challenge associated with this strategy is to validate the YAC clone structure and ensure that the deduced sequence organization reflects that of native genomic DNA. As with the cloning strategy itself, we took advantage of the unique properties of telomeres to assist with this problem. Because the telomere forms the end of a very large linear DNA molecule, a site-specific cleavage of DNA close to the end of the chromosome is expected to release a DNA fragment whose size corresponds to the distance between the cleavage site and the end of the chromosome. A site-specific DNA cleavage method developed in the early 1990s (Ferrin and Camerini-Otero 1991; Koob et al. 1992) was thus applied to this problem in order to validate the sequence organization of individual alleles of many telomeres. Similar mapping methods are turning out to be an important tool for analyzing large variant subtelomeric regions in human populations.

The elucidation of finished DNA sequence from each of these regions has taken several paths, each of which takes advantage of the telomere linkage of half-YAC sequences and/or the physical mapping experiments but which differ in the extent to which DNA sequences are derived from the half-YACs themselves. The phenomenal throughput of the public draft sequencing project allowed connection of most of the telomere regions with the human working draft sequence (Riethman et al. 2001), and the extension of this work to near completion is summarized in Table 1. Most recently, half-YACs have been gel-purified, shotgun-cloned, sequenced, and assembled directly without an intermediate cosmid subcloning step; although technically more difficult, this approach provides the most direct route to complete subtelomeric sequences, has now been used successfully by several sequencing centers, and will be the approach used to sequence large variant alleles of subtelomeric regions (see Fig. 2).

## ASSEMBLY AND ANALYSIS

Subtelomeric clones and sequence accessions that were identified and connected to telomeres previously (IHGSC 2001; Riethman et al. 2001) were used to nucleate the assembly of new and more complete subtelomeric draft/finished sequence contigs for these regions as the IHGSC has continued sequencing of these clones. The sequences used for the described assembly and analysis were those available in the public databases in August 2002.

The overall sequence organization of each subtelomeric assembly was evaluated initially in terms of subtelomeric repeats, segmental duplications, satellite sequences, and internal (TTAGGG)n-like sequence content. First, BLAST (Altschul et al. 1997) analysis of a database of the subtelomeric assemblies was done; Srpt sequences were defined as any non-self sequence match having >90% identity within the subtelomeric sequence database. Second, sequence comparisons between the subtelomeric assemblies and public databases were used to define additional homology segments; segmental duplications were defined as non-self sequence matches in the NR or HTGS databases, but absent in the Srpt database, that were greater than 90% identity and greater than 1 kb in length. Finally, satellite-related and (TTAGGG)n-related sequences were identified using the high-sensitivity RepeatMasker parameters (Smit and Green [http://ftp.genome.washington.edu/RM/RepeatMasker.html]).

Figure 1 illustrates a range of different types of subtelomeric repeat organization found in the reference sequences. The bulk of Srpt sequences are confined to the most distal regions of the subtelomere, although there are several examples (2p, 2q, 3q, 5p, 7p, 8p, and 12p) where, in addition to a terminal block of Srpt, there are additional smaller segments interspersed within the adjacent 1-copy DNA and segmentally duplicated DNA. Segmental duplication blocks were often found adjacent to Srpts but displayed a highly variable pattern of content and distribution at each chromosome end (Fig. 1). Overall, 12.5% of the 18.87 Mb of subtelomeric DNA analyzed was

**Table 1.** Progress in Completing Telomeric Reference Sequences

| Tel | Telomere gaps[a] | Evidence for Tel linkage[b] | Large-scale variation[c] |
|---|---|---|---|
| 1p | 0 | RARE, Srpt | P, Hi |
| 1q | 10 | RARE, Srpt | Lo |
| 2p | 0 | RARE, Srpt | Lo |
| 2q | 16 | RARE, Srpt | P, Lo |
| 3p | 35 | Srpt | unk |
| 3q | 55 | Srpt | unk |
| 4p | 0 | RARE, Srpt | unk |
| 4q | 10 | Srpt | P, Hi |
| 5p | 70 | RARE, Srpt | unk |
| 5q | 20 | RARE, Srpt | P, Lo |
| 6p | 5 | RARE, Srpt | P, Hi |
| 6q | 3 | RARE, Srpt | P, Lo |
| 7p | 34 | RARE, Srpt | P, Hi |
| 7q | 0 | RARE, Srpt | Lo |
| 8p | 0 | RARE, Srpt | P, Hi |
| 8q | 0 | RARE, Srpt | Lo |
| 9p | 0 | RARE, Srpt | unk |
| 9q | 0 | Srpt | P, Hi |
| 10p | 14 | Srpt | unk |
| 10q | 0 | RARE, Srpt | P, Hi |
| 11p | 0 | RARE, Srpt | P, Hi |
| 11q | 0 | Srpt | unk |
| 12p | 16 | Srpt | P, Hi |
| 12q | 60 | RARE, Srpt | Lo |
| 13q | 15 | RARE, Srpt | Lo |
| 14q | 7 | RARE, Srpt | P, Hi |
| 15q | 0 | Srpt | Lo |
| 16p | 0 | RARE, Srpt | P, Hi |
| 16q | 5 | Srpt | P, Hi |
| 17p | 38 | RARE, Srpt | P, Hi |
| 17q | 120 | RARE, Srpt | unk |
| 18p | 0 | RARE, Srpt | Lo |
| 18q | 0 | RARE, Srpt | Lo |
| 19p | 11 | RARE, Srpt | P, Lo |
| 19q | 5 | Srpt | P, Hi |
| 20p | 55 | Srpt | unk |
| 20q | unk (50) | Srpt | P, Hi |
| 21q | 0 | RARE, Srpt | Lo |
| 22q | 20 | Srpt | unk |
| Xp/Yp | 37 | PFGE, Srpt | unk |
| XqYq | 0 | PFGE, Srpt | unk |

[a] "Telomere gaps" are defined as the distance (in kb) between the telomeric end of DNA sequence on the reference allele and the beginning of the terminal (TTAGGG)n tract. Indicated sizes based on 7_3_03 AGPs.

[b]Where designated, mapping experiments using a site-specific cleavage method (RARE cleavage; Riethman et al. 1997) have been done to demonstrate colinearity of the half-YAC insert DNA with the cognate telomere. In the absence of RARE cleavage data, the presence of subtelomeric repeats adjacent to terminal (TTAGGG)n sequences in a half-YAC clone is taken as strong evidence for proximity to the telomere; this has been borne out by the RARE cleavage experiments carried out so far.

[c]Telomeres with a frequency of >10% large variant alleles in the small populations sampled are considered to have "Hi" polymorphism in the context of this proposal, and those with less than 10% large variant alleles are considered to have "Lo" polymorphism. For the telomeres not listed, no molecular data are available with respect to large-scale variations, and the available FISH data are inconclusive with respect to potential large-scale variation. The polymorphism frequencies detected by FISH are minimum numbers, since detection depends on the variable presence/absence of only one specific FISH probe at the telomere. The size(s) of the polymorphisms cannot be determined by FISH but are assumed to be at least the size of the probe used (based upon similar FISH signal intensities at all sites). Data on polymorphic telomeres are from Wilkie et al. (1991); Ijdo et al. (1992); Cook et al. (1994); Macina et al. (1994, 1995); Martin-Gallardo et al. (1995); Reston et al. (1995); Monfouilloux et al. (1998); Trask et al. (1998); van Overveld et al. (2000).

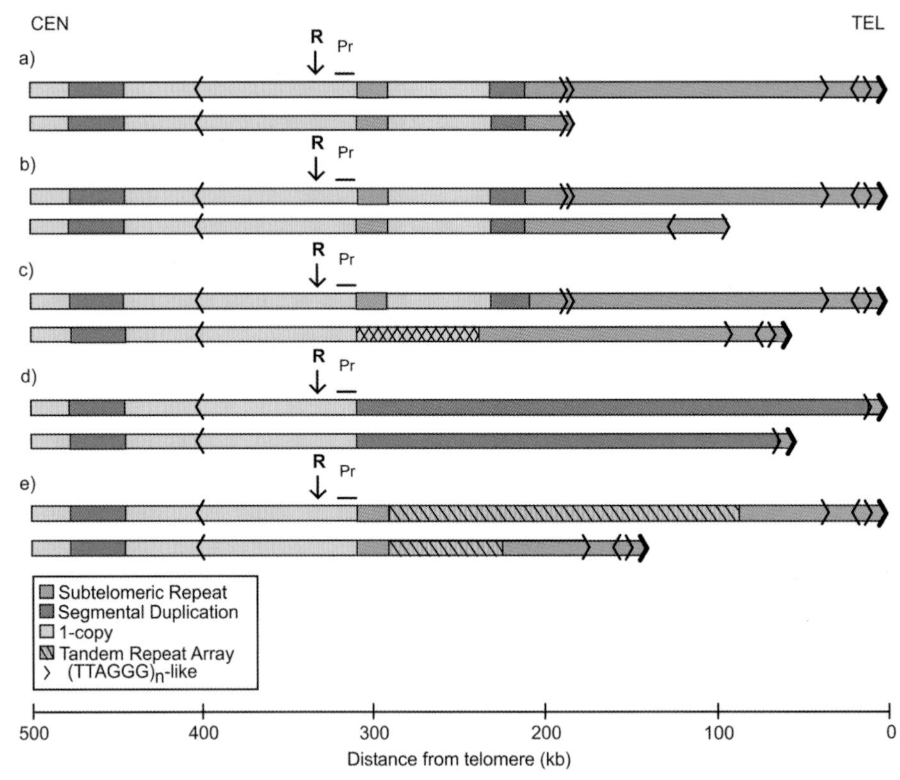


**Figure 2.** Models of large-scale variation leading to length polymorphisms at human telomeres. The models are based on published mapping studies (see Table 1) as well as our unpublished results. In *a*, the alleles vary by addition of a Srpt segment to an existing telomere. Some of the 11p, 16p, and 8p variants appear to have this structure. In *b*, the alleles vary in both Srpt content and Srpt organization, without a simple relationship to each other in the variant region. Some of the 6p and 8p variants appear to follow this model, as does a 19p variant. In *c*, the two alleles diverge just before the Srpt region begins, leading to a large insertion/deletion polymorphism in distal 1-copy DNA. Evidence for this sort of variation exists at the 2q telomere. In *d*, the two alleles are similar except for an insertion/deletion polymorphism contained entirely within the segmentally duplicated DNA. Evidence for this exists near the 14q telomere where duplicated IgG VH genes are located. In *e*, the alleles vary by insertion/deletion polymorphism of tandemly repeated DNA tracts located within the Srpt region. Evidence for this model comes from mapping the D4Z4 repeat tract near the 4q and 10q telomeres. Combinations of these five classes of alternative subtelomeric sequence organization are possible, as are additional types not yet characterized.

made up of Srpt and 8.3% of segmentally duplicated DNA, for a total of 20.8% segmental duplications of both types. Genome-wide, an estimated 5% of genomic DNA is believed to contain segmentally duplicated sequences (IHGSC 2001; Bailey et al. 2002).

Interstitial (TTAGGG)n-like sequence distribution was examined because of its potential roles in subtelomeric recombination and telomere healing (Mondello et al. 2000; Azzalin et al. 2001; Ruiz-Herrera et al. 2002) and its hypothesized role as a boundary element for subtelomeric DNA compartments (Flint et al. 1997b). The internal subtelomeric (TTAGGG)n-like sequence islands ranged in size from 24 bp to 823 bp; most were in a rather tight size range of 151–200 bp. Those shorter than this size tended to be in 1-copy sequence regions, those longer, in Srpt sequence. The boundary (TTAGGG)n islands ranged from 57 bp to 257 bp in size. There were 40 (TTAGGG)n-like sequence islands in Srpt, 0 in segmental duplications, and 26 in 1-copy regions. Nine (TTAGGG)n-like sequence islands were at boundaries (1 at SD/Srpt, 4 at Srpt/1-copy, and 4 at allele boundaries). The 4 (TTAGGG)n-like islands that occurred at the allele boundaries were within the internal Srpt regions of long subtelomeric alleles but mapped to the precise coordinates of the termini of shorter alleles for these same telomeres (8p, 9q, 11p, 16p). This suggests that the longer alleles of these telomeres might have been formed by simple addition of a terminal subtelomeric sequence segment to a preexisting telomere.

A comparison of the number of interstitial (TTAGGG)n-like islands found in subtelomeric DNA with those found genome-wide showed that (TTAGGG)n-like islands are highly enriched (>20-fold) in subtelomeric regions. In addition, they tend to be both longer and more similar to perfect (TTAGGG)n tracts in subtelomeric DNA compared to elsewhere in the genome. From an evolutionary perspective, this suggests that most subtelomeric interstitial (TTAGGG)n tracts have arisen more recently than those found elsewhere in the genome, have originated via a separate mechanism than (TTAGGG)n islands found elsewhere (see, e.g., Azzalin et al. 2001), or are under some selective pressure to maintain similarity to (TTAGGG)n (Flint et al. 1997b).

Transcripts were annotated in each subtelomeric assembly, noting their mapping coordinates relative to their respective telomere and whether they originate in duplicated DNA or single-copy DNA. We used a database of unique transcripts representing each Unigene cluster (Schuler 1997; ftp://ftp.ncbi.nih.gov/repository/UniGene/ ; Hs.seq.uniq.Z file available from the Unigene build available July 1, 2002, containing transcript sequences representing ~128,000 Unigene clusters). One thousand twelve subtelomeric transcripts were annotated in this manner, 732 from 1-copy genomic regions, and 280 from segmentally duplicated DNA and subtelomeric repeat DNA. Overall, the subtelomeric region is somewhat enriched in Unigene transcripts (54 transcripts per Mb) relative to the genome-wide average (43 transcripts per Mb). The enrichment of transcripts in subtelomeric DNA is consistent with earlier studies (see, e.g., Flint et al. 1997b), although there is a great deal of variation in gene concentration from telomere to telomere. Of the transcripts embedded within the segmental duplications and subtelomeric repeats, an unknown but significant fraction are likely to be pseudogenes (see, e.g., Flint et al. 1997a), whereas others are likely to be members of gene families with many closely related but nonidentical functional transcripts (see, e.g., Flint et al. 1997b; Mah et al. 2001; Fan et al. 2002). Cross-boundary transcripts contain part of a sequence from a duplicated genomic segment and part from a 1-copy segment, or parts from a segmental duplication and from a subtelomeric repeat. These transcripts might represent transcribed pseudogenes generated by juxtaposition of progenitor transcript segments, or they might generate new functionalities by virtue of exon shuffling upon duplication (Bailey et al. 2002; Fan et al. 2002); they include transcripts for an F-box protein, for a zinc finger-containing protein, and for many unknown potential proteins. It is essential to acquire complete finished sequences for each distinct allele of each subtelomeric region in order to identify and analyze these genes and gene families, and to de-convolute the many instances of over-clustered Unigenes and mRNAs derived from separate but highly similar duplicated genomic DNA fragments.

Subtelomeric gene families with members having nucleotide sequence similarity in the 70% to 90% level include the immunoglobulin heavy-chain genes (found at 14q), olfactory receptor genes (1-copy regions of 1q, 5q, 10q, and 15q as well as previously characterized subtelomeric repeat DNA at 1p, 6p, 8p, 11p, 15q, 19p, and 3q [Trask et al. 1998]), and zinc-finger genes (4p, 5q, 8p, 8q, 12q, and 19q). Transcripts for multiple members of these gene families were found within many of the individual subtelomeric regions. The abundance of gene families in subtelomeric regions is a common feature of most eukaryotes and may reflect a generally increased recombination and tolerance of subtelomeric DNA for rapid evolutionary change.

## VARIATION AND TELOMERIC CLOSURE

Large variant alleles of many human subtelomeric regions exist and are believed to consist mainly or entirely of subtelomeric repeats (Wilkie et al. 1991; Macina et al. 1995; Trask et al. 1998). For example, Wilkie et al. (1991) found 3 alleles varying in length up to 260 kb at the 16p telomere among the 47 chromosomes sampled. The variant DNA regions appeared to comprise low-copy subtelomeric repeat sequences, and each allele appeared to be in complete linkage disequilibrium with markers at both the proximal and the distal ends of the polymorphic segment of DNA; this suggested that the subtelomeric repeat segment contained in each allele behaved as a single block of DNA, with no detectable recombination within the block. Trask et al. (1998) examined the structure and genomic distribution of a cosmid-sized block of segmentally duplicated subtelomeric DNA. They found that this block was consistently present at the 3q, 15q, and 19p telomeres in humans, was variably distributed at an additional subset of human telomeres, but was present in a single copy in nonhuman primate genomes. More detailed analysis of a 12-kb segment of this block that encodes olfactory receptor genes revealed evidence for evolutionarily recent interchromosomal exchanges involving this segment, suggesting that the mosaic patchworks of duplications that comprise subtelomeric repeat regions are not merely linear descendants of the original elements, but are still evolving and exchanging with each other (Mefford and Trask 2002). Similar studies have more recently demonstrated that the evolution of most primate subtelomeric regions has involved multiple, lineage-dependent duplications in recent evolutionary time (Martin et al. 2002; van Geel et al. 2002). The duplications have colonized many individual human subtelomeric regions in a variable fashion since the divergence of human and primate lineages, and at least some of them are still capable of interacting and exchanging sequences interchromosomally.

A dramatic example of this is the interchromosomal exchange of the tandemly repeated D4Z4 sequence tract between human 4q and 10q telomere regions (van Deutekom et al. 1996). The size of the D4Z4 repeat tract at both 4qtel and 10qtel is highly variable in individuals, and it has been suggested that relatively frequent meiotic pairing interactions between subtelomeric regions of these nonhomologous chromosomes may contribute to their atypically high variation in D4Z4 tract length. The deletion of most of the D4Z4 tract on a particular 4q allele in the population causes FSHD, a type of muscular dystrophy (Lemmers et al. 2002). Interestingly, nucleotide sequence variation in the subtelomere region distal to the D4Z4 repeat between the 4qA allele and the 4qB allele is unusually high (Lemmers et al. 2002), suggesting unusual recombinational or selective pressures that keep these two 4qtel alleles distinct while still permitting promiscuous interchromosomal exchange of the D4Z4 repeat tracts between 4qtel and 10qtel. It is unclear how the deletion of most of the D4Z4 tract on the 4qA allele causes FSHD, but a widely discussed potential mechanism is a position effect on a relatively distant gene caused by disruption of subtelomeric heterochromatin in the vicinity of the D4Z4 repeats.

Characterization of large-scale human subtelomeric variation is still in its infancy, mainly because accurately mapped and assembled reference sequences for these regions were unavailable until now and because current methods for detecting and analyzing large-scale variation are limited. FISH-based detection of polymorphism is limited to DNA contained in the cloned probe and provides little information with respect to the real size of the polymorphic segment. Nonetheless, it has provided information on subtelomeric regions that are clearly variable (Ijdo et al. 1992; Trask et al. 1998). Conventional PFGE analysis of DNA fragments enabled the initial observations of large-scale variation at the 16p telomere (Wilkie et al. 1991), but this method is limited by the variable methylation of genomic DNA and the sensitivity of infrequently cutting restriction enzymes to CpG methylation.

We have used a site-specific DNA cleavage method (*RecA*-assisted *r*estriction *e*ndonuclease [RARE] cleavage; Riethman et al. 1997) combined with pulsed-field gel electrophoresis (PFGE) analysis of the cleaved large DNA fragments to analyze subtelomere structure. A RARE cleavage experiment targeting a single genomic site in a subtelomeric region is expected to release a telomere-terminal fragment of genomic DNA; the size of this fragment corresponds to the distance from the cleavage site to the end of the chromosome. This simple principle makes RARE cleavage mapping an ideal method for physically mapping telomere regions, simplifies validation of half-YAC clone structure, and enables the systematic analysis of large polymorphisms in human subtelomeric regions.

Table 1 summarizes the current state of the experimental characterization of human subtelomeric variation. On the basis of current data and the number of telomeres for which no large-scale variation data are yet available, ~2–4 large-scale length variants are expected to exist at each of 20–25 human telomeres. This is, however, a very preliminary and rough estimate, and a proper assessment of the occurrence and frequency of large subtelomeric variants awaits a more systematic analysis.

Figure 2 shows models of large-scale variation leading to large-length polymorphisms at human telomeres. The models are based on published mapping studies (see Table 1) as well as our unpublished results. In Figure 2a, the alleles vary by addition of a Srpt segment to an existing telomere. Some of the 11p, 16p, and 8p variants appear to have this structure. In Figure 2b, the alleles vary in both Srpt content and Srpt organization, without a simple relationship to each other in the variant region. Some of the 6p and 8p variants appear to follow this model, as does a19p variant. In Figure 2c, the two alleles diverge just before the Srpt region begins, leading to a large insertion/deletion polymorphism in distal 1-copy DNA. Evidence for this sort of variation exists at the 2q telomere. In Figure 2d, the two alleles are similar except for an insertion/deletion polymorphism contained entirely within the segmentally duplicated DNA. Evidence for this exists near the 14q telomere where duplicated IgG VH genes are located. In Figure 2e, the alleles vary by insertion/deletion polymorphism of tandemly repeated DNA tracts located within the Srpt region. Evidence for this model comes from mapping the D4Z4 repeat tract near the 4q and 10q telomeres. Combinations of these five classes of alternative subtelomeric sequence organization are possible, as are additional types not yet detected.

## HUMAN (TTAGGG)n-ADJACENT DNA

The critical DNA regions required for developing reagents for single-telomere (TTAGGG)n tract-length assays are those immediately adjacent to the terminal (TTAGGG)n tract (subterminal sequences). Within the past 6 months there has been a dramatic expansion in the availability of these sequences as work in our own lab and elsewhere has resulted in completion of telomere regions for many reference alleles (see Table 1). Sequences for multiple variant subtelomere regions will continue to be added to the databases in the near future as our own and other telomere mapping and sequencing projects progress, providing the raw material for PCR-based single telomere length assays (Forstemann et al. 2000; Baird et al. 2003). In these assays, a sequence for priming PCR toward the centromere from the natural (TTAGGG)n tract terminus is added (either by ligation or by a terminal transferase-based method), and a second primer derived from subterminal DNA and oriented toward the telomere is paired with the first primer for PCR. A hybridization probe derived from subterminal DNA distal to the telomere-oriented primer is used to detect the PCR product. The specificity of the PCR for a single telomere (or a specific allele of a telomere) relies on the selection of the telomere-oriented primer from what are often closely related subterminal sequences; thus, as a database of subtelomeric and subterminal sequences is expanded, the power to select specific primers for PCR-based telomere-length assays will increase.

Figure 3 illustrates the three most common types of subterminal sequence organizations found in fully sequenced telomere alleles so far. The sequence organization shown in Figure 3a is the simplest and most straightforward for PCR-based assay design. There is a short (2–10 kb) region of low-copy Srpt sequence immediately adjacent to the (TTAGGG)n tract, followed by 1-copy DNA from which unique probes and primers can be derived. The original STELA assay was derived from such a telomere (Xp/Yp telomere; Baird et al. 2003), and allele-specific STELA and telomere-PCR (Forstemann et al. 2000) assays can be prepared by using a telomere-oriented primer that is allele-specific (A_sp in Fig. 3a). Telomeres in this class include 7qtel, 8qtel, 11qtel, 14qtel, 18qtel, and Xp/Yptel, each of which already has a fully sequenced or nearly sequenced reference allele.

Figure 3b shows the TelBam11 class of subterminal low-copy repeat sequence (Brown et al. 1990) that is found on 10–20 telomeres in each individual, and is represented on fully sequenced reference alleles for 10qtel and 21qtel. In this case, a DNA hybridization probe we isolated (HC1208) that is specific for this low-copy Srpt sequence recognizes all members of this subtelomeric repeat family; the specificity of the telomere-length assay for single telomeres and for single alleles must therefore come entirely from the telomere-oriented primer (A_sp), which must therefore be carefully selected from divergent sequences within the adjacent subtelomeric DNA. Preliminary results indicate that TelBam11-associated sequences typically display 90–98% nucleotide sequence identity, permitting sufficient divergence for development of

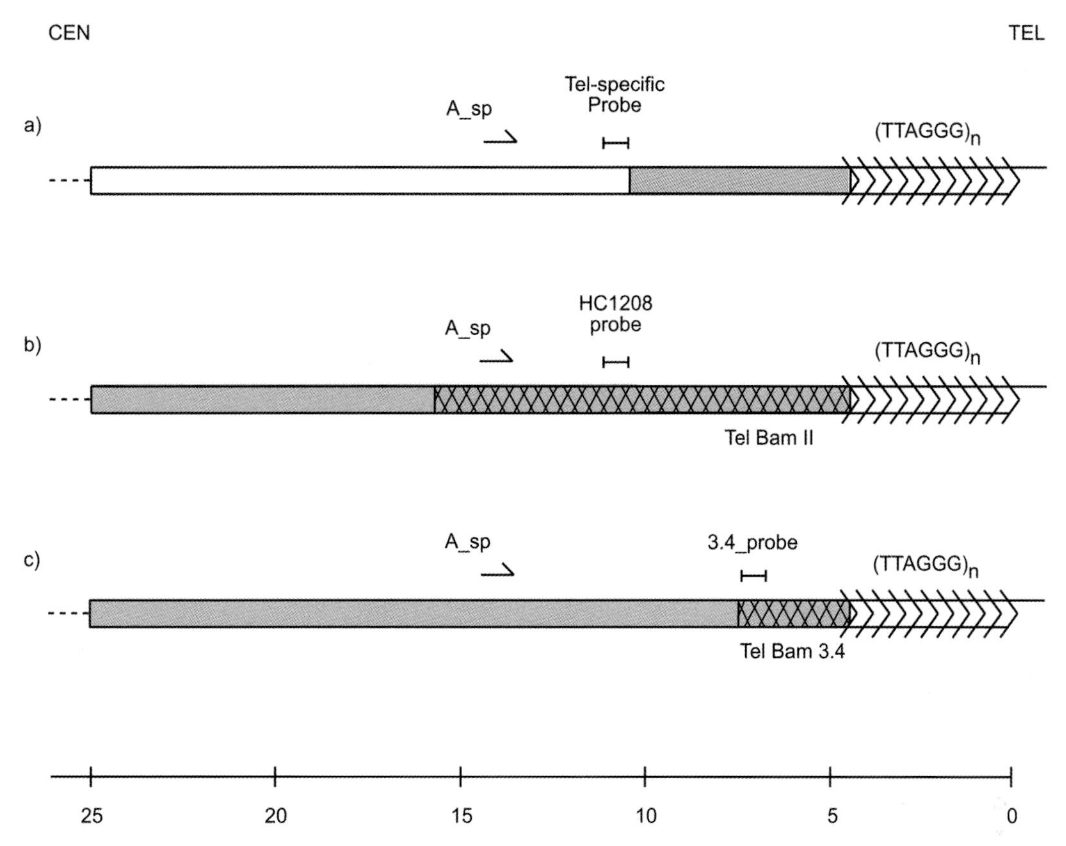


**Figure 3.** Human subterminal sequence organization. The shaded region is low-copy subtelomeric repeat (Srpt) DNA; unshaded is single-copy DNA. The terminal (TTAGGG)n tracts are depicted by the arrows, with a 3′-end overhang as shown. The cross-hatched portion of the Srpt regions in *b* and *c* are the TelBam11 and TelBam3.4 segments, respectively. The subterminal regions from which single-copy and low-copy hybridization probes can be derived are shown, as are regions from which allele-specific primers (A_sp) can be selected.

telomere-specific and allele-specifc primers. The TelBam3.4 class of subterminal low-copy repeat sequence (Brown et al. 1990) has properties similar to the TelBam11 class, including a low-copy DNA hybridization probe specific for this subtelomeric repeat and sufficient divergence for development of single-telomere-specific and allele-specific telomere-oriented primers (Fig. 3c). Completed reference telomeres that include TelBam3.4 family members are those for 9ptel, 15qtel, 16ptel, and Xq/Yqtel.

Individual-specific telomere-length typing of all 92 telomeres in a given genome would have a dramatic impact on basic studies of human telomere capping/uncapping, end-resection and 3′-overhang processing, and telomere replication and maintenance. In addition, this capability may reveal inborn individual-specific differences in telomere lengths that could affect susceptibility to aging and cancer phenotypes. The completion of the human genome telomere regions and the ongoing analysis of variation in human subtelomeric regions will enable progress toward this goal.

## CONCLUSIONS

As shown in Table 1 and Figure 2, there are still very large gaps in our understanding of human subtelomeric DNA sequence content, organization, and structure. The complete sequencing of these regions is absolutely essential for an accurate picture of the human genome. Even from the little that is currently known, it is clear that large-scale subtelomeric variation can have important functional consequences in terms of mutations that cause human disease (van Deutekom et al. 1996), relative susceptibility of telomere regions to chromosome rearrangement/instability (Bailey et al. 2002), and subtelomeric gene dosage, gene variation, and gene evolution (Trask et al. 1998; Mah et al. 2001; Bailey et al. 2002). It is very likely that the specific sequence content and organization of subtelomere regions will have an impact on telomere position effects, telomeric chromatin/heterochromatin structure, and DNA replication near telomeres. Telomere maintenance pathways, particularly those influenced by recombination, may likewise be affected by differential subtelomeric structure and sequence organization. Little is currently known of possible recombination within and adjacent to large variant regions, and how the haplotype structure of subtelomeric regions might be affected by specific subtelomeric sequence content and organization during human evolution.

Current "finished" reference sequences for most subtelomeric regions are amalgams of sequenced BAC, PAC, and half-YAC clones from separate alleles, and the only cases where some certainty exists with respect to the sequence organization of single, specific subtelomeric alleles are those where DNA from entire half-YACs has been sequenced. Most subtelomeric variants have yet to be characterized, cloned, and sequenced in their entirety, and perhaps many are still undetected. This gap in our understanding of the structure and sequence content of these dynamic and rapidly evolving chromosome regions must be filled in order to complete the human genome sequence and to develop reagents for understanding its function and evolution. Sequencing of representatives of all common large-scale variants of chromosome ends will provide a comprehensive picture of the occurrence and frequency of different subtelomeric length variants in human populations, and will permit development of sequence-based marker sets for more facile analysis of variant regions using PCR-based methods. These aims are achievable and essential for understanding the structure and function of the human genome.

## ACKNOWLEDGMENTS

We thank the members of the International Human Genome Sequencing Consortium, many of whom participated in the sequencing of subtelomeric regions. Bob Moyzis, Jonathan Flint, and William Brown collaborated or provided reagents for the earlier stages of this work, and Uma Mudunuri provided technical support. John Rux and the Wistar Bioinformatics Facility provided programming and computational support. Financial support was provided by National Institutes of Health grants HG-00567 and CA-25874, and by the Commonwealth Universal Research Enhancement Program, Pennsylvania Department of Health.

## REFERENCES

Altschul S.F., Madden T.L., Schäffer A.A., Zhang J., Zhang Z., Miller W., and Lipman D.J. 1997. Gapped BLAST and PSI-BLAST: A new generation of protein database search programs. *Nucleic Acids Res.* **25:** 3389.

Azzalin C.M., Nergadze S.G., and Giulotto E. 2001. Human intrachromosomal telomeric-like repeats: Sequence organization and mechanisms of origin. *Chromosoma* **110:** 75.

Bailey J.A., Gu Z., Clark R.A., Reinert K., Samonte R.V., Schwartz S., Adams M.D., Myers E.W., Li P.W., and Eichler E.E. 2002. Recent segmental duplications in the human genome. *Science* **297:** 1003.

Baird D.M., Rowson J., Wynford-Thomas D., and Kipling D. 2003. Extensive allelic variation and ultrashort telomeres in senescent human cells. *Nat. Genet.* **33:** 203.

Blasco M.A., Gasser S.M., and Lingner J. 1999. Telomeres and telomerase. *Genes Dev.* **13:** 2353.

Brown W.R., MacKinnon P.J., Villasante A., Spurr N., Buckle V.J., and Dobson M.J. 1990. Structure and polymorphism of human telomere-associated DNA. *Cell* **63:** 119.

Cook G.P., Tomlinson I.M., Walter G., Carter N.G., Riethman H.C., Winter G., and Rabbitts T.H. 1994. A map of the human immunoglobulin $V_H$ locus completed by analysis of the telomeric region of chromosome14q. *Nat. Genet.* **7:** 162.

Fan Y., Newman T., Linardopoulou E., and Trask B.J. 2002. Gene content and function of the ancestral chromosome fusion site in human chromosome 2q13-2q14.1 and paralogous regions. *Genome Res.* **12:** 1663.

Ferrin L.J. and Camerini-Otero R.D. 1991. Selective cleavage of human DNA: RecA-assisted restriction endonuclease (RARE) cleavage. *Science* **254:** 1494.

Feuerbach F., Galy V., Trelles-Sticken E., Fromont-Racine M., Jacquier A., Gilson E., Olivo-Marin J.C., Scherthan H., and Nehrbass U. 2002. Nuclear architecture and spatial positioning help establish transcriptional states of telomeres in yeast. *Nat. Cell Biol.* **4:** 214.

Flint J., Thomas K., Micklem G., Raynham H., Clark K., Doggett N.A., King A., and Higgs D.R. 1997a. The relationship between chromosome structure and function at a human telomeric region. *Nat. Genet.* **15:** 252.

Flint J., Bates G.P., Clark K., Dorman A., Willingham D., Roe B.A., Micklem G., Higgs D.R., and Louis E.J. 1997b. Sequence comparison of human and yeast telomeres identifies structurally distinct subtelomeric domains. *Hum. Mol. Genet.* **6:** 1305.

Forstemann K., Hoss M., and Lingner J. 2000. Telomerase-dependent repeat divergence at the 3′ ends of yeast telomeres. *Nucleic Acids Res.* **28:** 2690.

Griffith J.D., Comeau L., Rosenfield S., Stansel R.M., Bianchi A., Moss H., and de Lange T. 1999. Mammalian telomeres end in a large duplex loop. *Cell* **97:** 503.

Hemann M.T., Strong M.A., Hao L.Y., and Greider C.W. 2001. The shortest telomere, not average telomere length, is critical for cell viability and chromosome stability. *Cell* **107:** 67.

Ijdo J.W., Lindsay E.A., Wells R.A., and Baldini A. 1992. Multiple variants in subtelomeric regions of normal karyotypes. *Genomics* **14:** 1019.

International Human Genome Sequencing Consortium (IHGSC). 2001. Initial sequencing and analysis of the human genome. *Nature* **409:** 860.

Koob M., Burkiewicz A., Kur J., and Szybalski W. 1992. RecA-AC: Single-site cleavage of plasmids and chromosomes at any predetermined restriction site. *Nucleic Acids Res.* **20:** 5831.

Lansdorp P.M., Verwoerd N.P., van de Rijke F.M., Dragowska V., Little M.T., Dirks R.W., Raap A.K., and Tanke H.J. 1996. Heterogeneity in telomere length of human chromosomes. *Hum. Mol. Genet.* **5:** 685.

Lemmers R.J., de Kievit P., Sandkuijl L., Padberg G.W., van Ommen G.J., Frants R.R., and van der Maarel S.M. 2002. Facioscapulohumeral muscular dystrophy is uniquely associated with one of the two variants of the 4q subtelomere. *Nat. Genet.* **32:** 235.

Mah N., Stoehr H., Schulz H.L., White K., and Weber B.H. 2001. Identification of a novel retina-specific gene located in a subtelomeric region with polymorphic distribution among multiple human chromosomes. *Biochim. Biophys. Acta* **1522:** 167.

Macina R.A., Negorev D.G., Spais C., Ruthig L.A., Hu X-L., and Riethman H.C. 1994. Sequence organization of the human chromosome 2q telomere. *Hum. Mol. Genet.* **3:** 1847.

Macina R.A., Morii K., Hu X.-L., Negorev D.G., Spais C., Ruthig L.A., and Riethman H.C. 1995. Molecular cloning and RARE cleavage mapping of human 2p, 6q, 8q, 12q, and 18q telomeres. *Genome Res.* **5:** 225.

Martin C.L., Wong A., Gross A., Chung J., Fantes J.A., and Ledbetter D.H. 2002. The evolutionary origin of human subtelomeric homologies—Or where the ends begin. *Am. J. Hum. Genet.* **70:** 972.

Martin-Gallardo A., Lamerdin J., Sopapan P., Friedman C., Fertitta A.L., Garcia E., Carrano A., Negorev D., Macina R.A., Trask B.J., and Riethman H. 1995. Molecular analysis of a novel subtelomeric repeat with polymorphic chromosomal distribution. *Cytogenet. Cell Genet.* **71:** 289.

Mefford H.C. and Trask B.J. 2002. The complex structure and dynamic evolution of human subtelomeres. *Nat. Rev. Genet.* **3:** 91.

Mondello C., Pirzio L., Azzalin C.M., and Giulotto E. 2000. Instability of interstitial telomeric sequences in the human genome. *Genomics* **68:** 111.

Monfouilloux S., Avet-Loiseau H., Amarger V., Balazs I., Pourcel C., and Vergnaud G. 1998. Recent human-specific spreading of a subtelomeric domain. *Genomics* **51:** 165.

Moyzis R.K., Buckingham J.M., Cram S., Dani M., Deaven L.L., Jones M.D., Meyne J., Ratliff R. L., and Wu J.R. 1988. A highly conserved repetitive DNA sequence, $(TTAGGG)_n$, present at the telomeres of human chromosomes. *Proc. Natl. Acad. Sci.* **85:** 6622.

Reston J.T., Hu X.-L., Macina R.A., Spais C., and Riethman H. 1995. Structure of the terminal 300 kb of DNA from human chromosome 21q. *Genomics* **26:** 31.

Riethman H. 2003. Cloning, mapping, and sequencing telomeres. In *Genomic mapping and sequencing* (ed. I. Dunham), p. 257. Horizon Press, Wymondham, United Kingdom.

Riethman H., Birren B., and Gnirke A. 1997. Preparation, manipulation, and mapping of high molecular weight DNA. In *Genome analysis: A laboratory manual*, vol. 1. Analyzing DNA (ed. B. Birren, et al.), p. 83. Cold Spring Harbor Laboratory Press, Cold Spring Harbor, New York.

Riethman H.C., Moyzis R.K., Meyne J., Burke D.T., and Olson M.V. 1989. Cloning human telomeric DNA fragments into *Saccharomyces cerevisiae* using a yeast-artificial-chromosome vector. *Proc. Natl. Acad. Sci.* **86:** 6240.

Riethman H.C., Xiang Z., Paul S., Morse E., Hu X.L., Flint J., Chi H.C., Grady D.L., and Moyzis R.K. 2001. Integration of telomere sequences with the draft human genome sequence. *Nature* **409:** 948.

Ruiz-Herrera A., Garcia F., Azzalin C., Giulotto E., Egozcue J., Ponsa M., and Garcia M. 2002. Distribution of intrachromosomal telomeric sequences (ITS) on *Macaca fascicularis* (primates) chromosomes and their implication for chromosome evolution. *Hum. Genet.* **110:** 578.

Schuler G.D. 1997. Pieces of the puzzle: Expressed sequence tags and the catalog of human genes. *J. Mol. Med.* **75:** 694.

Smit A.F.A. and Green P. RepeatMasker Home page: http://ftp.genome.washington.edu/RM/RepeatMasker.html

Trask B.J., Friedman C., Martin-Gallardo A., Rowen L., Akinbami C., Blankenship J., Collins C., Giorgi D., Iadonato S., Johnson F., Kuo W.L., Massa H., Morrish T., Naylor S., Nguyen O.T., Rouquier S., Smith T., Wong D.J., Youngblom J., and van den Engh G. 1998. Members of the olfactory receptor gene family are contained in large blocks of DNA duplicated polymorphically near the ends of human chromosomes. *Hum. Mol. Genet.* **7:** 13.

van Deutekom J.C.T., Bakker E., Lemmers R.J.L.F., Van der Wielen M.J.R., Bik E., Hofker M.H., Padberg G.W., and Frants R.R. 1996. Evidence for subtelomeric exchange of 3.3 kb tandemly repeated units between chromosomes 4q35 and 10q26: Implications for genetic counseling and etiology of FSHD1. *Hum. Mol. Genet.* **5:** 1997.

van Geel M., Eichler E.E., Beck A.F., Shan Z., Haaf T., van der Maarel S.M., Frants R.R., and de Jong P.J. 2002. A cascade of complex subtelomeric duplications during the evolution of the hominoid and Old World monkey genomes. *Am. J. Hum. Genet.* **70:** 269.

van Overveld P.G., Lemmers R.J., Deidda G., Sandkuijl L., Padberg G.W., Frants R.R., and van Der Maarel S.M. 2000. Interchromosomal repeat array interactions between chromosomes 4 and 10: A model for subtelomeric plasticity. *Hum. Mol. Genet.* **9:** 2879.

Wilkie A.O.M., Higgs D.R., Rack K.A., Buckle V.J., Spurr N.K., Fischel-Ghodsian N., Ceccherini I., Brown W.R.A., and Harris P.C. 1991. Stable length polymorphism of up to 260 kb at the tip of the short arm of human chromosome 16. *Cell* **64:** 595.

Zijlmans J.M., Martens U.M., Poon S.S., Raap A.K., Tanke H.J., Ward R.K., and Lansdorp P.M. 1997. Telomeres in the mouse have large inter-chromosomal variations in the number of T2AG3 repeats. *Proc. Natl. Acad. Sci.* **94:** 7423.

# Genome Research: The Next Generation

F.S. Collins
*National Human Genome Research Institute, National Institutes of Health, Bethesda, Maryland 20892*

With the essential completion of the human genome sequence in April 2003, all of the original goals of the Human Genome Project have been achieved or surpassed. We are now at last in the genome era. Phenomenal opportunities now exist to build upon this foundation and to accelerate the application of this fundamental information to human health.

To develop a conceptual framework for research endeavors in the genome era, the National Human Genome Research Institute (NHGRI) hosted a landmark series of planning meetings from 2001 to 2003 involving more than 600 U.S. and foreign scientists representing government, industry, and academia. Ideas gathered from these and other scientific leaders were used by NHGRI to develop a bold vision for genomics research, which employs the visual metaphor of a three-story building resting firmly upon the foundation of the Human Genome Project (Collins et al. 2003). In this blueprint for the future, the first floor represents Genomics to Biology; the second floor, Genomics to Health; and the third floor, Genomics to Society (Fig. 1). The door of this metaphorical structure has been deliberately left open, providing free and easy access to any researcher seeking genomic data, or wishing to enter and work on any of the floors.

## GENOMICS TO BIOLOGY

The Human Genome Project represented biology's first major foray into discovery-driven, rather than hypothesis-driven, science. Despite concerns expressed by skeptics in its early years, the project has proved to be a boon to both kinds of biological research. Not only has this discovery-driven effort produced powerful new research tools and data sets that have transformed hypothesis-driven biological research, it has also paved the way for other large-scale genomics projects to develop additional innovative technologies and community resource data sets. Our vision for basic biology seeks to accelerate

**Figure 1.** The National Human Genome Research Institute's blueprint for the future of genome research. (Reprinted, with permission, from Collins et al. 2003 [copyright Macmillan].)

and expand the genomics revolution now under way in nearly all realms of biological science.

### Build a Human Haplotype Map

One of the greatest opportunities facing biological researchers in the genome era is defining and analyzing human genetic variation. Although humans are 99.9% genetically identical, the 0.1% variation in DNA sequences holds crucial clues to individual differences in susceptibility to disease and response to drugs.

In October 2003, a public–private research consortium consisting of Japan, the United Kingdom, Canada, China, and the United States launched the International HapMap Project (http://www.genome.gov/HapMap) with the goal of defining and tagging the most common blocks of genetic variation, or haplotype blocks, that cover 80–90% of the human genome. When completed, the human haplotype map (HapMap) will serve as a valuable tool for scientists searching for common genetic variations associated with complex diseases, as well as variations associated with differences in drug response.

Initially, researchers will study DNA from 270 people in widely distributed geographic regions, which were selected in an effort to find genetic variations that are common in most populations around the globe. Samples are being collected from the Yorubas in Nigeria, Japanese, Han Chinese, and U.S. residents of northern and western European descent.

Expected to take three years to complete, the HapMap will reduce from 10 million to roughly 300,000 the number of single-nucleotide polymorphisms (SNPs) required to examine the whole genome for association with a phenotype. This will make genome-scan approaches to finding genes that influence disease susceptibility much more efficient and comprehensive, since effort will not be wasted typing more SNPs than necessary, and all regions of the genome can be included. Ultimately, the HapMap with its select, gold-standard set of haplotype tag SNPs is expected to enable genome-wide association studies to be conducted at 30–40 times less cost than is currently possible.

In addition to speeding the hunt for genes associated with a wide range of complex diseases and conditions, the HapMap should prove a valuable resource for studying the genetic factors contributing to individual variation in response to drugs and vaccines, as well as those that influence response to environmental factors.

### Sequence Additional Genomes

Although the effort to produce a reference sequence of the *Homo sapiens* genome is essentially completed, there remains a compelling need to sequence the genomes of many more species, both vertebrate and invertebrate. The ability to conduct comparative analyses of genomic sequences of various species has provided biologists with an amazing new window into evolution and genomic function (Thomas et al. 2003). Given the power of comparative genomics, there exists an immense hunger among biologists for free and publicly available sequence data on a wide variety of organisms, and sequencers should feed that hunger.

### Develop New Technologies for Genomics Research

Quantum leaps in technology must be made if our genomics-based vision is to realize its full potential for biomedical research. One of the most ambitious goals is the development of technology that would allow researchers to sequence the genome of a human or other mammals for $1,000 or less—a cost at least four orders of magnitude lower than is possible with current technology.

Yet another major technological hurdle involves lowering the cost of synthesizing DNA, with the ultimate goal being the synthesis of any DNA molecule at high accuracy for $0.01 or less per base. This would transform many of the ways in which biological research is carried out, allowing researchers to efficiently "write" DNA sequences in much the same way they can now easily "read" sequence by DNA sequencing. Researchers also want and should have the technological capacity to identify and simultaneously analyze a wide array of genetic and protein elements within a single cell, including determining the methylation status of all DNA.

### Identify All Functional Elements of the Genome

Although biology has proven enormously successful in sequencing genomes, our experimental and computational methods are still primitive in identifying elements that are not involved in protein coding, which make up about two-thirds of the highly conserved DNA in the human genome. An NHGRI-led consortium made up of scientists in government, industry, and academia recently set out to develop efficient ways of identifying and precisely locating all of the functional elements contained in the human DNA sequence. The ultimate goal of the *ENC*yclopedia *O*f *D*NA Elements (ENCODE) project (http://www.genome.gov/ENCODE/) is to create a reference work that will help researchers mine and fully utilize the human sequence to gain a deeper understanding of human biology, as well as to develop new strategies for improving human health.

The first phase of ENCODE will focus on developing high-throughput methods for rigorously analyzing a defined set of DNA target regions comprising ~30 megabases, or 1%, of the human genome. This pilot project should lay the groundwork for a large-scale effort to characterize all of the protein-coding genes, non-protein-coding genes, and other sequence-based functional elements in the human genome. As has been the case with the Human Genome Project, data from the ENCODE project will be collected and stored in a database that will be freely available to the entire scientific community.

### Identify All Proteins in the Cell and Their Interactions; Develop a Computational Model of the Cell

All of the technological quantum leaps set forth in NHGRI's vision for genomics research can be considered ambitious and audacious, but two are even bolder than the rest. The ultimate challenges—and ultimate payoffs—in terms of advancing biological understanding are development of a technology or technologies for determining the abundance and modification state of *all* proteins in a single cell in a single experiment, and then coupling such experimental data with other data sets and the power of bioinformatics to create a computational model of the cell.

## GENOMICS TO HEALTH

The consensus is that the time is now right to move genomics in an intentional way toward the translational applications that have motivated the Human Genome Project from the very beginning and have inspired great hope for medical benefits among both the general public and the biomedical research community. Bearing in mind that the road from genomics to health is likely to contain many unexpected twists and turns, we should not expect these benefits overnight. But the stage is now set for a major push toward defining the causes and potential cure or prevention of a long list of human diseases.

### Identify Genetic and Environmental Risk Factors for All Common Diseases

Once the human HapMap is constructed and this community resource is made freely available to researchers worldwide, it will be possible to conduct whole-genome association studies of nearly all diseases in all populations, thereby helping to decipher the complicated interplay of multiple genes and multiple nongenetic factors in many common diseases. In addition, new computational and experimental methods will need to be devised to allow the better detection of gene–gene and gene–environment interactions in these often complex disorders.

### Develop Sentinel Systems for Disease Detection; Develop a Molecular Taxonomy of Illness

Genomic technologies possess enormous potential for establishing new approaches to the prediction and prevention of disease. Testing or screening asymptomatic or presymptomatic individuals for gene expression patterns associated with increased disease risk could be used to detect diseases far earlier than is now possible and/or to develop individual preventive medicine strategies. In addition, clinicians could use the detection of gene variants that correlate with drug response to better tailor prescribing patterns to the individual patient. Much remains to be done, however, before such strategies are widely implemented in the clinic. For example, the cost of genome analysis needs to be lowered, and greater care must be taken to ensure the clinical validity of genetic tests.

The field of genomics also stands poised to revolutionize clinical medicine's current taxonomy of disease states, which until now has been based largely on empirical classification schemes. Armed with systematic analyses of DNA sequence, somatic mutations, epigenetic modifications, gene expression, protein expression, and protein modification, we should begin the process of using detailed molecular characterizations to understand, and to potentially reclassify, all human illnesses. A molecular taxonomy of disease will provide a more precise, scientific approach to the diagnosis and prevention of disease, as well as predicting response to treatment.

### Develop and Deploy High-throughput Robotic Screening of Small Molecules for Academic Research

Our current toolbox for studying translation of the human genome is woefully understocked. Among the many community resources that needed to be built or expanded are full-length cDNA collections, siRNA collections, collections of knockout mice, and transcriptome reference data sets.

However, we believe there is another tool that will offer an unprecedented opportunity to establish a new paradigm for academic biological research: a publicly available library of small, organic drug-like molecules and the capacity to screen this library against any high-throughput assay (Austin 2003). Although small organic molecules have a good track record of modulating gene function and have been the target of intense efforts by the pharmaceutical industry, biologists in academia and the public sector have not had easy access to large libraries of these compounds and therefore have not taken full advantage of the tremendous power of small molecules to serve as probes to advance our understanding of biology. Further underscoring the need for a public small-molecule library is the fact that those libraries that exist within the pharmaceutical industry represent a very skewed distribution of genomic targets. More than 40% of small-molecule drugs now on the market target G-protein-coupled receptors, while vast areas of the genome exist for which no small-molecule probes have been identified at all.

Providing public-sector scientists with easy access to small-molecule tools on the scale now available to the pharmaceutical industry will greatly broaden the scope of biological inquiry and have a transformative effect on basic biomedical research, speeding functional studies of the genome and the development of novel therapeutics. This area of genomics research, often referred to as chemical genomics, will open the door to a new type of forward/reverse approach to understanding biological pathways in which small-molecule probes can be used to elucidate gene function by studying either the chemical's

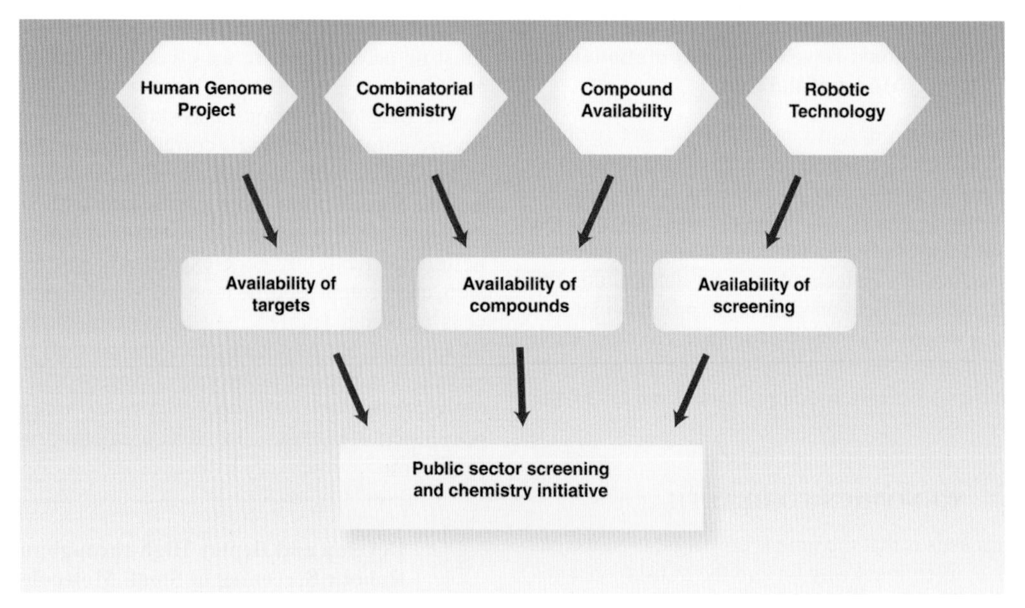


**Figure 2.** A public-sector small-molecule initiative is made possible by four convergent developments in biology and chemistry.

effect on phenotype or its impact on the protein product of a specific gene (Stockwell 2000).

Although major organizational and funding challenges are associated with any novel science effort of this scale, a public-sector initiative in chemical genomics is now within our reach because of convergent developments in biology and chemistry: the essential completion of the human genome sequence, advances in combinatorial chemistry, the wide availability of small-molecule compounds, and the development of high-throughput robotic technology for high-throughput screening (Fig. 2).

Although it will build upon what has been learned in similar efforts within the pharmaceutical industry, a public chemical genomics initiative should not be viewed as an effort to turn academic researchers into drug developers. Instead, we consider it an opportunity for academic and government biologists to contribute in a much more empowered way to the earliest stages of the drug development pipeline where there is reasonable certainty of success, namely the identification of biological targets, the development of biologically relevant assays, the screening of small-molecule libraries, and initial medicinal chemistry to convert screening hits into useful biological probes (Fig. 3). To realize this vision, we anticipate the establishment of centralized facilities for high-throughput screening of small molecules, as well as the creation of public-sector capacity to advise researchers on the development of assays and to provide assistance in medicinal chemistry to render the initial "hits" more useful. As has been the case with sequence data generated by the Human Genome Project, these resources would be freely and readily accessible to the entire scientific community, and a central database would allow potentially powerful inferences to be drawn about biological pathways and networks.

### Catalyze Development of Large Human Cohorts

If we are to fully discern genotype–phenotype correlations and environmental contributions to common illnesses, it is imperative that we assemble large cross-sectional human cohorts that encompass multiple populations. Given the time and expense needed to pull together such studies, we must embark on this effort very soon or else face the risk of not having such large cohorts in place when we need them to understand genotype–environment–phenotype correlations in an unbiased way. In addition to furnishing valuable clues about major causes of morbidity and mortality, large cohorts may also prove priceless in efforts to find alleles that protect against disease and help to promote health and longevity.

### Elucidate the Role That Genomics Can Play to Reduce Health Disparities and Improve Health in the Developing World

Disparities in health status constitute a significant problem in the United States and throughout the world. Although socioeconomic and other nongenetic factors are major contributors to such disparities, genomics may be able to contribute to the reduction of disparities by shedding light on the effects of disease-associated gene variants on the health status of certain populations. Likewise, many efforts to improve health in developing nations may seem to lie outside the direct realm of genomics. But genomics can offer significant contributions by improving understanding of the genetic factors that influence susceptibility and response to infectious diseases, as well as by reducing research and development costs of vaccines and pharmaceuticals.

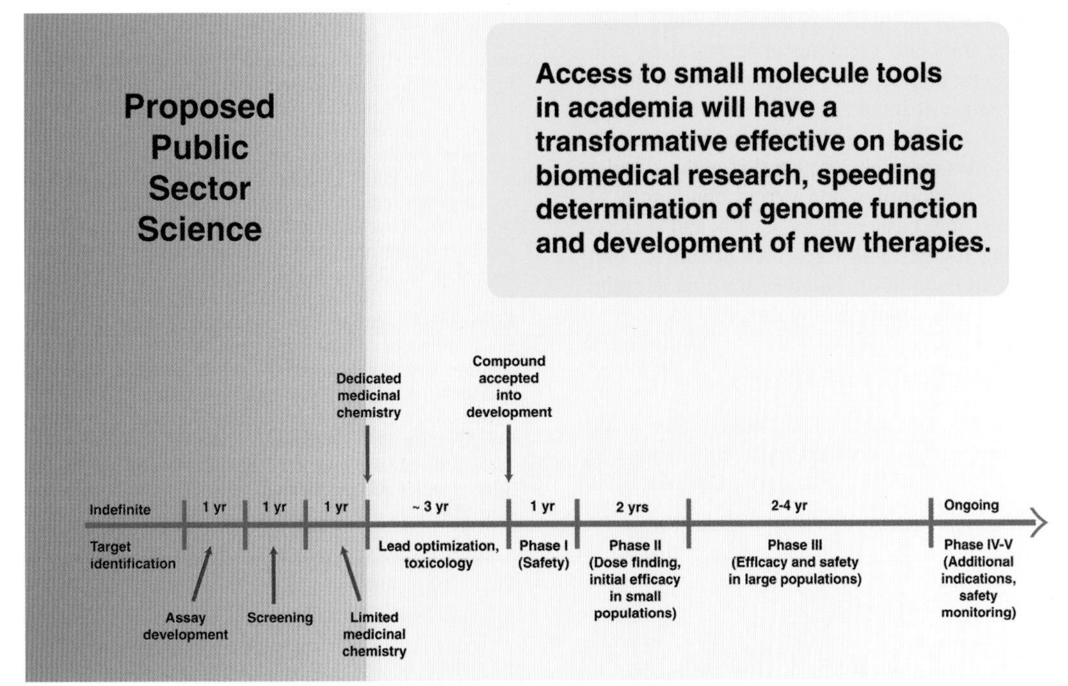


**Figure 3.** A proposed role for the public sector in the early stages of the drug development pipeline. Previously, public sector efforts have been largely limited to the first step of target identification. Providing access to high-throughput screening for academic investigators will empower a whole new approach to understanding biology, and also will encourage the exploration of small-molecule approaches to the treatment of rare diseases.

## GENOMICS TO SOCIETY

The third floor of NHGRI's blueprint for the future, Genomics to Society, is vital to realizing the visions of its other two floors. If we neglect to consider the implications of genome research for society as a whole, people may suffer unnecessary harm and society may not support the rest of these worthwhile efforts.

Among the many complex ethical, legal, and social issues that must be addressed are how to enhance genetic privacy and how to better protect against genetic discrimination. We also must gain a better understanding of the relationship of genomics to race and ethnicity and then bring this understanding to bear as a voice of scientific reason in the often contentious dialog about race and ethnicity.

One legal issue of particular interest to the biomedical research community revolves around intellectual property. Many new types of genomic data sets will be produced in the future. Consequently, we need to encourage the development of appropriate patenting and licensing practices to benefit the public without stifling the intellectual atmosphere that gives rise to new scientific ideas and the marketplace that develops and produces new therapeutics.

Another controversial area centers on the ramifications of advances in the understanding of genetic factors that may influence human behavior. A history of poorly designed association studies has tied certain genetic alleles to behavioral traits commonly perceived as negative, creating a potential for stigmatization of individuals with such alleles. One purpose of producing the human haplotype map is to help improve research design in behavioral genetics so that the role of genetic factors is not oversimplified or overstated as it often has been in the past. Still, we fully expect that genomics will shed new and scientifically valid light on the biological roots of many types of human behavior, and we as a society must develop strategies for integrating this newfound knowledge with existing views of who we are.

Applications of genomics that cross over the blurred line between treatment and enhancement will challenge our society, and the question must be asked (as it has been with human reproductive cloning) whether there are boundaries that should not be crossed. Furthermore, applications of genomics extend far beyond the fields of biology and medicine to the legal system, the insurance industry, the military, and even educational institutions. There are no easy answers as to who should set the boundaries for genomics applications or how these policies should be enforced. However, we need to start tackling these complex questions today because some of these uses could radically change the landscape of life for future generations.

## CONCLUSIONS

After operating in the "pre-genome" era for all of its existence, the science of biology officially entered the genome era on April 14, 2003, with the successful completion of the Human Genome Project. However, the challenge of truly understanding the human genome has just begun. A staggering amount of research and technological development remains to be done before we can define all the parts of the human genome and their functions, let alone translate genomic understanding into medical benefits for all people around the globe. With the insightful assistance of hundreds of the world's leading biologists, we have proposed a blueprint for the next generation of genomics research—a plan that gathers under one broad roof the complex and varied challenges of moving genomics into basic biology, health, and society. Now, it is time for genomics research to unroll that blueprint, load up its toolbox, and take the first exciting steps toward building tomorrow's reality.

## ACKNOWLEDGMENTS

I thank all of the thousands of researchers who made the dreams of the Human Genome Project come true, and the hundreds of scientists who participated in NHGRI's planning process, 2001–2003.

## REFERENCES

Austin C.P. 2003. The completed human genome: Implications for chemical biology. *Curr. Opin. Chem. Biol.* **7:** 511.

Collins F.S., Green E.D., Guttmacher A.E., and Guyer M.S. 2003. A vision for the future of genomics research: A blueprint for the genomic era. *Nature* **422:** 835.

Stockwell B.R. 2000. Chemical genetics: Ligand-based discovery of gene function. *Nat. Rev. Genet.* **1:** 116.

Thomas J.W., Touchman J.W., Blakesley R.W., Bouffard G.G., Beckstrom-Sternberg S.M., Margulies E.H., Blanchette M., Siepel A.C., Thomas P.J., McDowell J.C., Maskeri B., Hansen N.F., Schwartz M.S., Weber R.J., Kent W.J., Karolchik D., Bruen T.C., Beven R., Cutler D.J., Schwartz S., Elnitski, L., Idol J.R., Prasad A.B., Lee-Lin S.-Q., Maduro V.V.B., Summers T.J., Portnoy M.E., Dietrich N.L., Akhter N., Ayele K., Benjamin B., Cariaga K., Brinkley C.P., Brooks S.Y., Granite S., Guan X., Gupta J., Haghighi P., Ho S-L., Huang M.C., Karlins E., Laric P.L., Legaspi R., Lim M.J., and Maduro, et al. 2003. Comparative analyses of multi-species sequences from targeted genomic regions. *Nature* **424:** 788.

# DNA Sequence Variation of *Homo sapiens*

D.R. BENTLEY
*The Wellcome Trust Sanger Institute, Hinxton, Cambridge CB10 1SA, United Kingdom*

The finished genome sequence of *Homo sapiens* (Rogers, this volume) provides a starting point for the study of sequence variation in the human population. Every variant that is discovered can be mapped back to the human genome and correlated with genes, regulatory elements, and other functionally important sequences. As we gain a better understanding of the biological information encoded by the human genome sequence, we should aim to define the sequence variants that have biochemical and phenotypic consequences.

The genome sequence enables us to develop targeted strategies to search for disease-related sequence variants. This requires a better understanding of the patterns of variation and of the genetic variants that can be used as reference markers throughout the genome. Genome sequence information will also underpin future surveys of somatic variation and cancer. Studying the DNA sequence variation of *Homo sapiens* will enable us to understand our origins and evolution and will help characterize the genetic basis of our individuality—for example, in our susceptibility or resistance to disease, and our variable response to drugs, toxins, and other environmental factors.

## THE ORIGIN OF SEQUENCE VARIATION

Modern humans are believed to have migrated east out of Africa 50,000–60,000 years ago (Jobling et al. 2004) and subsequently spread across the world, replacing earlier *Homo* species. This pattern was originally deduced largely from archaeological and anthropological evidence (Stringer 2002) but received substantial reinforcement from DNA sequence information. For example, genetic variability is generally higher in Africa than on other continents, and phylogenetic reconstructions of non-recombining regions usually place the root in Africa (Cavalli-Sforza and Feldman 2003; Pääbo 2003). A subset of the genetic variants in Africa at the time were therefore present in the migrant founders of all later subsequent groups, while many variants remained only within population subgroups in Africa. More recent migrations between different parts of the globe have since contributed to admixture between multiple subgroups, and in the last few hundred years, this process has increased substantially.

Sequence variation arises as a result of new mutation and recombination (see Fig. 1). Based on observed variation and simulations, the average human mutation rate has been estimated to be 1–2 bases in 100 million per generation (Drake et al. 1998; Giannelli et al. 1999; Reich et al. 2002) which corresponds to around 30–60 new mutations per gamete. For variants that are neutral (not under selection), the allele frequency in the population will be affected by random drift. Many new mutations will disappear within a few generations; a few may become com-

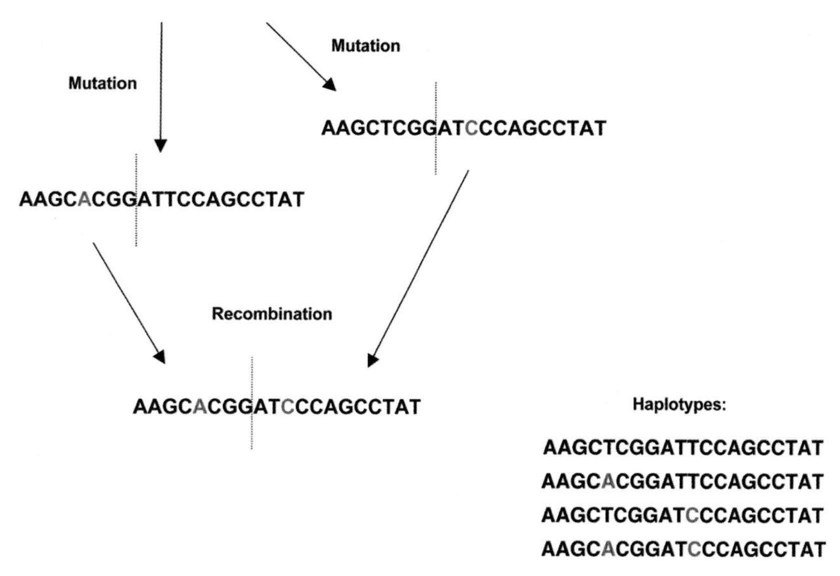


**Figure 1.** Origin of sequence variation. Sequence variation arises by mutation (*colored bases*) and by recombination (*dotted lines*). These processes give rise to individual haplotypes (listed on the right) that coexist in the population.

mon over many generations. Most high-frequency sequence variants therefore arose in Africa, were present in the migrant founders, and are common to multiple population groups around the world (Barbujani et al. 1997). New variants that are under positive selection—i.e., functional variants that confer a survival advantage—will increase in frequency more quickly than expected by random processes. A change in the environment (such as the appearance of a new virus, or release of a new toxin, or change in food source) may impose a new selective pressure on existing functional variants, leading to alterations in allele frequency in a particular population. Balancing or frequency-dependent selection may act to maintain both alleles of a polymorphism stably in the population.

Recombination during meiosis results in cross-overs between homologous chromosomes. This leads to reassortment of previously existing variants into new combinations in the haploid gametes. Assuming a recombination rate of 1 site per 100 million bases per generation (Yu et al. 2001) (i.e., around 30 recombination events per gamete), large segments of each parental homolog are passed on from one generation to the next. Over multiple generations, further recombination events result in progressive fragmentation of the original ancestral segments into more and more pieces. Individual present-day chromosomes are thus mosaics of segments of relatively few ancestral chromosomes, each segment having its own lineage history (Pääbo 2003). The sequence along a chromosome defines the complete set of variants in an individual haploid sequence and represents a particular "haplotype" (from "haploid genotype") (see Fig. 1).

Somatic mutation occurs in individual cells at any stage following the formation of a diploid zygote, during development and throughout adulthood. The mutation process may be enhanced by exposure to particular nongenetic factors such as radiation or toxins. Particular combinations of somatic mutations, sometimes in conjunction with germ-line variants, alter the normal program of cell proliferation, differentiation, and apoptosis, and lead to cancer.

## THE NATURE OF SEQUENCE VARIATION

Individual copies of the haploid human genome differ at approximately one site per kilobase on average (Li and Sadler 1991; Sachidanandam et al. 2001). Single-nucleotide polymorphisms (SNPs) account for around 90% of sequence variants in the human population, the remaining 10% being insertions or deletions ("indels"). It has been estimated that the world's human population contains over 10 million SNPs with a minor allele frequency (m.a.f.) of at least 1% (Kruglyak and Nickerson 2001). Several in-depth studies of sequencing in specific genes have demonstrated that within exons there is a slightly lower density of polymorphic sites (1 site per 2,000 bases), and also typically a lower allele frequency (Cargill et al. 1999; Halushka et al. 1999). Assuming that 1.5% of the genome sequence encodes protein (Lander et al. 2001), the global population contains around 90,000 variants (with m.a.f. >1%) in protein-coding sequence. From previous surveys, the protein-coding variants can be subdivided into ~50% synonymous (causing no amino acid change), 33% nonsynonymous and conservative (i.e., leading to conservative amino acid replacement), and 17% nonsynonymous, nonconservative changes (Cargill et al. 1999; Halushka et al. 1999).

Genetic variants that give rise to nonconservative amino acid changes are particularly good functional variant candidates, but a number of observations suggest that this generalization is likely to be inadequate. Not all nonconservative amino acid changes necessarily alter function. For example, in an exhaustive survey (Green et al. 1999), only half of all possible amino acid changes (233/454) in factor IX cause hemophilia B (Haemophilia B Mutation Database v12: http://www.kcl.ac.uk/ip/petergreen/haemBdatabase.html).

Functionally important amino acids in proteins tend to be conserved between species; on average, about one-third of amino acids are highly or absolutely conserved in vertebrates. Conversely, other types of changes (either at the amino acid or nucleotide level) can alter function. Conservative changes may affect protein function, and any changes may affect transcript function; for example, by introducing or abolishing splice sites. There is no good estimate of the number of functionally important changes outside protein-coding sequence. Changes in promoters or enhancers may affect transcription (Li et al. 1999; Saur et al. 2004). Comparison between multiple genomes suggests that around 5% of the human genome is under selection to conserve sequence (Waterston et al. 2002). These analyses suggest that, in addition to the protein-coding sequence, another 3.7% of the genome is equally strongly conserved and may be important for gene regulation or chromatin behavior (Waterston et al. 2002; Margulies et al. 2003). On this basis, an initial estimate of variants in functional sequence (with m.a.f. < 1%) would be nearer 300,000, and an unknown fraction of these will have phenotypic consequences.

The first genome-wide survey of human variation resulted in detection of 1.42 million candidate SNPs that were mapped to a unique position in the working draft sequence (Sachidanandam et al. 2001). This set was estimated to contain possibly 12% of all SNPs with a m.a.f. of 1% or more (Kruglyak and Nickerson 2001). More recently, the total number of SNPs available with a unique map position has increased to 7 million (dbSNP release 120, http://www.ncbi.nlm.nih.gov/SNP) following generation of additional shotgun data, and may now contain 40% of SNPs with m.a.f. <1% (estimate based on Kruglyak and Nickerson [2001]). A fully characterized catalog of human sequence variation may still be a long way off. In the meantime, however, efforts are focused on how to develop the resources that are available from the finished genome sequence, and how to use them to study the genetics of human disease.

## SEQUENCE VARIATION AND DISEASE

The genes and underlying sequence variants that cause over 1400 diseases have already been identified (data from Online Mendelian Inheritance in Man; http://www.ncbi.nlm.nih.gov). The majority of these have been rare single-gene disorders. In most cases, linkage studies in large affected families have been used to target a genomic region in order to screen for mutations that disrupt a candidate gene ("positional cloning"; Collins 1992). In a few studies, detection of a cytogenetic abnormality (e.g., a translocation) provided the necessary positional information, and a few genes were discovered using prior knowledge of the protein (e.g., amino acid sequence). The "parametric" linkage approach used in the single-gene disorders has been successful in only a few polygenic conditions, where the gene has a sufficiently strong effect to show a clear familial inheritance pattern. Examples include susceptibility loci for breast cancer (BRCA1 [Miki et al. 1994] and BRCA2 [Wooster et al. 1995]) and the initial study of maturity onset of diabetes of the young (MODY) (for review, see Bell and Polonsky 2001). However, in most studies, this approach has been largely unsuccessful as little or no convincing linkage has been found. Alternative linkage strategies that do not rely on specific transmission patterns initially held great promise for complex trait applications (as seen, for example, in the discovery of NOD2 with Crohn's disease [Hugot et al. 2001; Ogura et al. 2001] and CTLA-4 with Graves' disease [Ueda et al. 2003]), but have not yet yielded many consistent results. Population-based association studies are believed to offer greater statistical power in detecting genetic effects underlying complex traits (Risch 2000).

An association study is used to test for a positive correlation between a sequence variant in the genome and a disease or measurable phenotype (Risch and Merikangas 1996). For example, a genotyping assay is used to determine the frequency of each allele at a particular locus in separate, matched groups of patients and controls (a case-control study). If one of the alleles has a high frequency in cases compared to controls, this is evidence for association of that allele with the disease. A direct association study uses a set of candidate gene variants that are presumed to include the causal variant(s) (Risch and Merikangas 1996). It tests the hypothesis that a particular variant is directly involved in the phenotype. An indirect study uses anonymous variants (such as SNPs) as markers (Collins et al. 1997). It tests the hypothesis that a marker is closely linked to an unknown causative variant. A direct association study was used to demonstrate the association of a 32-base deletion (Δ32) in the cytokine receptor 5 gene (*CKR5*) and resistance to HIV. The CKR5 protein is a G-protein-coupled receptor on the surface of $CD4^+$ T-lymphocytes that appears to be an efficient coreceptor for HIV-1 viral strains. The deletion variant is nonfunctional with respect to both its natural function and its capacity to mediate HIV-1 infection. The homozygous form (Δ32/Δ32) was strongly associated with a protective effect against HIV-1 infection, and there was also evidence that the heterozygous form (+/Δ32) delayed progression to acquired immune deficiency syndrome in some cases (Dean et al. 1996). An indirect association study was used to follow up an earlier linkage to part of Chromosome 5q31 to inflammatory bowel disease (IBD) (Rioux et al. 2001). Using a dense panel of SNPs, the disease was associated with a common haplotype, and the critical region was reduced from ~1 Mb to 250 kb.

The power of a direct association study is influenced by the allele frequency and risk ratio of the causative variant in relationship to the sample size. Detecting causative variants in association studies requires a progressively larger sample size as the allele frequency or risk ratio of the variant decreases. For indirect association studies, an additional factor is the degree of correlation between the anonymous markers used in the study and the causative variant. If one of the markers is completely correlated with the causative variant, the power of the indirect study will match that of the direct approach. In most cases, however, we expect the available anonymous markers to be incompletely correlated, and the indirect approach will have less power. If the allele frequency of the anonymous markers is substantially different from the unknown functional variants, this will also reduce power (Risch 2000; Zondervan and Cardon 2004).

Causative variants that are known to contribute to common disease include examples of a range of allele frequencies (0.85–0.01 in the examples listed in Table 1). There may be diseases caused by multiple rare variants at the same locus (e.g., the NOD2 locus; Hugot et al. 2001; Ogura et al. 2001). In these cases, each variant is of independent origin and is therefore carried on a different haplotype. The ability to detect these variants by association will depend on the power of the study to detect each individual variant. The case-control association studies reported to date have demonstrated the ability to detect disease-associated variants with modest relative risks. For example, the association of LTA-3 with myocardial infarction was essentially hypothesis-free: 65,671 SNPs distributed in 13,738 genes throughout the genome were tested in 1,133 cases and 1,006 controls (in a two-tier study), and homozygosity with respect to each of two SNPs (both in the LTA-3 gene) showed significant association with the disease (Ozaki et al. 2002). Evidence was obtained suggesting a possible functional significance for each variant that might correlate with phenotype. If so, then this would be a direct association study. Studies such as this provide an indication of the future potential and limitations of the approach. For some of the other disease studies listed in Table 1, prior linkage to a chromosomal region was also detected, and this positional information was used to select genes or variants for use in the association study.

Before starting a direct association study, it is necessary to find the functional variants for the study by sequencing candidate genes in multiple individuals. For maximum sensitivity, the best approach would be to sequence a group of affected individuals (specific to each study) that should be enriched for disease-associated variants. Alternatively, a large number of control samples would need to be sequenced to capture the low-frequency variants that might be expected (for estimates, see Kruglyak and Nickerson 2001). Either option is limited by cost and by the fact that we do not yet know all the functional regions of each gene that would need to be se-

**Table 1.** Genetic Disease Variants

| Reference | Phenotype | Gene | Allele | Frequency (between 0 and 1) | Allelic relative risk |
|---|---|---|---|---|---|
| Bentley et al. (1986) | Hemophilia B | FIX | Arg -4 Gln | $1.7 \times 10^{-9}$ | ∞ |
| Bell et al. (1984) | IDDM | INS | VNTR | 0.7 | 2 |
| Altshuler et al. (2000) | NIDDM | PPARG | Pro 12 Ala | 0.8 | 1.25 |
| Ozaki et al. (2002) | MI | LTA-3 | Thr 26 Asn | 0.5 | 1.8 |
| Ueda et al. (2003) | Graves | CTLA4 | Thr 17 Ala | 0.35 | 1.7 |
| Palmer et al. (1991) | Creutzfeldt-Jacob | PRNP | Met 129 Val | 0.35 | 3 |
| Saunders et al. (1993) | Alzheimer's | APOE | Cys 112 Arg | 0.16 | 4 |
| Dean et al. (1996) | HIV resistance | CCR5 | del 32 | 0.10 | 7 |
| Bertina et al. (1994) | thrombosis | FV | Arg 506 Gln | 0.05 | 7 |
| Ogura et al. (2001) | Crohn's | NOD2 | 1007 fs | 0.04 | 2 |
| Hugot et al. (2001) | Crohn's | NOD2 | Arg 702 Trp | 0.04 | 3 |
| Hugot et al. (2001) | Crohn's | NOD2 | 980 fs | 0.02 | 6 |
| Hugot et al. (2001) | Crohn's | NOD2 | Gly 908 Arg | 0.01 | 6 |

quenced. However, it is important to pursue this strategy, as it will be very effective to search for rare variants, either in control groups or patient collections, in order to explore the full extent of functional sequence variation in the genome.

Before embarking on the indirect approach, we need to develop a comprehensive panel of anonymous markers (SNPs). This can be done for each gene or chromosomal region of interest, but it will be much more valuable to do it systematically across the whole genome and develop a freely available resource. Once created, this SNP panel can be applied universally to any search for associations with any disease or phenotype. It is critical that as near as possible, every part of the genome is tightly linked to at least one marker of the panel. Development of this panel requires a dense SNP map, technology to genotype a large number of SNPs in multiple DNA samples, an effective way to measure linkage between SNPs, a strategy that can be adjusted to characterize regions of high and low linkage disequilibrium (LD), and a way to assess the extent that common patterns of variation are captured in each population group. Recent research on a number of isolated genomic regions and whole chromosomes has led to evaluation of these requirements, as discussed below.

## LINKAGE DISEQUILIBRIUM AND HAPLOTYPES

The degree of correlation between two markers can be determined by population genetic analysis. If the alleles of two neighboring SNPs are in equilibrium in the population, the alleles at each SNP are independent from one another. If this is the case, each haplotype (each combination of two alleles) occurs at a frequency that is the product of the frequency of each individual allele. If, on the other hand, particular SNP alleles occur together more often than expected by the equilibrium model, they are said to be in association, or LD. LD between two loci ("pair-wise" LD) is determined empirically by carrying out genotyping experiments on a population sample and calculating the difference between expected and observed frequencies of each combination of alleles. The most commonly used measures of LD are D′ (Lewontin 1964) and $r^2$ (Hill and Robertson 1968; Ohta and Kimura 1969; for further discussion, see Wall and Pritchard 2003).

In general, pair-wise LD decreases with increasing physical distance between markers. This is because over longer distances there is a higher chance that ancestral recombination events have occurred between the markers. However, LD is also highly variable with respect to physical distance ( Clark et al. 1998; Abecasis et al. 2001; Reich et al. 2001; Dawson et al. 2002). Another observation is that variants which arose recently (for example, most rare variants) tend to exhibit LD over longer distances, reflecting the low probability of a nearby recombination in relatively few generations.

A whole-chromosome study enabled systematic correlation of LD with respect to physical distance and a wide range of structural and genetic features (Dawson et al. 2002). Individual genotypes were collected for 1,504 evenly spaced SNPs along Chromosome 22 (average spacing 1 SNP/20 kb) in a panel of DNAs of North European origin. The profile of LD along the chromosome was determined by averaging pair-wise LD measurements within 1.7-Mb sliding windows. This study revealed highly variable patterns of LD in different chromosome regions, and two notable regions where LD extended over hundreds of kilobases (Fig. 2). LD did not correlate with any features of the gene content, repetitive sequence, or other structural features such as base composition. However, there was a strong correlation between high LD and low recombination frequency (Fig. 2), indicating that historical and contemporary recombination are related, and nonrandom along the chromosome. Within the regions of high LD, it was also possible to observe that a limited number of haplotypes accounted for the majority in the population. For example, over a 700-kb region, 5 extended haplotypes together accounted for 76% of the individual chromosomes studied, the remainder being accounted for in 12 rare haplotypes (Dawson et al. 2002).

The results of several parallel studies revealed similar findings: that LD was highly variable, and that it was possible to define short regions where LD was consistently high between all markers and where there were a limited number of common haplotypes. Following the initial work of Daly et al. (2001), Gabriel et al. (2002) developed a method to define regions of high LD as occurring

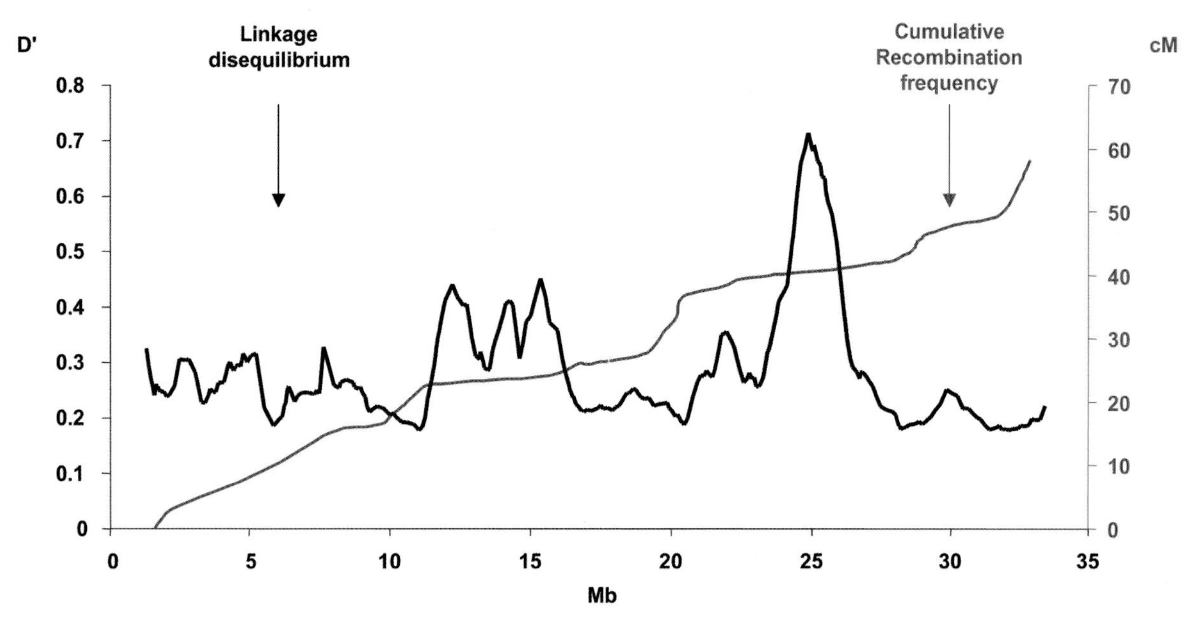


**Figure 2.** Linkage disequilibrium (LD) and meiotic recombination: Chromosome 22. The LD profile is based on average D′ values in sliding windows (see text). LD and cumulative recombination frequency are plotted relative to physical distance along the chromosome, with the telomere of the long arm on the right of the figure.

where pair-wise LD values (based on D′) between three or more adjacent SNPs exceeded a defined threshold (see Fig. 3a–c). Within regions of high LD, it was possible to define a limited number of common haplotypes that represented most of the chromosomes in the study (Fig. 3d). These regions were termed "haplotype blocks" (Gabriel et al. 2002). They were separated by regions where LD had not been detected, either because LD was low or because there were too few markers available. The block method did not fully take into account LD between blocks and did not provide a continuous view of the pattern of LD. Nevertheless, it would be possible to use it as a simple and reliable indicator of regions where data from more SNPs was needed. As expected from previous work

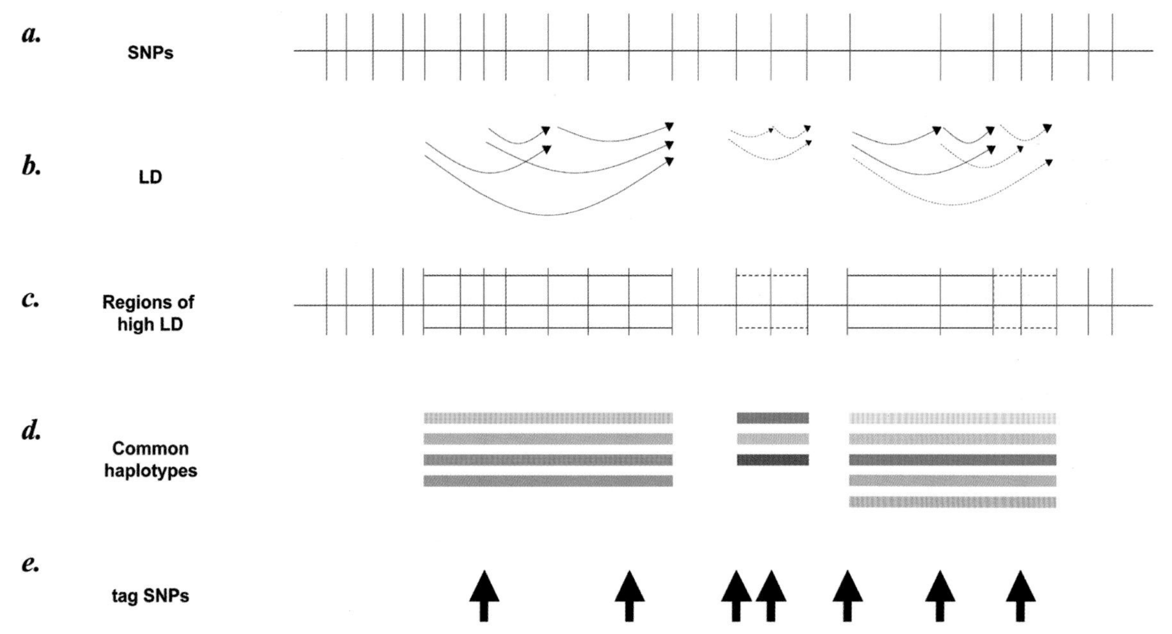


**Figure 3.** SNPs, linkage disequilibrium, and haplotypes. The schematic diagram depicts the course of an experiment carried out in two tiers as follows. *First tier:* From the SNP map of the genome, SNPs are selected (*blue vertical bars* in *a*) and used to genotype DNA samples. The results are used to calculate pair-wise LD (*curved arrows* in *b*), and the results are compiled to define regions of high LD (*diagonally shaded boxes* in *c*, drawn over the SNP map). At this point, overall progress in characterizing LD is assessed, and additional SNPs are selected where more data are needed (*red vertical bars* in *a* and *c*). Common haplotypes (*colored bars* in *d*) are determined from the LD data. A subset of the SNPs ("tag" SNPs; *arrows* in *e*) is selected that captures most or all of the common variation in the region (see text for details).

(Clark et al. 1998; Reich et al. 2001), this study also revealed greater diversity in African-Americans than in North Europeans with respect to ancestral recombination and polymorphism. A lower average block size was estimated in the African-Americans than in North Europeans (11 kb vs. 22 kb), and within blocks, the average haplotype diversity was higher in African-Americans than in North Europeans (5 vs. 4.2 haplotypes per block). The study by Patil et al. (2001) took a different approach. Oligonucleotide sequencing arrays representing the nonrepetitive sequence in Chromosome 21 were used to detect SNPs and common haplotypes by analysis of 20 individuals. 24,047 SNPs detected blocks of common haplotypes of average size 13 kb and 3.2 haplotypes per block covering 80% of the chromosome. (Blocks with 1 or 2 SNPs were excluded from these calculations.)

Comparison of the studies listed in Table 2 revealed that for most of them, detection of LD was limited by SNP availability. Caution should be exercised in this comparison, as some data sets were small and the criteria for measuring high LD were slightly different; but the overall trend was clear. As SNP spacing decreased from 20 kb to 8 kb, 5 kb, and 1 kb, the proportion of the studied genomic regions where high LD was detected increased from 20% to 58%, 78%, and 95%, respectively (see Table 2). Future studies of the genome therefore needed more SNPs, particularly in regions where little or no LD had been detected.

Another important observation from these studies is based on the idea that the variation data derived from the initial SNPs could be used to select a subset of the SNPs that distinguished the different haplotypes, and thus captured most or all of the information on common variation in the region. These "haplotype tag SNPs" (htSNPs) (see Fig. 3e) (Johnson et al. 2001; see also Fig. 1 in International HapMap Consortium 2003) would be maximally informative and could be used in future association studies, whereas the other SNPs could be discarded. In three of the studies listed in Table 1, htSNPs were identified comprising between 27% and 20% of the initial SNPs. Determining LD and haplotype patterns empirically could therefore enable fourfold savings in future association studies with little or no loss of power, assuming the patterns of haplotype diversity were the same in the initial population and the subsequent disease sample. The conclusions from these studies provided the motivation for planning a large-scale pilot study on Chromosome 20 (see below), and also the International HapMap Project, to determine common LD and haplotype patterns throughout the human genome, in multiple ethnic groups (International HapMap Consortium 2003).

## CHROMOSOME 20

The experience gained in the previous studies illustrated the need to produce genotype data with very high densities of SNP maps, and to evaluate how SNP density affects measurement of LD and haplotype analysis over large genomic regions. These conclusions formed the basis for a study of Chromosome 20. At the time, the existing map for this Chromosome contained a total of 46,000 candidate SNPs (1 SNP/1.3 kb on average). Further analysis of the SNP distribution along the chromosome revealed that only 30% of the sequence contained 10 or more SNPs per 10-kb window. Given that the density of SNPs used in the study of Jeffreys et al. (2001) (see Table 2) was higher than this, it was necessary to generate many more SNPs in order to obtain this density over the rest of Chromosome 20. Random shotgun sequencing of Chromosome 20 (purified by flow sorting) generated additional SNPs to obtain a minimum density of 10 SNPs per 10 kb (close to the density used in the Jeffreys study) for almost all the chromosome (P. Deloukas et al., unpubl.). The need to supplement the existing SNP map to this level was confirmed in the findings of the subsequent LD analysis and has since been adopted on a genome-wide basis (see above).

SNPs were selected from the new Chromosome 20 map at ~1-kb spacing and typed using the Golden Gate assay (Fan et al., this volume) in samples of North European, African-American, and East Asian origin. Analysis of a 10-Mb region (Ke et al. 2004) revealed variable LD along the chromosome and good correlation between high LD and low recombination frequency, as observed in the Chromosome 22 study. Progressive removal of subsets of the raw data did not alter the LD profile obtained by the sliding windows method (using 500-kb windows), illustrating that this view of LD is robust at SNP densities of 1 SNP/10 kb and above. The profiles were also consistent between all three population groups, although overall LD was lower in the African-Americans than in the other two groups.

In contrast to the LD profiles, it was found that haplotype block patterns were not stable. These differed depending on the SNP densities, and were particularly unstable at low densities both in regions of high and low LD (see Fig. 4) (Ke et al. 2004). This result suggested that future studies should be carried out at a minimum density of 1 SNP/5 kb. More SNPs were also required, particularly in areas of low LD, and more SNPs might also be required to confirm patterns of LD and haplotypes in areas of high LD (see SNPs marked in blue in Fig. 3a and c). At the highest SNP density, the overall coverage of the re-

**Table 2.** LD Studies

| Study | Region studied | kb studied | SNPs (total) | SNP spacing (kb) | LD (% region covered) | tag SNPs ( % of total) |
|---|---|---|---|---|---|---|
| Dawson et al. (2002) | Chromosome 22 | 30,000 | 1,504 | 20 | 20 | — |
| Gabriel et al. (2002) | 51 regions | 13,000 | 1,970 | 7.8 | 58 | — |
| Daly et al. (2001) | region 5q31 | 460 | 103 | 5 | 78 | 25 |
| Johnson et al. ( 2001) | selected genes | 135 | 122 | 1.1 | n.d. | 27 |
| Jeffreys et al. (2001) | human MHC | 216 | 179 | 1.2 | 95 | 20 |

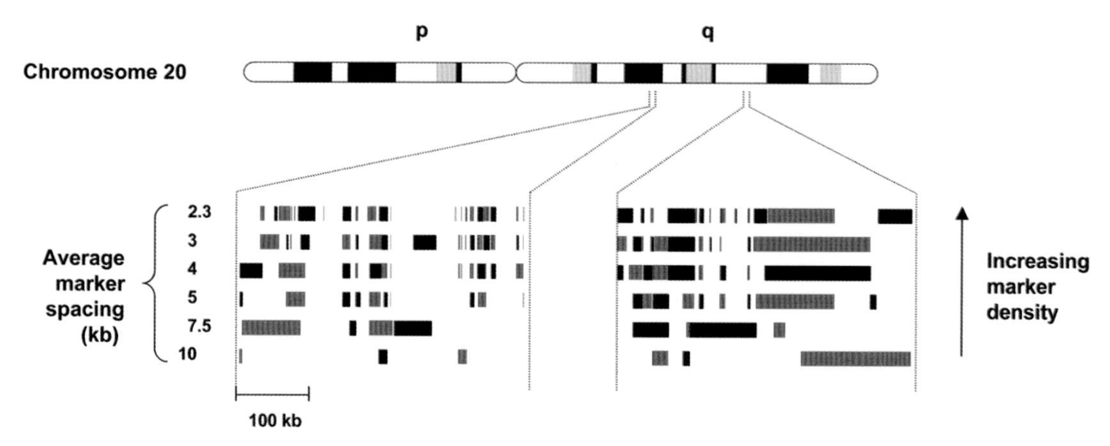


**Figure 4.** LD analysis of Chromosome 20. Haplotype blocks (*red* and *black boxes*) were computed from LD data on Chromosome 20 and are shown for two regions, one each of high and low overall LD. The analysis is repeated using data from different densities of SNPs (average marker spacings in each analysis are listed on the left of the figure, and increasing SNP density is indicated by the vertical arrow).

gion in high LD was estimated to be 65% in the North Europeans and 45% in the African-Americans (Ke et al. 2004). The coverage estimates for North Europeans fits the trend in Table 1. However, the lower coverage figure for the African-Americans suggests that more dense SNP sets may need to be used in these populations—echoing the earlier findings of the Gabriel study (Gabriel et al. 2002). The block patterns also varied significantly when using different analytical methods, indicating the need for other ways to describe LD patterns on a fine scale. For example, LD unit maps (Maniatis et al. 2002) reflect continuous views of LD, although they are to some extent sensitive to varying allele frequency. A promising approach is the development of methods to estimate recombination rates on a fine scale using the same data sets described above (McVean et al. 2004). These approaches could enable a precise assessment of progress in characterization of LD in the genome, identify and prioritize regions that require more data, and help assess the ability of panels of anonymous markers to detect unknown disease-related variants.

## IMPLICATIONS FOR FUTURE STUDIES

The first global study of patterns of common variation in the human genome has crystallized in the form of the International HapMap Project (International HapMap Consortium 2003). The density of SNPs that are now available in the public domain is likely to be sufficient for this initial analysis of LD and common haplotypes, although some targeted SNP discovery to fill gaps may be necessary. This project will produce a freely available genome-wide panel of tested SNPs that can be assessed for use in association studies. It is intended to reduce the need for researchers to search for their own SNPs, and to enable better searches for disease variants in genes, chromosomal regions, or ultimately, the whole genome using the indirect association approach. Evaluation of the parameters that govern association studies are still required, and the results of pilot studies might help calibrate the HapMap and other studies to fit more precisely with their anticipated applications. In parallel, it is also necessary to address the challenge of developing sufficiently large sample collections with accurate, accessible phenotype information. Without this, even a perfect SNP panel that covers 100% of the genome will lack the power to establish an association between genotype and phenotype.

We should continue sequence-based discovery of variants on a genome-wide basis; this would give us an unbiased and properly quantified picture of the extent of common sequence variation in the human population, replacing the current simulation-based estimates. Achieving the first step at this level would benefit from upward of 100 individual human genome sequences, a task that requires a greater than tenfold increase in sequence output and is likely to demand the successful implementation of new technologies for cheap, accurate, high-throughput resequencing. Discovering rare variants (m.a.f. < 1%) would require improvements of a further one or two orders of magnitude in sequencing. At present, therefore, pursuing rare variants remains in the domain of individual candidate genes and may be best carried out in conjunction with a genetic disease study. Both candidate gene and whole-genome sequencing will also be instrumental in characterizing somatic variants that cause cancer, as demonstrated, for example, by Davies et al. (2002). Systematic pursuit of these areas of research will help us gain a view of the complete spectrum of variant allele frequencies that contributes to human disease.

As we learn more about the full extent of functional sequences in the genome, our knowledge of the functionally important subset of variants will increase (for further discussion, see Bentley 2004). Developing a panel of functional variants for all genes would increase the power of association studies. However, we must be aware of the unknown; any association study based on the direct approach needs to take account of the possibility that the true causative variant is not included in the study. It is therefore necessary to establish the underlying mechanism and explain the functional significance of the variant. Within the medical field, understanding the mechanism that underlies disease will be the most effective way to make progress in diagnosis, alleviation, and the effective treatment of disease.

## ACKNOWLEDGMENTS

The author especially thanks Ines Barroso, Lon Cardon, Panos Deloukas, Peter Donnelly, Christine Rees, and others for discussion and assistance. The author gratefully acknowledges the support of the Wellcome Trust.

## REFERENCES


Abecasis G.R., Noguchi E., Heinzmann A., Traherne J.A., Bhattacharyya S., Leaves N.I., Anderson G.G., Zhang Y., Lench N.J., Carey A., Cardon L.R., Moffatt M.F., and Cookson W.O. 2001. Extent and distribution of linkage disequilibrium in three genomic regions. *Am. J. Hum. Genet.* **68:** 191.

Altshuler D., Hirschhorn J.N., Klannemark M., Lindgren C.M., Vohl M.C., Nemesh J., Lane C.R., Schaffner S.F., Bolk S., Brewer C., Tuomi T., Gaudet D., Hudson T.J., Daly M., Groop L., and Lander E.S. 2000. The common PPARgamma Pro12Ala polymorphism is associated with decreased risk of type 2 diabetes. *Nat. Genet.* **26:** 76.

Barbujani G., Magagni A., Minch E., and Cavalli-Sforza L.L. 1997. An apportionment of human DNA diversity. *Proc. Natl. Acad. Sci.* **94:** 4516.

Bell G. I. and Polonsky K.S. 2001. Diabetes mellitus and genetically programmed defects in beta-cell function. *Nature* **414:** 788.

Bell G.I., Horita S., and Karam J.H. 1984. A polymorphic locus near the human insulin gene is associated with insulin-dependent diabetes mellitus. *Diabetes* **33:** 176.

Bentley A.K., Rees D.J., Rizza C., and Brownlee G.G. 1986. Defective propeptide processing of blood clotting factor IX caused by mutation of arginine to glutamine at position -4. *Cell* **45:** 343.

Bentley D.R. 2004. Genome vision. *Nature* (in press).

Bertina R.M., Koeleman B.P., Koster T., Rosendaal F.R., Dirven R.J., de Ronde H., van der Velden P.A., and Reitsma P.H. 1994. Mutation in blood coagulation factor V associated with resistance to activated protein C. *Nature* **369:** 64.

Cargill M., Altshuler D., Ireland J., Sklar P., Ardlie K., Patil N., Shaw N., Lane C.R., Lim E.P., Kalyanaraman N., Nemesh J., Ziaugra L., Friedland L., Rolfe A., Warrington J., Lipshutz R., Daley G.Q., and Lander E.S. 1999. Characterization of single-nucleotide polymorphisms in coding regions of human genes. *Nat. Genet.* **22:** 231.

Cavalli-Sforza L.L. and Feldman M.W. 2003. The application of molecular genetic approaches to the study of human evolution. *Nat. Genet.* (suppl.) **33:** 266.

Clark A.G., Weiss K.M., Nickerson D.A., Taylor S.L., Buchanan A., Stengard J., Salomaa V., Vartiainen E., Perola M., Boerwinkle E., and Sing C.F. 1998. Haplotype structure and population genetic inferences from nucleotide-sequence variation in human lipoprotein lipase. *Am. J. Hum. Genet.* **63:** 595.

Collins F.S. 1992. Positional cloning: Let's not call it reverse anymore. *Nat. Genet.* **1:** 3.

Collins F.S., Guyer M.S., and Charkravarti A. 1997. Variations on a theme: Cataloging human DNA sequence variation. *Science* **278:** 1580.

Daly M.J., Rioux J.D., Schaffner S.F., Hudson T.J., and Lander E.S. 2001. High-resolution haplotype structure in the human genome. *Nat. Genet.* **29:** 229.

Davies H., Bignell G.R., Cox C., Stephens P., Edkins S., Clegg S., Teague J., Woffendin H., Garnett M.J., Bottomley W., Davis N., Dicks E., Ewing R., Floyd Y., Gray K., Hall S., Hawes R., Hughes J., Kosmidou V., Menzies A., Mould C., Parker A., Stevens C., Watt S., and Hooper S., et al. 2002. Mutations of the BRAF gene in human cancer. *Nature* **417:** 949.

Dawson E., Abecasis G.R., Bumpstead S., Chen Y., Hunt S., Beare D.M., Pabial J., Dibling T., Tinsley E., Kirby S., Carter D., Papaspyridonos M., Livingstone S., Ganske R., Lohmussaar E., Zernant J., Tonisson N., Remm M., Magi R., Puurand T., Vilo J., Kurg A., Rice K., Deloukas P., and Mott R., et al. 2002. A first-generation linkage disequilibrium map of human chromosome 22. *Nature* **418:** 544.

Dean M., Carrington M., Winkler C., Huttley G.A., Smith M.W., Allikmets R., Goedert J.J., Buchbinder S.P., Vittinghoff E., Gomperts E., Donfield S., Vlahov D., Kaslow R., Saah A., Rinaldo C., Detels R., and O'Brien S.J. 1996. Genetic restriction of HIV-1 infection and progression to AIDS by a deletion allele of the CKR5 structural gene. Hemophilia Growth and Development Study, Multicenter AIDS Cohort Study, Multicenter Hemophilia Cohort Study, San Francisco City Cohort, ALIVE Study. *Science* **273:** 1856.

Drake J.W., Charlesworth B., Charlesworth D., and Crow J.F. 1998. Rates of spontaneous mutation. *Genetics* **148:** 1667.

Gabriel S.B., Schaffner S.F., Nguyen H., Moore J.M., Roy J., Blumenstiel B., Higgins J., DeFelice M., Lochner A., Faggart M., Liu-Cordero S.N., Rotimi C., Adeyemo A., Cooper R., Ward R., Lander E.S., Daly M.J., and Altshuler D. 2002. The structure of haplotype blocks in the human genome. *Science* **296:** 2225.

Giannelli F., Anagnostopoulos T., and Green P.M. 1999. Mutation rates in humans. II. Sporadic mutation-specific rates and rate of detrimental human mutations inferred from hemophilia B. *Am. J. Hum. Genet.* **65:** 1580.

Green P.M., Saad S., Lewis C.M., and Giannelli F. 1999. Mutation rates in humans. I. Overall and sex-specific rates obtained from a population study of hemophilia B. *Am. J. Hum. Genet.* **65:** 1572.

Halushka M.K., Fan J.B., Bentley K., Hsie L., Shen N., Weder A., Cooper R., Lipshutz R., and Chakravarti A. 1999. Patterns of single-nucleotide polymorphisms in candidate genes for blood-pressure homeostasis. *Nat. Genet.* **22:** 239.

Hill W.G. and Robertson A. 1968. Linkage disequilibrium in finite populations. *Theor. Appl. Genet.* **38:** 226.

Hugot J.P., Chamaillard M., Zouali H., Lesage S., Cezard J.P., Belaiche J., Almer S., Tysk C., O'Morain C.A., Gassull M., Binder V., Finkel Y., Cortot A., Modigliani R., Laurent-Puig P., Gower-Rousseau C., Macry J., Colombel J.F., Sahbatou M., and Thomas G. 2001. Association of NOD2 leucine-rich repeat variants with susceptibility to Crohn's disease. *Nature* **411:** 599.

International HapMap Consortium. 2003. The International HapMap Project. *Nature* **426:** 789.

Jeffreys A.J., Kauppi L., and Neumann R. 2001. Intensely punctate meiotic recombination in the class II region of the major histocompatibility complex. *Nat. Genet.* **29:** 217.

Jobling M.A., Hurles M.E., and Tyler-Smith C. 2004. *Human evolutionary genetics*. Garland Science, New York.

Johnson G.C., Esposito L., Barratt B.J., Smith A.N., Heward J., Di Genova G., Ueda H., Cordell H.J., Eaves I.A., Dudbridge F., Twells R.C., Payne F., Hughes W., Nutland S., Stevens H., Carr P., Tuomilehto-Wolf E., Tuomilehto J., Gough S.C., Clayton D.G., and Todd J.A. 2001. Haplotype tagging for the identification of common disease genes. *Nat. Genet.* **29:** 233.

Ke X., Hunt S., Tapper W., Lawrence R., Stavrides G., Ghori J., Whittaker P., Collins A., Morris A.P., Bentley D., Cardon L.R., and Deloukas P. 2004. The impact of SNP density on fine-scale patterns of linkage disequilibrium. *Hum. Mol. Genet.* **13:** 577.

Kruglyak L. and Nickerson D.A. 2001. Variation is the spice of life. *Nat. Genet.* **27:** 234.

Lander E.S., Linton L.M., Birren B., Nusbaum C., Zody M.C., Baldwin J., Devon K., Dewar K., Doyle M., FitzHugh W., Funke R., Gage D., Harris K., Heaford A., Howland J., Kann L., Lehoczky J., LeVine R., McEwan P., McKernan K., Meldrim J., Mesirov J.P., Miranda C., Morris W., and Naylor J., et al. (International Human Genome Sequencing Consortium). 2001. Initial sequencing and analysis of the human genome. *Nature* **409:** 860.

Lewontin R.C. 1964. The interaction of selection and linkage. I. General considerations: Heterotic models. *Genetics* **49:** 49.

Li W.D., Reed D.R., Lee J.H., Xu W., Kilker R.L., Sodam B.R., and Price R.A. 1999. Sequence variants in the 5′ flanking region of the leptin gene are associated with obesity in women. *Ann. Hum. Genet.* **63:** 227.

Li W.H. and Sadler L.A. 1991. Low nucleotide diversity in man. *Genetics* **129:** 513.

Maniatis N., Collins A., Xu C.F., McCarthy L.C., Hewett D.R., Tapper W., Ennis S., Ke X., and Morton N.E. 2002. The first linkage disequilibrium (LD) maps: Delineation of hot and cold blocks by diplotype analysis. *Proc. Natl. Acad. Sci.* **99:** 2228.

Margulies E.H., Blanchette M., Haussler D., and Green E.D. 2003. Identification and characterization of multi-species conserved sequences. *Genome Res.* **13:** 2507.

McVean G.A.T., Myers S.R., Hunt S.E.H., Deloukas P., Bentley D.R., and Donnelly P. 2004. The fine-scale structure of recombination rate variation in the human genome. *Science* (in press).

Miki Y., Swensen J., Shattuck-Eidens D., Futreal P.A., Harshman K., Tavtigian S., Liu Q., Cochran C., Bennett L.M., and Ding W., et al. 1994. A strong candidate for the breast and ovarian cancer susceptibility gene BRCA1. *Science* **266:** 66.

Ogura Y., Bonen D.K., Inohara N., Nicolae D.L., Chen F.F., Ramos R., Britton H., Moran T., Karaliuskas R., Duerr R.H., Achkar J.P., Brant S.R., Bayless T.M., Kirschner B.S., Hanauer S.B., Nunez G., and Cho J.H. 2001. A frameshift mutation in NOD2 associated with susceptibility to Crohn's disease. *Nature* **411:** 603.

Ohta T. and Kimura M. 1969. Linkage disequilibrium due to random genetic drift. *Genet. Res.* **13:** 47.

Ozaki K., Ohnishi Y., Iida A., Sekine A., Yamada R., Tsunoda T., Sato H., Hori M., Nakamura Y., and Tanaka T. 2002. Functional SNPs in the lymphotoxin-alpha gene that are associated with susceptibility to myocardial infarction. *Nat. Genet.* **32:** 650.

Pääbo S. 2003. The mosaic that is our genome. *Nature* **421:** 409.

Palmer M.S., Dryden A.J., Hughes J.T., and Collinge J. 1991. Homozygous prion protein genotype predisposes to sporadic Creutzfeldt-Jakob disease. *Nature* **352:** 340.

Patil N., Berno A.J., Hinds D.A., Barrett W.A., Doshi J.M., Hacker C.R., Kautzer C.R., Lee D.H., Marjoribanks C., McDonough D.P., Nguyen B.T., Norris M.C., Sheehan J.B., Shen N., Stern D., Stokowski R.P., Thomas D.J., Trulson M.O., Vyas K.R., Frazer K.A., Fodor S.P., and Cox D.R. 2001. Blocks of limited haplotype diversity revealed by high-resolution scanning of human chromosome 21. *Science* **294:** 1719.

Reich D.E., Schaffner S.F., Daly M.J., McVean G., Mullikin J.C., Higgins J.M., Richter D.J., Lander E.S., and Altshuler D. 2002. Human genome sequence variation and the influence of gene history, mutation and recombination. *Nat. Genet.* **32:** 135.

Reich D.E., Cargill M., Bolk S., Ireland J., Sabeti P.C., Richter D.J., Lavery T., Kouyoumjian R., Farhadian S.F., Ward R., and Lander E.S. 2001. Linkage disequilibrium in the human genome. *Nature* **411:** 199.

Rioux J.D., Daly M.J., Silverberg M.S., Lindblad K., Steinhart H., Cohen Z., Delmonte T., Kocher K., Miller K., Guschwan S., Kulbokas E.J., O'Leary S., Winchester E., Dewar K., Green T., Stone V., Chow C., Cohen A., Langelier D., Lapointe G., Gaudet D., Faith J., Branco N., Bull S.B., and McLeod R.S., et al. 2001. Genetic variation in the 5q31 cytokine gene cluster confers susceptibility to Crohn disease. *Nat. Genet.* **29:** 223.

Risch N.J. 2000. Searching for genetic determinants in the new millennium. *Nature* **405:** 847.

Risch N. and Merikangas K. 1996. The future of genetic studies of complex human diseases. *Science* **273:** 1516.

Sachidanandam R., Weissman D., Schmidt S.C., Kakol J.M., Stein L.D., Marth G., Sherry S., Mullikin J.C., Mortimore B.J., Willey D.L., Hunt S.E., Cole C.G., Coggill P.C., Rice C.M., Ning Z., Rogers J., Bentley D.R., Kwok P.Y., Mardis E.R., Yeh R.T., Schultz B., Cook L., Davenport R., Dante M., and Fulton L., et al. 2001. A map of human genome sequence variation containing 1.42 million single nucleotide polymorphisms. *Nature* **409:** 928.

Saunders A.M., Strittmatter W.J., Schmechel D., George-Hyslop P.H., Pericak-Vance M.A., Joo S.H., Rosi B.L., Gusella J.F., Crapper-MacLachlan D.R., and Alberts M.J., et al. 1993. Association of apolipoprotein E allele epsilon 4 with late-onset familial and sporadic Alzheimer's disease. *Neurology* **43:** 1467.

Saur D., Vanderwinden J.M., Seidler B., Schmid R.M., De Laet M.H., and Allescher H.D. 2004. Single-nucleotide promoter polymorphism alters transcription of neuronal nitric oxide synthase exon 1c in infantile hypertrophic pyloric stenosis. *Proc. Natl. Acad. Sci.* **101:** 1662.

Stringer C. 2002. Modern human origins: Progress and prospects. *Philos. Trans. R. Soc. Lond. B Biol. Sci.* **357:** 563.

Ueda H., Howson J.M., Esposito L., Heward J., Snook H., Chamberlain G., Rainbow D.B., Hunter K.M., Smith A.N., Di Genova G., Herr M.H., Dahlman I., Payne F., Smyth D., Lowe C., Twells R.C., Howlett S., Healy B., Nutland S., Rance H.E., Everett V., Smink L.J., Lam A.C., Cordell H.J., and Walker N.M., et al. 2003. Association of the T-cell regulatory gene CTLA4 with susceptibility to autoimmune disease. *Nature* **423:** 506.

Wall J.D. and Pritchard J.K. 2003. Haplotype blocks and linkage disequilibrium in the human genome. *Nat. Rev. Genet.* **4:** 587.

Waterston R.H., Lindblad-Toh K., Birney E., Rogers J., Abril J.F., Agarwal P., Agarwala R., Ainscough R., Alexandersson M., An P., Antonarakis S.E., Attwood J., Baertsch R., Bailey J., Barlow K., Beck S., Berry E., Birren B., Bloom T., Bork P., Botcherby M., Bray N., Brent M.R., Brown D.G., and Brown S.D., et al. (Mouse Genome Sequencing Consortium). 2002. Initial sequencing and comparative analysis of the mouse genome. *Nature* **420:** 520.

Wooster R., Bignell G., Lancaster J., Swift S., Seal S., Mangion J., Collins N., Gregory S., Gumbs C., and Micklem G. 1995. Identification of the breast cancer susceptibility gene BRCA2. *Nature* **378:** 789.

Yu A., Zhao C., Fan Y., Jang W., Mungall A.J., Deloukas P., Olsen A., Doggett N.A., Ghebranious N., Broman K.W., and Weber J.L. 2001. Comparison of human genetic and sequence-based physical maps. *Nature* **409:** 951.

Zondervan K.T. and Cardon L.R. 2004. The complex interplay among factors that influence allelic association. *Nat. Rev. Genet.* **5:** 89.

# SNP Genotyping and Molecular Haplotyping of DNA Pools

P.-Y. Kwok and M. Xiao
*Department of Dermatology and Cardiovascular Research Institute, University of California, San Francisco, California 94114*

With the completion of a comprehensive sequence of the human genome in the 50th anniversary year of the discovery of the double-helical structure of DNA, we have entered the genome era. One of the promises of the genome era is that genomic tools and strategies will be available to further our understanding of the role of genetic factors in human health and disease (Collins et al. 2003). For common human disorders, genetic theory predicts that the search for genetic factors associated with these disorders requires tens to hundreds of thousands of single nucleotide polymorphism (SNP) markers typed on thousands of individuals, depending on whether one takes a gene-based or map-based approach (Botstein and Risch 2003).

As one considers how to utilize these tools in population studies, one quickly comes to the inevitable conclusion that a modest study (with 500 cases and 500 controls and 300,000 SNPs) will be too expensive and time-consuming to pursue with current technology. For example, at the current cost of $0.50 per genotype and the current rate of 20,000 genotypes per day, such a modest study will cost $150 million and take 41 years to do, even when one works non-stop 7 days a week. Before substantial technological advances to reduce the cost and increase the throughput of genotyping are achieved, a possible strategy for comprehensive genetic studies of common human disorders is to reduce the number of assays needed by pooling the DNA samples for analysis (Sham et al. 2002). In fact, the technologies are either already available or close to available for analyzing SNPs and haplotypes in pooled DNA samples.

The rationale for pooling DNA samples is that, in association studies, one is basically looking for differences in allele frequencies for SNPs or haplotype frequencies between those with the disorder and those without the disorder. If the cases and controls are ethnically matched, the vast majority of the SNPs and haplotypes in the genome will have similar allele frequencies and haplotype frequencies between the cases and controls. The handful of alleles and haplotypes associated with the disorder being studied will be the only ones with significant frequency differences between the two groups. If allele frequencies of SNPs and haplotype frequencies in a population can be determined accurately by analyzing the samples in pools rather than as individuals, association studies can be done very efficiently and cost-effectively. Using the same modest study described above as an example, pooling the 500 cases and 500 controls reduces the number of experiments to less than 1 million, as opposed to performing 300 million individual genotyping reactions. In addition to the 300-fold reduction in time needed for genotyping, substantial savings in cost are also realized, since the cost of genotyping pooled samples is not much higher than the cost of genotyping individual samples.

The question then is, How well does DNA pool analysis perform both technically and as a practical strategy for association studies? The technical limitations are real. For example, 1 ng of DNA contains ~300 copies of the haploid, double-stranded human genome. If the genotyping assay uses 5 ng of DNA, pooling DNA samples from more than 1,500 individuals means that some individuals' DNA will not be represented, by definition. In fact, with a random Poisson distribution, the optimal number of samples one can pool together to avoid bias is much smaller, with some authors advocating an optimal pool size of 50 (Barratt et al. 2002). Another technical limitation is that each individual in the pool must contribute equal amounts of DNA to it. Accurate DNA quantification is difficult, and some uncertainty is always there. Fortunately, simulation studies have shown that errors due to this concern are insignificant (Shifman et al. 2002). Without individual genotypes, there is a loss of power when the data are analyzed. However, several algorithms have been developed recently to extract haplotype information from pooled sample data, and these authors showed that the loss of power, although real, was manageable (Barratt et al. 2002; Ito et al. 2003; Wang et al. 2003; Yang et al. 2003).

## POOL GENOTYPING METHODS

Any DNA quantification method can be used to determine allele frequencies in pooled DNA samples. In the case of association studies, however, what one looks for are not absolute allele frequencies, but relative allele frequencies. Therefore, methods which can show that two pools have significantly different allele frequencies for a SNP are also useful.

There are many published allele frequency estimation methods. A partial list of the methods with acceptable accuracy include the 5′-nuclease assay (Breen et al. 2000), kinetic allele-specific PCR (Germer et al. 2000), denaturing high-performance liquid chromatography (Giordano et al. 2001), primer extension assay with MALDI-TOF mass spectrometry detection (Mohlke et al. 2002), invader assay (Lyamichev et al. 1999), and pyrosequencing (Wasson et al. 2002). In all these assays, one can detect 3% allele frequency differences between two pools in a reproducible and accurate way. These methods, however, suffer from relatively high assay development costs due to the use of modified/labeled olignonucleotide probes or the need for expensive equipment. Since the assays are used in a small number of reactions with the pooling strategy, high assay development cost imposes a substantial financial burden on the study.

Recently, we have developed a method for allele frequency estimation that has the lowest assay development cost and is very simple to implement. This method is based on real-time monitoring of fluorescence quenching in a primer extension reaction (Xiao and Kwok 2003). For each assay, three unlabeled, PCR-grade primers are designed for PCR amplification and the primer extension reaction. Dye-labeled terminators in standardized kits are incorporated onto the SNP primers allele-specifically by DNA polymerase as dictated by the allelic base(s) present in the target DNA. The fluorescent labels on the terminators lose their fluorescence intensity when they are incorporated onto the SNP primer due to the natural quenching properties of DNA. Since the rate of decrease in fluorescence intensity is proportional to the amount of DNA target present, real-time monitoring of fluorescence intensity of a fluorescent label yields the fluorescence intensity profile that reflects the amount of target DNA in the reaction. Comparing the rate of decrease in fluorescence intensity of the labels for the allelic nucleotides allows one to estimate the relative abundance of the two alleles in the DNA sample. If a reference DNA sample, such as one from a heterozygote, is used, one can estimate the allele frequencies of the two alleles in a pooled sample.

## POOL HAPLOTYPING METHODS

A major challenge in genetic analysis of common diseases is the determination of haplotypes in case-control studies. Without the benefit of genotypes from family members of the subjects, haplotypes are usually constructed with the use of computer algorithms (Excoffier and Slatkin 1995; Long et al. 1995). Based on simple assumptions of population genetics and taking into account allele frequencies of the SNPs involved, these algorithms perform quite well when the minor allele frequencies of the SNPs in the locus are relatively high. When the minor frequency of a SNP in the locus is low, however, the haplotype prediction programs do not work as well.

The current approaches to haplotyping involve the separation of the DNA from the parental chromosomes and determining the alleles present in these separated DNA fragments. This can be done by cloning or by allele-specific PCR (Sarkar and Sommer 1991; Patil et al. 2001; Eitan and Kashi 2002; McDonald et al. 2002; Tost et al. 2002). However, these approaches are costly and labor-intensive. It will be difficult to implement these approaches in a high-throughput way or to use them in pool haplotyping for DNA encompassing markers forming haplotype block that are typically 15–20 kb in length.

A possible solution to this problem is to analyze individual molecules. If one is able to determine the nature and the distance separating the alleles on single DNA molecules, one can obtain the haplotypes present in the DNA sample, since alleles found on the same piece of DNA are by definition part of the same haplotype. We have been exploring ways to achieve single-molecule detection for the purpose of molecular haplotyping, with the goal of estimating haplotype frequencies in pooled DNA samples.

In one approach, long-range PCR products 10–20 kb in size are labeled allele-specifically with fluorescent "padlock" ligation probes (Nilsson et al. 1994; Jarvius et al. 2003). The backbone of the double-stranded DNA is also labeled. The labeled PCR products are then uncoiled and threaded through a nanochannel with the fluorescence monitored by a confocal microscopic assembly. Coincidence of the allele-specific labels and DNA backbone label allows one to infer the haplotype present on the molecule. Preliminary studies showed that a two-SNP system on a 13-kb PCR product can be detected readily, with a read rate of 20,000 molecules per minute (P.-Y. Kwok and M. Xiao, unpubl.).

The current limitations of single-molecule detection are threefold. First, long-range PCR is not robust at >20 kb. Haplotypes at longer distances must be determined by combining overlapping PCR products. Second, allele-specific labeling using ligation probes is not 100% efficient. This becomes a significant concern as the number of SNPs in the haplotype block increases. However, because the number of molecules one can analyze is large, haplotypes of individuals can still be determined when a small but significant fraction of molecules are fully labeled. When pooled samples are analyzed, lack of efficient labeling will compromise the accuracy of haplotype frequency estimates. Third, distance resolution in the flow counting system is crude. Without distance information, two unique dye labels are needed for each SNP, which in turn leads to the need for a similar number of lasers. As the number of SNPs goes up, one soon runs out of fluorescent labels and/or lasers. With good distance information available, only two fluorescent dyes are needed for each SNP, and the haplotypes can be read out like barcodes: color1-distance1/2-color 2-distance 2/3-color 3-distance 3/4-...etc. Even if some of the molecules enter the nanochannel backwards, the inverse bar code is quickly deduced from the readout.

## CONCLUSIONS

Genetic analysis of pooled DNA samples has been used successfully since the early days of genetic mapping, first with microsatellite markers in family studies (Nystuen et al. 1996; Scott et al. 1996) and now with SNPs in case-control studies (Shaw et al. 1998; Mohlke et al. 2002). As the number of SNP markers one must use to study genetic factors associated with common human diseases increases, DNA pools will be used more widely and allele frequency and haplotype frequency estimation in population studies will be commonplace. One can imagine the two-stage mapping strategy of pool SNP genotyping on the genome scale to identify regions of the genome associated with a disorder followed by pooled haplotyping using single-molecule detection methods to narrow down the associated regions. Once the associated regions are mapped to 10-kb resolution, the regions can be analyzed further by comprehensive DNA sequencing of individuals with the disorder and the controls. As the technological and theoretical methods improve, pool DNA analysis will be a powerful tool for genetic study of common human disorders.

## ACKNOWLEDGMENTS

This work is funded by grants from the National Institutes of Health, the Smith Family Foundation, and the Sandler Family Foundation. The authors thank US Genomics, Inc. and Perkin-Elmer Life Sciences Inc. for instrument and reagent support.

## REFERENCES

Barratt B.J., Payne F., Rance H.E., Nutland S., Todd J.A., and Clayton D.G. 2002. Identification of the sources of error in allele frequency estimations from pooled DNA indicates an optimal experimental design. *Ann. Hum. Genet.* **66:** 393.

Botstein D. and Risch N. 2003. Discovering genotypes underlying human phenotypes: Past successes for Mendelian disease, future approaches for complex disease. *Nat. Genet.* (suppl.) **33:** 228.

Breen G., Harold D., Ralston S., Shaw D., and St Clair D. 2000. Determining SNP allele frequencies in DNA pools. *Biotechniques* **28:** 464.

Collins F.S., Green E.D., Guttmacher A.E., Guyer M.S., and U.S. National Human Genome Research Institute. 2003. A vision for the future of genomics research. *Nature* **422:** 835.

Eitan Y. and Kashi Y. 2002. Direct micro-haplotyping by multiple double PCR amplifications of specific alleles (MD-PASA). *Nucleic Acids Res.* **30:** e62.

Excoffier L. and Slatkin M. 1995. Maximum-likelihood estimation of molecular haplotype frequencies in a diploid population. *Mol. Biol. Evol.* **12:** 921.

Germer S., Holland M.J., and Higuchi R. 2000. High-throughput SNP allele-frequency determination in pooled DNA samples by kinetic PCR. *Genome Res.* **10:** 258.

Giordano M., Mellai M., Hoogendoorn B., and Momigliano-Richiardi P. 2001. Determination of SNP allele frequencies in pooled DNAs by primer extension genotyping and denaturing high-performance liquid chromatography. *J. Biochem. Biophys. Methods* **47:** 101.

Ito T., Chiku S., Inoue E., Tomita M., Morisaki T., Morisaki H., and Kamatani N. 2003. Estimation of haplotype frequencies, linkage-disequilibrium measures, and combination of haplotype copies in each pool by use of pooled DNA data. *Am. J. Hum. Genet.* **72:** 384.

Jarvius J., Nilsson M., and Landegren U. 2003. Oligonucleotide ligation assay. *Methods Mol. Biol.* **212:** 215.

Long J.C., Williams R.C., and Urbanek M. 1995. An E-M algorithm and testing strategy for multiple-locus haplotypes. *Am. J. Hum. Genet.* **56:** 799.

Lyamichev V., Mast A.L., Hall J.G., Prudent J.R., Kaiser M.W., Takova T., Kwiatkowski R.W., Sander T.J., de Arruda M., Arco D.A., Neri B.P., and Brow M.A. 1999. Polymorphism identification and quantitative detection of genomic DNA by invasive cleavage of oligonucleotide probes. *Nat. Biotechnol.* **17:** 292.

McDonald O.G., Krynetski E.Y., and Evans W.E. 2002. Molecular haplotyping of genomic DNA for multiple single-nucleotide polymorphisms located kilobases apart using long-range polymerase chain reaction and intramolecular ligation. *Pharmacogenetics* **12:** 93.

Mohlke K.L., Erdos M.R., Scott L.J., Fingerlin T.E., Jackson A.U., Silander K., Hollstein P., Boehnke M., and Collins F.S. 2002. High-throughput screening for evidence of association by using mass spectrometry genotyping on DNA pools. *Proc. Natl. Acad. Sci.* **99:** 16928.

Nilsson M., Malmgren H., Samiotaki M., Kwiatkowski M., Chowdhary B.P., and Landegren U. 1994. Padlock probes: Circularizing oligonucleotides for localized DNA detection. *Science* **265:** 2085.

Nystuen A., Benke P.J., Merren J., Stone E.M., and Sheffield V.C. 1996. A cerebellar ataxia locus identified by DNA pooling to search for linkage disequilibrium in an isolated population from the Cayman Islands. *Hum. Mol. Genet.* **5:** 525.

Patil N., Berno A.J., Hinds D.A., Barrett W.A., Doshi J.M., Hacker C.R., Kautzer C.R., Lee D.H., Marjoribanks C., McDonough D.P., Nguyen B.T., Norris M.C., Sheehan J.B., Shen N., Stern D., Stokowski R.P., Thomas D.J., Trulson M.O., Vyas K.R., Frazer K.A., Fodor S.P., and Cox D.R. 2001. Blocks of limited haplotype diversity revealed by high-resolution scanning of human chromosome 21. *Science* **294:** 1719.

Sarkar G. and Sommer S.S. 1991. Haplotyping by double PCR amplification of specific alleles. *Biotechniques* **10:** 436.

Scott D.A., Carmi R., Elbedour K., Yosefsberg S., Stone E.M., and Sheffield V.C. 1996. An autosomal recessive nonsyndromic-hearing-loss locus identified by DNA pooling using two inbred Bedouin kindreds. *Am. J. Hum. Genet.* **59:** 385.

Sham P., Bader J.S., Craig I., O'Donovan M., and Owen M. 2002. DNA Pooling: A tool for large-scale association studies. *Nat. Rev. Genet.* **3:** 862.

Shaw S.H., Carrasquillo M.M., Kashuk C., Puffenberger E.G., and Chakravarti A. 1998. Allele frequency distributions in pooled DNA samples: Applications to mapping complex disease genes. *Genome Res.* **8:** 111.

Shifman S., Pisante-Shalom A., Yakir B., and Darvasi A. 2002. Quantitative technologies for allele frequency estimation of SNPs in DNA pools. *Mol. Cell. Probes* **16:** 429.

Tost J., Brandt O., Boussicault F., Derbala D., Caloustian C., Lechner D., and Gut I.G. 2002. Molecular haplotyping at high throughput. *Nucleic Acids Res.* **30:** e96.

Wang S., Kidd K.K., and Zhao H. 2003. On the use of DNA pooling to estimate haplotype frequencies. *Genet. Epidemiol.* **24:** 74.

Wasson J., Skolnick G., Love-Gregory L., and Permutt M.A. 2002. Assessing allele frequencies of single nucleotide polymorphisms in DNA pools by pyrosequencing technology. *Biotechniques* **32:** 1144.

Xiao M. and Kwok P.Y. 2003. DNA analysis by fluorescence quenching detection. *Genome Res.* **13:** 932.

Yang Y., Zhang J., Hoh J., Matsuda F., Xu P., Lathrop M., and Ott J. 2003. Efficiency of single-nucleotide polymorphism haplotype estimation from pooled DNA. *Proc. Natl. Acad. Sci.* **100:** 7225.

# Highly Parallel SNP Genotyping

J.-B. Fan,* A. Oliphant,* R. Shen,* B.G. Kermani,* F. Garcia,* K.L. Gunderson,* M. Hansen,* F. Steemers,* S.L. Butler,*‡ P. Deloukas,† L. Galver,* S. Hunt,† C. McBride,* M. Bibikova,* T. Rubano,* J. Chen,* E. Wickham,* D. Doucet,* W. Chang,* D. Campbell,* B. Zhang,*¶ S. Kruglyak,* D. Bentley,† J. Haas,*§ P. Rigault,* L. Zhou,* J. Stuelpnagel,* and M.S. Chee*

**llumina, Inc., San Diego, California 92121; †The Wellcome Trust Sanger Institute, Hinxton, Cambridge CB10 1SA, United Kingdom*

The genetic factors underlying common disease are largely unknown. Discovery of disease-causing genes will transform our knowledge of the genetic contribution to human disease, lead to new genetic screens, and underpin research into new cures and improved lifestyles. The sequencing of the human genome has catalyzed efforts to search for disease genes by the strategy of associating sequence variants with measurable phenotypes. In particular, the Human Genome Project and follow-on efforts to characterize genetic variation have resulted in the discovery of millions of single-nucleotide polymorphisms (SNPs) (Patil et al. 2001; Sachidanandam et al. 2001; Reich et al. 2003). This represents a significant fraction of common genetic variation in the human genome and creates an unprecedented opportunity to associate genes with phenotypes via large-scale SNP genotyping studies.

To make use of this information, efficient and accurate SNP genotyping technologies are needed. However, most methods were designed to analyze only one or a few SNPs per assay, and are costly to scale up (Kwok 2001; Syvanen 2001). To help enable genome-wide association studies and other large-scale genetic analysis projects, we have developed an integrated SNP genotyping system that combines a highly multiplexed assay with an accurate readout technology based on random arrays of DNA-coated beads (Michael et al. 1998; Oliphant et al. 2002; Gunderson et al. 2004). Our aim was to reduce costs and increase productivity by ~2 orders of magnitude. We chose a multiplexed approach because it is more easily scalable and is intrinsically cost-efficient (Wang et al. 1998). Although existing multiplexed approaches lacked the combination of accuracy, robustness, scalability, and cost-effectiveness needed for truly large-scale endeavors (Wang et al. 1998; Ohnishi et al. 2001; Patil et al. 2001; Dawson et al. 2002; Gabriel et al. 2002), we hypothesized that some of these limitations could be overcome by designing an assay specifically for multiplexing.

To increase throughput and decrease costs by ~2 orders of magnitude, it was necessary to eliminate bottlenecks throughout the genotyping process. It was also desirable to minimize sources of variability and human error in order to ensure data quality and reproducibility. We therefore took a systems-level view to technology design, development, and integration. Although the focus of this paper is on a novel, highly multiplexed genotyping assay, the GoldenGate™ assay, four other key technologies that were developed in parallel, as part of the complete BeadLab system (Oliphant et al. 2002), are briefly described below.

## BEADARRAY™ PLATFORM

We developed an array technology based on random assembly of beads in micro-wells located at the end of an optical fiber bundle (Michael et al. 1998). This technology has advantages over conventional microarrays and is particularly suited to the needs of high-throughput genotyping (Oliphant et al. 2002; Gunderson et al. 2004). Arrays currently in use have up to 50,000 beads, each ~3 microns in diameter. The beads are distributed among 1,520 bead types, each bead type representing a different oligonucleotide probe sequence. This gives, on average, ~30 copies of each bead type, with the result that a genotype call is based on the average of many replicates. The inherent redundancy increases robustness and genotyping accuracy.

We took advantage of the fact that the arrays have a small footprint to design an *array matrix*, comprising 96 arrays arranged in an 8 x 12 matrix that matches the well spacing of a standard microtiter plate (Fig. 1). With this format, samples can be processed in standard microtiter plates, using standard laboratory equipment. The array

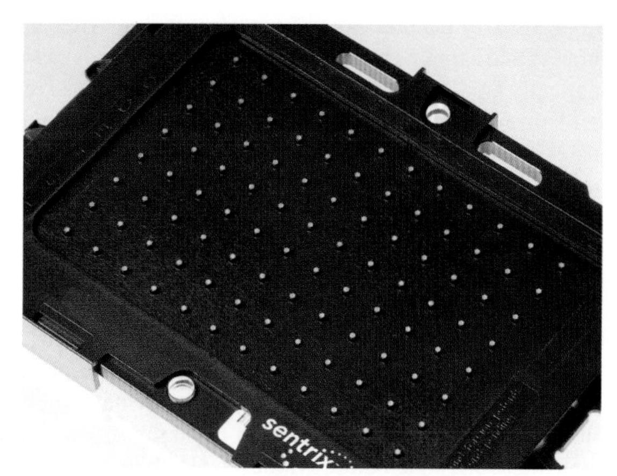


**Figure 1.** The Sentrix™ array matrix.

Present addresses: ‡Pfizer Global R&D, La Jolla Laboratories, 10777 Science Center Drive, San Diego, California 92121; ¶Activx Biosciences, 11025 N. Torrey Pines Road, La Jolla, California 92037; §13438 Russet Leaf Lane, San Diego, California 92129.

matrix is then mated to the microtiter plate, allowing 96 hybridizations to be carried out simultaneously. At the current multiplex level of 1,152, a total of ~110,000 genotypes can be obtained from each matrix of 96 arrays.

## BEADARRAY READER

Scanners for conventional microarrays typically have imaging spot sizes in the range of 3–5 microns, insufficient to resolve the ~5-micron-spaced features on the randomly assembled optical fiber-based arrays. We therefore developed a compact confocal-type imaging system with ~0.8-μm resolution and two-laser illumination (532 and 635 nm; Barker et al. 2003). This scanner is able to image a 96-array matrix in both color channels in about 1.5 hours, which allows a throughput of ~8–10 array matrices, corresponding to ~1 million genotypes, per scanner per day.

## AUTOMATION AND A LABORATORY INFORMATION MANAGEMENT SYSTEM

Automation of a process can be achieved by building a custom instrument that performs multiple functions without human intervention. However, mechanical integration tends to be inflexible: Even minor changes in the process might require a costly redesign of the instrument. An alternative strategy is to keep the system modular, and loosely but accurately coupled through a laboratory information management system (LIMS), designed hand-in-hand with automation (Oliphant et al. 2002). This flexible design philosophy is well-suited to molecular biology assays, which can be both complex and rapidly evolving.

In collaboration with Wildtype Informatics, we developed a LIMS that tracks objects as they are processed through the laboratory. Physical objects that contain samples and reagents, such as microtiter plates and array matrices, are bar-coded. The LIMS supervises each step where information is associated with a new object. For example, when samples are transferred from one plate to another, the robot application requests permission from LIMS to perform the process for the specified plate bar codes. Should LIMS approve the transaction, the robot proceeds with the process and sample transfer. After the process and sample transfers are complete, LIMS is automatically updated with the bar code information. This fail-safe approach, called positive sample tracking, eliminates common sources of human error, such as mislabeling and plate mix-ups.

## OLIGATOR® DNA SYNTHESIZER

SNP assays require one or more oligonucleotides (the GoldenGate assay requires $3n + 3$ oligonucleotides for $n$ SNPs), which are most efficiently generated by de novo chemical synthesis. Anticipating a need to create millions of SNP assays, we developed a high-throughput, low-cost centrifugal oligonucleotide synthesizer (Lebl et al. 2001). This LIMS-integrated automated instrument is able to produce hundreds to thousands of oligonucleotides per day. With this technology, we are able to develop SNP genotyping assays on a genome-wide scale, rapidly and cost-effectively.

## DESIGN OF A HIGHLY MULTIPLEXED SNP GENOTYPING ASSAY

The GoldenGate assay was developed specifically for multiplexing to high levels while retaining the flexibility to choose any SNPs of interest to assay. There are a number of key design elements. In particular, the assay performs allelic discrimination directly on genomic DNA (gDNA), generates a synthetic allele-specific PCR template afterward, then performs PCR on the artificial template. In contrast, conventional SNP genotyping assays typically use PCR to amplify a SNP of interest. Allelic discrimination is then carried out on the PCR product. By reversing the conventional order, we require only three universal primers for PCR, and eliminate primer sequence-related differences in amplification rates between SNPs. We also attach the gDNA to a solid support prior to the start of the assay proper. After attachment, assay oligonucleotides targeted to specific SNPs of interest are annealed to the gDNA (Fig. 2). This attachment step improves assay specificity by allowing unbound and nonspecifically hybridized oligonucleotides to be removed by stringency washes. Correctly hybridized oligonucleotides remain on the solid phase.

Two allele-specific oligonucleotides (ASOs) and one locus-specific oligonucleotide (LSO) are designed for each SNP (Fig. 2). Each ASO consists of a 3′ portion that hybridizes to gDNA at the SNP locus, with the 3′ base complementary to one of two SNP alleles, and a 5′ portion that incorporates a universal PCR primer sequence (P1 or P2, each associated with a different allele). The LSOs consist of three parts: At the 5′ end is a SNP locus-specific sequence; in the middle is an address sequence, complementary to one of 1,520 capture sequences on the array; and at the 3′ end is a universal PCR priming site (P3′). Currently, a typical multiplex pool is designed to assay 1,152 SNPs, and thus contains 2,304 ASOs and 1,152 LSOs. The additional capacity of the array provides some room for expansion of the multiplex pool.

After the annealing and washing steps, an allele-specific primer extension step is carried out. This employs DNA polymerase to extend ASOs if their 3′ base is complementary to their cognate SNP in the gDNA template (Pastinen et al. 2000). Allele-specific extension is followed by ligation of the extended ASOs to their corresponding LSOs, to create PCR templates. Requiring the joining of two fragments to create a PCR template provides an additional level of genomic specificity. Any residual incorrectly hybridized ASOs and LSOs are unlikely to be adjacent, and therefore should not be able to ligate.

Next, the primers P1, P2, and P3 are added. P1 and P2 are fluorescently labeled, each with a different dye. For the SNP illustrated in Figure 2, where A and G represent the two alleles, the expected products are P1-A-P3 in the case of an AA homozygote, P2-G-P3 in the case of a GG

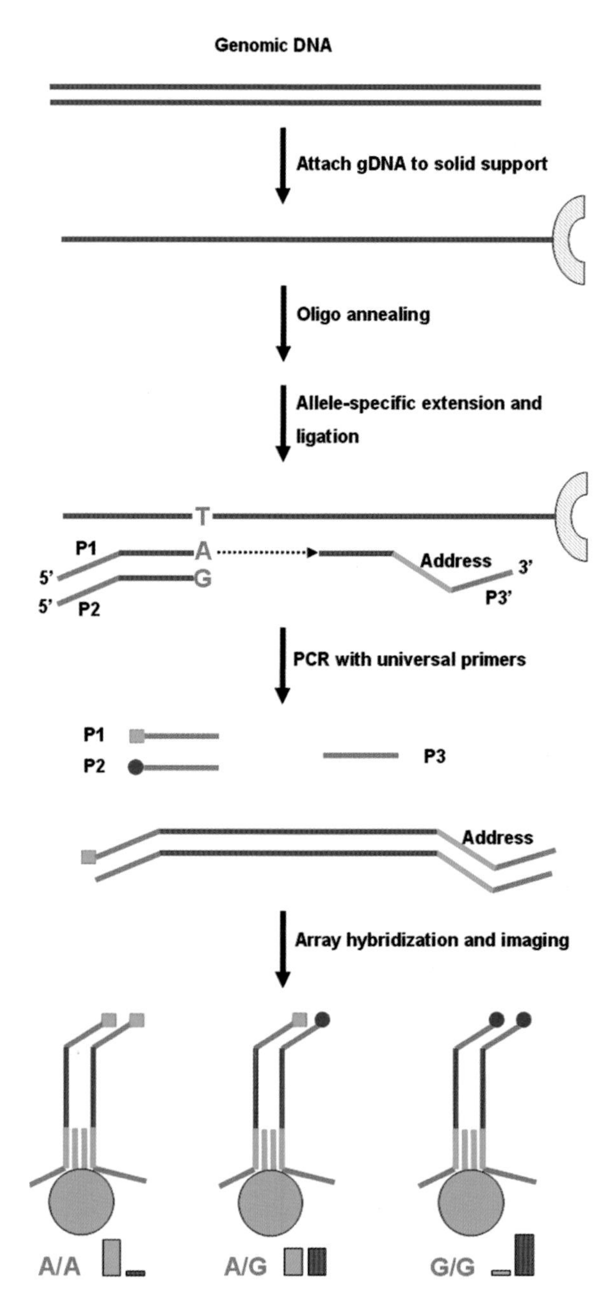


**Figure 2.** The GoldenGate SNP genotyping assay scheme. See Appendix for detailed procedures.

homozygote, or an equimolar mixture of P1-A-P3 and P2-G-P3 in the case of an AG heterozygote. Because P1 is associated with the A allele and P2 with the G allele, the ratio of the two primer-specific fluorescent signals identifies the genotype as AA, AG, or GG.

Each SNP is assigned a different address sequence, which is contained within the LSO. Each of these addresses is complementary to a unique capture sequence represented by one of the bead types in the array. Therefore, the products of the 1,152 assays hybridize to different bead types in the array, allowing all 1,152 genotypes to be read out simultaneously. This universal address system, consisting of artificial sequences that are not SNP specific, allows any set of SNPs to be read out on a common, standard array (Gerry et al. 1999; Cai et al. 2000; Chen et al. 2000; Fan et al. 2000; Iannone et al. 2000). This provides flexibility and reduces array manufacturing costs. Custom sets of assays can be made on demand, simply by building the address sequences into the SNP-specific assay oligonucleotides. The use of universal PCR primers to associate a fluorescent dye with each SNP allele also saves on costs. Because only three primers, two labeled and one unlabeled, are needed regardless of the number of SNPs to be assayed, the primer costs are negligible, as they are amortized over large numbers of assays.

The GoldenGate assay uses ~40 bp surrounding the SNP. Either strand can be chosen for the assay, but we use design rules to designate a preferred strand. The ASOs are designed to have a $T_m$ of 60°C (57–62°C), whereas the LSO has a $T_m$ of 57°C (54–60°C). It is possible to design a similar assay that omits the allele-specific polymerase extension step. However, we have found that allelic discrimination using polymerase, followed by ligation, increases the signal-to-noise ratio (data not shown) (Abravaya et al. 1995). Another advantage is that a variable gap between the ASOs and LSO (typically 1–20 bases) provides flexibility to position the LSO to avoid unfavorable sequences.

## GENOTYPING RESULTS

To date, we have developed well in excess of 100,000 SNP assays. Representative data are shown in Figure 3. Figure 3A shows a fluorescence image from a single array in a 96-array matrix. The image is a false-color composite of the Cy3 (green) and Cy5 (red) images collected in separate channels. A portion of the image is expanded in Figure 3B to show individual beads. Red and green beads are indicative of homozygous genotypes. Yellow indicates a heterozygous genotype, resulting from the presence of both Cy3 and Cy5 on the same bead (Fig. 2). In the next stage of data processing, a trimmed mean intensity is calculated for each bead type, for both Cy3 and Cy5 (on average, there are ~30 beads of each of the 1,520 bead types in an array). Figure 3C shows trimmed mean intensities for 96 DNA samples genotyped on one SNP. The SNP is one of 1,152 assayed in a multiplex pool. As shown, the 96 DNA samples cluster into three groups, showing that all three genotypes are represented in the sample set.

Actual genotype calls are made after transforming the intensity values into modified polar coordinates (Fig. 4). By taking into account the intensity distribution of beads, averaging, and rejecting outliers, measurement precision is improved (Fig. 4A vs. 4B and 4C vs. 4D). It is also shown in Figure 4 that occasional beads are outliers, and would, on their own, give inaccurate genotypes. Even though there are relatively few such beads, they could have a detrimental impact given the requirements for low error rates in large-scale genotyping studies. However, the redundancy in the system ensures a minimum of 5 beads of each type, greatly reducing the chance of an incorrect call.

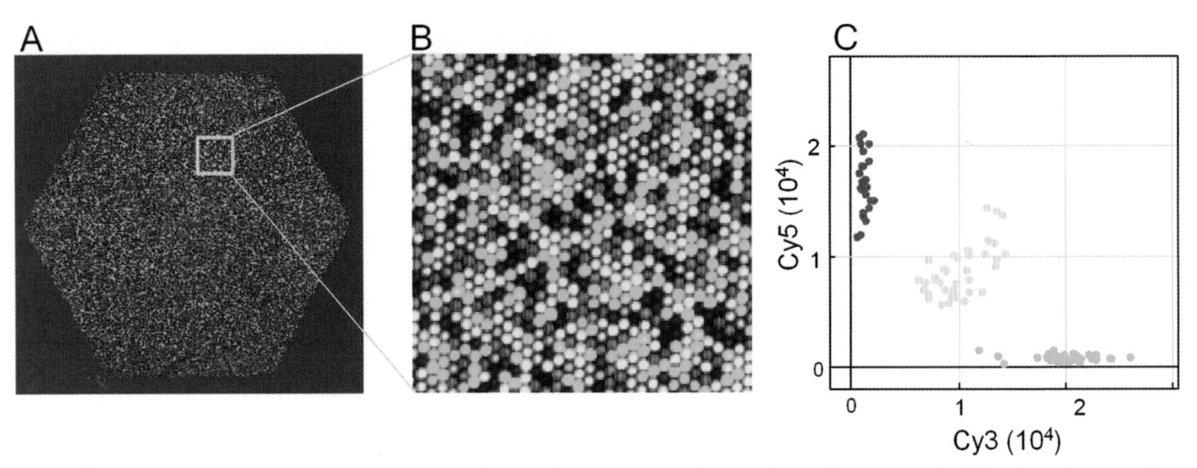


**Figure 3.** Views of genotyping data. (*A*) Fluorescence hybridization image of an ~1.4-mm-diameter optical fiber bundle containing 49,777 fibers in a monolithic, hexagonally packed array. (*B*) A portion of the hybridization image magnified to show individual beads. (*C*) Genotype calls for a single SNP on 96 DNA samples.

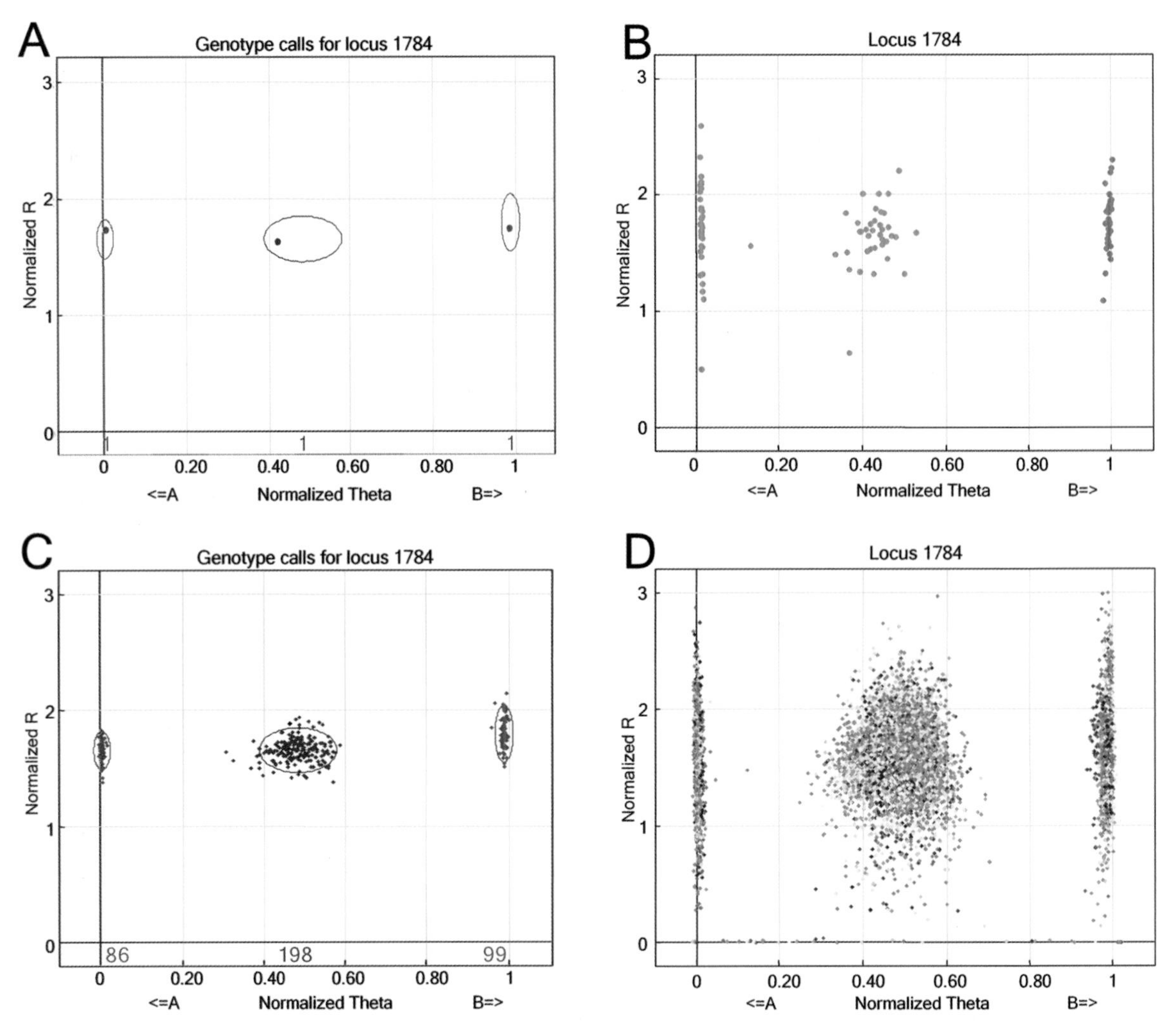


**Figure 4.** Genotyping data for one SNP in a 1,152 multiplex pool. Genotyping plots were created by graphing normalized Intensity (R = normalized X + normalized Y) vs. Theta = $^2/\pi$ $Tan^{-1}(Cy5/Cy3)$. (*A*) Bead type data for three DNAs representing three different genotypes. Each data point represents a trimmed mean intensity derived from a population of beads. (*B*) Data from individual beads corresponding to the three data points shown in *A*. (*C*) Same as *A*, for 383 of 384 DNAs (one DNA sample failed to yield data for this SNP). The number of DNAs in each genotype cluster is shown above the *x* axis. (*D*) Same as *B*, showing individual bead data for all 383 DNAs.

### ASSAY CONVERSION RATES

The fraction of SNPs that can be converted to working assays influences cost and the probability that any particular SNP can be genotyped. Currently, for most genetic studies, cost is of greater importance, and we maximize the assay conversion rate by using bioinformatic screens to rank SNPs according to likelihood of success. Sequences flanking the targeted SNP are evaluated on sequence composition and the presence of duplicated or more highly repetitive sequences, palindromes, and neighboring polymorphisms. The algorithm generates a quantitative score that reflects the likelihood of successfully developing an assay. Therefore, assay conversion rates depend on the quality of the set of input SNPs (e.g., some sets contain a higher fraction of sequencing errors and rare polymorphisms), whether or not we apply a bioinformatic screen, and the rigor of the screen. The quality of the oligonucleotides used in the assay is also an important factor. Given these variables, we have obtained assay conversion rates ranging from <50% to >97%. From a random sampling of dbSNP (http://www.ncbi.nlm.nih.gov/SNP/), attempting assays on both strands, we obtained an assay conversion rate of 81% (87% when excluding sequences that could not be assayed by other methods). This compares favorably to other technologies (Ohnishi et al. 2001; Gabriel et al. 2002; Hardenbol et al. 2003; Phillips et al. 2003).

Recently, many high-quality "double-hit" SNPs have been deposited in dbSNP (Reich et al. 2003; J. Mullikin, pers. comm.). These have each allele supported by two independent sequence reads, and therefore are more likely to be genuine SNPs and to have a relatively high minor allele frequency. To estimate the upper bound of our ability to convert candidate SNPs into assays, we selected 17,280 SNPs (corresponding to 15 x 1,152 multiplexes) that were predicted to have the highest likelihood of success from a collection of ~124,000 double hit SNPs (~14%). By choosing double-hit SNPs for this experiment, we effectively avoided the confounding variable of SNP quality (i.e., sequencing errors and rare SNPs). We developed assays for 10 x 1,152 SNPs on both strands, and assays for a further 5 x 1,152 SNPs on one strand only. This allowed us to compare success rates between the two levels of coverage. We obtained assay conversion rates of 96–99% when assaying both strands and using the best strand, and 94–96% when assaying a single strand (Table 1). Because these assays were carried out on the highest quality SNPs, we expect success rates to be more typically in the range of 90% when developing assays for double-hit SNPs on one strand, our standard approach.

Most of our assays have been developed at 1,152-plex, but we recently increased multiplexing levels to 1,536-plex. We have not yet determined the limits of multiplexing for this assay, but we have achieved excellent accuracy and call rates at multiplex levels of 1,152 to 1,536.

### ACCURACY

There are several ways to estimate accuracy, including reproducibility, strand correlation, concordance with other genotyping methods, and consistency of Mendelian inheritance. Each measure has strengths and weaknesses (Oliphant et al. 2002). Here we report on an analysis of 5,704 SNPs from human chromosome 20. In a study of a 10-Mb region of Chromosome 20 (Contig NT_011362.7; 3,726,000 - 13,824,000 bp), 11,328 SNPs were selected for assay development. All assays were developed on both strands at a multiplex level of 1,152 and used to genotype 384 samples, including 100 unrelated African-Americans, 191 Caucasians (95 individuals from 12 three-generation Utah CEPH families and 96 UK Caucasians), 32 Japanese and 10 Chinese DNAs and controls,

**Table 1.** Assay Development and Genotyping Results for 17,280 SNPs

| Bundle | Multiplex | Successful assays | Successful DNAs | Conversion rate (%) | Called genotypes | Possible genotypes | Call rate (%) |
|---|---|---|---|---|---|---|---|
| DS_1 | 1,152 | 1,136 | 95 | 99 | 107,882 | 107,920 | 99.96 |
| DS_2 | 1,152 | 1,129 | 95 | 98 | 107,232 | 107,255 | 99.98 |
| DS_3 | 1,152 | 1,132 | 95 | 98 | 107,515 | 107,540 | 99.98 |
| DS_4 | 1,152 | 1,134 | 95 | 98 | 107,704 | 107,730 | 99.98 |
| DS_5 | 1,152 | 1,121 | 95 | 97 | 106,464 | 106,495 | 99.97 |
| DS_6 | 1,152 | 1,112 | 95 | 97 | 105,619 | 105,640 | 99.98 |
| DS_7 | 1,152 | 1,112 | 95 | 97 | 105,575 | 105,640 | 99.94 |
| DS_8 | 1,152 | 1,120 | 95 | 97 | 106,376 | 106,400 | 99.98 |
| DS_9 | 1,152 | 1,114 | 95 | 97 | 105,794 | 105,830 | 99.97 |
| DS_10 | 1,152 | 1,107 | 95 | 96 | 105,139 | 105,165 | 99.98 |
| SS_1 | 1,152 | 1,101 | 95 | 96 | 104,540 | 104,595 | 99.95 |
| SS_2 | 1,152 | 1,107 | 95 | 96 | 105,111 | 105,165 | 99.95 |
| SS_3 | 1,152 | 1,097 | 95 | 95 | 104,189 | 104,215 | 99.96 |
| SS_4 | 1,152 | 1,086 | 95 | 94 | 103,132 | 103,170 | 99.96 |
| SS_5 | 1,152 | 1,084 | 95 | 94 | 102,947 | 102,980 | 99.97 |
| **Total** | **17,280** | **16,692** | **95** | **97** | **1,585,219** | **1,585,740** | **99.97** |

Each bundle corresponds to a SNP set assayed as a multiplex pool. All the assays are read out on a single array per sample (96 samples per 96-array matrix). Each sample plate contained 95 DNAs and a negative control. The first ten SNP sets, designated DS, were assayed on both strands. The remaining five SNP sets, designated SS, were assayed on one strand only. The DS and SS sets are sorted in the table by assay conversion rate.

including 32 duplicated DNAs. A total of 5,704 SNPs with a minor allele frequency of ≥4% in the combined set of DNA samples was selected for further analysis. Figure 4 shows genotyping data for one of the SNPs in this study. A linkage disequilibrium analysis of these data will be published elsewhere (P. Deloukas).

We used duplicate genotypes (from assays on both strands and DNA duplicates) and inheritance (CEPH family panel only) to identify discrepant genotypes. After removing 5 DNAs with poor results, the GenCall confidence score (provided with each genotype; see Analysis section below) was used as a threshold. The genotypes retained above the cutoff had a concordance rate of >99.7%. In addition, 566 of the 5,704 SNPs were also genotyped using the Homogeneous Mass Extend assay and MALDI-TOF mass spectrometry (CEPH panel only; P. Deloukas and colleagues, Wellcome Trust Sanger Institute). The concordance between the two different methods was 99.68%, based on a total of 27,901 genotypes.

We have also estimated accuracy from the sum of reproducibility and heritability errors for the data sets shown in Table 1. The accuracy of each of the 15 data sets shown in the table was ~99.9%. These results are consistent with a number of other studies (A. Oliphant, unpubl.) that have estimated the accuracy of genotypes after applying quality cutoffs to be in the range of 99.7–99.9%.

## CALL RATE

There is a tradeoff between accuracy and call rate, which is also an important genotyping performance metric. We define the call rate as the fraction of genotype calls that are made as a fraction of possible calls, excluding unsuccessful assays. In the Chromosome 20 study described above, the overall accuracy of ~99.7% was achieved with an average call rate of 91.7%. We have since achieved even higher call rates while maintaining high accuracy. The more recent data set shown in Table 1 had a total of 1,585,219 genotypes called of a possible 1,585,740, for a call rate of 99.97%.

## THE IMPORTANCE OF DNA QUALITY AND QUANTITY

We have analyzed the effects of key assay variables on data quality and found that gDNA concentration and purity are the most important variables in routine operation. Figure 5 shows the reproducibility of the GoldenGate assay as a function of the amount of input gDNA. A major advantage of a highly multiplexed assay is that relatively little DNA is consumed per genotype, assuming that many SNPs are assayed per sample. We routinely use 200 ng of gDNA for SNP assays multiplexed at 1,152-plex. At this level of multiplexing, DNA consumption is ~0.2 ng per genotype. Furthermore, we have shown that the assay works well with amplified gDNA, allowing large-scale genotyping from only ~ 10 ng of gDNA (D.L. Barker et al., in prep.).

## ASSAY CONTROLS

Internal controls are used to monitor key steps in the procedure. These include gDNA/oligo annealing, PCR, array hybridization, and imaging. For example, assay specificity is checked by assaying nonpolymorphic sites in the genome with an ASO pair, of which one is a perfect match and the other a 3′ end mismatch. To illustrate, a site containing a G base might be assayed with an ASO containing a 3′C and a mismatch ASO containing a 3′T. The ratio of signals from the two ASOs is a measure of specificity. Similarly, imbalances in the amplification from P1 and P2 can be detected by assaying a nonpolymorphic site in the genome with two ASOs that are identical except that one incorporates P1 and the other P2. A

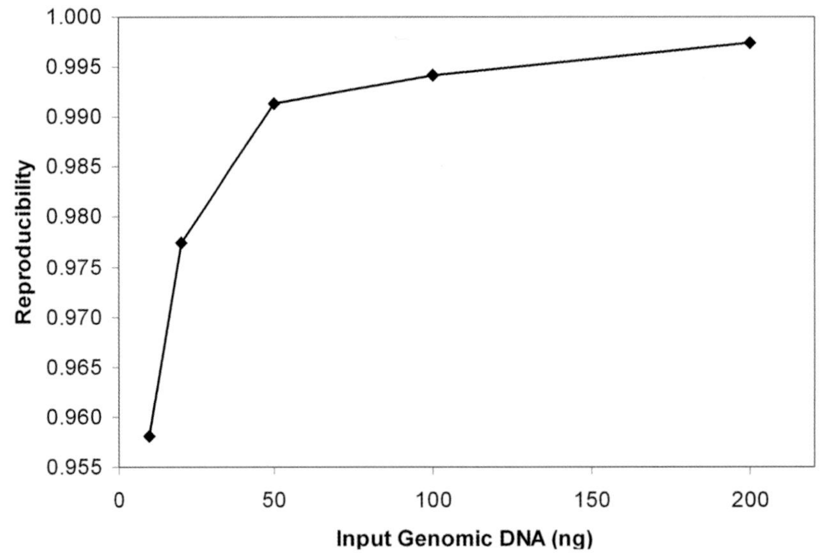


**Figure 5.** Relationship between the amount of input genomic DNA and genotyping reproducibility.

double-labeled control is used in array hybridization to check the optical balance of the Cy3 and Cy5 detection channels.

## ANALYSIS OF LARGE GENOTYPING DATA SETS

To cope with the large amount of data generated, we developed automated methods to extract and analyze data, and derived quantitative measures of data quality. To call genotypes, we developed GenCall, a software program that interprets sample data using a model based on reference data. GenCall is used in conjunction with GenTrain software, which applies a custom clustering algorithm to a reference data set to obtain a set of locus-specific variables for each SNP. This information is provided as input to GenCall.

GenCall also calculates a quality score for each genotype called, which has been shown to correlate with the accuracy of the genotyping call (Fig. 6) (Oliphant et al. 2002). GenCall scores are in the range of 0 to 1, with 1 indicating the highest probability of the score being accurate. The score reflects the degree of separation between homozygote and heterozygote clusters and the placement of the individual call within a cluster, which can be considered key measures of signal-to-noise in the assay data. Besides being important for quality control of complex lab processes, the ability to evaluate objectively the quality of large data sets also enables process improvement to be carried out in a systematic way.

As shown in Figure 6, lower GenCall scores reflect less correlation between strands. Although the relationship is not linear, it is nevertheless useful in helping to establish a threshold for ensuring data quality. Currently, the relationship between GenCall score and accuracy can only be interpreted within a given study. Given the relatively large sizes of studies being undertaken, and the completeness of the data obtained, this is a minor limitation. Nevertheless, we are working to establish a more uniform relationship between GenCall score and accuracy.

## ASSAY PANELS

Assay panels need to be readily available before any SNP genotyping system can be broadly useful for human genetic studies. We have developed a human linkage mapping panel, comprising ~4,600 highly informative SNPs, and a fine mapping panel comprising ~40,000 SNPs. We can also easily make custom panels on demand, with a high assay development success rate. The content for these panels can be chosen from the millions of available SNPs.

We are also involved in the International HapMap Project, which aims to create a haplotype map of the human genome and to make this information freely available in the public domain. This will enable genome-wide genetic association studies, potentially revolutionizing the search for the genetic basis of common diseases (http://hapmap.cshl.org). The SNPs are being genotyped in a set of samples representing African, Asian, and Caucasian populations. The data will be used to define haplotype patterns that are common in each population, and to identify a specific set of SNPs ("tag SNPs") that will be maximally informative for future genome-wide association studies. These genome-wide association studies will investigate the role of common variants in common diseases. Illumina is developing the haplotype map for Chromosomes 8q, 9, 18q, 22, and X, covering 15.5% of the genome, and the Wellcome Trust Sanger Institute is analyzing Chromosomes 1, 6, 10, 13, and 20, covering 24% of the genome. Other leading genome centers are also using the GoldenGate assay and the BeadLab system to develop HapMap assays for other regions of the genome, so far totaling an additional ~20%.

## GENOTYPING RNA

Inherited variation in allelic mRNA abundance provides a means of studying the genetic basis of gene regulation and may enable associations to be made between genetic variation and disease (Yan et al. 2002; Lo et al. 2003). Studies of this type would benefit from the ability to assay efficiently large sets of SNPs occurring in coding regions (cSNPs). In preliminary studies of the GoldenGate assay for allele-specific quantitative mRNA profiling, 32 cSNPs were genotyped on matched pairs of DNA and RNA samples isolated from an ovarian tissue (Fig. 7). The RNA samples were first converted to cDNA, then genotyped using the same procedures used for gDNA. Of the 32 cSNPs, two scored as heterozygous in DNA, but

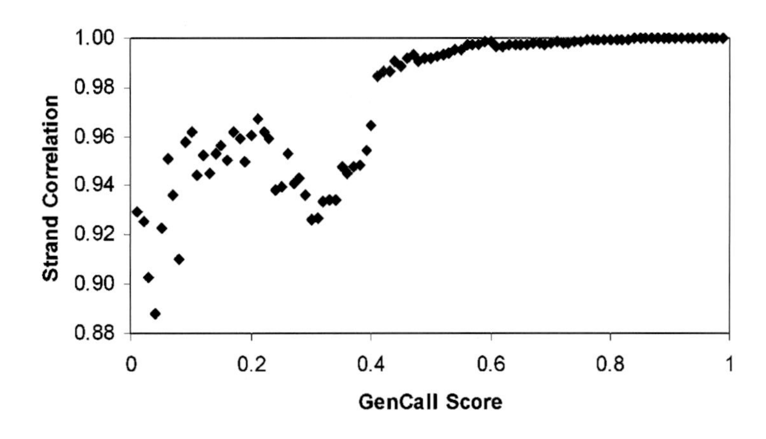


**Figure 6.** GenCall score predicts accuracy. The correlation between genotypes attempted on both strands of a SNP is a proxy for accuracy. A set of 2,916,654 genotype calls was analyzed. Only 99 data points representing the tail of the distribution are shown in this plot.

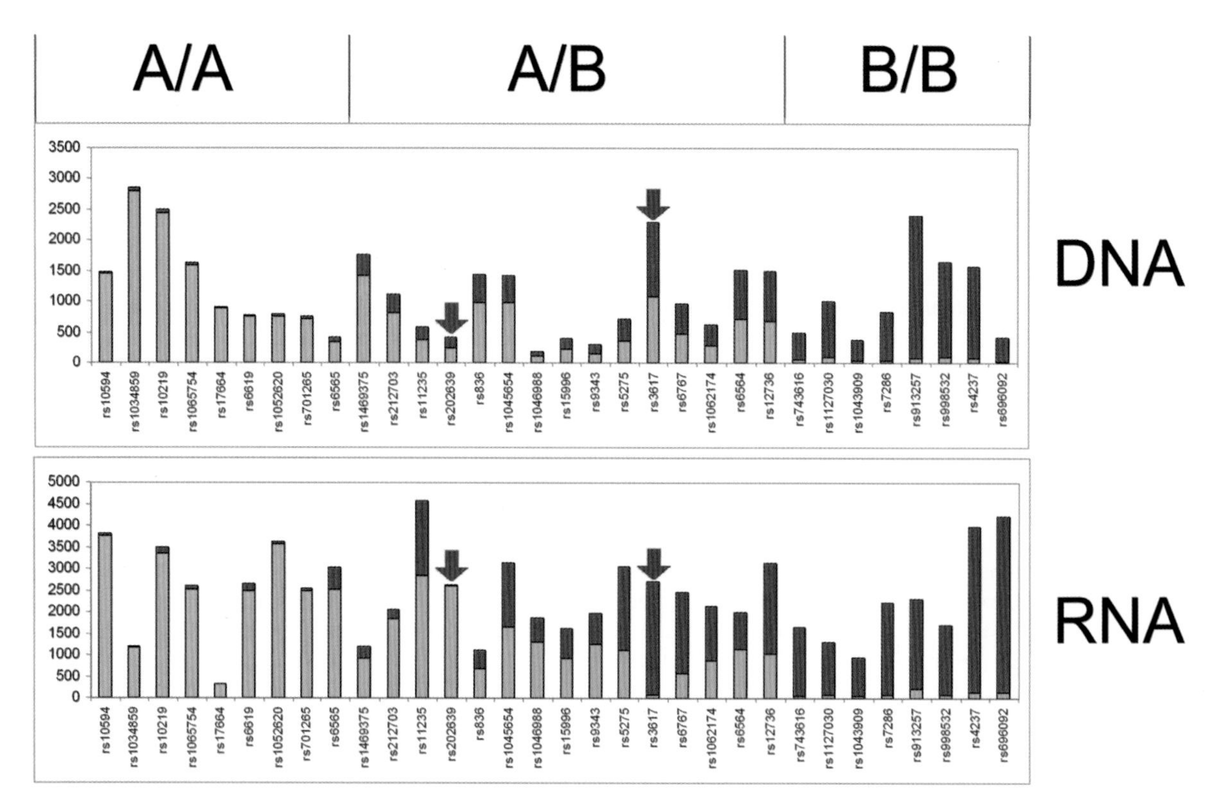


**Figure 7.** Allele-specific expression monitoring. Both genomic DNA and total RNA (converted to cDNA) from an ovarian tissue sample were genotyped for 32 SNPs in gene coding regions. Arrows indicate SNPs, called as heterozygous in gDNA, which show a clear allelic imbalance in mRNA abundance.

showed a strong allele-specific bias in transcript abundance. In addition, assays for 1,152 cSNPs from 450 human cancer-related genes were monitored in pairs of DNA/RNA samples isolated from different human tissues. Of the cSNPs (genes) that were expressed at detectable levels in the RNA and heterozygous in the corresponding genomic DNA, 20% showed differences in allelic abundance at 95% confidence; in many cases, only one allele from a heterozygous genomic locus was detected (J.-B. Fan, unpubl.). These preliminary studies show the feasibility of using the GoldenGate assay to genotype cDNA derived from mRNA populations.

## CONCLUSIONS

We have developed a flexible, accurate, and scalable genotyping system, and have achieved high accuracy together with high call rates. Conventional wisdom was that single-plex assays would be more accurate than highly multiplexed assays, and that it would be difficult to optimize assays in a multiplex format. In fact, our accuracy and call rates are similar to those obtained from single-plex genotyping systems, but at over 1,000 times the assay sequence complexity. These results demonstrate the specificity of the GoldenGate assay format, as well as the reproducibility and accuracy of the BeadArray platform and the BeadLab genotyping system as a whole.

We previously published a ligation-based RNA assay for analysis of mRNA splice variants (Yeakley et al. 2002). In the course of this work, others have also published genotyping assays that are conceptually similar (Schouten et al. 2002; Hardenbol et al. 2003), and others have also achieved high multiplexing levels using direct hybridization-based methods, albeit without the flexibility to assay any SNPs of interest (Kennedy et al. 2003). The assay format used by Schouten et al. (2002), which is read out using gel electrophoresis, is used for the analysis of DNA copy number variation, including the detection of deletions in gDNA.

The GoldenGate assay is versatile and can be adapted to a variety of other applications, such as methylation profiling (J.-B. Fan, unpubl.). In addition, since we use a universal address scheme, different address sequences can be assigned to the same SNP locus to interrogate the same SNP in different samples. This strategy can be quite useful for studies that involve many samples but relatively few SNPs, as in some plant and animal breeding applications. We tested this scheme with an experimental design in which a set of 96 SNPs was associated with 10 discrete sets of 96 address sequences and used to genotype 10 samples in parallel, with readout on one array. We obtained exactly the same genotyping results using this pooling scheme as with our standard approach.

In conclusion, we believe that the GoldenGate assay format is an exemplar for a new class of highly multiplexed assays that utilize parallel readout systems. It represents a significant departure from single and low-multiplex assays and is well suited for large-scale analysis of complex biological systems. We expect that large-scale genotyping studies using this approach will help elucidate the genetic basis of complex diseases, and we are also optimistic that many new applications will spring from the general approach we have developed.

## ACKNOWLEDGMENTS

The authors thank Steven Barnard, Diping Che, Todd Dickinson, Michael Graige, Robert Kain, Michal Lebl, and Chanfeng Zhao for their contributions to the development of the BeadArray and Oligator technologies that provided the foundation for this work. We also thank Steffen Oeser for assistance with graphics. The work described here was supported in part by grants R44 HG-02003, R43 CA-81952, and HG-002753 from the National Institutes of Health to M.S.C.

## APPENDIX

### GoldenGate™ Assay Procedures

Details may vary depending on the specifics of the genotyping system used. All robotic processes were performed on a Tecan Genesis Workstation 150 (Tecan). Up-to-date protocols are supplied with genotyping systems from Illumina, Inc.

***Immobilization of genomic DNA to streptavidin-coated magnetic beads.*** Genomic DNA (20 µl at 100 ng/µl) was mixed with 5 µl of photobiotin (0.2 µg/µl, Vector Laboratories) and 15 µl of mineral oil, and incubated at 95°C for 30 minutes. Trizma base (25 µl of 0.1 M) was added, followed by two extractions with 75 µl of Sec-butanol to remove unreacted photobiotin. The extracted gDNA (20 µl) was mixed with 34 µl of Paramagnetic Particle A Reagent (MPA; Illumina) and incubated at room temperature for 90 minutes. The immobilized gDNA was washed twice with DNA wash buffer (WD1) (Illumina) and resuspended at 10 ng/µl in WD1. In each subsequent reaction, 200 ng (10 µl) of DNA was used.

***Annealing of assay oligonucleotides to genomic DNA.*** Annealing reagent (MA1; Illumina; 30 µl) and SNP-specific oligonucleotides (10 µl containing 25 nM of each oligonucleotide) were combined with immobilized DNA (10 µl) to a final volume of 50 µl. LSOs were synthesized with a 5′ phosphate to enable ligation. Annealing was carried out by ramping temperature from 70°C to 30°C over ~8 hours, then holding at 30°C until the next processing step.

***Assay oligo extension and ligation.*** After annealing, excess and mishybridized oligonucleotides were washed away, and 37 µl of master mix for extension (MME; Illumina) was added to the beads. Extension was carried out at room temperature for 15 minutes. After washing, 37 µl of master mix for ligation (MML; Illumina) was added to the extension products, and incubated for 20 minutes at 57°C to allow the extended upstream oligo to ligate to the downstream oligo.

***PCR amplification.*** After extension and ligation, the beads were washed with universal buffer 1 (UB1; Illumina), resuspended in 35 µl of elution buffer (IP1; Illumina) and heated at 95°C for one minute to release the ligated products. The supernatant was then used in a 60-µl PCR. PCR reactions were thermocycled as follows: 10 seconds at 25°C; 34 cycles of (35 seconds at 95°C, 35 seconds at 56°C, 2 minutes at 72°C); 10 minutes at 72°C; and cooled to 4°C for 5 minutes. The three universal PCR primers (P1, P2, and P3) are labeled with Cy3, Cy5, and biotin, respectively.

***PCR product preparation.*** Double-stranded PCR products were immobilized onto paramagnetic particles by adding 20 µl of Paramagnetic Particle B Reagent (MPB; Illumina) to each 60-µl PCR, and incubated at room temperature for a minimum of 60 minutes. The bound PCR products were washed with universal buffer 2 (UB2; Illumina), and denatured by adding 30 µl of 0.1 N NaOH. After 1 minute at room temperature, 25 µl of the released ssDNAs was neutralized with 25 µl of hybridization reagent (MH1; Illumina) and hybridized to arrays.

***Array hybridization and imaging.*** Arrays were hydrated in UB2 for 3 minutes at room temperature, and then preconditioned in 0.1 N NaOH for 30 seconds. Arrays were returned to the UB2 reagent for at least 1 minute to neutralize the NaOH. The pretreated arrays were exposed to the labeled ssDNA samples described above. Hybridization was conducted under a temperature gradient program from 60°C to 45°C over ~12 hours. The hybridization was held at 45°C until the array was processed. After hybridization, the arrays were first rinsed twice in UB2 and once with IS1 (IS1; Illumina) at room temperature with mild agitation, and then imaged at a resolution of 0.8 microns using a BeadArray Reader (Illumina; Barker et al. 2003). PMT settings were optimized for dynamic range, channel balance, and signal-to-noise ratio. Cy3 and Cy5 dyes were excited by lasers emitting at 532 nm and 635 nm, respectively.

***Genotyping with RNA samples.*** A 20-µl reverse transcription reaction containing a reaction mix (MMC; Illumina) and total RNA (up to 1 µg), was incubated at room temperature for 10 minutes and then at 42°C for 1 hour. After cDNA synthesis, the remainder of the assay was identical to the GoldenGate assay described above.

## REFERENCES

Abravaya K., Carrino J.J., Muldoon S., and Lee H.H. 1995. Detection of point mutations with a modified ligase chain reaction (Gap-LCR). *Nucleic Acids Res.* **23:** 675.

Barker D.L., Therault G., Che D., Dickinson T., Shen R., and Kain R. 2003. Self-assembled random arrays: High-performance imaging and genomics applications on a high-density microarray platform. *Proc. SPIE* **4966:** 1.

Cai H., White P.S., Torney D., Deshpande A., Wang Z., Keller R.A., Marrone B., and Nolan J.P. 2000. Flow cytometry-based minisequencing: A new platform for high-throughput single-nucleotide polymorphism scoring. *Genomics* **66:** 135.

Chen J., Iannone M.A., Li M.S., Taylor J.D., Rivers P., Nelsen A.J., Slentz-Kesler K.A., Roses A., and Weiner M.P. 2000. A microsphere-based assay for multiplexed single nucleotide polymorphism analysis using single base chain extension. *Genome Res.* **10:** 549.

Dawson E., Abecasis G.R., Bumpstead S., Chen Y., Hunt S., Beare D.M., Pabial J., Dibling T., Tinsley E., Kirby S., Carter D., Papaspyridonos M., Livingstone S., Ganske R., Lohmussaar E., Zernant J., Tonisson N., Remm M., Magi R., Puurand T., Vilo J., Kurg A., Rice K., Deloukas P., and Mott R., et al. 2002. A first-generation linkage disequilibrium map of human chromosome 22. *Nature* **418:** 544.

Fan J.B., Chen X., Halushka M.K., Berno A., Huang X., Ryder T., Lipshutz R.J., Lockhart D.J., and Chakravarti A. 2000. Parallel genotyping of human SNPs using generic high-density oligonucleotide tag arrays. *Genome Res.* **10:** 853.

Gabriel S.B., Schaffner S.F., Nguyen H., Moore J.M., Roy J., Blumenstiel B., Higgins J., DeFelice M., Lochner A., Faggart M., Liu-Cordero S.N., Rotimi C., Adeyemo A., Cooper R., Ward R., Lander E.S., Daly M.J., and Altshuler D. 2002. The structure of haplotype blocks in the human genome. *Science* **296:** 2225.

Gerry N.P., Witowski N.E., Day J., Hammer R.P., Barany G., and Barany F. 1999. Universal DNA microarray method for multiplex detection of low abundance point mutations. *J. Mol. Biol.* **292:** 251.

Gunderson K.L., Kruglyak S., Graige M.S., Garcia F., Kermani B., Zhao C., Che D., Dickinson T., Wickham E., Bierle J., Doucet D., Milewski M., Yang R., Siegmund C., Haas J., Zhou L., Oliphant A., Fan J.-B., Barnard S., and Chee M.S. 2004. Decoding randomly ordered DNA arrays. *Genome Res.* (in press).

Hardenbol P., Baner J., Jain M., Nilsson M., Namsaraev E.A., Karlin-Neumann G.A., Fakhrai-Rad H., Ronaghi M., Willis T.D., Landegren U., and Davis R. W. 2003. Multiplexed genotyping with sequence-tagged molecular inversion probes. *Nat. Biotechnol.* **21:** 673.

Iannone M.A., Taylor J.D., Chen J., Li M.S., Rivers P., Slentz-Kesler K.A., and Weiner M.P. 2000. Multiplexed single nucleotide polymorphism genotyping by oligonucleotide ligation and flow cytometry. *Cytometry* **39:** 131.

Kennedy G.C., Matsuzaki H., Dong S., Liu W.M., Huang J., Liu G., Su X., Cao M., Chen W., Zhang J., Liu W., Yang G., Di X., Ryder T., He Z., Surti U., Phillips M.S., Boyce-Jacino M.T., Fodor S.P., and Jones K.W. 2003. Large-scale genotyping of complex DNA. *Nat. Biotechnol.* **21:** 1233.

Kwok P.Y. 2001. Methods for genotyping single nucleotide polymorphisms. *Annu. Rev. Genomics Hum. Genet.* **2:** 235.

Lebl M., Burger C., Ellman B., Fambro S., Hachmann J., Heiner D., Ibrahim G., Jones A., Kim S., Nibbe M., Pires J., Santos C., Touhy S., Mudra P., Pokorny V., Poncar P., and Zenisek K. 2001. Fully automated parallel oligonucleotide synthesizer. *Collect. Czech. Chem. Commun.* **66:** 1299.

Lo H.S., Wang Z., Hu Y., Yang H.H., Gere S., Buetow K.H., and Lee M. P. 2003. Allelic variation in gene expression is common in the human genome. *Genome Res.* **13:** 1855.

Michael K.L., Taylor L.C., Schultz S.L., and Walt D.R. 1998. Randomly ordered addressable high-density optical sensor arrays. *Anal. Chem.* **70:** 1242.

Ohnishi Y., Tanaka T., Ozaki K., Yamada R., Suzuki H., and Nakamura Y. 2001. A high-throughput SNP typing system for genome-wide association studies. *J. Hum. Genet.* **46:** 471.

Oliphant A., Barker D.L., Stuelpnagel J.R., and Chee M.S. 2002. BeadArray technology: Enabling an accurate, cost-effective approach to high-throughput genotyping. *Biotechniques* **32:** S56.

Pastinen T., Raitio M., Lindroos K., Tainola P., Peltonen L., and Syvanen A.C. 2000. A system for specific, high-throughput genotyping by allele-specific primer extension on microarrays. *Genome Res.* **10:** 1031.

Patil N., Berno A.J., Hinds D.A., Barrett W.A., Doshi J.M., Hacker C.R., Kautzer C.R., Lee D.H., Marjoribanks C., McDonough D.P., Nguyen B.T., Norris M.C., Sheehan J.B., Shen N., Stern D., Stokowski R.P., Thomas D.J., Trulson M.O., Vyas K.R., Frazer K.A., Fodor S.P., and Cox D.R. 2001. Blocks of limited haplotype diversity revealed by high-resolution scanning of human chromosome 21. *Science* **294:** 1719.

Phillips M.S., Lawrence R., Sachidanandam R., Morris A.P., Balding D.J., Donaldson M.A., Studebaker J.F., Ankener W.M., Alfisi S.V., Kuo F.S., Camisa A.L., Pazorov V., Scott K.E., Carey B.J., Faith J., Katari G., Bhatti H.A., Cyr J.M., Derohannessian V., Elosua C., Forman A.M., Grecco N.M., Hock C.R., Kuebler J.M., and Lathrop J.A., et al. 2003. Chromosome-wide distribution of haplotype blocks and the role of recombination hot spots. *Nat. Genet.* **33:** 382.

Reich D.E., Gabriel S.B. and Altshuler D. 2003. Quality and completeness of SNP databases. *Nat. Genet.* **33:** 457.

Sachidanandam R., Weissman D., Schmidt S.C., Kakol J.M., Stein L.D., Marth G., Sherry S., Mullikin J.C., Mortimore B.J., Willey D.L., Hunt S.E., Cole C.G., Coggill P.C., Rice C.M., Ning Z., Rogers J., Bentley D.R., Kwok P.Y., Mardis E.R., Yeh R.T., Schultz B., Cook L., Davenport R., Dante M., and Fulton L., et al. 2001. A map of human genome sequence variation containing 1.42 million single nucleotide polymorphisms. *Nature* **409:** 928.

Schouten J.P., McElgunn C.J., Waaijer R., Zwijnenburg D., Diepvens F., and Pals G. 2002. Relative quantification of 40 nucleic acid sequences by multiplex ligation-dependent probe amplification. *Nucleic Acids Res.* **30:** e57.

Syvanen A.C. 2001. Accessing genetic variation: Genotyping single nucleotide polymorphisms. *Nat. Rev. Genet.* **2:** 930.

Wang D.G., Fan J.-B., Siao C.-J., Berno A., Young P., Sapolsky R., Ghandour G., Perkins N., Winchester E., Spencer J., Kruglyak L., Stein L., Hsie L., Topaloglou T., Hubbell E., Robinson E., Mittmann M., Morris M.S., Shen N., Kilburn D., Rioux J., Nusbaum C., Rozen S., Hudson T.J., and Lipshutz R., et al. 1998. Large-scale identification, mapping, and genotyping of single-nucleotide polymorphisms in the human genome. *Science* **280:** 1077.

Yan H., Yuan W., Velculescu V.E., Vogelstein B., and Kinzler K. W. 2002. Allelic variation in human gene expression. *Science* **297:** 1143.

Yeakley J.M., Fan J.B., Doucet D., Luo L., Wickham E., Ye Z., Chee M.S., and Fu X.D. 2002. Profiling alternative splicing on fiber-optic arrays. *Nat. Biotechnol.* **20:** 353.

# Structure of Linkage Disequilibrium in Humans: Genome Factors and Population Stratification

J. Bertranpetit, F. Calafell, D. Comas, A. González-Neira, and A. Navarro
*Unitat de Biologia Evolutiva, Universitat Pompeu Fabra, 08003 Barcelona, Catalonia, Spain*

Human genomes differ at, on average, one per thousand bases; this statement, which has been repeated ad nauseam, takes on a whole new dimension when the physical organization of the human genome is taken into account. Thus, each allele at each polymorphic site does not exist independently, but is physically linked to the allele occurring at the next polymorphic site, and so on, all along each chromosome. A haplotype is any combination of allelic states at linked sites. Such physical links are quite strong and are only broken by recombination via crossing-over or similar DNA exchange processes such as gene conversion. Recombination, the shuffling of chromosome segments of maternal and paternal origin during gamete formation, happens at such a broad scale that relatively wide DNA portions tend to travel together from one generation to the next. In fact, the average chance of two adjacent bases being separated by recombination in one generation is in the order of one in a hundred million ($10^{-8}$), or, if considering polymorphic sites at a density of one per kilobase, their probability of not staying together in one generation would be roughly $10^{-5}$.

Frequently, haplotype frequencies do not derive from the random assortment of alleles at each locus; preferential associations do exist. The departure of haplotype frequencies from the expectation under random association of their integrating alleles is called linkage disequilibrium (LD). LD arises by several mechanisms, including random genetic drift, mutation, migration, changes in population size, and selection due to genome and population factors (Bertranpetit and Calafell 2001). All this newly created LD will be eroded by recombination, which will be more efficient in reshuffling the alleles at the most distant loci. Thus, LD is expected to decay with physical distance. In summary, sites that are physically close will tend to carry particular combinations of alleles.

From a purely numerical point of view, LD can be characterized in two ways: as a direct measure or as the result of a formal hypothesis test. For the sake of simplicity, we will refer to LD between a pair of polymorphisms, although the LD concept can be extended, not without some difficulty, to the relationships among a larger number of markers. LD ranges between two possible extremes: On one hand, knowledge of the allele in one locus may not convey any information about the content of the other locus; both are independent and LD is, in fact, nil. On the other hand, LD can be complete in the sense that knowing which allele is at one locus determines which allele is at the other. Measures of LD (derived from allele and haplotype frequencies) take extreme values (conveniently, 0 and 1) for these two extreme cases and the appropriate intermediate value for intermediate situations. The two most popular measures, $D'$ and $r^2$, are based on the departure from the expected haplotype frequencies under linkage equilibrium and on the correlation coefficient between haplotype frequencies, respectively. However, a given $D'$ or $r^2$ value does not reveal by itself the statistical significance of LD; for that, proper testing is needed, usually by means of a chi-square statistic or by Fisher's exact test.

LD has a clear role to play in biomedical research: The physical position of a gene contributing to a phenotype (usually a disease) can be deduced from the polymorphic markers with which it is found at high LD. Most genes involved in the causation of Mendelian disorders were not found via LD, but with linkage mapping, in which the location of a disease gene is investigated by observing, in families with affected and nonaffected individuals, the patterns of joint inheritance of the phenotype and each of the polymorphic markers in a genetic map. Accordingly, the events that are useful to reject a genome segment as containing the disease locus are recombinations happening within the families collected for the study. Alternatively, if LD is assessed in the whole population, the whole history of recombination between the disease locus and the map markers can be used to locate the gene. This approach was used first by Hastbacka et al. (1992) to refine the position of the gene for diastrophic dysplasia within the interval provided by linkage analysis.

LD is implicitly embedded in association mapping, one of the most popular approaches in the search for the genetic components of complex diseases. In association studies, the allelic frequencies in one or more polymorphisms are compared between affected individuals and healthy controls. If frequencies are significantly different, there are two possible explanations: Either the polymorphism itself contributes to the phenotype (which can often be ruled out given that most polymorphisms used are synonymous or otherwise neutral variants), or the polymorphism is in linkage disequilibrium with genetic variants actually involved in the disease phenotype.

Association studies often involve polymorphisms in or around candidate genes that, given their known biological function, may participate in the phenotype. This approach is as good as our a priori knowledge of the molecular biology of the disease, which, in many cases, may be scant, if not altogether nonexistent. A wholesale approach can be envisioned to overcome our lack of knowledge: Type a high density of polymorphisms all through the human genome and try to capture blindly the LD signal. The markers of choice are single nucleotide polymorphisms (SNPs), which are the most amenable to high-throughput typing. A very large number of markers are needed for this method to work. Their quantity is a function of a number of parameters such as the actual contribution of each unknown genetic variant to the disease, the frequency of the alleles involved, and, crucially, the physical extent of LD. The extent of LD is known to vary enormously between genome regions and, for the same region, between populations, although the amount and distribution of LD variation among human groups is still under debate. As discussed below, such variation depends on a number of genomic and demographic factors. All of these factors have to be considered before embarking on the daunting task of mass SNP typing.

## LD IN A GENOME CONTEXT

### The Great Variation of LD across the Genome

LD patterns are greatly variable across the genome (Reich et al. 2001; Wall and Pritchard 2003b). Indeed, it is a well-known fact that different genomic regions present enormous differences in their LD levels and distribution not only due to their different evolutionary histories or to the local differences in relevant genomic variables such as recombination, but also due to the great stochasticity associated with LD measurements (Pritchard and Przeworski 2001; Reich et al. 2002). In certain regions, strong association between pairs of markers can extend beyond one megabase, whereas in others, there is no trace of LD at very short distances. Within the same region, very close pairs of sites can be in strong LD whereas intermediate markers are independent. Moreover, LD patterns among close markers, intensively studied in medical genetics, only follow the expected decay of LD when a large number of pair-wise LD measures for different distances are observed, and a mean trend usually emerges.

As a result, patterns of LD have long been considered noisy and unpredictable (Wall and Pritchard 2003b). Furthermore, although the existing knowledge about the genomic determinants of LD patterns is incomplete, it reveals a complex scenario: LD can be affected by heterogeneity in recombination rates, several kinds of natural selection, and, of course, genetic drift (Hudson 2001; Nordborg and Tavare 2002). Naturally, these factors may influence the diversification of LD patterns among human groups.

### Selection and LD

Natural selection represents the interaction between a particular gene or genes and the environment, obviously mediated by the gene function and the phenotypic variation it might create. Natural selection, in its wide variety of forms, can have a great impact on LD patterns and can generate associations over long distances. Moreover, the effects of selection may be felt beyond the particular loci being selected and may influence LD patterns of linked neutral variants. It is unclear how directional selection (either positive or purifying) may affect LD patterns between high-frequency variants in linked regions. Positive selection may create an area of high LD through a selective sweep. Any genetic variant that is advantageous and selected will increase its frequency in the population and become eventually fixed. However, any such variant is embedded in a particular haplotype background; if the selective pressure is strong enough compared to recombination, a large flanking region will be driven to fixation as well. This means that, in that zone, polymorphism will basically be eliminated and will start from scratch by mutation. Any new mutation will appear on a blank slate and, when increasing its frequency, will create LD in the region as the mutation will lie in a specific background. This suggests an effect of positive selection on LD patterns between low-frequency alleles, which are usually overlooked in association and LD surveys. Purifying selection weeds out the much more frequent deleterious variants. With each gene that is selected against, a whole chromosome is taken down from the species pool. This reduces the effective population size, which leads to an expected increase in LD, as discussed below. Otherwise, the effect of this process in LD patterns should be small.

The more studied instance of selection is balancing selection, which leads to strong associations in neighboring regions (Hudson 1990). A paradigmatic example of the latter is balancing selection acting on the MHC. Yet, old polymorphisms under balancing selection are considered to be rare (Przeworski et al. 2000), and it is possible that this form of selection is just difficult to detect. Most of the knowledge about the relationship between selection and LD comes from single-locus models, and it is uncertain how their predictions can be applied at a whole-genome scale in which selection has many interacting targets. Thus, although for given genome regions the effect of selection may be recognized and even related to a given gene and its variants, on a genome scale the effect of selection may be impossible to elucidate among the variation induced by heterogeneity in recombination rates and stochastic processes. Theoretical models predict that certain kinds of epistatic interactions between sites and, in particular, between sites under balancing selection, can generate extremely strong linkage disequilibrium over long regions (Navarro and Barton 2002). Interestingly, however, epistatic interactions of opposite sign can reduce LD even between closely linked sites (Navarro and Barton 2002). It is also unclear how epistatic selection will interact with mutation, varying recombination rates, and/or population structure. Current models of human LD largely ignore multilocus selection but, given that complex diseases are the outcome of interactions among many sites, this may soon change. It would be interesting to identify complex human adaptations in response to different environments that would undoubtedly have an impact on LD.

### LD and Recombination Rates

Because LD is eroded by recombination, the general view is that variation in recombination rates, both at large and at fine scales, is the single most important determinant of LD patterns. Current estimates of recombination rates show, first, that their values in the genome change between regions by more than an order of magnitude, from 0.1 cM/Mb to more than 3 cM/Mb; and, second, that far from varying smoothly along a chromosome, recombination rates present sharp local peaks and valleys (Kong et al. 2002). Even at the scale of a few kilobases, variation in recombination rates is dramatic, and the existence of recombination hot spots is quickly gaining acceptance. However, only a few actual examples have been reported (Jeffreys and Newman 2002; Kauppi et al. 2003), and little is known about the underlying biological mechanisms causing hot spots or about their evolutionary dynamics. The sizes of hot spots seem to range from 1 to 2 kb (Jeffreys and Newman 2002) but can extend to several Mb (Arnheim et al. 2003). The existence of allele-specific hot spots (Jeffreys and Neumann 2002) suggests that recombination-associated motifs are the main cause of hot spots, but other mechanisms cannot be ruled out (Arnheim et al. 2003; Zhang et al. 2003).

Recombination rates can also be influenced by two other important factors that are largely ignored by most models: homologous gene conversion and chromosomal rearrangements. Homologous gene conversion is only a relevant recombination source at very short scales (~1 kb) and, thus, pedigree-based recombination estimates are thought to be underestimating recombination rates at small scales. Therefore, gene conversion has been proposed as an explanation for the short-range intragenic defect of linkage disequilibrium in humans (Przeworski and Wall 2001). Finally, polymorphic chromosomal rearrangements, such as inversions or translocations, can induce enormous changes in recombination rates. First, when rearrangements segregate in a population, they modify the recombinational context of whole regions; second, inversions are known to suppress recombination in heterokaryotypes, thus reducing recombination rates; third, and most importantly, the presence of rearrangements introduces a substructuration in the population, and strong association between alleles linked to different arrangements can easily be built by mutation and/or drift (Navarro et al. 1996). Polymorphisms for short inversions, whose origin would be nonhomologous recombination between duplicated segments, seem to be frequent in a genome as repetitive as the human and may have a huge impact on LD patterns.

## LD AS A FUNCTION OF POPULATION FACTORS

Population history has been shown to have important effects on LD patterns. Small population size, through genetic drift, results in an overall increase of LD levels (Pritchard and Przeworski 2001; Nordborg and Tavare 2002). In a smaller population, the number of possible targets for recombination is lower and fewer recombinant chromosomes arise; thus, LD tends to be preserved. High levels of LD are also generated in a population bottleneck: A small sample of chromosomes survives the bottleneck, and the haplotypes they carry may be found in proportions that deviate randomly but greatly from their previous equilibrium frequencies, thus generating LD. A similar effect is achieved when a small group founds a new population, an important process in the demographic history of humans that has left an enduring footprint both at the global structure of LD and at a microgeographic scale. LD levels are also increased by population substructure, because deme differentiation leads to strong associations even between unlinked sites. On the other hand, population growth decreases LD (Kruglyak 1999; Pritchard and Przeworski 2001).

### The Genetic History of Humans and Its Possible Consequences on LD

The current landscape of genetic diversity in humans is the product of population history. Its first and most prominent feature is its homogeneity: As explained above, the human genome is remarkably poor in polymorphism. It suffices, however, to allow the search of population substructure at several scales, from continental groups to divergence due to local historical processes of settlement and expansion. The main conclusion of this search is that the extant divisions in populations do not result in sharp genetic partitions. Although highly significant, population differentiation explains only 15% of the total genetic variation (Romualdi et al. 2002), whereas the remaining 85% is explained by differences among individuals. If the main continental divisions are considered, they explain 8%, and 7% of the total genetic diversity is accounted for by differences between populations of the same continent.

The human genetic landscape is not uniform. As a general trend, African populations carry the largest variation in terms of allele and nucleotide diversity, and, when gene genealogies are reconstructed, lineages found in Africa tend to have split first. Genetic data support higher population size in Africans and lack of recent bottlenecks that have been detected in non-African populations. This has been interpreted as the result of the "out of Africa" or replacement model of human origins. On the contrary, genetic evidence is far more difficult to accommodate into the alternative multiregional model. We humans are a young species. The time we have had to accumulate genetic diversity may be one-fourth to one-tenth of that which our closest relatives, chimpanzees and gorillas, have had (Bertranpetit and Calafell 2003). Current estimates place the origin of the species at ~150,000 years ago, with a probable bottleneck ~50,000 years ago when our direct ancestors first left Africa to populate the rest of the world. This explains the overall genetic homogeneity of the species and the pattern of geographic subdivision, with genetic diversity in non-Africans being a subset of that in Africa.

This genetic landscape (and the population history that shaped it) may have had profound implications for the geographical patterns of LD in humans. The low nucleotide diversity in humans means that the density of markers in a map may be restricted, and not all SNPs will be as abundant as required to pick up an LD signal with, for example, an infrequent variant with a small contribution to a complex disease.

As explained above, population size and its fluctuations over time have a clear effect on LD, with low effective population sizes and bottlenecks contributing to it. If the history of a population is sufficiently well known, this can be verified. For instance, Laan and Pääbo (1997) and Kaessmann et al. (2002) found that the Saami and Evenki, reindeer herders of presumably constant size, show higher levels of LD when compared to neighboring, expanding populations. This has two implications: (1) As explained in the next section, it is possible to attempt the reverse inference; that is, to infer (although in a rather vague and qualitative manner) population history from LD and (2) population history matters in biomedical research aimed at deciphering the genetic architecture of complex disease making successful gene hunting more likely in some populations. In particular, populations of constant size, such as the Saami and Evenki, may harbor extensive LD that may allow the detection of the relevant genetic associations in complex diseases. Moreover, these populations tend to be isolated, with lower genetic diversity, which may also be reflected in a lesser allelic and genic heterogeneity in complex diseases, thus facilitating the discovery of such variants.

Despite the hopes created by the isolated population approach, however, it is not clear what contribution it may have to the understanding of genetic disease. At worst, population history may be so complex and so poorly known that LD in a population cannot be predictable a priori and should be determined empirically for each population and genome region. The sources of paleodemographic data are scarce and are provided mainly by paleoanthropology, archaeology, and historical sources, which necessarily implies large uncertainties. The example of the Icelandic population is quite revealing; from the first settlement by Vikings in the Middle Ages, the demographic history of the island is known in exceptionally excruciating detail, given the Icelandic national passion for genealogy. However, the unknown diversity in the settlers has precluded the use of LD as a tool for association mapping in an efficient way, and the most valuable resource for biomedical research in Iceland remains the genealogical database.

The genome-wide structure of LD in humans needs to be tackled and understood if LD is to be used efficiently as a gene discovery tool. However, the population sampling density required to obtain a description of the required level of detail is not clearly known. A cynical view would be that only the populations that can afford the costly pharmaceutical diagnostics and treatments need to be sampled; in an ideal world of universal health care, the level of detail that should be both relevant and nonredundant may be just guessed if LD is to behave as genetic diversity. Then, there would be almost as much variation within as among continents, and all continents should be sampled, with much more variation within Africa than elsewhere. The old-fashioned vision of human diversity as made of monolithic Europeans, East Asians, and Africans still prevails in genetics without a solid basis, and there is resistance to the realization that although it accounts for a large fraction of the world population, it does not relate to the actual structure of LD and genome variation.

## WORLDWIDE VARIATION OF LD IN SPECIFIC GENE REGIONS

The usage of LD as a tool to unravel population history and to reconstruct demographic aspects of human evolution has been pioneered by Kenn Kidd at Yale University, and his collaborators. The first global analysis of the LD variation in human populations was undertaken in a region of around 9.8 kb within the CD4 locus (Tishkoff et al. 1996). These first results based on 42 worldwide populations showed extremely reduced LD in sub-Saharan African populations and significantly higher levels of haplotype diversity than in non-African populations. This fact has been explained by the larger effective population size maintained in sub-Saharan populations and the effect of a recent bottleneck in the expansion of modern humans out of Africa. The rationale of this conclusion is based on the fact that LD decays through time due to recombination, and therefore large and old populations tend to present reduced levels of LD, whereas not enough time has elapsed to reduce LD in small and recently formed populations that preserve the strong LD produced in a bottleneck. Thus, the global LD approach was consistent with the hypothesis of the "out of Africa" origin of modern humans, already supported by a large body of genetic evidence. However, these results were derived from a single locus and may be the spurious result of other processes such as differential selective pressures among human populations or simple stochastic variation. Population history, nonetheless, should affect all the genome with the same average strength, and this signal should be detected in other genome regions.

Subsequent worldwide analyses in several other gene regions did show a similar pattern of lower LD among sub-Saharan African populations and a LD increase outside Africa. However, there were a number of deviations from this overall theme. The extreme LD pattern shown by the CD4 locus (that is, low LD in Africa and much higher LD elsewhere) is also present in DRD2 (Kidd et al. 1998), DM (Tischkoff et al. 1998), PLAT (Tishkoff et al. 2000), and COMT (DeMille et al. 2002). Such LD geographical structure is moderate in PAH (Kidd et al. 2000) and PKLR-GBA region (Mateu et al. 2002); whereas LD in Africans is similar to that of non-Africans in CFTR (Mateu et al. 2001), and even higher in SLC6A4 (Gelernter et al. 1999). These different LD geographic structures may be explained by different recombination rates at each locus: Loci with higher recombination rates would have recovered faster from the bottleneck, reaching again the low African LD levels. Low-grained specific recombination estimates are available (Kong et al. 2002), but they do not seem to correlate with LD geographical patterns for the specific genome regions, as shown in Table 1. The increase in LD caused by a bottleneck in the migration

**Table 1.** Linkage Disequilibrium Analyses Performed in Worldwide Samples

| | N pop | N ind | Crom. band | Kb | Sex-average rate DeCode cM/Mb | Reference |
|---|---|---|---|---|---|---|
| CD4 | 42 | 1600 | 12p13.31 | 10 | 3.5 | Tishkoff et al. (1996) |
| DRD2 | 28 | 1342 | 11q23.2 | 25 | 0.8 | Kidd et al. (1998) |
| DM | 25 | 1231 | 19q13.32 | 20 | 1.8 | Tishkoff et al. (1998) |
| SLC6A4 | 7 | 242 | 17q11.2 | 15 | 0.3 | Gelernter et al. (1999) |
| PLAT | 30 | 1287–1420 | 8p11.21 | 22 | 0.6 | Tishkoff et al. (2000) |
| PAH | 29 | 1475 | 12q23.2 | 75 | 1.5 | Kidd et al. (2000) |
| CFTR | 18 | 972 | 7q31.2 | 163 | 0.4 | Mateu et al. (2001) |
| G6PD | 15 | 605 | Xq28 | 23 | 2.0 | Tishkoff et al. (2001) |
| DRD4 | 5 | 300 | 11p15.5 | 5 | 1.2 | Ding et al. (2002) |
| CAPN10 | 11 | 561 | 2q37.3 | 12 | 1.6 | Fullerton et al. (2002) |
| COMT | 32 | 1,701 | 22q11.21 | 28 | 2.6 | DeMille et al. (2002) |
| PKLR-GBA | 17 | 896 | 1q22 | 70 | 1.3 | Mateu et al. (2002) |
| ADH | 41 | 2,134 | 4q23 | 112 | 1.5 | Osier et al. (2002) |

out of Africa is not a fixed quantity but has a very large stochastic variation. Therefore, the different geographical patterns observed in different loci can be interpreted as the distribution of a stochastic variable, in which CD4 and CFTR would be opposite extremes. It is possible, when sufficient data from a larger number of regions have accumulated, that the size and duration of the bottleneck might be derived from the average and variance of the increase of LD in non-Africans compared to Africans. Certainly, a more detailed knowledge of locus-specific recombination rates should be factored in this calculation.

In addition to these different results in the analyzed loci due to genetic drift, several unusual LD patterns in global analyses have been interpreted as being the result of the effects of positive selection in specific loci. The high frequency of some haplotypes in groups of populations and the low LD decay over chromosomal distance cannot be explained by genetic drift alone. This is true in the case of G6PD (Tishkoff et al. 2001), DRD4 (Ding et al. 2002), CAPN10 (Fullerton et al. 2002), and ADH (Osier et al. 2002). Interestingly, the LD analysis and its decay along the chromosomal region have been shown to be a powerful approach to detect positive selection in human populations (Sabeti et al. 2002). Therefore, global LD surveys will allow detection of population differences in selective pressures due to different environmental factors, such as exposure to certain pathogens, without prior knowledge of a specific variant or selective advantage. No doubt this approach will lead to the understanding of differential adaptation among human populations.

Besides the different patterns observed in African compared to non-African populations, other subtle population aspects might be detected. The unusual LD pattern observed for the American populations (e.g., PLAT and PAH loci) might draw our attention to specific populations in order to score all the human LD diversity present beyond major continental groups. The founder effect experienced by the ancestors of Native American populations during the colonization of the continent might have left its effect in the LD pattern of extant populations. Strong founder effects are also detected in Pacific populations, such as the Micronesian Nasioi. Therefore, beyond the main demographic effects detected in the LD pattern of the human species, such as the expansion out of Africa of modern humans, other minor demographic processes that might have left their footprint in extant LD population patterns could be traced by LD analysis.

## BLOCK STRUCTURE OF HUMAN LD

Recently, several studies have proposed that, despite their obvious complexity and variability, LD patterns in the human genome can be described by means of a quite simple framework. According to these models, the genome is divided in a series of "haplotype blocks" (Daly et al. 2001; Patil et al. 2001; Dawson et al. 2002; Gabriel et al. 2002; Stumpf 2002), sets of consecutive sites in very strong LD between each other which only constitute a few haplotypes. Contiguous blocks are separated by regions showing strong evidence of historical recombination. These models are currently the subject of great interest, because, if proven true, it would become straightforward to choose SNPs in whole-genome association studies. Once the haplotypes forming a block region were defined, only a few haplotype-tagging SNPs would be necessary to screen that region for association (Zhang et al. 2002; Meng et al. 2003).

Different authors have detected haplotype blocks using different data sets, different methods, and, crucially, very different definitions of a block (Wall and Pritchard 2003b). In general, block definitions are operational and may change within the same study to fit a priori knowledge about the involved populations. For example, when studying African populations, Daly et al. (2001) decrease their LD threshold for a pair of markers to be considered part of the block, because of the known lower LD levels of these populations. Thus, assessing the real "blockiness" of our genome is a difficult issue. In a couple of recent works, Wall and Pritchard (2003a,b) tackle this problem. They reanalyze several data sets, both real and simulated, using a variation of the block definition of Gabriel et al. (2002), which defines blocks as long stretches of extremely high pair-wise LD (measured as $|D'|$) separated by regions of extensive recombination. The amount of blockiness of real and simulated data is assessed by means of three ad hoc statistics: the amount of sequence covered by blocks, the internal consistency of blocks, and the degree of block-boundary overlapping. Their results suggest high heterogeneity in recombination rates, the best fit of simulations to actual data being obtained with a strong hot spot model (in which 75% of recombination takes place in hot spots). And, of course, it is not surprising to see that block boundaries roughly coincide with low-recombination regions (Zhang et al. 2003). Still, they find that the actual data substantially depart from a simple haplotype block model. As seems to be a constant in anything that relates to LD, some regions of the genome do fit the haplotype block model, whereas others do not render themselves to such a convenient description.

Present data show that there are important population differences in block structure of the human genome. Differences do exist in the haplotype composition and in the amount of LD, affecting the length of the block. Differences seem to be due to population factors, with no genome factor having yet been described to have a geographic stratification that could have a different effect in different human groups.

### Maintenance of Blocks in the Human Genome: The Example of GBA Region

The postulated structure of the human genome in haplotype blocks raises the question of whether it can be explained by recombination alone. As most of these blocks contain many polymorphisms, many in intermediate frequencies, the block structure must have had to be maintained just because of historical lack of recombination. In fact, it is usually assumed that once produced, the dynamics of LD are mainly driven by its decay in time, which is a function of the recombination rate between the markers considered. Usually, when describing a given pattern of LD (or its absence), it is assumed that the observed LD is the remainder of a previously higher LD (either still present since the emergence of some of the polymorphisms studied, or having been produced by a population history event) that has not had enough time to decay. No doubt detailed knowledge of recombination rates may allow the elucidation of LD dynamics in a given genome region, but it is difficult to estimate the degree of detail needed to make this knowledge useful. The recombination-based genetic map of the human genome based on STRs has achieved high detail, but it may contain a large amount of heterogeneity between the markers used (Kong et al. 2002).

If general populations are considered, like Europeans or Asians, the last bottleneck that may have influenced LD is rather old, related to the out of Africa event(s) and the settlement of modern humans, more than 40,000 years or 2,000 generations ago; if Africans are considered, their LD pattern should be related to older events, not still pinpointed beyond the existence of a small population size for most of the ancient human history. Nonetheless, LD may have increased more recently due to certain historical events, but only for specific populations with a particular demographic history.

Haplotype blocks do exist in general, nonisolated populations. Whether in their maintenance, factors other than recombination rate play a main role is accepted, but its importance is not clear. Although for some specific regions recombination alone may explain the level of LD through hot spots (in the MHC class II region; Kauppi et al. 2003), it is unlikely that it will explain the general pattern in the genome. This is why the possible action of selection could be a factor in a more complex (and complete) way to explain the persistence of LD blocks.

An interesting case is the GBA region, where the existence of a long LD block was observed for markers situated 90 kb apart that comprises several genes known to be involved in genetic diseases (Mateu et al. 2001). The understanding of the dynamics of the region was possible thanks to the information provided by a resequencing study of a pseudogene (psGBA; Martínez-Arias et al. 2001); in fact, the detailed knowledge (at the nucleotide sequence level) of a random sample of chromosomes is the best approach to unravel the footprint of factors such as selection, extremely difficult to recognize only with SNP variation (but see Sabeti et al. 2002).

The allelic partition of the variants, with a high number of singletons and two haplotypes found at high frequency, did not fit the expected pattern under a neutral model. In addition, results of several statistical tests (Fu's Fs and Fay and Wu's test) clearly pointed to the existence of positive selection. The interesting point is that psGBA is well characterized as a pseudogene, without any biological function and therefore free from any direct selective pressure. At least one genetic hitchhiking event needs to be presumed to understand the variation pattern, which highlights the role of the genome context to understand the variation in the human genome.

In this case, it has been possible to show that the haplotype structure and the high level of LD are related to the existence of a selective force that has acted in a region of low recombination and has produced a pattern of variation clearly identified as a haplotype block but with a dynamics fundamentally shaped by positive selection in an unknown location within a large block with low recombination. Even if it would be of interest to point to a specific gene or a specific nucleotide as the main cause for positive selection (which is the goal of most efforts to find human-specific differences in relation to other species), it must be admitted that with the haplotype structure in blocks it is much more difficult to recognize the specific changes that may have brought about selection.

A last observation holds true for the future. We know that there are geographic differences in the haplotype composition when there is LD among markers, a fact independent of the LD structure. It could be fruitful to use, at the same time, the geographic variation in haplotype structure and the existence of LD to uncover geographic differences in the action of positive selection.

## POPULATION DIFFERENCES IN LD IN GENOME ANALYSIS

### Wide Analyses on a Small Number of Populations

Recently available technology for SNP typing is allowing the study of the genetic variation on a genomic scale and its use for large-scale mapping of genes contributing to common diseases. To design this analysis correctly, knowledge of the distribution of LD along the genome is crucial to estimate how many markers are needed to obtain adequate power in genomewide studies. In some areas, a few haplotypes may encompass a region of the order of megabases, whereas in others, SNPs a few hundred bases apart may be in equilibrium, and thus in-depth screening may require typing of each of them.

The goal of the International *HapMap* Project is to create a powerful shortcut to identify genes linked to complex disorders, identifying haplotype blocks in which a few tag SNPs may be defined in order to use them as surrogates for the whole haplotype block. The project will produce haplotype maps of the whole genome in four populations: Americans of European ancestry, Japanese, Chinese, and Yorubans. However, the successful application of *HapMap* in association studies hinges on two debated key assumptions. First, the basic assumption for LD-based mapping of genes contributing to complex diseases is the so-called common disease/common variant model. That is, genetic factors contributing to common disease are assumed to be relatively few for each disease and to comprise frequent alleles at each site. Actual evidence for this model is weak (Pritchard 2001; Reich and Lander 2001) and has raised some skepticism. Even worse, if this model is true, it may be more difficult to find those genes: Frequent alleles tend to be older and, therefore, less likely to remain in LD with their genomic background.

The second assumption is that the analysis of the four selected populations will provide valid results for all the human species, independently of the ethnic background. To be precise, it is assumed that the most prevalent haplotypes will be described with the four populations analyzed and that the block structure of these populations, including LD decay with distance, will give the framework for any other human group. The studies of geographic distribution of LD, both in single genes or regions and in whole-genome approaches, do not support this homogeneity in LD structure.

Gabriel et al. (2002) characterized the haplotype patterns across 51 autosomal regions in samples from Africa (African-American and sub-Saharan African samples), European-American individuals, and East Asians, to determine the structure of LD and its variation across populations. They provide strong evidence that most of the human genome is organized into haplotype blocks. The blocks found in African populations are smaller than those in European and Asian populations, estimating that half of the human genome exists in blocks of around 22 kb in Africans and African-Americans and in blocks of around 40 kb in Europeans and Asians. In addition, the boundaries of these blocks and the common haplotypes found within are extremely correlated across populations. Another large-scale study reported by Reich et al. (2001) examined 19 randomly selected genomic regions in a United States population of north-European descent and a Nigerian population in order to characterize population differences in LD patterns around genes. Again, vast differences in the extent of LD between populations were found, and block size was smaller in the African than in the non-African sample.

These LD studies have been used to justify the decision of analyzing four "representative" ethnic groups in the *HapMap* Project. Most of the existing studies described with large-scale data sets and different average distance between markers have been performed in a reduced number of populations, usually one European, one Asian, and one African. As a result of this, several questions are still unanswered: Are haplotype blocks and LD structure from different populations within continental groups similar, and what is the amount and grain of the substructure? Do other ethnic groups harbor a divergent structure from the simple European/Asian/African framework proposed by *HapMap*, for example, in Native Americans or Oceanians? Is the expected heterogeneity within Africa found in LD and block structure? These questions are, entirely, on the universality of the design of *HapMap*.

There is a further concern related to the population differences: the quality of the SNPs database used to select markers. There is a clear ascertainment bias of SNPs reported in databases, and additional SNP discovery is required in highly diverse populations, such as African populations, which will help to ensure that the LD map will be powerful enough in all ethnic populations. Current data from multiple loci show that major haplotypes in one population could be absent in others. These results illustrate that tag SNPs determined in one population may not necessarily be good tag SNPs in another if the populations are sufficiently differentiated. Therefore, the importance of taking into account the actual genetic population differentiation of humans has not been considered in the making of the LD studies design.

### A Pilot Study of LD in a Large Number of Populations

To gain insight on the amount of population structure in patterns of LD and haplotype composition, we have studied a region of chromosome 22, spanning 1.78 Mb, in which strong LD and clear haplotype blocks were described in an English population (Dawson et al. 2002). Twelve SNPs that were described as flanking haplotype blocks in a high LD region have been genotyped (A. González-Neira et al., unpubl.). The study has been carried out in 1110 individuals (2220 chromosomes) from 39 populations chosen to represent most of the human genetic variation (mainly from the HGDP-CEPH panel; Cann et al. 2002). When compared to the English population, the decay of LD with genetic distance is very heterogeneous, with some European populations showing a similar pattern, while other populations have a remarkably different pattern, either with high LD and sharp decrease or low LD and smooth decay (Fig. 1).

There is no single way of elucidating the structure of LD and haplotype composition for a large number of populations. We present here just a graphic overview of the pattern of variation (Fig. 2). Considering the well-described English population (Dawson et al. 2002), all the other populations are depicted according to their divergence from it in two different but complementary param-

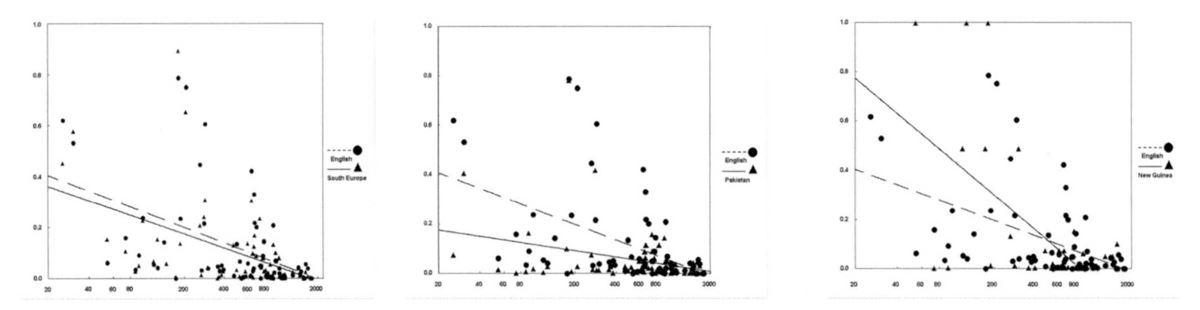


**Figure 1.** Decay of LD with physical distance. The pattern observed in the British population is compared, in each plot, to a population from South Europe, Pakistan, and New Guinea. LD has been measured by $r^2$ and physical distance in kilobases.

eters. (1) The *x* axis represents a measure of genetic differentiation in haplotype composition. Thus, each population is compared to the English by means of $F_{ST}$, the most appropriate genetic distance when differences have been produced by drift, the main force in extant human genome differentiation (Pérez-Lezaun et al. 1997). Higher values mean a strong difference in haplotype composition, either with the same haplotypes at very different frequency, or with different haplotype composition. (2) On the *y* axis there is a condensed measure of similarity between the pattern of LD of each population and that of the English. For each population, LD was calculated between all pairs of neighboring markers through $r^2$. Pairs of populations were compared by computing correlation coefficients between the LDs of adjacent SNPs.

The plot simultaneously informs about haplotype composition (of interest if tagSNPs have to be defined for a small set of main haplotypes) and LD structure (that is directly related to the block structure and LD decay with distance). Results (Fig. 2) are straightforward: There is a strong similarity among all European populations (so strong that they cannot be properly plotted due to their concentration near the English reference). Central Asian populations behave in a very similar way to the Europeans, and both groups may be embodied in a single pattern. East Asian populations are found in a large cluster, with a similar structure among them and well differentiated from the Europeans.

An interesting result is the structure of Africans, who despite not having a strong divergence in haplotype composition, present high heterogeneity in LD structure, confirming the idea from other genetic studies of its high heterogeneity. The most surprising result comes from the Native American populations, with strong differences in both parameters in relation to Europeans and a high level of heterogeneity among them; it is not possible to define a Native American pattern, because it varies from one population to the other.

## CONCLUSIONS

LD in human populations is shaped by a large number of factors that interact with each other in a hugely complex way. The population factors that shaped LD have been used to model expected LD patterns, but this is a challenging task, since we rarely have (and may never have) detailed information about the relevant population parameter values; additionally, there is a large amount of stochastic variation. A given pattern of LD may be used to infer population history events that could have caused them, but still the inference of the past from LD remains vaguely qualitative.

Genomic factors that could be expected to be well known are poorly understood in what refers to both their theoretical expectations and the application of these expectations to the real world. In particular, there is little knowledge about how interactions between loci shaping complex characters may affect LD patterns. Selection may have played a main role in shaping LD patterns across the genome and in generating differences between populations, but its footprint is extremely difficult to detect due to the lack of power of most statistical tests, and in most cases its action must be assumed, rather than demonstrated, to explain a given genetic pattern.

The distribution of recombination rates, which for a long time was assumed to be roughly constant along the genome, is not only the single most important parameter affecting LD, but also one that adds a great deal of complexity. It is now accepted that recombination rates vary enormously, and recombination hot spots are a factor of still unknown importance. We can guess about the variability of the recombination rate, but at a much too coarse scale. In general, there is plenty of room for improvement in the description of LD across the genome. Haplotype blocks may be a convenient and proper description of LD patterns in part of the genome, but other genomic regions do not present such a structure and, thus, it is possible that the easy SNP tagging methods suggested by haplotype blocks will not be of general application.

The main way to advance in the understanding of LD is through large amounts of data, an effort that is warranted by the promise that the knowledge of LD will facilitate the discovery of susceptibility alleles for complex diseases and will be useful for future pharmacogenomics. The *HapMap* project will produce a huge amount of SNP data for four selected populations. But, as we have seen, the geographic structure of LD is at least as complex as that of allele variation, if not more so. It is a vast oversimplification to assume that a few general populations can act as a broad umbrella for the characterization of global LD.

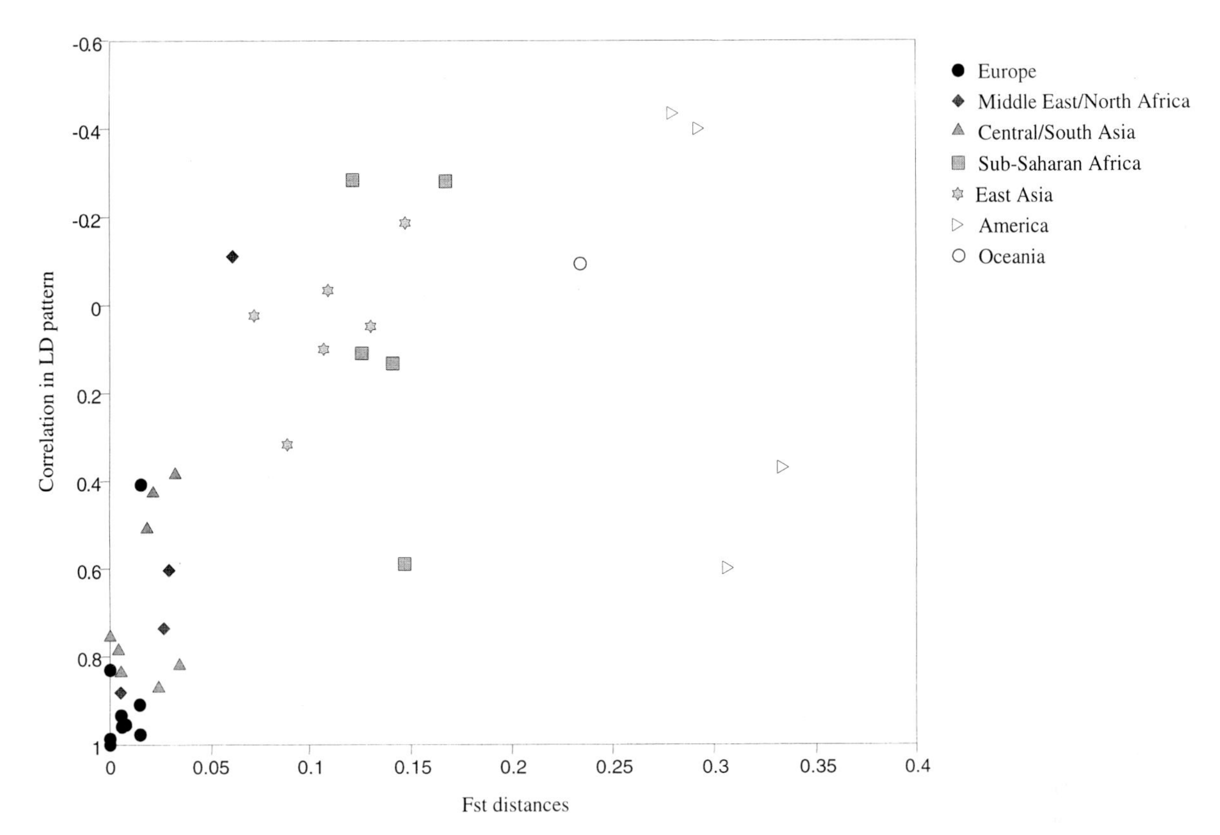


**Figure 2.** Relative positions of 39 worldwide populations in relation to the English for haplotype composition (measured as an Fst genetic distance) and for LD pattern (measured by the correlation coefficient between each population and the English of the LD measure for twelve neighboring markers). See text for details.


## ACKNOWLEDGMENTS

We thank Michelle Gardner for careful revision of the manuscript and Oscar Lao for his help with the analysis and figures. We also thank Aida Andrés and Tomàs Marquès for their helpful comments and assistance. This study was supported by the European Commission project QLG2-CT-2001-00916, by the Dirección General de Investigación (Spanish Government) grant BMC2001-0772, and Grup de Recerca Consolidat SGR2001-00285 from Generalitat de Catalunya.


## REFERENCES


Arnheim N., Calabrese P., and Nordborg M. 2003. Hot and cold spots of recombination in the human genome: The reason we should find them and how this can be achieved. *Am. J. Hum. Genet.* **73:** 5.

Bertranpetit J. and Calafell F. 2001. Genome versus population understanding in human population studies. In *Genes, fossils and behaviour* (ed. P. Donnelly and R.A. Foley), p. 49. IOS Press, Amsterdam, The Netherlands.

———. 2003. Genome views on human evolution. In *Evolution: From molecules to ecosystems* (ed. A. Moya and E. Font), p. 260. Oxford University Press, Oxford, United Kingdom.

Cann H.M., de Toma C., Cazes L., Legrand M.F., Morel V., Piouffre L., Bodmer J., Bodmer W.F., Bonne-Tamir B., Cambon-Thomsen A., Chen Z., Chu J., Carcassi C., Contu L., Du R., Excoffier L., Ferrara G.B., Friedlaender J.S., Groot H., Gurwitz D., Jenkins T., Herrera R.J., Huang X., Kidd J., and Kidd K.K., et al. 2002. A human genome diversity cell line panel. *Science* **296:** 261.

Daly M.J., Rioux J.D., Schaffner S.F., Hudson T.J., and Lander E.S. 2001. High-resolution haplotype structure in the human genome. *Nat. Genet.* **29:** 229.

Dawson E., Abecasis G.R., Bumpstead S., Chen Y., Hunt S., Beare D.M., Pabial J., Dibling T., Tinsley E., Kirby S., Carter D., Papaspyridonos M., Livingstone S., Ganske R., Lohmussaar E., Zernant J., Tonisson N., Remm M., Magi R., Puurand T., Vilo J., Kurg A., Rice K., Deloukas P., Mott R., Metspalu A., Bentley D.R., Cardon L.R., and Dunham I. 2002. A first-generation linkage disequilibrium map of human chromosome 22. *Nature* **418:** 544.

DeMille M.M., Kidd J.R., Ruggeri V., Palmatier M.A., Goldman D., Odunsi A., Okonofua F., Grigorenko E., Schulz L.O., Bonne-Tamir B., Lu R.B., Parnas J., Pakstis A.J., and Kidd K.K. 2002. Population variation in linkage disequilibrium across the COMT gene considering promoter region and coding region variation. *Hum. Genet.* **111:** 521.

Ding Y.C., Chi H.C., Grady D.L., Morishima A., Kidd J.R., Kidd K.K., Flodman P., Spence M.A., Schuck S., Swanson J.M., Zhang Y.P., and Moyzis R.K. 2002. Evidence of positive selection acting at the human dopamine receptor D4 gene locus. *Proc. Natl. Acad. Sci.* **99:** 309.

Fullerton S.M., Bartoszewicz A., Ybazeta G., Horikawa Y., Bell G.I., Kidd K.K., Cox N.J., Hudson R.R., and Di Rienzo A. 2002. Geographic and haplotype structure of candidate type 2 diabetes susceptibility variants at the calpain-10 locus. *Am. J. Hum. Genet.* **70:** 1096.

Gabriel S.B., Schaffner S.F., Nguyen H., Moore J.M., Roy J., Blumenstiel B., Higgins J., DeFelice M., Lochner A., Faggart M., Liu-Cordero S.N., Rotimi C., Adeyemo A., Cooper R., Ward R., Lander E.S., Daly M.J., and Altshuler D. 2002. The structure of haplotype blocks in the human genome. *Science* **296:** 2225.

Gelernter J., Cubells J.F., Kidd J.R., Pakstis A.J., and Kidd K.K.

1999. Population studies of polymorphisms of the serotonin transporter protein gene. *Am. J. Med. Genet.* **88:** 61.

Hastbacka J., de la Chapelle A., Kaitila I., Sistonen P., Weaver A., and Lander E. 1992. Linkage disequilibrium mapping in isolated founder populations: Diastrophic dysplasia in Finland. *Nat. Genet.* **23:** 204.

Hudson R. 1990. Gene genealogies and the coalescent process. *Oxf. Surv. Evol. Biol.* **7:** 1.

———. 2001. Linkage disequilibrium and recombination. In *Handbook of statistical genetics* (ed. D. Balding et al.), p. 309. Wiley & Sons, New York.

Jeffreys A.J. and Neumann R. 2002. Reciprocal crossover asymmetry and meiotic drive in a human recombination hot spot. *Nat. Genet.* **31:** 267.

Kaessmann H., Zollner S., Gustafsson A.C., Wiebe V., Laan M., Lundeberg J., Uhlen M., and Paabo S. 2002. Extensive linkage disequilibrium in small human populations in Eurasia. *Am. J. Hum. Genet.* **70:** 673.

Kauppi L., Sajantila A., and Jefrreys A. 2003. Recombination hotspots rather than population history dominate linkage disequilibrium in the MHC class II region. *Hum. Mol. Genet.* **121:** 33.

Kidd J.R., Pakstis A.J., Zhao H., Lu R.B., Okonofua F.E., Odunsi A., Grigorenko E., Tamir B.B., Friedlaender J., Schulz L.O., Parnas J., and Kidd K.K. 2000. Haplotypes and linkage disequilibrium at the phenylalanine hydroxylase locus, PAH, in a global representation of populations. *Am. J. Hum. Genet.* **66:** 1882.

Kidd K.K., Morar B., Castiglione C.M., Zhao H., Pakstis A.J., Speed W.C., Bonne-Tamir B., Lu R.B., Goldman D., Lee C., Nam Y.S., Grandy D.K., Jenkins T., and Kidd J.R. 1998. A global survey of haplotype frequencies and linkage disequilibrium at the DRD2 locus. *Hum. Genet.* **103:** 211.

Kong A., Gudbjartsson D.F., Sainz J., Jonsdottir G.M., Gudjonsson S.A., Richardsson B., Sigurdardottir S., Barnard J., Hallbeck B., Masson G., Shlien A., Palsson S.T., Frigge M.L., Thorgeirsson T.E., Gulcher J.R., and Stefansson K. 2002. A high-resolution recombination map of the human genome. *Nat. Genet.* **31:** 241.

Kruglyak L. 1999. Prospects for whole-genome linkage disequilibrium mapping of common disease genes. *Nat. Genet.* **22:** 139.

Laan M. and Pääbo S. 1997. Demographic history and linkage disequilibrium in human populations. *Nat. Genet.* **17:** 435.

Martinez-Arias R., Calafell F., Mateu E., Comas D., Andres A., and Bertranpetit J. 2001. Sequence variability of a human pseudogene. *Gen. Res.* **11:** 1071.

Mateu E., Perez-Lezaun A., Martinez-Arias R., Andres A., Valles M., Bertranpetit J., and Calafell F. 2002. PKLR-GBA region shows almost complete linkage disequilibrium over 70 kb in a set of worldwide populations. *Hum. Genet.* **110:** 532.

Mateu E., Calafell F., Lao O., Bonne-Tamir B., Kidd J.R., Pakstis A., Kidd K.K., and Bertranpetit J. 2001. Worldwide genetic analysis of the CFTR region. *Am. J. Hum. Genet.* **68:** 103.

Meng Z., Zaykin D.V., Xu C.F., Wagner M., and Ehm M.G. 2003. Selection of genetic markers for association analyses, using linkage disequilibrium and haplotypes. *Am. J. Hum. Genet.* **73:** 115.

Navarro A. and Barton N.H. 2002. The effect of multilocus balancing selection on neutral variability. *Genetics* **161:** 849.

Navarro A., Betran E., Zapata C., and Ruiz A. 1996. Dynamics of gametic disequilibria between loci linked to chromosome inversions: The recombination redistributing effect of inversions. *Genet. Res.* **67:** 67.

Nordborg M. and Tavare S. 2002. Linkage disequilibrium: What history has to tell us. *Trends Genet.* **18:** 83.

Osier M.V., Pakstis A.J., Soodyall H., Comas D., Goldman D., Odunsi A., Okonofua F., Parnas J., Schulz L.O., Bertranpetit J., Bonne-Tamir B., Lu R.B., Kidd J.R., and Kidd K.K. 2002. A global perspective on genetic variation at the ADH genes reveals unusual patterns of linkage disequilibrium and diversity. *Am. J. Hum. Genet.* **71:** 84.

Patil N., Berno A.J., Hinds D.A., Barrett W.A., Doshi J.M., Hacker C.R., Kautzer C.R., Lee D.H., Marjoribanks C., McDonough D.P., Nguyen B.T., Norris M.C., Sheehan J.B., Shen N., Stern D., Stokowski R.P., Thomas D.J., Trulson M.O., Vyas K.R., Frazer K.A., Fodor S.P., and Cox D.R. 2001. Blocks of limited haplotype diversity revealed by high-resolution scanning of human chromosome 21. *Science* **294:** 1719.

Pérez-Lezaun A., Calafell F., Mateu E., Comas D., Ruiz-Pacheco R., and Bertranpetit J. 1997. Microsatellite variation and the differentiation of modern humans. *Hum. Genet.* **99:** 1.

Pritchard J.K. 2001. Are rare variants responsible for susceptibility to complex diseases? *Am. J. Hum. Genet.* **69:** 124.

Pritchard J. and Przeworski M. 2001. Linkage disequilibrium in humans: Models and data. *Am. J. Hum. Genet.* **69:** 1.

Przeworski M. and Wall J.D. 2001. Why is there so little intragenic linkage disequilibrium in humans? *Genet. Res.* **77:** 143.

Przeworski M., Hudson R.R., and Di Rienzo A. 2000. Adjusting the focus on human variation. *Trends Genet.* **16:** 296.

Reich D.E. and Lander E.S. 2001. On the allelic spectrum of human disease. *Trends Genet.* **17:** 502.

Reich D.E., Schaffner S.F., Daly M.J., McVean G., Mullikin J.C., Higgins J.M., Richter D.J., Lander E.S., and Altshuler D. 2002. Human genome sequence variation and the influence of gene history, mutation and recombination. *Nat. Genet.* **32:** 135.

Reich D.E., Cargill M., Bolk S., Ireland J., Sabeti P.C., Richter D.J., Lavery T., Kouyoumjian R., Farhadian S.F., Ward R., and Lander E.S. 2001. Linkage disequilibrium in the human genome. *Nature* **411:** 199.

Romualdi C., Balding D., Nasidze I.S., Risch G., Robichaux M., Sherry S.T., Stoneking M., Batzer M.A., and Barbujani G. 2002. Patterns of human diversity, within and among continents, inferred from biallelic DNA polymorphisms. *Genome Res.* **12:** 602.

Sabeti P.C., Reich D.E., Higgins J.M., Levine H.Z., Richter D.J., Schaffner S.F., Gabriel S.B., Platko J.V., Patterson N.J., McDonald G.J., Ackerman H.C., Campbell S.J., Altshuler D., Cooper R., Kwiatkowski D., Ward R., and Lander E.S. 2002. Detecting recent positive selection in the human genome from haplotype structure. *Nature* **419:** 832.

Stumpf M.P. 2002. Haplotype diversity and the block structure of linkage disequilibrium. *Trends Genet.* **18:** 226.

Tishkoff S.A., Dietzsch E., Speed W., Pakstis A.J., Kidd J.R., Cheung K., Bonne-Tamir B., Santachiara-Benerecetti A.S., Moral P., and Krings M. 1996. Global patterns of linkage disequilibrium at the CD4 locus and modern human origins. *Science* **271:** 1380.

Tishkoff S.A., Goldman A., Calafell F., Speed W.C., Deinard A.S., Bonne-Tamir B., Kidd J.R., Pakstis A.J., Jenkins T., and Kidd K.K. 1998. A global haplotype analysis of the myotonic dystrophy locus: Implications for the evolution of modern humans and for the origin of myotonic dystrophy mutations. *Am. J. Hum. Genet.* **62:** 1389.

Tishkoff S.A., Pakstis A.J., Stoneking M., Kidd J.R., Destro-Bisol G., Sanjantila A., Lu R.B., Deinard A.S., Sirugo G., Jenkins T., Kidd K.K., and Clark A.G. 2000. Short tandem-repeat polymorphism/alu haplotype variation at the PLAT locus: Implications for modern human origins. *Am. J. Hum. Genet.* **67:** 901.

Tishkoff S.A., Varkonyi R., Cahinhinan N., Abbes S., Argyropoulos G., Destro-Bisol G., Drousiotou A., Dangerfield B., Lefranc G., Loiselet J., Piro A., Stoneking M., Tagarelli A., Tagarelli G., Touma E.H., Williams S.M., and Clark A.G. 2001. Haplotype diversity and linkage disequilibrium at human G6PD: Recent origin of alleles that confer malarial resistance. *Science* **293:** 455.

Wall J.D. and Pritchard J.K. 2003a. Assessing the performance of the haplotype block model of linkage disequilibrium. *Am. J. Hum. Genet.* **73:** 502.

—. 2003b. Haplotype blocks and linkage disequilibrium in the human genome. *Nat. Rev. Genet.* **4:** 587.

Zhang K., Calabrese P., Nordborg M., and Sun F. 2002. Haplotype block structure and its applications to association studies: Power and study designs. *Am. J. Hum. Genet.* **71:** 1386.

Zhang K., Akey J.M., Wang N., Xiong M., Chakraborty R., and Jin L. 2003. Randomly distributed crossovers may generate block-like patterns of linkage disequilibrium: An act of genetic drift. *Hum. Genet.* **113:** 51.

# Genome-wide Association of Haplotype Markers to Gene Expression Levels

A. WINDEMUTH, M. KUMAR, K. NANDABALAN, B. KOSHY, C. XU, M. PUNGLIYA, AND R. JUDSON
*Genaissance Pharmaceuticals, New Haven, Connecticut 06511*

Gene regulation has been studied intensely for decades. With the emergence of microarrays, there has been a spurt of literature looking at global patterns of gene expression. It has also been proposed that for genetic analysis, gene expression is a quantitative trait that is inherited (Cheung and Spielman 2002). To date, most of the genetic analysis of expression phenotypes has been analyzed in model organisms such as yeast (Brem et al. 2002; Cavalieri et al. 2000; Steinmetz et al. 2002), *Drosophila* (Jin et al. 2001), and mice (Sandberg et al. 2000). For the most part, these model animal studies used linkage analysis employing genome-wide linkage markers. A general scheme for performing such studies in plants was described by Jansen and Nap (2001). Schadt et al. (2003) performed extensive linkage analyses in corn, mice, and humans in which mRNA expression levels were treated as quantitative traits. They were able to use mRNA expression as an intermediate phenotype in order to discover two quantitative trait loci (QTL) for obesity in their mouse model. The study published by Cheung et al. (2003) using human lymphoblastoid cell lines confirmed that there is heritable natural variation of gene expression, based on the observation that variation is greater between unrelated individuals than it is within families. Yan et al. (2002), using gene expression measurements from 96 CEPH cell lines, demonstrated the existence of *cis*-acting SNPs, i.e., those that affect the expression of the gene in which they reside, in 6 genes. Rockman and Wray (2002) carried out an extensive literature survey of *cis*-acting polymorphisms and compiled a set of 106 genes showing significant experimental evidence of this effect. All of these studies confirm that expression is a highly heritable phenotype.

In this paper, we focus on genetic determinants of gene expression phenotypes in human B cells. We have discovered SNPs and gene-based haplotypes for 6,184 genes in a set of 93 human cell lines. Levels of mRNA expression were measured in these same cell lines, and a global analysis was performed to discover associations between genetic variation and expression levels. Because of the large number of comparisons carried out, we expect many false-positive associations. Several filters were used to minimize the number of false positives. The resulting associations were confirmed by a second round of cell growth, mRNA extraction, and expression measurement. We present two sets of associations for which the statistical evidence is strong, and which show a high rate of replication. The first is a set of *cis*-associations in 22 genes. Of these, 15 are significant in the original and replication set. In the second example, we show that two amino acid-changing SNPs in a zinc finger transcription factor are strongly associated with expression levels in a large number of other genes. From these data, we can infer that this protein (ZNF295) is involved in the transcription of these genes. These are only two of many interesting findings that have been gleaned from this data set. We believe that such experiments will greatly aid in the functional characterization of many genes.

## MATERIALS AND METHODS

***Samples.*** Blood samples were obtained from approximately 200 individuals recruited in two US locations (Miami, FL, and Anaheim, CA) and immortalized as B-cell lymphocyte cell lines by standard procedures (Neitzel 1986).

Data on family medical history and geographic origin of self, parents, and grandparents were obtained for each individual. From this collection, individuals were prioritized for sequencing according to the degree of homogeneous origin of their grandparents and to achieve an equal proportion of African-Americans, Asians, Caucasians, and Hispanic Latinos. Additionally, the medical data were examined so as not to introduce a bias for any known diseases. A set of 96 samples was constructed, containing DNA from 76 of these individuals, 10 individuals from a three-generation CEPH Caucasian family, 7 individuals from a two-generation African-American family, and one chimpanzee, as well as positive and negative DNA controls. Immortalized cell lines derived from the two families were obtained commercially. The sample set contained DNA from 82 unrelated individuals: 20 African-Americans, 20 Asians, 21 Caucasians, 18 Hispanic Latinos, and 3 Native Americans.

For the expression analysis, only 89 of the 93 cell lines produced data of sufficient quality for further analysis. This subset was made up of 75 unrelated subjects: 19 African-Americans, 19 Asians, 20 Caucasians, 14 Hispanic Latinos, and 3 Native Americans. There are 6 additional related African-Americans, 7 additional related Caucasians, and 1 additional related Hispanic Latino. Finally, 3 independent samples for one individual were measured to assess normalization consistency.

***Expression data.*** mRNA was extracted by standard protocols from snap-frozen pellets of B cells grown under standard conditions (Neitzel 1986) and harvested during the log phase of the growth when the cell density reached 500,000/ml. To isolate RNA, one hundred million cells from each cell line were used. mRNA samples were hybridized to the Affymetrix Human Genome U133 GeneChip (HG-U133A). The Affymetrix Microarray Suite software (MAS 5.0) was used to normalize the data. Data were extracted using the GeneLogic Express software suite. All association analyses were performed using custom software as described below. In order to replicate our findings, cell lines for a random subset of 67 of the 89 samples were independently grown up. We extracted mRNA and performed expression analysis independently using the Affymetrix protocol as described above, but using a different chip (HG-U95A).

***Genotype and haplotype data.*** Our database, the HAP database, draws on a large number of genes of pharmaceutical interest. These have been re-sequenced in the cell lines described above in selected regions of 6,184 genes. For each gene, each exon was sequenced along with 100 bp into the introns on each side of every exon. Up to 1,000 bp upstream from the first exon (the 5′ or promoter region) and 100 bp downstream of the last exon was sequenced. Within the sequenced regions, simple power calculations show that 99% of SNPs with a minor allele frequency of ≥10% in any of the major subpopulations will be seen at least once. More than 99% of SNPs with a minor allele frequency of 5% in the general population will also be seen. On average, 4,200 bp were sequenced for each gene. The database is further described elsewhere (Stephens et al. 2001).

***Association method.*** We have used custom software to screen the 6,184 gene loci vs. 22,242 mRNA expression levels for potential associations. The algorithm is described in more detail below. The calculation was performed on a network of 24 Windows PCs and took about one week to complete. The $p$ values for all potential associations and details on all associations with a $p$ value less than or equal to 0.001 were stored in an Oracle database for further analysis. This value was chosen so as to keep the use of storage space manageable and to select for the more promising candidate associations. The main features of our algorithm are the enumeration of all relevant markers for each gene locus and the use of permutation tests for accurate $p$ values. The latter is a well-known way to deal with multiple non-independent tests without making assumptions about variable distributions (Pitman 1937; Brown and Fears 1981; Heyse and Rom 1988).

To test for associations, all relevant single SNP and full haplotype markers were enumerated for each gene locus. A *marker* for the purpose of this discussion is a genetically determined rule dividing the sample population into two groups, called the Marker(+) and Marker(–) groups. A *single SNP marker* is based on the alleles of a single SNP, a *full haplotype marker* is based on the presence or absence of a particular haplotype. Each SNP and each haplotype gives rise to two markers: one following the dominant model and one following the recessive model. For a given SNP, a subject would be considered to be in the Marker(+) group of the recessive marker only if it was homozygous for the alternate allele. To belong to the Marker(+) group of the dominant SNP marker, the subject could either be heterozygous or homozygous for the alternate allele. Similarly, to be positive for a recessive full haplotype marker, a subject would have to have two copies of the haplotype, whereas one copy would be sufficient for the dominant marker. In some of our analyses, we also consider sub-haplotype markers, which are defined by a subset of SNPs and the presence or absence of a particular haplotype across the subset. Models other than recessive and dominant were not tested. A "heterozygous advantage" model was not considered because of its comparatively lower biological plausibility and the loss of sensitivity associated with the increase of the number of tests. An additive model would require more than two marker groups and complicate the statistical procedure. We have performed tests which show that an additive association will be detected by either of the recessive or dominant models with only a minor loss of sensitivity. A marker was considered *relevant* for the purpose of association screening if there were at least five subjects in each of the Marker(+) and Marker(–) groups. The reasons for not considering rarer markers are twofold: First, performance of the screening algorithm is greatly improved by looking only at the relevant markers, and second, any result with less than five subjects in one of the groups would be considered statistically questionable even if it stands up to rigorous significance testing.

To account for population structure, we used ethnicity and family membership as covariates in the statistical analysis. This reduces false positives caused by loci that act as surrogates for ethnicity, and at the same time improves sensitivity by eliminating that part of expression variation that is due to population structure. We included the covariates by calculating the residuals of the expression levels with respect to the covariates; i.e., the expression values were adjusted such that their averages within each of the four ethnic groups were the same. Similarly, the averages within each of three family groups (0: no family, 1: first family, 2: second family) were also adjusted to be the same.

For a given locus and a given expression level, a statistically rigorous significance level is calculated for the presence of an association between any of the enumerated markers and the expression level. Because there are potentially many markers in a gene locus, and many of these will be highly correlated due to marker overlap and linkage disequilibrium, we use a permutation test for multiple comparison adjustment (Pitman 1937; Brown and Fears 1981; Heyse and Rom 1988). First, a t-statistic is calculated for each marker to quantify the difference in expression between the Marker(+) and Marker(–) groups. Second, the most extreme of these statistics across all relevant markers is selected to serve as an overall indication of strength of association, the *omnibus* statistic. Third, a permutation test is applied to accurately quantify the significance of the result in the face of multiple testing of many markers. For this purpose, the omnibus statistic is recalculated for up to 6,250 randomly permuted data sets representing the null hypothesis of no true association. The permutation was done in such a way that the expression values were randomly redistributed among the subjects, whereas the genetic assignments were left alone. The fraction of these instances of the null hypothesis that result in a more extreme omnibus statistic is a direct measure for the probability of the observed statistic having arisen by chance, and is called the *p value*. More specifically, if $N$ random permutations are performed, and $M$ of the permutations yield an omnibus statistic more extreme than that of the original data, the probability of the null hypothesis is $p = M/N$. This procedure is nonparametric, meaning it makes no assumptions about the distribution of the expression values. In addition, because of the use of the omnibus statistic, the testing of multiple markers is fully accounted for. In fact, most other methods used to adjust for multiple comparisons would result in a large loss of sensitivity, since there is no way to accurately account for the high degree of correlation between the markers, many of which differ only slightly from each other. For efficiency, the permutation testing was done on successively larger sets of permutations $N$ = (50, 250, 1250, 6250), and the next higher set of permutations was skipped when the number of false positives $M$ was 10 or more in the previous set.

After the full association analysis was run, we further filtered the results based on the quality of the expression data. Each expression level for each sample is associated with both a quantitative expression value and binary present/absent label. This label is calculated by a statistical algorithm in the MAS 5.0 software, and a present label means that there is a probability of less than 5% that there is no signal and what is measured is only noise. For most analyses, we excluded fragments for which there were fewer than 10 present calls out of 89 total. Additionally, for some analyses, we excluded fragments for which the maximum fold change was less than two. This is calculated by taking the maximum and minimum expression values for the expression level in the subset of sample with "present" calls and taking the ratio. In addition, we excluded in the analysis the genes that showed more than twofold of mRNA expression levels within the three samples from the same individual.

## RESULTS

Association analysis was carried out examining 6,184 gene loci by 22,242 expression levels. At a significance level of $p \leq 0.001$, we found 66,141 significant associations between gene loci and expression levels. Because this is such a large and rich data set, we have concentrated our analysis on two interesting subsets of associations. The first of these analyses concentrates on a set of *cis*-acting SNPs; i.e., SNPs that affect the expression levels of the genes they reside in. The second example examines a gene coding for a largely uncharacterized zinc finger protein (ZNF295), which likely forms part of a transcription factor complex. Two amino acid-changing SNPs in this locus are strongly associated with expression levels in a large number of other genes. Finally, we examine the global statistics of the data set and attempt to compare our results to some reported previously.

### *cis*-Acting Genetic Variations

Not all of the gene loci for which we have haplotype data are represented by expression probes, and, vice versa, not all of the genes represented by expression probes have known haplotypes associated with them. In our experiment, there were 4,762 gene loci and 7,553 expression probes where each of the loci belonged to the same gene as one or more of the probes. We identified 22 genes from our initial screen that have candidate *cis*-associations significant at a level of $p<0.001$ and that have at least 10 present calls and at least twofold variation in expression. This is significantly more than would be expected by chance (see Discussion).

We initially searched for associations with individual SNPs and full haplotypes. Once this subset of genes was identified, we also looked for associations with sub-haplotypes defined by sets of 2, 3, 4, and 5 SNPs. Each of these genes showed a significant association ($p<0.05$) with one or more of these haplotype markers after corrections were made for multiple comparisons, as described in Methods. The identical calculations were run with the replication data set. Typically, we discovered a number of markers having significant associations to expression levels, all of which are highly correlated with one another. To simplify the analysis, we only examined the most significant multi-SNP marker and all significant single-SNP markers in the original data set, and compared them with the results of the replication experiment.

The most significant haplotype associations for these genes are summarized in Table 1. This table gives the gene symbol, a description of the haplotype marker, significance scores, and the counts and mean expression levels for the Marker(+) and Marker(–) groups. In all tables, significance scores are given as a logarithmic representation of the $p$ value (score = –10 log[$p$]). In the original data set, we present the raw significance and the significance after correction for multiple comparisons. In the replication set, only the raw significance scores are shown because we have already been given a marker from the first experiment and do not need to perform the multiple comparisons again.

To illustrate the interpretation of haplotype markers, let us consider the AIM1 marker, ([Rec 5] -28G, 38695G, 53185T, 53784C, 54768C). This is a recessive marker containing 5 SNPs. We examined the positions –28, 38695, 53185, 53784, and 54768, and looked for the haplotype GGTCC at those positions. The Marker(+) group is made up of those individuals having two copies of GGTCC at those sites. The Marker(–) group is everyone else. In this case, 16 individuals had two copies of GGTCC, and the mean expression level of AIM1 among those subjects was 529.5. The mean expression level of AIM1 among the remaining individuals was 365.5. The

**Table 1.** A Summary of the *cis*-Associations

| Gene | Best marker | Sig. (Orig.) | Adj. Sig. (Orig.) | Sig. (Rep.) | Sig. Best (Rep.) | Count marker(+) | Count marker(–) | Mean marker(+) | Mean marker(–) |
|---|---|---|---|---|---|---|---|---|---|
| AIM1 | (Rec 5) -28G, 38695G, 53185T, 53784C, 54768C | 75 | 26 | 43* | 54* | 16 | 73 | 529.5 | 365.3 |
| ARTS-1 | (Rec 3) -4495C, 380C, 9947T | 58 | 25 | x | 29* | 29 | 60 | 53.8 | 106.6 |
| ATP5O | (Rec 2) -491G, 6675C | 44 | 27 | 4 | 13* | 23 | 66 | 1788.9 | 1552.6 |
| CHKL | (Dom 2) 1142C, 3734C | 48 | 23 | 15* | 31* | 13 | 76 | 161.8 | 120.0 |
| DDX17 | (Dom 1) 11625G | 112 | 45 | 66* | 66* | 6 | 83 | 177.2 | 32.0 |
| DDX3 | (Dom 2) 9964A, 10171T | 44 | 33 | 20* | 34* | 52 | 37 | 812.2 | 616.6 |
| DNM1L | (Dom 1) -272A | 40 | 29 | 22* | 32* | 19 | 70 | 171.7 | 140.7 |
| ENTPD1 | (Dom 2) -860G, -409T | 44 | 26 | x | 15* | 66 | 23 | 585.6 | 694.7 |
| FLJ11342 | (Rec 3) -470T, -216A, 12232T | 37 | 25 | NA | NA | 53 | 36 | 226.4 | 201.3 |
| HBS1L | (Rec 1) 86620T | 57 | 33 | x | 30* | 53 | 36 | 255.0 | 210.2 |
| HMG14 | (Rec 3) 3366G, 5677T, 6047G | 144 | 45 | x | 44* | 33 | 56 | 818.4 | 1494.5 |
| HSPC023 | (Dom 1) 18T | 42 | 28 | NA | NA | 46 | 43 | 549.4 | 616.8 |
| LOC51082 | (Rec 2) -110C, 1316C | 65 | 45 | NA | NA | 35 | 54 | 513.8 | 425.4 |
| LOC51103 | (Dom 1) -5154A | 49 | 32 | 10 | 19* | 51 | 38 | 139.2 | 160.3 |
| MUT | (Dom 1) -407T | 40 | 38 | 15* | 19* | 50 | 39 | 43.6 | 35.1 |
| PEX6 | (Rec 2) 15386G, 16710G | 63 | 42 | x | 64* | 16 | 73 | 45.7 | 63.0 |
| PLEK | (Dom 2) 18267T, 31404G | 59 | 30 | 35* | 40* | 66 | 23 | 448.7 | 296.3 |
| PPP1R11 | (Rec 2) 1084A, 1148G | 46 | 31 | 3 | 18* | 73 | 16 | 371.3 | 313.1 |
| RNAHP | (Dom 2) 5159T, 23042G | 58 | 32 | 6 | 14* | 38 | 51 | 252.9 | 205.7 |
| RNPS1 | (Dom 2) -3132A, 9107T | 103 | 45 | x | 14* | 72 | 17 | 284.5 | 148.1 |
| UROD | (Rec 1) 1562G | 40 | 33 | 12 | 12 | 78 | 11 | 114.1 | 150.8 |
| VAMP8 | (Dom 1) 1497T | 127 | 45 | 10 | 15* | 52 | 37 | 434.0 | 619.1 |

For each gene, the best marker column gives a shorthand description of the most significant marker. "Dom" indicates a dominant marker and "Rec" a recessive marker. The following number is the number of SNPs used in the marker. The remaining information provides the position and allele of each SNP in the haplotype marker. Positions are relative to the transcription start site, with the A of the ATG at position 1. Sig. (Orig.) indicates the raw significance of the marker in the original data set. All significance scores are given as $-10 \log (p)$. A score >12 indicates a $p$ value of <0.05. Adj. Sig. (Orig.) gives the multiple comparisons corrected significance in the original set. Sig. (Rep.) is the raw significance of this particular marker in the replication set. An "x" in this column indicates that the marker was not seen in the replication set. Sig. Best (Rep.) gives the significance of the best marker in the replication set. In both of the replication columns, an asterisk indicates that the $p$ value was >0.05. The Count and Mean columns give the number of subjects and the mean values of the expression levels in the Marker(+) and Marker(–) groups.

multiple comparison corrected $p$ value for this association is 0.0025 (i.e., –10 log [0.0025] = 26).

A total of 19 of the 22 genes had data in the replication set. Of these, the most significant marker in the original data set was also significant in the replication set in seven cases (AIM1, CHKL, DDX17, DDX3, DNM1L, MUT, and PLEK). All but one of the genes (UROD) had at least one significant marker in the replication set. Most of these markers are simple (1 or 2 SNPs) and most have a relatively good balance between the number of individuals in the Marker(+) and Marker(–) groups.

Many individual SNPs in these genes displayed strong associations (raw $p \leq 0.001$) with the corresponding expression levels. These are summarized in Table 2. This table gives the gene, the location of the SNP, the allele frequency, and significance scores in the original and replication sets. It also indicates genes for which we observed a significant association with more than one of the probes for that gene on the Affymetrix HG-U133A chip. For SNPs, 12 genes showed significant associations with at least one SNP in both the original and replication sets (AIM1, DDX17, ENTPD1, HBS1L, HMG14, LOC51103, MUT, PEX6, PLEK, RNPS1, UROD, and VAMP8). When we combine the haplotype and single SNP markers, there is strong evidence for replicated significant associations in 15 of the 22 genes (AIM1, CHKL, DDX17, DDX3, DNM1L, ENTPD1, HBS1L, HMG14, LOC51103, MUT, PEX6, PLEK, RNPS1, UROD, and VAMP8), and there is moderate evidence in an additional 4 genes.

### *trans*-Acting Variation in a Transcription Factor: ZNF295

To find potential novel transcription factors, we looked for gene loci whose variants were associated with differential expression of a much larger number of genes than a typical locus. One of the most extreme cases we found was the gene ZNF295 (GenBank accession AP001745.1), which codes for a largely uncharacterized zinc finger protein. We saw 1,291 unique associations with ZNF295 with permutation-corrected $p$ values $<0.05$, out of the subset of expression levels with more than 10 present calls and more than a twofold change. Table 3 gives the distribution of observed associations as a function of significance score. Throughout this range, we see significantly more associations than would be expected by chance. We discovered a total of 12 SNPs in this gene, which are given in Table 4. Note that only 3 of these SNPs are seen in more than one sample. The 2 most common SNPs alter amino acids (Asn185Ser and Lys218Gln). Both of these coding SNPs have been reported previously in HGVBase (Brookes et al. 2000; Fredman et al. 2002). The annotated amino acid sequence of the gene is shown in Figure 1.

The protein coded for by ZNF295 contains a BTB/POZ protein-binding domain which is seen near the amino terminus of certain zinc finger proteins (Deweindt et al. 1995; Oyake et al. 1996; Li et al. 1997). This domain mediates homomeric dimerization and, in some instances, heterodimeric dimerization. The protein also contains a series of C2H2 zinc finger domains. The two amino acid changes that drive all of the associations occur in the linker region of the protein between the protein-binding domain and the DNA-binding zinc finger domains.

Recall that we test many markers within a gene locus for associations with each expression level. Because of overlap and linkage disequilibrium, markers are highly correlated, and we expect to see slightly different but related markers to be selected as the best marker for each associated expression level, even if the mechanism is exactly the same. A total of 6 different ZNF295 markers accounted for the 1,291 expression levels associated with ZNF295 haplotypes with a permutation-adjusted $p$ value $<0.05$. Table 5 gives the markers and the number of expression levels for which that marker gave the most significant association in the original and replicated data sets. All of these markers involve SNPs 1, 3, and 8, and most are similar in terms of how they divide the sample set into two groups. The Marker(+) group ranges in size from 7 to 28, with the exception of the 8T marker, which is inverse in the sense that its Marker(–) group of 10 samples might correspond to the Marker(+) groups of the other markers. The 7 samples in the Marker(+) group of the smallest and most often observed marker (1G, 8T) are all included in the Marker(+) groups of three of the remaining five markers (1G; 3C; 3C, 8T). The other two markers (1A, 3C; 8T) have almost no overlap with the most common marker, but are similar to each other. Since it is unlikely that more than one functional variation should exist in a single gene locus, it is satisfying to see that the unrelated markers (1A, 3C; 8T) in fact are not significantly confirmed in the replication data set. They are likely to be the expected false positives due to sampling error.

These 1,291 significant associations between expression levels and ZNF295 markers were next examined in the replication data set. A total of 248 associations were also significant in the replication set with an adjusted $p$ value $\leq 0.05$. This is a much larger number than would be expected by chance. Table 6 lists this set of expression levels that are replicated, along with adjusted significance scores, number of individuals in the Marker(+) and Marker(–) groups, and the group mean expression levels. Table 3 also gives the distribution of the number of replicated associations as a function of significance score. In 150 cases, the most significant marker in the original data set was also the most significant marker in the replication data set.

We further probed the reproducibility of these associations by calculating how well the marker could predict shifts in expression levels between the original and replication set. We calculated the fractional change in expression accounted for by the marker as (<Marker(-)>-<Marker(+)>)/(<Marker(–)>) where <...> indicates the mean expression level for the group. This was calculated for each of the 150 expression levels having significant associations in both the original and replication data sets for which the best marker was the same in both data sets. These data are illustrated in Figure 2. There is a strong correlation with $R^2 = 0.51$. Note that the slope and inter-

**Table 2.** Significant SNPs for the *cis*-Associations (Raw $p \leq 0.001$)

| Gene | Region | Position | Freq | Change | Probes w/ sig. assoc. (Orig.) | Sig (Orig.) | Sig (Rep.) |
|---|---|---|---|---|---|---|---|
| AIM1 | exon 2 | 7330 | 14 | T/C Cys 405 Arg | 1 | 48 | 39* |
| AIM1 | intron 14 | 40559 | 24 | C/G | 1 | 46 | 16* |
| AIM1 | intron 15 | 40925 | 28 | C/T | 1 | 31 | 15* |
| AIM1 | intron 15 | 43461 | 41 | C/T | 1 | 41 | 21* |
| AIM1 | intron 17 | 45931 | 26 | A/G | 1 | 35 | 13* |
| AIM1 | intron 17 | 45960 | 26 | G/A | 1 | 35 | 13* |
| ARTS-1 | exon 2 | 380 | 42 | C/G Pro 127 Arg | 1 | 57 | 7 |
| ARTS-1 | intron 2 | 564 | 15 | G/A | 1 | 74 | 1 |
| ARTS-1 | intron 5 | 9947 | 3 | T/C | 1 | 42 | 0 |
| ARTS-1 | exon 6 | 10095 | 14 | A/G Met 349 Val | 1 | 74 | 1 |
| ARTS-1 | exon 6 | 10118 | 35 | C/T Ala 356 Ala | 1 | 51 | 6 |
| ARTS-1 | intron 7 | 11725 | 39 | G/A | 1 | 56 | 4 |
| ARTS-1 | exon 9 | 13322 | 46 | C/T Ser 453 Ser | 1 | 56 | 5 |
| ARTS-1 | intron 9 | 13433 | 46 | C/T | 1 | 56 | 5 |
| ARTS-1 | intron 10 | 13720 | 46 | A/G | 1 | 56 | 5 |
| ARTS-1 | intron 10 | 15183 | 38 | G/C | 1 | 52 | 4 |
| ARTS-1 | exon 11 | 15300 | 46 | G/A Arg 528 Lys | 1 | 56 | 5 |
| ARTS-1 | intron 11 | 17349 | 46 | G/T | 1 | 56 | 5 |
| ARTS-1 | intron 11 | 17370 | 46 | C/A | 1 | 56 | 5 |
| ARTS-1 | exon 12 | 17420 | 15 | G/A Asp 575 Asn | 1 | 74 | 1 |
| ARTS-1 | intron 12 | 17636 | 46 | C/T | 1 | 56 | 5 |
| ARTS-1 | intron 12 | 17915 | 46 | A/G | 1 | 56 | 5 |
| ARTS-1 | exon 13 | 18106 | 50 | A/G Ala 637 Ala | 1 | 39 | 1 |
| ARTS-1 | intron 16 | 22273 | 1 | A/G | 1 | 54 | 0 |
| ATP5O | intron 2 | 3334 | 34 | C/T | 1 | 35 | 5 |
| ATP5O | exon 4 | 6675 | 34 | C/T Asn 107 Asn | 1 | 43 | 6 |
| CHKL | intron 10 | 3417 | 16 | T/C | 1 | 33 | 5 |
| CHKL | intron 10 | 3497 | 16 | C/T | 1 | 39 | 7 |
| CHKL | 3′UTR | 3734 | 14 | T/C | 1 | 46 | 8 |
| DDX17 | intron 8 | 11625 | 3 | A/G | 3 | 999 | 66* |
| DDX17 | intron 10 | 13859 | 3 | C/T | 3 | 999 | 66* |
| DDX17 | exon 13 | 20007 | 3 | G/A Pro 622 Pro | 3 | 999 | 66* |
| DNM1L | exon 3 | 24321 | 11 | G/A Gly 84 Gly | 1 | 41 | 9 |
| DNM1L | intron 14 | 52224 | 15 | C/T | 1 | 31 | 6 |
| DNM1L | intron 14 | 52406 | 11 | G/A | 1 | 39 | 10 |
| DNM1L | exon 20 | 58428 | 11 | G/C | 1 | 41 | 9 |
| DNM1L | exon 20 | 58429 | 11 | A/T | 1 | 41 | 9 |
| ENTPD1 | promoter | –860 | 47 | A/G | 3 | 64 | 15* |
| ENTPD1 | intron 5 | 15692 | 46 | G/C | 3 | 54 | 7 |
| ENTPD1 | intron 6b | 17844 | 21 | G/A | 3 | 56 | 7 |
| FLJ11342 | promoter | –735 | 12 | A/C | 1 | 42 | NA |
| FLJ11342 | promoter | –467 | 11 | A/G | 1 | 41 | NA |
| FLJ11342 | promoter | –42 | 12 | T/C | 1 | 42 | NA |
| FLJ11342 | intron 4 | 12232 | 13 | T/G | 1 | 45 | NA |
| FLJ11342 | exon 5 | 12774 | 13 | A/C | 1 | 45 | NA |
| HBS1L | promoter | –381 | 23 | A/C | 2 | 80 | 9 |
| HBS1L | intron 5 | 52051 | 10 | G/C | 1 | 46 | 3 |
| HBS1L | intron 10 | 66413 | 43 | A/G | 1 | 33 | 28* |
| HBS1L | intron 14 | 74194 | 21 | A/G | 2 | 67 | 7 |
| HBS1L | intron 15 | 74440 | 44 | G/C | 1 | 31 | 15* |

*(Continued on facing page.)*

cept in this plot are approximately 1 and 0, as expected for true replications. A total of 4 outlier points were removed for this analysis.

## DISCUSSION

### Statistical Significance

Naturally, with 6,184 times 22,242 possibilities, there will be a large number of very small *p* values simply by chance. Recall that the permutation adjustment for multiple comparisons was done to account for testing multiple markers in a single association, not for testing multiple associations. In fact, we do not find more positives in the overall set of associations than would be expected by chance. There are two ways to extract useful results from such a data set: (1) Corroborate the findings with results from an independent experiment or (2) select a subset of associations of interest without taking the results of the association test into account. We have employed the former method by performing a second experiment, and we have employed the second method when focusing on the *cis*-associations, among which the number of loci with *p* values of <0.001 is significantly greater than expected by chance.

Overall, at a significance level of $p < 0.001$, we find a total of 66,141 positive associations. Under the assumption of no true association and unbiased *p* values, we would expect approximately 137,545 false-positive associations. This indicates that there is a bias in the *p*-value distribution toward higher *p* values, or fewer significant results. We have determined that the source of this bias is

**Table 2.** (*Continued from facing page*)

| Gene | Region | Position | Freq | Change | Probes w/ sig. assoc. (Orig.) | Sig (Orig.) | Sig (Rep.) |
|---|---|---|---|---|---|---|---|
| HBS1L | intron 15 | 74471 | 24 | A/G | 2 | 89 | 10 |
| HBS1L | intron 16 | 83798 | 24 | A/G | 2 | 89 | 10 |
| HBS1L | exon 17 | 86620 | 24 | T/C Phe 659 Phe | 2 | 89 | 10 |
| HMG14 | promoter | –768 | 30 | T/G | 1 | 38 | 10 |
| HMG14 | intron 4 | 1603 | 32 | G/A | 1 | 92 | 25* |
| HMG14 | intron 4 | 3366 | 34 | G/T | 1 | 103 | 30* |
| HMG14 | intron 5 | 5669 | 21 | T/C | 1 | 74 | 19* |
| HMG14 | 3′UTR | 5982 | 6 | G/A | 1 | 30 | 7 |
| HSPC023 | promoter | –568 | 36 | G/C | 1 | 31 | NA |
| HSPC023 | promoter | –268 | 24 | T/C | 1 | 30 | NA |
| HSPC023 | exon 1 | 18 | 32 | C/T Arg 6 Arg | 1 | 40 | NA |
| HSPC023 | intron 1 | 146 | 32 | C/A | 1 | 40 | NA |
| HSPC023 | exon 2 | 193 | 32 | A/G Lys 39 Arg | 1 | 40 | NA |
| HSPC023 | intron 2 | 323 | 36 | A/G | 1 | 31 | NA |
| LOC51082 | promoter | –626 | 7 | C/T | 1 | 42 | NA |
| LOC51082 | promoter | –609 | 4 | C/T | 1 | 41 | NA |
| LOC51082 | exon 1 | –88 | 9 | G/C | 1 | 33 | NA |
| LOC51082 | exon 2 | 1316 | 32 | C/T | 1 | 66 | NA |
| LOC51103 | promoter | –5154 | 34 | G/A | 1 | 63 | 19* |
| LOC51103 | exon 2 | 26 | 21 | G/A Arg 9 His | 1 | 34 | 6 |
| LOC51103 | exon 2 | 92 | 17 | G/T Arg 31 Leu | 1 | 47 | 12 |
| MUT | promoter | –407 | 30 | A/T | 2 | 46 | 19* |
| PEX6 | 5′UTR | –55 | 38 | C/T | 2 | 82 | 64* |
| PEX6 | exon 1 | 399 | 38 | G/T Val 133 Val | 2 | 82 | 64* |
| PEX6 | intron 1b | 3390 | 33 | T/C | 2 | 68 | 35* |
| PEX6 | intron 7 | 12915 | 43 | T/C | 2 | 96 | 55* |
| PEX6 | intron 9 | 14412 | 46 | A/G | 2 | 83 | 46* |
| PEX6 | intron 15 | 16197 | 43 | A/G | 2 | 96 | 55* |
| PEX6 | exon 17 | 16710 | 46 | A/G Glu 938 Glu | 2 | 81 | 36* |
| PEX6 | exon 17 | 16712 | 38 | C/A Pro 939 Gln | 2 | 81 | 64* |
| PLEK | intron 1a | 101 | 12 | C/T | 1 | 30 | 16* |
| PLEK | exon 9 | 31404 | 45 | G/C | 1 | 70 | 24* |
| PPP1R11 | intron 1 | 1084 | 6 | A/G | 1 | 32 | 6 |
| PPP1R11 | intron 2 | 1385 | 6 | A/G | 1 | 32 | 6 |
| RNAHP | intron 1 | 5159 | 24 | G/T | 1 | 104 | 6 |
| RNPS1 | promoter | –3689 | 43 | T/C | 1 | 103 | 8 |
| RNPS1 | promoter | –3132 | 37 | A/C | 1 | 92 | 14* |
| RNPS1 | promoter | –3131 | 37 | G/C | 1 | 92 | 14* |
| RNPS1 | intron 7 | 9107 | 42 | T/C | 1 | 110 | 13* |
| UROD | intron 5 | 1562 | 6 | G/C | 1 | 54 | 19* |
| UROD | exon 6 | 1772 | 6 | A/G Pro 201 Pro | 1 | 54 | 19* |
| VAMP8 | intron 1 | 1299 | 49 | T/C | 1 | 70 | 15* |
| VAMP8 | exon 2 | 1497 | 38 | C/T Asn 46 Asn | 1 | 129 | 12 |
| VAMP8 | exon 3 | 3091 | 46 | C/T | 1 | 82 | 15* |

This table gives the gene, the region, the base-pair position relative to the transcription start site, and the alternative alleles for all SNPs that were individually significant before correction for multiple corrections. Significance scores are $-10 \log(p)$. In the replication set, all SNPs with a significance score >12 ($p \leq 0.05$) are indicated by an asterisk. The change column gives the alleles of the SNP, and (where appropriate) the alternative amino acid for SNPs in coding regions. The Probes column gives the number of expression probes examined for that gene which showed a significant association for that SNP (sig. >12). Where there are multiple associated probes, the best of their scores is given. A score of 999 indicates a $p$ value that is zero within the limits of the numerical precision of the $t$-test implementation.

the application of ethnicity as a covariate to our statistical procedure. The samples used are from several different ethnic populations, and there are many markers that differ in frequency between the different ethnic groups. Simultaneously, there are expression differences between the populations, due to either genetic or environmental factors. When the association tests are done without adjusting for ethnicity, any ethnicity-dependent expression level shows association with all ethnicity-dependent markers, regardless of any mechanistic connection between the associated genes. To counteract this, we have adjusted the expression data in such a way that the mean expression values are equal between different ethnic groups and between families. This eliminates the incidental associations based on ethnicity and increases our sensitivity for truly mechanistic associations. However, by performing the covariate adjustment before the permutation tests, we have biased the data set to show less association between ethnically dependent genes than a randomly permuted version of the adjusted data set. This leads to the bias in $p$ values. The correct way to get accurate $p$ values would be to perform the covariate adjustment separately on the permuted as well as the actual data set, which is an avenue for future work. Although this $p$-value bias does not permit us to treat the $p$ values as accurate probabilities, it does not diminish their usefulness in ranking and selecting results. Furthermore, the bias is entirely conservative, such that a significant biased $p$ value is certain to indicate at least that level of significance for a given result. In some of the following discussion, we will assume the $p$ values to be unbiased, and keep in mind that this will be a conservative assumption.

**Table 3.** Summary of the Number of Associations Seen and Confirmed with ZNF295 as a Function of Significance Score

| Sig. | Cumulative significant assoc. | Expected significant assoc. | Ratio (significant/ expected) | Cumulative replicated assoc. | Ratio (replicated/ significant) |
|---|---|---|---|---|---|
| 45 | 16 | 1 | 22.75 | 10 | 0.63 |
| 42 | 27 | 1 | 19.24 | 13 | 0.48 |
| 40 | 36 | 2 | 16.19 | 16 | 0.44 |
| 39 | 44 | 3 | 15.71 | 20 | 0.45 |
| 38 | 53 | 4 | 15.03 | 24 | 0.45 |
| 37 | 60 | 4 | 13.52 | 28 | 0.47 |
| 36 | 75 | 6 | 13.42 | 30 | 0.40 |
| 35 | 90 | 7 | 12.80 | 35 | 0.39 |
| 34 | 98 | 9 | 11.07 | 38 | 0.39 |
| 33 | 109 | 11 | 9.78 | 41 | 0.38 |
| 32 | 129 | 14 | 9.19 | 48 | 0.37 |
| 31 | 141 | 18 | 7.98 | 53 | 0.38 |
| 30 | 157 | 22 | 7.06 | 61 | 0.39 |
| 29 | 174 | 28 | 6.21 | 65 | 0.37 |
| 28 | 207 | 35 | 5.87 | 70 | 0.34 |
| 27 | 230 | 44 | 5.18 | 73 | 0.32 |
| 26 | 266 | 56 | 4.76 | 89 | 0.33 |
| 25 | 309 | 70 | 4.39 | 98 | 0.32 |
| 24 | 350 | 89 | 3.95 | 104 | 0.30 |
| 23 | 402 | 111 | 3.61 | 112 | 0.28 |
| 22 | 444 | 140 | 3.16 | 118 | 0.27 |
| 21 | 497 | 177 | 2.81 | 126 | 0.25 |
| 20 | 584 | 222 | 2.63 | 142 | 0.24 |
| 19 | 637 | 280 | 2.27 | 152 | 0.24 |
| 18 | 750 | 353 | 2.13 | 174 | 0.23 |
| 17 | 856 | 444 | 1.93 | 187 | 0.22 |
| 16 | 962 | 559 | 1.72 | 205 | 0.21 |
| 15 | 1066 | 703 | 1.52 | 220 | 0.21 |
| 14 | 1204 | 885 | 1.36 | 233 | 0.19 |
| 13 | 1291 | 1115 | 1.16 | 248 | 0.19 |

Cumulative associations are those with a score better than or equal to the one in the first column. The number of expected significant associations is based on the formula $N = 22{,}242 \times 10^{-sig/10}$.

### *cis*-Acting Genetic Variations

The statistical significance of our findings on *cis*-associations is readily estimated under the assumption that the $p$ values are a true representation of the probability of false positives. The number of false positives expected from testing 4,752 genes for *cis*-associations at a significance level of $p = 0.001$ is 4.752. Our finding of 22 *cis*-associations is inconsistent with this expectation. More generally, the number of false positives $n$ observes the binomial distribution

$$d(n) = \binom{n}{N} p^n (-p)^{N-n},$$

which puts the probability of finding 22 associations or more by chance at $p = 7.6 * 10^{-9}$. We conclude from this that most of the 22 associations are representative of actual genetic effects on gene expression.

Intriguingly, several of the genes have significant *cis*-associations with SNPs in the 5′ upstream region, which is implicated as a crucial regulatory region in a majority of genes studied. Some of these SNPs are found in known

**Table 4.** SNPs and Frequencies Seen in ZNF295

| SNP | ATG offset | Change | Grantham value | Grantham code | Common | Het. | Rare |
|---|---|---|---|---|---|---|---|
| **1** | **554** | **A/G Asn 185 Ser** | **46** | **C** | **76** | **18** | **0** |
| 2 | 574 | C/A Pro 192 Thr | 38 | C | 93 | 1 | 0 |
| **3** | **652** | **A/C Lys 218 Gln** | **53** | **MC** | **62** | **29** | **1** |
| 4 | 696 | C/A Ile 232 Ile | – | – | 91 | 1 | 0 |
| 5 | 846 | T/C Val 282 Val | – | – | 90 | 1 | 0 |
| 6 | 970 | T/C Ser 324 Pro | 74 | MC | 91 | 1 | 0 |
| 7 | 1311 | G/C Leu 437 Leu | – | – | 92 | 1 | 0 |
| 8 | 1500 | T/C Asn 500 Asn | – | – | 79 | 9 | 0 |
| 9 | 1538 | A/G Glu 513 Gly | 98 | MC | 87 | 1 | 0 |
| 10 | 1690 | C/G His 564 Asp | 81 | MC | 92 | 1 | 0 |
| 11 | 2434 | C/A Pro 812 Thr | 38 | C | 92 | 1 | 0 |
| 12 | 3229 | G/A | – | – | 82 | 1 | 0 |

Grantham values and codes indicate the degree of conservation (C: conservative, MC: moderately conservative) (Grantham 1974). The two amino acid changes that were found most responsible for the associations are indicated in bold.

```
   1 MEGLLHYINP AHAISLLSAL NEERLKGQLC DVLLIVGDQK FRAHKNVLAA SSEYFQSLFT
  61 NKENESQTVF QLDFCEPDAF DNVLNYIYSS SLFVEKSSLA AVQELGYSLG ISFLTNIVSK
 121 TPQAPFPTCP NRKKVFVEDD ENSSQKRSVI VCQSRNEAQG KTVSQNQPDV SHTSRPSPSI
 181 AVKANTNKPH VPKPIEPLHN LSLTEKSWPK DSSVVYAKSL EHSGSLDDPN RISLVKRNAV
 241 LPSKPLQDRE AMDDKPGVSG QLPKGKALEL ALKRPRPPVL SVCSSSETPY LLKETNKGNG
 301 QGEDRNLLYY SKLGLVIPSS GSGSGNQSID RSGPLVKSLL RRSLSMDSQV PVYSPSIDLK
 361 SSQGSSSVSS DAPGNVLCAL SQKSSLKDCS EKTALDDRPQ VLQPHRLRSF SASQSTDREG
 421 ASPVTEVRIK TEPSSPLSDP SDIIRVTVGD AATTAAASSS SVTRDLSLKT EDDQKDMSRL
 481 PAKRRFQADR RLPFKKLKVN EHGSPVSEDN FEEGSSPTLL DADFPDSDLN KDEFGELEGT
 541 RPNKKFKCKH CLKIFRSTAG LHRHVNMYHN PEKPYACDIC HKRFHTNFKV WTHCQTQHGI
 601 VKNPSPASSS HAVLDEKFQR KLIDIVRERE IKKALIIKLR RGKPGFQGQS SSQAQQVIKR
 661 NLRSRAKGAY ICTYCGKAYR FLSQFKQHIK MHPGEKPLGV NKVAKPKEHA PLASPVENKE
 721 VYQCRLCNAK LSSLLEQGSH ERLCRNAAVC PYCSLRFFSP ELKQEHESKC EYKKLTCLEC
 781 MRTFKSSFSI WRHQVEVHNQ NNMAPTENFS LPVLDHNGDV TGSSRPQSQP EPNKVNHIVT
 841 TKDDNVFSDS SEQVNFDSED SSCLPEDLSL SKQLKIQVKE EPVEEAEEEA PEASTAPKEA
 901 GPSKEASLWP CEKCGKMFTV HKQLERHQEL LCSVKPFICH VCNKAFRTNF RLWSHFQSHM
 961 SQASEESAHK ESEVCPVPTN SPSPPPLPPP PPLPKIQPLE PDSPTGLSEN PTPATEKLFV
1021 PQESDTLFYH APPLSAITFK RQFMCKLCHR TFKTAFSLWS HEQTHN
```

**Figure 1.** Annotated protein sequence of ZNF295. The BTB/POZ domain is underlined (bp 20–120), as are the zinc finger domains. The conserved residues of the domains are indicated in bold. The two residues associated with variable expression levels are circled.
Domain annotation:

1. 20-120 BTB/POZ domain
2. 185 SNP Asn / Ser [SNP1]
3. 218 SNP Lys / Gln [SNP3]
4. 546-569 $Cys_2$-$His_2$ type zinc finger domain
5. 575-998 $Cys_2$-$His_2$ type zinc finger domain
6. 670-692 $Cys_2$-$His_2$ type zinc finger domain
7. 777-798 $Cys_2$-$His_2$ type zinc finger domain
8. 911-927 potential $Cys_2$-$His_2$ type zinc finger domain
9. 937-959 $Cys_2$-$His_2$ type zinc finger domain
10. 1043-1065 $Cys_2$-$His_2$ type zinc finger domain

**Table 5.** Significant Markers for ZNF295 Associations in the Original and Replicated Data Sets

| Marker | Model | Original | Replicated | *N*(Marker+)/ *N*(Marker–) |
|---|---|---|---|---|
| 1G | dominant | 56 | 21 | 17/72 |
| 3C | dominant | 53 | 25 | 28/61 |
| 1A, 3C | dominant | 23 | 0 | 12/77 |
| 1G, 8T | dominant | 952 | 155 | 7/82 |
| 3C, 8T | dominant | 130 | 43 | 19/70 |
| 8T | recessive | 77 | 4 | 79/10 |

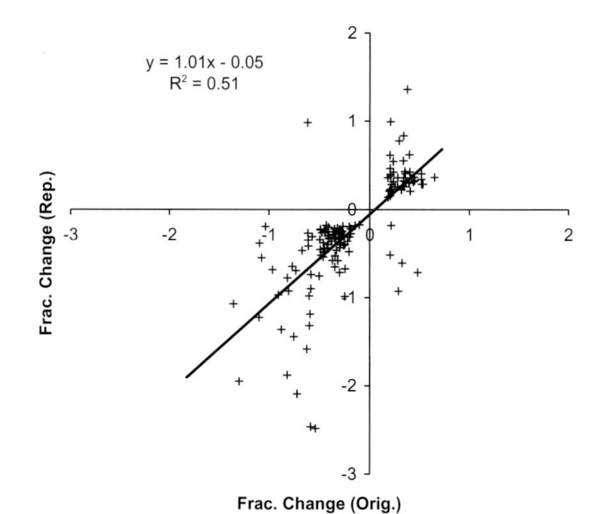


**Figure 2.** Plot showing the correlation between the fractional change in expression between the Marker(+) and Marker(–) groups in the original and replication sets. Only those expression levels for which the most significant marker was the same in the original and replication data sets are shown. Each point represents one of the expression levels associated with the ZNF295 marker. Note that the slope and intercept are approximately 1 and 0, as expected for true positives. A total of 4 extreme outliers have been removed for this analysis.

transcription-factor-binding motifs, but most are not. We also observed genes whose expression was tightly associated with polymorphisms occurring in exonic and intronic regions. More effort will be required to find which SNPs are functionally responsible for these expression differences and which are merely in linkage disequilibrium with nearby functional SNPs that are not observed directly. However, it is not entirely unexpected to find associations with SNPs in regions commonly thought of as nonfunctional, as there is ample experimental evidence (Hariharan et al. 1991; Mattick 1994; Ayoubi and Van De Ven 1996; Cramer et al. 1999; Majewski and Ott 2002; Dupont et al. 2003) suggesting that gene expression is modulated not only by the promoter, but also by elements in the introns and even exons of genes. Similarly, alternate promoters have been detected in introns in some genes (Lecanda et al. 2001; Lin et al. 2002). One class of intronic polymorphisms that affect expression levels were discovered in splice junctions (Beaufrere et al. 1998; Gabellini 2001). None of the intronic polymorphisms that we identified to have associations with the mRNA expression profile in the present study is situated in splice junctions.

There are several points of comparison between the experiments of Chueng et al. (2003) and those we report. In both cases, expression analysis was performed on a cohort of human B-cell lymphocyte cell lines that included both related and unrelated individuals. These authors note that the level of variability in expression for most genes is higher in unrelated individuals than within families. We also see this in our sample. This is shown in Figure 3 for the 3-generation Caucasian CEPH family included in our cohort. These authors also highlighted a set of 5 genes showing especially high levels of variation: ST3GALVi, ACTG2, GK, HNRPA2B1, and DHFR. We examined the

**Table 6.** List of All the Genes Whose Expression Is Associated with Markers in ZNF295

| RNA level | Adj P (Original) | N (Marker+) | N (Marker–) | Mean (Marker+) | Mean (Marker–) | Best marker (Original) | Adj P (Replication) | N (Marker+) | N (Marker–) | Mean (Marker+) | Mean (Marker–) | Best marker (Replication) |
|---|---|---|---|---|---|---|---|---|---|---|---|---|
| RAB1 | 45 | 7 | 82 | 277.53 | 176.68 | Dom 2 1G, 8T | 42 | 6 | 61 | 1164.42 | 890.94 | Dom 2 1G, 8T |
| HNRPD | 45 | 7 | 82 | 1136.38 | 816.36 | Dom 2 1G, 8T | 30 | 17 | 50 | 556.50 | 486.23 | Dom 1 3C |
| FLJ20274 | 45 | 7 | 82 | 151.30 | 84.38 | Dom 2 1G, 8T | 28 | 10 | 57 | 223.92 | 163.27 | Dom 2 3C, 8T |
| SCP2 | 45 | 7 | 82 | 343.93 | 241.59 | Dom 2 1G, 8T | 23 | 6 | 61 | 156.05 | 108.50 | Dom 2 1G, 8T |
| ARGBP2 | 45 | 7 | 82 | 205.35 | 144.48 | Dom 2 1G, 8T | 21 | 13 | 54 | 125.41 | 104.49 | Dom 1 1G |
| PCTK1 | 45 | 7 | 82 | 72.41 | 109.44 | Dom 2 1G, 8T | 21 | 6 | 61 | 9.72 | 58.65 | Dom 2 1G, 8T |
| SMARCE1 | 45 | 7 | 82 | 369.77 | 248.26 | Dom 2 1G, 8T | 17 | 6 | 61 | 309.66 | 253.57 | Dom 2 1G, 8T |
| MYO9B | 45 | 7 | 82 | 134.48 | 204.90 | Dom 2 1G, 8T | 15 | 6 | 61 | 178.87 | 241.45 | Dom 2 1G, 8T |
| CLECSF2 | 45 | 7 | 82 | 194.65 | 82.61 | Dom 2 1G, 8T | 15 | 6 | 61 | 130.60 | 63.03 | Dom 2 1G, 8T |
| ARHGEF1 | 45 | 7 | 82 | 119.01 | 183.18 | Dom 2 1G, 8T | 14 | 13 | 54 | 204.84 | 236.30 | Dom 1 1G |
| CD164 | 42 | 7 | 82 | 273.20 | 130.41 | Dom 2 1G, 8T | 42 | 6 | 61 | 456.14 | 329.72 | Dom 2 1G, 8T |
| MDS019 | 42 | 7 | 82 | 447.23 | 293.07 | Dom 2 1G, 8T | 15 | 17 | 50 | 343.85 | 286.77 | Dom 1 3C |
| MINK | 42 | 7 | 82 | 63.89 | 111.53 | Dom 2 1G, 8T | 14 | 10 | 57 | 39.53 | 80.52 | Dom 2 3C, 8T |
| IF2 | 40 | 7 | 82 | 35.85 | 17.15 | Dom 2 1G, 8T | 18 | 13 | 54 | 10.58 | –0.56 | Dom 1 1G |
| ARPC2 | 40 | 7 | 82 | 1261.94 | 956.15 | Dom 2 1G, 8T | 16 | 6 | 61 | 2446.83 | 2023.76 | Dom 2 1G, 8T |
| HNRPK | 40 | 7 | 82 | 891.32 | 686.02 | Dom 2 1G, 8T | 14 | 13 | 54 | 922.40 | 822.78 | Dom 1 1G |
| IFI16 | 39 | 7 | 82 | 1107.54 | 741.27 | Dom 2 1G, 8T | 35 | 6 | 61 | 527.08 | 393.54 | Dom 2 1G, 8T |
| USP1 | 39 | 7 | 82 | 139.85 | 67.37 | Dom 2 1G, 8T | 26 | 6 | 61 | 227.00 | 146.58 | Dom 2 1G, 8T |
| LOC55884 | 39 | 7 | 82 | 79.16 | 45.14 | Dom 2 1G, 8T | 17 | 6 | 61 | 11.72 | 4.80 | Dom 2 1G, 8T |
| SLU7 | 39 | 7 | 82 | 1380.91 | 766.52 | Dom 2 1G, 8T | 16 | 10 | 57 | 1227.24 | 858.95 | Dom 2 3C, 8T |
| DUT | 38 | 7 | 82 | 1215.06 | 900.21 | Dom 2 1G, 8T | 20 | 6 | 61 | 1022.79 | 838.79 | Dom 2 1G, 8T |
| ATP6H | 38 | 7 | 82 | 858.63 | 649.37 | Dom 2 1G, 8T | 18 | 10 | 57 | 590.27 | 512.21 | Dom 2 3C, 8T |
| LYPLA1 | 38 | 7 | 82 | 362.26 | 225.09 | Dom 2 1G, 8T | 17 | 6 | 61 | 246.28 | 182.87 | Dom 2 1G, 8T |
| STAT6 | 38 | 7 | 82 | 42.67 | 88.93 | Dom 2 1G, 8T | 13 | 6 | 61 | 112.18 | 157.37 | Dom 2 1G, 8T |
| AKAP2 | 37 | 7 | 82 | 357.73 | 187.68 | Dom 2 1G, 8T | 32 | 6 | 61 | 216.66 | 110.06 | Dom 2 1G, 8T |
| SMT3H2 | 37 | 7 | 82 | 1041.70 | 764.20 | Dom 2 1G, 8T | 31 | 6 | 61 | 1621.78 | 1296.81 | Dom 2 1G, 8T |
| DKFZp667O2416 | 37 | 19 | 70 | 116.85 | 148.77 | Dom 2 3C, 8T | 17 | 13 | 54 | 120.71 | 141.39 | Dom 1 1G |
| IRF4 | 37 | 79 | 10 | 2.23 | 8.14 | Rec 1 8T | 13 | 17 | 50 | 1036.69 | 923.93 | Dom 1 3C |
| ARHGDIA | 36 | 7 | 82 | 584.56 | 1051.93 | Dom 2 1G, 8T | 25 | 6 | 61 | 857.59 | 1354.29 | Dom 2 1G, 8T |
| YY1 | 36 | 7 | 82 | 387.73 | 285.23 | Dom 2 1G, 8T | 18 | 17 | 50 | 20.85 | 12.08 | Dom 1 3C |
| NKTR | 35 | 7 | 82 | 78.08 | 30.72 | Dom 2 1G, 8T | 39 | 10 | 57 | 31.47 | 10.33 | Dom 2 3C, 8T |
| EVI2B | 35 | 7 | 82 | 430.56 | 296.93 | Dom 2 1G, 8T | 19 | 6 | 61 | 281.65 | 195.42 | Dom 2 1G, 8T |
| MCL1 | 35 | 7 | 82 | 64.98 | 44.05 | Dom 2 1G, 8T | 18 | 6 | 61 | 607.50 | 481.50 | Dom 2 1G, 8T |
| GLRX | 35 | 7 | 82 | 1276.91 | 883.37 | Dom 2 1G, 8T | 16 | 6 | 61 | 479.49 | 386.76 | Dom 2 1G, 8T |
| SUPT6H | 35 | 7 | 82 | 105.38 | 157.70 | Dom 2 1G, 8T | 15 | 10 | 57 | 60.05 | 108.96 | Dom 2 3C, 8T |
| ATP6A1 | 34 | 7 | 82 | 66.73 | 29.01 | Dom 2 1G, 8T | 45 | 6 | 61 | 145.16 | 49.22 | Dom 2 1G, 8T |
| MLC-B | 34 | 7 | 82 | 1098.61 | 783.73 | Dom 2 1G, 8T | 40 | 6 | 61 | 969.78 | 728.68 | Dom 2 1G, 8T |
| YWHAB | 34 | 7 | 82 | 432.84 | 305.25 | Dom 2 1G, 8T | 19 | 17 | 50 | 974.76 | 895.09 | Dom 1 3C |
| ZNF91 | 33 | 7 | 82 | 147.19 | 84.93 | Dom 2 1G, 8T | 42 | 6 | 61 | 264.08 | 155.88 | Dom 2 1G, 8T |
| RAB2 | 33 | 7 | 82 | 273.15 | 212.31 | Dom 2 1G, 8T | 26 | 6 | 61 | 116.83 | 87.32 | Dom 2 1G, 8T |
| CARM1 | 33 | 7 | 82 | 149.63 | 219.56 | Dom 2 1G, 8T | 23 | 13 | 54 | 294.85 | 380.77 | Dom 1 1G |
| PSMA3 | 32 | 7 | 82 | 1146.16 | 890.30 | Dom 2 1G, 8T | 42 | 6 | 61 | 824.69 | 648.57 | Dom 2 1G, 8T |

| | | | | | | | | | | | | |
|---|---|---|---|---|---|---|---|---|---|---|---|---|
| PRKCL2 | 32 | 7 | 82 | 74.05 | 40.79 | Dom 2 1G, 8T | 34 | 6 | 61 | 86.39 | 48.64 | Dom 2 1G, 8T |
| PPARBP | 32 | 7 | 82 | 21.92 | 4.92 | Dom 2 1G, 8T | 32 | 6 | 61 | 78.35 | 37.37 | Dom 2 1G, 8T |
| IL2RG | 32 | 7 | 82 | 660.08 | 939.62 | Dom 2 1G, 8T | 29 | 13 | 54 | 634.88 | 748.08 | Dom 1 1G |
| PSMC2 | 32 | 7 | 82 | 122.21 | 75.76 | Dom 2 1G, 8T | 24 | 6 | 61 | 234.64 | 185.38 | Dom 2 1G, 8T |
| C6orf9 | 32 | 7 | 82 | 121.93 | 208.67 | Dom 2 1G, 8T | 19 | 6 | 61 | 315.94 | 490.48 | Dom 2 1G, 8T |
| GYPC | 32 | 19 | 70 | 385.62 | 507.71 | Dom 2 3C, 8T | 14 | 6 | 61 | 486.16 | 593.43 | Dom 2 1G, 8T |
| HRB2 | 31 | 7 | 82 | 98.73 | 61.38 | Dom 2 1G, 8T | 36 | 6 | 61 | 67.03 | 47.37 | Dom 2 1G, 8T |
| KRAS2 | 31 | 7 | 82 | 162.77 | 77.54 | Dom 2 1G, 8T | 31 | 6 | 61 | 57.82 | 26.00 | Dom 2 1G, 8T |
| DSS1 | 31 | 7 | 82 | 646.35 | 448.31 | Dom 2 1G, 8T | 29 | 6 | 61 | 388.65 | 308.23 | Dom 2 1G, 8T |
| PCMT1 | 31 | 7 | 82 | 236.24 | 158.35 | Dom 2 1G, 8T | 25 | 10 | 57 | 235.45 | 186.21 | Dom 2 3C, 8T |
| ACTR1B | 31 | 7 | 82 | 174.16 | 242.60 | Dom 2 1G, 8T | 19 | 17 | 50 | 90.53 | 118.65 | Dom 1 3C |
| KPNB2 | 30 | 7 | 82 | 587.97 | 367.05 | Dom 2 1G, 8T | 40 | 6 | 61 | 166.25 | 83.90 | Dom 2 1G, 8T |
| SLC16A7 | 30 | 7 | 82 | 37.22 | 16.87 | Dom 2 1G, 8T | 28 | 6 | 61 | 16.75 | 1.83 | Dom 2 1G, 8T |
| P5 | 30 | 7 | 82 | 671.78 | 455.97 | Dom 2 1G, 8T | 27 | 13 | 54 | 452.56 | 361.13 | Dom 1 1G |
| SLC16A6 | 30 | 7 | 82 | 58.84 | 33.28 | Dom 2 1G, 8T | 26 | 6 | 61 | 66.41 | 40.32 | Dom 2 1G, 8T |
| SLC25A1 | 30 | 7 | 82 | 111.19 | 197.20 | Dom 2 1G, 8T | 20 | 6 | 61 | 259.48 | 378.80 | Dom 2 1G, 8T |
| DJ328E19.C1.1 | 30 | 28 | 61 | 277.96 | 331.92 | Dom 1 3C | 18 | 6 | 61 | 148.19 | 91.27 | Dom 2 1G, 8T |
| SFRS10 | 30 | 7 | 82 | 604.57 | 467.15 | Dom 2 1G, 8T | 18 | 6 | 61 | 472.82 | 384.25 | Dom 2 1G, 8T |
| ATP1B3 | 30 | 7 | 82 | 1004.21 | 775.23 | Dom 2 1G, 8T | 14 | 17 | 50 | 728.92 | 650.23 | Dom 1 3C |
| UBE2L3 | 29 | 7 | 82 | 1199.02 | 957.32 | Dom 2 1G, 8T | 19 | 13 | 54 | 1122.47 | 985.06 | Dom 1 1G |
| SS18 | 29 | 7 | 82 | 64.13 | 32.66 | Dom 2 1G, 8T | 17 | 6 | 61 | 111.18 | 66.02 | Dom 2 1G, 8T |
| PARVB | 29 | 7 | 82 | 114.79 | 230.45 | Dom 2 1G, 8T | 17 | 10 | 57 | 379.37 | 517.53 | Dom 2 3C, 8T |
| RAD23B | 29 | 7 | 82 | 15.94 | 5.64 | Dom 2 1G, 8T | 17 | 10 | 57 | 231.69 | 199.80 | Dom 2 3C, 8T |
| SMARCD1 | 28 | 7 | 82 | 81.44 | 134.94 | Dom 2 1G, 8T | 18 | 6 | 61 | 37.41 | 98.57 | Dom 2 1G, 8T |
| SH2D2A | 28 | 7 | 82 | 48.31 | 78.71 | Dom 2 1G, 8T | 17 | 13 | 54 | 13.73 | 42.23 | Dom 1 1G |
| MBNL | 28 | 7 | 82 | 898.96 | 691.73 | Dom 2 1G, 8T | 17 | 6 | 61 | 415.22 | 339.20 | Dom 2 1G, 8T |
| CIRBP | 28 | 7 | 82 | 564.11 | 422.12 | Dom 2 1G, 8T | 16 | 6 | 61 | 575.14 | 449.70 | Dom 2 1G, 8T |
| GC20 | 28 | 7 | 82 | 414.00 | 292.41 | Dom 2 1G, 8T | 14 | 6 | 61 | 83.91 | 64.96 | Dom 2 1G, 8T |
| GPM6B | 27 | 12 | 77 | 17.28 | 7.76 | Dom 2 1A, 3C | 17 | 60 | 7 | –8.10 | –0.30 | Rec 1 8T |
| XRCC3 | 27 | 7 | 82 | 12.93 | 36.99 | Dom 2 1G, 8T | 16 | 6 | 61 | 41.24 | 64.86 | Dom 2 1G, 8T |
| JIK | 27 | 7 | 82 | 143.75 | 105.41 | Dom 2 1G, 8T | 13 | 6 | 61 | 114.53 | 89.47 | Dom 2 1G, 8T |
| YWHAQ | 26 | 7 | 82 | 1226.94 | 1031.40 | Dom 2 1G, 8T | 40 | 6 | 61 | 1018.92 | 837.02 | Dom 2 1G, 8T |
| KIAA0117 | 26 | 7 | 82 | 33.07 | 19.76 | Dom 2 1G, 8T | 32 | 17 | 50 | 41.52 | 27.90 | Dom 1 3C |
| NFKB2 | 26 | 7 | 82 | 47.08 | 86.31 | Dom 2 1G, 8T | 26 | 13 | 54 | 106.22 | 200.09 | Dom 1 1G |
| MAP4 | 26 | 7 | 82 | 86.89 | 136.33 | Dom 2 1G, 8T | 23 | 10 | 57 | 317.30 | 421.71 | Dom 2 3C, 8T |
| PPP1R2 | 26 | 7 | 82 | 291.03 | 213.13 | Dom 2 1G, 8T | 21 | 13 | 54 | 106.83 | 85.08 | Dom 1 1G |
| CAPZA1 | 26 | 7 | 82 | 1076.14 | 895.15 | Dom 2 1G, 8T | 21 | 6 | 61 | 586.85 | 491.75 | Dom 2 1G, 8T |
| ZYX | 26 | 7 | 82 | 172.16 | 312.17 | Dom 2 1G, 8T | 19 | 10 | 57 | 86.54 | 251.77 | Dom 2 3C, 8T |
| SLC1A5 | 26 | 7 | 82 | 106.13 | 166.05 | Dom 2 1G, 8T | 18 | 6 | 61 | 232.12 | 338.20 | Dom 2 1G, 8T |
| ARCN1 | 26 | 7 | 82 | 411.74 | 320.34 | Dom 2 1G, 8T | 18 | 6 | 61 | 187.45 | 148.79 | Dom 2 1G, 8T |
| TNFRSF8 | 26 | 19 | 70 | 169.88 | 246.26 | Dom 2 3C, 8T | 18 | 6 | 61 | 47.50 | 118.90 | Dom 2 1G, 8T |
| MT2A | 26 | 19 | 70 | 815.09 | 576.74 | Dom 2 3C, 8T | 18 | 10 | 57 | 222.86 | 159.34 | Dom 2 3C, 8T |
| DYRK4 | 26 | 17 | 72 | 91.96 | 121.00 | Dom 1 1G | 17 | 6 | 61 | 39.89 | 71.88 | Dom 2 1G, 8T |
| GYS1 | 26 | 7 | 82 | 169.91 | 246.08 | Dom 2 1G, 8T | 16 | 10 | 57 | 119.89 | 161.75 | Dom 2 3C, 8T |

*(Table continues on following pages.)*

**Table 6.** (*Continued*)

| RNA level | Adj P (Original) | N (Marker+) | N (Marker–) | Mean (Marker+) | Mean (Marker–) | Best marker (Original) | Adj P (Replication) | N (Marker+) | N (Marker–) | Mean (Marker+) | Mean (Marker–) | Best marker (Replication) |
|---|---|---|---|---|---|---|---|---|---|---|---|---|
| IFIT1 | 26 | 19 | 70 | 340.19 | 203.55 | Dom 2 3C, 8T | 16 | 10 | 57 | 390.99 | 266.46 | Dom 2 3C, 8T |
| CD36L1 | 26 | 7 | 82 | 162.60 | 238.20 | Dom 2 1G, 8T | 14 | 6 | 61 | 362.22 | 466.80 | Dom 2 1G, 8T |
| PC4 | 26 | 7 | 82 | 2060.14 | 1540.15 | Dom 2 1G, 8T | 13 | 13 | 54 | 1001.55 | 868.25 | Dom 1 1G |
| PIK3R1 | 25 | 7 | 82 | 56.89 | 31.30 | Dom 2 1G, 8T | 45 | 6 | 61 | 45.99 | 15.99 | Dom 2 1G, 8T |
| SP100 | 25 | 7 | 82 | 90.81 | 52.75 | Dom 2 1G, 8T | 38 | 6 | 61 | 65.62 | 21.22 | Dom 2 1G, 8T |
| PSMA4 | 25 | 7 | 82 | 793.65 | 624.81 | Dom 2 1G, 8T | 35 | 6 | 61 | 555.32 | 431.03 | Dom 2 1G, 8T |
| LY117 | 25 | 28 | 61 | 31.66 | 58.06 | Dom 1 3C | 31 | 6 | 61 | –105.65 | 49.89 | Dom 2 1G, 8T |
| LAMP2 | 25 | 7 | 82 | 122.08 | 83.86 | Dom 2 1G, 8T | 26 | 10 | 57 | 29.40 | 15.43 | Dom 2 3C, 8T |
| MAN1A1 | 25 | 7 | 82 | 203.31 | 135.54 | Dom 2 1G, 8T | 24 | 6 | 61 | 51.27 | 29.22 | Dom 2 1G, 8T |
| CBX3 | 25 | 7 | 82 | 879.37 | 694.79 | Dom 2 1G, 8T | 20 | 6 | 61 | 602.47 | 506.73 | Dom 2 1G, 8T |
| LSM2 | 25 | 7 | 82 | 295.26 | 371.56 | Dom 2 1G, 8T | 19 | 6 | 61 | 530.63 | 627.54 | Dom 2 1G, 8T |
| IMAGE145052 | 25 | 7 | 82 | 692.68 | 483.88 | Dom 2 1G, 8T | 16 | 6 | 61 | 423.96 | 360.08 | Dom 2 1G, 8T |
| M96 | 24 | 7 | 82 | 446.87 | 344.43 | Dom 2 1G, 8T | 45 | 6 | 61 | 139.07 | 81.06 | Dom 2 1G, 8T |
| TPD52 | 24 | 7 | 82 | 380.84 | 290.95 | Dom 2 1G, 8T | 34 | 13 | 54 | 33.66 | 21.92 | Dom 1 1G |
| HOXA7 | 24 | 7 | 82 | 73.36 | 45.90 | Dom 2 1G, 8T | 20 | 6 | 61 | 41.90 | 18.09 | Dom 2 1G, 8T |
| PTOV1 | 24 | 7 | 82 | 50.72 | 105.13 | Dom 2 1G, 8T | 18 | 6 | 61 | 100.32 | 168.54 | Dom 2 1G, 8T |
| SEC22L1 | 24 | 7 | 82 | 97.48 | 67.15 | Dom 2 1G, 8T | 14 | 6 | 61 | 382.09 | 313.41 | Dom 2 1G, 8T |
| ERAL1 | 24 | 7 | 82 | 126.03 | 167.87 | Dom 2 1G, 8T | 14 | 10 | 57 | 175.58 | 227.30 | Dom 2 3C, 8T |
| UBE1 | 23 | 7 | 82 | 872.23 | 1320.61 | Dom 2 1G, 8T | 33 | 6 | 61 | 743.45 | 1100.59 | Dom 2 1G, 8T |
| KIAA1096 | 23 | 7 | 82 | 204.63 | 113.40 | Dom 2 1G, 8T | 30 | 6 | 61 | 89.85 | 46.65 | Dom 2 1G, 8T |
| PTK2B | 23 | 19 | 70 | 72.90 | 108.74 | Dom 2 3C, 8T | 18 | 17 | 50 | 414.91 | 568.10 | Dom 1 3C |
| KLF12 | 23 | 7 | 82 | 19.08 | 10.06 | Dom 2 1G, 8T | 17 | 60 | 7 | 28.21 | 42.20 | Rec 1 8T |
| AGPAT1 | 23 | 7 | 82 | 263.05 | 332.71 | Dom 2 1G, 8T | 16 | 6 | 61 | 216.09 | 281.25 | Dom 2 1G, 8T |
| ATP2B1 | 23 | 7 | 82 | 122.97 | 82.57 | Dom 2 1G, 8T | 15 | 6 | 61 | 72.14 | 49.59 | Dom 2 1G, 8T |
| ACTN4 | 23 | 7 | 82 | 217.53 | 289.73 | Dom 2 1G, 8T | 15 | 10 | 57 | 400.26 | 473.67 | Dom 2 3C, 8T |
| SPTAN1 | 23 | 7 | 82 | 150.54 | 273.72 | Dom 2 1G, 8T | 13 | 6 | 61 | 184.11 | 274.55 | Dom 2 1G, 8T |
| B4GALT1 | 22 | 7 | 82 | 118.57 | 246.30 | Dom 2 1G, 8T | 38 | 6 | 61 | 345.09 | 525.29 | Dom 2 1G, 8T |
| FCER2 | 22 | 19 | 70 | 319.70 | 464.60 | Dom 2 3C, 8T | 26 | 17 | 50 | 352.03 | 504.46 | Dom 1 3C |
| SF3A2 | 22 | 7 | 82 | 145.97 | 243.76 | Dom 2 1G, 8T | 19 | 6 | 61 | 153.62 | 267.31 | Dom 2 1G, 8T |
| KIAA0790 | 22 | 28 | 61 | 52.93 | 36.26 | Dom 1 3C | 17 | 17 | 50 | 88.85 | 64.41 | Dom 1 3C |
| LYPLA2 | 22 | 7 | 82 | 256.49 | 326.80 | Dom 2 1G, 8T | 17 | 6 | 61 | 1064.54 | 902.02 | Dom 2 1G, 8T |
| WAS | 22 | 7 | 82 | 671.02 | 878.47 | Dom 2 1G, 8T | 14 | 6 | 61 | 297.13 | 515.22 | Dom 2 1G, 8T |
| SDHD | 21 | 7 | 82 | 312.15 | 226.04 | Dom 2 1G, 8T | 25 | 6 | 61 | 416.76 | 306.58 | Dom 2 1G, 8T |
| MED6 | 21 | 7 | 82 | 156.44 | 107.02 | Dom 2 1G, 8T | 24 | 6 | 61 | 80.95 | 52.61 | Dom 2 1G, 8T |
| TCF3 | 21 | 7 | 82 | 32.28 | 69.24 | Dom 2 1G, 8T | 17 | 6 | 61 | 135.62 | 190.39 | Dom 2 1G, 8T |
| B9 | 21 | 79 | 10 | 6.79 | 13.41 | Rec 1 8T | 17 | 6 | 61 | 28.51 | 58.27 | Dom 2 1G, 8T |
| TRAF1 | 21 | 19 | 70 | 323.82 | 409.29 | Dom 2 3C, 8T | 16 | 13 | 54 | 257.01 | 331.32 | Dom 1 1G |
| KIAA0102 | 21 | 7 | 82 | 1101.52 | 899.08 | Dom 2 1G, 8T | 15 | 17 | 50 | 493.29 | 428.92 | Dom 1 3C |
| ASNA1 | 21 | 7 | 82 | 217.08 | 270.80 | Dom 2 1G, 8T | 15 | 6 | 61 | 142.51 | 179.71 | Dom 2 1G, 8T |
| DXS9928E | 21 | 17 | 72 | 191.02 | 218.20 | Dom 1 1G | 14 | 17 | 50 | 0.94 | 36.33 | Dom 1 3C |
| LYN | 20 | 7 | 82 | 534.29 | 397.62 | Dom 2 1G, 8T | 37 | 6 | 61 | 278.78 | 182.39 | Dom 2 1G, 8T |
| KPNA3 | 20 | 28 | 61 | 151.68 | 123.57 | Dom 1 3C | 35 | 6 | 61 | 150.92 | 96.71 | Dom 2 1G, 8T |

| | | | | | | | | | | | | |
|---|---|---|---|---|---|---|---|---|---|---|---|---|
| ITM2B | 20 | 19 | 70 | 443.79 | 336.08 | Dom 2 3C, 8T | 26 | 17 | 50 | 309.22 | 214.92 | Dom 1 3C |
| H4FG | 20 | 7 | 82 | 516.92 | 275.65 | Dom 2 1G, 8T | 26 | 6 | 61 | 852.61 | 361.20 | Dom 2 1G, 8T |
| MAP2K2 | 20 | 7 | 82 | 100.60 | 139.64 | Dom 2 1G, 8T | 25 | 6 | 61 | 335.44 | 450.22 | Dom 2 1G, 8T |
| VBP1 | 20 | 7 | 82 | 607.83 | 505.78 | Dom 2 1G, 8T | 25 | 6 | 61 | 286.15 | 218.74 | Dom 2 1G, 8T |
| RANGAP1 | 20 | 7 | 82 | 100.16 | 152.11 | Dom 2 1G, 8T | 23 | 6 | 61 | 189.53 | 296.17 | Dom 2 1G, 8T |
| UBA2 | 20 | 7 | 82 | 345.53 | 239.56 | Dom 2 1G, 8T | 23 | 6 | 61 | 463.25 | 371.83 | Dom 2 1G, 8T |
| RAB14 | 20 | 7 | 82 | 223.42 | 149.81 | Dom 2 1G, 8T | 22 | 6 | 61 | 218.68 | 177.86 | Dom 2 1G, 8T |
| LRPAP1 | 20 | 28 | 61 | 132.32 | 101.84 | Dom 1 3C | 20 | 6 | 61 | 165.64 | 112.71 | Dom 2 1G, 8T |
| SURF5 | 20 | 7 | 82 | 78.95 | 103.43 | Dom 2 1G, 8T | 18 | 6 | 61 | 89.91 | 131.15 | Dom 2 1G, 8T |
| RECQL5 | 20 | 7 | 82 | 29.68 | 57.12 | Dom 2 1G, 8T | 18 | 6 | 61 | 45.25 | 26.36 | Dom 2 1G, 8T |
| UCP2 | 20 | 7 | 82 | 428.75 | 575.50 | Dom 2 1G, 8T | 17 | 6 | 61 | 633.05 | 850.88 | Dom 2 1G, 8T |
| SFRS2 | 20 | 7 | 82 | 1136.15 | 865.87 | Dom 2 1G, 8T | 15 | 60 | 7 | 1295.04 | 1507.22 | Rec 1 8T |
| KRT10 | 20 | 19 | 70 | 319.79 | 281.88 | Dom 2 3C, 8T | 14 | 13 | 54 | 271.26 | 232.89 | Dom 1 1G |
| RAB33A | 20 | 28 | 61 | 107.51 | 133.89 | Dom 1 3C | 13 | 17 | 50 | 60.06 | 75.90 | Dom 1 3C |
| KIAA0928 | 19 | 7 | 82 | 162.69 | 100.22 | Dom 2 1G, 8T | 42 | 6 | 61 | 52.17 | 20.21 | Dom 2 1G, 8T |
| DAD1 | 19 | 19 | 70 | 696.30 | 630.68 | Dom 2 3C, 8T | 38 | 10 | 57 | 540.73 | 461.09 | Dom 2 3C, 8T |
| ATP5D | 19 | 7 | 82 | 304.93 | 383.53 | Dom 2 1G, 8T | 30 | 6 | 61 | 105.39 | 196.80 | Dom 2 1G, 8T |
| SLC26A2 | 19 | 7 | 82 | 39.04 | 24.62 | Dom 2 1G, 8T | 20 | 6 | 61 | 14.17 | 4.09 | Dom 2 1G, 8T |
| CD22 | 19 | 19 | 70 | 359.09 | 448.57 | Dom 2 3C, 8T | 20 | 10 | 57 | 538.52 | 686.63 | Dom 2 3C, 8T |
| SON | 19 | 7 | 82 | 641.44 | 532.51 | Dom 2 1G, 8T | 17 | 6 | 61 | 50.18 | 33.90 | Dom 2 1G, 8T |
| RBBP2 | 19 | 7 | 82 | 149.69 | 120.13 | Dom 2 1G, 8T | 14 | 6 | 61 | 68.20 | 48.85 | Dom 2 1G, 8T |
| MAP4K1 | 19 | 7 | 82 | 329.62 | 465.81 | Dom 2 1G, 8T | 14 | 10 | 57 | 519.03 | 630.81 | Dom 2 3C, 8T |
| IRAK1 | 19 | 79 | 10 | 491.71 | 654.36 | Rec 1 8T | 14 | 10 | 57 | 174.85 | 222.16 | Dom 2 3C, 8T |
| TLOC1 | 19 | 7 | 82 | 485.95 | 375.58 | Dom 2 1G, 8T | 13 | 6 | 61 | 52.54 | 37.48 | Dom 2 1G, 8T |
| GOSR1 | 18 | 7 | 82 | 212.78 | 155.35 | Dom 2 1G, 8T | 39 | 6 | 61 | 255.02 | 161.45 | Dom 2 1G, 8T |
| PROSC | 18 | 7 | 82 | 74.62 | 51.72 | Dom 2 1G, 8T | 38 | 6 | 61 | 156.54 | 105.52 | Dom 2 1G, 8T |
| BAK1 | 18 | 17 | 72 | 127.30 | 151.78 | Dom 1 1G | 28 | 6 | 61 | 30.63 | 89.82 | Dom 2 1G, 8T |
| FALZ | 18 | 7 | 82 | 89.01 | 124.94 | Dom 2 1G, 8T | 28 | 6 | 61 | 176.46 | 91.58 | Dom 2 1G, 8T |
| KIAA1054 | 18 | 7 | 82 | 9.10 | 18.14 | Dom 2 1G, 8T | 27 | 13 | 54 | –14.39 | –6.77 | Dom 1 1G |
| PDXK | 18 | 19 | 70 | 279.66 | 327.25 | Dom 2 3C, 8T | 27 | 6 | 61 | 222.18 | 332.08 | Dom 2 1G, 8T |
| ARF6 | 18 | 7 | 82 | 678.91 | 498.41 | Dom 2 1G, 8T | 27 | 10 | 57 | 356.45 | 249.41 | Dom 2 3C, 8T |
| TBPL1 | 18 | 7 | 82 | 180.77 | 138.48 | Dom 2 1G, 8T | 25 | 6 | 61 | 28.81 | 3.25 | Dom 2 1G, 8T |
| MPST | 18 | 7 | 82 | 151.82 | 213.02 | Dom 2 1G, 8T | 24 | 10 | 57 | 165.74 | 222.37 | Dom 2 3C, 8T |
| TBCC | 18 | 7 | 82 | 121.67 | 94.31 | Dom 2 1G, 8T | 23 | 6 | 61 | 141.00 | 102.35 | Dom 2 1G, 8T |
| MAN2A1 | 18 | 7 | 82 | 60.96 | 38.51 | Dom 2 1G, 8T | 23 | 6 | 61 | 83.63 | 48.06 | Dom 2 1G, 8T |
| SLC29A1 | 18 | 7 | 82 | 102.91 | 143.02 | Dom 2 1G, 8T | 22 | 10 | 57 | 150.41 | 209.23 | Dom 2 3C, 8T |
| LSP1 | 18 | 7 | 82 | 303.66 | 501.44 | Dom 2 1G, 8T | 21 | 6 | 61 | 550.89 | 786.44 | Dom 2 1G, 8T |
| DJ971N18.2 | 18 | 7 | 82 | 135.19 | 84.94 | Dom 2 1G, 8T | 21 | 6 | 61 | 49.10 | 22.45 | Dom 2 1G, 8T |
| HMG14 | 18 | 7 | 82 | 1688.03 | 1437.02 | Dom 2 1G, 8T | 20 | 17 | 50 | 1256.86 | 1109.66 | Dom 1 3C |
| NME4 | 18 | 7 | 82 | 217.22 | 337.61 | Dom 2 1G, 8T | 20 | 6 | 61 | 218.16 | 368.71 | Dom 2 1G, 8T |
| GRN | 18 | 7 | 82 | 376.82 | 484.85 | Dom 2 1G, 8T | 17 | 17 | 50 | 285.33 | 374.83 | Dom 1 3C |
| ITGB1 | 18 | 7 | 82 | 261.37 | 176.26 | Dom 2 1G, 8T | 17 | 6 | 61 | 153.24 | 100.87 | Dom 2 1G, 8T |
| RAB5C | 18 | 7 | 82 | 112.38 | 169.24 | Dom 2 1G, 8T | 15 | 6 | 61 | 82.04 | 182.88 | Dom 2 1G, 8T |
| ING3 | 18 | 7 | 82 | 166.68 | 125.50 | Dom 2 1G, 8T | 14 | 6 | 61 | 39.70 | 27.10 | Dom 2 1G, 8T |
| KIAA0964 | 18 | 7 | 82 | 44.48 | 66.51 | Dom 2 1G, 8T | 13 | 6 | 61 | 143.41 | 195.55 | Dom 2 1G, 8T |

*(Table continues on following pages.)*

**Table 6.** (*Continued*)

| RNA level | Adj P (Original) | N (Marker+) | N (Marker–) | Mean (Marker+) | Mean (Marker–) | Best marker (Original) | Adj P (Replication) | N (Marker+) | N (Marker–) | Mean (Marker+) | Mean (Marker–) | Best marker (Replication) |
|---|---|---|---|---|---|---|---|---|---|---|---|---|
| ARF4 | 18 | 7 | 82 | 578.19 | 454.61 | Dom 2 1G, 8T | 13 | 10 | 57 | 331.13 | 273.62 | Dom 2 3C, 8T |
| TANK | 17 | 7 | 82 | 451.57 | 363.01 | Dom 2 1G, 8T | 26 | 6 | 61 | 55.95 | 33.42 | Dom 2 1G, 8T |
| DKFZP564B167 | 17 | 7 | 82 | 673.11 | 535.40 | Dom 2 1G, 8T | 25 | 6 | 61 | 384.51 | 272.96 | Dom 2 1G, 8T |
| LAMP1 | 17 | 7 | 82 | 18.18 | 8.93 | Dom 2 1G, 8T | 22 | 6 | 61 | 723.98 | 607.85 | Dom 2 1G, 8T |
| CHD1 | 17 | 7 | 82 | 176.71 | 132.19 | Dom 2 1G, 8T | 20 | 6 | 61 | 172.78 | 131.06 | Dom 2 1G, 8T |
| TYK2 | 17 | 7 | 82 | 184.07 | 246.01 | Dom 2 1G, 8T | 18 | 10 | 57 | 183.82 | 253.74 | Dom 2 3C, 8T |
| HNRPH1 | 17 | 7 | 82 | 1756.50 | 1339.86 | Dom 2 1G, 8T | 17 | 6 | 61 | 985.84 | 782.27 | Dom 2 1G, 8T |
| PRKCL1 | 17 | 19 | 70 | 88.61 | 113.32 | Dom 2 3C, 8T | 17 | 10 | 57 | 114.48 | 160.22 | Dom 2 3C, 8T |
| SSR2 | 17 | 17 | 72 | 1085.48 | 949.54 | Dom 1 1G | 16 | 17 | 50 | 813.77 | 722.99 | Dom 1 3C |
| EVI2A | 17 | 7 | 82 | 266.61 | 193.69 | Dom 2 1G, 8T | 16 | 6 | 61 | 92.14 | 67.80 | Dom 2 1G, 8T |
| TAZ | 17 | 17 | 72 | 81.75 | 99.89 | Dom 1 1G | 16 | 10 | 57 | 132.83 | 155.94 | Dom 2 3C, 8T |
| STAT5B | 17 | 19 | 70 | 46.34 | 58.18 | Dom 2 3C, 8T | 16 | 10 | 57 | 10.98 | 28.48 | Dom 2 3C, 8T |
| HSXIAPAF1 | 17 | 7 | 82 | 133.63 | 82.68 | Dom 2 1G, 8T | 15 | 6 | 61 | –0.50 | –31.49 | Dom 2 1G, 8T |
| XBP1 | 17 | 7 | 82 | 721.58 | 475.25 | Dom 2 1G, 8T | 13 | 17 | 50 | 1442.19 | 1163.47 | Dom 1 3C |
| PCK2 | 16 | 7 | 82 | 167.11 | 211.02 | Dom 2 1G, 8T | 35 | 6 | 61 | 116.06 | 187.19 | Dom 2 1G, 8T |
| FLJ12619 | 16 | 7 | 82 | 287.71 | 213.72 | Dom 2 1G, 8T | 34 | 6 | 61 | 88.91 | 53.81 | Dom 2 1G, 8T |
| NDUFB5 | 16 | 7 | 82 | 618.27 | 506.63 | Dom 2 1G, 8T | 33 | 6 | 61 | 392.25 | 288.34 | Dom 2 1G, 8T |
| MBD3 | 16 | 7 | 82 | 181.65 | 237.30 | Dom 2 1G, 8T | 25 | 6 | 61 | 93.82 | 205.93 | Dom 2 1G, 8T |
| KIAA0878 | 16 | 7 | 82 | 81.90 | 53.25 | Dom 2 1G, 8T | 25 | 6 | 61 | 11.22 | 3.22 | Dom 2 1G, 8T |
| SHOC2 | 16 | 7 | 82 | 344.65 | 289.88 | Dom 2 1G, 8T | 24 | 6 | 61 | 319.69 | 251.14 | Dom 2 1G, 8T |
| PTP4A1 | 16 | 7 | 82 | 208.64 | 159.79 | Dom 2 1G, 8T | 22 | 6 | 61 | 195.26 | 146.41 | Dom 2 1G, 8T |
| VARS2 | 16 | 19 | 70 | 26.53 | 44.58 | Dom 2 3C, 8T | 19 | 10 | 57 | 288.38 | 364.01 | Dom 2 3C, 8T |
| MVD | 16 | 7 | 82 | 22.57 | 48.30 | Dom 2 1G, 8T | 18 | 13 | 54 | –27.17 | 25.28 | Dom 1 1G |
| KIAA0101 | 16 | 7 | 82 | 1603.67 | 1312.44 | Dom 2 1G, 8T | 17 | 13 | 54 | 662.64 | 572.18 | Dom 1 1G |
| HMG20B | 16 | 79 | 10 | 67.70 | 91.99 | Rec 1 8T | 17 | 6 | 61 | 84.49 | 134.31 | Dom 2 1G, 8T |
| TGFB1 | 16 | 7 | 82 | 207.69 | 322.11 | Dom 2 1G, 8T | 17 | 10 | 57 | 55.99 | 75.83 | Dom 2 3C, 8T |
| MSH5 | 16 | 19 | 70 | 43.94 | 55.50 | Dom 2 3C, 8T | 17 | 10 | 57 | 0.17 | 25.62 | Dom 2 3C, 8T |
| DNMT2 | 16 | 28 | 61 | 27.01 | 21.05 | Dom 1 3C | 16 | 13 | 54 | 14.89 | 6.41 | Dom 1 1G |
| HMG2 | 16 | 7 | 82 | 1304.32 | 1074.41 | Dom 2 1G, 8T | 16 | 17 | 50 | 766.05 | 668.21 | Dom 1 3C |
| SLC35A3 | 16 | 7 | 82 | 47.16 | 33.64 | Dom 2 1G, 8T | 16 | 6 | 61 | 122.14 | 92.50 | Dom 2 1G, 8T |
| DGKA | 16 | 28 | 61 | 29.61 | 36.27 | Dom 1 3C | 13 | 17 | 50 | 157.17 | 183.05 | Dom 1 3C |
| FXR2 | 16 | 7 | 82 | 73.05 | 95.12 | Dom 2 1G, 8T | 13 | 10 | 57 | 89.47 | 118.81 | Dom 2 3C, 8T |
| MADH1 | 15 | 7 | 82 | 196.98 | 124.35 | Dom 2 1G, 8T | 39 | 6 | 61 | 128.59 | 67.71 | Dom 2 1G, 8T |
| CCR6 | 15 | 19 | 70 | 76.12 | 53.54 | Dom 2 3C, 8T | 31 | 6 | 61 | 56.52 | 25.55 | Dom 2 1G, 8T |
| PMS2L11 | 15 | 7 | 82 | 203.14 | 294.03 | Dom 2 1G, 8T | 23 | 10 | 57 | 143.71 | 207.23 | Dom 2 3C, 8T |
| GCN5L2 | 15 | 17 | 72 | 102.21 | 120.70 | Dom 1 1G | 22 | 6 | 61 | 187.60 | 377.80 | Dom 2 1G, 8T |
| KIAA0607 | 15 | 7 | 82 | 46.53 | 68.84 | Dom 2 1G, 8T | 22 | 6 | 61 | –318.90 | –198.25 | Dom 2 1G, 8T |
| JUND | 15 | 7 | 82 | 13.60 | 37.32 | Dom 2 1G, 8T | 21 | 13 | 54 | 115.86 | 216.32 | Dom 1 1G |

|  |  |  |  |  |  |  |  |  |  |  |  |  |
|---|---|---|---|---|---|---|---|---|---|---|---|---|
| KIAA1049 | 15 | 7 | 82 | 379.11 | 463.96 | Dom 2 1G, 8T | 20 | 6 | 61 | 176.30 | 275.71 | Dom 2 1G, 8T |
| EPOR | 15 | 12 | 77 | 416.46 | 321.89 | Dom 2 1A, 3C | 19 | 6 | 61 | 404.34 | 308.82 | Dom 2 1G, 8T |
| CACNB1 | 15 | 7 | 82 | 21.99 | 38.44 | Dom 2 1G, 8T | 18 | 17 | 50 | 53.50 | 68.54 | Dom 1 3C |
| C1D | 15 | 7 | 82 | 151.22 | 107.29 | Dom 2 1G, 8T | 16 | 6 | 61 | 549.27 | 457.70 | Dom 2 1G, 8T |
| BTBD2 | 15 | 7 | 82 | 17.09 | 48.73 | Dom 2 1G, 8T | 15 | 10 | 57 | 37.28 | 67.06 | Dom 2 3C, 8T |
| BATF | 15 | 19 | 70 | 348.94 | 423.90 | Dom 2 3C, 8T | 14 | 17 | 50 | 126.86 | 155.49 | Dom 1 3C |
| DKFZP586D0623 | 15 | 7 | 82 | 186.44 | 145.08 | Dom 2 1G, 8T | 14 | 6 | 61 | 108.92 | 82.07 | Dom 2 1G, 8T |
| ARNT | 15 | 79 | 10 | 28.52 | 20.70 | Rec 1 8T | 14 | 6 | 61 | 49.23 | 70.47 | Dom 2 1G, 8T |
| FANCA | 15 | 79 | 10 | 35.14 | 49.17 | Rec 1 8T | 13 | 6 | 61 | −15.53 | 11.76 | Dom 2 1G, 8T |
| ZNF161 | 14 | 7 | 82 | 97.51 | 78.17 | Dom 2 1G, 8T | 40 | 6 | 61 | 59.51 | 29.97 | Dom 2 1G, 8T |
| MAP3K14 | 14 | 7 | 82 | 81.86 | 107.29 | Dom 2 1G, 8T | 26 | 10 | 57 | −25.28 | 14.64 | Dom 2 3C, 8T |
| ENIGMA | 14 | 7 | 82 | 34.24 | 52.59 | Dom 2 1G, 8T | 24 | 10 | 57 | 33.67 | 59.47 | Dom 2 3C, 8T |
| TMOD | 14 | 28 | 61 | 82.15 | 116.55 | Dom 1 3C | 21 | 17 | 50 | 10.39 | 46.70 | Dom 1 3C |
| KIAA0870 | 14 | 19 | 70 | 59.79 | 83.20 | Dom 2 3C, 8T | 21 | 10 | 57 | 71.66 | 111.95 | Dom 2 3C, 8T |
| MGC10433 | 14 | 7 | 82 | 146.09 | 184.46 | Dom 2 1G, 8T | 20 | 6 | 61 | 100.89 | 168.08 | Dom 2 1G, 8T |
| MAPKAPK2 | 14 | 7 | 82 | 45.00 | 69.34 | Dom 2 1G, 8T | 19 | 6 | 61 | 70.95 | 124.58 | Dom 2 1G, 8T |
| FKBP1A | 14 | 7 | 82 | 3669.97 | 4704.30 | Dom 2 1G, 8T | 19 | 6 | 61 | 663.89 | 825.78 | Dom 2 1G, 8T |
| PPP3CC | 14 | 7 | 82 | 47.96 | 37.67 | Dom 2 1G, 8T | 18 | 6 | 61 | 131.09 | 105.32 | Dom 2 1G, 8T |
| CCND2 | 14 | 17 | 72 | 398.55 | 326.07 | Dom 1 1G | 17 | 6 | 61 | 499.56 | 390.86 | Dom 2 1G, 8T |
| GALNT1 | 14 | 7 | 82 | 186.33 | 143.75 | Dom 2 1G, 8T | 17 | 6 | 61 | 108.96 | 75.90 | Dom 2 1G, 8T |
| CEP2 | 14 | 79 | 10 | 39.46 | 57.50 | Rec 1 8T | 15 | 10 | 57 | 33.34 | 55.78 | Dom 2 3C, 8T |
| SLC7A8 | 14 | 7 | 82 | 28.83 | 46.08 | Dom 2 1G, 8T | 13 | 6 | 61 | 7.37 | −20.32 | Dom 2 1G, 8T |
| HSPCB | 13 | 7 | 82 | 2931.52 | 3686.73 | Dom 2 1G, 8T | 34 | 6 | 61 | 147.21 | 97.04 | Dom 2 1G, 8T |
| NCOA3 | 13 | 7 | 82 | 157.80 | 119.82 | Dom 2 1G, 8T | 28 | 6 | 61 | 181.13 | 114.56 | Dom 2 1G, 8T |
| FLJ20244 | 13 | 7 | 82 | 50.46 | 78.81 | Dom 2 1G, 8T | 27 | 10 | 57 | −34.51 | 45.97 | Dom 2 3C, 8T |
| TIEG | 13 | 7 | 82 | 251.69 | 172.33 | Dom 2 1G, 8T | 25 | 6 | 61 | 156.50 | 105.64 | Dom 2 1G, 8T |
| COL5A2 | 13 | 79 | 10 | 6.09 | 10.72 | Rec 1 8T | 20 | 17 | 50 | 15.87 | 4.98 | Dom 1 3C |
| CDC2 | 13 | 7 | 82 | 347.54 | 258.69 | Dom 2 1G, 8T | 19 | 6 | 61 | 242.03 | 181.95 | Dom 2 1G, 8T |
| ARPP-19 | 13 | 7 | 82 | 127.34 | 104.48 | Dom 2 1G, 8T | 18 | 6 | 61 | 111.14 | 84.98 | Dom 2 1G, 8T |
| HNRPH2 | 13 | 7 | 82 | 173.24 | 128.27 | Dom 2 1G, 8T | 18 | 6 | 61 | 316.11 | 240.91 | Dom 2 1G, 8T |
| SIL | 13 | 79 | 10 | 93.53 | 74.48 | Rec 1 8T | 18 | 6 | 61 | 70.93 | 50.22 | Dom 2 1G, 8T |
| PTCH | 13 | 7 | 82 | 73.76 | 55.25 | Dom 2 1G, 8T | 18 | 10 | 57 | −11.54 | 10.06 | Dom 2 3C, 8T |
| PPP2R4 | 13 | 7 | 82 | 132.82 | 182.13 | Dom 2 1G, 8T | 17 | 10 | 57 | 196.72 | 330.30 | Dom 2 3C, 8T |
| AIP | 13 | 19 | 70 | 202.03 | 236.73 | Dom 2 3C, 8T | 15 | 6 | 61 | 23.01 | 42.73 | Dom 2 1G, 8T |
| CASP1 | 13 | 7 | 82 | 77.50 | 52.80 | Dom 2 1G, 8T | 14 | 10 | 57 | 67.73 | 46.30 | Dom 2 3C, 8T |
| CHS1 | 13 | 79 | 10 | 41.34 | 26.76 | Rec 1 8T | 14 | 60 | 7 | 17.31 | 2.38 | Rec 1 8T |
| PSMB1 | 13 | 7 | 82 | 1443.91 | 1261.37 | Dom 2 1G, 8T | 13 | 6 | 61 | 1694.28 | 1432.72 | Dom 2 1G, 8T |

For each gene, the adjusted significance score, the sizes of the Marker(+) and Marker(−) groups, their mean expression values, and the description of the best marker are given in both the original and the replication dat aset.

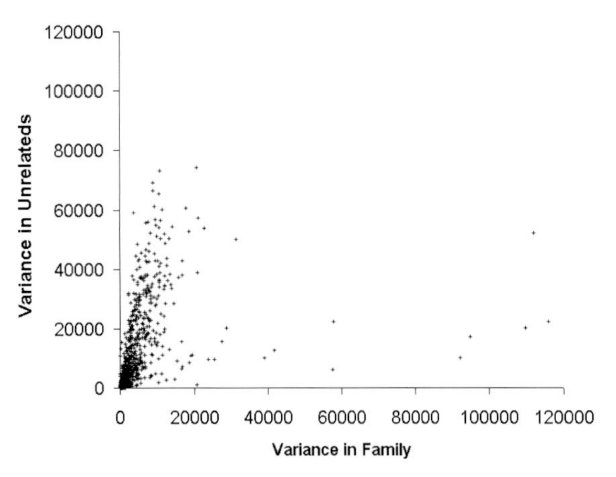


**Figure 3.** Variance comparison between one CEPH family and the unrelated individuals.

average levels of expression and variance for these 5 genes in our panel. We did not find these genes to have significantly higher variance compared to all the other genes.

We also looked for correspondences in two recent studies that examined *cis*-associations. Yan et al. (2002) described 6 genes with *cis*-acting associations. Five of these were found in our database: CAPN10, CARD15, CAT, SP100, and TP73. Of these, we find significant (adjusted $p < 0.05$) *cis*-association in CAT and SP100, and only CAT is reproduced in our confirmation data set. These results are shown in Table 7. There is good evidence for CARD15, but it is just below our significance threshold of 13 ($p = 0.05$). Overall, to find 2 out of 5 genes confirmed in one experiment and find 1 of these also confirmed in another constitutes significant concordance between their study and ours.

Rockman and Wray (2002) performed an extensive literature survey of *cis*-associations derived using a variety of experimental techniques and cell types. They identified 106 genes showing reproducible *cis*-effects. Of these genes, 93 have genotypes and expression data in our data set. Only four genes show significant *cis*-associations in our data set: CAT, IL1A, LPL, and MPO. Only CAT, which has been discussed above, can also be confirmed in our second experiment. Because of the number of tests, none of our results can be interpreted to confirm the findings of Rockman and Wray, with the exception of CAT, which has evidence from multiple sources. This is not much of a surprise, nor should it diminish their work, since their findings were based on a variety of experiments, most of which differ greatly from ours in terms of cell type, experimental conditions, and the particular polymorphisms examined.

**Table 7.** *cis*-Association Significance Scores for the Genes Studied by Yan et al. (2002)

| Gene | Adj. Sig. (Orig) | Adj. Sig. (Rep) | Raw Sig. (Orig) | Raw Sig. (Rep) |
|---|---|---|---|---|
| CAPN10 | 2 | - | 10 | - |
| CARD15 | 12 | - | 26 | - |
| CAT | 25 | 17 | 40 | 29 |
| SP100 | 15 | 1 | 28 | 11 |
| TP73 | 0 | - | 4 | - |

Adjusted and raw significance scores are given for both the primary and the replication data sets. Only CAT and SP100 were tested in the replication data set.

## ZNF 295 as a Potential General Transcription Factor

Judging the significance of the findings on ZNF295 is more difficult than in the case of *cis*-associations, because ZNF295 has been selected out of a larger set partly on the merit of its significance scores, biasing the $p$ values. However, comparison of the results with the independent replication experiment in Table 3 shows that many more positives are confirmed than would be expected if they were false. Figure 4 shows how the percentage of confirmed positives increases with the significance score. The expected confirmation rate for the null hypothesis of no association (at $p \leq 0.05$) is also shown (dashed line).

One possible explanation of the association of a large number of expression values to a single gene is that there is some underlying population substructure confounding the observation. As described previously, this problem has been addressed by using ethnicity and family membership as covariates, but it is possible that hidden substructure remains. We examined the individuals who were in the Marker(+) and Marker(–) groups and found no obvious patterns. For instance, the marker seen most often (GT at SNPs 1 and 8) had 7 individuals in the Marker(+) group. Of these, 2 are African-American, 2 are Asian, 1 is Caucasian, and 2 are Hispanic Latino. Five are male and 2 are female. None of these individuals is related to one another. There is no age bias between the individuals with and without the markers. The other markers for which there were associations also showed no obvious bias due to population substructure.

ZNF295 is a BTB/POZ domain zinc finger protein in humans with hitherto unassigned function. The BTB/POZ domain has very high levels of homology with other BTB class proteins such as PZLF, BAB1, and BCL6, and the zinc finger domains are of the C2H2 type found in glucocorticoid hormones (Iuchi 2001). The BTB domain is an evolutionarily conserved protein–protein interaction domain present in the amino terminus of several transcription factors involved in development, chromatin modeling, and human cancers (Melnick et al. 2002; T'Jampens et al. 2002). This domain is involved in a number of biological functions such as DNA binding, regulation of gene transcription, and organization of macromolecular structures such as chromatin and cell adhesion matrices. The BTB domain of the BCL6 oncoprotein binds to SMRT, the corepressor of unliganded retinoic acid receptors and thyroid receptors (Dhordain et al. 1997). The *Drosophila* BAB1 and BAB2 genes are developmentally regulated transcription factors that have been found to interact with dTAF(II)155, a TATA-box protein-associated factor (Pointud et al. 2001; Couderc et al. 2002). Residues in the BTB/POZ domain of the Tramtrack 69 (Ttk69) protein are important for Ttk69 activity; single amino acid changes in the domain abolish transcriptional repression activity (Wen et al. 2000). Furthermore, the zinc fingers in the carboxyl terminus of the Ttk69 protein were also important for DNA binding and function. ZNF295 is expressed in a wide variety of tissues (Unigene cluster Hs.157079) and is also conserved in mouse (ZNP295, 83.7% identity, Homologene) and in rat (LOC304056, 85.1% identity, Homologene).

One interesting observation (Fig. 2) is that the direction of change of the expression levels is not the same for all genes. In some cases, the Marker(+) group has higher expression levels than the Marker(–) group, whereas in others, the opposite is seen. One can speculate that the coding changes seen in this marker are directly responsible for altering the activity of a transcription complex that includes ZNF295. If this is the case, one would assume that the sign of this activity change would be consistent across transcripts regulated by the complex. However, it could also be true that some of these associations are second-order effects; i.e., ZNF295 is responsible for the transcription of some number of the associated genes, but the remaining associated genes are themselves induced by the directly transcribed genes. We prefer the second explanation because, as discussed below, many associations of ZNF295 haplotypes are with transcription factors, coactivators and genes involved in transcriptional regulation. This induction could be positively or negatively affected by changes in expression levels of the original genes.

There are several interesting observations to be made in the set of genes whose expression levels are associated with markers in ZNF295. There is a statistically significant excess of observed genes on Chromosomes 17 and 22. Of the transcripts on Chromosome 22, several cluster on sub-band 22q13.1. As shown in Table 6, ZNF295 haplotype markers are associated with 36 genes that are involved in transcriptional regulation; have DNA-binding properties; or are involved in DNA repair and metabolism. Expression levels of 14 genes that are involved in RNA processing or that have RNA-binding properties are also associated to markers within the ZNF295 gene. Furthermore, expression levels of several members of very important classes of genes such as GTPases (10 genes), immune response genes (10 genes), ATPase or ATP-binding (9 genes), genes involved in cell cycle or cell division or cell–cell signaling (10 genes), protein kinases (13 genes), protein trafficking (11 genes), genes involved in protein binding and folding (13 genes), genes involved in signal transduction (11 genes), and genes involved in apoptosis (5 genes) are highly significantly and reproducibly associated with markers within the ZNF295 gene. One example that illustrates the impact of ZNF295 haplotypes on expression levels of potential target genes is the association that links ZNF295 with expression levels of the transcription factor SP100. The 7 individuals in the Marker(+) group have a mean expression level for SP100 of 90.8, whereas the 82 individuals in the Marker(–) group have a mean expression level of 52.8. The raw and permutation-adjusted $p$ values for the association are 0.00024 and 0.0034, respectively. In the replication experiment, the Marker(+) and Marker(–) groups had 6 and 61 individuals, respectively. The respective means were 65.6 and 21.2, and the adjusted $p$ value was 0.00016. Three other instances are noteworthy because of the involvement of genes that have the POZ domain; the association between markers in the ZNF295 gene and expression levels in the ENIGMA, BTBD2, and TAZ genes (Guy et al. 1999; Kanai et al. 2000; Carim-Todd et al. 2001). The BTBD2 and the TAZ genes are involved in transcriptional regulation, whereas the ENIGMA gene is involved in receptor-mediated endocytosis.

Another noteworthy feature of our data is that the two amino acid changes (Asn181Ser and Lys218Gln) fall outside of the POZ and the zinc finger domains. Classic mutational approaches (such as the changes in the POZ domain of the Tramtrack protein mentioned above) usually abolish transcriptional activity such that subtle relationships between members of the transcriptional machinery are not revealed. By exploiting the naturally occurring variants in ZNF295, we have not only obtained clues to the biological function of ZNF295 itself, but also to how it is networked into the larger transcriptional machinery and the extent of its influence on the different cellular processes. These results demonstrate the utility of combining expression analyses with genetic analyses (such as the use of haplotype markers) to unravel gene networks and discover gene functions.

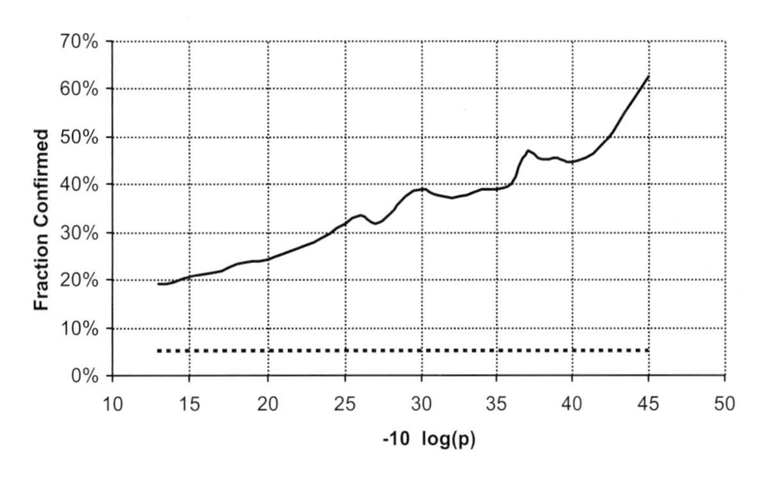


**Figure 4.** Fraction of associations with ZNF295 that are confirmed ($p \leq 0.05$ in the replication experiment) as a function of $p$ value (adjusted) in the original experiment. The dashed line represents the expectation under the null hypothesis of no confirmation.

Given our results and all the evidence discussed above, it is highly likely that ZNF295 is a general transcription factor and is a very good candidate for more detailed experimental approaches to investigate its transcriptional activity, target genes, and possible role in cellular growth and differentiation.

### Use of Genetic Associations for the Inference of Regulatory Networks

There have been many attempts to deduce regulatory networks from expression data (Gat-Viks and Shamir 2003; Johansson et al. 2003; Kwon et al. 2003; Segal et al. 2003). A major reason for this being such a hard task is that correlated expression levels do not provide information on the causality of the correlation. If gene A and gene B are found to be correlated, this could be due to any of the five following mechanisms: (1) A regulates B, (2) B regulates A, (3) A regulates X which regulates B, (4) B regulates X which regulates A, (5) X regulates both A and B. To resolve the resulting ambiguities, it is necessary to collect expression data as time series, on a sample that is not in a steady state. Alternatively, causality information can be obtained by actively disturbing the system; e.g., by knocking out genes or interfering with transcription in other ways. Both of these methods require many measurements under carefully controlled conditions, followed by complex analysis.

Genetic associations with expression provide a much more direct picture of the causality of regulation, since causation has to go from genotype to phenotype. In the same terms as above, if the genotype of gene A is found associated with the expression of gene B, there are now only two possibilities: (1) A regulates B and (2) A regulates X regulates B. Furthermore, only one measurement is needed, and the sample can be in a steady state. One way of looking at it is that the role of the perturbation necessary for deducing causality is played by the naturally occurring variations in the genome. This opens the possibility that studies such as the one described here will be a useful tool for the discovery of gene–gene interactions in the emerging field of Systems Biology.

## ACKNOWLEDGMENTS

We thank the entire team at Genaissance Pharmaceuticals for their valuable contributions to building the Hap database that made this work possible.

## REFERENCES

Ayoubi T.A. and Van De Ven W.J. 1996. Regulation of gene expression by alternative promoters. *FASEB J.* **10:** 453.

Beaufrere L., Rieu S., Hache J.C., Dumur V., Claustres M., and Tuffery S. 1998. Altered rep-1 expression due to substitution at position +3 of the IVS13 splice-donor site of the choroideremia (CHM) gene. *Curr. Eye Res.* **17:** 726.

Brem R.B., Yvert G., Clinton R., and Kruglyak L. 2002. Genetic dissection of transcriptional regulation in budding yeast. *Science* **296:** 752.

Brookes A.J., Lehvaslaiho H., Siegfried M., Boehm J.G., Yuan Y.P., Sarkar C.M., Bork P., and Ortigao F. 2000. HGBASE: A database of SNPs and other variations in and around human genes. *Nucleic Acids Res.* **28:** 356.

Brown C.C. and Fears T.R. 1981. Exact significance levels for multiple binomial testing with application to carcinogenicity screens. *Biometrics* **37:** 763.

Carim-Todd L., Sumoy L., Andreu N., Estivill X., and Escarceller M. 2001. Identification and characterization of BTBD1, a novel BTB domain containing gene on human chromosome 15q24. *Gene* **262:** 275.

Cavalieri D., Townsend J.P., and Hartl D.L. 2000. Manifold anomalies in gene expression in a vineyard isolate of *Saccharomyces cerevisiae* revealed by DNA microarray analysis. *Proc. Natl. Acad. Sci.* **97:** 12369.

Cheung V.G. and Spielman R.S. 2002. The genetics of variation in gene expression. *Nat. Genet.* (suppl.) **32:** 522.

Cheung V.G., Conlin L.K., Weber T.M., Arcaro M., Jen K.Y., Morley M., and Spielman R.S. 2003. Natural variation in human gene expression assessed in lymphoblastoid cells. *Nat. Genet.* **33:** 422.

Couderc J.L., Godt D., Zollman S., Chen J., Li M., Tiong S., Cramton S.E., Sahut-Barnola I., and Laski F.A. 2002. The bric a brac locus consists of two paralogous genes encoding BTB/POZ domain proteins and acts as a homeotic and morphogenetic regulator of imaginal development in *Drosophila. Development* **129:** 2419.

Cramer P., Caceres J.F., Cazalla D., Kadener S., Muro A.F., Baralle F.E., and Kornblihtt A.R. 1999. Coupling of transcription with alternative splicing: RNA pol II promoters modulate SF2/ASF and 9G8 effects on an exonic splicing enhancer. *Mol. Cell* **4:** 251.

Deweindt C., Albagli O., Bernardin F., Dhordain P., Quief S., Lantoine D., Kerckaert J.P., and Leprince D. 1995. The LAZ3/BCL6 oncogene encodes a sequence-specific transcriptional inhibitor: A novel function for the BTB/POZ domain as an autonomous repressing domain. *Cell Growth Differ.* **6:** 1495.

Dhordain P., Albagli O., Lin R.J., Ansieau S., Quief S., Leutz A., Kerckaert J.P., Evans R.M., and Leprince D. 1997. Corepressor SMRT binds the BTB/POZ repressing domain of the LAZ3/BCL6 oncoprotein. *Proc. Natl. Acad. Sci.* **94:** 10762.

Dupont A., Fontana P., Bachelot-Loza C., Reny J.L., Bieche I., Desvard F., Aiach M., and Gaussem P. 2003. An intronic polymorphism in the PAR-1 gene is associated with platelet receptor density and the response to SFLLRN. *Blood* **101:** 1833.

Fredman D., Siegfried M., Yuan Y.P., Bork P., Lehvaslaiho H., and Brookes A.J. 2002. HGVbase: A human sequence variation database emphasizing data quality and a broad spectrum of data sources. *Nucleic Acids Res.* **30:** 387.

Gabellini N. 2001. A polymorphic GT repeat from the human cardiac Na+Ca2+ exchanger intron 2 activates splicing. *Eur. J. Biochem.* **268:** 1076.

Gat-Viks I. and Shamir R. 2003. Chain functions and scoring functions in genetic networks. *Bioinformatics* (suppl. 1) *19:* I108.

Grantham R. 1974. Amino acid difference formula to help explain protein evolution. *Science* **185:** 862.

Guy P.M., Kenny D.A., and Gill G.N. 1999. The PDZ domain of the LIM protein enigma binds to beta-tropomyosin. *Mol. Biol. Cell* **10:** 1973.

Hariharan N., Kelley D.E., and Perry R.P. 1991. Delta, a transcription factor that binds to downstream elements in several polymerase II promoters, is a functionally versatile zinc finger protein. *Proc. Natl. Acad. Sci.* **88:** 9799.

Heyse J. and Rom D. 1988. Adjusting for multiplicity of statistical tests in the analysis of carcinogenicity studies. *Biom. J.* **30:** 883.

Iuchi S. 2001. Three classes of C2H2 zinc finger proteins. *Cell Mol. Life Sci.* **58:** 625.

Jansen R.C. and Nap J.P. 2001. Genetical genomics: The added value from segregation. *Trends Genet.* **17:** 388.

Jin W., Riley R.M., Wolfinger R.D., White K.P., Passador-Gurgel G., and Gibson G. 2001. The contributions of sex,

genotype and age to transcriptional variance in *Drosophila melanogaster*. *Nat. Genet.* **29:** 389.
Johansson O., Alkema W., Wasserman W.W., and Lagergren J. 2003. Identification of functional clusters of transcription factor binding motifs in genome sequences: The MSCAN algorithm. *Bioinformatics* (suppl. 1) **19:** I169.
Kanai F., Marignani P.A., Sarbassova D., Yagi R., Hall R.A., Donowitz M., Hisaminato A., Fujiwara T., Ito Y., Cantley L.C., and Yaffe M.B. 2000. TAZ: A novel transcriptional co-activator regulated by interactions with 14-3-3 and PDZ domain proteins. *EMBO J.* **19:** 6778.
Kwon A.T., Hoos H.H., and Ng R. 2003. Inference of transcriptional regulation relationships from gene expression data. *Bioinformatics* **19:** 905.
Lecanda J., Urtasun R., Recalde S., Prieto J., and Medina J.F. 2001. A novel polymorphism IVS2+843C>T in the alternate promoter b1 of the human AE2 anion exchanger gene. *Hum. Mutat.* **17:** 82.
Li X., Lopez-Guisa J.M., Ninan N., Weiner E.J., Rauscher F.J., III, and Marmorstein R. 1997. Overexpression, purification, characterization, and crystallization of the BTB/POZ domain from the PLZF oncoprotein. *J. Biol. Chem.* **272:** 27324.
Lin C.S., Chow S., Lau A., Tu R., and Lue T.F. 2002. Human PDE5A gene encodes three PDE5 isoforms from two alternate promoters. *Int. J. Impot. Res.* **14:** 15.
Majewski J. and Ott J. 2002. Distribution and characterization of regulatory elements in the human genome. *Genome Res.* **12:** 1827.
Mattick J.S. 1994. Introns: Evolution and function. *Curr. Opin. Genet. Dev.* **4:** 823.
Melnick A., Carlile G., Ahmad K.F., Kiang C.L., Corcoran C., Bardwell V., Prive G.G., and Licht J.D. 2002. Critical residues within the BTB domain of PLZF and Bcl-6 modulate interaction with corepressors. *Mol. Cell. Biol.* **22:** 1804.
Neitzel H. 1986. A routine method for the establishment of permanent growing lymphoblastoid cell lines. *Hum. Genet.* **73:** 320.
Oyake T., Itoh K., Motohashi H., Hayashi N., Hoshino H., Nishizawa M., Yamamoto M., and Igarashi K. 1996. Bach proteins belong to a novel family of BTB-basic leucine zipper transcription factors that interact with MafK and regulate transcription through the NF-E2 site. *Mol. Cell. Biol.* **16:** 6083.
Pitman E.J.G. 1937. Significance tests which may be applied to samples from any populations. *J. Roy. Stat. Soc. Series B* **4:** 119.
Pointud J.C., Larsson J., Dastugue B., and Couderc J.L. 2001. The BTB/POZ domain of the regulatory proteins Bric a brac 1 (BAB1) and Bric a brac 2 (BAB2) interacts with the novel *Drosophila* TAF(II) factor BIP2/dTAF(II)155. *Dev. Biol.* **237:** 368.
Rockman M.V. and Wray G.A. 2002. Abundant raw material for *cis*-regulatory evolution in humans. *Mol. Biol. Evol.* **19:** 1991.
Sandberg R., Yasuda R., Pankratz D.G., Carter T.A., Del Rio J.A., Wodicka L., Mayford M., Lockhart D.J., and Barlow C. 2000. Regional and strain-specific gene expression mapping in the adult mouse brain. *Proc. Natl. Acad. Sci.* **97:** 11038.
Schadt E.E., Monks S.A., Drake T.A., Lusis A.J., Che N., Colinayo V., Ruff T.G., Milligan S.B., Lamb J.R., Cavet G., Linsley P.S., Mao M., Stoughton R.B., and Friend S.H. 2003. Genetics of gene expression surveyed in maize, mouse and man. *Nature* **422:** 297.
Segal E., Shapira M., Regev A., Pe'er D., Botstein D., Koller D., and Friedman N. 2003. Module networks: Identifying regulatory modules and their condition-specific regulators from gene expression data. *Nat. Genet.* **34:** 166.
Steinmetz L.M., Sinha H., Richards D.R., Spiegelman J.I., Oefner P.J., McCusker J.H., and Davis R.W. 2002. Dissecting the architecture of a quantitative trait locus in yeast. *Nature* **416:** 326.
Stephens J.C., Schneider J.A., Tanguay D.A., Choi J., Acharya T., Stanley S.E., Jiang R., Messer C.J., Chew A., Han J.H., Duan J., Carr J.L., Lee M.S., Koshy B., Kumar A.M., Zhang G., Newell W.R., Windemuth A., Xu C., Kalbfleisch T.S., Shaner S.L., Arnold K., Schulz V., Drysdale C.M., Nandabalan K., Judson R.S., Ruano G., and Vovis G.F. 2001. Haplotype variation and linkage disequilibrium in 313 human genes. *Science* **293:** 489.
T'Jampens D., Devriendt L., De C., V, Vandekerckhove J., and Gettemans J. 2002. Selected BTB/POZ-kelch proteins bind ATP. *FEBS Lett.* **516:** 20.
Wen Y., Nguyen D., Li Y., and Lai Z.C. 2000. The N-terminal BTB/POZ domain and C-terminal sequences are essential for Tramtrack69 to specify cell fate in the developing *Drosophila* eye. *Genetics* **156:** 195.
Yan H., Yuan W., Velculescu V.E., Vogelstein B., and Kinzler K.W. 2002. Allelic variation in human gene expression. *Science* **297:** 1143.

# Genetic Variation and the Control of Transcription

C. COTSAPAS, E. CHAN, M. KIRK, M. TANAKA, AND P. LITTLE
*School of Biotechnology and Biomolecular Sciences, University of New South Wales, Sydney, New South Wales 2052, Australia*

Identifying and understanding genetic variation is a key driver of agricultural, biotechnological, and biomedical research and commercialization; the major focus of genetic variation research has until now been on changes to protein-coding sequences because these are computationally and experimentally accessible. In contrast, the contribution of genetic variation to the temporal or spatial control of transcription is not well understood, in part because we have no simple technologies to identify functional variants in control regions, nor do we, for the majority of genes, have a comprehensive understanding of the proteins that control gene expression.

Only relatively recently (Jin et al. 2001; Schadt et al. 2003) has it become apparent that levels of mRNA in a cell are influenced by genetic variation (for review, see Cheung and Spielman 2002). Such quantitative variation in mRNA levels has two distinct origins, which we define as *cis*- or *trans*-acting (Fig. 1).

- *Cis*-acting variants: The underlying sequence change(s) is in DNA that controls expression of a nearby gene, including promoter/enhancers, splice determinants, poly(A) addition sequences, mRNA stability, and transport signals.
- *Trans*-acting variants: These are located in genes whose products, protein or RNA, in the broadest sense, control mRNA levels of genes. Thus, variation in a distant gene is responsible for alteration in the transcript levels of a target gene—hence the description of "*trans*" variation. The most obvious candidates for *trans*-acting effectors are transcription factors.

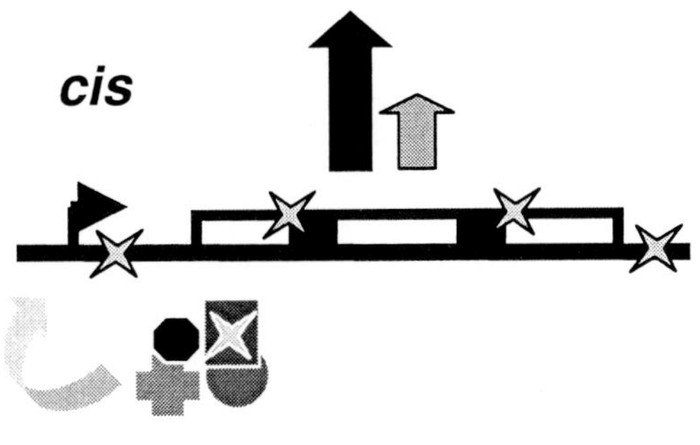


**Figure 1.** Two types of genetic variation influencing amounts of mRNA in a cell or tissue (shown by *large filled arrows*). Sites of variations are marked by a star: those within the gene are acting *in cis*, those within effector proteins of any type are *in trans*.

*Cis*-acting variation: Variation in mRNA levels has been studied in a genetic context by Yan et al. (2002), who demonstrated that allelic variation in expression of 6 human genes was *cis*-acting and this was "relatively common among normal individuals." Sandberg et al. (2000) used cDNA microarrays to show that 24 genes were expressed at differing levels in the brains of 129SVEv and C57BL/6 mice, and there have been reports of interstrain differences in mRNA levels of individual genes, including *Gas5* (Muller et al. 1998). In humans, variations near the *INS* and *IGF2* genes contribute to alteration in mRNA levels, and these are associated with susceptibility to type 1 diabetes (Bennett et al. 1995), and variations in the promoter region of the presenilin 1 gene are associated with an increased risk of early-onset Alzheimer's disease (Theuns et al. 2000). More generally, Cargill et al. (1999) and Halushka et al. (1999) estimated the frequency of single-nucleotide polymorphisms (SNPs) near, but not within, coding sequences—and consequently candidates for promoter polymorphisms—as 1/354 bp, and Yamada et al. (2000) found a value of 1/562 bases. These data show that quantitative variation is relatively common and contributes to significant differences in phenotype.

*Trans*-acting effects on transcript levels: This class of quantitative variation is not well studied. Hustert et al. (2001) showed that variations in the pregnane X receptor, a transcription factor (TF) for the CYP3A4 gene, influenced expression of the target gene. Schadt et al. (2003) used a mouse backcross approach to study the extent to which variation occurred with two inbred strains of mouse, and showed that 33% of all genes exhibited alterations in the level of mRNA in at least 10% of the backcross mice analyzed; they treated mRNA levels as classic quantitative trait loci and showed that 34–71% of those influences were in *cis*. These limited data suggest that *in trans* genetic variation occurs and is of significant physiological consequence.

For the sake of inclusiveness, we will refer to TFs, splice components, and stability determinant proteins as simply mRNA level "effectors," since all might contribute to quantitative variation. We suggest that genetic variation of *trans*-acting control of gene expression potentially may have more profound effects on cells than variations in other proteins, simply because such effectors influence multiple genes and whole pathways, suggesting a potentially important role in evolution and in the generation of species diversity. For example, Enard et al. (2002) have suggested that amounts of mRNA expressed within the cells of the human and chimpanzee are altered more significantly in the brain than in the liver and argued that this might therefore be a key component of the differences between the cognitive abilities of our respective brains.

It is probably simplest (but far from being the exclusive mechanism) to think of the influence of protein sequence variations within TFs in this respect. As a class, functional changes in TF action might be expected to have very significant effects on cells because many TFs have multiple target genes and some, such as NF-κB (for review, see Valen et al. 2001), regulate pathways of genes whose proteins have connected functions. We note that mutations in TFs have been identified in 17 human genetic conditions (Human Gene Mutation Database, ver. 14/01/2003 at www.hgmd.org). Far less is known about the specificity of non-TF effectors such as iron regulatory proteins (IRP) (for review, see Eisenstein 2000) or proteins that bind AU-rich mRNA regions (Laroia et al. 2002). These regulate mRNA stability, and therefore, variants in such proteins might have a similar pleiotropic effect. The influence, if any, of sequence variations on RNAi (see, e.g., Shi 2003) remains unknown, but of course these also might make a contribution.

We conclude from this brief review that detecting variation of mRNA levels, distinguishing between *cis* and *trans* effects, and identifying *trans* effectors is an important biological objective that could identify a significant molecular mechanism contributing to variation in phenotype in all living creatures.

## RESULTS

### Genetic Variation of Transcription Factors

The first question we addressed was the extent to which genetic variation influenced TFs. Ramensky et al. (2002) analyzed data from HGVBase (http://hgvbase.cgb.ki.se/) to show that protein-coding variations in human TFs are relatively less common than similar variations in other classes of proteins. This suggests that they are likely to be under negative selection and implies a strong influence of purifying selection. However, the data in HGVBase are susceptible to bias in the candidate genes that had been selected for study, and the only case in which the genome of multiple individuals has been systematically sequenced is that of the mouse. The complete sequence of the C57BL/6 mouse has been established by the Public Consortium (Waterston et al. 2002) and those of 129S1/SvImJ (0.256 coverage) 129X1/SvJ (0.691 coverage), A/J (0.899 coverage), and DBA/2J (0.789 coverage) have been established by Celera (Kerlavage et al. 2002). These data allow us to compare systematically the genetic variation with whole classes of proteins. We therefore set out to answer two questions to allow us to establish the extent and nature of genetic variation in the transcriptional machinery and so identify potential *trans*-acting contributions to differences in mRNA level. Is functional variation in transcription factors in inbred mice similar in extent to variations in other proteins? Is the silent/missense ratio of coding sequence variations similar in TFs? Differences in this ratio are in part due to differing selection pressures on protein

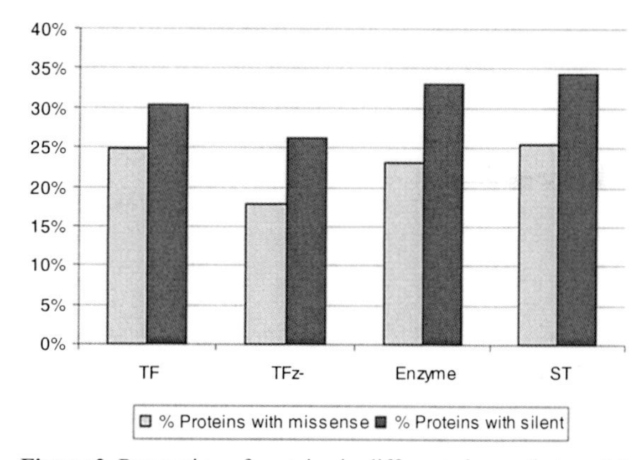


**Figure 2.** Proportion of proteins in different classes that contain missense or silent substitutions: TF are transcription factors, TFz⁻ are transcription factors excluding zinc finger proteins, Enzyme are proteins with an enzymatic function, ST are proteins involved in signal transduction.

function and can provide evidence for functional selection of variations. We used the Celera gene function ontology to define 1549 TFs (containing 695 TFs classified as "zinc finger" proteins), 1465 enzymes, and 1519 signal-transducing proteins and retrieved missense and sense coding sequence variations within these using the Celera Mouse Reference SNP database. As shown in Figure 2, 30.24% and 24.76% of TFs contained silent and missense variations, respectively (reduced to 26.20% and 17.82%, if only non-zinc-finger proteins are included), whereas 32.97% and 23.06% of enzymes and 34.38% and 25.33% of signal-transducing proteins contained silent and missense variations, respectively. The number of silent and missense variations can only be compared directly if there is no more than one nucleotide difference between each homologous pair of codons (Nei and Gojobori 1986). This is because there is no way of knowing the order of mutation where multiple nucleotide differences have occurred, and this can preclude classification into silent or missense variation. A total of 17 TFs, 12 enzymes, and 15 STs were excluded from these analyses because they contained two sequence differences.

The ratio of silent to missense variations in TFs was 1.32 (1.68 if zinc fingers are excluded), 2.11 for enzymes, and 1.68 for signal-transducing proteins, giving a chi-squared test probability of $10^{-12}$ (Fig. 3). The Celera database does not allow us to calculate a per-site rate of variation, which is the classic method (Nei and Gojobori 1986) of establishing the influence of selection. Thus, the highly statistically significant increase in missense variations in TFs compared to enzymes is compatible with either TFs being under selection or with TFs being under weaker structural constraints than other proteins. However, the observations of Ramensky et al. (2002) argue that the latter explanation cannot be correct, since they observe that genetic variation in human TFs is less frequent than in any other class of protein.

We conclude that genetic variation is as common in TFs as it is in any other class of proteins. We therefore set out to establish the extent of mRNA variation between

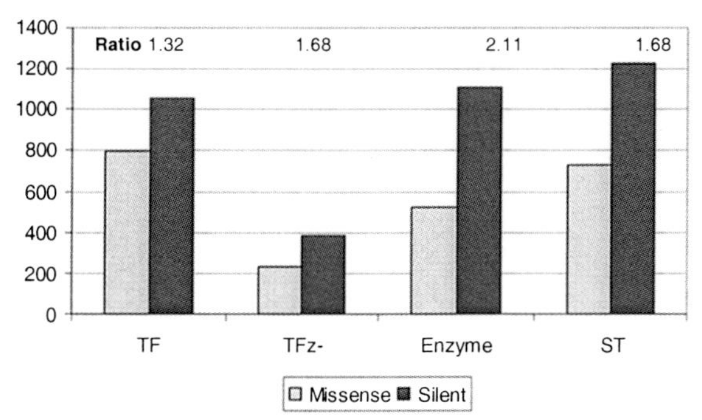


**Figure 3.** Numbers of missense or silent variations in the protein classes detailed in Fig. 1. Ratios are of silent/missense variations in the protein class.

two of the inbred mouse strains, C57BL/6 and DBA/2J, whose DNA sequences differ on average at one base in every 1660.

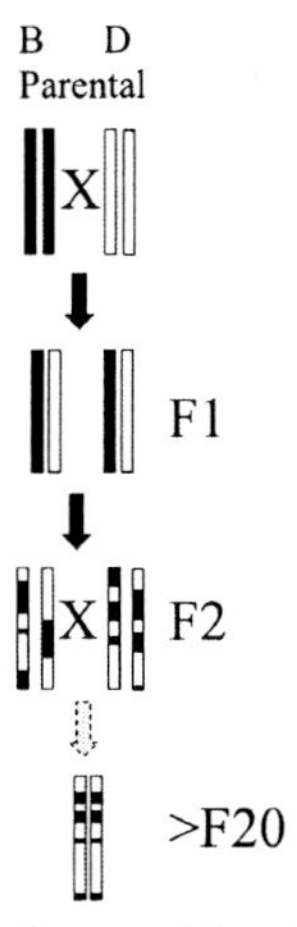


**Figure 4.** The creation of a recombinant inbred line of mice. Inbreeding of $F_2$ mice will ultimately give rise to a homozygous mouse with genes derived from either parental strain: This is diagrammed as a black or white chromosome, representative of the whole mouse genome.

## Experimental Design for Detecting mRNA Variation

The design of an experimental system to study genetic variation in mRNA levels is taxing. We require, at a DNA sequence level, knowledge of the genetic variation between two individuals, which can also be bred to produce progeny of defined genotype for mRNA level assessment. Ideally, we would like these individuals to be homozygous, to minimize noise associated with heterozygosity. An organism that fulfils all these criteria is the laboratory mouse, *Mus musculus*, whose genetics are relatively well understood and for which there exist a number of inbred, homozygous lines with significant phenotypic differences. The short breeding cycle also allows us to breed custom mice of defined genotypes for specific hypothesis testing.

Analyzing the levels of mRNA in mice should allow a rigorous distinction between genetic and nongenetic sources of variation—essentially a genetic filter. We have created this experimental analysis using the recombinant inbred (RI) lines of mice developed from the inbred strains C57BL/6 and DBA/2J and using cDNA microarrays to analyze the mRNA levels of many genes simultaneously.

## Recombinant Inbred Mice

The key component of the experiment is to use a series of RI strains of mice. RI mice are made by crossing two different inbred "parental" strains, in our case C57BL/6 ("B" mice) and DBA/2J ("D" mice); these are termed BxD strains. The properties of an RI strain are illustrated in Figure 4. The first generation ($F_1$) contains pairs of B and D chromosomes, which recombine to make chromosomes in the $F_2$ that are derived from a mix of B and D alleles. Repeated brother/sister mating in subsequent generations will result, from about $F_{20}$ onward, in a homozygous mouse with the allele at any locus derived from either the B or D parental strain, depending solely on the recombinations that occurred early in the breeding history of the individual line. To make the BxD RI strains, 36 pairs of $F_2$ brother/sister mice were selected at random to make 36 "lines" of mice by repeated brother/sister matings for 117 generations to date. RI lines therefore contain homozygous loci derived from an arbitrary mixture of the alleles in the parental strains. Any single gene that is variant between the parental strains will be homozygous D or B in different lines; the final pattern of gene origin in all 36 lines is called the strain distribution pattern or SDP—simply a string representing the homozygous genotype in the 36 lines; e.g., DBBDB-DDB, etc. to 36 letters.

## Computing the Parental Origin of Genes in the RIs

In practice, SDPs are known for ~1600 markers, polymorphic between the parental strains, typed in 26 RI strains—the remaining 10 having been typed for a subset (Tanaka et al. 2000). By mapping these markers to the whole-genome mouse assembly (02/02 freeze, http://genome.ucsc.edu/goldenPath/mmFeb2002/bigZips/), we can infer SDPs, with varying degrees of accuracy, for any gene lying between two markers. Using the 800 markers we have mapped to date, we have inferred SDPs for 10,832 cDNAs from the NIA 15K set.

## The Experimental Design

The underlying principle is illustrated in Figure 5: In 5 RI lines, chromosomal regions derived from B are black, those from D are gray. For a gene whose mRNA level is different in B and D (the "cognate" gene) due to a *cis* effect, the mRNA level (gray/black arrows) will correlate with the actual parental origin of the gene. In contrast, if the variation is due to a *trans*-acting effector, TF, the lev-

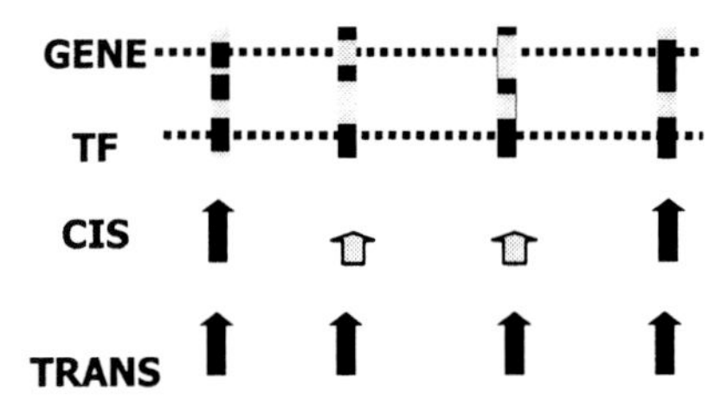


**Figure 5.** The detection of *cis* or *trans* origins of difference in mRNA levels of a gene. The location of the gene is indicated by a dashed line: If the amounts of mRNA are controlled *in cis*, the mRNA levels (*arrows* labeled *cis*) are concordant with the origin of the cognate gene. Conversely, if the levels are controlled *in trans* (*arrows* labeled *trans*), then they are concordant with the origin of the transcription factor (TF) and not with the cognate gene itself.

els of mRNA (TRANS in figure) will correlate with the parental origin of TF, NOT the gene: In this case, TF is B-derived in all five lines. In practice, the level of mRNA is measured with microarrays.

Our experimental approach was to create pools of total RNA derived from brain tissue from three animals of each strain. Each pool was reverse-transcribed, fluorescently labeled, and competitively hybridized with a reference cDNA sample from a pool of three C57BL/6J animals. All animals were age- and sex-matched. The arrays, produced by the Ramaciotti Centre for Gene Function Analysis (UNSW, Sydney, Australia), were spotted with the 15K NIA cDNA clone set.

### Statistical Analysis of *cis* Variation

We have analyzed these array hybridizations using a statistical approach implemented with custom-written software. In outline, we perform lowess print tip normalization of background-subtracted fluorescence intensity ratios (RI/B) to define relative hybridization for each cDNA (Yang et al. 2002).

Using the computed SDP for the relevant gene, M values from each RI are assigned to B or D as appropriate, and a $t$ statistic is calculated to assess the difference between the two groups. To assess significance, we permute all possible combinations of M values for the number of Bs and Ds in the SDP, and calculate $t$ statistics for each. By placing the experimental $t$ statistic onto this distribution, we can derive a percentile (P) value: essentially, a measure of probability that the fit between the data and the SDP is derived by chance. The importance of this approach is that it requires no a priori assumptions of difference levels between samples and can be used to assess differences between the parental strains or within the RI lines. Using a variety of data based on observation, we have modeled this process and conclude that we have ~40% power (with confidence interval, C.I., of 99.8%) to detect 1.5-fold variation of signal with a variance of 1, which increases to >68% (with C.I. of 99.8%) as the fold difference is increased to >2. These observations agree with anecdotal evidence as to the resolution power of microarrays (Claverie 1999).

*Trans* variation is identified by a similar process, but in calculating $t$-statistics we input all SDPs known to occur within the RI lines (there are 807 of these), rather than using the single cognate gene SDP, and searching for a significant match. In other words, we ask whether *any* gene's strain distribution pattern correlates with the fluctuation of transcript level from the RI lines, with the expectation that a *trans* effector will do so. This approach generates substantial problems of multiple sampling and computation time, and we necessarily have to preselect the cDNAs to reduce the confounding multiple testing, using a number of criteria. We are currently limiting this analysis to cDNAs we believe to be differentially expressed between the parental strains themselves; concurrently, we are developing statistical approaches to analyze the complete, non-selected data set.

### Experimental Analysis of the Origin of Quantitative Variation

We have used three approaches to determine the number of genes contained within the NIA 15K set that are differentially expressed between the parental C57BL/6J and DBA/2J strains. Eight replicates of B/D microarray data were analyzed using the three different methods to identify candidate differentially expressed genes in the parental lines: by a simple heuristic (8 replicates > 2-fold [2/8 min]), 36 genes (0.3%), by a $t$-statistic with a 0.1% cutoff, 241 genes (1.7%), and by B-statistic (Lonnstedt and Speed 2002), 13 genes (0.09%) are differentially expressed between C57BL/6J and DBA/2J. There is overlap between these analyses yielding 279 unique candidate genes. Each of these genes was examined in the RI series and expression patterns were assessed against the known SDP of the gene. Of the 192 genes that could be analyzed, only 5 (2.6%) had *cis* determinants with a 99.8% confidence.

Strikingly, 8 additional genes, not differentially expressed between the parental strains, could be shown in the RI strains to exhibit *cis* variation at 99.8% confidence, which suggests to us that genetic background is critical in determining the mRNA levels.

*Trans* analysis: The 279 genes were analyzed using the $t$-statistic approach by matching the individual RI strains' M values against all 808 SDPs contained within the RI set. Of 210,000 matches, 412 were significant at the 99.8% level, and these contained 56% of the differentially expressed genes.

Our interpretation of these preliminary data is that of the genes exhibiting differential expression, ~60% are susceptible to monogenic *cis* or *trans* influences, whereas ~40% are due to oligogenic, polygenic, or other influences that cannot readily be mapped in these analyses. We also detect, but cannot presently quantitate, significant epistatic influences in those genes that are differentially expressed in the RI lines but not the parental strains.

### Identifying a Regulon

A regulon is a group of genes that are coordinately expressed because they share a common control mechanism; a variant TF, for example, might influence the expression of all genes within the regulon. These can be detected in our experiments if we observe an SDP in the *trans* analysis which appears to be associated with variation in the levels of mRNA in multiple genes that do not map to this SDP.

We identified two SDPs that appeared to be influencing 12 different genes. The target genes of one such case include Slc25a1 (solute carrier family 25 member 1), Xrn1 (5′-3′ exoribonuclease 1), Xpnpep1 (X-prolyl aminopeptidase P1, soluble), Ctps2 (cytidine 5′-triphosphate synthase 2), BG065446 (homology with RNA polymerase 1–3 16-kD subunit), C77518 (homology with PTPL-1-associated Rho-GAP), D7Ertd59e (an expressed marker), as well as five proteins of unknown function. Our working hypothesis is that the shared SDP defines a region that contains a variant transcription effector that is modulating the behavior of SETS of genes that comprise a regulon (a set of coregulated genes). The expectation, given that our data set contains 412 matches, is that any one SDP will appear 412/800 times: Thus, a single SDP appearing 12 times is unusual. The SDPs define a region of chromosome 14 that spans 7 Mb. Using the UCSC genome browser, we have identified two potential transcription factors within the region, Dnase1l3 and 2610511E03Rik (Rnase P-related). 2610511E03Rik contains one replacement variation (Met132Val) in C57BL/6 compared to DBA/2J in the Celera database: The other protein contains only silent/noncoding variants. This leads us to the working hypothesis that variation in 2610511E03Rik is the cause of variation in the target genes. In addition, we have identified one case of an SDP influencing 9 genes, an SDP influencing 7 genes, two cases of an SDP influencing 6 genes, three cases of an SDP influencing 5 genes, fourteen cases of an SDP influencing 3 genes, and one hundred and twelve cases of an SDP influencing 2 genes, and these are being analyzed intensively. The existence of identifiable regulons partially explains the frequency of *trans* influence discussed above. We believe these preliminary data are graphic illustrations of the potential of our analytic and experimental analyses.

## CONCLUSION

We have developed a powerful system for analyzing genetic variation and its influence on mRNA levels. Our approach is readily comparable to that of Schadt et al. (2003), who used a backcross between C57BL/6 and DBA/2J to analyze mRNA level variation. Such animals are either homozygous or heterozygous for variations, in contrast to RI strains which are homozygous, and this may account for the greater variability seen in mRNA levels in their analyses, where 33% of genes appeared to be differentially expressed within the progeny. Further analysis will resolve this issue.

An important distinction between the use of RI or backcross mice is that the RI lines are genetically stable and can be bred at will. We believe that ability to tailor genotypes by selection of RI lines and other strains reinforces, once again, the great power of the mouse as a genetic model for human variation, because establishing the relationship of quantitative mRNA variation to ultimate phenotype is not simple. The parental C57BL/6 and DBA/2J mice differ significantly in many physical, biochemical, and behavioral respects (Festig 1998), and these data in principle can be related to underlying genetic variations. In practice, until we have a better idea of the specific effectors and genes, it is difficult to define testable hypotheses.

It is possible to extend the approach we have developed here to humans. Unlike RI lines, humans are frequently heterozygous for variations, outbred, susceptible to environmental influence, and not a ready source of tissue. Despite these reservations, we believe it will be possible to carry out preliminary experiments on tissue mRNAs isolated from extended human families; specifically, the three-generation "reference" CEPH families that have been very extensively typed using microsatellite markers as part of the Human Genome Project. These data, publicly available at http://lpg.nci.nih.gov/CHLC/, enable us to calculate the parental origin of any genomic region, which in turn enables us to construct the equivalent of an SDP. This will be the "parent of origin distribution pattern" or PODP of the gene in the family. Analogous to our mice experiments, concordance or discordance of expression levels with the PODP will indicate *cis* or *trans* influence on expression levels.

Confounding this experiment are numerous nongenetic factors associated with the intrinsic variability of transformed lymphoblastoid cell lines, but these influences are not expected to be identically distributed to Mendelian patterns of inheritance, and frank genetic signal should in principle be isolable by the analysis we have proposed.

Our observation of a significant amount of variation within the machinery that controls transcription, allied with our preliminary data and that of Schadt et al. (2003), leads us to propose a new class of project. We suggest that identifying sequence variation in this machinery, in any organism, will provide more significant insights into the molecular basis of phenotypic variation than conventional candidate gene approaches based on more limited physiological function.

## ACKNOWLEDGMENTS

This work was carried out with the support of a start-up grant from the University of New South Wales and with the aid of Australian Postgraduate Award scholarships to E.C. and M.K. We are indebted to the staff of the Clive and Vera Ramaciotti Centre for Gene Function Analysis for provision of mouse microarrays and to Matt Wand, Willam Dunsmuir, and David Nott (School of Mathematics at UNSW) for a continuing collaboration on statistical analysis of our data.

## REFERENCES

Bennett S.T., Lucassen A.M., Gough S.C., Powell E.E., Undlien D.E., Pritchard L.E., Merriman M.E., Kawaguchi Y., Dronsfield M.J., and Pociot F. 1995. Susceptibility to human type 1 diabetes at IDDM2 is determined by tandem repeat variation at the insulin gene minisatellite locus. *Nat. Genet.* **9:** 284.

Cargill M., Altshuler D., Ireland J., Sklar P., Ardlie K., Patil N., Shaw N., Lane C.R., Lim E.P., Kalyanaraman N., Nemesh J., Ziaugra L., Friedland L., Rolfe A., Warrington J., Lipshutz R., Daley G.Q., and Lander E.S. 1999. Characterization of single-nucleotide polymorphisms in coding regions of human genes. *Nat. Genet.* **22:** 231.

Cheung V.G. and Spielman R.S. 2002. The genetics of variation in gene expression. *Nat. Genet.* (suppl.) **32:** 522.

Claverie J.M. 1999. Computational methods for the identification of differential and coordinated gene expression. *Hum. Mol. Genet.* **8:** 1821.

Eisenstein R.S. 2000. Iron regulatory proteins and the molecular control of mammalian iron metabolism. *Annu. Rev. Nutr.* **20:** 627.

Enard W., Khaitovich P., Klose J., Zollner S., Heissig F., Giavalisco P., Nieselt-Struwe K., Muchmore E., Varki A., Ravid R., Doxiadis G.M., Bontrop R.E., and Paabo S. 2002. Intra- and interspecific variation in primate gene expression patterns. *Science* **296:** 340.

Festing M.W. 1998. Inbred strains if mice. (http://www.informatics.jax.org/external/festing/mouse/STRAINS.shtml).

Halushka M.K., Fan J.B., Bentley K., Hsie L., Shen N., Weder A., Cooper R., Lipshutz R., and Chakravarti A. 1999. Patterns of single-nucleotide polymorphisms in candidate genes for blood-pressure homeostasis. *Nat. Genet.* **22:** 239.

Hustert E., Zibat A., Presecan-Siedel E., Eiselt R., Mueller R., Fuss C., Brehm I., Brinkmann U., Eichelbaum M., Wojnowski L., and Burk O. 2001. Natural protein variants of pregnane X receptor with altered transactivation activity toward CYP3A4. *Drug Metab. Dispos.* **29:** 1454.

Jin W., Riley R.M., Wolfinger R.D., White K.P., Passador-Gurgel G., and Gibson G. 2001. The contributions of sex, genotype and age to transcriptional variance in *Drosophila melanogaster*. *Nat. Genet.* **29:** 389.

Kerlavage A., Bonazzi V., di Tommaso M., Lawrence C., Li P., Mayberry F., Mural R., Nodell M., Yandell M., Zhang J., and Thomas P. 2002. The Celera Discovery System. *Nucleic Acids Res.* **30:** 129.

Laroia G., Sarkar B., and Schneider R.J. 2002. Ubiquitin-dependent mechanism regulates rapid turnover of AU-rich cytokine mRNAs. *Proc. Natl. Acad. Sci.* **99:** 1842.

Lonnstedt I. and Speed T. 2002. Replicated microarray data. *Statistica Sinica* **12:** 31.

Muller A.J., Chatterjee S., Teresky A., and Levine A.J. 1998. The gas5 gene is disrupted by a frameshift mutation within its longest open reading frame in several inbred mouse strains and maps to murine chromosome 1. *Mamm. Genome.* **9:** 773.

Nei M. and Gojobori T. 1986. Simple methods for estimating the numbers of synonymous and non-synonymous nucleotide substitutions. *Mol. Biol. Evol.* **3:** 418.

Ramensky V., Bork P., and Sunyaev S. 2002. Human non-synonymous SNPs: Server and survey. *Nucleic Acids Res.* **30:** 3894.

Sandberg R., Yasuda R., Pankratz D.G., Carter T.A., Del Rio J.A., Wodicka L., Mayford M., Lockhart D.J., and Barlow C. 2000. Regional and strain-specific gene expression mapping in the adult mouse brain. *Proc. Natl. Acad. Sci.* **97:** 11038.

Schadt E.E., Monks S.A., Drake T.A., Lusis A.J., Che N., Colinayo V., Ruff T.G., Milligan S.B., Lamb J.R., Cavet G., Linsley P.S., Mao M., Stoughton R.B., and Friend S.H. 2003. Genetics of gene expression surveyed in maize, mouse and man. *Nature* **422:** 297.

Shi Y. 2003. Mammalian RNAi for the masses. *Trends Genet.* **19:** 9.

Tanaka T.S., Jaradat S.A., Lim M.K., Kargul G.J., Wang X., Grahovac M.J., Pantano S., Sano Y., Piao Y., Nagaraja R., Doi H., Wood W.H., III, Becker K.G., and Ko M.S. 2000. Genome-wide expression profiling of mid-gestation placenta and embryo using 15k mouse developmental cDNA microarray. *Proc. Natl. Acad. Sci.* **97:** 9127.

Theuns J., Del-Favero J., Dermaut B., van Duijn C.M., Backhovens H., Van den Broeck M.V., Serneels S., Corsmit E., Van Broeckhoven C.V., and Cruts M. 2000. Genetic variability in the regulatory region of presenilin 1 associated with risk for Alzheimer's disease and variable expression. *Hum. Mol. Genet.* **9:** 325.

Valen G., Yan Z.Q., and Hansson G.K. 2001. Nuclear factor kappa-B and the heart. *J. Am. Coll. Cardiol.* **38:** 307.

Waterston R.H., Lindblad-Toh K., Birney E., Rogers J., Abril J.F., Agarwal P., Agarwala R., Ainscough R., Alexandersson M., An P., Antonarakis S.E., Attwood J., Baertsch R., Bailey J., Barlow K., Beck S., Berry E., Birren B., Bloom T., Bork P., Botcherby M., Bray N., Brent M.R., Brown D.G., and Brown S.D., et al. (Mouse Genome Sequencing Consortium). 2002. Initial sequencing and comparative analysis of the mouse genome. *Nature* **420:** 520.

Yamada R., Tanaka T., Ohnishi Y., Suematsu K., Minami M., Seki T., Yukioka M., Maeda A., Murata N., Saiki O., Teshima R., Kudo O., Ishikawa K., Ueyosi A., Tateishi H., Inaba M., Goto H., Nishizawa Y., Tohma S., Ochi T., Yamamoto K., and Nakamura Y. 2000. Identification of 142 single nucleotide polymorphisms in 41 candidate genes for rheumatoid arthritis in the Japanese population. *Hum. Genet.* **106:** 293.

Yan H., Yuan W., Velculescu V.E., Vogelstein B., and Kinzler K.W. 2002. Allelic variation in human gene expression. *Science* **297:** 1143.

Yang Y.H., Dudoit S., Luu P., Lin D.M., Peng V., Ngai J., and Speed T.P. 2002. Normalization for cDNA microarray data: A robust composite method addressing single and multiple slide systematic variation. *Nucleic Acids Res.* **30:** e15.

# Genome-wide Detection and Analysis of Recent Segmental Duplications within Mammalian Organisms

J.A. BAILEY AND E.E. EICHLER
*Department of Genetics, Center for Computational Genomics, Comprehensive Cancer Center and Center for Human Genetics, Case Western Reserve University School of Medicine and University Hospitals of Cleveland, Cleveland, Ohio 44106*

Protein encoding DNA sequence represents a paltry fraction of the human genome—an estimated 1.5% of the ~3 billion base pairs. In stark contrast, more than half of the genome is found to be repetitive, and, if the origin of all sequences could be discerned, the true fraction of repetitive DNA would be much higher. The highly repetitive nature of the genome was initially recognized through DNA reassociation studies ("Cot analysis") (Schmid and Deininger 1975; Deininger and Schmid 1976). More recent molecular analyses have confirmed the multitude of different high-copy or common repeats that constitute this repetitive fraction for many organisms. Among vertebrate species, five categories of high-copy-repeat elements may be distinguished (Table 1). For the majority of these repeats, studies have yielded descriptions of the consensus sequence, transposition/replication mechanism, and basic phylogeny. These properties have been used to classify them. Although a few repeats, such as α-satellite in humans, have been assigned a function, the majority are commonly referred to as "junk DNA" or "selfish DNA" because these repeats are considered to have no benefit to the "host" organism (Ohno 1972).

Unlike high-copy-repeat elements, segmental duplications represent portions or "segments" that occur multiple times within the genome and have arisen by genomic duplication as opposed to retrotransposition processes. These duplicated segments include both protein-encoding regions and common repeat sequences such as LINEs, SINEs, etc. From the perspective of genome assembly, both common interspersed repeats and segmental duplications may be classified as repetitive in nature. The distinction between the two, however, is critical. Segmental duplications may include genic material and therefore provide potential redundancy and ultimately diversity of genes within an evolving species. With the exception of encoding proteins responsible for their own propagation (autonomous high-copy repeats), high-copy repeats have limited capacity to expand the host gene repertoire. Furthermore, high-copy repeats may be easily recognized as such by virtue of defined sequence properties such as the presence of direct repeats or genes responsible for their self-propagation. Thus, high-copy repeats can easily be identified using consensus by programs such as REPEATMASKER (A.F.A. Smit and P. Green; http://repeatmasker.genome.washington.edu). In this context, "high copy (or common) repeats" classify the numerous aforementioned interspersed and tandem elements, and "segmental duplication" will only be used when referring to the duplication of genomic DNA.

Even though segmental duplications have long been implicated in the processes of gene evolution (Muller 1936; Ohno et al. 1968), their prevalence, complex architecture and high degree of sequence identity within the human genome may be considered one of the most surprising and unanticipated findings of the Human Genome Project (Collins et al. 1998). The discovery and sequence properties of these duplications suggest a level of genome dynamism during evolution hitherto unrealized. This "surprise" is due in large part to the conclusions that have been drawn from mammalian and vertebrate comparative studies over the past 30 years (Ohno et al. 1968). These studies showed gross conservation of genome and chromosomal structure, creating the impression of static genome evolution. Genome sizes within mammals appear conserved, hovering around 3 billion base pairs (although

**Table 1.** Major High-Copy (Common) Repeat Classes and Characteristics

| Element | Replication method | Organization | Location | Copies | Unit length | % of genome |
|---|---|---|---|---|---|---|
| LINE | autonomous retrotransposition | interspersed | dispersed (AT-rich regions) | 868 k | 6–8 kb | 21 |
| SINE | non-autonomous retrotransposition | interspersed | dispersed (GC-rich regions) | 1.5 M | 100–300 bp | 13 |
| LTR | Retrotransposition non and autonomous | interspersed | dispersed | 450 k | 1.5–11 kb | 8 |
| Satellite[a] | unequal homologous recombination | tandem | heterochromatin, centromeres | | 100–2000 bp | 4–9[b] |
| DNA elements | DNA transposition | interspersed | dispersed | 300 k | 100–3000 bp | 3 |

[a]Other tandem repeats include minisatellites and microsatellites.
[b]Estimate of 4% is lower limit based only on euchromatic sequence.

the extremes, 1.7 Gb and 6.7 Gb, differ by roughly 4-fold) (Graur and Li 2000). High-resolution G-banding techniques detected only a few differences between human and great ape karyotypes (Yunis 1976; Yunis and Prakash 1982). More recent comparative studies, using chromosomal paints and radiation-hybrid panels, have been able to detect more subtle rearrangements; however, these studies have confirmed the conservation of large-scale mammalian chromosome structure (one rearrangement per 10 million years, on average) (Wienberg et al. 1997; Murphy et al. 2001). In contrast to this static view of genome evolution, our analysis of the architecture of human segmental duplications implies a large number of radical smaller-scale changes (<4 Mb) over the extremely short period of primate evolution (<40 million years).

### General Characteristics of Segmental Duplications

Our analysis of segmental duplication focused on regions between 1 and 1000 kb in length that were found in more than one place in the genome (Table 2). In addition, we restricted our analysis to duplications exhibiting ≥90% sequence identity (Bailey et al. 2001). Our rationale was to exclude potentially uncharacterized high-copy repeats that had emerged by retrotransposition (the effective size of most of these is less than 1 kb) while identifying larger duplications specific to the primate lineage. Approximately 10% sequence divergence would identify duplications that had emerged 30–40 million years, based on a neutral model for genome evolution (Nei 1968) and excluding large-scale gene conversion events. In many respects, segmental duplication is an umbrella term meant to exclude the duplication of an entire chromosome (aneuploidy) or the entire genome (tetraploidy). In contrast to satellite and interspersed high-copy repeats, segmental duplications represent a different hierarchy, since they usually contain high-copy repeats. Segmental duplications do not have any known transposase, long-terminal repeat, or any other defining sequence characteristic that allows for their a priori designation as duplicated material. In other words, these sequences are normal genomic DNA. Duplication boundaries or junctions are not always conserved from one duplication event to the next. Concomitantly, segmental duplications with multiple copies in the genome may have different boundaries at each site. The only way to truly define a segmental duplication is by discovery that a sequence is represented more than once in the haploid genome.

### Computational Analysis of Segmental Duplications

To understand their organization and impact, we developed two independent computational approaches to identify segmental duplications within the human genome assembly. The first was a whole-genome analysis comparison (WGAC) method using BLAST as the underlying search tool to identify seed regions for further analysis (Bailey et al. 2001). This approach was completely dependent on the accuracy of the sequence assembly. Duplications, particularly those that were large and approached allelic levels of overlap, however, were frequently and incorrectly collapsed as allelic overlaps. This led to false negatives in the detection of segmental duplication and created concomitant gaps within the assembly. Alternatively, if two unique regions were not correctly assembled such that overlaps within the BACs could not be joined, they appeared as very highly homologous segmental duplications and therefore false positives in the analysis. Both misassembly and duplication collapse led to unreliable estimates within the early versions of the draft assembly (Bailey et al. 2001; Lander et al. 2001).

To circumvent these limitations, we developed a second computational approach that was independent of the assembly. This alternate approach compared the primary data sets from both the public and private efforts to sequence the human genome (Bailey et al. 2002b). Each sequence from 32,610 BACs (public initiative sequencing substrate) was compared to the 26.5 million human sequence reads (Celera initiative). The number and the average percent sequence identity of these reads in 5-kb windows was used as a metric to determine the extent, homology, and approximate copy number of the duplications within each BAC (Fig. 1). Based on this whole-genome shotgun sequence detection (WSSD) strategy,

**Table 2.** Comparison of Segmental Duplications with Other Repeat Classes

| Element | Sequence | End points | # of copies per element | Location | Transposition mechanism | Size (kb) |
|---|---|---|---|---|---|---|
| Segmental duplications | normal genomic DNA that can contain genes and other repeats | variable | 2–50 | clustered: chromosome-specific, pericentromeric regions, subtelomeric regions | unknown/unequal homologous recombination | 1–500+ |
| High copy interspersed elements | well-defined consensi | well defined | 100–millions | dispersed | DNA and retro-transposition | 0.1–8 |
| High copy tandem elements | well-defined consensi | well defined | 100–millions | centromeres, telomeres and hetrochromatic regions | unequal homologous recombination | 0.1–1 |
| Processed pseudogenes | reverse transcribed mRNA | well defined | 1–100 | dispersed | retrotransposition | <6 |

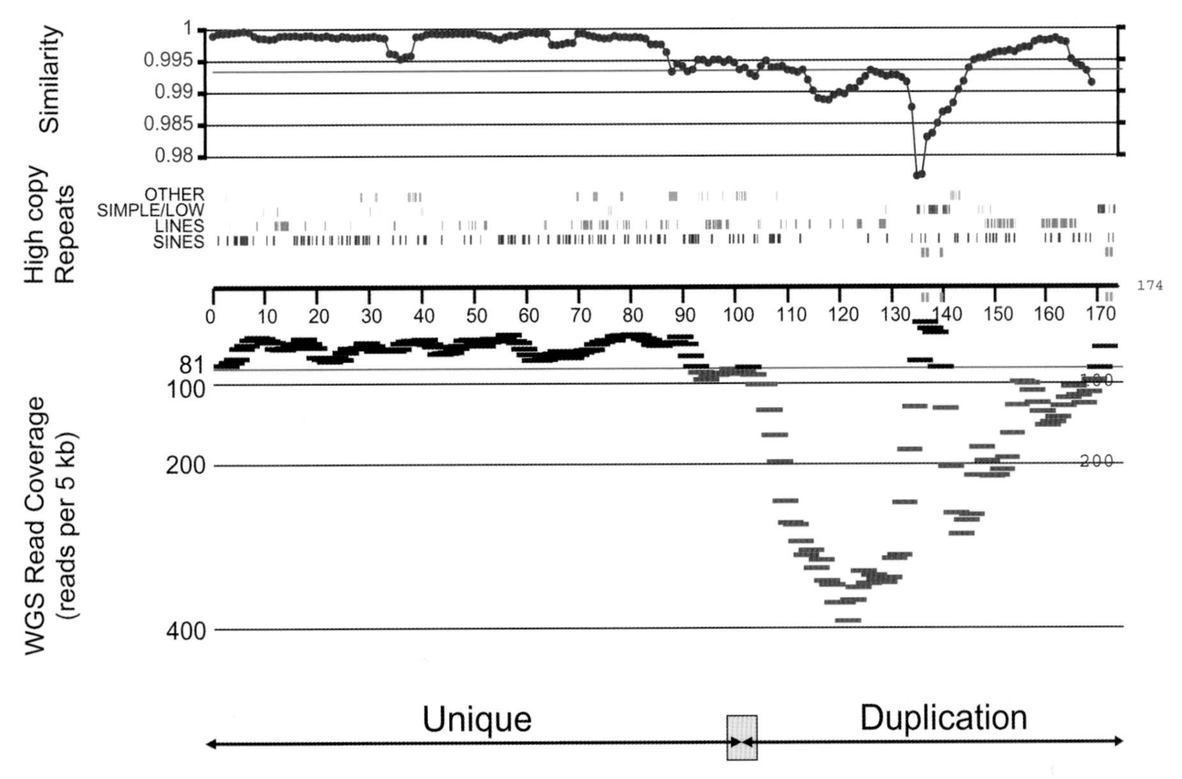


**Figure 1.** Detection of segmental duplications using whole-genome shotgun (WGS) sequence. We compared the sequence of all 32,610 publicly available clones to 27.3 million Celera WGS reads (Bailey et al. 2002b). Reads were recruited to a public sequence if the similarity was >94%. We then calculated the number of reads per 5 kb (bottom *black* and *red* bars). Based on a set of unique clones, windows that surpassed a threshold of 81 reads per 5 kb (3 standard deviations) were considered duplicated (*red bars*). For this example (AC008079.2), a large duplicated region is evident on the right. This clone represents a transition point between unique sequence and the segmental duplications on chromosome 22 involved in velocardiofacial syndrome. Note the tight distribution of read depth within unique sequence (*small black bars*). Also, for divergent duplications, the average fraction similarity (*top*) drops below allelic levels, providing confirmation for all but the most recent duplications.

we created a database of segmental duplication. A total of 8,595 positive duplicated segments were detected within 2,972 clones corresponding to a total of 130.4 Mb of duplicated sequence. This in silico assay was designed to detect duplications between 94% and 100% sequence identity. Experimental analyses confirmed that the strategy had ~99% sensitivity to detect duplications ≥2 copies and ≥15 kb in length. More importantly, it allowed the facile identification of duplications approaching allelic levels of variation. This database was subsequently used to aid in the assembly of the human genome, particularly within duplicated regions. Using this database combined with whole-genome BLAST analysis, we studied the pattern and nature of duplications both on a genome scale and at the chromosome level.

The combination of these computational analyses showed that the human genome is composed of a relatively large percentage of highly similar duplications (5–7%) which appear to have arisen discontinuously over the past 40 million years of primate evolution. The pattern of duplication varies greatly between chromosomes (Figs. 2 and 3). These duplications are distributed nonrandomly within the genome. Duplications showed 3- to 5-fold enrichment within pericentromeric and subtelo meric regions—the pericentromeric and subtelomeric regions of the human genome account for ~60% of all segmental duplications (Bailey et al. 2002b). Each of these regions consists of juxtaposed mosaic sequence blocks that in turn are found on multiple chromosomes. Detailed comparative FISH studies reveal that the pericentromeric region of chromosome 22 is particularly evolutionarily unstable. It varies greatly within the great apes, and contains a human-specific duplication event of at least 600 kb (Bailey et al. 2002a). Intrachromosomal duplications also demonstrate interspersed clustering unlike traditional views of tandemly repeated clusters. Such regions often show a mosaic pattern of duplicated modules, suggesting a complex history of transposition likely due to the predisposition of duplicated material to initiate further duplications and rearrangements. Through detailed analyses of chromosome 22, we identified 11 novel or modified human transcripts created as a result of segmental duplications (Bailey et al. 2002a). Whole-genome analyses show that duplicated genes on average are less conserved when compared to mouse than genes in unique regions (Bailey et al. 2002a). Together, these studies suggest segmental duplication has been an ongoing process of primate evolution contributing to gene evolution and to the rapid remodeling and transformation of genome architecture among closely related species.

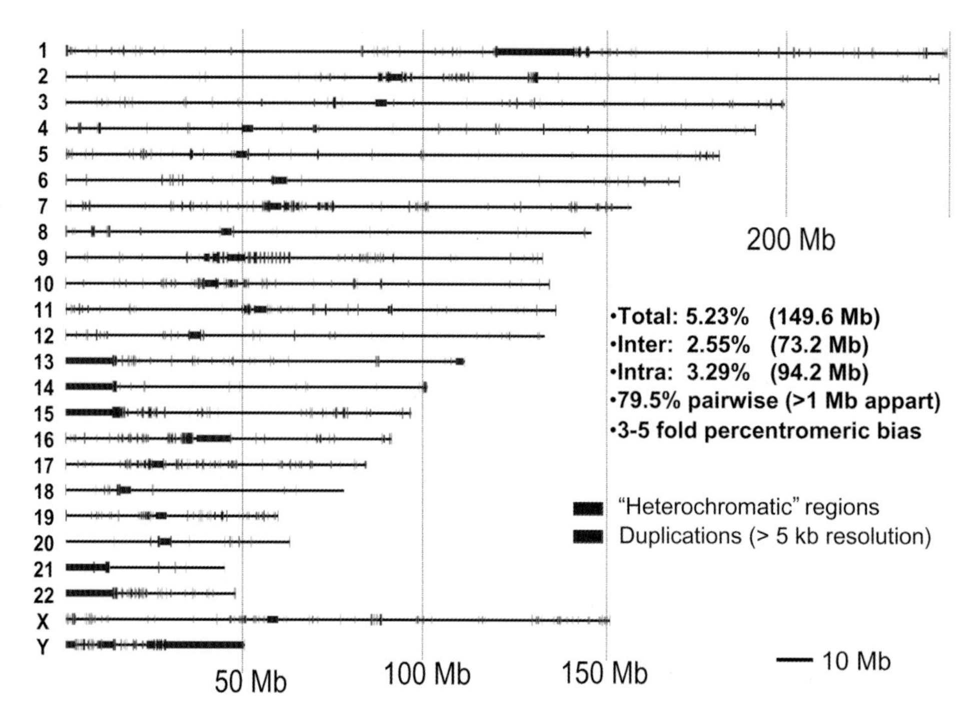


**Figure 2.** Overview of human segmental duplications (build 31). The genome-wide distribution of human segmental duplications (≥5 kb for clarity) are shown with both inter- and intrachromosomal duplications colored red. Purple bars are heterochromatic regions not targeted by the Human Genome Project. Overall, 5.23% (149.6 Mb) of sequence is involved in segmental duplication (≥90% identity and ≥1 kb length). Pericentromeric enrichment (3- to 5-fold on average) is evident for many chromosome arms including all acrocentric chromosomes. One major facet that cannot be appreciated from this picture is that 3 out of 4 alignments are separated by over 1 Mb.

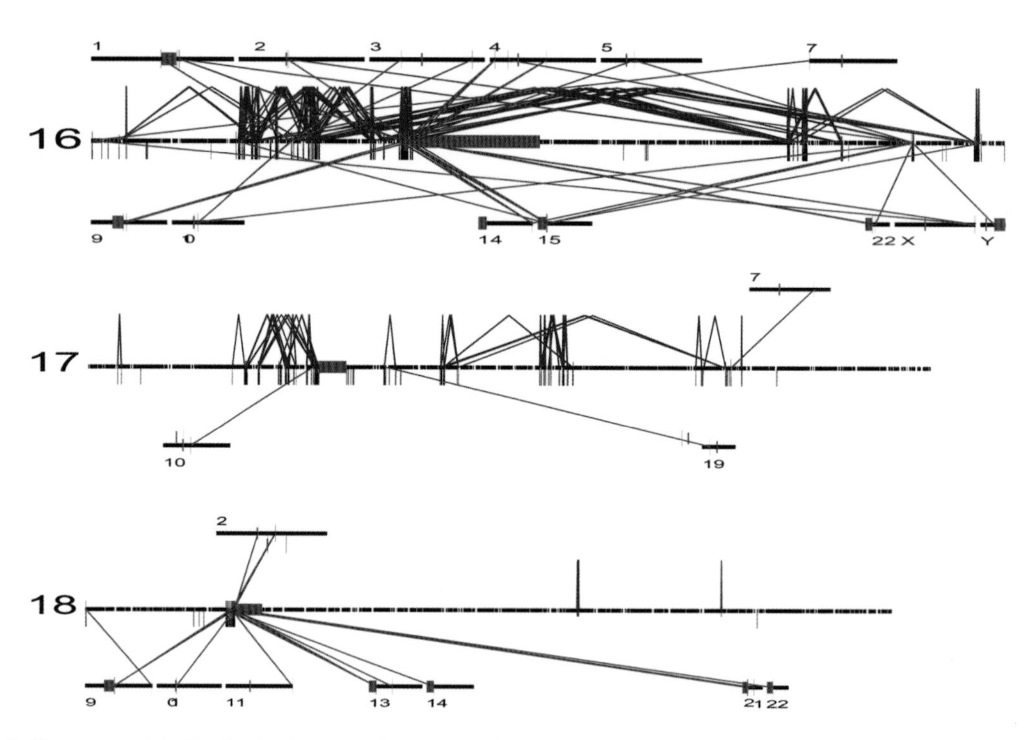


**Figure 3.** Chromosomal duplication landscapes. The patterns of large (≥20 kb) intrachromosomal (*blue*) and interchromosomal (*red*) duplications are shown for human chromosomes 16, 17, and 18 (build 31). Purple bars are heterochromatic regions not targeted by the Human Genome Project. Each chromosome has a distinctive amount and pattern of duplication. Chromosome 16 is one of the most highly duplicated chromosomes with large amounts of both inter- and intrachromosomal duplications. In contrast, chromosome 17 is limited mainly to intrachromosomal duplication, and chromosome 18 is limited to interchromosomal duplication. Note again the pericentromeric localization for the vast majority of interchromosomal duplications.

### Structure and Dynamics of Pericentromeric Duplicated Regions

Concurrent analyses of pericentromeric regions, particularly chromosomes 2 and 10, confirmed our in silico results and yielded further insight into the nature of pericentromeric duplication (Guy et al. 2000, 2003; Horvath et al. 2000, 2001, 2003; Bailey et al. 2002a; Crosier et al. 2002). These analyses suggest that pericentromeric regions are often solely composed of large juxtaposed modules of duplicated sequence (Jackson et al. 1999; Loftus et al. 1999; Ruault et al. 1999; Guy et al. 2000; Horvath et al. 2000; McPherson et al. 2001). There is also evidence for preferential duplication among pericentromeric regions. First, for any given module, there are often large numbers of pericentromeric copies, and the least similar copy (presumably the ancestral source locus) is usually found outside of the pericentromeric region. For instance, the 4 exons comprising the module from ALD (Xq28) are now found in pericentromeric regions at 2p11, 10p11, 16p11, and 22q11, and the most divergent copy is the Xq28 source locus (Eichler et al. 1997). In addition, a partial gene duplication of *NF1*-related sequences is even more dramatic—pericentromeric copies are found on chromosomes 2,14 (x2), 15 (x2), 22 (Regnier et al. 1997). Additional evidence for duplication between pericentromeric regions is that juxtaposed duplicated modules within one pericentromeric region are often found in the same order and orientation within other pericentromeric regions. This argues against independent duplication events of an individual module to the multiple pericentromeric regions, since the more parsimonious explanation is that a larger duplication event between pericentromeric regions moved multiple adjacent modules together, effectively preserving their order and orientation.

These observations combined with in silico data have led to the proposal of a pericentromeric two-step model (Fig. 4) (Eichler et al. 1997; Horvath et al. 2000). The first step is the seeding event, an initial duplication event bringing material from elsewhere in the genome to a pericentromeric region. The second step is subsequent duplication events that ensue between pericentromeric regions. These events usually move larger blocks that encompass the module plus other flanking modules and thus create large regions of shared paralogy between pericentromeric regions. This model is now supported by more extensive studies (Luijten et al. 2000; Crosier et al. 2002; Guy et al. 2003; Horvath et al. 2003).

The close estimate of seeding events, as well their subsequent duplications to other pericentromeric regions, suggests that these duplications have occurred in a punctuated manner during primate evolution, intriguingly near the trichotomization of the African apes (Eichler 2001). Detailed comparative studies of several pericentromeric regions on 2p11 and 16p11 support this thesis. Although such suggestions await larger-scale analysis of multiple pericentromeric regions, these conclusions are not necessarily supported by all duplications characterized to date. Pericentromeric duplications exist that have been seeded before the divergence of great apes and have spread within the pericentromeric regions over long periods of time. The *NF1*-related duplications occurred 22–23 million years ago before the divergence of the great apes, and unlike sequences such as *ALD*, have continued to spread between pericentromeric regions until the present (Regnier et al. 1997; Luijten et al. 2000). In the case of keratinocyte growth factor, interchromosomal duplications are found within orangutan, suggesting that duplications occurred before the divergence of orangutan and the other great apes (Zimonjic et al. 1997). These observations do not preclude increased amounts of preferential pericentromeric duplications occurring around the trichotomization of human, chimp, and gorilla. These contrasting observations of seeding and exchange events at different times can be explained only if certain locations within pericentromeric regions are active at any given time, in terms of susceptibility both to initial seeding events and to subsequent duplications to other pericentromeric regions. However, answering these questions will require a global analysis to parsimoniously reconstruct the evolution of these regions. At present, genome-wide analyses can merely establish whether pericentromeric regions are enriched for duplications and provide some estimate of the degree of sequence similarity.

Two other points regarding pericentromeric regions are worth noting. First, examination by comparative FISH in the great apes, chimpanzees, gorillas, and orangutan often reveals variation in the number and position of duplication modules. Typically, FISH signals are lost or gained in a lineage-specific manner, suggesting that pericentromeric duplications continued after the divergence of the African apes from each other (Zimonjic et al. 1997; Horvath et al. 2000). In other words, these regions appear to be continuously evolving and thus have the potential to underlie phenotypic differences between these species. Second, these regions appear to be gene-poor. Jackson

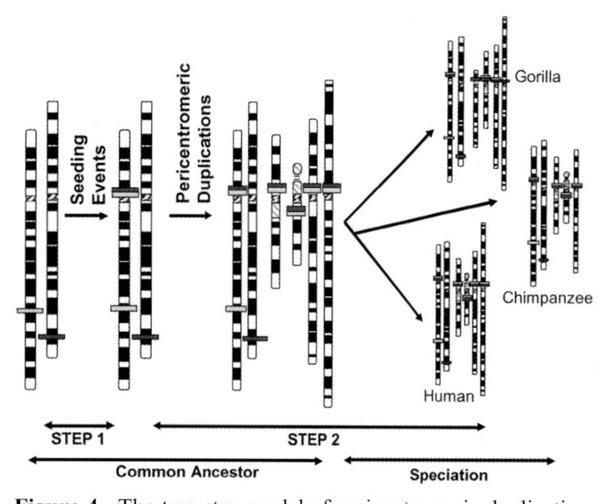


**Figure 4.** The two-step model of pericentromeric duplication. Initial seeding events (step *1*) are followed by duplications of larger blocks of sequence between pericentromeric regions (step *2*). When speciation events occur, further duplications, deletions, and other rearrangements create variation between species. Figure based on duplications of ALD and 4q24 evolution (Horvath et al. 2000).

and other investigators have suggested that the perceived enrichment of duplications within the pericentromeric regions may simply be due to the lack of deleterious consequences that would select against their fixation (Jackson et al. 1999; Guy et al. 2000). Other gene-poor regions of the genome that are non-pericentromeric and lack duplications do not support this claim (Hattori et al. 2000). It is likely that reduced selection and increased generation of duplications act in concert to create the large blocks of duplications within pericentromeric regions.

### Structure and Dynamics of Subtelomeric Regions

Telomeres are essential for chromosome stability and consist of a TTAGGG repeat that is created and maintained by the enzyme telomerase. Between these "stabilizing repeats" and the unique euchromatic sequences of the chromosome arms lie the subtelomeric regions that consist of divergent telomeric sequence interspersed with segmental duplications (Riethman et al. 2001; Mefford and Trask 2002). Like pericentromeric regions, subtelomeric regions appear to preferentially duplicate material to other subtelomeric regions (Monfouilloux et al. 1998; Grewal et al. 1999; van Geel et al. 2002). Many of these regions, when examined by FISH, differ both between species and also within the human population (Monfouilloux et al. 1998; Trask et al. 1998a; Park et al. 2000), suggesting that gene differences caused by the presence or absence of these regions could explain phenotypic differences both within and between species. Unlike pericentromeric regions, subtelomeric regions are gene-rich, and the duplications often contain expressed genes (Riethman et al. 2001). Examples of transcribed genes that have been duplicated include *RAB*-like genes, *TUB4Q* members, and myosin light-chain kinase (Brand-Arpon et al. 1999; Wong et al. 1999; van Geel et al. 2000). The most striking gene duplications found in the subtelomeric regions involve the olfactory receptors. Olfactory receptors are found throughout the genome (Glusman et al. 2001), but many appear to have spread recently through subtelomeric segmental duplication (Rouquier et al. 1998; Trask et al. 1998b; Glusman et al. 2001; Mefford et al. 2001). The evolutionary instability of subtelomeric regions and their propensity to accumulate segmental duplications may have been critical in the expansion of this gene family.

### Segmental Duplications and Human Disease

It has long been recognized that repetitive DNA can mediate misalignment resulting in aberrant homologous recombination that creates rearrangements—mainly deletions and duplications. Clinical diseases caused by this mechanism have been labeled genomic disorders and have been reviewed extensively (Mazzarella and Schlessinger 1997, 1998; Lupski 1998; Ji et al. 2000; Stankiewicz and Lupski 2002). The thalassemias were the first genomic disorders to be characterized as an event occurring as a result of unequal-homologous recombination. For both α and β thalassemia, misalignment and recombination arise due to the highly similar tandemly repeated sequence within and between the genes (Kunkel et al. 1969; Phillips et al. 1980). In the case of β-globin, recombination takes place within the 5′ situated γ-globin and the 3′ β-globin, resulting in a γ–β deletion fusion called Hb Lepore. The corresponding meiotic homolog is a chromosome with 3 genes: γ-globin, β–γ fusion, and β-globin.

In addition to tandemly duplicated genes, another class of genomic disorders is due to interspersed duplications. Since the identification of the CMT-1A REPs (Pentao et al. 1992), a growing list of genomic disorders have been characterized that involve rearrangements between distantly spaced, highly similar (>98%) sequences. Such interspersed segmental duplications are often referred to as low-copy repeats (LCRs) or duplicons and are also associated with nonallelic homologous recombination (NAHR) (Stankiewicz and Lupski 2002) between copies. When NAHR occurs between these distantly spaced copies, it results in rearrangements not only of the duplicons, but also of the intervening sequence, often resulting in gains and losses of blocks of intervening unique sequence (segmental aneusomy). If these unique regions contain genes that are triplosensitive, haploinsufficient, or imprinted, disease may occur. Characterized interspersed genomic disorders now include the 17p11 duplication of Charcot-Marie-Tooth disease 1A (Reiter et al. 1996); the 17p11 deletion of hereditary neuropathy with liability to pressure palsies (HNPP) (Chen et al. 1997); the 22q11 deletion in velocardiofacial syndrome (VCFS), the most common microdeletion syndrome (Morrow et al. 1997; Shaikh et al. 2000); and the 15q11-q13 deletion in Prader-Willi/Angelman syndrome (PW/AS) (Amos-Landgraf et al. 1999). An extensive list of genomic disorders mediated by nonallelic homologous recombination has been described previously (Stankiewicz and Lupski 2002).

From the examination of the sequence properties of segmental duplications associated with genetic disease, it becomes apparent that the size, relative orientation, degree of sequence identity, and length of intervening sequence are key factors in determining the frequency and type of rearrangement. In general, as segmental duplications become larger and share higher sequence identity, there is an increased propensity for rearrangement (Reiter et al. 1996, 1998; Edelmann et al. 1999; Lopez-Correa et al. 2001). We examined the human genome for all occurrences where duplications had >95% sequence identity and were larger than 10 kb in length, and where the intervening sequence separating two duplications was between 50 kb and 5 Mb in length (Bailey et al. 2002b). This analysis identified 163 potential regions of the human genome that would be predicted to be sites of structural rearrangement—23 of these regions were already associated with some form of genomic disease. A detailed assessment, thus, offers considerable clinical value in the sense that it provides a road map for the identification and characterization of sites of recurrent structural rearrangement within the human population associated with both genetic disease and large-scale polymorphism.

### Comparative Analyses

Are recent interspersed duplications of genomic sequence a common property of all genomes, or is their occurrence largely restricted to the primate genome? Our analysis indicates two types of recent interspersed duplications exist in the human genome, the chromosome-specific and interchromosomal repeats. Similar duplications, at least in apparent size or frequency, have not been reported as of yet for any other organism. Differences in methods of genomic and genetic characterization in these species, however, could largely have explained this effect. With the advent of whole-genome sequencing, we sought to address this question in an unbiased fashion, by direct examination of nucleotide sequence within other species. An identical analysis was performed for the recently published genomes of *Caenorhabditis elegans* (Consortium 1998), *Drosophila melanogaster* (Adams et al. 2000), *Takifugu rubripes* (Aparicio et al. 2002), and *Mus musculus* (Waterston et al. 2002). Very little evidence for large (≥20 kb), highly homologous (≥90%) duplications could be found within these species (Table 3). Although retroposon accumulation biases have been documented for *Drosophila*, no subtelomeric or pericentromeric clustering of duplicated segments could be ascertained. Furthermore, these comparisons indicate that the human genome is enriched 5- to 100-fold for such duplications when compared to genomes of model organisms (Table 3). Although there may be several explanations for this effect (differences in genome size, differential rates of recombination, methodological differences in sequence assembly, etc.), the structure of the human genome appears structurally distinct with respect to recent large-scale interspersed duplications.

**Table 3.** Segmental Duplications in Other Sequenced Organisms

| Size | Fly | Worm | Fugu[a] | Mouse | Human[b] |
|---|---|---|---|---|---|
| ≥1 kb | 1.20% | 4.25% | 2.18% | ND | 5.23% |
| ≥5 kb | 0.37% | 1.50% | 0.03% | 1.95% | 4.78% |
| ≥10 kb | 0.08% | 0.66% | 0.00% | 0.70% | 4.52% |
| ≥20 kb | 0.00% | ND | 0.00% | 0.11% | 4.06% |

[a]*Takifugu rubripes* build 3.0.
[b]Build 31, Nov. 2002.

## CONCLUSIONS

Based on our current analysis of genomes, the genome of *Homo sapiens* is unique in the abundance and distribution of large (≥20 kb), highly homologous (≥95%) segmental duplications. This unusual architecture of the human genome has important practical and biological implications. In the late 1990s, there was considerable debate between advocates of the whole-genome shotgun and those of the clone-ordered approaches for human genome sequence and assembly (Green 1997; Weber and Myers 1997). Our analysis of the initial private and public genome assemblies (Lander et al. 2001; Venter et al. 2001) indicated that neither effectively resolved the organization and sequence of regions containing large segmental duplications. Ironically, combining both of these approaches did provide the most effective means for the identification, characterization, and subsequent resolution of many of these regions. These data suggest that a combined whole-genome shotgun and clone-ordered approach may be the best strategy for the completion of complex genomes that are laden with large segmental duplications. Given the fact that the human genome harbors nearly 100 such sites, each greater than 300 kb, high-sequence-identity duplications continue to impede gene annotation and SNP characterization of the human genome. Emerging evidence that such regions vary structurally depending on the human haplotype further complicates their sequence and assembly. Taken in this light, it is perhaps not surprising that the majority of the large gaps that remain within the human genome project as of April 2003 are flanked by such duplications.

From the biological perspective, this architecture has two important implications—one functional and the other structural. The ability to juxtapose segments that would have never shared proximity in the genome of an ancestral species offers tremendous potential for exon shuffling and domain accretion of the proteome. Many such chimeric transcripts have now been documented (Courseaux and Nahon 2001; Bailey et al. 2002a; Stankiewicz and Lupski 2002; Bridgland et al. 2003) where different portions of the transcript originate from diverse regions. Few of these "fusions" appear to produce functional proteins (Hillier et al. 2003). Rare exceptions have been noted, such as the emergence of the *TRE2* (also known as *USP6*) gene specifically within the hominoid lineage. In this case, approximately half of its 30 exons arose from a segmental duplication of the *USP32* ancestral gene, whereas the amino-terminal portion of this oncogene originated as a duplication of the *TBC1D3* ancestral gene. The fused transcript emerged during the radiation of the great apes, producing a gene with tissue specificity different from either of the progenitor genes (Paulding et al. 2003). In addition to gene innovation through exon shuffling, segmental duplications have the potential to lead to the emergence of novel genes through adaptive evolution. The remarkable positive selection of the *morpheus* gene family on chromosome 16, where the genes show accelerated amino acid replacement an order of magnitude above the neutral expectation, may be an example of such an effect (Johnson et al. 2001).

From the structural perspective, the current architecture of the human genome suggests that subtle remodulation of many specific chromosomal regions has occurred over short periods of primate evolution. This observation challenges the rather static notion of genome evolution that has emerged from early karyotype and chromosome-painting studies. Whereas the majority of the human genome seems to fit well with this model of conservation, the duplicated regions, in contrast, have been particularly prone to multiple, independent occurrences of rearrangement through duplication. In humans, the majority of intrachromosomally duplicated copies are separated by more than a megabase of intervening sequence. Such architecture has been rarely observed in other model organisms. The mechanism responsible for this event is unknown, but it is noteworthy that Alu repeat elements are frequently observed at the breakpoints of segmental duplications (Bailey et al. 2003). It is possible that the unusual phylogeny of the Alu repeat family may have particularly predisposed the primate genome to segmental duplication. In addition to subtle structural events, it is also becoming apparent that segmental duplications are associated with large-scale structural changes originally observed in the early karyotype studies of human and great-ape chromosomes (Yunis and Prakash 1982). To date, four of five large-scale chromosomal rearrangements that have been characterized at the molecular level show the presence of segmental duplications precisely at the breakpoint (Nickerson et al. 1999; Stankiewicz et al. 2001; Kehrer-Sawatzki et al. 2002; Eder et al. 2003). Although the cause and consequence relationship has not been defined, the data suggest that an understanding of the evolution and origin of segmental duplications will be critical to realizing the nature and pattern of primate chromosome variation. Such information is crucial in our complete reconstruction of the evolutionary history of the genome of *Homo sapiens*.

## REFERENCES


Adams M.D., Celniker S.E., Holt R.A., Evans C.A., Gocayne J.D., Amanatides P.G., Scherer S.E., Li P.W., Hoskins R.A., Galle R.F., George R.A., Lewis S.E., Richards S., Ashburner M., Henderson S.N., Sutton G.G., Wortman J.R., Yandell M.D., Zhang Q., Chen L.X., Brandon R.C., Rogers Y.H., Blazej R.G., Champe M., and Pfeiffer B.D., et al. 2000. The genome sequence of *Drosophila melanogaster*. *Science* **287:** 2185.

Amos-Landgraf J.M., Ji Y., Gottlieb W., Depinet T., Wandstrat A.E., Cassidy S.B., Driscoll D.J., Rogan P.K., Schwartz S., and Nicholls R.D. 1999. Chromosome breakage in the Prader-Willi and Angelman syndromes involves recombination between large, transcribed repeats at proximal and distal breakpoints. *Am. J. Hum. Genet.* **65:** 370.

Aparicio S., Chapman J., Stupka E., Putnam N., Chia J.M., Dehal P., Christoffels A., Rash S., Hoon S., Smit A., Gelpke M.D., Roach J., Oh T., Ho I.Y., Wong M., Detter C., Verhoef F., Predki P., Tay A., Lucas S., Richardson P., Smith S.F., Clark M.S., Edwards Y.J., and Doggett N., et al. 2002. Whole-genome shotgun assembly and analysis of the genome of *Fugu rubripes*. *Science* **297:** 1301.

Bailey J.A., Liu G., and Eichler EE. 2003. An Alu transposition model for the origin and expansion of human segmental duplications. *Am. J. Hum. Genet.* **73:** 823.

Bailey J.A., Yavor A.M., Massa H.F., Trask B.J., and Eichler E.E. 2001. Segmental duplications: Organization and impact within the current human genome project assembly. *Genome Res.* **11:** 1005.

Bailey J.A., Yavor A.M., Viggiano L., Misceo D., Horvath J.E., Archidiacono N., Schwartz S., Rocchi M., and Eichler E.E. 2002a. Human-specific duplication and mosaic transcripts: The recent paralogous structure of chromosome 22. *Am. J. Hum. Genet.* **70:** 83.

Bailey J.A., Gu Z., Clark R.A., Reinert K., Samonte R.V., Schwartz S., Adams M.D., Myers E.W., Li P.W., and Eichler E.E. 2002b. Recent segmental duplications in the human genome. *Science* **297:** 1003.

Brand-Arpon V., Rouquier S., Massa H., de Jong P.J., Ferraz C., Ioannou P.A., Demaille J.G., Trask B.J., and Giorgi D. 1999. A genomic region encompassing a cluster of olfactory receptor genes and a myosin light chain kinase (MYLK) gene is duplicated on human chromosome regions 3q13-q21 and 3p13. *Genomics* **56:** 98.

Bridgland L., Footz T.K., Kardel M.D., Riazi M.A., and Mc Dermid H.E.. 2003. Three duplicons form a novel chimeric transcription unit in the pericentromeric region of chromosome 22q11. *Hum. Genet.* **112:** 57.

Chen K., Manian P., Koeuth T., Potocki L., Zhao Q., Chinault A., Lee C., and Lupski J. 1997. Homologous recombination of a flanking repeat gene cluster is a mechanism for a common contiguous gene deletion syndrome. *Nat. Genet.* **17:** 154.

Collins F.S., Patrinos A., Jordan E., Chakravarti A., Gesteland R., and Walters L. 1998. New goals for the U.S. Human Genome Project: 1998–2003. *Science* **282:** 682.

Consortium (The *C. elegans* Sequencing Consortium). 1998. Genome sequence of the nematode *C. elegans:* A platform for investigating biology. *Science* **282:** 2012.

Courseaux A. and Nahon J.L. 2001. Birth of two chimeric genes in the Hominidae lineage. *Science* **291:** 1293.

Crosier M., Viggiano L., Guy J., Misceo D., Stones R., Wei W., Hearn T., Ventura M., Archidiacono N., Rocchi M., and Jackson M.S. 2002. Human paralogs of KIAA0187 were created through independent pericentromeric-directed and chromosome-specific duplication mechanisms. *Genome Res.* **12:** 67.

Deininger P.L. and Schmid C.W. 1976. Thermal stability of human DNA and chimpanzee DNA heteroduplexes. *Science* **194:** 846.

Edelmann L., Pandita R.K., and Morrow B.E. 1999. Low-copy repeats mediate the common 3-Mb deletion in patients with velo-cardio-facial syndrome. *Am. J. Hum. Genet.* **64:** 1076.

Eder A., Mario V., Ianigro M., Teti M., Rocchi M., and Archidiacono N. 2003. Chromosome 6 phylogeny in primates and centromere repositioning. *Mol. Biol. Evol.* **20:** 1506.

Eichler E.E. 2001. Recent duplication, domain accretion and the dynamic mutation of the human genome. *Trends Genet.* **17:** 661.

Eichler E.E., Budarf M.L., Rocchi M., Deaven L.L., Doggett N.A., Baldini A., Nelson D.L., and Mohrenweiser H.W. 1997. Interchromosomal duplications of the adrenoleukodystrophy locus: A phenomenon of pericentromeric plasticity. *Hum. Mol. Genet.* **6:** 991.

Glusman G., Yanai I., Rubin I., and Lancet D. 2001. The complete human olfactory subgenome. *Genome Res.* **11:** 685.

Graur D. and Li W.-H. 2000. *Fundamentals of molecular evolution*. Sinauer, Sunderland, Massachusetts.

Green P. 1997. Against a whole-genome shotgun. *Genome Res.* **7:** 410.

Grewal P.K., van Geel M., Frants R.R., de Jong P., and Hewitt J.E. 1999. Recent amplification of the human FRG1 gene during primate evolution. *Gene* **227:** 79.

Guy J., Spalluto C., McMurray A., Hearn T., Crosier M., Viggiano L., Miolla V., Archidiacono N., Rocchi M., Scott C., Lee P.A., Sulston J., Rogers J., Bentley D., and Jackson M.S. 2000. Genomic sequence and transcriptional profile of the boundary between pericentromeric satellites and genes on human chromosome arm 10q. *Hum. Mol. Genet.* **9:** 2029.

Guy J., Hearn T., Crosier M., Mudge J., Viggiano L., Koczan D., Thiesen H.J., Bailey J.A., Horvath J.E., Eichler E.E., Earthrowl M.E., Deloukas P., French L., Rogers J., Bentley D., and Jackson M.S. 2003. Genomic sequence and transcriptional profile of the boundary between pericentromeric satellites and genes on human chromosome arm 10p. *Genome Res.* **13:** 159.

Hattori M., Fujiyama A., Taylor T.D., Watanabe H., Yada T., Park H.S., Toyoda A., Ishii K., Totoki Y., Choi D.K., Soeda E., Ohki M., Takagi T., Sakaki Y., Taudien S., Blechschmidt K., Polley A., Menzel U., Delabar J., Kumpf K., Lehmann R., Patterson D., Reichwald K., Rump A., and Schillhabel M., et al. 2000. The DNA sequence of human chromosome 21. The chromosome 21 mapping and sequencing consortium. *Nature* **405:** 311.

Hillier L.W., Fulton R.S., Fulton L.A., Graves T.A., Pepin K.H., Wagner-McPherson C., Layman D., Maas J., Jaeger S., Walker R., Wylie K., Sekhon M., Becker M.C., O'Laughlin M.D., Schaller M.E., Fewell G.A., Delehaunty K.D., Miner T.L., Nash W.E., Cordes M., Du H., Sun H., Edwards J., Bradshaw-Cordum H., and Ali J., et al. 2003. The DNA sequence of human chromosome 7. *Nature* **424:** 157.

Horvath J., Schwartz S., and Eichler E. 2000. The mosaic structure of a 2p11 pericentromeric segment: A strategy for characterizing complex regions of the human genome. *Genome Res.* **10:** 839.

Horvath J.E., Bailey J.A., Locke D.P., and Eichler E.E. 2001. Lessons from the human genome: Transitions between euchromatin and heterochromatin. *Hum. Mol. Genet.* **10:** 2215.

Horvath J.E., Gulden C.L., Bailey J.A., Yohn C., McPherson J.D., Prescott A., Roe B.A., De Jong P.J., Ventura M., Misceo D., Archidiacono N., Zhao S., Schwartz S., Rocchi M., and Eichler E.E. 2003. Using a pericentromeric interspersed repeat to recapitulate the phylogeny and expansion of human centromeric segmental duplications. *Mol. Biol. Evol.* **20:** 1463.

Jackson M.S., Rocchi M., Thompson G., Hearn T., Crosier M., Guy J., Kirk D., Mulligan L., Ricco A., Piccininni S., Marzella R., Viggiano L., and Archidiacono N. 1999. Sequences flanking the centromere of human chromosome 10 are a complex patchwork of arm-specific sequences, stable duplications, and unstable sequences with homologies to telomeric and other centromeric locations. *Hum. Mol. Genet.* **8:** 205.

Ji Y., Eichler E.E., Schwartz S., and Nicholls R.D. 2000. Structure of chromosomal duplicons and their role in mediating human genomic disorders. *Genome Res.* **10:** 597.

Johnson M.E., Viggiano L., Bailey J.A., Abdul-Rauf M., Goodwin G., Rocchi M., and Eichler E.E. 2001. Positive selection of a gene family during the emergence of humans and African apes. *Nature* **413:** 514.

Kehrer-Sawatzki H., Schreiner B., Tanzer S., Platzer M., Muller S., and Hameister H. 2002. Molecular characterization of the pericentric inversion that causes differences between chimpanzee chromosome 19 and human chromosome 17. *Am. J. Hum. Genet.* **71:** 375.

Kunkel H.G., Natvig J.B., and Joslin F.G. 1969. A "Lepore" type of hybrid gamma-globulin. *Proc. Natl. Acad. Sci.* **62:** 144.

Lander E.S., Linton L.M., Birren B., Nusbaum C., Zody M.C., Baldwin J., Devon K., Dewar K., Doyle M., FitzHugh W., Funke R., Gage D., Harris K., Heaford A., Howland J., Kann L., Lehoczky J., LeVine R., McEwan P., McKernan K., Meldrim J., Mesirov J.P., Miranda C., Morris W., and Naylor J., et al. (International Human Genome Sequencing Consortium). 2001. Initial sequencing and analysis of the human genome. *Nature* **409:** 860.

Loftus B.J., Kim U.J., Sneddon V.P., Kalush F., Brandon R., Fuhrmann J., Mason T., Crosby M.L., Barnstead M., Cronin L., Deslattes Mays A., Cao Y., Xu R.X., Kang H.L., Mitchell S., Eichler E.E., Harris P.C., Venter J.C., and Adams M.D. 1999. Genome duplications and other features in 12 Mbp of DNA sequence from human chromosome 16p and 16q. *Genomics* **60:** 295.

Lopez-Correa C., Dorschner M., Brems H., Lazaro C., Clementi M., Upadhyaya M., Dooijes D., Moog U., Kehrer-Sawatzki H., Fryns J.P., Rutkowski J.L., Marynen P., Stephens K., and Legius E. 2001. Recombination hotspot in NF1 microdeletion patients. *Hum. Mol. Genet.* **10:** 1387.

Luijten M., Wang Y., Smith B.T., Westerveld A., Smink L.J., Dunham I., Roe B.A., and Hulsebos T.J. 2000. Mechanism of spreading of the highly related neurofibromatosis type 1 (NF1) pseudogenes on chromosomes 2, 14 and 22. *Eur. J. Hum. Genet.* **8:** 209.

Lupski J.R. 1998. Genomic disorders: Structural features of the genome can lead to DNA rearrangements and human disease traits. *Trends Genet.* **14:** 417.

Mazzarella R. and Schlessinger D. 1997. Duplication and distribution of repetitive elements and non-unique regions in the human genome. *Gene* **205:** 29.

———. 1998. Pathological consequences of sequence duplications in the human genome. *Genome Res.* **8:** 1007.

McPherson J.D., Marra M., Hillier L., Waterston R.H., Chinwalla A., Wallis J., Sekhon M., Wylie K., Mardis E.R., Wilson R.K., Fulton R., Kucaba T.A., Wagner-McPherson C., Barbazuk W.B., Gregory S.G., Humphray S.J., French L., Evans R.S., Bethel G., Whittaker A., Holden J.L., McCann O.T., Dunham A., Soderlund C., and Scott C.E., et al. (International Human Genome Mapping Consortium). 2001. A physical map of the human genome. *Nature* **409:** 934.

Mefford H.C. and Trask B.J. 2002. The complex structure and dynamic evolution of human subtelomeres. *Nat. Rev. Genet.* **3:** 91.

Mefford H.C., Linardopoulou E., Coil D., van den Engh G., and Trask B.J. 2001. Comparative sequencing of a multicopy subtelomeric region containing olfactory receptor genes reveals multiple interactions between non-homologous chromosomes. *Hum. Mol. Genet.* **10:** 2363.

Monfouilloux S., Avet-Loiseau H., Amarger V., Balazs I., Pourcel C., and Vergnaud G. 1998. Recent human-specific spreading of a subtelomeric domain. *Genomics* **51:** 165.

Morrow B., Edelmann L., Ferreira J., Pandita R., Carlson C., Procter J., Jackson M., Wilson D., Goldberg R., Shprintzen R., and Kucherlapati R. 1997. A duplication of chromosome 22q11 is the basis for the common deletion that occurs in velo-cardiofacial syndrome patients. *Am. J. Hum. Genet.* (suppl.) **61:** A7.

Muller H.J. 1936. Bar duplication. *Science* **83:** 528.

Murphy W.J., Stanyon R., and O'Brien S.J. 2001. Evolution of mammalian genome organization inferred from comparative gene mapping. *Genome Biol.* **2:** REVIEWS0005.

Nei M. 1968. The frequency distribution of lethal chromosomes in finite populations. *Proc. Natl. Acad. Sci.* **60:** 517.

Nickerson E., Gibbs R.A., and Nelson D.L. 1999. Sequence analysis of the breakpoints of a pericentric inversion distinguishing the human and chimpanzee chromosomes 12. *Am. J. Hum. Genet.* **65:** A291.

Ohno S. 1972. So much "junk" DNA in our genome. *Brookhaven Symp. Biol.* **23:** 366.

Ohno S., Wolf U., and Atkin N. 1968. Evolution from fish to mammals by gene duplication. *Hereditas* **59:** 169.

Park H.S., Nogami M., Okumura K., Hattori M., Sakaki Y., and Fujiyama A. 2000. Newly identified repeat sequences, derived from human chromosome 21qter, are also localized in the subtelomeric region of particular chromosomes and 2q13, and are conserved in the chimpanzee genome. *FEBS Lett.* **475:** 167.

Paulding C.A., Ruvolo M., and Haber D.A. 2003. The Tre2 (USP6) oncogene is a hominoid-specific gene. *Proc. Natl. Acad. Sci.* **100:** 2507.

Pentao L., Wise C., Chinault A., Patel P., and Lupski J. 1992. Charcot-Marie-Tooth type 1A duplication appears to arise from recombination at repeat sequences flanking the 1.5 Mb monomer unit. *Nat. Genet.* **2:** 292.

Phillips J.A., III, Vik T.A., Scott A.F., Young K.E., Kazazian H.H., Jr., Smith K.D., Fairbanks V.F., and Koenig H.M. 1980. Unequal crossing-over: A common basis of single alpha-globin genes in Asians and American blacks with hemoglobin-H disease. *Blood* **55:** 1066.

Regnier V., Meddeb M., Lecointre G., Richard F., Duverger A., Nguyen V.C., Dutrillaux B., Bernheim A., and Danglot G. 1997. Emergence and scattering of multiple neurofibromatosis (NF1)-related sequences during hominoid evolution suggest a process of pericentromeric interchromosomal transposition. *Hum. Mol. Genet.* **6:** 9.

Reiter L.T., Hastings P.J., Nelis E., De Jonghe P., Van Broeckhoven C., and Lupski J.R. 1998. Human meiotic recombination products revealed by sequencing a hotspot for homologous strand exchange in multiple HNPP deletion patients. *Am. J. Hum. Genet.* **62:** 1023.

Reiter L., Murakami T., Koeuth T., Pentao L., Muzny D., Gibbs R., and Lupski J. 1996. A recombination hotspot responsible for two inherited peripheral neuropathies is located near a mariner transposon-like element. *Nat. Genet.* **12:** 288.

Riethman H.C., Xiang Z., Paul S., Morse E., Hu X.L., Flint J., Chi H.C., Grady D.L., and Moyzis R.K. 2001. Integration of telomere sequences with the draft human genome sequence. *Nature* **409:** 948.

Rouquier S., Taviaux S., Trask B.J., Brand-Arpon V., van den Engh G., Demaille J., and Giorgi D. 1998. Distribution of olfactory receptor genes in the human genome. *Nat. Genet.* **18:** 243.

Ruault M., Trichet V., Gimenez S., Boyle S., Gardiner K., Rolland M., Roizes G., and De Sario A. 1999. Juxta-centromeric region of human chromosome 21 is enriched for pseudogenes and gene fragments. *Gene* **239:** 55.

Schmid C.W. and Deininger P.L. 1975. Sequence organization of the human genome. *Cell* **6:** 345.

Shaikh T.H., Kurahashi H., Saitta S.C., O'Hare A.M., Hu P., Roe B.A., Driscoll D.A., McDonald-McGinn D.M., Zackai E.H., Budarf M.L., and Emanuel B.S. 2000. Chromosome 22-specific low copy repeats and the 22q11.2 deletion syndrome: Genomic organization and deletion endpoint analysis. *Hum. Mol. Genet.* **9:** 489.

Stankiewicz P. and Lupski J.R. 2002. Genome architecture, rearrangements and genomic disorders. *Trends Genet.* **18:** 74.

Stankiewicz P., Park S.S., Inoue K., and Lupski J.R. 2001. The evolutionary chromosome translocation 4;19 in Gorilla gorilla is associated with microduplication of the chromosome fragment syntenic to sequences surrounding the human proximal CMT1A-REP. *Genome Res.* **11:** 1205.

Trask B.J., Massa H., Brand-Arpon V., Chan K., Friedman C., Nguyen O.T., Eichler E.E., van den Engh G., Rouquier S., Shizuya H., and Giorgi D. 1998a. Large multi-chromosomal duplications encompass many members of the olfactory receptor gene family in the human genome. *Hum Mol. Genet.* **7:** 2007.

Trask B., Friedman C., Martin-Gallardo A., Rowen L., Akinbami C., Blankenship J., Collins C., Giorgi D., Iadonato S., Johnson F., Kuo W., Massa H., Morrish T., Naylor S., Nguyen O., Rouquier S., Smith T., Wong D., Younglbom J., and van den Engh G. 1998b. Members of the olfactory receptor gene family are contained in large blocks of DNA duplicated polymorphically near the ends of human chromosomes. *Hum. Mol. Genet.* **7:** 13.

van Geel M., Eichler E.E., Beck A.F., Shan Z., Haaf T., van Der Maarel S.M., Frants R.R., and de Jong P.J. 2002. A cascade of complex subtelomeric duplications during the evolution of the hominoid and Old World monkey genomes. *Am. J. Hum. Genet.* **70:** 269.

van Geel M., van Deutekom J.C., van Staalduinen A., Lemmers R.J., Dickson M.C., Hofker M.H., Padberg G.W., Hewitt J.E., de Jong P.J., and Frants R.R. 2000. Identification of a novel beta-tubulin subfamily with one member (TUBB4Q) located near the telomere of chromosome region 4q35. *Cytogenet. Cell Genet.* **88:** 316.

Venter J.C., Adams M.D., Myers E.W., Li P.W., Mural R.J., Sutton G.G., Smith H.O., Yandell M., Evans C.A., Holt R.A., Gocayne J.D., Amanatides P., Ballew R.M., Huson D.H., Wortman J.R., Zhang Q., Kodira C.D., Zheng X.H., Chen L., Skupski M., Subramanian G., Thomas P.D., Zhang J., Gabor Miklos G.L., and Nelson C., et al. 2001. The sequence of the human genome. *Science* **291:** 1304.

Waterston R.H., Lindblad-Toh K., Birney E., Rogers J., Abril J.F., Agarwal P., Agarwala R., Ainscough R., Alexandersson M., An P., Antonarakis S.E., Attwood J., Baertsch R., Bailey J., Barlow K., Beck S., Berry E., Birren B., Bloom T., Bork P., Botcherby M., Bray N., Brent M.R., Brown D.G., and Brown S.D., et al. 2002. Initial sequencing and comparative analysis of the mouse genome. *Nature* **420:** 520.

Weber J.L. and Myers E.W. 1997. Human whole-genome shotgun sequencing. *Genome Res.* **7:** 401.

Wienberg J., Stanyon R., Nash W.G., O'Brien P.C., Yang F., O'Brien S.J., and Ferguson-Smith M.A. 1997. Conservation of human vs. feline genome organization revealed by reciprocal chromosome painting. *Cytogenet. Cell Genet.* **77:** 211.

Wong A.C., Shkolny D., Dorman A., Willingham D., Roe B.A., and McDermid H.E. 1999. Two novel human RAB genes with near identical sequence each map to a telomere-associated region: The subtelomeric region of 22q13.3 and the ancestral telomere band 2q13. *Genomics* **59:** 326.

Yunis J.J. 1976. High resolution of human chromosomes. *Science* **191:** 1268.

Yunis J.J. and Prakash O. 1982. The origin of man: A chromosomal pictorial legacy. *Science* **215:** 1525.

Zimonjic D., Kelley M., Rubin J., Aaronson S., and Popescu N. 1997. Fluorescence in situ hybridization analysis of keratinocyte growth factor gene amplification and dispersion in evolution of great apes and humans. *Proc. Natl. Acad. Sci.* **94:** 11461.

# The Effects of Evolutionary Distance on TWINSCAN, an Algorithm for Pair-wise Comparative Gene Prediction

M. WANG, J. BUHLER, AND M.R. BRENT

*Department of Computer Science and Engineering, Washington University, St. Louis, Missouri 63130*

Although the human genome sequence is finished, complete delineation of all human protein-coding genes remains a distant prospect. There currently are only about 13,000 genes (loci) for which at least one complete open reading frame is known with high confidence (http://mgc.nci.nih.gov/, http://www.ncbi.nlm.nih.gov/LocusLink/RSstatistics.html) (Pruitt and Maglott 2001; Strausberg et al. 2002), out of an estimated total of at least 20,000 (Roest Crollius et al. 2000; Waterston et al. 2002). Thus, we are in urgent need of improved techniques for delineating complete gene structures. One way in which such improvements have come about in the last few years is through comparison of the human genome to other sequenced vertebrate genomes. New gene modeling programs were developed to exploit information in alignments between the mouse and human genomes (Bafna and Huson 2000; Korf et al. 2001; Alexandersson et al. 2003; Flicek et al. 2003; Parra et al. 2003), and these are now being used to obtain cDNA sequence via hypothesis-driven RT-PCR and sequencing experiments (Guigó et al. 2003; Wu et al. 2004). One of the first comparative gene modeling programs to achieve substantial improvements over the previous state of the art was TWINSCAN, which can annotate a *target* genome by exploiting alignments from an *informant* genome even if the informant sequences are unassembled whole-genome shotgun reads (Flicek et al. 2003).

The rapid pace at which new vertebrate genomes are being sequenced can be expected to lead to further improvements in the accuracy of gene modeling systems. We now have complete, published draft sequences of the mouse (Waterston et al. 2002) and pufferfish (Aparicio et al. 2002) genomes, an unpublished assembly of the rat, and five- to sixfold coverage of the chicken and dog in whole-genome shotgun reads (also see Kirkness et al. 2003). Furthermore, regions orthologous to a 1.8-Mb segment of human Chromosome 7 containing CFTR and 9 other genes (the "greater CFTR region") have now been fully sequenced in a number of vertebrate species (Thomas et al. 2003). It is therefore possible, for the first time, to evaluate the evolutionary distance at which pair-wise comparison of vertebrate genomes is most useful for improving the accuracy of gene modeling.

In this paper, we explore the effects of evolutionary distance on gene prediction using TWINSCAN. To gain insight into our observations about gene prediction accuracy, we investigate the characteristics of BLASTN alignments in coding and noncoding sequence as a function of evolutionary distance. Mouse sequence is used as the target in order to maximize the range of evolutionary distances at which whole-genome informant sequences are available. We first focus on the previously studied CFTR region, then test the generality of our CFTR results using complete mouse chromosomes.

TWINSCAN takes as input local alignments between a target genome and a database of sequences from an informant genome. For each nucleotide of the target genome, only the highest-scoring local alignment overlapping that nucleotide is used. These alignments are converted into a representation called *conservation sequence*, which assigns one of three symbols to each nucleotide of the target genome (Fig. 1). Each target nucleotide is paired with "|" if the alignment contains a match, ":" if the alignment contains a gap or mismatch, and "." if there is no overlapping alignment. TWINSCAN has separate probability models for the likelihood that each conservation sequence pattern will occur in coding regions, UTRs, splice signals, and translation initiation and termination signals. Given a target DNA sequence and its conservation sequence, TWINSCAN predicts the most likely gene structures according to its probability model.

## RESULTS

### Mouse CFTR Region

We ran TWINSCAN on the mouse CFTR region using the CFTR regions of *Fugu, Tetraodon*, chicken, human, cat, and rat as the informant databases. TWINSCAN performs best with the chicken informant. When TWIN-

```
                 Coding region                                 Intron

Human   ACCAGACCAGATAGATACTTGTCTGCCACCCTC    AGATGCAAAAGAAACAGGTACCGCAGTG---CCCT
        |||||||||||||| || ||||| || || |||          |||   | |||||||||   ||||   ||||
Mouse   ACCAGACCAGATAGGTATTTGTCAGCTACTCTC          AAAAGAAACAGGTACCGCAGTGTCTCCCT

Human   ACCAGACCAGATAGATACTTGTCTGCCACCCTC    AGATGCAAAAGAAACAGGTACCGCAGTGCCCT
Conseq  ||||||||||||||:||:|||||:||:||:|||    ......|||::|:||||||||||::||:|||||
```

**Figure 1.** Conversion of the best local alignment in each region of the target genome (*top*) into the conservation sequence representation used by TWINSCAN (*bottom*). (*Left*) A typical coding region, in which there are no unaligned bases or gaps, and the distances between mismatches tend to be multiples of three. (*Right*) A typical intron, in which there are unaligned regions, gaps and adjacent mismatches.

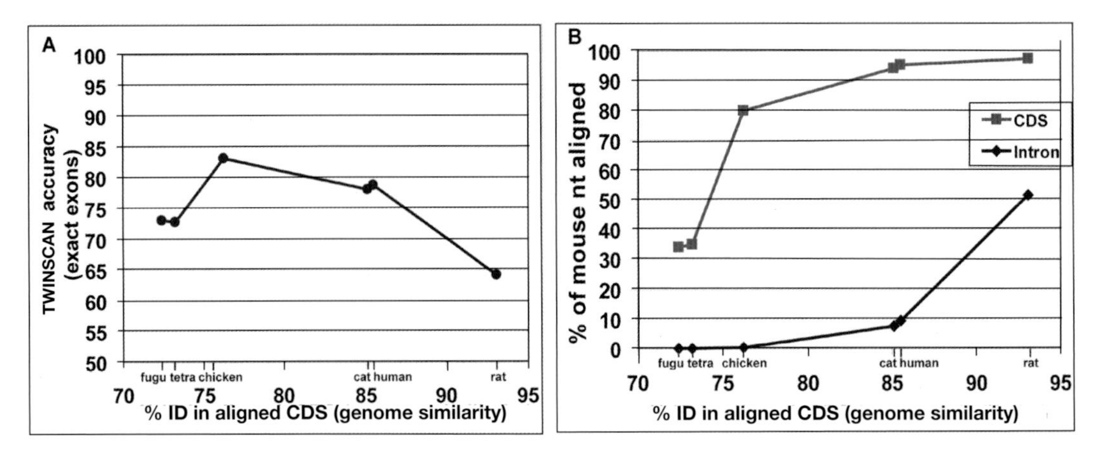


**Figure 2.** (*A*) TWINSCAN performance on the greater CFTR region of mouse with informant sequence from various organisms, plotted by accuracy of exact exon prediction (Y-axis) versus percent identity in aligned coding regions (a proxy for evolutionary distance, X-axis). (*B*) BLASTN alignments between the greater CFTR region of mouse and informant sequence from various organisms. For each pair of species, both the percentage of CDS sequence that aligns and the percentage of intron sequence that aligns (excluding splice sites) are plotted against percent identity in aligned coding regions.

SCAN accuracy using each informant database is plotted against the percent identity in aligned mouse coding sequences (a proxy for evolutionary divergence), the result is a unimodal curve peaking at chicken (Fig. 2A). Human is the second best informant for mouse, then *Fugu*, then rat. Accuracy with cat as the informant is very similar to accuracy with human, as expected given their similar divergence from mouse; likewise, the *Tetraodon* and *Fugu* informant sequences yield similar accuracy.

To gain insight into these accuracy results, we analyzed the BLASTN alignments that were used to create the conservation sequences for TWINSCAN (see Methods). For each informant database, we compared the percentage of mouse coding sequence (CDS) that aligns with the informant to the percentage of mouse intron sequence that aligns with the informant (excluding splice site regions). This analysis provides a compelling explanation for the observed differences in gene prediction accuracy (Fig. 2B). The comparison between mouse and *Fugu* exhibits essentially no alignment in the introns, but less than 40% of the CDS aligns. Moving closer in divergence, intron alignment remains very low in the mouse–chicken comparison (0.3%), but CDS alignment jumps to more than 80%. Thus, chicken alignments appear to have great power to discriminate between coding and noncoding sequence. Moving even closer, the human alignments cover ten times more of the mouse introns than do the chicken alignments, but only about 1.2 times more of the CDS. Since there is 48 times more noncoding sequence than exon sequence in this region, about 25 times more noncoding bases than CDS bases align to rat. Intuitively, the fact that many more of the aligned bases are noncoding than coding would seem to yield little discriminative power, even though a higher percentage of coding bases are aligned than noncoding bases. The results for cat and human are very similar to one another, as are those for *Fugu* and *Tetraodon*, suggesting that most of the observed differences are due to evolutionary distance.

Close examination of individual genes provides additional insight into how alignments affect gene-structure prediction. For example, the CFTR gene itself contains 27 exons and spans more than 150 kb of genomic sequence. Alignments of four informant databases from different lineages to the mouse CFTR gene are shown in Figure 3A. Clearly, chicken alignments correspond very closely to the mouse exons, whereas many exons are missed by the fish

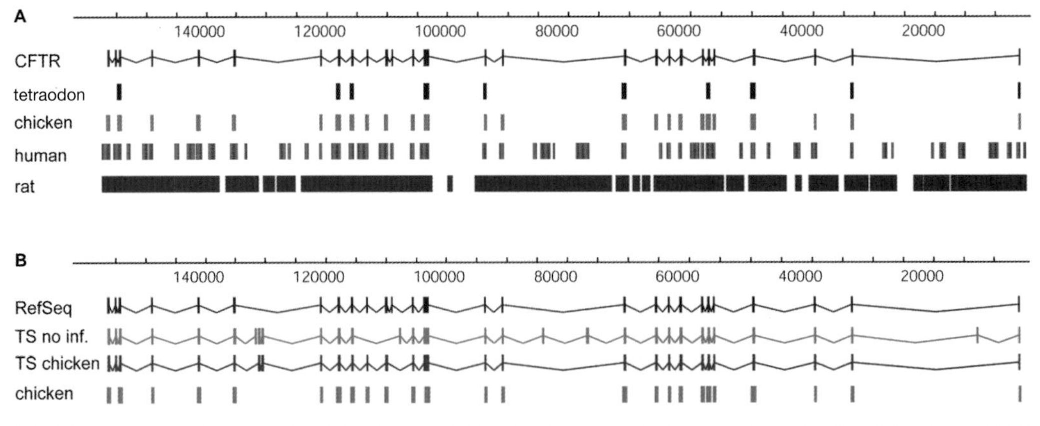


**Figure 3.** (*A*) The mouse CFTR gene (*red*) together with BLASTN alignments from the "greater CFTR" regions of *Tetraodon*, chicken, human, and rat. (*B*) The mouse CFTR gene (*red*), the corresponding TWINSCAN prediction without using any informant (*green*), the TWINSCAN prediction using chicken as the informant (*blue*), and blocks of mouse–chicken alignment (*black*).

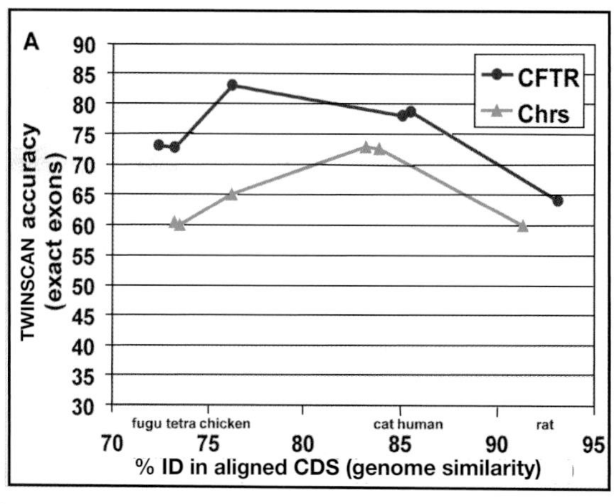


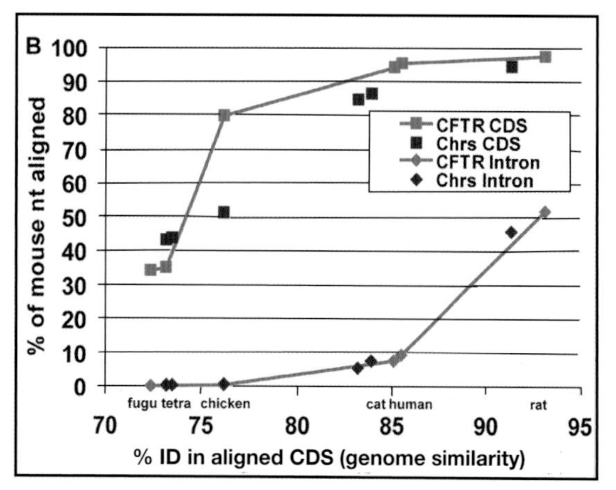


**Figure 4.** (*A*) TWINSCAN performance on both the greater CFTR region of mouse (*purple*) and chromosomes 11, 17, and 19 of mouse (*turquoise*) with informant sequence from various species. (*B*) BLASTN alignments of both the greater CFTR region of mouse (*teal*) and chromosomes 11, 17, and 19 of mouse (*burgundy*) with informant sequence from various species. For each pair of species, both the percentage of CDS sequence that aligns (*squares*) and the percentage of intron sequence that aligns (excluding splice sites, *diamonds*) are plotted against percent identity in aligned coding regions.

alignments, and many intronic segments align to human. (Note that exon 17 of human CFTR is masked for low complexity and hence does not align to mouse.) Finally, most of this genomic region aligns to rat. The benefits of exploiting chicken alignments are shown in Figure 3B. TWINSCAN without any informant alignments, which is very similar to GENSCAN (Burge and Karlin 1997), calls seven extra exons and misses two. TWINSCAN using chicken alignments calls only two extra exons (compared to seven) and misses only one exon (compared to two).

### Mouse Chromosomes 11, 17, and 19

To see how the results from the CFTR region generalize, we applied the same analysis to mouse chromosomes 11, 17, and 19. These chromosomes were selected because they have high densities of aligned mouse MGC (Strausberg et al. 2002) transcripts. Aligned MGC transcript sequences were chosen as the gold standard annotation in this study because this collection of transcript sequences is the most accurate one we know of. The informant databases consisted of an estimated 5x shotgun coverage of *Fugu*, *Tetraodon*, chicken, human, dog, and rat (see Methods for fold-coverage calculations). Dog replaced cat because there was not sufficient coverage of cat in whole-genome shotgun reads. TWINSCAN accuracy with each informant database is plotted in Figure 4A (see Methods for accuracy calculations). Whereas the chicken alignments were best in the CFTR region, the human alignments are best in this broader study. The difference is explained by the alignment characteristics (Fig. 4B). Whereas 80% of the CDS in the mouse CFTR region aligns to chicken, only 50% of the CDS in the three chromosomes aligns. This major drop, which is not seen in the other comparisons, erases the jump in percentage of CDS aligned as one moves from fish to chicken in the CFTR alignments. In the whole-chromosome alignments of Figure 4B, chicken lies on a straight line running from fish to human. This difference between CFTR and the broader genome is unlikely to be due to incomplete shotgun coverage of the informant genomes because the percentage of mouse CDS aligning to each informant database is near saturation at 5x coverage (Fig. 5). It is not due to other quirks in the shotgun databases, because aligning these databases to the CFTR region yields the same pattern of results as aligning the orthologous sequences from BACs (data not shown).

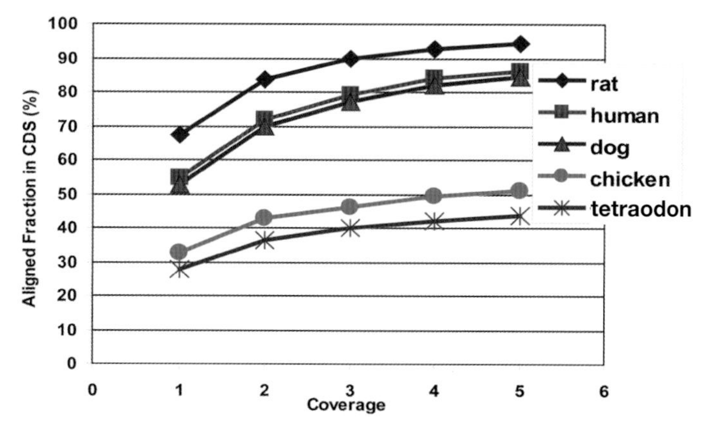


**Figure 5.** Percent identity in CDS for alignments between mouse chromosomes 11, 17, and 19 and whole-genome shotgun reads from various species at one- to fivefold redundancy. The 1x data points are averages over five independent 1x databases, and the 2x points are averages over two independent databases.

## METHODS

***Sequences: CFTR study.*** The sequence of the mouse CFTR region was downloaded from the NISC Web site (http://www.nisc.nih.gov/data/20020612_Target1_0051/mouse_T1.fasta).

***Sequences: Mouse chromosomes study.*** The sequences of mouse chromosomes 11, 17, and 19 were downloaded from (http://genome.ucsc.edu/golden Path/mmFeb2003/chromosomes/). The chromosomes were divided into nonoverlapping 1-Mb segments for both the BLAST and TWINSCAN portions of the analysis. Shotgun reads from *Fugu* (*Takifugu rubripes*), chicken (*Gallus gallus*), dog (*Canis familiaris*), human (*Homo sapiens*), and rat (*Rattus norvegicus*) were downloaded from the NCBI trace archive (ftp://ftp.ncbi.nih.gov/pub/TraceDB/).

***Fold coverage calculations.*** Each 1x database consists of clipped reads whose total length is equal to the estimated size of the source genome. The genome sizes for *Tetraodon* (0.31 Gb), *Fugu* (0.31 Gb), and rat (2.6 Gb) were estimated as the number of bases in the assembly; the size for human (2.9 Gb) was taken from the mouse genome paper (Waterston et al. 2002), the size for dog (2.4 GB) was taken from the recent shotgun sequencing survey paper (Kirkness et al. 2003), and the size for chicken (1.2 Gb) was taken from the proposal to sequence the chicken genome (McPherson et al. 2002). An alternative definition of 1x is the number of raw reads needed to achieve an N-fold redundant assembly, divided by N (Waterston et al. 2002; Flicek et al. 2003). This alternative yields substantially larger 1x databases because assemblers typically discard many of the raw input reads.

***BLAST.*** To prepare BLAST databases, we masked repeats in the informant genome sequences with Repeat Masker (Smit and Green, http://ftp.genome.washington.edu/RM/RepeatMasker.html), performed an additional round of low-complexity masking with nseg (Wootton and Federhen 1996) using default parameters, and removed all strings of 15 or more consecutive Ns in order to speed processing. All BLAST jobs were run using WUBLAST 2.0, 18-Jan-2003, running under x86 Linux. The analysis reported here uses the following BLAST parameters: M = 1, N = –1, Q = 5, R = 1, Z = 3,000,000,000, Y = 3,000,000,000, B = 10,000, V = 100, W = 10, X = 30, S = 30, S2 = 30, and gapS2 = 30. The seg and dust filter options were used.

***Sequence annotation.*** The annotation of the mouse CFTR region created by the NISC was downloaded from http://www.nisc.nih.gov/data/20020612_Target1_0051/mouse_T1.annot.ff (Genbank accession AE017189). For the mouse chromosomes, we downloaded the MGC transcripts aligned to the genome by BLAT from http://genome.ucsc.edu . This annotation contains 8,547 known genes for the entire genome.

***TWINSCAN.*** We used TWINSCAN version 1.3. Both TWINSCAN source code and a Web server are available at http://genes.cse.wustl.edu.

***Accuracy calculations.*** The predictions were compared to the MGC annotations using the Eval software package (Keibler and Brent 2003; http://genes.cse.wustl.edu/eval/). The accuracy measure plotted in Figure 2A is the average of exact exon sensitivity (the fraction of annotated exons for which TWINSCAN predicts both splice sites correctly) and exon specificity (the fraction of exons predicted by TWINSCAN that exactly match annotated exons). Because the MGC transcripts include only a fraction of the exons in the mouse genome, specificity measured against MGC is a systematic underestimate of actual specificity—all predicted exons that are not in the MGC transcripts are counted as wrong. In order to make the results on the chromosome comparable to those on the well-annotated CFTR region, the accuracy measure plotted in Figure 4A is the average of exon sensitivity and 3.5 times the exon specificity.

## DISCUSSION

The availability of extensive genome sequence from a variety of vertebrate lineages has enabled the first direct investigation of how evolutionary distance affects comparative gene prediction in the vertebrates. Initially, we studied the greater CFTR region using the sequences of BACs that were selected for orthology to the human CFTR region. A previous analysis of this region (Thomas et al. 2003) based on different alignment methods (Schwartz et al. 2003) reported that chicken alignments covered a large fraction of coding sequence (CDS) in the mammalian CFTR region but only a small fraction of noncoding sequence. Our analysis confirms this and shows that it has the expected positive effect on the accuracy of TWINSCAN, a state-of-the-art gene prediction system. In addition to improving coding versus noncoding predictions, chicken alignments yielded the greatest accuracy as measured by prediction of exact exon boundaries.

Application of the same analysis to a much broader sample of the mouse genome (chromosomes 11, 17, and 19) told a very different story. The CDS of the greater CFTR region is exceptionally well conserved between the mammalian and avian lineages, relative to other portions of the genome. In the broader survey, there was no sudden jump in the percentage of mouse CDS that aligns as one moves from fish comparisons to chicken comparisons. Instead, the aligned percentage seems to increase linearly with the percent identity in aligned regions as one moves from fish to chicken to human (Fig. 4). This difference was reflected directly in the TWINSCAN performance, which peaked at human rather than chicken. The curves in Figure 5, as well as previously reported results (Flicek et al. 2003), suggest that using complete, assembled informant genomes in place of the 5x shotgun coverage is unlikely to have any qualitative effect on the relative utility of the comparisons. Likewise, changing alignment parameters or algorithms affects the absolute fraction of intron and CDS that aligns in each genome pair, but it does not change the relative values of comparisons at the distances studied here (data not shown). Using translated alignments (TBLASTX) also has little qualitative effect and does not improve TWINSCAN performance. Finally, our results on CFTR agree qualitatively with those that Thomas et al. obtained with BLASTZ (Schwartz et al. 2003), an algorithm that produces a very different type of alignment than BLASTN.

The accuracy curve shown in Figure 4A reflects comparisons at only four significantly different evolutionary distances. The alignment curves in Figure 4B suggest that filling in the intermediate distances will yield an accuracy curve with a single peak at the evolutionary distance that is optimal for informing gene modeling. If so, the peak is at a distance farther than that between mouse and rat but closer than that between mouse and chicken. The peak at the mouse–human comparison in Figure 4A is consistent with a highly idealized theoretical analysis suggesting a peak somewhere in the vicinity of mouse–human, or possibly somewhat farther out (Zhang et al. 2003). To determine the peak location more precisely, we will need the sequences of more genomes. The most immediate possibility for a comparison at a distance intermediate between mouse–human and mouse–chicken will come from the opossum, *Monodelphis domestica*, a marsupial that has been designated as a high-priority sequencing target by the National Human Genome Research Institute (http://www.genome.gov/page.cfm?pageID=10002154).

Although the mouse–rat divergence is much too close for optimal annotation of mammalian genomes using a single genome pair, preliminary data suggest that the situation may be quite different when the target is a genome with very short introns. For example, gene prediction in *Cryptococcus neoformans* serotype D benefits from alignments to serotype A. These alignments cover 100% of CDS and 87% of intron bases, excluding splice sites (A. Tenney, unpubl.). Visual inspection suggests that when TWINSCAN is trained on these alignments, it predicts introns or intergenic regions that include most unaligned regions; conversely, it rarely predicts an intron that does not overlap an unaligned region. Because the introns are very short (68 bp on average), the locations of the intron boundaries are quite constrained relative to those of mammalian introns. Apparently, TWINSCAN uses these alignments to find the general locations of introns, rather than their precise boundaries.

The research presented here is a significant step toward determining the optimal distance for gene modeling using pair-wise genome alignments. Looking ahead to the next step, we and other investigators are developing methods that use information from multiple genome alignments rather than choosing a single best alignment (Boffelli et al. 2003; Siepel and Haussler 2003). When it has been fully developed, the multi-genome approach is expected to yield real breakthroughs in the accuracy of gene modeling.

## ACKNOWLEDGMENTS

We are grateful to the centers that produced the genome sequences used in this study. Special mention is due to the Washington University Genome Sequencing Center for producing the chicken whole-genome shotgun sequence, the National Institutes of Health Intramural Sequencing Center for producing BAC-based sequences of the greater CFTR regions, and the Whitehead Institute Genome Research Center for producing the dog whole-genome shotgun sequence. The authors were supported in part by grant HG-02278 from the National Human Genome Research Institute.

## REFERENCES

Alexandersson M., Cawley S., and Pachter L. 2003. SLAM: Cross-species gene finding and alignment with a generalized pair hidden Markov model. *Genome Res.* **13:** 496.

Aparicio S., Chapman J., Stupka E., Putnam N., Chia J.M., Dehal P., Christoffels A., Rash S., Hoon S., Smit A., Gelpke M.D., Roach J., Oh T., Ho I.Y., Wong M., Detter C., Verhoef F., Predki P., Tay A., Lucas S., Richardson P., Smith S.F., Clark M.S., Edwards Y.J., and Doggett N., et al. 2002. Whole-genome shotgun assembly and analysis of the genome of *Fugu rubripes*. *Science* **297:** 1301.

Bafna V. and Huson D.H. 2000. The conserved exon method for gene finding. *Proc. Int. Conf. Intell. Syst. Mol. Biol.* **8:** 3.

Boffelli D., McAuliffe J., Ovcharenko D., Lewis K.D., Ovcharenko I., Pachter L., and Rubin E.M. 2003. Phylogenetic shadowing of primate sequences to find functional regions of the human genome. *Science* **299:** 1391.

Burge C. and Karlin S. 1997. Prediction of complete gene structures in human genomic DNA. *J. Mol. Biol.* **268:** 78.

Flicek P., Keibler E., Hu P., Korf I., and Brent M.R. 2003. Leveraging the mouse genome for gene prediction in human: From whole-genome shotgun reads to a global synteny map. *Genome Res.* **13:** 46.

Guigó R., Dermitzakis E.T., Agarwal P., Ponting C., Parra G., Reymond A., Abril J.F., Keibler E., Lyle R., Ucla C., Antonarakis S.E., and Brent M.R. 2003. Comparison of mouse and human genomes followed by experimental verification yields an estimated 1,019 additional genes. *Proc. Natl. Acad. Sci.* **100:** 1140.

Keibler E. and Brent M.R. 2003. Eval: A software package for analysis of genome annotations. *BMC Bioinformatics* **4:** 50.

Kirkness E.F., Bafna V., Halpern A.L., Levy S., Remington K., Rusch D.B., Delcher A.L., Pop M., Wang W., Fraser C.M., and Venter J.C. 2003. The dog genome: Survey sequencing and comparative analysis. *Science* **301:** 1898.

Korf I., Flicek P., Duan D., and Brent M.R. 2001. Integrating genomic homology into gene structure prediction. *Bioinformatics* (suppl. 1) **17:** S140.

McPherson J.D., Dodson J., Krumlauf R., and Olivier P. 2002. Proposal to sequence the genome of the chicken. (http://www.wattnet.com/library/DownLoad/PD12 genome.pdf)

Parra G., Agarwal P., Abril J.F., Wiehe T., Fickett J.W., and Guigó R. 2003. Comparative gene prediction in human and mouse. *Genome Res.* **13:** 108.

Pruitt K.D. and Maglott D.R. 2001. RefSeq and LocusLink: NCBI gene-centered resources. *Nucleic Acids Res.* **29:** 137.

Roest Crollius H., Jaillon O., Bernot A., Dasilva C., Bouneau L., Fischer C., Fizames C., Wincker P., Brottier P., Quetier F., Saurin W., and Weissenbach J. 2000. Estimate of human gene number provided by genome-wide analysis using *Tetraodon nigroviridis* DNA sequence. *Nat. Genet.* **25:** 235.

Schwartz S., Kent W.J., Smit A., Zhang Z., Baertsch R., Hardison R.C., Haussler D., and Miller W. 2003. Human-mouse alignments with BLASTZ. *Genome Res.* **13:** 103.

Siepel A.C. and Haussler D. 2003. Combining phylogenetic and hidden Markov models in biosequence analysis. In RECOMB 2003 (ed. W. Miller et al.), p. 277. ACM Press (ACM Digital Library), New York.

Strausberg R.L., Feingold E.A., Grouse L.H., Derge J.G., Klausner R.D., Collins F.S., Wagner L., Shenmen C.M., Schuler G.D., Altschul S.F., Zeeberg B., Buetow K.H., Schaefer C.F., Bhat N.K., Hopkins R.F., Jordan H., Moore T., Max S.I., Wang J., Hsieh F., Diatchenko L., Marusina K., Farmer A.A., Rubin G.M., and Hong L., et al. 2002. Generation and initial analysis of more than 15,000 full-length human and mouse cDNA sequences. *Proc. Natl. Acad. Sci.* **99:** 16899.

Thomas J.W., Touchman J.W., Blakesley R.W., Bouffard G.G., Beckstrom-Sternberg S.M., Margulies E.H., Blanchette M., Siepel A.C., Thomas P.J., McDowell J.C., Maskeri B., Hansen N.F., Schwartz M.S., Weber R.J., Kent W.J., Karolchik D., Bruen T.C., Bevan R., Cutler D.J., Schwartz S., Elnitski L., Idol J.R., Prasad A.B., Lee-Lin S.Q., and Maduro V.V., et al. 2003. Comparative analyses of multi-species sequences from targeted genomic regions. *Nature* **424:** 788.

Waterston R.H., Lindblad-Toh K., Birney E., Rogers J., Abril J.F., Agarwal P., Agarwala R., Ainscough R., Alexandersson M., An P., Antonarakis S.E., Attwood J., Baertsch R., Bailey J., Barlow K., Beck S., Berry E., Birren B., Bloom T., Bork P., Botcherby M., Bray N., Brent M.R., Brown D.G., and Brown S.D., et al. (Mouse Genome Sequencing Consortium). 2002. Initial sequencing and comparative analysis of the mouse genome. *Nature* **420:** 520.

Wootton J.C. and Federhen S. 1996. Analysis of compositionally biased regions in sequence databases. *Methods Enzymol.* **266:** 554.

Wu J.Q., Shteynberg D., Arumugam M., Gibbs R.A., and Brent M.R. 2004. Identification of rat genes by TWINSCAN gene prediction, RT-PCR, and direct sequencing. *Genome Res.* **14:** 665.

Zhang L., Pavlovic V., Cantor C.R., and Kasif S. 2003. Human-mouse gene identification by comparative evidence integration and evolutionary analysis. *Genome Res.* **13:** 1190.

## WEBSITE REFERENCES

Mammalian Gene Collection (MGC)
http://mgc.nci.nih.gov/

NCBI LocusLink
http://www.ncbi.nlm.nih.gov/LocusLink/RSstatistics.html

NCBI trace archive
ftp://ftp.ncbi.nih.gov/pub/TraceDB/

NHGRI's sequencing priority list
http://www.genome.gov/page.cfm?pageID=10002154

NISC's annotation of the greater CFTR region of mouse
http://www.nisc.nih.gov/data/20020612_Target1_0051/mouse_T1.fasta

Annotation of the CFTR region
http://www.nisc.gov/data/20020612_Target1_0051/mouse_T1.annot.ff

NISC's map of orthologous vertebrate BAC clones
http://www.nisc.nih.gov/projects/zooseq/pubmap/PubM.cgi

UCSC's repeat masked mouse genome sequence
http://genome.ucsc.edu/goldenPath/mmFeb2003/chromosomes/chr*.fa.zip

RepeatMasker software and documentation
http://ftp.genome.Washington.edu/RM/RepeatMasker.html

Eval software and documentation
http://genes.cse.wustl.edu/eval/

TWINSCAN software, documentation, and web server
http://genes.cse.wustl.edu

# Lineage-specific Expansion of KRAB Zinc-finger Transcription Factor Genes: Implications for the Evolution of Vertebrate Regulatory Networks

A.T. Hamilton,* S. Huntley,* J. Kim,* E. Branscomb,† and L. Stubbs*†
*Genome Biology Division, Lawrence Livermore National Laboratory, Livermore, California 94550;
†DOE Joint Genome Institute, Walnut Creek, California 94598

A substantial fraction of the vertebrate gene repertoire is conserved across the animal kingdom and beyond (Lander et al. 2001; Aparicio et al. 2002; Lespinet et al. 2002). In addition, unique 1:1 ortholog pairings reveal substantial domains of syntenic conservation when closely related genomes are compared (see, e.g., Dehal et al. 2001; Waterston et al. 2002). However, for certain gene types, 1:1 orthologous relationships are the exception and not the rule, apparently because duplicate copies of these genes have been generated and fixed at unusually high rates. These ongoing duplication events have yielded substantial numbers of lineage-specific genes, giving rise to gene-repertoire differences that distinguish even very closely related species. Examples include genes encoding olfactory receptors (OR) and *Krüppel*-type zinc finger (KZNF) proteins, which together comprise 2–5% of known mammalian genes. Importantly, most of the genes in these rapidly expanding families exist in contiguous familial clusters, consistent with the model that they have arisen through a process of repeated tandem duplications (Ohno 1970; Huntley et al. 2003). Although the mechanisms driving these in situ duplications require more study, the evolutionary generation of such tandem gene arrays is responsible for the overwhelming majority of gene repertoire expansion in metazoans (Friedman and Hughes 2003). What underlies this dramatic difference in evolutionary fate between these "fast-lane" genes and those in the conserved core genome?

Available evidence supports the view that gene duplication is a common event affecting all genes more or less uniformly (Nei et al. 1997). For most gene types, the fraction of new duplicates that survive as novel functional genes is exceedingly low; the vast majority of copies are lost as pseudogenes within a few million years (Lynch and Conery 2000). However, the ratio of intact genes to pseudogenes in many tandem gene clusters is much higher than would be expected from the genome-average rates of gene copy fixation (see, e.g., Xie et al. 1997; Trowsdale et al. 2001; Vanhalst et al. 2001; Shannon et al. 2003). These and other observations suggest that exceptional evolutionary mechanisms may operate in these cases to dramatically increase the likelihood that a new gene duplicate will acquire a preserved novel function rather than experience mutational inactivation or selective rejection.

Traditionally, studies of gene family evolution have focused on genes encoding receptors such as those involved in immune function and olfaction. These examples represent biological systems in which the pressure to detect a constantly changing set of molecules can theoretically select for the retention of new gene copies (or alleles) due to the advantages of greater diversity in reacting to novel challenges, e.g., antigens or chemical scents. However, the KZNF gene family, one of the largest families of mammalian genes with an estimated 800 human members (Lander et al. 2001; Aparicio et al. 2002), encodes not receptor proteins but highly specific DNA-binding proteins that are typically involved in regulating the expression of other genes. One prominent KZNF subfamily, in which the DNA-binding KZNF domains are attached to a strong repressor motif called the *Krüppel*-associated box (KRAB), has been particularly prolific in vertebrates. Interestingly, genes containing both KRAB and KZNF motifs have not been identified in the sequenced genomes of invertebrates or fish. Instead, the KRAB-KZNF (KK) combination has only been found in tetrapods (perhaps first occurring in lobe-finned fish or early amphibians). After its origin there apparently was a rapid expansion, producing a family of hundreds of genes in mammals.

Like the similarly numerous OR genes, KK gene clusters have expanded independently in different vertebrate lineages yielding, for example, distinct repertoires of related genes in primates and rodents (Dehal et al. 2001). Recent studies have indicated that new KK gene copies have diverged through a unique strategy, combining redirection of paralog expression sites, positive selection for non-synonymous change at critical DNA-interaction residues, and deletions and duplications within the KZNF repeats (Shannon et al. 2003). By these mechanisms this large gene family is evolving rapidly, creating large numbers of novel transcription factors (TFs) with altered DNA-binding properties and expression sites independently in each mammalian lineage.

The modular structure of KZNF genes, their unique strategy for structural divergence, and the possibility that KK proteins are coevolving with the *cis*-acting regulatory sequences with which they interact adds a unique aspect of complexity to the investigation of their origins and evolution. What selective drive was behind the retention of hundreds of new repressively acting regulatory genes in higher vertebrates relative to fish and to the dramatic differentiation in KZNF gene repertoire in different mammalian lineages? How has the rapid rise of the KK TF family affected expression-site specificity of the ancient, core set of conserved vertebrate genes? This profligate elaboration of repressive transcription factors has presumably led to a widespread modifcation of gene regulation with correspondingly important biological effects. The impact of this on the evolution of pathways, genomes, and species is potentially vast and needs to be considered.

Here we explore the divergence strategy of the mammalian KK gene family in further detail and discuss the potential role that rapid lineage-specific diversification of this class of TFs may have played in remodeling gene regulatory networks in vertebrate evolution.

## STRUCTURE AND FUNCTION OF *KRÜPPEL*-TYPE ZINC FINGER GENES

### Zinc-finger Structure and DNA Binding

Several types of eukaryotic protein motifs contain specific arrangements of cysteine and sometimes histidine amino acids that form zinc-binding structures (or "fingers") (Krishna et al. 2003). However, KZNF proteins, in which the standard finger unit contains two cysteines and two histidines arranged around the zinc ion, are the predominant type of zinc finger motif in metazoan genomes. In most KZNF proteins, multiple zinc-finger motifs are arranged in tandem arrays, with the repeated motifs typically encoded together on a single exon (Fig. 1). The KZNF proteins studied to date function primarily as transcription factors with the fingers binding target DNA sites in a sequence-specific manner (see, e.g., Pieler and Bellefroid 1994; Turner and Crossley 1999; Tanaka et al. 2002). Some KZNF proteins function as transcriptional activators, whereas others serve as repressors; certain proteins can assume either role (Dang et al. 2000). In addition to DNA binding, other functions have also been proposed for some KZNF motifs, including double-stranded RNA binding and mRNA splicing (Ladomery and Dellaire 2002) and protein interaction (Zheng et al. 2000).

The KZNF finger motif is a ~21 residue ββ-α globular domain stabilized at its base by zinc coordination to the paired cysteine and histidine residues (Fig. 1D). The α-helical structure of each finger fits into and interacts with the major groove as the KZNF finger polymer wraps around the DNA double helix (Jacobs 1992; Choo and Klug 1994; Laity et al. 2000). Analyses of crystal structures of certain KZNF proteins bound to target DNA have identified specific amino acid positions within each finger that appear generally to be most critical for DNA sequence recognition; each finger potentially recognizes a 2- to 5-bp DNA sequence, and adjacent fingers are in contact with contiguous, overlapping sets of neighboring nucleotides at the target site (Choo and Klug 1994; Pabo et al. 2001; Wolfe et al. 2001; Huntley et al. 2003). This relationship implies colinearity between protein and DNA-target sequence (Fig. 2), a notion that suggests a simple method might be found for predicting target sites from protein sequence once the putative "code" is understood.

More complex factors also influence target site choice for KZNF domains, however, including cooperative interactions between adjacent fingers (Isalan et al. 1997; Laity et al. 2000). The influence of such inter-finger interactions makes the computer-modeling and code-resolution problem much more difficult (Miller and Pabo 2001; Wolfe et al. 2001). This complexity of the interactions, especially when multiple fingers are involved, may prohibit the generation of a simple "sequence code," and it may not be possible to extrapolate directly rules derived from experiments with specific KZNF proteins to DNA-binding properties of other zinc-finger arrays. In particular, it appears that there may be alternate amino acid sequence combinations that can recognize the same core target DNA site (Peng et al. 2002), and some finger arrays may be capable of recognizing more than one target sequence (Filippova et al. 1996; Ohlsson et al. 2001; Daniel et al. 2002).

### The *Krüppel*-associated Box: A Mechanism of Repression

In most KZNF proteins studied to date, the DNA-binding domain is attached to a protein-interaction or "effector" motif, which translates the signal of DNA binding into specific transcriptional outcomes. Several different effector motifs have become associated with KZNF

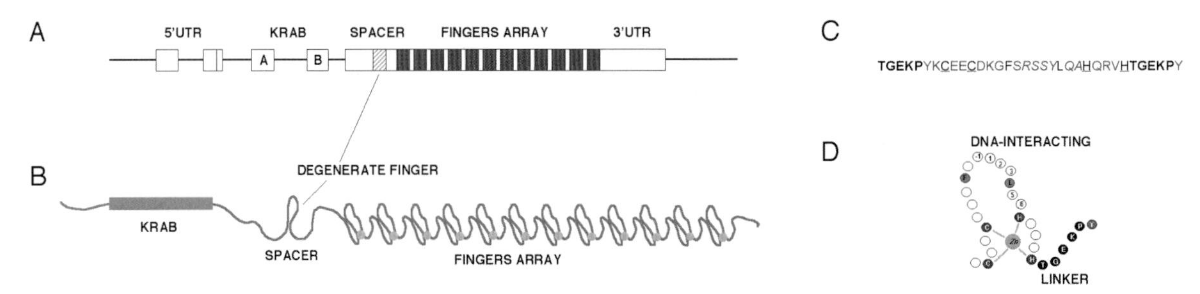


**Figure 1.** (*A,B*) Diagram of the typical exon structure for a KRAB-containing zinc-finger gene and the resulting protein. The KRAB consists of a primary A module and a secondary B module, which is often absent. The black boxes in the largest exon represent individual zinc-finger motifs. The striped box in the spacer indicates a "degenerate" zinc finger. (*C*) Amino acid sequence of a typical *Krüppel*-type zinc finger. The paired cysteines and histidines (C2H2) necessary for zinc binding are underlined; the conserved "linker" sequence between fingers is in bold, and the most variable positions (including those thought to be critical for DNA sequence recognition) are in italics. (*D*) Simple depiction of the amino acid positions in a zinc finger, with the C2H2, linker, and other conserved amino acid positions darkened. The more variable positions are numbered (in relation to the start of an α-helix).

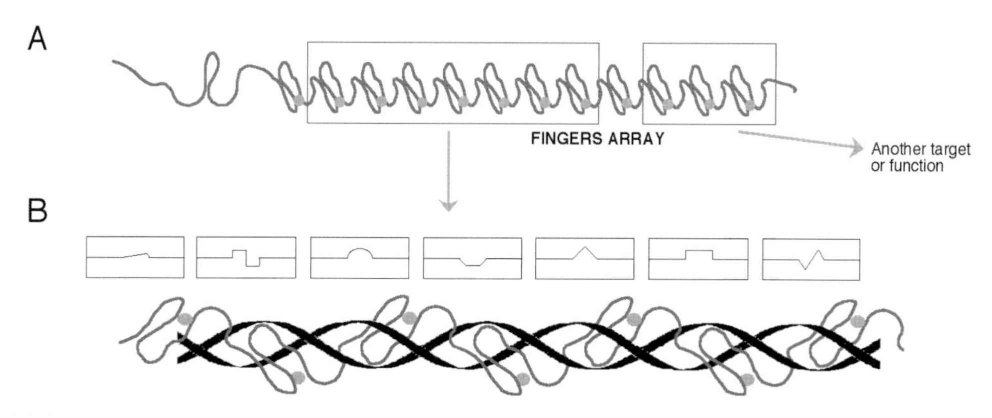


**Figure 2.** (*A*) Generic polydactyl zinc-finger protein, in this example depicted with subsets of its zinc-finger motifs having two separate targets or functions. (*B*) Schematic of a set of zinc fingers bound to their target, with a "code" simulated by the interacting shapes (protein on top, DNA below) to show the colinearity of the amino acid sequence in the fingers and the target stretch of DNA.

DNA-binding domains during the course of eukaryotic evolution, and each confers a particular type of activity to the DNA-binding KZNF domains. The effectors function by mediating interactions between the KZNF proteins and other protein cofactors, including some involved directly or indirectly in chromatin remodeling (Collins et al. 2001; Schultz et al. 2002). The most prevalent effector motif in mammalian KZNF proteins, the KRAB domain, is predicted to include charged amphipathic helices that bind to the RBCC region of the transcriptional corepressor TRIM28 (a.k.a. KAP-1, TIF1-β) (Collins et al. 2001). TRIM28, in turn, recruits other proteins such as HP1 heterochromatin proteins, histone methyltransferases, and histone deacetylases. The complex of proteins attached to the TRIM28 "scaffold" is thought to alter chromatin structure via histone modification to repress the transcription of the targeted gene (Schultz et al. 2002). The KRAB can be divided into subregions called A and B, which are typically encoded on separate exons; several lines of evidence have indicated that the A box is necessary for repression, whereas the B box contributes but is less critical for TF function, and many KK proteins do not contain the KRAB B motif (Mark et al. 1999).

The KRAB A motifs of KK proteins are highly conserved, both within and between species; this strict conservation is likely to reflect constraints imposed by the requirement for TRIM28 binding. In contrast, the DNA-binding KZNF domains can be radically different in length, due to the variable number of zinc fingers they can include. Although the individual finger motifs have a common structure, diversity is allowed both in the sequence of the DNA contact region and in the number of repeats, as discussed below.

## DUPLICATION AND DIVERGENCE OF THE KRAB-KZNF GENE FAMILY IN MAMMALS

### Cluster Evolution

With only three mammalian genomes sequenced (human, mouse, rat), our picture of KK gene evolution in this lineage is still incomplete. Nonetheless, many new insights have arisen from comparisons between those genomes. First, unlike other types of vertebrate KZNF loci, most human and rodent KK genes are found in large tandem clusters containing up to 40 related genes. Although the majority of human KK gene clusters are represented by related families in syntenically homologous regions of the mouse genome, mouse and human KK clusters contain strikingly different numbers of genes. Indeed, sequence comparisons between homologous clusters point to active gene gain and loss since divergence of the primate and rodent lineages (Dehal et al. 2001; Shannon et al. 2003). Second, most mouse and human KK loci contain open reading frames capable of encoding fully functional proteins (Dehal et a. 2001; E. Branscomb et al., unpubl.), suggesting that the differential duplications have yielded substantial numbers of novel lineage-specific proteins. Finally, KK clusters contain few pseudogenes even compared to other types of familial gene clusters (see, e.g., Gaudieri et al. 1999; Young and Trask 2002). Therefore, tandemly clustered ZNF genes appear to be subject to unusual selective pressures that actively favor the maintenance of duplicated copies as functional genes.

### Lessons from a Differentially Expanded Cluster in Humans and Mice

We have recently described the structure and evolutionary history of a pair of homologous KK gene clusters located in human chromosome 19q13.2 (Hsa19q13.2) and mouse chromosome 7 (Mmu7), respectively (Shannon et al. 2003). Twenty-one human genes and 10 mouse genes are found in these homologous gene clusters, although only three sets of 1:1 orthologous pairs exist. The clusters also include two cases in which a single gene in one species is related to multiple genes in the other (Fig.3). The relative orders of genes comprising these five sets of homologs are maintained in human and mouse, suggesting that this arrangement of genes was present in a common ancestor of primates and rodents. These findings are consistent with the idea that the differences between mouse and human clusters reflect independent histories of duplication and loss starting from a basic set of as few as five ancestral genes (Shannon et al. 2003).

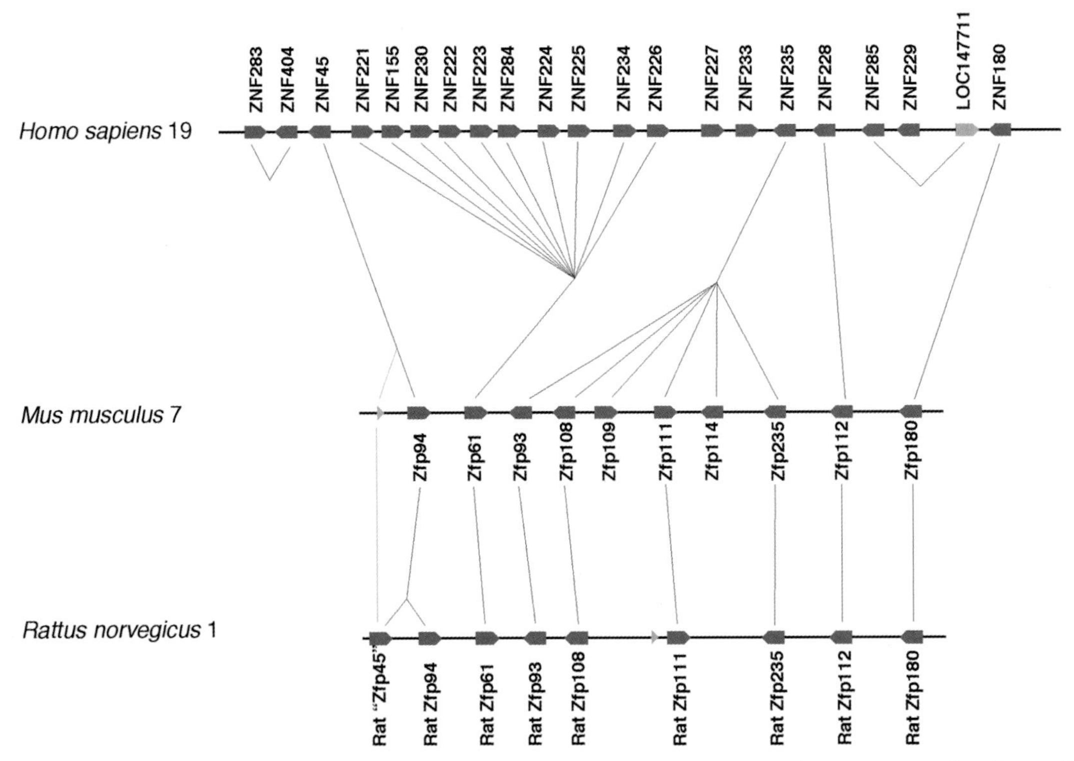


**Figure 3.** Three-species comparison of the zinc-finger gene cluster on Hsa19q13.2 and the homologous mouse and rat clusters. Block-arrows represent orientations and relative positions of genes; lines connecting genes indicate close relationships based on the sequences of the zinc-finger arrays (relationships can include 1:1 orthology or, in several cases, clades where a single gene in one species is related to an expanded group of lineage-specific paralogs in the other). Human *ZNF45* is related to mouse *Zfp94* and two rat genes; the alternate gray line between mouse and rat maps indicates the position of an apparent remnant of the second *ZNF45*-like gene in mouse (indicated by a triangle).

Sequence alignments of the lineage-specific KK duplicates have revealed mechanisms by which new KK duplicates may be rapidly diverging in function. One mechanism involves selection for nonconservative single-nucleotide substitutions at critical DNA-recognition sites. Most amino acid residues in KZNF domains are required for zinc binding and structural integrity, and these are typically highly conserved (Jacobs 1992; Mascle et al. 2003). Therefore, as expected, general surveys of nonsynonymous vs. synonymous mutations for complete KZNF-domain sequences reveal a strong signature of purifying selection. However, when only those amino acids predicted to be key to DNA binding are considered, a trend toward positive selection is indicated in some comparisons between paralogs, whereas orthologs often conserve these amino acid positions. Such evidence was recently documented in alignments of lineage-specific duplicates in the Hsa19q13.2/Mmu7 clusters (Shannon et al. 2003) plus selected additional KK genes (Looman et al. 2002). A higher amino acid replacement rate for the DNA-interacting sites has also been seen between other types of closely related KZNF genes (Sander et al. 2003). These data suggest that, in some cases at least, selection may be operating to favor diversity in the target recognition of paralogous proteins.

Paralog alignments have also revealed a second mechanism for divergence in KZNF arrays that may have even more striking impact on DNA-binding properties of the duplicated genes. Not all genes have the same number of functional zinc-finger motifs, even if they are closely related. For example, alignment of the six mouse homologs of the singleton human gene, *ZNF235*, revealed a pattern of finger deletions and duplications that alters the arrangement and number of DNA-binding motifs. In most cases, the size and linker-spacing of the surrounding fingers were maintained, indicating that the deletions and duplications are driven by recombination between the tandem finger repeats (Shannon and Stubbs 1998; Shannon et al. 2003). In the case of *ZNF235* and its six mouse relatives, one mouse gene (*Zfp235*) has retained the same finger number and arrangement as in the human gene, hinting at a shared functional homology, but all other mouse *Zfp235* paralogs carry differently altered arrangements of fingers.

A third mode of KK gene divergence involves redirection of expression patterns for the duplicated genes, such that new gene copies are active in tissues or cell types in which ancestral genes or other paralogs are not expressed. For example, in the Hsa19q13.2 gene cluster, some of the human genes are ubiquitously expressed, but closely related paralogs *ZNF225* and *ZNF284* are expressed in limited, nonoverlapping tissue types (Fig. 4). Therefore, KK paralog diversification may also arise by acquisition of altered expression patterns, presumably

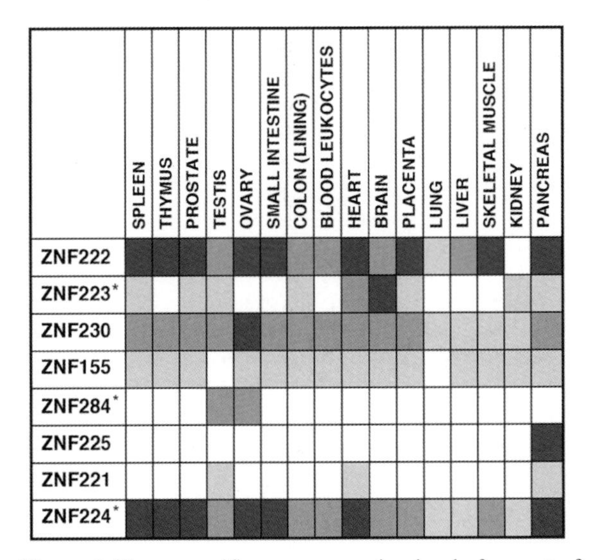


**Figure 4.** Tissue-specific gene expression levels for a set of closely related primate-specific paralogs including *ZNF224*. (Based on data in Shannon et al. 2003). Darker colors represent higher expression levels based on northern blot data. Genes are listed in groups of closest relatives. Asterisks indicate genes in which there is evidence of alternate splicing products.

due to the evolution of the promoter and enhancer sequences that are duplicated along with the genes. Diversification of mRNA splicing patterns also appears to play a role in paralog divergence; for instance, *ZNF284* is alternatively spliced to yield multiple mRNA isoforms, whereas *ZNF225* produces a single mRNA species (Shannon et al. 2003). Alternative splicing has been shown to generate KK protein isoforms with truncated or even missing KRAB A and KRAB B domains (Odeberg et al. 1998; Takashima et al. 2001), and may in some cases alter finger-domain structure through the use of cryptic splice sites within the KZNF-encoding exon (Peng et al. 2002; Hennemann et al. 2003). Both types of alternative splicing events have potential to produce KK protein isoforms with altered functional properties.

### Recent History: Comparing KK Gene and Cluster Structure in Different Rodents

To gain additional insights into the evolutionary history of the Hsa19q13.2/Mmu7 KK gene cluster, we analyzed sequence of the homologous cluster in rat. The rat and mouse gene clusters are structurally similar, but changes have clearly taken place since the split of the two rodent lineages 15–20 million years ago (Mya) (Fig. 3). For instance, the rat genome carries only four genes related to human gene *ZNF235*, whereas mouse carries six paralogous copies. In another example of change, the rat cluster includes two genes related to human *ZNF45* and its single mouse homolog, *Zfp94* (Fig. 5). This duplication leaves only two of the five sets of predicted primate–rodent homologs within this cluster as conserved 1:1:1 orthologs in human, rat, and mouse. In both cases (the two extra mouse *ZNF235* duplicates and the extra rat *ZNF45*-like locus), the duplication events probably predated the mouse/rat split and lineage-specific losses followed, as indicated by a preliminary analysis of sequence divergence in the KZNF exons and duplicated repetitive elements (A. Hamilton, unpubl.). The case for a mouse-specific loss of the *ZNF45*-like gene is supported by an isolated pseudo-KRAB A sequence that is similar to the (also disrupted) KRAB A in the rat locus and in the same relative position (Fig. 3).

Besides the change in gene number, structural changes in finger repeat arrangements have also taken place in orthologous pairs of mouse and rat genes. For example, rat *Zfp111* carries two additional fingers inserted in the middle of the KZNF domain as compared to the same gene in the mouse (Fig. 5), a type of change that potentially generates a major change in the DNA-binding "code" of the protein (see below) (Fig. 6). Mouse and rat *Zfp93, Zfp108*, and *Zfp112* differ by loss or gain of one finger per pair, but in each case these changes have occurred at the ends of the arrays (not shown). The gene-structure differences between mouse and rat genes suggest that KZNF array deletions and duplications are relatively frequent evolutionary events.

### Implications of the Modes of Evolution for Zinc-finger Arrays

KZNF domains can therefore be altered in several ways that can potentially affect their ability to bind to a specific DNA target sequence. Although typically each motif acts as a discrete DNA-binding element (Pabo et al. 2001), adjacent finger repeats cooperate in determining target site recognition and binding stability (Isalan et al.

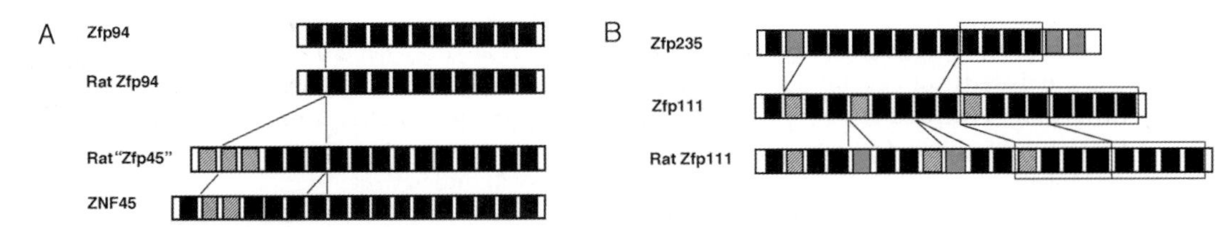


**Figure 5.** Examples of changes in the number of finger motifs in sets of related ZNF orthologs and paralogs. (*A*) Human *ZNF45* with mouse and rat copies of *Zfp94* (GenBank accession XM_218438 for the predicted rat sequence) and the second rat locus (fingers represented in XM_218439) (*B*) Mouse *Zfp111* and rat *Zfp111* (also known as *rkr2*; NM_133323) showing the addition of two fingers in the rat array. The *Zfp235* diagram represents mouse *Zfp235*, rat *Zfp235*, and human *ZNF235*, all of which have the same array structure; this arrangement represents the closest human relative of the rodent Zfp111 protein. Rat and mouse *Zfp111* copies carry duplications of a block of fingers relative to *Zfp235* (designated in boxes).

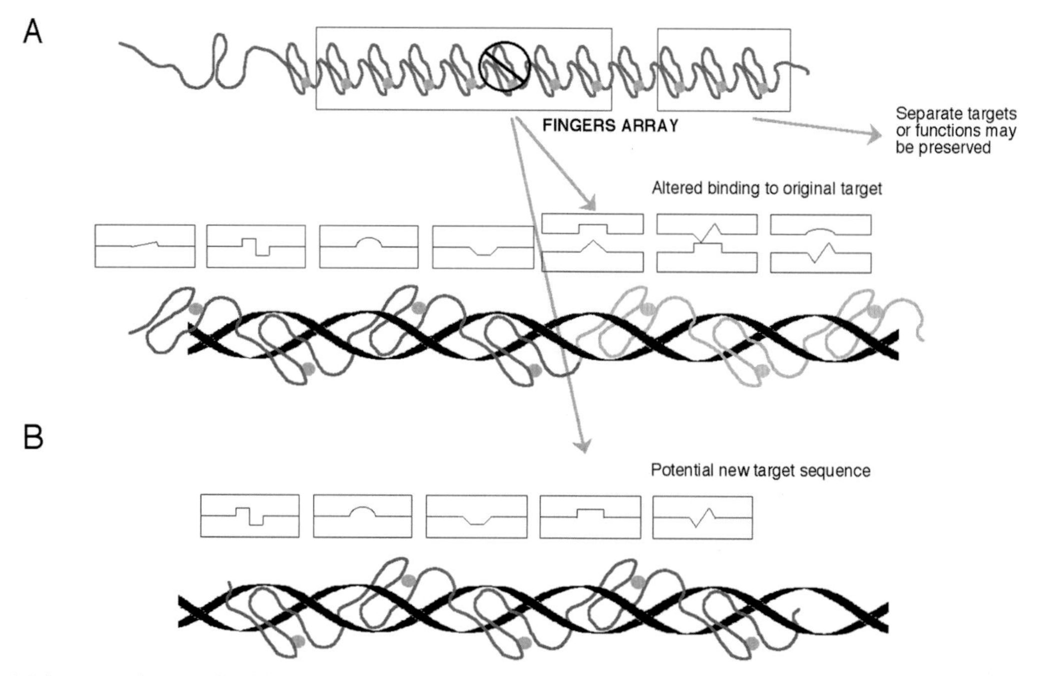


**Figure 6.** Diagrammatic example of the potential dramatic effects of zinc-finger motif gain or loss on the DNA sequence recognition of a hypothetical "code." (*A*) Loss of a finger from the middle of the array affects several other fingers on one side of it, causing a "frameshift"-like disruption of any colinear amino acid-nucleotide recognition code and potentially reducing binding affinity for its original target. Note that a deletion on the ends of the array (not shown) may not have had as much effect on binding this target (but the remaining fingers may have alternate functions which would be disrupted). (*B*) The same mutant protein shown with an increased affinity for a different target DNA sequence. This is one proposed way in which duplicated zinc-finger genes may find a role that allows them to be retained by selection.

1997; Laity et al. 2000), so a change in the code of a single finger may have wider effects on protein function. However, the gain or loss of whole fingers may have much greater potential to drive paralog divergence by altering arrangement of the KZNF motifs and interrupting the matched geometries of fingers and nucleotides at binding sites (Fig. 6). Considering the mode of interaction between DNA and KZNF domains, recombination between the adjacent finger-repeat structures may provide an unusually rapid path to functional divergence for KZNF proteins, producing more dramatic alterations in DNA-binding structure than could possibly be generated through any other type of single mutational event.

How might such changes in finger organization affect the function of KZNF proteins? Experiments in which different fingers of specific proteins are mutated using recombinant DNA techniques often produce proteins with altered DNA-binding specificity (see, e.g., Isalan et al. 1997; Ohlsson et al 2001; Hennemann et al. 2003; Mascle et al. 2003). However, several lines of data have suggested that all fingers in a polydactyl protein do not contribute equally to DNA binding (Filippova et al. 1996; Obata et al. 1999; Quitschke et al. 2000; Ohlsson et al. 2001; Daniel et al. 2002; Peng et al. 2002; Hennemann et al. 2003). Longer or shorter versions of a duplicated KZNF array in which the core-binding set of fingers is retained may potentially bind the same gene targets with altered stabilities or specificities. Such subtly altered TF proteins may be useful in fine-tuning gene regulation; e.g., in different tissues or at specific times in development. Indeed, one set of rodent-specific KK duplicates has recently been shown to be involved in regulation of different genes in a gender-specific manner in mice (Krebs et al. 2003).

The functional impact of deletions or duplications of finger motifs from the middle of a KZNF array has not been studied extensively and is therefore a matter of conjecture. The answer may be complex, depending on details of structure and function of a particular KZNF domain. Internal rearrangements of finger motifs may have a dramatic effect on DNA binding, analogous to a "frameshift" in the target-recognition code. If such a disruption occurs within a core set of fingers, KZNF motifs on one side or the other of the change could be mismatched to the DNA sequence (Fig. 6). However, internal duplications and deletions may also generate novel DNA-binding arrays with affinities for new target sites; by opening new opportunities for gene regulation, even radical changes in KZNF array structure might sometimes be adaptive. Finally, not all changes in the physical center of an array need be disruptive; it is possible that when core subsets of fingers at either end of a long or divided array bind different targets (Morris et al. 1994), a duplication or deletion between these subsets may be allowed with little effect.

The importance of individual fingers may also shift over evolutionary time in coordination with mutations in the target sequence. For example, human and chicken variants of the highly conserved 11-finger KZNF protein, CTCF, use different (but overlapping) sets of fingers to bind the diverging promoter sequences of the target gene, c-*myc* (Filippova et al. 1996). These intriguing data highlight the power of evolutionary comparisons in understanding TF protein function. They also provide evidence for the notion that KZNF arrays are coevolving with their regulatory binding sites and demonstrate that patterns of conservation and divergence in related proteins may reveal which fingers are most crucial for core target-sequence recognition. For KK proteins, alignments between lineage-specific paralogs can be especially illuminating. For instance, the 5′-most finger arrays in *ZNF224* and its human paralogs are relatively highly conserved, whereas the 3′ fingers are most divergent (Huntley et al. 2003). This suggests that the amino-terminal fingers of the paralogous proteins have retained the ability to bind to a common core target sequence, with carboxy-terminal regions possibly influencing stability or adding specificity. Indeed, finger-deletion studies with ZNF224 have confirmed that the amino-terminal fingers are most critical to DNA target site binding (Medugno et al. 2003).

### Known Regulatory Targets of KZNF Transcription Factor Proteins

Regulatory targets have been identified for conserved KZNF transcription factors in several different species (see, e.g., Pavletich and Pabo 1993; Ohlsson et al. 2001; Waltzer et al. 2001; Cowden and Levine 2002; Yoon et al. 2002; Mascle et al. 2003), providing the basis for understanding KZNF protein function including target-site recognition and binding. The best-studied KZNF proteins are highly conserved with 3–4 finger arrays (see, e.g., Dang et al. 2000), but targets have also been determined for several longer polydactyl proteins. For example, more than 30 targets have been identified for the 8-finger neuron-restrictive silencing factor (NRSF) protein, which serves to silence expression of neuron-specific genes in nonneural tissues (Chen et al. 1998). ZNF202, a SCAN-KRAB-KZNF protein, plays a central role in repression of genes involved in lipid metabolism pathways (Wagner et al. 2000). Zfp263, a mouse KK protein, has been implicated in negative regulation of *Col11a2*, a highly conserved collagen gene involved in skeletal development (Tanaka et al. 2002). ZNF239 and its rodent counterpart, Zfp239, bind a common 18-bp recognition sequence within the gene encoding the interphotoreceptor retinoid-binding protein, IRBP (Arranz et al. 2001). Each of these polydactyl TF proteins is encoded by unique, singleton genes with clear 1:1 orthologs in other vertebrate species.

In contrast, regulatory targets are known for only two members of the evolutionarily labile fraction of the KK protein family, and both proteins regulate ancient, conserved genes. One well-studied 8-fingered KK protein, ZBRK1, binds to a 15-bp sequence located in an intron of the DNA damage-inducible gene, *GADD45*, repressing its expression. Interestingly, the *GADD45*-repressor activity of ZBRK1 appears to be mediated by the product of the breast cancer gene, *BRCA1* (Zheng et al. 2000). *ZBRK1* is intriguing because it is a member of an expanded cluster of KK genes in Hsa19q13.3 and does not have a clear mouse ortholog in the syntenically homologous region of Mmu17 or any other genomic region (L. Stubbs, unpubl.). Although it is possible that another mouse protein has arisen to regulate *Gadd45* in a similar BRCA1-dependent manner, no obvious rodent candidate for this function has been identified.

In addition, the protein encoded by *ZNF224*, a member of the 10-gene human expansion in the Hsa19q13.2 KK cluster (Fig. 3), has been implicated in repression of *ALDOA*, another ancient and unique human gene. ZNF224 binds to a negative regulatory element, called AldA-NRE, which is highly conserved in rodents (Medugno et al. 2003). Several of ZNF224's ten human paralogs share conserved versions of the amino-terminal fingers implicated in AldA-NRE binding, making it possible that different human proteins are capable of repressing *ALDOA* activity. However, curiously, the single rodent representative of this clade, Zfp61, has lost some of the amino-terminal fingers predicted to be critical to AldA-NRE binding in ZNF224 and has undergone other major changes, including a multi-finger deletion toward the carboxy-terminal end of the array compared to its closest human relatives (Shannon et al. 2003). The high degree of divergence between the finger arrays in ZNF224 and Zfp61 makes it unlikely that these human and mouse proteins bind similar targets. The protein partner for the highly conserved AldA-NRE sequence, and the regulation of *ALDOA* expression in rats and mice, therefore remain open mysteries.

## CONCLUSION: KK PROTEIN DIVERGENCE AND THE EVOLUTION OF VERTEBRATE REGULATORY NETWORKS

KK protein repertoires vary significantly in different vertebrates, with a considerable amount of overlap but also a substantial level of lineage-specific novelty, and the change in gene number and type has been particularly dramatic in mammals. How have vertebrate genomes tolerated such dramatic gene repertoire change? What role might the KK family's ongoing divergence play in intraspecies variation and speciation? What do these data imply about the constancy of gene regulation patterns in vertebrates and, for example, assumptions regarding regulatory network conservation in humans and mice? What is the fate of conserved regulatory targets, like *AldoA* and *Gadd45*, in species without KK proteins that regulate those genes in other lineages?

In the case of other large and rapidly changing gene families, the answers to similar questions are beginning to emerge. For example, gene gain and loss in the olfactory receptor gene family correlates clearly with species-specific differences in the acuity and range in the sense of smell (Mombaerts 1999). These differences may have been exploited by certain lineages in identification of new food sources, in predator detection, and in social interactions, including mate choice. The dramatic loss of functional OR genes in recent primate evolution and particularly in humans may relate to the reduced dependence on olfaction with the development of and increased reliance upon vision (Gilad et al. 2003). Gene gain and loss in gene families associated with immune function translates clearly into differences in disease resistance and susceptibility; for example, haplotypes containing different numbers and types of encoding killer cell immunoglobulin-like receptor (KIR) genes are associated clearly with susceptibility to arthritis and with differences in response to human immunodeficiency virus (HIV) infection (Hsu et al. 2002).

The biological drive for—and impact of—the massive gains and losses and pervasive subtle changes that are observed in most vertebrate tandem gene families is not yet clearly understood, and KZNF proteins are no exception. By participating in the regulation of multiple downstream target genes, some of which may themselves encode TF proteins or other types of regulators, each new or lost transcription factor has potential to directly or indirectly affect the function of multiple biological pathways and processes. However, as illustrated by examples above and a limited number of other cases, the downstream regulatory targets are yet known for only a handful of vertebrate KK genes.

Has KK gene family expansion coincided with the rise of other gene families, such that new functional genes have acquired individually tailored transcription factors? In this model, the duplication and divergence of *cis*-regulatory sequences would provide a driver for KK gene amplification and finger-domain change. The subtle changes that arise in duplicated promoters and enhancers —frequently consisting of small deletions and insertions of sequence —might indeed provide ideal new targets for duplicated KK proteins with subtly altered arrangements of fingers. This model seems plausible in light of the fact that modification of gene expression represents a major mode of divergence for vertebrate gene duplicates in certain gene families (Bird 1995; Makova and Li 2003). However, several KK proteins have been shown to regulate unique genes, and therefore at least some of these novel transcriptional repressors have evolved to impose new layers of control over the genome's ancient, conserved core gene set. In this model, the rise of the KK gene family may have played a significant role in mammalian evolution by bringing different sets of unique genes under TRIM28-mediated negative control, either replacing or modifying the earlier regulatory pathways that existed before the rise of KK genes.

Although far from providing complete information, current research has shed some interesting light on the function of polydactyl KZNF proteins and raises new questions regarding the biological impact of the KK family's striking evolutionary expansion. Clearly, given the massive differences in KK gene content and structure observed in different vertebrate lineages, no single change in a gene or its targets can be meaningfully studied in isolation. Unraveling the impact of evolutionary change on regulatory circuits will instead require a system-wide approach, incorporating data regarding target-site choice and target-gene fate in closely related lineages. Without such data, our picture of vertebrate gene regulation will remain incomplete. The modification of the regulatory networks that control gene expression patterns could influence evolution of multiple aspects of biology, including physical body plans (Carroll 1995), and a diversity of available transcriptional repressors may be vital in allowing both major and minor changes in developmental pathways. Although, given their evolutionary fluidity, it seems likely that the impact of adding, gaining, or altering each new protein may be exceedingly subtle, we predict that the cumulative effects of the massive KK gene expansion and their rapid evolutionary divergence have contributed substantially to shaping biological differences between species.

## ACKNOWLEDGMENTS

The authors thank Colleen Elso and Ivan Ovcharenko for critical comments on the manuscript. This work was supported by grants from the U.S. Department of Energy, Office of Biological and Environmental Research, under contract no. W-7405-ENG-48 with the University of California, Lawrence Livermore National Laboratory.

## REFERENCES

Aparicio S., Chapman J., Stupka E., Putnam N., Chia J.M., Dehal P., Christoffels A., Rash S., Hoon S., Smit A., Gelpke M.D., Roach J., Oh T., Ho I.Y., Wong M., Detter C., Verhoef F., Predki P., Tay A., Lucas S., Richardson P., Smith S.F., Clark M.S., Edwards Y.J., and Doggett N., et al. 2002. Whole-genome shotgun assembly and analysis of the genome of *Fugu rubripes. Science* **297:** 1301.

Arranz V., Dreuillet C., Crisanti P., Tillit J., Kress M., and Ernoult-Lange M. 2001. The zinc finger transcription factor, MOK2, negatively modulates expression of the interphotoreceptor retinoid-binding protein gene, IRBP. *J. Biol. Chem.* **276:** 11963.

Bird A.P. 1995. Gene number, noise reduction and biological complexity. *Trends Genet* **11:** 94.

Carroll S.B. 1995. Homeotic genes and the evolution of arthropods and chordates. *Nature* **376:** 479.

Chen Z.F., Paquette A.J., and Anderson D.J. 1998. NRSF/REST is required in vivo for repression of multiple neuronal target genes during embryogenesis. *Nat. Genet.* **20:** 136.

Choo Y. and Klug A. 1994. Toward a code for the interactions of zinc fingers with DNA: Selection of randomized fingers displayed on phage. *Proc. Natl. Acad. Sci.* **91:** 11163.

Collins T., Stone J.R., and Williams A.J. 2001. All in the family: The BTB/POZ, KRAB, and SCAN domains. *Mol. Cell. Biol.* **21:** 3609.

Cowden J. and Levine M. 2002. The Snail repressor positions Notch signaling in the *Drosophila* embryo. *Development* **129:** 1785.

Dang D.T., Pevsner J., and Yang V.W. 2000. The biology of the mammalian Kruppel-like family of transcription factors. *Int. J. Biochem. Cell Biol.* **32:** 1103.

Daniel J.M., Spring C.M., Crawford H.C., Reynolds A.B., and Baig A. 2002. The p120(ctn)-binding partner Kaiso is a bimodal DNA-binding protein that recognizes both a sequence-specific consensus and methylated CpG dinucleotides. *Nucleic Acids Res.* **30:** 2911.

Dehal P., Predki P., Olsen A.S., Kobayashi A., Folta P., Lucas S., Land M., Terry A., Ecale Zhou C.L., Rash S., Zhang Q., Gordon L., Kim J., Elkin C., Pollard M.J., Richardson P., Rokhsar D., Uberbacher E., Hawkins T., Branscomb E., and Stubbs L. 2001. Human chromosome 19 and related regions in mouse: Conservative and lineage-specific evolution. *Science* **293:** 104.

Filippova G.N., Fagerlie S., Klenova E.M., Myers C., Dehner

Y., Goodwin G., Neiman P.E., Collins S.J., and Lobanenkov V.V. 1996. An exceptionally conserved transcriptional repressor, CTCF, employs different combinations of zinc fingers to bind diverged promoter sequences of avian and mammalian c-myc oncogenes. *Mol. Cell. Biol.* **16:** 2802.

Friedman R. and Hughes A.L. 2003. The temporal distribution of gene duplication events in a set of highly conserved human gene families. *Mol. Biol. Evol.* **20:** 154.

Gaudieri S., Kulski J.K., Dawkins R.L., and Gojobori T. 1999. Different evolutionary histories in two subgenomic regions of the major histocompatibility complex. *Genome Res.* **9:** 541.

Gilad Y., Man O., Paabo S., and Lancet D. 2003. Human specific loss of olfactory receptor genes. *Proc. Natl. Acad. Sci.* **100:** 3324.

Hennemann H., Vassen L., Geisen C., Eilers M., and Moroy T. 2003. Identification of a novel Kruppel-associated box domain protein, Krim-1, that interacts with c-Myc and inhibits its oncogenic activity. *J. Biol. Chem.* **278:** 28799.

Hsu K.C., Chida S., Geraghty D.E., and Dupont B. 2002. The killer cell immunoglobulin-like receptor (KIR) genomic region: Gene-order, haplotypes and allelic polymorphism. *Immunol. Rev.* **190:** 40.

Huntley S., Hamilton A.T., Kim J., Branscomb E., and Stubbs L. 2003. Tandem gene family expansion and genomic diversity In *Comparative genomics: A guide to the analysis of eukaryotic genomes* (ed. M.D. Adams). Humana Press, Totowa, New Jersey. (In press.)

Isalan M., Choo Y., and Klug A. 1997. Synergy between adjacent zinc fingers in sequence-specific DNA recognition. *Proc. Natl. Acad. Sci.* **94:** 5617.

Jacobs G.H. 1992. Determination of the base recognition positions of zinc fingers from sequence analysis. *EMBO J.* **11:** 4507.

Krebs C.J., Larkins L.K., Price R., Tullis K.M., Miller R.D., and Robins D.M. 2003. Regulation of sex-limitation (Rsl) encodes a pair of KRAB zinc-finger genes that control sexually dimorphic liver gene expression. *Genes. Dev.* **17:** 2664.

Krishna S.S., Majumdar I., and Grishin N.V. 2003. Structural classification of zinc fingers: Survey and summary. *Nucleic Acids Res.* **31:** 532.

Ladomery M. and Dellaire G. 2002. Multifunctional zinc finger proteins in development and disease. *Ann. Hum. Genet.* **66:** 331.

Laity J.H., Dyson H.J., and Wright P.E. 2000. DNA-induced alpha-helix capping in conserved linker sequences is a determinant of binding affinity in Cys(2)-His(2) zinc fingers. *J. Mol. Biol.* **295:** 719.

Lander E.S., Linton L.M., Birren B., Nusbaum C., Zody M.C., Baldwin J., Devon K., Dewar K., Doyal M., FitzHugh W., Funke R., Gage D., Harris K., Heaford A., Howland J., Kann L., Lehoczky J., LeVine R., McEwan P., McKernan K., Meldrim J., Mesirov J.P., Miranda C., Morris W., and Naylor J., et al. (International Human Genome Sequencing Consortium). 2001. Initial sequencing and analysis of the human genome. *Nature* **409:** 860.

Lespinet O., Wolf Y.I., Koonin E.V., and Aravind L. 2002. The role of lineage-specific gene family expansion in the evolution of eukaryotes. *Genome Res.* **12:** 1048.

Looman C., Abrink M., Mark C., and Hellman L. 2002. KRAB zinc finger proteins: An analysis of the molecular mechanisms governing their increase in numbers and complexity during evolution. *Mol. Biol. Evol.* **19:** 2118.

Lynch M. and Conery J.S. 2000. The evolutionary fate and consequences of duplicate genes. *Science* **290:** 1151.

Makova K.D. and Li W.-H. 2003. Divergence in the spatial pattern of gene expression between human duplicate genes. *Genome Res.* **13:** 1638.

Mark C., Abrink M., and Hellman L. 1999. Comparative analysis of KRAB zinc finger proteins in rodents and man: Evidence for several evolutionarily distinct subfamilies of KRAB zinc finger genes. *DNA Cell Biol.* **18:** 381.

Mascle X., Albagli O., and Lemercier C. 2003. Point mutations in BCL6 DNA-binding domain reveal distinct roles for the six zinc fingers. *Biochem. Biophys. Res. Commun.* **300:** 391.

Medugno L., Costanzo P., Lupo A., Monti M., Florio F., Pucci P., and Izzo P. 2003. A novel zinc finger transcriptional repressor, ZNF224, interacts with the negative regulatory element (AldA-NRE) and inhibits gene expression. *FEBS Lett.* **534:** 93.

Miller J.C. and Pabo C.O. 2001. Rearrangement of side-chains in a Zif268 mutant highlights the complexities of zinc finger-DNA recognition. *J. Mol. Biol.* **313:** 309.

Mombaerts P. 1999. Seven-transmembrane proteins as odorant and chemosensory receptors. *Science* **286:** 707.

Morris J.F., Hromas R., and Rauscher F.J., III. 1994. Characterization of the DNA-binding properties of the myeloid zinc finger protein MZF1: Two independent DNA-binding domains recognize two DNA consensus sequences with a common G-rich core. *Mol. Cell. Biol.* **14:** 1786.

Nei M., Gu X., and Sitnikova T. 1997. Evolution by the birth-and-death process in multigene families of the vertebrate immune system. *Proc. Natl. Acad. Sci.* **94:** 7799.

Obata T., Yanagidani A., Yokoro K., Numoto M., and Yamamoto S. 1999. Analysis of the consensus binding sequence and the DNA-binding domain of ZF5. *Biochem. Biophys. Res. Commun.* **255:** 528.

Odeberg J., Rosok O., Gudmundsson G.H., Ahmadian A., Roshani L., Williams C., Larsson C., Ponten F., Uhlen M., Asheim H.C., and Lundeberg J. 1998. Cloning and characterization of ZNF189, a novel human Kruppel-like zinc finger gene localized to chromosome 9q22-q31. *Genomics* **50:** 213.

Ohlsson R., Renkawitz R., and Lobanenkov V. 2001. CTCF is a uniquely versatile transcription regulator linked to epigenetics and disease. *Trends Genet.* **17:** 520.

Ohno S. 1970. *Evolution by gene duplication.* Springer-Verlag, Berlin, New York,.

Pabo C.O., Peisach E., and Grant R.A. 2001. Design and selection of novel Cys2His2 zinc finger proteins. *Annu. Rev. Biochem.* **70:** 313.

Pavletich N.P. and Pabo C.O. 1993. Crystal structure of a five-finger GLI-DNA complex: New perspectives on zinc fingers. *Science* **261:** 1701.

Peng H., Zheng L., Lee W.H., Rux J.J., and Rauscher F.J., III. 2002. A common DNA-binding site for SZF1 and the BRCA1-associated zinc finger protein, ZBRK1. *Cancer Res.* **62:** 3773.

Pieler T. and Bellefroid E. 1994. Perspectives on zinc finger protein function and evolution: An update. *Mol. Biol. Rep.* **20:** 1.

Quitschke W.W., Taheny M.J., Fochtmann L.J., and Vostrov A.A. 2000. Differential effect of zinc finger deletions on the binding of CTCF to the promoter of the amyloid precursor protein gene. *Nucleic Acids Res.* **28:** 3370.

Sander T.L., Stringer K.F., Maki J.L., Szauter P., Stone J.R., and Collins T. 2003. The SCAN domain defines a large family of zinc finger transcription factors. *Gene* **310:** 29.

Schultz D.C., Ayyanathan K., Negorev D., Maul G.G., and Rauscher F.J., III. 2002. SETDB1: A novel KAP-1-associated histone H3, lysine 9-specific methyltransferase that contributes to HP1-mediated silencing of euchromatic genes by KRAB zinc-finger proteins. *Genes Dev.* **16:** 919.

Shannon M. and Stubbs L. 1998. Analysis of homologous XRCC1-linked zinc-finger gene families in human and mouse: Evidence for orthologous genes. *Genomics* **49:** 112.

Shannon M., Hamilton A.T., Gordon L., Branscomb E., and Stubbs L. 2003. Differential expansion of zinc-finger transcription factor loci in homologous human and mouse gene clusters. *Genome Res.* **13:** 1097.

Takashima H., Nishio H., Wakao H., Nishio M., Koizumi K., Oda A., Koike T., and Sawada K. 2001. Molecular cloning and characterization of a KRAB-containing zinc finger protein, ZNF317, and its isoforms. *Biochem. Biophys. Res. Commun.* **288:** 771.

Tanaka K., Tsumaki N., Kozak C.A., Matsumoto Y., Nakatani F., Iwamoto Y., and Yamada Y. 2002. A Kruppel-associated box-zinc finger protein, NT2, represses cell-type-specific promoter activity of the alpha 2(XI) collagen gene. *Mol. Cell. Biol.* **22:** 4256.

Trowsdale J., Barten R., Haude A., Stewart C.A., Beck S., and

Wilson M.J. 2001. The genomic context of natural killer receptor extended gene families. *Immunol. Rev.* **181:** 20.

Turner J. and Crossley M. 1999. Mammalian Kruppel-like transcription factors: More than just a pretty finger. *Trends Biochem. Sci.* **24:** 236.

Vanhalst K., Kools P., Vanden Eynde E., and van Roy F. 2001. The human and murine protocadherin-beta one-exon gene families show high evolutionary conservation, despite the difference in gene number. *FEBS Lett.* **495:** 120.

Wagner S., Hess M.A., Ormonde-Hanson P., Malandro J., Hu H., Chen M., Kehrer R., Frodsham M., Schumacher C., Beluch M., Honer C., Skolnick M., Ballinger D., and Bowen B.R. 2000. A broad role for the zinc finger protein ZNF202 in human lipid metabolism. *J. Biol. Chem.* **275:** 15685.

Waltzer L., Vandel L., and Bienz M. 2001. Teashirt is required for transcriptional repression mediated by high Wingless levels. *EMBO J.* **20:** 137.

Waterston R.H., Lindblad-Toh K., Birney E., Rogers J., Abril J.F., Agarwal P., Agarwala R., Ainscough R., Alexandersson M., An P., Antonarakis S.E., Attwood J., Baertsch R., Bailey J., Barlow K., Beck S., Berry E., Birren B., Bloom T., Bork P., Botcherby M., Bray N., Brent M.R., Brown D.G., and Brown S.D., et al. (Mouse Genome Sequencing Consortium). 2002. Initial sequencing and comparative analysis of the mouse genome. *Nature* **420:** 520.

Wolfe S.A., Grant R.A., Elrod-Erickson M., and Pabo C.O. 2001. Beyond the "recognition code:" Structures of two Cys2His2 zinc finger/TATA box complexes. *Structure* **9:** 717.

Xie S., Green J., Bixby J.B., Szafranska B., DeMartini J.C., Hecht S., and Roberts R.M. 1997. The diversity and evolutionary relationships of the pregnancy-associated glycoproteins, an aspartic proteinase subfamily consisting of many trophoblast-expressed genes. *Proc. Natl. Acad. Sci.* **94:** 12809.

Yoon J.W., Kita Y., Frank D.J., Majewski R.R., Konicek B.A., Nobrega M.A., Jacob H., Walterhouse D., and Iannaccone P. 2002. Gene expression profiling leads to identification of GLI1-binding elements in target genes and a role for multiple downstream pathways in GLI1-induced cell transformation. *J. Biol. Chem.* **277:** 5548.

Young J.M. and Trask B.J. 2002. The sense of smell: Genomics of vertebrate odorant receptors. *Hum. Mol. Genet.* **11:** 1153.

Zheng L., Pan H., Li S., Flesken-Nikitin A., Chen P.L., Boyer T.G., and Lee W.H. 2000. Sequence-specific transcriptional corepressor function for BRCA1 through a novel zinc finger protein, ZBRK1. *Mol. Cell* **6:** 757.

# Sequence Organization and Functional Annotation of Human Centromeres

M.K. Rudd,*†, M.G. Schueler,†‡ and H.F. Willard*
*Institute for Genome Sciences & Policy, Department of Molecular Genetics and Microbiology, Duke University, Durham, North Carolina 27710; †Department of Genetics, Case Western Reserve University, Cleveland, Ohio 44106

With the near completion of the archival human genome sequence, attention has turned to defining the attributes of the sequence that are responsible for its many functions, including determination of gene content, functional identification of both short- and long-range regulatory sequences, and characterization of structural elements of the genome responsible for maintaining genome stability and integrity (Collins et al. 2003). In addition, the task remains to close the several hundred gaps that disrupt the landscape of an otherwise contiguous sequence for each of the 24 human chromosomes. The largest of these gaps correspond to regions of known or presumed heterochromatin at or near the centromeres of each chromosome, regions that were largely excluded from both the public (Lander et al. 2001) and private (Venter et al. 2001) sequencing efforts due in part to their highly repetitive DNA content. Although there are several large regions of noncentromeric heterochromatin in the human karyotype (Trask 2002), the repetitive DNA associated with the primary constriction of each chromosome has been strongly implicated in centromere function (Willard 1998; Lamb and Birchler 2003) and thus remains to be fully characterized at the genomic and functional levels.

The centromere is essential for normal segregation of chromosomes in both mitotic and meiotic cells. Paradoxically, although this role is conserved throughout eukaryotic evolution, the sequences that accomplish centromere function in different organisms are not (Henikoff et al. 2001). In part reflecting the different perspectives brought to chromosome structure and function by cell biologists on the one hand and by geneticists (and now genomicists) on the other, two fundamentally different approaches have contributed to the current understanding of centromere biology. The top-down, extrinsic approach, classically adopted by cytologists and cell biologists, has focused predominantly on the functions of the kinetochore, the proteinaceous complex that assembles at the primary constriction and mediates both attachment of chromosomes to microtubules and movement of chromosomes along the mitotic and meiotic spindles (Rieder and Salmon 1998; Tanaka 2002). This approach has been highly successful in identifying a range of proteins that are critical for establishing the specialized chromatin that marks the position of the centromere on complex eukaryotic chromosomes and directs assembly of the kinetochore complex (Sullivan et al. 2001; Cleveland et al. 2003).

The complementary bottom-up approach, with its fundamentally more intrinsic focus, seeks to explore the underlying genomic basis for centromere function and to define the *cis*-acting DNA sequences that must, at some level, specify where a centromere forms (Nicklas 1971; Willard 1998). Influenced by the success of elegant experiments in budding yeast that defined a minimal ~120-bp sequence which directed centromere function (Clarke 1998), the large blocks of heterochromatin that dominate the pericentric regions of human (and other mammalian) genomes were initially discounted as serious candidates for anything remotely functional. However, both physical and genetic mapping studies over the past 15 years have given credibility to the possibility that the satellite DNA sequences themselves do indeed play a role in the structural integrity of the centromere and/or directly in specifying kinetochore assembly. In many complex eukaryotic genomes, including plants (Dong et al. 1998; Copenhaver et al. 1999; Zhong et al. 2002; Nagaki et al. 2003), fission yeast (Kniola et al. 2001), *Drosophila* (Sun et al. 2003), as well as human (Lander et al. 2001; Venter et al. 2001) and mouse (Waterston et al. 2002), a variety of repetitive DNAs have now been characterized genomically to varying degrees at or near the centromere. In an increasing number of these cases, direct functional studies now support a role for repetitive sequences in centromere specification and function (for review, see Cleveland et al. 2003). In the human genome, both in-depth studies of centromeric chromatin (Vafa and Sullivan 1997; Ando et al. 2002; Blower et al. 2002) and the formation of human artificial chromosomes with de novo centromeres (Harrington et al. 1997; Ikeno et al. 1998) provide direct evidence of a functional role for centromere-associated satellite DNA.

The functional importance of centromeric sequences notwithstanding, these regions of the human genome remain poorly understood at the level of sequence assembly and annotation. Indeed, most reported contigs of pericentromeric regions in the human genome terminate at clones containing just a few heterogeneous satellite

‡Present address: Genome Technology Branch, National Human Genome Research Institute, Bethesda, Maryland 20892.

DNA repeats without reaching the extensive homogeneous arrays that characterize all normal human centromeres (Schueler et al. 2001). The exact role of primary DNA sequence in centromere function continues to be a matter of some debate, and likely both genomic and epigenetic factors are involved (Choo 2001; Sullivan et al. 2001; Cleveland et al. 2003). With growing interest in the elements of a functioning centromere, complete sequence assemblies of human centromeres, combined with their functional annotation, will be an essential resource.

## GENOMIC ORGANIZATION OF THE α-SATELLITE FAMILY

Originally identified in the human genome over 25 years ago (Manuelidis 1976; Manuelidis and Wu 1978), α-satellite DNA, defined by a diverged 171-bp motif repeated in a tandem head-to-tail fashion, has been identified at the centromeres of all normal human chromosomes studied to date (Willard and Waye 1987; Alexandrov et al. 2001). Human chromosomes (as well as those characterized in great apes) contain α satellite organized hierarchically into multimeric, higher-order repeat arrays in which a defined number of monomers have been homogenized as a unit (Fig. 1). On at least the majority of human chromosomes, these large arrays span several megabases of DNA (Wevrick and Willard 1989) and are coincident with the centromere as defined by genetic mapping (Mahtani and Willard 1998; Laurent et al. 2003) and with the cytogenetically visible primary constriction and site of a number of centromere and kinetochore proteins that have been implicated in centromere function (Tyler-Smith et al. 1993; Blower et al. 2002; Spence et al. 2002).

In addition to these highly homogeneous arrays of multimeric α satellite, many centromeric regions also contain a variable amount of heterogeneous α-satellite monomers that, unlike the higher-order repeats, fail to show any evidence of hierarchical organization (Fig. 1B,C). Where examined at the level of genome maps, these stretches of monomeric α satellite flank the higher-order arrays and have been, on some chromosomes, linked on sequence contigs with the euchromatin of the chromosome arms (Horvath et al. 2001). Only on the X chromosome (see below) has a sequence contig successfully bridged from a chromosome arm, through monomeric α satellite, to the higher-order repeats of the functionally defined centromere (Schueler et al. 2001). However, as the precise functional determinants of the centromere on different chromosomes remain undefined and as different classes of α-satellite DNA (and other pericentromeric satellite DNAs) show substantial sequence heterogeneity both inter- and intrachromosomally (Lee et al. 1997; Alexandrov et al. 2001), it becomes important to complete sequence contigs of each chromosome arm/centromere junction and to functionally annotate the different types of sequence element present throughout pericentromeric regions.

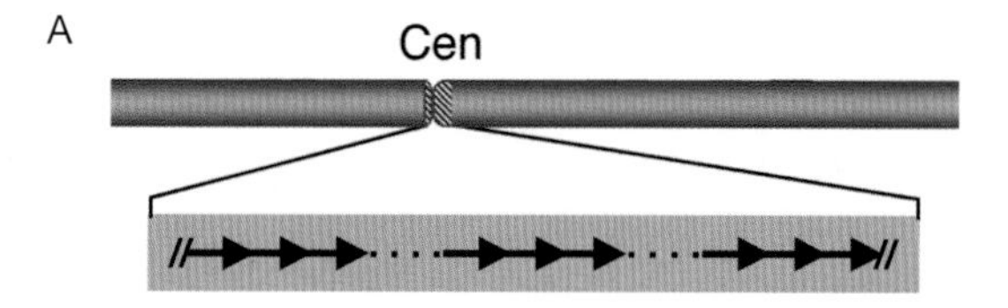


B Monomeric alpha satellite

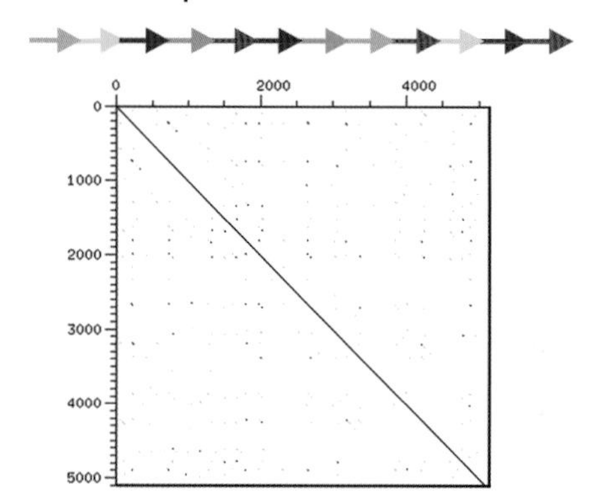


C Higher-order alpha satellite

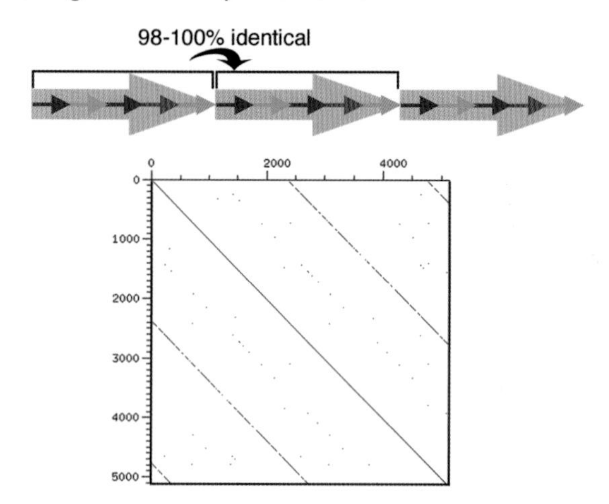


**Figure 1.** α-Satellite organization in the human genome. (*A*) The centromere region of human chromosomes comprises megabases of repetitive α-satellite DNA organized largely in tandem arrays that localize to the primary constriction. (*B*) Monomeric α satellite lacks a higher-order periodicity, and individual monomers are ~65–80% identical in sequence. Each arrow represents a ~171-bp monomer. A dotter plot (Sonnhammer and Durbin1995) (~95–99% stringency over 100 bp) of 5 kb of monomeric α satellite from the chromosome 17 pericentromeric region. High identity is seen only along the diagonal, indicating a lack of close sequence relationships among the ~30 monomers whose sequence is illustrated. (*C*) Higher-order arrays of α satellite consist of tandem multimeric units (shown as 5-monomer units in the schematic) that are nearly identical in sequence. D17Z1-B α satellite is based on a 2.4-kb higher-order repeat unit, and higher-order repeats (consisting of 14 adjacent monomers each) are ~98–99% identical to each other (see text). This multimeric, higher-order relationship is apparent from the dotter plot as diagonals at 2.4-kb intervals parallel to the self-diagonal.

## THE CENTROMERES OF CHROMOSOMES X AND 17

As models to explore the feasibility of detailed functional and genomic mapping of human centromeres, we initially selected chromosomes 17 and X (Willard et al. 1983; Waye and Willard 1986). α Satellite from both chromosomes has been characterized extensively; each belongs to the pentameric subfamily of human α satellite in which the homogeneous higher-order repeats (12 and 16 monomers long for the DXZ1 and D17Z1 arrays on the X and 17, respectively) are based on an underlying five-monomer unit. Both arrays span ~2–4 Mb on most copies of these chromosomes in the human population, and detailed long-range restriction maps have been determined (Wevrick et al. 1990; Mahtani and Willard 1998). Furthermore, study of a variety of both naturally occurring and engineered human chromosome abnormalities had demonstrated correspondence between the DXZ1 and D17Z1 loci and several kinetochore proteins associated specifically with functionally active centromeres (Wevrick et al. 1990; Sullivan and Willard 1998; Higgins et al. 1999; Mills et al. 1999; Lee et al. 2000). Notwithstanding this background of information and resources, the centromeric region of these two chromosomes remained poorly represented. Although the available sequence contigs of both the X chromosome and chromosome 17 are marked by a number of gaps, the largest and most notable of these lie at the centromeres (Fig. 2).

We focused our initial efforts on the short arm of the X chromosome (Xp), as both the public and private draft sequence assemblies terminated in clones that contained small amounts of monomeric α satellite (Schueler et al. 2001). As reported by Schueler et al. (2001), the final clone assembly, based on both in silico strategies and screening of multiple large-clone libraries, spanned an abrupt transition (the "satellite junction," Fig. 3) from the euchromatin of proximal Xp to the first satellite sequences, encountered less than 150 kb from the most proximal fully annotated gene, *ZXDA*, on Xp (Schueler et al. 2001). Although the most distal portion of the pericentromeric heterochromatin on Xp contains representatives of several different satellite DNA families (α satellite, γ satellite, and a 35-bp satellite; see Fig. 3), it transitions proximally through ~200 kb of α satellite until the contig reaches a junction (the "array junction," Fig. 3) with members of the higher-order repeat array DXZ1. Notably, repeats from the DXZ1 locus have been annotated as the functional centromere on the X by deletion mapping using chromosome variants (Higgins et al. 1999; Lee et al. 2000; Schueler et al. 2001; Spence et al. 2002), by formation of de novo centromeres in human artificial chromosome assays (see below and Schueler et al. 2001; Rudd et al. 2003), and by mapping a domain of topoisomerase II activity associated with centromere function within the DXZ1 array near the Xp array junction (Spence et al. 2002). Thus, these studies provide a proof of principle for assembly and functional annotation of a human centromere (Henikoff 2002).

To extend these studies and to determine how general the strategy employed on Xp might be for examining other centromeres, we selected chromosome 17 and the D17Z1 locus, which previous studies had demonstrated had the functional attributes of the centromere (Wevrick et al. 1990; Haaf et al. 1992; Harrington et al. 1997). In the course of developing a contig between D17Z1 and the short arm of chromosome 17 (17p) (M.K. Rudd et al., unpubl.), we discovered a novel type of higher-order repeat, D17Z1-B, that adds to the complexity of this region (Fig. 4A). Most centromere regions that have been examined in detail have revealed a single, highly homogeneous higher-order repeat array, typically several megabases in length. Studies of the DXZ1 array have both strengthened this concept (Schueler et al. 2001; Schindelhauer and Schwarz 2002) and extended it, with the description of a set of localized, diverged copies of the DXZ1 repeat at the junction between DXZ1 and monomeric α satellite (Schueler et al. 2001). In contrast, a few chromosomes (such as chromosome 7) are characterized by two, otherwise unrelated, higher-order repeat arrays (Wevrick and Willard 1991). Chromosome 17 appears to represent a third type of centromere organization in that there are two physically distinct higher-order repeat arrays that are clearly related to each other evolutionarily.

Whereas D17Z1 comprises 16 monomers to make up a 2.7-kb higher-order repeat (Waye and Willard 1986), D17Z1-B is 2.4 kb long and made of 14 monomers. D17Z1 and D17Z1-B are both made up of monomers arranged in a pentameric fashion with corresponding monomers in the same order; however, D17Z1-B is missing two monomers present in D17Z1 (Fig. 4B), likely reflecting a deletion or duplication event mediated by unequal crossing-over. We identified two overlapping clones in this contig that contain D17Z1-B α satellite; BAC RPCI-11 285M22 has been completely sequenced and BAC RPCI-11 18L18 is currently partially se-

A X chromosome

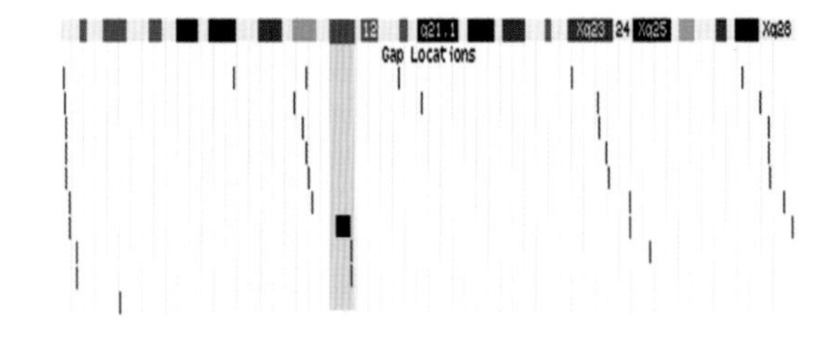


B chromosome 17

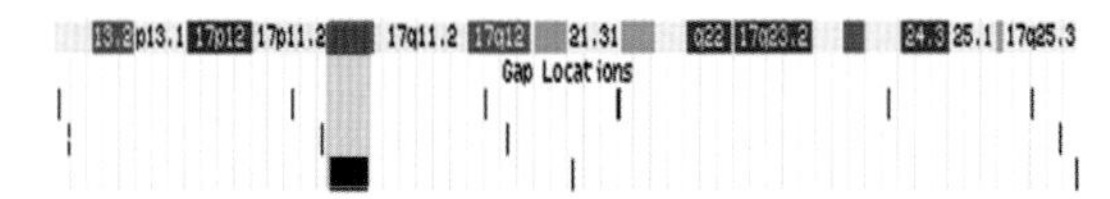


**Figure 2.** Gaps in the public genome assembly of chromosomes X and 17 (April 2003 freeze http://genome.ucsc.edu/). The chromosome is shown at the top of the figure and the centromere region is indicated in red. Gaps in the sequence assembly are illustrated as black vertical lines.

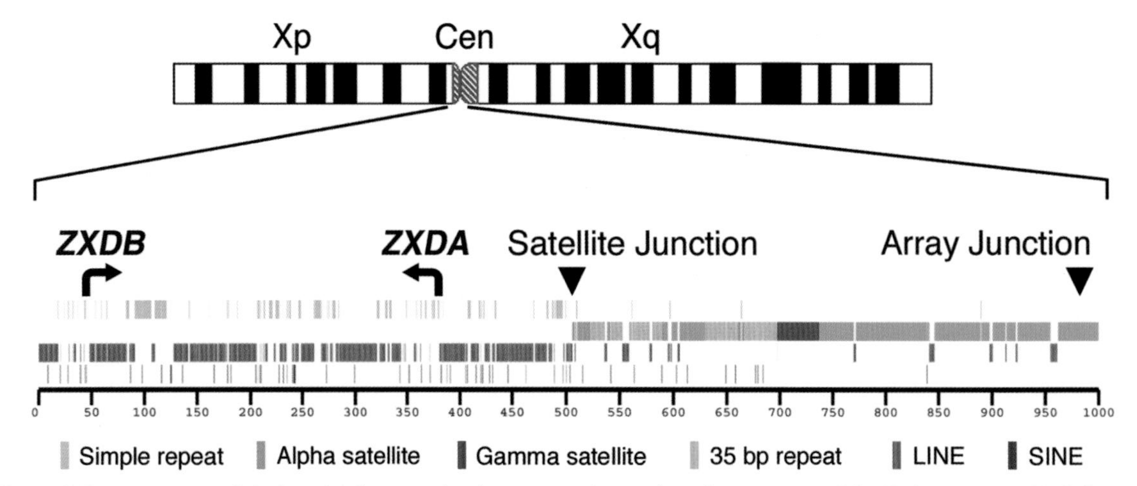


**Figure 3.** Repeat content of the junction between the short-arm euchromatin and centromere of the X chromosome. Each line and color illustrates a different type of repeat family present in the most proximal 1 Mb on Xp (drawn using rm2parasight; D. Locke, unpubl.). The "satellite junction" indicates an abrupt transition from the euchromatin of Xp and the first pericentromeric satellite sequences. At least three types of satellites are present here: monomeric α satellite, γ satellite, and a 35-bp satellite (HSAT4). The most proximal annotated genes on Xp (*ZXDA* and *ZXDB*) are shown, with the direction of transcription indicated by the arrows. The "array junction" indicates the transition between monomeric α satellite and DXZ1 higher-order repeats, which extend a further ~3 Mb at the centromere. (Based on Schueler et al. 2001.)

quenced (low-pass sequence sampling). BAC 285M22 bridges the junction between the D17Z1-B array and monomeric α satellite; fluorescence in situ hybridization studies demonstrated that this junction lies to the short arm side of D17Z1 (Fig. 4A,C). When we compared the higher-order repeats in 285M22 and 18L18 ($n = 7$), all were 98–99% identical, indicating that they are part of a highly homogeneous array of α satellite. Notably, however, whereas D17Z1 and D17Z1-B share a similar multimeric structure, they are only 92% identical, demonstrating that they are distinct yet related higher-order repeats. (Such a relationship differs from a number of variants of D17Z1 that, although distinctive in their higher-order structure and their genomic localization within the large D17Z1 locus, are *not* distinguished by levels of overall sequence relatedness [Warburton and Willard 1990, 1995].)

Although a contig extending across the full D17Z1 and D17Z1-B arrays remains to be achieved, we have established a complete contig linking D17Z1-B to the euchromatin of 17p (M.K. Rudd, unpubl.), essentially using the strategy developed as part of our Xp work (Schueler et al. 2001). The genomic content of this part of chromosome 17 is, however, somewhat more complex than that of the X, in that there are three different blocks of monomeric α satellite located within ~500 kb of D17Z1-B, separated by regions of genomic sequence populated by a number of different repeat families (both satellite and non-satellite) and at least some transcribed elements. A similar picture appears to be emerging on the long arm (17q) side of the centromere, although a contig containing stretches of monomeric α satellite and proximal 17q sequences has not yet been linked to the large D17Z1 array at the centromere (M.K. Rudd, unpubl.).

The coexistence within pericentromeric regions of both higher-order repeat arrays of α satellite and more limited stretches of monomeric α satellite without any detectable higher-order structure appears to be a consistent theme of a number of human chromosomes (Horvath et al. 2001). Although the organizational distinction between these types of α satellites (i.e., multimeric vs. monomeric) is apparent at short range with standard homology-finding programs (e.g., Fig. 1B,C), it is also important to address the phylogenetic sequence relationships among the individual monomers populating these different classes of the repeat family.

To illustrate this point, we examined ~100 monomers of monomeric α satellite from each of the Xp and 17q contigs, together with 28 monomers making up the DXZ1 and D17Z1 higher-order repeats. As seen in Figure 5, the sequences fall into two distinct phylogenetic clades, corresponding precisely to their monomeric or multimeric origin. Within the multimeric clade, the DXZ1 and D17Z1 monomers exhibit close sequence relatedness, as established previously (Waye and Willard 1986). Notably, the monomeric clade includes monomers from both chromosomes (Fig. 5); in other words, clusters of monomeric α satellite from one chromosome are more closely related to monomeric α satellite from the *other* chromosome than they are to the multimeric α satellite that maps only a few hundred kilobases away on the *same* chromosome.

These relationships presumably reflect the highly efficient homogenization mechanisms that drive and maintain the high degree of sequence homogeneity *within* higher-order repeat arrays, even across several megabases of genomic DNA (Willard and Waye 1987; Durfy and Willard 1989; Schueler et al. 2001; Schindelhauer and Schwarz 2002). However, the phylogenetic data argue that homogenization mechanisms *between* monomeric and multimeric α satellite are poorly efficient or nonexistent, even over much shorter genomic dis-

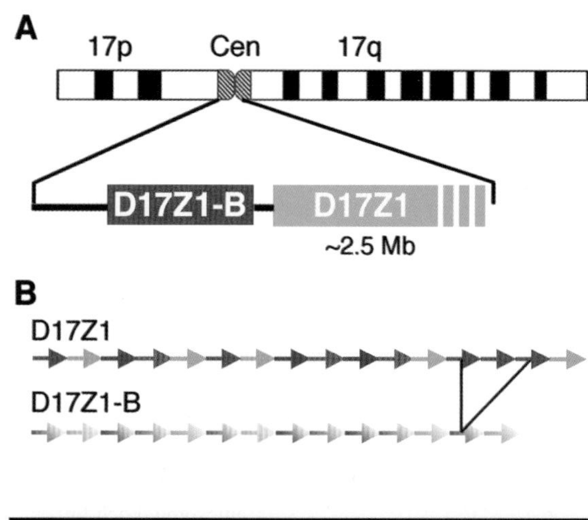


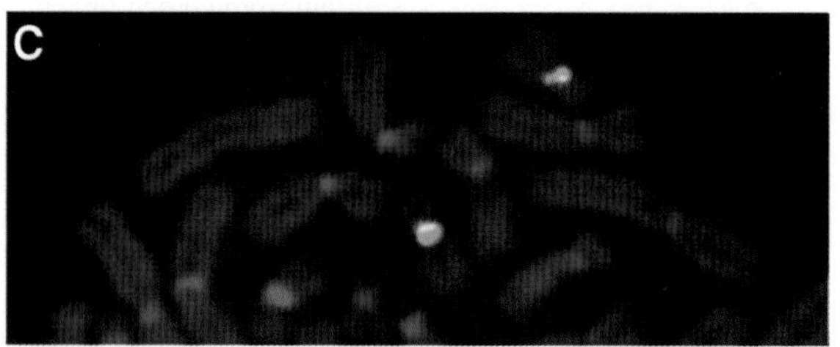


**Figure 4.** Organization of D17Z1 and D17Z1-B higher-order repeats at the centromere of chromosome 17. (*A*) The array of D17Z1 repeats spans ~ 2.5 Mb and lies adjacent to D17Z1-B, estimated to be ~500 kb. (*B*) Monomer organization of D17Z1 (16-mer) and D17Z1-B (14-mer). The two higher-order repeats share a similar monomer arrangement and are 92% identical to one another. (*C*) Chromosomal orientation of D17Z1 and D17Z1-B, determined by fluorescence in situ hybridization. D17Z1-B (*red*) hybridizes to the 17p side of D17Z1 (*green*). Chromosomes are counterstained with DAPI in blue.

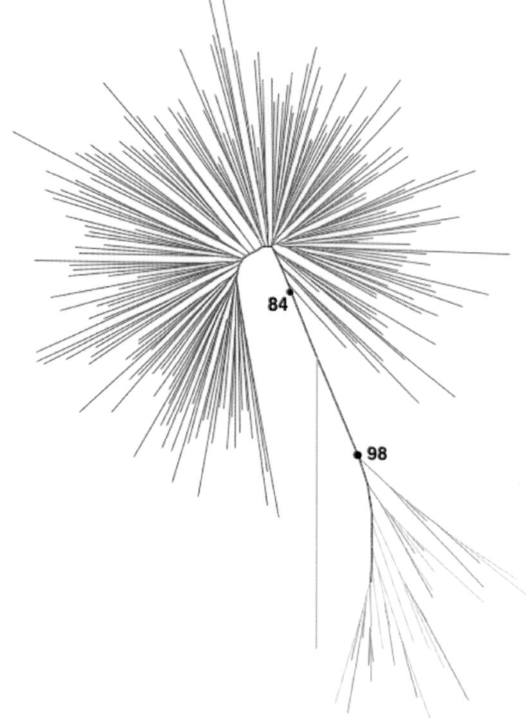


**Figure 5.** Phylogenetic analysis of 230 α-satellite monomers from the X chromosome and chromosome 17. Monomers from DXZ1 (*light blue*), D17Z1 (*pink*), monomeric X α satellite (*blue*, from Schueler et al. 2001), and monomeric 17 α satellite (*red*, from M.K. Rudd, unpubl.) were analyzed using neighbor-joining methods (http://www.megasoftware.net/). One α-satellite monomer from African green monkey (*green*) was used as an outgroup. 1000 bootstrap replicates were performed, and bootstrap values for the well-supported monomeric and higher-order nodes are shown in bold type.

tances. Plausibly, this could reflect the inhibitory effect of local discontinuities (i.e., the introduction of non-α-satellite sequences interrupting an otherwise continuous array of tandem repeats) on homogenization mechanisms. Alternatively (or in addition), it may indicate that the initial establishment of a multimeric repeat (presumably by unequal crossing-over mediated by pairing of monomers out of register) is relatively inefficient and represents the rate-limiting step evolutionarily in generation of a highly homogeneous array (Smith 1976). This inefficiency may be a by-product of the average sequence divergence between any two monomers within monomeric α satellite; even neighboring monomers differ in their sequence by ~20–35%, presumably quite a bit less than the levels of identity usually associated with unequal pairing and recombination (see, e.g., Stankiewicz and Lupski 2002). In contrast, even the very first two multimeric repeats (no matter how unlikely their formation in the first place) are virtually 100% identical in sequence, providing a highly efficient substrate for additional rounds of unequal crossing-over and thus driving formation of homogeneous higher-order repeat arrays (Warburton and Willard 1996).

The relationships among classes of α satellites both within a given centromeric region and between different chromosomes across the genome provide illustrative examples of a range of evolutionary mechanisms that are almost certainly at play elsewhere in the genome as well. The potential consequences of unequal crossing-over, sequence conversion, and sequence homogenization mechanisms for both human disease and genome evolution have been amply demonstrated. Notable examples include the clusters of highly homogeneous, low-copy repeats associated with genomic rearrangements (Inoue and Lupski 2002; Stankiewicz and Lupski 2002), the high level of intra- and interchromosomal duplications in the human genome (Bailey et al. 2002; Eichler and Sankoff 2003), and the unusually high level of palindromic sequences on the human Y chromosome associated with recurrent sequence conversion (Rozen et al. 2003; Skaletsky et al. 2003).

## FUNCTIONAL GENOME ANNOTATION WITH HUMAN ARTIFICIAL CHROMOSOMES

To fully understand the nature of sequences associated with the centromeric regions of the human genome requires not only complete sequence annotation, but also detailed functional annotation in terms of the impact different sequence elements have on chromosome structure and function. Although some sequences in the vicinity of centromeres may indeed be without demonstrable function (and likely no different in that respect from the bulk of human genome sequences), others are plausible candidates for defining boundaries between euchromatin and heterochromatin, for establishing chromosomal regions with characteristic levels of gene expression, for influencing (i.e., blocking or mediating) potential position effects on gene function, for anchoring chromosomes within preferred nuclear territories, and for contributing to sister chromatid cohesion during cell division, in addition to being responsible for centromere specification and function.

As a step toward establishing an experimental approach suitable for functional genome annotation, we and others developed an assay based on formation of human artificial chromosomes (Harrington et al. 1997; Ikeno et al. 1998), building on the success and impact of yeast artificial chromosome technology for understanding the function of components of the budding yeast genome (Murray and Szostak 1983). The development of an efficient and tractable human artificial chromosome system would involve assembly of required chromosomal elements (centromere, telomeres, and origins of DNA replication), together with genomic fragments whose genic or other functions one wished to examine (Larin and Mejia 2002). Progress toward this goal has been made, and several different approaches have been used or are under development, based on cotransfection of candidate genomic sequences (Grimes et al. 2001), ligation of synthetic centromere and telomere components (Harrington et al. 1997), or modification of human sequences isolated in yeast (Henning et al. 1999; Kouprina et al. 2003) and bacterial artificial chromosome constructs (Ebersole et al. 2000; Mejia et al. 2001; Schueler et al. 2001; Grimes et al. 2002; Rudd et al. 2003). Whereas much of the early focus of this technology has been on optimization of de novo centromere formation (Harrington et al. 1997; Ohzeki et al. 2002; Rudd et al. 2003), proof-of-principle experiments have shown that genes containing large fragments of chromosomal DNA from the human genome can also be expressed and thus are amenable to study using such assays (Grimes et al. 2001; Mejia et al. 2001; Ikeno et al. 2002).

The most straightforward assay is illustrated in Figure 6. In this approach, BACs containing ~30–100 kb of α satellite are modified to introduce a drug-resistance marker for selection in mammalian cells and, if desired, additional fragments from the human genome for functional testing. The BAC is then transfected (or microinjected) into cells in culture, and the resulting drug-resistant colonies are screened for the presence of a cytogenetically visible human artificial chromosome (Fig. 6). In ~5–50% of colonies (depending in part on the particular α-satellite sequences used), a mitotically stable human artificial chromosome is detected in a high proportion of cells, containing both vector and input sequences and colocalizing with kinetochore proteins detected by indirect immunofluorescence. Importantly, a number of non-α-satellite and noncentromeric control genomic fragments are incapable of de novo centromere formation using this assay (Ebersole et al. 2000; Grimes et al. 2002), indicating that the assay is specific for functional centromeric sequences.

Using such an assay, we have demonstrated that both DXZ1 and D17Z1 sequences are capable of generating de novo centromeres in human artificial chromosomes (Harrington et al. 1997; Schueler et al. 2001; Grimes et al. 2002; Rudd et al. 2003). This provides functional annotation for at least part of the pericentromeric contigs described earlier. However, it should be emphasized that the ability of sequences adjacent to these higher-order repeat arrays to function as centromeres in this assay has not yet been evaluated. Furthermore, the α-satellite sequences alone do not completely recapitulate mitotic centromere function, as human artificial chromosomes show a level of chromosome nondisjunction and anaphase lag that is significantly higher than that of intact, endogenous centromeres (Rudd et al. 2003). This suggests that other sequences in these centromeric contigs may be necessary for at least some aspect of faithful chromosome segregation and stability. Thus, to completely annotate these regions will require testing both monomeric α satellite and the D17Z1-B sequences, using both the human artificial chromosome assay and extended chromatin studies to identify which DNA is involved in the assembly of the specialized chromatin that underlies centromere function (Blower et al. 2002; Cleveland et al. 2003).

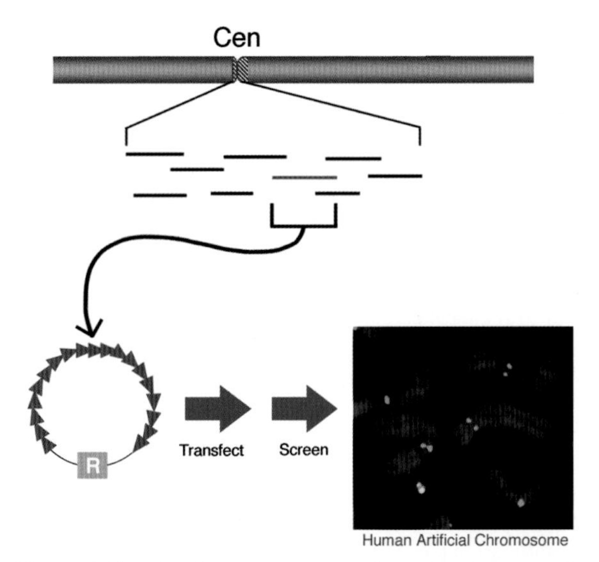


**Figure 6.** Functional centromere annotation using a human artificial chromosome assay (Grimes et al. 2002; Rudd et al. 2003). α-Satellite DNA hypothesized to play a role in centromere function (*purple arrows*) is cloned into a BAC vector containing a drug-resistance gene (R) and transfected into human tissue culture cells. Drug-resistant clones are screened for the presence of an artificial chromosome. The artificial chromosome can be identified by hybridization to a red α-satellite probe. Like normal chromosomes, artificial chromosomes bind antibodies to centromere and kinetochore proteins (*green*).

## CONCLUSIONS

Current studies of the genomic organization of the centromeric regions of human chromosomes reveal several features, despite the large gaps that are apparent in the current genome sequence assemblies (e.g., Fig. 2). The most proximal contigs on several chromosome arms in the genome have reached α-satellite DNA, including those on chromosomes 7 (Hillier et al. 2003; Scherer et al. 2003), 16 (Horvath et al. 2000), 21 (Brun et al. 2003), 22 (Dunham et al. 1999), and the Y chromosome (Skaletsky et al. 2003), in addition to our work on the X chromosome and chromosome 17, as summarized here. Other fully sequenced contigs (chromosome 10, Guy et al. 2003) terminate in other types of satellite DNA, short of connecting to α satellite at the centromere. However, only the contigs on Xp and 17p span from euchromatin of the chromosome arm to higher-order α-satellite repeat arrays that have been annotated functionally with centromere assays (Fig. 7). Others (like the chromosome 21q junction illustrated in Fig. 7) terminate in monomeric α satellite, but are separated by a gap of undetermined size from the higher-order sequences of the functional centromere. Thus, chromosome arm/centromere junctions remain important goals for future research, requiring a combination of directed efforts to extend existing contigs and suitable functional assays to provide validation and functional annotation.

As we move toward an understanding of the organization, function, and evolution of the human genome, any claims of a "complete" sequence will need to include full analysis of the pericentromeric and other heterochromatic regions of our chromosomes. The data presented here and elsewhere (Schueler et al. 2001) suggest that, notwithstanding their repetitive content, the satellite-containing centromeric regions of human chromosomes can be mapped, sequenced, assembled, and annotated functionally. Complete assembly of centromere contigs should, therefore, be feasible and will provide an important source of genomic and functional data for studies of chromosome biology, as well as genome evolution.

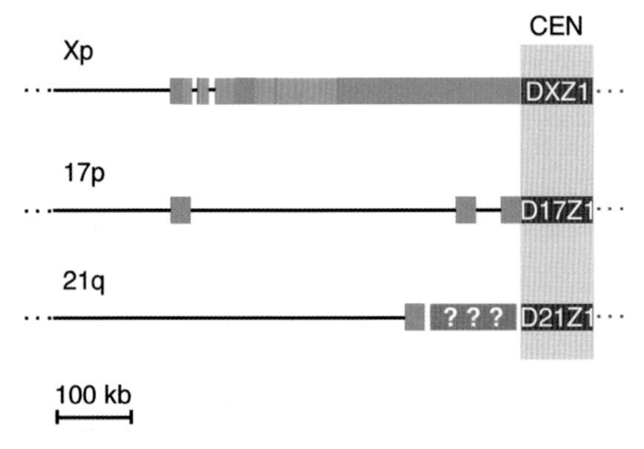


**Figure 7.** Genome assembly of the centromeric regions of the X chromosome, and chromosomes 17 and 21. Both the X and 17 centromeres have contiguous sequence on the short-arm sides, connecting euchromatin to monomeric α satellite (*green*) to higher-order repeat α satellite (*red*). Orange indicates other satellite sequences. A chromosome 21q contig has reached monomeric α satellite but has not connected to higher-order α satellite (gap indicated by question marks). For these three human chromosomes, the centromere activity of their respective higher-order repeat α satellites has been functionally annotated using a human artificial chromosome assay (Harrington et al. 1997; Ikeno et al. 1998; Schueler et al. 2001).

Scientific arguments aside, there is also a strong historical and philosophical imperative for including centromeres in the final stages of gap closure in the archival, truly complete sequence of the genome of *Homo sapiens*. After all, which part of the Rosetta Stone would one choose to omit?

## ACKNOWLEDGMENTS

We thank Evan Eichler, Jeff Bailey, Devin Locke, and Eric Green for helpful discussions and assistance. Work in the authors' lab has been supported by research grants from the National Institutes of Health and the March of Dimes Birth Defects Foundation.

## REFERENCES

Alexandrov I., Kazakov A., Tumeneva I., Shepelev V., and Yurov Y. 2001. Alpha-satellite DNA of primates: Old and new families. *Chromosoma* **110:** 253.

Ando S., Yang H., Nozaki N., Okazaki T., and Yoda K. 2002. CENP-A, -B, and -C chromatin complex that contains the I-type alpha-satellite array constitutes the prekinetochore in HeLa cells. *Mol. Cell. Biol.* **22:** 2229.

Bailey J.A., Yavor A.M., Viggiano L., Misceo D., Horvath J.E., Archidiacono N., Schwartz S., Rocchi M., and Eichler E.E. 2002. Human-specific duplication and mosaic transcripts: The recent paralogous structure of chromosome 22. *Am. J. Hum. Genet.* **70:** 83.

Blower M.D., Sullivan B.A., and Karpen G.H. 2002. Conserved organization of centromeric chromatin in flies and humans. *Dev. Cell* **2:** 319.

Brun M., Ruault M., Ventura M., Roizes G., and DeSario A. 2003. Juxtacentromeric region of chromosome 21: A boundary between centromeric heterochromatin and euchromatic chromosome arms. *Gene* **312:** 41.

Choo K.H. 2001. Domain organization at the centromere and neocentromere. *Dev. Cell* **1:** 165.

Clarke L. 1998. Centromeres: Proteins, protein complexes, and repeated domains at centromeres of simple eukaryotes. *Curr. Opin. Genet. Dev.* **8:** 212.

Cleveland D.W., Mao Y., and Sullivan K.F. 2003. Centromeres and kinetochores: From epigenetics to mitotic checkpoint signaling. *Cell* **112:** 407.

Collins F.S., Green E.D., Guttmacher A.E., and Guyer M.S. 2003. A vision for the future of genomics research. *Nature* **422:** 835.

Copenhaver G.P., Nickel K., Kuromori T., Benito M.I., Kaul S., Lin X., Bevan M., Murphy G., Harris B., Parnell L.D., McCombie W.R., Martienssen R.A., Marra M., and Preuss D. 1999. Genetic definition and sequence analysis of *Arabidopsis* centromeres. *Science* **286:** 2468.

Dong F., Miller J.T., Jackson S.A., Wang G.L., Ronald P.C., and Jiang J. 1998. Rice (Oryza sativa) centromeric regions consist of complex DNA. *Proc. Natl. Acad. Sci.* **95:** 8135.

Dunham I., Shimizu N., Roe B.A., Chissoe S., Hunt A.R., Collins J.E., Bruskiewich R., Beare D.M., Clamp M., Smink L.J., Ainscough R., Almeida J.P., Babbage A., Bagguley C., Bailey J., Barlow K., Bates K.N., Beasley O., Bird C.P., Blakey S., Bridgeman A.M., Buck D., Burgess J., Burrill W.D., and O'Brien K.P., et al. 1999. The DNA sequence of human chromosome 22. *Nature* **402:** 489.

Durfy S.J. and Willard H.F. 1989. Patterns of intra- and interarray sequence variation in alpha satellite from the human X chromosome: Evidence for short range homogenization of tandemly repeated DNA sequences. *Genomics* **5:** 810.

Ebersole T.A., Ross A., Clark E., McGill N., Schindelhauer D., Cooke H., and Grimes B. 2000. Mammalian artificial chromosome formation from circular alphoid input DNA does not require telomere repeats. *Hum. Mol. Genet.* **9:** 1623.

Eichler E.E. and Sankoff D. 2003. Structural dynamics of eukaryotic chromosome evolution. *Science* **301:** 793.

Grimes B.R., Rhoades A.A., and Willard H.F. 2002. Alpha-satellite DNA and vector composition influence rates of human artificial chromosome formation. *Mol. Ther.* **5:** 798.

Grimes B.R., Schindelhauer D., McGill N.I., Ross A., Ebersole T.A., and Cooke H.J. 2001. Stable gene expression from a mammalian artificial chromosome. *EMBO Rep.* **2:** 910.

Guy J., Hearn T., Crosier M., Mudge J., Viggiano L., Koczan D., Thiesen H.J., Bailey J.A., Horvath J.E., Eichler E.E., Earthrowl M.E., Deloukas P., French L., Rogers J., Bentley D., and Jackson M.S. 2003. Genomic sequence and transcriptional profile of the boundary between pericentromeric satellites and genes on human chromosome arm 10p. *Genome Res.* **13:** 159.

Haaf T., Warburton P.E., and Willard H.F. 1992. Integration of human alpha satellite DNA into simian chromosomes: Centromere protein binding and disruption of normal chromosome segregation. *Cell* **70:** 681.

Harrington J.J., Van Bokkelen G., Mays R.W., Gustashaw K., and Willard H.F. 1997. Formation of de novo centromeres and construction of first-generation human artificial microchromosomes. *Nat. Genet.* **15:** 345.

Henikoff S. 2002. Near the edge of a chromosome's "black hole". *Trends Genet.* **18:** 165.

Henikoff S., Ahmad K., and Malik H.S. 2001. The centromere paradox: Stable inheritance with rapidly evolving DNA. *Science* **293:** 1098.

Henning K.A., Novotny E.A., Compton S.T., Guan X.Y., Liu P.P., and Ashlock M.A. 1999. Human artificial chromosomes generated by modification of a yeast artificial chromosome containing both human alpha satellite and single-copy DNA sequences. *Proc. Natl. Acad. Sci.* **96:** 592.

Higgins A.W., Schueler M.G., and Willard H.F. 1999. Chromosome engineering: Generation of mono- and dicentric isochromosomes in a somatic cell hybrid system. *Chromosoma* **108:** 256.

Hillier L.W., Fulton R.S., Fulton L.A., Graves T.A., Pepin K.H., Wagner-McPherson C., Layman D., Maas J., Jaeger S., Walker R., Wylie K., Sekhon M., Becker M.C., O'Laughlin M.D., Schaller M.E., Fewell G.A., Delehaunty K.D., Miner T.L., Nash W.E., Cordes M., Du H., Sun H., Edwards J., Bradshaw-Cordum H., and Ali J., et al. 2003. The DNA sequence of human chromosome 7. *Nature* **424:** 157.

Horvath J.E., Bailey J.A., Locke D.P., and Eichler E.E. 2001. Lessons from the human genome: Transitions between euchromatin and heterochromatin. *Hum. Mol. Genet.* **10:** 2215.

Horvath J.E., Viggiano L., Loftus B.J., Adams M.D., Archidiacono N., Rocchi M., and Eichler E.E. 2000. Molecular structure and evolution of an alpha satellite/non-alpha satellite junction at 16p11. *Hum. Mol. Genet.* **9:** 113.

Ikeno M., Inagaki H., Nagata K., Morita M., Ichinose H., and Okazaki T. 2002. Generation of human artificial chromosomes expressing naturally controlled guanosine triphosphate cyclohydrolase I gene. *Genes Cells* **7:** 1021.

Ikeno M., Grimes B., Okazaki T, Nakano M., Saitoh K., Hoshino H., McGill N.I., Cooke H., and Masumoto H. 1998. Construction of YAC-based mammalian artificial chromosomes. *Nat. Biotechnol.* **16:** 431.

Inoue K. and Lupski J.R. 2002. Molecular mechanisms for genomic disorders. *Annu. Rev. Genomics Hum. Genet.* **3:** 199.

Kniola B., O'Toole E., McIntosh J.R., Mellone B., Allshire R., Mengarelli S., Hultenby K., and Ekwall K. 2001. The domain structure of centromeres is conserved from fission yeast to humans. *Mol. Biol. Cell* **12:** 2767.

Kouprina N., Ebersole T., Koriabine M., Pak E., Rogozin I. B., Katoh M., Oshimura M., Ogi K., Peredelchuk M., Solomon G., Brown W., Barrett J. C., and Larionov V. 2003. Cloning of human centromeres by transformation-associated recombination in yeast and generation of functional human artificial chromosomes. *Nucleic Acids Res.* **31:** 922.

Lamb J.C. and Birchler J.A. 2003. The role of DNA sequence in centromere formation. *Genome Biol.* **4:** 214.

Lander E.S., Linton L.M., Birren B., Nusbaum C., Zody M.C., Baldwin J., Devon K., Dewar K., Doyle M., FitzHugh W., Funke R., Gage D., Harris K., Heaford A., Howland J., Kann L., Lehoczky J., LeVine R., McEwan P., McKernan K., Meldrim J., Mesirov J.P., Miranda C., Morris W., and Naylor J., et al. (International Human Genome Sequencing Consortium). 2001. Initial sequencing and analysis of the human genome. *Nature* **409:** 860.

Larin Z. and Mejia J.E. 2002. Advances in human artificial chromosome technology. *Trends Genet.* **18:** 313.

Laurent A.M., Li M., Sherman S., Roizes G., and Buard J. 2003. Recombination across the centromere of disjoined and non-disjoined chromosome 21. *Hum. Mol. Genet.* **12:** 2229.

Lee C., Critcher R., Zhang J.G., Mills W., and Farr C.J. 2000. Distribution of gamma satellite DNA on the human X and Y chromosomes suggests that it is not required for mitotic centromere function. *Chromosoma* **109:** 381.

Lee C., Wevrick R., Fisher R.B., Ferguson-Smith M.A., and Lin C.C. 1997. Human centromeric DNAs. *Hum. Genet.* **100:** 291.

Mahtani M.M. and Willard H.F. 1998. Physical and genetic mapping of the human X chromosome centromere: Repression of recombination. *Genome Res.* **8:** 100.

Manuelidis L. 1976. Repeating restriction fragments of human DNA. *Nucleic Acids Res.* **3:** 3063.

Manuelidis L. and Wu J.C. 1978. Homology between human and simian repeated DNA. *Nature* **276:** 92.

Mejia J.E., Willmott A., Levy E., Earnshaw W.C., and Larin Z. 2001. Functional complementation of a genetic deficiency with human artificial chromosomes. *Am. J. Hum. Genet.* **69:** 315.

Mills W., Critcher R., Lee C., and Farr C. J. 1999. Generation of an approximately 2.4 Mb human X centromere-based minichromosome by targeted telomere-associated chromosome fragmentation in DT40. *Hum. Mol. Genet.* **8:** 751.

Murray A.W. and Szostak J.W. 1983. Construction of artificial chromosomes in yeast. *Nature* **305:** 189.

Nagaki K., Song J., Stupar R.M., Parokonny A.S., Yuan Q., Ouyang S., Liu J., Hsiao J., Jones K.M., Dawe R.K., Buell C.R., and Jiang J. 2003. Molecular and cytological analyses of large tracks of centromeric DNA reveal the structure and evolutionary dynamics of maize centromeres. *Genetics* **163:** 759.

Nicklas R. B. 1971. Mitosis. *Adv. Cell Biol.* **2:** 225.

Ohzeki J., Nakano M., Okada T., and Masumoto H. 2002. CENP-B box is required for de novo centromere chromatin assembly on human alphoid DNA. *J. Cell Biol.* **159:** 765.

Rieder C.L. and Salmon E.D. 1998. The vertebrate cell kinetochore and its roles during mitosis. *Trends Cell Biol.* **8:** 310.

Rozen S., Skaletsky H., Marszalek J.D., Minx P.J., Cordum H.S., Waterston R.H., Wilson R.K., and Page D.C. 2003. Abundant gene conversion between arms of palindromes in human and ape Y chromosomes. *Nature* **423:** 873.

Rudd M.K., Mays R.W., Schwartz S., and Willard H.F. 2003. Human artificial chromosomes with alpha satellite-based de novo centromeres show increased frequency of nondisjunction and anaphase lag. *Mol. Cell. Biol.* **23:** 7689.

Scherer S.W., Cheung J., MacDonald J.R., Osborne L.R., Nakabayashi K., Herbrick J.A., Carson A.R., Parker-Katiraee L., Skaug J., Khaja R., Zhang J., Hudek A.K., Li M., Haddad M., Duggan G.E., Fernandez B.A., Kanematsu E., Gentles S., Christopoulos C.C., Choufani S., Kwasnicka D., Zheng X.H., Lai Z., Nusskern D., and Zhang Q., et al. 2003. Human chromosome 7: DNA sequence and biology. *Science* **300:** 767.

Schindelhauer D. and Schwarz T. 2002. Evidence for a fast, intrachromosomal conversion mechanism from mapping of nucleotide variants within a homogeneous alpha-satellite DNA array. *Genome Res.* **12:** 1815.

Schueler M.G., Higgins A.W., Rudd M.K., Gustashaw K., and Willard H.F. 2001. Genomic and genetic definition of a functional human centromere. *Science* **294:** 109.

Skaletsky H., Kuroda-Kawaguchi T., Minx P.J., Cordum H.S., Hillier L., Brown L.G., Repping S., Pyntikova T., Ali J., Bieri T., Chinwalla A., Delehaunty A., Delehaunty K., Du H., Fewell G., Fulton L., Fulton R., Graves T., Hou S.F., Latrielle P., Leonard S., Mardis E., Maupin R., McPherson J., and Miner T., et al. 2003. The male-specific region of the human Y chromosome is a mosaic of discrete sequence classes. *Nature* **423:** 825.

Smith G.P. 1976. Evolution of repeated DNA sequences by unequal crossover. *Science* **191:** 528.

Sonnhammer E.L. and Durbin R. 1995. A dot-matrix program with dynamic threshold control suited for genomic DNA and protein sequence analysis. *Gene* **167:** 1.

Spence J.M., Critcher R., Ebersole T.A., Valdivia M.M., Earnshaw W.C., Fukagawa T., and Farr C.J. 2002. Co-localization of centromere activity, proteins and topoisomerase II within a subdomain of the major human X alpha-satellite array. *EMBO J.* **21:** 5269.

Stankiewicz P. and Lupski J.R. 2002. Molecular-evolutionary mechanisms for genomic disorders. *Curr. Opin. Genet. Dev.* **12:** 312.

Sullivan B.A. and Willard H.F. 1998. Stable dicentric X chromsomes with two functional centromeres. *Nat. Genet.* **20:** 227.

Sullivan B.A., Blower M.D., and Karpen G.H. 2001. Determining centromere identity: Cyclical stories and forking paths. *Nat. Rev. Genet.* **2:** 584.

Sun X., Le H.D., Wahlstrom J.M., and Karpen G.H. 2003. Sequence analysis of a functional *Drosophila* centromere. *Genome Res.* **13:** 182.

Tanaka T.U. 2002. Bi-orienting chromosomes on the mitotic spindle. *Curr. Opin. Cell Biol.* **14:** 365.

Trask B.J. 2002. Human cytogenetics: 46 chromosomes, 46 years and counting. *Nat. Rev. Genet.* **3:** 769.

Tyler-Smith C., Oakey R.J., Larin Z., Fisher R.B., Crocker M., Affara N.A., Ferguson-Smith M.A., Muenke M., Zuffardi O., Jobling M.A. 1993. Localization of DNA sequences required for human centromere function through an analysis of rearranged Y chromosomes. *Nat. Genet.* **5:** 368.

Vafa O. and Sullivan K.F. 1997. Chromatin containing CENP-A and alpha satellite DNA is a major component of the inner kinetochore plate. *Curr. Biol.* **7:** 897.

Venter J.C., Adams M.D., Myers E.W., Li P.W., Mural R.J., Sutton G.G., Smith H.O., Yandell M., Evans C.A., Holt R.A., Gocayne J.D., Amanatides P., Ballew R.M., Huson D.H., Wortman J.R., Zhang Q., Kodira C.D., Zheng X.H., Chen L., Skupski M., Subramanian G., Thomas P.D., Zhang J., Gabor Miklos G.L., and Nelson C., et al. 2001. The sequence of the human genome. *Science* **291:** 1304.

Warburton P.E. and Willard H.F. 1990. Genomic analysis of sequence variation in tandemly repeated DNA. Evidence for localized homogenous sequence domains within arrays of alpha-satellite DNA. *J. Mol. Biol.* **216:** 3.

———. 1995. Interhomologue sequence variation of alpha satellite DNA from human chromosome 17: Evidence for concerted evolution along haplotypic lineages. *J. Mol. Evol.* **41:** 1006.

———. 1996. Evolution of centromeric alpha satellite DNA: Molecular organization within and between human and primate chromosomes. In *Human genome evolution* (ed. S.T. Jackson and G. Dover), p. 121. BIOS Scientific Publishers, Oxford, United Kingdom.

Waterston R.H., Lindblad-Toh K., Birney E., Rogers J., Abril J.F., Agarwal P., Agarwala R., Ainscough R., Alexandersson M., An P., Antonarakis S.E., Attwood J., Baertsch R., Bailey J., Barlow K., Beck S., Berry E., Birren B., Bloom T., Bork P., Botcherby M., Bray N., Brent M.R., Brown D.G., and Brown S.D., et al. (Mouse Genome Sequencing Consortium). 2002. Initial sequencing and comparative analysis of the mouse genome. *Nature* **420:** 520.

Waye J.S. and Willard H.F. 1986. Structure, organization, and sequence of alpha satellite DNA from human chromosome 17: Evidence for evolution by unequal crossing-over and an ancestral pentamer repeat shared with the human X chromosome. *Mol. Cell. Biol.* **6:** 3156.

Wevrick R. and Willard H.F. 1989. Long-range organization of tandem arrays of alpha satellite DNA at the centromeres of human chromosomes: High frequency array-length polymorphism and meiotic stability. *Proc. Natl. Acad. Sci.* **86:** 9394.

———. 1991. Physical map of the centromeric region of human chromosome 7: Relationship between two distinct alpha satellite arrays. *Nucleic Acids Res.* **19:** 2295.

Wevrick R., Earnshaw W.C., Howard-Peebles P.N., and Willard H.F. 1990. Partial deletion of alpha satellite DNA associated with reduced amounts of the centromere protein CENP-B in a mitotically stable human chromosome rearrangement. *Mol. Cell. Biol.* **10:** 6374.

Willard H.F. 1998. Centromeres: The missing link in the development of human artificial chromosomes. *Curr. Opin. Genet. Dev.* **8:** 219.

Willard H.F. and Waye J.S. 1987. Hierarchical order in chromosome-specific human alpha satellite DNA. *Trends Genet.* **3:** 192.

Willard H.F., Smith K.D., and Sutherland J. 1983. Isolation and characterization of a major tandem repeat family from the human X chromosome. *Nucleic Acids Res.* **11:** 2017.

Zhong C.X., Marshall J.B., Topp C., Mroczek R., Kato A., Nagaki K., Birchler J. A., Jiang J., and Dawe R.K. 2002. Centromeric retroelements and satellites interact with maize kinetochore protein CENH3. *Plant Cell* **14:** 2825.

# Evolutionary Strategies of Human Pathogens

J. PARKHILL AND N. THOMSON
*The Sanger Institute, Wellcome Trust Genome Campus, Hinxton, Cambridge, CB10 1SA, United Kingdom*

Over the last few thousand years, the human population has changed dramatically in size and density. Such changes have altered the niches which are available for bacteria that interact with human beings. As a group, the bacterial pathogens of humans are highly diverse, in terms of both their phylogeny and their strategies for taking advantage of the opportunities offered by the human host. The large population sizes and rapid growth rates of many bacteria, coupled with their ability to exchange genes with close and distant relatives, mean that they are extremely well adapted to take advantage of these new environmental and host niches as they become available. The genomes of several of the most feared human pathogens reveal the signatures of rapid evolutionary change, in terms of both gene acquisition and loss, implying that these pathogens have recently changed their niche within the host. Prime examples of these pathogens include *Salmonella enterica* serovar Typhi (*S.* Typhi), and *Yersinia pestis*, both members of the enteric group of pathogens, which cause typhoid fever and plague, respectively, and *Bordetella pertussis*, the causative agent of whooping cough.

## *S. ENTERICA* SEROVAR TYPHI, AN OBLIGATE HUMAN PATHOGEN

*S.* Typhi is a member of the enterobacteriaceae, a group of pathogens that also includes such organisms as *Escherichia* and *Yersinia*. Enteric pathogens tend to reside in the gut of animals, hence their family designation. Some of the enteric pathogens have a broad host range and can cause a variety of diseases; *S. enterica* serovar Typhimurium (*S.* Typhimurium), for example, can infect, among others, humans, causing a mild gastroenteritis, and mice, where it causes a systemic disease similar to human typhoid. Others, such as *S.* Typhi, are host restricted, only infecting a single host. For *S.* Typhi this is *Homo sapiens*, in which it causes the acute systemic disease typhoid fever. Comparison of the genome of *S.* Typhi (Parkhill et al. 2001a) with those of close relatives such as *S.* Typhimurium (McClelland et al. 2001) and more distant relatives such as *Escherichia coli* (Blattner et al. 1997; Perna et al. 2001) reveals a pattern of long-term gene acquisition and recent loss that has shaped the evolution of this pathogen, and can be used to identify the genetic explanations for these phenotypic differences.

### Large Insertions (Pathogenicity Islands)

Previous comparisons of pathogenic and nonpathogenic members of the same species have revealed that the former often carry large chromosomal insertions encoding pathogenicity determinants. To date, many such regions which are linked to virulence have been characterized both in gram-negative and gram-positive bacteria (for review, see Hacker et al. 1997). This led to the concept of pathogenicity islands (PAI), which appear to be mobile, or mobilizable, elements that can be exchanged between strains or species. Characteristically, PAIs are inserted adjacent to stable RNA genes and have atypical G+C content, or dinucleotide composition. In addition to virulence-related functions, these regions often carry genes encoding transposase or integrase-like proteins and are unstable or self-mobilizable (Blum et al. 1994; Hacker et al. 1997). The salmonellae are known to possess large inserts when compared to *E. coli* (*Salmonella* pathogenicity islands or SPIs) which are thought to have been acquired laterally. The gene products encoded by pathogenicity islands SPI-1 (Mills et al. 1995; Galan 1996) and SPI-2 (Ochman et al. 1996; Shea et al. 1996) have been shown to be important for different stages of the infection process. Both of these islands carry type III secretion systems (specialized organelles for secreting proteins into the cytoplasm of host cells) and their associated secreted protein effectors. SPI-1 confers the ability to invade epithelial cells on to all salmonellae. SPI-2 has been shown to be important for various aspects of the systemic infection allowing *Salmonella* to spread from the intestinal tissue into the blood and eventually to infect, and survive within, macrophages of the liver and spleen (for review, see Kingsley and Baumler 2002). Comparison of the genomes of *E. coli*, *S.* Typhimurium, and *S.* Typhi shows many such large differences (Fig. 1), many of which are candidate PAIs. The genome of *S.* Typhi contains 10 readily identifiable SPIs, of which three (SPI-7, SPI-8, and SPI-10) are not present in *S.* Typhimurium; SPI-7 displays many of the elements of a classical pathogenicity island (Fig. 2). It encodes *S.* Typhi-specific virulence factors such as the Vi extracellular polysaccharide, as well as an additional super-insertion of a phage carrying the *sopE* gene which encodes a type III secreted effector molecule. SPI-7 is inserted adjacent to a tRNA gene and encodes a phage integrase that is likely to be responsible for this insertion. Also present is a large operon encoding a type IVB pilus, which has been implicated in interaction with host cells (Zhang et al. 2000) and may have been involved in the acquisition of the island by DNA transfer (similar pili are used for conjugation by plasmids such as R64; Zhang et al. 1997). Of the two other *S.* Typhi-specific islands, SPI-8 encodes an apparently nonfunctional bacteriocin biosynthesis system, and SPI-10 carries a bacteriophage and a chaperone-usher fimbrial operon.

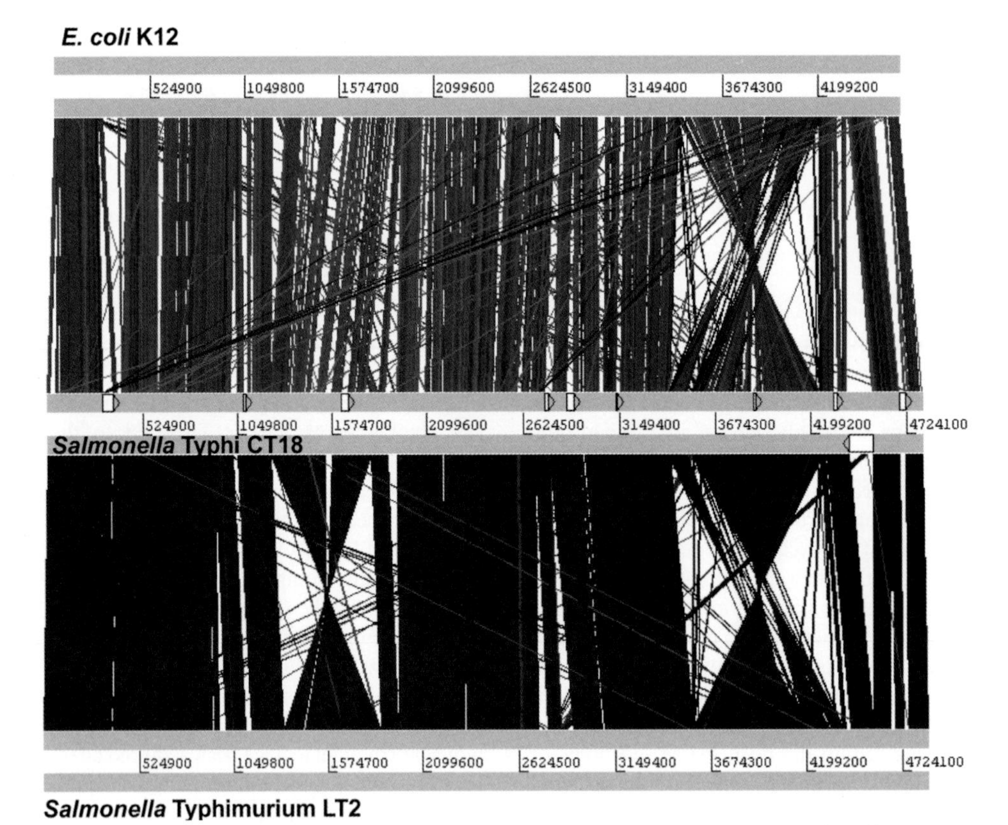


**Figure 1.** The figure shows DNA:DNA matches (computed using BLASTN and displayed using ACT http://www.sanger.ac.uk/Software/ACT) between the genomes of *E. coli* K12 (*top*) (Blattner et al. 1997), *S.* Typhi (*middle*) (Parkhill et al. 2001a), and *S.* Typhimurium (*bottom*) (McClelland et al. 2001). The gray bars between the genomes represent individual BLASTN matches. Some of the shorter and weaker BLASTN matches have been removed to show the overall structure of the comparison. The open boxes on the *S.* Typhi genome represent the ten pathogenicity islands identified in the genome. The open box on the lower strand represents SPI-7, shown in more detail in Fig. 2.

## Small Insertions and Replacements

In addition to the groups of genes exchanged as part of large pathogenicity islands, very many smaller differences exist between these genomes. A gene-by-gene analysis of the differences between *S.* Typhi and *S.* Typhimurium, and of those between *S.* Typhi and *E. coli* K12, clearly shows that the large majority of insertions or deletion events between these pairs of organisms are in fact small (Fig. 3A). The majority of separate insertion or deletion events involve just a few genes. Even taking into account the fact that insertion or deletion of the larger islands involves many more genes per event, it is clear that nearly equivalent numbers of species-specific genes are attributable to insertion or deletion events involving 10 genes or fewer, as are due to events involving 20 genes or more (Fig. 3B).

It should be clear, therefore, that the acquisition and

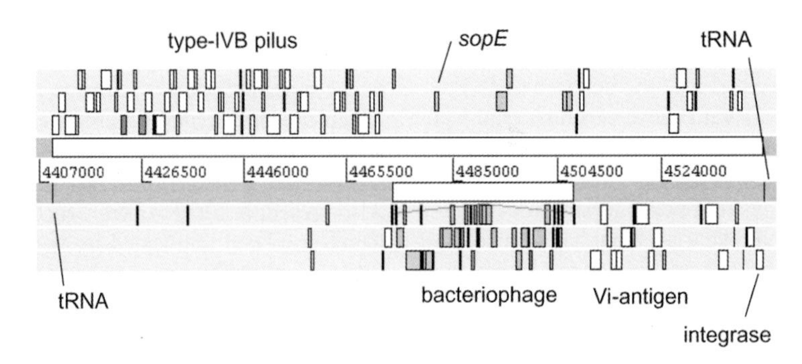


**Figure 2.** The *S.* Typhi SPI-7 pathogenicity island. The dark gray bars in the center represent the DNA strands, and the island is indicated by a large white box on the upper strand. The numbers are chromosomal coordinates in base pairs. The tRNA genes, indicating the point of insertion of the island, are shown on the lower DNA strand. Genes are represented by small open boxes above and below the DNA strands, with the various levels representing the six translational reading frames. The inserted bacteriophage is shown as a large white box on the lower strand, and its genes are shown as a darker gray, as are two pseudogenes in the type IVB pilus region.

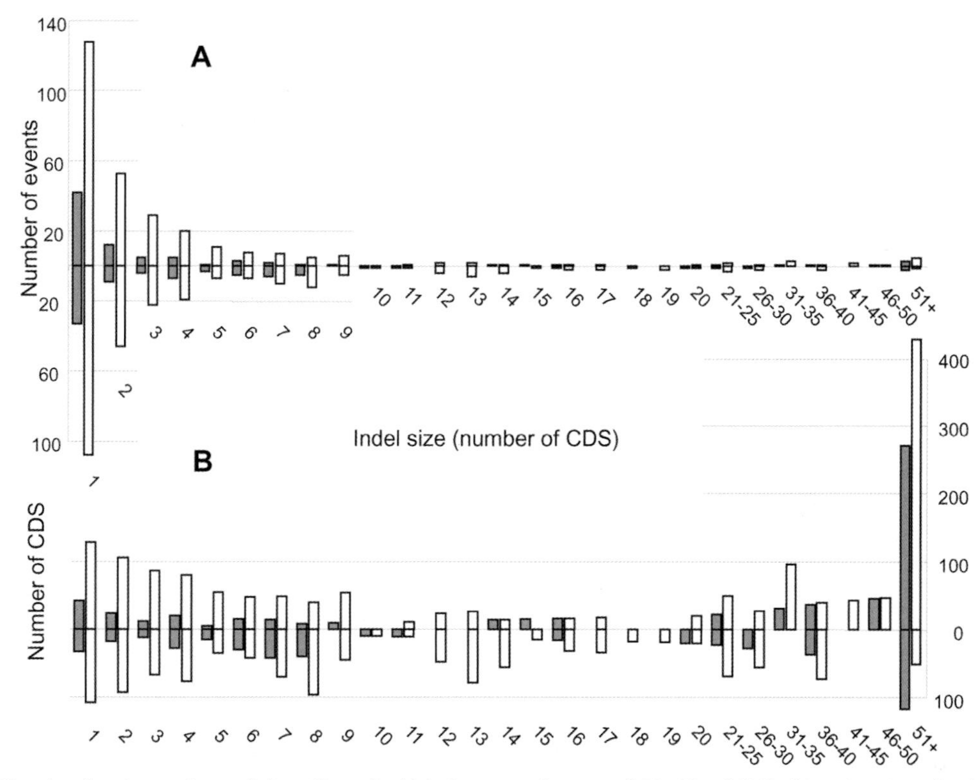


**Figure 3.** Plot showing the number and size of insertion/deletion events between *S.* Typhi and *S.* Typhimurium (gray bars) and *S.* Typhi and *E. coli* (*white bars*). The X axis shows the size of the insertion/deletion events expressed as number of coding sequences. Bars above the axis show insertions in *S.* Typhi relative to *S.* Typhimurium or *E. coli*. Bars below the axis show deletions in *S.* Typhi relative to *S.* Typhimurium or *E. coli*. The Y axis in plot A shows the total number of insertion/deletion events, and in plot B shows the total number of CDSs in all the events of that size. (Adapted from Parkhill et al. 2001a.)

exchange of small islands is likely to be of significant importance to the overall phenotype of the organisms; a few examples will underline this point. SspH2 is a leucine-rich protein that is secreted by the *Salmonella* type III secretion system encoded by SPI-2. In *S.* Typhi, SspH2 is encoded by a unique region relative to *E. coli* K12, which contains just one other intact gene and three pseudogenes of phage origin. The *S.* Typhimurium equivalent of this island contains additional genes of phage origin, suggesting the entire region is a prophage remnant. Elsewhere, the single *S.* Typhimurium gene *envF* (encoding a putative lipoprotein) is replaced in *S.* Typhi by a five-gene block encoding distant homologs of the *Campylobacter jejuni* toxin subunit CdtB and the *Bordetella pertussis* toxin subunits PtxA and PtxB. Many other such examples exist for other organisms and in other comparisons.

In the majority of these small islands, no direct evidence exists for genes that might allow them to be self-mobile. However, it is likely that small islands of this type are indeed exchanged between members of a species and constitute part of the species gene pool. It is easy to see that, once present in one member of the species, they can be easily exchanged by generalized transduction mechanisms (Neidhardt and Curtiss 1996), followed by homologous recombination between the near-identical flanking genes to allow integration into the chromosome.

This type of mechanism of genetic exchange would also allow for nonorthologous gene replacement, involving the exchange of related genes at identical regions in the backbone. We could also consider these alternative gene sets to be equivalent in some ways to the eukaryotic concept of alternative alleles at a single locus. A specific example of this can be seen in the capsular switching of *Neisseria meningitidis* (Swartley et al. 1997) and *Streptococcus pneumoniae* (Dillard and Yother 1994; Dillard et al. 1995), where different sets of genes responsible for the biosynthesis of different capsular polysaccharides are found at identical regions in the chromosome, flanked by conserved genes. The implied mechanism for capsular switching involves replacement of the polysaccharide-specific gene clusters by homologous recombination between the chromosome and exogenous DNA within conserved flanking genes.

This type of gene exchange may well be more general, and good candidates for this occur among the numerous chaperone-usher fimbrial systems of *E. coli* and *Salmonella*. These systems have been seen to be variable in number and occurrence in both *E. coli* (Perna et al. 2001) and *S. enterica* (Townsend et al. 2001). Perna et al. (2001) noted that fimbrial operons were among the most variable sequences that were present in both *E. coli* K12 and O157:H7, an observation that also extends to the comparison between *S.* Typhi and *E. coli* K12. For example, the *yadC*-containing operon in *E. coli*, and the apparently orthologous *sta* fimbrial operon in *S.* Typhi, reside in the same chromosomal context, and the flanking genes show very high levels of similarity. This contrasts with the fimbrial genes themselves, which are only weakly conserved. Again, it seems likely that related gene sets have been exchanged at the same chromosomal location by homologous recombination between the conserved flanking genes and exogenously acquired DNA.

### Pseudogenes and Gene Loss

This long-term acquisition and exchange of genes has been offset by recent gene loss, in the form of gene inactivation or pseudogene formation. The genome of *S.* Typhi was predicted to contain over 200 pseudogenes, and this number is almost certainly an underestimate, as the criteria used to identify them were fairly stringent. Genes were only suggested to be pseudogenes where they had a mutation that would prevent correct translation, such as in-frame stop codons, frameshifts, deletions, or IS-element insertions. Clearly, genes can be functionally inactivated in other ways, including promoter mutations and mis-sense coding changes, and these would probably not be identified from sequencing alone.

As described above, *S.* Typhi is host-restricted, and appears to be only capable of infection of a human host, whereas *S.* Typhimurium, which causes a milder disease in humans, has a much broader host range. After a careful comparison, *S.* Typhimurium was predicted to contain only around 39 pseudogenes (McClelland et al. 2001), compared to *S.* Typhi's 204. It is apparent that the pseudogenes in *S.* Typhi are not randomly spread throughout the genome: They are overrepresented in genes that are unique to *S.* Typhi when compared to *E. coli* (59% of the pseudogenes lie in the unique regions, compared to 33% of all *S.* Typhi genes being unique), and many of the pseudogenes in *S.* Typhi have intact counterparts in *S.* Typhimurium that have been shown to be involved in aspects of virulence and host interaction. Specific examples of this include the leucine-rich repeat protein *slrP* (involved in host-range specificity in *S.* Typhimurium [Tsolis et al. 1999] and secreted through a type III system), other type-III-secreted effector proteins including *sseJ* (Miao and Miller 2000), *sopE2* (Bakshi et al. 2000), and *sopA* (Wood et al. 2000; Zhang et al. 2002), and the genes *shdA, ratA,* and *sivH*, which are present in an island unique to *Salmonellae* infecting warm-blooded vertebrates (Kingsley and Baumler 2000). Many other inactivated genes may also have been involved in virulence or host interaction, including components of seven of the twelve chaperone-usher fimbrial systems. Given this distribution of pseudogenes, it is possible that the host specificity of *S.* Typhi may be due to the loss of an ability to interact with a broader host range due to functional inactivation of the necessary genes. In contrast to other organisms containing multiple pseudogenes, such as *Mycobacterium leprae* (Cole et al. 2001), most of the pseudogenes in *S.* Typhi are caused by a single mutation, suggesting that they have been inactivated relatively recently. This is consistent with the fact that worldwide, *S.* Typhi is seen to be clonal (Reeves et al. 1989), and the serovar may be only a few tens of thousands of years old.

It is apparent, therefore, that *S.* Typhi has recently changed its niche, from a gut organism with a broad host range to a systemic pathogen with a restricted host range. The ability to exploit this new niche is likely to have arisen through acquisition of novel genetic material, as part of the long-term ebb and flow of genes that occurs between members of the same or related species within this group of organisms. Such a niche change would, of necessity, have involved a small population of organisms, causing an evolutionary bottleneck. Such bottlenecks reduce the ability of competition and purifying selection to remove mutations from the population (Andersson and Hughes 1996), and hence lead to an apparent rise in mutation rate, leaving the derived strain with many pseudogenes. The loss of these genes may have a short-term selective advantage, or it may be selectively neutral. It is also possible that some of the genes lost might have been advantageous in the longer term, but cannot now be recovered by the organism.

## *YERSINIA PESTIS*, A VECTOR-ADAPTED MAMMALIAN PATHOGEN

*Y. pestis*, another member of the *Enterobacteriaceae*, is the causative agent of plague, and as such has been responsible for an enormous amount of human mortality over the last 1,500 years. As with *S.* Typhi, *Y. pestis* is also near clonal in its world-wide spread, and thus has apparently very recently emerged as a species. It has been estimated that *Y. pestis* evolved from *Y. pseudotuberculosis*, which causes gastroenteritis, between 1,500 and 20,000 years ago (Achtman et al. 1999). Therefore, *Y. pestis* has changed from being a gut bacterium (*Y. pseudotuberculosis*), transmitting via the fecal–oral route, to an organism capable of utilizing a flea vector for systemic infection of a mammalian host (Perry and Fetherston 1997; Achtman et al. 1999). As with *S.* Typhi, this change is thought to be the result of a series of long-term gene acquisitions, culminating in the organism being in the position to move into a new niche. Again, in taking this route, the organism appears to have gone through an evolutionary bottleneck, leading to pseudogene formation and, in this case, IS element expansion.

### Gene Acquisition by *Yersinia*

It has been known for some time that *Y. pestis* has very recently acquired novel DNA in the form of plasmids. *Y. pestis, Y. pseudotuberculosis,* and *Y. enterocolitica* possess a 70-kb plasmid (pYV/pCD1 in *Y. pestis* and pIB1 in *Y. pseudotuberculosis*) that encodes the Yop virulon, a type III secretion system essential for virulence in both organisms. Subsequently, *Y. pestis* has acquired two unique plasmids that encode a variety of other virulence determinants. The 9.5-kb plasmid (pPST/pPCP1) encodes the plasminogen activator and putative invasin Pla (Cowan et al. 2000) which is essential for virulence by the subcutaneous route. The 100- to 110-kb plasmid (pFra/pMT1) encodes the murine toxin (Ymt) and the $F_1$ capsular protein, which have both been shown to play a role in the transmission of plague (Perry and Fetherston 1997). In addition to these well-studied plasmids, however, *Y. pestis* appears to have substantial numbers of recently acquired genes inserted into the backbone of the chromosome (Parkhill et al. 2001b).

As described above, the hallmarks of horizontally acquired DNA are often differential nucleotide composition and/or location and gene content. Using these measures, a number of apparently horizontally acquired regions within the *Y. pestis* genome could be delineated (Table 1). As would be expected, many of these genes are likely to be involved in interaction with, and survival inside, the host. Examples of these include adhesins, novel type III secretion genes, and iron-sequestration and uptake genes. The best-studied of these is the high-pathogenicity island (HPI) (Buchrieser et al. 1998), which directs the production and uptake of the iron-chelating siderophore yersiniabactin. HPI-like elements are widely distributed in enterobacteria including *E. coli, Klebsiella, Enterobacter,* and *Citrobacter* spp (Bach et al. 2000; Schubert et al. 2000; Hayashi et al. 2001) and like many prophage, HPIs are found adjacent to *asn*-tRNA genes. (tRNA genes are common sites for bacteriophage integration into genomes [Reiter et al. 1989].) The *Yersinia* HPI has been shown to move and integrate in a similar manner to bacteriophage and can be found integrated at multiple *asn*-tRNA loci in *Y. pestis* (Hare et al. 1999; Rakin et al. 2001). Also prominent within these gene sets are those that appear to be involved in pathogenicity within an insect host. One such locus encodes homologs of the high-molecular-weight insecticidal toxin complexes (Tcs) from *Photorhabdus luminescens, Serratia entomophila,* and *Xenorhabdus nematophilus* (Waterfield et al. 2001). The characterized toxins have been shown to be complexes of the products of three different gene families: *tcaB/tcdA, tcaC/tcaB*, and *tccC*. Analysis of the histopathological effects of purified Tca toxin on an insect midgut has shown that it causes the lining epithelium to bleb into the midgut lumen and eventually disintegrate (Blackburn et al. 1998). In *Y. pestis*, three adjacent genes encoding homologs of *P. luminescens* TcaA, TcaB, and TcaC were identified, separated from nearby homologs of TccC by phage-like genes. These genes are, however, not unique to *Y. pestis*; it has been shown that orthologs of the *Y. pestis tca* insecticidal toxin genes were also present within a strain of *Y. pseudotuberculosis* (Parkhill et al. 2001b).

Other insect pathogen-related islands included a low G+C region encoding a protein similar to viral enhancins. This island is inserted alongside a tRNA gene and carries several transposase fragments. Like the Tca toxin, viral enhancins also attack the insect midgut. The peritrophic membrane, a noncellular matrix composed of chitin, proteins, and glycoproteins, lines the insect midgut and is thought to act as a barrier for pathogenic microorganisms (Wang and Granados 1997). The proteolytic activity of enhancins has been shown to degrade the peritrophic membrane and to allow the escape of viral pathogens from the gut into the deeper tissues.

It seems clear, then, that recent gene acquisition by *Y. pestis* and its immediate ancestor was involved in enhancing its ability to interact with, and to parasitize, both mammalian and insect hosts, probably through the fecal–oral route in each case. However, in order to get a fuller understanding of the changes that occurred in the speciation of *Y. pestis*, we must look at the gene loss that occurred coincident with this event.

**Table 1.** Putative Horizontally Acquired Islands of *Y. pestis*

| Range[a] | G+C | Insertion site | Function |
|---|---|---|---|
| YPO0255–YPO0273 | 49.1% | — | type III secretion system |
| YPO0335–YPO0340 | 36.6% | tRNA-Phe | insect viral enhancing factor |
| YPO0590–YPO0642 | 50.2% | tRNA-Met | contains adhesin, autotransporter, protein kinase, resistance proteins, and secreted proteins |
| YPO0684–YPO0697 | 36.6% | IS1541 | adherence proteins |
| YPO0770–YPO0778 | 49.2% | IS100 | high pathogenicity island 2 (HPI-2)—siderophore biosynthesis |
| YPO0803–YPO0818 | 32.8% | — | type-II-related secretion system (GSP) |
| YPO0871–YPO0887 | 46.5% | tRNA-Gly | Bacteriocin?, phage remnant |
| YPO0961–YPO0995 | 48.3% | tRNA-Phe | ClpB homolog, quorum sensing genes, siderophore biosynthesis |
| YPO1083–YPO1098 | 45.8% | tRNA-Asp/tmRNA | phage remnant |
| YPO1224–YPO1259 | 47.3% | tRNA-Thr/tRNA-Pro | virulence related proteins—outer membrane protease, *Salmonella msgA* homolog; phage remnant |
| YPO1448–YPO1480 | 45.8% | tRNA-Ser | *E. coli* cytotoxic nectotizing factor homolog, fatty acid metabolism genes |
| YPO1900–YPO1917 | 56.4% | tRNA-Asn | high pathogenicity island—yersiniabactin biosynthesis |
| YPO1951–YPO1954 | 46.7% | — | Hms hemin storage (pigmentation) locus |
| YPO2084–YPO2140 | 47.2% | — | prophage |
| YPO2274–YPO2280 | 41.8% | — | phage remnant |
| YPO2311–YPO2321 | 46.9% | IS1541 | insecticidal toxin complex *tccC* gene |
| YPO2434–YPO2443 | 44.0% | — | multiple antibiotic resistance protein homolog, iron transport |
| YPO2863–YPO2868 | 31.6% | IS285/IS1541 | membrane proteins |
| YPO2934–YPO2948 | 45.4% | IS100 | chaperone-usher fimbrial system, *clpB* homolog |
| YPO3673–YPO3682 | 45.1% | — | insecticidal toxin complex genes *tccC, tcaABC* |
| YPO4014–YPO4033 | 44.4% | tRNA-SeC | iron transport genes |

Adapted from Parkhill et al. (2001b).

[a]These are regions that appear to be insertions relative to *E. coli* or *S.* Typhi, and have one or more of the following characteristics: virulence determinants, unusual % G+C, flanking tRNA, IS elements or integrases, phage proteins.

**Gene Loss and IS-element Expansion in *Y. pestis***

At the time of publication, *Y. pestis* CO92 was predicted to encode 149 pseudogenes, although for the reasons described above, this is likely to be an underestimate. In addition to pseudogene formation by point mutation, many genes have been inactivated by IS-element insertion. IS (*i*nsertion *s*equence) elements are small mobile elements that carry genes only for their own transposition, and can thus be considered to be selfish parasitic elements within the genome. *Y. pestis* has undergone a tenfold expansion in IS elements since separating from *Y. pseudotuberculosis* (Odaert et al. 1996; McDonough and Hare 1997), and it appears that this expansion of IS elements can be attributed to the same evolutionary processes as the increase in point mutations. During normal bacterial growth, IS elements will be actively transposing at a low rate and inserting into new chromosomal loci. In a large, freely recombining population, most of these events will be removed by competition and selection. However, as described above, during an evolutionary bottleneck many of these lesions will not be removed, due to the reduced level of purifying selection, and will become fixed. The result is an apparent increase in IS elements in the daughter populations.

Analysis of the pattern of gene loss revealed that it was not spread randomly among all classes of genes, but that pseudogenes were specifically over- and underrepresented in particular functional classes (Fig. 4). Pseudogenes are underrepresented in classes involved in central and intermediary metabolism, indicating that the core functions of the organism have been maintained. Conversely, many genes involved in pathogenicity and host interaction have been inactivated. This may seem counterintuitive, given that *Y. pestis* is a more virulent pathogen than its immediate ancestor, but closer examination reveals that many of these genes were specifically involved in pathogenicity via the fecal–oral route and are likely to be unnecessary, or indeed a hindrance, for systemic infection and spread. An example of this is the *yadA* and *inv* genes, the products of which encode an adhesin and an invasin, which are important for adherence to surfaces of the gut and the invasion of the cells lining it (Pepe and Miller 1993; El Tahir and Skurnik 2001). Replacement of the mutated *yadA* gene with an intact version in *Y. pestis* has been shown to result in a significant decrease in virulence by the subcutaneous infection route (Rosqvist et al. 1988). In addition to *yadA* and *inv*, many of the pseudogenes were also predicted to encode other adhesin-like molecules that may also have been important for enteropathogenicity.

The second largest group of pseudogenes was predicted to encode surface-exposed or exported proteins. Five pseudogenes were in the lipopolysaccharide (LPS) biosynthesis operon. The LPS O-side chain plays a key role in the virulence of a range of pathogens including *Y. enterocolitica* (Darwin and Miller 1999). The O-side chain is known to mediate resistance to complement-mediated and phagocyte killing, and it might also play a role

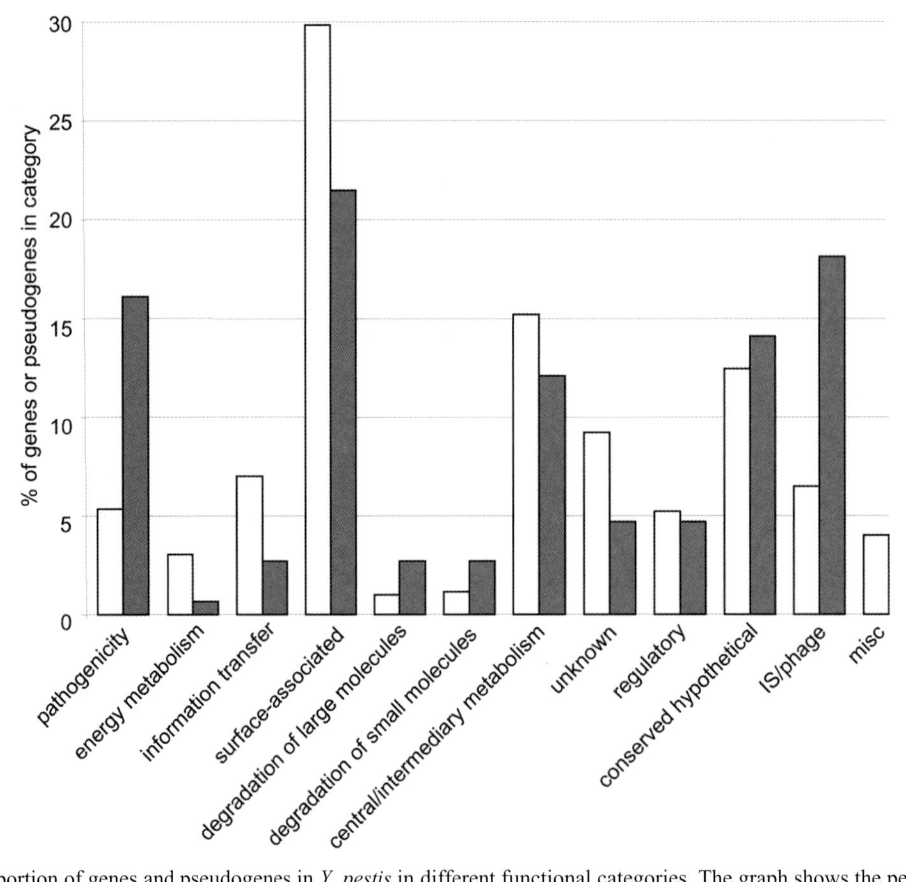


**Figure 4.** Proportion of genes and pseudogenes in *Y. pestis* in different functional categories. The graph shows the percentage of all genes (in *white*), and pseudogenes (in *gray*) in each of the functional categories. (Adapted from Parkhill et al. 2001b.)

in survival in the gut by protecting the bacterium from bile salts (Nesper et al. 2001). *Y. pestis* produces a rough LPS, lacking an O-antigen as a consequence of these mutations within the O-antigen biosynthesis cluster (Skurnik et al. 2000). Motility is required for efficient host-cell invasion by *Y. enterocolitica* (Young et al. 2000) and *Y. pseudotuberculosis; Y. pestis* strains are uniformly nonmotile, but the genome sequence showed that they possess two separate clusters of flagellar genes as well as a chemotaxis operon, which total more than 80 genes (~2% of the total genes). Consistent with the failure to observe motility in *Y. pestis*, the flagellar and chemotaxis gene clusters contain several mutations (Parkhill et al. 2001b; Deng et al. 2002). The most important appears to be the frameshift in the regulator *flhD*, the effect of which seems to be the silencing of the entire flagellar biosynthesis system.

In addition to mutations in the genes required for enteropathogenicity, several of the newly discovered genes that appear to specify pathogenicity for an insect host have also been inactivated. These include the insecticidal toxin genes *tcaB*, which contained a frameshift mutation, and *tcaC*, which possessed an internal deletion. As described above, many of these genes are present in the ancestor, *Y. pseudotuberculosis*, and it may be that the disruption of these genes is necessary for the new lifestyle of *Y. pestis*, which persists in the flea gut for relatively long periods of time.

The change in niche accomplished by *Y. pestis* thus is not from a mammalian enteric pathogen to a systemic pathogen using an entirely new insect vector for infection. Rather, it appears that the ancestral organism had reached the point of maintaining separate fecal–oral infections in both mammalian and insect hosts. The specific change in niche was to connect the two hosts by a novel transmission route, spreading directly from the flea gut to the mammalian host via a subcutaneous injection. During, or immediately following, this change, the organism appears to have inactivated genes that were unnecessary for, or would hinder, this new lifestyle, specifically those involved in interacting with the mammalian gastrointestinal tract, or those causing unwanted toxicity in the insect host. The evolutionary bottleneck associated with this change may also have allowed the introduction of mutations into genes, such as those encoding metabolic enzymes, that were neutral or only marginally beneficial to the organism.

## CONCLUSIONS

We can see that the evolution of these pathogens is opportunistic, taking advantage of new niches as they arise, and losing the ability to occupy the niches previously held. These examples are by no means unique. Preliminary data from the sequencing of *Bordetella pertussis*, a human-specific pathogen that is the causative agent of whooping cough, indicates that it is a recently derived degenerate clone of the broad host-range mammalian pathogen *Bordetella bronchiseptica*. Many of the same processes, including IS element expansion, pseudogene formation, and genome rearrangement, have occurred in the derivation of this host-specific pathogen. In cases like these, and others, it appears that the rapid changes that have occurred in the human population may have created these new niches, and we can expect bacterial pathogens to rapidly exploit the further opportunities afforded them in the future as the human population continues to expand and diversify.

## ACKNOWLEDGMENTS

We are very grateful to the sequencing and annotation teams of the Sanger Institute Pathogen Sequencing Unit for their considerable efforts in generating the data described in this review.

## REFERENCES

Achtman M., Zurth K., Morelli G., Torrea G., Guiyoule A., and Carniel E. 1999. *Yersinia pestis*, the cause of plague, is a recently emerged clone of *Yersinia pseudotuberculosis*. *Proc. Natl. Acad. Sci.* **96:** 14043.

Andersson D.I. and Hughes D. 1996. Muller's ratchet decreases fitness of a DNA-based microbe. *Proc. Natl. Acad. Sci.* **93:** 906.

Bach S., de Almeida A., and Carniel E. 2000. The *Yersinia* high-pathogenicity island is present in different members of the family *Enterobacteriaceae*. *FEMS Microbiol. Lett.* **183:** 289.

Bakshi C.S., Singh V.P., Wood M.W., Jones P.W., Wallis T.S., and Galyov E.E. 2000. Identification of SopE2, a *Salmonella* secreted protein which is highly homologous to SopE and involved in bacterial invasion of epithelial cells. *J. Bacteriol.* **182:** 2341.

Blackburn M., Golubeva E., Bowen D., and French-Constant R.H. 1998. A novel insecticidal toxin from *Photorhabdus luminescens*, toxin complex A (Tca), and its histopathological effects on the midgut of *Manduca sexta*. *Appl. Environ. Microbiol.* **64:** 3036.

Blattner F.R., Plunkett G., Bloch C.A., Perna N.T., Burland V., Riley M., Collado-Vides J., Glasner J.D., Rode C.K., Mayhew G.F., Gregor J., Davis N.W., Kirkpatrick H.A., Goeden M.A., Rose D.J., Mau B., and Shao Y. 1997. The complete genome sequence of *Escherichia coli* K-12. *Science* **277:** 1453.

Blum G., Ott M., Lischewski A., Ritter A., Imrich H., Tschape H., and Hacker J. 1994. Excision of large DNA regions termed pathogenicity islands from tRNA- specific loci in the chromosome of an *Escherichia coli* wild-type pathogen. *Infect. Immun.* **62:** 606.

Buchrieser C., Prentice M., and Carniel E. 1998. The 102-kilobase unstable region of *Yersinia pestis* comprises a high-pathogenicity island linked to a pigmentation segment which undergoes internal rearrangement. *J. Bacteriol.* **180:** 2321.

Cole S.T., Eiglmeier K., Parkhill J., James K.D., Thomson N.R., Wheeler P.R., Honore N., Garnier T., Churcher C., Harris D., Mungall K., Basham D., Brown D., Chillingworth T., Connor R., Davies R.M., Devlin K., Duthoy S., Feltwell T., Fraser A., Hamlin N., Holroyd S., Hornsby T., Jagels K., and Lacroix C., et al. 2001. Massive gene decay in the leprosy bacillus. *Nature* **409:** 1007.

Cowan C., Jones H.A., Kaya Y.H., Perry R.D., and Straley S.C. 2000. Invasion of epithelial cells by *Yersinia pestis*: Evidence for a *Y. pestis*-specific invasin. *Infect. Immun.* **68:** 4523.

Darwin A.J. and Miller V.L. 1999. Identification of *Yersinia enterocolitica* genes affecting survival in an animal host using signature-tagged transposon mutagenesis. *Mol. Microbiol.* **32:** 51.

Deng W., Burland V., Plunkett G., III, Boutin A., Mayhew G.F., Liss P., Perna N.T., Rose D.J., Mau B., Zhou S., Schwartz D.C., Fetherston J.D., Lindler L.E., Brubaker R.R., Plano G.V., Straley S.C., McDonough K.A., Nilles M.L., Matson J.S., Blattner F.R., and Perry R.D. 2002. Genome sequence of *Yersinia pestis* KIM. *J. Bacteriol.* **184:** 4601.

Dillard J.P. and Yother J. 1994. Genetic and molecular characterization of capsular polysaccharide biosynthesis in *Streptococcus pneumoniae* type 3. *Mol. Microbiol.* **12:** 959.

Dillard J.P., Caimano M., Kelly T., and Yother J. 1995. Capsules and cassettes: Genetic organization of the capsule locus of *Streptococcus pneumoniae*. *Dev. Biol. Stand.* **85:** 261.

El Tahir Y. and Skurnik M. 2001. YadA, the multifaceted

*Yersinia* adhesin. *Int. J. Med. Microbiol.* **291:** 209.

Galan J.E. 1996. Molecular genetic bases of *Salmonella* entry into host cells. *Mol. Microbiol.* **20:** 263.

Hacker J., Blum-Oehler G., Muhldorfer I., and Tschape H. 1997. Pathogenicity islands of virulent bacteria: Structure, function and impact on microbial evolution. *Mol. Microbiol.* **23:** 1089.

Hare J.M., Wagner A.K., and McDonough K.A. 1999. Independent acquisition and insertion into different chromosomal locations of the same pathogenicity island in *Yersinia pestis* and *Yersinia pseudotuberculosis*. *Mol. Microbiol.* **31:** 291.

Hayashi T., Makino K., Ohnishi M., Kurokawa K., Ishii K., Yokoyama K., Han C.G., Ohtsubo E., Nakayama K., Murata T., Tanaka M., Tobe T., Iida T., Takami H., Honda T., Sasakawa C., Ogasawara N., Yasunaga T., Kuhara S., Shiba T., Hattori M., and Shinagawa H. 2001. Complete genome sequence of enterohemorrhagic *Escherichia coli* O157:H7 and genomic comparison with a laboratory strain K-12. *DNA Res.* **8:** 11.

Kingsley R.A. and Baumler A.J. 2000. Host adaptation and the emergence of infectious disease: The *Salmonella* paradigm. *Mol. Microbiol.* **36:** 1006.

———. 2002. Pathogenicity islands and host adaptation of *Salmonella* serovars. *Curr. Top. Microbiol. Immunol.* **264:** 67.

McClelland M., Sanderson K.E., Spieth J., Clifton S.W., Latreille P., Courtney L., Porwollik S., Ali J., Dante M., Du F., Hou S., Layman D., Leonard S., Nguyen C., Scott K., Holmes A., Grewal N., Mulvaney E., Ryan E., Sun H., Florea L., Miller W., Stoneking T., Nhan M., Waterston R., and Wilson R.K. 2001. Complete genome sequence of *Salmonella enterica* serovar Typhimurium LT2. *Nature* **413:** 852.

McDonough K.A. and Hare J.M. 1997. Homology with a repeated *Yersinia pestis* DNA sequence IS100 correlates with pesticin sensitivity in *Yersinia pseudotuberculosis*. *J. Bacteriol.* **179:** 2081.

Miao E.A. and Miller S.I. 2000. A conserved amino acid sequence directing intracellular type III secretion by *Salmonella typhimurium*. *Proc. Natl. Acad. Sci.* **97:** 7539.

Mills D.M., Bajaj V., and Lee C.A. 1995. A 40 kb chromosomal fragment encoding *Salmonella typhimurium* invasion genes is absent from the corresponding region of the *Escherichia coli* K- 12 chromosome. *Mol. Microbiol.* **15:** 749.

Neidhardt F.C. and Curtiss R., Eds. 1996. Escherichia coli *and* Salmonella: *Cellular and molecular biology*. ASM Press, Washington, D.C.

Nesper J., Lauriano C.M., Klose K.E., Kapfhammer D., Kraiss A., and Reidl J. 2001. Characterization of *Vibrio cholerae* O1 El tor *galU* and *galE* mutants: Influence on lipopolysaccharide structure, colonization, and biofilm formation. *Infect. Immun.* **69:** 435.

Ochman H., Soncini F.C., Solomon F., and Groisman E.A. 1996. Identification of a pathogenicity island required for *Salmonella* survival in host cells. *Proc. Natl. Acad. Sci.* **93:** 7800.

Odaert M., Berche P., and Simonet M. 1996. Molecular typing of *Yersinia pseudotuberculosis* by using an IS*200*-like element. *J. Clin. Microbiol.* **34:** 2231.

Parkhill J., Dougan G., James K.D., Thomson N.R., Pickard D., Wain J., Churcher C., Mungall K.L., Bentley S.D., Holden M.T., Sebaihia M., Baker S., Basham D., Brooks K., Chillingworth T., Connerton P., Cronin A., Davis P., Davies R.M., Dowd L., White N., Farrar J., Feltwell T., Hamlin N., and Haque A., et al. 2001a. Complete genome sequence of a multiple drug resistant *Salmonella enterica* serovar Typhi CT18. *Nature* **413:** 848.

Parkhill J., Wren B.W., Thomson N.R., Titball R.W., Holden M.T., Prentice M.B., Sebaihia M., James K.D., Churcher C., Mungall K.L., Baker S., Basham D., Bentley S.D., Brooks K., Cerdeno-Tarraga A.M., Chillingworth T., Cronin A., Davies R.M., Davis P., Dougan G., Feltwell T., Hamlin N., Holroyd S., Jagels K., and Karlyshev A.V., et al. 2001b. Genome sequence of *Yersinia pestis*, the causative agent of plague. *Nature* **413:** 523.

Pepe J.C. and Miller V.L. 1993. *Yersinia enterocolitica* invasin: A primary role in the initiation of infection. *Proc. Natl. Acad. Sci.* **90:** 6473.

Perna N.T., Plunkett G., III, Burland V., Mau B., Glasner J.D., Rose D.J., Mayhew G.F., Evans P.S., Gregor J., Kirkpatrick H.A., Posfai G., Hackett J., Klink S., Boutin A., Shao Y., Miller L., Grotbeck E.J., Davis N.W., Lim A., Dimalanta E.T., Potamousis K.D., Apodaca J., Anantharaman T.S., Lin J., and Yen G., et al. 2001. Genome sequence of enterohaemorrhagic *Escherichia coli* O157:H7. *Nature* **409:** 529.

Perry R.D. and Fetherston J.D. 1997. *Yersinia pestis*—Etiologic agent of plague. *Clin. Microbiol. Rev.* **10:** 35.

Rakin A., Noelting C., Schropp P., and Heesemann J. 2001. Integrative module of the high-pathogenicity island of *Yersinia*. *Mol. Microbiol.* **39:** 407.

Reeves M.W., Evins G.M., Heiba A.A., Plikaytis B.D., and Farmer J.J., III. 1989. Clonal nature of *Salmonella typhi* and its genetic relatedness to other salmonellae as shown by multilocus enzyme electrophoresis, and proposal of *Salmonella bongori* comb. nov. *J. Clin. Microbiol.* **27:** 313.

Reiter W.D., Palm P., and Yeats S. 1989. Transfer RNA genes frequently serve as integration sites for prokaryotic genetic elements. *Nucleic Acids Res.* **17:** 1907.

Rosqvist R., Skurnik M., and Wolf-Watz H. 1988. Increased virulence of *Yersinia pseudotuberculosis* by two independent mutations. *Nature* **334:** 522.

Schubert S., Cuenca S., Fischer D., and Heesemann J. 2000. High-pathogenicity island of *Yersinia pestis* in enterobacteriaceae isolated from blood cultures and urine samples: Prevalence and functional expression. *J. Infect. Dis.* **182:** 1268.

Shea J.E., Hensel M., Gleeson C., and Holden D.W. 1996. Identification of a virulence locus encoding a second type III secretion system in *Salmonella typhimurium*. *Proc. Natl. Acad. Sci.* **93:** 2593.

Skurnik M., Peippo A., and Ervela E. 2000. Characterization of the O-antigen gene clusters of *Yersinia pseudotuberculosis* and the cryptic O-antigen gene cluster of *Yersinia pestis* shows that the plague bacillus is most closely related to and has evolved from *Y. pseudotuberculosis* serotype O:1b. *Mol. Microbiol.* **37:** 316.

Swartley J.S., Marfin A.A., Edupuganti S., Liu L.J., Cieslak P., Perkins B., Wenger J.D., and Stephens D.S. 1997. Capsule switching of *Neisseria meningitidis*. *Proc. Natl. Acad. Sci.* **94:** 271.

Townsend S.M., Kramer N.E., Edwards R., Baker S., Hamlin N., Simmonds M., Stevens K., Maloy S., Parkhill J., Dougan G., and Baumler A.J. 2001. *Salmonella enterica* serovar Typhi possesses a unique repertoire of fimbrial gene sequences. *Infect. Immun.* **69:** 2894.

Tsolis R.M., Townsend S.M., Miao E.A., Miller S.I., Ficht T.A., Adams L.G., and Baumler A.J. 1999. Identification of a putative *Salmonella enterica* serotype typhimurium host range factor with homology to IpaH and YopM by signature-tagged mutagenesis. *Infect. Immun.* **67:** 6385.

Wang P. and Granados R.R. 1997. An intestinal mucin is the target substrate for a baculovirus enhancin. *Proc. Natl. Acad. Sci.* **94:** 6977-6982.

Waterfield N.R., Bowen D.J., Fetherston J.D., Perry R.D., and French-Constant R.H. 2001. The tc genes of *Photorhabdus:* A growing family. *Trends Microbiol.* **9:** 185.

Wood M.W., Jones M.A., Watson P.R., Siber A.M., McCormick B.A., Hedges S., Rosqvist R., Wallis T.S., and Galyov E.E. 2000. The secreted effector protein of *Salmonella dublin*, SopA, is translocated into eukaryotic cells and influences the induction of enteritis. *Cell. Microbiol.* **2:** 293.

Young G.M., Badger J.L., and Miller V.L. 2000. Motility is required to initiate host cell invasion by *Yersinia enterocolitica*. *Infect. Immun.* **68:** 4323.

Zhang S., Santos R.L., Tsolis R.M., Stender S., Hardt W.D., Baumler A.J., and Adams L.G. 2002. The *Salmonella enterica* serotype typhimurium effector proteins SipA, SopA, SopB, SopD, and SopE2 act in concert to induce diarrhea in calves. *Infect. Immun.* **70:** 3843.

Zhang X.L., Morris C., and Hackett J. 1997. Molecular cloning, nucleotide sequence, and function of a site-specific recombinase encoded in the major 'pathogenicity island' of Salmonella typhi. *Gene* **202:** 139.

Zhang X.L., Tsui I.S., Yip C.M., Fung A.W., Wong D.K., Dai X., Yang Y., Hackett J., and Morris C. 2000. *Salmonella enterica* serovar typhi uses type IVB pili to enter human intestinal epithelial cells. *Infect. Immun.* **68:** 3067.

# Gene Expression Profiling of Cells, Tissues, and Developmental Stages of the Nematode *C. elegans*

S.J. McKay,* R. Johnsen,† J. Khattra,* J. Asano,* D.L. Baillie,† S. Chan,* N. Dube,¶ L. Fang,† B. Goszczynski,‡ E. Ha,† E. Halfnight,¶ R. Hollebakken,† P. Huang,* K. Hung,† V. Jensen,† S.J.M. Jones,* H. Kai,¶ D. Li,† A. Mah,† M. Marra,* J. McGhee,‡ R. Newbury,¶ A. Pouzyrev,§ D.L. Riddle,§ E. Sonnhammer,** H. Tian, ‡ D. Tu,† J.R. Tyson,† G. Vatcher,* A. Warner,¶ K. Wong,* Z. Zhao,† and D.G. Moerman¶

**Genome Sciences Centre, BC Cancer Agency, Vancouver, B.C., Canada, V6T 1Z4; †Department of Molecular Biology and Biochemistry, Simon Fraser University, Burnaby, B.C., Canada, V5A 1S6; ‡Department of Biochemistry and Molecular Biology, University of Calgary, Calgary, Alberta, Canada T2N 4N1; ¶Department of Zoology, University of British Columbia, Vancouver, B.C., Canada V6T 1Z4; §Division of Biological Sciences, University of Missouri, Columbia, Missouri 65211-7400; **Karolinska Institute, Stockholm, Sweden*

Completion of the DNA sequences of the human genome and that of the nematode *Caenorhabditis elegans* allows the large-scale identification and analysis of orthologs of human genes in an organism amenable to detailed genetic and molecular analyses. We are determining gene expression profiles in specific cells, tissues, and developmental stages in *C. elegans*. Our ultimate goal is not only to describe detailed gene expression profiles, but also to gain a greater understanding of the organization of gene regulatory networks and to determine how they control cell function during development and differentiation.

The use of *C. elegans* as a platform to investigate the details of gene regulatory networks has several major advantages. Two key advantages are that it is the simplest multicellular organism for which there is a complete sequence (*C. elegans* Sequencing Consortium 1998), and it is the only multicellular organism for which there is a completely documented cell lineage (Sulston and Horvitz 1977; Sulston et al. 1983). *C. elegans* is amenable to both forward and reverse genetics (for review, see Riddle et al. 1997). A 2-week life span and generation time of just 3 days for *C. elegans* allows experimental procedures to be much shorter, more flexible, and more cost-effective compared to the use of mouse or zebrafish models for genomic analyses. Finally, the small size, transparency, and limited cell number of the worm make it possible to observe many complex cellular and developmental processes that cannot easily be observed in more complex organisms. Morphogenesis of organs and tissues can be observed at the level of a single cell (White et al. 1986). As events have shown, investigating the details of *C. elegans* biology can lead to fundamental observations about human health and biology (Sulston 1976; Hedgecock et al. 1983; Ellis and Horvitz 1986).

We are using complementary approaches to examine gene expression in *C. elegans*. We are constructing transgenic animals containing promoter green fluorescent protein (GFP) fusions of nematode orthologs of human genes. These transgenic animals are examined to determine the time and tissue expression pattern of the promoter::GFP constructs. Concurrently, we are undertaking serial analysis of gene expression (SAGE) on all developmental stages of intact animals and on selected purified cells. Tissues and selected cells are isolated using a fluorescence activated cell sorter (FACS) to sort promoter::GFP marked cell populations. To date we have purified to near homogeneity cell populations for embryonic muscle, gut, and a subset of neurons. The SAGE and promoter::GFP expression data are publicly available at http://elegans.bcgsc.bc.ca.

## PROMOTER::GFP FUSIONS AS INDICATORS OF SPECIFIC TISSUE AND TEMPORAL GENE EXPRESSION

Our ultimate goal is to examine the in vivo spatial and temporal expression profiles of as many genes in the *C. elegans* genome as possible. Presently, the most effective methods for determining expression patterns in the worm are either antibodies or reporter fusion constructs. We have opted to use the more cost-effective promoter::GFP fusion technique. GFP reporter constructs are exquisitely sensitive and can detect expression at the resolution of a single cell (Chalfie et al. 1994; Chalfie 1995). The *C. elegans* community is fortunate to have an excellent GFP insertion vector kit available (developed by Dr. Andrew Fire, Carnegie Institution, http://www.ciwemb.edu/pages/firelab.html). We have been preceded in our approach by others, in particular, the laboratory of Ian Hope (Hope 1991; Lynch et al. 1995), where 350 expressing reporter gene fusions have been constructed (http://bgypc086.leeds.ac.uk/). Although the use of GFP fusions as expression reporters is not novel, the scale of our project is unprecedented.

To make a viable high-throughput approach for GFP fusion constructs, we needed a method that was both fast and efficient. Over the past year, we have demonstrated that fusion-PCR, also known as "stitching," enables construction of GFP fusions on a genome-wide scale. This PCR stitching technique has been used successfully by at least two groups (Cassata, Kagoshima et al. 1998; Hobert 2002) and we demonstrate here that it is scalable.

## Choosing Candidate *C. elegans* Genes for Promoter GFP Analysis

Our study focuses on nematode homologs of human genes. A comparison of the two predicted proteomes with INPARANOID (Remm et al. 2001) identified 4367 *C. elegans* proteins with probable human orthologs (http://inparanoid.cgb.ki.se). This list of genes provides an excellent opportunity to use the worm to infer biological information for genes potentially relevant to human biology and health care. Of particular interest are predicted worm/human homologs for which there are no data concerning function; more than half of the worm orthologs have no functional annotation associated with them. These are particularly important gene targets, as they may form a new set of "Rosetta stone" proteins.

Most of the genome annotations used in the selection of our list of target genes were obtained from WormBase (www.wormbase.org; Stein et al. 2001; Harris et al. 2003). The list was filtered to remove rRNA genes and genes with SL2 trans-splice acceptor sites, which are associated with operons (Blumenthal 1995; Blumenthal et al. 2002). Also removed were genes with characterized mRNAs, an indication that the gene was already well studied. Preference was given to genes with EST-confirmed 5′ ends and those identified as embryonically expressed in Intronerator (Kent and Zahler 2000). We did not remove genes for which other researchers have constructed reporter fusions, because such genes act as a control set for our work. Indeed, thus far, at least four examples of expression patterns we have observed with our promoter::GFP constructs are identical to those observed by other investigators using either antibodies or functional GFP fusions.

The PCR stitching technique uses a two-step approach. First, the promoterless GFP gene and the putative *C. elegans* promoter region are PCR-amplified separately. In a second-round PCR, a complementary region engineered into the 3′ primer of the promoter amplicon and the 5′ primer of the GFP amplicon allows them to prime each other to form a chimeric amplicon containing a complete expression cassette. The PCR experiments were designed to capture putative promoter regions by amplifying about 3 kb of genomic DNA sequence immediately upstream of the predicted ATG initiator site. When an upstream gene was within 3 kb, the size of the amplicon was adjusted downward. Early PCR experiments were designed semi-manually with the aid of primer3 (Rozen and Skaletsky 2000). To facilitate scale-up, we took advantage of the excellent *C. elegans* genome informatics resources to automate the PCR experimental design process. We used Perl and AcePerl (Stein and Thierry-Mieg 1998) to extract *C. elegans* genomic DNA sequence and annotations from wormbase to tie them together with the primer design and validation programs primer3 and e-PCR (Schuler 1997). To provide flexible, real-time design of PCR-based GFP fusion experiments, an interactive Web version of the program is available (http://elegans.bcgsc.bc.ca; S. McKay et al., in prep,).

## Constructing Transgenic Animals with Heritable Promoter::GFP Constructs

Transgenic worms were generated by a modification of the method described by Mello et al. (1991). Promoter::GFP constructs and *dpy-5*(+) plasmid (pCeh-361) (kindly provided by C. Thacker and A. Rose) were used to construct transgenic strains. Transformants were identified by rescue of the Dpy-5 mutant phenotype. In *C. elegans*, transgene constructs usually form large extrachromosomal arrays. Due to the holokinetic nature of *C. elegans* chromosomes, these arrays can be partitioned during mitosis as though they were small chromosomes. However, extrachromosomal arrays must be large to be heritable (Stinchcomb et al. 1985; Clark et al. 1990; Mello and Fire 1995; Mello et al. 1991). Heritability of the GFP transgene construct is of considerable importance here, as somatic mosaicism or loss of the construct during gametogenesis could confound inferred gene expression patterns.

To determine whether our GFP transgenes form sufficiently large concatemeric arrays in vivo, we used quantitative PCR to estimate the copy number of the promoter::GFP constructs and plasmids in 20 different transgenic strains. We estimate that there are about 5–10 copies of promoter::GFP and 100–600 copies of the *dpy-5* plasmid in the heritable arrays. Although linear GFP DNA appears to be incorporated into arrays an order of magnitude less efficiently than circular plasmids, the sensitivity of the GFP assay does not require high copy numbers. To date, we have generated transgenic lines representing more than 1000 genes.

The ultimate in stable inheritance is ensured by chromosomal integration of the transgene, a process that can be induced by creating double-stranded breaks in chromosomes with ionizing radiation. Although the necessary handling and strain cleanup steps make this process less amenable to scale-up, we are constructing chromosomal integrant strains for a subset of the GFP constructs using low-dose X-ray irradiation (1500R). To date, such strains have been constructed for 80 genes. All of the strains generated from this study will be made available through the *Caenorhabditis* Genetics Center (biosci.umn.edu/CGC/CGChomepage.htm).

## Expression Analysis of Promoter::GFP Constructs

As transformants carrying GFP fusions become available, they are subjected to a detailed in vivo analysis. As a first pass, we determine the developmental stage, tissues, and where possible, the individual cells where GFP expression is observed (Table 1). To date, we have observed GFP expression for 450 (56%) of 802 different transgenic lines. Possible reasons why no GFP expression was observed in the remaining lines include (1) germ-line silencing (Kelly et al. 1997; for review, see Seydoux and Schedl 2001), (2) absence of promoter::GFP in the heritable arrays, (3) conditional gene expression, or (4) failure to capture the entire transcription control element. The PCR experiments were designed to amplify as much

**Table 1.** Temporal and Tissue-specific Expression of Promoter::GFP Fusions

| Tissue | Larval exclusive | Adult exclusive | Both larval and adult stages |
|---|---|---|---|
| Pharynx | 2 | 11 | 59 |
| Intestinal | 23 | 3 | 66 |
| Vulval | 0 | 33 | 1 |
| Spermatheca | 0 | 6 | 1 |
| Body wall muscle | 3 | 6 | 40 |
| Hypodermis | 3 | 1 | 17 |
| Seam cells | 0 | 0 | 2 |
| Anal sphincter and depressor muscle | 0 | 9 | 12 |
| Excretory cell | 0 | 3 | 7 |
| Nerve ring | 5 | 0 | 36 |
| Ventral nerve cord | 6 | 1 | 23 |
| Dorsal nerve cord | 0 | 1 | 3 |
| Head neurons | 6 | 2 | 45 |
| Tail neurons | 5 | 5 | 48 |
| Body neurons | 2 | 2 | 9 |

of the potential promoter region as possible (up to 3 kb). Although it is rare in *C. elegans*, there are cases where important transcription control elements lie outside this 3-kb range and therefore preclude expression of the GFP construct.

Preliminary classification of GFP expression is done using a low-power GFP dissecting microscope. More detailed follow-up is done using a standard or confocal microscope equipped with epifluorescence and Nomarski optics. Ultimately, detailed expression patterns and gene activation in embryos are captured with live, two-channel four-dimensional microscopy. The fourth dimension is time; Z-stacks of developing embryos are recorded using Nomarski microscopy every 30–45 seconds. Interspersed with the normal Z-stacks we record GFP fluorescence in specific cells, which are then mapped and identified relative to the Nomarksi images. Software that supports this recording and analysis has been developed (Schnabel et al. 1997; also see Fire 1994; Thomas and White 1998; Bürglin 2000), and we are using programs derived from the study by Schnabel et al. (1997).

A survey of temporal GFP expression patterns is shown in Table 1, and some illustrative examples are displayed in Figure 1. We have detected GFP at all developmental stages and have identified expressed GFP in all major tissues except the germinal gonad. We did not expect to observe germ-line expression with any of our extrachromosomal array constructs, because germ-line silencing affects genes in extrachromosomal arrays (for review, see Seydoux and Schedl 2001). So far, we have not observed germ-line expression in any of the integrated lines derived from extrachromosomal arrays. A majority of the promoters we have examined thus far drive GFP expression in the intestine (92) and the nervous system (70), many exclusively in one of these tissues (Table 1). The large number of genes expressed in the intestine, the functional equivalent of the human stomach, intestine, and liver, agrees with our findings using SAGE on adult dissected intestine (see below and Table 2). Besides the intestine and nervous system, other major tissues including muscle and hypodermis are well represented in our data set. Subsets of cells and tissues within these broad categories are also delineated; we have observed GFP expression specific to the nerve ring, sensory neurons, ventral nerve cord, pharynx, seam cells, the excretory canal, the spermatheca, and anal sphincter muscles (Table 1). Our single biggest challenge in determining cell identity concerns the 302 cells that comprise the nematode nervous system (White et al. 1986). Neural expression patterns display a myriad of combinatorial possibilities, a fraction of which are represented in Table 1.

**Table 2.** SAGE Libraries

| Stage | Tissue | Tags | | Genes |
|---|---|---|---|---|
| | | total | unique | |
| Embryo 14-bp tags | whole | 133,825 | 25,885 | 8,187 |
| Embryo 21-bp tags | whole | 220,032 | 44,992 | 8,929 |
| L1 larvae starved | whole | 116,363 | 19,494 | 6,429 |
| L1 larvae normal | whole | 109,994 | 17,532 | 6,705 |
| L2 larvae | whole | 130,209 | 24,658 | 7,264 |
| L3 larvae | whole | 127,924 | 24,039 | 7,667 |
| L4 larvae | whole | 141,878 | 25,701 | 8,046 |
| Young adult | whole | 119,222 | 23,128 | 6,302 |
| Adult (glp-4) | dissected gut | 138,346 | 14,386 | 4,892 |
| Adult (glp-4) | whole | 117,529 | 19,140 | 6,974 |
| Embryo (myo-3::GFP) | FACS sorted muscle | 58,147 | 16,967 | 4,850 |
| 6-day adult (fer-15) | whole | 110,306 | 19,861 | 6,758 |
| 1-day adult (fer-15;daf-2) | whole | 101,939 | 16,960 | 5,159 |
| 6-day adult (fer-15;daf-2) | whole | 100,737 | 14,004 | 4,687 |
| 10-day adult (fer-15;daf-2) | whole | 116,336 | 19,183 | 5,594 |
| Mixed stage[a] | whole | 175,995 | 37,894 | 9,222 |
| Dauer larvae[a] | whole | 65,828 | 18,136 | 5,373 |
| Meta library (14-bp tags) | | 1,806,431 | 130,112 | 14,661 |
| Meta library (21-bp tags) | | 278,179 | 53,738 | 10,896 |

The total and unique tag numbers are for the raw tag collection prior to filtering. Libraries were filtered to remove tags with low sequence quality (below phred20) and tags originating from duplicate ditags (possible PCR artifacts). The number of genes refers to genes whose expression was detected by the presence of one or more tags mapped unambiguously to a single mRNA.

[a]From Jones et al. (2001). Only tags with known sequence quality were considered.

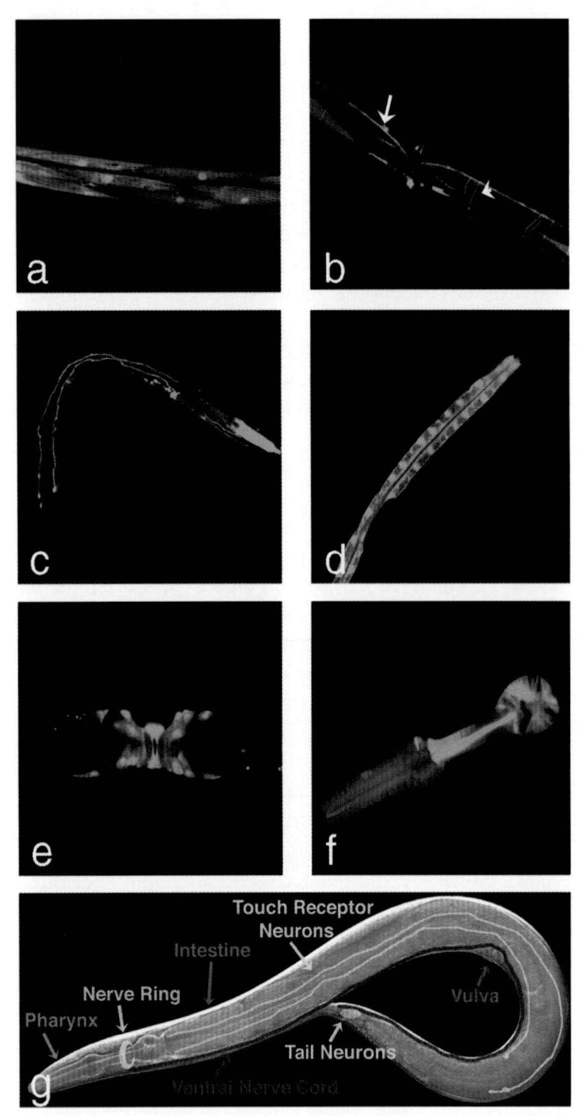


**Figure 1.** A gallery of promoter::GFP expression patterns for *C. elegans*. (*a*) Body wall muscle—gene B0228.4. (*b*) Ventral cord neurons and commissures—gene Y102A11A.2. Arrow indicates ventral cord and arrowhead points to a commissure. (*c*) Touch cells and pharynx—gene F32F2.1. (*d*) Intestine—gene Y102A11A.2. (*e*) vulval cells—gene Y47G6A.7. (*f*) Pharynx—gene C09G4.1. (*g*) Diagram illustrating location of some of the cells shown in panels *a–f*.

Because neuronal cells have been assigned to 118 classes (White et al. 1986), it is perhaps not surprising that there are many different neuronal gene expression patterns.

The GFP expression strains developed as part of this project are intended to become useful reagents for the biomedical research community. If enough promoters from different genes can be analyzed, we hope to be able to deduce logical rules regulating gene expression. Certainly this goal will be achievable if we combine this GFP expression set with the SAGE studies described below. A first step toward understanding the coordinate regulation of genes whose promoters drive similar expression patterns is to both computationally and biologically dissect the function of the promoter region. To facilitate this, a new comparative tool has emerged in the form of DNA sequence alignments between *C. elegans* and *C. briggsae* genomes (L. Stein et al. 2001). These two species are sufficiently diverged (80 to 100 million years) that noncoding sequences have diverged, but coding and other functional sequences remain conserved. This property can be exploited to refine existing gene models and create new ones, as well as to help identify potentially important *cis*-regulatory elements upstream of conserved genes. The latter could prove invaluable for dissecting the function of promoter regions of genes that we select for further study.

## SERIAL ANALYSIS OF GENE EXPRESSION: TEMPORAL AND TISSUE-SPECIFIC EXPRESSION PROFILING

Several studies using either DNA microarray analysis or serial analysis of gene expression (SAGE) have been done to examine the expression of *C. elegans* genes in the whole organism (see, e.g., Hill et al. 2000; Reinke et al. 2000; Jones et al. 2001: Kim et al. 2001). SAGE is complementary to microarray analysis and, at present, is the most sensitive and specific method for obtaining qualitative and quantitative information on expressed RNAs (Velculescu et al. 1995). Using this approach, we can establish the portion of the genome that is transcribed and contributes to the protein profile at various time points during growth and development. Within the RNA profile, we can also identify many genes that do not encode proteins, but produce only RNA products. Finally, we can gain useful insight into alternatively spliced mRNA isoforms, their changes over time, and relative abundance. Ours is the first study to use SAGE to examine all the developmental stages of *C. elegans*. Thus far, we have constructed 17 libraries spanning all developmental stages from embryo to adult and also representing tissue, cell-type, and mutation-specific populations (Table 2). Taken together, these libraries include ~1.8 million observed tags.

### Tag to Gene Mapping: Building the "Conceptual" Transcriptome

For a SAGE tag to be associated with a specific gene, it is first necessary to build a conceptual transcriptome representing the processed transcripts of all known genes and predicted gene models. Although tags corresponding to the mitochondrial and noncoding transcriptomes are also represented in our *C. elegans* SAGE libraries, most tags correspond to the predicted nuclear transcriptome. The process we used to build the conceptual transcriptome is illustrated in Figure 2. An examination of WormBase (www.wormbase.org; release WS110) , the public repository of information on the biology and genome of *C. elegans*, reveals that nearly 40% of the 22,156 *C. elegans* gene models have no EST evidence to confirm gene structure or expression. About 44% of WormBase genes have sufficient EST coverage to determine the extent of 3′UTRs (untranslated regions) in the processed transcripts.

Because the SAGE technique captures transcripts by their poly-A tail, and the tags are usually anchored at the

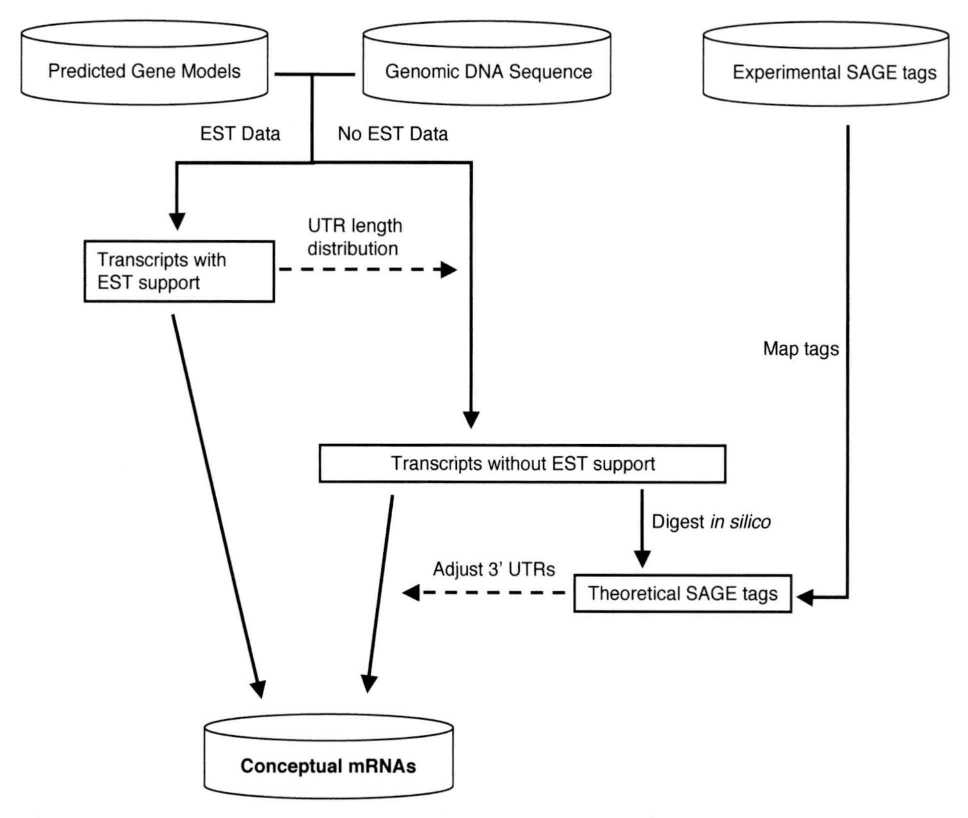


**Figure 2.** Building the conceptual transcriptome. Conceptual transcripts were assembled with known UTRs for genes with EST coverage and predicted UTRs for other genes based on the distribution of known UTR lengths. Introns were excised from the coding DNA and UTRs. Predicted 3′UTRs were adjusted according to potential polyadenylation signals, and both 3′ and 5′UTRs were truncated where required to avoid overlapping other genes. In some cases, overestimated 3′UTR lengths were detected by abundant experimentally observed SAGE tags occurring at the penultimate *Nla*III site (position 2). These predicted UTRs were truncated accordingly.

3′-most *Nla*III site, mRNAs with a cut site in their 3′UTR would be missed if coding sequences alone were used to map tags. For the 12,272 gene models lacking confirmed 3′UTRs, the untranslated regions of processed transcripts were predicted using a method modified from that of Pleasance et al. (2003). UTR lengths were estimated based on size distributions that cover 95% of known UTRs. About 5,550 of the predicted 3′UTRs include a *Nla*III site. Because the highest frequency SAGE tag for a transcript occurs at the first tag position, we used pooled SAGE data from more than a million SAGE tags to further refine the 3′UTR predictions for 1,449 gene models.

To determine how many transcripts we can identify, a meta-library of ~1.8 million tags was constructed by pooling all of the SAGE libraries (excluding longSAGE) in Table 2. A "specific" tag is defined as a tag that uniquely matches to a single gene or that can be resolved to a single gene by taking the lowest position match. To minimize the potential impact of sequencing errors, only tags with a cumulative phred score of 20 (Ewing and Green 1998) were considered. A score of Phred20 corresponds to a 99% probability that a base is called correctly. In this case, the score represents the average sequence quality of the entire tag sequence. A total of 26,682 specific tags corresponding to mRNAs for nuclear genes were observed. The total number of genes whose expression was detected by a SAGE tag for at least one transcript was 14,661. A distinct advantage of the SAGE technique is its ability to discriminate between alternative splice variants. Indeed, 7,073 (49%) of the detected genes are represented by two or more tags. A subset of just 1,126 (8%) of these genes have previously observed alternative splice variants documented in WormBase. Even among these previously well-studied genes, over 800 have multiple tags, potentially representing previously unobserved splice variants.

### A Comparison of Short (14 bp) Versus Long (21 bp) SAGE Tags

Until very recently, all SAGE libraries were constructed using the tagging enzyme *Bsm*FI, which generates a 14-bp tag. Theoretically, a 14-bp tag is sufficient to unambiguously identify any gene in the *C. elegans* genome. In practice, not all tags map unambiguously to a single location. Two factors contribute to this ambiguity. First, there are multigene families stemming from ancestral sequence duplications; these related genes can share similar 3′ ends. Second, there appears to be some sequence compositional bias in 3′UTRs that tend to be AT-rich. Based on a theoretical analysis of the *C. elegans* transcriptome, Pleasance et al. (2003) observed that, of all *C. elegans* genes that have an *Nla*III site in their conceptual transcript, about 12% would not be unambiguously identified by 14-bp tags. With an additional three nucleotides, they predicted that this number could be reduced to about 6% but, beyond 17 bp, there was no substantial reduction in ambiguity. Although 17-bp SAGE tags are not currently available, the need for longer tags has recently been addressed by the new longSAGE technique, which uses the enzyme *Mme*I to generate 21-bp tags (Saha et al. 2002). Now that there is a means for generating 21-bp tags, why would one not always use it? It comes down to cost, the bulk of which is in sequencing. It is possible to obtain greater sample depth of sequencing with 14-bp tags than for the same amount of sequencing with 21-bp tags. There is a trade-off between sampling deep enough to detect low-abundance transcripts and sequencing longer tags to reduce ambiguity.

To empirically determine the benefits of longer tags, we examined the same embryonic mRNA sample with both normal SAGE and longSAGE (Table 2). The two approaches identified 6,118 common genes (Fig. 3A) but also identified a substantial number of specific tags unique to one library. Since the majority of the nonoverlapping transcripts were of low abundance (Fig. 3B), it appears that such transcripts were detected stochastically at the sampling depth used. LongSAGE emerged as the method of choice, however, as it met the theoretical expectation of a twofold reduction in the ambiguities observed in assigning 14-bp tags to genes. Using longSAGE, it was possible to infer 2,896 specific genomic sites for tags shared by both libraries, but for which no unambiguous single sight could be assigned using a 14-bp tag. On the basis of the differences in tag ambiguity between the two protocols, we conclude that it is necessary to utilize longSAGE if resolving power is of utmost importance.

### A Comparison of DNA Microarrays and SAGE

There are close to ten published studies using DNA microarray analysis to profile *C. elegans* gene expression (for review, see Reinke 2002; also see Stuart et al. 2003), but so far only two SAGE studies (Jones et al. 2001; Holt and Riddle 2003). It is therefore important to compare the two approaches, as they both have advantages and pitfalls. The Affymetrix GeneChip™ array for *C. elegans* was designed to represent 22,500 *C. elegans* transcripts or EST clusters. Sequence information for probe design came from the December 05, 2000 Sanger Center ACeDB database release and GenBank release 121, and was re-annotated by Affymetrix. We remapped the Affymetrix probe sets to our current conceptual transcriptome to allow direct comparison of transcript profiles with SAGE. Because of changing gene models and genomic DNA annotations, not all transcripts predicted in 2000 can be compared directly to the 2003 version. However, ~ 90% of the Affymetrix probe sets can poten-

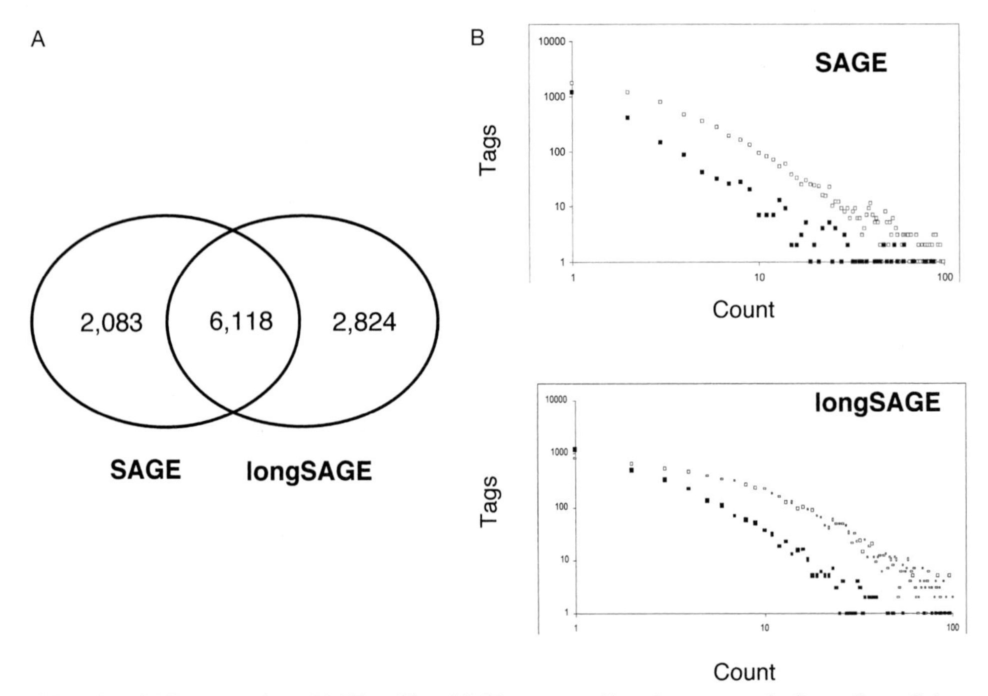


**Figure 3.** SAGE vs. longSAGE comparison. SAGE and longSAGE tags were filtered to remove duplicate ditags, linker tags, or tags with low-quality sequence. Only tags mapping unambiguously to the positive strand of a single transcript, or that could be resolved to a single sequence by taking the lowest position match (specific tags), were considered. Both libraries were constructed from the same mRNA sample, extracted from a synchronized embryonic population (Table 2). (*A*) A Venn diagram comparison of genes identified by SAGE and longSAGE. The region of overlap indicates genes for which specific tags were observed in both libraries. (*B*) Log X log plots of tag count distributions of shared tags (with the same 14-nt 5′ end) and unique tags. Due to the logarithmic scale, tag counts of 0 and 1 are not distinguishable. Note that the tags unique to the SAGE or longSAGE libraries (*filled squares*) are much less abundant.

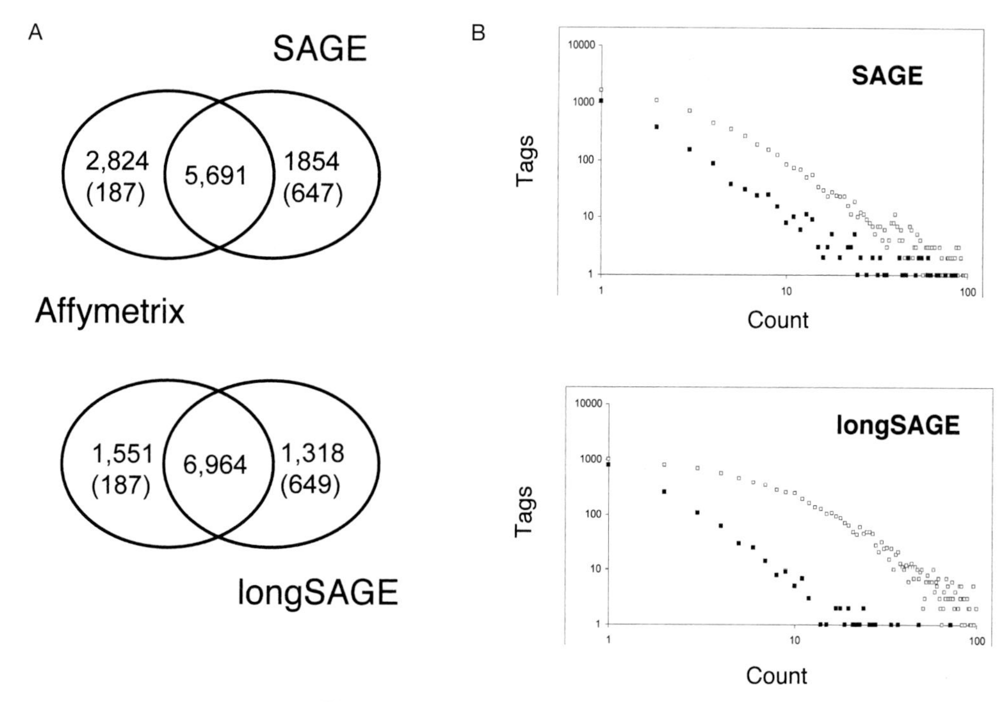


**Figure 4.** SAGE vs. AffyMetrix GeneChip™. Each 25mer probe in an Affymetrix probe set (obtained from www.affymetrix.com) was compared to all transcripts in the conceptual transcriptome. There are 20,291 probe sets that map specifically to 17,147 conceptual mRNAs. The remaining 2,257 probe sets mapped to multiple transcripts, or did not map to any. Only specific SAGE or longSAGE tags were considered. Mitochondrial transcripts, which are absent from the Affymetrix chip, were not considered. Oligonucleotide sequences from each probe set on the Affymetrix chip that was called as "present" in three replicate Affymetrix GeneChip experiments were mapped to transcripts in our virtual transcriptome. (*A*) Comparison of expressed genes detected by SAGE and longSAGE vs. the Affymetrix chip. For the SAGE methods, the number in parentheses represents transcripts not detectable by the chip. For Affymetrix, the number in parentheses represents transcripts without *Nla*III sites (undetected by SAGE). (*B*) Log x log plots of SAGE tag count distributions of shared and unique tags. Transcripts unique to the SAGE methods (*filled squares*) are of lower abundance. Only transcripts for which a direct SAGE/Affymetrix comparison was possible were considered.

tially be identified by SAGE, and 17,123 map unambiguously to single conceptual transcripts. To compare the chip and the two SAGE methods empirically, we hybridized the same mRNA samples we used for the whole-embryo SAGE and longSAGE libraries to Affymetrix chips. The pool of transcripts detected by both methods is limited: (1) Transcripts lacking an *Nla*III site or for which no specific tags were observed are excluded from SAGE and (2) probe sets that do not unambiguously detect a single transcript (or set of alternatively spliced transcripts) from our conceptual transcriptome are excluded from Affymetrix. Even with these caveats, a comparison of transcription profiles of embryonic libraries derived from SAGE and DNA chip analyses reveals a great deal of concordance (Fig. 4A), with more than half of all transcripts detected present in both data sets. As with the comparison between SAGE and longSAGE, most of the disagreement between methods is due to low-abundance transcripts (Figs. 3B and 4B). What is clear from the SAGE/Affymetrix comparison is that, because of the improved specificity in tag-to-gene mapping, longSAGE improves correspondence to the Affymetrix chip data (Fig. 4A). Given appropriate filtering to avoid sequencing errors and other artifacts, a rare, positive observation of a specific tag indicates that a transcript is likely present, albeit at low abundance. However, failing to observe a specific SAGE tag does not unequivocally demonstrate the absence of a transcript, as very rare transcripts would not be consistently detectable at normal sampling depths. The fact that rare transcripts can be observed at all is an important advantage of SAGE over microarrays because the signal from such low-abundance transcripts would be difficult to distinguish from background noise. An equally important advantage is that SAGE does not require a priori understanding of the transcriptome in order to detect transcripts. Among the sets of transcripts found only by SAGE are those transcription units or alternative splice variants that are not currently represented on the Affymetrix chip because they are novel. Finally, an important advantage that microarray analysis has over SAGE is that microarrays are less costly. The next generation of chips for transcription profiling stands to be greatly improved by the addition of novel transcripts identified by SAGE. Bearing in mind that not all genes are currently suitable for direct comparison between SAGE and Affymetrix analyses, the intersect between chip and longSAGE (Fig. 4) sets a conservative estimate of at least 7,000 as the minimal number of different transcripts expressed during embryogenesis, a full third of *C. elegans* predicted genes.

### Embryonic Muscle and Adult Intestine: Examples of Tissue-specific Expression Profiling

For analysis of cellular function during development and growth, we are performing SAGE analysis on purified or enriched samples from specific cell populations and tissues. We use two different protocols depending on whether we are purifying tissue from embryos or from adult animals. To examine developing embryos, we used enzymatic digestion and mechanical shearing to free individual cells (Christensen et al. 2002). If cells of interest are labeled with a GFP marker, they can be isolated using a fluorescence activated cell sorter (FACS). Depending on when the GFP tag is expressed, labeled cells can be isolated directly from a fragmented embryo, or the cells can be plated and allowed to differentiate further prior to sorting. There are GFP tags available for every major tissue during development (hypodermis, nervous system, intestine, and muscle), and for subpopulations of those tissues. The isolation of developing gut cells after sorting is shown in Figure 5. We now have the means to isolate and purify analyzable quantities of specific cell populations from *C. elegans* embryos.

Our first embryonic tissue-specific SAGE library was for embryonic muscle using *myo-3*::GFP as a tissue-specific marker (Okkema et al. 1993). The *myo-3* myosin heavy-chain gene is expressed during late embryogenesis in nascent body wall muscle cells. It is first detected as the cells migrate from a lateral position to muscle quadrants located on the dorsal and ventral sides of the embryo (Epstein et al. 1993). We were able to fragment embryos and obtain individual muscle cells via FACS in sufficient quantities to extract mRNA and construct a longSAGE library. This library allowed us to detect 4,850 different genes (Table 2). Among this set of transcripts were many of the genes one expects to find, including body wall myosins, actins, and several components associated with sarcomere assembly. Although it is not surprising to detect mRNA for major structural proteins because they are expected to be relatively abundant, the sensitivity of tissue enrichment and SAGE is demonstrated by detection of the relatively rare mRNA for *hlh-1*, the nematode homolog of myoD (Krause et al. 1990). We currently have tissue-specific embryonic libraries under construction for all the major germ layers including embryonic gut, the developing nervous system, and the hypodermis.

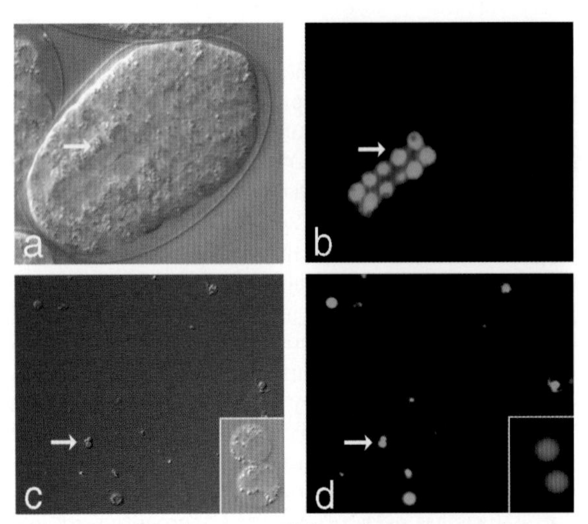


**Figure 5.** FACS of promoter::GFP marked embryonic intestine cells. (*a*) Late-stage living embryo viewed using Nomarski optics. Arrow points to double row of intestinal cells. This is a dorsal view, and anterior is to the top right corner. (*b*) Same embryo viewed using fluorescence microscopy. The promoter for the Elt-2 transcription factor is fused to GFP and acts as a marker for developing intestinal cells. (*c,d*) Disaggregated embryo cells enriched for GFP expression after FAC sorting. *c* shows cells viewed using Nomarski optics and *d* shows the same field of cells viewed using fluorescence optics. Arrow points to identical cells in both views.

Although isolation of most tissues or organs from adult worms has not been possible, hand dissection of a few adult intestines has been used to study vitellogenin synthesis (Kimble and Sharrock 1983). We used the temperature-sensitive *glp-4* mutant, *bn2*, which lacks a gonad when raised at 25°C (Beanan and Strome 1992), thereby removing one of the major internal organs of the worm and making gut dissection much easier. We constructed a SAGE library from 1,863 dissected adult intestines. As a control, a library was also made from whole *glp-4(bn2)* worms grown under identical conditions. Both show the expected distribution of transcripts, with a few transcripts present at very high levels (1,000–2,000 tags per library) and many transcripts present at 1 tag per library (Table 2) (S. McKay et al., unpubl.). The quality of the dissection is judged to be good based on the low-to-undetectable level of tags in the gut library corresponding to transcripts that are known to be expressed outside the gut (e.g., cuticular collagens, major sperm proteins, muscle proteins). A preliminary estimate of the number of different transcripts detected in the adult intestine is about 4,900 (Table 2).

There are two ways in which to view SAGE profiles from specific tissues. The first view is that it provides an enduring archive, an inventory of genes that are needed to make embryonic muscle or an adult worm gut. The second view derives from our interest in gene regulation, where there is significant value in knowing whether a particular gene is expressed only in a particular tissue. As an example of this approach, we have used the gut SAGE data. By comparing the number of tags to the gut-specific vitellogenin genes in the gut library and in the intact *glp-4(bn2)* library, we estimate that 1,000–2,000 of the genes expressed in the adult gut are gut-specific. Even at this early stage, several conclusions can be drawn. Perhaps not surprisingly, many of the genes expressed at the highest level only in the gut are digestive enzymes, in particular aspartic proteases. The *asp-1* gene encodes such a protease and has previously been demonstrated to be expressed strongly and specifically in the intestine (Tcherepanova et al. 2000); the current results certainly confirm this at >2,000 tags in the gut library. Other members of the same protease family are expressed at comparable or even higher levels. We have a special interest in gut transcription factors, and here the SAGE list is proving invaluable. As expected, transcripts for transcription factors are present at reasonably low levels (a few dozen or fewer tags per library). Tags corresponding to the gut-specific GATA-type zinc-finger factor *elt-2* (Hawkins and McGhee 1995; Fukushige et al. 1998) are present at the highest level of any recognizable transcription factor in the gut library. The library provides an intriguing list of a dozen or more transcription factors that, by the level of transcript enrichment, are judged to be gut-specific, and yet nothing is yet known about them.

Perhaps the most intriguing finding in either tissue-specific library is the presence of experimentally unverified gene models derived from computational analysis for which there is no functional annotation. These genes promise much new territory to explore. As many of these predicted genes have human homologs, they are of particular relevance to the themes of this symposium.

### Exploiting the SAGE Data Sets: Developmental Profiling and Gene Discovery

Throughout the life cycle of any organism there are dynamic changes in the expression profile of the genome. Tracking and displaying these changes in a way that leads to further understanding of individual gene function and overall gene regulation is one of the most significant challenges of 21st-century biology. We are exploring how best to mine the SAGE data for information pertaining to gene regulation and gene pathways and also to explore how best to present the data for exploitation by others.

The following are three examples of how one might track and display a large gene family through development (Fig. 6). In the first example, we examined the large zinc-finger gene family. WormBase identifies 785 potential zinc-finger-encoding transcripts. From our studies of all developmental stages, we identified 1,299 specific SAGE tags corresponding to 625 genes. Their expression profile is illustrated in Figure 6A. We have done a similar study of cuticle collagens (SAGE identifies 167 of 206 potential collagen genes annotated in WormBase, Fig. 6B) and kinases (SAGE identifies 652 of 734 potential kinase-encoding genes annotated in WormBase, Fig. 6B). In each of these large families, we were able to track at least two-thirds of the genes. Although this is already a significant achievement, we should be able to detect even more members of these large families if we use enriched tissues.

Tracking known genes is an important use of expression profiling data, but SAGE also enables gene discovery. In the embryonic SAGE and longSAGE libraries, a total of 1,070 14-bp tags and 2,730 21-bp tags map to unique locations in the genomic DNA but do not map to any known nuclear, mitochondrial, or rRNA transcript. The larger number of unambiguous long tags demonstrates the resolving power of longSAGE, which resolved the ambiguity of one-third of the ambiguous 14-bp tags in this class. The SAGE protocol involves DNase treatment of an RNA sample, so tags that map to genomic DNA but not to predicted transcripts can be used to infer novel transcribed sequences, or undocumented alternative splice variants, or UTRs of known genes. For example, Jones et al. (2001) identified two novel transcribed sequences possessing telomeric repeat-like sequences and no obvious open reading frame that are present at high abundance in dauer larvae but not in other life stages. Among the 14-bp "genomic" tags, 445 map to introns of known genes; this could be explained in large part by previously unknown exons or, more rarely, by small genes nested entirely within the intron. The remaining 625 genomic tags map to regions for which there are no annotated transcribed sequences, and likely represent novel transcription units or, if they are near known genes, alter-

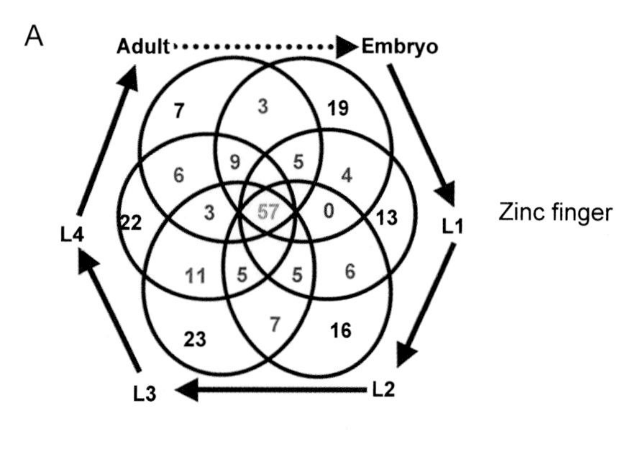


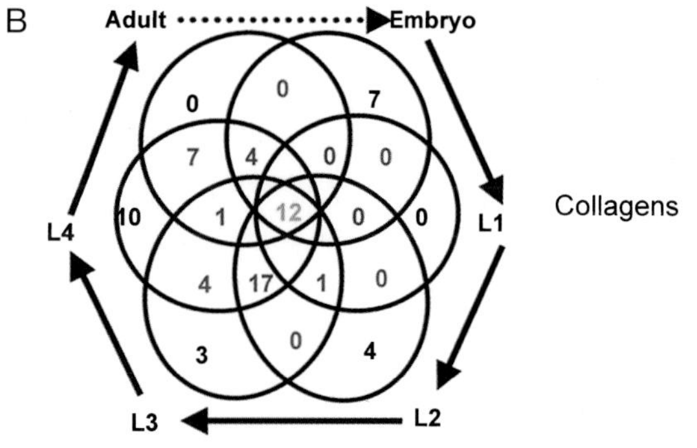


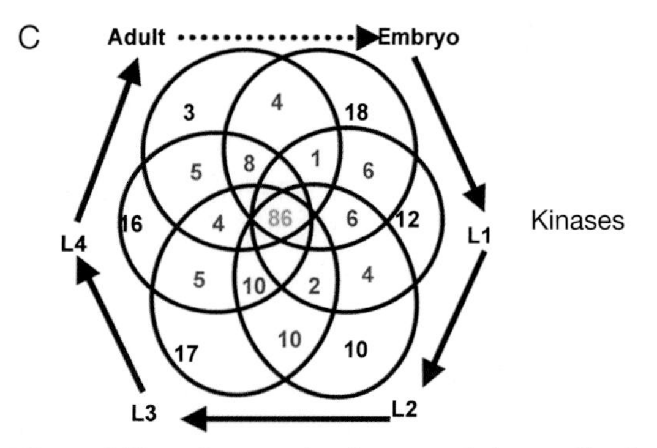


**Figure 6.** Venn diagrams showing transcription profiles for six stages of *C. elegans* development. Only specific SAGE tags were considered. Transcripts are counted by virtue of presence or absence rather than relative abundance. Putative alternative splice variants identified by different SAGE tags for the same gene are counted separately. The criterion for presence of a transcript is the observation of at least one tag of high-quality sequence (phred40). Regions of overlap indicate the number of transcripts common to all affected stages. Numbers in black, red, blue, and green correspond to the count of transcripts observed in one, two, three, or all developmental stages. (*A*) Zinc-finger genes. (*B*) Collagen genes. (*C*) Kinase genes.

native polyadenylation signals or UTRs not fully represented by ESTs. The improved resolution of longSAGE allowed us to identify 1,257 tags that map within introns of known genes and 1,473 that occur in intergenic regions. Many of the novel tags overlap with regions of sequence similarity in orthologous regions of the genome of the related nematode *C. briggsae,* suggesting conserved regions may have functional significance. Comparative genomics and directed RT-PCR experiments will be required to characterize the novel transcribed sequences whose presence we infer with genomic SAGE tags.

## CONCLUSIONS

1. The *C. elegans* and human genomes are estimated to have at least 4,300 orthologous gene pairs.
2. The function of many of these genes is unknown in either organism, but the simplicity of nematode anatomy coupled with powerful genetic tools should contribute to the understanding of their function.
3. There are temporal and tissue-specific promoters, but few or no single-cell promoters. Individual cell identity within a tissue would then appear to result from combinatorial overlaps.
4. SAGE technology confirms the expression of at least 14,600 genes in the nematode. This number will increase as the technology is refined.
5. SAGE reveals multiple different tags for half of the genes in *C. elegans*, suggesting that alternative splicing of genes is common in this organism.
6. Many of the SAGE tags map to unannotated regions of the nematode genome and thus may identify new genes.
7. The studies outlined here using promoter::GFP constructs and SAGE will lead to the establishment of a gene expression database that can be interrogated to understand temporal and spatial patterning during development.
8. We have established the minimum number of genes required for nematode embryogenesis, 7,000, and the number of genes required to determine and maintain the function of a specific tissue, about 5,000.

## ACKNOWLEDGMENTS

We thank David Miller III and his colleagues at Vanderbilt for providing us with their protocol for fragmenting embryos. Funding for this project was provided by Genome British Columbia and Genome Canada to D.L.B., S.J.M.J., M.M., and D.G.M. J.T. is the recipient of an International Prize Travelling Research Fellowship from the Wellcome Trust. S.J. and M.M. are Michael Smith Foundation Health Research Scholars. Additional funding was provided by National Institutes of Health operating grants AG-12689 and GM-60151 to D.L.R. and a Canadian Institute of Health Research grant to J.M.

## REFERENCES

Beanan M.J. and Strome S. 1992. Characterization of a germ-line proliferation mutation in *C. elegans*. *Development* **116:** 755.

Blumenthal T. 1995. Trans-splicing and polycistronic transcription in *Caenorhabditis elegans*. *Trends Genet.* **11:** 132.

Blumenthal T., Evans D., Link C.D., Guffanti A., Lawson D., Thierry-Mieg J., Thierry-Mieg D., Chiu W.L., Duke K., Kiraly M., and Kim S.K. 2002. A global analysis of *Caenorhabditis elegans* operons. *Nature* **417:** 851.

Bürglin T.R. 2000. A two-channel four-dimensional image recording and viewing system with automatic drift correction. *J. Microsc.* **200:** 75.

Cassata G., Kagoshima H., Pretot R.F., Aspock G., Niklaus G., and Bürglin T.R. 1998. Rapid expression screening of *Caenorhabditis elegans* homeobox open reading frames using a two-step polymerase chain reaction promoter-gfp reporter construction technique. *Gene* **212:** 127.

*C. elegans* Sequencing Consortium. 1998. Genome sequence of the nematode *C. elegans:* A platform for investigating biology. *Science* **282:** 2012.

Chalfie M. 1995. Green fluorescent protein. *Photochem. Photobiol.* **62:** 651.

Chalfie M., Tu Y., Euskirchen G., Ward W.W., and Prasher D.C. 1994. Green fluorescent protein as a marker for gene expression. *Science* **263:** 802.

Christensen M., Estevez A., Yin X., Fox R., Morrison R., McDonnell M., Gleason C., Miller D.M., III, and Strange K. 2002. A primary culture system for functional analysis of *C. elegans* neurons and muscle cells. *Neuron* **33:** 503.

Clark D.V., Johnsen R.C., McKim K.S., and Baillie D.L. 1990. Analysis of lethal mutations induced in a mutator strain that activates transposable elements in *Caehorhabditis elegans*. *Genome* **33:** 109.

Ellis H.M. and Horvitz H.R. 1986. Genetic control of programmed cell death in the nematode *C. elegans*. *Cell* **44:** 817.

Epstein H.F., Casey D.L., and Ortiz I. 1993. Myosin and paramyosin of *Caenorhabditis elegans* embryos assemble into nascent structures distinct from thick filaments and multifilament assemblages. *J. Cell Biol.* **122:** 845.

Ewing B. and Green P. 1998. Base-calling of automated sequencer traces using phred. II. Error probabilities. *Genome Res.* **8:** 186.

Fire A. 1994. A four dimensional digital archiving system for cell lineage tracing and retrospective embryology. *Comput. Appl. Biosci.* **10:** 443.

Fukushige T., Hawkins M.G., and McGhee J.D. 1998. The GATA-factor elt-2 is essential for formation of the *Caenorhabditis elegans* intestine. *Dev. Biol.* **198:** 286.

Harris T.W., Lee R., Schwarz E., Bradnam K., Lawson D., Chen W., Blasier D., Kenny E., Cunningham F., Kishore R., Chan J., Muller H.M., Petcherski A., Thorisson G., Day A., Bieri T., Rogers A., Chen C.K., Spieth J., Sternberg P., Durbin R., and Stein L.D. 2003. WormBase: A cross-species database for comparative genomics. *Nucleic Acids Res.* **31:** 133.

Hawkins M.G. and McGhee J.D. 1995. elt-2, a second Gata factor from the nematode *Caenorhabditis elegans*. *J. Biol. Chem.* **270:** 14666.

Hedgecock E.M., Sulston J.E., and Thomson J.N. 1983. Mutations affecting programmed cell deaths in the nematode *Caenorhabditis elegans*. *Science* **220:** 1277.

Hill A.A., Hunter C.P., Tsung B.T., Tucker-Kellogg G., and Brown E.L. 2000. Genomic analysis of gene expression in *C. elegans*. *Science* **290:** 809.

Hobert O. 2002. PCR fusion-based approach to create reporter gene constructs for expression analysis in transgenic *C. elegans*. *Biotechniques* **32:** 728.

Holt S.J. and Riddle D.L. 2003. SAGE surveys *C. elegans* carbohydrate metabolism: Evidence for an anaerobic shift in the long-lived dauer larva. *Mech. Ageing Dev.* **124:** 779.

Hope I.A. 1991. "Promoter trapping" in *Caenorhabditis elegans*. *Development* **113:** 399.

Jones S.J.M., Riddle D.L., Pouzyrev A.T., Velculescu V.E., Hillier L., Eddy S.R., Stricklin S.L., Baillie D.L., Waterston R., and Marra M.A. 2001. Changes in gene expression associated with developmental arrest and longevity in *Caenorhabditis elegans*. *Genome Res.* **11:** 1346.

Kelly W.G., Xu S., Montgomery M.K., and Fire A. 1997. Distinct requirements for somatic and germline expression of a generally expressed *Caenorhabditis elegans* gene. *Genetics* **146:** 227.

Kent W.J. and Zahler A.M. 2000. The intronerator: Exploring introns and alternative splicing in *Caenorhabditis elegans*. *Nucleic Acids Res.* **28:** 91.

Kim S.K., Lund J., Kiraly M., Duke K., Jiang M., Stuart J.M., Eizinger A., Wylie B.N., and Davidson G.S. 2001. A gene expression map for *Caenorhabditis elegans*. *Science* **293:** 2087.

Kimble J. and Sharrock W.J. 1983. Tissue-specific synthesis of yolk proteins in *Caenorhabditis elegans*. *Dev. Biol.* **96:** 189.

Krause M., Fire A., Harrison S.W., Priess J., and Weintraub H. 1990. CeMyoD accumulation defines the body wall muscle cell fate during *C. elegans* embryogenesis. *Cell* **63:** 907.

Lynch A.S., Briggs D., and Hope I.A. 1995. Developmental expression pattern screen for genes predicted in the *C. elegans* genome sequencing project. *Nat. Genet.* **11:** 309.

Mello C. and Fire A. 1995. DNA transformation. *Methods Cell Biol.* **48:** 451.

Mello C.C., Kramer J.M., Stincomb D., and Ambros V. 1991. Efficient gene transfer in *C. elegans:* Extrachromosomal maintenance and integration of transforming sequences. *EMBO J.* **10:** 3959.

Okkema P.G., Harrison S.W., Plunger V., Aryana A., and Fire A. 1993. Sequence requirements for myosin gene expression and regulation in *Caenorhabditis elegans*. *Genetics* **135:** 385.

Pleasance E.D., Marra M.A., and Jones S.J. 2003. Assessment of SAGE in transcript identification. *Genome Res.* **13:** 1203.

Reinke V. 2002. Functional exploration of the *C. elegans* genome using DNA microarrays. *Nat. Genet.* **32:** 541.

Reinke V., Smith H.E., Nance J., Wang J., Van Doren C., Begley R., Jones S.J., Davis E.B., Scherer S., Ward S., and Kim S.K. 2000. A global profile of germline gene expression in *C. elegans*. *Mol. Cell* **6:** 605.

Remm M., Storm C.E., and Sonnhammer E.L. 2001. Automatic clustering of orthologs and in-paralogs from pairwise species comparisons. *J. Mol. Biol.* **314:** 1041.

Riddle D.L., Blumenthal T., Meyer B.J., and Preiss J.R. 1997. *C. elegans II*. Cold Spring Harbor Laboratory Press, Cold Spring Harbor, New York.

Rozen S. and Skaletsky H. 2000. Primer3 on the WWW for general users and for biologist programmers. *Methods Mol. Biol.* **132:** 365.

Saha S., Sparks A.B., Rago C., Akmaev V., Wang C.J., Vogelstein B., Kinzler K.W., and Velculescu V.E. 2002. Using the transcriptome to annotate the genome. *Nat. Biotechnol.* **20:** 508.

Schnabel R., Hutter H., Moerman D.G., and Schnabel H. 1997. Assessing normal embryogenesis in *Caenorhabditis elegans* using a 4D microscope: Variability of development and regional specification. *Dev. Biol.* **184:** 234.

Schuler G.D. 1997. Sequence mapping by electronic PCR. *Genome Res.* **7:** 541.

Seydoux G. and Schedl T. 2001. The germline in *C. elegans:* Origins, proliferation, and silencing. *Int. Rev. Cytol.* **203:** 139.

Stein L.D. and Thierry-Mieg J. 1998. Scriptable access to the *Caenorhabditis elegans* genome sequence and other ACEDB databases. *Genome Res.* **8:** 1308.

Stein L., Sternberg P., Durbin R., Thierry-Mieg J., and Spieth J. 2001. WormBase: Network access to the genome and biology of *Caenorhabditis elegans*. *Nucleic Acids Res.* **29:** 82.

Stein L.D., Bao Z., Blasiar D., Blumenthal T., Brent M.R., Chen N., Chinwalla A., Clarke L., Clee C., Coghlan A., Coulson A., D'Eustachio P., Fitch D.H., Fulton L.A., Fulton R.E., Griffiths-Jones S., Harris T.W., Hillier L.W., Kamath R., Kuwabara P.E., Mardis E.R., Marra M.A., Miner T.L., Minx P., and Mullikin J.C., et al. 2003. The genome sequence of *Caenorhabditis briggsae:* A platform for comparative genomics. *PLoS Biol.* **1:** E45.

Stinchcomb D.T., Shaw J.E., Carr S.H., and Hirsh D. 1985. Extrachromosomal DNA transformation of *Caenorhabditis elegans*. *Mol. Cell. Biol.* **5:** 3484.

Stuart J.M., Segal E., Koller D., and Kim S.K. 2003. A gene-coexpression network for global discovery of conserved genetic modules. *Science* **302:** 249.

Sulston J.E. 1976. Post-embryonic development in the ventral cord of *Caenorhabditis elegans*. *Philos. Trans. R. Soc. Lond. B Biol. Sci.* **275:** 287.

Sulston J.E. and Horvitz H.R. 1977. Postembryonic cell lineages of the nematode *Caenorhabditis elegans*. *Dev. Biol.* **82:** 41.

Sulston J.E., Schierenberg E., White J.G., and Thomson J.N. 1983. The embryonic cell lineage of the nematode *Caenorhabditis elegans*. *Dev. Biol.* **100:** 64.

Tcherepanova I., Bhattacharyya L., Rubin C.S., and Freedman J.H. 2000. Aspartic proteases from the nematode *Caenorhabditis elegans*. Structural organization and developmental and cell-specific expression of asp-1. *J. Biol. Chem.* **275:** 26359.

Thomas C.F. and White J.G. 1998. Four dimensional imaging: The exploration of space and time. *Trends Biotechnol.* **16:** 175.

Velculescu V.E., Zhang L., Vogelstein B., and Kinzler K.W. 1995. Serial analysis of gene expression. *Science* **270:** 484.

White J.G., Southgate E., Thomson J.N., and Brenner S. 1986. The structure of the nervous system of the nematode *Caenorhabditis elegans*. *Philos. Trans. R. Soc. Lond. B Biol. Sci.* **314:** 1.

# Building Comparative Maps Using 1.5x Sequence Coverage: Human Chromosome 1p and the Canine Genome

R. GUYON,* E.F. KIRKNESS,† T.D. LORENTZEN,‡ C. HITTE,* K.E. COMSTOCK, ‡ P. QUIGNON,* T. DERRIEN,* C. ANDRÉ,* C.M. FRASER, † F. GALIBERT,* AND E.A. OSTRANDER‡

*UMR 6061 CNRS, Génétique et Développement, Faculté de Médecine, 35043 Rennes Cédex, France; †The Institute for Genomic Research, Rockville, Maryland 20850; ‡Clinical and Human Biology Divisions, Fred Hutchinson Cancer Research Center, D4-100, Seattle, Washington 98109-1024

The ability to test the association between phenotype and genotype within biological systems of interest is limited by the degree to which the genome map of any model system can be rigorously aligned with the reference human and mouse maps. Although extensive reciprocal chromosome paint studies have outlined the general evolutionary relationships between the chromosomes of dog and other mammals, details of the conserved synteny that exist between the human and dog genomes are still lacking. Similarly, there is a paucity of information regarding the relationship between model genomes.

In the case of the canine genome, reciprocal chromosome painting has enabled investigators to broadly establish the evolutionary relationship between canine chromosomes and cytogenetic bands defining human chromosome arms (Breen et al. 1999b; Yang et al. 1999). These data suggest the existence of 68–73 conserved regions (Breen et al. 1999b; Yang et al. 1999; Sargan et al. 2000), and radiation hybrid data suggest a total of 76 conserved segments between human and dog (Guyon et al. 2003). Radiation hybrid data demonstrate that several canine chromosomes, such as *Canis familiaris* chromosomes (CFA) 8, 12, 22–24, and most of the smaller canine chromosomes are apparently composed of a single continuous section of the human genome; others, such as CFA1 through CFA7, retain two to four portions of several distinct human chromosomes, and still others, such as CFA15, correspond to as many as five HSA fragments (Guyon et al. 2003). This fact, combined with the nearly 1600 microsatellite markers now ordered on the canine radiation hybrid (RH) map (Parker et al. 2001; Guyon et al. 2003), ensures that genome-wide scans on informative canine families can be carried out with relative ease, and the corresponding chromosome arm in the human genome can be quickly and correctly identified.

However, within the large syntenic regions that define each canine chromosome, there is a paucity of mapped genes that severely limits the ability to move from a general region of interest to selection of specific candidate genes. Indeed, only 900 canine-specific genes have been placed on the most recent version of the canine RH map (Guyon et al. 2003), and still fewer on the meiotic linkage map (Parker et al. 2001). Overall, the distribution of gene-based markers averages only one per 3 Mb. Although these data support the hypothesis that blocks of several megabases are well conserved throughout the canine genome, the number of mapped genes within any single block is insufficient for assigning breakpoints of conserved synteny. This limits the degree to which initial findings of linkage in canine families can be followed by successful positional cloning efforts, and reduces the utility of the human genome sequence for tackling problems of interest in other mammalian systems. Thus, it remains a priority of the canine genome-mapping community to add more gene-based markers to the canine map.

In this study, we have tested the hypothesis that 1x sequence coverage of the canine genome is sufficient to permit identification and mapping of the canine orthologs of most human genes. Toward this aim, we focused on the human chromosome 1p arm (HSA1p), which is known to contain several disease-associated genes of interest. The 1x sequence was used to identify canine orthologs of 158 genes from HSA1p, and RH mapping of 120 of them allowed production of a dense comparative map between human and dog in this region of interest. Human HSA1p corresponds to seven conserved segments within five chromosomal regions (CFA 2, 5, 6, 15, and 17) with gene orders and limits well defined. The study presented here, therefore, illustrates the power of combining 1.5x shotgun sequence data and a one megabase resolution RH map for building a comparative map with the human sequence. The net result is a unified resource suitable for studies aimed at positional cloning of mapped loci, candidate gene assessment, and evolutionary analyses. We suggest that, for many additional genomes, this will be a powerful and economical approach for characterizing genome structure and evolutionary relationships.

## RESULTS

Starting from a set of 187 HSA1p genes, fragments of 158 putative orthologs were retrieved from the canine genomic sequence data. For 144 genes, canine-specific primers were designed and 126 were successfully typed on the RHDF5000-2 canine RH panel (Vignaux et al. 1999). RH data from the markers were computed with the latest 3270-marker RH map (Guyon et al. 2003) using the MultiMap and Traveling Salesman (TSP)/CONCORDE softwares (Matise et al.1994; Agarwala et al. 2000). Of these 126 gene markers, 120 could be incorporated in 5 of

**Table 1.** Map Statistics of Conserved Segments between HSA1p and Canine Chromosomes

| HSA1p | CFA | Canine CS size (TSP units) | Number of canine markers | Number of mapped positions | Number of canine gene-based markers | Human CS limits (Mb) | Human CS size (Mb) | Total number of human genes in the CS[a] | Number of anchor sites | Average distance between anchor sites in human (Mb) |
|---|---|---|---|---|---|---|---|---|---|---|
| CS I | 5 | 567 | 19 | 14 | 13 | 1–9.8 | 8.8 | 123 | 13 | 0.7 |
| CS V | | 1886 | 36 | 34 | 21 | 52.6–66.4 | 13.8 | 91 | 24 | 0.6 |
| CS II | 2 | 1691 | 48 | 39 | 35 | 10.4–31.2 | 20.8 | 332 | 35 | 0.6 |
| CS III | 15 | 1737 | 20 | 20 | 12 | 35–42.2 | 7.2 | 97 | 10 | 0.7 |
| CS IV | | 1424 | 22 | 18 | 14 | 42.6–51.1 | 8.7 | 107 | 14 | 0.6 |
| CS VI | 6 | 1320 | 78 | 41 | 52 | 67.1–110.4 | 43.3 | 273 | 54 | 0.8 |
| CS VII[b] | 17 | 293 | 18 | 10 | 14 | 110.9–123.5 | 12.6 | 104 | 13 | 1.0 |
| TOTAL | | 8918 | 241 | 176 | 161 | | 115.2 | 1127 | 163 | 0.7 |

[a]Numbers were retrieved from the NCBI Web site: http://www.ncbi.nih.gov/mapview/maps.cgi?org=hum&chr=1.
[b]The numbers corresponding to CS VII only refers to HSA1p orthologous markers.

the 38 canine autosomes, and 6 markers remained unlinked (see http://www-recomgen.univ-rennes1.fr/doggy.html and http://www.fhcrc.org/science/dog_genome/dog.html). The five chromosomes that are shown by this analysis to correspond to HSA1p are: CFA 2, 5, 6, 15, and 17. Markers were then ordered on each chromosome using the TSP/CONCORDE software. These 120 gene markers were added to the 41 previously mapped (Guyon et al. 2003), bringing the total number of canine gene markers to 161 in these five chromosomal regions (Table 1). However, for four of those previously mapped genes, no human counterparts could be found on the NCBI Build 31 database of the human sequence, thus 157 gene markers constitute informative anchor sites between the human and dog genomes. The current comparative map between HSA1p and the canine orthologous regions is shown on Figure 1.

The total number of canine gene-based markers assigned to each of the five chromosomal regions orthologous to HSA1p ranges from 52 on CFA6 to 14 on CFA17, whereas the total number of canine markers ranges from 78 on CFA6 to 18 on CFA17 (Table 1). On CFA6, however, the 78 markers are mapped to 41 unique positions, and on CFA17 the 18 markers are mapped to 10 unique positions. The increase in the number of markers that are co-positioned when the total number of markers mapped to some regions increases reflects the limited resolution of the RHDF5000-2 panel. As indicated in previous studies (Breen et al. 2001; Guyon et al. 2003), the panel resolution has been estimated to be ~600 kb; therefore, markers present in an interval of 1 Mb cannot be accurately ordered with respect to their immediate neighbors (Priat et al. 1998).

As shown in Figure 1, we noted different degrees of conserved homologous gene association as defined by the First International Workshop on Comparative Genome Organization (Andersson et al. 1996). Conserved synteny, defined as the association of two or more homologous genes in two species, regardless of gene order or interspacing of noncontiguous segments, has been observed for CFA 2, 5, 6, 15, and 17. Conserved segments are defined as the syntenic association of two or more homologous and contiguous genes not interrupted by different chromosome segments in either species. The HSA1p orthologous region of CFA 15 is split in two conserved segments by an asyntenic fragment. Conversely, the CFA 5 orthologous region of HSA1p is made of two conserved segments separated by an asyntenic fragment. The HSA1p orthologous regions of CFA 2, 6, and 17 are made of only one conserved segment, with a part of the CFA17 conserved segment being inverted. Conserved order is defined as the demonstration that three or more homologous genes lie on one chromosome in the same order in two species. In this study, conserved orders were observed in the conserved segments of CFA 5, 6, and 15. Indeed, a detailed screen for gene orders allowed us to split both CFA 2 (CS II ) and CFA17 (CS VII) conserved segments into two distinct blocks where gene order is head to head.

A detailed analysis of CFA 5 and 15 shows that the syntenic association of homologous genes is contiguous

**Figure 1.** Comparative map of HSA1p and CFA 2, 5, 6, 15, and 17. HSA1p and part of HSA1q are symbolized by a vertical bar, graduated every 10 Mb. The anchor sites, indicated by lines between HSA1p /HSA1q and markers placed on the canine RH map, allow one to define conserved segments between human and dog (CS I to CS VII). The entire CFAs are symbolized by vertical bars in which colored boxes delineate the human evolutionarily conserved segments determined by reciprocal chromosome painting (Breen et al. 1999a,b; Yang et al. 1999). Numbers indicate HSA origin of the conserved segments. The orthologous position of the HSA1p/1q chromosome on RH maps and CFAs (*red-colored boxes*) is indicated by brackets. Note that canine maps are inverted with respect to their chromosomal positions. For each CFA, the RH map shows the statistical support symbolized by horizontal bars of variable lengths reflecting the five maps automatically delivered by TSP/CONCORDE. In blue at the top of the RH map, a scale of 0 to 100% reflects the confidence level for the position of each marker. In scrambled regions, markers occupying several positions are bracketed in order to narrow the problematic region into smaller intervals. Cumulated distances between RH markers are reported in TSP units at the end of the horizontal bars. Marker names indicated in red correspond to gene-based markers (Type I); other markers are colored black. Markers in bold indicate genes or noncoding markers that constitute anchor sites. Markers in italics and outside brackets belong to other HSA orthologous regions. Characteristics of all markers are available at (http://www-recomgen.univ-rennes1.fr/doggy.html) (http://www.fhcrc.org/science/dog_genome/dog.html).

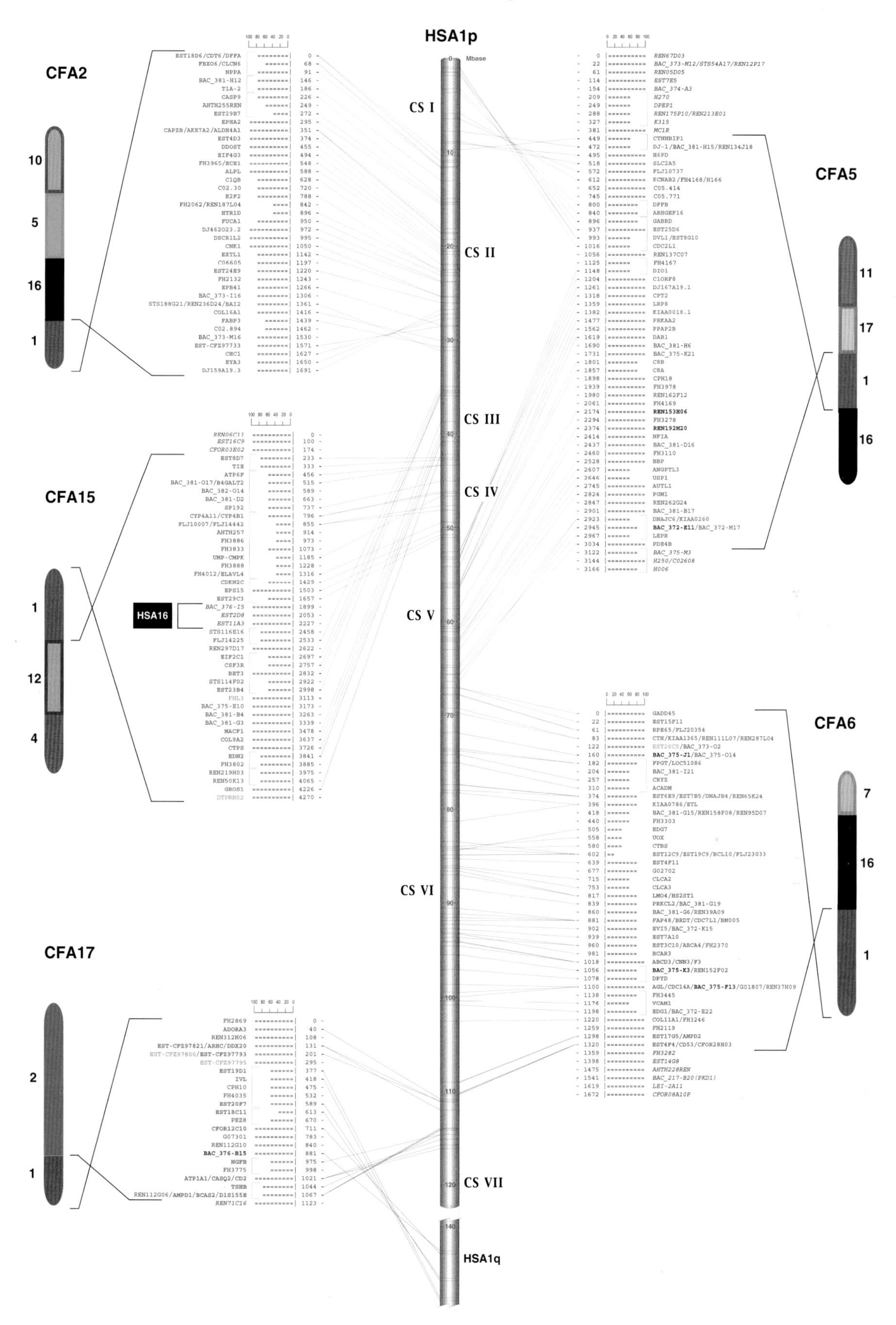


**Figure 1.** (*See facing page for legend.*)

only in one of the two species (Fig. 1). For CFA5, CS I and CS V, separated by 42.6 Mb in the human sequence, appear as a single contiguous block in dog. CS I is identified by 13 anchor sites and spans 8.8 Mb on HSA1p. CS V is identified by 24 anchor sites and spans 13.8 Mb (Table 1). The gene order inside each block is conserved between human and dog, but CS I is inverted relative to HSA1p. For CFA15, the HSA1p orthologous region is split into two conserved segments of 14 and 10 anchor sites by an asyntenic region 328 TSP units long, and composed of three markers orthologous to a 0.2-Mb interval of HSA16. The gene orders in these two conserved segments are in the same orientation, but their relative positions reveal a transposition event in the dog genome (Fig. 1).

CS VII on CFA17 is composed of two blocks of 5 and 14 anchor sites, respectively, the second one overlapping the centromeric region of HSA1 and being in inverted orientation, includes 6 genes of HSA1q. Although gene orders within each of these blocks cannot be assessed with certainty because of the RH panel resolution limit, this organization reveals an inversion event including the centromeric region of HSA1 or in the CFA17 orthologous region. The whole conserved segment between CFA17 and HSA1 represents 1067 TSP units in the canine map. In human, it includes the 21 Mb of the centromere of HSA1 (HSA1p11.1-q12) and spans 39.8 Mb.

In contrast to the above, we see little variation associated with CFA2 and CFA6, where 35 and 54 anchor sites have been mapped, respectively. The association of homologous genes in human and dog is contiguous and in accordance with the definition of conserved segments. Finally, we note that the gene order between HSA1p and CFA6 (CS VI) is entirely conserved. However, CS II is subdivided in two blocks of conserved order, harboring 23 and 7 anchor sites, respectively, highlighting an additional rearrangement in one of the two species.

## DISCUSSION

Using a simple statistical model, it can be estimated that the 1.5x coverage of the dog genome will provide 70% of the genome sequence, with an average gap length of ~480 bases (Lander and Waterman 1988). It can also be estimated that sequences containing 100 bp of an exon (or exon fragment) will be sufficient to identify dog orthologs of most human genes. The probability of a specific 100-base fragment of the genome occurring entirely within a single sequence read of 576 bases is only 0.58. However, most human genes appear to be composed of at least 4 exons (Venter et al. 2001), and given the known similarity in gene structure between humans and canids (see, e.g., Szabo et al. 1996; Credille et al. 2001; Haworth et al. 2001), the same is likely to be true for dog. Consequently, the probability that at least one 100-base exon fragment from a gene is contained within the genomic sequence data can be estimated as >0.95. It has not yet been determined for what proportion of human genes 1:1 orthology can be detected in the dog genome. For mouse, this value has been estimated to be ~80% (Okazaki et al. 2002; Waterston et al. 2002). If dogs and humans share a similar number of orthologous genes, we can estimate that the dog genomic sequence data will yield at least one orthologous exon fragment for ~80% of human genes. In this study, fragments of putative dog orthologs were identified for 158 of 187 (84%) selected human genes. Recently, a more comprehensive analysis has indicated that 79% of all annotated human genes (and 96% of those that have detectable orthologs in mouse) are represented by orthologous dog sequences in the 1.5x coverage (Kirkness et al. 2003).

If the primary objective of a sequencing project is to generate gene-based markers for RH mapping,1.5x sequence coverage of a genome offers several advantages over large collections of ESTs. Unlike cDNA libraries, the representation of genes is unaffected by cellular expression levels, and identification of orthologous exons is not biased by the length of 3′-untranslated mRNA. In addition, the low but significant conservation of intronic sequences between species is useful for distinguishing between paralogous sequences that share substantial sequence identity within exons.

The most recent iteration of the canine RH map (Guyon et al. 2003) featured 870 markers for which orthologous sequences have been identified on the human genome. Although the HSA1p orthologous regions were shown to correspond to five canine chromosomes, CFA 2, 5, 6, 15, and 17 in the previous RH map, by increasing more than fourfold the number of markers, this work clearly delineates the gene order and breakpoints for seven conserved segments. The largest increase in resolution is in the HSA1p orthologous region of CFA5 (CS I and V), which now contains 34 genes compared to 4 in the previous version of the canine map.

This comparative map allows us to characterize more precisely the conserved segments orthologous to HSA1p that were previously identified by reciprocal chromosome painting studies (Breen et al. 1999a; Yang et al. 1999; Sargan et al. 2000) on CFA5, CFA15, and CFA17 (Fig. 1). On CFA5, although contiguous in dog, two conserved segments (I and V) are split in human. On CFA15, the region is split by a novel asyntenic fragment of HSA16, as previously shown by Guyon (Guyon et al. 2003), leaving two conserved segments (III and IV) harboring inverted positions in human versus dog. Finally, on CFA17, the region is split into blocks but constitutes a unique conserved segment. The two remaining canine chromosomal regions (CFA2 and CFA6) each constitute a unique conserved segment.

The seven regions span 115 Mb of the 123-Mb HSA1p arm, indicating that for roughly 8 Mb of HSA1p, no canine counterparts have yet been mapped. Between those conserved segments, six regions that contain the breakpoints of interest range in length from 0.3 to 3.8 Mb, and represent a total of 7.3 Mb. Together, those intervals contain 113 human genes (http://www.ncbi.nih.nlm.gov/mapview) ranging from 58 genes in the 3.8-Mb region between CS II and III to 6 genes in the 0.4-Mb region between CS III and IV. In addition, 42 genes are present in the 1 Mb most telomeric region of HSA1p above CS I. Despite the high density of anchor sites along HSA1p

(1/700 kb), eight intervals greater than two megabases with no mapped genes in dog still remain inside conserved segments. The two largest span 7.3 Mb in CS VII and 6.5 Mb in CS VI and contain 40 and 50 genes, respectively. The other six intervals spanning less than 3 Mb contain from 3 to 35 genes. These intervals are likely to contain additional conserved segments that will be resolved by RH mapping additional genes retrieved from the 1.5x canine sequence. Additional sequencing can be done to more clearly delineate the breakpoints.

This comparative map has allowed us to compare gene orders in human and dog, and to comment on possible intrachromosomal rearrangements. In five of the seven conserved segments, the gene orders are strictly conserved, whereas CS II on CFA2 contains a small inverted segment. Despite the fact that gene order cannot be precisely assessed in CS VII, the two blocks probably harbor inverted gene order as a consequence of the chromosomal inversion that brought HSA1q orthologous genes between HSA1p orthologous blocks.

On the current canine RH map, some local discrepancies leading to an artifactual inversion of local orders are observed. This is likely due to the resolution limit of the RHDF5000-2 panel, estimated to be about 600 kb (Vignaux et al. 1999). A related problem, the high number of colocalized anchor sites, especially on CFA6 (CS VI), highlights the saturation of the canine HSA1p orthologous map in discrete regions. The use of a higher-resolution canine RH panel would allow us to circumvent both problems. Some local discrepancies in the comparative map are, however, likely due to slight distortions in the human sequence assembly, typically observed when updating the human localization of anchor sites from one NCBI Build to the next. Indeed, according to NCBI build 31, the HSA1p arm is still composed of at least 56 contigs, separated by gaps of unknown size and sequence.

To date evolutionary breakpoints between human and dog identified in HSA1p and to establish in which lineages such events happened, the comparison of the conservation between HSA1p and various mammals is very instructive. The ancestral genome of primates and carnivores was likely a low-numbered, largely metacentric genome that evolved at a slow rate to human (11 steps), cat (6 steps), mink (10 steps), and seal (8 steps) (O'Brien et al. 1999). Chromosome rearrangements can be used as characters for phylogenetic reconstruction following the principle of outgroup comparison (Yang et al. 2000). The HSA1p region appears to be entirely syntenic between human and cat (Murphy et al. 2000). This indicates that the split into five chromosomal segments in dog occurred in the Canoidea lineage following the Canoidea and Feloidea radiation, some 60 million years ago (Wayne 1993). Yang and colleagues (Yang et al. 1999) showed by reciprocal chromosome painting that HSA1p is also split in five chromosomal segments in the red fox, indicating that these evolutionary events occurred before the dog and red fox divergence, some ten million years ago (Wayne 1993). This time estimate could be refined by the comparison of genomic rearrangements between human and other Canoidea superfamily members, provided an appropriate comparative map with human is well established.

Whereas the mammal radiations generally display a slow rate of chromosome exchange, ~1–2 exchanges per 10 million years, certain lineages show a more rapid pattern of chromosome change. Consider, for example, the primate lineage, in which the genome is mostly conserved between human, chimpanzee, and macaque, while it is dramatically shuffled in the gibbon lineage (O'Brien et al. 1999; O'Brien and Stanyon 1999). Similarly, in the carnivore lineage, the dog, as well as other canids, has an appreciably rearranged genome relative to the ancestral carnivore organization, indicating a high rate of chromosome exchange (Wayne et al. 1987; Wayne 1993; Yang et al. 1999). Although only HSA1p orthologous regions are considered here, this study suggests similar findings.

In this comparative map, an HSA16 orthologous region is found contiguous to or within HSA1p orthologous regions in four of five instances. The HSA16 conserved segments are found contiguous to HSA1p in CFA2, 5, and 6, whereas a small conserved segment is found inside the HSA1p region in CFA15. In Carnivora, this association is not found in cat, arguably because its genome is less rearranged and very close to human in this region (Murphy et al. 2000; Yang et al. 2000). We have no explanation for this association; it may be a consequence of poorly understood evolutionary forces, or merely a coincidence.

Detailed comparative maps between closely and distantly related species are of great interest in understanding the evolutionary relationships between species, families, and orders. The study presented here illustrates the joint utility of the 1.5x shotgun sequence approach and a relatively dense RH map for building a comparative map with the human genome. The net result is a unified resource that can facilitate studies aimed at genetic mapping, positional cloning of mapped loci, and evolutionary studies of species of interest.

## METHODS

### Selection of Orthologs Derived from Canine 1.5x Sequence

Sequence from the canine genome was derived as follows: Genomic DNA from a male standard poodle was used to prepare plasmid libraries of small- and medium-sized inserts (~2 kb and ~10 kb, respectively). End-sequencing of clones from each library was conducted at Celera Genomics as described previously (Venter et al. 2001) and yielded 3.42 million reads (86.7% paired) from 2 kb clones, and 2.81 million reads (86.4% paired) from 10 kb clones. Read quality was evaluated in 50-bp windows using Paracel's TraceTuner, with each read trimmed to include only those consecutive 50-bp segments with a minimum mean accuracy of 97%. End windows (both ends of the trace) of 1, 5, 10, 25, and 50 bases were trimmed to a mean accuracy of 98%. Every read was checked further for vector and contaminant matches of 50 bases or more. The finished sequence data consist of 6.22

million reads (mean read length, 576 bases), representing ~1.5x coverage of the 3-Gb haploid canine genome (Vinogradov 1998).

For 187 genes known to span HSA1p, the associated peptide sequence was searched against the complete collection of dog reads using tblastn. For each peptide, all homologous dog reads that were identified by the blast searches were assembled at high stringency (99% nucleotide identity) using TIGR Assembler (http://www.tigr.org/softlab/assembler/). Each assembly, or unassembled read, was then searched back against the Ensembl (release 1.1) collection of confirmed cDNAs and peptides (using blastn and blastx, respectively). If the highest scoring hits (for both the DNA- and protein-sequence searches) were to the gene that was used originally for searching, the assembly was considered a fragment of a putative ortholog. The coordinates of each human gene on HSA1p were obtained from NCBI build 31 of the human genome (http://genome.ucsc.edu/).

### Radiation Hybrid Mapping

Genes were mapped on the 118 cell lines of the RHDF5000-2 panel described previously (Vignaux et al. 1999). In brief, PCR primers were selected for mapping using a standard selection program, i.e., Primer3 (http://www-genome.wi.mit.edu/cgi-bin/primer/primer3_www.cgi). Whenever possible, both primers were selected in the two introns flanking the annotated exon sequence. Alternatively, to better ensure amplification of the correct gene, in some cases one primer was selected from a flanking intron and the other from a corresponding exon. Primers were preferentially selected to be 25 bp in length and to work under a single optimal set of PCR conditions (salt, Tm, $Mg^{++}$, etc.) generating PCR products of 200–250 bp.

Typing of markers was done using existing infrastructure described previously (Priat et al. 1998; Mellersh et al. 2000; Breen et al. 2001; Guyon et al. 2003). In brief, all reactions are done using a 96-well or 384-well format in a volume of 10–15 µl. An initial screen using 50 ng of dog DNA, 50 ng of hamster DNA, and a 1:3 mix of dog/hamster DNA (50 ng) is used to select primers suitable to be placed across the entire panel. PCRs were done with 50 ng of RH DNA, and products were resolved on 1.8% or 2% agarose gels, electrophoresed for 30 minutes as described previously (Priat et al. 1998; Mellersh et al. 2000; Breen et al. 2001). Bands were viewed under UV light after ethidium bromide staining, and an image was recorded.

All markers were typed in duplicate and were considered consistent when the two vectors for each marker had a discrepancy value <16%, calculated for each marker based on its retention value within the panel. This threshold of 16% was determined to correspond to a distance lower than the resolution limit of the RHDF5000-2 panel (600 kb). Details and PCR conditions for all markers are available in Table A at:

http://www-recomgen.univ-rennes1.fr/doggy.html
http://www.fhcrc.org/science/dog_genome/dog.html

### Analysis and Map Construction

Novel markers were incorporated into the latest 3270 marker RH data set (Guyon et al. 2003). The corresponding RH groups were computed by pair-wise calculations using the MultiMap software (Matise et al. 1994) at a LOD threshold ≥8.0, thus allowing HSA1p orthologous canine gene markers to be assigned to specific chromosomes. To refine the region of interest containing orthologous HSA1p genes, the relevant chromosomes were split into smaller RH groups using the MultiMap algorithm and a LOD threshold of >9.0. Contiguous groups of the same chromosome origin were computed together. RH groups containing at least one HSA1p orthologous marker were then ordered using the TSP approach as specified by the CONCORDE computer package (http://www.math.princeton.edu/tsp/concorde.html) (Agarwala et al. 2000). TSP/CONCORDE computes five independent RH maps, and the resulting maps were subsequently evaluated to produce a consensus map using a method developed by us (Hitte et al. 2003). Intermarker distances were determined with the rh_tsp_map1.0 version of TSP/CONCORDE, which produces map positions in arbitrary TSP units.

## ACKNOWLEDGMENTS

We acknowledge the American Kennel Club Canine Health Foundation, U.S. Army Grant DAAD19-01-1-0658 (E.A.O. and F.G.), and National Institutes of Health R01CA-92167 (E.A.O, E.K., and F.G.). In addition, E.A.O is supported by K05 CA-90754 and is the recipient of a Burroughs Wellcome Award in Functional Genomics. R.G. is partly supported by an AKC and CNRS fellowship, and P.Q. by a Conseil Regional de Bretagne fellowship.

## REFERENCES

Agarwala R., Applegate D.L., Maglott D., Schuler G.D., and Schaffer A.A. 2000. A fast and scalable radiation hybrid map construction and integration strategy. *Genome Res.* **10:** 350.

Andersson L., Archibald A., Ashburner M., Audun S., Barendse W., Bitgood J., Bottema C., Broad T., Brown S., Burt D., Charlier C., Copeland N., Davis S., Davisson M., Edwards J., Eggen A., Elgar G., Eppig J.T., Franklin I., Grewe P., Gill T., III, Graves J.A., Hawken R., Hetzel J., and Womack J., et al. 1996. Comparative genome organization of vertebrates. The First International Workshop on Comparative Genome Organization. *Mamm. Genome* **7:** 717.

Breen M., Thomas R., Binns M.M., Carter N.P., and Langford C.F. 1999a. Reciprocal chromosome painting reveals detailed regions of conserved synteny between the karyotypes of the domestic dog (*Canis familiaris*) and human. *Genomics* **61:** 145.

Breen M., Langford C.F., Carter N.P., Holmes N.G., Dickens H.F., Thomas R., Suter N., Ryder E.J., Pope M., and Binns M.M. 1999b. FISH mapping and identification of canine chromosomes. *J. Hered.* **90:** 27.

Breen M., Jouquand S., Renier C., Mellersh C.S., Hitte C., Holmes N.G., Cheron A., Suter N., Vignaux F., Bristow A.E., Priat C., McCann E., André C., Boundy S., Gitsham P., Thomas R., Bridge W.L., Spriggs H.F., Ryder E.J., Curson A., Sampson J., Ostrander E.A., Binns M.M, and Galibert F. 2001. Chromosome-specific single-locus FISH probes allow anchorage of an 1800-marker integrated radiation-hybrid/

linkage map of the domestic dog genome to all chromosomes. *Genome Res.* **11:** 1784.

Credille K.M., Venta P.J., Breen M., Lowe J.K., Murphy K.E., Ostrander E.A., Galibert F., and Dunstan R.W. 2001. DNA sequence and physical mapping of the canine transglutaminase 1 gene. *Cytogenet. Cell Genet.* **93:** 73.

Guyon R., Lorentzen T.D., Hitte C., Kim L., Cadieu E., Parker H.G., Quignon P., Lowe J.K., Renier C., Gelfenbeyn B., Vignaux G., DeFrance H.B., Gloux S., Mahairas G.G., André C., Galibert F., and Ostrander E.A. 2003. A 1 Mb resolution radiation hybrid map of the canine genome. *Proc. Natl. Acad. Sci.* **100:** 5296.

Haworth K.E., Islam I., Breen M., Putt W., Makrinou E., Binns M., Hopkinson D., and Edwards Y. 2001. Canine TCOF1; cloning, chromosome assignment and genetic analysis in dogs with different head types. *Mamm. Genome* **12:** 622.

Hitte C., Lorentzen T., Guyon R., Kim L., Cadieu E., Parker H., Quignon P., Lowe J., Gelfenbeyn B., André C., Ostrander E.A., and Galibert F. 2003. Comparison of the MultiMap and TSP/CONCORDE packages for constructing radiation hybrid maps. *J. Hered.* **94:** 9.

Kirkness E.F., Bafna V., Halpern A.L., Levy S., Remington K., Rusch D.B., Delcher A.L., Pop M., Wang W., Fraser C.M., and Venter J.C. 2003. The dog genome: Survey sequencing and comparative analysis. *Science* **301:** 1854.

Lander E. and Waterman M. 1988. Genomic mapping by fingerprinting random clones: A mathematical analysis. *Genomics* **2:** 231.

Matise T.C., Perlin M., and Chakravarti A. 1994. Automated construction of genetic linkage maps using an expert system (MultiMap): A human genome linkage map. *Nat. Genet.* **6:** 384.

Mellersh C.S., Hitte C., Richman M., Vignaux F., Priat C., Jouquand S., Werner P., André C., DeRose S., Patterson D.F., Ostrander E.A., and Galibert F. 2000. An integrated linkage-radiation hybrid map of the canine genome. *Mamm. Genome* **11:** 120.

Murphy W.J., Sun S., Chen Z., Yuhki N., Hirschmann D., Menotti-Raymond M., and O'Brien S.J. 2000. A radiation hybrid map of the cat genome: Implications for comparative mapping. *Genome Res.* **10:** 691.

O'Brien S.J. and Stanyon R. 1999. Phylogenomics. Ancestral primate viewed. *Nature* **402:** 365.

O'Brien S.J., Eisenberg J.F., Miyamoto M., Hedges S.B., Kumar S., Wilson D.E., Menotti-Raymond M., Murphy W.J., Nash W.G., Lyons L.A., Menninger, J.C., Stanyon R., Wienberg J., Copeland N.G., Jenkins N.A., Gellin J., Yerle M., Andersson L., Womack J., Broad T., Postlethwait J., Serov O., Bailey E., James M.R., and Marshall Graves J.A., et al. 1999. Genome maps 10. Comparative genomics. Mammalian radiations. Wall chart. *Science* **286:** 463.

Okazaki Y., Furuno M., Kasukawa T., Adachi J., Bono H., Kondo S., Nikaido I., Osato N., Saito R., Suzuki H., Yamanaka I., Kiyosawa H., Yagi K., Tomaru Y., Hasegawa Y., Nogami A., Schonbach C., Gojobori T., Baldarelli R., Hill D.P., Bult C., Hume D.A., Quackenbush J., Schriml L.M., and Kanapin A., et al. (FANTOM Consortium; RIKEN Genome Exploration Research Group Phase I & II Team). 2002. Analysis of the mouse transcriptome based on functional annotation of 60,770 full-length cDNAs. *Nature* **420:** 563.

Parker H.G., Yuhua X., Mellersh C.S., Khan S., Shibuya H., Johnson G.S., and Ostrander E.A. 2001. Meiotic linkage mapping of 52 genes onto the canine map does not identify significant levels of microrearrangement. *Mamm. Genome* **12:** 713.

Priat C., Hitte C., Vignaux F., Renier C., Jiang Z., Jouquand S., Cheron A., André C., and Galibert F. 1998. A whole-genome radiation hybrid map of the dog genome. *Genomics* **54:** 361.

Sargan D.R., Yang F., Squire M., Milne B.S., O'Brien P.C., and Ferguson-Smith M.A. 2000. Use of flow-sorted canine chromosomes in the assignment of canine linkage, radiation hybrid, and syntenic groups to chromosomes: Refinement and verification of the comparative chromosome map for dog and human. *Genomics* **69:** 182.

Szabo C.I., Wagner LA., Francisco L.V., Roach J.C., Argonza R., King M.C., and Ostrander E.A. 1996. Human, canine and murine BRCA1 genes: Sequence comparison among species. *Hum. Mol. Genet.* **5:** 1289.

Venter J.C., Adams M.D., Myers E.W., Li P.W., Mural R.J., Sutton G.G., Smith H.O., Yandell M., Evans C.A., Holt R.A., Gocayne J.D., Amanatides P., Ballew R.M., Huson D.H., Wortman J.R., Zhang Q., Kodira C.D., Zheng X.H., Chen L., Skupski M., Subramanian G., Thomas P.D., Zhang J., Gabor Miklos G.L., and Nelson C., et al. 2001. The sequence of the human genome. *Science* **291:** 1304.

Vignaux F., Hitte C., Priat C., Chuat J.C., André C., and Galibert F. 1999. Construction and optimization of a dog whole-genome radiation hybrid panel. *Mamm. Genome* **10:** 888.

Vinogradov A.E. 1998. Genome size and GC-percent in vertebrates as determined by flow cytometry: The triangular relationship. *Cytometry* **31:** 100.

Waterston R.H., Lindblad-Toh K., Birney E., Rogers J., Abril J.F., Agarwal P., Agarwala R., Ainscough R., Alexandersson M., An P., Antonarakis S.E., Attwood J., Baertsch R., Bailey J., Barlow K., Beck S., Berry E., Birren B., Bloom T., Bork P., Botcherby M., Bray N., Brent M.R., Brown D.G., and Brown S.D., et al. (Mouse Genome Sequencing Consortium). 2002. Initial sequencing and comparative analysis of the mouse genome. *Nature* **420:** 520.

Wayne R.K. 1993. Molecular evolution of the dog family. *Trends Genet.* **9:** 218.

Wayne R.K., Nash W.G., and O'Brien S.J. 1987. Chromosomal evolution of the Canidae. II. Divergence from the primitive carnivore karyotype. *Cytogenet. Cell Genet.* **44:** 134.

Yang F., Graphodatsky A.S., O'Brien P.C., Colabella A., Solanky N., Squire M., Sargan D.R., and Ferguson-Smith M.A. 2000. Reciprocal chromosome painting illuminates the history of genome evolution of the domestic cat, dog and human. *Chromosome Res.* **8:** 393-404.

Yang F., O'Brien P.C., Milne B.S., Graphodatsky A.S., Solanky N., Trifonov V., Rens W., Sargan D., and Ferguson-Smith M.A. 1999. A complete comparative chromosome map for the dog, red fox, and human and its integration with canine genetic maps. *Genomics* **62:** 189.

## WEB SITE REFERENCES

http://www.tigr.org/softlab/assembler/(The Institute for Genomic Research TIGR Assembler 2.0 website).

http://genome.ucsc.edu/ (University of California, Santa Cruz Genome Bioinformatics website).

http://www-genome.wi.mit.edu/cgi-bin/primer/primer3_www.cgi (Whitehead Institute/MIT Center for Genome Research Primer 3 website).

http://www-recomgen.univ-rennes1.fr/doggy.html (UMR-CNRS 6061, Universite de Rennes Canine Radiation Hybrid Project website).

http://www.fhcrc.org/science/dog_genome/dog.html (Fred Hutchinson Cancer Research Center Dog Genome Project website).

http://www.math.princeton.edu/tsp/concorde.html (TSP/CONCORDE Home webpage).

http://www.ncbi.nih.nlm.gov/blast (NCBI web site).

# Positional Identification of Structural and Regulatory Quantitative Trait Nucleotides in Domestic Animal Species

M. GEORGES* AND L. ANDERSSON†

*Department of Genetics, Faculty of Veterinary Medicine, University of Liege (B43), 4000 Liege, Belgium;
†Department of Medical Biochemistry and Microbiology, Uppsala University, BMC, S-751 24 Uppsala, Sweden

Although the identification of mutations underlying monogenic traits has become nearly trivial, the molecular dissection of multifactorial traits—which include the majority of medically and agronomically important phenotypes—remains a major challenge. Despite the substantial resources that have been allocated to such efforts—particularly in human—a recent survey reported only 30 successful outcomes, all organisms confounded (Glazier et al. 2002). This number has to be compared with the more than 1,600 Mendelian traits for which the causal mutation has been identified in the human only. As recently stated by Rutherford and Henikoff (2003), "The nature of quantitative-trait variation is one of the last unexplored frontiers in genetics, awaiting the future cloning and definitive identification of quantitative-trait determinants, whether they be genetic or epigenetic." Indeed, it remains largely unknown how many mutations underlie the genetic variation for a typical quantitative trait, what the nature of these mutations is (structural or regulatory; genetic or epigenetic), what the actual distribution of effect size is, what the importance of dominance, epistatic and gene-by-environment interactions is, how genetic variation is maintained, etc.

## LIVESTOCK POPULATIONS OFFER UNIQUE ADVANTAGES FOR THE MOLECULAR DISSECTION OF MULTIFACTORIAL TRAITS

Herein, we argue that livestock populations have a number of features which make them particularly suited for the identification of genes and mutations underlying complex traits and that, in this regard, this very valuable resource has been underutilized. Among these features are the following:

1. At least in the developed world, **a growing list of phenotypes is being systematically recorded on a substantial proportion of the breeding population** as part of ongoing management and selection programs. The list of phenotypes includes production traits (such as milk yield and composition in dairy cattle, carcass composition and quality in pigs and poultry, and number and quality of the eggs in poultry), "type" traits describing individual morphology, health-related traits including, e.g., udder health, traits relating to both male and female fertility, and even behavioral traits.
2. **Individual phenotypes are being processed** in conjunction with detailed information about living conditions and pedigree relationships **to compute environmental and genetic variance components, from which estimates of heritability can be derived.** Because related animals are being raised across multiple environments, there is much less risk of confounding environment and genetics when compared to most studies on human subjects. The relative contribution of genetic factors to the phenotypic variation is therefore known with high accuracy, which is obviously paramount when aiming at mapping and identifying the underlying genetic determinants. Related biometric methods allow the generation of individual phenotypes that are pre-corrected for the identified environmental effects. The use of such "residuals" potentially increases the power to detect the underlying genes.
3. **Extant pedigrees with a suitable structure for the mapping of quantitative trait loci (QTL) underlying the *within-population* genetic variance can often be readily collected.** This is in large part due to the frequent occurrence of large paternal half-sib pedigrees ("harems"), particularly in species where artificial insemination (AI) is common practice. As an example, bulls having more than 50,000 daughters are common in dairy cattle.
4. As a result of the unusual population structure (e.g., the often very small breeding male-to-female ratio) used to maximize genetic response, **the effective population size is typically restricted,** even for the most widespread breeds. As an example, the effective population size of the black and white Holstein-Friesian dairy cattle breed has been estimated at less than 200, despite an actual, worldwide population of several tens of millions. This has two important consequences for the identification of genes underlying complex traits. First, **the genetic complexity of the trait in terms of allelic and locus heterogeneity is bound to be reduced.** In this sense, livestock breeds resemble isolated populations that are eagerly sought by human geneticists. Second, it is now becoming apparent that population-wide **linkage disequilibrium (LD) in livestock extends over tens of centimorgans** rather than subcentimorgan regions as typically observed in human. Capturing this LD signal can therefore be achieved with much sparser marker maps than in human. As a result, locating genetic determinants by whole-genome association studies may be considerably easier in livestock than in man.

5. There is growing evidence that present-day livestock species derive from multiple independent domestication events that in some instances involved founders from distinct subspecies. As a result of subsequent admixture, some **livestock populations now resemble advanced inter-cross lines between highly divergent genomes exhibiting nucleotide diversities as high as 1/200.** For instance, it is well documented that Asian pigs, which derive from the Asian wild boar, were imported in Europe during the 19th century and repeatedly crossed with European pigs derived from the European wild boar (Giuffra et al. 2000). Likewise, there is strong evidence that "... African cattle are the products of a progressive, male-driven admixture between the indigenous taurine breeds (*Bos taurus*) and immigrating zebu cattle (*Bos indicus*) ..." (Bradley and Cunningham 1999). This situation is therefore reminiscent of the mosaic structure of variation that was recently demonstrated for laboratory mice (Wade et al. 2002). Below, we demonstrate how this may contribute to ultra-fine-mapping QTLs in these populations.
6. Although more laborious and time-consuming than with conventional model organisms, **experimental livestock inter- and back-cross pedigrees can be generated at will to map QTLs that underlie the *between-population* genetic variance.** Numerous such crosses have been generated between parental "lines" that are highly divergent for the traits of interest. Examples include the $F_2$ pedigrees that were generated between Meishan and Yorkshire pigs to study prolificacy and other traits (de Koning et al. 1999; Rohrer et al. 1999; Bidanel et al. 2001), between Boran and N'Dama cattle to study trypanotolerance (see, e.g., Hanotte et al. 2003), and between layer and broiler chickens to study growth and egg production (Sewalem et al. 2002). Unique versions of this approach are the experimental crosses that were established by crossing domestic animals with extant representatives of their wild-type ancestor species. Such crosses have been generated between the large white and European wild boar (Andersson et al. 1994), as well as between white leghorn chickens and the red jungle fowl (see, e.g., Kerje et al. 2003). The latter experiments seek to identify the molecular basis of the spectacular metamorphoses obtained by less than 10,000 years of artificial selection. It is important to realize that, contrary to genuine model organisms, the parental livestock "lines" that are used to generate such experimental crosses are never inbred. Mapping methods have to be adjusted for these idiosyncrasies and results interpreted with caution (see below).

Over the last 15 years, hundreds of QTLs affecting a broad range of traits have been mapped in livestock (see, e.g., Andersson 2001). Herein, we describe the major steps that led to the first two successful identifications of the causal quantitative trait nucleotides (QTNs), one underlying a QTL influencing milk composition in cattle, the other a QTL influencing muscle mass and fat deposition in the pig.

## A NONCONSERVATIVE LYSINE TO ALANINE SUBSTITUTION IN THE BOVINE *DGAT1* GENE HAS A MAJOR EFFECT ON MILK YIELD AND COMPOSITION

### Exploiting Progeny Testing to Map a QTL with Major Effect on Milk Yield and Composition to Proximal BTA14

The spectacular genetic progress that has been achieved in dairy cattle over the last 50 years is mainly due to the combination of (1) the systematic use of AI, allowing the widespread dissemination of genetically superior germ plasm, and (2) the selection of genetically superior bulls for AI by means of progeny testing. Candidate AI bulls are continuously being produced by mating so-called "bull-sires" and "bull-dams" selected on their respective breeding values (BVs). The actual genetic merit of resulting bulls is measured with high accuracy on the basis of the average "performance" of 50–100 daughters: the progeny test. Note that progeny testing a bull takes 5–6 years and costs an estimated $30,000 per animal.

As a result of this systematic reliance on AI and progeny testing, pedigrees composed of a sire and 50–100 of its progeny-tested sons can readily be sampled from AI companies and used to map QTLs. The phenotypes used for QTL mapping are the BVs of the sons estimated from their respective daughters. The reliabilities (correlation between true and estimated BVs) of these estimated BVs are a function of the number of daughters and the heritability of the trait but are typically on the order of 70–90%. This experimental design for the mapping of QTLs was initially proposed by Weller et al. (1990) and referred to as the "granddaughter design" (GDD). The GDD requires 3–4 times less genotyping when compared to the alternative "daughter design" (DD) that relies on the use of paternal half-sister pedigrees (Georges et al. 1995). Note that in both the GDD and DD, only the paternal chromosome provides linkage information unless one uses more sophisticated "whole pedigree analysis" methods (see, e.g., Hoeschele et al. 1997).

Like many others, our laboratory performed a whole-genome scan using a GDD counting 29 paternal half-brother pedigrees for a total of 1,200 black-and-white Holstein-Friesian (HF) bulls having BVs for 45 traits. The genome was covered with a battery of 350 autosomal microsatellite markers. Because only male meioses are informative, the X chromosome escapes exploration when relying on the basic GDD. This experiment led to the identification of numerous QTLs influencing a broad range of traits (see, e.g., Spelman et al. 1996, 1999; Arranz et al. 1998; Schrooten et al. 2000; W. Coppieters et al., unpubl.). The QTL yielding the most significant signal in this experiment influenced milk fat percentage and, to a lesser degree, milk yield and milk protein percentage, and mapped to proximal BTA14 (Coppieters et al. 1998).

### Exploiting Linkage Disequilibrium to Refine the Map Position of the QTL to a 3-cM Interval

Initial mapping experiments using the GDD typically led to QTL locations with confidence intervals of 20 cM or more. These are too large to envisage either positional cloning or the efficient utilization of the identified QTL for marker-assisted selection (MAS). The ability to refine the map position of the identified QTL is therefore crucial. Three factors determine the achievable mapping resolution: (1) the marker density in the chromosome region of interest, (2) the cross-over density in the chromosome region of interest among the available informative chromosomes, and (3) the "detectance" or probability of a given QTL genotype, given the phenotype.

Increasing the marker density, even if still laborious in most livestock species, is conceptually the simplest bottleneck to resolve. To increase the cross-over density, one can either increase the number of "current" recombinants by sampling or producing additional offspring, or attempt to exploit the "historical" recombination events that occurred in the ancestors of the available chromosome panel, i.e., exploit linkage disequilibrium (LD). However, because LD typically extends over shorter chromosome regions than linkage, this approach requires a commensurate increase in marker density. For instance, it has been estimated that on the order of 1 marker per 3–10 kb may be needed to map genes by means of LD in the human (see, e.g., Wall and Pritchard 2003). To evaluate the feasibility of an LD-based approach in livestock, we measured the degree of genome-wide LD in the HF dairy cattle population. To our surprise, we found that LD extends over several tens of centimorgans and that gametic association is even common between nonsyntenic loci (Farnir et al. 2000). We showed that drift caused by the small effective population size accounts for most of the observed LD. As a consequence, it seemed that LD could be useful for fine-mapping purposes even with the medium density marker maps available in livestock. One major advantage of using LD in the context of the GDD is that it allows the extraction of information from the maternally inherited chromosomes of the sons. As noted before, when performing linkage analysis, the only informative chromosomes in the GDD are the sires' chromosomes. Exploiting LD therefore potentially doubles the amount of information that can be extracted from the GDD.

The GDD also offers interesting features in terms of QTL "detectance." As noted before, the use of the estimated BVs of the sons reduces the environmental noise, thereby improving the QTL detectance when compared to the use of the daughters' phenotypes in the DD. In addition, the evidence for segregation or absence of segregation in each paternal half-sib pedigree provides considerable information about the QTL genotype of the sires. If a pedigree shows clear evidence for the segregation of a QTL linked to a given marker, the sire is deemed to be heterozygous at that QTL or *Qq* under a biallelic QTL model. If we make the assumption that all heterozygous "*Qq*" sires carry an identical-by-descent *Q* allele that appeared by mutation or migration on a founder chromosome and swept through the population as a result of its favorable effect on milk composition, all the *Q*-bearing chromosomes of the heterozygous sires are predicted to share a marker haplotype tracing back to the founder, and whose detection leads to the localization of the QTL. We initially used this biallelic QTL model to refine the map location of the BTA14 QTL to a 3-cM chromosome interval, bounded by microsatellite markers *BULGE30–BULGE9* (Riquet et al. 1999; Farnir et al. 2002). Interestingly, the *Q* allele was shown to be associated with different marker haplotypes in the Dutch and the New Zealand (NZ) HF dairy cattle populations (respectively referred to as marker haplotypes $\mu H^{Q\text{-}D}$ and $\mu H^{Q\text{-}NZ}$), suggesting that the *Q* alleles might be distinct in these two populations. As expected, the *q* alleles were associated with multiple marker haplotypes in both populations (collectively referred to as $\mu h^{q}$).

Less constraining polyallelic QTL models have since been developed and have confirmed these initial results (Kim and Georges 2002). The latter approach, which was first articulated by Meuwissen and Goddard (2001a,b), models the phenotype by means of a linear model including fixed environmental effects, a random QTL effect, a random polygenic effect, and a random error term. The covariances between the QTL effects of individual chromosomes are derived from the pair-wise identity-by-descent (IBD) probabilities conditional on marker data computed using a combination of conventional linkage analysis and a simplified coalescent model to extract the LD signal. In the Kim and Georges (2002) version, the chromosomes are hierarchically clustered on the basis of the pair-wise IBD probabilities. The covariances between individual polygenic effects are computed in a standard way according to the individual animal model (see, e.g., Lynch and Walsh 1997). Variance components and individual effects are estimated by classical restricted maximum likelihood (REML) techniques. This approach has proven highly effective in the dissection of at least two more QTLs, one influencing twinning (Meuwissen et al. 2002), and the other milk composition (Blott et al. 2003). Despite these initial successes, QTL fine-mapping remains the major hurdle when attempting to clone the underlying genes. This is primarily due to the fact that many of the identified QTLs appear to reflect the combined action of an undetermined number of linked QTNs, whereas most QTL mapping methods erroneously assume a single, or at best two, QTLs per chromosome. It remains to be determined whether emerging, more sophisticated statistical models will be able to satisfactorily deal with more complicated situations (see, e.g., T.H. Meuwissen and M.E. Goddard, in prep.).

Note that because of the common occurrence of very large paternal half-sister pedigrees, an alternative fine-mapping method using "current" recombinants to increase cross-over density combined with selective genotyping of extremes to increase detectance is possible but has not been utilized so far. Daughters with extreme phenotypes (e.g., the 1000 bottom 5% and 1000 top 5% of 20,000 daughters) could be genotyped for a pair of markers flanking a QTL for which the sire is known to segregate, based on prior knowledge. Daughters that have inherited a recombinant paternal chromosome could then easily be singled out and genotyped for a battery of markers spanning the interval of interest. Methods akin to marker difference regression (MDR) could then be applied to refine the location of the one or multiple QTLs in the region (see, e.g., Lynch and Walsh 1997).

### Association Studies Point toward *DGAT1* as the Causal Gene

A BAC contig spanning the ~1.4-Mb *BULGE30–BULGE9* interval was constructed using a combination of chromosome walking and STS content mapping. STSs for contig construction were mainly cDNA probes corresponding to genes predicted to map to the interval of interest based on human–bovine comparative mapping information, and STS derived from BAC end sequences (Grisart et al. 2002 and in prep.).

The BAC contig was shown to contain a very strong positional candidate, diacylglycerol acyl transferase (*DGAT1*). As its name implies, *DGAT1* indeed catalyzes the final step in triglyceride (TG) synthesis. TGs make up 98% of milk fat—the trait most profoundly affected by the BTA14 QTL. In addition, constitutive inactivation of *DGAT1* in knockout mice was shown to completely abrogate lactation, pointing toward a unique function of *DGAT1* in lactational physiology (Smith et al. 2000).

The *DGAT1* gene was completely sequenced from a small number of chromosomes carrying the $\mu H^{Q\text{-}D}$, $\mu H^{Q\text{-}NZ}$, and $\mu h^{q}$ marker haplotypes. This led to the identification of eight SNPs: six in introns, one in the 3′UTR, and one in exon VIII. The latter (referred to as *K232A*) was particularly intriguing because (1) it resulted in the nonconservative substitution of a lysine by an alanine, (2) the lysine residue was perfectly associated with both the $\mu H^{Q\text{-}D}$ and $\mu H^{Q\text{-}NZ}$ haplotypes, whereas the alanine residue was found on all analyzed $\mu h^{q}$ haplotypes, and (3) all eight sequenced mammals carried a lysine residue at the corresponding position except *Cercopithecus aethiops*, which nevertheless had a basic amino acid, arginine, in position 232.

Cohorts of respectively 1,800 bulls and 500 cows were genotyped for the *K232A* mutation, and its effect on milk yield and composition was estimated using a linear model including a random polygenic component accounting for the relatedness between individuals. The *K232A* mutation was shown to have a major effect on milk fat percentage ($p < 10^{-122}$ in the bulls!) explaining as much as 30% of the phenotypic variation in the cows. All other traits measuring milk yield and milk composition were very significantly affected as well. These results were perfectly compatible with the *DGAT1 K232A* mutation's being directly responsible for the BTA14 QTL effect.

It is worthwhile noting that, contrary to expectation, the *K* allele corresponding to the *Q* QTL allele is in fact the ancestral state shared with the other mammalian species, the *A* allele corresponding to the *q* allele being the derived state. However, the fact that in both the Dutch and the New Zealand (NZ) population the *K* allele is associated with a unique albeit distinct marker haplotype testifies for the selective sweep that it has undergone independently in both populations, probably following the implementation of systematic selection for increased fat yield in the 1960s.

One could argue that most other polymorphisms in the *BULGE9–BULGE30* interval would have shown strong effects on milk yield and composition as a result of extensive LD in this region. At least three other *DGAT1* SNPs (in virtually perfect association with *K232A*) indeed yielded equally strong effects. To test the unique status of *DGAT1* within this interval, we developed a panel of 20 SNP markers spanning the *BULGE9–BULGE30* interval and genotyped a cohort of 1,818 Dutch and 227 NZ HF bulls using a high-throughput oligonucleotide ligation assay (OLA). The effect of each SNP on milk fat percentage was estimated using the previously described linear model. In the Dutch population, the most significant effects (LOD scores > 200) were obtained for an entire block of nine SNPs (including the *DGAT1* SNPs) located at the centromeric end of the *BULGE9–BULGE30* interval. All nine SNPs yielded essentially equally significant effects. This suggested that the QTL was indeed located within the portion of the BAC contig encompassing *DGAT1*, but it did not allow us to exclude the other genes located in this segment, including the *CHRP* gene that was previously incriminated by Looft et al. (2001). In the NZ population, however, where *DGAT1* was not as strongly associated with the remainder of the proximal SNP block, the effect on milk fat percentage was clearly much stronger with the *DGAT1* SNPs than with any other polymorphism in the region. This provided very strong evidence that *DGAT1* is indeed the gene responsible for the BTA14 QTL effect.

### Functional Tests Incriminate the *K232A* Mutation as the Causative QTN

The previously described results identified *DGAT1* as the gene underlying the BTA14 QTL, and identified *K232A* as the most likely causal mutation. To provide additional evidence for the causality of the K232A mutation, we expressed bovine *DGAT1* alleles that differed only at the *K232A* position in Sf9 insect cells using a baculovirus expression system. We measured *DGAT1* activity in the corresponding microsome preparations by measuring the rate of incorporation of $^{14}$C-labeled oleoyl-CoA in diacylglycerol, using thin-layer chromatography and phosphor imaging to separate and quantify the synthesized TG (B. Grisart et al., in prep.). The activity of the *K* allele at apparent $V_{max}$ concentrations was shown to be 1.5 times higher when compared to that of the *A* allele. This is in perfect agreement with the phenotypic effect of the *K232A* mutation, the *A* to *K* substitution causing an increase in milk fat percentage in the live animal.

These results therefore strongly supported the causality of the *K232A* mutation, the only structural mutation differentiating the *Q* and *q DGAT1* alleles. Any other mutation would have to be a regulatory mutation controlling the expression level of *DGAT1*. We measured the relative amounts of *K* and *A* mRNA in the mammary glands of heterozygous *K/A* cows and showed these to be equivalent. Therefore, this does not support the alternative hypothesis of an as yet unidentified regulatory mutation controlling *DGAT1* expression levels that would be in strong association with the *K232A* SNP.

This combination of genetic and functional evidence made a very compelling case that the *DGAT1 K232A* mutation is indeed single-handedly responsible for the BTA14 QTL effect on milk yield and composition.

## A G TO A SUBSTITUTION IN A SILENCER ELEMENT CONTROLLING *IGF2* TRANSCRIPTION HAS A MAJOR EFFECT ON SKELETAL AND CARDIAC MUSCLE MASS IN PIGS

### Using Experimental Crosses to Map an Imprinted QTL with Major Effect on Muscle Mass to Proximal SSC2

As mentioned previously, when working with livestock species experimental crosses can be designed at will to search for QTLs that underlie the phenotypic differences between highly divergent parental lines or breeds. We herein describe results obtained on two such pedigrees: (1) a European wild boar (EWB) x large white (LW) intercross counting 200 $F_2$ offspring generated to identify the genetic determinants accounting for the spectacular phenotypic evolution accrued through 10,000 years of domestication (Andersson et al. 1994) and (2) a Piétrain (P) x LW intercross counting 650 $F_2$ offspring, generated to identify QTLs underlying the differences in growth and carcass composition between two breeds that are being extensively used in commercial pig production worldwide (Nezer et al. 2002). Whole-genome scans were performed in both instances using microsatellite markers spanning the entire genome.

Both studies revealed a QTL with major effect on muscle mass and fat deposition mapping to the distal end of the short arm of pig chromosome SSC2 (Andersson-Eklund et al. 1998; Nezer et al. 2002). SSC2p was known from comparative mapping data to be orthologous to HSA11pter-q13, harboring at least two intriguing candidate genes, *MyoD* and *IGF2*. *MyoD* was shown to map to SSC2, however, at more than 60 cM from the most likely position of the QTL. The position of *IGF2*, on the contrary, was shown to coincide exactly with that of the QTL in both studies. *IGF2* is known to be imprinted and expressed exclusively from the paternal allele in man, mouse, rabbit, and sheep. We demonstrated that *IGF2* is likewise imprinted in the pig. To test whether *IGF2* might underlie the observed QTL effect, we applied an imprinting model to our pedigree data. We clearly showed in both studies that only the paternal SSC2 QTL allele influences the phenotype, the maternal allele having very little if any phenotypic effect at all. This strongly supported the involvement of an imprinted gene with exclusive paternal expression, possibly *IGF2* (Nezer et al. 1999; Jeon et al. 1999).

The QTL allele substitution effect amounted to 2–3% of muscle mass, underpinning the potential economic importance of this QTL. It is worthwhile noting that the QTL allele increasing muscle mass originated from the LW parent in the EWB x LW cross, and from the P parent in the P x LW cross, suggesting that multiple QTL alleles might exist. Subsequent studies performed in Meishan (M) x LW and Berkshire (B) x LW intercrosses detected the same imprinted QTL effect, with the allele increasing muscle mass originating in both cases from the LW parent (de Koning et al. 2000; Thomsen et al. 2002).

It is noteworthy that imprinting can normally not be tested in a standard $F_2$ pedigree. Indeed, testing imprinting boils down to evaluating the contrast, $\Delta_{IMP}$, between (1) $F_2$ individuals sorted by paternally inherited chromosome, with the maternal chromosome held fixed, and (2) $F_2$ individuals sorted by maternally inherited chromosome, with the paternal chromosome held fixed:

$$\Delta_{IMP} = {}^1/_2[(\overline{11}-\overline{21}) + (\overline{21}-\overline{22})] - {}^1/_2(\overline{11}-\overline{12}) + (\overline{21}-\overline{22})] = (\overline{12}-\overline{21})$$

in which $\overline{XX}$ corresponds to the phenotypic average of the *XX* genotypic class where the first digit refers to the paternal chromosome, and the second to the maternal chromosome. Because it is impossible to distinguish "12" from "21" individuals in a standard $F_2$, the imprinting hypothesis cannot be tested. Contrary to the situation applying to genuine model organisms in which the parental lines are inbred, in livestock, parental lines are never inbred and segregate for multiple marker alleles. The alternate $F_2$ heterozygotes can therefore often be differentiated, allowing one to test the imprinting hypothesis in a typical livestock $F_2$ design. This has been extensively applied in animal genetics following our initial report and has led to a flurry of so-called imprinting effects in the animal genetics literature (see, e.g., de Koning et al. 2000). There is a caveat, however. Just as parental lines in livestock are not fixed for alternate marker alleles, they cannot simply be assumed to have fixed alternate QTL alleles. As a consequence, all $F_1$ individuals cannot be assumed to be systematically heterozygous for the same QTL and/or same QTL alleles for a given QTL. The phenotypic contrast between the alleles of different $F_1$ individuals at a given locus is therefore expected to vary. Because the number of $F_1$ parents is typically very small (especially for the males), it cannot be expected that the average contrast between chromosomes inherited from $F_1$ sires will be the same as the average contrast between chromosomes inherited from the $F_1$ dams, even in the absence of true parental imprinting (de Koning et al. 2002). This problem will be exacerbated in the likely case of LD between marker and QTL alleles in the paternal lines. Indeed, by definition it will be possible to test imprinting only for chromosome regions for which the $F_1$ sire and dam have inherited distinct marker alleles from one or both $F_0$ parents. This will increase the probability that the $F_1$ sire and dam have a distinct QTL genotype thus yielding a parent-of-origin-specific contrast between $F_1$ chromosomes even in the absence of parental imprinting. It is crucial to be very cautious before concluding that a parent-of-origin effect detected in an $F_2$ design is due to genuine parental imprinting.

### Using LD and Marker-assisted Segregation Analysis to Refine the Map Position of the SSC2 QTL to a 250-kb Segment Containing *IGF2*

Measuring LD in European pig populations demonstrated that it was essentially as widespread in pigs as previously recognized in cattle and sheep. LD was found to be common between syntenic markers several tens of centimorgans apart, and gametic association was even found between nonsyntenic markers, especially in synthetic populations derived by the recent admixture of distinct breeds (N. Harmegnies and M. Georges, unpubl.). To provide additional support for the direct involvement of *IGF2* in the SSC2 QTL, we therefore used an LD-based IBD mapping method to refine the map position of the QTL. The approach chosen was very similar to the one previously described for the fine-mapping of the BTA14 QTL in dairy cattle. To increase the QTL detectance, the QTL genotype of a series of boars sampled in multiple populations was determined with high accuracy by marker-assisted segregation analysis using large cohorts of offspring. Heterozygous *Qq* sires were predicted to carry an IBD *Q* QTL allele having appeared by mutation or migration on a founder chromosome *g* generations ago and having undergone a selective sweep due to its favorable effect on muscle mass. As a consequence, a shared marker haplotype spanning the QTL was predicted for all *Q*-bearing chromosomes. We developed a panel of 51 SNPs, and 5 microsatellite markers uniformly spanning the *KVLQT1–H19* imprinted domain and genotyped heterozygous *Qq* boars for all of these including sufficient offspring to establish marker phase. As predicted, all *Q*-bearing chromosomes were indeed sharing a haplotype spanning 250 kb and containing *INS* and *IGF2* as only known paternally expressed genes. Such a shared haplotype was not found among *q* chromosomes, which were showing a much higher degree of genetic diversity, as expected. This provided strong additional support for a direct involvement of *IGF2* (Nezer et al. 2003). Contrary to the *DGAT1* story, however, sequencing the *Q* and *q* alleles for the coding portions of *IGF2* did not reveal any structural difference. This suggested that the QTL effect was most likely due to a regulatory rather than a structural mutation.

### Extensive Resequencing of a 28.5-kb Chromosome Segment Encompassing the *IGF2* Gene Identifies the QTN

To identify the hypothetic regulatory mutation, we sequenced 28.5 kb (encompassing the last exon of the *TH* gene, the complete *INS* gene, and the complete *IGF2* gene up to the *SWC9* microsatellite marker in its 3′UTR corresponding to the distal end of the marker haplotype shared by all *Q* alleles) for three *Q* chromosomes and seven *q* chromosomes, each carrying a distinct marker haplotype across the *KVLQT1–H19* domain (Van Laere et al. 2003). We identified a staggering 258 SNPs, corresponding to one polymorphic base pair every 111 nucleotides. A phylogenetic tree describing the relationship between the ten analyzed chromosomes was constructed by neighbor-joining analysis. It showed that all three *Q* chromosomes were virtually identical to each other, whereas the *q* chromosomes essentially fell into two distinct clades. One of these included the EWB, suggesting that the chromosomes in this cluster were of European descent. Chromosomes in this cluster exhibited an average nucleotide diversity of 0.0007 among themselves and of 0.0039 with the chromosomes from the *Q* clade. The other *q* clade comprised two more distantly related chromosomes exhibiting a nucleotide diversity of 0.0025 among themselves, of 0.0032 with the *Q* clade, and 0.0034 with the other *q* clade.

For 33 of the 258 identified SNP positions, the three *Q* chromosomes shared a nucleotide not encountered on any of the analyzed *q* chromosomes. These would most likely correspond to mutations having occurred on the branch separating the *Q* from the other chromosomes and would therefore be prime candidates for the QTN. Two of these 33 SNPs reside in evolutionary footprints, regions exhibiting a high degree of conservation between human, mouse, and pig, which were detected in a preliminary sequence exploration of the porcine *IGF2* and *H19* genes (Amarger et al. 2002).

We then screened a large number of sire families using one of the 33 candidate SNPs (located between *TH* and *INS*) as being diagnostic of the *Q* versus *q* status. We identified two sire families that were not showing any evidence for the segregation of a QTL despite the fact that the corresponding boars were clearly heterozygous for the SNP: one Hampshire boar and one $F_1$ M x LW boar. We completely sequenced the 28.5-kb segment for the four chromosomes of the corresponding boars. The Hampshire boar proved to carry one chromosome that belonged to the *Q* clade and one chromosome that was clearly a recombinant chromosome, being from the *q* clade between *TH* and the middle of the *IGF2* intron 1 and from the *Q* clade for the remainder of the interval. This allowed us to exclude the entire segment for which this boar was heterozygous, eliminating 9 of the 33 candidate SNPs. The $F_1$ M x LW boar proved to carry one chromosome—of Meishan origin—belonging apparently to the *Q* clade over its entire 28.5-kb length, the other—of large white origin—belonging to the *q* clade over the entire 28.5 kb. Intriguingly, however, the Meishan chromosome differed from all other known *Q* chromosomes at one position, corresponding to one of the two candidate SNPs located in an evolutionary footprint: a CpG island in the third intron of *IGF2*. The Meishan chromosome was sharing an *A* residue with all *q* chromosomes at that position whereas all other *Q* chromosomes had a *G* residue. The *q* LW chromosome, on the contrary, shared the same residue with all other *q* chromosomes at all the 33 QTN candidate positions. The easiest interpretation of these results is that the chromosome of the M x LW $F_1$ boar clustering within the *Q* clade was in fact functionally of *q* type differing with the genuine *Q* chromosomes at the actual "holy grail": the QTN. The functional *q* status of this Meishan chromosome was in agreement with a previous publication detecting the imprinted SSC2 QTL in a M x LW intercross, and in which the Meishan chromosome was associated with a decrease in muscle mass and concomitant increase in fat deposition (de Koning et al. 2000). Extending our screen, we identified one boar that was heterozygous at the QTN only: The QTL proved to segregate in its offspring, providing very strong support that the *G* residue at the QTN was necessary for a chromosome to be functionally *Q*.

The fact that the *q* chromosome of Meishan origin clusters with the *Q* clade indicates that the latter is of Asian origin. The *G* to *A* transition either could have been sampled as such during pig domestication in Asia, or could have occurred after domestication either in Asia or in Europe after the importation of Asian germ plasm in the 19th century.

### Functional Analyses Confirm That the QTN Is Not Only Necessary but Also Sufficient

Given the phenotypic effect of the QTL, and the lack of a structural *IGF2* difference between the *Q* and *q* alleles, we predicted that the QTN would be a regulatory mutation causing a postnatal increase in *IGF2* levels in striated muscle (A.-S.Van Laere et al., in prep.). Indeed, the major effect of the QTL was on muscle mass, fat deposition, and heart size, measured at six months of age. There was no evidence for a perinatal growth effect or a postnatal effect on the size of any other organs (Jeon et al. 1999). Note that in the mouse a deletion involving DMR1, an associated block of directly repeated sequences and the CpG island containing the QTN, leads to derepression of the maternal allele during fetal development as well as continued postnatal expression from the maternal and paternal alleles in mesodermal tissue (Constancia et al. 2000). To test the effect of the QTL on *IGF2* expression in vivo, we produced an $F_2$ population segregating for the QTN. $F_2$ offspring were slaughtered at five stages of development: fetal, one, two, four, and six months of age. We first demonstrated that the QTL genotype had no effect on the imprinting status of the *IGF2* gene. For all analyzable genotypes (i.e., *qq*, *Qq*, and *qQ*), we found that (1) before birth, *IGF2* was expressed exclusively from the paternal allele in all examined tissues and (2) after birth, *IGF2* remained preferentially expressed from the paternal allele with, however, progressive reactivation of the maternal allele in some tissues, including skeletal muscle. This might explain why in one of the initial QTL mapping experiments muscle growth was found to be slightly superior in $q^{pat}/Q^{mat}$ versus *qq* animals, and in *QQ* versus $Q^{pat}/q^{mat}$ animals (Jeon et al. 1999). We then used northern blot analysis as well as RT-PCR to quantify the amounts of *IGF2* mRNA in different tissues of $F_2$ animals sorted according to the *IGF2* allele inherited from their sire. Before birth, the QTN genotype had no detectable effect on *IGF2* levels in any of the analyzed tissues. This is in agreement with the lack of detectable phenotypic effect of the QTL at birth. After birth, the QTN genotype had a very significant effect on *IGF2* levels in skeletal muscle (~threefold increase) and heart, although not in liver. This is again in agreement with the postnatal effect of the QTL on skeletal muscle mass and heart size—resulting probably from a paracrine *IGF2* effect—but lack of effect on the size of any other organs. These results thus provided very strong additional support for a direct role of *IGF2*. To check whether the identified QTN was sufficient to confer the functional *Q* status to a chromosome, we performed a series of in vitro experiments. The QTN is the antepenultimate nucleotide of a 16-bp motif that is perfectly conserved among all tested mammals (eight species) and which contains an 8-bp palindromic sequence susceptible to adopt a hairpin structure. Despite the fact that it does not present any obvious similarity with known *cis*-regulatory elements, its remarkable conservation suggested that it might bind a *trans*-acting factor. We thus performed electrophoretic mobility shift assays using 27-mers including the QTN that were incubated with nuclear extracts from C2C12 myoblast cells. The *q* allele was clearly able to bind a factor not bound by the *Q* allele. Competition with cold *q* but not *Q* oligonucleotide would effectively displace the *q* probe. These results suggest that the QTN indeed affects a *cis*-acting silencer element binding in its wild-type conformation to a *trans*-acting suppressor. This interaction would be abrogated in the *Q* allele, leading to an overexpression of *IGF2* in skeletal muscle and, hence, increased muscle mass.

To further test this hypothesis, we cloned a 578-bp fragment containing the QTN in front of the luciferase reporter gene driven by the *IGF2* P3 promoter. We demonstrated in vivo that most of the *IGF2* transcript in postnatal skeletal muscle derives from the P3 promoter and that the QTN genotype influences the amount of P3-driven *IGF2* transcripts. We clearly demonstrated by transient transfection in C2C12 cells that including the *q* fragment led to a fourfold reduction in expression level, whereas the *Q* fragment reduced the expression level by $^1/_4$ only. This confirmed that the evolutionarily conserved segment containing the QTN indeed functions as a silencer element in myoblastic cells, and that the QTN interferes with this function leading to a threefold increase in expression levels from the *Q* when compared to the *q* allele (i.e., noticeably very similar to the expression ratios observed in vivo). Altogether, these results strongly suggest that the QTN is not only necessary, but also sufficient to determine the QTL status of the corresponding chromosome (Van Laere et al. 2003).

## ON THE IMPORTANCE OF SEQUENCING LIVESTOCK GENOMES

In this paper, we have described the strategies that were used to successfully identify the causal mutations underlying two QTL in livestock. Given the limited resources that are available in the field of livestock genomics, these results are quite remarkable and demonstrate the value of livestock populations for the molecular dissection of complex traits. We strongly believe, therefore, that a more systematic use of livestock populations could very significantly contribute to a fundamental understanding of the molecular architecture of complex inherited traits, as well as to the identification of biochemical pathways affecting phenotypes that are not only of importance for agriculture but have relevance to human health as well. This is well illustrated by the recent identification of a missense mutation in the porcine *PRKAG3* gene causing the RN phenotype associated with a large effect on glycogen storage in skeletal muscle and direct relevance for the pathogenesis of non-insulin-dependent diabetes mellitus in humans (Milan et al. 2000). Other examples abound: The differences in skeletal muscle mass and fat deposition among pig and poultry breeds as described in this paper can be viewed as models of mild obesity; understanding the differences in calcium metabolism between jungle fowl and leghorn chickens linked to egg production could reveal pathways relevant to osteoporosis; cystic ovarian disease is a very common complex inherited disease in dairy cattle and potentially a good model for polycystic ovarian syndrome, one of the most common endocrine disorders of women.

Contrary to previous considerations (Nadeau and Frankel 2000), we believe that QTL mapping and subsequent QTN identification have the potential to make a significant contribution to narrow the "phenotype gap"; i.e., the lack of functional information from mutation-induced phenotypes for most mammalian genes. The detection of the QTN underlying the SSC2 QTL illustrates this vividly. The *Q* to *q* substitution effect corresponds to a difference of 2–3% in muscle mass, which would have been virtually impossible to detect in a phenotype-driven mutagenesis screen. Yet this mutation accounts for 25% of the phenotypic difference in the $F_2$ generation, and its identification revealed a novel *cis*-acting regulatory element in *IGF2*, a gene that has been extensively studied using standard molecular biology.

The genome sequences of a number of livestock species are expected to become available in the not too distant future. The poultry genome should be completed, at least at eightfold coverage, by the end of 2003 (http://genomewustl.edu/projects/chicken/), and the sequencing of the bovine genome should be initiated before the end of the year. A onefold coverage of the porcine genome will be generated by the end of 2003 (M. Fredholm, pers. comm.), and more extensive sequencing of the pig genome will hopefully commence soon after the completion of the poultry and bovine genomes.

So far, the motivation to sequence the genomes of livestock species has mainly come from the realization that comparison of genome sequences from evolutionarily diverse species is a powerful approach to identify functionally important genetic elements (see, e.g., Collins et al. 2003). Having the sequence of the livestock genomes at hand will also immensely facilitate the identification of additional QTNs in these species in which—contrary to the status in the human—large numbers of very convincing QTLs have already been mapped. The potential value of understanding the molecular architecture of QTLs segregating in livestock populations not only for agriculture but, equally importantly, for fundamental biology and the biomedical science, is an additional reason to ensure that livestock genomes are rapidly and completely sequenced.

## ACKNOWLEDGMENTS

Results described in this paper were made possible thanks to the financial support of the Belgian and Walloon Ministries of Agriculture, the EU, Cr-Delta (Arnhem, The Netherlands), LIC (Hamilton, New Zealand), VLB (Auckland, New Zealand), Gentec (Buggenhout, Belgium), and the Swedish Research Council for Environment, Agricultural Sciences and Spatial Planning. We are grateful to Dr. Fredholm for critically reviewing this manuscript.

## REFERENCES

Amarger V., Nguyen M., Van Laere A.S., Nezer C., Georges M., Andersson L. 2002. Comparative sequence analysis of the INS-*IGF2*-H19 gene cluster in pigs. *Mamm. Genome* **13:** 388.

Andersson L. 2001. Genetic dissection of phenotypic diversity in farm animals. *Nat. Rev. Genet.* **2:**130.

Andersson L., Haley C.S., Ellegren H., Knott S.A., Johansson M., Andersson K., Andersson-Eklund L., Edfors-Lilja I., Fredholm M., and Hansson I., et al. 1994. Genetic mapping of quantitative trait loci for growth and fatness in pigs. *Science* **263:**1771.

Andersson-Eklund L., Marklund L., Lundstrom K., Haley C.S., Andersson K., Hansson I., Moller M., and Andersson L. 1998. Mapping quantitative trait loci for carcass and meat quality traits in a wild boar x Large White intercross. *J. Anim. Sci.* **76:** 694.

Arranz J.-J., Coppieters W., Berzi P., Cambisano N., Grisart B., Karim L., Marcq F., Riquet J., Simon P., Vanmanshoven P., Wagenaar D., and Georges M. 1998. A QTL affecting milk yield and composition maps to bovine chromosome 20: A confirmation. *Anim. Genet.* **29:** 107.

Bidanel J.P., Milan D., Iannuccelli N., Amigues Y., Boscher M.Y., Bourgeois F., Caritez J.C., Gruand J., Le Roy P., Lagant H., Quintanilla R., Renard C., Gellin J., Ollivier L., and Chevalet C. 2001. Detection of quantitative trait loci for growth and fatness in pigs. *Genet. Sel. Evol.* **33:** 289.

Blott S., Kim J.-J., Moisio S., Schmidt-Küntzel A., Cornet A., Berzi P., Cambisano N., Ford C., Grisart B., Johnson D., Karim L., Simon P., Snell R., Spelman R., Wong J., Vilkki J., Georges M., Farnir F., and Coppieters W. 2003. Molecular dissection of a QTL: A phenylalanine to tyrosine substitution in the transmembrane domain of the bovine growth hormone receptor is associated with a major effect on milk yield and composition. *Genetics* **163:** 253.

Bradley D.G. and Cunningham E.P. 1999. Genetic aspects of domestication. In *The genetics of cattle* (ed. R. Fries and A. Ruvinsky), p. 15. CAB International, Wallingford, Oxon, United Kingdom.

Collins F.S., Green E.D., Guttmacher A.E., and Guyer M.S. 2003. A vision for the future of genomics research. *Nature* **422:** 835.

Constancia M., Dean W., Lopes S., Moore T., Kelsey G., and Reik W. 2000. Deletion of a silencer element in *IGF2* results in loss of imprinting independent of H19. *Nat. Genet.* **26:** 203.

Coppieters W., Riquet J., Arranz J.-J., Berzi P., Cambisano N., Grisart B., Karim L., Marcq F., Simon P., Vanmanshoven P., Wagenaar D., and Georges M. 1998. A QTL with major effect on milk yield and composition maps to bovine chromosome 14. *Mamm. Genome* **9:** 540.

de Koning D.J., Bovenhuis H., and van Arendonk J.A. 2002. On the detection of imprinted quantitative trait loci in experimental crosses of outbred species. *Genetics* **161:** 931.

de Koning D.J., Rattink A.P., Harlizius B., van Arendonk J.A., Brascamp E.W., and Groenen M.A. 2000. Genome-wide scan for body composition in pigs reveals important role of imprinting. *Proc. Natl. Acad. Sci.* **97:** 7947.

de Koning D.J., Janss L.L., Rattink A.P., van Oers P.A., de Vries B.J., Groenen M.A., van der Poel J.J., de Groot P.N., Brascamp E.W., and van Arendonk J.A. 1999. Detection of quantitative trait loci for backfat thickness and intramuscular fat content in pigs (*Sus scrofa*). *Genetics* **152:** 1679.

Farnir F., Grisart B., Coppieters W., Riquet J., Berzi P., Cambisano N., Karim L., Mni M., Moisio S., Simon P., Wagenaar D., Vilkki J., and Georges M. 2002. Simultaneous mining of linkage and linkage disequilibrium to fine-map QTL in outbred half-sib pedigrees: Revisiting the location of a QTL with major effect on milk production on bovine chromosome 14. *Genetics* **161:** 275.

Farnir F., Coppieters W., Arranz J.-J., Berzi P., Cambisano N., Grisart B., Karim L., Marcq F., Moreau L., Mni M., Nezer C., Simon P., Vanmanshoven P., Wagenaar D., and Georges M. 2000. Extensive genome-wide linkage disequilibrium in cattle. *Genome Res.* **10:** 220.

Georges M., Nielsen D., Mackinnon M., Mishra A., Okimoto R., Pasquino A.T., Sargeant L.S., Sorensen A., Steele M.R., Zhao X., Womack J.E., and Hoeschele I. 1995. Mapping quantitative trait loci controlling milk production by exploiting progeny testing. *Genetics* **139:** 907.

Giuffra E., Kijas J.M.H., Amarger V., Carlborg Ö., Jeon J.-T., and Andersson L. 2000. The origin of the domestic pig: Independent domestication and subsequent introgression. *Genetics* **154:** 1785.

Glazier A.M., Nadeau J.H., and Aitman T.J. 2002. Finding genes that underlie complex traits. *Science* **298:** 2345.

Grisart B., Coppieters W., Farnir F., Karim L., Ford C., Cambisano N., Mni M., Reid S., Spelman R., Georges M., and Snell R. 2002. Positional candidate cloning of a QTL in dairy cattle: Identification of a missense mutation in the bovine DGAT gene with major effect on milk yield and composition. *Genome Res.* **12:** 222.

Hanotte O., Ronin Y., Agaba M., Nilsson P., Gelhaus A., Horstmann R., Sugimoto Y., Kemp S,. Gibson J., Korol A., Soller M., and Teale A. 2003. Mapping of quantitative trait loci controlling trypanotolerance in a cross of tolerant West African N'Dama and susceptible East African Boran cattle. *Proc. Natl. Acad. Sci.* **100:** 7443.

Hoeschele I., Uimari P., Grignola F.E., Zhang Q., and Gage K.M. 1997. Advances in statistical methods to map quantitative trait loci in outbred populations. *Genetics* **147:**1445.

Jeon J.T., Carlborg O., Törnsten A., Giuffra E., Amarger V., Chardon P., Andersson-Eklund L., Andersson K., Hansson I., Lundström K., and Andersson L. 1999. A paternally expressed QTL affecting skeletal and cardiac muscle mass in pigs maps to the *IGF2* locus. *Nat. Genet.* **21:** 157.

Kerje S., Carlborg O., Jacobsson L., Schutz K., Hartmann C., Jensen P., and Andersson L. 2003. The twofold difference in adult size between the red junglefowl and White Leghorn chickens is largely explained by a limited number of QTLs. *Anim. Genet.* **34:** 264.

Kim J.J. and Georges M. 2002. Evaluation of a new fine-mapping method exploiting linkage disequilibrium: A case study analysing a QTL with major effect on milk composition on bovine chromosome 14. *Asian-Australas. J. Anim. Sci.* **15:** 1250.

Looft C., Reinsch N., Karall-Albrecht C., Paul S., Brink M., Thomsen H., Brockmann G., Kuhn C., Schwerin M., and Kalm E. 2001. A mammary gland EST showing linkage disequilibrium to a milk production QTL on bovine chromosome 14. *Mamm. Genome* **12:** 646.

Lynch M. and Walsh B. 1997. *Genetics and analysis of quantitative traits.* Sinauer, Sunderland, Massachusetts.

Meuwissen T.H. and Goddard M.E. 2001a. Fine mapping of quantitative trait loci using linkage disequilibria with closely linked marker loci. *Genetics* **155:** 421.

———. 2001b. Prediction of identity by descent probabilities from marker-haplotypes. *Genet. Sel. Evol.* **33:** 605.

Meuwissen T.H., Karlsen A., Lien S., Olsaker I., and Goddard M.E. 2002. Fine mapping of a quantitative trait locus for twinning rate using combined linkage and linkage disequilibrium mapping. *Genetics* **161:** 373.

Milan D., Jeon J.T., Looft C., Amarger V., Robic A., Thelander M., Rogel-Gaillard C., Paul S., Iannuccelli N., Rask L., Ronne H., Lundstrom K., Reinsch N., Gellin J., Kalm E., Roy P.L., Chardon P., and Andersson L. 2000. A mutation in PRKAG3 associated with excess glycogen content in pig skeletal muscle. *Science* **288:**1248.

Nadeau J.H. and Frankel W.N. 2000. The roads from phenotypic variation to gene discovery: Mutagenesis versus QTLs. *Nat. Genet.***25:** 381.

Nezer C., Moreau L., Wagenaar D., and Georges M. 2002. Results of a whole genome scan targeting QTL for growth and carcass characteristics in a Piétrain X Large White intercross. *Gen. Sel. Evol.* **34:** 371.

Nezer C., Collette C., Moreau L., Brouwers B., Kim J.J., Giuffra E., Buys N., Andersson L., and Georges M. 2003. Haplotype sharing refines the location of an imprinted QTL with major effect on muscle mass to a 250 Kb chromosome segment containing the porcine *IGF2* gene. *Genetics* **165:** 277.

Nezer C., Moreau L., Brouwers B., Coppieters W., Detilleux J., Hanset R., Karim L., Kvasz A., Leroy P., and Georges M. 1999. An imprinted QTL with major effect on muscle mass and fat deposition maps to the IgfII locus in pigs. *Nat. Genet.* **21:** 155.

Riquet J., Coppieters W., Cambisano N., Arranz J.-J., Berzi P., Davis S., Grisart B., Farnir F., Karim L., Mni M., Simon P., Taylor J., Vanmanshoven P., Wagenaar D., Womack J.E., and Georges M. 1999. Identity-by-descent fine-mapping of QTL in outbred populations: Application to milk production in dairy cattle. *Proc. Natl. Acad. Sci.* **96:** 9252.

Rohrer G.A., Ford J.J., Wise T.H., Vallet J.L., and Christenson R.K. 1999. Identification of quantitative trait loci affecting female reproductive traits in a multigeneration Meishan-White composite swine population. *J. Anim. Sci.* **77:** 1385.

Rutherford S.L. and Henikoff S. 2003. Quantitative epigenetics. *Nat. Genet.* **33:** 6.

Schrooten C., Bovenhuis H., Coppieters W., and Van Arendonk J.A.M. 2000. Whole genome scan to detect quantitative trait loci for conformation and functional traits in dairy cattle. *J. Dairy Sci.* **83:** 795.

Sewalem A., Morrice D.M., Law A., Windsor D., Haley C.S., Ikeobi C.O., Burt D.W., and Hocking P.M. 2002. Mapping of quantitative trait loci for body weight at three, six, and nine weeks of age in a broiler layer cross. *Poult. Sci.* **81:** 1775.

Smith S.J., Cases S., Jensen D.R., Chen H.C., Sande E., Tow B., Sanan D.A., Raber J., Eckel R.H., and Farese R.V., Jr. 2000. Obesity resistance and multiple mechanisms of triglyceride synthesis in mice lacking Dgat. *Nat. Genet.* **25:** 87.

Spelman R.L., Coppieters W., Karim L., Van Arendonk J.A.M., and Bovenhuis H. 1996. Quantitative trait loci analysis for five milk production traits on chromosome six in the dutch Holstein-Friesian population. *Genetics* **144:**1799.

Spelman R.J., Huisman A.E., Singireddy S.R., Coppieters W., Arranz J., Georges M., and Garrick D.J. 1999. Short communication: Quantitative trait loci analysis on 17 nonproduction traits in the New Zealand dairy population. *J. Dairy Sci.* **82:** 2514.

Thomsen H., Dekkers J.C.M., Lee H.K., and Rothschild M. 2002. Characterisation of quantitative trait loci for growth and meat quality in a breed cross in swine. In *7th World Congress of Genetics Applied to Livestock Production*, Montpellier, France.

Van Laere A.-S., Nguyen M., Braunschweig M., Nezer C., Collette C., Moreau L., Archibald A.L., Haley C.S., Buys N., Tally M., Andersson G., Georges M., and Andersson L. 2003. Positional identification of a regulatory mutation in *IGF2* causing a major QTL effect on muscle growth in the pig. *Nature* **425:** 832.

Wade C.M., Kulbokas E.J. III, Kirby A.W., Zody M.C., Mullikin J.C., Lander E.S., Lindblad-Toh K., and Daly M.J. 2002. The mosaic structure of variation in the laboratory mouse genome. *Nature* **420:** 574.

Wall J.D. and Pritchard J.K. 2003. Haplotype blocks and linkage disequilibrium in the human genome. *Nat. Rev. Genet.* **4:** 587.

Weller J.I., Kashi I., and Soller M. 1990. Power of daughter and granddaughter designs for determining linkage between marker loci and quantitative trait loci in dairy cattle. *J. Dairy Sci.* **73:** 2525.

# Evolving Methods for the Assembly of Large Genomes

R.A. Gibbs and G.M. Weinstock
*Human Genome Sequencing Center, Baylor College of Medicine, Houston, Texas 77030*

Several large genomes have been sequenced in recent years, leading to the general perception that genome assembly is a solved problem, and, aside from the considerable expense of generating the DNA sequence reads, the pathway to completing future projects is straightforward. Closer examination reveals that this is not the case, and the complexity of large genomes and the relative newness of the tools available for piecing them together, plus the rapid evolution of related technologies, make this an active area for scientific development. A particular concern is the ongoing role of mapping and utilization of large insert bacterial artificial chromosome clones (BACs) when tackling new species for which resources and reagents are scarce.

Current approaches for assembling large genomes include the "hierarchical" strategy, used for the publicly funded nematode consortium (The *C. elegans* Sequencing Consortium 1998) and human genome projects (Lander et al. 2001), and the whole-genome shotgun (WGS) methods (Weber and Myers 1997) used for the privately funded human sequence (Venter et al. 2001), *Drosophila melanogaster* (Adams et al. 2000), mouse (Waterston et al. 2002b), and, more recently, ciona (Dehal et al. 2002), zebrafish (http://www.sanger.ac.uk/Projects/D_rerio/), *Drosophila pseudoobscura* (http://www.hgsc.bcm.tmc.edu/projects/drosophila/), the chimp (Pennisi 2003), pufferfish (Aparicio et al. 2002), dog (Kirkness et al. 2003), and honeybee (http://www.hgsc.bcm.tmc.edu/projects/honeybee/). Conceptually, these methods represent opposite extremes (Fig. 1). In the hierarchical strategy, considerable effort is required at the beginning of the project and generally includes the generation, manipulation, and analysis of BACs. Ideally, the precise positional relationship between the BACs is established prior to sequencing. Next, each individual ~200-kb clone is treated as an individual, localized DNA sequence project. This includes random sequencing as well as a more painstaking finishing phase. Finally, the BAC sequences are joined at the end of the project, to reconstruct the complete genome.

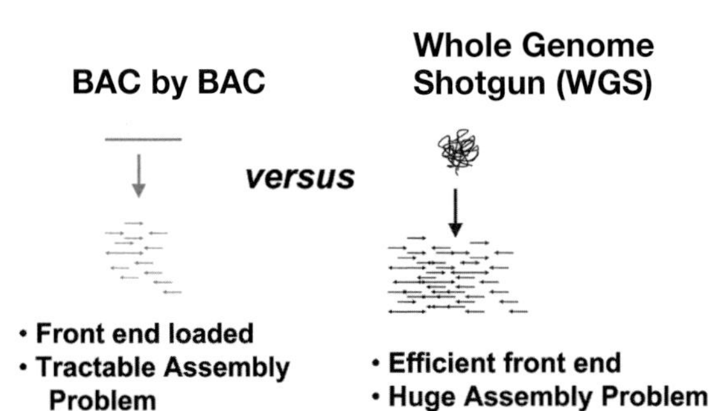


**Figure 1.** The hierarchical strategy versus whole-genome sequencing.

In contrast, in a WGS project, sequences are generated from the ends of inserts of randomly selected (shotgunned) subclones of the whole genome. When a sufficient average depth of sequence coverage is achieved, the genome is assembled by advanced software that primarily functions by recognizing individual sequence overlaps. To maintain assembly accuracy over long genome distances, the sequences are generated from subclones with different insert sizes. The assembly is subsequently constrained by requiring that sequence reads from the ends of the same subclone (mate-pairs) are at their expected relative positions in the final sequence (Edwards et al. 1990). In many instances, these clone-end sequences include at least a small fraction of BAC ends, since those clones are easily manipulated and their inserts span hundreds of kilobases.

Depending on the depth of coverage, a randomly sequenced genome can be referred to as a preliminary draft (<4x coverage), a draft (4–6x), or a deep draft (≥7x). The overall quality of the draft improves steadily with the average sequence coverage, regardless of whether the sequence unit is a single BAC or a whole genome (Bouck et al. 1998).

The distinction between the hierarchical approach and the WGS method has often focused on the overall quality of the final sequence. This is an inappropriate comparison, since in either case the termination of effort at the draft stage will result in an incomplete sequence. Following the random sequencing and assembly, the current technologies require a careful hand-curation phase to achieve the most accurate product. Thus far, no WGS assemblies of eukaryotic genomes have been directly raised to the highest standards, although a *D. melanogaster* WGS assembly has been subsequently finished on a BAC-by-BAC basis (Celniker et al. 2002), and a similar process is under way in the mouse.

## COMBINED METHOD FOR GENOME SEQUENCING:

The relative performance of the hierarchical strategy when extensively using BAC clones, versus the WGS method with minimal use of BAC data, has been indirectly debated in the context of the different ways that competing groups have assembled the human DNA sequence (Myers et al. 2002; Waterston et al. 2002a, 2003). Even apart from these debates, the shortcomings of each approach are apparent. On the one hand, the hierarchical approach is laborious and requires the development of an intermediate product (the clone map) that is ultimately abandoned. On the other hand, WGS sequencing requires extensive work to piece together the segments of genomes that are repeated over distances longer than one kilobase.

To improve the sequencing of large genomes, we therefore developed a new "Combined Strategy" and applied it to the Rat Genome Sequencing Project (RGSP) (Rat Genome Sequencing Project Consortium 2004). The underlying principle is to combine the precision and orientation afforded by BAC clones with the ease and scalability of a WGS approach (Fig. 2). As a consequence, the RGSP was designed to include low-coverage (1.5x) sequence from a full set of BAC clones that covered the entire rat genome, as well as an abundance of WGS reads that would provide deep overall sequence coverage. The rationale was that specific WGS reads could be recruited to join the sequences that were derived from within BAC clones. When brought together, these mixed reads could be assembled using our familiar software for handling BAC-sized sequencing projects.

This was the first, real-time combined use of BACs and WGS data to generate a complete genome assembly, since, although the previous *Drosophila* and concurrent murine genome projects both used WGS and BACs, these were employed in sequential and separate phases. We predicted the BAC component of the Combined Strategy would be particularly important in the rat genome sequencing program, as it aimed to generate a "draft" of about 7x sequence coverage, but not finished sequence. In this case, the BACs were likely to confer an overall higher quality on the eventual assembly.

Figure 3 illustrates the process of *BAC Fishing*, the first step and a key element of the Combined Strategy. At this stage, the small numbers of "bait" sequence reads from individual BACs are used to identify larger numbers of matching "catch" DNA sequence reads from WGS libraries. Typically 500–800 BAC reads "fish" 2,500–3,000 reads from a 6x WGS pool. We developed *BAC-fisher* (Havlak et al. 2004) software that enabled this fishing process, and PHRAP was used for the subsequent stringent step of local assembly to form the "*enriched BACs*" (*eBACs*) containing the two kinds of reads.

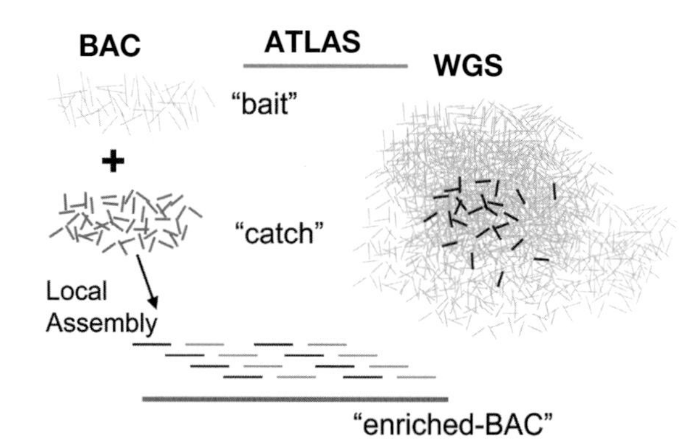


**Figure 3.** Formation of eBACs by BAC-Fishing. Sequencing reads from a lightly sequenced (~1x coverage) BAC are used to identify reads from WGS sequencing (5–6x coverage) that overlap and thus come from the genomic region of the BAC. The relevant WGS reads are co-assembled with the BAC reads using Phrap to produce an eBAC, an assembly with much greater coverage (6–7x) than the original skimmed BAC sequence.

The high quality of the data contained within each *eBAC* validated the basic principle of the Combined Strategy (Fig. 4). The *eBACs* were made publicly available as a useful resource while the project progressed, and formed the foundation for the remainder of the rat assembly. To generate a complete rat genome draft, ~19,000 *eBACs* were assembled into strings of *BACtigs* on the basis of their sequence overlaps. The *BACtigs* were next joined into *super-BACtigs* by large clone insert mate-pair information and ultimately aligned to genome-mapping data to form the complete assembly (Fig. 5). More details of this assembly procedure are discussed elsewhere (Havlak et al. 2004).

The overall quality of the rat assembly was very high, and the project therefore managed to capture both the economic and genome coverage advantages of WGS while retaining the ease and precision of assembly afforded by the BAC components. The additional advantages of the BACs were illustrated by a comparison with

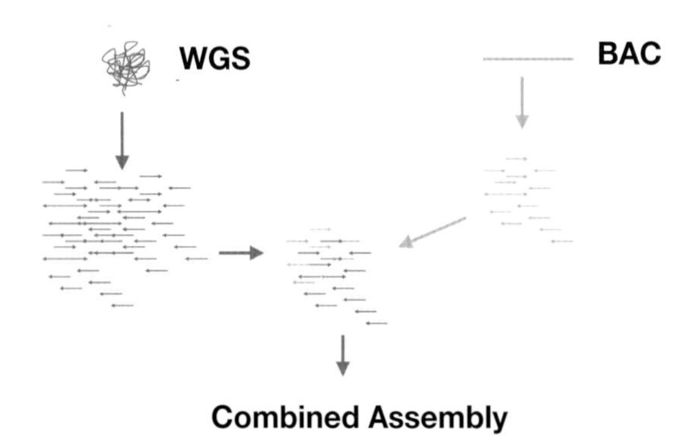


**Figure 2.** The Combined Strategy for sequencing large genomes.

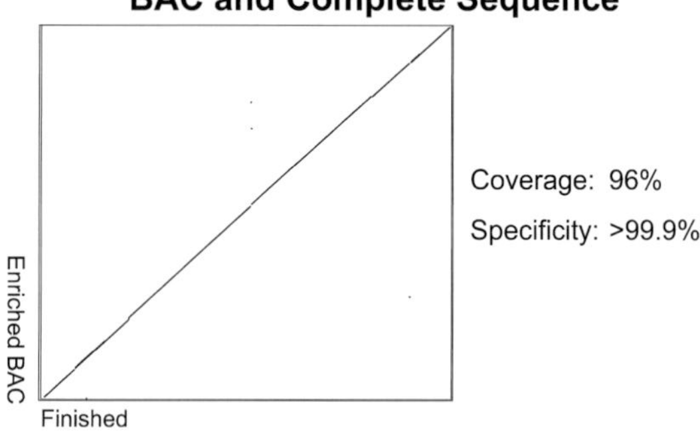


**Figure 4.** High quality of eBACs. An example of a base-by-base comparison between an eBAC (*ordinate*) and the same region of finished sequence (*abscissa*). The colinearity and a few gaps are evident. No repeat filtering was performed, resulting in the large amount of signal off the diagonal throughout the graph.

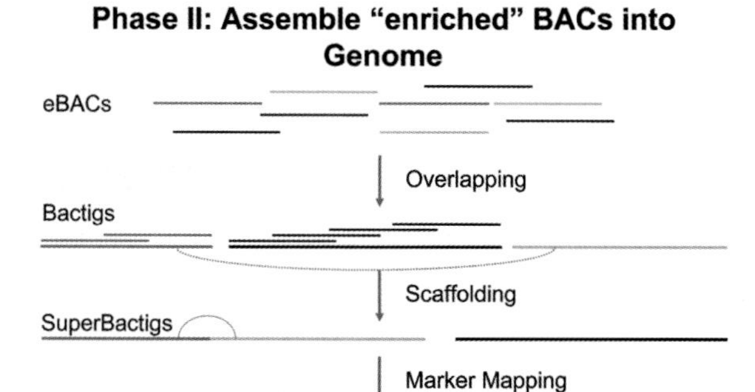


**Figure 5.** Assembly of eBACs. Overlaps between eBACs are identified by BAC end sequences and other methods (Chen et al. 2004), and sequences of overlapping eBACs are assembled into Bactigs. Bactigs are ordered and oriented with respect to each other, first using read pair information (superbactigs), and then based on marker, fingerprint, and other information (ultrabactigs). Sequences of chromosomes are thus based on the arrangement of ultrabactig units.

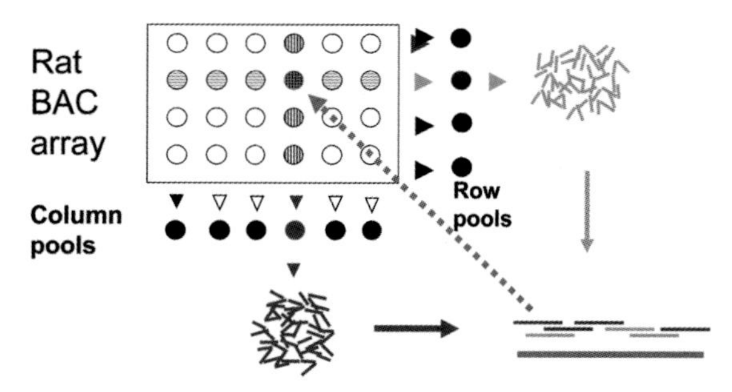


**Figure 6.** Clone Array Pooled Shotgun Sequencing (CAPSS). A 4 x 6 array of BACs is shown. For each column, the BACs are pooled and a shotgun library is made and sequenced (*hatched reads*). The same process is repeated for each row (*black reads*). Then each column pool is co-assembled with each row pool, and contigs composed of both row and column reads are identified. These contigs belong to the BAC at the intersection of the row and column, and this deconvolutes a set of contigs for each BAC. These contigs can then be used for *BAC-Fishing* with WGS reads as described above to build a more complete sequence of the BAC. (Reprinted, with permission, from Cai et al. 2001.)

the contemporaneous murine WGS draft assembly (see Table 1) (Waterston et al. 2002b). Among other favorable statistics, the rat data showed the combined method had captured many more regions of genomic duplication (2.9% vs. 0.01%). The overall genome size in the assembly was also larger for the rat than the initial mouse assembly (by 250 Mb), which we attribute in part to the BAC component enabling the recovery of repetitive regions of the rat genome. These were two subtle but significant distinctions between the results of the two assembly approaches.

## CLONE-ARRAY POOLED SHOTGUN SEQUENCING STRATEGY

The experiences above indicated the value of a BAC-based component in large genome products and suggested that their continued inclusion would be advantageous. This is especially useful if there is to be no further effort to "finish" the genome. A significant challenge was introduced, however, due to the large number of individual BAC preparations and subclone libraries that are required: For a mammalian-sized project, this involves the preparation of 20,000 BAC plasmids and shotgun libraries. In the rat genome project, these were generated over a period of 15 months in one of the most challenging and least automated parts of the project. Development of more efficient and inexpensive methods for the use of BAC clones in large genome projects was therefore highly desirable.

To address this need, the Baylor College of Medicine–Human Genome Sequencing Center (BCM-HGSC) further improved whole-genome sequencing by adopting the approach of Cai et al. (2001). *Clone-Array Pooled Shotgun Sequencing* (CAPSS) (Fig. 6) utilizes physical arrays of BAC clone DNAs that are sequenced as shotgun libraries constructed from row or column pools. The sequence data are then assembled by iterative computational comparison of the row and column data. Sequence contigs containing mixtures of reads from a row and a column map back to the individual well (and BAC) at the intersecting point in the array. The simple logic of CAPSS has instant appeal, especially since it obviates the need for many of the individual shotgun libraries, reducing them to as low as a "$2\sqrt{N}$" requirement ($N$ = the number of BACs). Other, more subtle advantages of CAPSS include drastically reduced computational requirements and the opportunity to provide more genome coverage with BAC clones in each genome project. This latter benefit may eliminate the need for BAC fingerprinting in future projects.

**Table 1.** Rat, Mouse, and Human Genome Assemblies

| | Rat | Mouse | Human |
|---|---|---|---|
| Size | 2.75 Gb | 2.5 Gb | 2.9 Gb |
| Segmental duplications | 2.9% | 0.01% | 5% |

Assemblies: rat (Rat Genome Sequencing Project Consortium 2004), mouse (Waterston et al. 2002b), and human (Lander et al. 2001). Segmental duplications from Bailey et al. (2002, 2004) and Waterston et al. (2002b). The mouse fraction represents that in the initial high-coverage WGS assembly and increased to 1–1.2% in subsequent assemblies.

## CAPSS SIMULATIONS

Experiments employing CAPSS in both simulations and arrays (10 x 10, 24 x 24, 48 x 48) with real sequence reads showed that reads can be successfully assigned to over 80% of BACs in an array and can be assembled into contigs (Csuros and Milosavljevic 2002, 2003, 2004; Milosavljevic et al. 2003). This can be most robustly accomplished when transverse pooling is used, with the BACs in two arrays, each containing the same BACs but in a shuffled order so that no two BACs are in a row or column together in both arrays. This allows various issues to be effectively handled, such as the presence of overlapping BACs, which can give misleading results with assignment from a single array, or unequal representation of BACs in a pool, which is generally found sporadically and is of limited impact when multiple pools are used. Typically, a total sequence coverage of about 1× per BAC is distributed among the rows and columns, and this is sufficient for deconvolution of most reads to individual BACs. The contigs that result from assembly of the deconvoluted reads, while representing partial coverage of a BAC, can be used as bait in *BAC-Fishing* to add WGS reads which, when assembled, result in full BAC coverage comparable to eBACs, described above. Moreover, since the overhead associated with construction of individual eBAC sequences is reduced dramatically by pooling, it is possible to employ deeper tiling paths, improving the quality of draft sequences by reducing gaps and resolving duplications on a finer scale.

This same approach can be used with much lower sequence coverage to map BACs to a reference sequence and build a BAC map. *Pooled Genome Indexing* (PGI) provides cross-species alignments by comparing reads against the genome of a closely related organism (Csuros and Milosavljevic 2002, 2004). Where a row read and a column read align close to each other in the genome (within a BAC insert length), the reads are deconvoluted to the BAC at the row–column intersection, and the BAC is simultaneously mapped to the region between the two matches. PGI pooling schemes are being applied to map rhesus macaque BAC clones to the human genome. This sequence-based mapping integrates the mapping and sequencing components of a genome project. Although it is possible that a PGI approach could replace BAC fingerprinting, in practice, fingerprinting still provides an independent data set that is valuable in quality assessment. However, the ability to map high clone coverage with PGI suggests that an effective strategy is to limit future fingerprinting to those BACs on a tiling path derived from PGI. These various BAC pooling methods are on firm theoretical and technical footing and are currently being tested in the genome projects at the BCM-HGSC (Table 2). They are expected to lead to cost reductions of the order of $1.5–2.0 million for a full draft sequence of a mammalian genome.

## OTHER COMPONENTS OF GENOME PROJECTS

Additional data, beyond the final draft DNA sequence, are required to realize the full utility of a genome project. Three avenues of additive data that increase value of a genome project are finishing, characterization of polymorphisms, and sequencing cDNAs. Not surprisingly, the use of BACs in the genome assemblies has ramification for the development of these additional component data.

*Finishing* encompasses a range of activities that produce differing degrees of polishing of the draft sequence. At one end extreme are gap-filling and quality improvements that can be driven by automated sequence analysis and produce a useful but not complete improvement in the sequence. Alternatively, to reach the highest grade of sequence quality requires much more sophisticated and labor-intensive efforts for resolving difficult regions. These latter approaches employ a variety of techniques such as different sequencing chemistries, specialized shotgun library techniques, or transposon-based methods.

A reasonable compromise at the genome scale is for a draft sequence product to be greatly improved by more automated approaches, whereas selected regions are finished to a higher grade. This makes practical the possibility of finishing essentially all coding regions as well as targeted features, such as genes of interest recommended by the research community (e.g., QTLs); genes related by homology with disease or important models in other species; genes of interest identified by differential rates of evolution; presence/absence in closely related genomes; boundaries of syntenic regions with human or closely related genomes; difficult to assemble regions (e.g., due to high repeat content); and members of new repeat classes.

As noted above in the rat project, finished sequence also provides a gold standard against which the draft assembly can be compared to quantitatively assess quality (Fig. 4). BAC-based projects provide a convenient way to define regions for finishing as well as the ideal reagent to use for these directed finishing approaches.

*Exploring polymorphism* in a genome is also an impor-

**Table 2.** Genome Sequencing Projects Under Way at the BCM-HGSC

| Species | Dates | Size | Goal | Status | Methods[a] | %BCM |
|---|---|---|---|---|---|---|
| Human (*Homo sapiens*) | 1990–2003 | 2.9 Gb | finished | complete | clone by clone | 10.5 |
| Mouse (*Mus musculus*) | 1997–1999 | 2.6 Gb | finished | N/A | clone by clone | <1.0 |
| Rat (*Rattus norvegicus*) | 1999–2003 | 2.8 Gb | draft | complete | combined | 50 |
| Fruit fly (*D. melanogaster*) | 1998–2003 | 0.13 Gb | finished | complete | mixed-sequential | 35 |
| Fruit fly (*D. pseudoobscura*) | 2002–2003 | 0.13 Gb | draft | complete | WGS | 100 |
| Honeybee (*Apis mellifera*) | 2002–2003 | 0.27 Gb ? | draft | in progress | CAPSS-Combined | 100 |
| Social amoeba (*D. discoideum*) | 1998–2003 | 34 Mb | finished | complete | chromosome shotgun | 20 |
| cDNAs (*Homo sapiens*) | 1995–2003 | 7,000 | finished | N/A | CCS | 30 |
| cDNAs (*Mus musculus*) | 1999–2003 | 7,000 | finished | N/A | CCS | 30 |
| Human re-sequencing | 2003–2004 | 2.9 Gb | reads | in progress | shotgun | 50 |
| Sea urchin (*S. purpuratus*) | 2003– | 0.8 Gb | draft | in progress | CAPSS-Combined | 100 |
| Bovine (*Bos taurus*) | 2003– | 2.9 Gb | draft | in progress | CAPSS-Combined | >80[a] |
| Rhesus monkey (*M. mulatta*) | 2003– | 2.9 Gb | draft | in progress | CAPSS/PGI-Combined | >60[a] |
| Red flour beetle (*T. castaneum*) | 2004– | 0.2 Gb | draft | in progress | WGS | 100 |
| Bacteria (numerous) | ongoing | <5 Mb | finished | in progress | WGS | 100 |

[a](WGS) Whole-genome shotgun; (CAPSS) clone-array pooled shotgun sequencing (Cai et al. 2001); (CAPSS/PGI) CAPSS/pooled genomic indexing.

tant adjunct to a draft reference sequence. Mapping such sequence variation has high value in genetic analysis, as well as for insights into population structures and natural variation. Where structured inbred laboratory populations or well-studied outbred populations are available, there is a clear justification for SNP discovery within genome sequencing efforts. In other organisms, SNP discovery is harder to justify—for example, when there is little genetic analysis to exploit such markers.

The range of genomes currently under study at the BCM-HGSC illustrates the variable value in studying polymorphism (Table 2). The human studies justified targeted SNP discovery by sequencing additional human genomes to light coverage. Similar data generation would be of utility in the rat and bovine genome projects, as each has interesting and mapped quantitative traits. In another project, the rhesus macaque, SNP discovery is harder to justify, but the degree of natural diversity of these populations is an intriguing question. Finally, in the sea urchin project, the most polymorphic outbred genome we are studying, there is no obvious application for a map of polymorphisms.

Polymorphism is also problematic for genome assemblies, as it results in a high level of mismatched bases between overlapping sequence reads. In the extreme, a very highly diverged polymorphism can require the independent assembly of each separate haplotype—effectively doubling the effort required for the genome sequence. However, BACs offer a pathway to resolve this difficulty, as each BAC clone comes from a single chromosome and allows greater definition of its haplotype than would be possible with purely WGS reads.

*Sequencing cDNAs* is also an important part of the menu of items that can contribute to a completed genome. The BCM-HGSC has previously developed a novel method for rapid full-length cDNA sequencing and applied it as part of the Mammalian Genome Collection project (http://mgc.nci.nih.gov/) to sequence over 15,000 individual cDNAs. These data have enormous utility in annotation as well as providing a data set for measuring completeness of the draft sequence, and therefore cDNA generation has been included in each subsequent project where possible.

In the case of cDNA sequencing, BACs do not contribute a direct benefit in the generation of the primary data. Nevertheless, in BAC-based projects, there are frequent occasions where exons of interesting cDNAs are observed in the genome, and the BAC clones from which they are derived are scrutinized in order to give the most accurate genomic structure and complete the gene model.

## OTHER KINDS OF DATA ON THE HORIZON

Fluorescent Sanger DNA sequencing technology has now been used for ~20 years, and has so far provided a superior performance to alternative procedures. Individual sequence determinations typically span >700 contiguous high-quality bases, creating a high standard for comparison to other technologies. Nevertheless, there are many innovative alternative strategies under development and considerable interest in improving on the cumbersome gel-matrix separation phase of the Sanger approach. Much of this effort is in the industrial sector and is directed at the enablement of highly parallelized procedures, so many more reactions can occur simultaneously.

A common feature of each new method is therefore that the likely length of each single contiguous sequence determination will be much less than the existing technology. In some cases, the effective length of new sequence read types will measure in the range of 100 bases or so, but techniques that produce very short reads (10–20 bases) cannot be discounted. It is also likely that the base reads generated will have an overall lower quality and reproducibility. Nevertheless, these new methods will very possibly provide superior performance in terms of the total number of newly sequenced bases for a given time or cost. Therefore, assembly strategies and algorithms will be required to use these data in an efficient manner.

The ability to confidently align sequences diminishes as the read length and quality decrease (reduced signal) and the total sequence target complexity increases (increased noise). In addition, shorter reads will not have the same ability to span many classes of genome repeats. Hence, the likely influx of these new data will provide a much greater challenge for WGS projects than for more localized BAC alignments. This is a strong additional argument for a current focus on refining methods for assembly based on BAC clones.

## SUMMARY OF STRATEGY FOR SEQUENCING NEW GENOMES

The balance of available resources and scientific priorities dictates that most new genomes will be "drafted" and not finished as in the human and mouse genome projects. A practical strategy therefore aims to maximize the yield of biological information that can be realized within this "draft sequencing" model by generating sequence coverage that is as complete, contiguous, and high-quality as possible. Achieving this requires coordination of many elements. On the basis of our previous experience, we recommend exploiting components of BAC-constrained sequences, clone pooling, WGS reads, and a modest amount of physical mapping. The overall concert that is orchestrated is complex, and each component has features that need to be tuned to take advantage of the individual contributions and characteristics of the particular genome being sequenced, to optimize the overall result. For example, the balance between the depth of coverage of BAC reads and WGS reads, and the genome size, needs careful choice. Similarly, the role of clones with different insert sizes in each of the categories is important in order to ensure precise joining of "contigs" into scaffolds. Each of these needs to be considered in the context of the ease and expense of generating that class of data and the history of resources available for that organism. The precise strategy for each individual organism is slightly different. A global goal is to both standardize the approaches that can be used for future genomes and to better understand which elements of the characterization of individual genomes must be most precisely defined at the outset, to ensure smooth execution of a genome project.

## CONCLUSIONS

Extensive BAC-based sequence data remain an important component of sequencing efforts directed at large complex genomes, where there are likely to be significant regions of duplication. The BACs can also aid the resolution of regions of extensive heterozygosity during genome assembly. These advantages are particularly important for the analysis of genomes that are expected to remain as "draft" sequences and not be finished to high quality by expensive and largely manual finishing efforts. The use of individual BACs adds costs, however, and this has driven our development of innovative methods including BAC pooling protocols for streamlined clone handling. The ongoing use of BACs will likely be of particular importance in the future. New sequencing technologies will produce copious quantities of inexpensive data that will be of lower quality than the current expensive fluorescent Sanger sequencing reads, and the BAC-enabled constraints will assist in their correct assembly.

## ACKNOWLEDGMENTS

We thank the many staff members of the BCM-HGSC who have contributed effort and ideas over the 10-year history of the program. We thank the members of the Rat Genome Sequencing Consortium for their dedication to that project. Special thanks to the National Human Genome Research Institute for support, as well as the National Heart, Lung, and Blood Institute and the Department of Energy.

## REFERENCES

Adams M.D., Celniker S.E., Holt R.A., Evans C.A., Gocayne J.D., Amanatides P.G., Scherer S.E., Li P.W., Hoskins R.A., Galle R.F., George R.A., Lewis S.E., Richards S., Ashburner M., Henderson S.N., Sutton G.G., Wortman J.R., Yandell M.D., Zhang Q., Chen L.X., Brandon R.C., Rogers Y.H., Blazej R.G., Champe M., and Pfeiffer B.D., et al. 2000. The genome sequence of *Drosophila melanogaster*. *Science* **287:** 2185.

Aparicio S., Chapman J., Stupka E., Putnam N., Chia J.M., Dehal P., Christoffels A., Rash S., Hoon S., Smit A., Gelpke M.D., Roach J., Oh T., Ho I.Y., Wong M., Detter C., Verhoef F., Predki P., Tay A., Lucas S., Richardson P., Smith S.F., Clark M.S., Edwards Y.J., and Doggett N., et al. 2002. Whole-genome shotgun assembly and analysis of the genome of *Fugu rubripes*. *Science* **297:** 1301.

Bailey J.A., Church D.M., Ventura M., Rocchi M., and Eichler E.E. 2004. An analysis of segmental duplications and genome assembly in the mouse. *Genome Res.* (in press).

Bailey J.A., Gu Z., Clark R.A., Reinert K., Samonte R.V., Schwartz S., Adams M.D., Myers E.W., Li P.W., and Eichler E.E. 2002. Recent segmental duplications in the human genome. *Science* **297:** 1003.

Bouck J., Miller W., Gorrell J.H., Muzny D., and Gibbs R.A. 1998. Analysis of the quality and utility of random shotgun sequencing at low redundancies. *Genome Res.* **8:** 1074.

Cai W.W., Chen R., Gibbs R.A., and Bradley A. 2001. A clone-array pooled shotgun strategy for sequencing large genomes. *Genome Res.* **11:** 1619.

Celniker S.E., Wheeler D.A., Kronmiller B., Carlson J.W., Halpern A., Patel S., Adams M., Champe M., Dugan S.P., Frise E., Hodgson A., George R.A., Hoskins R.A., Laverty T., Muzny D.M., Nelson C.R., Pacleb J.M., Park S., Pfeiffer B.D., Richards S., Sodergren E.J., Svirskas R., Tabor P.E., Wan K., and Stapleton M., et al. 2002. Finishing a whole-genome shotgun: Release 3 of the *Drosophila melanogaster* euchromatic genome sequence. *Genome Biol.* **3:** RESEARCH0079.

Chen R., Sodergren E., Gibbs R., and Weinstock G.M. 2004. Dynamic clone tiling path building in the rat genome sequencing project. *Genome Res.* (in press).

Consortium (The *C. elegans* Sequencing Consortium). 1998. Genome sequence of the nematode *C. elegans:* A platform for investigating biology. *Science* **282:** 2012.

Csuros M. and Milosavljevic A. 2002. Pooled genomic indexing (PGI): Mathematical analysis and experiment design. In *Algorithms in bioinformatics: Second International Workshop* (ed. R. Guigo and D. Gusfield), vol. 2452, p. 10. Springer Verlag, Heidleberg, Germany.

———. 2003. Clone-array pooled shotgun mapping and sequencing: Design and analysis of experiments. *Genome Informatics* **14:** 186.

———. 2004. Pooled genomic indexing (PGI): Analysis and design of experiments. *J. Comput. Biol.* (in press).

Dehal P., Satou Y., Campbell R.K., Chapman J., Degnan B., De Tomaso A., Davidson B., Di Gregorio A., Gelpke M., Goodstein D.M., Harafuji N., Hastings K.E., Ho I., Hotta K., Huang W., Kawashima T., Lemaire P., Martinez D., Meinertzhagen I.A., Necula S., Nonaka M., Putnam N., Rash S., Saiga H., and Satake M., et al. 2002. The draft genome of *Ciona intestinalis:* Insights into chordate and vertebrate origins. *Science* **298:** 2157.

Edwards A., Voss H., Rice P., Civitello A., Stegemann J., Schwager C., Zimmermann J., Erfle H., Caskey C.T., and Ansorge W. 1990. Automated DNA sequencing of the human HPRT locus. *Genomics* **6:** 593.

Havlak P., Chen R., Durbin K.J., Egan A., Ren Y.S.X., Weinstock G., and Gibbs R. 2004. The Atlas genome assembly system. *Genome Res.* (in press).

Kirkness E.F., Bafna V., Halpern A.L., Levy S., Remington K., Rusch D.B., Delcher A.L., Pop M., Wang W., Fraser C.M., and Venter J.C. 2003. The dog genome: Survey sequencing and comparative analysis. *Science* **301:** 1898.

Lander E.S., Linton L.M., Birren B., Nusbaum C., Zody M.C., Baldwin J., Devon K., Dewar K., Doyle M., FitzHugh W., Funke R., Gage D., Harris K., Heaford A., Howland J., Kann L., Lehoczky J., LeVine R., McEwan P., McKernan K., Meldrim J., Mesirov J.P., Miranda C., Morris W., and Naylor J., et al. (International Human Genome Sequencing Consortium). 2001. Initial sequencing and analysis of the human genome. *Nature* **409:** 860.

Milosavljevic A.M., Csuros M., Weinstock G., and Gibbs R. 2003. Shotgun sequencing, clone pooling, and comparative strategies for mapping and sequencing. *Targets* **2:** 245.

Myers E.W., Sutton G.G., Smith H.O., Adams M.D., and Venter J.C. 2002. On the sequencing and assembly of the human genome. *Proc. Natl. Acad. Sci.* **99:** 4145.

Pennisi E. 2003. Evolution. Chimp genome draft online. *Science* **302:** 1876.

Rat Genome Sequencing Project Consortium. 2004. Evolution of the mammalian genome: Sequence of the genome of the brown Norway rat. *Nature* (in press).

Venter J.C., Adams M.D., Myers E.W., Li P.W., Mural R.J., Sutton G.G., Smith H.O., Yandell M., Evans C.A., Holt R.A., Gocayne J.D., Amanatides P., Ballew R.M., Huson D.H., Wortman J.R., Zhang Q., Kodira C.D., Zheng X.H., Chen L., Skupski M., Subramanian G., Thomas P.D., Zhang J., Gabor Miklos G.L., and Nelson C., et al. 2001. The sequence of the human genome. *Science* **291:** 1304.

Waterston R.H., Lander E.S., and Sulston J.E. 2002a. On the sequencing of the human genome. *Proc. Natl. Acad. Sci.* **99:** 3712.

—. 2003. More on the sequencing of the human genome. *Proc. Natl. Acad. Sci.* **100:** 3022.

Waterston R.H., Lindblad-Toh K., Birney E., Rogers J., Abril J.F., Agarwal P., Agarwala R., Ainscough R., Alexandersson M., An P., Antonarakis S.E., Attwood J., Baertsch R., Bailey J., Barlow K., Beck S., Berry E., Birren B., Bloom T., Bork P., Botcherby M., Bray N., Brent M.R., Brown D.G., and Brown S.D., et al. (Mouse Genome Sequencing Consortium). 2002b. Initial sequencing and comparative analysis of the mouse genome. *Nature* **420:** 520.

Weber J.L. and Myers E.W. 1997. Human whole-genome shotgun sequencing. *Genome Res.* **7:** 401.

# Mouse Genome Encyclopedia Project

Y. Hayashizaki

*Laboratory for Genome Exploration Research Group, Riken Genomic Sciences Center (GSC), Riken Yokohama Institute; 1-7-22 Suehiro-cho, Tsurumi-ku, Yokohama, Kanagawa, 230-0045 Japan; and Genome Science Laboratory, Discovery and Research Institute, RIKEN Wako Main Campus, 2-1 Hirosawa, Wako, Saitama, 351-0198 Japan*

The large-scale approach in life science was begun to cover various "omics," resulting in the sketching out of a whole view of genetic information. On April 14, 2003, the completion of the sequencing of the human genome was announced by the six countries involved in the International Human Genome Project. This essential platform for life science in the 21st century was achieved through enormous effort as a complete international collaboration. The genome sequence is clearly the gold standard for the next generation of life science; however, it is not enough merely to understand what is written in the genome sequence. Two independent approaches are essential. One is genome sequence analysis and the other is transcriptome analysis; without this, we have no direct proof of the existence of the actual gene. Prediction of a gene from the genome sequence by computer software could not determine which exons are connected together, in which tissue(s), and at which stage(s). In this sense, comprehensive transcriptome analysis is essential to understanding the content of the genome, and especially in understanding the actual transcripts. To achieve this goal, transcriptome analysis should be based on the principle of physical full-length cDNA (FL cDNA) clones being collected, clustered, and sequenced. In 1995, the Human Genome Consortium, led by the US and UK, decided to go into the sequence stage for the human genome (Gibbs 1995) and to complete the full sequence by 2003. In the same year, our small group started the Mouse Genome Encyclopedia Project to fulfil the other essential part of the genome project by analyzing the transcriptome. In this article, I recount the history of the Riken Mouse Genome Encyclopedia Project and mention the results of the project and future prospects.

## THE BACKGROUND AND THE CONCEPT OF THE MOUSE GENOME ENCYCLOPEDIA PROJECT

The basic idea for the Riken Mouse Genome Encyclopedia originated in my mind at the beginning of the 1990s or the end of the 1980s, much earlier than the project's actual start in 1995. In 1995, the situation was that (1) the sequence facility was operated using a slab-type automated sequencer, (2) viral and bacterial genome sequences (*H. influenzae* was the first report) started to be completed, and (3) the US and UK declared that they planned to go into the actual sequencing stage of the human genome. Under these circumstances, I considered what would be an appropriate target for Japan to contribute to this community in a complementary way. FL cDNA was one possible target, as FL cDNA is proof of the existence of real transcripts, not just the prediction of transcripts. In addition, it provides us with the full sequence information of transcripts; expression of proteins and final protein products including bioactive RNA molecules could be detected using FL cDNA. Furthermore, the promoter region could be analyzed in combination with the genome sequence; FL cDNA is a good tool for the assembly of a shotgun genome sequence. Although the physical collection of FL cDNA required the development of higher technology, I considered it a worthwhile attempt. Thus, two goals were proposed for the Riken Mouse Encyclopedia Project. One was the development of a series of technologies required for the collection of FL cDNA and sequence analysis and the other was the establishment of the Mouse Genome Encyclopedia, the comprehensive collection of FL cDNA. Two key technologies required development: FL cDNA technologies and high-speed sequencing techniques. Our Mouse Genome Encyclopedia was defined as consisting of five components: the cDNA clone bank, the cDNA sequence database, the mapping of cDNA on the genome sequence, the expression profile of the comprehensive transcriptome, and the protein–protein interaction database.

## STRATEGY TO COLLECT THE FL cDNA CLONES

Figure 1 shows the strategy of collection, clustering, and sequencing of the complete FL cDNA. Using our FL cDNA technologies, which are described below, clones in the FL-enriched library were spread onto agar plates. Clones were picked randomly and end-sequenced. Using these end sequences, all clones were clustered to make a nonredundant cDNA bank. A representative clone from each cluster was picked, and the entire sequence was determined. The complete cDNA sequence was subjected to subsequent transcriptome analysis, using FANTOM (functional annotation of mouse cDNA) analysis. In this process, the required capacity for sequencing was calculated. The collection and determination of FL cDNAs covering a significant proportion of the transcriptome was achieved in a few years. We estimated that at least 40,000 sequencing passes would be required. In 1995,

§Corresponding author.

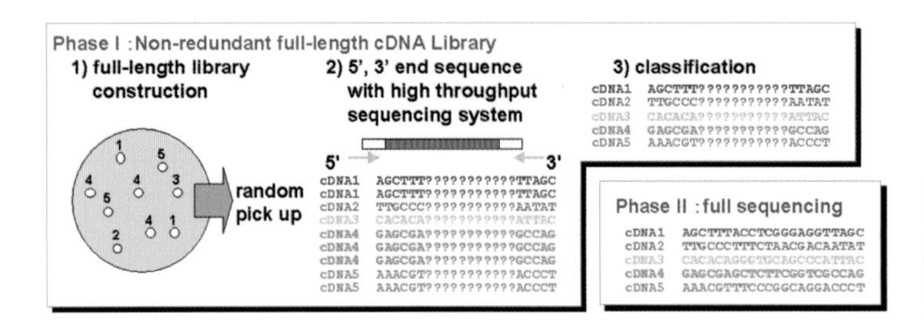


**Figure 1.** Strategy for constructing the mouse genome encyclopedia. This consists of two phases, the first for the construction of the nonredundant FL cDNA library, and the second for determination and analysis of the FL sequence.

neither a capillary sequencer nor a template-DNA preparation system with a sufficiently large capacity had been developed. In this project, we decided to construct a system to fulfil the required capacity and named it the Riken integrated sequence analysis (RISA) system (Itoh et al. 1997, 1999; Shibata et al. 2000). RISA is designed as a pipeline for pumping out FL cDNA data from mouse house to database. RISA consists of several key components for the collection and analysis of FL cDNA (Carninci et al. 1996, 1997, 1998, 2000, 2001), such as a plasmid preparator (Itoh et al. 1997, 1999), large-scale cycle sequence system (Sasaki et al. 1997) or transcriptional sequence system based on RNA polymerase (Izawa et al. 1998; Sasaki et al. 1998a,b), RISA 384 capillary sequencer (Shibata et al. 2000), and FANTOM system (Kasukawa et al. 2003). In the following sections, the key technologies developed in this project are described briefly.

## FL cDNA TECHNOLOGY SYSTEM

An ideal FL cDNA is a complete copy of the mRNA population in all its length variations. The mRNA samples prepared from tissues in the conventional methodologies always include truncated and partial mRNAs. Furthermore, reverse transcriptase is unable to make copies of the mRNA template with high fidelity. Differential expression levels of genes are responsible for creating undesirable bias in the frequency of cDNAs in the library. Furthermore, cloning bias on the size of mRNA creates difficulty in isolating FL cDNAs of a large size. To overcome these problems, our group has developed a series of new technologies described below.

### Elongation Technology (Trehalose Method) (Carninci et al. 1998)

Reverse transcriptase stops at the bottom of the stem loop of mRNA. In other words, the major cause of partial cDNAs is the secondary structure of mRNA. To synthesize the first-strand FL cDNA, the secondary structure of mRNA needs to be broken. A higher temperature for reverse transcription is the easiest way to make this possible. However, reverse transcriptase itself is inactivated by heat. Thus, the development of a heat-stable reverse transcriptase would resolve this problem. We proposed the following working hypothesis. Chaperonin functions to refold denatured protein or fold peptides which are in the process of synthesis in the ribosome. If a chaperonin-like substance is present in the reverse transcription reaction, it may protect the enzyme. For this, the chaperonin-like substance must itself be heat stable. We found two references (Attfield 1987; Hottiger et al. 1987) indicating that the disaccharide trehalose is induced after heat shock in *Saccharomyces cerevisiae* and that a disruptive yeast mutant of trehalose synthesis enzyme is very weak against heat shock (De Virgilio et al. 1994). This suggested to us that trehalose itself is a chaperonin-like substance and that the disaccharide must be heat stable. This hypothesis was tested.

As shown in Figure 2a, the 5-kb in-vitro-transcribed RNA is used as the template RNA for first-strand synthesis by reverse transcriptase. The products of reverse transcription are internally labeled using [$\alpha$-$^{32}$P]rUTP. In lane 1, the traditional protocol for reverse transcription at 42°C is used and partial cDNA is synthesized as well as the 5 kb of the full-length product. At 60°C, a much higher temperature than usual, the reverse transcriptase is inactivated and no signal is detected. However, in the presence of trehalose, the 5-kb FL cDNA product is visible in lanes 2 and 3 without any partial products, showing low electrophoretic mobility. In Figure 2b, natural

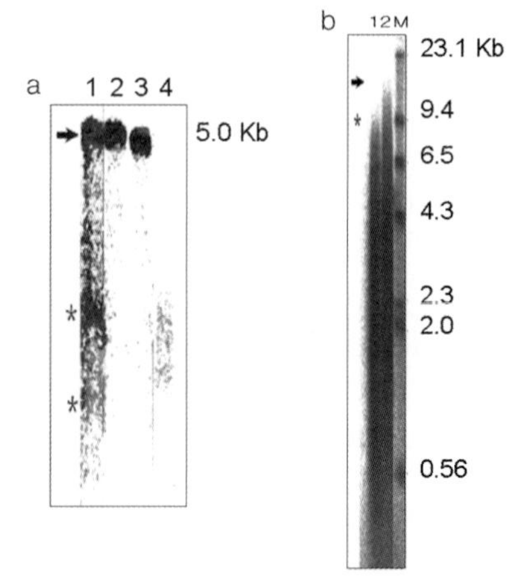


**Figure 2.** Improvement of full-length cDNA yield by thermostabilization of reverse transcriptase by trehalose. (*a*) 5-kb cDNAs were synthesized in the absence of trehalose at 42°C (lane *1*), 60°C (lane *4*); synthesized in the presence of 0.6 M trehalose at 55°C (lane *2*), 60°C (lane *3*). (*Arrow*) Full-length cDNA; (asterisks) truncated cDNAs. (*b*) Reverse transcription of mouse brain mRNA: (Lane *1*) Standard optimized reaction; (lane *2*) reaction in the presence of trehalose at standard temperature; M, $\lambda$-*Hind*III markers. (*Arrow* and *asterisk*) Longest full-length cDNA detectable in the absence and the presence of trehalose, respectively.

mRNA was used. Using the traditional protocol, the longest size of the first-strand cDNA is around 9 kb. However, in the presence of trehalose, it is much improved to 14–16 kb, covering almost all of the transcripts present.

### Selection Technology (CAP Trapper Technology)

Figure 3a shows the structure of mRNA. The CAP site should be labeled to select FL cDNAs. We exploited the "diol" structure shown in this figure, which is a specific feature of the CAP site. The CAP and poly(A) sites are the only two sites where diol groups occur. The diol group can be oxidized by sodium periodine ($NaIO_4$) to produce a dialdehyde group that could be connected to a hydrazide group. Biotin-hydrazide, which is commercially available, is very convenient for direct biotinylation in this reaction. This procedure is based on the chemical reaction, and the efficiency is almost perfect (Carninci et al. 1996, 1997).

Figure 3b shows the whole procedure of CAP trapping. After single-stranded cDNA synthesis, the CAP and poly(A) sites are labeled using the protocol shown in Figure 3a. RNase I is used to cleave remaining single-stranded RNA, but does not act on RNA–DNA hybrids. Single-stranded RNA is cleaved by RNase I at a 5′ site. Cleavage also occurs at a poly(A) site (Fig. 3b). The non-poly(T) complementary sequence is designed as a primer for the first cDNA. Hybridization of mRNA with FL cDNA preserves the biotinylated sites, which would otherwise be lost following RNase I digestion. Therefore, only mRNA hybridized with full-length single-stranded cDNA is trapped by avidin beads. The first-strand cDNA is then isolated and subjected to the subsequent cloning process.

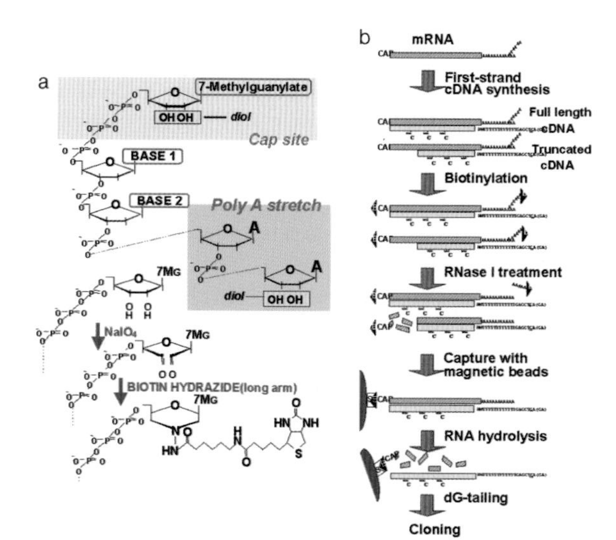


**Figure 3.** (*a*) Structure of mRNA and the procedure of its biotinylation. Two diol groups are oxidized by $NaIO_4$, and then coupled with biotin hydrazide. (*b*) Scheme of CAP Trapper for FL cDNA selection.

### Normalization and Subtraction Technology

A standard library reflecting the original mRNA population is biased, causing unpermissible redundant clones in the process of collecting FL cDNAs. Three populations are defined within the collected cDNAs. These are "abundant cDNA," "cDNAs already collected," and "rare and new cDNA." We used the biotinylated RNAs of the starting material, and biotinylated RNAs transcribed in vitro from cDNA already collected, as subtraction drivers (Carninci et al. 2000). The first-strand cDNA and these drivers are mixed, and hybrids are trapped using avidin beads. Thus, the undesirable populations of first-strand cDNA are eliminated, and only rare and new cDNAs are collected. In this protocol, we did not use PCR because this produces a large cloning bias. This resulted in a good efficiency of new gene discovery in our library.

### New Cloning Vector

We developed cloning vectors λFLC1, λFLC2, λFLC3, and λFLC4, based on infection after packaging and in vivo circularization, using the Cre-lox recombination system (Carninci et al. 2001). λFLC1 is the prototype of our vector, which is based on the insertional λ vector. λFLC2 is the substitution vector for longer cDNAs, of more than 3.5 kb on average. The super-rare cutter sites in λFLC3 are designed to transfer the cDNA to the expression vector. Finally, λFLC4 was designed to clone superlong cDNA, with an average length of around 10 kb. λFLC4 carries the F-factor replication origin, which is very stable even when carrying very long insert DNA; for example, that used in BAC vectors for genomic DNA cloning.

### Summary of the Riken cDNA Library

Riken cDNA technology created a FL cDNA library of good quality. The average length of cDNA is 2.0–2.5 kb in a standard library, and more than 3.5 kb in λFLC2 and λFLC3. Using λFLC4, cDNAs greater than 10 kb are easily cloned. The full-length coverage of the Riken library is such that 90–95% of clones in the standard library encode at least initiation codon (ATG), and 60% of clones in the accumulated cDNA bank cover whole coding sequences (CDS). In the accumulated bank, the full-length rate is lower than that of the single library, because we compromised and incorporated partial cDNAs into the bank if the sequence was novel.

## RISA SYSTEM

As explained above, a high-throughput sequence system with the capacity to finalize the collection of FL cDNA within a few years had not been developed when we started. The required capacity was 40,000 sequence passes per day. Therefore, we decided to develop it in house. A detailed description is given elsewhere (Shibata et al. 2000), but the following points deserve brief mention.

After construction of the FL cDNA library, the clones spread onto agar plates should be picked up by a Qbot system (Genetix Ltd., UK) to make a master plate. A replica plate should be made from this master plate as a backup (Itoh et al. 1999; Shibata et al. 2000). The plated clones should then undergo inoculation, cultivation, plasmid preparation, cycle sequencing reaction or transcriptional sequencing, and sequencing steps. This was set up in the form of a pipeline through our RISA Inoculator, RISA Filtrator and Disitometer, RISA plasmid preparator, GS384 thermal cycler, and RISA 384 capillary sequencer. A basic concept in the development of the RISA system was that error in the identification of clones should be eradicated. A significant difference between the requirements for large-scale cDNA sequencing and large-scale genome sequencing is that the correspondence of the sequence data to clones on master plates has to be perfect in the case of cDNA cloning. We always have to go back to the master plate to recover the FL cDNA clones; therefore, the identification of clones is essential for analysis after sequencing. However, with genome shotgun sequencing, only trace sequences are assembled after sequencing, and the shotgun clones are not used. Thus, in the RISA system, the relative location of the clones is not changed until the data are obtained. Therefore, for processing by the RISA system the samples are transferred on 384-format plates (or 96-format plates for the plasmid preparatory step). The RISA sequencer is also designed to inject all samples directly into 384-format capillary arrays from 384-format plates.

## COLLECTION OF FL cDNA CLONES (PHASE I)

We started to collect the samples from various organs of mice at varying stages in development (Carninci et al. 2003). In total, 267 tissues were collected, including primordial germ cells, fertilized eggs, two-cell and four-cell zygotes, and so on. After construction of FL cDNA, we picked up clones randomly from each library and subjected the first 5000 clones to end sequencing. If the library seems to be saturated, we stop to sequence more clones from it. If it is not saturated, we continue to sequence. In Figure 4, the horizontal axis represents the date from June 1996 to July 2002, and the vertical axis represents the number of clusters. The abundant cDNAs (more than 10 appearances) and the middle abundant cDNAs (6–10 appearances) have already been collected, as shown in this figure. Middle rare cDNAs (2–5 appearances) are also saturated. Only new cDNAs (single appearance) collected as singletons are still increasing.

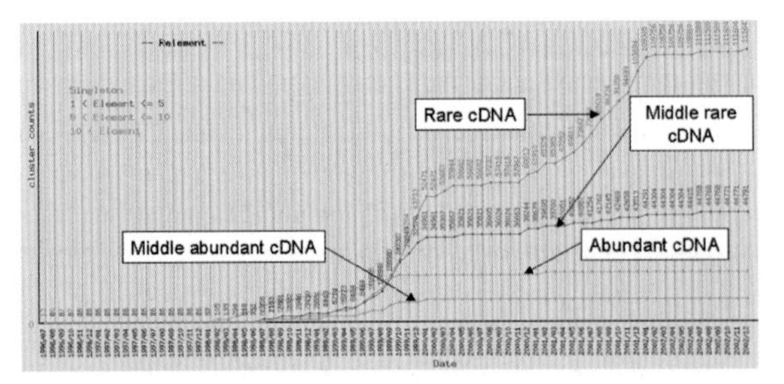


**Figure 4.** The number of genes discovered in the course of this project. The discoveries of new singleton cDNAs are increasing, whereas abundant and middle abundant cDNAs have already become saturated.

In total, 1,916,592 clones from normalized and subtracted libraries were picked up. Calculation of the enrichment efficiency for normalization and subtraction suggests this is equivalent to 14,400,000 clones in a standard library. All of these clones were subjected to end-sequencing from the 3′ site, and these were classified into 188,000 clusters. The data from the clustering and mapping onto the genome sequence of all 520,311 5′-end sequences were used to select 60,770 clones to subject to full sequencing analysis.

## EXPRESSION PROFILES AND PROTEIN–PROTEIN INTERACTION

Expression profiles were analyzed using Stanford-type cDNA microarrays (Miki et al. 2001; Bono et al. 2003). We used mRNA prepared from the whole body of a mouse C57BL/6 embryo at 17.5 days. We produced expression profiles using 20,000 genes, with redundancy in 49 tissues, at FANTOM1. Additionally, we produced 60,000 genes with redundancy in 21 tissues at FANTOM2. To analyze protein–protein interactions, we have developed a high-throughput, mammalian, two-cell hybrid system. The system has a capacity of 20,000 wells/day and operates as one assay system, consisting of one robot and one fluorescent reader. As a pilot study, we have produced 6000 x 6000 protein–protein interactions (Suzuki et al. 2001, 2003). These two systems are very good platforms for subsequent analysis and annotation.

## FANTOM AND INTERNATIONAL FANTOM CONSORTIUM

To annotate the full-length sequencing data, we organized an international meeting named FANTOM (*F*unctional *AN*no*T*ation *o*f *M*ouse cDNA). Over a hundred first-class scientists working in various life science fields were invited to be members of the international FANTOM consortium to annotate the function of each gene. We had two meetings, FANTOM 1 (from August 29 to September 5, 2000) (Kawai et al. 2001) and FANTOM2 (FANTOM2 Typhoon meeting; October 2001, FANTOM2 Cherry blossom meeting; April 2002) (Okazaki et al. 2002). For FANTOM2, we developed a teleconference system (named MATRICS) allowing on-line annotation to be carried out. Every member can annotate in detail the function of genes. FANTOM members revised 25% of the annotation database made by computer software. The conclusion was that 60,770 FANTOM2 sequences still had redundancy and that the database contained 33,994 unique sequences. We had intentionally avoided redundant efforts to sequence the known genes present in public databases, so by adding 44,122 redundant sequences of cDNA known in public databases, we established a database carrying 36,830 unique sequences. This we named Representative Transcript and Protein Set (RTPS). RTPSs are the unique FL cDNA sequences encoding protein coding or protein noncoding transcripts supported by physical clones.

In the process of selecting the 60,770 sequences to subject to full-stretch sequencing, many unique sequences, contained by 1,916,592 clones of the original master bank, were missed. Using the new version (MGSCv3) of the mouse genome sequence database that was provided by the Mouse Genome Sequence Consortium (MGSC) as a part of the collaboration, it was concluded that at least 43,000 unique transcripts are encoded in the mouse genome. It was estimated that the total number of "genes," a word that should be used very carefully (see the next section), is around 60,000.

## TRANSCRIPTIONAL UNIT

To analyze the total number of genes for subsequent research, words with ambiguous definitions, such as "gene," "locus," and so on, should be avoided. In the genome community, the word "gene" is used for "the protein-coding genes (or loci)". This was tacitly defined due to the convenience of computer-assisted exon prediction, but genome sequences do not give any proof of the existence of transcripts. Computer prediction of exons is based on open reading frames (ORF). However, in transcriptome analysis, we meet a lot of "noncoding genes," which should also be incorporated into the transcriptome. Thus, we should not use the word "gene" or "locus" to discuss a genomic region which is transcribed, the so-called "gene" in the past definition.

For this reason, we coined a new term, transcriptional unit (TU). TU is defined as a segment of the genome flanked by the most distal exons from which transcripts are generated. This term can be used as purely computational and unequivocal. The transcripts sharing any exon are encoded by a single TU. If two transcripts do not share any single exon, these two transcripts are in different TUs, even if one is localized in the intron of the other. Where two transcripts encode the sense and antisense strands, respectively, in the same region of genomic DNA, these two genomic segments are different TUs. Thus, by the definition of the new term "TU," we could count the number of so-called "genes" (Okazaki et al. 2002).

## CONTRIBUTION OF RIKEN FANTOM2 CLONES INTERNATIONALLY (APRIL 2003)

After our contribution to the world-wide mouse gene (TU) database, the number in the RTPS has increased to 37,086. As shown in Figure 5, the Riken FANTOM2 collection covers 90% of the total TUs (33,459/37,086); although the remaining 10%, including many known TUs, were covered by our master bank, they were not sequenced to avoid the redundant efforts. A significant proportion (24% of TUs; 8,898/37,086) of the RTPS is also covered by the Mammalian Gene Collection (MGC) supported by the National Institutes of Health (Strausberg et al. 2002). From this analysis, we estimate that a significant proportion of mouse TUs have still not been covered, much less the whole transcriptome, which includes variant transcripts with alternative splicing and differential 5′ and 3′ sites.

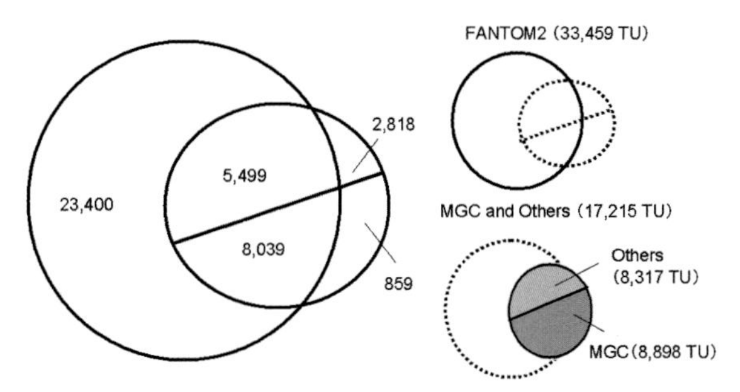


**Figure 5.** Contribution of RIKEN FANTOM2 clones in the world. The FANTOM2 collection covers 90% of all TU published in the world.

## FUNCTIONAL CLASSIFICATION OF MOUSE TRANSCRIPTOME AND NONPROTEIN CODING TRANSCRIPTS

All sequence data were classified by level of homology to known genes. As shown in Figure 6, 20% of all FANTOM2 FL cDNA sequences are identical to known mouse transcripts, and 14% show various levels (85%, 70%, and 50%) of homology with protein sequences from

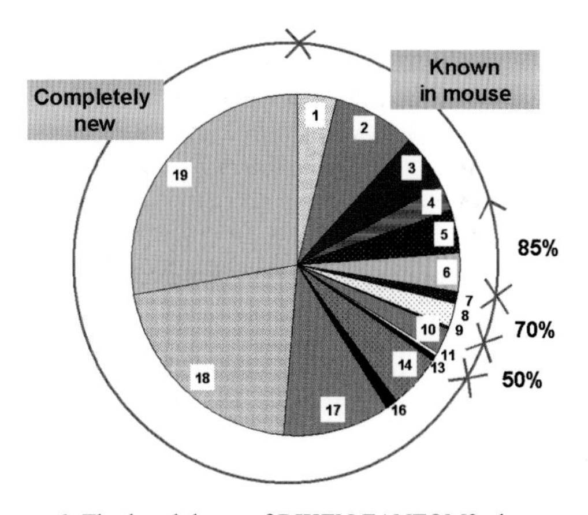


**Figure 6.** The breakdown of RIKEN FANTOM2 clones to the categories defined below. (*1*) Directly assigned by MGI, (*2*) Mouse DNA Hit (>98% ID, >100 bp) Complete, (*3*) Mouse DNA Hit (>98% ID, >100 bp) Partial, (*4*) Protein Hit (>98% ID, >100% length, mouse) Complete, (*5*) Protein Hit (>85% ID, >90% length) Complete, (*6*) Protein Hit (>85% ID, >90% length) Partial, (*7*) Protein Hit (>70% ID, >70% length) Complete, (*8*) Protein Hit (>70% D, >70% length) Partial, (*9*) Protein Hit (>50% ID, >50% length) Complete, (*10*) Protein Hit (>50% ID, >50% length) Partial, (*11*) Inferred from TIGR/UniGene Cluster, (*12*) Inferred from UniGene Cluster, (*13*) Inferred from TIGR Cluster, (*14*) InterPro domain/motif containing, (*15*) MDS domain/motif containing, (*16*) SCOP structural domain/motif containing, (*17*) hypothetical protein, (*18*) EST hit, (*19*) unclassifiable.

other organisms. Surprisingly, the remaining FANTOM2 sequences are novel. These novel sequences can be classified into four categories. The first category consists of sequences containing InterPro motifs (Apweiler et al. 2001), MDS domains (Kawaji et al. 2002), or SCOP structural domains (Gough et al. 2001). The second category constitutes sequences with ORFs of significant size (more than 100 amino acids) that code for hypothetical proteins but contain no recognizable motifs or domains. The sequences in the third category have no ORFs but hybridize to ESTs reported in public databases. The final category contains totally unknown sequences with no ORFs, known motifs, or homology with ESTs.

The first step in FANTOM annotation was identification of ORFs in each sequence. However, sequences with no ORFs were found more frequently than expected. FANTOM2 cDNAs contained 20,487 protein-coding sequences and 16,599 noncoding sequences; this is surprising, given that only around 100 non-coding RNAs (other than tRNAs) were known before FANTOM2 was developed (Fig. 6). A new "continent" of functional noncoding RNAs has been discovered; previously the protein continent has been better explored in the examination of final gene expression. Artifacts including unspliced cDNAs (despite efforts to prepare cytoplasmic RNA only) and genomic contamination were present among 16,599 noncoding RNAs. A significant population of these noncoding RNAs are likely not to be "junk": On average, each noncoding RNA is spliced from three exons, and CpG islands or CG-rich sequences are preferentially located at the 5′ ends of the noncoding RNAs. Since 32,000 protein-coding sequences are predicted from the completed human genome sequence, more than several thousand protein-coding sequences are missing from FANTOM2. However, the prediction of so-called "genes" from the human genome sequence missed a major population (at least several thousand) of noncoding RNAs. Analysis of the functions of these noncoding RNAs is an important task for future research.

## LARGE-SCALE MAPPING OF FL cDNA ONTO THE MOUSE GENOME SEQUENCE

To clarify the chromosomal distribution of the transcripts, we mapped the FL cDNAs onto the mouse genome draft sequence (MGSCv3) (Waterston et al. 2002). Mapping was possible in 32,568 out of 33,047 cDNA sequences. The remaining 479 sequences did not show any homology with mouse genome sequences. One possible explanation for this is the exclusive use of the female mouse in generating the genome database, thus the database does not contain the Y chromosome sequence. Additionally, the mouse genome sequence database is still only in draft form, with 3% of the genome currently unsequenced. The data on the chromosomal locations of the FL cDNAs constitute a powerful tool that will facilitate the positional candidate approach to identifying the genes responsible for specific phenotypes in mutant mice and in human disease. We developed a computer software package, GENOMAPPER, that can list candidate genes when given the flanking marker, expression sites, and, if possible, protein–protein interactions.

## SENSE AND ANTISENSE RNA PAIRS

In the FANTOM2 set, we found an unexpectedly large number of pairs of sense and antisense transcripts. These were discovered through the mapping of cDNA onto the genomic sequence. The pairs include all combinations of coding and noncoding RNAs, and spliced and unspliced sequences. Natural sense and antisense transcripts may regulate gene expression in various ways. For example, the antisense imprinter RNA (AIR) was reported as the noncoding, intronless transcript controlling the imprinted expression of Igf2r (Sleutels et al. 2002). When AIR is disrupted, imprinted expression is heavily perturbed. In some of the sense–antisense pairs, antisense-strand RNA may function to repress the function of the sense-strand RNA by RNAi, the suppression of translation, or other mechanisms. Investigation of the functions of sense–antisense pairs should be an interesting direction for future research.

## CDS ANNOTATION AND ANALYSIS OF PROTEIN-CODING SEQUENCE

The first step of annotation, the identification of the coding sequence, addresses many questions. Is there a significant ORF? Is this ORF protein coding or noncoding? If it is protein coding, is the transcript spliced? Which ORF is the real one? Are there any frameshift errors, or initiation and termination codon errors? Is the cDNA full-length or truncated? Which ATG is the initiation codon? We developed a computer software package, Protein Coding Region Estimator (ProCrest), to answer these questions (Y. Hayashizaki et al., unpubl.). This software calculates the amino acid frequency, tandem amino acid weight matrix, degeneracy of genetic code, tRNA anticodon usage (wobble rules: [U, C], A, G), bias of base contents (G,C and A,G), Kozak consensus, and polyadenylation sites for a given sequence.

Using this software, we identified sequences as protein coding or noncoding and identified CDSs. With the resulting CDS database, functional analysis of the protein sequences predicted by ProCrest was possible. Gene ontology (GO) analysis, motif analysis with Inter Pro and/or Pfam, and gene family analysis were also performed on the CDSs. The software developed by Matsuda et al. (Kawaji et al. 2002) was used to find new motifs in FANTOM2 clones, resulting in the discovery of 10 new putative motifs (Okazaki et al. 2002).

## DYNAMIC VARIATION OF TRANSCRIPTS PRODUCED FROM STATIC GENOME SEQUENCE

One remarkable finding of the FANTOM2 research is that the genome of higher organisms, including humans and mice, can encode much more information than suggested by the number of TUs (genes). To control such enormously complex biological systems, dynamic variation is employed at the transcriptional level, since genomic information itself is very static. In our database, we found an unexpectedly large number of 5′ and 3′ end variations and alternatively spliced sequences. Variation at the 5′ end is mainly due to differences in gene promoters, which are regulated in a tissue- and stage-specific manner. 3′-end variation is so common that it is also very likely to be functional. When classified by 3′-end sequence, our original 1,916,592 clones fell into 188,000 clusters. When analyzed in terms of complexity, these 188,000 3′-end variations are equivalent to ~60,000 TUs. Each TU therefore has more than three 3′-end variations on average. In the 3′-end untranslated region (UTR), many consensus sequences with clear functions are known. One good example is the small stretch "UUAUUUAUU." When included in a 3′-end variation, this motif appears to be the target attacked by endonucleases. mRNAs with the "UUAUUUAUU" motif are degraded very rapidly, and the half-life of such transcripts is very short, whereas mRNAs without this motif have a long half-life. 3′-end variation, therefore, is clearly functional.

Living cells control the selection and ordering of exons using tissue- and stage-specific splicing machinery. Our clones were collected using normalization and subtraction methods to cover a greater variety of TUs. If our subtraction system had worked perfectly, we would not have collected any alternatively spliced transcripts. However, the system is leaky; given the large-scale collection of FL cDNA, plenty of alternatively spliced transcripts were isolated in FANTOM2 analysis. Surprisingly, more than 41% of all TUs encode alternatively spliced forms, and 79% of alternatively spliced TUs alter the amino acid sequences of CDSs. Thus, the number of transcripts and proteins is much larger than the number of TUs (genes). This mechanism allows living cells to expand their genomic information to finely control their life processes.

## THE DISTRIBUTION SYSTEM FOR THE POST-GENOME RESOURCE (DNABOOK)

The Riken Mouse Genome Encyclopedia Project produced two major resources: the transcriptome database and the mouse FL cDNA clone bank. These two resources are key platforms for use in post-genome and post-transcriptome life science. The distribution of these resources is therefore very important in facilitating research in the 21st century. The FANTOM2 database was published on the Internet on December 5, 2002. However, at that point, the clone bank of 60,770 FL cDNA clones still had to be shipped in a box with 100 kg of dry ice, a very inconvenient and tedious process.

We have solved this problem by developing a new technology for the distribution of FL cDNA clones (Hayashizaki and Kawai 2003; Hayashizaki et al. 2003; Kawai and Hayashizaki 2003). The DNA is printed onto paper sheets and may be shipped as a bound "DNABook." This idea was first conceived at the beginning of this project. Figure 7 shows the cover of the first edition of the *Riken Mouse Encyclopedia DNABook* and has been

**Figure 7.** The cover design represents full-length cDNA analyzed by RIKEN Integrated Sequence Analyzer (RISA) and FANTOM computer system. Designed by Dr. Chie Owa.

used from the start of this project as the title slide in our presentations. This design illustrates the following: All of the FL cDNAs were fed into the RISA system, many sequences have come out of RISA, and the database was subsequently established on computer hard disk. An image of the hard-copy encyclopedia "book" is also included in this figure.

We tested the performance of the DNA printed sheet under various conditions of temperature, humidity, pressure, and other environmental factors, and its resistance to scratching or touching. Under normal conditions, we believe it possible for cDNA clones to be maintained on the paper sheet for at least several years.

The DNABook has the potential to significantly improve the processes of storing and shipping genome resources. Traditionally, a large dry-ice box for shipping and a large freezer at –80°C for storage are required. Shipping and handling times are typically onc or two weeks. The DNABook, however, can be shipped as easily as printed material. Once a user has the encyclopedia DNABook, clones may be obtained by simply punching out the relevant spots and subjecting them to PCR. Clones are available for experiments within two hours. One book can replace a freezer full of clones.

In the pre-genome era (before about 1995), we managed small amounts of sequence information and small numbers of clones. Sequence information and DNA clones were transferred using printed matter and dry-ice boxes. During the post-genome era, up until the present day, we used a huge amount of sequence information but a small number of clones. Information technology (IT) was replacing printed material for the transfer of sequence information. Since only relatively few DNA clones needed to be shipped, dry-ice boxes could still be used. The post-transcriptome era has now begun. The quantities of both sequence information and cDNA clones are huge. DNA printing technology will replace the dry-ice box for the distribution of clones.

The printing technology developed by Johannes Gutenberg in the 15th century is viewed as one of the most significant inventions of the last millennium because it enabled the sharing of information on a worldwide scale. However, the role of printing in information transfer is rapidly being supplanted by IT. Printing technology has found a new role in DNABook as a vehicle for the distribution of biological information. The resources produced by genome and transcriptome research can now be efficiently distributed by DNABook; this may become the basic platform for the next generation of life science research.

## CONCLUSIONS

This paper summarizes the history and future prospects of the Riken Mouse Genome Encyclopedia Project. Transcriptome analysis has been useful in elucidating many aspects of genetic information. The basic strategies used to encode the additional genetic information required for the regulation of complex life differ among higher animals, plants, and other organisms. In higher animals, the additional genetic information is encoded using unexpectedly high levels of variation, created through alternative splicing and 5′- and 3′-end variation in transcripts originating from the same TU (gene). Plants, in contrast, use a larger number of TUs rather than transcriptional variants to encode the additional information. More than 41% of all animal genes are alternatively spliced, with 79% of alternatively spliced TUs containing altered amino acid sequences. In rice, by contrast, only 13% of genes are alternatively spliced, but 78% of the alternatively spliced transcripts of these TUs change the protein sequences (Kikuchi et al. 2003).

Higher animals appear to use noncoding sequences to achieve more complex control of gene expression. For example, 45% (16,599/37,086 TUs) of the mouse transcriptome consists of noncoding sequences (although a significant proportion of these are leaky transcripts), whereas the rice transcriptome contains a much smaller proportion of noncoding sequence (13%; Y. Hayashizaki et al., unpubl.).

Noncoding RNA is much more significant than we expected in the mouse transcriptome. Clearly, proteins are the major final biologically active products encoded by the genome. However, the significant population of noncoding RNA functions directly by various mechanisms. The geographical analogy of a second "RNA continent" separate from the "continent" of expressed proteins aids the visualization of this concept. Noncoding RNA will be an essential component in genome network analysis.

Large-scale life science is going to cover the entire spectrum of "Omics". Generally, research activities directed toward exploring entire "Omics" require interdisciplinary activity achieved by many scientists, technicians, and other supporters, such as administrators. In this sense, the Riken Mouse Encyclopedia Project could be realized by real international collaboration via the international FANTOM consortium and many other collaborators. The RISA system could be developed and data produced by factorial operation. In addition, large-scale life science is closely connected to industry in terms of the supply of technology, collaboration for the development of technology, and even the use and/or distribution of the results.

The platform for transcriptome analysis developed by the Riken Mouse Genome Encyclopedia Project could become an indispensable tool in understanding life as a molecular system.

## ACKNOWLEDGMENTS

I thank all the members of RIKEN Genome Research Exploration Group and Genome Science Laboratory for dedicated work. I especially thank the members of FANTOM consortium for cooperation, and I also thank RIKEN executives and the members of Yokohama Research Promotion Division for supporting and encouraging this project. This work was mainly supported by research grants for the RIKEN Genome Exploration Research project and for the National Project on Protein Structural and Functional Analysis from MEXT of the Japanese government to Y.H., and by CREST to Y.H. from JST. It was further supported by The Rice Genome FL cDNA Library Construction Project from BRAIN to Y.H.

## REFERENCES

Apweiler R., Attwood T.K., Bairoch A., Bateman A., Birney E., Biswas M., Bucher P., Cerutti L., Corpet F., Croning M.D., Durbin R., Falquet L., Fleischmann W., Gouzy J., Hermjakob H., Hulo N., Jonassen I., Kahn D., Kanapin A., Karavidopoulou Y., Lopez R., Marx B., Mulder N.J., Oinn T.M., and Pagni M., et al. 2001. The InterPro database, an integrated documentation resource for protein families, domains and functional sites. *Nucleic Acids Res.* **29:** 37.

Attfield P.V. 1987. Trehalose accumulates in *Saccharomyces cerevisiae* during exposure to agents that induce heat shock response. *FEBS Lett.* **225:** 259.

Bono H., Yagi K., Kasukawa T., Nikaido I., Tominaga N., Miki R., Mizuno Y., Tomaru Y., Goto H., Nitanda H., Shimizu D., Makino H., Morita T., Fujiyama J., Sakai T., Shimoji T., Hume D.A., Hayashizaki Y., and Okazaki Y. 2003. Systematic expression profiling of the mouse transcriptome using RIKEN cDNA microarrays. *Genome Res.* **13:** 1318.

Carninci P., Nishiyama Y., Westover A., Itoh M., Nagaoka S., Sasaki N., Okazaki Y., Muramatsu M., and Hayashizaki Y. 1998. Thermostabilization and thermoactivation of thermolabile enzymes by trehalose and its application for the synthesis of full length cDNA. *Proc. Natl. Acad. Sci.* **95:** 520.

Carninci P., Shibata Y., Hayatsu N., Sugahara Y., Shibata K., Itoh M., Konno H., Okazaki Y., Muramatsu M., and Hayashizaki Y. 2000. Normalization and subtraction of cap-trapper-selected cDNAs to prepare full-length cDNA libraries for rapid discovery of new genes. *Genome Res.* **10:** 1617.

Carninci P., Shibata Y., Hayatsu N., Itoh M., Shiraki T., Hirozane T., Watahiki A., Shibata K., Konno H., Muramatsu M., and Hayashizaki Y. 2001. Balanced-size and long-size cloning of full-length, cap-trapped cDNAs into vectors of the novel lambda-FLC family allows enhanced gene discovery rate and functional analysis. *Genomics* **77:** 79.

Carninci P., Westover A., Nishiyama Y., Ohsumi T., Itoh M., Nagaoka S., Sasaki N., Okazaki Y., Muramatsu M., Schneider C., and Hayashizaki Y. 1997. High efficiency selection of full-length cDNA by improved biotinylated cap trapper. *DNA Res.* **4:** 61.

Carninci P., Kvam C., Kitamura A., Ohsumi T., Okazaki Y., Itoh M., Kamiya M., Shibata K., Sasaki N., Izawa M., Muramatsu M., Hayashizaki Y., and Schneider C. 1996. High-efficiency full-length cDNA cloning by biotinylated CAP trapper. *Genomics* **37:** 327.

Carninci P., Waki K., Shiraki T., Konno H., Shibata K., Itoh M., Aizawa K., Arakawa T., Ishii Y., Sasaki D., Bono H., Kondo S., Sugahara Y., Saito R., Osato N., Fukuda S., Sato K., Watahiki A., Hirozane-Kishikawa T., Nakamura M., Shibata Y., Yasunishi A., Kikuchi N., Yoshiki A., and Kusakabe M., et al. 2003. Targeting a complex transcriptome: The construction of the mouse full-length cDNA encyclopedia. *Genome Res.* **13:** 1273.

De Virgilio C., Hottiger T., Dominguez J., Boller T., and Wiemken A. 1994. The role of trehalose synthesis for the acquisition of thermotolerance in yeast. I. Genetic evidence that trehalose is a thermoprotectant. *Eur. J. Biochem.* **219:** 179.

Gibbs R.A. 1995. Pressing ahead with human genome sequencing. *Nat. Genet.* **11:** 121.

Gough J., Karplus K., Hughey R., and Chothia C. 2001. Assignment of homology to genome sequences using a library of hidden Markov models that represent all proteins of known structure. *J. Mol. Biol.* **313:** 903.

Hayashizaki Y. and Kawai J. 2003. *RIKEN mouse genome encyclopedia DNA book.* RIKEN Genomic Sciences Center, Yokohama, Japan.

Hayashizaki Y., Kawai J., and Kanehisa M. 2003. *RIKEN human cDNA encyclopedia metabolome DNA book.* RIKEN Genomic Sciences Center, Yokohama, Japan.

Hottiger T., Boller T., and Wiemken A. 1987. Rapid changes of heat and desiccation tolerance correlated with changes of trehalose content in *Saccharomyces cerevisiae* cells subjected to temperature shifts. *FEBS Lett.* **220:** 113.

Itoh M., Carninci P., Nagaoka S., Sasaki N., Okazaki Y., Ohsumi T., Muramatsu M., and Hayashizaki Y. 1997. Simple and rapid preparation of plasmid template by a filtration method using microtiter filter plates. *Nucleic Acids Res.* **25:** 1315.

Itoh M., Kitsunai T., Akiyama J., Shibata K., Izawa M., Kawai J., Tomaru Y., Carninci P., Shibata Y., Ozawa Y., Muramatsu M., Okazaki Y., and Hayashizaki Y. 1999. Automated filtration-based high-throughput plasmid preparation system. *Genome Res.* **9:** 463.

Izawa M., Sasaki N., Watahiki M., Ohara E., Yoneda Y., Muramatsu M., Okazaki Y., and Hayashizaki Y. 1998. Recognition sites of 3′-OH group by T7 RNA polymerase and its application to transcriptional sequencing. *J. Biol. Chem.* **273:** 14242.

Kasukawa T., Furuno M., Nikaido I., Bono H., Hume D.A., Bult C., Hill D.P., Baldarelli R., Gough J., Kanapin A., Matsuda H., Schriml L.M., Hayashizaki Y., Okazaki Y., and Quackenbush J. 2003. Development and evaluation of an automated annotation pipeline and cDNA annotation system. *Genome Res.* **13:** 1542.

Kawai J. and Hayashizaki Y. 2003. DNA book. *Genome Res.* **13:** 1488.

Kawai J., Shinagawa A., Shibata K., Yoshino M., Itoh M., Ishii Y., Arakawa T., Hara A., Fukunishi Y., Konno H., Adachi J., Fukuda S., Aizawa K., Izawa M., Nishi K., Kiyosawa H., Kondo S., Yamanaka I., Saito T., Okazaki Y., Gojobori T., Bono H., Kasukawa T., Saito R., and Kadota K., et al. (RIKEN Genome Exploration Research Group Phase II Team and the FANTOM Consortium). 2001. Functional annotation of a full-length mouse cDNA collection. *Nature* **409:** 685.

Kawaji H., Schonbach C., Matsuo Y., Kawai J., Okazaki Y., Hayashizaki Y., and Matsuda H. 2002. Exploration of novel motifs derived from mouse cDNA sequences. *Genome Res.* **12:** 367.

Kikuchi S., Satoh K., Nagata T., Kawagashira N., Doi K., Kishimoto N., Yazaki J., Ishikawa M., Yamada H., Ooka H., Hotta I., Kojima K., Namiki T., Ohneda E., Yahagi W., Susuki K., Li C.J., Ohtsuki K., Shishiki T., Otomo Y., Murakami K., Iida Y., Sugano S., Fujimura T., and Suzuki Y., et al. 2003. Collection, mapping, and annotation of over 28,000 cDNA clones from japonica rice. *Science* **301:** 376.

Miki R., Kadota K., Bono H., Mizuno Y., Tomaru Y., Carninci P., Itoh M., Shibata K., Kawai J., Konno H., Watanabe S., Sato K., Tokusumi Y., Kikuchi N., Ishii Y., Hamaguchi Y., Nishizuka I., Goto H., Nitanda H., Satomi S., Yoshiki A., Kusakabe M., DeRisi J.L., Eisen M.B., and Iyer V.R., et al. 2001. Delineating developmental and metabolic pathways in vivo by expression profiling using the RIKEN set of 18,816 full-length enriched mouse cDNA arrays. *Proc. Natl. Acad. Sci.* **98:** 2199.

Okazaki Y., Furuno M., Kasukawa T., Adachi J., Bono H., Kondo S., Nikaido I., Osato N., Saito R., Suzuki H., Yamanaka I., Kiyosawa H., Yagi K., Tomaru Y., Hasegawa Y., Nogami A., Schonbach C., Gojobori T., Baldarelli R., Hill D.P., Bult C., Hume D.A., Quackenbush J., Schriml L.M., and Kanapin A., et al. (FANTOM Consortium; RIKEN Genome Exploration Research Group Phase I & II Team). 2002. Analysis of the mouse transcriptome based on functional annotation of 60,770 full-length cDNAs. *Nature* **420:** 563.

Sasaki N., Izawa M., Watahiki M., Ozawa K., Tanaka T., Yoneda Y., Matsuura S., Carninci P., Muramatsu M., Okazaki Y., and Hayashizaki Y. 1998a. Transcriptional sequencing: A method for DNA sequencing using RNA polymerase. *Proc. Natl. Acad. Sci.* **95:** 3455.

Sasaki N., Izawa M., Shimojo M., Shibata K., Akiyama J., Itoh M., Nagaoka S., Carninci P., Okazaki Y., Moriuchi T., Muramatsu M., Watanabe S., and Hayashizaki Y. 1997. A novel control system for polymerase chain reaction using a RIKEN GS384 thermalcycler. *DNA Res.* **4:** 387.

Sasaki N., Izawa M., Sugahara Y., Tanaka T., Watahiki M., Ozawa K., Ohara E., Funaki H., Yoneda Y., Matsuura S., Muramatsu M., Okazaki Y., and Hayashizaki Y. 1998b. Identification of stable RNA hairpins causing band compression in transcriptional sequencing and their elimination by use of inosine triphosphate. *Gene* **222:** 17.

Shibata K., Itoh M., Aizawa K., Nagaoka S., Sasaki N., Carninci P., Konno H., Akiyama J., Nishi K., Kitsunai T., Tashiro H., Itoh M., Sumi N., Ishii Y., Nakamura S., Hazama M., Nishine T., Harada A., Yamamoto R., Matsumoto H., Sakaguchi S., Ikegami T., Kashiwagi K., Fujiwake S., Inoue K., and Togawa Y., et al. 2000. RIKEN integrated sequence analysis (RISA) system–384-format sequencing pipeline with 384 multicapillary sequencer. *Genome Res.* **10:** 1757.

Sleutels F., Zwart R., and Barlow D.P. 2002. The non-coding Air RNA is required for silencing autosomal imprinted genes. *Nature* **415:** 810.

Strausberg R.L., Feingold E.A., Grouse L.H., Derge J.G., Klausner R.D., Collins F.S., Wagner L., Shenmen C.M., Schuler G.D., Altschul S.F., Zeeberg B., Buetow K.H., Schaefer C.F., Bhat N.K., Hopkins R.F., Jordan H., Moore T., Max S.I., Wang J., Hsieh F., Diatchenko L., Marusina K., Farmer A.A., Rubin G.M., and Hong L., et al. 2002. Generation and initial analysis of more than 15,000 full-length human and mouse cDNA sequences. *Proc. Natl. Acad. Sci.* **99:** 16899.

Suzuki H., Saito R., Kanamori M., Kai C., Schonbach C., Nagashima T., Hosaka J., and Hayashizaki Y. 2003. The mammalian protein-protein interaction database and its viewing system that is linked to the main FANTOM2 viewer. *Genome Res.* **13:** 1534.

Suzuki H., Fukunishi Y., Kagawa I., Saito R., Oda H., Endo T., Kondo S., Bono H., Okazaki Y., and Hayashizaki Y. 2001. Protein-protein interaction panel using mouse full-length cDNAs. *Genome Res.* **11:** 1758.

Waterston R.H., Lindblad-Toh K., Birney E., Rogers J., Abril J.F., Agarwal P., Agarwala R., Ainscough R., Alexandersson M., An P., Antonarakis S.E., Attwood J., Baertsch R., Bailey J., Barlow K., Beck S., Berry E., Birren B., Bloom T., Bork P., Botcherby M., Bray N., Brent M.R., Brown D.G., and Brown S.D., et al. (Mouse Genome Sequencing Consortium). 2002. Initial sequencing and comparative analysis of the mouse genome. *Nature* **420:** 520.

# DNA Sequence Assembly and Multiple Sequence Alignment by an Eulerian Path Approach

Y. ZHANG* AND M.S.WATERMAN†
*Departments of *Mathematics and †Biological Sciences, University of Southern California, Los Angeles, California 90089*

## DNA SEQUENCE ASSEMBLY

Assembly of short DNA fragments (500–1000 bp) generated by shotgun sequencing is a widely used technique for sequencing large genomes, including the human genome. The most popular framework of DNA fragment assembly algorithms in the past 25 years is the "overlap-layout-consensus" approach. All high-quality DNA fragments are first compared to each other for possible overlaps, then a layout is made by arranging all DNA fragments into relative positions and orientations according to the overlap information, and finally, a multiple alignment is computed to obtain a consensus sequence that will be used as the genomic sequence. The main difficulty with this framework, in addition to the computation required, comes from the fact that genomic sequences always contain a large amount of repeat regions accumulated along their evolutionary history. In particular, repeats that are longer than the fragment length and have >98% identity are hard to distinguish from true overlaps, and hence finding a correct path in the layout step is difficult.

Surprisingly, a 15-year-old computational model from DNA arrays provides the basis for a novel approach to assembly. Sequencing by hybridization (SBH) is conceptually analogous to DNA fragment assembly by regarding each DNA fragment as a $k$-tuple. Idury and Waterman (1995) mimicked SBH procedure by breaking each DNA fragment of length $n$ into $n–k+1$ overlapping $k$-tuples and hybridized all $k$-tuples in silicon such that a DNA fragment assembly problem is mapped into a SBH problem. A de Bruijn graph is constructed in their SBH approach where each edge represents a $k$-tuple from fragments; two edges share a common vertex if they share a common ($k$–1)-tuple; identical $k$-tuples share the same edge. Repeats and sequencing errors make the simple de Bruijn graph extremely tangled, and the solution of DNA fragment assembly from such a complex graph remains challenging. Pevzner et al. (2001b) took the SBH idea and provided some additional features. Instead of building an overlap graph in the traditional "overlap-layout-consensus" framework, the Eulerian path approach builds a de Bruijn graph that represents all fragments and their relationships in a much simpler way. In addition, the difficulty resulting from the repeats and sequencing errors often is much easier to conquer in the de Bruijn graph structure.

Assume a DNA sequence consists of four unique subsequences that are separated by one triple repeat, as shown in Figure 1a. The traditional overlap layout consensus approach builds an overlap graph by regarding every fragment as a vertex, and two vertices are connected if two corresponding fragments overlap. Figure 1b shows the overlap graph in our example. The assembly problem is thus finding a path in the overlap graph that visits every vertex exactly once, a Hamilton path problem that is well known NP-complete. On the contrary, the Eulerian path approach breaks all fragments into $k$-tuples and builds a de Bruijn graph as described above (Fig. 1c). A conceptually ideal de Bruijn graph is shown in Figure 1d, a much simpler representation of repeats than the overlap graph. The most important advantage of this representation is that the assembly problem is now finding a path that visits every edge exactly once, an Eulerian path problem that has linear-time solutions (Fleischner 1990).

The implementation of the Eulerian path approach to DNA fragment assembly problems is called EULER. Traditional algorithms postpone the error correction until the last consensus step, whereas EULER applies error correction at the beginning of assembly (this is the innovation of Pevzner et al. 2001b). Without knowing the finished genomic sequence or even the layout of fragments, error correction is still possible by approximating the $k$-tuple spectrum, and the result is usually accurate enough. After error correction, EULER constructs a de Bruijn graph and stores the fragment information in corresponding edges. This fragment information is the fundamental difference between EULER and SBH where such information is unavailable. EULER's superpath idea can successfully solve many repeats by using fragment information. Finally, EULER outputs an Eulerian path from the de Bruijn graph that represents the finished genomic sequence.

### Error Correction

Sequencing errors make the de Bruijn graph a tangle of erroneous edges, thus very difficult to solve. EULER's error correction procedure reduces 97% of errors from the original DNA fragments and makes the data almost error-free. The idea is based on an approximation to the spectrum of real genomic sequence, i.e., to find a collection of $k$-tuples all of which are from the genomic sequence instead of sequencing errors. With sequencing errors, many $k$-tuples are erroneous and many "true" $k$-tuples are missing. EULER calls a $k$-tuple solid if it appears in more than

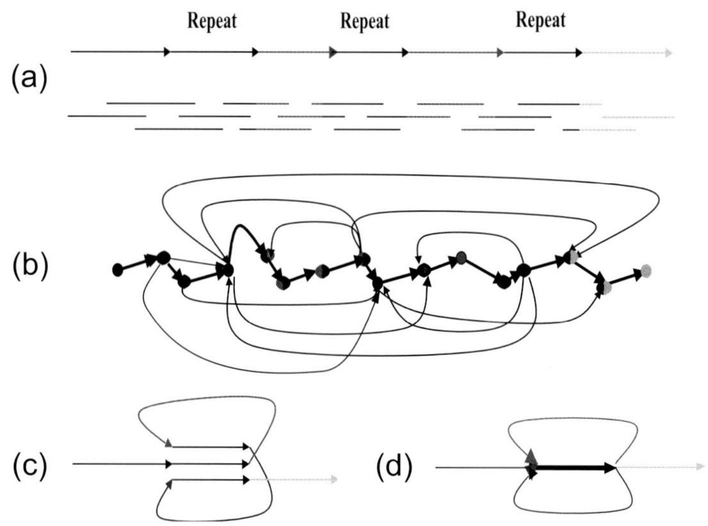


**Figure 1.** (*a*) DNA sequence with a triple repeat R; (*b*) the overlap graph; (*c*) construction of the de Bruijn graph by gluing repeats; (*d*) de Bruijn graph. (Reprinted, with permission, from Pevzner et al. 2001b.)

*M* fragments, otherwise it is a weak *k*-tuple. The error correction problem is thus to transform the spectrum of the original DNA fragments into the spectrum of a genomic sequence by changing weak *k*-tuples to solid *k*-tuples. Without knowing the real genomic sequence, one natural criterion of error correction is to minimize the total number of distinct *k*-tuples in the spectrum. One error in a fragment will create at most $2k$ (including the reverse complement part) erroneous *k*-tuples in the spectrum, or $2d$ ($d<k$) if the error appears near either end of a fragment so that at most $2d$ *k*-tuples can be affected. EULER uses a greedy approach to look for error corrections that reduce the number of weak *k*-tuples by $2k$ or $2d$.

The *Neisseria meningitidis* (NM) sequencing project (Parkhill et al. 2000), one of the most difficult-to-assemble and repeat-rich bacterial genomes completed so far, was used to demonstrate the efficiency of EULER's error correction method. The NM genome contains 2,184,406 nucleotides, with 126 long nearly perfect repeats up to 3,832 bp in length. The sequencing project resulted in 53,263 fragments (coverage 9.7), with 255,631 sequencing errors in total. EULER corrected 97.7% errors and made the original sequencing data almost error-free with 0.11 errors per fragment (Pevzner et al. 2001a).

## Construction of de Bruijn Graph

EULER constructs a de Bruijn graph from the corrected DNA fragments in the following way: Given a set of DNA fragments, EULER breaks each fragment and its reverse complement into overlapping *k*-tuples; each *k*-tuple represents a directed edge in the graph, and the direction of an edge is the direction of reading a *k*-tuple; tuple positions and fragment indices are stored in the corresponding edges; each *k*-tuple contains two ($k$–1)-tuples that will represent vertices connected at two ends of an edge; all edges and vertices are "glued" together if they correspond to identical *k*-tuples and ($k$–1)-tuples. A de Bruijn graph of two fragments, ATGC and ATGT, is shown in Figure 2.

An edge is called single if the edge represents a single *k*-tuple in the genomic sequence (but this *k*-tuple may appear in many fragments that cover it), otherwise the edge is called multiple. By regarding each multiple edge as *m* parallel single edges if the edge represents *m* occurrences of the *k*-tuple in the genomic sequence, a fragment assembly problem is then to find a Eulerian path that visits each edge exactly once. The algorithm for finding a Eulerian path costs linear time and can detect erroneous edges that will then be discarded. In addition, without knowing orientations, EULER builds all fragments and their reverse complements into the de Bruijn graph, and expects that the graph can be partitioned into two complementary subgraphs, corresponding to reading the sequence in each direction.

## Superpath Transformation

A vertex $v$ is called a source if indegree($v$) = 0, or a sink if outdegree($v$) = 0. A branching vertex is a vertex that has indegree($v$) x outdegree($v$) > 1. The de Bruijn graph corresponding to the original fragments of the NM genome

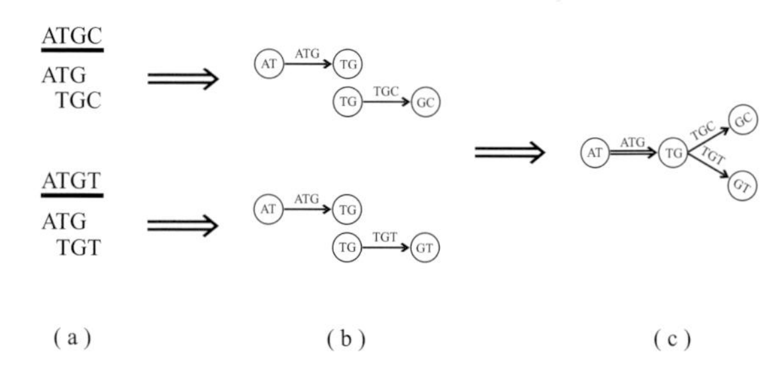


**Figure 2.** (*a*) Two DNA fragments and their 3-tuples (for simplicity, their reverse complements are not included); (*b*) edge and vertex presentation of those 3-tuples; (*c*) a de Bruijn graph made by "gluing" identical edges and vertices.

has 502,843 branching vertices ($k = 20$), and this number is reduced to 12,175 after error correction. Even with error-free data, however, the de Bruijn graph is still very complicated for shotgun sequencing projects. EULER uses the fragment information stored in edges to handle this difficulty.

Define a repeat structure a path $P(v1 \rightarrow vn) = v1...vn$ in the graph, where indegree($v1$) > 1, outdegree($vn$) > 1, indegree($vi$) = outdegree($vi$) = 1, for $1 < i < n$, and outdegree($v1$) = indegree($vn$) = 1 if $n > 1$. A repeat structure represents a possible repeat in the assembled sequence. If indegree($v1$) = $p$ and outdegree($vn$) = $q$, then there will be $p \times q$ possible pairings of edges that a path enters the repeat structure from one edge and exits from the other, while the correct pairings corresponding to the assembled sequence are unknown. If a fragment covers the entire repeat, the correct pairings can be detected by following the fragment path. If no fragments covered the entire repeat, the correct pairings will remain unclear and the repeat structure will form a tangle.

A superpath transformation of the de Bruijn graph is then introduced to solve repeat structures. The goal of superpath transformations is to do a series of transformations to the graph so that the final graph contains no multiple edges. Transformations on multiple edges should be performed with caution, because fragment information stored in multiple edges must be partitioned and stored separately into superpaths. For a detailed discussion, please refer to the original paper by Pevzner et al. (2001b).

One feature of the Eulerian superpath approach to the DNA sequence assembly problem is that it uses, rather than struggles with, the imperfect repeats. By superpath transformation, most imperfect repeats will eventually be separated into different paths. By using the linear-time Eulerian path algorithm, the Eulerian approach has the potential to assemble larger eukaryotic genomes in the future.

## MULTIPLE DNA SEQUENCE ALIGNMENT

The linear-time Eulerian path algorithm has the potential to assemble a large eukaryotic genome of length up to gigabases. A closely related but complementary question is now asked: Can the Eulerian path idea be applied to the multiple sequence alignment problem? That is, instead of assembling a long genomic sequence, can we use this idea in aligning a large number of short specific sequences?

Many MSA algorithms have been developed in the past decades. One bottleneck for all of them is the expensive computational cost when aligning extremely long sequences or a huge number of sequences simultaneously. We demonstrate that the Eulerian path idea can provide almost linear time MSA with accurate solution of this problem. In particular, we present a program called EulerAlign (Zhang and Waterman 2003) that uses the Eulerian path method to solve the global MSA of a large number of DNA sequences. An application on a genome resequencing project is then presented to demonstrate its performance.

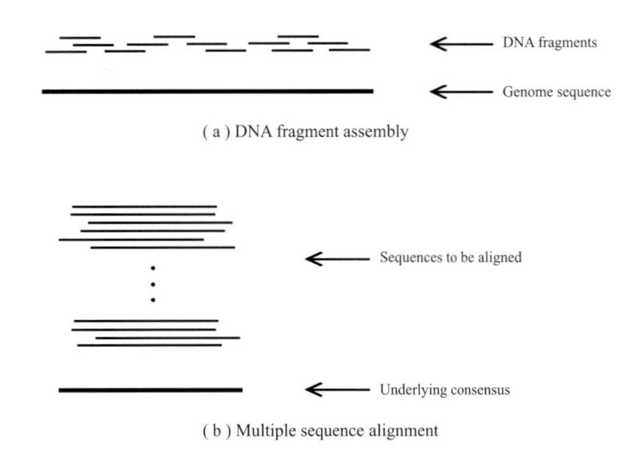


**Figure 3.** Analogy between (*a*) DNA fragment assembly problem and (*b*) multiple sequence alignment problem. *a* and *b* are similar except for the positional distribution of fragments.

### Motivation

EulerAlign takes the Eulerian path idea that builds all sequences into a de Bruijn graph, and then extracts the graph information to do the multiple sequence alignment. Recall the process of a sequence assembly project, where genomic sequences are broken into short pieces and these pieces are then randomly cloned to build a library of short DNA fragments. All fragments are sequenced and input into a DNA fragment assembly program to reconstruct the original genomic sequence. A global MSA problem is a special DNA sequencing project (Fig. 3), where the "genome" is short enough to be sequenced directly, but the "sequencing machine" makes a very high rate of errors.

The Eulerian path approach for the DNA assembly problem requires the input fragment to be error-free, whereas EULER did this by an error-correction procedure. In a MSA problem, however, there could be thousands of sequences to be aligned, and the genetic differences among the sequences result in many errors compared with sequencing errors. Thus, the error correction as used in EULER will be helpful but not likely to succeed in correcting most of them. A MSA problem is to recover the underlying consensus sequence from a large number of divergent sequences, where mutations do not pose a barrier because each sequence covers (almost) the entire consensus sequence. This difference allows us to construct an accurate alignment even with the presence of many mutations.

A consensus sequence is typically obtained from a given alignment by extracting the majority letters (in the simplest case) in each column of the alignment. EulerAlign reverses this procedure by first obtaining a consensus sequence from the graph and then builds the alignment from the consensus. In the de Bruijn graph, each sequence represents a sequence path, and each multiple edge represents a $k$-tuple shared by many sequence paths (see Fig. 4). The letters that are common to a majority of sequences are thus determined by multiple edges visited by many sequence paths. The multiplicity for an edge is the number of sequence paths visiting the edge. The higher the multiplicity is, the larger the chance that the edge represents a $k$-tuple in the consensus sequence.

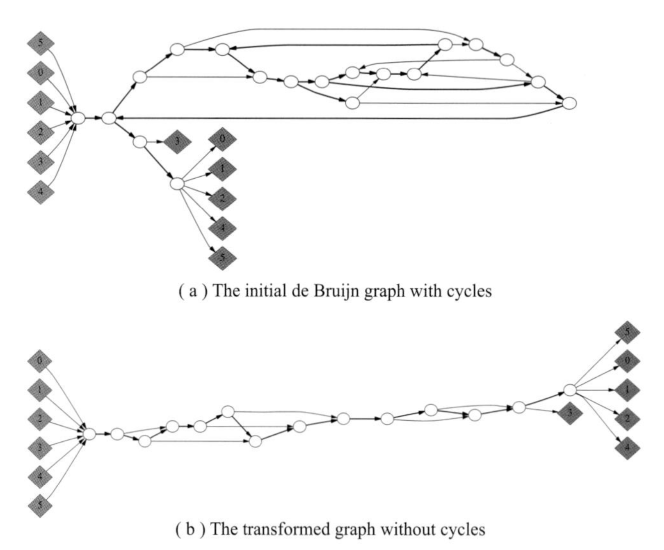

( a ) The initial de Bruijn graph with cycles

( b ) The transformed graph without cycles

**Figure 4.** Example of de Bruijn graph constructed from 6 sequences of 100 bp each. (*a*) The initially constructed graph has many cycles. (*b*) After removing self-cycles, the heaviest acyclic path appears as thick edges that are visited by many sequence paths. Diamond vertices correspond to the sequence ends.

Based on this idea, EulerAlign assigns weight to each edge as a function of the edge's multiplicity and length, then uses a heaviest-path algorithm to extract a path with the largest sum of weights. The heaviest-path problem (identical to the shortest-path problem with negative weights) has linear time solutions for a directed acyclic graph (DAG), and EulerAlign uses this heaviest path as the consensus to construct the final alignment. The pipeline for EulerAlign is as follows: (1) construct a de Bruijn graph; (2) transform the graph to a DAG; (3) extract a heaviest path as the consensus sequence; (4) do consensus alignment.

### Graph Transformation and Consensus Alignment

The initially constructed graph contains cycles due to repeats and random matches, but an optimal heaviest path is acyclic in the graph. If the true consensus includes repeat regions, an acyclic heaviest path will not accurately represent the true consensus unless the cycles (repeats) are solved. EulerAlign uses superpath transformations defined in EULER to solve these cycles. A superpath transformation captures the sequence path information, and hence can minimize the loss of similarity information stored in edges (multiplicities). When the number of sequences is large, however, it could be very time-consuming to remove all cycles by superpath transformations. In addition, an equivalent superpath transformation may not exist in some situations, and thus, removing all cycles may cause significant information loss. EulerAlign compromises by applying superpath transformations only on a particular subset of cycles. A cycle is easily detected when a sequence path visits a vertex more than once, and such cycles are called self-cycles because they are entirely involved in one sequence path. EulerAlign uses the superpath transformation to remove a cycle if, and only if, the cycle is a self-cycle, and all sequence paths will be individually acyclic after removing all self-cycles. Because the heaviest path itself is acyclic, this procedure is designed to allow the heaviest path to capture most similarity information. A depth first search in the graph is then performed to eliminate all remaining cycles. Although losing all information stored in the removed cycles, this procedure enables a rigorous heaviest-path algorithm to be applied. The time-cost for removing all remaining cycles is linear in the size of the graph.

After obtaining an acyclic graph, EulerAlign extracts the heaviest path according to the weight function defined for edges. The algorithm for finding the heaviest path costs linear time proportional to the size of the graph. Finally, EulerAlign applies the classical banded dynamic programming algorithm to align each sequence with the consensus and builds the final multiple sequence alignment. Positional specific scoring functions derived from "heavy" edges and large potential indels indicated by the position shifts of $k$-tuples can be used during this process.

### Performance

We have tested EulerAlign on both simulated and real DNA sequences. For simulated sequences, we randomly add substitutions and indels according to two models: the equidistance model and the evolutionary model. In the equidistance model, each sequence is independently mutated with a common mutation distribution, and hence on average all sequences are equally similar to each other. In the evolutionary model, all sequences are related to a common ancestral sequence along an evolutionary tree. Mutations in each sequence are generated along the tree and hence are correlated among sequences. The equidistance model perfectly fits the requirement of a consensus alignment, but the evolutionary model reflects a more realistic situation.

Two scoring systems are used to evaluate the perfor-

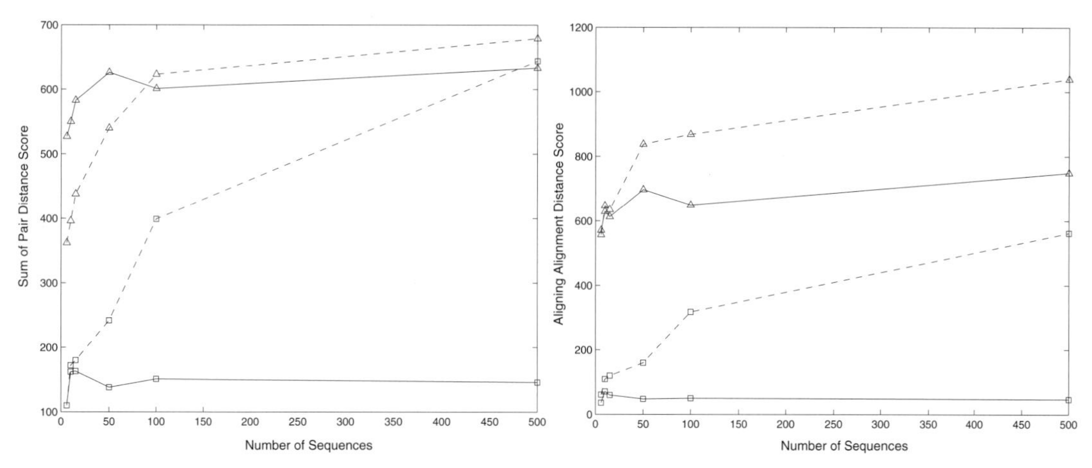


**Figure 5.** SP and AA scores (distance score) with respect to the number of sequences. The squares and triangles indicate different pair-wise similarities (square, 90%, triangle, 70%). Both SP and AA scores are computed from a single alignment test. Solid lines connect points from EulerAlign, and dashed lines connect points from ClustalW.

mance of EulerAlign: (1) sum of pairs (SP) score, a popular and simple measure, and (2) aligning alignment (AA) score, comparison of an alignment to the true alignment; by simulating sequences, the true alignment (rather than the mathematically optimal alignment) is known. We used ClustalW (Higgins and Sharp 1989; Thompson et al. 1994), a well-studied and popular MSA software, as the reference. Figure 5 shows the comparison between EulerAlign and ClustalW on sequence sets generated by the evolutionary model with different mutation rates: 5.2% and 16.4%, respectively, corresponding to 90% and 70% pair-wise sequence similarities. The comparison on the equidistance model is not shown, simply because EulerAlign is designed for that model and hence achieves a better result. The linear growth of the computational time with respect to the number of aligned sequences by EulerAlign is shown in Figure 6a, and a significant comparison to the quadratic growth by ClustalW is shown in Figure 6b. We used distance scores; hence, the smaller the score the better the result. All tests are done on a SUN UltraSPARC 750MHz workstation.

## Application on *Arabidopsis* Sequences

*Arabidopsis thaliana* is widely used as a model organism for genetic study in plant biology. As an application on real genomic sequences, we used EulerAlign to construct alignments for several sets of short specific sequences sampled from 96 *Arabidopsis* individuals by PCR experiments with certain primers. These alignments are then used to study the genetic variations and hence evolutionary relationships in the *Arabidopsis* population. Sequence data are kindly provided by M. Nordborg at USC. Presented with base-calling errors, an accurate multiple alignment is crucial for efficiently detecting real genetic variations other than sequencing errors. The main difference between genetic variations and sequencing errors is that sequencing errors are more independently and randomly distributed (although of course a function of position in the sequence fragment).

To reduce base-calling errors, each individual is sequenced from both forward and backward strands, and each base-call has a quality value assigned by Phred (Ew-

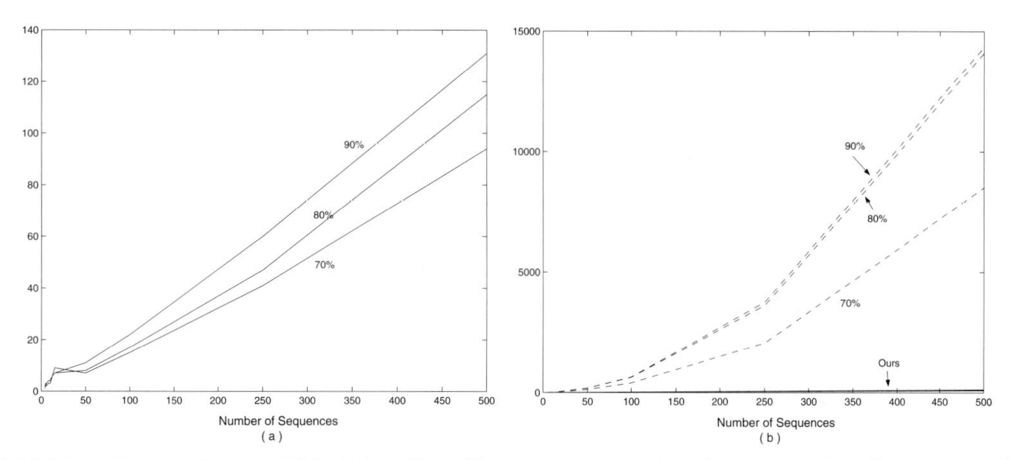


**Figure 6.** (*a*) Linear time cost (in seconds) by EulerAlign with respect to the number of sequences; three lines correspond to 90%, 80%, and 70% pair-wise similarities. (*b*) Comparison to the quadratic time cost by ClustalW (*dashed lines*). The tested numbers of sequences are 10, 15, 50, 100, 250, and 500.

ing and Green 1998). Because the forward and backward pairs represent the same genome region, any discrepancy between them indicates that at least one of two base-calls is wrong. Our experience shows that the base-calls in two strands are asymmetric; that is, base-calls in one strand tend to make certain errors more frequently than in the other strand. All sequence pairs (forward and backward strands) are combined by Phrap (Green 1994) according to their quality values before doing multiple alignment.

Poor-quality values are often assigned as the base-call approaches the end of sequences, and occasionally outlier sequences are generated by wrong PCR amplification. Regions with low-quality values in each sequence are not trimmed or discarded before doing alignment; hence, informative segments can be found even though they have low qualities. Since the quality values within and among sequences are highly variant even after the combination of two strands, we must incorporate the quality values into the computation to avoid misalignments due to low-quality regions.

We use ClustalW version 1.83 as the reference program. For ways in which quality values are used in EulerAlign and ClustalW, refer to our previous paper (Zhang and Waterman 2003). We tested 20 sequence sets, including both good sequence sets and bad sequence sets, where up to 18% of sequences are either of very poor quality (<20 in average after combining both strands) or outliers. The parameters for both EulerAlign and ClustalW are tuned "optimal" by human effort. By optimal we mean the best-scoring functions and utilization of quality values for all 20 sequence sets, not individually. All alignments using ClustalW with quality values are done by Tina Hu, at USC.

We used a modified version of sum-of-pair scores to evaluate alignments from both programs. The scores are adjusted according to the quality value of each letter. For example, the mismatch penalty of two low-quality letters is smaller than the penalty of two high-quality letters. To test the robustness of EulerAlign, we also computed the alignments by both programs without using quality values; i.e., poor-quality sequences and outliers are equally considered with high-quality sequences. The sum-of-pair scores for these alignments are, of course, not adjusted by quality values. Table 1 shows the comparison. We use similarity scores so that the level of pair-wise identities of each sequence set can be inferred from the relationship between sequence lengths and scores.

We found that the performance of both programs is comparable after tuning the alignment parameters and using quality values when doing alignment. ClustalW wins in 11 sequence sets whereas EulerAlign wins in 9. By checking the alignments, we found the major alignment difference by two programs are for the outliers and at either end of sequences which have low quality and less identities. This result is possibly due to the different utilizations of quality values by EulerAlign and ClustalW when doing multiple alignment. On the other hand, without using quality values, both programs computed alignments by their default parameters, and EulerAlign outperforms ClustalW in all 20 sequence sets.

It is known that the sum-of-pair scores or other scoring schema cannot always reflect the correctness of a biologically meaningful alignment. Examples of misalignments, one by ClustalW and one by EulerAlign, are shown in Figure 7. We argue that these misalignments are due to the improper scoring functions used by each program. Since the alignment parameters have been tuned optimal in both programs (for all 20 sequence sets instead of for each set individually), we conclude that EulerAlign made a more reasonable alignment than ClustalW in this case. Because the real alignments for DNA sequences are unknown, we hope the sum-of-pair scoring scheme

**Table 1.** Comparison of Alignments by EulerAlign and ClustalW on *Arabidopsis* Sequences

| | | | With quality | | Without quality | |
|---|---|---|---|---|---|---|
| Set | N | L | EulerAlign | ClustalW | EulerAlign | ClustalW |
| At_000000166 | 96 | 665 | 355.8 | 356.4 | 396.7 | 390.5 |
| At_000000244 | 96 | 677 | 588.4 | 588.5 | 598.6 | 597.9 |
| At_000000245 | 84 | 677 | 263.6 | 262.6 | 322.2 | 301.1 |
| At_000000296 | 94 | 687 | 480.2 | 480.1 | 508.0 | 504.1 |
| At_000000300 | 96 | 541 | 478.7 | 477.2 | 483.6 | 481.0 |
| At_000000308 | 95 | 623 | 502.9 | 503.5 | 524.4 | 520.4 |
| At_000000325 | 87 | 676 | 262.4 | 264.6 | 348.6 | 330.5 |
| At_000000331 | 96 | 716 | 455.8 | 455.2 | 480.2 | 461.8 |
| At_000000383 | 93 | 639 | 364.6 | 368.5 | 410.2 | 398.7 |
| At_000000397 | 92 | 680 | 277.3 | 279.2 | 353.9 | 327.0 |
| At_000000403 | 69 | 644 | 384.7 | 381.9 | 440.7 | 416.2 |
| At_000000454 | 95 | 556 | 499.3 | 500.0 | 504.0 | 502.7 |
| At_000000459 | 95 | 731 | 630.5 | 632.7 | 648.3 | 643.7 |
| At_000000466 | 87 | 719 | 586.3 | 585.6 | 621.8 | 614.4 |
| At_000000504 | 95 | 575 | 530.8 | 532.2 | 536.5 | 535.3 |
| At_000000541 | 95 | 760 | 413.9 | 414.8 | 456.4 | 447.3 |
| At_000000550 | 95 | 760 | 358.1 | 358.3 | 391.8 | 375.0 |
| At_000000584 | 87 | 715 | 259.7 | 250.5 | 344.7 | 313.6 |
| At_000000603 | 83 | 695 | 338.6 | 337.7 | 410.5 | 405.4 |
| At_000000689 | 92 | 724 | 564.6 | 560.9 | 592.6 | 588.0 |

N is the number of sequences. L is the average sequence length. Scoring functions are (match, mismatch, gapopen, gapextention) = (1, 0, –4, –1). All scores are normalized by N(N–1)/2.

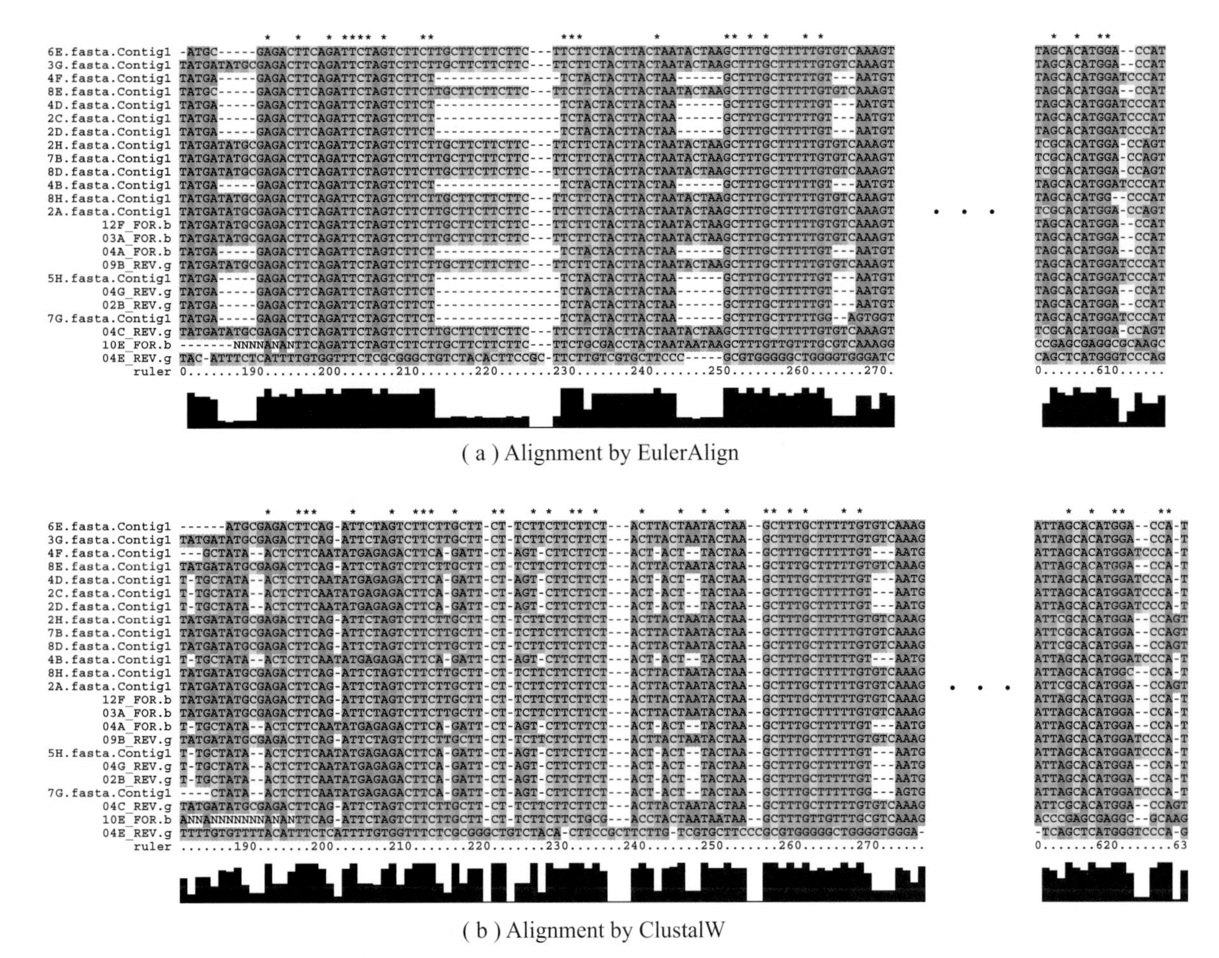


**Figure 7.** Part alignments of sequence set At_000000541 by (*a*) EulerAlign and (*b*) ClustalW. In the first part, EulerAlign made a correct alignment whereas ClustalW did not. Alignments by two programs look very different, but they are indeed the same region. In the second part, ClustalW made a correct alignment whereas EulerAlign did not. Sequence order in *a* is adjusted to the same order in *b*. The bottom sequence is an outlier. The visualization tool is ClustalX (Thompson et al. 1997).

demonstrates, to some extent, the robustness and high performance of EulerAlign. More significantly, EulerAlign computed each alignment of the test sets within 20 seconds and obtained a comparable result to ClustalW, which used 10 minutes on the same machine.

EulerAlign is an efficient alignment tool for multiple DNA sequence alignment. By incorporating additional information, such as quality values, EulerAlign is able to provide a fast and accurate way for automated resequencing analysis in many biological applications.

## DISCUSSION

The Eulerian path approach is an efficient and accurate solution to both the DNA fragment assembly problem and the multiple sequence alignment problem. It applies the de Bruijn graph structure that stores the sequence and similarity information simultaneously with economic memory requirements. By breaking sequences into overlapping $k$-tuples and tracing sequence paths in the corresponding de Bruijn graph, similarity among sequences, or even a rough multiple sequence alignment, is readily available in a resolution of $k$ consecutive matches. All repeat regions shorter than sequence length are easily detectable by following sequence paths, and the superpath transformation effectively solves these repeat structures. In an assembly problem, a Eulerian path visiting each edge exactly once is extracted in a linear time with respect to the size of fragment sets. In a MSA problem, a heaviest path representing the consensus sequence is obtained in a linear time with respect to the size of aligned sequences. This linearity makes the Eulerian path approach extremely efficient in dealing with large data sets. Although ad hoc in nature, the Eulerian path approach to the DNA fragment assembly outperforms many other assembly algorithms with respect to accuracy. As for the multiple sequence alignment, the Eulerian path approach is at least as good as ClustalW in both simulated and real DNA sequence sets but requires very much less time in computation. Although best fitted to the alignment of sequences within one family, EulerAlign can be extended to align sequences from different families (the algorithm is in progress). In conclusion, we believe that the Eulerian path idea and its combinatorial framework can be fit into many practical problems in computational biology.

## ACKNOWLEDGMENTS

We thank Professor Magnus Nordborg at USC for kindly providing the *Arabidopsis* sequencing data, and we are grateful to Professor Lei Li at USC, Professor Pavel Pevzner, and Dr. Haixu Tang at UCSD for many helpful discussions. This research was supported by National Institutes of Health grant R01 HG-02360-01.

## REFERENCES

Ewing B. and Green P. 1998. Base-calling of automated sequencer traces using phred. II. Error probabilities. *Genome Res.* **8:** 186.

Fleischner H. 1990. *Eulerian graphs and related topics*. Elsevier Science, London, United Kingdom.

Green P. 1994. Documentation for Phrap. (http://bozeman.mbt.washington.edu/phrap.docs/phrap.html)

Higgins D.G. and Sharp P.M. 1989. Fast and sensitive multiple sequence alignments on a microcomputer. *CABIOS* **5:** 151.

Idury R. and Waterman M.S. 1995. A new algorithm for DNA sequence assembly. *J. Comput. Biol.* **2:** 291.

Parkhill J., Achtman M., James K.D., Bentley S.D., Churcher C., Klee S.R., Morelli G., Basham D., Brown D., Chillingworth T., Davies R.M., Davis P., Devlin K., Feltwell T., Hamlin N., Holroyd S., Jagels K., Leather S., Moule S., Mungall K., Quail M.A., Rajandream M.A., Rutherford K.M., Simmonds M., and Skelton J., et al. 2000. Complete DNA sequence of a serogroup A strain of *Neisseria meningitidis* Z2491. *Nature* **404:** 502.

Pevzner P.A., Tang H., and Waterman M.S. 2001a. A new approach to fragment assembly in DNA sequencing. In *Proceedings of the Fifth International Conference on Computational Biology*, Montreal, Quebec, Canada (RECOMB 2001), p. 256.

———. 2001b. An Eulerian path approach to DNA fragment assembly. *Proc. Natl. Acad. Sci.* **98:** 9748.

Thompson J.D., Higgins D.G., and Gibson T.J. 1994. CLUSTAL W: Improving the sensitivity of progressive multiple sequence alignment through sequence weighting, position-specific gap penalties and weight matrix choice. *Nucleic Acids Res.* **22:** 4673.

Thompson J.D., Gibson T.J., Plewniak F., Jeanmougin F., and Higgins D.G. 1997. The ClustalX windows interface: Flexible strategies for multiple sequence alignment aided by quality analysis tools. *Nucleic Acids Res.* **24:** 4876.

Zhang Y. and Waterman M.S. 2003. An Eulerian path approach to global multiple alignment for DNA sequences. *J. Comput. Biol.* **10:** 803.

# Ensembl: A Genome Infrastructure

E. BIRNEY AND THE ENSEMBL TEAM
*EBI, Wellcome Trust Genome Campus, Hinxton, Cambridge CB10 1SD; and Sanger Institute, Wellcome Trust Genome Campus, Hinxton, Cambridge CB10 1SA, United Kingdom*

The genome sequence of any organism is an invaluable resource for molecular biologists. Experiments are either trivial to design or greatly enhanced by the knowledge of the genome sequence; linkage analysis leads directly to a set of genes in the critical region, and association studies can be designed to any region (including, in theory, the entire genome). At least as important as the ease of experimental design is the fact that the genome is essentially complete, and therefore, researchers can be confident that the aspects of biology they are studying must be present in the genome sequence in some manner. The fact that all biology is somehow associated with some aspect of the genome, that the genome is complete and essentially unchanging, means that it is a foundational resource for biology. The generation of the human (Landfer et al. 2001) and mouse (Waterston et al. 2002) genome sequences provided landmarks for the understanding of human biology.

There is a catch. The genome sequence of large organisms is itself a large, unwieldy data set and is opaque to analysis: For all of the arguments of completeness of the genome, if we can't ascribe at least some function to parts of the sequence, it becomes effectively unusable. In addition to this scientific problem of knowing which sequences are functional or not, there is the somewhat mundane problem of simply handling the data size. This engineering problem is compounded due to the number of genomes now sequenced and the churn rate of genome and cDNA sequence.

Ensembl is designed to overcome these problems and therefore to make genomes far more useful to a broad range of audiences (Clamp et al. 2003). Ensembl focuses on providing information to three classes of users: (1) Bench biologists, who generally are focused on one or two genes and want a user-friendly, graphical web-based system to access the genome. The ensembl web site, www.ensembl.org, is focused on this user. (2) Mid-scale functional genomics users, who are working with sets of genes, either due to positional cloning or expression analysis. These users often need their own "slice and dice" data-extraction routines. The EnsMart data-mining system (described below) is focused on this user. (3) Large-scale genomics groups and other bioinformatics groups. We have found that by simply being open in terms of both software and data, we are able to satisfy most of this group's needs.

Ensembl is not the only group analyzing and displaying these genomes. The UCSC group under David Haussler and the NCBI group are both very active in this area. We enjoy healthy competition with these groups and collaborate on the underlying data resources, ensuring, for example, that the assembly is common between all three sites.

## RESULTS

Table 1 outlines the genomes which Ensembl displays. We make a distinction between genomes where we predict genes and genomes where the gene structures are provided by another group. Notice that for all genomes there have been multiple gene builds, in each case marshaling a set of data resources (e.g., cDNA and EST data sets) and tuning the gene prediction process for each genome. A gene build takes information from three sources. cDNAs generated from the target genome are reconciled back onto the genome sequence by a specific "best in genome" process. Then, remaining pieces of the genome which show strong protein similarity to genes in other organsisms are used to generate "novel" genes via the program genewise. Finally, EST sequences are mapped back to the genome, clustered, and then a minimum set of transcripts which represent the clustered ESTs are generated. Depending on the species, the EST data are sometimes merged with the main cDNA and protein build (e.g., in *Anopheles gambiae*) and sometimes are kept separate (e.g., in *Homo sapiens*). This is due to the difference in EST quality in different genomes, with the large, error-prone human EST set proving the hardest to use.

The Ensembl web site (www.ensembl.org) is designed for biological researchers to quickly orient themselves on the genome and design experiments on the basis of the genome information. Our two main displays are focused on genomic sequence and gene products. Increasingly, we have discovered that more and more people want to work with subsets of gene products encoded by a particular genome. This "set" working behavior is catered to by the EnsMart data-mining tool available both through the Web and as a downloadable command-line tool.

Assessing the quality of our gene prediction is hard because we use all available data at any point in our build process, and new experimental cDNA approaches tend to focus on currently undiscovered cases. Assessments via overlap to dual-genome predictors, which use only the human and mouse genome as input (and therefore should be unbiased toward other cDNA or EST evidence), suggest that there are around another 1,000 protein-coding genes outside of Ensembl to predict (Guigó et al. 2003).

**Table 1.** General Information on Species Activities

| Species | Genome size | Gene number | Exons/ gene | Builds to date | Assembly authority | %Web hits |
|---|---|---|---|---|---|---|
| In-house genebuilds | | | | | | |
| *Homo sapiens* | 3.23 Gb | 24037 | 8.7 | 11 | NCBI | 57 |
| *Mus musculus* | 2.50 Gb | 24948 | 8.7 | 4 | NCBI | 20 |
| *Rattus norvegicus* | 2.56 Gb | 23751 | 7.9 | 3 | RGSC[a] | 9 |
| *Danio rerio* | 1.57 Gb | 20062 | 7.9 | 5 | Sanger Institute | 7 |
| *Anopheles gambiae* | 278 Mb | 14707 | 4.0 | 4 | IAGP[b] | 1 |
| *Caenorhabditis briggsae* | 106 Mb | 11884 | 7.2 | 1 | Sanger Institute | 1 |
| Genebuilds performed outside Ensembl | | | | | | |
| *Fugu rubripes* | 390 Mb | 35180 | 4.7 | 2 | IFGC[c] | 1 |
| *Drosophila melanogaster* | 128 Mb | 13525 | 4.6 | N/A | FlyBase | 1 |
| *Caenorhabditis elegans* | 103 Mb | 19988 | 6.2 | N/A | WormBase | 3 |

[a]RGSC = Rat Genome Sequencing Consortium: Baylor College of Medicine, Celera Genomics, Genome Therapeutics, The Institute for Genome Research, The University of British Columbia.
[b]IAGP = International Anopheles Genome Project: EBI/Sanger Institute, Celera Genomics, Genoscope, University of Notre Dame, EMBL, Institut Pasteur, IMBB, TIGR.
[c]IFGC = International *Fugu* Genome Consortium: Institute of Molecular and Cell Biology (Singapore), Joint Genome Institute, Human Genome Mapping Project Resource Centre, Institute for Systems Biology.

Other estimates have run closer to 4,000 new genes. There has been a healthy debate on the number of final genes in the human genome, which I was foolish enough to open a book on in 2000, taking bets for the number of genes in the human genome. The rules of the bet (written in 2000) were that we would declare a winner in 2003. Sadly, our estimates of the gene number do not have a close margin of error to come to any firm number; however, I was saved by the fact that the distribution of bets is centered around 50,000 (median number in the betting pool was 52,689). This large overbetting has meant that, although we are not certain of the precise number, there were only three individuals who placed even close to estimated bets below 30,000 protein-coding genes. We duly split the betting pool between these three participants without having to fix on a particular number.

Ensembl also calculates orthology relationships between the protein-coding gene sets from the different genomes. Each genome pair has a number of unmatched genes that have no obvious orthologs in the other species (complex lineage-specific duplications are allowed). We call these unmatched genes "orphans." Table 2 shows the number of orphans found between different comparisons, and the total number of orphans across all comparisons in each species. The presence of orphans is a combination of fast-evolving genes that have lost the clear protein similarity signatures to assign orthology solely on protein sequence; the fact that none of the genomes is finished; errors in the gene prediction process on genomes, in particular, the presence of pseudogenes; and finally, erronously submitted cDNA information (for example, genomic contamination that has a significant open reading frame). By sampling random sets of orphan sequences, we believe that nearly all orphans are due to either misappropiate pseudogene classification (classifying a pseudogene as a real gene) or cloning artifacts from library sequencing (for example, cloning projects that used differential display to clone cancer-specific genes). These

**Table 2.** Comparative Analysis of Ensembl Gene Structure Predictions

| | | Number of orphans per species | | | | | | | | |
|---|---|---|---|---|---|---|---|---|---|---|
| | | Hs | Mm | Rn | Dr | Fr[a] | Dm[b] | Ag | Ce[c] | Cb |
| Pairwise comparison | Hs:Mm | 2723 | 2237 | | | | | | | |
| | Hs:Rn | 2845 | | 1261 | | | | | | |
| | Hs:Dr | 5465 | | | 1333 | | | | | |
| | Hs:Fr | 4835 | | | | 13016 | | | | |
| | Mm:Rn | 1929 | | 717 | | | | | | |
| | Mm:Dr | 5544 | | | 1510 | | | | | |
| | Mm:Fr | 4696 | | | | 13271 | | | | |
| | Rn:Dr | 4340 | | | 1613 | | | | | |
| | Rn:Fr | 3147 | | | | 13376 | | | | |
| | Dm:Ag | | | | | | 3043 | 3747 | | |
| | Ce:Cb | | | | | | | | 4035 | 22 |
| Orphans across all comparisons | | 2416 | 1423 | 375 | 719 | 10397 | 3043 | 3747 | 4035 | 22 |

Numbers represent the number of orphans both in pairwise comparisons and in terms of comparison to all closely related species.
[a]*Fugu rubripes*, [b]*Drosophila melanogaster*, and [c]*Caenorhabditis elegans* gene structures are imported from external sources.

projects often submit their sequence to the main mRNA section of the EMBL/GenBank/DDBJ repository, and also, some projects have an appreciable level of genomic contamination or other small ORF-containing transcripts that have no similarity to known genes in other organisms.

## DISCUSSION

Ensembl is one of the main genome-management and display systems available for users. We deliver genome information in a variety of ways, from website access through to complete downloads and local installations. In doing so, we are accelerating molecular biology research worldwide. For example, the Diabetes and Inflammation Laboratory (DIL) group at the Cambridge Institute for Medical Research uses Ensembl data access to dramatically simplify their in-house gene and SNP discovery processes. Another example is the in-house installation of Ensembl at the Center for Human Genetics at Duke University, which allows those researchers to integrate their work on genetic disease discovery. Ensembl's role here is to provide a good infrastructure for genetic and molecular biology-based research and so free up in-house time and effort that would otherwise be duplicated across many laboratories worldwide.

Ensembl also remains at the forefront of trying to understand more aspects of the information stored in genomes. Of particular interest are comparative genomics approaches which leverage the action of evolution on genomes. We have worked extensively with Chris Ponting's group from Oxford to provide an insight into positive selection of particular genes in mouse and rat. In the future, we hope to contribute to the understanding of mammalian *cis*-regulation and other features.

## ACKNOWLEDGMENTS

Ensembl is a large multidisciplinary team, and I thank the entire team for their effort and enthusiasm over the last three years. Ensembl is a joint project between the EBI and the Sanger Insitute and is predominantly funded by the Wellcome Trust, with additional funding from the National Institutes of Health–National Institute of Allergic and Infectious Diseases and EMBL.

## REFERENCES

Clamp M., Andrews D., Barker D., Bevan P., Cameron G., Chen Y., Clark L., Cox T., Cuff J., Curwen V., Down T., Durbin R., Eyras E., Gilbert J., Hammond M., Hubbard T., Kasprzyk A., Keefe D., Lehvaslaiho H., Iyer V., Melsopp C., Mongin E., Pettett R., Potter S., and Rust A., et al. 2003. Ensembl 2002: Accommodating comparative genomics. *Nucleic Acids Res.* **31:** 38.

Guigó R., Dermitzakis E.T., Agarwal P., Ponting C.P., Parra G., Reymond A., Abril J.F., Keibler E., Lyle R., Ucla C., Antonarakis S.E., and Brent M.R. 2003. Comparison of mouse and human genomes followed by experimental verification yields an estimated 1,019 additional genes. *Proc. Natl. Acad. Sci.* **100:** 1140.

Lander E.S., Linton L.M., Birren B., Nusbaum C., Zody M.C., Baldwin J., Devon K., Dewar K., Doyle M., FitzHugh W., Funke R., Gage D., Harris K., Heaford A., Howland J., Kann L., Lehoczky J., LeVine R., McEwan P., McKernan K., Meldrim J., Mesirov J.P., Miranda C., Morris W., and Naylor J., et al. (International Human Genome Sequencing Consortium). 2001. Initial sequencing and analysis of the human genome. *Nature* **409:** 860.

Waterston R.H., Lindblad-Toh K., Birney E., Rogers J., Abril J.F., Agarwal P., Agarwala R., Ainscough R., Alexandersson M., An P., Antonarakis S.E., Attwood J., Baertsch R., Bailey J., Barlow K., Beck S., Berry E., Birren B., Bloom T., Bork P., Botcherby M., Bray N., Brent M.R., Brown D.G., and Brown S.D., et al. (Mouse Genome Sequencing Consortium). 2002. Initial sequencing and comparative analysis of the mouse genome. *Nature* **420:** 520.

# Prediction, Annotation, and Analysis of Human Promoters

M.Q. Zhang
*Cold Spring Harbor Laboratory, Cold Spring Harbor, New York 11724*

Since the celebrated discovery of the Watson-Crick double-helix structure of DNA, it has taken 50 years for the human genome to be sequenced. It may very well take another 50 years for the functional information to be fully decoded. Until recently, genome research has mainly focused on coding regions, and the immediate questions have been, "Where are the protein coding regions?" and "What are the functions of the gene products?" Increasingly, the field is advancing toward noncoding regions, where the central questions become, "Where are the regulatory regions?" and "How do they control gene expressions?" In 1961, Jacob and Monod published "On the regulation of gene activity" at the 26th Cold Spring Harbor Symposium on Quantitative Biology, in which some of the fundamental concepts of gene regulation were first elegantly formulated. Regulatory regions are most fundamental, because all the gene structures are defined by and recognized through the *cis* elements in such regions; furthermore, what a gene does in vivo is intimately related to when, where, and how much it is expressed. A phenotype, upon which the selection force is acting, is the integrated result of gene function and regulation. It is argued that animal diversity is mainly due to evolutionary expansion in regulatory complexity (Levine and Tjian 2003). Most regulations occur at the transcriptional level, and the initiation of transcription is largely determined by the promoter located at the beginning of each gene; identification of promoters and *cis* regulatory elements within them has become the prerequisite for understanding of gene regulation. For a few model organisms with compact genomes (such as phage, bacteria, and yeast), many of the gene regulatory pathways or networks have been worked out. But for mammalian systems, such as human, systematic identification of regulatory regions and gene networks has turned out to be extremely difficult, largely due to the size and complexity of the genomes (hence, as a result, the diversity of the cell/tissue types and the complication of developmental stages).

Here I outline our approaches to this problem. Because genome research is data and technology driven, many approaches in the field can soon become obsolete once new or additional data or technologies become available. Here, I try to state generic ideas and methods that may be evolving with or refined by new data or technologies. I also try to point out open problems and to suggest new experiments to attack them.

## IN SILICO PREDICTION OF MAMMALIAN PROMOTERS

Transcription of a eukaryotic protein-coding gene is preceded by multiple events; these include decondensation of the locus, nucleosome remodeling, histone modifications, binding of transcriptional activators (or derepressors) and coactivators to enhancers and promoters, and recruitment of the basal transcription machinery to form the preinitiation complex (PIC) at the core promoter. A core promoter is defined approximately as the DNA region (–40, +40) with respect to the transcriptional start site (TSS). It may contain the TFIIB recognition element (BRE) and the TATA-box at the 5′ end, the initiator (Inr) around the TSS, and the downstream promoter element (DPE) at the 3′ end (see, e.g., Smale and Kadonaga 2003). Although, in a mammalian genome, distal enhancers/silencers can be 10–100 kb away from the target gene, most of the *cis*-regulatory elements are contained in a proximal promoter region of 0.5–2 kb in size. Putative mapping of known transcription factor-binding site (TFBS) density profiles was originally used to develop the first computational promoter prediction program, called *Promoterscan* (Prestridge 1995; for survey and evaluation of earlier promoter prediction programs, see Fickett and Hatzigeorgiou 1997), later discriminative oligonucleotide-based algorithms, such as *PromoterInspector* (Scherf et al. 2000), showed much improved performance.

We hypothesized that molecular pattern recognition may be achieved by different molecular machinery with different resolutions at different scales (Zhang 1998a). An analogy would be that, if one tries to locate a landmark on earth from an airplane, one could use a coarse-grained tool to locate a regional landscape before zooming in with a finer mapping tool. Ideally, a coarse-grained promoter finder should be able to detect a chromatin and/or epigenetic landscape at the proximal promoter level (resolution <2 kb). It could be an easier problem if one had 3D structural images (and this could happen within the next 10 years). With only the primary DNA sequences, one would have to use large-scale statistical features of those length characteristics. Fortunately, for human (or vertebrate), CpG islands can provide one such discriminative feature for at least 50% of genes (Antequera and Bird 1993)! The human genome contains ~50,000 CpG islands, ~30,000 after repeatmasking, and the majority of these are near promoters. Computationally, a CpG island is defined (Gardiner-Garden and Frommer 1987) by a DNA region >200 bp that has >50% GC content and >0.6 ratio of CpG over expected CpG. Using this criterion alone, one would find ~345,000 CpG islands in the human genome. By detecting promoter-associated CpG islands, we have developed an algorithm (called *CpG_Promoter*) for coarse-grained promoter mapping (Ioshikhes and Zhang 2000). Promoter-associated CpG islands tend to be larger (0.5–2 kb), with higher GC content and CpG ratio; other CpG islands are mostly associated with Alu repeats. Takai and Jones (2002) proposed a new definition: size >500 bp that has >55% GC content and >0.65 CpG ratio. Using this new criterion, one would find ~37,000 CpG islands. Other CpG island-based promoter prediction algorithms, such as *CpG+* (Hannenhalli and Levy 2001) and *CpGProD* (Ponger and Mouchiroud 2002), have also become available. We would like to see more large-scale experimental data, such as chromosome bandings, methylation patterns, histon modification profiles, DNase hypersensitive sites, ChIP profiles, and genomewide transcription reporter constructs. Integrating these data will allow better promoter landscape-mapping algorithms to be developed.

For finer promoter mapping, aiming at predicting the TSS with a resolution <100 bp, we developed an algorithm, called *CorePromoter* (Zhang 1998b), based on quadratic discrimination analysis (QDA) using position-dependent oligonucleotide features (these positions are designed to capture the known core-promoter elements). By combining coarse-grained and fine prediction tools, I demonstrated how the TSS could be precisely located for the App gene (a 300-kb gene in Chromosome 21) encoding amyloid precursor (Zhang 2000). Instead of oligomers, *Eponine* uses a set of weight matrices in a hybrid machine-learning approach (Down and Hubbard 2002) to identify TSS. *Dragon Promoter Finder* (*DPF*) is an *a*rtificial *n*eural *n*etwork (ANN)-based algorithm that uses multiple sensors (promoters, exons, and introns) to predict TSS (Bajic et al. 2002).

Because gene structures are often correlated (i.e., neighboring introns or exons can help predict promoters as demonstrated in *DPF* above; for review, see Zhang 2002a), we have developed *FirstEF* that integrates promoter, 5´UTR, and first-intron information for predicting human first exons and promoters simultaneously (Davuluri et al. 2001). There is increasing evidence that transcription and splicing are coupled; we expect that the promoter may influence the first donor site selection. Recently, *DPF* output and CpG islands were integrated into a larger ANN program called *Gene Start Finder* (*DGSF*) to achieve a comparable promoter prediction in a test using Chromosomes 4, 21, and 22 (Bajic and Seah 2003). Although modern gene prediction programs, such as *Genscan* (Burge and Karlin 1997), try to predict first coding exons, *FirstEF* is the only program that is capable of predicting noncoding (untranslated) first exons. The most important open problems in promoter prediction are (1) how to improve accuracy on predicting promoters (or first exons) that are not CpG island-associated; (2) how to predict alternative promoters and to predict multiple TSSs (a single promoter can regulate multiple start sites, especially when the multiple start site downstream element (MED-1) is present (Ince and Scotto 1995).

## AUTOMATIC CONSTRUCTION OF CSHL MAMMALIAN PROMOTER DATABASE REFERENCE SYSTEM

As microarray expression data become prevalent, biologists often need to extract various sets of promoter sequences from clustered genes (Zhang 1999a). Originally, we developed *PEG* (*p*romoter *e*xtraction from *G*enBank) using a set of accession numbers or ESTs as the input to facilitate the extraction of large sets of promoters (Zhang and Zhang 2001). When the nearly finished genome became available in April 2003 (Human built 33), we developed our automated annotation pipeline (an expert system) called *FexAnnotator* (*f*irst *e*xon *a*nnotator; R.V. Davuluri et al., in prep.), which can reduce false positives and false negatives from the *FirstEF* predictions by using existing knowledge in the public sequence database annotations (mRNA/EST and ENSEMBL genes). In this first-pass annotation, we have ~53,000 first exons (including ~8,000 alternative first exons only annotated for the Refseq genes). The accuracy check shows that among ~10,000 experimentally verified first exons (such as those in *EPD* and in *DBTSS*), ~80% were found within 500 bp of our pipeline predictions. Another check uses known TFBSs in *TRANSFAC*; the density of these TFBSs is indeed concentrating within the vicinity of annotated core promoters.

For genome scale regulation studies, building a high-quality promoter database, which allows easy and flexible data query or retrieval as well as on-the-fly analysis, is essential. Our *Saccharomyces cerevisiae* Promoter Database (*SCPD;* Zhu and Zhang 1999) has proved to be very useful for the yeast community. To better annotate human promoters, we are currently building the *CSHL Mammalian Promoter Database,* which initially includes Homo sapiens Promoter Database (HsPD, based on Human built 33, April 2003), Mus musculus Promoter Database (MmPD, based on Mouse release Feb. 2003), and Rattus norvegicus Promoter Database (RnPD, based on Rat release Jan. 2003). A new pipeline has been developed (Z.Y. Xuan et al., ubpubl.), in addition to ENSEMBL (Hubbard et al. 2002), it also makes use of results from GenomeScan (Yeh et al. 2001), *Fgeneh+* (Solovyev 2002), and TwinScan (Korf et al. 2001) in order to annotate promoters for potential novel genes (for further experimental validations). The new pipeline, taking advantage of cross-species comparisons, can automatically annotate multiple genomes in parallel on a Linux cluster (Fig. 1) and has been used to create the initial reference system for the *CSHL Mammalian Promoter Database* (Z.Y. Xuan et al., unpubl.). In this database, orthologous promoters will be linked so that a user can input a list of UnigeneIDs or Accession Numbers (from a clustered microarray data, say), specify the range of the promoter region, extract orthologous promoter sequences, do motif finding on the fly, or select a gene of interest, do orthologous promoter alignment on the fly and look for conserved motifs (Fig. 2). Maintaining com-

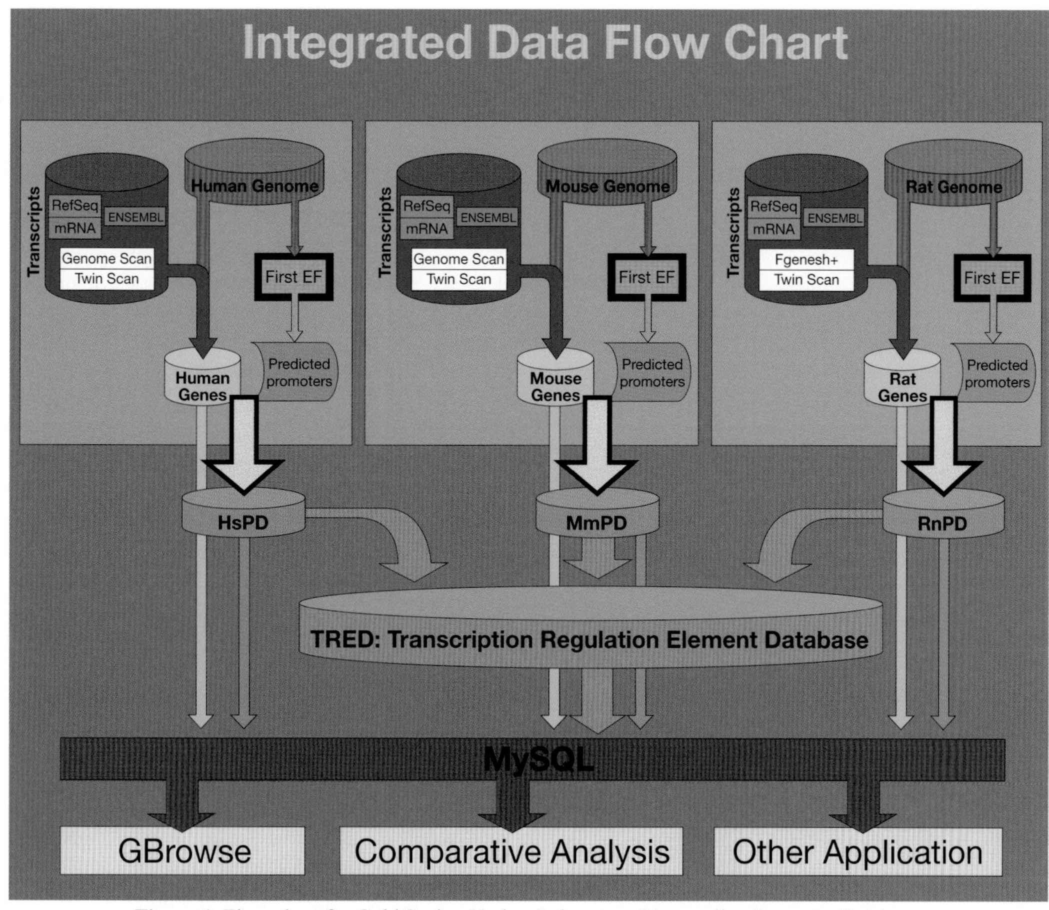


**Figure 1.** Flow chart for Cold Spring Harbor Laboratory Mammalian Promoter Database.

putability in addition to manual browsibility will serve well both computational and experimental biologists.

## FUNCTIONAL CURATION OF CELL CYCLE TRANSCRIPTION FACTORS AND THEIR TARGET GENES

A promoter reference system created by automatic pipeline can ensure completeness; it is consistent with most of the known information and also has reasonable accuracy. It must contain rich functional information (TFs, TFBSs, TSS, CpG islands) and links to other related databases and literature reference in order to be useful. Therefore, we are adding to *HsPD/MmPD/RnPD*, *TRED* (Fig. 1) (*t*ranscription *r*egulatory *e*lement *d*atabase; F. Zhao et al., unpubl.), which allows semi-automated or even hand-curated information to be entered. The three most important issues every useful database must address are (1) to assign quality value to the raw record; (2) to ensure accuracy and usefulness; and (3) to open data disseminations. For issue 1, we have assigned different quality values to promoters and TFBSs according to how they were derived. For issue 3, we are discussing with NCBI (D. Lipman, pers. comm.) and EBI (E. Birney, pers. comm.) ways to incorporate our results into public databases. The most difficult and time-consuming task is issue 2, which involves hand-curation and outreach to transcription expert labs. We are initially focusing on cell cycle and cancer-related TFs including their target genes, and we will give authorship to related transcription labs that contribute data or expertise. Currently, of 60,519 promoters (40,658 genes) in the human part of *TRED*, only 2,003 promoters (1,853 genes) are in the best-quality class (known and curated class). Other classes are (1) known but not curated, (2) predicted based on Refseq, (3) predicted based on other mRNAs, (4) predicted based on other ESTs, and (5) purely predicted. As an example, for human E2F targets, *TRED* contains 233 promoters (182 genes) in the best-quality class.

## HIGH-THROUGHPUT EXPERIMENTAL VALIDATIONS

All computational predictions must be subjected to experimental verification, and both positive and negative results are crucial feedbacks for further database and algorithm improvement. A lack of high-throughput experimental validation has become the bottleneck in this feedback loop. As cDNA libraries become more saturating, novel gene finding has gradually shifted its paradigm from EST sequencing to computational prediction plus experimental validation (Das et al. 2001; Guigo et al. 2003). To validate first exons and TSSs, getting 5′ complete cDNAs is essential (Davuluri et al. 2000; Suzuki et

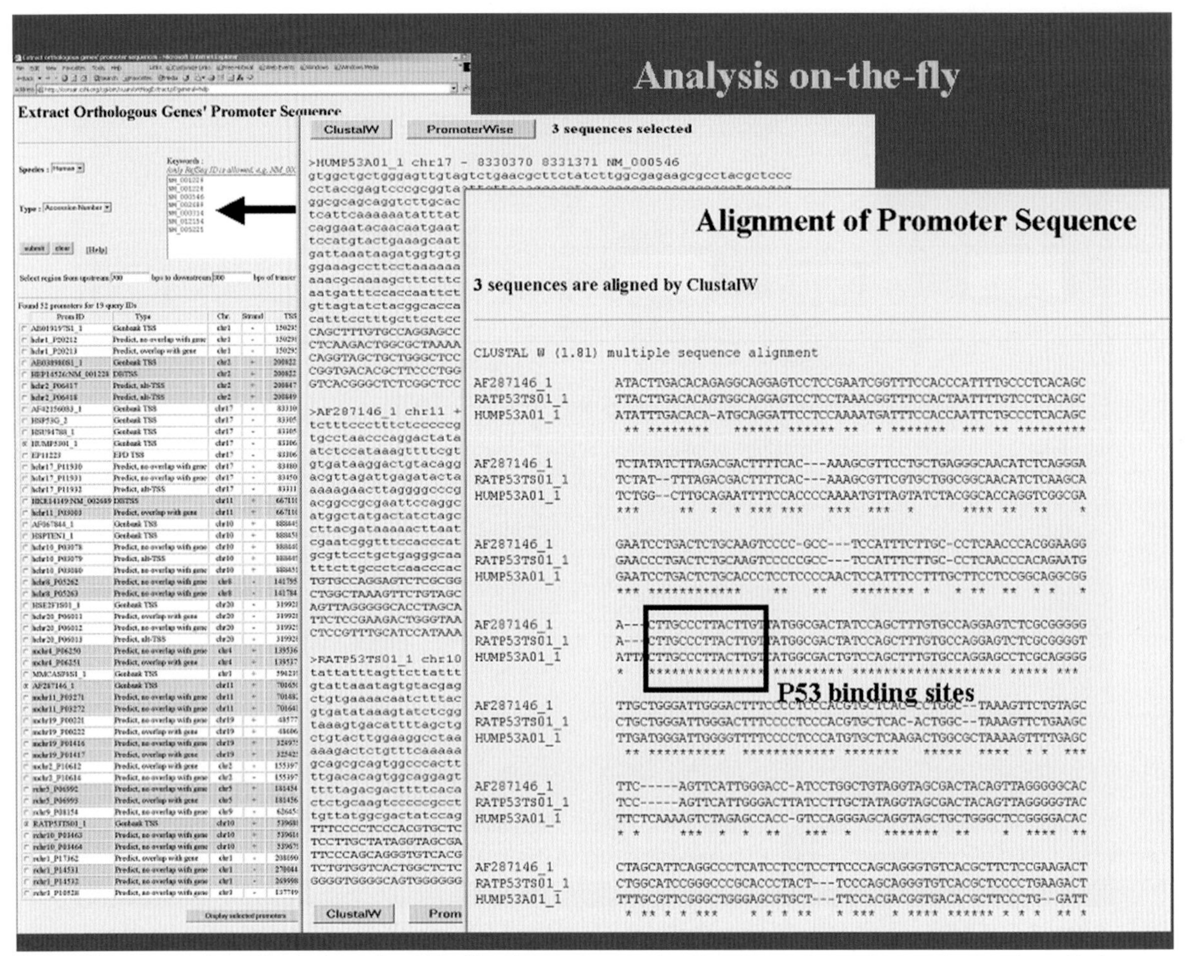


Figure 2. Demonstration of "analysis on-the-fly" utility: A user can paste in a list of genes (accession numbers) and specify the range (–700, +300), extract promoter sequences (including orthologous sequences), select P53 gene promoters, do promoter alignment, and identify the motif in conserved regions.

al. 2000). Recently, using reporter construct, the 5′ quality of Refseq and MGC clones has been randomly assayed for transcriptional activity of the upstream sequences (Trinklein et al. 2003).

In collaboration with the McCombie and Hannon labs at Cold Spring Harbor Laboratory on developing high-throughput experimental genome annotation technologies, we have performed systematic 5′-RACE-PCR validation of 300 first-exon predictions in 15 mouse tissue libraries (S. Dike et al., in prep.). We have selected the predicted exons in five categories according to support evidence from (1) EPD (this serves as the positive control), (2) Refseq, (3) more than two ESTs, (4) only one EST, and (5) pure prediction. The success rates are 12/13, 17/27, 18/23, 28/169, and 16/68, respectively. Here ~25% of predicted novel genes are likely to be real.

Working with the Wang lab at the University of Chicago on developing GLGI (*g*eneration of *l*onger 3′cDNA from SAGE tag for *g*ene *i*dentification; see J.J. Chen et al. 2003)-based genome annotation technology, we also obtained 57 positives from a test of 104 first-exon predictions in human tissues, and 15 full-length cDNAs were sequenced from 47 novel exon/SAGE-tag clones (S.M. Wang, pers. comm.).

To test promoter activities, we have collaborated with the Stubbs lab at Lawrence Livermore National Laboratory in annotating predicted genes/promoters in an 800-kb region (containing 48 genes) of human ch19q13 using the luciferase report system in addition to RT-PCR. Of 38 tested predictions, 26 tested positive (see Fig. 3) (L. Stubbs, pers. comm.).

These experimental exercises have demonstrated the validity of the large-scale computational prediction plus experimental verification approach for accurate genome annotation. It is also alarming that many previous false positives can be turned into true positives after more issues are tested or more sensitive experimental techniques are used (Kapranov et al. 2003). The new challenge for computational biologists is how to recover false negatives; for experimental biologists the challenge is how to prove a false positive!

## COMPUTATIONAL CHALLENGES IN IDENTIFICATION OF *cis*-REGULATORY ELEMENTS AND TRANSCRIPTIONAL NETWORKS

Although most TFBSs are in the promoter region, many may be in the first intron (which can also be located by *FirstEF* prediction), and some may be in the 3′-flank-

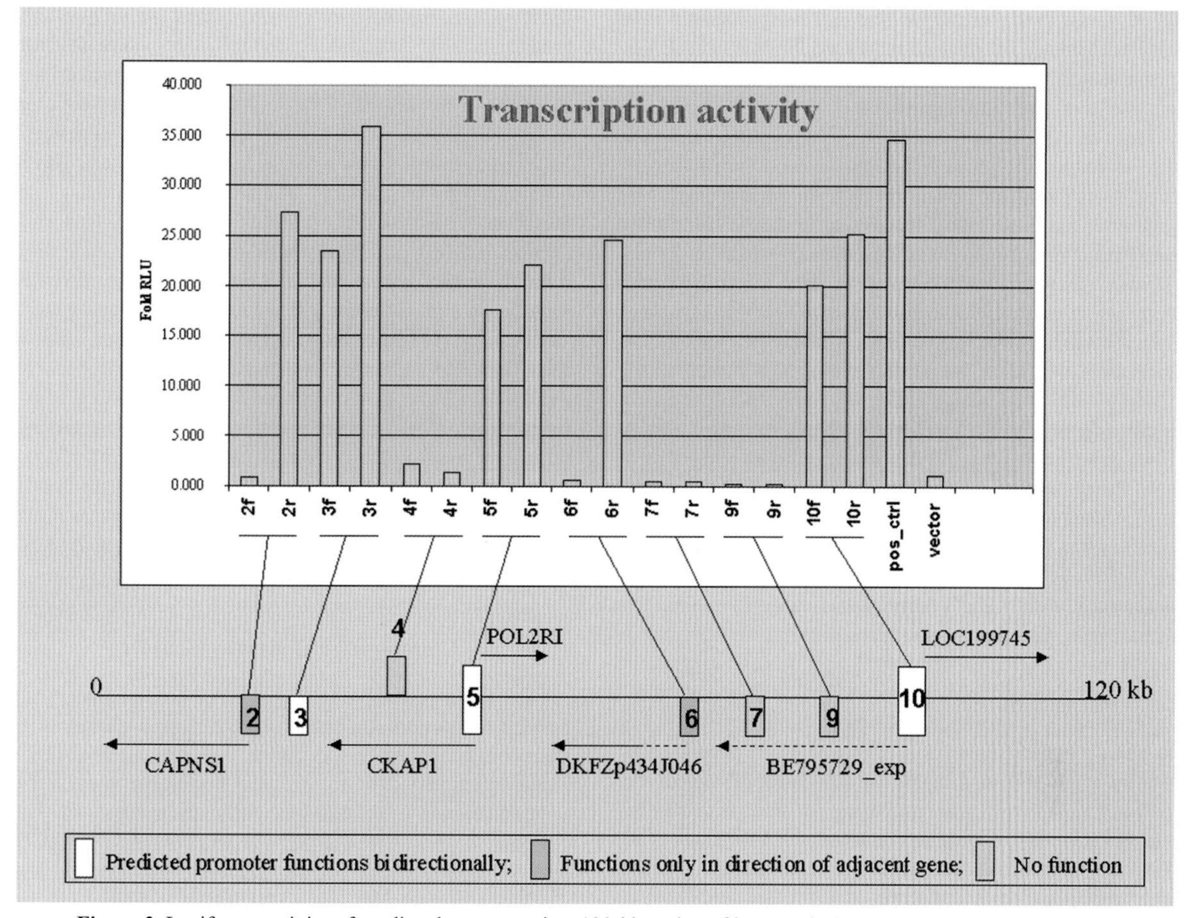


**Figure 3.** Luciferase activity of predicted promoters in a 120-kb region of human ch19q13 (provided by Lisa Stubbs).

ing region (which can be located by EST/poly(A) mapping). There are also many distal enhancers/silencers/boundary elements that are the most difficult to find because they are so far away from the target genes. Even if they are found, linking to the correct target genes is still no easy task. We are focusing on the proximal promoter region for *cis*-regulatory element discovery; many of the methods may also be applied to other regulatory regions once they are approximately localized (e.g., by comparative genomic analysis, by DNase hypersensitivity mapping, or by enhancer trapping technologies).

### Computation-Then-Validation Paradigm

Traditionally, identification of a *cis*-regulatory element is very laborious: collect known binding sites, build consensus or weight matrix, and search for new loci. One cannot discover novel sites in this way. To study human cell cycle regulation, we have developed E2F *SiteScan* based on a genetic algorithm trained on known sites in *TRANSFAC* and have scanned ~5,000 promoters in the public database to identify more than 300 E2F targets, many of which were also validated by the ChIP-PCR method (Kel et al. 2001). Since the E2F motif was built mainly from known cell cycle genes, they may be biased, as E2F also plays important roles in other biological pathways (such as apoptosis and DNA repair). By analyzing promoters from ChIP-PCR top candidates, we were able to identify novel E2F targets that do not have the conventional binding motif (Weinmann et al. 2001). However, the scope with PCR is still very limited. When large-scale genome-wide data and technologies become available, one will be able to study TFBSs in the whole genome together with their transcriptional readouts. It is expected that computational approaches will become more indispensable and will play more important roles in the future (Zhang 2002b,c).

### Large-scale Gene Expression Analysis

DNA microarray gene expression has become the widely used method for studying gene regulation. It provides a direct readout of cellular transcriptional programs. Interpretation of gene expression patterns by *cis* elements and *trans* factors, or conversely, by reconstruction of regulatory circuits from transcriptional responses, is the main challenge in the 21st century (Zhang 1999a; Banerjee and Zhang 2002). Using cluster analysis followed by motif searching of promoters of coregulated genes, we were quite successful in identification of *cis* elements involved in yeast cell cycle regulation (Spellman et al. 1998; Zhang 1999b). By combining functional information, such as *MIPS* (Zhu and Zhang 2000) or *GO*

(G.X. Chen et al., in prep.), one can further select gene clusters that are not only coexpressed, but also share significant number of genes involved in similar functional pathways or structural complexes.

Human *cis*-element detection is much more difficult due to much smaller signal-to-noise ratio (promoter region is much larger and uncertain, motifs are more degenerate, there are many repeats, etc.). Most commonly used motif finders, such as *Consensus* (Hertz et al. 1990), *MEME* (Bailey and Elkan 1994), and *Gibbs sampler* (Lawrence et al. 1993; Neuwald et al. 1995), assume a specific background model (e.g., Markov of order $k$). To increase specificity, we have developed a novel motif-finding software package called *BEAST* (*b*inding *e*lement *a*naly*s*is *t*ools; N. Hata and M.Q. Zhang, in prep.) that allows arbitrary background sequences to be the control set. The algorithm is based on an exhaustive word-counting strategy (allowing gap and reverse-complement, overlapping word is treated similarly, as in van Helden et al. 2000). For each motif, the Fisher exact test (or chi-square test with Yates's correction) is used to evaluate $p$-value (with multiplicity correction) for the significance of motif association to the target (promoter) sequences against the background control. *BEAST* has been applied to microarray expression data from transcription factor knockout experiments (G.X. Chen et al., in prep.), using the up-regulated promoters, the down-regulated, or the combination as the target and using the unchanged as the control. Combined with *GO* annotation (Ashburner et al. 2000), results agree well with the corresponding ChIP-chip analysis (data not shown).

*BEAST* was tested in detecting liver-specific promoter elements when a set of 35 proximal promoters of known liver-specific genes was used for the targets and the pool of 1800 EPD promoters was used as the control. The HNF-1 motif YAMT..TTRA ($p = 6.1 \times 10^{-12}$) was clearly identified on top of other putative motifs (N. Hata and M.Q. Zhang, in prep.). The new challenge is to apply *BEAST* systematically to mammalian tissue expression data, using tissue-specific gene promoters as the targets and using the pool as the control, for discovering tissue-specific promoter elements. Future adaptation of *BEAST* with weight matrices should further improve its sensitivity for degenerate motifs or motif combinations.

### Large-scale Chromatin Localization Analysis

Unlike the indirect coregulation strategy above, the ChIP-chip assay allows detection of TF-binding targets in the whole genome by cross-linking protein to chromatin DNA in vivo. The first two human ChIP-chip experiments were done using a CpG island DNA chip (Weinmann et al. 2002) or using a Refseq gene promoter chip (Ren et al. 2002) to map E2F4 target genes.

In collaboration with the Ren lab, we have used the ChIP-chip assay to discover a global transcriptional regulatory role for c-myc in Burkitt's lymphoma cells (Li et al. 2003). We find that c-myc and its heterodimeric partner, Max, occupy more than 15% of the gene promoters tested, and they colocalize with TFIID in these cells, indicating a general role for overexpressed c-myc in global gene regulation of some cancer cells. One surprise from the promoter analysis is that many of the targets do not have the conventional E-box; instead, we find a novel motif **CGGAAG** by *BEAST* which is the most significant *cis* element shared by a large number of c-myc/Max-binding target promoters (N. Hata et al., unpubl.). Furthermore, most of the elements are located near TSS (within 100 bp), and their positions are conserved among human, mouse, and rat (data not shown). We are currently seeking an experimental test for its functional relevance.

Recently, two other motif detection algorithms suitable for ChIP-chip and expression data analysis have become available. One is a word-based linear regression algorithm called *REDUCE* (Bussemaker et al. 2001), and another is a hybrid (word enumeration and weight matrix) greedy search algorithm called *MDscan* (Liu et al. 2002). Compared to these, *BEAST* conveniently provides motif $p$-values and is more discriminative against a given background control set.

### Comparative Genomic Analysis

Increasingly, comparative genomics has become a very powerful method for detecting functional elements in noncoding regions. We began with a comparative DNA sequence analysis of mouse and human protocadherin gene clusters in collaboration with experimentalists. The genomic organization of the human protocadherin $\alpha$, $\beta$, and $\gamma$ gene clusters (designated *Pcdh*$\alpha$, *Pcdh*$\beta$, and *Pcdh*$\gamma$) is remarkably similar to that of immunoglobulin and T-cell-receptor genes. The extracellular and transmembrane domains of each protocadherin protein are encoded by an unusually large "variable" region exon, whereas the intracellular domains are encoded by three small "constant" region exons located downstream from a tandem array of variable region exons. By comparing human draft and mouse BAC sequences, we were able to identify an alternative CpG-island-associated promoter in front of each variable exon in the $\alpha$ and $\gamma$ gene clusters, as well as a highly conserved *cis*-regulatory element within the promoter (Wu et al. 2001). Later, it was further confirmed that these *cis* elements are functionally important (Wang et al. 2002) and alternative promoter choice determines first-intron splice-site selection (Tasic et al. 2002).

To build our comparative genomics infrastructure, we carried out a whole-genome comparison between human and both (Celera and public) versions of mouse assemblies and published our *CSEdb* (*Conserved Sequence Element*) (Xuan et al. 2002). CSEs cover ~3% of the human genome. One-third of these CSEs are related to known genes, some are related to other functional elements (such as RNA genes and antisense genes); but more than half are still functionally unknown. Unknown CSEs provide excellent candidates for discovering novel genes or *cis*-regulatory regions. CSEs also allow us to arrive at another independent estimate of the number of human genes (~40,000).

Although comparative genomics has proved to be promising for discovering *cis*-regulatory regions (Pennacchio and Rubin 2001), because different promoters evolve at different rates, multiple species would be needed for narrowing down to short TFBSs. Initial success in yeast (Cliften et al. 2003; Kellis et al. 2003) may not directly translate to human; novel integrated approaches would be required to find functional *cis* elements even if the number of mammalian genomes were doubled.

### Integration, Combinatorial Analysis, and Network Reconstruction

Genomic data is noisy; the best weapon for combating noise is signal correlation analysis. Combinatorial interaction among TFs introduces correlation among their binding sites. Recently, there have been new motif-finding algorithms, such as *CO-Bind* (GuhaThakurta and Stormo 2001), that are designed specifically for detecting correlated motifs. Integration of evolutionary conservation with word-pair analysis can yield a better regression to expression data (Chiang et al. 2003).

Integrating ChIP-chip and expression data at the single motif level has recently been attempted (Conlon et al. 2003). We have developed two methods for studying cooperativity by integrating ChIP-chip data and microarray expression data. For a given pair of TFs, A and B, the first method compares expression patterns of the targets of both TFs to that of A or B alone. If the former is more coherent (correlated), it is more likely that the two TFs are interacting in the transcription regulation of their common targets (Banerjee and Zhang 2003). The second method further integrates with promoter sequence analysis in order not only to infer the interacting TFs, but also to assign their corresponding binding sites by iteratively and exhaustively searching for significant TF combinations and motif combinations up to the triplet level (M. Kato et al., in prep.). After analyzing over one hundred TF ChIP-chip data (Lee et al. 2002), we were able to reconstruct the yeast cell cycle transcriptional regulation network so that (1) it extends the previous chain of single regulators to an expanded chain of regulatory modules; (2) modules at adjacent phases often share a common component that can bridge the continuity of the cycle; (3) there are modules at specific checkpoints (branchpoints) that allow cell entry or exit of the cycle according to external signals (Fig. 4). Experimental verification is necessary to confirm any network predictions (Segal et al. 2003).

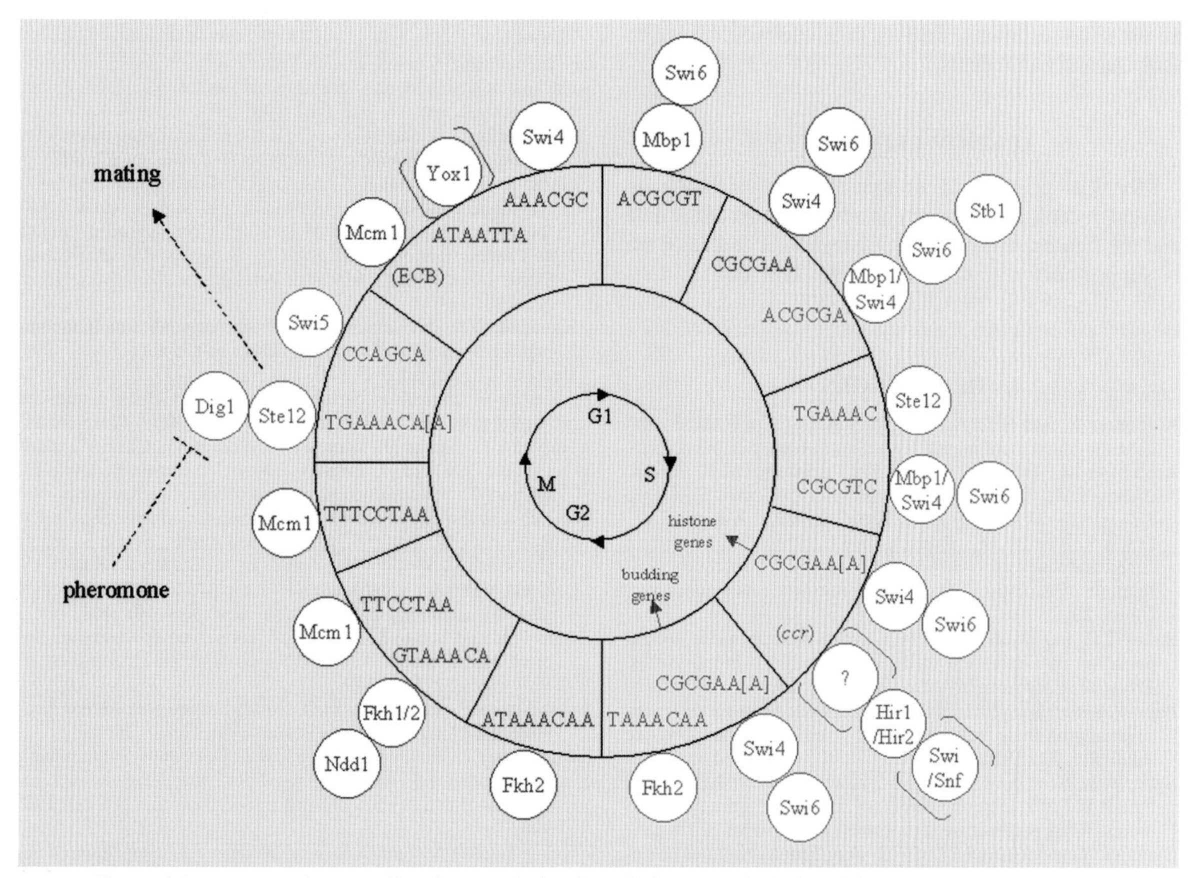


**Figure 4.** Reconstructed yeast cell cycle transcriptional regulation network. (Adapted from M. Kato et al., in prep.)

We are waiting for experimentalists to generate good-quality data of ChIP-chip and expression from the same sample preparations for mammalian systems, as well as to sequence multiple vertebrate genomes. Mammals alone are not enough for *cis*-element studies about human; one needs distant organisms (such as chicken, for phylogenetic footprinting) as well as close ones (such as chimpanzee, for phylogenetic shadowing).

## CONCLUSIONS

It is clear now that having a "periodic table" of genes is not enough; we also need a network diagram telling us how the genes are connected, and for this, we are going to need another "periodic table" of gene regulatory elements and a "wiring diagram" that connects each regulator to its targets. A combination of computational and functional genomics will help us to fill up these tables quickly. Infrastructure such as promoter databases and *cis*-element/*trans*-factor databases is urgently needed. New technologies that can provide a different genome-wide view of the regulatory networks and new algorithms that integrate various large-scale data will be the keys for attacking human gene regulation problems (Banerjee and Zhang 2002). Conservation is important for revealing function; non-conservation can be even more important for understanding evolution (Wray et al. 2003). The recent discovery of a promoter that acquired p53 responsiveness during primate evolution through microsatellite expansion of weak binding sites (Contente et al. 2003) is an amazing testimony, and for this, one would have to look beyond just rodents.

## ACKNOWLEDGMENTS

I thank all present and previous members of the Zhang lab and my collaborators for contributing most of the data and the figures, many before publication. The Zhang lab is supported by grants (HG-01696, GM-60513, CA-81152, CA-88351) from the National Institutes of Health.

## REFERENCES

Antequera F. and Bird A. 1993. Number of CpG islands and genes in human and mouse. *Proc. Natl. Acad. Sci.* **90:** 11995.

Ashburner M., Ball C.A., Blake J.A., Botstein D., Butler H., Cherry J.M., Davis A.P., Dolinski K., Dwight S.S., Eppig J.T., Harris M.A., Hill D.P., Issel-Tarver L., Kasarskis A., Lewis S., Matese J.C., Richardson J.E., Ringwald M., Rubin G.M., and Sherlock G. 2000. Gene ontology: Tool for the unification of biology. The Gene Ontology Consortium. *Nat. Genet.* **25:** 25.

Bailey T.L. and Elkan C.P. 1994. Fitting a mixture model by expectation maximization to discover motifs in biopolymers. *Proc. Int. Conf. Intell. Syst. Mol. Biol.* **2:** 28.

Bajic V.B. and Seah S.H. 2003. Dragon Gene Start Finder identifies approximate locations of the 5′ ends of genes. *Nucleic Acids Res.* **31:** 3560.

Bajic V.B., Seah S.H., Chong A., Zhang G., Koh J.L.Y., and Brusic V. 2002. Dragon Promoter Finder: Recognition of vertebrate RNA polymerase II promoter. *Bioinformatics* **18:** 198.

Banerjee N. and Zhang M.Q. 2002. Functional genomics as applied to mapping transcription regulatory networks. *Curr. Opin. Microbiol.* **5:** 313.

———. 2003. Identification of cooperativity among transcription factors controlling cell cycle in yeast. *Nucleic Acids Res.* **31:** 7024.

Burge C. and Karlin S. 1997. Prediction of complete gene structures in human genomic DNA. *J. Mol. Biol.* **268:** 78.

Bussemaker H.J., Li H., and Siggia E.D. 2001. Regulatory element detection using correlation with expression. *Nat. Genet.* **27:** 167.

Chen J.J., Lee S., Zhou G., Rowley J.D., and Wang S.M. 2003. Generation of longer cDNA fragments from SAGE tags for gene identification. *Methods Mol. Biol.* **221:** 207.

Chiang D.Y., Moses A.M., Kellis M., Lander E.S., and Eisen M. 2003. Phylogenetically and spatially conserved word pairs associated with gene-expression changes in yeasts. *Genome Biol.* **4:** R43.

Cliften P., Sudarsanam P., Desikan A., Fulton L., Fulton B., Majors J., Waterson R., Cohen B.A., and Johnston M. 2003. Finding functional features in *Saccharomyces* genomes by phylogenetic footprinting. *Science* **301:** 71.

Conlon E.M., Liu X.S., Lieb J.D., and Liu J.S. 2003. Integrating regulatory motif discovery and genome-wide expression analysis. *Proc. Natl. Acad. Sci.* **100:** 3339.

Contente A., Zischler H., Einspanier A., and Dobbelstein M. 2003. A promoter that acquired p53 responsiveness during primate evolution. *Cancer Res.* **63:** 1756.

Das M., Burge C.B., Park E., Colinas J., and Pelletier J. 2001. Assessment of the total number of human transcription units. *Genomics* **77:** 71

Davuluri R.V., Grosse I., and Zhang M.Q. 2001. Computational identification of promoters and first exons in the human genome. *Nat. Genet.* **29:** 412.

Davuluri R.V., Suzuki Y., Sugano S., and Zhang M.Q. 2000. CART classification of 5′UTR sequences. *Genome Res.* **10:** 1807.

Down T.A. and Hubbard T.J. 2002. Computational detection and location of transcription start sites in mammalian genomic DNA. *Genome Res.* **12:** 458.

Fickett J.W. and Hatzigeorgiou A.G. 1997. Eukaryotic promoter recognition. *Genome Res.* **7:** 861.

Gardiner-Garden M. and Frommer M. 1987. CpG islands in vertebrate genomes. *J. Mol. Biol.* **196:** 261.

GuhaThakurta D. and Stormo G.D. 2001. Identifying target sites for cooperatively binding factors. *Bioinformatics* **17:** 608.

Guigo R., Dermitzakis E.T., Agarwal P., Ponting C.P., Parra G., Reymond A., Abril J.F., Keibler E., Lyle R., Ucla C., Antonarakis S.E., and Brent M.R. 2003. Comparison of mouse and human genomes followed by experimental verification yields an estimated 1,019 additional genes. *Proc. Natl. Acad. Sci.* **100:** 1140.

Hannenhalli S. and Levy S. 2001. Promoter prediction in the human genome. *Bioinformatics* (suppl. 1) **17:** S90.

Hertz G.Z., Hartzell G.W., III, and Stormo G.D. 1990. Identification of consensus patterns in unaligned DNA sequences known to be functionally related. *Comput. Appl. Biosci.* **6:** 81.

Hubbard T., Barker D., Birney E., Cameron G., Chen Y., Clark L., Cox T., Cuff J., Curwen V., Down T., Durbin R., Eyras E., Gilbert J., Hammond M., Huminiecki L., Kasprzyk A., Lehvaslaiho H., Lijnzaad P., Melsopp C., Mongin E., Pettett R., Pocock M., Potter S., Rust A., Schmidt E., Searle S., Slater G., Smith J., Spooner W., Stabenau A., Stalker J., Stupka E., Ureta-Vidal A., Vastrik I., and Clamp M. 2002. The Ensembl genome database project. *Nucleic Acids Res.* **30:** 38.

Ince T.A. and Scotto K.W. 1995. A conserved downstream element defines a new class of RNA polymerase II promoters. *J. Biol. Chem.* **270:** 30249.

Ioshikhes I.P. and Zhang M.Q. 2000. Large-scale human promoter mapping using CpG islands. *Nat. Genet.* **26:** 61.

Kapranov P., Cawley S.E., Drenkow J., Bekiranov S. Strausberg R.L., Fodor S.P., and Gingeras T.R. 2003. Large-scale transcriptional activity in chromosomes 21 and 22. *Science* **296:** 916.

Kel A.E., Kel-Margoulis O.V., Farnham P.J., Stephanie M.B., Wingender. E., and Zhang M.Q. 2001. Computer-assisted identification of cell cycle-related genes—New targets for

E2F transcription factors. *J. Mol. Biol.* **309:** 99.

Kellis M., Patterson N., Endrizzi M., Birren B., and Lander E.S. 2003. Sequencing and comparison of yeast species to identify genes and regulatory elements. *Nature* **423:** 241.

Korf I., Flicek P., Duan D., and Brent M.R. 2001. Integrating genomic homology into gene structure prediction. *Bioinformatics* (suppl. 11) **17:** S140.

Lawrence C.E., Altschul S.F., Bogouski M.S., Liu J.S., Neuwald A.F., and Wooten J.C. 1993. Detecting subtle sequence signals: A Gibbs sampler strategy for multiple alignment. *Science* **262:** 208.

Lee T.I., Rinaldi N.J., Robert F., Odom D.T., Bar-Joseph Z., Gerber G.K., Hannett N.M., Harbison C.T., Thompson C.M., Simon I., Zeitlinger J., Jennings E.G., Murray H.L., Gordon D.B., Ren B., Wyrick J.J., Tagne J.B., Volkert T.L., Fraenkel E., Gifford D.K., and Young R.A. 2002. Transcriptional regulatory networks in *Saccharomyces cerevisiae. Science* **298:** 799.

Levine M. and Tjian R. 2003. Transcription regulation and animal diversity. *Nature* **424:** 147.

Li Z., van Calcar S., Qu C., Kolodner R., Cavenee W.K., Zhang M.Q., and Ren B. 2003. A global transcriptional regulatory role for c-myc in Burkitt's lymphoma cells. *Proc. Natl. Acad. Sci.* **100:** 8164.

Liu X.S., Brutlag D.L., and Liu J.S. 2002. An algorithm for finding protein-DNA binding sites with application to chromatin-immunoprecipitation microarray experiments. *Nat. Biotechnol.* **20:** 835.

Neuwald A.F., Liu J.S., and Lawrence C.E. 1995. Gibbs motif sampling: Detection of bacterial outer membrane repeats. *Protein Sci.* **4:** 1618.

Pennacchio L.A. and Rubin E.M. 2001. Genomic strategies to identify mammalian regulatory sequences. *Nat. Rev. Genet.* **2:** 100.

Ponger L. and Mouchiroud D. 2002. CpGProD: Identifying CpG islands associated with transcription start sites in large genomic mammalian sequences. *Bioinformatics* **18:** 631.

Prestridge D.S. 1995. Predicting Pol II promoter sequences using transcription factor binding sites. *J. Mol. Biol.* **249:** 923.

Ren B., Cam H., Takahashi Y., Volkert T., Terragni J., Young R.A., and Dynlacht B.D. 2002. E2F integrates cell cycle progression with DNA repair, replication, and G(2)/M checkpoints. *Genes Dev.* **16:** 245.

Scherf M., Klingenhoff A., and Werner T. 2000. Highly specific location of promoter regions in large genomic sequences by PromoterInspector: A novel context analysis approach. *J. Mol. Biol.* **297:** 599.

Segal E., Shapira M., Regev A., Pe'er D., Botstein D, Koller D., and Friedman N. 2003. Module networks: Identifying regulatory modules and their condition-specific regulators from gene expression data. *Nat. Genet.* **34:** 166.

Smale S.T. and Kadonaga J.T. 2003. The RNA polymerase II core promoter. *Annu. Rev. Biochem.* **72:** 449.

Solovyev V.V. 2002. Finding genes by computer: Probabilistic and discriminative approaches. In *Current topics in computational molecular biology* (ed. T. Jiang et al.), p. 201. The MIT Press, Cambridge, Massachusetts.

Spellman P.T., Sherlock G., Zhang M.Q., Iyer V.R., Anders K., Eisen M.B., Brown P.O., Botstein D., and Futcher B. 1998. Comprehensive identification of cell cycle-regulated genes of the yeast *Saccharomyces cerevisiae* by microarray hybridization. *Mol. Biol. Cell* **9:** 3273.

Suzuki Y., Ishihara D., Sasaki M., Nakagawa H., Hata H, Tsunoda T., Watanabe M., Komatsu T., Ota T., and Isogai T. 2000. Statistical analysis of the 5′ untranslated region of human mRNA using "Oligo-Capped" cDNA libraries. *Genomics* **64:** 286.

Takai D. and Jones P.A. 2002. Comprehensive analysis of CpG islands in human chromosomes 21 and 22. *Proc. Natl. Acad. Sci.* **99:** 3740.

Tasic B., Nabholz C.E., Baldwin K.K., Kim Y., Rueckert E.H., Ribich S.A. Cramer P., Wu Q., Axel R., and Maniatis T. 2002. Promoter choice determines splice site selection in protocadherin alpha and gamma pre-mRNA splicing. *Mol. Cell* **10:** 21.

Trinklein N.D., Aldred S.J., Saldanha A.J., and Myers R.M. 2003. Identification and functional analysis of human transcriptional promoters. *Genome Res.* **13:** 308.

van Helden J., Rios A.F., and Collado-Vides J. 2000. Discovering regulatory elements in non-coding sequences by analysis of spaced dyads. *Nucleic Acids Res.* **28:** 1808.

Wang X., Su H., and Bradley A. 2002. Molecular mechanisms governing Pcdh-gamma gene expression: Evidence for a multiple promoter and cis-alternative splicing model. *Genes Dev.* **16:** 1890.

Weinmann A.S., Bartley S.M., Zhang T., Zhang M.Q., and Farnham P.J. 2001. The use of chromatin immunoprecipitation to clone novel E2F target promoters. *Mol. Cell. Biol.* **21:** 6820.

Weinmann A.S., Yan P.S., Oberley M.J., Huang T.H., and Farnham P.J. 2002. Isolating human transcription factor targets by coupling chromatin immunoprecipitation and CpG island microarray analysis. *Genes Dev.* **16:** 235.

Wray G.A., Hahn M.W., Abouheif E., Balhoff J.P., Pizer M., Rockman M.V., and Romano L. 2003. The evolution of transcriptional regulation in eukaryotes. *Mol. Biol. Evol.* **20:** 1377.

Wu Q., Zhang T., Cheng J.-F., Kim Y., Grimwood J., Schmulz J., Dickson M., Noonan J.P., Zhang M.Q., Myers R.M., and Maniatis T. 2001. Comparative DNA sequence analysis of mouse and human protocadherin gene clusters. *Genome Res.* **11:** 389.

Xuan Z.Y., Wang J.H., and Zhang M.Q. 2002. Computational comparison of two mouse draft genomes and the human goldenpath. *Genome Biol.* **4:** R1

Yeh R.F., Lim L.P., and Burge C.B. 2001. Computational inference of homologous gene structures in the human genome. *Genome Res.* **11:** 803.

Zhang M.Q. 1998a. A discrimination study of human core-promoters. *Pac. Symp. Biocomput. 1998*, p. 240.

———. 1998b. Identification of human gene core promoters in silico. *Genome Res.* **8:** 319.

———. 1999a. Large scale gene expression data analysis: A new challenge to computational biologists. *Genome Res.* **9:** 681.

———. 1999b. Promoter analysis of co-regulated genes in the yeast genome. *Comput. Chem.* **23:** 233.

———. 2000. Discriminant analysis and its application in DNA sequence motif recognition. *Brief. Bioinform.* **1:** 331.

———. 2002a. Computational prediction of eukaryotic protein coding genes. *Nat. Rev. Genet.* **3:** 698.

———. 2002b. Computational methods for promoter recognition. In *Current topics in computational molecular biology* (ed. T. Jiang et al.), p. 201. The MIT Press. Cambridge, Massachusetts.

———. 2002c. Extracting functional information from microarrays: A challenge for functional genomics. *Proc. Natl. Acad. Sci.* **99:** 12509.

Zhang T. and Zhang M.Q. 2001. Promoter extraction from GenBank (PEG): Automatic extraction of eukaryotic promoter sequences in large sets of genes. *Bioinformatics* **17:** 1232.

Zhu J. and Zhang M.Q. 1999. SCPD: A promoter database of yeast *Saccharomyces cerevisiae. Bioinformatics* **15:** 607.

———. 2000. Cluster, function and promoter: Analysis of yeast expression array. *Pac. Symp. Biocomput. 2000*, p. 479.

# Ontologies for Biologists: A Community Model for the Annotation of Genomic Data

M. ASHBURNER,* C.J. MUNGALL,[†‡] AND S.E. LEWIS[‡]
*Department of Genetics, University of Cambridge & EMBL - EBI, Hinxton, Cambridge, United Kingdom; [†]Howard Hughes Medical Institute, and [‡]University of California, Berkeley, California

*Genomics has made biology into an information science.*

David Botstein, June, 2003

We celebrate the 50th anniversary of the discovery of the structure of DNA in 1953. The nature of the genetic code became *the* theoretical preoccupation of the ensuing decade, and immediately the language of information science entered biology (see Kay 2000). The discoveries of the genetic code, both its structure (syntax) and content (semantics), and of mechanisms for error recovery during its transmission, were followed, in the next two decades, by the development of methods to readily determine DNA sequences, to clone specific DNA sequences in bacterial hosts, and to amplify sequences in vitro by the polymerase chain reaction. Within 40 years or so of Watson and Crick's discovery, the first complete sequences of bacterial genomes were determined; today we celebrate the completion of the human genome. These advances brought a revolution that has affected even the most conservative fields of biology (see, e.g., Hebert et al. 2003).

None of these advances, except the very first, would have been possible without the application of methods from computer science, a field whose growth and maturity have closely paralleled that of molecular biology. Computational methods were introduced to biology in the early 1950s, for the calculation of Fourier summations in protein crystallography (Bennett and Kendrew 1952; Huxley 1990). However, it was the need to capture, store, assemble, and analyze DNA sequence data (Staden 1980) and to correlate them with other biological knowledge that motivated a new field of science—bioinformatics. The first attempts to collect protein and nucleic acid sequence data were published as slim printed books (Dayhoff et al. 1965; Croft 1973; Barrell and Clark 1974). The development of public computer files of sequence data soon followed, with the establishment of the PIR Database in 1980, the EMBL Data Library and Genbank in 1982, and the DDBJ in 1986.

Few can doubt that, without the international nucleic sequence data library as a common, freely available, depository of all public sequence data, neither the achievement of sequencing the human genome, nor its analysis, would have been possible. MEDLINE, Swiss-Prot, and PDB are equally core databases that also are very broad in content ("horizontal") and play an absolutely central role enabling genomic research. In addition, there are an increasing number of "vertical" databases, narrow in content, but covering this content in great detail. These include the model organism databases and databases for a restricted class of objects; for example, transcription factors or eukaryotic promoters. In the last decade or so, model organism databases have been developed as community projects for all of the biological organisms commonly used in research, and the number of specialist databases for biological objects has mushroomed. This is illustrated by the growth, from 24 in 1993 to 129 in 2003, in the number of databases that have contributed to the annual database issue of *Nucleic Acids Research*.

The proliferation of databases in this general field led many, in the early 1990s, to agonize about questions of database "interoperability" and to attempt to develop "federations" of databases (see, e.g., Fasman 1994) or database warehouses (e.g., the Integrated Genome Database [Ritter 1994]; see Davidson et al. 1995) or to dictate to database builders a common technical solution to database design. These attempts failed, with the only possible exception being the success of ACeDB in a number of communities (Durbin and Thierry-Mieg 1991). Yet the problems that the multiplicity of databases posed to both bench biologists and computational biologists have not gone away; indeed, they have worsened. They have worsened for two reasons: One is that the number of databases (and the amount of data) has simply increased, the other is that biologists have increasingly realized the importance of knowledge from a variety of organisms other than their own favorite model. Mouse biologists found that they needed information about genes from *Drosophila* or yeast; human geneticists needed information about genes in *Caenorhabditis elegans* or zebrafish.

One strategy to overcome some of the problems resulting from the dispersion of biological knowledge is to exploit insights from related fields. This is the approach taken by the many different databases participating in the Gene Ontology project. This project began in mid-1998 as a collaboration of three model organism databases (those for *Saccharomyces cerevisiae, Drosophila*, and mouse) to solve a very well defined problem: "How can gene products be described in a biologically meaningful way?" We realized that solving this problem would be a major task, hence one far better done collaboratively than competitively, and that, were we to succeed, we would—within this domain—have solved one of the major barriers to integration; that is, inconsistent semantics. Indeed, we envisioned in 1998 that, were we to succeed, we would have achieved de facto a degree of database integration, without having the considerable technical and social upheaval of redesigning our individual databases.

## STRUCTURE OF THE GENE ONTOLOGY

The Gene Ontology (GO) project's twin aims are to conceptualize biological knowledge within the limited domain of gene product attributes and then to describe gene products in their respective distributed databases by applying these conceptualizations (see Table 1). For the purposes of GO, gene products are considered to have three attributes: one or more molecular functions, one or more cellular locations, and one or more roles in one or more biological processes. GO builds three independent (in principle, orthogonal) graphs of concepts of these three classes of attributes. The GO graphs are not strict hierarchies, but rather are directed acyclic graphs (DAGs) in which multiple parentage is permitted. A typical strict hierarchy, familiar to all biologists, is the Linnean classification of phyla, kingdoms, classes, orders, families, and species. In the Linnean classification system, there is a strict cardinality of one between a child concept and a parent concept (a species is a member of one genus and one genus only), whereas in a DAG a child concept can have one or more than one parent.

Another contrast to be made between the GO graph and a Linnean classification is in the types of relationships that may be formed between concepts. The relationship in a Linnean classification is one of subsumption, each child concept is narrower in its scope than its parent concept; for example, a species is a narrower concept than a genus. In the GO graphs the relationship between parent and child concept may be one of subsumption—then this relationship is described as an *isa* relationship (also known as a hyponomy; Fellbaum 1998). Thus, the concept DNA binding is a hyponym of its parent concept nucleic acid binding. Alternatively, GO concepts may be related by a *partof* relationship (meronomy; see Miller 1998). Thus, the concept mitochondrion is considered a meronym of the concept cell; more simply, the mitochondrion is *partof* the cell. Both of these relationships are transitive. If *a* isa *b*, and *b* isa *c*, then *a* isa *c*; *x* is partof *y*, and *y* is partof *z*, then *x* is partof *z*. Concepts may have different relationships with their different parent concepts: For example, the concept nuclear membrane *isa* membrane, but is also *partof* the nucleus.

It is recognized that both the *isa* and *partof* relationships are complex (see, e.g., Brachman 1983; Winston et al. 1987; Rogers and Rector 2000). Indeed, the precise meaning of these relationships is, presently, context-dependent within GO. For example, within the context of the GO graph of biological processes, the *partof* relationship between the concept process *b* and its parent concept process *a* is that *b* is a subprocess of *a*. In principle, GO could have a far richer taxonomy of relationships between concepts (as an example, the UMLS Metathesaurus has over 50 different relationships between parent and child concepts [UMLS 2003]; for mappings between GO and UMLS, see McCray et al. [2002] and Sarkar et al. [2003]). The GO community is not convinced that to increase the complexity of relationships would increase functionality; it would, however, very considerably increase the complexity of maintaining the Gene Ontology itself.

Many concepts in biology may be described in several different ways; for this reason, GO concepts may have one or more synonyms. For the purposes of GO, a synonym and a concept have exactly the same meaning: These terms can be exchanged without altering the truth value of a sentence that contains them. However, it is also very useful within GO (e.g., for searching), to relate terms that have a less strict relationship than synonymy; such terms then have the relationship *is_ related*. Thus the term "antibody" is not a GO concept, yet it *is_related* to

**Table 1.** Gene Products, Including EST Collections, That Are Annotated with GO Terms

| | |
|---|---|
| Viruses | *Herpes viruses |
| Microbes[a] | **Shewanella oneidensis* |
| | **Vibrio cholerae* |
| | **Bacillus anthracis* |
| | *Brucella suis* |
| | *Enterococcus faecalis* |
| | *Pseudomonas putida* |
| | *Pseudomonas syringae* |
| | **Coxiella burnetii* |
| | *Treponema pallidum* |
| | *Dictyostelium discoideum* |
| Protozoa | **Plasmodium falciparum* |
| | *Plasmodium yoelii* |
| | **Trypanosoma brucei* |
| | **Leishmania major* |
| | *Tetrahymena*[b] |
| Algae | *Chlamydomonas rheinharti*[b] |
| Fungi | **Saccharomyces cerevisiae* |
| | **Schizosaccharomyces pombe* |
| | *Paracoccidioides brasiliensis* |
| Nematodes | **Caenorhabditis elegans* |
| | *Meloidogyne incognita* |
| Insects/arachnids | *Amblyomma variegatum* |
| | *Anopheles gambiae* |
| | **Drosophila melanogaster* |
| | **Glossina morsitans* |
| | *Apis mellifera* |
| Vertebrates | **Danio rerio* |
| | *Xenopus laevis*[b] |
| | **Mus musculus* |
| | **Rattus norvegicus* |
| | *Bos taurus* |
| | *Sus scrofa* |
| | *Canis familiaris* |
| | **Homo sapiens* |
| Plants | **Arabidopsis thaliana* |
| | **Oryza sativa* |
| | *Zea mays* |

Items marked with an asterisk are available from the GO database and its search engines, such as AmiGO.

[a] Over 30 other microbial genomes have been automatically annotated with GO terms at TIGR (http://www.tigr.org/tigr-scripts/CMR2/GO_terms.spl?db=CMR).

[b]In preparation.

the GO concept "antigen binding activity" in the GO database for utilitarian reasons.

All GO concepts have a unique identifier number and a definition that explicitly states their precise meaning. There is a strict one-to-one correspondence between a GO_identifier and its definition (rather than to the lexical string used to refer to the concept). Thus, a change in a concept's lexical string (but not its definition) will not alter its GO_identifier; conversely, if there is a change in the definition that alters the meaning of a GO concept then, even if its lexical string remains identical, this will result in the new concept having a new GO_identifier. If a GO concept is found to be incorrect or irrelevant it is not simply discarded, it is marked with the attribute *is_obsolete* and it, its GO_identifier, and definition remain in the database.

Finally, GO concepts may well be equivalent or have a close relationship with a term in some other database. For example, a GO concept describing the activity of an enzyme can be considered to have a close relationship with the name of that enzyme within the Enzyme Commission's database. This relationship may be expressed as a database cross-reference. GO also maintains a number of tables that represent semantic mappings of GO concepts to concepts in other databases; e.g., between GO concepts and Swiss-Prot keywords and between GO concepts and the functional catalog of MIPS (MIPS 2003).

## LIMITATIONS OF THE GO MODEL

During the early stages of the development of GO, a decision was made to restrict GO to using only the relationships of subsumption and meronomy. This decision was made for pragmatic reasons, and it was accepted that at some time in the future the implicit limitation in the expressive power of GO might become limiting. That future has arrived: GO, and similar artifacts, need to be property-based ontologies to be sufficiently expressive.

Currently, GO includes a large number of terms that can be exemplified by "biosynthesis of interleukin-13." It is obvious that this is a compound term, composed of a noun describing the action "biosynthesis" and a prepositional phrase "of interleukin-13," describing what is being biosynthesized. The same idea could also be expressed using the verb "biosynthesizes" and the direct object "interleukin-13." In either expression there is a central topic, "biosynthesis" or "biosynthesizes," that requires an additional supporting term, "interleukin-13," to completely convey the entire concept.

At present, GO terms are indivisible; there is no notion of the decomposition of phrases into either individual words or concepts and properties. If GO continues to introduce compound terms, the system will become progressively more redundant, less flexible, and increasingly difficult to manage. This is apparent in the parts of GO that involve implicit cross-products (e.g., between a process like "transport" and an ontology of chemical compounds). We must therefore provide a solution that both disambiguates fundamental concepts and also enables accurate annotation. One approach is to migrate to a property-based ontology (a *Description Logic*; see Baader et al. 2003). Another way to view this is a progression beyond a vocabulary of fixed phrases toward a combination of a vocabulary and a grammar for the creation of phrases (see, e.g., WordNet 2003).

GO will use a Description Logic (DL) to allow certain GO terms to have *properties* (i.e., attributes or *slots*). Properties provide a formalism for dealing with the classification of finely granular terms by allowing the flexible creation of phrases. This approach will move GO away from being merely a word-turned-phrase-based ontology toward a property-based ontology. This will offer the biological curators a structured means of composing phrases to use during annotation. A phrase is composed of a primary term (e.g., "biosynthesis') which dictates what other types of terms must interact with it in supporting or secondary roles (e.g., "interleukin-13"). The term "biosynthesis" would include a property for a required term indicating the thing-being-synthesized and thus an explicit GO term for "interleukin-13 biosynthesis" would no longer be necessary. When annotating, if the curator selected "biosynthesis," she would then be obligated to fill in the "thing-being-synthesized" property, in this case, "interleukin-13" or another identifier from a biochemical-moiety ontology. From the gene product curator's perspective, the annotations become phrases, composed of a primary term and some number of other terms that are necessary to complete the explanation. The number of properties required for a term may vary from zero to many. As an example to illustrate this, consider the term "localization." A curator would never use this term without indicating both the object and a prepositional phrase. Neither "localization," nor "localization of mRNA," nor "localization to the cell membrane" suffices; to make sense one must say "localization of mRNA to the cell membrane." From the ontology curator's perspective, defining a term will now include specifying the required properties for that term, to indicate whether or not, and what type of, supporting terms are required when a gene product is actually being annotated. The use of properties appears robust and extensible, and gives both ontology curators and biological curators flexibility, while retaining semantic rigor. It is entirely consistent with both Description Logics and frame-based ontologies, both of which are widely used in the knowledge representation–artificial intelligence (AI) community. This approach is also very modular because, as phrases are constructed, they may be used recursively to create more complex phrases.

Terms that are modifiers also may be added to the phrase, the difference being that modifying terms are not demanded by the primary term, but simply provide additional information. So, for the term "biosynthesis," although it is obligatory that a curator supply a term for the thing-being-synthesized property, they would not be obliged to fill in the property for a mediator of this biosynthesis. An annotation for a protein could then be made for biosynthesis (primary term) *of* GPI anchor (mandatory completion property) *via* N-glycyl-glycosylphosphatidylinositolethanolamine (optional informational property).

To illustrate the difference of this approach from that now used, this is a current entry from Swiss-Prot for a human protein annotation (some columns removed for clarity):

| PRODUCT ID | PRODUCT SYMBOL | GO ID |
|---|---|---|
| Q14116 | IL13_HUMAN | GO:0042231 interleukin-13 biosynthesis |

This would be replaced with:

| PRODUCT ID | PRODUCT SYMBOL | GO ID | PROPERTY (VALUE) |
|---|---|---|---|
| Q14116 | IL13_HUMAN | GO:0042089 cytokine biosynthesis | biosynthesizes (interleukin-13) |

Of course, the property "biosynthesizes" will be restricted to specific terms, and it will be the task of the ontology curators to specify the allowable properties for any given term. Any child term connected purely by *isa* links (but not *partof*) will inherit the properties of its parent. We anticipate that the immediate need is for a small number of properties applicable to a limited number of terms. Despite the small number, the list will be maintained in a computable form and thus allow software to automatically check the gene association input files for invalid property data and avoid nonsensical annotations (e.g., "prolactin receptor synthesizes pyrimidines"). This list of primary terms and their supporting role-playing properties can be represented as a 4-column table:

| TERM | PRODUCT | PROPERTY-FILLER-TERM | Cardinality |
|---|---|---|---|
| transport | transports | compound | 1 |
| biosynthesis | synthesizes | compound | 1 |
| biosynthesis | mediated-by | compound | 0 or 1 |
| protein-binding | binds | protein-or-protein-family | 1 |
| transcription factor | regulates | gene | 1 |

The property-filler-class is used to restrict the vocabulary from which the terms may be drawn for the property. If, for example, a gene product is annotated with the term "protein-binding" then the permissible value for the "binds" property must always be a protein family or a certain protein. The protein family would be represented with an ID from an appropriate database; for example, UniProt.

As shown in the table above, some terms can have more than one property type (e.g., "biosynthesis"). Properties can have different cardinalities (e.g., the mediated-by property is optional). More specific terms can have more specific property-filler terms (e.g., "protein biosynthesis" terms can only synthesize proteins as opposed to the more general "compound").

One disadvantage of this system is that we no longer will represent annotations as a simple 2-dimensional matrix of gene product and GO term; rather, we will have a matrix of gene product and "annotation phrase," which can have unbounded dimensionality. This has implications for everyone who writes software that uses GO. For example, in implementing searches, e.g., a search for "pyrimidine biosynthesis," the software would need to decompose the phrase and traverse both the function graph and the protein family/compound ontology (pyrimidine *isa* nucleotide). It is obvious, however, that one can simply temporarily instantiate the phrases by using the existing annotations. That is, for every leading term T, find all the supporting values, V1, V2...Vn, in the current annotations and generate "temporary phrase terms" TV1, TV2 ... TVn. In an AmiGO type view (Lewis et al. 2002), one would then see these phrases appearing beneath the primary term. The existence of simple phrase-construction rules is assumed for clarity—for example, any dynamically constructed "biosynthesis" phrase could be written by prefixing the word "biosynthesis" with the value filled in the "synthesizes" property.

```
[is-a] metabolism ; GO:0008152
[is-a] biosynthesis ; GO:0009058 (2000 products)
 [is-a] cytokine biosynthesis ; GO:0042089 (500 products)
 [is-a] interleukin-1 biosynthesis (100 products)
 [is-a] interleukin-2 biosynthesis (120 products)
 [is-a] interleukin-3 biosynthesis (140 products)
 [is-a] interleukin-4 biosynthesis (200 products)
```

This approach is both simple enough to use and powerful enough to solve many related problems and also opens the door to other uses. As an example, consider the following term now present in the GO: "positive regulation of vulval development." This is clearly a compound term composed of terms from many separate vocabularies; it also illustrates the recursive use of other compound terms. The linguistic composition of this term is shown in Figure 1.

When the worm gene CE08399 is to be annotated to "positive regulation of vulval development," the underlying association data would look like this:

| PRODUCT | GO ID | COMPLETION PROPERTY | INFORMATIONAL PROPERTY |
|---|---|---|---|
| CE08399 | GO:nnnnnnn (regulation) | regulates (development of (vulva)) | type(+) |

This would apply in other situations as well:

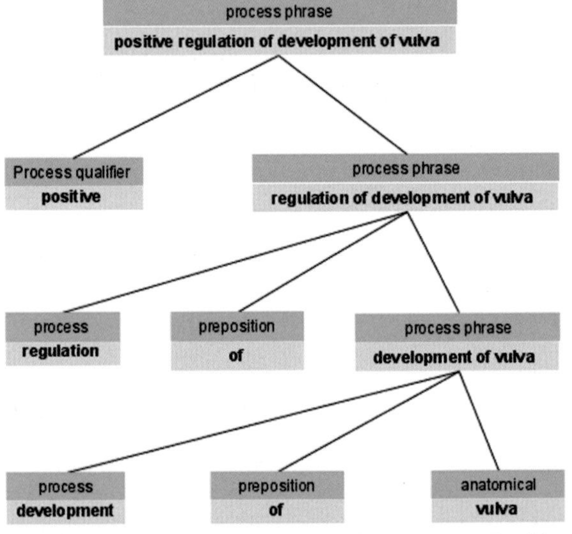


**Figure 1.** The linguistic composition of the GO term "positive regulation of vulval development."

| PRODUCT | TOPIC TERM | COMPLETION PROPERTY | INFORMATIONAL PROPERTY |
|---|---|---|---|
| Gene 1 | regulation | of(growth of (eye)) | type(+) |
| Gene 2 | regulation | of(growth of (leg)) | type(+) |
| Gene 3 | regulation | of(development of (B800-850 antenna complex)) | type(-) |

The last entry would appear in text (including browsers such as AmiGO) as "negative regulation of development of B800-850 antenna complex." Although this appears as an English-like phrase, internally this structured grammar is maintained as an inverted tree.

One unanswered issue is how we will attach definitions and synonyms to these composed phrases. We believe that this can be solved by having the phrase inherit the union of the definitions from the terms that it comprises, as has been described by Hill et al. (2002). The use of properties to indicate prepositions and direct objects will enable other compound phrases to indicate, *with, from, by*, and so forth. For example, a phenotypic observation such as, "black-marking *on*(ventral-surface *of* abdomen))," is composed of the topic term ("marking"), an optional descriptive term ("black"), and the sub-phrase "ventral-surface of abdomen," which in turn is composed of a position term with a property for an anatomical term.

Although we do not intend to make the proposed solution dependent on any particular technology or tool, or to force any complicated recasting of GO, it is worth noting that our approach is very compatible with that being developed by the W3C to support the "SemanticWeb," i.e., Ontology Web Language (OWL) (OWL 2003). In the Semantic Web Activity Statement, the W3C states "For the Web to reach its full potential, it must evolve into a Semantic Web, providing a universally accessible platform that allows data to be shared and processed by automated tools as well as by people" (Semantic Web 2003). We cannot guarantee when, if ever, the Semantic Web will become more than a dream, but the Ontology Web Language that is being developed is proving useful as a Web-worthy syntax able to fully describe the complex semantic content in a computable form.

We have developed tools for converting the GO format into a description logic format, DAML+OIL the predecessor of OIL. This has also been done independently by Wroe et al. (2003). The mechanics of the conversion are simple, *isa* relationships are converted into a *subClassOf* relationship, and all other relationship types are converted using *has-type* restrictions on the relationship. Here are two examples of GO information in a precursor to OWL, the Ontology Inference Layer (OIL), which we use here because it is slightly more human-readable:

```
class-def defined biosynthesis subclass-of metabolism
    property-constraint synthesizes has-type
biological_entity
    class-def defined cytokine_biosynthesis subclass-of
biosynthesis
    property-constraint synthesizes has-type cytokine
class-def defined regulation subclass-of
biological_process
    property-constraint regulates has-type biological_process
    property-constraint has-regulation-type has-type
regulation_type
    property-constraint has-body-part has-type body_part
class-def defined regulation_type
class-def defined positive_regulation_type
class-def defined negative_regulation_type
```

## THE CONTENT, USE, AND AVAILABILITY OF GO

The core of GO is represented by the three graphs of concepts for the annotation of the function, biological role, and cellular component of gene products. These now include about 14,000 terms, of which about 80% are defined. These are maintained by the GO editorial team and their close collaborators working as curators for the model organism databases. However, anyone can suggest new GO terms, or point out errors in the GO graphs, through a site maintained at SourceForge (http://sourceforge.net/projects/geneontology).

The primary use of GO is for the annotation of gene products within the context of model organism or protein databases. There are two major modes of annotation. The first of these is by literature curation, a model used by most databases for the capture of information of many classes. When annotating a gene product, a database curator will relate one or more GO concepts to that gene product with an attribution to a particular publication and with an indication of the evidence used by that publication for the assertion for the relationship. This evidence may be the result of an experiment ("inferred from direct assay") or may be an inference from sequence comparison ("inferred from sequence similarity"). If the latter evidence code is used, then the object to which the annotated sequence is similar is recorded, e.g., "inferred from sequence similarity to Swiss-Prot:P12345." This allows GO curators (and others) to detect transitive errors of annotation (see Gilks et al. 2002).

GO concepts may also be inferred by electronic annotation. For example, the GO concepts relevant to a particular gene product may be inferred automatically by a program that compares its sequence with a set of protein sequences that have previously been annotated with GO concepts (other than those whose annotations are themselves "inferred from electronic annotation"). M. Yandell's LOVEATFIRSTSIGHT program was developed for the *Drosophila* genome annotation (Adams et al. 2000) for this purpose, and others have subsequently been developed (see, e.g., Pouliot et al. 2001; Xie et al. 2002; Mi et al. 2003). The GO Consortium maintains a database of annotated proteins that can be used by automatic prediction programs (ftp://ftp.geneontology.org/pub/go/gp2protein/).

Methods other than direct sequence comparison have been used to predict GO terms associated with gene products. Some of these are indirect, that is via the literature or microarray data (see below); others use properties of proteins other than their primary sequence (e.g., post-translational modifications; see, e.g., Jensen et al. 2003) or recognized protein domains (Schug et al. 2002). King et al. (2003; see also Berriz et al. 2003) used the GO annotations of both FlyBase and the *Saccharomyces* Genome Database (SGD) to model relationships among GO terms with decision trees and Bayesian networks to predict GO terms that had been missed by the curators of these databases.

Members of the GO Consortium contribute tables of

associations between gene products and GO concepts to the GO database (see ftp://ftp.geneontology.org/pub/go/gene-associations/). It is these data that provide the de facto integration of many disparate databases. With a suitable tool, for example the GO AmiGO browser http://www.godatabase.org/cgi-bin/go.cgi), users can now query the model organism databases for almost 20 different species, plus Swiss-Prot, PDB, and the TIGR gene indices for all gene products annotated with any particular gene concept (Fig. 2). Thus, a user can query AmiGO for all gene products annotated as playing a role in "O-linked glycosylation," to discover 41 gene products from different organisms. GO is also being used by the Genome Knowledge Base (GK 2003), which represents richly annotated biological "pathway" data.

The automatic extraction of "knowledge" from the biomedical literature (see Hirschman et al. 2002) is a field to which the work of the Gene Ontology Consortium can make a major contribution. The gene_association tables in the GO database are easily parsible to form links between PUBMED abstracts and GO terms. Indeed, these have been used as a training set to predict GO terms associated with the complete PUBMED corpus of 12 million or so abstracts, by a commercial concern using an undisclosed algorithm (ReelTwo 2003). Others, e.g., Raychaudhuri et al. (2002), have used a variety of statistical methods to automatically assign GO terms to PUBMED abstracts and then, by transitivity, to the genes referenced therein (see also Jenssen et al. 2001). Supervised learning algorithms and other methods are also being used to predict GO terms to gene products from high-throughput gene expression data (Hvidsten et al. 2001; Khatri et al. 2002; Draghici et al. 2003; Lagreid et al. 2003).

## obo: OPEN BIOLOGICAL ONTOLOGIES

The experience of the GO Consortium, as an open community effort both in building common ontologies for biologists and in implementing these ontologies within the context of model organism and other community databases, has encouraged efforts to extend this concept to other domains in biology (see, e.g., Bard and Winter 2001). These efforts are being driven by two needs. The first is that of the GO Consortium for ontologies of, for example, anatomies and chemical compounds, so as to allow the implementation of a richer formalism for GO; that is, the values of properties (see above). The other is the need of the community in general for semantic integration, allowing both a more rigorous annotation of objects within the context of a single database and the development of tools for cross-database querying.

obo is a clearinghouse for ontologies for biology. The obo site at SourceForge (obo 2003) provides a single point of access for many different ontologies. These ontologies are available if their developers agree to a few simple principles: that the ontologies are freely available to all without license, that they are instantiated within an agreed syntax (so as to encourage re-use of common software tools), that they are orthogonal to other obo ontologies (to encourage cooperation rather than competition within common domains), that they use unique identifiers for their concepts, and that their concepts are accompanied by text definitions.

**Table 2.** The Current Content of obo

- *OBO relationship types*
- *MESH terms*
- Genomic & proteomic
  - *gene structure and variation*
  - gene product
    - gene product name
    - *molecular function*
    - *biological process*
    - *cellular component*
    - protein
      - *protein domain*
      - *protein covalent bond*
      - *protein-protein interaction*
- Biochemical
  - biochemical substance*
  - cell signaling
  - *physical-chemical methods and properties*
- Developmental time line
  - plant development
    - *Arabidopsis development*
    - *Rice development*
  - animal development
    - human development*
    - *Mus anatomy and development*
    - *zebrafish anatomy and development*
    - *medaka fish anatomy and development*
    - *Drosophila development*
    - *Caenorhabditis development*
    - *Plasmodium development*
- Anatomy
  - gross anatomy
    - *microbial structure*
    - *Dictyostelium* anatomy*
    - plant gross anatomy
      - *Arabidopsis gross anatomy*
      - *Cereal gross anatomy*
    - animal gross anatomy
      - *Caenorhabditis* gross anatomy*
      - *Drosophila* gross anatomy
      - *zebrafish anatomy and development*
      - *medaka fish anatomy and development*
      - Mus gross anatomy
        - *Mus gross anatomy and development*
        - *Mus adult gross anatomy*
  - organ
  - tissue
  - cell type
    - *generalized cell types*
    - *microbial structure*
- Phenotype
  - mutant phenotype*
  - *mammalian phenotype*
  - *mouse pathology*
  - *plant trait*
  - human disease*
- Ethology
  - *Loggerhead nesting*
  - *Habronattus courtship*
  - *Mus* behavior*
- Experimental conditions
  - *microarray experimental conditions*
  - *plant environmental conditions*
  - *biological imaging methods*
  - *physical-chemical methods and properties*
- Taxonomic classification
  - *NCBI organismal classification*
  - *Swiss-Prot organismal classification*

Ontologies in italics are now available, at least in a form available for discussion. The others are either desiderata or (marked with an asterisk) under active development.

Table 2 gives an indication of the current content of obo. The Sequence Ontology (SO) is one example of an obo ontology (SONG 2003). This is being developed by

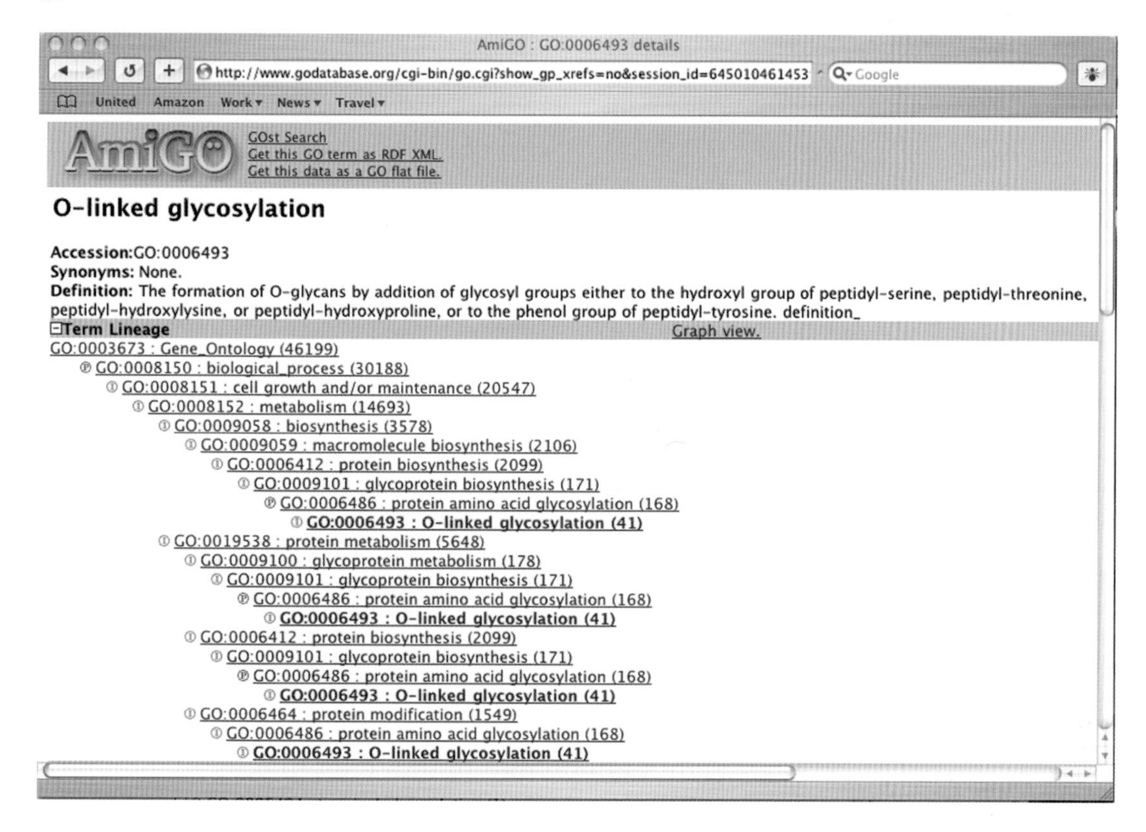


**b.**

| | | | |
|---|---|---|---|
| PAGT_HUMAN | SPTr | TAS E | Polypeptide N-acetylgalactosaminyltransferase |
| O95395 | SPTr | TAS E | Beta-1,6-N-acetylglucosaminyltransferase |
| OGT_CAEEL | SPTr | ISS | UDP-N-acetylglucosamine--peptide N-acetylglucosaminyltransferase |
| OFU1_CAEEL | SPTr | ISS | Putative GDP-fucose protein O-fucosyltransferase 1 (Fragment) |
| OFU1_MOUSE | SPTr | ISS | GDP-fucose protein O-fucosyltransferase 1 precursor |
| OFU1_DROME | SPTr | ISS | Putative GDP-fucose protein O-fucosyltransferase 1 precursor |
| Galnt1 | MGI | IDA | UDP-N-acetyl-alpha-D-galactosamine:polypeptide N-acetylgalactosaminyltra |
| Galnt7 | MGI | IDA | UDP-N-acetyl-alpha-D-galactosamine: polypeptide N-acetylgalactosaminyltr |
| Galnt9 | MGI | IDA | UDP-N-acetyl-alpha-D-galactosamine:polypeptide N-acetylgalactosaminyltra |
| Ogt | MGI | TAS | O-linked N-acetylglucosamine (GlcNAc) transferase (UDP-N-acetylglucosami |
| Ogtl | MGI | ISS | O linked N-acetylglucosamine transferase like |
| Siat8f-pending | MGI | IDA | sialyltransferase 8 (alpha-2, 8-sialyltransferase) F |
| DPM1 | SGD | IMP15056 | dolichol phosphate mannose synthase |
| KRE2 | SGD | IMP13239 | alpha-1,2-mannosyltransferase |
| KTR1 | SGD | IMP17219 | type II transmembrane protein |
| KTR3 | SGD | IMP17219 | alpha-1,2-mannosyltransferase (putative) |
| MNN1 | SGD | IDA309 | alpha-1,3-mannosyltransferase |
| MNT2 | SGD | IDA3857 | alpha-1,3-mannosyltransferase |
| MNT3 | SGD | IDA3857 | alpha-1,3-mannosyltransferase |

**Figure 2.** Screen shots from the Amigo browser of the Gene Ontology. (*a*) The result of a search for "O-linked glycosylation." The numbers in parentheses after each GO term are the numbers of gene products annotated with each term. (*b*) A small part of the output, showing the gene products annotated with this GO term (column *1*); column *2* indicates the database responsible for the annotation. (SPTr) Swiss-Prot/TrEMBL; (MGI) Mouse Genome Informatics; (SGD) *Saccharomyces* Genome Database; column *3* is the evidence for the annotation (TAS) traceable author statement; (ISS) inferred from sequence similarity; (ISA) inferred from direct assay; (IMP) inferred from mutant phenotype; column *4* shows the protein names corresponding to these gene products.

a group that includes the authors, Lincoln Stein and Richard Durbin. It has several obvious uses: It provides a structured controlled vocabulary for the description of primary annotations of nucleic acid sequence, e.g., the annotations shared by a DAS server or distributed in GFF or GAMEML formats; it provides a structured representation of these annotations within genomic databases. Were genes within model organism databases to be annotated with these terms, it would be possible to query all these databases for, for example, all genes whose transcripts are edited, or trans-spliced, or bound by a particular protein. SO also provides a structured controlled vocabulary for the description of mutations at both sequence and grosser levels in the context of genomic databases. The four major nodes of the complete Sequence Ontology allow for the annotation of chromosome mutations, the consequences of mutation, objects that can be located (in base coordinates) on a sequence, and general attributes of genes.

## CONCLUSION

The GO is not a mere exercise in knowledge representation; rather, its development is driven by a purely practical need—the need to provide semantic standards so that biological data can be more usefully stored and queried. The task is nontrivial. It is to create a language for biological research that is both computationally tractable and humanly understandable.

The technical aspect of the development of GO and related ontologies is, however, relatively trivial compared to the social aspect of the problem. It is predicated on community agreement. The community of developers and researchers must agree to agree. Achieving resolution requires a great deal of patience, as reaching a common understanding is often a lengthy process. It is this acceptance, that reaching community consensus is essential, that is the most notable aspect of the GO Consortium. But this is not itself sufficient; building ontologies is not an end in itself, the ontologies must be implemented, which requires their acceptance by the broader community of bioinformaticists and biologists. A common computable biological language is essential in order to communicate, share, and exploit information from the past and present. Without a strong semantic foundation, we will be living in a tower of Babel. With a common language, the community is connected and will benefit both individually from the ready availability of information and collectively by sharing and coordinating the exchange of information. Although the challenge is formidable, the benefits are even greater.

## ACKNOWLEDGMENTS

The Gene Ontology Consortium is supported by a grant from the National Institutes of Health (HG-02273, PI Judy Blake), and a grant from the Medical Research Council, UK to M.A. (G-9827766). The early development of GO was supported by donations from Astra Zeneca and Incyte Genomics. C.J.M. is supported by the Howard Hughes Medical Institute. We thank our many colleagues in the GO Consortium, without whom this project would have been impossible. We also thank the many users of GO for their feedback. Our use of description logic formats has benefited greatly from discussions with Drs. Robert Stevens and Chris Wroe of the University of Manchester.

## REFERENCES

Adams M.D., Celniker S.E., Holt R.A., Evans C.A., Gocayne J.D., Amanatides P.G., Scherer S.E., Li P.W., Hoskins R.A., Galle R.F., George R.A., Lewis S.E., Richards S., Ashburner M., Henderson S.N., Sutton G.G., Wortman J.R., Yandell M.D., Zhang Q., Chen L.X., Brandon R.C., Rogers Y.H., Blazej R.G., Champe M., and Pfeiffer B.D., et al. 2000. The genome sequence of *Drosophila melanogaster. Science* **287:** 2185.

Baader F., McGuinness D., Nardi D., and Patel-Schneider P., Eds. 2003. *The Description logic handbook.* Cambridge University Press, Cambridge, United Kingdom.

Bard J. and Winter R. 2001. Ontologies of developmental anatomy: Their current and future roles. *Brief. Bioinform.* **2:** 289.

Barrell B.G. and Clark B.F.C. 1974. *Handbook of nucleic acid sequences.* Joynson-Bruvvers, Eynsham, United Kingdom.

Bennett J.M. and Kendrew J.C. 1952. Computation of Fourier syntheses with a digital electronic calculating machine. *Acta Crystallogr.* **5:** 109.

Berriz G.F., White J.V., King O.D., and Roth, F.P. 2003. GoFish finds genes with combinations of Gene Ontology attributes. *Bioinformatics* **19:** 788.

Brachman R.J. 1983. What ISA-A is and isn't: An analysis of links in semantic networks. *Computer* **16:** 30.

Croft L.R. 1973. *Handbook of protein sequences.* Joynson-Bruvvers, Eynsham, United Kingdom.

Davidson S.B., Overton C., and Buneman P. 1995. Challenges in integrating biological data sources. *J. Comput. Biol.* **2:** 557.

Dayhoff M.O., Eck R.V., Chang Y., and Sochard S.R. 1965. *Atlas of protein sequence and structure.* National Biomedical Research Foundation, Washington, D.C.

Draghici S., Khatri P., Martins R.P., Ostermeier G.C., and Krawetz S.A. 2003. Global profiling of gene expression. *Genomics* **81:** 98.

Durbin R. and Thierry-Mieg J. 1991-. A *C. elegans* database. (http://www.faqs.org/faqs/acedb-faq/).

Fasman K. 1994. Restructuring the genome data base: A model for a federation of biological databases. *J. Comput. Biol.* **1:** 165.

Fellbaum C., Ed. 1998. *WordNet. An electronic lexical database.* MIT Press, Cambridge, Massachusetts.

Gilks W.R., Audit B., De Angelis D., Tsoka S., and Ouzounis C.A. 2002. Modeling the percolation of annotation errors in a database of protein sequences. *Bioinformatics* **18:** 1641.

GK (Genome KnowledgeBase). 2003. (http://www.genomeknowledge.org)

Hebert P.D.N., Cywinska A., Ball S.L., and deWaard J.R. 2003. Biological identification through DNA barcodes. *Proc. Roy. Soc. Lond. B* **270:** 312.

Hill D.P., Blake J.A., Richardson J.E., and Ringwald M. 2002. Extension and integration of the gene ontology (GO): Combining GO vocabularies with external vocabularies. *Genome Res.* **12:** 1982.

Hirschman L., Park J.C., Tsujii, J., Wong L., and Wu C.H. 2002. Accomplishments and challenges in literature data mining for biology. *Bioinformatics* **18:** 1553.

Huxley H.E. 1990. An early adventure in crystallographic computing. In *Selections and reflections: The legacy of Sir L. Bragg* (ed. J.M. Thomas and D. Phillips), p.133. Science Reviews, Northwood, United Kingdom.

Hvidsten T.R., Komorowski J., Sandvik A.K., and Laegreid A. 2001. Predicting gene function from gene expressions and ontologies. *Pac. Symp. Biocomput.* **2001:** 299.

Jensen L.J., Gupta R., Staerfeldt H.H., and Brunak S. 2003. Prediction of human protein function according to Gene Ontology categories. *Bioinformatics* **19:** 635.

Jenssen T.K., Laegreid A., Komorowski J., and Hovig E. 2001. A literature network of human genes for high-throughput analysis of gene expression. *Nat. Genet.* **28:** 21.

Kay L.E. 2000. *Who wrote the book of life? A history of the genetic code.* Stanford University Press, Stanford, California.

Khatri P., Draghici S., Ostermeier G.C., and Krawetz S.A. 2002. Profiling gene expression using onto-express. *Genomics* **79:** 266.

King O.D., Foulger R.E., Dwight S.S., White J.V., and Roth F.P.

2003. Predicting gene function from patterns of annotation. *Genome Res.* **13:** 896.

Lagreid A., Hvidsten T.R., Midelfart H., Komorowski J., and Sandvik A.K. 2003. Predicting gene ontology biological process from temporal gene expression patterns. *Genome Res.* **13:** 965.

Lewis S.E., Searle S.M., Harris N., Gibson M., Lyer V., Richter J., Wiel C., Bayraktaroglir L., Birney E., Crosby M.A., Kaminker J.S., Matthews B.B., Prochnik S.E., Smithy C.D., Tupy J.L., Rubin G.M., Misra S., Mungall C.J., and Clamp M.E. 2002. Apollo: A sequence annotation editor. *Genome Biol.* **3:** RESEARCH0082.

McCray A.T., Browne A.C., and Bodenreider O. 2002. The lexical properties of the Gene Ontology (GO). *Proceedings of the American Medical Informatics Association Symposium*, p. 504.

Mi H., Vandergriff J., Campbell M., Narechania A., Majoros W., Lewis S.E., Thomas P.D., and Ashburner M. 2003. Assessment of genome-wide function classification for *Drosophila melanogaster*. *Genome Res.* **13:** 2118.

Miller G.A. 1998. Nouns in WordNet. In *WordNet. An electronic lexical database* (ed. C. Fellbaum), ch. 1. MIT Press, Cambridge, Massachusetts.

MIPS (Munich Information Center for Protein Sequences). 2003. Comprehensive yeast genome database (http://mips.gsf.de/proj/yeast/CYGD/db/index.html).

obo (Open Biological Ontologies). 2003. (http://obo.sf.net/).

OWL (Ontology Web Language). 2003. (http://www.w3.org/2001/sw/WebOnt/).

Pouliot Y., Gao J., Su Q.J., Liu G.G., and Ling X.B. 2001. DIAN: A novel algorithm for genome ontological classification. *Genome Res.* **11:** 1766.

Raychaudhuri S., Chang J.T., Sutphin P.D., and Altman R.B. 2002. Associating genes with gene ontology codes using a maximum entropy analysis of biomedical literature. *Genome Res.* **12:** 203.

ReelTwo. 2003. (http:/www.reeltwo.com/).

Ritter O. 1994. The integrated genomic database. In *Computational methods in genome research* (ed. S. Suhai), p. 57. Plenum Press, New York.

Rogers J. and Rector A. 2000. GALEN's model of parts and wholes: Experiences and comparisons. In *Proceedingsof the American Medical Informatics Association Symposium*, p. 714.

Sarkar I.N., Cantor M.N., Gelman R., Hartel F., and Lussier Y.A. 2003. Linking biomedical language information and knowledge resources in the 21st century: GO and UMLS. *Pac. Symp. Biocomput.* **8:** 427.

Schug J., Diskin S., Mazzarelli S. Brunk B.P., and Stoeckert C.J., Jr. 2002. Predicting gene ontology functions from ProDom and CDD protein domains. *Genome Res.* **12:** 648.

Semantic Web. 2003. (http://www.w3.org/2001/sw/).

SONG (Sequence Ontology Project). 2003. (http://song.sf.net/).

Staden R. 1980. A new computer method for the storage and manipulation of DNA gel reading data. *Nucleic Acids Res.* **8:** 3673.

UMLS (Unified Medical Language System). 2003. (http://www.nlm.nih.gov/research/umls/).

Winston M.E., Chaffin R., and Herrmann D. 1987. A taxonomy of part-whole relations. *Cognitive Sci.* **11:** 417.

WordNet. 2003. WordNet: A lexical database for the English language (http://www.cogsci.princeton.edu/~wn/).

Wroe C.J., Stevens R., Goble C.A., and Ashburner M. 2003. A methodology to migrate the gene ontology to a description logic environment using DAML+OIL. *Pac. Symp. Biocomput.* **8:** 624.

Xie H., Wasserman A., Levine Z., Novik A., Grebinskiy V., Shoshan A., and Mintz L. 2002. Large scale protein annotation through gene ontology. *Genome Res.* **12:** 785.

# The Genome Knowledgebase: A Resource for Biologists and Bioinformaticists

G. Joshi-Tope,* I. Vastrik,† G.R. Gopinath,* L. Matthews,* E. Schmidt,†
M. Gillespie,‡ P. D'Eustachio,¶ B. Jassal,† S. Lewis,§ G. Wu,* E. Birney,† and L. Stein*
*Cold Spring Harbor Laboratory, Cold Spring Harbor, New York 11790; †European Bioinformatics Institute, Hinxton Outstation, Hinxton, Cambridge, United Kingdom; ‡Department of Biology, St. Johns University, Queens, New York 11439; ¶New York University School of Medicine, New York, New York 10016; §Berkeley Drosophila Genome Project, University of California, Berkeley, California 94720

Biological science now has access to the sequenced genomes of dozens of organisms spanning the phylogenetic gamut from prokaryotes (Fleischmann et al. 1995) to people (Lander et al. 2001). We have reasonable estimates on the number and nature of most of the protein-coding genes, a fact that has radically changed the nature of gene hunting from an activity that is primarily done at the bench to one that is mostly done using the computer. Our knowledge of the genome contents is augmented by high-throughput experimental techniques such as expression chips (DeRisi et al. 1997) and yeast two-hybrid studies (Fields and Song 1989) for probing protein interactions and regulatory networks.

The amount of genomic information is rising exponentially. To manage this onslaught, the bioinformatics community has created a series of gene-centric "catalogs" such as RefSeq (Pruitt et al. 2000; Pruitt and Maglott 2001), GeneCards (Safran et al. 2002), KEGG (Kanehisa and Goto 2000), and YPD (Constanzo et al. 2000). These databases provide a gene-by-gene view of the genome, presenting all the information known about the structure and function of a gene and its protein product(s) in a single record. These gene catalog efforts have been greatly aided by the Gene Ontology (GO; Ashburner et al. 2000), which provides a detailed controlled vocabulary for describing the subcellular location, molecular function, and associated biological process of a gene product.

However, biological researchers rarely think a gene at a time. Instead, they are concerned with the complex interactions among proteins, protein complexes, nucleic acids, and small molecules that carry out a complex biological process. A great preponderance of papers in the field of molecular biology deal with dissecting the choreographed interactions between ensembles of macromolecules, or with the function of a particular gene product within the larger context of a biological pathway. Hence, there is a mismatch between the "gene at a time" design of the current generation of genome databases, and the whole-pathway approach of much of the scientific literature. The online gene catalogs cannot express the pathway concepts embodied in the literature, and reciprocally, the research community can only obtain fragmentary and sometimes misleading information about pathways from the gene catalogs.

To narrow this gap, we created the Genome Knowledgebase (GK; http://www.genomeknowledge.org), an open, online database of fundamental human biological processes. This paper describes GK, and the lessons we have learned in the two years since we first began the project.

## INTENDED AUDIENCE

GK has two target audiences. One is the wet-bench researcher who has stumbled onto an unfamiliar gene product and wants to get a quick overview of what the product is and what it does. The GK Web site allows such a researcher to quickly locate a gene product, to find the reactions and processes it participates in, and to learn about the role the gene product plays in the larger context of a biological pathway. Included within this target audience are undergraduates, graduate students, and postdocs, who can use GK as a "Cliff Notes™" of biological pathways, to rapidly bring themselves up to speed on the foundations of a biological process and its core literature. To this target audience, GK appears as a Web-based publication, an online textbook of molecular biology.

The second target audience is the bioinformaticist who is trying to draw conclusions from a large data set like a series of spotted cDNA expression chip experiments. To this type of researcher, GK appears as a database that can be queried interactively or downloaded in bulk form. The pathway information contained in GK can be superimposed on top of the experimental data set, revealing information about pathways already known and suggesting new relationships.

## THE GK WEB SITE

The front page of the GK Web site is shown in Figure 1. From this perspective, GK appears like a top-down textbook of biological processes. The knowledge contained within GK is organized into modules such as "DNA Replication." Each module has one or more primary authors who are typically experts recruited from the research community, an editor who is a full-time member of the GK staff, and one or more peer reviewers who are also recruited from the community. A module has an ini-

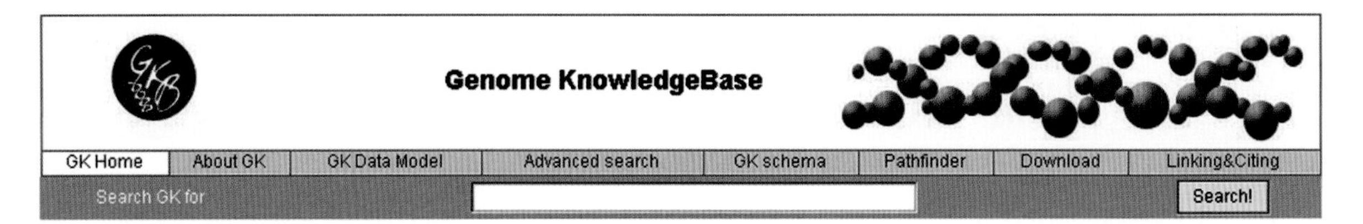


**Figure 1.** The GK front page.

tial release date and possibly a more recent revision date. These features allow the module to be cited like a publication.

Figure 2 shows what the user sees as he moves down following a pathway, in this case a step in the $G_1$ phase of the mitotic cell cycle. The navigation bar on the left shows an expandable outline of the current pathway, keeping track of the context of the current step in the context of the larger pathway. The detail panel on the right features a human-readable summary of the step, an optional diagram, and a summary of all complexes, protein products, and other macromolecules that participate in this step. The detail panel also lists orthologous reactions that occur or are predicted to occur in other model organisms/ species, and literature citations directly relevant to the step. (The citations and orthologous reactions are scrolled out of view in Fig. 2.) All gene product names currently link out to SwissProt. Links to EnsEMBL, RefSeq, and model organism databases will be phased in during the autumn of 2003/winter of 2004.

Naturally, pathways do not stand alone but are interconnected. When a user is viewing the details of a molecular reaction that participates in multiple pathways, the navigation bar at the left expands to show him each of the pathways that the reaction participates in.

The bioinformaticists' view of GK, using an experimental "pathfinder" viewer, is shown in Figure 3. Here we have just executed a query to find the shortest path between D-fructose and pyruvate. GK finds a path through the reactions that define intermediate metabolism and displays them in an interactive format. Although this representation is easiest to understand in the context of classical biochemical pathways, it can be used to explore any biological process; for example, to find the steps that connect a primary RNA transcript located in the nucleus to a capped, spliced, polyadenylated messenger RNA located in the cytoplasm.

From the bioinformaticist's point of view, GK is a database that organizes proteins and other macromolecules into reactions and organizes these in turn into pathways. This information is invaluable when searching for the biological signatures hidden in large-scale data sets such as microarray expression studies and protein-interaction screens.

## THE GK MISSION

GK's central mission is to provide a nexus for connecting the online gene-centric databases to the research literature through sets of curated reactions and pathways. Since this is a monumental task, we have limited the scope of our work in the following ways:

- The focus of GK is on biological pathways that are thought to operate in humans. Photosynthesis, the yeast mating type switch, and the iron/sulfur biochemistry of the inhabitants of thermal vents can all be safely ignored.
- GK deliberately avoids "cutting edge" research. We ask our authors to stay about a year behind the current

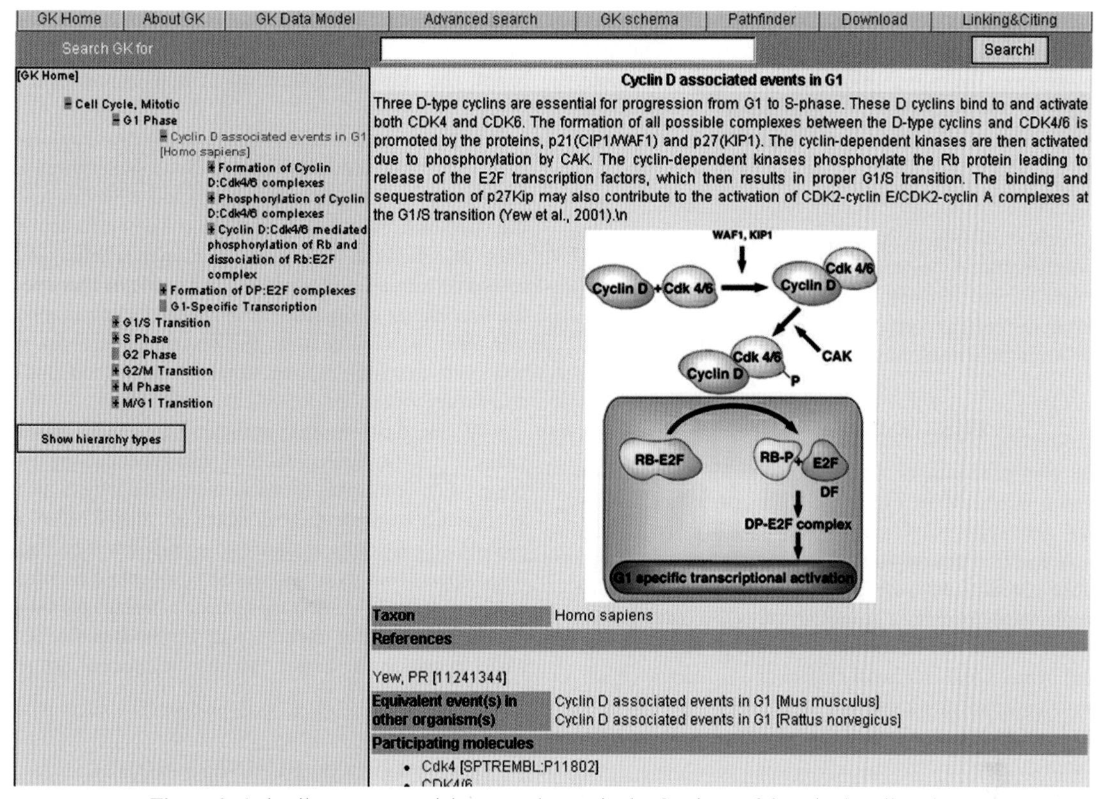


**Figure 2.** A detail page summarizing an early step in the $G_1$ phase of the mitotic cell cycle.

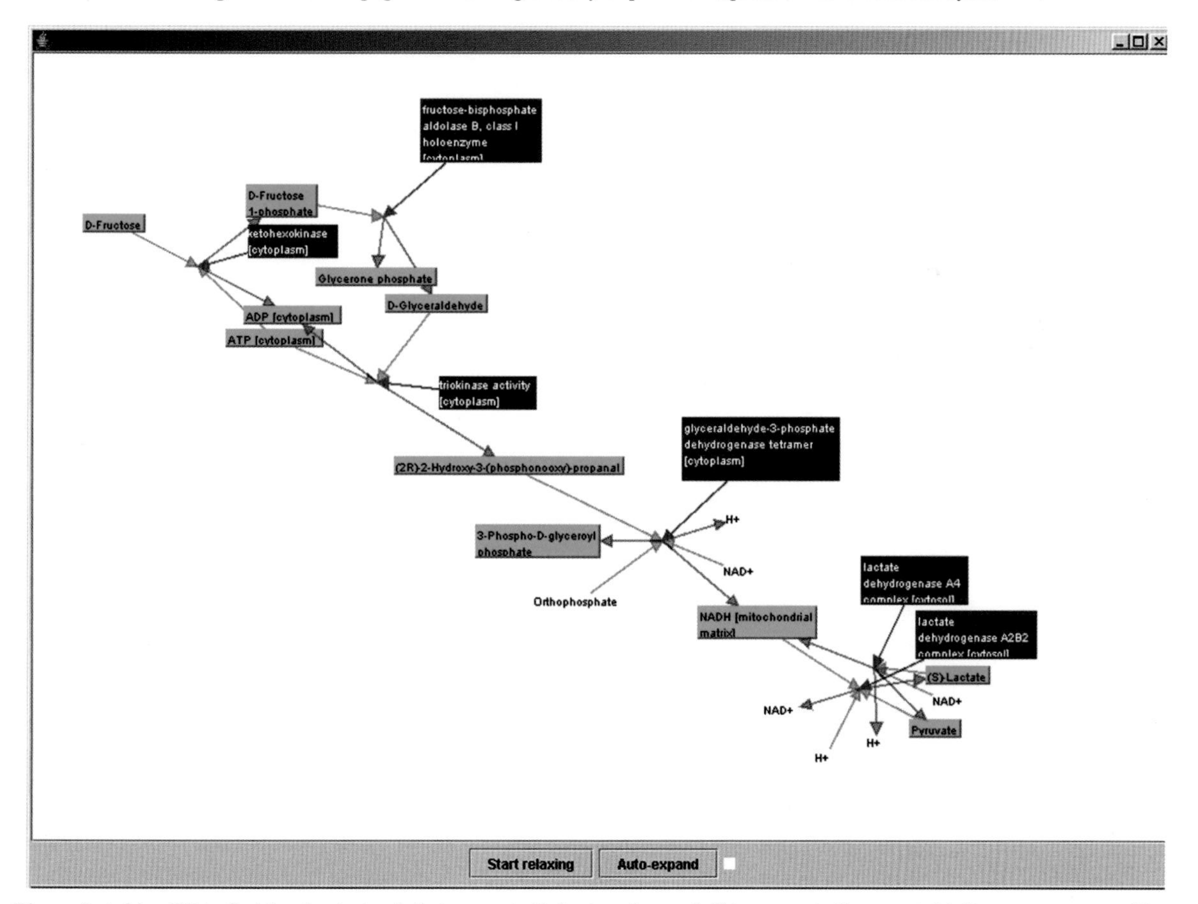


**Figure 3.** Asking GK to find the shortest path that connects D-fructose (*upper left*) to pyruvate (*bottom right*). Boxes represent entities: encoded entities (proteins and nucleic acids, *dark blue*) or non-encoded entities (small molecules, *light blue*). Entities are connected to the events in which they participate by arrows whose colors denote the entities' roles: (*red*) catalyst; (*green*) input; (*blue*) output.

state of knowledge. In terms of currency, GK tries to be more of a textbook than a review article. This policy reduces the rate at which modules go out of date, and also allows authors to defer controversial or uncertain assertions. This is partially offset by an update policy which calls for each module to be reviewed and updated once every three years.

- As discussed below, our data model makes deliberate simplifications. For example, we do not attempt to account for rate constants, concentration gradients, or other numerical values that would be key for true physiological modeling.

## THE GK DATA MODEL

The core of GK is its data model, a simplified version of which is shown in Figure 4. There are three fundamental data types: Physical Entity, Reaction, and Pathway. A reaction is a biological event that consumes one or more physical entities as input, and produces one or more physical entities as output. A physical entity, as its name implies, is any molecule that participates in a biological process and can be as simple as a molecule of water or as complex as a multi-subunit receptor complex. A Pathway is group of reactions that together drive forward a biological process.

This paradigm turns out to be quite general as demonstrated by examples from two very different fields—classic biochemistry and the more modern study of signal transduction. The phosphorylation of glucose to form glucose-6-phosphate is a classic reaction in intermediate metabolism, and it is easy to see how it is represented in GK: The inputs are D-glucose and ATP, and the outputs are glucose-6-phosphate and ADP. Downstream reactions, such as those of glycolysis, are dependent on the availability of glucose-6-phosphate as input.

Now let us look at the conformational change that occurs when calcium binds to calmodulin, thereby putting calmodulin into its "active" form, and relaying the signal to downstream effector molecules. GK represents the inputs as calmodulin and inorganic calcium, and the outputs as the calmodulin–calcium complex. Downstream reactions, such as those mediated by protein kinase A, are dependent on the calmodulin–calcium complex. We have thus neatly disposed of the need to represent elusive ideas like "activation" and "transduction," and have enabled

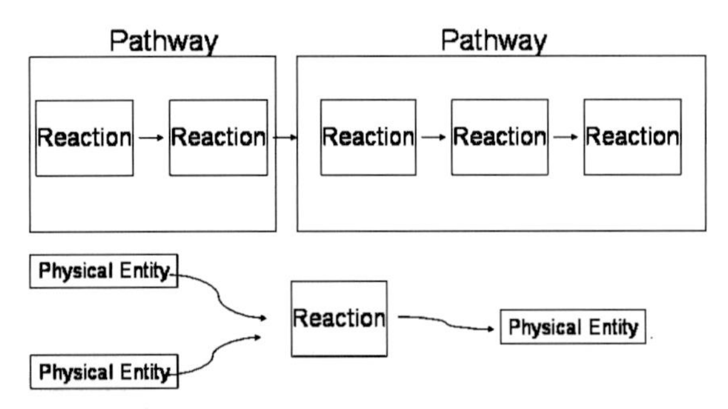


**Figure 4.** The GK data model is built around the Reaction and Physical Entity. Pathways are used to organize reactions in a traditional goal-oriented manner.

pathways to be built by reasoning, based on the presence of required inputs. Reactions are also used in GK to represent the stepwise assembly of macromolecular complexes.

Reactions include a number of attributes in addition to their inputs and outputs. These include such things as catalyst and catalyst activities associated with the reaction, the species in which the reaction is known to occur (human, or, when the existence of a human reaction must be inferred from another species, the name of that species), the subcellular compartment in which the reaction occurs, GO terms associated with the reaction, the phase of the cell cycle to which the reaction may be limited, and literature references that document this reaction.

The GK data model includes many subclasses and variants of the basic reaction and physical entity data types. For example, there is a generic physical entity type used to represent the concept of "all mRNAs" as opposed to a particular instance of an mRNA. The Pathway data type allows us to group reactions into a hierarchical series of steps and sub-steps. This is crucial for GK's didactic function and needed to accommodate the frequent cases in which the exact reactions that occur are speculative.

## THE GK BUSINESS MODEL

GK is organized along the lines of an electronic publication. The GK editor-in-chief, in consultation with the principal investigators and an editorial advisory board, annually maps out a calendar of modules and a rough release schedule. We then solicit experts in the field to author modules. We have had most success when we have identified a single module "lead author" who then helps develop the structure of the module, recruits additional authors, and supervises the completion of the annotation.

The mechanics of authoring a GK module are simple but effective. We give each author a Microsoft PowerPoint™ presentation, which contains a series of blank diagrams in a standardized format, and ask the author to fill in the blanks. For example, one diagram shows a generic binding reaction in which two molecules interact to form a complex. The author copies this diagram and changes the labels as appropriate, to indicate a specific reaction. The author then fills in fields that indicate the species, phase of the cell cycle, subcellular location, and the citations that describe this reaction.

When the PowerPoint™ representation is complete, the author sends it to a GK curator, who extracts the information and enters it into the GK database using a Knowledgebase editor tool called Protégé (Noy et al. 2001). The draft module is reviewed internally by the GK staff and is subsequently made available on a prepublication Web site for the author to review and approve. After the author approves the draft, it is made available to one or more peer reviewers, who are asked to verify the completeness and correctness of the module. We supplement the human peer review with a series of software checks to verify the integrity of the described pathways, the validity of protein and citation accession numbers, and concordance between the molecular functions assigned to gene products by the human authors and those recorded in the Gene Ontology databases. Several iterations of the draft module may be needed before the software validity checks, GK curator, and peer reviewers all agree that a module is ready for release.

Recently, we have made increasing use of the "mini-jamboree," a kickoff meeting in which the authors for a related set of modules get together in the same locale to hash out the scope of the modules and work out the relationships among them. This gives authors the chance to meet with the GK staff and to receive individualized help with the GK data model and PowerPoint™ templates. The first series of mini-jamborees have been piggybacked on relevant Cold Spring Harbor Laboratory scientific meetings, thereby saving participants time and cost. However, given authors' enthusiastic response to the mini-jamboree format, we are looking to increase the number and broaden the location of these meetings.

## EVIDENCE TRACKING

In principle, GK is a database of human biological processes. In practice, many of the processes that we understand come from experiments that have been performed in bacteria, yeast, fruit fly, or other model organisms. We use a strict form of evidence tracking, which distinguishes between reactions that have been documented by direct experimental evidence (e.g., on human tissue culture cells) and those that are inferred indirectly from work with model organisms. In the former case, we create a GK Reaction whose taxon is *H. sapiens*, and document it with citations which demonstrate that the reaction occurs in humans. In the latter case, we create two Reactions: a reaction occurring in the model organism documented with the appropriate direct literature citations, and an inferred human reaction that is indirectly supported by what is known to occur in the model organism.

The key reason for creating dual reactions in the model organism system and humans is to prevent their genomes from becoming "entangled" in the database. In a reaction directly confirmed in yeast, the Physical Entities involved will be yeast proteins accessioned in SwissProt or Swiss-Prot/TREMBL. Likewise, a reaction directly confirmed in human cells will refer to the relevant human proteins. When a reaction in yeast is used to indirectly support a putative reaction in human, we populate the human reaction with the human orthologs of the yeast proteins when the correct orthologs are known, or create placeholder physical entities for the human proteins when the correct orthologs are not known. In either case, we strictly avoid creating human reactions that involve non-human proteins. This allows pathway builders and other reasoning algorithms to traverse the reactions without risking nonsensical jumps from one species to another.

Our current approach is to fill out pathways completely in humans, and to bring in reactions from model organisms only when necessary to span a gap in knowledge. However, this data model leaves open the possibility of filling out pathways completely in several of the model organism species as well, allowing meaningful comparisons among them.

## GK SOFTWARE

Internally, GK runs on top of a MySQL database (DuBois 2003), a popular open source relational database management system. The data model was tested with Protégé-2000 (Noy et al. 2001), an open source editor for frame-based knowledge systems, and implemented using a Protégé to MySQL input/output adapter that we developed for the project. The Web pages are driven by Perl scripts running under the open source Apache Web server and using a Perl middleware layer that translates high-level requests into reactions and physical entities into low-level SQL queries. We have also developed a Java interface to the database and are in the process of developing a Java-based authoring and editing tool for GK.

All software developed for GK is available for use under an open source license.

## DISCUSSION

GK is one of a small but growing number of pathway-oriented databases. The oldest and most mature entrant in this field is EcoCyc (Karp et al. 2002), a database of *E. coli* metabolic pathways. Recently, the EcoCyc data model has been successfully used to represent pathways in *Arabidopsis*, yeast, and human. Like GK, the Cyc databases use a frame-based representation of biological pathways that are centered around the concept of a reaction, although the technical details differ considerably, including Cyc's use of the LISP programming language. However, the chief difference between the systems is their scope: Cyc focuses on intermediate metabolism, whereas GK emphasizes higher-order interactions among macromolecules such as transcription and translation.

The KEGG database of genomes contains considerable information on pathways, but, like Cyc, its focus is on intermediate metabolism. A fundamental difference between the GK and KEGG data models is that in KEGG pathways the catalyst is represented as an enzymatic activity, or EC number. However, the relationship between an enzymatic activity and a protein is not as simple as it might seem: Proteins that are only distantly related may have the same EC number if they share a catalytic activity, and the same protein subunit may carry different enzymatic activities depending on other subunits with which it is currently complexed. For this reason, it is not straightforward to connect a KEGG reaction to a human gene product.

The BioCarta project (http://www.biocarta.com) is oriented toward high-level processes such as signal transduction, much as GK is. However, its primary product is a series of diagrams illustrating biological pathways, and the project as a whole is a proprietary commercial enterprise.

Last, there are BIND (Bader and Hogue 2000; Bader et al. 2001) and MINT (Zanzoni et al. 2002), two databases of protein–protein interaction data, as well as such project-specific databases as the Alliance for Cell Signalling (AfCS; Li et al. 2002). These databases emphasize the management of large amounts of raw interaction data from high-throughput data sets, such as yeast two-hybrid and co-immunoprecipitation studies. Because they contain the cutting-edge, unverified information that GK explicitly excludes, they are entirely complementary to GK. In the near future, we hope to establish reciprocal links with BIND, AfCS, and other such resources.

As noted earlier, GK has deliberately limited its scope in several areas in order to make the project manageable. One of the more controversial of these decisions was to omit tissue-specific expression information from the description of proteins. In the GK world, all proteins are expressed in all cells of the human body. This decision makes it difficult to express, for example, differences in the glycolytic pathways in liver and skeletal muscle, and ultimately interferes with the ability of pathway traversal algorithms to correctly detect impossible pathways. Our reasoning is that not too far into the future there will be online atlases of human tissue-specific protein expression patterns derived from oligonucleotide arrays and other high-throughput technologies. This information will close the gap in GK's data set. In the meantime, we are using a variety of workarounds to describe tissue-specific differences in GK pathways.

How far has GK gone toward our ambitious goal of covering all essential topics in human biology? It has been roughly one year since GK went from development mode into full production. During that time, we have released ten modules spanning such topics as DNA replication, mRNA processing, RNA transcription, cell cycle checkpoints, and a fair chunk of intermediate metabolism. Roughly speaking, we complete a new module each month, although the size and scope of each module varies tremendously depending on its content.

One crude measure is to count the number of SwissProt and SwissProt/TREMBL proteins that we have accessioned in GK and then to divide by the total number of human proteins in SwissProt and SwissProt/TREMBL. By this measure, GK covers slightly more than 6% of the information space. However, this procedure is biased downward because it fails to account for the fact that a considerable number of the proteins in TREMBL come from gene predictions and have no known function. A better measure might involve estimating the number of reactions described in medical textbooks or the review literature. We are currently debating how best to accomplish this.

Looking ahead, our biggest priority for the next year of the project is to develop more tools for mining and visualizing GK data. The database is only as good as the tools it offers to bioinformaticists and other researchers for organizing and interpreting large-scale data sets. One proposed visualization tool is a zoomable "starry sky" view of the human physiome in which each protein occupies a fixed position in the night sky and connecting lines define the "constellation" pathways. By superimposing microarray expression data on this view, researchers could see at a glance which pathways are involved by a set of up- or down-regulated genes.

A second priority is to speed the authoring and curation process. As noted above, we have developed and are now testing an authoring tool that allows authors to interact directly with the GK database. If this tool is successful, we will no longer have to rely on PowerPoint™ templates as the primary authoring tool. This will free up curators from the task of transferring information from PowerPoint™ into Protégé and allow us to recruit more authors.

Finally, we need to bring more automation into the curatorial process. GK curators spend a considerable amount of time looking up literature references, matching gene names to SwissProt accession numbers, and checking the descriptions of protein functions against Gene Ontology terms. This process would be greatly aided by a set of software tools to generate a "pre-authored" document that would contain proposed Gene Ontology terms, protein accession numbers, and literature references for the topic currently under review. Over the next year, we will explore a number of approaches to such pre-authoring tools.

## ACKNOWLEDGMENTS

The development of Genome KnowledgeBase is supported by grant R01 HG-002639 from the National Human Genome Research Institute at the National Institutes of Health and a subcontract from the NIH-funded Cell Migration Consortium.

## REFERENCES

Ashburner M., Ball C.A., Blake J.A., Botstein D., Butler H., Cherry J.M., Davis A.P., Dolinski K., Dwight S.S., Eppig J.T., Harris M.A., Hill D.P., Issel-Tarver L., Kasarskis A., Lewis S., Matese J.C., Richardson J.E., Ringwald M., Rubin G.M., and Sherlock G. (The Gene Ontology Consortium). 2000. Gene ontology: Tool for the unification of biology. *Nat. Genet.* **25:** 25.

Bader G.D.and Hogue C.W. 2000. BIND: A data specification for storing and describing biomolecular interactions, molecular complexes and pathways. *Bioinformatics* **16:** 465.

Bader G.D., Donaldson I., Wolting C., Ouellette B.F., Pawson T., and Hogue C.W. 2001. BIND: The Biomolecular Interaction Network Database. *Nucleic Acids Res.* **29:** 242.

Costanzo M.C., Hogan J.D., Cusick M.E., Davis B.P., Fancher A.M., Hodges P.E., Kondu P., Lengieza C., Lew-Smith J.E., Lingner C., Roberg-Perez K.J., Tillberg M., Brooks J.E., and Garrels J.I. 2000. The yeast proteome database (YPD) and *Caenorhabditis elegans* proteome database (WormPD): Comprehensive resources for the organization and comparison of model organism protein information. *Nucleic Acids Res.* **28:** 73.

DeRisi J.L., Iyer V.R., and Brown P.O. 1997. Exploring the metabolic and genetic control of gene expression on a genomic scale. *Science* **278:** 680.

DuBois P. 2003. *MySQL cookbook.* O'Reilly and Associates, Sebastopol, California.

Fields S. and Song O. 1989. A novel genetic system to detect protein-protein interactions. *Nature* **340:** 245.

Fleischmann R.D., Adams M.D., White O., Clayton R.A., Kirkness E.F., Kerlavage A.R., Bult C.J., Tomb J.F., Dougherty B.A., Merrick J.M., McKenney K., Sutton G.G., FitzHugh W., Fields C.A., Gocayne J.D., Scott J.D., Shirley R., Liu L.I., Glodek A., Kelley J.M., Weidman J.F., Phillips C.A., Spriggs T., Hedblom E., and Cotton M.D., et al. 1995. Whole-genome random sequencing and assembly of *Haemophilus influenzae* Rd. *Science* **269:** 496.

Kanehisa M. and Goto S. 2000. KEGG: Kyoto encyclopedia of genes and genomes. *Nucleic Acids Res.* **28:** 27.

Karp P.D., Riley M., Saier M., Paulsen I.T., Collado-Vides J., Paley S.M., Pellegrini-Toole A., Bonavides C., and Gama-Castro S. 2002. The EcoCyc database. *Nucleic Acids Res.* **30:** 56.

Lander E.S., Linton L.M., Birren B., Nusbaum C., Zody M.C., Baldwin J., Devon K., Dewar K., Doyle M., FitzHugh W., Funke R., Gage D., Harris K., Heaford A., Howland J., Kann L., Lehoczky J., LeVine R., McEwan P., McKernan K., Meldrim J., Mesirov J.P., Miranda C., Morris W., and Naylor J., et al. (International Human Genome Sequencing Consortium). 2001. Initial sequencing and analysis of the human genome. *Nature* **409:** 860.

Li J., Ning Y., Hedley W., Saunders B., Chen Y., Tindill N., Hannay T., and Subramanian S. 2002. The Molecule Pages database. *Nature* **420:** 716.

Noy N.F., Sintek M., Decker S., Crubezy M., Fergerson R.W., and Musen M.A. 2001. Creating semantic web contents with Protégé-2000. *IEEE Intelligent Systems* **16:** 60.

Pruitt K.D. and Maglott D.R. 2001. RefSeq and LocusLink: NCBI gene-centered resources. *Nucleic Acid Res.* **29:** 137.

Pruitt K.D., Katz K.S., Sicotte H., and Maglott D.R. 2000. Introducing RefSeq and LocusLink: Curated human genome resources at the NCBI. *Trends Genet.* **16:** 44.

Safran M., Solomon I., Shmueli O., Lapidot M., Shen-Orr S., Adato A., Ben-Dor U., Esterman N., Rosen N., Peter I., Olender T., Chalifa-Caspi V., and Lancet D. 2002. GeneCards 2002: Towards a complete, object-oriented, human gene compendium. *Bioinformatics* **18:** 1542.

Zanzoni A., Montecchi-Palazzi L., Quondam M., Ausiello G., Helmer-Citterich M., Cesareni G. 2002. MINT: A Molecular INTeraction database. *FEBS Lett.* **513:** 135.

# The Share of Human Genomic DNA under Selection Estimated from Human–Mouse Genomic Alignments

F. CHIAROMONTE,* R.J. WEBER,[†] K.M. ROSKIN,[†] M. DIEKHANS,[†] W.J. KENT,[†] AND D. HAUSSLER[‡]

** Department of Statistics and Department of Health Evaluation Sciences, Pennsylvania State University, University Park, Pennsylvania 16803; [†]Center for Biomolecular Science and Engineering, University of California, Santa Cruz, California 95064; [‡]Howard Hughes Medical Institute, University of California, Santa Cruz, California 95064*

Draft sequences covering most euchromatic parts have recently become available for two mammalian genomes, human (Lander et al. 2001; Venter et al. 2001) and mouse (Waterston et al. 2002). This raises the possibility of using comparative genomics to estimate what fraction of the human genome evolves under purifying selection. Lacking genomes of other mammals, this comparative exercise is still in its preliminary stages. However, a rough estimate has been made that ~5% of the human genome is in short segments that appear to be under selection based on comparison with mouse (Waterston et al. 2002). Here, as a basis for future refinements, we present the computational strategy that led to this estimate, providing details on scoring functions, data preparation, and statistical techniques. We also describe stability analyses, control experiments, and tests for the effects of artifacts that were performed to establish robustness of our results, and discuss possible alternate interpretations.

Our strategy hinges on three elements: (1) the construction of various collections of short aligned windows of the human genome (e.g., 50 bp)—in particular, a large collection of such windows that are very likely to have evolved neutrally since the divergence of human and mouse ("ancestral repeats," relics of transposons that were present in the genome of our common ancestor with mouse); (2) the development of a score function quantifying conservation in short aligned windows, and providing a satisfactory "template" for neutral behavior when computed on windows in ancestral repeats; and (3) statistical techniques to estimate and compare the score distributions for genome-wide and ancestral repeat windows, and thus infer an upper bound on the share of genome-wide windows that are compatible with the neutral template. The remaining share of the genome is populated by windows that are too conserved to be modeled by the neutral template, and hence are either evolving under purifying selection, or are evolving neutrally but are experiencing fewer substitutions than nearby windows in ancestral repeats for some unknown reasons.

Because ancestral transposons have been inactive since their insertion in the genome of the common ancestor of human and mouse, they are one type of human DNA that is most likely to have evolved free of any selective pressure. The rate of substitution in these sites between human and mouse is similar to, but slightly less than, that observed in fourfold degenerate sites from codons, and covaries regionally with that rate (Waterston et al. 2002; Hardison et al. 2003). This suggests that both of these types of sites provide reasonable models to evaluate the rate of neutral substitution, and that this rate depends on some local properties of the chromosome where it is measured, as was known from previous studies on the effects of GC content on substitution rates (Bernardi 1993). Because ancestral repeats constitute 22% of the human genome and are still reliably alignable to mouse, they allow us to construct a very large number of short aligned windows of neutrally evolving DNA (Waterston et al. 2002; Schwartz et al. 2003).

There are many ways to measure conservation in short aligned windows, even with just two species. The aim here is to provide a simple template for neutral behavior that allows, in comparison, a satisfactory separation of aligned sequence that is undergoing purifying selection. To this end, we further explore the *normalized percent identity* score introduced in Roskin et al. (2002), and used in Waterston et al. (2002). Definitions of several variants of this score, and a brief discussion of the first crude estimates of the share under selection, which were made with one of these variants, can be found in Roskin et al. (2002, 2003). Yet more possible scores are analyzed in Elnitski et al. (2003), with a particular focus on separating regulatory elements from neutral DNA.

The normalized percent identity score involves no assumptions on the characterization of DNA functions that might be under purifying selection, except that they result in a higher degree of conservation. It has a straightforward definition, and the advantage of being relatively easy to compute. The score is obtained by calculating the fraction of aligned bases in the window that are identical between human and mouse, and then subtracting a mean and dividing by a standard deviation estimated under neutrality. The only subtlety comes from estimating the neutral mean and standard deviation. The neutral mean for a window is estimated locally, using only aligned bases from ancestral repeats in a region surrounding the window. Local estimation of the neutral mean percent identity allows the conservation score to compensate for regional variations in the rate of neutral evolution (Roskin et. al. 2002, 2003; Waterston et al. 2002; Hardison et al. 2003). This includes variations induced by changes in GC content and other features. The standard deviation estimate is derived from the mean estimate using a simple binomial model, and thus is also local.

We make no parametric assumptions in estimating the normalized percent identity score distribution for either genome-wide or ancestral repeat windows. With approximately two million data points in the smaller data set of ancestral repeat windows, there is no need for such assumptions. However, we do use Gaussian kernel smoothing to estimate a continuous nonparametric score distribution from these empirical data. We decompose the continuous genome-wide distribution as a mixture of a neutral component and a component that appears to be under selection.

## METHODS

***Data preparation.*** Our collections of short aligned windows were constructed using a fixed grid of locations along the human sequence. The grid is such as to always guarantee nonoverlapping windows for the sizes we consider. For a given window size ($W$) and alignment filtering threshold ($T$), the genome-wide collection is constructed first extending windows of $W$ bases at each location, and then discarding all windows with less than $T$ bases aligned with mouse. For the same window size and filtering threshold, the collection of windows relative to a particular feature type (ancestral repeats, coding regions) is constructed in a similar fashion, first extending windows of size $W$ at grid locations, and then discarding windows whose overlap with aligned features of that type is less than $T$ bases. Table 1 gives coverage provided by genome-wide windows for the $W = 50$, $T = 40$ case presented in our main analysis, as well as other combinations of window size and filtering threshold.

Ancestral repeats were repeats identified by RepeatMasker (available at http://ftp.genome.washington.edu/RM/RepeatMasker.html; Smit and Green 1999) and present at orthologous sites. A list of specific families of ancestral repeats is given in the Methods web-available compendium to Waterston et al. (2002).

Known coding region annotation was obtained by aligning the RefSeq (Pruitt and Maglott 2001) human mRNAs from GenBank release 130.0 to the human genome with BLAT (Kent 2002; Kent et al. 2002). We selected annotations that had an aligned mouse position and met the following criteria: (1) CDS appeared complete in both human and mouse, beginning with a start codon, and ending with a stop codon. The mouse stop codon was allowed up to 20 codons before the human stop codon. (2) There were no in-frame stop codons. (3) Introns in human CDS had splice sites in the form *GT..AG*, *GC..AG*, or *AT..AC*. This resulted in 11,718 gene alignments.

Further details on data preparation can be found in the Methods web-available compendium to Waterston et al. (2002), and in Schwartz et al. (2003).

***Eliminating pseudogenes.*** The initial BLASTZ alignment contained numerous processed and nonprocessed pseudogenes that could artificially inflate our estimate of the share under selection. To remove these pseudogenes, we apply a filter that only keeps each reciprocal best pair of alignments between human and mouse: If a segment of mouse sequence aligns to multiple human genome locations, we only keep the region that aligns back to that same region in mouse and gives the highest alignment score. Pseudogenes are clearly under different selective pressure than the genes they are duplicated from, so they should not align as well in both directions as the genes themselves. Applying this filter removes ~14% of the initial alignment, and whereas the initial alignment covers 89% of RefSeq genes, the filtered one only covers 83%. Therefore, our filter errs on the side of caution, likely removing more highly conserved sequence than needed to eliminate pseudogenes' effects, but this is acceptable in an attempt to produce a conservative, lower bound estimate of the share under selection. In other experiments, we used the chaining method described in Kent et al. (2003) in place of this reciprocal best filtering method and obtained similar results, with slightly higher estimates of the share under selection (not shown in this paper).

***Normalized percent identity.*** The normalization presented in Equation 1 centers the fraction of aligned bases in a window ($m(w)$) by an estimated regional expectation under neutrality ($m_o$), given by the average fraction of identical aligned base pairs in ancestral repeats in a region surrounding the window, but not containing it. The region is chosen to contain $K = 6{,}000$ aligned bases that are believed not to be under selective pressure, including those in the window itself (for instance, when creating the neighborhood of an ancestral repeat window of size $W = 50$ with at least $T = 40$ aligned bases, this corresponds to between 5,950 and 5,960 bases once the window itself is removed). The average size of the regions constructed in this way is 379,079 bp. The parameter 6,000 was chosen to reduce the variance among normalized scores of ancestral repeat windows. The results are not very sensitive to this parameter: For instance, using $K = 600$ leads to an estimate of 5.11% for the share under selection, $K = 3{,}000$ gives 5.19%, and $K = 12{,}000$ gives 5.08%. As $K$ grows, the estimated local mean $m_o$ approaches the global mean. In the limit, for infinitely large $K$, we obtain an estimate 4.84%. This shows that we apparently do lose a bit in the estimate of the share under selection if we do not try to account for local evolutionary rate variation, but the numbers we obtain are still in the same ballpark.

***Gaussian kernel density estimation.*** Gaussian kernel smoothing (see Eq. 2) was implemented using the *R* language (Ihaka and Gentlman 1996) routine *density(x, n, window, bw, na.rm=T, from, to)* where

- *x* is the vector of observations (e.g., the vector of S-scores for 50-bp ancestral repeat WA-windows for estimating the neutral density, and that of 50-bp genome-wide WA-windows for estimating the genome-wide density).
- *n* determines the number of equispaced abscissa values between *from* and *to* on which the smooth curve ordinate values are computed. We fixed the same *n* (10,000) *from* and *to* (minimum and maximum observed scores for genome-wide windows) for all estimations, to have density values on exactly the same abscissa grid.
- *window* determines the type of kernel to be employed. We used "g" for Gaussian.

- $bw$ is the parameter defining the degree of smoothing. We used $bw = 0.5$ (which according to the routine specifications corresponds to a Gaussian kernel standard deviation of 0.5) when considering 50-,100-, and 200-bp WA-windows, and $bw = 0.75$ (kernel standard deviation = 0.75) when considering 30-bp WA-windows.

The $na.rm = T$ is a technical argument ensuring that missing values, if any, be discarded from the calculation.

***Mixture estimation.*** Based on the upper bound expressed by Equation 5, we approximate the neutral weight from above using the empirical minimum of a ratio: $p_o = \min_S[f_{genome}(S)/f_{neutral}(S)]$. Under-smoothing in the density estimates may translate in ragged fluctuations of this ratio, especially for extreme values of $S$ where very few observations are available and thus both densities are very close to 0. These fluctuations complicate a reliable assessment of the empirical minimum. This problem, whose potential effect was evaluated through some control experiments, can be satisfactorily mitigated by selecting an appropriate degree of smoothing in the Gaussian kernel procedure, and implementing an additional "trimming" procedure for ratio fluctuations on extreme $S$ values.

***Trimming the neutral density.*** The small fluctuations in the estimated neutral density cause $p_o f_{neutral}(S) > f_{genome}(S)$ for some values of $S$, but according to the mixture in Equation 3, this cannot happen. As a consequence, the estimate for $p_o$ must be decreased until all the fluctuations are below the genome-wide density. This causes $(1 - p_o) f_{genome}(S)$ to increase and makes some known neutral windows appear selected. Alternatively, we can explicitly model the error in the neutral density so

$$f_{neutral}(x) = f^*_{neutral}(x) + \varepsilon$$

where $f_{neutral}(S)$ is the density estimated from the data, $f^*_{neutral}(S)$ is the true neutral density, and $\varepsilon$ is a positive constant error term. The amount of trimming is set to $\alpha = 0.01$ where $\int \varepsilon \, dx \leq \alpha$ and therefore $\int f^*_{neutral} \, dx > 1 - \alpha$. With this error term the estimate of $p_o$ becomes

$$p_o = \frac{f_{genome}(x)}{\max(0, f_{neutral}(x) - \varepsilon)}$$

Even this simple constant error model has a dramatic effect in reducing the number of neutral windows incorrectly labeled selected, as results of the control experiment described below illustrate (see Fig. 5).

***Probability of selection estimation.*** The equality in Equation 5 is derived from the mixture in Equation 3 as follows:

$$\Pr(w \text{ selected}|S(w) = S) = 1 - \Pr(w \text{ } neutral|S(w) = S)$$

$$= 1 - \frac{\Pr(w \text{ neutral} \cap S(w) = S)}{\Pr(S(w) = S)}$$

$$= 1 - \frac{\Pr(w \text{ neutral})\Pr(S(w) = S|w \text{ neutral})}{f_{genome}(S)}$$

$$= 1 - p_o \frac{f_{neutral}(S)}{f_{genome}(S)}$$

Thus, the probability of selection as a function of normalized percent identity is conservatively estimated with the curve $1 - p_o \, [f_{neutral}(S)/f_{genome}(S)]$.

## RESULTS

### Main Analysis

Our main analysis uses the collection of all 50-bp nonoverlapping windows of the human genome with at least 40 bases aligned to mouse, referred to as "well-aligned windows" or *WA-windows* below, plus the subset of these windows within aligned ancestral repeat sequence (see Methods, and Table 1 for coverage statistics). The average numbers of bases aligned in these WA-windows are 47.5 and 46.94, respectively. To score a window $w$, we compute the fraction $m(w)$ of aligned base pairs in $w$ that are identical between human and mouse and subtract from it an estimated regional expectation under neutrality, $m_o$. This estimate is the average fraction of identical aligned base pairs in $K = 6{,}000$ aligned ancestral repeat sites in a region surrounding the window $w$, a regional size of about 400 kb, determined so as to optimize a tradeoff between sample variance in the estimate of $m_o$ and regional fluctuations in $m_o$. The results are not greatly sensitive to the choice of $K$ (see Methods). We then rescale $(m(w) - m_o)$ to take into account differences in fluctuation magnitude due to $m_o$ and to the number of aligned positions in the window, $n(w)$. This results in the normalized percent identity score

$$S(w) = \frac{(m(w)\, n(w) - m_o\, n(w))}{\sqrt{m_o(1 - m_o)n(w)}} = \sqrt{\frac{n(w)}{m_o(1 - m_o)}}(m(w) - m_o) \quad (1)$$

As shown in the red curve in Figure 1, for ancestral repeat WA-windows, the empirical distribution of $S$ is tight and symmetric about 0 (mean = – 0.119, S.D. = 1.208, median = – 0.126). It is bell-shaped, but its tails are too heavy for a Gaussian. On the other hand, for genome-wide WA-windows (blue curve in Fig. 1), the empirical distribution is broader and asymmetric, with a heavier right tail (mean = 0.367, S.D. = 1.541, median = 0.239).

**Table 1.** Estimates of the Share of the Human Genome under Selection for Different Window Sizes ($W$) and Required Number of Aligned Bases ($T$)

| W | T | $p_1 = (1 - p_O)$ | Coverage | $a_{sel}$ (%) |
|---|---|---|---|---|
| 30 | 20 | 0.15 | 846472K (30.4%) | 4.51 |
| | 25 | 0.17 | 743308K (26.7%) | 4.50 |
| | 30 | 0.23 | 439501K (15.8%) | 3.65 |
| 50 | 40 | 0.19 | 756051K (27.1%) | 5.19 |
| | 45 | 0.22 | 623286K (22.4%) | 4.90 |
| | 50 | 0.31 | 292506K (10.5%) | 3.31 |
| 100 | 80 | 0.23 | 739836K (26.6%) | 6.15 |
| | 90 | 0.29 | 550530K (19.8%) | 5.8 |
| | 100 | 0.52 | 122437K (4.4%) | 2.29 |
| 200 | 160 | 0.31 | 708701K (25.4%) | 7.92 |
| | 180 | 0.40 | 467954K (16.8%) | 6.68 |
| | 200 | 0.81 | 328668K (1.2%) | 0.96 |

The table reports the estimated mixture coefficient for the selected component, $p_1 = 1 - p_O$, coverage of the human genome (in terms of number of bases and percentage), and estimated share of the genome contained in windows under selection, $a_{sel}$.

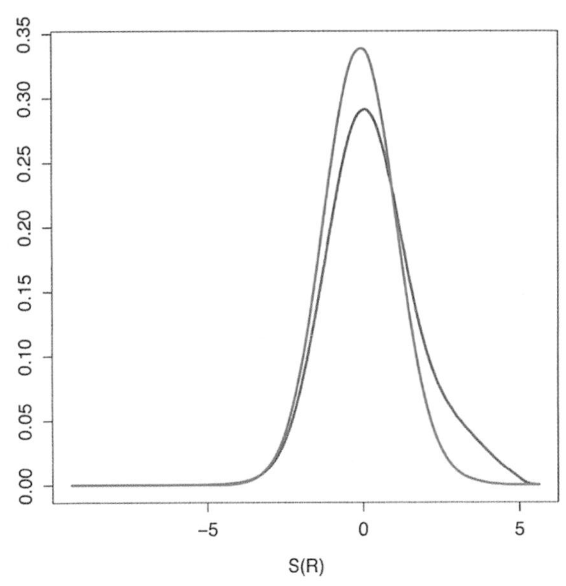


**Figure 1.** Smoothed densities of normalized percent identity for ancestral repeats and genome-wide WA-windows (50 bp, at least 40 aligned). $f_{neutral}(S)$ and $f_{genome}(S)$ are depicted in red and blue, respectively. They are obtained through Gaussian kernel smoothing, a technique that employs the convolution of a Gaussian density with the discrete distribution placing equal mass on each observed value.

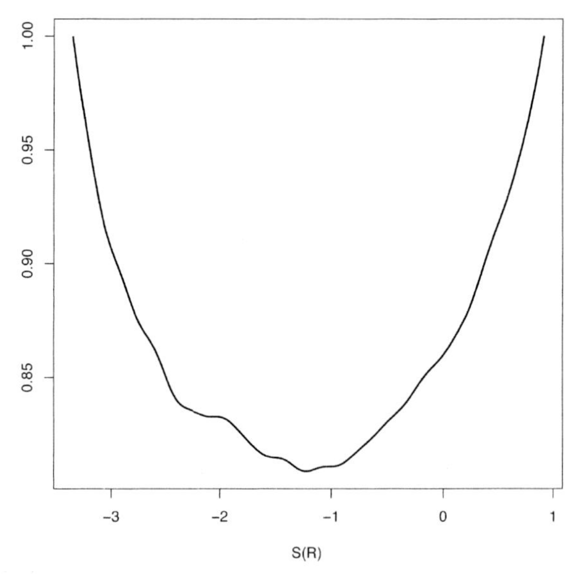


**Figure 2.** Ratio between the smoothed densities of normalized percent identity for genome-wide and ancestral repeat WA-windows (50 bp, at least 40 aligned). The minimum of this curve, $p_o$ = 0.808, estimates an upper bound for the neutral weight in the mixture (i.e., the share of genome-wide windows compatible with the neutral template provided by ancestral repeats).

We employed Gaussian kernel smoothers to produce the estimated density functions $f_{neutral}(S)$ and $f_{genome}(S)$ depicted by the blue and red curves in Figure 1. A Gaussian kernel smoother (Wegman 1972; Silverman 1986) estimates the density of a variable $X$, for which observations $\{x_1,...x_N\}$ are available, by convolving the density of a normal $N(0,\sigma^2)$ with a distribution placing mass $1/N$ on each observed value:

$$f(X) = \frac{1}{N}\sum_{i=1}^{N}\frac{1}{\sqrt{2\pi}\sigma}\exp\left\{-\frac{(X-x_i)^2}{2\sigma^2}\right\} \quad (2)$$

We decompose the distribution of $S$ for genome-wide WA windows as a mixture of a neutral component (the score distribution for WA-windows in ancestral repeats) and a component that appears to be under selection, with weights $p_o$, and $p_1 = (1 - p_o)$, respectively:

$$f_{genome}(S) = p_o f_{neutral}(S) + (1 - p_o) f_{selected}(S) \quad (3)$$

(For background on mixtures, see Lindsay 1995; McLachlan and Peel 2000; for an approach similar to the one used here, see Efron et al. 2001.) Thus, a WA-window is assumed to be neutral (have conservation consistent with $f_{neutral}$) with probability $p_o$, and undergoing selection (have conservation consistent with $f_{selected}$) with probability $(1 - p_o)$.

We have estimated $f_{neutral}$ and $f_{genome}$ from our data, and will use Equation 3 to estimate $p_o$, which will then determine $f_{selected}$. Although the parameter $p_o$ is not univocally determined by Equation 3, non-negativity of densities implies that

$$p_o \leq \frac{f_{genome}(S)}{f_{neutral}(S)} \quad (4)$$

for all scores $S$. Thus, we estimate an *upper bound* to the neutral weight as $p_o = min_S[f_{genome}(S)/f_{neutral}(S)]$, which gives a value of 0.808 . This is illustrated in Figure 2. (In practice, additional steps are taken to ensure that inaccuracies in the estimated density ratio $f_{genome}(S)/f_{neutral}(S)$ do not affect the result; see Control Experiments below and Methods.) Figure 3 summarizes the corresponding "conservative" mixture decomposition: the blue curve de-

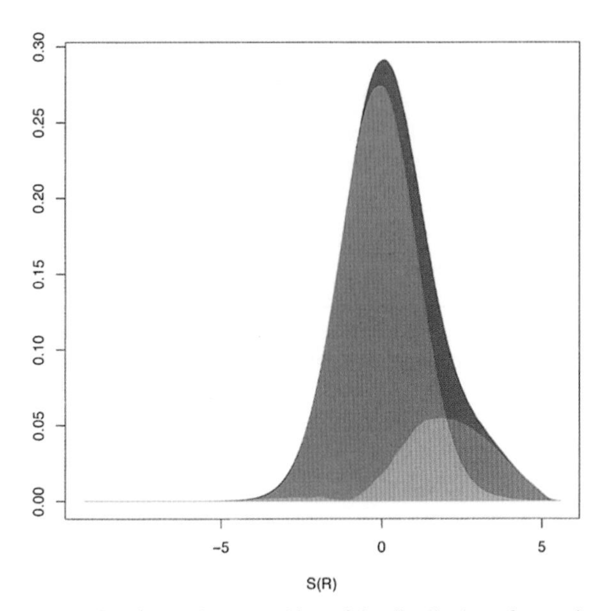


**Figure 3.** Mixture decomposition of the distribution of normalized percent identity for genome-wide WA-windows (50 bp, at least 40 aligned) into a neutral component and a component under selection. This is a "conservative" decomposition that uses the estimated upper bound $p_o$. The blue curve depicts $f_{genome}(S)$, the red curve depicts $p_o f_{neutral}(S)$, and the green curve depicts the difference $f_{genome}(S) - p_o f_{neutral}(S) = (1 - p_o) f_{selected}(S)$.

picts $f_{genome}(S)$, the red curve depicts $p_o f_{neutral}(S)$, and the green curve depicts the difference $f_{genome}(S) - p_o f_{neutral}(S) = (1 - p_o) f_{selected}(S)$ (the estimated score distribution for WA-windows under selection, rescaled by its weight). Note that no parametric assumptions are used in this decomposition. The density of the scores in the selected component captures the empirical structure of all observed conservation levels in 50-base windows beyond those that can be explained by the neutral model; we don't assume that the amount of "selection" follows any particular parametric model.

This calculation suggests that, at most, 80.8% of the genome-wide WA-windows are consistent with neutral evolution, with the remainder (at least 19.2%) appearing to be under selection, or neutral but accumulating substitutions at a slower rate than those in ancestral repeats. Because ~27.1% of all human bases are covered by WA-windows, under the additional conservative assumption that no regions outside these well-aligned windows are under selection, this result implies that a fraction $a_{selected}$ of at least 0.192*0.271 = 0.0520 (about 5.2%) of the human genome is contained in 50-bp windows that appear to be under selection by this test.

## Window Size and Alignment Threshold: Separating Selected and Neutral Behaviors

Fully investigating stability of the above results with respect to different choices of alignment and score functions is beyond the scope of this paper. However, we note that very similar results were obtained on another alignment, using a somewhat different score function (taking into account base composition and adjacent bases' effects on neutral evolution), and a cruder mixture modeling method (Roskin et al. 2002, 2003). In addition to the fact that ancestral repeats accumulate substitutions slower than fourfold degenerate sites (Waterston et al. 2002; Hardison et al. 2003), this is further evidence against the hypothesis that DNA in ancestral repeats accumulates substitutions faster than other types of neutral DNA, perhaps due to some lingering base-compositional property of the ancient relics, and hence that the analysis above overestimates the share of the genome under selection. We discuss other tests for this type of "biased neutral model" effect below. However, to get a general feel for the stability of the results, we first investigate the effect of window size ($W$) and threshold number of aligned bases ($T$) on our "conservative" estimate of the mixture coefficient $p_o$, and subsequent lower bound estimate of the share under selection.

Outcomes for various choices of $W$ and $T$ are reported in Table 1. The estimated fraction of WA-windows under selection increases with increasing window size and required number of aligned bases, while the total fraction of the genome covered by WA-windows decreases. The variation in the estimated share under selection, $a_{selected}$, reflects a tradeoff between these two effects.

As the window size and/or the alignment threshold decrease, neutral and genome-wide distributions of normalized percent identity become more similar, making it more difficult to statistically separate neutral and selected components. This is reflected in the results given in Table 1. When the neutral and selected distributions are highly overlapping, and thus the neutral and genome-wide distributions more similar, the lower bound we produce is very weak, which in turn leaves room for gross underestimation of the apparent share under selection (in the extreme case of two identical score distributions for neutral and selected windows, our conservative estimation of $p_o$ would be 1, and thus our lower bound estimate of the share under selection 0, although the actual neutral weight and share under selection could be anywhere between 0 and 1). This effect becomes more severe for smaller window sizes; the smaller the size, the less neutral and selected windows separate in terms of normalized percent identity.

To see this, we considered windows we have good reason to believe are under selection; namely, windows entirely contained in the coding regions of known genes in the RefSeq database (Pruitt and Maglott 2001). For window sizes $W$ = 30, 50, 100, and 200 bp, we set the alignment threshold to $T$ = 25, 40, 80, and 160, respectively, and compared the score distribution for well-aligned coding windows (*WAC-windows*) to the distribution for neutral WA-windows, i.e., WA-windows in ancestral repeats. We found a substantial overlap for 30-bp windows, but much less overlap for windows of 50 bp and larger (Fig. 4).

Using the mixture decomposition for a fixed window size, say $W$ = 50 bp, we can estimate the probability that a generic 50-bp window $w$ is under selection given its normalized percent identity:

$$\Pr(w \text{ selected } | S(w) = S) = 1 - p_o \frac{f_{neutral}(S)}{f_{genome}(S)} \quad (5)$$

For any collection $C$ containing $N$ 50-bp windows, we

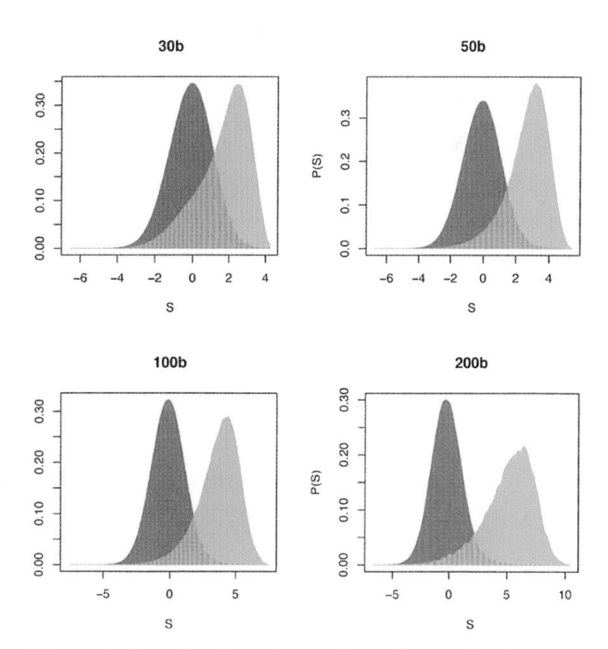


**Figure 4.** Gaussian Kernel smoothing of normalized percent identity distributions for WAC (well-aligned coding) windows (*green*) and WA-windows in ancestral repeats (red), for window sizes 30 bp (alignment threshold = 25), 50 bp (alignment threshold = 40), 100 bp (alignment threshold = 80), and 200 bp (alignment threshold = 160).

can use this formula to calculate the expected fraction of windows in $C$ that are under selection as

$$\frac{1}{N}\sum_{w\in C}\Pr(w \text{ selected}|S(w)) \qquad (6)$$

If, for example, we apply Equation 6 with $C$ defined as all $W$ = 50-bp WA-windows (alignment threshold $T$ = 40), we recover the mixture coefficient $p_1$ = 0.192 discussed above, because $p_1$ is the fraction of these WA-windows that are estimated by the mixture decomposition to be under selection, and this must be the same as the expected fraction of windows under selection. Here Equation 6 merely provides another way of calculating the same number, and hence a nice test for our software. However, if we apply Equation 6 with $C$ defined as $W$ = 50-bp WAC-windows, then we can calculate something more interesting; namely, the expected fraction of well-aligned coding windows that are under selection. We performed this calculation for various window sizes.

For 200-bp windows we obtained 86%, for 100-bp windows 78%, for 50-bp windows 65%, but for 30-bp windows, we obtained only 48%. This further indicates how our mixture decomposition method produces a very conservative lower bound for the share under selection when applied to the normalized percent identity distribution of small windows.

## A Tighter Lower Bound: Splitting Well-aligned Windows

Our computational strategy requires enough separation between the neutral and selected distribution of normalized percent identity for the mixture to reliably detect the difference. In fact, the definition of well-aligned windows ($T$ = 25 for $W$ = 30 bp, $T$ = 40 for $W$ = 50 bp, $T$ = 80 for $W$ = 100 bp, $T$ = 160 for $W$ = 200 bp) and choice of window size for the main analysis ($W$ = 50 bp) stemmed from separation considerations; see also the Discussion section below. However, if ancestral transposon relics are a good neutral model, our figure of 5.2% may still represent a fairly conservative lower bound for the share under selection. As a means to tighten this lower bound, we can further isolate extremely well-aligned genome-wide and neutral windows, splitting WA-windows into a high and a low alignment range. We tried, respectively, 20–24 and 25–30 aligned bases for $W$ = 30, 40–44 and 45–50 for $W$ = 50, 80–94, and 95–100 for $W$ = 100, and 160–194 and 195–200 for $W$ = 200.

We repeated our calculations (estimating smooth densities for neutral and genome-wide scores, decomposing the genome-wide score distribution into a neutral and a selected component, computing a share under selection based on the mixture weight estimate and coverage) separately for high- and low-range WA-windows, and added the results. As shown in Table 2, this consistently produces slightly higher share figures.

The reason for the tighter lower bound is that neutral and genome-wide normalized percent identity distributions are more dissimilar within each of the two groups than they are for WA-windows as a whole; that is, the split increases separation between neutral and selected behavior. From a purely theoretical point of view, splitting could either increase or decrease separation (this represents an interesting area for further theoretical study), but if it increases separation, then still finer partitions of WA-windows may lead to even higher share estimates. However, finer partitions lead to the compounding of errors in the calculations performed for each group, and this limits their utility. We address the issue of statistical error next.

## Control Experiments

As a control for the error associated with our Gaussian smoothing and mixture decomposition, using 50-bp windows with a threshold of 40 aligned bases, we divided the WA-windows in ancestral repeats into two sets, A and B, at random. Set A was used to estimate the neutral score distribution. Set B was used to estimate a genome-wide distribution under a "null" scenario of no selection. Since both data sets contain neutral windows, one expects a near 0 estimate for the fraction under selection: If $f_{neutral}(S) = f_{genome}(S)$ exactly for all scores $S$, we would have $p_o = \min_S[f_{genome}(S)/f_{neutral}(S)] = 1$, and hence $1 - p_o = 0$. However, random differences between $f_{\text{neutral}}(S)$ and $f_{genome}(S)$ do occur, especially for extreme values of $S$ where very few observations are available and thus both densities are very close to 0. These differences between small density values can generate fairly wide fluctuations in the ratio, resulting in a minimum sizably smaller than 1 (on some control experiments the minimum was $<0.9$).

The magnitude of this error can be greatly reduced by selecting an appropriate degree of smoothing in the Gaussian kernel procedure and implementing an additional "trimming" procedure for ratio fluctuations on extreme $S$ values (see Methods). With these steps, the control experiments resulted in ratio minima above 0.985. Figure 5

**Table 2.** Estimates of the Share of the Human Genome under Selection Obtained Splitting WA-windows into a High and a Low Alignment Range, for Various Window Sizes ($W$)

| | | Low | | High | | Summed | WA-windows |
|---|---|---|---|---|---|---|---|
| $W$ | $T$ | range | $a_{sel,L}$ (%) | range | $a_{sel,H}$ (%) | $a_{sel,+}$ (%) | $a_{sel}$ (%) |
| 30 | 20 | 20–24 | 0.22 | 25–30 | 4.5 | 4.72 | 4.51 |
| 50 | 40 | 40–44 | 0.344 | 45–50 | 4.955 | 5.30 | 5.15 |
| 100 | 80 | 80–94 | 1.53 | 95–100 | 4.9 | 6.43 | 6.15 |
| 200 | 160 | 160–194 | 4.7 | 195–200 | 3.45 | 8.15 | 7.92 |

The table reports estimated share of the genome contained in windows under selection for low range ($a_{sel,L}$) and high range ($a_{sel,H}$), and the overall estimate obtained as their sum ($a_{sel,+}$). The last column contains the estimate obtained without partitioning WA-windows ($a_{sel}$).

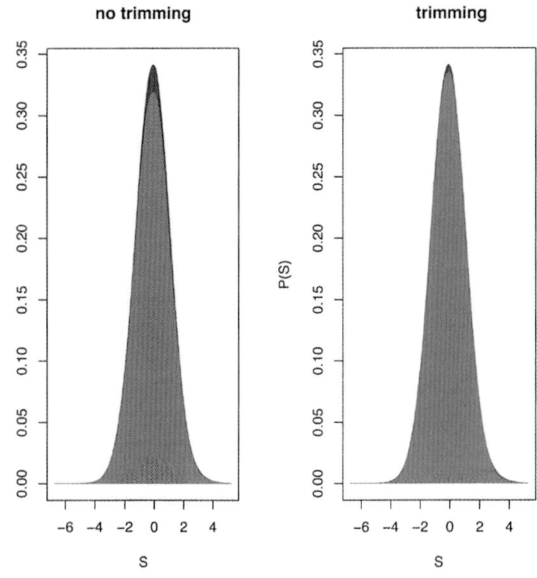


**Figure 5.** Results of a control experiment for 50-bp WA-windows (alignment threshold = 40). The set of ancient repeat windows is randomly divided into two subsets, A and B, of equal size. A is used to estimate the neutral density (*red*), B is used to estimate a genome-wide density (*blue*) under a "null" scenario of no selection, and the usual procedure is applied to estimate $p_0$ and the density under selection (*green*). Except for error affecting the Gaussian kernel estimation and mixture decomposition, the red and blue curves should be almost coincident, and the green curve negligible. The two panels show the decompositions obtained without (*left*) and with (*right*) "trimming."

illustrates the effectiveness of trimming on a control experiment.

### Tests for Alignment Artifacts

A concern with the use of ancestral repeats as a model of neutral substitutions between human and mouse is the reliability of their cross-species alignments. One problem is the possibility that nonorthologous repeats are aligned. This risk is effectively minimized by the BLASTZ alignment procedure used to obtain human–mouse whole-genome alignments: The procedure very carefully first seeds all alignments off unique DNA matches between the two genomes, and only after this extends these matches into the adjacent repetitive regions (Schwartz et al. 2003). Estimates of the amount of nonorthologous DNA that was aligned by this method are quite small (Waterston et al. 2002).

To further ensure that we were not getting nonorthologous alignment, we additionally refined these alignments using a reciprocal filtering method. This method removes nonorthologous alignments by selecting, among BLASTZ alignments, those that can be aligned in both directions (human to mouse and mouse to human) with the highest score—reciprocal best alignments. In our earlier analyses, alignments of mouse genes to human processed pseudogenes had been occasionally included, and as a result, human pseudogenes had appeared largely as if they were under selection. When we switched to reciprocal best alignments, we saw a reduction in our share under selection estimate of about 0.33% (e.g., from 5.33% to 5.0%)—mostly because of the removal of the alignments to processed pseudogenes. Consequently, all the results in this paper use reciprocal best alignments. We note that this filtering eliminates more alignments than the recently proposed chaining (Kent et al. 2003), thereby leading to potentially more conservative lower bounds on the share under selection (see Methods). In fact, we did some experiments recomputing our estimate on chained, syntenic alignments, and obtained results very similar to those obtained on reciprocal best alignments.

Another artifact could derive from failure to correctly align at the base-by-base level some human–mouse orthologous pairs of ancestral repeats. A bias in our estimates could be introduced by an inability to find the most diverged pairs of orthologous repeats, causing them to be absent from our data set, or because after finding these diverged pairs, the relatively large distance between them tempts the optimization method used in the detailed pair-wise alignment to find more base identities between them than are actually there in the evolutionarily correct alignment. Similar effects have been observed in simulation studies of pair-wise alignments on synthetic sequences derived by too many substitutions from a synthetic ancestral sequence (Holmes and Durbin 1998). Both of these types of bias would cause the observed level of conservation in ancestral repeats to be greater than it should be, and hence make the true share under selection even larger than what we are estimating. This analysis further reinforces our claim that our estimates produce a lower bound on the share under selection, provided the model of neutral DNA by ancestral repeats is adequate. The additional weakness in the lower bound caused by these potential alignment artifacts is not great because, as mentioned above, the observed levels of conservation between human and mouse in fourfold degenerate sites of codons is similar to that in ancestral repeats, and the former are not subject to alignment artifacts.

### Inadequacies of Ancestral Repeats as a Model of Neutral Evolution

The final issue is whether or not the relics of ancestral transposons provide an unbiased model of neutral evolution. We cannot completely resolve this with the data we have, but we have done some tests, in addition to the use of alternate score functions that compensate for base compositional biases, and effects of flanking bases, including dinucleotide effects like bias in substitution rates for CpGs (Roskin et al. 2003).

First, any property of genomic DNA that causes rates of neutral substitution to vary from region to region, such as GC content, should be compensated for by the way we compute our score function relative to the neutral rate estimated from a surrounding window of DNA, rather than on an absolute scale. This prevents one class of DNA from standing out as apparently richer in elements under selection just because it is in a general region of the genome that is accumulating changes at a slower rate.

However, we could still have a bias in the estimate of the share under selection if there are properties that cannot be corrected for by a score function that accounts for compositional biases, and that affect the neutral substitution rate differently in relics of ancestral transposons than they do in other types of neutral DNA.

One possibility is that relics of ancestral transposons, because of their similarity to each other, are more apt to undergo gene conversion (or ectopic conversion) events than other neutral DNA. If we are looking at a pair of orthologously placed transposon relics in the human and mouse genomes, but one of them, say the one in the mouse, has undergone a gene conversion in the rodent lineage, then additional substitutions may have been introduced by that event, making the human–mouse aligned elements less conserved than would be a pair that did not undergo lineage-specific gene conversion. If this were quite common, it would cause us to overestimate the share under selection using our neutral model. However, we note that all gene conversion events that occurred to transposon relics before the human–mouse split would have no effect: They would merely replace the DNA that is inherited by both species. Only primate or rodent lineage-specific gene conversion events can introduce bias. Since triggering a gene conversion event requires a reasonably high degree of sequence identity between the two copies, this means that relics from ancestral transposon families that were only active long before the human–mouse split, and hence whose copies were highly dissimilar to each other at the time of the split, are much less likely to have had a lineage-specific gene conversion event after the split. Basing the neutral model on these "most ancient" relics would then remove the bias.

We divided ancestral repeats into 130 subfamilies as defined by RepeatMasker (Smit and Green 1999), computed the average number of mismatches between each repeat and the consensus sequence for its subfamily, and eliminated from the analysis all repeats belonging to the 65 least diverged subfamilies, representing those transposon relics that were present in the ancestral genome the least amount of time before the human–mouse split. This eliminated roughly 60% of the bases in the original collection of ancestral repeats. For $W$ = 50-bp WA-windows, the estimated share under selection obtained with this restricted set of ancestral repeats was very similar to that obtained with the full set of ancestral repeats as a neutral model (ratio of the two estimates was between 0.95 and 1 in different tests of this type with various data sets and alignments). This argues against gene conversion being a source of bias.

Another possibility is that after transposons are inserted they undergo a more rapid substitution rate. Possible causes for this may include mechanisms to suppress transposon transcription, or some holdover from another ancient cellular defense mechanism against insertions of transposons, as well as rapid adaptation of their GC content to the GC content of the surrounding DNA (Bernardi 1993). This would also cause an overestimate in the share under selection using transposon relics as neutral model. However, it seems plausible that such an increased rate of substitutions would diminish as the transposon relics age, so that a very old transposon relic which has been accumulating substitutions in the genome for 50 to 100 million years would behave more like typical neutral DNA. Consequently, we would have expected to see more of a change in the estimate of the share under selection in the above-mentioned experiments, in which we eliminated "younger" ancestral transposon relics from the neutral model. The largest effect we saw when experimenting with different subsets of ancestral transposons to define the neutral model was when we used only SINEs (both "young" and "old" ancestral SINEs). Here the estimate dropped to 4.67%, possibly indicating some bias, but still not as big a fluctuation as we saw by varying the window size and alignment threshold (Table 1).

Finally, one type of neutral DNA that may affect our results because it evolves in a distinct way, different from that of transposon relics, is DNA in simple repeats. However, we found that there are only 2 million bases of human simple repeats aligned with mouse, less than 1/1000th of the genome, so these by themselves could not substantially affect our estimate of the share under selection. Essentially, even rejecting all simple repeats as a priori not being under selection, we would not reduce our estimate of the share of the genome under selection.

## DISCUSSION

Ultimately, one would like to identify individual bases of the human genome that are under selection. However, with only one alignment available (to the mouse genome), there is insufficient information to do so at this time. Even a global statistical estimate of the share under selection cannot be made from data relative to single bases, because the neutral and genome-wide distributions will be very similar for any reasonable conservation score computed on two-species comparisons of individual bases.

To illustrate this, consider the simple conservation function that has score = 1 if the aligned bases are identical in human and mouse, and score = 0 otherwise. The estimated probability of score = 1 is 0.667 for aligned neutral bases (from ancestral repeats) and 0.699 for aligned bases genome-wide. Applying our basic mixture estimation method to these data gives an upper-bound estimate of $p_o$ = 0.904. This is the maximum fraction of the genome-wide aligned sites that is compatible with the neutral score distribution, and is (implicitly) obtained by assuming that bases under selection are *always* identical between human and mouse. We cannot do any better without assuming prior knowledge of the score distribution for selected sites, something we have avoided in our approach. When we then convert $p_o$ into a lower bound on the fraction of sites in the human genome undergoing selection, we obtain $a_{selected}$ = 0.35*(1–0.904) = 0.0336, or about 3.4% (35% of the bases in the human genome are aligned to mouse). In light of the implicit assumption that all selected bases are exactly conserved, this is clearly a very weak lower bound. The problem here is the strong overlap between the neutral distribution and the (unobserved) selected distribution of the score, which makes them hard to separate. As we have seen, this strong overlap is also present in conservation scores computed on windows when those windows are small, e.g., only 30 bp. This casts doubt on the stringency of lower bounds obtained from very small windows: Technically, they are valid lower bounds, but they might yield significant underestimates of the share under selection.

Using larger windows (instead of single-base positions or very small windows) to estimate the share under selection carries other limitations. We can produce a more stringent lower bound, and thus a better estimate for the fraction of windows that appear to be under selection and the fraction of the human genome that is contained in such windows. However, because the score applies to each window as a whole, we need to restrict attention to well-aligned windows, and recall that our share estimate does not automatically equate to an estimate of the fraction of individual bases under selection. This may not be a severe limitation, because bases are not selected entirely independently from their neighbors; altogether, it may make more sense to consider small regions under selection than individual bases under selection.

From a certain point onward, increasing the window size appears to cause an inflation in the total estimated share of the genome under selection beyond what we can attribute to better separation of the neutral and selected score distributions; see Table 1. However, this is a misinterpretation of the results. For instance, in attempting to compare "on a base-by-base level" the estimated fraction of the genome in 50-bp windows under selection to the estimated fraction in 100-bp windows, we are implicitly converting the probability that a window is selected, Pr($w$ selected | $S(w) = S$) defined above, into the expected number of bases in the window $w$ that are under selection, which is not legitimate. In fact, the estimates for 50-bp windows and for 100-bp windows are not directly comparable in this fashion: They are estimates of two different underlying quantities, one measuring evolution of smaller (50 bp) segments and the other larger (100 bp) segments.

From a biological perspective, we would like to reliably detect the effects of purifying selection on as small a unit as makes sense; ideally at most a few tens of bases. Given the limitations posed by employing only the human–mouse alignment, the best we can do at this point is to use 50-bp WA-windows: About 5% of the human genome is contained in 50-bp WA-windows that are more conserved than neighboring neutral windows (modeled by ancestral repeats) and thus appear to be under selection. As discussed above, we cannot eliminate the possibility that mechanisms other than purifying selection explain the data we see. In particular, some unidentified specialized types of molecular evolution within ancestral repeats could be causing some kinds of neutral windows to be significantly more conserved than neighboring neutral windows from ancestral repeats, which would artificially inflate our estimate of the share under selection. Tests with alternate score functions that compensate for compositional effects (Roskin et al. 2002, 2003) and tests with different subsets of ancestral repeats as neutral models provide some evidence against the existence of an extreme bias of this type, but cannot eliminate this possibility. Additional evidence will be required to positively prove that the effect we are seeing is due to selection.

The estimate, if valid, leads to the question of what function these elements under selection may possess. 5% is considerably more than can be accounted for by the estimated fraction of the genome that is coding, which is about 1.5%. Note that including all 50-bp windows that contain any coding bases typically adds only about 25 bp on either end of a 200-bp coding exon, increasing the 1.5 coding percentage by a factor of only 5/4. Moreover, this is a considerable overassessment of the effect of coding bases on our estimate of the share under selection because, as we have seen, only about 70% of fully coding 50-bp WA-windows (WAC-windows) are contributing to the estimate as it is, and we expect the fraction to be less for partially coding 50-bp WA-windows. Hence, the bulk of the "selection signal" we are detecting is likely to be coming from noncoding bases, possibly performing regulatory or other important functions.

With multiple alignments to several mammals, it should be possible to develop better score functions based on more accurate models of molecular evolution. These will allow us to separate neutral and selected windows more effectively, and thus to further investigate the properties of small regions of the human genome that are under selection.

## ACKNOWLEDGMENTS

We thank all researchers involved in the International Mouse Genome Sequencing Consortium for help and data sharing. In particular, we are grateful to E. Lander, M. Zody, R. Waterston, F. Collins, P. Green, N. Goldman, A. Smit, and W. Miller for their suggestions and comments. We also thank T. Pringle for comments on the research. F.C. was supported by National Human Genome Research Institute grant HG-02238. R.J.W., K.M.R., M.D., W.J.K., and D.H. were supported by NHGRI grant 1P41HG-02371. D.H. was also supported by the Howard Hughes Medical Institute.

## REFERENCES

Bernardi G. 1993. The isochore organization of the human genome and its evolutionary history: A review. *Gene* **135:** 57.

Efron B., Tibshirani R., Storey J., and Tusher V. 2001. Empirical Bayes analysis of a microarray experiment. *J. Am. Stat. Assoc.* **96:** 1151.

Elnitski L., Hardison R., Li J., Yang S., Kolbe D., Eswara P., O'Connor M., Schwartz S., Miller W., and Chiaromonte F. 2003. Distinguishing regulatory DNA from neutral sites. *Genome Res.* **13:** 64.

Hardison R., Roskin K., Yang S., Diekhans M., Kent W., Weber R., Elnitski L., Li J., O'Connor M., Kolbe D., Schwartz S., Furey T.S., Whelan S., Goldman N., Smit A., Miller W., Chiaromonte F., and Haussler D. 2003. Co-variation in frequencies of substitution, deletion, transposition and recombination during eutherian evolution. *Genome Res.* **13:** 13.

Holmes I. and Durbin R. 1998. Dynamic programming alignment accuracy. *J. Comput. Biol.* **5:** 493.

Ihaka R. and Gentlman R. 1996. R: A language for data analysis and graphics. *J. Comput. Graph. Stat.* **5:** 299.
Kent W.J. 2002. BLAT: The BLAST-like alignment tool. *Genome Res.* **12:** 656.
Kent W.J., Baertsch R., Hinrichs A., Miller W., and Haussler D. 2003. Evolution's cauldron: Duplication, deletion, and rearrangement in the mouse and human genomes. *Proc. Natl. Acad. Sci.* **100:** 11484.
Kent W.J., Sugnet C., Furey T., Roskin K., Pringle T., Zahler A., and Haussler D. 2002. The human genome browser at UCSC. *Genome Res.* **12:** 996.
Lander E.S., Linton L.M., Birren B., Nusbaum C., Zody M.C., Baldwin J., Devon K., Dewar K., Doyle M., FitzHugh W., Funke R., Gage D., Harris K., Heaford A., Howland J., Kann L., Lehoczky J., LeVine R., McEwan P., McKernan K., Meldrim J., Mesirov J.P., Miranda C., Morris W., and Naylor J., et al. (International Human Genome Sequencing Consortium). 2001. Initial sequencing and analysis of the human genome. *Nature* **409:** 860.
Lindsay B.G. 1995. *Mixture models: Theory, geometry and applications* (NFS-CBMS Regional Conference Series in Probability and Statistics), vol. 5. American Statistical Association, Alexandria, Virginia.
McLachlan G.J. and Peel D. 2000. *Finite mixture models*. Wiley, New York.
Pruitt K. and Maglott D. 2001. RefSeq and LocusLink: NCBI gene-centered resources. *Nucleic Acids Res.* **29:** 137.
Roskin K.M., Diekhans M., and Haussler D. 2003. Scoring two-species local alignments to try to statistically separate neutrally evolving from selected DNA segments. In *Proceedings of the 7th Annual International Conference on Research in Computational Molecular Biology* (RECOMB 2003), p. 257.
Roskin K., Diekhans M., Kent W., and Haussler D. 2002. Score functions for assessing conservation in locally aligned regions of DNA from two species. UCSC Technical Report CRL-02-30, September 14, 2002. Center for Biomolecular Science and Engineering, Baskin Engineering, University of California, Santa Cruz.
Schwartz S., Kent W., Smit A., Zhang Z., Baertsch R., Hardison R., Haussler D., and Miller W. 2003. Human-mouse alignments with Blastz. *Genome Res.* **13:**103.
Silverman B. 1986. *Density estimation for statistics and data analysis*. Chapman and Hall, London, United Kingdom.
Venter J.C., Adams M.D., Myers E.W., Li P.W., Mural R.J., Sutton G.G., Smith H.O., Yandell M., Evans C.A., Holt R.A., Gocayne J.D., Amanatides P., Ballew R.M., Huson D.H., Wortman J.R., Zhang Q., Kodira C.D., Zheng X.H., Chen L., Skupski M., Subramanian G., Thomas P.D., Zhang J., Gabor Miklos G.L., and Nelson C., et al. 2001. The sequence of the human genome. *Science* **291:** 1304.
Waterston R.H., Lindblad-Toh K., Birney E., Rogers J., Abril J.F., Agarwal P., Agarwala R., Ainscough R., Alexandersson M., An P., Antonarakis S.E., Attwood J., Baertsch R., Bailey J., Barlow K., Beck S., Berry E., Birren B., Bloom T., Bork P., Botcherby M., Bray N., Brent M.R., Brown D.G., and Brown S.D., et al. (Mouse Genome Sequencing Consortium). 2002. Initial sequencing and comparative analysis of the mouse genome. *Nature* **420:** 520.
Wegman E. 1972. Nonparametric probability density estimation. *Technometrics* **14:** 533.

## WEB SITE REFERENCE

Smit A.F. and Green P. 1999. RepeatMasker. (http://ftp.genome.washington.edu/RM/RepeatMasker.html)

# Detecting Highly Conserved Regions of the Human Genome by Multispecies Sequence Comparisons

E.H. MARGULIES,* NISC COMPARATIVE SEQUENCING PROGRAM,*† AND E.D. GREEN*†

*Genome Technology Branch and †NIH Intramural Sequencing Center (NISC), National Human Genome Research Institute, National Institutes of Health, Bethesda, Maryland 20892

The Human Genome Project's recent completion of a high-quality sequence of the human genome represents a landmark scientific accomplishment of great historic significance. It also signifies a critical transition for the field of genomics, as the focus now shifts from elucidating the human genome sequence to establishing the functional information that it encodes. To capitalize on this new and powerful informational resource as well as the ever-advancing technologies for performing genetic and genomic studies, plans for compelling and ambitious research programs have been formulated (Collins et al. 2003).

One of the highest-priority areas is the systematic analysis of the completed human genome sequence, including the identification of all the functional elements that it contains. Such elements fall into two major classes: sequences that reflect "classically" defined protein-encoding genes and sequences that confer function in other ways (i.e., functional noncoding elements). Significant advances have been made in generating a complete inventory of human genes (Hubbard et al. 2002; also see http://www.ensembl.org/Homo_sapiens/), and these have been aided by the availability of complementary data sets (actual gene sequences, such as expressed-sequence tags [Boguski et al. 1994; also see http://www.ncbi.nlm.nih.gov/dbEST] and full-length cDNA sequences [Strausberg et al. 2002; also see http://mgc.nci.nih.gov/]), ever-improving computational tools for performing gene predictions (Kulp et al. 1996; Burge and Karlin 1997; Batzoglou et al. 2000; Korf et al. 2001; Rogic et al. 2001; Solovyev 2001; Alexandersson et al. 2003; Flicek et al. 2003), and genomic sequence data from distantly related vertebrates (e.g., pufferfish; Aparicio et al. 2002) that can be used for detecting coding sequences. In short, it seems likely that the detection of most protein-encoding sequences in the human genome should ultimately prove to be relatively straightforward. In contrast, the identification of functional noncoding sequences is profoundly more challenging. There are many reasons for this, including the primitive state of knowledge about the compositional and functional characteristics of such sequences, the lack of available complementary data sets to aid in their detection, the rudimentary state of computational tools for predicting the existence of such elements, and the paucity of robust and efficient approaches for experimentally validating their presence. In short, the comprehensive identification of functional noncoding sequences in the human genome represents one of the most difficult tasks currently being faced in genomics.

The comparison of sequences derived from different species has emerged as a powerful strategy for identifying functionally important genomic elements (Pennacchio and Rubin 2001). The traditional approach for this involves comparing orthologous sequences from species separated by large evolutionary distances, with particular attention paid to sequences found to be highly similar. Such conserved sequences, which presumably were selectively retained throughout evolution, represent candidates for being functionally important. Often referred to as "phylogenetic footprinting" (Tagle et al. 1988; Duret and Bucher 1997; Blanchette and Tompa 2002), such an approach has been used to detect both coding regions (Pennacchio et al. 2001) and functional noncoding regions, including those involved in gene regulation (Hardison 2000). A variant approach for comparative sequence analysis has recently emerged that involves examining sequences from multiple, closely related species (Boffelli et al. 2003). Called "phylogenetic shadowing," this complementary strategy measures interspecies differences (rather than similarities) found at individual base positions to reveal the functional portion of the region under study.

Here we provide an overview of our efforts to generate and compare genomic sequences derived from multiple vertebrate species. In particular, we describe how we are using multispecies sequence comparisons for detecting small, discrete regions of highly conserved sequence, and demonstrate why such analyses are likely to become an important component of the armamentarium required to identify all the functional elements in the human genome.

## GENOME-WIDE VERSUS TARGETED SEQUENCING

The importance of sequence comparisons for understanding genome function was appreciated at the onset of the Human Genome Project, serving as a key rationale for the early generation of genome sequences of the nematode worm (*C. elegans* Sequencing Consortium 1998) and fruit fly (Adams et al. 2000) in conjunction with the sequencing of the human genome (Lander et al. 2001).

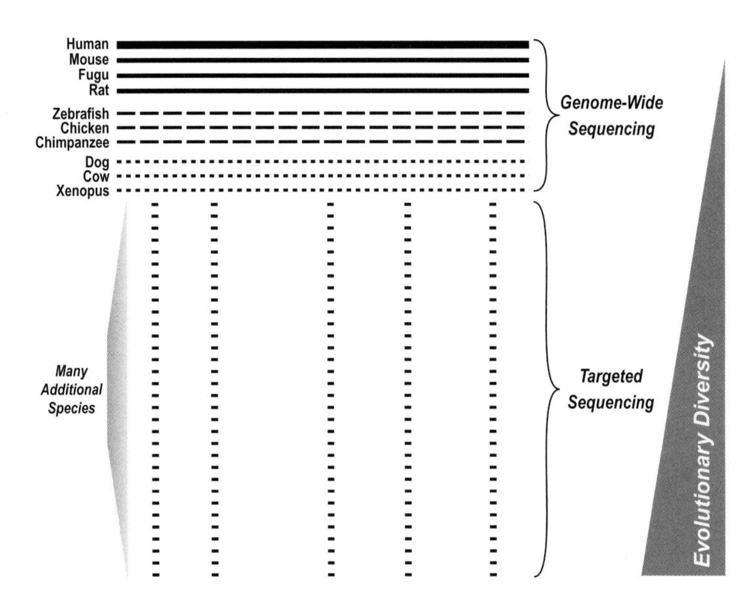


**Figure 1.** Genome-wide and targeted sequencing approaches. Genome-wide sequencing results in the generation of sequence data for an entire genome of a species. Indicated are various completed, ongoing, or planned genome-wide sequencing efforts involving vertebrates, with the extent of completion schematically indicated by the different line formats (see text for details). Targeted sequencing (Thomas et al. 2003), which involves the generation of sequence from the same genomic regions in multiple species, allows for the comparative analyses of sequences from a more evolutionarily diverse set of species.

The insights gained from subsequent comparisons of the resulting data have catalyzed numerous additional genome-sequencing projects, with the largest efforts focusing on the generation of sequence for the entire genome of a species. Toward that end, there are now mature draft sequences of the mouse (Waterston et al. 2002), rat (http://genome.ucsc.edu/cgi-bin/hgGateway?org=rat), and fugu (Aparicio et al. 2002) genomes, rapidly maturing sequences of the zebrafish, chicken, and chimpanzee genomes, and commitments to sequence the dog, cow, and *Xenopus* genomes (Fig. 1). Such genome-wide sequencing projects provide complete insight about the genetic architecture of the corresponding species and allow detailed comparisons to be performed with other available whole-genome sequences (Ureta-Vidal et al. 2003). However, the current costs of genome sequencing (upward of $50–100 million for generating a high-quality draft sequence of a mammalian genome) limit the number of vertebrate genomes that can be sequenced in a genome-wide fashion, at least for the foreseeable future.

As a complement to genome-wide sequencing efforts, targeted sequencing of small, defined genomic regions can be performed for many additional species, allowing comparative analyses of larger, more evolutionarily diverse sets of sequences (Fig. 1). Specifically, this involves first identifying and isolating a genomic region(s) of interest from a series of species (typically using bacterial artificial chromosome [BAC] clones; Shizuya et al. 1992; Birren et al. 1997). This process has been aided by the establishment of an ever-growing repertoire of available vertebrate BAC libraries (see http://www.genome.gov/10001852 and http://bacpac.chori.org) and the development of methods for efficiently isolating target-specific BACs from multiple species in parallel (Thomas et al. 2002). Sequencing of the isolated BACs can then be readily accomplished using standard shotgun-sequencing methods (Wilson and Mardis 1997; Green 2001). Although such a targeted sequencing approach only yields data for small genomic regions, it inherently provides the versatility for comparing sequences from many more species than can be readily accomplished by genome-wide approaches. Furthermore, by only generating data for delimited genomic regions, targeted sequencing strategies are significantly less costly than their whole-genome counterparts.

In short, genome-wide sequencing produces data that are comprehensive at the individual species level, but are more limited in terms of evolutionary diversity. In contrast, targeted sequencing produces data that are more limited at the individual species level, but provide the opportunity to perform comparative analyses with more evolutionarily diverse sequences.

## NISC COMPARATIVE SEQUENCING PROGRAM

The National Institutes of Health Intramural Sequencing Center (NISC) Comparative Sequencing Program is a large, multigroup effort that aims to generate and analyze targeted genomic sequences from multiple vertebrate species (see http://www.nisc.nih.gov). The major elements of this effort involve the (1) selection of targeted genomic regions of interest (already totaling >100 and comprising >60 Mb [or ~2% of the human genome]); (2) design and use of suitable "universal" hybridization probes for the large-scale isolation of BACs from phylogenetically diverse species (Thomas et al. 2002); (3) mapping of isolated clones, construction of BAC contig maps, and identification of minimally overlapping clones that together span the targeted genomic region in each species (Thomas et al. 2003); (4) shotgun sequencing of each selected BAC (Wilson and Mardis 1997; Green 2001); and (5) assimilation, annotation, and comparative analyses of the generated sequences (Thomas et al. 2003).

Our efforts to date have resulted in the generation of a large amount of multispecies sequence data (~350 Mb from >30 different species). The most complete analyses have been performed for two targeted regions on human

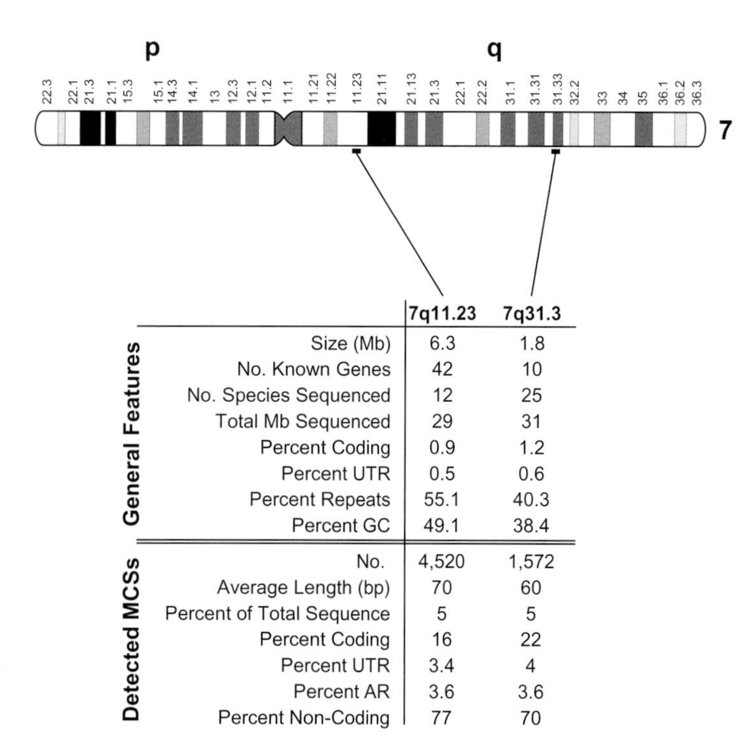


| | | 7q11.23 | 7q31.3 |
|---|---|---|---|
| General Features | Size (Mb) | 6.3 | 1.8 |
| | No. Known Genes | 42 | 10 |
| | No. Species Sequenced | 12 | 25 |
| | Total Mb Sequenced | 29 | 31 |
| | Percent Coding | 0.9 | 1.2 |
| | Percent UTR | 0.5 | 0.6 |
| | Percent Repeats | 55.1 | 40.3 |
| | Percent GC | 49.1 | 38.4 |
| Detected MCSs | No. | 4,520 | 1,572 |
| | Average Length (bp) | 70 | 60 |
| | Percent of Total Sequence | 5 | 5 |
| | Percent Coding | 16 | 22 |
| | Percent UTR | 3.4 | 4 |
| | Percent AR | 3.6 | 3.6 |
| | Percent Non-Coding | 77 | 70 |

**Figure 2.** Features of two targeted genomic regions sequenced in multiple species. The two indicated regions of human chromosome 7 were subjected to multispecies comparative sequencing (see http://www.nisc.nih.gov for details). The general features of each region are listed, with the indicated numbers corresponding to the human reference sequence. "No. Known Genes" is the number of annotated genes with defined coding sequences in each region. "Total Mb Sequenced" reflects the amount of sequence data generated in aggregate from the indicated number of different species. "Percent GC" is the average GC content (calculated in nonoverlapping 1-kb windows across each region). Also indicated is information about the MCSs detected in each region (see text). For the 7q31.3 region, the MCSs were identified using the sequences from a subset of 13 species (among those excluded were all primates, tetraodon, and zebrafish); for the 7q11.23 region, sequences from all 12 species were used to detect MCSs. The last four rows reflect the percent of MCS bases that overlap coding, UTRs, ARs, and noncoding sequence, respectively.

chromosome 7q31.3 (Thomas et al. 2003) and 7q11.23, respectively (Fig. 2; also see http://www.nisc.nih.gov/data). Together these regions span >8 Mb in the human genome and contain 52 known genes; they are also notably different with respect to their general sequence composition (e.g., GC content, amount of repetitive sequences, and proportion of coding sequence).

The sequences and associated analyses emanating from the NISC Comparative Sequencing Program reflect increasingly complex data sets, especially with respect to the number of vertebrate species represented and the various comparisons being performed. For visualizing and disseminating these data, we have collaborated with our colleagues at the University of California, Santa Cruz (UCSC) to facilitate their development of a "zoo browser" component of the UCSC Genome Browser (Kent et al. 2002; Karolchik et al. 2003; Thomas et al. 2003; also see http://genome.ucsc.edu). A sample display of this browser for a small portion of the 7q31.3 target is shown in Figure 3. In addition to providing convenient access to the assembled sequences for each species, the browser has been customized to display the results of pair-wise sequence alignments (which can be performed using any species' sequence as the reference) and other comparative analyses (see below). By direct integration with the standard UCSC Genome Browser environment, such data can then be viewed within the context of the ever-growing collection of annotations and associated information about the human genome sequence.

## MULTISPECIES CONSERVED SEQUENCES

The recent comparative analysis between the human and mouse genome sequences (Waterston et al. 2002) represents the most detailed comparison of vertebrate genomes performed to date. Several findings from this effort are particularly relevant to our multispecies sequencing program. First, roughly 40% of the mouse genome forms sequence alignments with the human genome using established alignment methods (e.g., blastz [Schwartz et al. 2003a]; see the human–mouse alignments depicted in Fig. 3). Second, only about 5% of the mammalian genome is estimated to be actively conserved (Waterston et al. 2002; Roskin et al. 2003), and this consists of roughly 1.5% that is protein coding and 3.5% that is noncoding. At present, the specific bases that constitute this 5% are not known. Thus, a critical challenge is to develop strategies for identifying this small fraction of actively conserved sequence, since it presumably reflects the bulk of the functionally important portion of the human genome. However, simple pair-wise comparisons of human and mouse sequences are ineffective at identifying the correct subset of actively conserved sequence in the mammalian genome (Thomas et al. 2003).

To broaden the above human–mouse sequence comparisons, we have performed similar pair-wise alignments between human and various other species' sequences (Fig. 3), revealing a number of interesting findings. First, primate sequences are highly similar to the human sequence, as expected. Second, alignments between the human sequence and that of the other mammals show patterns similar to that seen with mouse; however, the sequence conservation with human is generally higher for most nonrodent placental mammals compared to rodents (Thomas et al. 2003). Alignments between human and fish sequences are almost exclusively confined to coding regions. Interestingly, marsupial, monotreme, and chicken sequences show significantly fewer align-

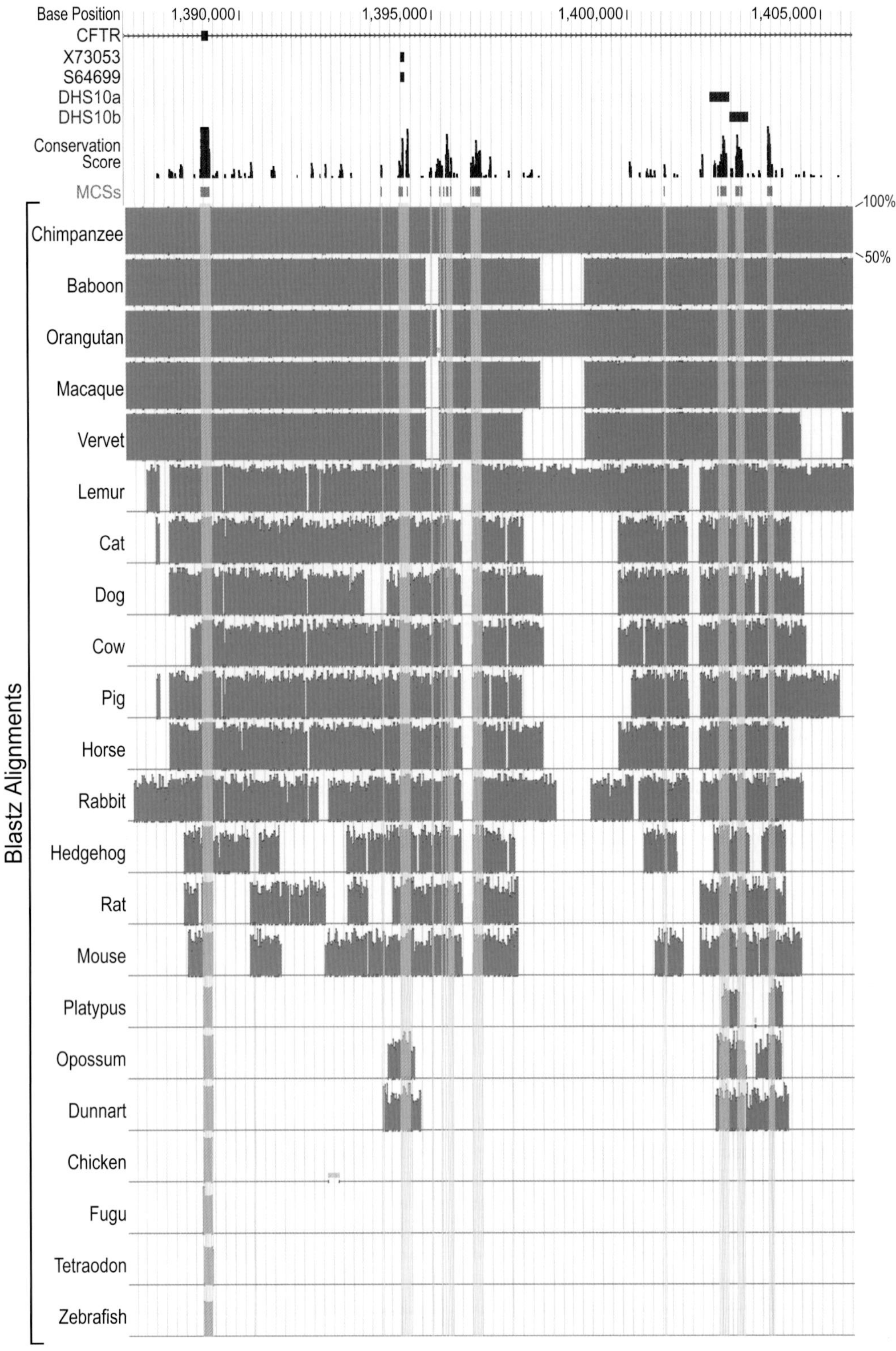


**Figure 3.** UCSC Genome Browser view of multispecies sequences and their comparative analyses. Depicted is an adapted view of the UCSC Genome Browser (see http://genome.ucsc.edu) showing the data for a ~20-kb region within the *CFTR* gene (7q31.3 target; see Fig. 2), with sequence generated from 22 species. Along the top are shown standard UCSC Genome Browser tracks. The tracks that are unique to the "zoo browser" are as follows: (1) Sites sensitive to cleavage by DNase I (DNase I hypersensitive sites DHS10a and DHS10b; Smith et al. 2000); (2) "Conservation Score," which depicts the conservation score calculated at each position using the multispecies sequence data (see text and Fig. 4); (3) "MCSs," which are regions exceeding an established conservation score threshold that together represent the top 5% most conserved sequence (Margulies et al. 2003); and (4) "Blastz Alignments," which depict the pair-wise alignments across the region between the indicated species' sequence and the human reference sequence, with the *y* axis indicating the percent sequence identity (50–100%) within the alignments (if present). The small horizontal gray bar depicts a region where there is currently no sequence for that species. Note that the vertical yellow lines, which are not present on the actual browser, highlight the positions of MCSs.

ments with the human sequence than that seen with the classic placental mammals, but notably more than that seen with the fish species.

The findings highlighted in Figure 3 illustrate how simple pair-wise alignments of vertebrate sequences, which reveal the chaotic-appearing patterns of sequence conservation, do not readily uncover the small fraction of the mammalian genome that is likely to serve a functional role. We have thus sought to use our multispecies sequence data to develop approaches for identifying those regions that are highly conserved across multiple species; we call such sequences MCSs—for Multispecies Conserved Sequences. In general, our methods involve analyzing the multispecies sequences as a group, in a fashion that allows the "signal" of actively conserved sequence to be discriminated from the "noise" inherent to pair-wise alignments.

One strategy we have developed for detecting MCSs (Margulies et al. 2003) is illustrated in Figure 4. In brief, this involves calculating a "conservation score" for each base across a sequenced region. Starting with human-referenced multisequence alignments (Fig. 4A) (e.g., derived from MultiPipMaker; Schwartz et al. 2003b), overlapping 25-base windows are used to calculate the cumulative binomial probability of encountering the observed number of base identities, given the neutral substitution rate calculated from fourfold degenerate sites (Margulies et al. 2003). The latter are the third positions of codons for which any base codes for the same amino acid; this is calculated separately for each targeted genomic region (e.g., shown in Fig. 4B for the 7q31.3 region). Thus, conserved sequences from more diverged species (e.g., platypus) make a larger relative contribution to the conservation score than do those from less diverged species (e.g., cat) (Fig. 4B). In calculating a final conservation score, and to normalize for any phylogenetic biases (e.g., the use of two carnivores but only one marsupial species), the data are then "phylogenetically averaged" by first averaging the conservation scores within each represented clade and then averaging across all the clades.

In light of the estimate that ~5% of the mammalian genome is actively conserved (Waterston et al. 2002; Roskin et al. 2003), we typically use a conservation score threshold such that the top 5% most conserved bases fall within MCSs (shaded in red in Fig. 4C). To monitor the effectiveness of the established thresholds at identifying the most conserved sequence, the distributions of coding bases (a "marker" for actively conserved sequence) versus noncoding bases (most of which are presumably not actively conserved) are examined. Representative distributions are schematically illustrated in Figure 4C. Note that the noncoding (purple) and coding (yellow) bases reflect ~99% and ~1% of the total sequence, respectively. Using a threshold that identifies the top 5% most conserved sequence typically results in the inclusion of >98% of coding bases within MCSs, along with a very small fraction of the noncoding bases.

Using the above methods, analyses of the multispecies sequences identified 1572 and 4520 MCSs in the 7q31.3 and 7q11.23 regions highlighted in Figure 2, respectively. Overall, the MCSs average 60–70 bp in size; however, those overlapping coding sequence tend to be larger than those in noncoding sequence (Margulies et al. 2003). By definition, the MCSs together comprise 5% of the total sequence, but there are notable differences in the composition of the MCSs relative to known annotated features in each region. For example, the percent of MCS bases residing within apparently noncoding sequence is higher for the 7q11.23 region than for the 7q31.3 region (Fig. 2). Remarkably, for both regions, this constitutes ≥70% of the MCS bases, providing a sizable amount of highly conserved sequence that might reflect functional noncoding elements. Finally, only a small fraction of MCS bases (3.6% for both regions) reside within ancestral repeats (ARs; relics of transposons inserted prior to the eutherian radiation and presumed not to be actively conserved); thus, our MCS-detection scheme detects very little sequence that is thought to be neutrally evolving.

More careful inspection reveals important insights about the locations of MCSs relative to known genomic features. This is nicely illustrated in the adapted view of the UCSC Genome Browser in Figure 3. Note the tracks (near the top) displaying the calculated conservation scores and corresponding positions of MCSs (which are further highlighted by vertical yellow lines) across the depicted region. As expected, an MCS overlaps an exon present in the RefSeq record for the *CFTR* gene; interestingly, another MCS overlaps a rarely expressed, alternately spliced *CFTR* exon (represented by GenBank records X73053 and S64699) (Melo et al. 1993; Will et al. 1993). Other MCSs overlap two previously detected DNase I hypersensitive sites (Smith et al. 2000), which are presumed to be associated with regions of transcription factor binding. The remaining MCSs overlap apparently noncoding sequences, for which no functional significance has yet been established; however, the highly conserved nature of these sequences makes them attractive candidates for further study. Finally, it is striking that many of the MCSs reside at the few genomic locations where there are alignments between the human sequence and marsupial (opossum and dunnart) or monotreme (platypus) sequences (Fig. 3); these findings illustrate why sequences from these species, which occupy unique positions on the evolutionary tree (Graves and Westerman 2002; Wakefield and Graves 2003), will likely be a valuable resource for identifying the most conserved sequences in the human genome.

## PERFORMANCE OF DIFFERENT SPECIES

The relative contribution of different species' sequences toward the identification of MCSs is of interest, in part because such information might help guide decisions about future genome-wide sequencing projects. We thus investigated the effectiveness of various species' sequences, both individually and in groups, at detecting MCSs.

For this analysis, we systematically recomputed the conservation scores using all possible subsets of species, and then assessed the ability of each subset to detect a reference set of MCSs (those detected using all species' se-

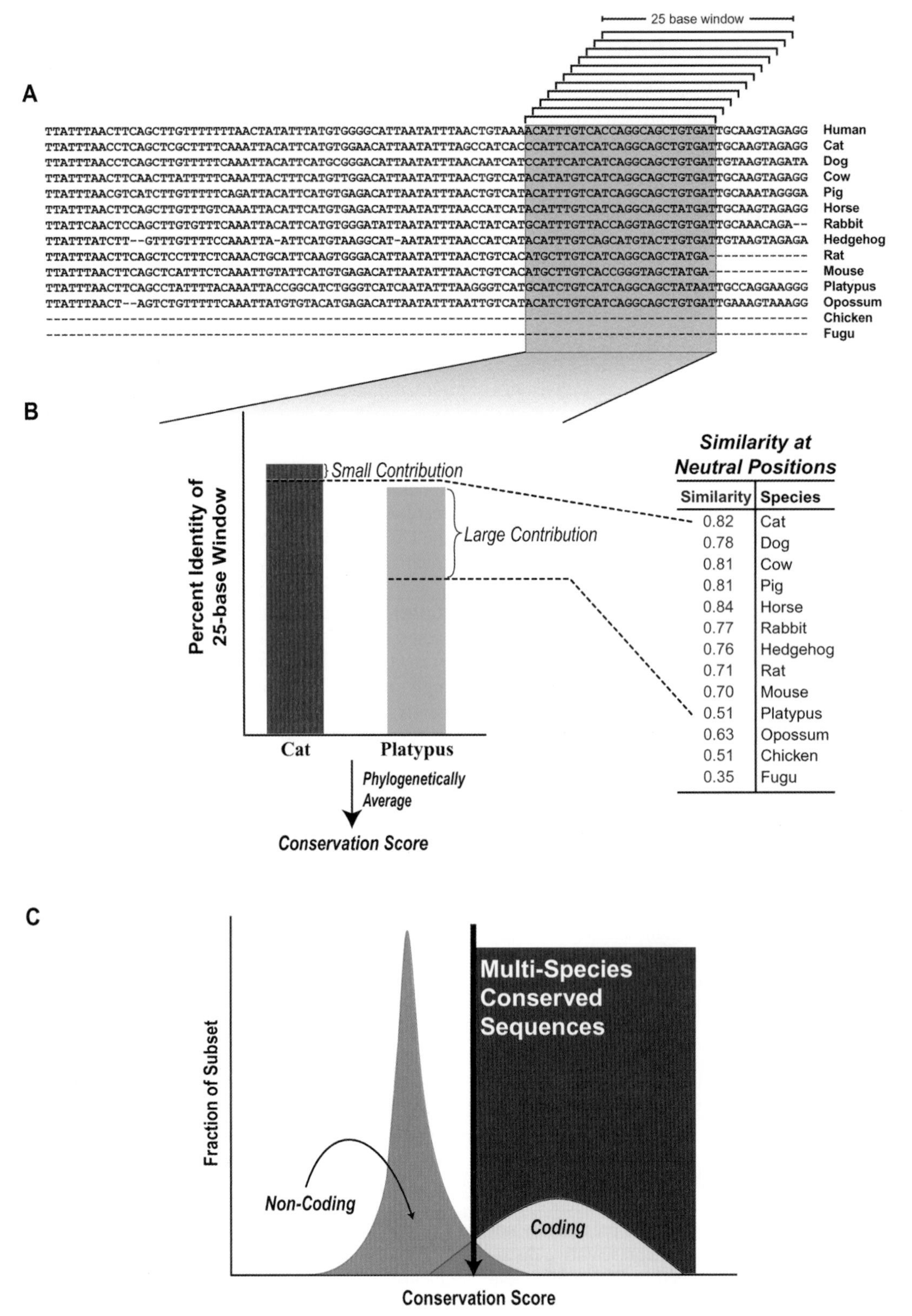


**Figure 4.** General strategy for detecting MCSs. (*A*) The amount of sequence conservation (with the human reference sequence) is calculated for each overlapping 25-base window of a multispecies sequence alignment. (*B*) The contribution of each species' sequence to the conservation score is weighted based on the neutral substitution rate between that species and human (indicated are the calculated similarities at neutral positions, with the neutral substitution rates being 1 minus these values); in this fashion, the ultimate contribution of an alignment to the resulting conservation score depends on the difference between the percent identity of that alignment and the corresponding neutral rate for that species (as illustrated for cat and platypus). (*C*) In general, coding regions (*yellow*) have higher conservation scores than noncoding regions (*purple*). A conservation score threshold (*vertical arrow*) is selected such that 5% of the total sequence resides within the detected MCSs (*red*); importantly, this results in virtually all coding sequence and a very small amount of noncoding sequence falling within MCSs. For additional details, see Margulies et al. (2003).

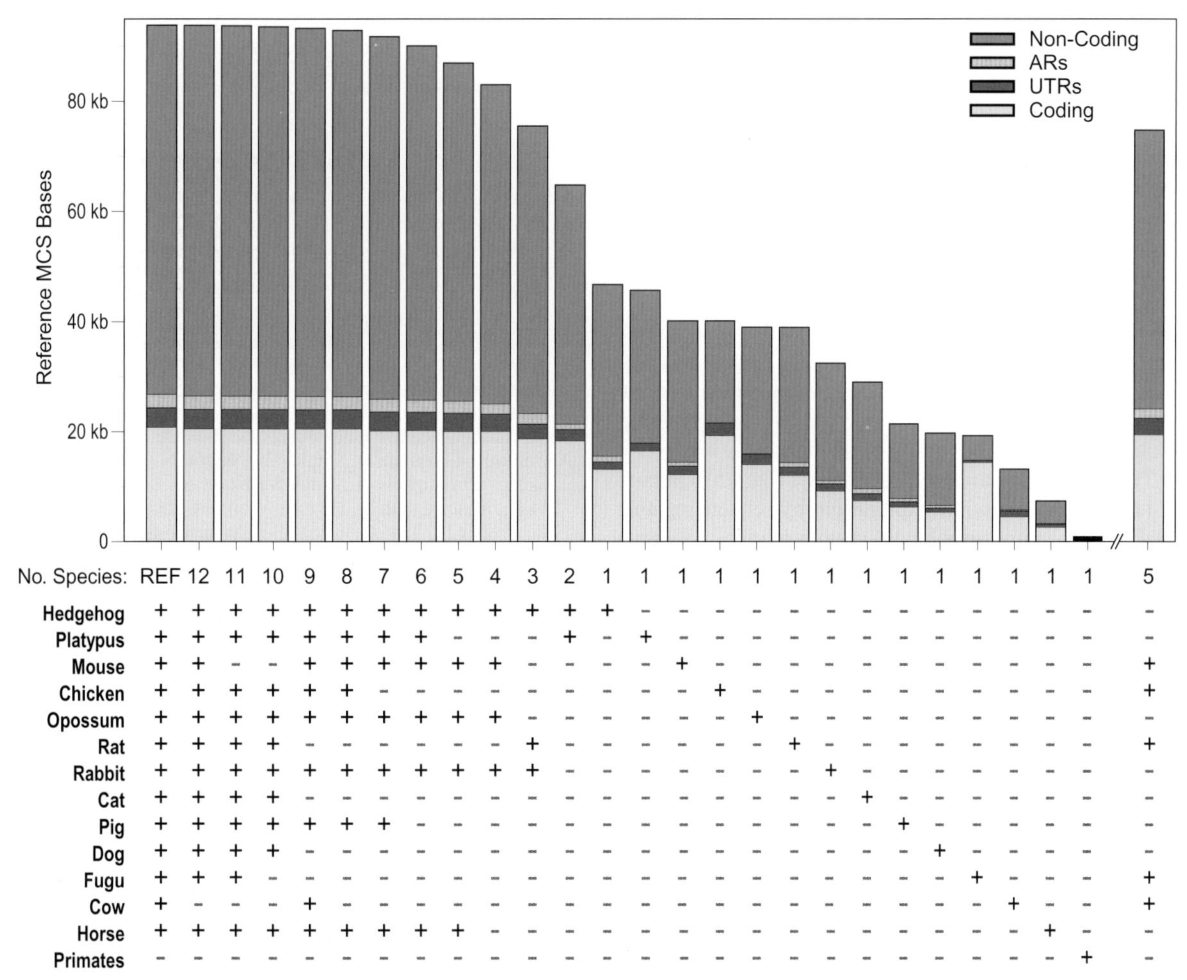


**Figure 5.** Performance of different combinations of species at detecting MCSs. Sequences from every combination of 13 nonhuman species were analyzed for their ability to detect the reference set of MCSs (far-left bar labeled "REF," which reflects those MCSs detected using all 13 of the indicated species' sequences). Results for the subset of each possible number of species (from 1 to 12, in addition to human) yielding the greatest overlap with the reference set of MCSs (at a 75% specificity) is also shown. The results for each individual species (including that of a typical primate) and the group of mouse, chicken, rat, fugu, and cow (for which whole-genome sequences are or will soon be available; see far-right bar) are also indicated. Note that the bars depict the overlap between the MCSs detected with the indicated species' sequences and the reference set of MCS bases, broken down for coding regions (*yellow*), UTRs (*blue*), ARs (*gray*), and noncoding sequence (*green*). For additional details, see Margulies et al. (2003).

quences). To ensure equivalent comparisons, in each case a conservation score threshold was selected such that 75% of the detected MCS bases overlapped the reference set of MCSs (i.e., resulted in a specificity of 75%; also see Margulies et al. 2003). In addition to examining each individual species, we also identified and compared the best-performing subsets of two species, three species, four species, and so forth, with the results highlighted in Figure 5.

A number of interesting findings emerge from these analyses. First, when used alone, some of the species (e.g., chicken, platypus, and fugu) are effective at detecting coding sequence (yellow portions of the bars in Fig. 5); indeed, only three or four species in combination are needed to detect virtually all of the coding MCS bases. In contrast, a larger group of species is needed for detecting virtually all of the noncoding MCS bases (green portions of the bars in Fig. 5). There are also notable differences in the types of sequences detected by individual species. For example, fugu does well at detecting coding MCS bases, but little else; hedgehog detects fewer coding MCS bases, but is the best single-species performer in terms of detecting noncoding MCS bases. In this regard, hedgehog and platypus appear to be the best-performing single species, with the rodents, chicken, and opossum representing the next tier in terms of single-species performance. Although some individual species can detect ~50% of the reference set of MCS bases, more than six species are required as a group to detect >96%; indeed, the additional species are particularly needed for detecting the noncoding MCS bases. Finally, note that MCS detection by the group of five species for which whole-genome sequences are (or will soon be) available (mouse, chicken, rat, fugu, and cow; see the far-right bar in Fig. 5) is roughly equivalent to that of the best-performing subset of three species (hedgehog, rat, and rabbit). Thus, the actual species present in a given subset greatly influence MCS detection.

## CONCLUSIONS

With the recent completion of the human genome sequence, attention turns to its interpretation. Important priorities now include identifying all functional elements in the human genome and acquiring a detailed understanding of its evolutionary origins. A powerful approach that greatly facilitates such endeavors involves comparing the sequences of the human and other extant vertebrate genomes—all of which reflect products of persistent evolutionary "experimentation" that has occurred over the past tens of millions of years. Indeed, cataloguing the genomic similarities and differences among closely and distantly related species will provide tremendous insight about common features of genome function as well as the molecular basis for the observed phenotypic diversity.

The NISC Comparative Sequencing Program is pursuing multispecies genome explorations by sequencing and analyzing the same orthologous regions in many different vertebrates. By limiting the genomic territory under study, data sets of unprecedented phylogenetic diversity are being generated. Above we describe one of our initial areas of comparative analyses—developing strategies for identifying regions that are highly conserved across multiple species (called MCSs). In short, our MCS-detection methods effectively distill the often extensive number of pair-wise sequence alignments to reveal discrete stretches of sequence that are conserved across multiple species (see Fig. 3). Importantly, many of the detected MCSs appear to contain functionally important sequences (Margulies et al. 2003; Thomas et al. 2003).

The detection of MCSs can help to establish prioritized lists of genomic regions that might harbor biologically important elements. Such sequences, especially those residing within noncoding regions, merit more careful scrutiny as well as follow-up experimental and computational examination. On a practical level, the detection of MCSs can be used as one metric for guiding the selection of genomes to sequence in the future, with the analyses highlighted here illustrating the analytical framework we have established for evaluating how different species' sequences contribute to MCS detection. Together, our studies, in conjunction with ongoing and future whole-genome sequencing efforts, should elevate the utility of comparative sequencing and make important contributions toward unraveling the functional and evolutionary complexities of the human genome.

## ACKNOWLEDGMENTS

This work was supported by funds provided by the National Human Genome Research Institute. We thank all participants of the NISC Comparative Sequencing Program for their contributions, in particular those performing BAC mapping (Jackie Idol, Valerie Maduro, Shih-Queen Lee-Lin, Matt Portnoy, Arjun Prasad, and Jim Thomas), BAC sequencing (Bob Blakesley, Alice Young, Gerry Bouffard, Jenny McDowell, Baishali Maskeri, Nancy Hansen, and Jeff Touchman), and sequence annotation (Pam Thomas and Laura Elnitski). We also thank the UCSC Genome Bioinformatics Group, in particular David Haussler and Mathieu Blanchette, as well as Webb Miller, and Phil Green for extensive collaborative interactions.

## REFERENCES

Adams M.D., Celniker S.E., Holt R.A., Evans C.A., Gocayne J.D., Amanatides P.G., Scherer S.E., Li P.W., Hoskins R.A., Galle R.F., George R.A., Lewis S.E., Richards S., Ashburner M., Henderson S.N., Sutton G.G., Wortman J.R., Yandell M.D., Zhang Q., Chen L.X., Brandon R.C., Rogers Y.H., Blazej R.G., Champe M., and Pfeiffer B.D., et al. 2000. The genome sequence of *Drosophila melanogaster*. *Science* **287:** 2185.

Alexandersson M., Cawley S., and Pachter L. 2003. SLAM: Cross-species gene finding and alignment with a generalized pair hidden Markov model. *Genome Res.* **13:** 496.

Aparicio S., Chapman J., Stupka E., Putnam N., Chia J.M., Dehal P., Christoffels A., Rash S., Hoon S., Smit A., Gelpke M.D., Roach J., Oh T., Ho I.Y., Wong M., Detter C., Verhoef F., Predki P., Tay A., Lucas S., Richardson P., Smith S.F., Clark M.S., Edwards Y.J., and Doggett N., et al. 2002. Whole-genome shotgun assembly and analysis of the genome of *Fugu rubripes*. *Science* **297:** 1301.

Batzoglou S., Pachter L., Mesirov J.P., Berger B., and Lander E.S. 2000. Human and mouse gene structure: Comparative analysis and application to exon prediction. *Genome Res.* **10:** 950.

Birren B., Mancino V., and Shizuya H. 1997. Bacterial artificial chromosomes. In *Genome analysis: A laboratory manual*, vol. 2: *Cloning systems* (ed. B. Birren et al.), p. 241. Cold Spring Harbor Laboratory Press, Cold Spring Harbor, New York.

Blanchette M. and Tompa M. 2002. Discovery of regulatory elements by a computational method for phylogenetic footprinting. *Genome Res.* **12:** 739.

Boffelli D., McAuliffe J., Ovcharenko D., Lewis K.D., Ovcharenko I., Pachter L., and Rubin E.M. 2003. Phylogenetic shadowing of primate sequences to find functional regions of the human genome. *Science* **299:** 1391.

Boguski M.S., Tolstoshev C.M., and Bassett D.E., Jr. 1994. Gene discovery in dbEST. *Science* **265:** 1993.

Burge C. and Karlin S. 1997. Prediction of complete gene structures in human genomic DNA. *J. Mol. Biol.* **268:** 78.

*C. elegans* Sequencing Consortium. 1998. Genome sequence of the nematode *C. elegans:* A platform for investigating biology. *Science* **282:** 2012.

Collins F.S., Green E.D., Guttmacher A.E., and Guyer M.S. 2003. A vision for the future of genomics research: A blueprint for the genomic era. *Nature* **422:** 835.

Duret L. and Bucher P. 1997. Searching for regulatory elements in human noncoding sequences. *Curr. Opin. Struct. Biol.* **7:** 399.

Flicek P., Keibler E., Hu P., Korf I., and Brent M.R. 2003. Leveraging the mouse genome for gene prediction in human: From whole-genome shotgun reads to a global synteny map. *Genome Res.* **13:** 46.

Graves J.A. and Westerman M. 2002. Marsupial genetics and genomics. *Trends Genet.* **18:** 517.

Green E.D. 2001. Strategies for the systematic sequencing of complex genomes. *Nat. Rev. Genet.* **2:** 573.

Hardison R.C. 2000. Conserved noncoding sequences are reliable guides to regulatory elements. *Trends Genet.* **16:** 369.

Hubbard T., Barker D., Birney E., Cameron G., Chen Y., Clark L., Cox T., Cuff J., Curwen V., Down T., Durbin R., Eyras E., Gilbert J., Hammond M., Huminiecki L., Kasprzyk A., Lehvaslaiho H., Lijnzaad P., Melsopp C., Mongin E., Pettett R., Pocock M., Potter S., Rust A., and Schmidt E., et al. 2002. The Ensembl genome database project. *Nucleic Acids Res.* **30:** 38.

Karolchik D., Baertsch R., Diekhans M., Furey T.S., Hinrichs A., Lu Y.T., Roskin K.M., Schwartz M., Sugnet C.W., Thomas D.J., Weber R.J., Haussler D., and Kent W.J. 2003.

The UCSC Genome Browser Database. *Nucleic Acids Res.* **31:** 51.

Kent W.J., Sugnet C.W., Furey T.S., Roskin K.M., Pringle T.H., Zahler A.M., and Haussler D. 2002. The human genome browser at UCSC. *Genome Res.* **12:** 996.

Korf I., Flicek P., Duan D., and Brent M.R. 2001. Integrating genomic homology into gene structure prediction. *Bioinformatics* (suppl. 1) **17:** S140.

Kulp D., Haussler D., Reese M.G., and Eeckman F.H. 1996. A generalized hidden Markov model for the recognition of human genes in DNA. *Proc. Int. Conf. Intell. Syst. Mol. Biol.* **4:** 134.

Lander E.S., Linton L.M., Birren B., Nusbaum C., Zody M.C., Baldwin J., Devon K., Dewar K., Doyle M., FitzHugh W., Funke R., Gage D., Harris K., Heaford A., Howland J., Kann L., Lehoczky J., LeVine R., McEwan P., McKernan K., Meldrim J., Mesirov J.P., Miranda C., Morris W., and Naylor J., et al. (International Human Genome Sequencing Consortium). 2001. Initial sequencing and analysis of the human genome. *Nature* **409:** 860.

Margulies E.H., Blanchette M., NISC Comparative Sequencing Program, Haussler D., and Green E.D. 2003. Identification and characterization of multi-species conserved sequences. *Genome Res.* **13:** 2507.

Melo C.A., Serra C., Stoyanova V., Aguzzoli C., Faraguna D., Tamanini A., Berton G., Cabrini G., and Baralle F.E. 1993. Alternative splicing of a previously unidentified CFTR exon introduces an in-frame stop codon 5′ of the R region. *FEBS Lett.* **329:** 159.

Pennacchio L.A. and Rubin E.M. 2001. Genomic strategies to identify mammalian regulatory sequences. *Nat. Rev. Genet.* **2:** 100.

Pennacchio L.A., Olivier M., Hubacek J.A., Cohen J.C., Cox D.R., Fruchart J.C., Krauss R.M., and Rubin E.M. 2001. An apolipoprotein influencing triglycerides in humans and mice revealed by comparative sequencing. *Science* **294:** 169.

Rogic S., Mackworth A.K., and Ouellette F.B. 2001. Evaluation of gene-finding programs on mammalian sequences. *Genome Res.* **11:** 817.

Roskin K., Diekhans M., and Haussler D. 2003. Scoring two-species local alignments to try to statistically separate neutrally evolving from selected DNA segments. In *The 7th Annual International Conference on Research in Computational Molecular Biology, Berlin* **RECOMB 2003:** 257.

Schwartz S., Kent W.J., Smit A., Zhang Z., Baertsch R., Hardison R.C., Haussler D., and Miller W. 2003a. Human-mouse alignments with BLASTZ. *Genome Res.* **13:** 103.

Schwartz S., Elnitski L., Li M., Weirauch M., Riemer C., Smit A., Green E.D., Hardison R.C., and Miller W. (NISC Comparative Sequencing Program). 2003b. MultiPipMaker and supporting tools: Alignments and analysis of multiple genomic DNA sequences. *Nucleic Acids Res.* **31:** 3518.

Shizuya H., Birren B., Kim U.J., Mancino V., Slepak T., Tachiiri Y., and Simon M. 1992. Cloning and stable maintenance of 300-kilobase-pair fragments of human DNA in *Escherichia coli* using an F-factor-based vector. *Proc. Natl. Acad. Sci.* **89:** 8794.

Smith D.J., Nuthall H.N., Majetti M.E., and Harris A. 2000. Multiple potential intragenic regulatory elements in the CFTR gene. *Genomics* **64:** 90.

Solovyev V.V. 2001. Statistical approaches in eukaryotic gene prediction. In *Handbook of statistical genetics* (ed. D.J. Balding et al.), p. 83. John Wiley & Sons, New York.

Strausberg R.L., Feingold E.A., Grouse L.H., Derge J.G., Klausner R.D., Collins F.S., Wagner L., Shenmen C.M., Schuler G.D., Altschul S.F., Zeeberg B., Buetow K.H., Schaefer C.F., Bhat N.K., Hopkins R.F., Jordan H., Moore T., Max S.I., Wang J., Hsieh F., Diatchenko L., Marusina K., Farmer A.A., Rubin G.M., and Hong L., et al. 2002. Generation and initial analysis of more than 15,000 full-length human and mouse cDNA sequences. *Proc. Natl. Acad. Sci.* **99:** 16899.

Tagle D.A., Koop B.F., Goodman M., Slightom J.L., Hess D.L., and Jones R.T. 1988. Embryonic ε and γ globin genes of a prosimian primate (*Galago crassicaudatus*). Nucleotide and amino acid sequences, developmental regulation and phylogenetic footprints. *J. Mol. Biol.* **203:** 439.

Thomas J.W., Prasad A.B., Summers T.J., Lee-Lin S.Q., Maduro V.V., Idol J.R., Ryan J.F., Thomas P.J., McDowell J.C., and Green E.D. 2002. Parallel construction of orthologous sequence-ready clone contig maps in multiple species. *Genome Res.* **12:** 1277.

Thomas J.W., Touchman J.W., Blakesley R.W., Bouffard G.G., Beckstrom-Sternberg S.M., Margulies E.H., Blanchette M., Siepel A.C., Thomas P.J., McDowell J.C., Maskeri B., Hansen N.F., Schwartz M.S., Weber R.J., Kent W.J., Karolchik D., Bruen T.C., Bevan R., Cutler D.J., Schwartz S., Elnitski L., Idol J.R., Prasad A.B., Lee-Lin S.Q., Maduro V.V., and Summers T.J., et al. 2003. Comparative analyses of multi-species sequences from targeted genomic regions. *Nature* **424:** 788.

Ureta-Vidal A., Ettwiller L., and Birney E. 2003. Comparative genomics: Genome-wide analysis in metazoan eukaryotes. *Nat. Rev. Genet.* **4:** 251.

Wakefield M.J. and Graves J.A. 2003. The kangaroo genome. Leaps and bounds in comparative genomics. *EMBO Rep.* **4:** 143.

Waterston R.H., Lindblad-Toh K., Birney E., Rogers J., Abril J.F., Agarwal P., Agarwala R., Ainscough R., Alexandersson M., An P., Antonarakis S.E., Attwood J., Baertsch R., Bailey J., Barlow K., Beck S., Berry E., Birren B., Bloom T., Bork P., Botcherby M., Bray N., Brent M.R., Brown D.G., and Brown S.D., et al. (Mouse Genome Sequencing Consortium). 2002. Initial sequencing and comparative analysis of the mouse genome. *Nature* **420:** 520.

Will K., Stuhrmann M., Dean M., and Schmidtke J. 1993. Alternative splicing in the first nucleotide binding fold of CFTR. *Hum. Mol. Genet.* **2:** 231.

Wilson R.K. and Mardis E.R. 1997. Analyzing DNA. In *Genome analysis: A laboratory manual*, vol. 1: *Shotgun sequencing* (ed. B. Birren et al.), p. 397. Cold Spring Harbor Laboratory Press, Cold Spring Harbor, New York.

# Comparative Analysis of Human Chromosome 22q11.1-q12.3 with Syntenic Regions in the Chimpanzee, Baboon, Bovine, Mouse, Pufferfish, and Zebrafish Genomes

B.A. Roe,* C. Lau,* S. Oommen,* J. Li,* A. Hua,* H.S. Lai,* S. Kenton,* J. White,* and H. Wang[†]

*Department of Chemistry and Biochemistry and [†]Department of Zoology, University of Oklahoma, Norman, Oklahoma 73019

It has been over three decades since the initial publication of the methods for DNA sequencing that were developed independently by Sanger, Nicklen, and Coulson at the Medical Research Council Laboratory of Molecular Biology in Cambridge, England (Sanger et al. 1977) and by Maxam and Gilbert (1977) in the biology department at Harvard University in Cambridge, Massachusetts. Since then, the Sanger dideoxynucleotide sequencing method has become the method of choice for large-scale DNA sequencing, has been automated by Lee Hood's group (Smith et al. 1986) and others (Ansorge et al. 1987; Prober et al. 1987; Brumbaugh et al. 1988; Kambara and Takahahi 1993), and commercially developed into a stable platform for large-scale, highly accurate, and rapid sequence data collection with robust sequencing chemistry. Today, because of these improvements, DNA sequencing almost is taken for granted. In this volume, we celebrate the completion of the entire human genomic DNA sequence, almost 50 years after the initial publication of the double-helix structure of DNA and its implications (Watson and Crick 1953a,b), and we look forward to the completion of the genomes of other organisms that either presently are in working draft form (Waterston et al. 2002) or whose sequence is being contemplated (J.E. Collins et al. 2003).

Although there is an ever-growing compilation of DNA sequence data, fewer than half of the predicted human genes, and ironically, a similar percentage of genes from bacteria and other model organisms, have been assigned well-defined functions (Lander et al. 2001; Shoemaker et al. 2001; Zhang 2002; Waterston et al. 2002; F.S. Collins et al. 2003). The challenge now is to develop methods and approaches that will result in detailed understanding of the regions contained within the human genome, whether they be exons coding for functional proteins, regions encoding the genes for stable RNAs, regulatory regions, or regions involved in regulating gene expression and other genome-specific functions. One approach that has emerged as a powerful tool for genomic functional annotation is to compare the sequences of orthologous regions that are syntenic to segments of the human genome (Jan et al. 1999; Lund et al. 2000; Footz et al. 2001; Ji et al. 2001; Deschamps et al. 2003; Thomas et al. 2003; Ureta-Vidal et al. 2003). Then, based on the hypothesis that if a region has been conserved over evolutionary time its structure contains important features required to maintain biological function, it is possible to illuminate these genomic features; e.g., genes, regulatory regions including antisense RNAs, and conserved evolutionary breakpoints which can be related directly to conserved biological functions that occur in evolutionarily distant organisms (Thomas et al. 2003; Ureta-Vidal et al. 2003). Once discovered, the function of these conserved predicted features must be confirmed, through gene expression studies to determine when during development and in what cells the gene is transcribed, to infer when and where its biological function(s) is performed. This knowledge is a natural prelude to the eventual detailed analysis of both the biochemical and biophysical properties of the genomic region and its final gene product, be it a stable RNA or functional protein.

Our laboratory has played a role in the development and implementation of DNA sequencing methods over the past 30 years (Chissoe et al. 1991; Bodenteich et al. 1994). This stems from my (B.A.R.) initial introduction to DNA sequencing as a member of the human mitochondrial genome sequencing project during my first sabbatical in Fred Sanger's laboratory (Anderson et al. 1981, 1982), through sequencing the genomic regions involved in the Philadelphia chromosomal translocation (Chissoe et al. 1995), and culminating in sequencing a significant portion of the upper half of human Chromosome 22 (Dunham et al. 1999) as a member of the international consortium that reported the sequence of this first completed human chromosome. More recently, we have focused our efforts on comparative genomics, prompted by the observation that although 598 protein-coding genes and 228 pseudogenes have been predicted on human Chromosome 22, only about half of the protein-coding genes have both a known function and expression profiles, one-quarter of the predicted genes have an unknown function but are expressed in one or more tissues and developmental stages, and one-quarter have neither a known function nor a known expression profile (Dunham et al. 1999; Shoemaker et al. 2001; J.E. Collins et al. 2003). These studies are revealing the identity of evolutionarily conserved structures in the upper third of human Chromosome 22 genes corresponding to both coding and noncoding regions. On the basis of these observations, we have begun examining the gene expression profiles of these conserved coding regions by whole-mount in situ hybridization in the zebrafish model system. Thus, through these two approaches, we will have taken the first steps to determine the genomic features of the full repertoire of higher eukaryote genes, their expression profiles, and eventually the function of their gene products.

## COMPARATIVE GENOMIC SEQUENCING

Much of our comparative genomic sequencing has been concentrated on regions of the chimpanzee, baboon, cow, and mouse genomes orthologous to the upper half of human Chromosome 22. The regions of sequence orthologous vertebrate clones and the chimp Chromosome 23 whole-genome sequence (WGS) reads that have homology with human Chromosome 22 and are aligned with the orthologous regions of human Chromosome 22 are shown below in Figure 1. At present, our combined assembly of 208,000 WGS reads and our BAC-based sequences results in a 33,077,854-bp assembly of chimpanzee Chromosome 23 with ~1,305 gaps and over 98% identity to human Chromosome 22, while similar regions of the baboon are ~92% identical.

This observation is confirmed by the percentage identity plot (PIP) (Schwartz et al. 2000, 2003) shown in Figure 2, where it is also shown that human Chromosome 22 contains retroviral and alu sequences that are not present in the orthologous regions of the baboon and chimpanzee. The completed sequences of these two primate chromosomes will allow us to discover additional differences that will further illuminate human-specific and chimp-specific genetic features that likely will include additional insertions, deletions, and even possible gene duplications.

These results are in agreement with those reported in the excellent comparative sequencing study from Eric Green's laboratory (Thomas et al. 2003) and a recent publication from the laboratory of Sakaki and Okada (Ohshima et al. 2003). However, at present, only relatively small regions from selected species have been compared. Thus, additional comparative sequencing from numerous species is needed to obtain the full picture of these observations on the broader genomic scale.

## WHOLE-MOUNT IN SITU HYBRIDIZATION STUDIES IN THE ZEBRAFISH SYSTEM

Because several hundred genes predicted on human Chromosome 22 have an unknown function but are expressed in one or more tissues and developmental stages, or have neither a known function nor known expression profile, it was critical to develop rapid, high-throughput methods to investigate the expression profiles of the zebrafish orthologs of these classes of human genes. To facilitate the automation of these analyses, we recently have developed optimized conditions for 96-well microtiter-based whole-mount in situ hybridization in zebrafish using the digoxigenin-labeled (DIG-labeled), small (~100 base), ssDNA unidirectional PCR probes (Kitazawa et al. 1999; Knuchel et al. 2000) produced from exon-specific PCR products similar to those used for high-throughput in situ hybridization in *Drosophila* (Tautz and Pfeifle 1989; Patel and Goodman 1992) and *Caenorhabditis elegans* (Seydoux and Fire 1995). In the experiment shown in Figure 3, a cDNA clone corresponding to dJ508I15.C22.5, a human Chromosome-22-

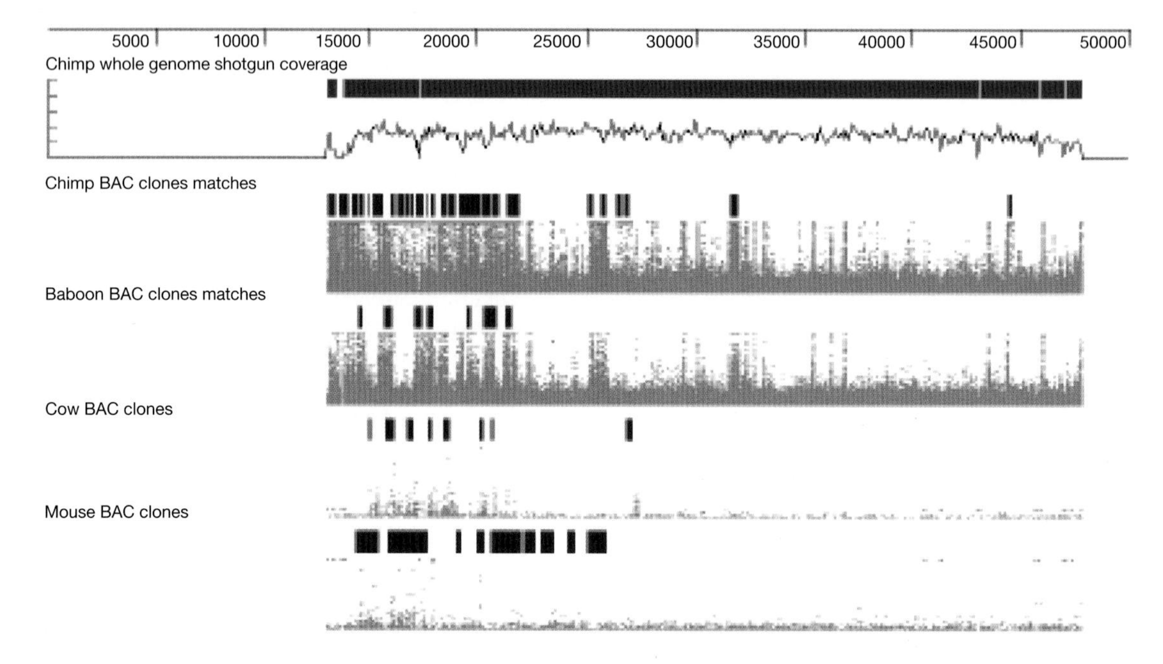


**Figure 1.** Schematic representation of the chimpanzee WGS and our sequenced chimpanzee, baboon, cow, and mouse BACs aligned with the orthologous regions in human Chromosome 22.

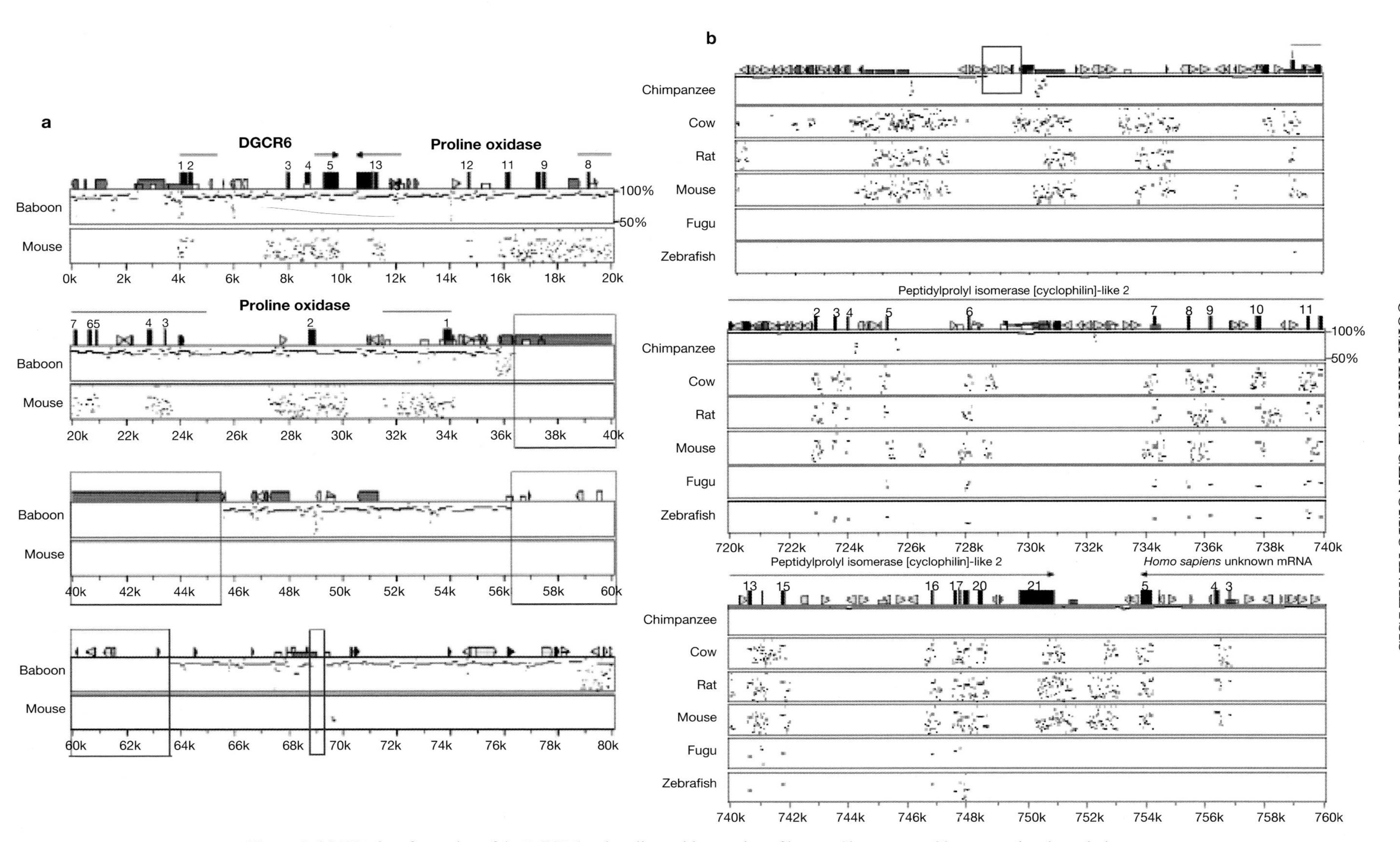


**Figure 2.** (*a*) PIP plot of a portion of the DGCR6 and proline oxidase region of human Chromosome 22 compared to the orthologous regions of baboon and mouse showing human-specific retroviral insertions (*red boxes*) that occurred in the primate lineage. (*b*) PIP plot of a portion of the peptidylprolyl isomerase (cyclophilin)-like 2 region of human Chromosome 22 compared to the orthologous regions of chimpanzee, cow, rat, mouse, fugu, and zebrafish showing a human-specific alu sequence (*red box*) that occurs only in the human lineage.

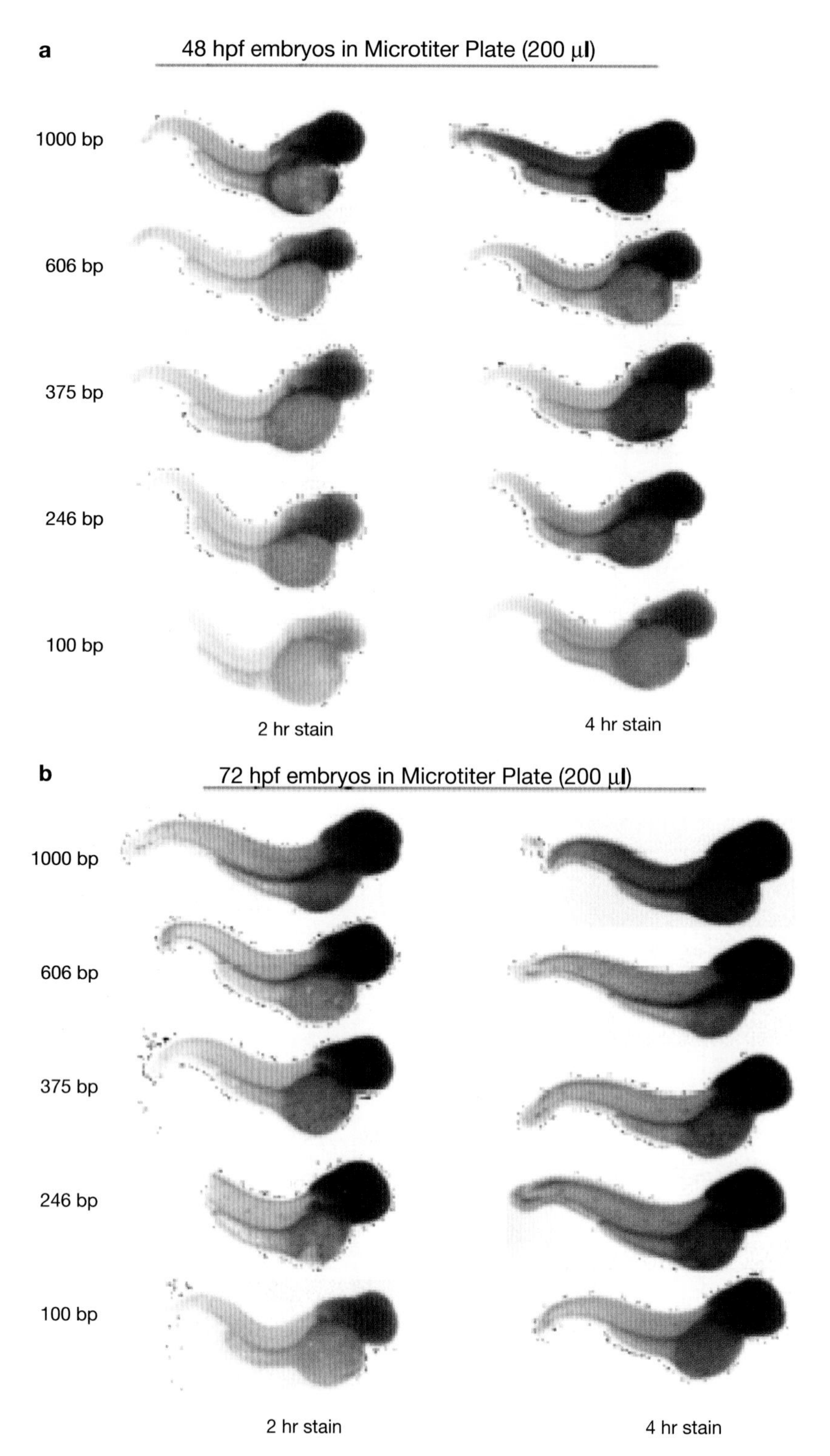


**Figure 3.** A 96-well microtiter-based whole-mount in situ hybridization in zebrafish using the DIG-labeled, small (~100 base), ssDNA asymmetric PCR probes (32,33) produced from an exon-specific PCR product corresponding to a brain-expressed at 48 (*a*) and 72 hours (*b*) postfertilization (hpf).

predicted gene with a previously unknown expression profile, was PCR-amplified using primers spaced between 1000 and 100 nucleotides apart, which subsequently were used as templates for unidirectional amplification in the presence of either the above sense or antisense primer to produce the single-stranded DNA probes. After the single-stranded probes were labeled using the Roche DIG-labeled DNA labeling kit (Roche Molecular Biochemicals, cat. #1277065), they were incubated for 18 hours at 48°C in 200 μl of hybridization buffer (hybridization buffer contained 50% formamide, 5x SSc [standard saline citrate], 50 μg/ml heparin, 500 μg/ml *E. coli* genomic DNA, 0.2% Tween 20, and 10 mM citric acid, pH 6.0, prepared as described at http://zfin.org/zf_info/zfbook/chapt9/9.82.html) with 48 and 72 hours postfertilization (hpf) zebrafish embryos in a 96-well microtiter plate (VWR Scientific, cat. #62402-933), stained for 4 or 8 hours to develop the characteristic dark color, washed, and photographed by a CCD-camera-equipped Leica MZFLIII fluorescence stereo-microscope. These results clearly indicate that single-stranded DNA probes ranging in size from 1 kb to 100 bases can be used for successful in situ hybridizations to zebrafish embryos in a 96-well format.

This robust protocol for zebrafish embryo whole-mount in situ hybridization using single-stranded DNA probes has allowed us to multiplex our experiments efficiently and thus begin to investigate several additional predicted human Chromosome 22 genes based on this exon-specific PCR, unidirectional amplification approach.

As another example of our optimized protocols, we investigated the expression of a human Chromosome 22 gene which has an unknown function, but whose expression has been observed through EST studies. This predicted gene corresponds to an EST, KIAA0819 (Nagase et al. 1998), and the results of our analysis are shown in Figure 4. Hybridization of a single-stranded unidirectional DNA probe (ssDNA) corresponding to exon 1 in one copy of the dual-copy zebrafish ortholog of mouse KIAA0819, as revealed by the PIP, is shown in Figure 4a. The hybridization results presented in Figure 4b indicate that only the antisense probe gives a positive hybridization signal to the otic vesicle at 24, 48, and 120 hpf.

In contrast to the above results, in the two experiments described below we have observed hybridization not only with the antisense probes corresponding to the respective mRNAs, but also with the sense probes corresponding to the antisense RNAs for two genes in the zebrafish system that correspond to two predicted human Chromosome 22 genes AP000553.6 and AP000354.2, as shown in Figures 5 and 6, respectively.

### Human AP000553.6

Ensembl GeneID ENSG00000161179, located on human Chromosome 22 from 20306935 to 20308868 bp, is a novel predicted gene that has no known function or expression data. This unknown gene contains five predicted exons in humans and its orthologs in mouse, rat, fugu, and zebrafish that were found by a reciprocal blast search. The PIP shown in Figure 5a indicates that this gene has highly conserved exonic sequences in mammals but shows a lower level of sequence conservation in fish, where several exons are less than 50% conserved.

To study the expression profile in zebrafish, exon-specific primers were selected to PCR-amplify the exon region from zebrafish genomic DNA. For the expression data shown in Figure 5b–d, the 337-bp 3′-most exon (exon 5, Ensembl ExonID ENSDARE0000013070) was amplified using a sense primer 5′TGAGCACCATGGG TAAGAAC3′ beginning at the 24th base of the exon and an antisense primer 5′GGATCCCTTGTTGCCTG TAA3′ beginning at the 317th bp of the exon. These primers generated a 293-bp PCR product that then was used to generate ssDNA sense and antisense probes in the presence of the DIG-labeled dNTP mix.

As shown in Figure 5b, no background hybridization was observed in the absence of probes when the antisense and sense probes shown in Figure 5c and d were incubated in parallel. As indicated in Figure 5c, the antisense hybridization probe reveals the presence of hybridizing RNA at 12 hpf in the mesoderm layer, at 24 hpf in notochord, at 48 hpf in both notochord and otic vesicle, and at 72 hpf in the otic vesicle and liver. In contrast, Figure 5d shows that the ssDNA sense probe hybridizes to RNA in the notochord at 24 hpf through 72 hpf, indicating the likely presence of an antisense regulatory RNA for this gene product expressed in the notochord that may play a role in suppressing the production of the AP00553.6 protein at the translational level.

### Human AP000354.2

Ensembl GeneID ENSG00000100014, located on Chromosome 22 at positions 23041871–23137773 bp, is a novel predicted gene that has no known function or expression data. This unknown gene contains 13 predicted exons, and a reciprocal blast search revealed orthologs in mouse, rat, fugu, and zebrafish. The PIP shown in Figure 6a reveals the regions of sequence similarity that are 50% or greater, and, as described above for the AP000553.6 gene, the exons of this AP000354.2 gene also are highly conserved in mammals but less conserved in fugu and zebrafish. Interestingly, of the 12 exons in mammals, only 5 are conserved recognizably in zebrafish, with 4 being more than 50% conserved.

To investigate the expression profile of the AP-000354.2 gene, PCR primers were picked to amplify the second, 331-bp zebrafish orthologous exon. Here the sense primer, 5′CAAACGGACAACATCCATGA3′, was 11 bp into the exon, and the antisense primer was 5’CCTCATGTCCCTCAGCTCAC3′. These two primers generated a 300-bp PCR product that served as a template for the unidirectional PCR to generate sense and antisense probes.

As shown in Figure 6b, although there is not much difference between the antisense and sense probe hybridization in the 12 hpf embryos, in the 24 hpf embryos, hybridization in the telencephalon portion of the forebrain, the midbrain, and the hindbrain is not visible with the sense probe, whereas the notochord and myotomes show strong hybridization with both antisense and sense probes. In the 48 hpf embryos, the myotomes and telencephalon portion of the forebrain seem to hybridize more strongly with the sense probe. In the 72 hpf embryos, the telencephalon portion of forebrain is more intensely hybridized with the sense probe, whereas the lateral line ganglia show similar hybridization to both the antisense and sense probes. Other features, including the disappearance of staining in notochord and myotomes, are similar with both probes. This complex expression pattern, observed for both antisense and sense probes, indicates that the AP000354.2 gene is expressed in multiple

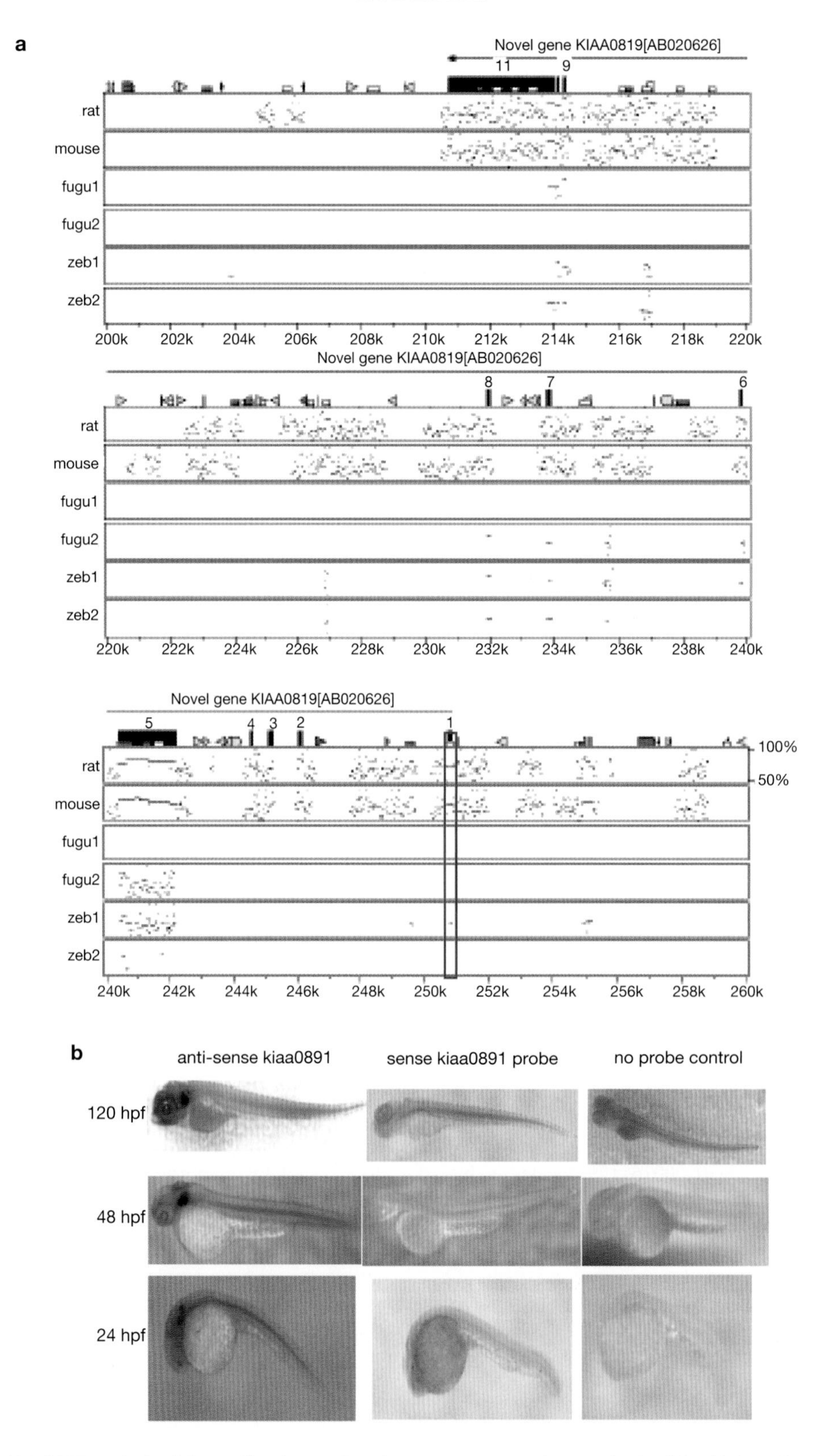


**Figure 4.** (*a*) MultiPIP analysis of the predicted genes from human, rat, mouse, fugu, and zebrafish with homology to cDNA probe KIAA0819. The unique unidirectional 708-base ssDNA probes from exon 1 of zebrafish 1 were DIG-labeled after PCR amplification of the 708-bp region of zebrafish genomic DNA indicated by the red rectangle. This probe is only a 74-base region with a percent identity of 79.7 out of 708 bases in zeb2 with an overall identity of less than 10%. (*b*) Whole-mount in situ hybridization of the ssDNA antisense, ssDNA sense, and no-probe control described in Fig. 7 to 24 hpf, 48 hpf, and 120 hpf zebrafish embryos indicating hybridization to the otic placode region only with the DIG-labeled antisense probe.

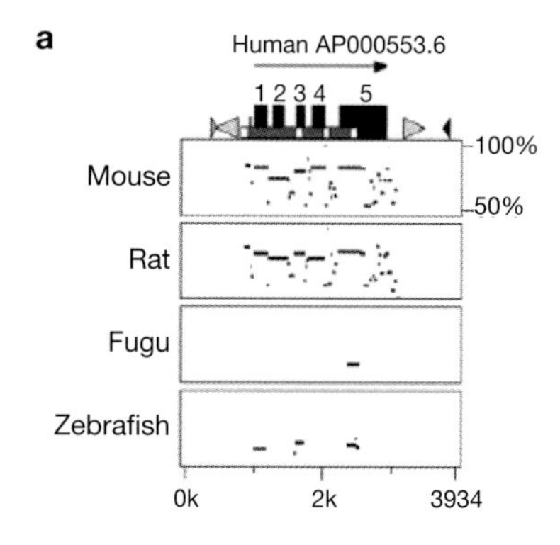


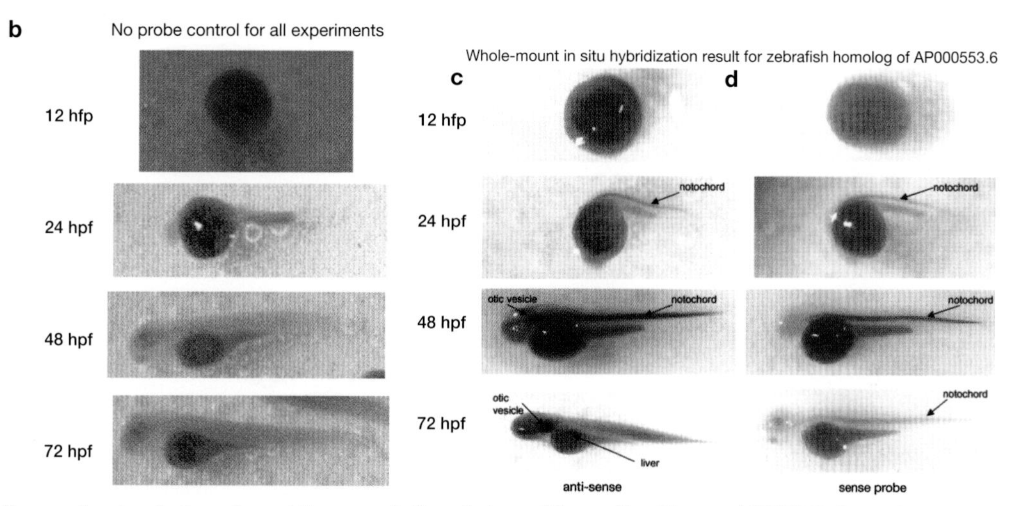


**Figure 5.** (*a*) PIP diagram showing the homology at the genomic for orthologs of the predicted human AP000553.6 gene from mouse, rat, fugu, and zebrafish. Whole-mount in situ hybridization for probes corresponding to zebrafish ortholog of the AP000553.6 gene (*b*) no probe controls, (*c*) ssDNA anti-sense probe, and (*d*) ssDNA sense probe to 12, 24, 48, and 72 hpf zebrafish embryos.

tissues in the developing embryo. If indeed the sense probe hybridizations indicate binding to siRNAs or other small antisense RNAs, the regulation of this gene is not only intriguing, but quite complex.

Finally, we recently obtained a zebrafish mutant from Dr. Nancy Hopkins at MIT that was created during her large-scale, insertional mutagenesis screen (Golling et al. 2002). This mutant has a proviral insertion in intron 1 of a phf5a-like Zn-finger gene (Trappe et al. 2002a,b), the human ortholog of which is predicted on human Chromosome 22 (AL008582.11.1.98758) at position 40,098,761 through 40,107,730. Figure 7 shows the results of our whole-mount in situ hybridization experiments using a single-stranded 280-base RNA probe that was produced from pooled 61 hpf and 72 hpf zebrafish mRNA after RT-PCR first with primers 43F1-GTTCTCTGCGCTCACTCTTT and 418R1-ACTCCTTCCGTTTCTTTCTT corresponding to the 5′ and 3′ UTR regions of the full-length mRNA, and then with a nested primer pair 65F2-TAATG GCAAAACATCACCCA and 358R2-GATCAGTCTTT GAGCTGCCC corresponding to regions of exon 1 and exon 4 of this phf5a-like gene, respectively. The final PCR product was cloned into the *Sma*I site of pUC-18 (Yanisch-Perron et al. 1985), excised with *Eco*RI and *Bam*HI, and then subcloned into the *Eco*RI and *Bam*HI sites of pBlue-Script SK- (Alting-Mees and Short 1989). The insert was sequenced to confirm its sequence, and subsequent in vitro transcription produced 280-base DIG-labeled RNA probes.

In the results presented in Figure 7, it can be clearly seen that since the inserted provirus disrupts intron 1, the expression of the phf5a-like gene is reduced significantly at 96, 120, and 168 hpf. Although both the mutant and wild-type zebrafish show phf5a-like gene expression in the brain past 48 hpf, the mutant does not show the characteristic downward tip of the tail, although it does have a slightly stunted growth and abnormally shaped yolk sac. As additional mutants become available, these will be targets for additional comparative in situ hybridization experiments.

## CONCLUSIONS

Through the above experiments, we clearly have demonstrated that combining highly accurate, high-throughput comparative genomic sequencing and subsequent computer-based analysis with a high-throughput approach to whole-mount in situ hybridization studies in

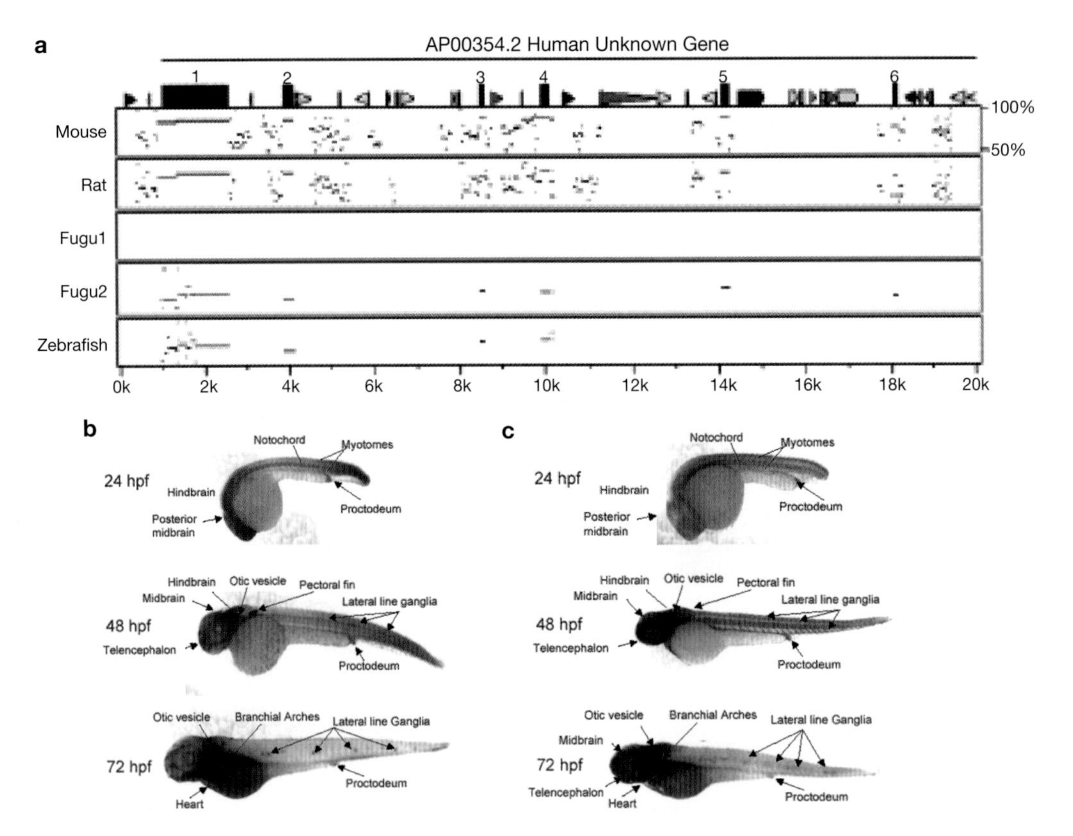


**Figure 6.** (*a*) PIP diagram showing the homology at the genomic level for orthologs of the predicted human AP000354.2 gene from mouse, rat, fugu, and zebrafish. (*b*) Detailed view of hybridization of AP000354.2 ssDNA antisense probe to 24 hpf embryo showing hybridization to the forebrain, posterior midbrain, hindbrain, notochord, pectoral fin, myotomes, tail bud, and the proctodeum, to 48 hpf zebrafish embryo showing hybridization to the telencephalon portion of the forebrain, the midbrain, hindbrain, otic vesicle, pectoral fin, lateral line ganglia, and proctodeum, and to a 72 hpf zebrafish embryo to the heart, the otic vesicle, branchial arches, lateral line ganglia, and the proctodeum. (*c*) Detailed view of hybridization of the AP000354.2 ssDNA sense probe to 24, 48, and 72 hpf zebrafish embryos showing the expression profile of antisense RNAs.

zebrafish embryos provides a powerful tool to investigate genes for which no expression evidence had yet been obtained. We anticipate that in the future these studies will uncover the expression profiles of the remaining predicted but heretofore undetected several hundred genes predicted on human Chromosome 22 and the many thousands of similar genes throughout the human genome. These results will serve as a prelude to future experiments aimed at determining the function(s) of these genes on phenotype (Nasevicius and Ekker 2000; Draper et al. 2001; Ekker and Larson 2001) and, where observed, their antisense regulatory RNAs.

18 hpf 24 hpf 48 hpf 55 hpf
mut mut mut
wt wt wt
72 hpf 96 hpf 120 hpf 168hrs

**Figure 7.** Whole-mount in situ hybridization of a DIG-labeled RNA antisense probe to the phf5a-like gene in zebrafish that corresponds to a predicted, but unknown, gene from human Chromosome 22.

## ACKNOWLEDGMENTS

We acknowledge the members of the sequencing teams at our Advanced Center for Genome Technology for their outstanding contributions to our comparative genomic sequencing studies, with a special note of thanks to Stephane Deschamps, Shaoping Lin, Fu Ying, Trang Do, Angela Prescott, Limei Yang, Ziyun Yao, and Sulan Qi. This work was supported by a grant from the National Human Genome Research Institute of the National Institutes of Health.

## REFERENCES

Alting-Mees M.A. and Short J.M. 1989. pBluescript II: Gene mapping vectors. *Nucleic Acids Res.* **17:** 9494.

Anderson S., Bankier A.T., Barrell B.G., de Bruijn M.H.L., Coulson A.R., Drouin J., Eperon I.C., Nierlich D.P., Roe B.A., Sanger F., Schreier P.H., Smith A.J.H., Staden R., and Young I.G. 1981. Sequence and organization of the human mitochondrial genome. *Nature* **290:** 457.

———. 1982. Comparison of the human and bovine mitochondrial genomes. In *Mitochondrial genes* (ed. P. Slonimski et al.), p. 5. Cold Spring Harbor Laboratory, Cold Spring Harbor, New York.

Ansorge W., Sproat B., Stegemann J., Schwager C., and Zenke M. 1987. Automated DNA sequencing: Ultrasensitive detection of fluorescent bands during electrophoresis. *Nucleic Acids Res.* **15:** 4593.

Bodenteich A., Chissoe S., Wang Y.F., and Roe B.A. 1994. Shotgun cloning as the strategy of choice to generate templates for high throughput dideoxynucleotide sequencing. In *Automated DNA sequencing and analysis techniques* (ed. M.D. Adams et al.), p. 42. Academic Press, London, United Kingdom.

Brumbaugh J.A., Middendorf L.R., Grone D.L., and Ruth J.L. 1988. Continuous, on-line DNA sequencing using oligodeoxynucleotide primers with multiple fluorophores. *Proc. Natl. Acad. Sci.* **85:** 5610.

Chissoe S.L., Wang Y.F., Clifton S.W., Ma N., Sun H.J., Lobsinger J.S., Kenton S.M., White J.D., and Roe B.A. 1991. Strategies for rapid and accurate DNA sequencing. *Methods* (Companion to *Methods Enzymol.*) **3:** 55.

Chissoe S.L., Bodenteich A., Wang Y.F., Wang Y.P., Burian D., Clifton S.W., Crabtree J., Freeman A., Iyer K., Jian L., Ma Y., McLaury H.J., Pan H.Q., Sarhan O.H., Toth S., Wang Z., Zhang G., Heisterkamp N., Groffen J., and Roe B.A. 1995. Sequence and analysis of the human c-abl gene, the bcr gene, and regions involved in the Philadelphia chromosomal translocation. *Genomics* **27:** 67.

Collins J.E., Goward M.E., Cole C.G., Smink L.J., Huckle E.J., Knowles S., Bye J.M., Beare D.M., and Dunham I. 2003. Reevaluating human gene annotation: A second-generation analysis of chromosome 22. *Genome Res.* **13:** 27.

Collins F.S., Green E.D., Guttmacher A.E., and Guyer M.S. 2003. A vision for the future of genomics research. *Nature* **422:** 835.

Deschamps S., Meyer J., Chatterjee G., Wang H., Lengyel P., and Roe B.A. 2003. The mouse Ifi200 gene cluster: Genomic sequence, analysis, and comparison with the human HIN-200 gene cluster. *Genomics* **82:** 34.

Draper B.W., Morcos P.A, and Kimmel C.B. 2001. Inhibition of zebrafish fgf8 pre-mRNA splicing with morpholino oligos: A quantifiable method for gene knockdown. *Genesis* **30:** 154.

Dunham I., Shimizu N., Roe B.A., Chissoe S., Hunt A.R., Collins J.E., Bruskiewich R., Beare D.M., Clamp M., Smink L.J., Ainscough R., Almeida J.P., Babbage A., Bagguley C., Bailey J., Barlow K., Bates K.N., Beasley O., Bird C.P., Blakey S., Bridgeman A.M., Buck D., Burgess J., Burrill W.D., and O'Brien K.P., et al. 1999. The DNA sequence of human chromosome 22. *Nature* **402:** 489.

Ekker S.C. and Larson J.D. 2001. Morphant technology in model developmental systems. *Genesis* **30:** 89.

Footz T.K., Brinkman-Mills P., Banting G.S., Maier S.A., Riazi M.A., Bridgland L., Hu S., Birren B., Minoshima S., Shimizu N., Pan H.Q., Nguyen T., Fang F., Fu Y., Ray L., Wu H., Shaull S., Phan S., Yao Z., Chen F., Hua A., Hu P., Wang Q., Loh P., Qi S., Roe B.A., and H.E. McDermid. 2001. Analysis of the cat eye syndrome critical region in humans and the region of conserved synteny in mice: A search for candidate genes at or near the human chromosome 22 pericentromere. *Genome Res.* **11:** 1053.

Golling G., Amsterdam A., Sun Z., Antonelli M., Maldonado E., Chen W., Burgess S., Haldi M., Artzt K., Farrington S., Lin S.Y., Nissen R.M., and Hopkins N. 2002. Insertional mutagenesis in zebrafish rapidly identifies genes essential for early vertebrate development. *Nat. Genet.* **31:** 135.

Jang W., Hua A., Spilson S.V., Miller W., Roe B.A., and Meisler M.H. 1999. Comparative sequencing of human and mouse BAC clones from the mnd2 region of chromosome 2p13. *Genome Res.* **9:** 53.

Ji W., Chen F., Do T., Do A., Roe B.A., and Meisler M.H. 2001. DQX1, an RNA dependent ATPase homolog with a novel DEAQ box: Expression pattern and genomic sequence comparison of the human and mouse genes. *Mamm. Genome* **12:** 456.

Kambara H. and Takahahi S. 1993. Multiple-sheath flow capillary array DNA analyzer. *Nature* **361:** 565.

Kitazawa S., Kitazawa R., and Maeda S. 1999. *In situ* hybridization with polymerase chain reaction-derived single-stranded DNA probe and S1 nuclease. *Histochem. Cell Biol.* **111:** 7.

Knuchel M.C., Graf B., Schlaepfer E., Kuster H., Fischer M., Weber R., and Cone R.W. 2000. PCR-derived ssDNA probes for fluorescent in situ hybridization to HIV-1 RNA. *J. Histochem. Cytochem.* **48:** 285.

Lander E.S., Linton L.M., Birren B., Nusbaum C., Zody M.C., Baldwin J., Devon K., Dewar K., Doyle M., FitzHugh W., Funke R., Gage D., Harris K., Heaford A., Howland J., Kann L., Lehoczky J., LeVine R., McEwan P., McKernan K., Meldrim J., Mesirov J.P., Miranda C., Morris W., and Naylor J., et al. (International Human Genome Sequencing Consortium). 2001. Initial sequencing and analysis of the human genome. *Nature* **409:** 860.

Lund J., Chen F., Hua A., Roe B., Budarf M., Emanuel B.S., and Reeves R.H. 2000. Comparative sequence analysis of 634Kbp of the mouse chromosome 16 region of conserved synteny with the human velocardiofacial region of Chr 22q11.2. *Genomics* **63:** 374.

Maxam A.M. and Gilbert W. 1977. A new method for sequencing DNA. *Proc. Natl. Acad. Sci.* **74:**560.

Nagase T., Ishikawa K., Suyama M., Kikuno R., Hirosawa M., Miyajima N., Tanaka A., Kotani H., Nomura N., and Ohara O. 1998. Prediction of the coding sequences of unidentified human genes. XII. The complete sequences of 100 new cDNA clones from brain which code for large proteins *in vitro. DNA Res.* **5:** 355.

Nasevicius A. and Ekker S.C. 2000. Effective targeted gene 'knockdown' in zebrafish. *Nat. Genet.* **26:** 216.

Ohshima K., Hattori M., Yada T., Gojobori T., Sakaki Y., and Okada N. 2003. Whole-genome screening indicates a possible burst of formation of processed pseudogenes and Alu repeats by particular L1 subfamilies in ancestral primates. *Genome Biol.* **4:** R74.

Patel N.H. and Goodman C.S. 1992. Preparation of digoxigenin-labeled single-stranded DNA probes. In *Non-radioactive labeling and detection of biomolecules* (ed. C. Kessler), p. 377. Springer-Verlag, Berlin, Germany.

Prober J.M., Trainor G.L., Dam R.J., Hobbs F.W., Robertson C.W., Zagursky R.J., Cocuzza A.J., Jensen M.A., and Baumeister K. 1987. A system for rapid DNA sequencing with fluorescent chain-terminating dideoxynucleotides. *Science* **238:** 336.

Sanger F., Nicklen S., and Coulson A.R. 1977. DNA sequencing with chain-terminating inhibitors. *Proc. Natl. Acad. Sci.* **74:** 5463.

Schwartz S., Zhang Z., Frazer K.A., Smit A., Riemer C., Bouck J., Gibbs R., Hardison R., and Miller W. 2000. PipMaker—A web server for aligning two genomic DNA sequences. *Genome Res.* **10:** 577.

Schwartz S., Elnitski L., Li M., Weirauch M., Riemer C., Smit A., Green E.D., Hardison R.C., Miller W., and the NISC Comparative Sequencing Program. 2003. MultiPipMaker and supporting tools: Alignments and analysis of multiple genomic DNA sequences. *Nucleic Acids Res.* **31:** 3518.

Seydoux G. and Fire A. 1995. Whole-mount in situ hybridization for the detection of RNA in *Caenorhabditis elegans* embryos. *Methods Cell Biol.* **48:** 323.

Shoemaker D.D., Schadt E.E., Armour C.D., He Y.D., Garrett-Engele P., McDonagh P.D., Loerch P.M., Leonardson A., Lum P.Y., Cavet G., Wu L.F., Altschuler S.J., Edwards S., King J., Tsang J.S., Schimmack G., Schelter J.M., Koch J., Ziman M., Marton M.J., Li B., Cundiff P., Ward T., Castle J., and Krolewski M., et al. 2001. Experimental annotation of the human genome using microarray technology. *Nature* **409:** 922.

Smith L.M., Sanders J.Z., Kaiser R.J., Hughes P., Dodd C., Connell C.R., Heiner C., Kent S.B., and Hood L.E. 1986. Fluorescence detection in automated DNA sequence analysis. *Nature* **321:** 674.

Tautz D. and Pfeifle C. 1989. A nonradioactive in situ hybridization method for the localization of specific RNAs in *Drosophila* embryos reveals translational control of the segmentation gene hunchback. *Chromosoma* **98:** 81.

Thomas J.W., Touchman J.W., Blakesley R.W., Bouffard G.G., Beckstrom-Sternberg S.M., Margulies E.H., Blanchette M., Siepel A.C., Thomas P.J., McDowell J.C., Maskeri B., Hansen N.F., Schwartz M.S., Weber R.J., Kent W.J., Karolchik D., Bruen T.C., Bevan R., Cutler D.J., Schwartz S., Elnitski L., Idol J.R., Prasad A.B., Lee-Lin S.Q., and Maduro V.V., et al. 2003. Comparative analyses of multi-species sequences from targeted genomic regions. *Nature* **424:** 788.

Trappe R., Schulze E., Rzymski T., Frode S., and Engel W. 2002a. The *Caenorhabditis elegans* ortholog of human PHF5a shows a muscle-specific expression domain and is essential for *C. elegans* morphogenetic development. *Biochem. Biophys. Res. Commun.* **297:** 1049.

Trappe R., Ahmed M., Glaser B., Vogel C., Tascou S., Burfeind P., and Engel W. 2002b. Identification and characterization of a novel murine multigene family containing a PHD-finger-like motif. *Biochem. Biophys. Res. Commun.* **293:** 816.

Ureta-Vidal A, Ettwiller L, and Birney E. 2003. Comparative genomics: Genome-wide analysis in metazoan eukaryotes. *Nat. Rev. Genet.* **4:** 251.

Waterston R.H., Lindblad-Toh K., Birney E., Rogers J., Abril J.F., Agarwal P., Agarwala R., Ainscough R., Alexandersson M., An P., Antonarakis S.E., Attwood J., Baertsch R., Bailey J., Barlow K., Beck S., Berry E., Birren B., Bloom T., Bork P., Botcherby M., Bray N., Brent M.R., Brown D.G., and Brown S.D., et al. (Mouse Genome Sequencing Consortium). 2002. Initial sequencing and comparative analysis of the mouse genome. *Nature* **420:** 520.

Watson J.D. and Crick F.H. 1953a. Molecular structure of nucleic acids; a structure for deoxyribose nucleic acid. *Nature* **171:** 737.

———. 1953b. Genetical implications of the structure of deoxyribonucleic acid. *Nature* **171:** 964.

Yanisch-Perron C., Vieira J., and Messing J. 1985. Improved M13 phage cloning vectors and host strains: Nucleotide sequences of the M13mp18 and pUC19 vectors. *Gene* **33:** 103.

Zhang M.Q. 2002. Computational prediction of eukaryotic protein-coding genes. *Nat. Rev. Genet.* **3:** 698.

# Genome-wide Analyses Based on Comparative Genomics

O. JAILLON, J.-M. AURY, H. ROEST CROLLIUS, M. SALANOUBAT, P. WINCKER, C. DOSSAT, V. CASTELLI, N. BOUDET, S. SAMAIR, R. ECKENBERG, S. BONNEVAL, W. SAURIN, C. SCARPELLI, V. SCHÄCHTER, AND J. WEISSENBACH
*CNRS UMR8030, Genoscope and University of Evry, Evry, France*

The establishment of an exhaustive inventory of genes is the primary goal of genome sequencing projects. When looking at multicellular genome annotations that are available in sequence data banks or on other sites, the level of available information is quite variable from genome to genome, and the degree of completion that has been reached among the gene inventories of the genomes sequenced to date is very difficult to assess. These gene inventories are typically carried out in an automated fashion by the annotation platforms of the major data banks, and rely mainly on two types of predictions: ab initio predictions and those based on sequence comparisons. A genomic DNA sequence can be subjected to direct or indirect comparisons. In direct comparisons the genomic DNA sequence is aligned with sequences of expression products, namely ESTs, cDNAs, or proteins from the same species. In indirect comparisons the genomic sequence is aligned with genomic or expressed sequences from other organisms.

To date, whole-genome comparisons have not been used extensively for a variety of reasons: They are rather demanding in computing capacity and, until recently, the number of complete or draft genome sequences was very limited. In addition, such comparisons generate massive data sets, from which the significant fraction is sometimes difficult to extract. Such massive sequence alignments can be performed at both DNA and protein levels.

Whole-genome comparisons between genomes from multicellular organisms can be used to detect sequence conservation in both coding and noncoding regions. Whereas conservation of coding regions can be detected between species separated by large evolutionary distances (e.g., between mammals and fish), the conservation of noncoding regions is usually much weaker and mainly detected between species that are separated by shorter evolutionary distances (e.g., within mammals). In the work described here, our primary interest was to identify yet undetected coding sequences belonging to unknown or already identified genes. For that purpose, we have developed a two-level procedure aimed at predicting (1) exons and (2) genes: (1) The first level, dubbed "Exofish," searches for ecores (*E*volutionary *CO*nserved *RE*gions) between pairs of genomes that can be very specifically ascribed to protein-coding DNA segments when using appropriate parameters determined on training sets of exhaustively annotated known genes. (2) Sets of ecores satisfying a certain colinearity property are then assembled into candidate gene models called ecotigs (*ECO*re con*TIGS*). Since Exofish (the first level) was first introduced and applied to the comparison of a set of annotated human genes and a collection of random shotgun sequence reads from *Tetraodon nigroviridis* (Roest Crollius et al. 2000), several comparative gene prediction methods have been introduced (for review, see Ureta-Vidal et al. 2003). These "dual-gene predictors" take one of the two following approaches: Asymmetrical "informant" methods (Korf et al. 2001; Wiehe et al. 2001; Yeh et al. 2001) complement traditional hidden Markov models (HMM) of gene structure on the target sequence with information from its alignment with another sequence, whereas symmetrical "Pair-HMM" methods (Meyer and Durbin 2002; Pachter et al. 2002) rely on a single HMM of a pair of sequences joined by orthology links to predict gene structure on both sequences simultaneously.

The "Exofish + ecotigs" approach takes a slightly different route. Its aim is not to perform ab initio gene prediction alone, but to provide a highly reliable set of core predictions that can be used jointly with other methods or resources for an initial annotation, or to improve existing annotations. Therefore, Exofish clearly favors specificity over sensitivity.

In addition, whereas the two levels of our method correspond roughly to the steps traditionally found in ab initio gene prediction algorithms (exon prediction, gene model construction), it is purely comparative—i.e., it does not rely on an a priori model of gene structure—and therefore quite simple; its levels are uncoupled; i.e., the gene prediction method builds on, but is independent of, the exon prediction method.

In this paper, we briefly describe our comparative procedure, show how Exofish comparisons and construction of derived gene models (ecotigs) have been applied to several pairs of multicellular eukaryotic genomes, and discuss how these comparisons can be used to reevaluate the degree of completion and accuracy of gene and exon identification in these genomes.

## METHODOLOGY

Our aim is to annotate a target genome using comparisons with a query genome. The results of sequence alignments are positioned on the target genome.

## Detection of Evolutionary Conserved Sequences (Ecores)

The methodology we are currently using is based on comparisons of the translated phases between two DNA sequences using TBLASTX (Altschul et al. 1990) as an engine. Although TBLASTX computation generates a huge number of sequence matches called HSPs (high scoring pairs), these can be categorized as true positives (matches that are located within coding exons) or false positives (located elsewhere).

We showed that it was possible to minimize the background of false-positive matches (Roest Crollius et al. 2000) to about 1% by (1) applying extensive and stringent masking of low-complexity repeats, and (2) filtering HSPs using appropriate combinations of threshold values for sequence identity rate and length of sequence alignments.

For a given length of sequence match, we use a specific percentage of sequence identity. These settings were initially determined as a broken line manually, i.e., the frontier between putative true positives and false positives was empirically approximated by a piecewise linear function, using reference annotations. Since the number of properly annotated genes is well above 1000 in several genomes, we are now using polynomial approximations both for practical reasons and to improve accuracy.

Finally, sets of overlapping true positive matches are assembled, and the resulting genomic regions are called ecores. Although an ecore designates a pair of corresponding regions, one on the target genome and one on the query genome, the pair of regions may be composed of HSPs that are not in strict correspondence. Depending on the context, we will employ the term "ecore" indifferently for the pair or for one of its components.

An example of such an optimization is shown in Figure 1 for a set of 1589 *Arabidopsis thaliana* genes that were compared to the Syngenta (Goff et al. 2002) sequence draft of the rice genome. The fraction of false positives in the test set on the right part of the curve (alignments above thresholds for length and sequence identity) is below 0.2%. However, this high specificity is at the expense of sensitivity which is limited to ~64% of exons of the set of tested genes. These settings have to be defined for each pair of genomes compared.

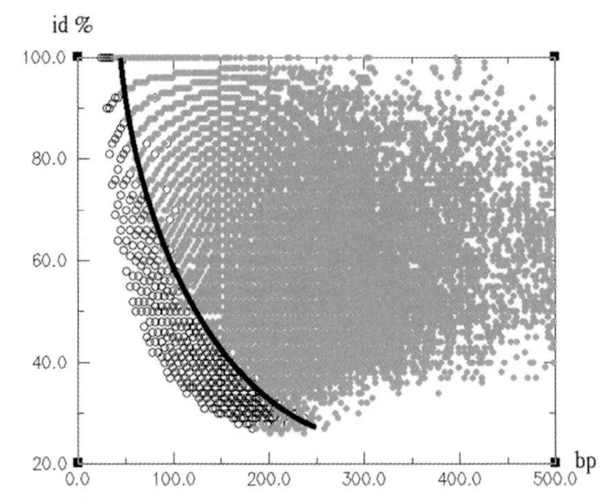


**Figure 1.** The conditions that generate optimal alignment in coding regions were first tested using a large range of TBLASTX (Altschul et al. 1990) conditions (W, X, scoring matrix) between a well-annotated set of 1589 genes including introns, exons, and 100 bp of intergenic region at both ends of each gene (P. Rouzé and S. Aubourg , pers. comm.) and the Syngenta rice draft sequence (Goff et al. 2002). The BLAST settings providing the highest sensitivity are: match score = 15, mismatch score = –3, W = 4, X = 13. All sequences were masked against known repeats from rice and *Arabidopsis*. For each condition, a filter was applied based on the length and percent identity of alignments. The gray dots correspond to HSPs that overlapped exons in 100% of cases. The circles correspond to HSPs that overlapped exons in less than 100% of cases. HSPs situated to the left of the curve were discarded.

## Detection of Conserved Contiguity of Ecores (Ecotigs)

Sequence conservation between genomic sequences of species that have diverged for 100 Myr or more is essentially restricted to coding segments. In a large majority of cases, there is only a single ecore match in exons that are matched. In addition, if two ecores remain consecutive (contiguous) in both genomes A and B, they are highly likely to belong to the same gene. In order to identify ecores that conserve such colinearity, named hereafter ecotigs (*ECO*res con*TIG*uous), we designed an algorithm that identifies conserved contiguity of ecores between two genomes A and B by constructing well-chosen paths in the joint "contiguity-similarity" graph. Contiguity edges were induced by exon neigborhood relationships of ecores from both genomes and similarity edges by correspondence between ecores (J.M. Aury et al., in prep.).

Figure 2 outlines the rationale of the procedure we use to identify and construct ecotigs. Ecotigs group ecores together as long as colinearity is preserved, up to a fixed tolerated "gap" in ecore succession. In other words, ecores that are consecutive in genome A will be included in an ecotig if they consist of at least two HSPs that are consecutive or separated by one (the chosen gap value) additional HSP at most on genome B (distance 1 and 2, respectively, in Fig. 2). When constructing ecotigs on genome A, we first consider the ecore pairs i and ii that are consecutive on genome A (ia and iia). Ecore ia is linked to ecores ib1 and ib2 on genome B. These latter are separated by a distance ≤2 from ecore iib. According to the chosen rules, i and ii will be grouped in the same ecotig. The ecotig is then tentatively extended to ecore iii. The distance separating iib and iiib on genome B is 2. Consequently, iii will be incorporated into the existing ecotig. By iterating of the process, we can group ecores i, ii, iii, iv, and v in a single ecotig. Ecores v and vi are contiguous on genome A (va and via) but are not syntenic on genome B, and vb and vib are hence separated by an infinite distance. This will stop extension of the ecotig. A new tentative ecotig is then initiated with ecore vi and the process is iterated.

A single ecotig on one genome may be related to multiple ecotigs on the other one. In addition, ecotigs may be composed of sequences from more than one gene, since

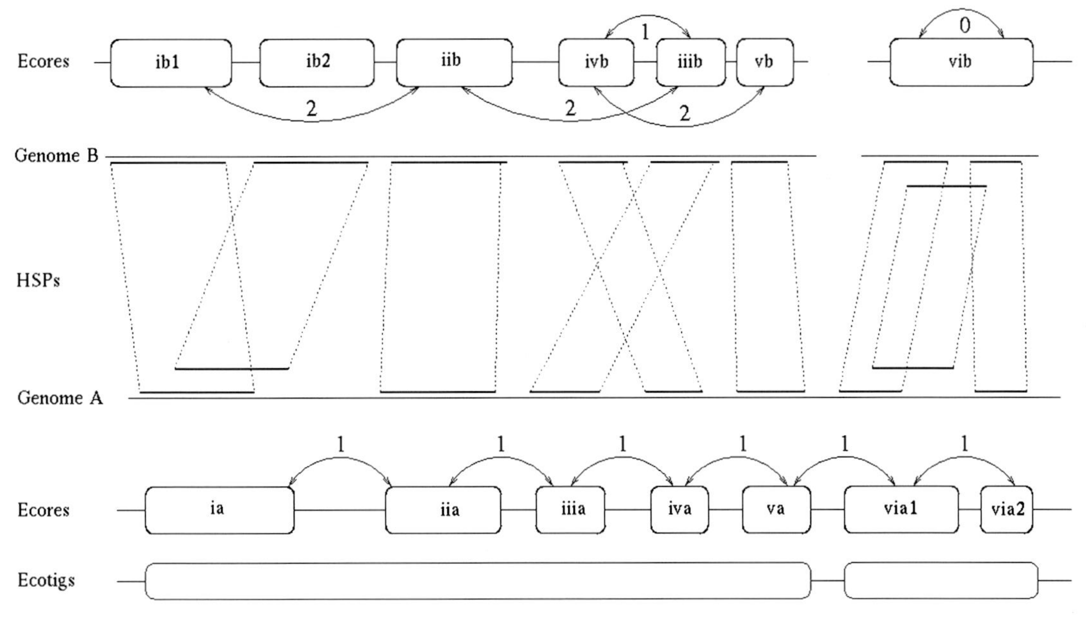


**Figure 2.** Construction of ecotigs. The two first lines represent, respectively, the ecores (*boxes*) and the HSPs (*segments*) detected on genome B, using genome A as a query. The two following lines represent, respectively, the HSPs and the ecores detected on genome A, using genome B as a query. The bottom line represents the ecotig gene models constructed on genome A. Matching HSPs are linked by dotted lines. Matching ecores are identified by the same prefix (i, ii, etc.). Numbers over (or under) arrows represent distances separating ecores that are consecutive on genome A (number of consecutive ecores minus one).

genes may remain colinear on the pair of compared genomes, depending on the degree of conserved synteny of the genome pair analyzed. In the rest of this paper, we designate the pair of matched genomes with the following notation: (query/target).

## RESULTS

The ecore detection and ecotig construction procedures have been applied to compare draft or complete genome sequences from various multicellular organisms. We applied Exofish comparisons to several genome pairs between plant, insect, and vertebrate genomes, namely, mammals (mouse/human) with pufferfish (*Tetraodon/Takifugu*), the *Drosophila melanogaster* sequence with the *Anopheles gambiae* draft, and the *Arabidopsis* sequence with the rice genome draft. Such comparisons were used (1) to detect gene models or exons that have not yet been identified in one or both genomes, (2) to extend existing gene models, and (3) to determine the degree of completion of existing annotations.

### Plant Genomes

The recent availability of a draft sequence of the rice genome with sufficient coverage (Goff et al. 2002) has opened the possibility of comparing whole plant genomes for the first time. In addition, the genome of *Arabidopsis* has been the focus of several extensive annotation projects that make this genome one of the best documented to date. This situation enabled us to use *Arabidopsis* as a reference on which Exofish performances can be evaluated. Such an evaluation benefited also from the availability of a source of new cDNA sequences that have not yet been used in the *Arabidopsis* genome analyses but have served as a support for experimental validation of comparison-based predictions.

Exofish was first calibrated using a set of 1589 *Arabidopsis* genes that had been manually annotated. The optimal conditions we determined produce a specificity above 99% and a sensitivity at the exon and gene level of 64% and 93%, respectively. The global Exofish comparison was performed between the finished *Arabidopsis* genome sequence The *Arabidopsis* Genome Initiative (2000) used as a target and the BAC-based sequence draft established by the International Rice Genome Sequencing Program used as a query. Ecores were mapped relative to the *Arabidopsis* genome annotation.

Statistics on ecores detected within and outside annotations are shown in Table 1. A total of 74% of the annotated genes (MIPS annotation) included one ecore at least, and 47% of the annotated exons are matched by one or more ecores. Conversely, 91% of the ecores are localized within the boundaries of annotated genes, and only about 1% of these ecores do not match an annotated exon. In a subset of 60 nonmatching ecores, experimental evidence based on new cDNAs showed that 59 cases correspond to novel exons. We thus estimate that about 98% of the ecores within gene annotations, but which do not match annotated exons, correspond to real exons that were missed during the annotation process. Taking into account that only one exon out of two is detected as an ecore (Table 1), an extrapolation of this analysis suggests that about 900 internal exons are still missing in the set of

**Table 1.** Distribution of (Rice/*Arabidopsis*) Ecores in the Sequence of *Arabidopsis thaliana*

| | Ecores | Genes | Genes detected | Ecores within genes | Exons | Exons detected | Ecores overlapping exons | Ecores in genes not in exons |
|---|---|---|---|---|---|---|---|---|
| Numbers | 80,010 | 26,027 | 19,235 | 73,119 | 135,318 | 64,032 | 72,396 | 723 |
| (%) | (100) | (100) | (74) | (91) | (100) | (47) | (90) | (1) |

11,620 annotated *Arabidopsis* genes for which no corresponding full-length cDNA is available.

A total of 6891 ecores were found to be located outside gene annotations. Of these, 2980 were found in other annotated features such as transposons, tRNAs, or pseudogenes. The presence of ecores in pseudogenes is expected and difficult to avoid. Transposons that were matched by ecores correspond to cases that escaped masking. However, we expect that a substantial fraction of the 3911 remaining un-annotated ecores correspond to gene extensions or to undetected genes. Again, experimental evidence based on new cDNAs confirmed that 150 ecores could be included in 93 gene extensions. It is, however, impossible to estimate the fraction of genes that could be extended, since we cannot determine the fraction of truly full length cDNAs in the collection of novel cDNAs that is being used for experimental validation.

To analyze these gene extensions further, we constructed ecotig gene models (see Methodology). Of the 80,010 ecores, 70,847 were incorporated in 15,311 ecotigs and 9,163 remained as singletons. A total of 14,308 ecotigs (67,607 ecores) matched 15,496 genes in *Arabidopsis*; 712 gene models are overlapped by two or more ecotigs (1,433 ecotigs). Conversely, 1,413 ecotigs led to the fusion of 3,307 annotated genes. It remains to be seen whether these fusions are correlated with conservation of synteny between genes from both plants. This is an obvious drawback of the ecotig method that may nevertheless be of interest in the identification of conservation of synteny between two genomes, for which it could even provide a measurement.

The construction of ecotigs can first be applied to extend gene models, and in their present stage, 697 annotated *Arabidopsis* genes could be potentially extended on the basis of the ecotigs (914 ecores). Of the 93 gene extensions that were experimentally supported by cDNA sequences described above, 64 could be included in ecotigs.

Among the 1,003 ecotigs (3,240 ecores) located in regions with no gene annotation, 619 match a transposon, a tRNA, or a pseudogene. Of the 384 remaining ecotigs, 245 were subjected to manual curation, which selected 98 (255 ecores) as potential gene candidates. Experimental evidence based on cDNAs was available for 19 of these candidates. In addition, singleton ecores may also indicate the existence of additional genes, since about 40 such singletons were confirmed by novel cDNAs. Interestingly, many of these novel gene candidates (55%) encode small open reading frames (smORFs) with a CDS <100 amino acids, suggesting that many such smORFs remain undetected and should become a major focus for future systematic searches (Kessler et al. 2003). The possibility that a fraction of them correspond to pseudogenes cannot be excluded, but non-detection of a corresponding real gene in the *Arabidopsis* genome using those novel cDNAs argues against this hypothesis. An analysis of the Exofish comparison and ecotig construction applied to the rice genome is in progress.

## Insect Genomes

Comparisons between insect genomes have also become possible with the recent availability of a sequence draft from the malaria vector mosquito *Anopheles gambiae* (Holt et al. 2002). The *D. melanogaster* genome has been the focus of several major annotation efforts and can be considered as one of the most exhaustively annotated genomes of a model organism. This situation is thus very similar to that of the pair of plant genomes compared as described above and has been exploited for the same purposes.

The available sequence assembly of *A. gambiae* was compared to the last two versions of the genome sequence of the fruit fly using Exofish with conditions determined on a set of reference genes (Jaillon et al. 2003). The global ecore counts obtained for releases 2 and 3 of the fly genome (Table 2) show a very slight decrease from 47,134 to 46,742 ecores, respectively, probably reflecting some minor changes in the sequence assembly (Celniker et al. 2002). More importantly, we observed a significant increase in the fraction of ecores mapping inside gene models (Misra et al. 2002), from 90.5% to 93.5% between the two releases. This illustrates how Exofish can provide a quantitative evaluation of the improvement of two successive versions of a whole-genome analysis of gene content.

Mosquito/fly genome comparisons reveal the presence of 4,063 ecores outside of annotated exons in the *D. melanogaster* genome. Since the mean ecore number in the *D. melanogaster* Gene Collection (used as a reference set) is higher than in other annotated genes, we expect that some gene models are still incomplete or fragmented. We expect that most of these ecores would correspond to additional exons of partially annotated genes. Conversely, it is not expected that the 4,063 ecores will contribute to a substantial increase in the total gene number of *D. melanogaster*. A verified example of a modification of a predicted gene indicated by Exofish is shown in Figure 3. In this case, a series of additional exons in the annotation of release 2 was predicted by Exofish, suggesting that a significant number of exons were missed in this region (Fig. 3, top). We reexamined the same region in re-

**Table 2.** Distributions of (*Anopheles/D.melanogaster*) Ecores in the Sequence of *Drosophila* in Two Successive Annotations

| BDGP annotation | Ecores | Genes | Genes detected | Ecores within genes | Exons | Exons detected | Ecores overlapping exons | Ecores in genes not in exons |
|---|---|---|---|---|---|---|---|---|
| Release 2 | 47,134 | 13,468 | 11,147 | 42,633 | 54,771 | 31,751 | 41,332 | 1072 |
| (%) | (100) | (100) | (83) | (90.5) | (100) | (58) | (88) | (2.5) |
| Release 3 | 46,742 | 13,666 | 11,167 | 43,705 | 61,085 | 33,996 | 42,679 | 1026 |
| (%) | (100) | (100) | (82) | (93.5) | (100) | (56) | (91.5) | (2) |

lease 3 and observed that, at present, all ecores have been placed in two gene models (Fig. 3, bottom).

Based on remote protein sequence or structure homologies, an additional set of 1,042 *D. melanogaster* candidate genes has been proposed (Gopal et al. 2001) (http://genomes.rockefeller.edu/dm). Ecores could be found in 18.7% of these new gene models (the list of the matches can be found at www.genoscope.cns/externe/Fly). This low fraction of matches could either result from a very low conservation of these genes between *A. gambiae* and *D. melanogaster*, possibly representing a subset of rapidly evolving genes, or indicate that a large fraction of these hypothetical genes should be dismissed. However, Exofish can also serve to validate a number of these potential genes. A genome-wide analysis was also performed on the assembly of the *A. gambiae* genome sequence draft (Holt et al. 2002). We found more ecores in the *Anopheles* assembly (54,069 in release 6.01a) than in the *D. melanogaster* genome (ratio = 1.16). Several explanations that are not mutually exclusive may account for this observation. The high number of ecores could reflect (1) an increased coding capacity in the genome of *Anopheles* or (2) a larger number of pseudogenes or unmasked transposable elements in *Anopheles* or (3) problems in the sequence assembly. The presence of at least two different haplotypes in the *A. gambiae* strain sequenced is known to have introduced a number of redundancies in the assembly, essentially as linked artefactual duplications and unanchored duplicated scaffolds (Holt et al. 2002). Work is in progress to test these hypotheses. We compared the

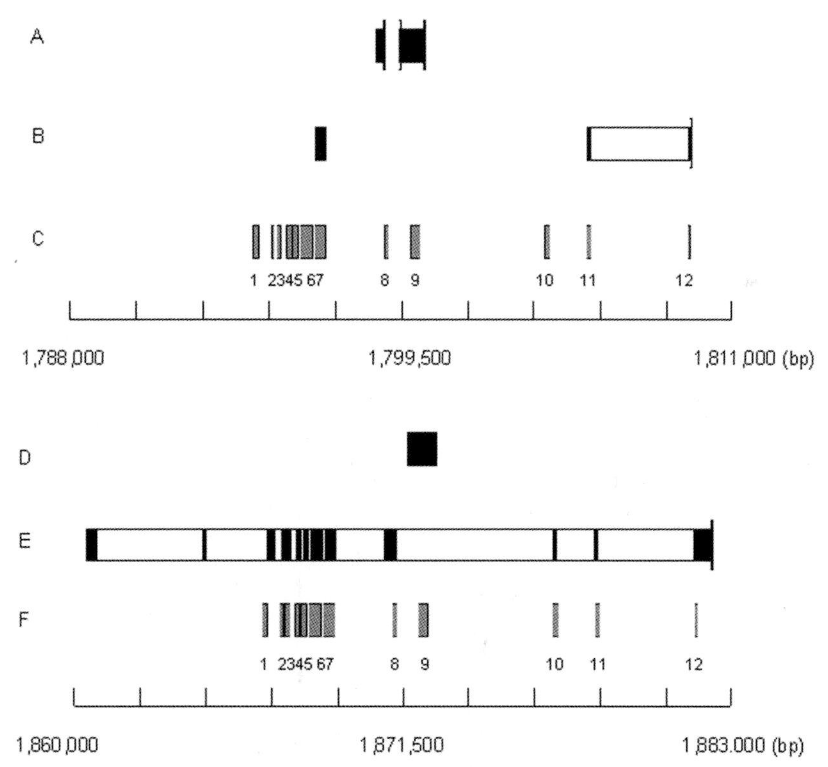


**Figure 3.** Exofish analysis on a region on arm 2L of the genome of *D. melanogaster* from 2 different releases of annotations, and around the same ecores. (*Top*) Results from release 2 of BDGP. (*Bottom*) Results from release 3 of BDGP. (*A,D*) BDGP annotations on the 5′–3′ strand. (*B,E*) BDGP annotations on the 3′–5′ strand. The genes are represented by boxes, with exons in black and introns in white. (*C,F*) Ecores (*gray boxes*). In release 2 (*top*), 5 ecores (numbers 7, 8, 9, 11, 12) overlap 4 gene models, and 7 ecores (numbers 1, 2, 3, 4, 5, 6, 10) do not overlap any annotation. In release 3, a large gene model overlaps all the ecores that fall exclusively in exons except ecore number 9. This ecore is part of a gene model on the 5′–3′ strand, which is predicted inside an intron on the 3′–5′ strand.

**Table 3.** Distribution of (*D.melanogaster/Anopheles*) Ecores on the Assembly of *Anopheles* in Two Successive Ensembl Annotations

| Ensembl annotation | Ecores | Genes | Genes detected | Ecores overlapping genes | Exons | Exons detected | Ecores overlapping exons | Ecores in genes not in exons |
|---|---|---|---|---|---|---|---|---|
| Release 6.1a | 54,069 | 15,088 | 11,929 | 42,693 | 53,693 | 32,553 | 40,278 | 2,415 |
| (%) | (100) | (100) | (79) | (79) | (100) | (60.5) | (74.5) | (4.5) |
| Release 10.2.1 | 53,132 | 14,658 | 10,759 | 39,749 | 56,573 | 32,610 | 39,247 | 502 |
| (%) | (100) | (100) | (73.5) | (75) | (100) | (57.5) | (74) | (1) |

54,069 ecores from the assembly of *Anopheles* to release 6.1a of the Celera-Ensembl joint annotations of *Anopheles* (http://www.ensembl.org/Anophelesgambiae). We found that 79% of the ecores matched 79.1% of the gene candidates (Table 3). The fraction of annotated *Anopheles* genes that is detected by Exofish is thus slightly lower than in *D. melanogaster*. Conversely, a large fraction (21%) of *Anopheles* ecores map outside of annotations. These observations indicate that a substantial fraction of exons were not annotated and that a number of gene models should be revised.

A more recent version of the *Anopheles* assembly and annotation has been released (version 10.2.1). Surprisingly, the percentage of ecores outside of annotations increased from 21% to 25.6% (Table 3). We found that a significant fraction of the duplicated ecores that were present in release 6.1a have been discarded as haplotype variants. This explained in large part the net disappearance of 937 ecores between the two versions.

### Vertebrate Genomes

The increase in the number of vertebrate genome drafts amplifies dramatically the number of possible genome-wide comparisons. Exofish is particularly suited to genomes that are separated by substantial evolutionary distances. We therefore applied Exofish to compare the available pufferfish genome drafts from *Takifugu* (Aparicio et al. 2002) and *Tetraodon* (recently assembled at the Whitehead Institute using the Arachne genome assembler) with the mammalian genomes from human and mouse (Lander et al. 2001; Waterston et al. 2002). The main difficulty with vertebrate genomes remains in the definition of a set of exhaustively annotated genes. Because alternative splicing involves about half of mammalian genes (Dunham et al. 1999; Heilig et al. 2003) and has only been exhaustively studied for a handful of genes, the estimates given hereafter remain tentative. Nevertheless, the additional ecores or gene models proposed by Exofish are worth taking into account and deserve further experimental investigations. The overall numbers of ecores observed in pairwise comparisons (one fish versus one mammal) are shown in Table 4. Both mammalian genomes show a very similar number of ecores regardless of the query fish genome. The difference between pufferfish ecores detected in human and mouse genomes is very low and may just reflect a difference in the degree of completion of each genome. Conversely, mammalian sequences match about 10% more ecores in *Tetraodon* than in *Takifugu*. However, the *Tetraodon* alignments on the mammalian genomes do not detect more ecores than *Takifugu* alignments. This suggests that the additional *Tetraodon* ecores do not correspond to sequences that are absent in *Takifugu* but rather to some redundancy in the *Tetraodon* assembly. It remains to be seen whether or not such redundancy is artefactual, possibly resulting from diverged haplotypes that were considered as distinct loci in the assembly process.

Combining ecores matched by both pufferfishes results in an overall increase in the number of ecores observed in mammals: Each fish draft sequence contains 7–8% ecores not found in the other fish genome. We conclude that these drafts each cover ~92–93% of the complete sequence. The overall ecore content in *Takifugu* and mammals is very close. However, this estimate remains global and cannot, for instance, detect gene families that would have undergone an increase or reduction in gene number in some species or groups of species. A similar number of ecores in pufferfish and mammals argues against a whole-genome duplication event in teleosts as has been proposed (Amores et al. 1998; Wittbrodt et al. 1998).

To evaluate the degree of completion of various sets of data or annotations of the human genome, we estimated the number of ecores located within and outside gene

**Table 4.** Ecores Determined by Pairwise Genome-wide Gxofish Comparisons

| Query genome[a] | Target genome[b]: human | mouse | *Tetraodon* | *Takifugu* |
|---|---|---|---|---|
| *Tetraodon* | 141,623 | 135,924 | NA | ND |
| *Takifugu* | 142,052 | 137,635 | ND | NA |
| Combined pufferfish | 152,208 | 146,714 | NA | NA |
| Human | NA | ND | 151,791 | 135,031 |
| Mouse | ND | NA | 150,444 | 133,537 |
| Combined mammals | NA | NA | 161,623 | 143,555 |

(NA) Not applicable. (ND) Not determined.
[a]Genome sequence that is compared to the target genome using Exofish.
[b]Genome on which the matching ecores are localized.

models defined by Ensembl or by a RefSeq cDNA aligned on the human genome assembly. Some 28,400 (*Tetraodon*/human) ecores did not overlap Ensembl annotations and corresponded to 32,500 (human/*Tetraodon*) ecores. Of these, 29,200 turned out to have a (*Tetraodon*/mouse) counterpart, of which 16,600 mapped within Ensembl mouse annotations and 12,600 did not. This indicates that a very large fraction of (*Tetraodon*/human) ecores (90%) that are not in human annotations are conserved in mouse and are related to (*Tetraodon*/mouse) ecores that are frequently (57%) located in mouse gene models. Work is in progress to determine the fraction of possible background due to pseudogenes or simple sequences and to check whether such ecores are in conserved synteny in mammals. About 6% of (*Tetraodon*/human) ecores that mapped within a gene model did not match an exon defined by these resources (Table 5). Ecotigs can be used to evaluate the fraction of ecores that extend annotated gene models. We found 3,416 ecotigs that extended existing Ensembl annotations by one ecore at least. A RefSeq sequence exists for a subset of 2,701 extending ecotigs, suggesting that a substantial number of these cDNAs are not yet complete.

## CONCLUSIONS

The studies summarized above show how whole-genome comparisons based on a tool like Exofish can be an efficient method to evaluate quality and to improve existing annotations of genomes as diverse as plants, insects, and vertebrates. Exofish has certain limitations. Because it was designed to maximize specificity, its sensitivity is rather low, probably reflecting an uneven rate of evolution between different genes, and even within coding regions of a single gene, that depend on nonuniform selective constraints imposed on the various proteins of an organism. Despite such limitations, Exofish data and ecotigs can be used at both genome-wide and gene-specific levels.

At the genome-wide level, ecores and ecotigs provide an independent assessment of the quality and improvement of analyses across successive annotation rounds for a given genome. The fact that a significant number of ecores do not overlap annotated exons, even for extensively studied species such as *D. melanogaster, A. thaliana, H. sapiens*, and *M. musculus*, illustrates the potential for interspecies comparisons. It also provides a measure of the amount of work and the type of additional experimental results that will be needed to improve existing annotations, especially for species like *A. gambiae*, for which it helps to set priorities.

At a gene-specific level, potential additional internal exons or gene extensions can be pinpointed, and Exofish predictions can be used for targeted experimental work, such as searches for splicing variants and/or 5′UTRs using RT-PCR. Exofish comparisons also provide highly reliable confirmation of ab initio gene and exon predictions. A web tool specifically designed to display pairwise genome comparisons and derived ecotig construction is available at www.genoscope.cns.fr/comparative.

Exofish was designed to address a precise type of problem, the identification of exons. Each setting is a compromise between sensitivity and reduction of background noise and has to be adjusted to the pair of genomes that are being compared. Other comparative studies should be able to highlight other essential features such as regulatory regions at the DNA level or at the transcript level, especially in UTRs.

Comparative studies have also shown that small open reading frames (smORFs) are conserved during evolution (Kessler et al. 2003). These studies, however, are just beginning, and most smORFs remain to be detected. Their discovery can greatly benefit from renewed analyses based on EST and cDNA collections in combination with comparative studies.

We have shown here that the use of conservation of contiguity between ecores in the construction of ecotigs is a valuable procedure for attempting to extend genes. This procedure is, however, not sufficient per se, since it may link ecores from consecutive genes together and should therefore be experimentally verified. It is hypothesized that the fusion of consecutive gene models into a single

**Table 5.** Whole-genome Comparison between Target, Human Genome Assembly (Build 31) and Query, *Tetraodon* Genome Assembly (Whitehead Institute/Genoscope)

| | Ecotigs | Ecores | Ensembl genes | Ensembl exons | RefSeq genes | RefSeq exons |
|---|---|---|---|---|---|---|
| Total 34,057 | 141,623 | 23,436 | 190,869 | 15,383 | | 145,416 |
| Ecotigs overlapping Ensembl genes | 23,298 | 113,240 | 17,294 | 126,202 | | |
| Ecotigs overlapping Ensembl exons | 19,160 | 106,557 | 16,180 | 91,868 | | |
| Ecores outside Ecotigs/Ensembl exons | | 6,090 | 16,180 | | | |
| Ecotigs overlapping RefSeq genes | 18,474 | 92,015 | | | 12,604 | 100,429 |
| Ecotigs overlapping RefSeq exons | 15,227 | 86,226 | | | 12,130 | 75,236 |

ecotig is a result of conservation of synteny between the pair of genomes analyzed. Ecotigs could therefore also be used to evaluate the degree of such conserved synteny.

Comparative genomics is a challenging new field of research that is still in its infancy. Its future will require the development of analysis tools which depend on a better understanding of the nature and action of evolutionary forces that shape the genomes of multicellular organisms.

## ACKNOWLEDGMENTS

This work was supported by Consortium national de recherche en génomique.

## REFERENCES


Altschul S.F., Gish W., Miller W., Myers E.W., and Lipman D.J. 1990. Basic local alignment search tool. *J. Mol. Biol.* **215:** 403.

Amores A., Force A., Yan Y.L., Joly L., Amemiya C., Fritz A., Ho R.K., Langeland J., Prince V., Wang Y.L., Westerfield M., Ekker M., and Postlethwait J.H. 1998. Zebrafish hox clusters and vertebrate genome evolution. *Science* **282:** 1711.

Aparicio S., Chapman J., Stupka E., Putnam N., Chia J.M., Dehal P., Christoffels A., Rash S., Hoon S., Smit A., Gelpke M.D., Roach J., Oh T., Ho I.Y., Wong M., Detter C., Verhoef F., Predki P., Tay A., Lucas S., Richardson P., Smith S.F., Clark M.S., Edwards Y.J., and Doggett N., et al. 2002. Whole-genome shotgun assembly and analysis of the genome of *Fugu rubripes*. *Science* **297:** 1301.

*Arabidopsis* Genome Initiative. 2000. Analysis of the genome sequence of the flowering plant *Arabidopsis thaliana*. *Nature* **408:** 796.

Celniker S.E., Wheeler D.A., Kronmiller B., Carlson J.W., Halpern A., Patel S., Adams M., Champe M., Dugan S.P., Frise E., Hodgson A., George R.A., Hoskins R.A., Laverty T., Muzny D.M., Nelson C.R., Pacleb J.M., Park S., Pfeiffer B.D., Richards S., Sodergren E.J., Svirskas R., Tabor P.E., Wan K., and Stapleton M., et al. 2002. Finishing a whole-genome shotgun: Release 3 of the *Drosophila melanogaster* euchromatic genome sequence. *Genome Biol.* **3:** RESEARCH0079.

Dunham I., Shimizu N., Roe B.A., and Chissoe S. 1999. The DNA sequence of human chromosome 22. *Nature* **402:** 489.

Goff S.A., Ricke D., Lan T.H., Presting G., Wang R., Dunn M., Glazebrook J., Sessions A., Oeller P., Varma H., Hadley D., Hutchison D., Martin C., Katagiri F., Lange B.M., Moughamer T., Xia Y., Budworth P., Zhong J., Miguel T., Paszkowski U., Zhang S., Colbert M., Sun W.L., and Chen L., et al. 2002. A draft sequence of the rice genome (*Oryza sativa* L. ssp. *japonica*). *Science* **296:** 92.

Gopal S., Schroeder M., Pieper U., Sczyrba A., Aytekin-Kurban G., Bekiranov S., Fajardo J.E., Eswar N., Sanchez R., Sali A., and Gaasterland T. 2001. Homology-based annotation yields 1,042 new candidate genes in the *Drosophila melanogaster* genome. *Nat. Genet.* **27:** 337.

Heilig R., Eckenberg R., Petit J.L., Fonknechten N., Da Silva C., Cattolico L., Levy M., Barbe V., de Berardinis V., Ureta-Vidal A., Pelletier E., Vico V., Anthouard V., Rowen L., Madan A., Qin S., Sun H., Du H., Pepin K., Artiguenave F., Robert C., Cruaud C., Bruls T., Jaillon O., and Friedlander L., et al. 2003. The DNA sequence and analysis of human chromosome 14. *Nature* **421:** 601.

Holt R.A., Subramanian G.M., Halpern A., Sutton G.G., Charlab R., Nusskern D.R., Wincker P., Clark A.G., Ribeiro J.M., Wides R., Salzberg S.L., Loftus B., Yandell M., Majoros W.H., Rusch D.B., Lai Z., Kraft C.L., Abril J.F., Anthouard V., Arensburger P., Atkinson P.W., Baden H., de Berardinis V., Baldwin D., and Benes V., et al. 2002. The genome sequence of the malaria mosquito *Anopheles gambiae*. *Science* **298:** 129.

Jaillon O., Dossat C., Eckenberg R., Eiglmeier K., Segurens B., Aury J.M., Roth C.W., Scarpelli C., Brey P.T., Weissenbach J., and Wincker P. 2003. Assessing the *Drosophila melanogaster* and *Anopheles gambiae* genome annotations using genome-wide sequence comparisons. *Genome Res.* **13:** 1595.

Kessler M.M., Zeng Q., Hogan S., Cook R., Morales A.J., and Cottarel G. 2003. Systematic discovery of new genes in the *Saccharomyces cerevisiae* genome. *Genome Res.* **13:** 264.

Korf I., Flicek P., Duan D., and Brent M.R. 2001. Integrating genomic homology into gene structure prediction. *Bioinformatics* (suppl. 1) **17:** S140.

Lander E.S., Linton L.M., Birren B., Nusbaum C., Zody M.C., Baldwin J., Devon K., Dewar K., Doyle M., FitzHugh W., Funke R., Gage D., Harris K., Heaford A., Howland J., Kann L., Lehoczky J., LeVine R., McEwan P., McKernan K., Meldrim J., Mesirov J.P., Miranda C., Morris W., and Naylor J., et al. (International Human Genome Sequencing Consortium). 2001. Initial sequencing and analysis of the human genome. *Nature* **409:** 860.

Meyer I.M. and Durbin R. 2002. Comparative ab initio prediction of gene structures using pair HMMs. *Bioinformatics* **18:** 1309.

Misra S., Crosby M.A., Mungall C.J., Matthews B.B., Campbell K.S., Hradecky P., Huang Y., Kaminker J.S., Millburn G.H., Prochnik S.E., Smith C.D., Tupy J.L., Whitfied E.J., Bayraktaroglu L., Berman B.P., Bettencourt B.R., Celniker S.E., de Grey A.D., Drysdale R.A., Harris N.L., Richter J., Russo S., Schroeder A.J., Shu S.Q., and Stapleton M., et al. 2002. Annotation of the *Drosophila melanogaster* euchromatic genome: A systematic review. *Genome Biol.* **3:** RESEARCH0083.

Pachter L., Alexandersson M., and Cawley S. 2002. Applications of generalized pair hidden Markov models to alignment and gene finding problems. *J. Comput. Biol.* **9:** 389.

Roest Crollius H., Jaillon O., Bernot A., Dasilva C., Bouneau L., Fischer C., Fizames C., Wincker P., Brottier P., Quetier F., Saurin W., and Weissenbach J. 2000. Estimate of human gene number provided by genome-wide analysis using *Tetraodon nigroviridis* DNA sequence. *Nat. Genet.* **25:** 235.

Ureta-Vidal A., Ettwiller L., and Birney E. 2003. Comparative genomics: Genome-wide analysis in metazoan eukaryotes. *Nat. Rev. Genet.* **4:** 251.

Waterston R.H., Lindblad-Toh K., Birney E., Rogers J., Abril J.F., Agarwal P., Agarwala R., Ainscough R., Alexandersson M., An P., Antonarakis S.E., Attwood J., Baertsch R., Bailey J., Barlow K., Beck S., Berry E., Birren B., Bloom T., Bork P., Botcherby M., Bray N., Brent M.R., Brown D.G., and Brown S.D., et al. (Mouse Genome Sequencing Consortium). 2002. Initial sequencing and comparative analysis of the mouse genome. *Nature* **420:** 520.

Wiehe T., Gebauer-Jung S., Mitchell-Olds T., and Guigo R. 2001. SGP-1: Prediction and validation of homologous genes based on sequence alignments. *Genome Res.* **11:** 1574.

Wittbrodt J., Meyer A., and Schartl M. 1998. More genes in fish? *Bioessays* **20:** 511.

Yeh R.F., Lim L.P., and Burge C.B. 2001. Computational inference of homologous gene structures in the human genome. *Genome Res.* **11:** 803.

# Comparative Genomic Tools for Exploring the Human Genome

I. OVCHARENKO* AND G.G. LOOTS†
*EEBI Computing Division, †Genome Biology Division
Lawrence Livermore National Laboratory, Livermore, California 94550

The human genome has been sequenced, but understanding its coding potential will take much longer than previously anticipated. As we play the numbers game in vain, we are left wondering how many genes are really encrypted in its sequence, and what the biological roles of the rest of the genome, comprising mostly noncoding sequences, may be. Only ~5% of the genome is predicted to encode proteins and ~45% is considered "junk DNA" due to old retroviral contaminations. In general, we have minimal substantiated evidence on the biological functions of the remaining ~50% of the genome that consists of noncoding sequences. Identifying all the protein-coding sequences, as well as decoding the noncoding portion of the human genome, is an enormous obstacle to overcome.

One possible solution to this problem has become evident in the recent years: We need to take advantage of the evolutionary forces that have separated us from our metazoan relatives through the use of comparative sequence analysis. Comparing multiple genomes provides insights into deeply conserved regions of the human genome that are ultimately important and furthers our understanding of the biology of *Homo sapiens*. Despite tremendous technological advances in engineering mutations in experimental animal models (particularly the laboratory mouse), assigning biological functions to newly identified genes and their associated regulatory elements still remains a daunting and time-consuming effort, dependent on highly skilled scientists who can generate such genetic manipulations. To minimize the search space within genomic sequences under biological investigation, we have created a suite of comparative genomic tools: the *ECR Browser*, *rVista*, and *eShadow* to assist in the analysis of the human genome. These tools are designed to perform genome-scale comparisons of multiple vertebrate genomes in order to prioritize genic and nongenic candidate regions that can be pursued experimentally. They are capable of detecting similar DNA patterns in highly divergent species, and of amplifying the biological signal of conservation in a limited set of very closely related organisms. In this paper, we summarize the utility and applications of these tools and briefly describe the fundamental principles behind the core algorithms that are implemented by these programs.

## COMPARING VERTEBRATE GENOMES: THE ECR BROWSER

Comparing the genomes of multiple organisms selected to represent a wide range of evolutionary distances is an extremely powerful approach for identifying functional coding and noncoding sequences—mostly because biologically relevant sequences tend to evolve at slower rates than nonfunctional DNA regions (Loots et al. 2000; Pennacchio et al. 2001; Lim et al. 2003). To create a resource that assists biologists in extracting phylogenomic information, we have aligned the genomes of two rodents (mouse and rat), three fishes (zebrafish and two pufferfish: *Fugu rubripes* and *Tetraodon nigroviridis*), an amphibian (the frog *Xenopus tropicalis*), and human, and displayed the pair-wise comparisons in a publicly available genome browser of evolutionary conservation, the *ECR Browser* (http://ecrbrowser.dcode.org/).

Genome browsers are databases designed to navigate a completely sequenced genome by scrolling and zooming through any DNA region and visualizing all the known sequence features or annotations. Available annotations include mRNAs, expressed sequence tags (ESTs), gene predictions, single nucleotide polymorphisms (SNPs), repeats, conserved elements, as well as many other additional features. *Ensembl* (http://www.ensembl.org/) and the *UCSC Genome Browser Database* (http://genome.ucsc.edu/) are two representative browsers that were originally designed to support the assembly and annotation needs for the Human Genome Project by creating an efficient, user-friendly data storage and retrieval system with a compact visual display (Hubbard et al. 2002; Karolchik et al. 2003; Ureta-Vidal et al. 2003). Both of these browsers have expanded rapidly to provide access to additional available assembled genomes and their accompanying annotations. The *ECR Browser* is designed as an extension to the genome resources compiled in the *Ensembl* and *UCSC Genome* databases, providing detailed comparative genomic information for distantly related vertebrate organisms that have been fully sequenced. The main purpose of the *ECR Browser* tool is to supply a framework of vertebrate evolutionary and phylogenetic relationships that can be easily manipulated and customized by users.

Aligning the human genome to another genome is by far a nontrivial task. Finding true orthologous matches, as well as connecting these matches to generate contiguous whole-genome sequence alignments, is a computationally challenging task that requires innovative approaches to increase the accuracy and efficiency for the currently established comparative tools (Schwartz et al. 2003). Identifying synteny of sequences from closely related organisms, such as primates, is a straightforward process. In contrast, determining sequence orthology between highly divergent regions from distantly related genomes, such as humans and rodents, is considerably more difficult, due to significant DNA rearrangements that have taken place over the course of evolution. Chromosomal segments have been reshuffled significantly throughout evolution; therefore, correct alignments of completely sequenced genomes require the recreation of the pathways that lead to these rearrangements. Challenges in generating accurate alignments persist even when performing comparisons of relatively small orthologous intervals that are only a few megabases (Mb) in length. This is due to the fact that even well conserved regions may contain insertions and deletions (indels), as well as single base-pair mutations. Indels represent sequences that lack orthologous counterparts and are indicated by gaps in alignments (Tatusova and Madden 1999; Bailey et al. 2002; Kent 2002), and even highly orthologous regions in closely related species are rich in indels and mutations (Frazer et al. 2003).

Another potential difficulty in obtaining accurate syntenic sequence alignments is created by the large numbers of tandem and segmental duplications found in the human genome (Bailey et al. 2002). Current assembly strategies are unable to accurately differentiate highly homologous duplicons from true overlapping sequences, resulting in erroneous genomic assemblies with underrepresented paralogs. In addition, lineage-specific segmental duplications may lack truly orthologous sequences (Shannon et al. 2003). These caveats diminish the power of comparative sequence analysis to highlight potentially functional regions in lineage-specific genomic intervals. The overwhelming number of highly similar paralogous regions in most vertebrate genomes leads to major difficulties in determining true orthology and synteny in the absence of a one-to-one sequence match and the presence of numerous short alignments with combinatorial structure of matches, mismatches, and gaps (Ureta-Vidal et al. 2003).

To overcome some of the problems faced while determining true synteny in alignments, we have scored all the similar matches and allowed multiple orthologous/paralogous regions from the queried species to align to a single location in the reference genome. As a result, the *ECR Browser* (http://ecrbrowser.dcode.org/) reveals the complete pattern of conservation for the human genome while comparing it with all other available genomes (Fig. 1B). This approach, which conceptually coalesces local and global syntenic principles, detects sequence similarities even when highly divergent genomes are being compared, where the gene structure and order are not well maintained over such large evolutionary distances (Schwartz et al. 2000; Kent 2002).

Users can enter a region of the human genome in the *ECR Browser* by searching for a landmark such as the name or acronym of a known gene, the NCBI accession number of an mRNA or DNA sequence, the numeric positions within a human chromosome (Fig. 1A), or by homology searches if a piece of sequence is provided. To search by homology (<GENOME ALIGNMENT> Feature of the Browser), users can type in a sequence, upload a sequence file in FASTA format, or provide an NCBI accession number for a piece of DNA from any organism (not necessarily from the available genomes) and search the three available genomes. Homologous regions will be displayed in the *ECR Browser* and shaded in different colors depending on the available gene annotations. Exons are depicted in blue, 3′ and 5′ untranslated regions (UTR) are displayed in yellow, repetitive elements are in green, intronic elements are shown in pink, and intergenic noncoding elements are depicted in red shadings (Fig. 1D). If a sequence file is used to search the *ECR Browser*, the human/input-sequence pair-wise comparisons can be visualized as an additional alignment track labeled with a question mark (Fig. 1E). In addition to the ready-made alignments, the *ECR Browser* can also align the sequences provided by the user to the available base genomes, and incorporate any available annotation into the dynamic visual display of the alignments as a *zPicture* (data not shown; http:///zPicture/dcode.org).

Although several genome browsers provide easy access to the underlying sequence data, virtually none provides detailed analysis of intergenic and intronic noncoding regions. Most available genome browsers mainly focus on providing annotations for protein-coding genes. In sharp contrast, the *ECR Browser*'s main purpose is to assist during the process of deciphering the noncoding portion of the human genome, especially the evolutionarily conserved regions (ECRs) that potentially code for transcriptional regulatory elements. The <Grab ECR> feature provides effortless access to any conserved DNA element, while allowing users to analyze the sequence structure of these specific regions. Three options are available: (1) graphical and textual alignment visualization, (2) statistical evaluation for the likelihood of an ECR element to resemble noncoding RNA gene structures (through the use of QRNA tool) (Rivas and Eddy 2001), and (3) transcription factor-binding site analysis via the *rVista* tool portal (Loots et al. 2002).

The *ECR Browser* is a highly versatile computational tool for exploring the phylogenetic relationships of the human genome with the other sequenced genomes. We plan to expand the current alignments to include pairwise alignments for all available vertebrate genomes, particularly for the dog, cow, and chicken, when assemblies for these genomes become available. We are currently using the gene annotation information available at the *UCSC Genome Database*, but we are also establishing alternative options that will incorporate user-submitted annotations such as coding exons, regulatory elements, and other biologically meaningful genomic features. Our goal is to provide a robust comparative genomics infrastructure that can be easily accessed and manipulated over the Web to facilitate the retrieval and analysis of phylogenomic information.

## ANALYZING NONCODING REGULATORY REGIONS: THE rVISTA TOOL

Decoding transcriptional regulatory networks controlling gene expression remains one of the most challenging endeavors of sequence-based research. In vertebrates,

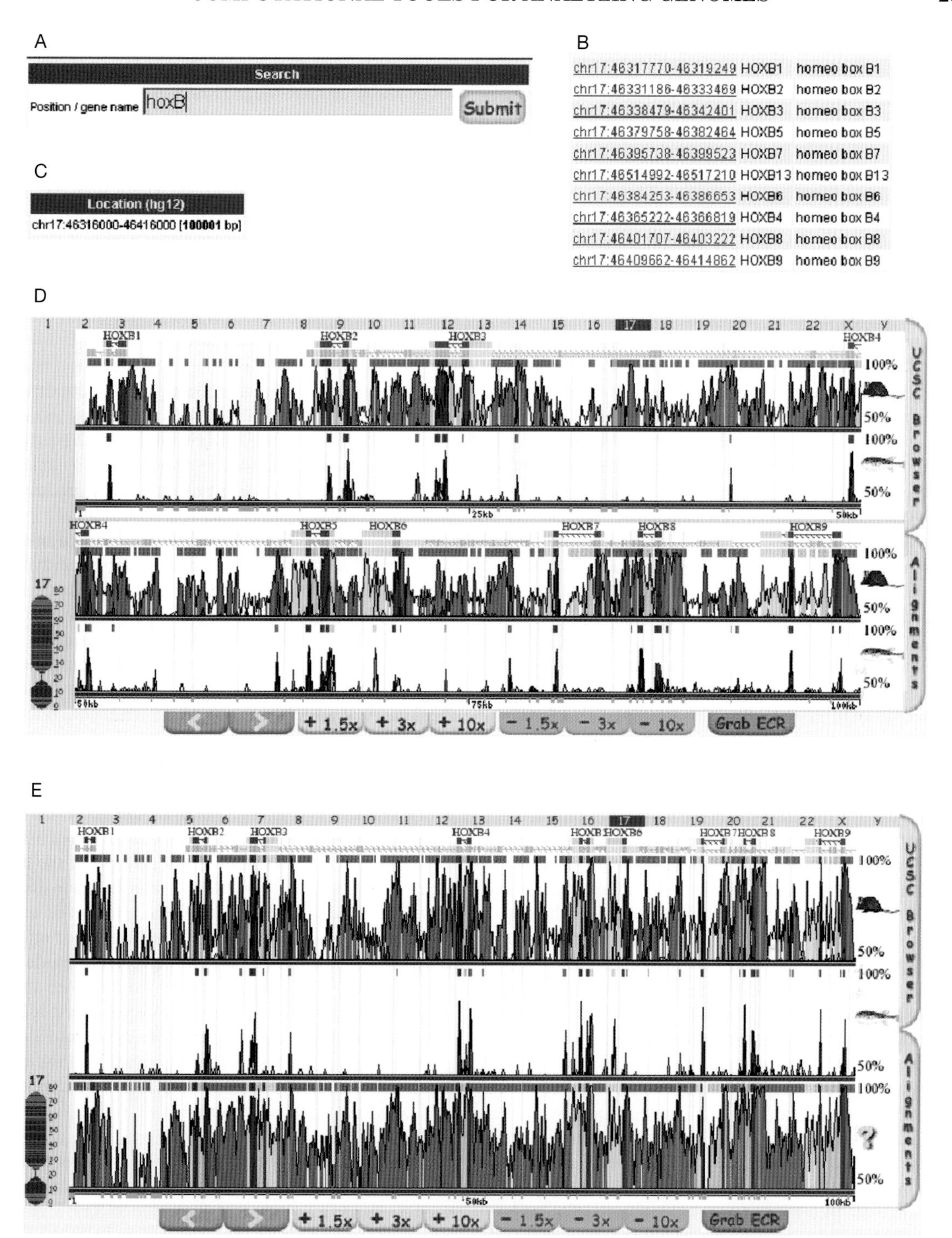


**Figure 1.** HoxB cluster viewed in the *ECR Browser*. To start using the *ECR Browser*, it is possible to begin by searching for a gene name or a specific location within the human genome. For example, to locate the HoxB cluster, the user should input "hoxb" in the search box and click on the submit button (*A*). The database will be searched and it will generate a list with all the regions that include *HoxB*-related genes (*B*). Since all the *HoxB* genes are adjacent to each other, any gene can be selected and it will direct to alignments on human chromosome 17. The window size can be adjusted to include all the *HoxB* genes present in this cluster (*C*), and the human/mouse and human/*Fugu* fish pair-wise alignments can be visualized as conservation plots (*D*). Annotated exons are colored in blue, untranslated regions in yellow, repetitive elements in green, intronic noncoding elements in pink, and intergenic noncoding elements in red (*D,E*). By selecting <UCSC Browser> and <Alignments> tabs, users can cross-reference the information displayed in the *ECR Browser* with the information available in the *UCSC* database, and can download sequence and annotation files, or customize alignment plots using the *zPicture* option (data not shown). Any sequence can be compared to the human, mouse, or *Fugu* genomes (upload sequences as FASTA files; or NCBI accession numbers). When a cow HoxB contig (AC129959.6) is compared and aligned to the human genome, the *ECR Browser* allows the visualization of the human/cow pair-wise alignment as an independent track (labeled by a question mark) (*E*).

modulation of gene expression is achieved through the complex interaction of regulatory proteins (*trans*-factors) with specific DNA regions (*cis*-acting regulatory sequences) (Krivan and Wasserman 2001). Intensive experimental efforts over several decades have identified numerous regulatory proteins, transcription factors (TF), and their DNA-binding specificities. The sequence-specific DNA-binding activity of TFs is central to transcriptional regulation and regulatory networks. For most of the known TFs, the DNA-binding sequence motifs were found to be short (6–12 bp) and highly degenerate, and such specificities are cataloged in the *TRANSFAC* databases (http://www.biobase.de/) (Wingender et al. 2001; Matys et al. 2003). Pattern-recognition programs (*MATCH* or *MatInspector*) (Quandt et al. 1995) use this database to carry out the reverse-engineering task of predicting significant motif matches in DNA sequences, which could serve as an in silico strategy for detecting transcription factor binding sites (TFBSs). Due to their highly degenerate nature, TF motifs occur very frequently in short genomic intervals, and only a very small fraction of the predicted sites are biologically significant. This fact limits the use of TFBS databases for sequence-based discovery of transcriptional regulatory elements (Fickett and Wasserman 2000).

Multispecies comparative sequence analysis or *phylogenetic footprinting* has been suggested as a strategy to counter the large numbers of TFBS false positives derived from the analysis of a single sequence (Gumucio et al. 1996; Duret and Bucher 1997; Levy et al. 2001). In general, it is believed that TFBSs are the building blocks of gene regulation. Several studies have carefully shown that transcriptional regulatory elements are evolutionarily conserved, supporting the use of genomic comparisons for the de novo discovery of gene regulatory elements (Hardison et al. 1997; Oeltjen et al. 1997; Loots et al. 2000). In addition, in complex organisms, gene expression results from the cooperative action of many different proteins simultaneously required to cooperatively activate and modulate gene expression (Berman et al. 2002). Therefore, a potential avenue for improving the discovery of functional regulatory elements is to identify multiple TFBSs that are specifically clustered together (Kel et al. 1999; Zhu et al. 2002). This strategy has been implemented successfully in the analysis of regulatory regions involved in muscle (Wasserman and Fickett 1998) and liver-specific gene expression (Krivan and Wasserman 2001).

To facilitate the efficient and accurate prediction of biologically functional regulatory sequences present in large genomic intervals, we have developed a computational tool, *Regulatory VISTA* or *rVISTA* (http://rvista.dcode.org/) that enriches for functional TFBSs using evolutionary conservation (Loots et al. 2002). The *rVISTA* tool combines TFBS motif recognition, orthologous sequence alignments, and TFBS cluster analysis to overcome some of the limitations associated with TFBS predictions of sequences derived from a single organism. The analysis proceeds in four steps: (1) identification of TFBS matches in the individual sequences, (2) identification of locally aligned noncoding TFBSs, (3) calculation of local conservation extending upstream and downstream from each orthologous TFBS, and (4) visualization of individual or clustered noncoding TFBSs. Alignments generated by the *zPicture* (http:zpicture.dcode. org/) program can be processed by *rVISTA* by following the *rVISTA* link provided in the results page. The user can also submit alignment files generated by *PipMaker* along with the corresponding sequence annotations (optional) to identify conserved TFBS matches present only in noncoding genomic intervals. Pre-computed matrices imported from the *TRANSFAC* database or user-defined consensus sequences can be used to identify TFBS motifs in the input sequences. The alignment and annotation files are used next to identify all the aligned TFBSs present in noncoding DNA and to calculate the degree of DNA conservation encompassing each TFBS. *rVISTA* calculates the maximum DNA conservation surrounding each aligned binding site in a dynamically shifting window, 20 bp in length, and filters out the sites present in regions that are less than 80% conserved (Fig. 2) (Loots et al. 2002).

The data generated by *rVISTA* are compiled into two types of outputs: (1) static data files and (2) a dynamic Web-interactive graphical user interface that maps TFBS on top of the conservation plot. The static text files include data tables with detailed statistics for all aligned and conserved TFBSs, alignment files depicting the text for each TFBS match in a different color, and files with the numerical position of each TFBS within the reference sequence. The visualization module allows the user to customize the data and graphically visualize TFBS together with the alignment conservation plot. Since regulatory regions in higher eukaryotes are represented by conglomerates of multiple TFBSs that act in concordance to directly modulate the expression patterns of the linked genes (Pilpel et al. 2001), *rVISTA* calculates the distance between all neighboring TFBSs and allows the user to perform customized clustering of individual or multiple unique transcription factors. One clustering module allows the user to selectively cluster two or more sites of the same TF present in regions of user-defined lengths, and a second clustering module allows the user to identify groups of multiple sites specific for different TFs, to predict DNA regions with unique regulatory signatures (Fig. 2) (Loots et al. 2002).

Recently, the *ECR Browser* has included an *rVISTA* portal that takes advantage of the available precomputed whole-genome pair-wise alignments, eliminating the need to create an alignment file prior to TFBS analysis while using the *rVISTA* tool. Users now have the option to browse the human genome and to perform *rVISTA* analysis on individual highly conserved noncoding elements by using the <Grab ECR> button, or on long genomic intervals containing several blocks of conservation by using the <Alignments> link. *rVISTA* analysis can then be conducted on any available pair-wise alignments by pushing the <run rVISTA> button. From this point on, TFBS analysis proceeds as described in Figure 2B–E, and is processed by the *rVISTA* software.

Annotating the noncoding portion of the human genome still remains one of the greatest challenges post-

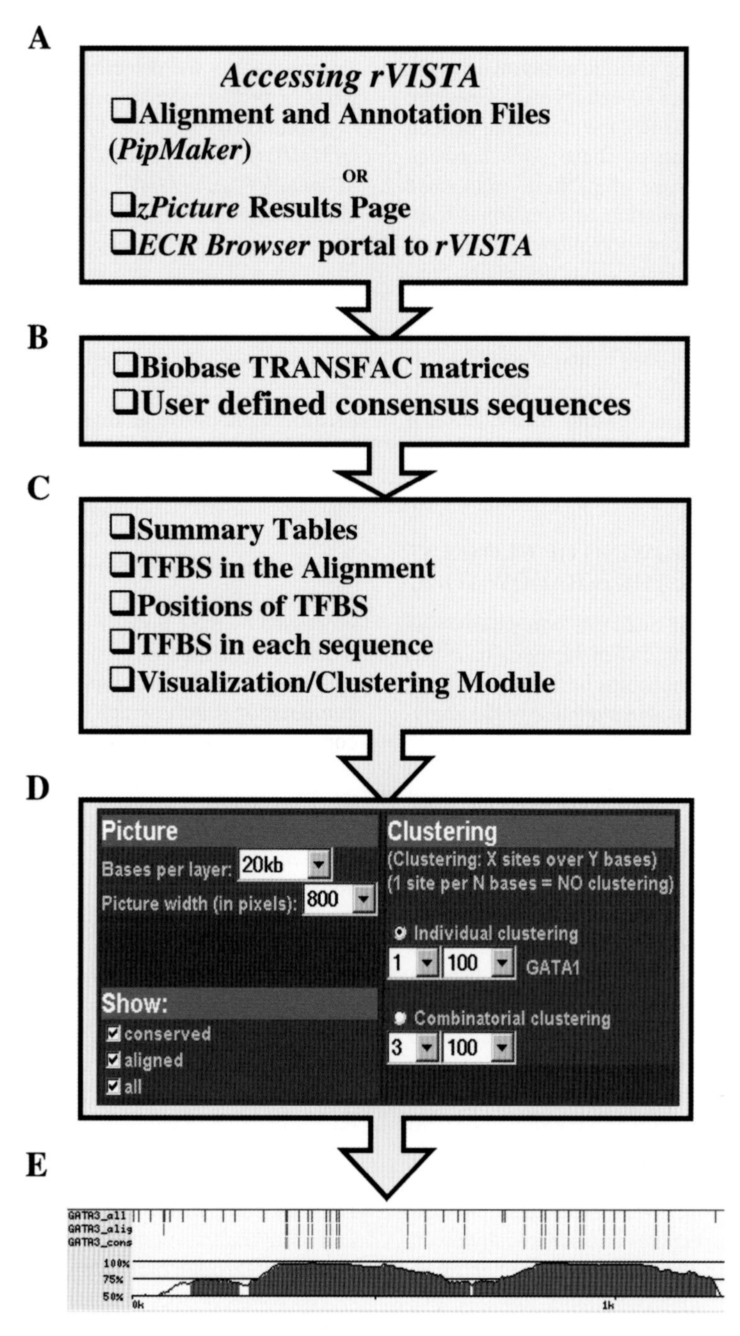


**Figure 2.** Analyzing TFBSs using the *rVISTA* tool. *rVISTA* can be accessed from its own home page (http://rvista.dcode.org/), or through the *zPicture* alignment program (http://zpicture.dcode.org/) and the *ECR Browser* (http:ecrbrowser.dcode.org/) by using the <Run rVISTA> option. *rVISTA* requires the user to submit an alignment file generated by *PipMaker* and optional annotation files (*A*). The imported *TRANSFAC* matrix library and the *MATCH* program are consequently used to identify all TFBS matches in each individual sequence and to generate a file with all TFBS matches in the reference sequence (used as baseline for visualization). The user can also provide customized consensus binding sites that *rVISTA* can use to search each individual sequence (*B*). The alignment and the annotation files are used next to identify all aligned TFBSs present in noncoding regions (in the absence of annotation, the program will identify all aligned sites across the entire alignment). DNA sequence conservation is determined by the hula-hoop module (Loots et al. 2002), which identifies TFBSs located inside islands of local conservation and generates a data table with detailed statistics. *rVISTA* will also return to the user a link to a Web-interactive visualization module, and an alignment file with TFBS highlighted in different colors (*C*). Using the interactive visualization component of *rVISTA*, the user can customize the data by choosing which TF sites to visualize, what *TRANSFAC* parameters to use for TF matches (*rVISTA* default 0.9/0.85), and by selectively clustering individual or combinatorial TFBS (*D*). The user can customize the clustering of the data sets (all matches in the reference sequence are depicted as *blue* tick marks, aligned TFBS matches are in *red*, and conserved TFBS matches are in *green*). Binding-site matches with the user-defined characteristics will be illustrated as tick marks above the alignment plot (*E*).

sequencing complete vertebrate and mammalian genomes. Clues for identifying sequences involved in the complex regulatory networks of eukaryotic genes are provided by the presence of TFBS, the clustering of such motifs, and the conservation of these sites between species. *rVISTA* takes advantage of all these established strategies to enhance the detection of functional transcriptional regulatory sequences controlling gene expression through its ability to identify evolutionarily conserved and clustered sites. *rVISTA*'s ability to use comparative data and clustering options in a user-friendly manner makes it particularly suited to assist investigators focused on biologically defined genomic intervals, as well as those interested in performing whole-genome analyses to identify functional TFBSs and transcriptional regulatory elements.

## PHYLOGENETIC SHADOWING OF CLOSELY RELATED SEQUENCES: THE eSHADOW TOOL

Despite the wide utility of available comparative sequence analysis tools and established methods based on the distant evolutionary relationships of mouse and man, their uses become constrained when applied to the analysis of recent segmental duplications (Bailey et al. 2002), actively evolving gene clusters of paralogous genes (Shannon et al. 2003), slowly diverging genomic intervals (Balavoine et al. 2002), or genomic regions specific to a lineage of closely related species (Boffelli et al. 2003). To obtain a deep understanding of the biology of *Homo sapiens*, as well as to gain valuable insights into the origins of hominoids, we need to be able to compare our genomic sequences not only to the genomes of distantly related organisms (rodents and lower vertebrates), but also to genomes closely related to us (apes and monkeys). Extracting biologically meaningful information from such comparisons is very challenging, since the genomes of humans and our closest relative, the chimpanzee, are almost indistinguishable, with greater than 98% nucleotide sequence identity.

Recently, a novel approach was introduced, *phylogenetic shadowing*, which computes and statistically evaluates conservation profiles of multiple sequence alignments from closely related species. Using this statistical method, it was demonstrated that exons and transcriptional regulatory elements can be accurately predicted, validating the use of this approach for deciphering primate-specific functional DNA sequences (Boffelli et al. 2003). We have created a publicly accessible automated tool based on the principle of *phylogenetic shadowing*, the *eShadow* tool (http://eshadow.dcode.org/) (I. Ovcharenko et al., in prep.) and broadened its applications to include protein sequence analysis in addition to multiple nucleotide sequence comparisons. We have also expanded the limits of the *phylogenetic shadowing* method to generate meaningful statistical data while analyzing a limited number of closely related sequences, with a minimum amount of information content, and found *eShadow* to be able to identify functional elements even from human/chimp/baboon multiple sequence alignments.

The *eShadow* program combines three different statistical approaches for detecting highly conserved DNA regions or protein domains that are actively being selected by evolutionary forces: (1) *Hidden Markov Model Islands* (*HMMI*), (2) *Divergence Threshold* (*DT*), and (3) *Match/Mismatch Chain* (*MMC*) (Fig. 3B). This tool then allows Web-interactive visualization of nucleotide or protein multiple sequence alignments to create conservation profiles that will define regions with the smallest amount of collective variation or *evolutionary Shadows* (*eShadow*) (Fig. 3A). In addition to being able to reliably identify known human exons and predict primate-specific noncoding regulatory elements, the *eShadow* tool is capable of performing amino acid alignments for a set of closely related orthologous or paralogous peptide sequences to correctly predict protein domains of known critical function (data not shown) (I. Ovcharenko et al., in prep.).

The user is required to submit two or more sequences (nucleotide or amino acid) in the standard FASTA format (Pearson 1990; Pearson and Lipman 1988) that will be processed by the *ClustalW* alignment program to create a multiple sequence alignment (Thompson et al. 1994). The alignment is next forwarded to the main *eShadow* computational unit, which converts the alignment into a conservation plot (Fig. 3A) and contains the three analytical components for detecting slow-mutating regions (highly conserved elements) within alignments (Fig. 3B). The *Hidden Markov Model Islands* (*HMMI*) component is the most robust module of the three different statistical approaches implemented by the *eShadow* tool. *HMMI* scans multiple sequence alignments, examines the distribution of gaps, and evaluates the conservation profile to generate statistically significant prediction for protein and DNA elements that are under selective pressure. It performs extremely well in identifying known coding exons and putative regulatory elements in comparisons of multiple closely related primate organisms and can deduce elements accurately even from a single pair-wise alignment between human and baboon genomic sequences (I. Ovcharenko et al., in prep.). The regions defined by comparisons between closely related species in *eShadow* analysis compare well with data obtained independently through human/mouse genome comparisons.

The *eShadow* tool is able to predict exonic sequences and determine approximate exon–intron boundaries of transcripts without any a priori knowledge of the underlying gene structure. Exons and exon–intron boundary predictions are determined by identifying regions of overlap between *HMMI* calculations, open reading frame (ORF), and splice site deductions. ORFs are visualized alongside *HHMI*, *DT*, and *MMC* predictions as bars of different color shadings, with yellow bars indicating regions likely to be exonic in nature, where ORFs superimposed with the *HMMI* blocks (Fig. 3C). An optimization module has also been included in the *eShadow* software and is able to train the underlying HMMI unit to differentially predict functional elements based on annotations provided by the user.

The greatest strength of the *eShadow* tool is its ability to identify slow-mutating conserved regions in align-

A

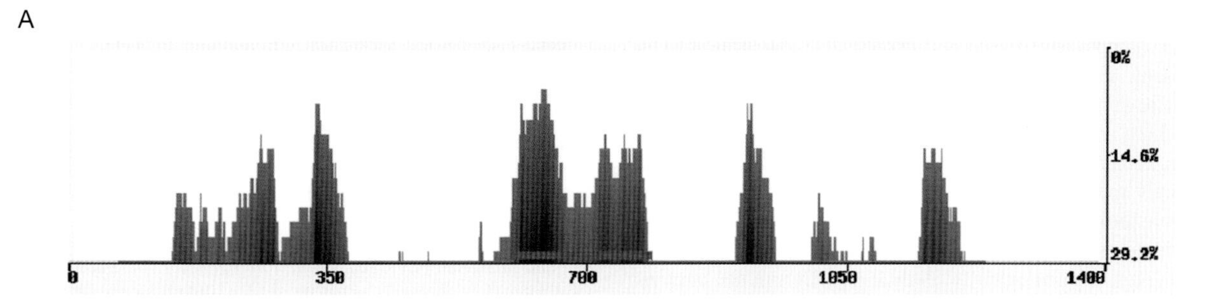


B

SUBMIT

request id:6815210

Picture settings

Sliding window size: 50 Bases per layer: 1400 Layer width (pixels): 150 Base organism: human

Mapping conserved regions

HMM Islands (Orange Bars) | Divergence Threshold (Green Bars) | Match/Mismatch Chain (Blue Bars)

Probabilities:
eS: 0.85 ; eF: 0.77 ; T: 0.1

Show ORFs
Parameters optimization
Golden section search (FA is required)
Baum-Welch (FA is useful)
Maximum Likelihood estimators (FA is required)

Maximum percent variation: 20
Minimal length: 80

1. Collapse matches: 2
2. Collapse mismatches: 5
3. Length filter: 80

Features annotation (FA)

602 784

Output data

Show eShadow text output below
Files: input sequences, clustalW alignment and output, evolutionary tree parameters

SUBMIT

C

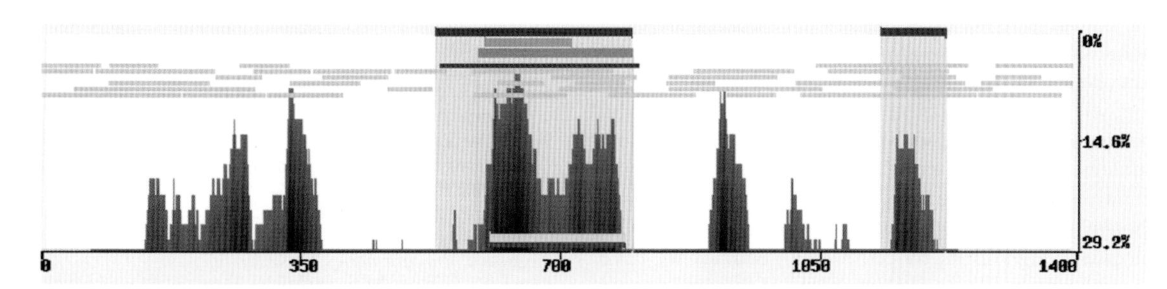


**Figure 3.** ApoB multiple primate sequence alignment visualized using the *eShadow* tool. Genomic ApoB DNA from the following species: *Homo sapiens*, *Macaca mulatta*, *Macaca nemestrina*, *Pongo pygmaeus*, *Presbytis entellus*, *Gorilla gorilla*, *Pan troglodytes*, *Pan paniscus*, *Saguinus labiatus*, *Saimiri sciureus*, *Nasalis larvatus*, *Alouatta seniculus*, *Hylobates lar*, and *Callicebus moloch* were aligned and visualized by the *eShadow* tool (*A*). Most highly conserved region in the alignment (*A*) corresponds with the *ApoB* exon (annotated by a *red* bar). Image settings, parameter values, and annotations can be adjusted and customized by the user (*B*). *HMM Islands* (*beige*), *Divergence Threshold* (*green*), and *Match/Mismatch Chain* (*blue*) predictions highly correlate with the known *ApoB* exon (*C*). The longest (*brown*) open reading frame (*gray*) overlapping *HMMI* (*beige*) predictions is used to deduce the location of putative exons (*yellow*). The predicted exon (*yellow*) exactly coincides with the *ApoB* known exon (*red*). (Percent variation) *y* axis, (size in bp) *x* axis (*A,C*).

ments of closely related species, making it very useful for the comparative analysis of organisms most similar to humans. This tool has the potential to identify primate-specific sequences, particularly gene regulatory elements, which are believed to be responsible for the majority of molecular differences that set primates and other mammals apart. *eShadow* is currently the only publicly available tool that can assist in the analysis of primate-specific elements, or indeed elements specific to any evolutionary lineage, by statistically evaluating alignments between highly similar protein and nucleotide sequences.

## CONCLUSIONS

In general, computational predictions have strongly correlated with functionally characterized coding and noncoding elements mostly because the training sets used for these analyses have been carefully chosen from biologically validated data sets. Thus, the practical limitations of computationally derived de novo predictions have not been avidly assessed through biological experimentation. In-depth characterizations of DNA sequences of unknown biological function represent a major bottleneck in expanding human genome sequence annotation to include noncoding sequences and novel genes. Comparative genomic tools and strategies have been used effectively to prioritize candidate regions (coding for novel genes and regulatory elements) to be tested in biological experiments, and have the potential to be extremely useful during the analysis of fully sequenced genomes.

The strong suite of bioinformatic tools we have presented here take advantage of phylogenomic relationships between different species and create a framework of evolutionary conservation between the human genome and the genomes of other fully or partially sequenced organisms. The precomputed data available at the *ECR Browser* as well as the data that can be generated using both the *rVista* and the *eShadow* tools, have the potential to computationally detect the most essential DNA elements present in a complex eukaryotic genome. These tools should therefore facilitate the discovery of novel genes and biologically important noncoding elements in the human genome. Bioinformatics and comparative genomics are emerging as strong disciplines presenting researchers with new in silico solutions for extracting biological information from sequence data. Currently, comparative sequence analysis is one of the best available strategies for computationally predicting and prioritizing putative functional regions in the human genome. Although not all computer-generated predictions may be biologically significant, by comparing genomes we have found a way to effectively filter some of the randomness produced by evolutionary forces to focus on the most likely functional elements for experimental studies.

## ACKNOWLEDGMENTS

The authors thank L. Stubbs and M. Nobrega for critical reading of the manuscript. This work was performed under the auspices of the U.S. Department of Energy by the University of California, Lawrence Livermore National Laboratory, Contract No. W-7405-Eng-48.

## REFERENCES

Bailey J.A., Gu Z., Clark R.A., Reinert K., Samonte R.V., Schwartz S., Adams M.D., Myers E.W., Li P.W., and Eichler E.E. 2002. Recent segmental duplications in the human genome. *Science* **297:** 1003.

Balavoine G., de Rosa R., and Adoutte A. 2002. Hox clusters and bilaterian phylogeny. *Mol. Phylogenet. Evol.* **24:** 366.

Berman B.P., Nibu Y., Pfeiffer B.D., Tomancak P., Celniker S.E., Levine M., Rubin G.M., and Eisen M.B. 2002. Exploiting transcription factor binding site clustering to identify cis-regulatory modules involved in pattern formation in the *Drosophila* genome. *Proc. Natl. Acad. Sci.* **99:** 757.

Boffelli D., McAuliffe J., Ovcharenko D., Lewis K.D., Ovcharenko I., Pachter L., and Rubin E.M. 2003. Phylogenetic shadowing of primate sequences to find functional regions of the human genome. *Science* **299:** 1391.

Duret L. and Bucher P. 1997. Searching for regulatory elements in human noncoding sequences. *Curr. Opin. Struct. Biol.* **7:** 399.

Fickett J.W. and Wasserman W.W. 2000. Discovery and modeling of transcriptional regulatory regions. *Curr. Opin. Biotechnol.* **11:** 19.

Frazer K.A., Chen X., Hinds D.A., Pant P.V., Patil N., and Cox D.R. 2003. Genomic DNA insertions and deletions occur frequently between humans and nonhuman primates. *Genome Res.* **13:** 341.

Gumucio D.L., Shelton D.A., Zhu W., Millinoff D., Gray T., Bock J.H., Slightom J.L., and Goodman M. 1996. Evolutionary strategies for the elucidation of cis and trans factors that regulate the developmental switching programs of the beta-like globin genes. *Mol. Phylogenet. Evol.* **5:** 18.

Hardison R.C., Oeltjen J., and Miller W. 1997. Long human-mouse sequence alignments reveal novel regulatory elements: A reason to sequence the mouse genome. *Genome Res.* **7:** 959.

Hubbard T., Barker D., Birney E., Cameron G., Chen Y., Clark L., Cox T., Cuff J., Curwen V., Down T., Durbin R., Eyras E., Gilbert J., Hammond M., Huminiecki L., Kasprzyk A., Lehvaslaiho H., Lijnzaad P., Melsopp C., Mongin E., Pettett R., Pocock M., Potter S., Rust A., and Schmidt E., et al. 2002. The Ensembl genome database project. *Nucleic Acids Res.* **30:** 38.

Karolchik D., Baertsch R., Diekhans M., Furey T.S., Hinrichs A., Lu Y.T., Roskin K.M., Schwartz M., Sugnet C.W., Thomas D.J., Weber R.J., Haussler D., and Kent W.J. 2003. The UCSC Genome Browser Database. *Nucleic Acids Res.* **31:** 51.

Kel A., Kel-Margoulis O., Babenko V., and Wingender E. 1999. Recognition of NFATp/AP-1 composite elements within genes induced upon the activation of immune cells. *J. Mol. Biol.* **288:** 353.

Kent W.J. 2002. BLAT—The BLAST-like alignment tool. *Genome Res.* **12:** 656.

Krivan W. and Wasserman W.W. 2001. A predictive model for regulatory sequences directing liver-specific transcription. *Genome Res.* **11:** 1559.

Levy S., Hannenhalli S., and Workman C. 2001. Enrichment of regulatory signals in conserved non-coding genomic sequence. *Bioinformatics* **17:** 871.

Lim L.P., Glasner M.E., Yekta S., Burge C.B., and Bartel D.P. 2003. Vertebrate microRNA genes. *Science* **299:** 1540.

Loots G.G., Ovcharenko I., Pachter L., Dubchak I., and Rubin E.M. 2002. rVista for comparative sequence-based discovery of functional transcription factor binding sites. *Genome Res.* **12:** 832.

Loots G.G., Locksley R.M., Blankespoor C.M., Wang Z.E., Miller W., Rubin E.M., and Frazer K.A. 2000. Identification of a coordinate regulator of interleukins 4, 13, and 5 by cross-species sequence comparisons. *Science* **288:** 136.

Matys V., Fricke E., Geffers R., Gossling E., Haubrock M., Hehl

R., Hornischer K., Karas D., Kel A.E., Kel-Margoulis O.V., Kloos D.U., Land S., Lewicki-Potapov B., Michael H., Munch R., Reuter I., Rotert S., Saxel H., Scheer M., Thiele S., and Wingender E. 2003. TRANSFAC: Transcriptional regulation, from patterns to profiles. *Nucleic Acids Res.* **31:** 374.

Oeltjen J.C., Malley T.M., Muzny D.M., Miller W., Gibbs R.A., and Belmont J.W. 1997. Large-scale comparative sequence analysis of the human and murine Bruton's tyrosine kinase loci reveals conserved regulatory domains. *Genome Res.* **7:** 315.

Pearson W.R. 1990. Rapid and sensitive sequence comparison with FASTP and FASTA. *Methods Enzymol.* **183:** 63.

Pearson W.R. and Lipman D.J. 1988. Improved tools for biological sequence comparison. *Proc. Natl. Acad. Sci.* **85:** 2444.

Pennacchio L.A., Olivier M., Hubacek J.A., Cohen J.C., Cox D.R., Fruchart J.C., Krauss R.M., and Rubin E.M. 2001. An apolipoprotein influencing triglycerides in humans and mice revealed by comparative sequencing. *Science* **294:** 169.

Pilpel Y., Sudarsanam P., and Church G.M. 2001. Identifying regulatory networks by combinatorial analysis of promoter elements. *Nat. Genet.* **29:** 153.

Quandt K., Frech K., Karas H., Wingender E., and Werner T. 1995. MatInd and MatInspector: New fast and versatile tools for detection of consensus matches in nucleotide sequence data. *Nucleic Acids Res.* **23:** 4878.

Rivas E. and Eddy S.R. 2001. Noncoding RNA gene detection using comparative sequence analysis. *BMC Bioinformatics* **2:** 8.

Schwartz S., Kent W.J., Smit A., Zhang Z., Baertsch R., Hardison R.C., Haussler D., and Miller W. 2003. Human-mouse alignments with BLASTZ. *Genome Res* **13:** 103.

Schwartz S., Zhang Z., Frazer K.A., Smit A., Riemer C., Bouck J., Gibbs R., Hardison R., and Miller W. 2000. PipMaker—A web server for aligning two genomic DNA sequences. *Genome Res.* **10:** 577.

Shannon M., Hamilton A.T., Gordon L., Branscomb E., and Stubbs L. 2003. Differential expansion of zinc-finger transcription factor loci in homologous human and mouse gene clusters. *Genome Res.* **13:** 1097.

Tatusova T.A. and Madden T.L. 1999. BLAST 2 Sequences, a new tool for comparing protein and nucleotide sequences. *FEMS Microbiol. Lett.* **174:** 247.

Thompson J.D., Higgins D.G., and Gibson T.J. 1994. CLUSTAL W: Improving the sensitivity of progressive multiple sequence alignment through sequence weighting, position-specific gap penalties and weight matrix choice. *Nucleic Acids Res.* **22:** 4673.

Ureta-Vidal A., Ettwiller L., and Birney E. 2003. Comparative genomics: Genome-wide analysis in metazoan eukaryotes. *Nat. Rev. Genet.* **4:** 251.

Wasserman W.W. and Fickett J.W. 1998. Identification of regulatory regions which confer muscle-specific gene expression. *J. Mol. Biol.* **278:** 167.

Wingender E., Chen X., Fricke E., Geffers R., Hehl R., Liebich I., Krull M., Matys V., Michael H., Ohnhauser R., Pruss M., Schacherer F., Thiele S., and Urbach S. 2001. The TRANSFAC system on gene expression regulation. *Nucleic Acids Res.* **29:** 281.

Zhu Z., Pilpel Y., and Church G.M. 2002. Computational identification of transcription factor binding sites via a transcription-factor-centric clustering (TFCC) algorithm. *J. Mol. Biol.* **318:** 71.

# Evolution of Eukaryotic Gene Repertoire and Gene Structure: Discovering the Unexpected Dynamics of Genome Evolution

I.B. Rogozin,* V.N. Babenko,* N.D. Fedorova,* J. D. Jackson,* A.R. Jacobs,* D.M. Krylov,* K.S. Makarova,* R. Mazumder,*†¶ S.L. Mekhedov,* B.G. Mirkin,† A.N. Nikolskaya,*¶ B.S. Rao,* S. Smirnov,* A.V. Sorokin,* A.V. Sverdlov,* S. Vasudevan,* Y.I. Wolf,* J.J. Yin,* D.A. Natale,*¶ and E.V. Koonin*

*National Center for Biotechnology Information, National Library of Medicine, National Institutes of Health, Bethesda, Maryland and †School of Information Systems and Computer Science, Birkbeck College, University of London, London, WC1E 7HX, United Kingdom*

## COMPARATIVE GENOMICS, EVOLUTIONARY CLASSIFICATION OF GENES, AND PHYLETIC PATTERNS

Comparative genomics has already changed our understanding of genome evolution. In what might amount to a paradigm shift in evolutionary biology, genome comparisons have shown that lineage-specific gene loss and horizontal gene transfer (HGT) are not freak incidents of evolution but extremely common phenomena that, to a large degree, have shaped the extant genomes, at least those of prokaryotes (Doolittle 1999; Gogarten et al. 2002; Snel et al. 2002; Mirkin et al. 2003). The extent of gene loss occurring in certain lineages of prokaryotes, particularly parasites, is astonishing: In some cases, >80% genes in the genome have been lost over ~200 million years of evolution (Moran 2002). Horizontal gene transfer is harder to document, but a strong case has been made for its extensive contribution to the evolution of prokaryotes (Ochman et al. 2000; Koonin et al. 2001; Mirkin et al. 2003).

Gene exchange between phylogenetically distant eukaryotes does not appear to be an important evolutionary phenomenon. In contrast, the contribution of gene loss to the evolution of eukaryotic genomes was probably substantial, although the level of genome fluidity observed in prokaryotes is unlikely to have been attained in eukaryotic evolution. A comparison of the genomes of two yeasts, *Saccharomyces cerevisiae* and *Schizosaccharomyces pombe*, showed that, in the *S. cerevisiae* lineage, up to 10% of genes have been lost since the divergence of the two species (Aravind et al. 2000). It appears likely that, in eukaryotic parasites with small genomes, e.g., the microsporidia, much more massive gene elimination has occurred (Katinka et al. 2001). In contrast, the extent of gene loss in complex, multicellular eukaryotes remains unclear, although the small number of unique genes in the human genome when compared to the mouse genome (and vice versa) suggests considerable stability of the gene repertoire (Waterston et al. 2002).

We are interested in quantitative analysis of the dynamics of genome evolution. A prerequisite for such studies is a classification of the genes from the sequenced genomes based on homologous relationships. The two principal categories of homologs are orthologs and paralogs (Fitch 1970; Sonnhammer and Koonin 2002). Orthologs are homologous genes that evolved via vertical descent from a single ancestral gene in the last common ancestor of the compared species. Paralogs are homologous genes, which, at some stage of evolution, have evolved by duplication of an ancestral gene. Orthology and paralogy are two sides of the same coin because, when a duplication (or a series of duplications) occurs after the speciation event that separated the compared species, orthology becomes a relationship between sets of paralogs, rather than individual genes (genes that belong to orthologous sets are sometimes called co-orthologs) (Sonnhammer and Koonin 2002).

Robust identification of orthologs and paralogs is critical for the construction of evolutionary scenarios, which include, along with vertical inheritance, lineage-specific gene loss and, possibly, HGT (Snel et al. 2002; Mirkin et al. 2003). The algorithms for the construction of these scenarios involve, in one form or another, tracing the fates of individual genes, which is feasible only when orthologs (including co-orthologs) are known. In principle, orthologs, including co-orthologs, should be identified by phylogenetic analysis of entire families of homologous proteins, which is expected to define orthologous protein sets as clades (e.g., Sicheritz-Ponten and Andersson 2001). However, for genome-wide protein sets, such analysis remains labor-intensive and error-prone. Thus, procedures have been developed for identification of sets of probable orthologs without explicit use of phylogenetic methods. Generally, these approaches are based on the notion of a genome-specific best hit (BeT), i.e., the protein from a target genome, which has the greatest sequence similarity to a given protein from the query genome (Tatusov et al. 1997; Huynen and Bork 1998). The central assumption here is that orthologs have a greater similarity to each other than to any other protein from the respective genomes due to the conservation of functional constraints. When multiple genomes are analyzed, pairs of probable orthologs detected on the basis of

¶Present address: Protein Identification Resource, Georgetown University Medical Center, 3900 Reservoir Road, NW, Washington, DC 20007.

BeTs are combined into orthologous clusters represented in all or a subset of the analyzed genomes (Tatusov et al. 1997; Montague and Hutchison 2000). This approach, amended with procedures for detecting co-orthologous protein sets and for treating multidomain proteins, was implemented in the database of *c*lusters of *o*rthologous *g*roups (COGs) of proteins (Tatusov et al. 1997, 2001). The current COG set includes ~70% of the proteins encoded in 69 genomes of prokaryotes and unicellular eukaryotes (Tatusov et al. 2003). The COGs have been extensively employed for genome-wide evolutionary studies, functional annotation of new genomes, and target selection in structural genomics (Koonin and Galperin 2002 and references therein).

A simple but critically important concept that was introduced in the context of the COG analysis is a phyletic (phylogenetic) pattern, which is the pattern of representation (presence–absence) of the analyzed species in each COG (Tatusov et al. 1997; Koonin and Galperin 2002). Similar notions have been independently developed and applied by others (Gaasterland and Ragan 1998; Pellegrini et al. 1999). The COGs show a wide scatter of phyletic patterns, with only a small minority (~1%) represented in all included genomes. Similarity and complementarity among the phyletic patterns of COGs have been successfully employed for prediction of gene functions (Galperin and Koonin 2000; Koonin and Galperin 2002; Myllykallio et al. 2002; Levesque et al. 2003). Phyletic patterns can be formally represented as strings of "1"s (for presence of a species) and "0"s (for absence of a species), which can be easily input to a variety of algorithms. The evolutionary parsimony methods are among those that naturally apply to these types of data. We recently showed that parsimonious evolutionary scenarios for most COGs involve multiple events of gene loss and/or HGT (Mirkin et al. 2003).

Recently, we extended the system of orthologous protein clusters to complex, multicellular eukaryotes by constructing clusters of eu*k*aryotic *o*rthologous *g*roups (KOGs) for seven sequenced genomes of animals, fungi, microsporidia, and plants (Tatusov et al. 2003). Here, we analyze the phyletic patterns of KOGs to extract the hidden evolutionary signals. In particular, we reconstruct the parsimonious scenario of evolution of the crown-group eukaryotes by assigning the loss of genes (KOGs) and emergence of new genes to the branches of the phylogenetic tree, and delineate the minimal gene sets for various ancestral forms. We then shift the study from the level of gene sets to the level of gene structure, construct the evolutionary scenario for intron positions in highly conserved genes, and compare the dynamics of evolution of the gene repertoire and gene structure.

## KOGS FOR SEVEN SEQUENCED EUKARYOTIC GENOMES: MAJOR TRENDS IN GENOME EVOLUTION

Eukaryotic KOGs were constructed by comparing the sequences of the (predicted) proteins encoded in the genomes of three animals (*Homo sapiens*, the fruit fly *Drosophila melanogaster*, and the nematode *Caenorhabditis elegans*), the flowering plant *Arabidopsis thaliana*, two fungi (budding yeast *S. cerevisiae* and fission yeast *S. pombe*), and the microsporidian *Encephalitozoon cuniculi*. The procedure for KOG construction was a modification of the one previously used for COGs (Tatusov et al. 1997, 2001) and is described in greater detail elsewhere (Tatusov et al. 2003). Unlike in the previous COG analyses, we strived to produce a complete evolutionary classification of eukaryotic genes. The original COGs consisted, at a minimum, of proteins from three species, which enhanced the power of the analysis and allowed incorporation of even those orthologs that showed low sequence similarity to each other. In the present analysis, we also identified clusters of putative orthologs from two species (TWOGs) and lineage-specific expansions of paralogs from each of the analyzed genomes (Lespinet et al. 2002; Tatusov et al. 2003). Thus, at least in principle, each gene in the seven analyzed eukaryotic genomes is accounted for in the emerging evolutionary classification.

Of the 112,920 analyzed proteins, 65,170 belonged to KOGs (including TWOGs), 23,436 belonged to LSEs, and 24,314 remain singletons ("orfans"). Both the considerable level of evolutionary conservation—given the wide phylogenetic span of the analyzed genomes, each KOG should be considered a highly conserved family—and the major contribution of LSEs are notable. Furthermore, the number of orfans is likely to be inflated as some of these undoubtedly are gene prediction artifacts. Figure 1 shows the assignment of the proteins from each of the analyzed eukaryotes to KOGs with different numbers of species (including TWOGs) and LSEs. The fraction of proteins assigned to KOGs generally decreases with the increasing genome size, from 81% for the fission yeast *S. pombe* to 51% for the largest, the human genome. The contribution of LSEs shows the opposite trend, being the greatest in the largest genomes, i.e., human and *Ara-*

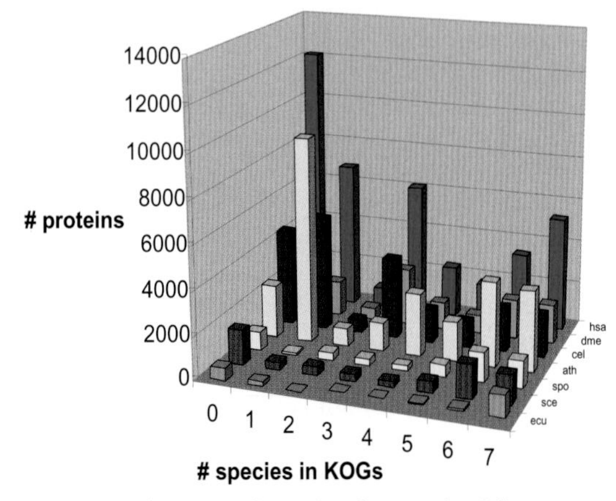


**Figure 1.** Assignment of proteins from each of the seven analyzed eukaryotic genomes to KOGs of different size and to LSEs. "0" indicates proteins without detectable homologs (orfans) and "1" indicates LSEs. Species abbreviations: (ath) *Arabidopsis thaliana*, (cel) *Caenorhabditis elegans*, (dme) *Drosophila melanogaster*, (ecu) *Encephalitozoon cuniculi*, (has) *Homo sapiens*, (sce) *Saccharomyces cerevisiae*, (spo,) *Schizosaccharomyces pombe*.

*bidopsis*, and minimal in the microsporidian (Fig. 1). A notable difference was observed in the representation of eukaryotic genomes in KOGs with different numbers of species. The three unicellular organisms are represented largely in highly conserved seven- or six-species KOGs; in contrast, in animals and *Arabidopsis*, a much larger fraction of the genes is accounted for by LSEs and by KOGs that include three or four genomes, e.g., animal-specific ones (Fig. 1).

The phyletic patterns of KOGs reveal a substantial, conserved eukaryotic gene core, but also notable diversity (Fig. 2). The "pan-eukaryotic" genes, which are represented in each of the seven analyzed genomes, comprise ~20% of the KOGs, and approximately the same number of KOGs include all species except for the microsporidian, an intracellular parasite with a highly degraded genome (Katinka et al. 2001). Among the remaining KOGs, a large fraction consists of representatives of the three analyzed animal species (worm, fly, and human), but ~30% are KOGs with unexpected patterns, e.g., one animal, one plant, and one fungal species (see http://www.ncbi.nlm.nih.gov/COG/new/shokog.cgi). During manual curation of the KOGs, those with unexpected patterns were specifically scrutinized to identify potential highly diverged members from one or more of the analyzed genomes, and the KOGs were modified when such diverged orthologs were detected. Some of these unexpected patterns probably indicate that a gene is still missing in the analyzed set of protein sequences from one or more of the included eukaryotic species. Largely, however, the unexpected phyletic patterns seem to reflect the extensive, lineage-specific gene loss that is apparently characteristic of eukaryotic evolution.

## A PARSIMONIOUS SCENARIO OF GENE LOSS AND EMERGENCE IN EUKARYOTIC EVOLUTION AND RECONSTRUCTION OF ANCESTRAL EUKARYOTIC GENE SETS

Assuming a particular species tree topology, methods of evolutionary parsimony analysis can be employed to construct a parsimonious scenario of evolution; i.e., mapping of different types of evolutionary events onto the branches of the tree such that the total number of the events is minimal. As discussed above, with prokaryotes, the problem is confounded by the fact that both lineage-specific gene loss and HGT apparently have made major contributions to genome evolution, with the relative rates of these processes remaining unknown (Snel et al. 2002; Mirkin et al. 2003). Since HGT between major lineages of eukaryotes apparently can be safely disregarded, an unambiguous parsimonious scenario including only gene loss and emergence of new genes as elementary events can be constructed. The phylogenetic tree of the eukaryotic crown group seems to have been established with near complete confidence. In particular, some conflicting observations notwithstanding, the majority of phylogenetic studies point to an animal–fungal clade, grouping of microsporidia with the fungi, and a coelomate (chordate–arthropod) clade among the animals (Blair et al. 2002; Hedges 2002; Y.I. Wolf et al., unpubl.). Assuming this tree topology and treating the phyletic pattern of each KOG as a string of binary characters (1 for the presence of the given species and 0 for its absence in the given KOG), the parsimonious scenario of gene loss and emergence during the evolution of the eukaryotic crown group was constructed. For this reconstruction, the Dollo parsimony approach was adopted (Farris 1977). Under this approach, gene loss is considered irreversible; i.e., a gene (a KOG member) can be lost independently in several evolutionary lineages but cannot be regained. The use of this assumption is justified by the negligible rate of HGT between eukaryotes (the Dollo approach is not valid for the reconstruction of prokaryotic ancestors).

In the resulting scenario, each branch was associated with a unique number of losses and gains, with the exception of the plant branch and the branch leading to the common ancestor of the fungi–microsporidial clade and the animals, to which gene losses could not be assigned with the current set of genomes (Fig. 3). Undoubtedly, once genomes of early-branching eukaryotes are included, gene loss associated with these branches will become apparent. The reconstructed scenario includes massive gene loss in the fungal clade, with additional

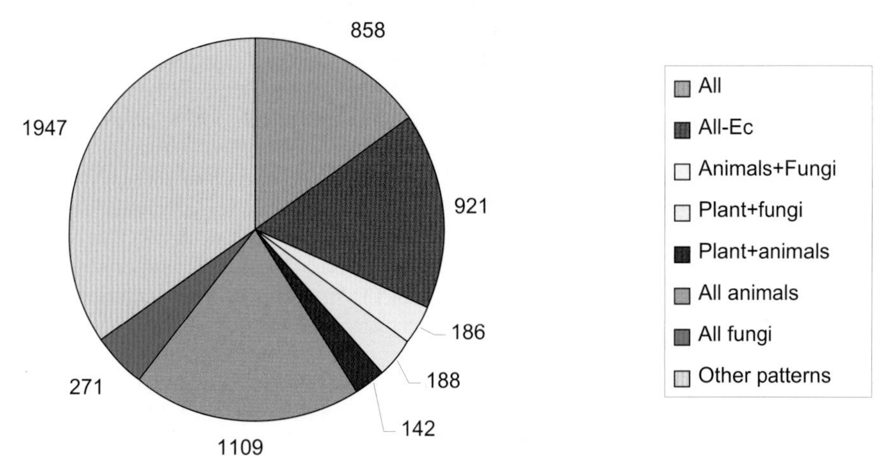


**Figure 2.** The phyletic patterns of KOGs. (All-Ec) KOGs represented in all analyzed genomes with the exception of *Encephalitozoon cuniculi*; (All animals, All fungi) KOGs represented exclusively in the three animal species or the two fungal species, respectively.

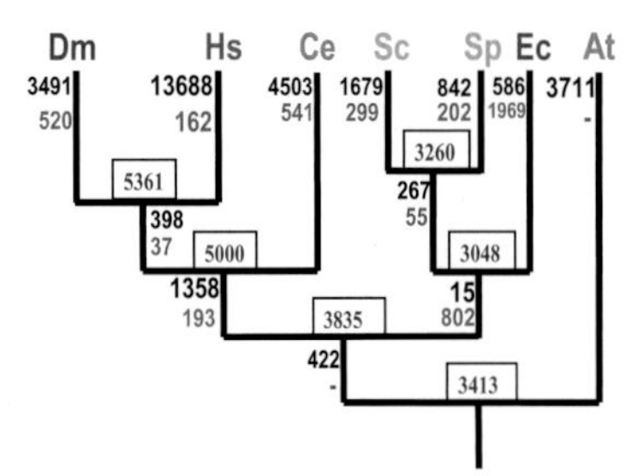


**Figure 3.** Parsimonious scenario of loss and emergence of genes (KOGs) for the most likely topology of the eukaryotic phylogenetic tree. Numbers in boxes indicate the inferred number of KOGs in the respective ancestral forms. Numbers next to branches indicate the number of gene gains (emergence of KOGs) (*top, black*) and gene (KOG) losses (*bottom, red*) associated with the respective branches; a dash indicates that the number of losses for a given branch could not be determined. Proteins from each genome that did not belong to KOGs as well as LSEs were counted as gains on the terminal branches. The species abbreviations are as in Fig. 1.

elimination of numerous genes in the microsporidian; emergence of a large set of new genes at the onset of the animal clade; and subsequent substantial gene loss in each of the animal lineages, particularly in the nematodes and arthropods (Fig. 3). The estimated number of genes lost in *S. cerevisiae* after the divergence from the common ancestor with the other yeast species, *S. pombe*, closely agreed with a previous estimate produced using a different approach (Aravind et al. 2000).

The parsimony analysis described here includes reconstruction of the gene sets of ancestral eukaryotic genomes. Under the Dollo parsimony model, an ancestral gene (KOG) set for a given clade is the union of the KOGs that are shared by the respective outgroup and each of the remaining species in the clade. Thus, the gene set for the common ancestor of the crown group includes all the KOGs in which *Arabidopsis* co-occurs with any of the other analyzed species. Similarly, the reconstructed gene set for the common ancestor of fungi and animals consists of all KOGs in which at least one fungal species co-occurs with at least one animal species. These are conservative reconstructions of ancestral gene sets because, as indicated above, gene losses in the lineages branching off the deepest bifurcation could not be detected. Under this conservative approach, 3,365 genes (KOGs) were assigned to the last common ancestor of the crown group (Fig. 3). Most likely, a certain number of ancestral genes have been lost in all, or all but one, of the analyzed lineages during subsequent evolution such that the gene set of the eukaryotic crown group ancestor might have been close in size to those of modern yeasts.

## EVOLUTION OF EUKARYOTIC GENE STRUCTURE

Most of the eukaryotic protein-coding genes contain multiple introns that are spliced out of the pre-mRNA by a distinct, large RNA–protein complex, the spliceosome, which is conserved in all eukaryotes (Dacks and Doolittle 2001). The positions of some spliceosomal introns are conserved in orthologous genes from plants and animals (Marchionni and Gilbert 1986; Logsdon et al. 1995; Boudet et al. 2001). A recent systematic analysis of pairwise alignments of homologous proteins from animals, fungi, and plants suggested that 10–15% of the introns are ancient (Fedorov et al. 2002). However, intron densities in different eukaryotic species differ widely, the location of introns in orthologous genes does not always coincide even in closely related species (Logsdon 1998), likely cases of intron insertion and loss have been described (see, e.g., Rzhetsky et al. 1997; Logsdon et al. 1998), and indications of a high intron turnover rate have been obtained (Lynch and Richardson 2002). It has been suggested that the proportion of shared intron positions decreased with increasing evolutionary distance and, accordingly, intron conservation could be a useful phylogenetic marker (Stoltzfus et al. 1997). However, the evolutionary history of introns and the selective forces that shape intron evolution remain mysterious. Although recent comparisons reveal the existence of many ancient introns shared by animals, plants, and fungi (Fedorov et al. 2002), the point(s) of origin of these introns in eukaryotic evolution and the relative contributions of intron loss and intron insertion in the evolution of eukaryotic genes remain unknown.

We used the KOG data set for analysis of evolution of intron–exon structure of eukaryotic genes on the scale of complete genomes. KOGs that are represented in all analyzed species, with a possible exception of *E. cuniculi*, were selected, and orthologs from two more eukaryotic species, the mosquito *Anopheles gambiae* and the apicomplexan malarial parasite *Plasmodium falciparum*, were added to these KOGs using the COGNITOR method (Tatusov et al. 1997). Many of the KOGs include multiple paralogs from one or more of the constituent species, due to lineage-specific duplications (see above); among these paralogs, the one showing the greatest evolutionary conservation (defined as the mean similarity to KOG members from other species) was selected. For a pair of introns to be considered orthologous, they were required to occur in exactly the same position in the aligned sequences of KOG members. Given the well-known problems in the annotation of gene structure and difficulties in aligning poorly conserved regions of protein sequences, we used two approaches to the analysis of evolutionary conservation of intron positions. Under the first schema, all intron positions were extracted from automatically produced alignments, whereas under the second schema, only positions surrounded by well-conserved, unambiguous portions of the alignment were analyzed. Altogether, 684 KOGs were examined for intron conservation; these comprised the great majority, if not the entirety, of highly conserved eukaryotic genes that are amenable for an analysis of the exon–intron structure over the entire span of crown group evolution. The analyzed KOGs contained 21,434 introns in 16,577 unique positions (10,066 introns in 7,236 positions when only the conserved portions of alignments were analyzed); 5,981 introns were conserved in two or more genomes (4,619 in conserved regions). Most of the con-

**Table 1.** Conservation of Intron Positions in Orthologous Gene Sets from Eight Eukaryotic Species

| Number of species | | 1 | 2 | 3 | 4 | 5 | 6 | 7 | 8 |
|---|---|---|---|---|---|---|---|---|---|
| Number of introns - total | observed[a] | 13,406 | 2,047 | 719 | 275 | 104 | 25 | 1 | 0 |
| | expected | 21,368 | 33 | 0 | 0 | 0 | 0 | 0 | 0 |
| | expected - 10% | 20,083 | 662 | 8 | 0 | 0 | 0 | 0 | 0 |
| Number of introns - conserved blocks | observed[a] | 5,446 | 1,122 | 411 | 163 | 74 | 19 | 1 | 0 |
| | expected | 9,982 | 42 | 0 | 0 | 0 | 0 | 0 | 0 |
| | expected - 10% | 8,613 | 689 | 25 | 0 | 0 | 0 | 0 | 0 |

[a]The probability that intron sharing in different species was due to chance, $p$(Monte Carlo) <0.0001 (applies both to the analysis of all alignment positions and to the test with 10% of the positions allowed for intron insertion).

served introns were present in only two species, but a considerable number were found in three genomes and several introns were shared by four to seven species (Table 1). A simulation of the intron distribution in the analyzed sample of orthologous gene sets by random shuffling of the intron positions showed that ~1% of the observed number of introns shared by two species was expected to occur by chance but none was expected to be shared by three or more species (Table 1). It has been proposed that introns insert into coding sequences not randomly but primarily into "proto-splice sites" (Dibb 1991). Although the proto-splice model has been questioned as inconsistent with the observed distribution of intron phases (Long et al. 1995), we considered the potential effect of nonrandom intron insertion on the apparent evolutionary conservation of intron positions. For this purpose, random simulation was repeated with intron insertion allowed in 10% of the positions in the analyzed genes. Obviously, this led to an increase in the expected number of shared introns in two or more species, but the excess of introns found in the same position remained substantial and highly statistically significant (Table 1). These observations show that the great majority of introns located in the same position in orthologous genes from different eukaryotic lineages are indeed orthologous; i.e., originate from an ancestral intron in the same position in the respective gene of the last common ancestor of the compared species.

The matrix of shared introns in all pairs of analyzed eukaryotic genomes revealed a decidedly unexpected pattern (Table 2). The number of conserved introns did not drop monotonically with the increase of the evolutionary distance between the compared organisms. On the contrary, human genes shared the greatest number of introns not with any of the three animals but with the plant *Arabidopsis*. In the conserved regions (the more accurate results, given the uncertainties in alignment in other parts of genes), 24% of the intron positions in the analyzed human genes were shared with *Arabidopsis* (these comprised ~27% of the *Arabidopsis* introns) compared to ~12–17% of intron positions shared by humans with the fly, mosquito, or worm (Table 2). The difference becomes even more dramatic when the numbers of introns conserved in *Arabidopsis* and each of the three animal species are compared: approximately three times more plant introns have a counterpart in the same position in orthologous human genes than in the fly or worm orthologs (Table 2). Although yeast *S. pombe* and the apicomplexan protist *Plasmodium* have few introns compared to plants or animals, the same asymmetry was observed for these organisms: The numbers of introns shared with *Arabidopsis* and humans are close and are 2–3 times greater than the number of introns shared with the insects or the worm (Table 2).

*Plasmodium*, a repesentative of the alveolates, is believed to have branched off the trunk of the eukaryotic tree almost 2 billion years ago, prior to the divergence of the crown-group lineages, including animals, plants, and fungi (Hedges 2002). Thus, it is particularly notable that *Plasmodium* shares 143 (nearly one-third) of the 450 introns present in the conserved regions of the analyzed genes with at least one crown-group species (Table 2). Thus, a substantial fraction of the introns in extant eukaryotic genomes seem to be inherited from a common ancestor of the crown group and the alveolates; i.e., almost from the onset of eukaryotic evolution. Furthermore, the common ancestor of the crown group apparently had an intron-rich genome; the majority of the

**Table 2.** Conservation of Intron Positions in Eukaryotic Orthologous Gene Sets: The Matrix of Pair-wise Interspecies Comparisons

| | Pf | Sc | Sp | At | Ce | Dm | Ag | Hs |
|---|---|---|---|---|---|---|---|---|
| Pf | **450/971** | 2 | 48 | 137 | 50 | 46 | 54 | 145 |
| Sc | 1 | **22/46** | 7 | 3 | 3 | 3 | 4 | 6 |
| Sp | 34 | 6 | **450/839** | 209 | 98 | 114 | 111 | 308 |
| At | 97 | 2 | 147 | **2933/5589** | 353 | 255 | 254 | 1148 |
| Ce | 33 | 2 | 63 | 240 | **1468/3465** | 315 | 312 | 948 |
| Dm | 32 | 1 | 72 | 161 | 179 | **723/1826** | 787 | 802 |
| Ag | 36 | 1 | 62 | 158 | 176 | 382 | **675/1768** | 771 |
| Hs | 104 | 3 | 207 | 787 | 557 | 433 | 403 | **3345/6930** |

The diagonal shows the total number of introns in the 684 analyzed genes (denominator) from the given species and the number of introns in conserved regions of alignments (numerator). For each pair of species, the total number of shared introns is shown above the diagonal and the number of introns in conserved regions is shown below the diagonal. Species abbreviations: (At) *Arabidopsis thaliana,* (Ce) *Caenorhabditis elegans*, (Dm) *Drosophila melanogaster*, (Hs) *Homo sapiens*, (Ag) *Anopheles gambiae*, (Pf) *Plasmodium falciparum*, (Sc) *Saccharomyces cerevisiae*, (Sp) *Schizosaccharomyces pombe*.

ancestral introns seem to have survived in plants and animals, but have been lost in yeasts, nematodes, and arthropods.

We examined the evolutionary dynamics of introns in greater detail by using phylogenetic analysis. To this end, intron positions were represented as a data matrix of intron presence/absence (encoded as 1/0) and the matrices for all analyzed KOGs were concatenated to produce one alignment, which consisted of 16,577 columns of "ones" and "zeroes" (7,236 columns in conserved portions of alignments). In this form, the intron presence/absence data are conducive to evolutionary parsimony analysis and, of the existing parsimony approaches, Dollo parsimony seems to be most appropriate in this case (Nei and Kumar 2000). The Dollo parsimony tree that we reconstructed from the matrix of intron presence/absence obviously did not mimic the species tree, with humans and *Arabidopsis* forming a strongly supported cluster embedded within the metazoan clade, and another anomalous cluster formed by yeast *S. pombe* and *Plasmodium* (not shown). Other phylogenetic approaches, including unweighted maximum parsimony and several distance methods, which were applied to the data of Table 2, reproduced the same, incorrect topology (not shown). The topology of these trees supported the notion, already suggested by the pair-wise comparisons summarized in Table 2, that ancestral introns have been, to a large extent, conserved in plants and vertebrates, but have been extensively eliminated in fungi, nematodes, and arthropods. Clustering of *Plasmodium* with *S. pombe*, to the exclusion of the other yeast species, *S. cerevisiae*, seems to be due to the fact that *Plasmodium* shared approximately as many introns with *S. pombe* as with worm and insects, but the total number of introns in *S. pombe* was substantially less than in animals (Table 2). Thus, conservaton of intron positions does not seem to be a good source of information for inferring phylogenetic relationships at long evolutionary distances.

Having shown that evolution of introns in eukaryotic genes did not follow the species tree, we applied Dollo parsimony in the opposite direction: Given a species tree topology, construct the most parsimonious scenario for the evolution of gene structure, i.e., the distribution of intron gain and loss events over the tree branches. This approach is completely analogous to the construction of the scenario for gene gain and loss described above.

The resulting scenario suggests an intron-rich ancestor for the crown group, with limited intron loss in the animal ancestor, but massive losses in yeasts (particularly, *S. cerevisiae*), worm, and insects (Fig. 4). The differences in the relative rates of intron gain and loss in the terminal branches are remarkable: There is a huge excess of gains over losses in humans and *S. pombe*, and equally obvious excess of losses in insects and *S. cerevisiae*, whereas *C. elegans* shows nearly identical numbers of gains and losses. All introns shared by *Plasmodium* and any of the crown-group species (at least 210 as shown by analysis of complete alignments) are assigned to the last common ancestor of alveolates and the crown group, which lived some 1.5–2.0 billion years ago (Hedges 2002). At present, loss of ancestral introns in *Plasmodium* cannot be documented because *Plasmodium* is the outgroup to the crown-group species; neither can losses be assigned to the internal branch that leads to the ancestor of the crown group. Hence, we produced a conservative estimate of the number of the most ancient introns in the analyzed gene set, which is likely to be a substantial underestimate, given that *Plasmodium* is a parasite with a highly degraded genome and low intron density. Sequencing and analysis of genomes of other early branching eukaryotes is expected to substantially increase the number of introns that have survived since the dawn of eukaryotic evolution.

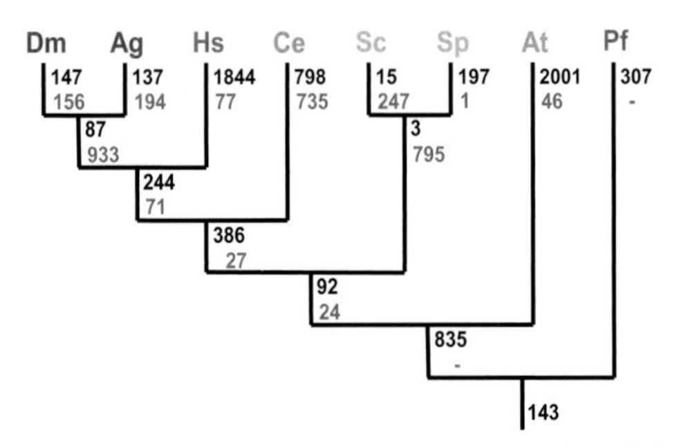


**Figure 4.** Parsimonious evolutionary scenario of intron gain/loss for the most likely topology of the eukaryotic phylogenetic tree. Intron gains and losses are mapped to each species and each internal branch. The designations are as in Fig. 3. The species abbreviations are as in Fig. 1, with the addition of *Anopheles gambiae* (Ag) and *Plasmodium falciparum* (Pf).

The present analysis pushes the origin of numerous spliceosomal introns back to the stage of eukaryotic evolution, 1.5–2.0 billion years ago, which precedes the origin of multicellularity, perhaps to the very point of origin of eukaryotes. Furthermore, as many as 25–30% of the introns in vertebrates and plants are apparently inherited from the common ancestor of the crown group. Other crown-group lineages experienced massive loss of introns. Intron elimination had been extensive already in the common ancestor of yeasts but was particularly catastrophic in *S. cerevisiae*, which apparently lost all ancestral introns. Those few introns that this organism does have are not shared with other species and probably have been regained secondarily (Table 2 and Fig. 4). More unexpectedly, insects and, particularly, the nematode have also lost the great majority of the ancestral introns, with the latter having secondarily acquired numerous new introns (Fig. 4).

Why have so many ancestral introns survived nearly 2 billion years of evolution? One explanation would involve functional roles of introns in specific positions, but there seems to be no evidence in support of such a possibility. The second and more likely explanation is that elimination of introns, particularly in highly conserved genes, which are often essential (such as those included in the present study), would often lead to gene inactivation and lethality. We found that ancient introns tend to be located in more highly conserved portions of genes than recently inserted ones (data not shown), which is compati-

**Table 3.** Dynamics of Intron Evolution in Lineage-specific Gene Expansions

| | LSEs | Intron losses | Intron gains | Intron-conserving duplications[a] | Conserved intron sites | Intron sites with gains/losses |
|---|---|---|---|---|---|---|
| *H. sapiens* | 2010 | 1492 | 4288 | 7078 | 5149 | 26110 |
| *A. thaliana* | 1842 | 1572 | 3158 | 7293 | 4927 | 16781 |
| *D. melanogaster* | 493 | 424 | 879 | 1648 | 424 | 3521 |
| *C. elegans* | 1009 | 1265 | 2398 | 3759 | 1345 | 15729 |
| *S. cerevisiae* | 44 | 47 | 38 | 194 | 41 | 46 |
| *S. pombe* | 45 | 33 | 47 | 123 | 45 | 80 |

[a]Number of duplications that have not been accompanied by loss or gain of introns.

ble with the above hypothesis. In some lineages, particularly the yeasts, but also insects and nematodes, this deleterious effect of intron loss apparently had been overcome by selective pressure for genome compaction or as a result of retrotransposition sweeps, or both. Absent such specific circumstances favoring intron elimination, many ancestral introns might have survived simply because losing them is costly.

Evolution of spliceosomal introns was long considered in the context of the "intron-early" versus "intron-late" debate. The intron-early hypothesis of Walter Gilbert posits that introns existed before the divergence of prokaryotes and eukaryotes and had an important role in the evolution of the very first functional proteins (Gilbert 1987; Gilbert and Glynias 1993). In contrast, the intron-late hypothesis holds that introns were inserted into eukaryotic genes after this primary divergence (Logsdon 1998; Lynch and Richardson 2002). Our present observations do not bear directly on the outcome of this debate, but they do show that many introns not only emerged shortly after or perhaps concomitantly with the origin of eukaryotes, but retained their positions at least in some eukaryotic lineages. Thus, the evidence presented here and elsewhere (Fedorov et al. 2002; Lynch and Richardson 2002) appears to be compatible with an "introns extremely early in eukaryotic evolution" view. It is even tempting to speculate that invasion of protein-coding genes by ancestors of introns was part of the dramatic and still mysterious series of events that led to the origin of the eukaryotic cell.

## DYNAMICS OF INTRON EVOLUTION IN PARALOGOUS GENE FAMILIES

In the previous section, we analyzed evolution of introns in highly conserved sets of orthologous eukaryotic genes and observed remarkable conservation of many intron positions, but also substantial lineage-specific loss of introns. Of no lesser interest is the dynamics of gene structure evolution in paralogous gene families. Lineage-specific expansions of paralogs exist in every branch on the tree of life, but they are particularly prominent in multicellular eukaryotes and, on many occasions, can be linked to the specific biology of the respective groups of organisms (Lespinet et al. 2002). In the process of constructing the KOG system, we delineated numerous LSEs, some of which are parts of KOGs, whereas others have no orthologs in other analyzed species (see above). For all these LSEs, we performed an analysis of the conservation of intron positions using the same strategy as described above for the KOG analysis. The preliminary results summarized in Table 3 reveal unexpected fluidity of the gene structure in LSEs, with numerous intron losses and an even substantially greater number of intron gains observed during the evolution of LSEs in each lineage. The great majority of intron sites in the LSEs have experienced at least one gain or loss event (Table 3). It has been predicted and subsequently supported by empirical analysis that gene evolution tends to accelerate after duplication (Ohno 1970; Lynch and Conery 2000; Kondrashov et al. 2002). Our current observations suggest that gene duplication also triggers mobilization of introns. Additionally, the results are compatible with the hypothesis that many duplications occurred via reverse transcription, with introns reinserting into new sites.

## CONCLUSIONS: CONGRUENT TRENDS IN THE EVOLUTION OF GENE REPERTOIRE AND GENE STRUCTURE

In the previous sections, we described the parsimonious scenarios of evolution for the eukaryotic gene repertoire and, separately, for the exon–intron structure of a set of highly conserved eukaryotic genes. Distinct trends of gene and intron loss and gain were detected for different eukaryotic lineages. To determine whether or not the loss and gain of genes and introns occurred in parallel, we plotted the respective values for each species and for the internal branches against each other. Extremely strong correlations were observed between the numbers of gene and intron gains (Fig. 5A) and gene and intron losses (Fig. 5B). To avoid confusion, it should be emphasized that the gain and loss of introns were measured for a set of highly conserved genes (KOGs), which have not experienced any gene loss or gain. Accordingly, the observed correlations were not trivial and appeared to indicate that genes and introns are gained and lost in parallel, reflecting the same, lineage-specific trends in genome evolution. Some eukaryotic lineages, such as fungi and, to a lesser extent, arthropods and nematodes, seem to be under strong selection for genome compaction, which is achieved through loss of dispensable genes, introns (even in the most conserved, housekeeping genes), and, probably, intergenic DNA. Other lineages,

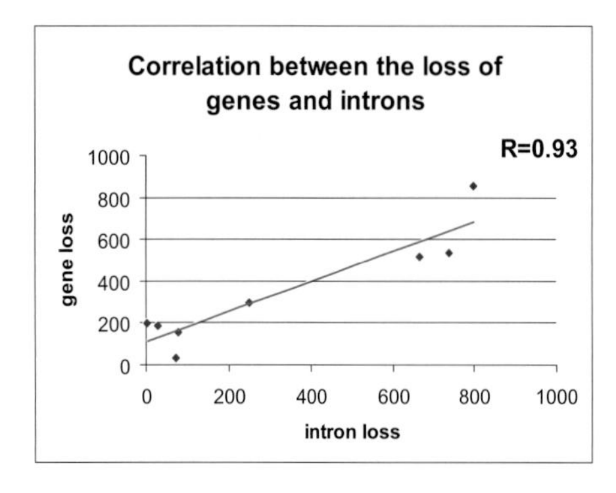


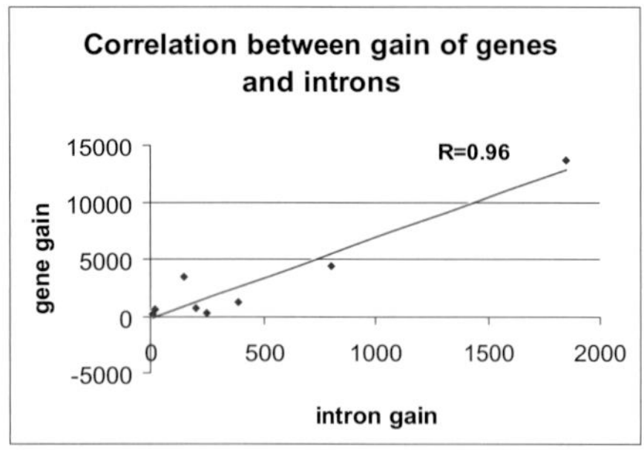


**Figure 5.** Comparative dynamics of evoluton of gene repertoires and gene structure. (*Top*) Gene (KOG) gain vs. intron gain. (*Bottom*) Gene (KOG) loss vs. intron loss. Data for gene and intron gain and loss are for each species, with the exception of *Encephalitozoon cuniculi* (intron data not analyzed since this genome contains practically no introns), *Anopheles gambiae*, and *Plasmodium falciparum* (two species not included in the KOGs), and for internal branches of the phylogenetic tree (Figs. 3 and 4).

such as vertebrates and plants, seem not to experience this selective pressure and are much more tolerant to various forms of apparently dispensable DNA, or even tend to expand their genomes. Understanding the deeper biological reasons behind these lineage-specific evolutionary trends is a fascinating task for the next generation of comparative-genomic studies.

## ACKNOWLEDGMENT

We thank Roman Tatusov for his contribution to the construction of the KOGs.

## REFERENCES

Aravind L., Watanabe H., Lipman D.J., and Koonin E.V. 2000. Lineage-specific loss and divergence of functionally linked genes in eukaryotes. *Proc. Natl. Acad. Sci.* **97:** 11319.

Blair J.E., Ikeo K., Gojobori T., and Hedges S.B. 2002. The evolutionary position of nematodes. *BMC Evol. Biol.* **2:** 7.

Boudet N., Aubourg S., Toffano-Nioche C., Kreis M., and Lecharny A. 2001. Evolution of intron/exon structure of DEAD helicase family genes in *Arabidopsis, Caenorhabditis,* and *Drosophila*. *Genome Res.* **11:** 2101.

Dacks J.B. and Doolittle W.F. 2001. Reconstructing/deconstructing the earliest eukaryotes: How comparative genomics can help. *Cell* **107:** 419.

Dibb N.J. 1991. Proto-splice site model of intron origin. *J. Theor. Biol.* **151:** 405.

Doolittle W.F. 1999. Lateral genomics. *Trends Cell Biol.* **9:** M5.

Farris J.S. 1977. Phylogenetic analysis under Dollo's Law. *Syst. Zool.* **26:** 77.

Fedorov A., Merican A.F., and Gilbert W. 2002. Large-scale comparison of intron positions among animal, plant, and fungal genes. *Proc. Natl. Acad. Sci.* **99:** 16128.

Fitch W.M. 1970. Distinguishing homologous from analogous proteins. *Syst. Zool.* **19:** 99.

Gaasterland T. and Ragan M.A. 1998. Microbial genescapes: Phyletic and functional patterns of ORF distribution among prokaryotes. *Microb. Comp. Genomics* **3:** 199.

Galperin M.Y. and Koonin E.V. 2000. Who's your neighbor? New computational approaches for functional genomics. *Nat. Biotechnol.* **18:** 609.

Gilbert W. 1987. The exon theory of genes. *Cold Spring Harbor Symp. Quant. Biol.* **52:** 901.

Gilbert W. and Glynias M. 1993. On the ancient nature of introns. *Gene* **135:** 137.

Gogarten J.P., Doolittle W.F., and Lawrence J.G. 2002. Prokaryotic evolution in light of gene transfer. *Mol. Biol. Evol.* **19:** 2226.

Hedges S.B. 2002. The origin and evolution of model organisms. *Nat. Rev. Genet.* **3:** 838.

Huynen M.A. and Bork P. 1998. Measuring genome evolution. *Proc. Natl. Acad. Sci.* **95:** 5849.

Katinka M.D., Duprat S., Cornillot E., Metenier G., Thomarat F., Prensier G., Barbe V., Peyretaillade E., Brottier P., Wincker P., Delbac F., El Alaoui H., Peyret P., Saurin W., Gouy M., Weissenbach J., and Vivares C.P. 2001. Genome sequence and gene compaction of the eukaryote parasite *Encephalitozoon cuniculi*. *Nature* **414:** 450.

Kondrashov F.A., Rogozin I.B., Wolf Y.I., and Koonin E.V. 2002. Selection in the evolution of gene duplications. *Genome Biol.* **3:** RESEARCH0008.

Koonin E.V. and Galperin M.Y. 2002. *Sequence - evolution - function: Computational approaches in comparative genomics*. Kluwer Academic, Boston, Massachusetts.

Koonin E.V., Makarova K.S., and Aravind L. 2001. Horizontal gene transfer in prokaryotes: Quantification and classification. *Annu. Rev. Microbiol.* **55:** 709.

Lespinet O., Wolf Y.I., Koonin E.V., and Aravind L. 2002. The role of lineage-specific gene family expansion in the evolution of eukaryotes. *Genome Res.* **12:** 1048.

Levesque M., Shasha D., Kim W., Surette M.G., and Benfey P.N. 2003. Trait-to-gene. A computational method for predicting the function of uncharacterized genes. *Curr. Biol.* **13:** 129.

Logsdon J.M., Jr. 1998. The recent origins of spliceosomal introns revisited. *Curr. Opin. Genet. Dev.* **8:** 637.

Logsdon J.M., Jr., Stoltzfus A., and Doolittle W.F. 1998. Molecular evolution: Recent cases of spliceosomal intron gain? *Curr. Biol.* **8:** R560.

Logsdon J.M., Jr., Tyshenko M.G., Dixon C., D-Jafari J., Walker V.K., and Palmer J.D. 1995. Seven newly discovered intron positions in the triose-phosphate isomerase gene: Evidence for the introns-late theory. *Proc. Natl. Acad. Sci.* **92:** 8507.

Long M., de Souza S.J., and Gilbert W. 1995. Evolution of the intron-exon structure of eukaryotic genes. *Curr. Opin. Genet. Dev.* **5:** 774.

Lynch M. and Conery J.S. 2000. The evolutionary fate and consequences of duplicate genes. *Science* **290:** 1151.

Lynch M. and Richardson A.O. 2002. The evolution of spliceosomal introns. *Curr. Opin. Genet. Dev.* **12:** 701.

Marchionni M. and Gilbert W. 1986. The triosephosphate isomerase gene from maize: Introns antedate the plant-animal divergence. *Cell* **46:** 133.

Mirkin B.G., Fenner T.I., Galperin M.Y., and Koonin E.V. 2003. Algorithms for computing parsimonious evolutionary scenarios for genome evolution, the last universal common ancestor and dominance of horizontal gene transfer in the evolution of prokaryotes. *BMC Evol. Biol.* **3:** 2.

Montague M.G. and Hutchison C.A., III. 2000. Gene content phylogeny of herpesviruses. *Proc. Natl. Acad. Sci.* **97:** 5334.

Moran N.A. 2002. Microbial minimalism: Genome reduction in bacterial pathogens. *Cell* **108:** 583.

Myllykallio H., Lipowski G., Leduc D., Filee J., Forterre P., and

Liebl U. 2002. An alternative flavin-dependent mechanism for thymidylate synthesis. *Science* **297:** 105.

Nei M. and Kumar S. 2000. *Molecular evolution and phylogenetics*. Oxford University Press, Oxford, United Kingdom.

Ochman H., Lawrence J.G., and Groisman E.A. 2000. Lateral gene transfer and the nature of bacterial innovation. *Nature* **405:** 299.

Ohno S. 1970. *Evolution by gene duplication*. Springer-Verlag, Berlin, Germany.

Pellegrini M., Marcotte E.M., Thompson M.J., Eisenberg D., and Yeates T.O. 1999. Assigning protein functions by comparative genome analysis: Protein phylogenetic profiles. *Proc. Natl. Acad. Sci.* **96:** 4285.

Rzhetsky A., Ayala F.J., Hsu L.C., Chang C., and Yoshida A. 1997. Exon/intron structure of aldehyde dehydrogenase genes supports the "introns-late" theory. *Proc. Natl. Acad. Sci.* **94:** 6820.

Sicheritz-Ponten T. and Andersson S.G. 2001. A phylogenomic approach to microbial evolution. *Nucleic Acids Res.* **29:** 545.

Snel B., Bork P., and Huynen M.A. 2002. Genomes in flux: The evolution of archaeal and proteobacterial gene content. *Genome Res.* **12:** 17.

Sonnhammer E.L. and Koonin E.V. 2002. Orthology, paralogy and proposed classification for paralog subtypes. *Trends Genet.* **18:** 619.

Stoltzfus A., Logsdon J.M., Jr., Palmer J.D., and Doolittle W.F. 1997. Intron "sliding" and the diversity of intron positions. *Proc. Natl. Acad. Sci.* **94:** 10739.

Tatusov R.L., Koonin E.V., and Lipman D.J. 1997. A genomic perspective on protein families. *Science* **278:** 631.

Tatusov R.L., Natale D.A., Garkavtsev I.V., Tatusova T.A., Shankavaram U.T., Rao B.S., Kiryutin B., Galperin M.Y., Fedorova N.D., and Koonin E.V. 2001. The COG database: New developments in phylogenetic classification of proteins from complete genomes. *Nucleic Acids Res.* **29:** 22.

Tatusov R.L., Fedorova N.D., Jackson J.D., Jacobs A.R., Kiryutin B., Koonin E.V., Krylov D.M., Mazumder R., Mekhedov S.L., Nikolskaya A.N., Rao B.S., Smirnov S., Sverdlov A.V., Vasudevan S., Wolf Y.I., Yin J.J., and Natale D.A. 2003. The COG database: An updated version includes eukaryotes. *BMC Bioinformatics* **4:** 41.

Waterston R.H., Lindblad-Toh K., Birney E., Rogers J., Abril J.F., Agarwal P., Agarwala R., Ainscough R., Alexandersson M., An P., Antonarakis S.E., Attwood J., Baertsch R., Bailey J., Barlow K., Beck S., Berry E., Birren B., Bloom T., Bork P., Botcherby M., Bray N., Brent M.R., Brown D.G., and Brown S.D., et al. (Mouse Genome Sequencing Consortium). 2002. Initial sequencing and comparative analysis of the mouse genome. *Nature* **420:** 520.

# Human–Mouse Comparative Genomics: Successes and Failures to Reveal Functional Regions of the Human Genome

L.A. PENNACCHIO,*† N. BAROUKH,* AND E.M. RUBIN*†
*Genome Sciences Department, MS 84-171, Lawrence Berkeley National Laboratory, Berkeley, California 94720;
†U.S. Department of Energy Joint Genome Institute, Walnut Creek, California 94598

Deciphering the genetic code embedded within the human genome remains a significant challenge despite the human genome consortium's recent success at defining its linear sequence (Lander et al. 2001; Venter et al. 2001). Although useful strategies exist to identify a large percentage of protein-encoding regions, efforts to accurately define functional sequences in the remaining ~97% of the genome lag. Our primary interest has been to utilize the evolutionary relationship and the universal nature of genomic sequence information in vertebrates to reveal functional elements in the human genome. This has been achieved through the combined use of vertebrate comparative genomics to pinpoint highly conserved sequences as candidates for biological activity and transgenic mouse studies to address the functionality of defined human DNA fragments. Accordingly, we describe strategies and insights into functional sequences in the human genome through the use of comparative genomics coupled with functional studies in the mouse.

## BACKGROUND

Mouse transgenesis experiments have constantly provided support for the universality of sequence-based regulatory information across vertebrates. Numerous examples exist where genes from a variety of vertebrates when introduced into mice as genomic transgenes express in a manner mimicking their expression in the natural host. One example of this is the human apolipoprotein A1 gene (*APOA1*), which has a well-described pattern of expression in the liver and intestines of both humans and mice. Indeed, in human *APOA1* transgenic mice, robust expression of the human gene in mouse liver and intestines was observed, consistent with the mouse's being able to recognize the regulatory sequences embedded within the human genomic transgene (Rubin et al. 1991). This *APOA1* study reflects data from a large number of mouse transgenesis experiments over the past 15 years which have repeatedly supported the idea that despite the ~80 million years since the last common ancestor of humans and mice, regulatory information has been conserved, and this supports the existence of a common gene regulatory vocabulary residing in the mammalian genome.

A particularly revealing mouse transgenesis study involved the generation and analysis of transgenic mice for a human gene for which there is no ortholog in the mouse genome. The human apolipoprotein (a) gene (*apo(a)*) recently arose in old-world monkeys, and when a large human genomic transgene (250 kb) containing *apo(a)* and flanking sequence was introduced into the mouse genome, its tissue expression pattern and components of its expression response to environmental factors mimicked that found in humans (Frazer et al. 1995). This study again highlights the existence of a highly conserved gene regulatory genetic code embedded in the noncoding sequence of mammals that determines neighboring gene expression characteristics.

## IDENTIFICATION OF A GENE REGULATORY ELEMENT THROUGH HUMAN–MOUSE COMPARATIVE GENOMICS

One challenge following traditional mouse transgenesis experiments and the many reports of successful recapitulation of human gene expression in the mouse is the downstream determination of the precise *cis*-regulatory sequences responsible for this activity. The recent availability of several vertebate genome sequences (human, mouse, rat, fugu, zebrafish) (Lander et al. 2001; Venter et al. 2001; Aparicio et al. 2002; Dehal et al. 2002; Waterston et al. 2002) has allowed the exploitation of comparative sequence analysis to reveal conserved intervals in the human genome as candidates for explaining this biological activity (Duret and Bucher 1997; Hardison et al. 1997; Hardison 2000; Pennacchio and Rubin 2001; Pennacchio et al. 2003). Since whole-genome sequence data sets for human and mouse are the most advanced (Lander et al. 2001; Venter et al. 2001; Waterston et al. 2002), we discuss the current power of comparing these two genomes, as well as the potential limitations of this single pair-wise comparison.

As an example of how comparative genomics can be used as a starting point to lead biological experimentation, we previously compared human–mouse sequence in an approximately one-megabase region (Mb) of human chromosome 5q31 (including five interleukins [IL] and 18 other genes) and its orthologous mouse region (Loots et al. 2000). This cross-species annotation of sequence identified 90 elements ≥100 bp that were conserved between human and mouse at a level of ≥70% identity (Fig. 1A). Within this data set, several previously characterized gene regulatory elements known to reside within this interval were readily identified by human–mouse sequence conservation, supporting the possibility of using such a strategy to identify gene regulatory elements.

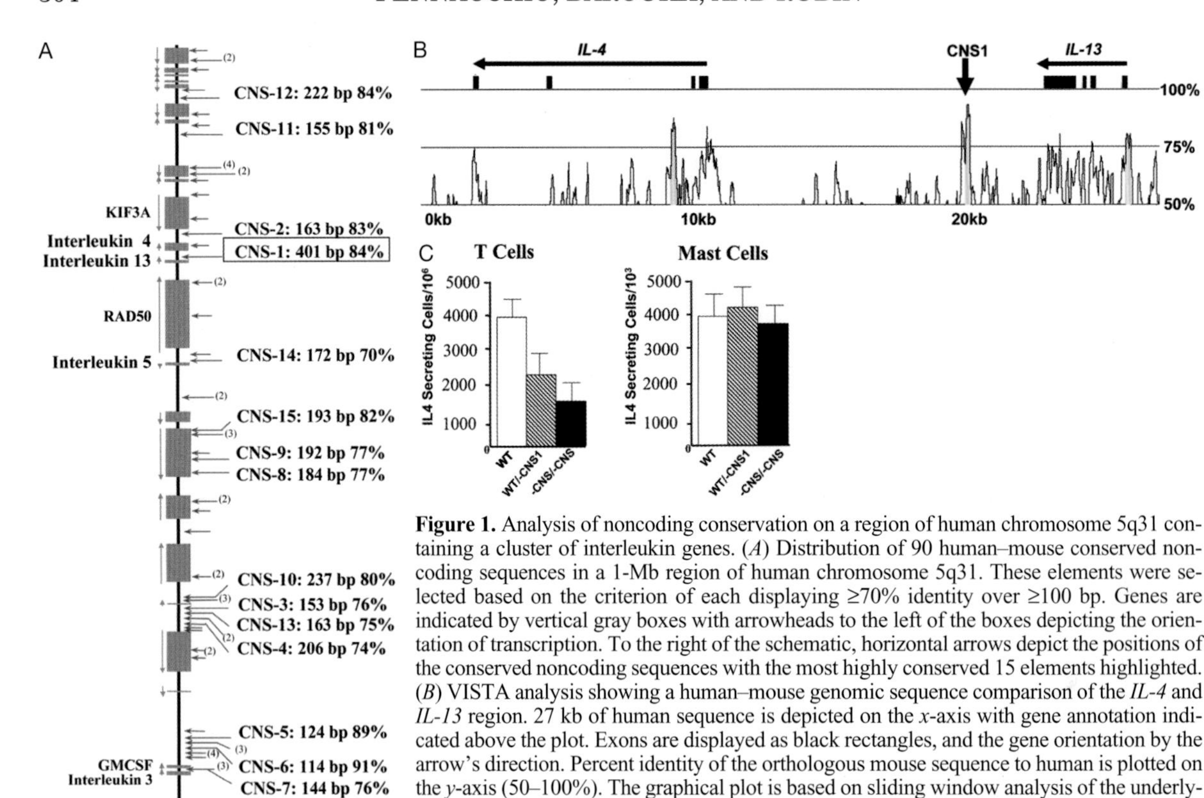


**Figure 1.** Analysis of noncoding conservation on a region of human chromosome 5q31 containing a cluster of interleukin genes. (*A*) Distribution of 90 human–mouse conserved noncoding sequences in a 1-Mb region of human chromosome 5q31. These elements were selected based on the criterion of each displaying ≥70% identity over ≥100 bp. Genes are indicated by vertical gray boxes with arrowheads to the left of the boxes depicting the orientation of transcription. To the right of the schematic, horizontal arrows depict the positions of the conserved noncoding sequences with the most highly conserved 15 elements highlighted. (*B*) VISTA analysis showing a human–mouse genomic sequence comparison of the *IL-4* and *IL-13* region. 27 kb of human sequence is depicted on the *x*-axis with gene annotation indicated above the plot. Exons are displayed as black rectangles, and the gene orientation by the arrow's direction. Percent identity of the orthologous mouse sequence to human is plotted on the *y*-axis (50–100%). The graphical plot is based on sliding window analysis of the underlying genomic alignment; in this illustration a 100-bp window is used which slides at 40-bp nucleotide increments. The vertical arrow indicates the location of CNS1, which was identified by its high degree of conservation between human and mouse (VISTA peak). (*C*) Expression analysis of mice targeted for a deletion of CNS1. Mast and T cells were isolated from wild-type, heterozygous, and homozygous mice for the deletion. In this example, T cells were stimulated with PMA, and the number of IL-4-secreting cells was determined.

To test the utility of comparative genomics to identify previously unknown gene regulatory elements, we studied the properties of a single conserved noncoding sequence (CNS1) located within the 15-kb interval between *IL-4* and *IL-13* (Fig. 1B). This single element was chosen for detailed characterization because of its large size and high percent identity in the 1-Mb interval (400 bp at ~87% identity between human and mouse). Furthermore, previous studies have suggested that *IL-4* and *IL-13* are coregulated in $T_h2$ cells, raising the possibility that this single element might explain the coregulation of these two genes. To characterize the function of CNS1, both transgenic and knockout mouse studies were performed (Loots et al. 2000; Mohrs et al. 2001). These independent in vivo strategies both revealed that CNS1 dramatically affected the expression of three human cytokine genes (*IL-4*, *IL-5*, and *IL-13*) separated by more than 120 kb of sequence. For instance, in mice engineered to lack CNS1, a significant reduction in the number of T cells secreting IL-4 was found, and this effect was not seen in Mast cells (Fig. 1C) (Mohrs et al. 2001). Thus, conservation of sequence alone led to the identification of a novel gene regulatory element that acts over long distances. Subsequent studies on CNS1 have further supported that this 400-bp element contains transcription factor-binding sites that co-activate *IL-4*, *IL-5*, and *IL-13* (Lee et al. 2001; Mohrs et al. 2001). It is interesting to note that although additional genes are found interspersed within these interleukin clusters, only the three interleukin genes appear to have altered expression when CNS1 is deleted in vivo.

This single study illustrated the complexity of long-range gene regulatory elements and the power of comparative biology to discover them. In the case of CNS1, as well as numerous other previously identified gene enhancers, these elements are found within highly conserved human–mouse intervals that are devoid of flanking noncoding conservation, making their identification straightforward. These findings implied that the rapid scanning of the human genome for noncoding conservation with mouse should reveal a large number of human gene regulatory elements, but how well does this hold true?

## PITFALLS: AN EXAMPLE WHERE COMPARATIVE GENOMICS FAILS TO REVEAL A GENE REGULATORY ELEMENT

In the field of science, hypotheses are put forth, and those that withstand rigorous testing are commonly reported as positive findings. Unfortunately, in addition to these positive stories, there are numerous failed experiments that more often than not go unreported. Although an increasing number of discoveries have been made using comparative genomics as a starting point with the hypothesis that *conserved sequences are functionally important*, failures have also occurred. One detailed example is provided below.

To identify gene regulatory elements within a four-gene apolipoprotein cluster on human chromosome 11q23 (Karathanasis 1985; Pennacchio et al. 2001), we performed human–mouse comparative analysis as a follow-up to the successful discovery of CNS1 within the interleukin gene cluster on human chromosome 5. Once again the goal was to find highly conserved human–mouse noncoding elements within this interval that could be tested for biological activity in vivo. Toward this goal, we chose to explore in further detail a ~600-bp human noncoding fragment that displayed ~70% identity with mouse (Fig. 2A). Similar to CNS1 in the interleukin cluster, this conserved sequence stood out discretely within a larger interval devoid of other noncoding conservation. This single finding supported that this human–mouse element has resisted "genetic drift," presumably due to functional constraints. The fact that it existed so prominently in a large interval containing four apolipoproteins with a complex expression pattern, and based on our previous experience with CNS1, suggested it too was a gene regulatory element.

To test this hypothesis, we engineered a bacterial artificial chromosome (BAC) containing the entire human apolipoprotein gene cluster with loxP sites flanking the

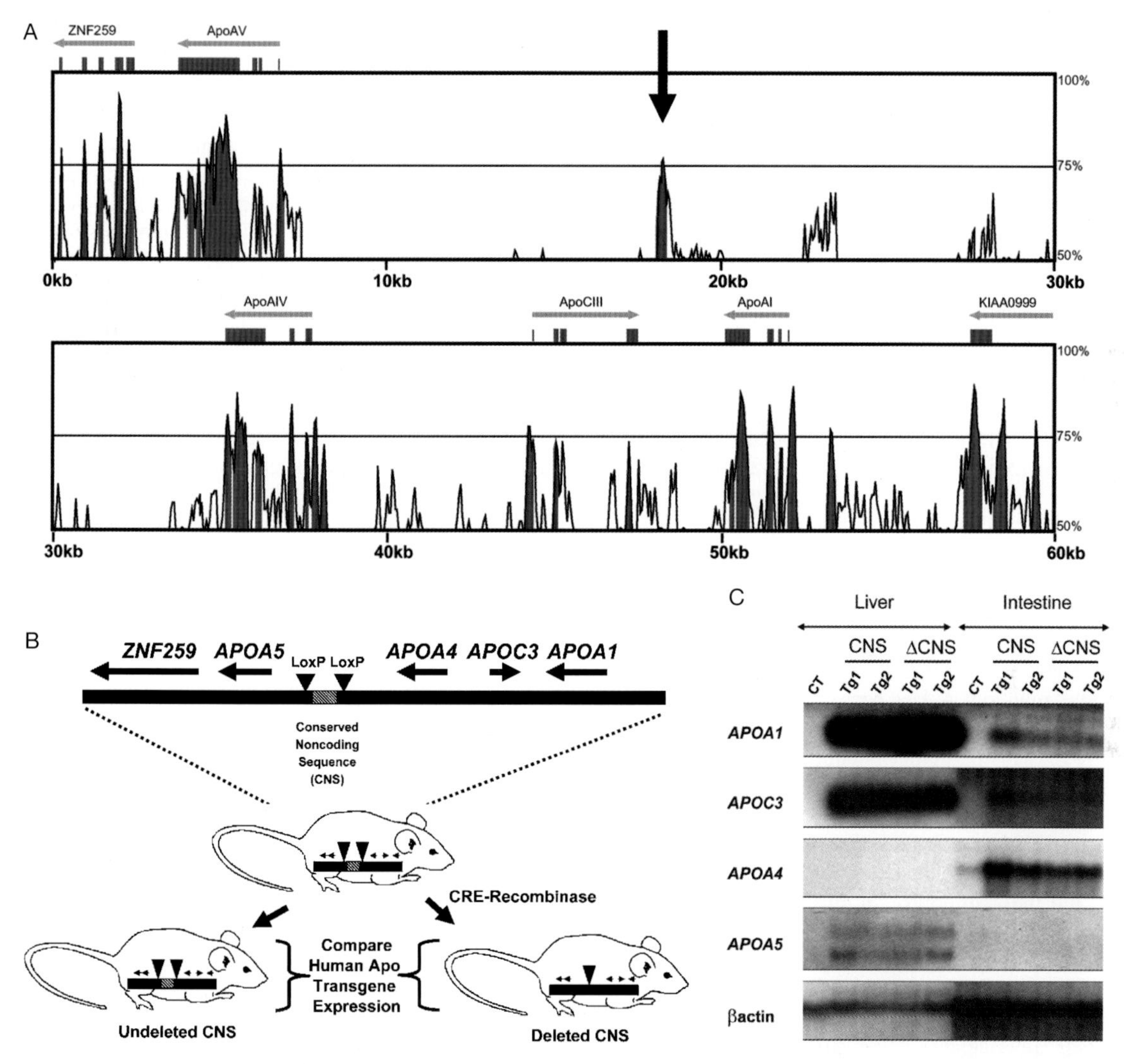


**Figure 2.** Identification and analysis of a highly conserved noncoding sequence in the *APOA1/C3/A4/A5* gene cluster. (*A*) A human–mouse VISTA plot displaying the level of genomic sequence conservation. In each panel, 30 kbp of contiguous human sequence is depicted on the *x*-axis. Above each panel, horizontal arrows indicate known genes and their orientation with each exon depicted by a box (gene names are indicated above each arrow). The VISTA graphical plot displays the level of homology between human and the orthologous mouse sequence. Human sequence is represented horizontally, and the percent similarity with the mouse sequence is plotted vertically (ranging from 50 to 100% identity). The vertical arrow indicates a highly conserved noncoding sequence. (*B*) Strategy for studying conserved noncoding sequences in vivo. A human BAC containing the apolipoprotein gene cluster is engineered to contain loxP sites flanking the conserved sequence of interest. Following the generation of a founder mouse, breeding experiments to Cre-recombinase-expressing mice generate a second line of animals with deletions for the conserved element of interest. (*C*) RNA analysis of transgenic mice (Tg) containing the conserved element (CNS) compared to transgenic mice lacking the element (ΔCNS). A wild-type control mouse is also provided (CT). Liver and intestine total RNA were prepared and hybridized with human-specific probes for *APOA1*, *APOC3*, *APOA4*, and *APOA5*. Mouse β-actin was used as an internal control. No differences were detected in transcript levels from transgenic animals containing the conserved element compared to animals lacking it.

highly conserved sequence (Fig. 2B). Our goal was to create two lines of transgenic mice; one that contained the human BAC plus the conserved element and a second that contained the human BAC minus the conserved element. This strategy was selected because it allowed us to compare the expression pattern of the human genes within the BAC in the two lines of transgenic mice in a position- and copy-number-independent manner. This was achieved by breeding the original transgenics for the BAC plus the conserved element flanked by loxP sites with Cre-Recombinase-expressing animals that produce a second line of mice where the conserved element was deleted (Fig. 2B).

Examination of mice with the conserved element compared to mice lacking the element revealed no detectable difference in any of the neighboring apolipoproteins' known expression pattern, despite extensive RNA analysis (Fig. 2C). In addition, determination of the protein levels of these genes in plasma also indicated no differences despite the deletion of the conserved element. These studies indicate that under the in vivo conditions in which these elements were assessed, no function could be assigned to this conserved sequence. Whether it functions in gene regulation at another time point or environmental condition or performs non-gene regulatory roles remains unclear. Alternatively, the element could be functionless. As a second approach to test for gene regulatory properties, we fused this conserved sequence to a minimal reporter vector and generated transgenic mice. Again, this 600-bp fragment was found to lack enhancer activity, in this case in 13.5-day embryos (data not shown). This example highlights the complexity of assigning function to highly conserved DNA elements and the determination of what assays are the best to capture the endless number of functional possibilities. Although human–mouse comparative genomics have provided the identity of numerous conserved elements with gene regulatory properties, many conserved elements are unlikely to be easily assigned a function. A key part of the interleukin CNS1 study was the detailed phenotypic analysis of *IL-4*, *IL-5*, and *IL-13*. This particular phenotype was only found in stimulated $T_h2$ cells that were analyzed by flow cytometry. Had a less sensitive phenotypic assay been performed, CNS1 would also appear nonfunctional. These two examples, the interleukin cluster on chromosome 5q31 and the apolipoprotein cluster on 11q23, provide early insights into the types of data expected to result from comparative genomic-driven studies. Having a strong understanding of a given gene's complex expression pattern and phenotypic assays to assess this complexity is anticipated to greatly aid in the identification of gene regulatory elements.

## EXTRAPOLATING GENE REGULATORY SCANS TO THE WHOLE GENOME

The recent completion and comparison of a draft genome sequence of mouse to that of human revealed a striking amount of DNA conservation. In one study, it was found that ~40% of the human genome could be aligned to the mouse genome at the nucleotide level (Waterston et al. 2002). In a second study, separate analysis uncovered the identity of over one million discrete human–mouse conserved elements across the human genome (≥70% identity over ≥100 bp) (Couronne et al. 2003). Further extrapolations from these studies strongly support that the vast majority of human–mouse conservation is found in noncoding DNA. For instance, if 40% of the human genome can be aligned to mouse and yet only ~5% of the genome is found in mature mRNA transcripts, most human–mouse conservation cannot be explained by this category of expressed DNA. In addition, of the more than one million discrete human–mouse conserved elements, current estimates suggest that only ~200,000 of these elements are conserved as the result of exons. Thus again, current predictions suggest that a significant amount of conservation exists in noncoding human DNA (Waterston et al. 2002; Couronne et al. 2003). A key question that remains is, What fraction of this noncoding DNA is functional, and what biological processes do they perform? High-throughput strategies are currently needed to categorize this large number of human–mouse conserved noncoding sequences.

One strategy to reduce the large number of human–mouse noncoding sequence elements for functional studies is to perform additional multispecies sequencing and analysis (Mayor et al. 2000; Schwartz et al. 2000; Pennacchio and Rubin 2001, 2003; Frazer et al. 2003). This can be achieved through the addition of a small number of more distantly related species (such as fish, bird, or amphibian), or through the use of a larger number of similarly distanced species (such as several additional mammals) (Bagheri-Fam et al. 2001; Gilligan et al. 2002; Gottgens et al. 2002; Cooper et al. 2003; Ureta-Vidal et al. 2003).

## DEEP PRIMATE SEQUENCE COMPARISONS TO REVEAL "PHYLOGENETIC SHADOWS"

In contrast to distant cross-vertebrate sequence comparisons, a recently developed strategy for annotating genomes has been to perform deep sequence comparisons of evolutionarily closely related species (Boffelli et al. 2003). The general goal previously described for cross-species sequence comparisons is to use species of relatively distant phylogenetic positions to maximize the identification of functionally conserved sequences. However, this strategy fails in the search for species-specific genes and regulatory sequences such as those unique to primates. Recent comparison of the human and mouse genomes indicates that only ~80% of human–mouse genes have a 1:1 orthologous relationship (Waterston et al. 2002). Thus, there is a need for strategies to characterize the 20% of genes and regulatory elements that do not have a true ortholog in both humans and mice. For these studies, comparing human sequences to that of closer evolutionary species is warranted. Yet, the use of primate sequences for cross-species sequence comparisons is limited due to the high level of homology between these species.

"Phylogenetic shadowing" was developed to overcome the excessive sequence identity shared between two primates, making their use in cross-species sequence comparisons possible (Boffelli et al. 2003). The principle be-

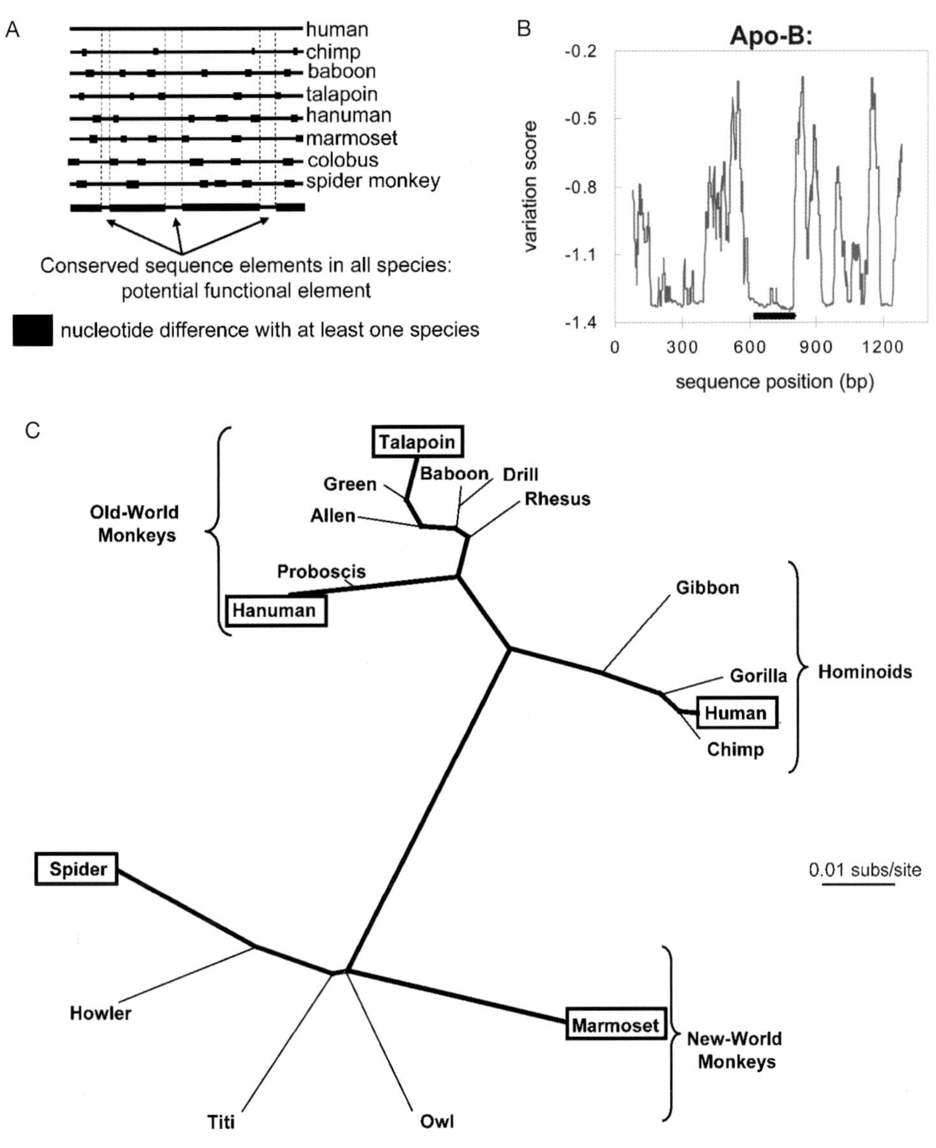


**Figure 3.** "Phylogenetic shadowing" of closely related species. (*A*) The alignment and comparison of sequences from multiple primate species reveal sequences that have been conserved across most species, making them candidates for being functionally relevant due to presumed evolutionary constraint at these sites. (*B*) Primate-specific phylogenetic shadowing reveals a previously defined exon for the apolipoprotein B gene (*APOB*). On the *x*-axis, a variation score is provided with more negative scores indicating less variable regions, and on the *y*-axis, 1500 bp of human sequence is displayed. The known *APOB* exon in this interval is depicted by a solid black line within the plot. Note the decreased amount of primate variation in regions corresponding to the exon. (*C*) Primate phylogenetic tree based on a single genomic interval. A carefully selected set of species that maximize phylogenetic distance (*boxed*) can capture the majority of the phylogenetic shadows and thus can reduce the amount of genomic sequence information required.

hind this strategy is to analyze orthologous sequence from numerous primate species to increase the sum of the evolutionary distance being compared. Rather than performing only pair-wise comparisons between human and mouse, phylogenetic shadowing compares a dozen or more different primate species. The additivity of these primate differences robustly defines regions of increased variation and "shadows" representing conserved segments (Fig. 3A).

In its first use, phylogenetic shadowing proved successful in the identification of both exons and putative gene regulatory elements (Boffelli et al. 2003). This work generated and analyzed 13–17 primate species for several orthologous genomic segments. Examination of a single exon from four independent genes revealed highly conserved shadows that overlapped with these functionally important regions (one example is provided in Fig. 3B for an exon of the apolipoprotein B gene). Further analysis of the human apolipoprotein (a) gene (*apo(a)*) revealed highly conserved motifs embedded within the upstream promoter region. Functional characterization of these phylogenetic shadows compared to more variable flanking DNA supported their role in regulating *apo(a)* expression (Boffelli et al. 2003).

Additional analyses of these data sets suggest that less than a dozen primate sequence comparisons can suffice to detect functional sequences, provided they maximize the phylogenetic distance. In fact, in Figure 3B, only five primates were examined, and they proved successful in identifying an exon of the apolipoprotein B gene based on conservation. These species included human, talapoin, hanuman, spider monkey, and marmoset, which represent the most diverse primates within the large primate sequence data set (Fig. 3C). This initial success warrants further examination of this technique in other genomic intervals to determine its overall utility and suggests that this approach on a genome-wide scale will aid in the identification of both human exons and gene regulatory elements.

## CONCLUSIONS

We have entered an era in which the entire genomes of an increasing number of vertebrates have been sequenced and human–mouse whole-genome comparisons are providing early insights into the realm of possible discoveries. The single human–mouse pair-wise comparison has revealed new genes, regulatory elements, and an entire catalog of highly conserved sequences with putative functionality. However, with this data set, it has become clear that no single pair-wise comparison is suited to capture all biological activity. Current efforts to sequence a wide variety of species, both evolutionarily closer and more distant from humans, are warranted (Boguski 2002; Sidow 2002). Clear examples exist of how human–mouse comparisons fail to capture known human functional elements and support closer sequence comparisons. In addition, certain regions of the human and mouse genomes are highly similar over long lengths, thereby shielding the identification of highly conserved motifs for functional studies. Thus, the generation of a wide-ranging sequence data set from a variety of vertebrates and beyond will provide useful information as to the genetic changes that have occurred over the evolutionary process and resulted in present-day *Homo sapiens*.

## ACKNOWLEDGMENTS

This work was supported in part by the National Institutes of Health–National Heart, Lung, and Blood Institute, programs for Genomic Application (grant HL-66681), and National Institutes of Health grant HL-071954A through the U.S. Department of Energy under contract no. DE-AC03-76SF00098.

## REFERENCES

Aparicio S., Chapman J., Stupka E., Putnam N., Chia J.M., Dehal P., Christoffels A., Rash S., Hoon S., Smit A., Gelpke M.D., Roach J., Oh T., Ho I.Y., Wong M., Detter C., Verhoef F., Predki P., Tay A., Lucas S., Richardson P., Smith S.F., Clark M.S., Edwards Y.J., and Doggett N., et al. 2002. Whole-genome shotgun assembly and analysis of the genome of *Fugu rubripes*. *Science* **297:** 1301.

Bagheri-Fam S., Ferraz C., Demaille J., Scherer G., and Pfeifer D. 2001. Comparative genomics of the SOX9 region in human and *Fugu rubripes:* Conservation of short regulatory sequence elements within large intergenic regions. *Genomics* **78:** 73.

Boffelli D., McAuliffe J., Ovcharenko D., Lewis K.D., Ovcharenko I., Pachter L., and Rubin E.M. 2003. Phylogenetic shadowing of primate sequences to find functional regions of the human genome. *Science* **299:** 1391.

Boguski M.S. 2002. Comparative genomics: The mouse that roared. *Nature* **420:** 515.

Cooper G.M., Brudno M., Green E.D., Batzoglou S., and Sidow A. 2003. Quantitative estimates of sequence divergence for comparative analyses of mammalian genomes. *Genome Res.* **13:** 813.

Couronne O., Poliakov A., Bray N., Ishkhanov T., Ryaboy D., Rubin E.M., Pachter L., and Dubchak I. 2003. Strategies and tools for whole-genome alignments. *Genome Res.* **13:** 73.

Dehal P., Satou Y., Campbell R.K., Chapman J., Degnan B., De Tomaso A., Davidson B., Di Gregorio A., Gelpke M., Goodstein D.M., Harafuji N., Hastings K.E., Ho I., Hotta K., Huang W., Kawashima T., Lemaire P., Martinez D., Meinertzhagen I.A., Necula S., Nonaka M., Putnam N., Rash S., Saiga H., and Satake M., et al. 2002. The draft genome of *Ciona intestinalis:* Insights into chordate and vertebrate origins. *Science* **298:** 2157.

Duret L. and Bucher P. 1997. Searching for regulatory elements in human noncoding sequences. *Curr. Opin. Struct. Biol.* **7:** 399.

Frazer K.A., Narla G., Zhang J.L., and Rubin E.M. 1995. The apolipoprotein(a) gene is regulated by sex hormones and acute-phase inducers in YAC transgenic mice. *Nat. Genet.* **9:** 424.

Frazer K.A., Elnitski L., Church D.M., Dubchak I., and Hardison R.C. 2003. Cross-species sequence comparisons: A review of methods and available resources. *Genome Res.* **13:** 1.

Gilligan P., Brenner S., and Venkatesh B. 2002. Fugu and human sequence comparison identifies novel human genes and conserved non-coding sequences. *Gene* **294:** 35.

Gottgens B., Barton L.M., Chapman M.A., Sinclair A.M., Knudsen B., Grafham D., Gilbert J.G., Rogers J., Bentley D.R., and Green A.R. 2002. Transcriptional regulation of the stem cell leukemia gene (SCL)—Comparative analysis of five vertebrate SCL loci. *Genome Res.* **12:** 749.

Hardison R.C. 2000. Conserved noncoding sequences are reliable guides to regulatory elements. *Trends Genet.* **16:** 369.

Hardison R.C., Oeltjen J., and Miller W. 1997. Long human-mouse sequence alignments reveal novel regulatory elements: A reason to sequence the mouse genome. *Genome Res.* **7:** 959.

Karathanasis S.K. 1985. Apolipoprotein multigene family: Tandem organization of human apolipoprotein AI, CIII, and AIV genes. *Proc. Natl. Acad. Sci.* **82:** 6374.

Lander E.S., Linton L.M., Birren B., Nusbaum C., Zody M.C., Baldwin J., Devon K., Dewar K., Doyle M., FitzHugh W., Funke R., Gage D., Harris K., Heaford A., Howland J., Kann L., Lehoczky J., LeVine R., McEwan P., McKernan K., Meldrim J., Mesirov J.P., Miranda C., Morris W., and Naylor J., et al. (International Human Genome Sequencing Consortium). 2001. Initial sequencing and analysis of the human genome. *Nature* **409:** 860.

Lee G.R., Fields P.E., and Flavell R.A. 2001. Regulation of IL-4 gene expression by distal regulatory elements and GATA-3 at the chromatin level. *Immunity* **14:** 447.

Loots G.G., Locksley R.M., Blankespoor C.M., Wang Z.E., Miller W., Rubin E.M., and Frazer K.A. 2000. Identification of a coordinate regulator of interleukins 4, 13, and 5 by cross-species sequence comparisons. *Science* **288:** 136.

Mayor C., Brudno M., Schwartz J.R., Poliakov A., Rubin E.M., Frazer K.A., Pachter L.S., and Dubchak I. 2000. VISTA: Visualizing global DNA sequence alignments of arbitrary length. *Bioinformatics* **16:** 1046.

Mohrs M., Blankespoor C.M., Wang Z.E., Loots G.G., Afzal V., Hadeiba H., Shinkai K., Rubin E.M., and Locksley R.M. 2001. Deletion of a coordinate regulator of type 2 cytokine expression in mice. *Nat. Immunol.* **2:** 842.

Pennacchio L.A. and Rubin E.M. 2001. Genomic strategies to identify mammalian regulatory sequences. *Nat. Rev. Genet.* **2:** 100.

———. 2003. Comparative genomic tools and databases: Providing insights into the human genome. *J. Clin. Invest.* **111:** 1099.

Pennacchio L.A., Olivier M., Hubacek J.A., Cohen J.C., Cox D.R., Fruchart J.C., Krauss R.M., and Rubin E.M. 2001. An apolipoprotein influencing triglycerides in humans and mice revealed by comparative sequencing. *Science* **294:** 169.

Rubin E.M., Ishida B.Y., Clift S.M., and Krauss R.M. 1991. Expression of human apolipoprotein A-I in transgenic mice results in reduced plasma levels of murine apolipoprotein A-I and the appearance of two new high density lipoprotein size subclasses. *Proc. Natl. Acad. Sci.* **88:** 434.

Schwartz S., Zhang Z., Frazer K.A., Smit A., Riemer C., Bouck J., Gibbs R., Hardison R., and Miller W. 2000. PipMaker—A web server for aligning two genomic DNA sequences. *Genome Res.* **10:** 577.

Sidow A. 2002. Sequence first. Ask questions later. *Cell* **111:** 13.

Ureta-Vidal A., Ettwiller L., and Birney E. 2003. Comparative genomics: Genome-wide analysis in metazoan eukaryotes. *Nat. Rev. Genet.* **4:** 251.

Venter J.C., Adams M.D., Myers E.W., Li P.W., Mural R.J., Sutton G.G., Smith H.O., Yandell M., Evans C.A., Holt R.A., Gocayne J.D., Amanatides P., Ballew R.M., Huson D.H., Wortman J.R., Zhang Q., Kodira C.D., Zheng X.H., Chen L., Skupski M., Subramanian G., Thomas P.D., Zhang J., Gabor Miklos G.L., and Nelson C., et al. 2001. The sequence of the human genome. *Science* **291:** 1304.

Waterston R.H., Lindblad-Toh K., Birney E., Rogers J., Abril J.F., Agarwal P., Agarwala R., Ainscough R., Alexandersson M., An P., Antonarakis S.E., Attwood J., Baertsch R., Bailey J., Barlow K., Beck S., Berry E., Birren B., Bloom T., Bork P., Botcherby M., Bray N., Brent M.R., Brown D.G., and Brown S.D., et al. (Mouse Genome Sequencing Consortium). 2002. Initial sequencing and comparative analysis of the mouse genome. *Nature* **420:** 520.

# High-throughput Mouse Knockouts Provide a Functional Analysis of the Genome

C.J. Friddle, A. Abuin, R. Ramirez-Solis, L.J. Richter, E.C. Buxton, J. Edwards, R.A. Finch, A. Gupta, G. Hansen, K.H. Holt, Y. Hu, W. Huang, C. Jaing, B.W. Key, Jr., P. Kipp, B. Kohlhauff, Z.-Q. Ma, D. Markesich, M. Newhouse, T. Perry, K.A. Platt, D.G. Potter, N. Qian, J. Shaw, J. Schrick, Z.-Z. Shi, M.J. Sparks, D. Tran, E.R. Wann, W. Walke, J.D. Wallace, N. Xu, Q. Zhu, C. Person, A.T. Sands, and B.P. Zambrowicz

*Lexicon Genetics Incorporated, The Woodlands, Texas 77381-1160*

One of the most laborious, costly, and time-consuming phases in the mouse knockout process is the generation of embryonic stem (ES) cell clones carrying the desired mutation. OmniBank is a library of more than 270,000 mouse ES cell clones, each containing a gene trap insertion event in a single gene. The trapped gene is identified by a sequence tag referred to as an OmniBank Sequence Tag (OST). The OSTs identify exons of the trapped genes and are stored in a searchable database in relational format (www.mouseknockout.com).

The recent completion and assembly of the human (Lander et al. 2001) and mouse (Waterston et al. 2002) genome sequences allows an unprecedented level of characterization of OmniBank gene trap events. First, the coding sequence immediately downstream of the retroviral insertion has been sequenced, identifying a set of preexisting mutations that covers well over half of the mouse genome. This database of trapped genes can be searched by BLASTN using the target gene of interest as the query. Clones for the gene of interest are confirmed using inverse PCR to determine the exact genomic site of retroviral integration. Analysis of hundreds of OmniBank lines using this genomic confirmation method shows that intragenic insertions lead to disruption of the endogenous transcript in all cases, as determined by RT-PCR analysis with primers to exons flanking the insertion site.

This resource has been used by Lexicon Genetics to generate hundreds of mouse lines that are knocked out for genes of both academic and pharmaceutical interest. The mouse genome was the second mammalian genome sequenced because of the power of mouse genetics for interpreting the biochemical, cellular, and physiological functions of genes (Bradley 2002; Sands 2003; Zambrowicz et al. 2003). This allows us to rapidly and efficiently predict the effect of a hypothetical drug that inhibits the targeted protein. Mouse lines that exhibit phenotypes of pharmaceutical interest, scored by the presence of beneficial traits and the absence of negative side effects, point to attractive candidates for high-throughput screening and lead compound discovery.

## A MAMMALIAN GENE FUNCTION SCREEN TO IDENTIFY NOVEL THERAPEUTIC TARGETS

The recently released mouse and human genome sequences represent new tools to employ in the pursuit of both the understanding and the treatment of human disease. To take full advantage of these tools, efficient methods are needed to determine mammalian gene function and to identify novel therapeutic targets. Indirect sequence-based methods, such as mRNA expression analysis, have proven effective as diagnostic tools (Sorlie et al. 2001) and in generating hypotheses (Rabitsch et al. 2001), but do not themselves provide definitive conclusions about the role of specific genes. In contrast, genetic screens carried out in the fly (Adams and Sekelsky 2002), worm (Jorgensen and Mango 2002), and other model organisms (Golling et al. 2002; Sessions et al. 2002) are more laborious, but have been invaluable for the identification of individual genes that play a role in any given process. Every gene in yeast has been knocked out to generate a finite set of yeast mutants covering the entire genome (Giaever et al. 2002). Transposon insertional mutagenesis in *Drosophila* has proven very efficient in both the generation of valuable phenotypes and also the mutation characterization that allows genotype–phenotype correlation (Oh et al. 2003). The implementation of a functional screen in the mouse leverages mammalian physiology to more directly probe the key genetic switches that maintain physiological homeostasis in mammals in general and humans in particular. The goal is not to recapitulate human disease in the mouse, whether acquired or genetic. Rather, the goal is to identify those proteins that regulate physiological processes of medical importance. These are not proteins that are commonly mutated to cause disease, such as hypercholesterolemia or schizophrenia, but rather these are the proteins that when inhibited by a pharmaceutical compound may yield medical benefit. The utility of this approach for drug discovery is supported by the strong correlation between the knockout phenotypes and the pharmacological effects of drugs against the major protein targets of the pharmaceutical industry (Zambrowicz and Sands 2003).

Three methods of mutagenesis illustrate the importance of tailoring the choice of method to the goals of the study. For researchers interested in a specific phenotype with no prior interest in the class of genes identified, chemical mutagenesis (Balling 2001) is an attractive choice. This strategy requires minimal initial investment and rapidly generates mutant mice that can be screened for the phenotype of interest. The difficulty lies in identifying the causative mutations. Since the process is random, and the mutated gene is not identified until well after the phenotype is characterized, this method is not appropriate for researchers with an interest in specific genes. Furthermore, this method provides no information about genes that show no phenotype in the selected screen. Such information can be exceedingly valuable when indirect methods of analysis have led a research group to focus on a particular target.

Researchers who are interested in a specific gene often choose gene targeting by homologous recombination (Thomas and Capecchi 1989). The labor and time required to generate these mutations makes homologous recombination most attractive for generating mutations that are not available by other means, such as those described below.

A third strategy utilizes significant initial effort to provide substantial rewards in downstream efficiency. Gene trapping by insertional mutagenesis (Friedrich and Soriano 1991; Skarnes et al. 1992; Zambrowicz and Friedrich 1998) has been proven to be scalable to the level of whole-genome mutagenesis, providing real-time correlation between genotype and phenotype. Gene trapping is a method of random mutagenesis in which the insertion of a DNA element into endogenous genes leads to their transcriptional disruption. Large numbers of ES cell fusion transcripts are sequenced and deposited into a searchable database. Because these sequences are exonic, they are informative with respect to the identity of the trapped gene (Fig. 1). This method is scalable across the entire genome because a single gene trap vector can be used to mutate thousands of individual genes as well as efficiently produce sequence tags for the rapid identification of altered alleles (Zambrowicz et al. 1998). This yields a bank of ES cells with unique predefined mutations that are suitable for the production of mouse knockouts for a large number of known genes. Prior knowledge of the site of each insertion mutation allows the preselection of genes in specific families or members of a common biochemical pathway, enabling a reverse genetic screen in mammals focused on candidate drug targets. This mutagenesis strategy is best suited for generating large numbers of mutations for a wide range of gene classes. It is not practical for use in targeting small numbers of specific genes.

We have previously described the development and automation of a high-throughput gene trapping technology (Zambrowicz et al. 1998, 2003) that has subsequently yielded well over 270,000 independent insertion events. Here we describe the application of that technology to a large-scale mammalian reverse genetics operation in mouse for the creation and analysis of thousands of mutant phenotypes of pharmaceutical relevance.

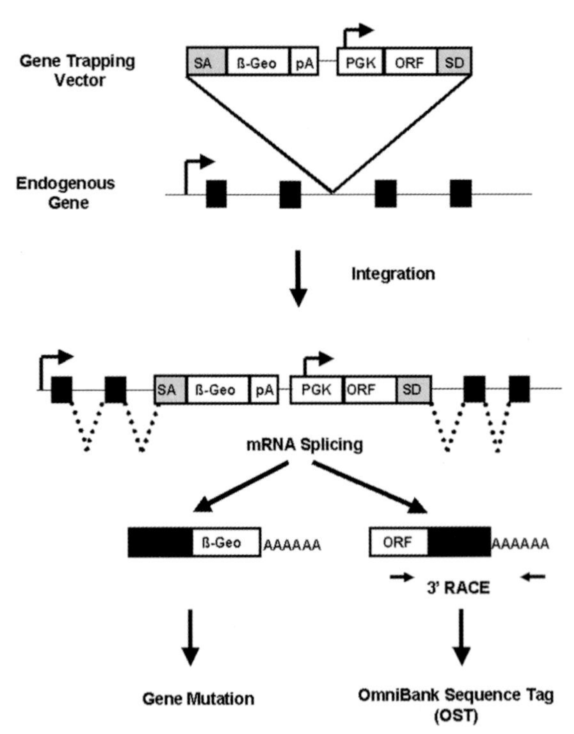


**Figure 1.** Simultaneous gene mutation and identification by gene trapping in mouse ES cells. A retroviral vector contains a splice acceptor sequence (SA) followed by a promoterless selectable marker such as betageo (β-geo), a functional fusion between the β-galactosidase and neomycin-resistance genes, with a polyadenylation signal (pA). Insertion of the retroviral vector into an expressed gene leads to the splicing of the endogenous upstream exons (*solid boxes*) into this cassette to generate a fusion transcript ending with β-geo. The vector also contains a promoter that is active in ES cells, the mouse phosphoglycerate kinase (PGK) promoter, followed by a synthetic single exon ORF upstream of a splice donor (SD) signal. Splicing from the PGK promoter-driven transcript to the exons downstream of the insertion gives rise to a fusion transcript that can be used to generate a sequence tag (OST) of the trapped gene by rapid amplification of cDNA ends (RACE)(Zambrowicz et al. 1998). The synthetic ORF contains a Kozak sequence and termination codons in all reading frames to prevent translation of downstream fusion transcripts.

## GENERATION OF THE OMNIBANK GENE TRAP LIBRARY

We have used high-throughput gene trapping with retroviral vectors in mouse ES cells to generate OmniBank, a library of over 270,000 mutated and sequenced ES cell clones. Each clone is frozen in duplicate in liquid nitrogen. We refer to the sequence trace generated from each clone as an OmniBank Sequence Tag (OST), which corresponds to endogenous exonic sequence immediately downstream of the genomic insertion site (Fig. 1).

## GENE COVERAGE IN OMNIBANK

Extrapolating from a set of 3,904 full-length mouse genes used to BLASTN (Altschul et al. 1997) against the OST collection, we estimate that the OmniBank library contains sequence-tagged clones representing gene trap mutations in ~60% of mouse genes. This reference set of

3,904 full-length mouse cDNAs was generated from the Ensembl database (www.ensembl.org) using three criteria: The mouse gene must have an identified human ortholog, it must be mapped to a specific chromosomal location in the mouse genome, and it must be represented in the RefSeq database at the National Center for Biotechnology Information (www.ncbi.nlm.nih.gov/LocusLink/refseq.html). These criteria ensure that the set is nonredundant and composed of full-length genes. BLASTN analysis of OmniBank with this reference gene list as a query, followed by manual review of the results, demonstrated that 2,170 (55.6%) were represented by OSTs. Although OSTs generally represent exonic sequence, a minority represent intronic sequence, presumably due to splicing into cryptic splice acceptor sites. We repeated the BLASTN query using the intronic sequences from this reference set, and an additional 148 genes were represented by OSTs, bringing the OmniBank coverage to 59.4%.

## GENERATION AND ANALYSIS OF OMNIBANK MOUSE LINES

The OmniBank ES cell clones of interest are chosen for mouse production based on the sequence identity between the corresponding OST and a gene of biological or pharmaceutical interest. Even though the OST identifies the trapped gene, it does not define the precise genomic insertion site in each OmniBank clone. This site is identified using inverse genomic PCR after the thawing of each clone (Fig. 2) (Silver and Keerikatte 1989). This insertional information provides direct evidence of vector insertion within the desired transcript. This genomic sequence is also used to design primers for genotyping of the resulting mice (Fig. 3).

To date, over 1,000 lines of knockout mice from both the OmniBank resource and homologous recombination efforts have been produced and phenotypically studied. We are currently analyzing the phenotypes associated with ~18 candidate drug targets per week, in an effort to rapidly identify genes of both biological importance and pharmaceutical tractability. To confirm the disruption of the endogenous transcript in homozygous mutants, RNA is isolated from selected tissues known to express the transcript and subjected to RT-PCR using primers complementary to exons flanking the insertion site (Fig. 4). Analysis of non-embryonic-lethal mouse lines demonstrates that gene trap insertions within both exons and introns of the gene of interest lead to the disruption of the endogenous mRNA transcript in all cases. Of these, over 96% show complete absence of wild-type message, with the remaining lines showing an average reduction in

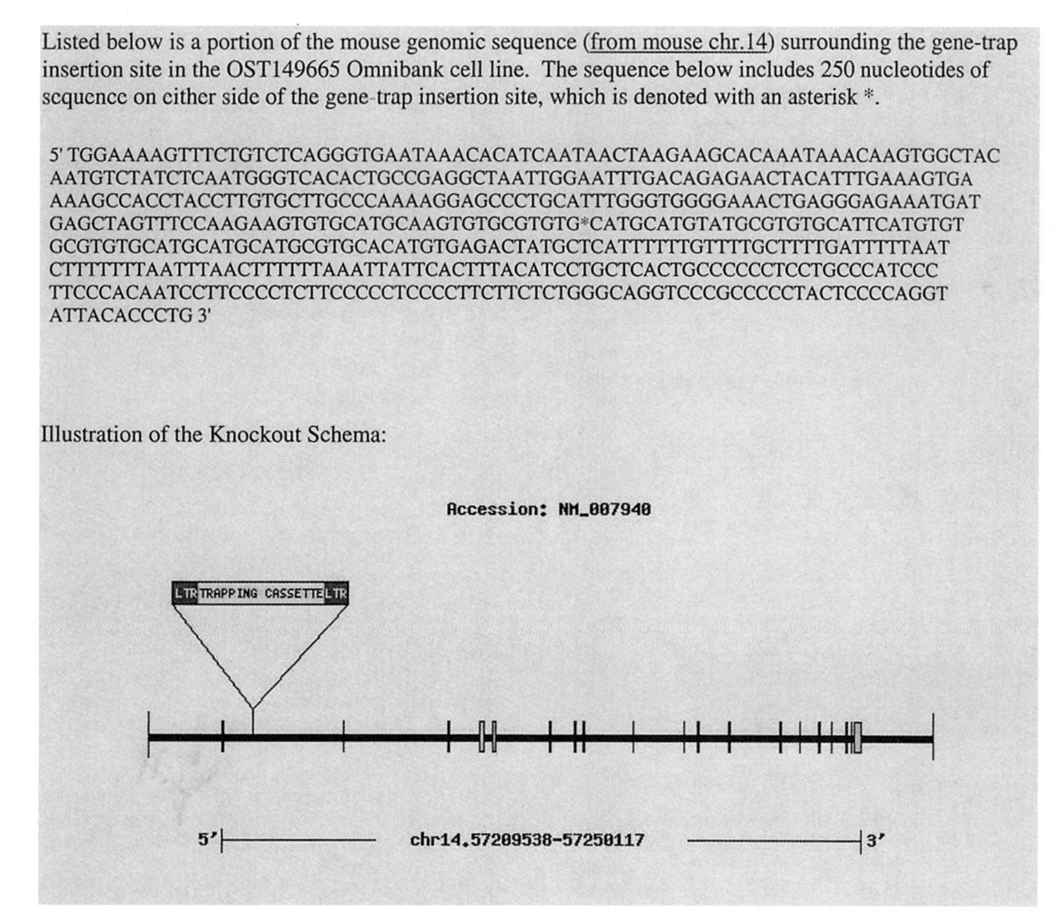


**Figure 2.** Sequence confirmation of OmniBank clones. OST149665 targets the epoxide hydrolase 2 gene (*Eh2*, NM_007940) as confirmed by genomic sequencing. The schema illustrates the insertion of the vector in relation to the intron/exon structure of the gene. LTR refers to the long terminal repeat of the gene trapping vector.

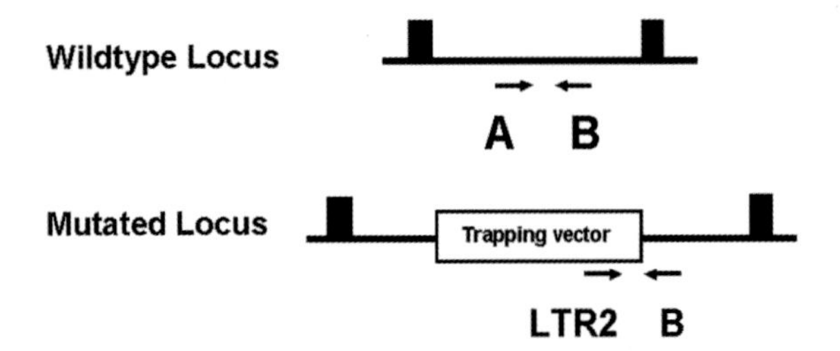


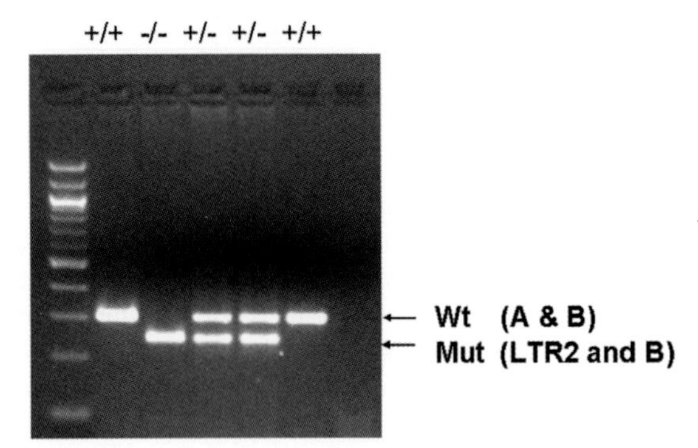


**Figure 3.** Genotyping OmniBank mice. Primers A and B flank the genomic insertion site in the epoxide hydrolase 2 gene (*Eh2,* NM_007940) and amplify a product for the wild-type allele. The LTR2 primer, complementary to OmniBank vectors, can be used in conjunction with a flanking primer to amplify the mutated allele.

mRNA levels of 91.6% as analyzed by quantitative PCR. These data demonstrate that intragenic insertion efficiently disrupts gene transcription in vivo and can be used to reliably predict mutagenicity prior to mouse production.

All lines of mice derived from this resource are subjected to a thorough phenotypic screen designed to identify potential drug targets for the treatment of major human diseases. This screen incorporates a variety of diagnostic tests, advanced medical technologies, imaging equipment, and challenge assays (BeltrandelRio et al. 2003; Zambrowicz and Sands 2003). Applying the same

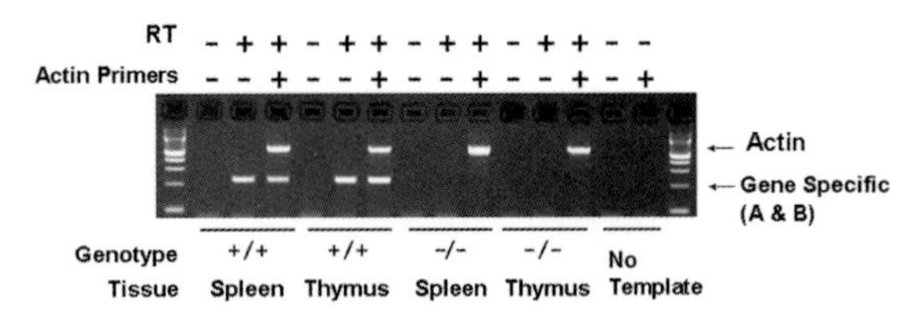


**Figure 4.** Assessing mutagenicity of OmniBank gene trap insertions by RT-PCR. Primers A and B are designed to exons flanking the insertion site in the *PolH* gene in mouse chromosome 17 (NM_030715). RT-PCR using primers A and B shows absence of endogenous message in the spleen and thymus of homozygous animals. Control primers to the murine β-actin gene were used (M12481). (RT) Reverse transcriptase.

screen to lines of mice produced by both gene trapping and gene targeting, we have determined that similar rates and types of phenotypes are observed regardless of the technology used to alter the gene. In fact, we have produced knockout lines for over 20 genes by both gene targeting and gene trapping and observed the same resulting phenotype.

## CONCLUSION

We describe strategies for systematically determining the function of mammalian genes on a genome-wide scale, including OmniBank, a unique sequence-tagged gene trap library in mouse ES cells. The OmniBank resource is enabling a mammalian reverse genetic screen of 5,000 candidate drug targets to determine novel points of potential therapeutic intervention. Bioinformatics mining of public sequence data has allowed the prioritization of all genes of unknown function within pharmaceutically important families, including GPCRs, kinases, transporters, ion channels, proteases, secreted proteins, and other key enzymes. Identification of corresponding gene trap sequence tags in preexisting mutant ES cell clones allows the rapid production of mouse knockouts targeting these gene classes. Resulting mouse mutant phenotypes are screened for direct effects on physiology that model potential therapeutic intervention within the context of mammalian physiology. This systematic approach is generating medically important discoveries for the development of novel therapeutics.

The majority of genes fall outside the gene classes of commercial interest. OmniBank clones that target genes outside these classes are available for use by academic research groups. In addition, the vast majority of mouse lines produced at Lexicon Genetics are not pursued for drug development. These mouse lines, whether generated through OmniBank or through gene targeting by homologous recombination, are also available for academic use. Information on how to search OmniBank and obtain clones for academic use is available at http://www.mouseknockout.com.

## REFERENCES

Adams M.D. and Sekelsky J.J. 2002. From sequence to phenotype: Reverse genetics in *Drosophila melanogaster*. *Nat. Rev. Genet.* **3:** 189.

Altschul S.F., Madden T.L., Schaffer A.A., Zhang J., Zhang Z., Miller W., and Lipman D.J. 1997. Gapped BLAST and PSI-BLAST: A new generation of protein database search programs. *Nucleic Acids Res.* **25:** 3389.

Balling R. 2001. ENU mutagenesis: Analyzing gene function in mice. *Annu. Rev. Genomics Hum. Genet.* **2:** 463.

Bradley A. 2002. Mining the mouse genome. *Nature* **420:** 512.

BeltrandelRio H., Kern F., Lanthorn T., Oravecz T., Piggott J., Powell D., Ramirez-Solis R., Sands A.T., and Zambrowicz B. 2003. Saturation screening of the druggable mammalian genome. In *Model organisms in drug discovery* (ed. P. Carroll and K. Fitzgerald), p. 251. John Wiley and Sons, Chichester, United Kingdom.

Friedrich G. and Soriano P. 1991. Promoter traps in embryonic stem cells: A genetic screen to identify and mutate developmental genes in mice. *Genes Dev.* **5:** 1513.

Giaever G., Chu A.M., Ni L., Connelly C., Riles L., Veronneau

S., Dow S., Lucau-Danila A., Anderson K., Andre B., Arkin A.P., Astromoff A., El-Bakkoury M., Bangham R., Benito R., Brachat S., Campanaro S., Curtiss M., Davis K., Deutschbauer A., Entian K.D., Flaherty P., Foury F., Garfinkel D.J., and Gerstein M., et al. 2002. Functional profiling of the *Saccharomyces cerevisiae* genome. *Nature* **418:** 387.

Golling G., Amsterdam A., Sun Z., Antonelli M., Maldonado E., Chen W., Burgess S., Haldi M., Artzt K., Farrington S., Lin S.Y., Nissen R.M., and Hopkins N. 2002. Insertional mutagenesis in zebrafish rapidly identifies genes essential for early vertebrate development. *Nat. Genet.* **31:** 135.

Jorgensen E.M. and Mango S.E. 2002. The art and design of genetic screens: *Caenorhabditis elegans*. *Nat. Rev. Genet.* **3:** 356.

Lander E.S., Linton L.M., Birren B., Nusbaum C., Zody M.C., Baldwin J., Devon K., Dewar K., Doyle M., FitzHugh W., Funke R., Gage D., Harris K., Heaford A., Howland J., Kann L., Lehoczky J., LeVine R., McEwan P., McKernan K., Meldrim J., Mesirov J.P., Miranda C., Morris W., and Naylor J., et al. (International Human Genome Sequencing Consortium). 2001. Initial sequencing and analysis of the human genome. *Nature* **409:** 860.

Oh S.W., Kingsley T., Shin H.H., Zheng Z., Chen H.W., Chen X., Wang H., Ruan P., Moody M., and Hou S.X. 2003. A P-element insertion screen identified mutations in 455 novel essential genes in *Drosophila*. *Genetics* **163:** 195.

Rabitsch K.P., Toth A., Galova M., Schleiffer A., Schaffner G., Aigner E., Rupp C., Penkner A.M., Moreno-Borchart A.C., Primig M., Esposito R.E., Klein F., Knop M., and Nasmyth K. 2001. A screen for genes required for meiosis and spore formation based on whole-genome expression. *Curr. Biol.* **11:** 1001.

Sands A.T. 2003. The master mammal. *Nat. Biotechnol.* **21:** 31.

Sessions A., Burke E., Presting G., Aux G., McElver J., Patton D., Dietrich B., Ho P., Bacwaden J., Ko C., Clarke J.D., Cotton D., Bullis D., Snell J., Miguel T., Hutchison D., Kimmerly B., Mitzel T., Katagiri F., Glazebrook J., Law M., and Goff S.A. 2002. A high-throughput *Arabidopsis* reverse genetics system. *Plant Cell* **14:** 2985.

Silver J. and Keerikatte V. 1989. Novel use of polymerase chain reaction to amplify cellular DNA adjacent to an integrated provirus. *J. Virol.* **63:** 1924.

Skarnes W.C., Auerbach B.A., and Joyner A.L. 1992. A gene trap approach in mouse embryonic stem cells: The lacZ reported is activated by splicing, reflects endogenous gene expression, and is mutagenic in mice. *Genes Dev.* **6:** 903.

Sorlie T., Perou C.M., Tibshirani R., Aas T., Geisler S., Johnsen H., Hastie T., Eisen M.B., van de Rijn M., Jeffrey S.S., Thorsen T., Quist H., Matese J.C., Brown P.O., Botstein D., Eystein Lonning P., and Borresen-Dale A.L. 2001. Gene expression patterns of breast carcinomas distinguish tumor subclasses with clinical implications. *Proc. Natl. Acad. Sci.* **98:** 10869.

Thomas K.R. and Capecchi M.R. 1989. Site-directed mutagenesis by gene targeting in mouse embryo-derived stem cells. *Cell* **51:** 503.

Waterston R.H., Lindblad-Toh K., Birney E., Rogers J., Abril J.F., Agarwal P., Agarwala R., Ainscough R., Alexandersson M., An P., Antonarakis S.E., Attwood J., Baertsch R., Bailey J., Barlow K., Beck S., Berry E., Birren B., Bloom T., Bork P., Botcherby M., Bray N., Brent M.R., Brown D.G., and Brown S.D., et al. (Mouse Genome Sequencing Consortium). 2002. Initial sequencing and comparative analysis of the mouse genome. *Nature* **420:** 520.

Zambrowicz B.P. and Friedrich G.A. 1998. Comprehensive mammalian genetics: History and future prospects of gene trapping in the mouse. *Int. J. Dev. Biol.* **42:** 1025.

Zambrowicz B.P. and Sands A.T. 2003. Knockouts model the 100 best-selling drugs—Will they model the next 100? *Nat. Rev. Drug Discov.* **2:** 38.

Zambrowicz B.P., Friedrich G.A., Buxton E.C., Lilleberg S.L., Person C., and Sands A.T. 1998. Disruption and sequence identification of 2,000 genes in mouse embryonic stem cells. *Nature* **392:** 608.

Zambrowicz B.P., Abuin A., Ramirez-Solis R., Richter L.J., Piggott J., BeltrandelRio H., Buxton E.C., Edwards J., Finch R.A., Friddle C.J., Gupta A., Hansen G., Hu Y., Huang W., Jaing C., Key B.W., Jr., Kipp P., Kohlhauff B., Ma X.Q., Markesich D., Payne R., Potter D.G., Qian N., Shaw J., and Schrick H.J., et al. 2003. Wnk1 kinase deficiency lowers blood pressure in mice: A gene-trap screen to identify potential targets for therapeutic intervention. *Proc. Natl. Acad. Sci.* **100:** 14109.

# Identification of Novel Functional Elements in the Human Genome

Z. LIAN,*† G. EUSKIRCHEN,*‡ J. RINN,*¶ R. MARTONE,*‡ P. BERTONE,*‡ S. HARTMAN,‡ T. ROYCE,¶ K. NELSON,‡ F. SAYWARD,§ N. LUSCOMBE,¶ J. YANG,§ J.-L. LI,§ P. MILLER,§ A.E. URBAN,‡ M. GERSTEIN,¶ S. WEISSMAN,† AND M. SNYDER‡

*†Department of Genetics, ‡Department of Molecular, Cellular and Developmental Biology, ¶Department of Molecular Biophysics and Biochemistry, and §Department of Anesthesiology, Yale University, New Haven, Connecticut 06520*

Recently, a nearly complete draft of the human genome has been determined, producing an enormous wealth of information (Olivier et al. 2001). However, the sequence by itself reveals little about the critical elements encoded in the DNA, and consequently, it is paramount to identify the functional elements encoded in the 3 billion base pairs and to determine how they work together to mediate complex processes such as development and responses to environmental alterations. Two essential tasks toward this goal are the identification of coding and transcriptionally active regions in the human genome and determining how they are regulated. The identification of these regions is an essential first step for the comprehensive and systematic analysis of gene and protein function. Thus far, a variety of different approaches have been used for identification of coding sequences and other functional elements in genomic DNA (Snyder and Gerstein 2003). Genes have been identified by generating and sequencing of cDNAs, expressed sequence tags (ESTs), and related approaches, and then mapping the mRNA coding sequences onto genomics DNA (Lander et al. 2001). Genes have also been identified by computational methods such as motif searches, identification of long open reading frames, and comparative genomic studies to identify conserved sequences, particularly those predicted to encode proteins (Lander et al. 2001; Venter et al. 2001; Waterston et al. 2002). The availability of the full genomic DNA sequence allows the direct identification of transcribed sequences by globally interrogating all regions of the genome using genomic DNA microarrays.

In addition to identification of genes, it is also of high interest to identify the elements that regulate their expression. Such information is crucial for understanding how the activity of genes is controlled and thereby what is essential for understanding cell proliferation and differentiation. Approaches to analyze gene regulation in the past have been hampered by the fact that the approaches are either not comprehensive or are indirect. For example, comparative analysis of gene expression using DNA microarrays in lines expressing or lacking a factor of interest is indirect—changes in gene expression may be due to downstream effects of the factor.

Recently, we have developed an approach for identifying the binding sites of transcription factors on a global scale (Iyer et al. 2001; Horak et al. 2002). This procedure involves immunoprecipitation of chromatin (ChIP) associated with a transcription factor of interest and using the associated DNA to probe a genomic DNA array containing the regulatory sequences or large segments of the genome. Thus, in one experiment, many binding sites for a transcription factor can be identified.

Below we describe the comprehensive analysis of transcribed regions and transcription factor-binding sites on a global scale. We have constructed an array containing most of the sequences of human Chromosome 22 and used it to identify novel transcribed regions and transcription factor-binding sites. These approaches are expected to be of broad utility for understanding the function and regulation of the human genome.

## CONSTRUCTION OF A HUMAN CHROMOSOME 22 ARRAY

We have prepared an array containing nearly all of the unique sequences of human Chromosome 22 (Dunham et al. 1999; Rinn et al. 2003). Chromosome 22 contains 35.X Mbp of DNA; approximately one-half comprises repetitive DNA, and it contains 545 annotated genes (Sanger release 2.3). To prepare the array, the repetitive DNA of human chromosome was detected computationally and subtracted from the total sequence. The remaining single-copy DNA sequence was amplified in 0.3- to 1.4-kb segments (mean size 820 bp) using oligonucleotide primers and PCR; 21, 024 PCR products were attempted and 93% were successful. The DNA products were printed in duplicate onto 2.5 slides. Sequencing of several hundred products has revealed that 95% of the sequences are identical or close matches to the expected fragments.

## IDENTIFICATION OF NOVEL TRANSCRIBED REGIONS

The chromosome 22 array was probed with a cDNA probe prepared from placental poly(A)$^+$ RNA (Rinn et al. 2003). The RNA had been purified three times using oligo(dT) cellulose. Labeled single-stranded cDNA was synthesized using a 50:50 mixture of oligo(dT) and ran-

*These authors contributed equally to this work.

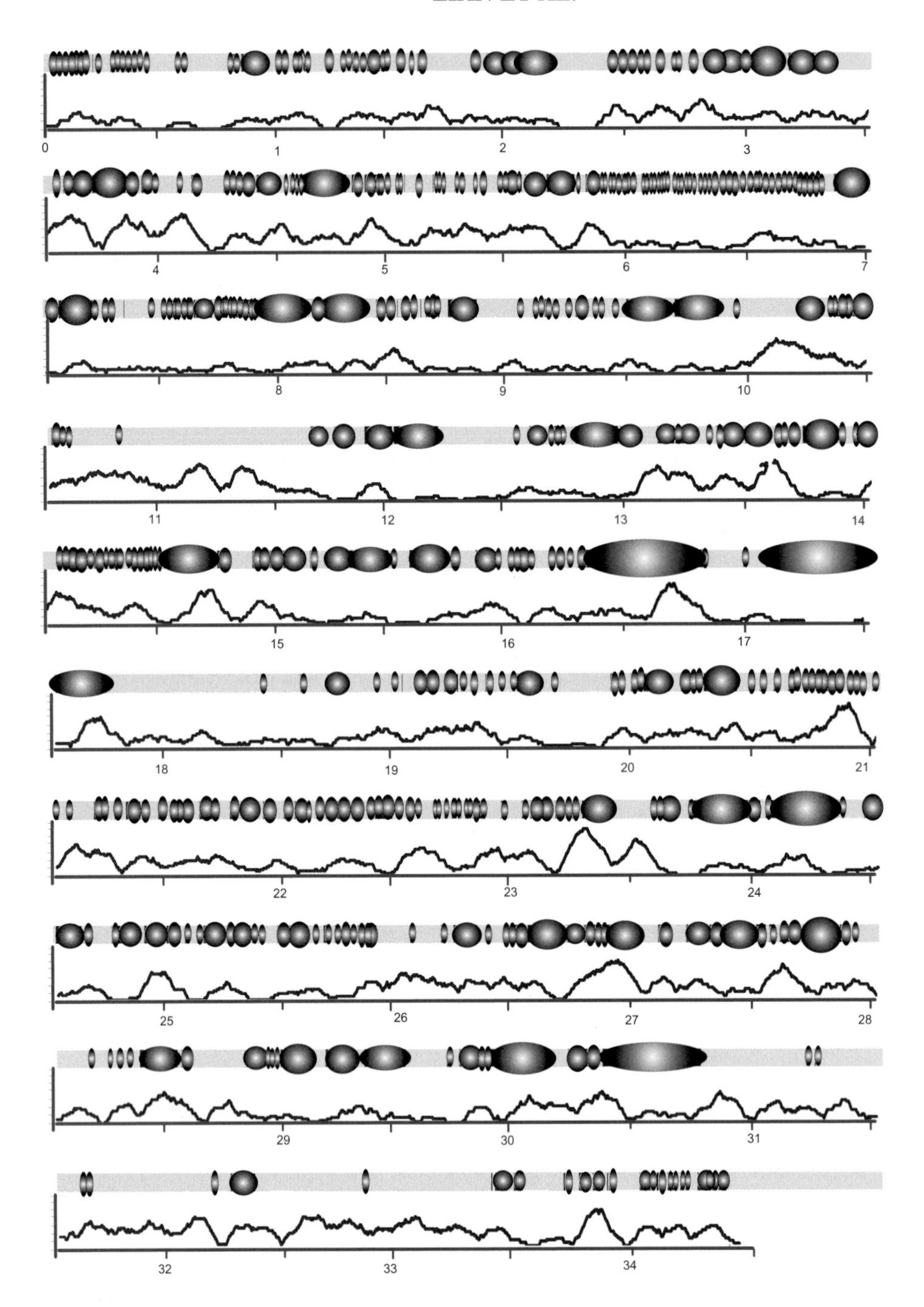


The top gray bar represents the sequence on Chromosome 22, ordered from centromere to telomere (unit: Mega bp).

Sanger Centre release 2.3 annotated genes in human Chromosome 22.

the value of the number of positive hybridizing fragments divided by the total number of fragments in a 100-kb window

**Figure 1.** The human Chromosome 22 placental transcriptome.

dom primers. Six sets of slides were probed, and 2470 fragments exhibited a significant signal in five or more replicas. The results are shown in Figure 1. Matching of the Sanger 2.3 annotation to the arrays revealed that 946 fragments corresponded to known exons; ~60% of known genes on Chromosome 22 were expressed in placental RNA. Importantly, more than half (1307 fragments) did not correspond to known exons or other annotations. These novel transcribed regions are designated TARs, for transcriptionally active regions.

To determine whether the novel TARs encoded discrete transcripts, 118 fragments were labeled and used to probe RNA blots of placental poly(A)$^+$ RNA (Rinn et al. 2003). Thirty (27%) reacted primarily with one RNA band, indi-

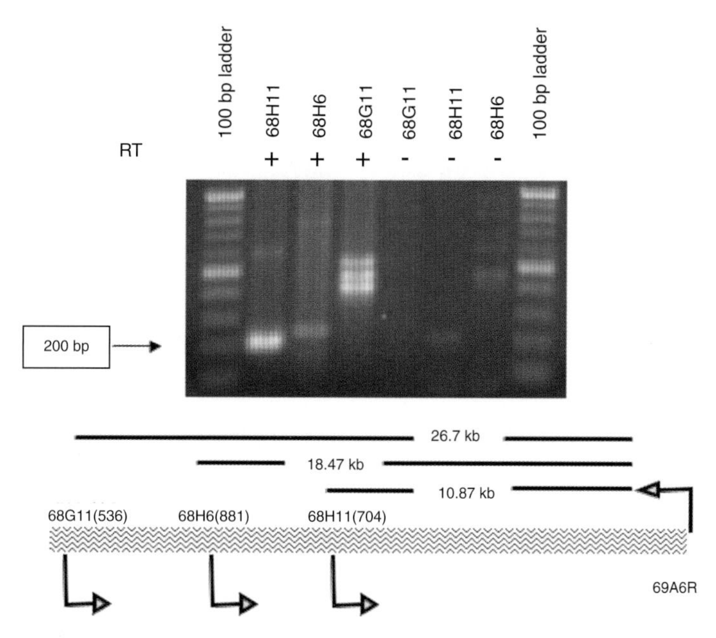


**Figure 2.** Characterization of a novel expressed coding fragment.

cating that they encode discrete transcripts. We further used BLAST to assure these probes could not be cross-hybridizing anywhere else in the genome. Indeed, 27 of the 30 matched only to Chromosome 22 sequence. The other 3 had weak homology with other regions in addition to a strong Chromosome 22 match. In one instance, two probes located 36 kb apart hybridized to the same 6-kb transcript, suggesting that they encoded different parts of the same message. We further investigated this region by a primer-walking experiment using a placenta cDNA library. Figure 2 demonstrates that sequential primers along this region produce increasingly larger amounts of transcript information, as evidenced by larger RT-PCR products. In each case, the increase in coding information is significantly smaller than the genomic sequence spanned by the primers, indicating the presence of a single mRNA coding region containing introns. Together the RNA blot analysis and RT-PCR primer-walking experiment indicate that many TARs encode discrete messages.

To further understand the nature of the TARs, 60-bp oligonucleotides were prepared to the regions of fragments from outside annotated genes or within introns. The sequences of the oligonucleotide were selected from the regions of the fragment that exhibited the highest score using gene prediction programs, e.g., GeneScan, Engrailed. Reverse complementary probes were also prepared. The oligonucleotides were spotted onto an array and probed with placental poly(A)$^+$ RNA. Fifty-three oligonucleotides showed a significant differential signal over the reverse complement oligo; interestingly, the proportion of signals from predicted coding versus reverse complementary oligonucleotides was identical, indicating that the gene prediction programs are not suitable for predicting novel TARs. We also found that, within introns, hybridization occurred nearly as often to the noncoding strand as to the coding strand. Thus, the transcription is unlikely to be residual unspliced messages, but rather most likely corresponds to novel transcribed regions.

We also explored the conservation of the TARs and their potential to encode protein. One third of the TARs are highly conserved with mouse sequences, indicating that they are functionally important. Approximately 8% of the hybridizing fragments with no prior annotation were found to have a homologous mouse protein. These are likely to represent novel exons associated with known annotated genes, as well as novel genes.

Recently, Karponov et al. (2002) mapped transcribed regions on human Chromosomes 21 and 22 using oligonucleotide arrays. A comparison of our data with theirs reveals that 90% of our hybridization results corresponded well with their results found for RNAs common to 6 of 11 cell lines. Thus, the independent methods each found common novel transcribed regions.

## MAPPING OF NF-κB SITES ALONG CHROMOSOME 22

In addition to mapping transcribed regions, we also have mapped potential regulatory regions by identifying the binding sites of several human transcription factors along human Chromosome 22 (Martone et al. 2003; and our unpublished data). Our study was initiated by analyzing the binding-site distribution of the NF-κB family member, p65, which has been implicated in a variety of cellular responses, including inflammation and apoptosis. The number of binding sites for NF-κB along an entire chromosome was unknown and difficult to predict. Few known targets for this factor resided on this chromosome.

We analyzed the distribution of NF-κB in HeLa cells in response to TNF-α stimulation using chIP chip and our Chromosome 22 genomic DNA array (Martone et al. 2003). Briefly, HeLa cells were treated with TNF-α for

90 minutes and the cells were treated with 1% formaldehyde which crosslinks protein to DNA. The cells were then lysed and chromatin sheared to ~500–600 bp final DNA size by sonication. Anti-p65 antibodies were used to immunoprecipitate the p65-bound chromatin, the crosslinks were reversed by heating, and the DNA was purified and labeled with Cy5. As a control, cells that were not incubated with TNF-α (in which NF-κB remains in the cytoplasm) were treated in an identical fashion, and nonspecific DNA that was precipitated by the antibodies was labeled with Cy3. The labeled probes were mixed and hybridized to the Chromosome 22 array; three separate experiments were performed. Using the ExpressYourself program (Luscombe et al. 2003), we found 209 binding sites of p65 along Chromosome 22 (Fig. 3). Verification of 75 was confirmed using PCR with primers

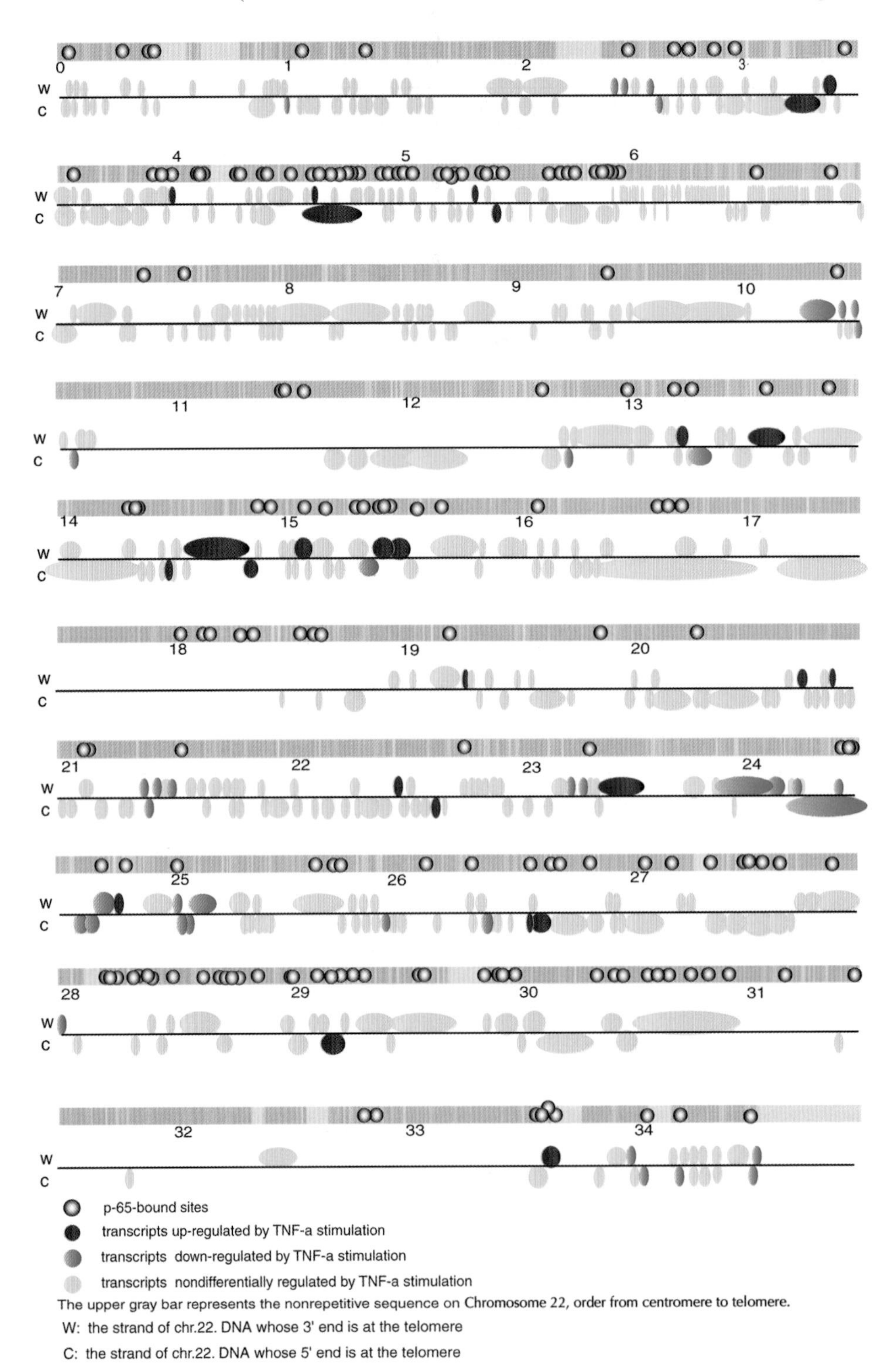


**Figure 3.** Chromosome 22q binding profile for p65.

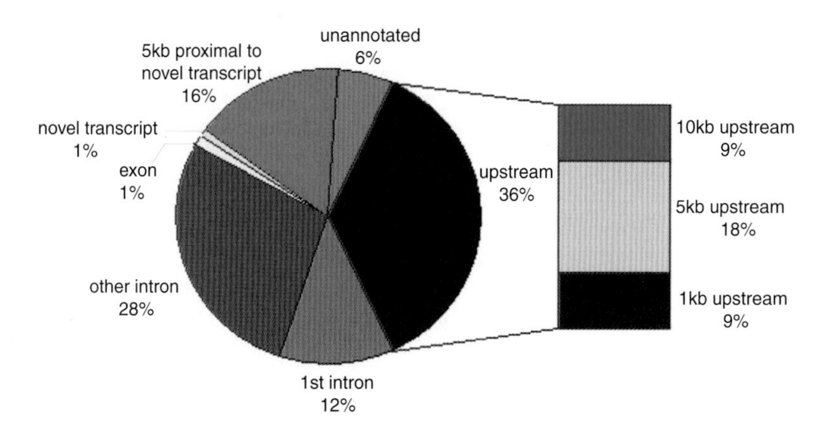


**Figure 4.** Distribution of p65-binding sites on Chromosome 22.

located within the hybridizing array fragment, and either gel electrophoresis and/or quantitative PCR. Approximately 80% of the sites confirm using these procedures.

The binding sites were mapped relative to known genes (Fig. 4). 77% of the sites are within 10 kb of annotated genes and 17% lie exclusively near a TAR. Only 6% do not lie near any annotated region or TAR. Interestingly, for annotated genes only a relatively small fraction (36%) of the total binding sites lie within 10 kb of the 5′ end of the gene. A large fraction lies in either the first intron (12%) or other introns (28%). Only 1% of p65-binding sites lie exclusively within an exon. The presence of NF-κB sites within introns is consistent with its initial discovery within introns (Baeuerle and Baltimore 1996). Nonetheless, these results suggest that many potential regulatory sites lie throughout a locus, not simply near the 5′ ends.

We also examined the binding distribution of p65 relative to its consensus binding site or that of a closely related factor, c-Rel, with which it is known to partner (Martone et al. 2003). 52% of the p65 binding fragments have identical matches to p65 or c-Rel consensus sites—the remainder have near matches. These results indicate that transcription factors are associated with both consensus and nonconsensus sites, indicating the importance of using experimental approaches for the detection of transcription factor-binding sites.

We also analyzed the types of genes that have p65-binding sites near or within them and found that the target genes often have biological functions that are consistent with those of p65. These include genes for PDGF, TIMP3, ATF4, EWSR1, IL2R-β, and PPAR. The binding of p65 suggests that NF-κB may be mediating some of its diverse effects through these gene targets.

Transcription factor binding does not always indicate regulation of gene expression. To correlate binding with gene expression, we examined the expression of HeLa cells upon treatment with TNF-α relative to untreated cells (Martone et al. 2003). By examining the median values of signals from exons, we found that 28 of human Chromosome 22 genes are up-regulated and 39 genes are down-regulated. Mapping of the p65-binding sites relative to expressed regions revealed that 12 lay near genes whose expression is induced by TNF-α and 6 lay near genes whose expression is repressed, indicating that TNF-α may have a previously unappreciated role as a transcriptional repressor. We also find that many p65-binding sites lie near genes whose expression is not affected by TNF-α. An example of the latter category is the λ-light chain genes; these genes may be regulated by p65 in B cells, but are not expected to be expressed in HeLa cells. These results indicate that both the gene and cellular context of binding are likely to be crucial for the regulation of gene expression; binding per se is not definitive for the regulation of gene expression. It is likely the p65 activity is modulated by functioning in concert with other transcription factors or modes of regulation (e.g., protein modification).

## CONCLUSION

These results demonstrate that it is possible to map novel transcribed regions and the binding sites of transcription factors over an entire chromosome using genomic tiling arrays. Extension of these types of technologies would allow mapping over the entire human genome, allowing a comprehensive analysis of both transcribed regions and binding of regulatory factors. Given that there are ~1000–2000 transcription factors in humans (http://www.godatabase.org/dev/database/) and over 250 different cell types, it should be possible to deduce the binding sites for all factors in all possible cell types, thereby revealing the entirely transcriptional circuitry for a human being.

## REFERENCES

Baeuerle P.A. and Baltimore D. 1996. NF-kappa B: Ten years after. *Cell* **87:** 13.

Dunham I., Shimizu N., Roe B.A., Chissoe S., Hunt A.R., Collins J.E., Bruskiewich R., Beare D.M., Clamp M., Smink L.J., Ainscough R., Almeida J.P., Babbage A., Bagguley C., Bailey J., Barlow K., Bates K.N., Beasley O., Bird C.P., Blakey S., Bridgeman A.M., Buck D., Burgess J., Burrill W.D., and O'Brien K.P., et al. 1999. The DNA sequence of human chromosome 22. *Nature* **402:** 489.

Horak C.E., Mahajan M.C., Luscombe N.M., Gerstein M., Weissman S.M., and Snyder M. 2002. GATA-1 binding sites mapped in the beta-globin locus by using mammalian chIp-chip analysis. *Proc. Natl. Acad. Sci.* **99:** 2924.

Iyer V.I., Horak C.A, Scafe C.S., Botstein D., Snyder M., and Brown P.O. 2001. Genomic binding distribution of the yeast cell-cycle transcription factors SBF and MBF. *Nature* **409:** 533.

Kapranov P., Cawley S.E., Drenkow J., Bekiranov S., Strausberg R.L., Fodor S.P., and Gingeras T.R. 2002. Large-scale transcriptional activity in chromosomes 21 and 22. *Science* **296:** 916.

Lander E.S., Linton L.M., Birren B., Nusbaum C., Zody M.C., Baldwin J., Devon K., Dewar K., Doyle M., FitzHugh W., Funke R., Gage D., Harris K., Heaford A., Howland J., Kann L., Lehoczky J., LeVine R., McEwan P., McKernan K., Meldrim J., Mesirov J.P., Miranda C., Morris W., and Naylor J., et al. (International Human Genome Sequencing Consortium). 2001. Initial sequencing and analysis of the human genome. *Nature* **409:** 860.

Luscombe N.M., Royce T.E., Bertone P., Echols N., Horak C.E., Chang J.T., Snyder M., and Gerstein M. 2003. ExpressYourself: A modular platform for processing and visualizing microarray data. *Nucleic Acids Res.* **31:** 3477.

Martone R., Euskirchen G., Bertone P., Hartman S., Royce T.E., Luscombe N.M., Rinn J.L., Nelson F.K., Miller P., Gerstein M., Weissman S., and Snyder M. 2003. Distribution of NF-kappaB-binding sites across human chromosome 22. *Proc. Natl. Acad. Sci.* **100:** 12247.

Olivier M., Aggarwal A., Allen J., Almendras A.A., Bajorek E.S., Beasley E.M., Brady S.D., Bushard J.M., Bustos V.I., Chu A., Chung T.R., De Witte A., Denys M.E., Dominguez R., Fang N.Y., Foster B.D., Freudenberg R.W., Hadley D., Hamilton L.R., Jeffrey T.J., Kelly L., Lazzeroni L., Levy M.R., Lewis S.C., and Liu X., et al. 2001. A high-resolution radiation hybrid map of the human genome draft sequence. *Science* **291:** 1298.

Rinn J.L., Euskirchen G., Bertone P., Martone R., Luscombe N.M., Hartman S., Harrison P.M., Nelson F.K., Miller P., Gerstein M., Weissman S., and Snyder M. 2003. The transcriptional activity of human chromosome 22. *Genes Dev.* **17:** 529.

Snyder M. and Gerstein M. 2003. Defining genes in the genomics era. *Science* **300:** 258.

Venter J.C., Adams M.D., Myers E.W., Li P.W., Mural R.J., Sutton G.G., Smith H.O., Yandell M., Evans C.A., Holt R.A., Gocayne J.D., Amanatides P., Ballew R.M., Huson D.H., Wortman J.R., Zhang Q., Kodira C.D., Zheng X.H., Chen L., Skupski M., Subramanian G., Thomas P.D., Zhang J., Gabor Miklos G.L., and Nelson C., et al. 2001. The sequence of the human genome. *Science* **291:** 1304.

Waterston R.H., Lindblad-Toh K., Birney E., Rogers J., Abril J.F., Agarwal P., Agarwala R., Ainscough R., Alexandersson M., An P., Antonarakis S.E., Attwood J., Baertsch R., Bailey J., Barlow K., Beck S., Berry E., Birren B., Bloom T., Bork P., Botcherby M., Bray N., Brent M.R., Brown D.G., and Brown S.D., et al. (Mouse Genome Sequencing Consortium). 2002. Initial sequencing and comparative analysis of the mouse genome. *Nature* **420:** 520.

# High-resolution Human Genome Scanning Using Whole-genome BAC Arrays

J. Li,* T. Jiang,* B. Bejjani,† E. Rajcan-Separovic,‡ and W.-W. Cai*

*Department of Molecular and Human Genetics, Baylor College of Medicine, Houston, Texas 77030; †Basic Medical Sciences Program, Washington State University Spokane Health Research and Education Center, Spokane, Washington 99210-1495; ‡Department of Pathology, University of British Columbia, Vancouver, V6H 3V4, British Columbia

Constitutional chromosome abnormalities are a frequent cause of many human syndromes, such as infertility, congenital anomalies, and mental retardation (Gardner and Sutherland 1996). Cytogenetic analysis of chromosomal integrity in patients with mental retardation (MR) indicates that 40% of severe (IQ<55) and 10–20% of mild (IQ = 55–70) MR is caused by chromosomal anomalies (Flint et al. 1995). However, due to the poor resolution of conventional cytogenetic analysis, subtle chromosomal rearrangements (<3 Mb) may be missed in a significant proportion of mild MR cases. Recent studies focusing on the subtelomeric regions of patients with idiopathic MR indeed indicated that the prevalence of subtle subtelomeric rearrangements could be as high as 6% in these patients (Flint et al. 1995; Knight et al. 1999), and it is now widely accepted that submicroscopic telomeric rearrangements are a significant cause of MR. It is quite natural to hypothesize that a substantial proportion of idiopathic MR may be caused by subtle rearrangements in other parts of the genome that were never detected due to a low resolution of cytogenetic analysis and the unavailability of high-resolution genome-scanning techniques. The fact that chromosomal rearrangements can occur in almost any region of the genome (Brewer et al. 1998, 1999), and the existence of subtle microdeletions and microduplications along the chromosomal arms in over 10 clinical syndromes associated with MR (Shapira 1998), support this hypothesis.

To discover subtle chromosomal rearrangements, the ideal approach is to perform a comprehensive genome scan to detect imbalances at the highest resolution without any assumption about the genomic structures and the underlying mechanisms causing the abnormalities. In the present study, we have demonstrated that construction of bacterial artificial chromosome (BAC) arrays covering the human genome at a 0.2-Mb resolution is possible and have analyzed eight patients with mild to moderate MR. Our preliminary study identified small chromosomal gains and losses across the genome in patients with idiopathic MR. Their validation and clinical relevance need further investigation, and larger population studies need to be initiated in the future to identify regions of polymorphisms that will help in distinguishing true changes in the genome from normal variation.

## ARRAY-BASED COMPARATIVE GENOMIC HYBRIDIZATION

The cytogenetic analysis of metaphase chromosomes to detect visible abnormalities can be considered as the first whole-genome scanning method. Since its introduction in the early 1970s, numerous chromosome abnormalities have been documented in various human syndromes (Gardner and Sutherland 1996). The development of comparative genomic hybridization (CGH) (Kallioniemi et al. 1992) was a significant technical advance because it obviated the requirement for preparation of metaphase chromosomes from patient samples, which is difficult for some types of tissues (e.g., solid tumors, tissues obtained from spontaneous abortions). However, the resolution of CGH is not much higher than obtained by routine cytogenetic analysis and is currently limited to 10 Mb in routine analysis. This resolution limit is not likely to be improved within the current CGH technical format because it relies on hybridization of differentially labeled patient (test) and control (reference) DNA to normal metaphase chromosomes.

DNA microarray-based comparative genomic hybridization (aCGH) emerged as a straightforward strategy to overcome the limitations of conventional cytogenetic and chromosomal CGH analysis (Fig. 1). In aCGH, cloned DNA fragments from known genomic regions are arrayed on a surface and CGH is performed on the array, allowing the whole genome to be scanned in a single experiment at a resolution only limited by the size of the cloned DNA fragments. However, to exploit the full potential of this array-based approach, many technical challenges need to be resolved. The technical aspects of aCGH, as outlined in Figure 1, include three important components: (1) a collection of mapped clones, (2) construction of arrays, and (3) development of quantitative hybridization methods.

## MAPPED DNA CLONES AS BASIC REAGENTS FOR aCGH

The human genome sequencing project and other genome-wide efforts have generated many mapped clone resources (Cheung et al. 1999, 2001; Korenberg et al., 1999; McPherson et al. 2001) that can be used for aCGH.

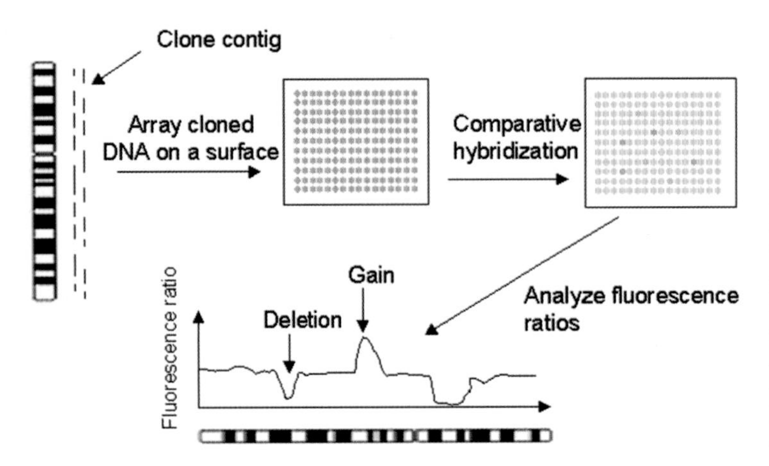


**Figure 1.** Array-based CGH to detect copy number variations. Cloned DNA samples representing different genomic regions are printed on glass surfaces to make microarrays. Test and normal control genomic DNA is differentially labeled with dye-labeled nucleotides and co-hybridized to the arrays. The copy number variations along a chromosome are represented by the fluorescence ratios of test DNA to control DNA different from 1 (usually 1.2 is considered as a cutoff for gain, and 0.8 as a cutoff for loss).

Large clones, mostly available as BACs, are preferred for making CGH arrays because they give good sensitivity, and information on mapped clones is easily accessible in the human genome databases (http://genome.ucsc.edu/; http://genome.wustl.edu/projects/human/). cDNA arrays have been demonstrated to be useful for CGH (Pollack et al. 1999; Heiskanen et al. 2000; Beheshti et al. 2003) but have several limitations which make them less attractive for detection of DNA copy number changes. First, they do not generate sufficient sensitivity for complex genomic probes because of their small size (0.5–3 kb). cDNA clones normally generate hybridization signals 30–80 times weaker than BAC clones for human or mouse genomic probes. This lack of sensitivity requires an additional signal amplification step (Heiskanen et al. 2000). Second, large genomic regions sparsely covered by cDNA clones will give poor resolution, and those genomic regions lacking genes will not be probed at all. Our analysis of the complete genomic sequence indicates that a total of 620-Mb euchromatic genomic sequences belong in regions larger than 500 kb without any annotated gene. These gene-barren regions may contain long-range genomic regulatory elements such as locus control regions, the deletion of which will affect multiple genes. Third, for applications that require high-level amplification of genomic DNA probes, amplification bias will more likely cause inaccurate copy number detection with cDNA arrays than BAC arrays because the larger size of BACs is more effective in averaging out amplification bias.

One apparent disadvantage of using BAC clones for CGH-based detection of DNA copy number changes is that some BACs contain significant amounts of repetitive sequences. Highly repetitive sequences in BACs can usually be effectively suppressed by Cot I DNA blocking. The most problematic BAC clones, however, are those containing high levels of low-copy repeats or similar cross-hybridizing sequences. These clones, in theory, will show false-positive or false-negative gains and losses, respectively (Fig. 2). If, for example, a genomic region is gained in the patient, false-positive gains are also caused by cross-hybridization of DNA from this region to other homologous regions in the genome. On the other hand, cross-hybridization may generate false-negative but not false-positive loss detection. False-negative loss detection is a major limitation that reduces the real resolution of BAC arrays, making some small genomic regions noninformative. A complete solution for the

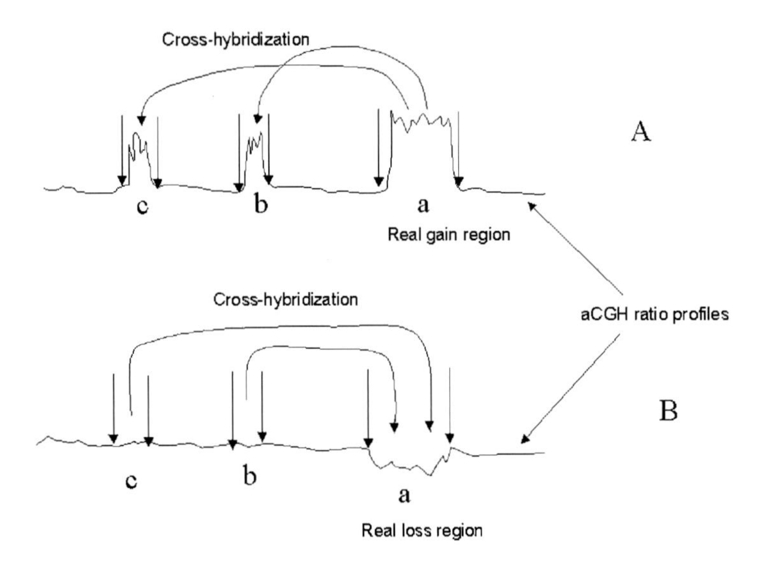


**Figure 2.** Models of false-positive and -negative detection caused by cross-hybridization. (*A*) Cross-hybridization causing false-positive gains. Gain of sequences in region **a** in the test sample can spill to cross-hybridizing regions **b** and **c**, causing false-positive gain detection in **b** and **c**. (*B*) Although real loss of sequences in region **a** exists, it is not detected due to partial or complete cross-hybridization of sequences from regions **b** and **c**, causing false-negative loss detection in region **a**.

false-positive and -negative problem with BAC clones is to use unique single-copy sequence fragments such as PCR-amplified genomic fragments as hybridization substrates on the arrays. Currently, this is not practical because the reagent cost is high, and effort involved in amplifying a large number of genomic fragments devoid of repeats is considerable.

## ARRAY PRODUCTION: A NOVEL APPROACH FOR MAKING DNA ARRAYS

Array production and hybridization are the most critical aspects of array CGH technology. Printing and subsequently crosslinking DNA fragments onto positively charged or reactive surfaces has now become a standard method for making cDNA arrays for gene expression analyses (Schena et al. 1995). However, this method does not work well for making CGH arrays because positively charged or reactive surfaces, even after extensive chemical inactivation blocking, tend to generate high hybridization background relative to the weak CGH signals. Entrapment of purified DNA in polymer matrix (Pinkel et al. 1998) does not require positively charged or reactive surfaces for DNA attachment, and higher amounts of target DNA can be confined in a spot area to increase hybridization efficiency. However, this "gluing down" method requires a very high concentration of highly purified BAC DNA, and the attached DNA is barely accessible for hybridization to generate directly visible spot signals (Pinkel et al. 1998). The requirement for a high amount of highly purified DNA is a significant disadvantage for this method for making high-density arrays.

To effectively minimize the hybridization background and maximize hybridization signals, we developed a novel approach for making DNA arrays (Cai et al. 2002) in which the BAC DNA is chemically modified using a crosslinker-like silane compound with an epoxide group that can covalently attach to DNA at slightly alkaline pH. The modified DNA retains normal hybridization specificity and could be purified after modification using standard procedures. Since the silanol groups at the other end of the silane molecule are only specific to glass surfaces (Fig. 3), DNA arrays can be conveniently made from modified DNA by simple deposition of the silanized DNA samples onto clean natural glass surfaces. Once contact is made with the glass surfaces, covalent attachment occurs at neutral pH after curing treatment in ethanol at an elevated temperature (Cai et al. 2002). DNA arrays made this way should have low background noise after hybridization washes because natural glass surfaces are slightly negatively charged and thus repulsive to DNA.

A distinct advantage of our DNA pre-modification method over all other methods using activated surfaces is that all activated surfaces are unstable, difficult to prepare reproducibly, and sensitive to impurities in the samples. Because of limited DNA-binding capacity, activated surfaces require careful calibration of DNA concentrations before printing to avoid overspill of excess DNA. In contrast, when DNA is modified before printing, it is not necessary to calibrate DNA concentrations because modified DNA molecules can cross-link with each other after drying to form a three-dimensional matrix, which has improved hybridization efficiency due to more available probe targets within spots. Modified DNA can be printed at any concentration without saturating the surface and causing spills, and even at concentrations of 2 μg/μl, all the spots are perfectly confined. On the other hand, printing the same DNA samples onto commercial slides without modification with a crosslinker-like silane compound at a concentration exceeding 300 ng/μl results in significant DNA spills. Because of this crosslinking property of silanized DNA, it can be arrayed on other surfaces such as metal and plastics, which are difficult to activate for DNA binding.

## QUANTITATIVE HYBRIDIZATION IN aCGH

One major challenge in aCGH is to attain quantitative hybridization on BAC arrays with unfractionated total genomic probes. Because both the targets and probes contain repetitive or duplicated sequences, hybridization signals contributed from these sequences must be suppressed to a sufficiently low level. Hybridization in aCGH is much more challenging than in conventional metaphase CGH for two reasons. First, excess of target DNA in spots demands highly effective suppression of repetitive sequence signals. On a BAC array, each spot has ~$10^5$–$10^6$ copies of 100- to 250-kb target DNA

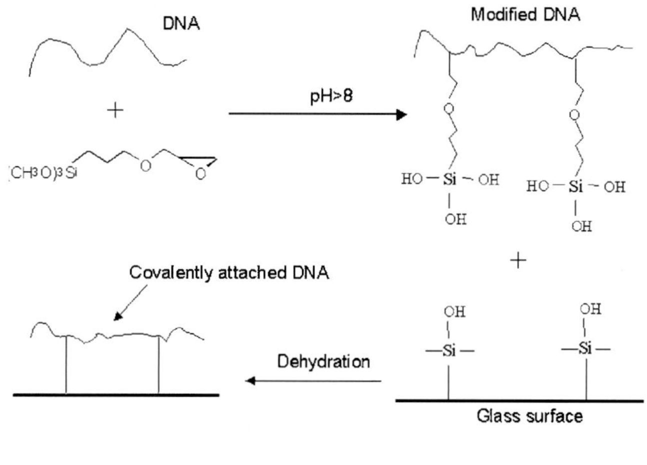

**Figure 3.** Chemical modification of DNA for making DNA microarrays. DNA can be covalently modified with a silane compound containing a reactive epoxide group in slightly basic solution. Upon drying, silanized DNA readily adheres to glass surfaces forming stable spots that can withstand hybridization and high stringent posthybridization washes.

molecules. Less than 1% of these target molecules are hybridized to unique sequences because of the limited specific sequences in genomic probes. Thus, even when 99% of the repetitive sequences in the arrayed clones are blocked, the contribution from the 1% repetitive sequences is still significant because any residual unblocked repetitive sequences will be saturated by the large amount of repetitive sequences in genomic probes. In contrast, in conventional metaphase CGH, the hybridization signals are the result of the hybridization of genomic DNA to a single DNA molecule on the metaphase chromosome, which is readily saturated by both unique and repetitive sequences. Moderate suppression of repetitive sequence signal is sufficient to intensify the unique sequence contribution to the total hybridization signal. Second, in metaphase CGH, the resolution is normally 10 Mb. Within such a large region, the effect of duplicated sequences or repetitive sequences is averaged out to the chromosome average level. However, for BACs, the unique sequence content fluctuates widely, making some of the BAC clones useless for aCGH.

Quantitative array hybridization depends on many variables. The commonly known variables are temperature, hybridization buffer composition, and conditions in posthybridization washes. Since there is little study regarding the hybridization behavior of cy3- or cy5-labeled DNA under different experimental conditions, optimal quantitative hybridization conditions must be found empirically. The general considerations in designing aCGH hybridization protocols include: (1) Blocking DNA should be enriched in all repeat sequence families including low-copy repeats. Current commercially available Cot I DNA mostly contains highly repetitive sequences. Although it is commonly used in aCGH, it may not be the optimal blocking DNA for aCGH. (2) If Cot DNA is used to block the array before hybridization, the fragment size of Cot DNA should not exceed 300 bp. Cot DNA with fragment size longer that 400 bp tends to introduce nonspecific unique sequences on the blocked BAC spots, subsequently introducing cross-hybridization signals. (3) The labeled patient and reference DNA probes should be fragmented to the size range of 200–400 bp to avoid cross hybridization. (4) A 30-fold or more excess amount of blocking DNA relative to the labeled probes should be used and prehybridized for 1–2 hours before applying to the arrays. (5) Long hybridization improves hybridization signals but lowers the sensitivity in detecting copy number differences because deterioration of cy5 and cy3 signals over prolonged hybridization occurs at different rates depending on as-yet-unknown variables. In hybridization buffers containing volume exclusion agents such as dextran sulfate or PEG, hybridization time of 16–24 hours usually generates sufficient fluorescent signals on arrays produced using the chemical premodification method (Cai et al. 2002).

We normally use test arrays containing a small number of BACs from chromosome X, chromosome Y, and some autosome clones as controls to test hybridization conditions. Female and male genomic DNA is co-hybridized to these test arrays. Under optimal conditions, the chromosome X copy-number difference between male and female DNA is clearly detected (Fig. 4). This test system is very stringent for optimization of aCGH hybridization conditions because most BAC clones on chromosome X contain a higher level of repetitive and duplicated sequences compared to the genome average (Lander et al. 2001; Venter et al. 2001). Success in achieving quantitative hybridization with these arrays indicates that at least repetitive sequences in probes and BAC targets on arrays are sufficiently blocked.

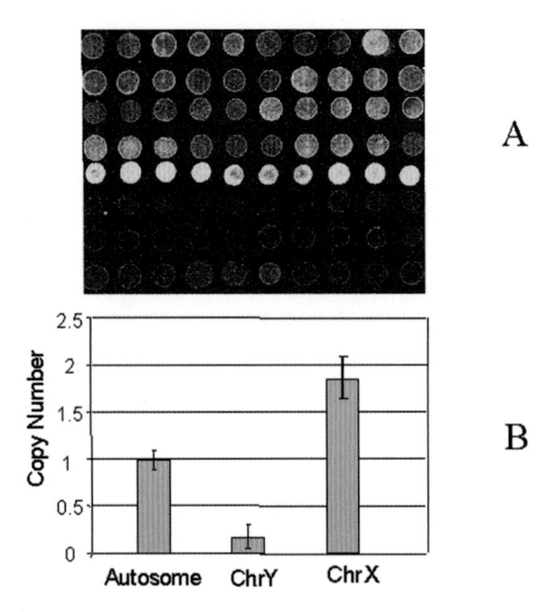


**Figure 4.** Test of quantitative aCGH using ChrXY arrays. (*A*) Portion of hybridization image from ChrXY test arrays. Spots in yellow are autosome BACs, in red, Chr. Y BACs, and in green, Chr.X BACs. The lower three rows are pUC clones without insert representing negative controls. (*B*) Results of copy number detection for normal female versus male experiments. Error bars represent standard deviations for different BACs on the arrays.

## FLUORESCENCE RATIO VARIABILITY IN aCGH

When equal amounts of control genomic probes labeled with cy5 and cy3 were hybridized to the BAC arrays, the variation of the fluorescence ratio for the two colors is usually within 10% for different BACs without concentration calibration before printing. This variation reflects the fluctuation and errors in hybridization and array spot quantitation. Additional sources of ratio variability may be the type of image quantification program used, which may cause differences in fluorescence ratios for the same spots up to 4%, and variability between replicate spots. Another significant source of ratio variation between BACs is the ratio drift toward one color (cy3 or cy5). This drift is spot-specific and depends on the fluorescence intensity. Some possible causes for this drift may be unequal labeling efficiency for cy3- and cy5-labeled probes, unequal accessibility of the differentially labeled probes to BAC DNA, or unequal dye stability and quenching. A recent study (Fare et al. 2003) indicated that the ozone level in the atmosphere is a significant factor affecting the fluorescence ratio in array experiments because of rapid cy5 signal degradation in posthybridization processing.

This intensity-dependent fluorescence ratio drift is one cause of false-positive detection in aCGH. Dye-reversal experiments are very effective in eliminating false positives caused by ratio drifts. Calibrating the modified DNA concentration before printing and taking fluorescence intensity into account in ratio normalization largely eliminated the ratio drifts, and the normalized ratio variation decreased from the initial 10% to an average of 6.7% across different BACs in experiments co-hybridizing normal to normal DNA.

## SENSITIVITY OF aCGH

Although the BAC CGH arrays promise to have wide applications in the study of the integrity of almost any genome, their current application is limited to samples with sufficient quantity of DNA isolated from relatively pure cells. In applying aCGH to analyze tumor samples, contamination of normal tissues in heterogeneous tumor samples will greatly reduce the detection rate for gain or loss. Recently, a laser-assisted microdissection technology called laser-capture microdissection has been developed for procurement of pure cell population from heterogeneous tissue (Emmert-Buck et al. 1996). This technology allows convenient extraction of any microscopic homogeneous cellular subpopulation from its complex surroundings. However, it is difficult to purify a large number of cells for array CGH, which requires at least 50,000 cells (minimally, 300 ng of genomic DNA is usually required). For conventional metaphase CGH analysis the problem of limited amounts of DNA was resolved by whole-genome amplification (WGA) (Zhang et al. 1992) using degenerate oligonucleotide-primed polymerase reaction (DOP-PCR) (Telenius et al. 1992).

The WGA technology has been improved and optimized recently for array CGH applications as well (Klein et al. 1999; Wells et al. 1999). However, incorporation of the WGA into the BAC CGH system will not be as straightforward as it might seem. The detection unit or resolution of BAC array CGH is defined as the size of the genomic region being sampled, and is at least 100-fold smaller than in conventional CGH (100–200 kb vs.10 Mb). At such high resolution, skewed amplification undetectable in a 10-Mb range can become prominent. In our hands, only the WGA technique based on the strand-displacement reaction (Dean et al. 2002; Lage et al. 2003) produced high levels of unbiased amplification. Using this method, we were able to perform aCGH with as little as 12 ng of genomic DNA. This sensitivity level should satisfy the sensitivity requirement for most applications. In addition, we have determined that the minimal amount of labeled probe within a 100-μm spot that can achieve the signal to noise (S/N) ratio of 10 was 0.01 pg. This amount of DNA is equivalent to ~60 copies of 150 kb of BAC molecules, which is equivalent to 150 kb of specific sequences from 0.2 ng of genomic DNA.

Although strand-displacement amplification can achieve a high level of amplification (Dean et al. 2002), its current limitation is the presence of contaminating DNA in the polymerases that will compete with specific amplification of the input DNA. This background amplification could generate significant amounts of labeled sequences that can hybridize to either the vector sequence on the BAC DNA or contaminating *Escherchia coli* DNA on the array spots.

## HUMAN WHOLE-GENOME BAC ARRAYS

To make human whole-genome (HWG) arrays, we selected 21,500 sequenced human BAC/PAC clones, which cover the human genome at an average resolution of 7 clones/MB. Most of the chromosomes are covered with BAC contigs with very small gaps. Chromosomes with a high level of repetitive sequences such as chromosome X and 19 were represented by fewer clones.

We printed all the clones on single slides and used a set of samples with known chromosome abnormalities to validate the HWG arrays and performed array analysis of five patients with a range of congenital anomalies with or without MR previously analyzed using conventional CGH techniques. We found that the results from HWG arrays agreed well with those of conventional CGH except in a case of a terminal deletion of chromosome 2q, which revealed less than the expected number of clones for a loss of a 10-Mb region. An example of the chromosomal CGH and aCGH detection of a gain of the terminal end of chromosome 5q is shown in Figure 5. In this patient the array data also indicated a small region of gain in the 5pter region.

Smaller regions of gains and losses that were not detected by conventional CGH were also seen in other patients. For example, in two patients we found a complex pattern of gains and losses of clones in the terminal end region of 1p spanning about 20 Mb. We used the chromosome 1p contig array to verify our WGA results. The 1p arrays were constructed using completely different sets of FISH-verified overlapping BAC clones, and the experiments were performed independently using a different hybridization protocol. In both cases, we found that the results of 1p arrays were consistent with our observations using HWG arrays.

The clinical relevance of these cytogenetically undetectable gains of terminal 5p and the complex pattern of gains and losses of terminal 1p remain unknown. Since they were detected in patients already carrying a significant and cytogenetically detectable chromosomal abnormality of a different chromosomal region, their contribution to the observed phenotypes may be minimal. Nevertheless, the detected copy number changes may help us identify regions with a high level of instability due to a variable number of repeats that consequently represent hot spots for unequal crossing-over.

## ANALYSIS OF IDIOPATHIC MR USING WHOLE-GENOME BAC ARRAYS

MR is one of the most common human congenital anomalies, and a large proportion of MR cases are idiopathic (Flint et al. 1995). To investigate whether subtle

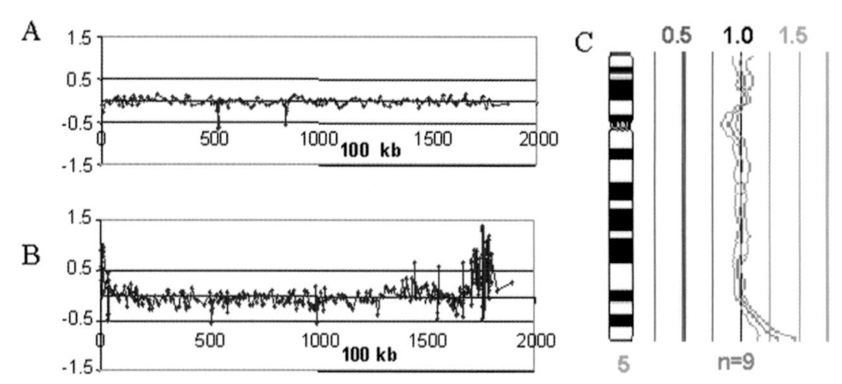


**Figure 5.** Comparison of CGH and aCGH profiles for chromosome 5. *A* and *B* are aCGH profiles of normal vs. normal hybridization and patient vs. normal control hybridization, respectively. The X axis represents linear positions of clones in 100 kb along the chromosome from p to q arm. Y axis is a normalized $\log_2$ fluorescence ratio calculated from a pair of dye-reversal experiments. (*C*) Metaphase CGH profile of chromosome 5 of the same patient. Both the aCGH and chromosomal CGH show gain of the terminal end of 5q. In addition, several clones along the terminal end of 5p show a gain.

genomic rearrangements exist in idiopathic MR, we initiated the genetic analysis of patients with idiopathic MR using human WGAs. All the recruited patients had mild MR as determined by formal testing (IQ 50–70). Following a thorough genetic evaluation at Baylor College of Medicine, none of the patients had detectable genetic etiology such as fragile X, microdeletion syndromes, subtelomere rearrangements, serious neurologic impairments (such as structural brain anomaly on imaging studies), or severe dysmorphic features.

We used the whole-genome BAC arrays for primary screening of potential regions of subtle chromosomal gain or loss in eight patients. For each patient sample we performed two hybridizations using the principle of dye reversal (Cai et al. 2002). The false-positive predictions due to random fluorescence ratio drifts or ratio normalization artifacts are greatly reduced by dye-reversal experiments. However, because the number of hybridization targets is high (over 20,000 spots), the number of gains and losses was also high in each of the eight patients, suggesting that a number of them are false-positive gains or losses. Since the verification of every clone showing a copy number change using conventional techniques such as FISH would be very time-consuming, we created subarrays containing clones from areas of loss or gain. We selected clones that were balanced (control clones) as well as clones showing a gain or loss. These small subarrays had five replicate spots for each of the selected clones. Dye-reversal hybridizations were repeated for subarrays for each sample, resulting in five pairs of data points for each positive and control clone. Arbitrarily, we defined a very stringent cutoff in which a clone must have at least four pairs of ratio data points outside the 2.5 S.D. units from the normalized average ratio of control clones to qualify for a true gain or loss. Among the 959 control BAC clones that did not show gain or loss in the HWG arrays, no clone was found to pass this cutoff and was therefore truly balanced. On the other hand, and as expected, only a fraction of the candidate positive BAC clones were verified in the secondary screen using these verification subarrays (Table 1). It is obvious that a verification step is necessary in order to eliminate false positives due to experimental variations on the HWG arrays. The results of our preliminary studies showed that gains of clones are much more frequent than losses. As previously discussed, consistent gains can be false positives caused by cross-hybridization from a region containing low copy repeats. Thus, further study is required to verify whether all regions showing gains are true positives.

## CONCLUSION

We demonstrate that high-resolution array-based comparative genomic hybridization can serve as a powerful approach for discovering small chromosomal rearrangements in a genome as complex as the human genome. Preliminary analysis of patients with known chromosomal abnormalities, as well as patients with normal karyotypes and idiopathic MR, were encouraging and identified areas of gain and loss. The biggest challenge in the future will be to distinguish between the changes due to normal variability in the genome versus gains and losses as the cause of the disorders. Population studies involving larger numbers of normal individuals are therefore required to establish the baseline variability and to identify regions most prone to polymorphisms. Large-scale studies of patients with mild MR are also required to resolve the issue of whether a combination of multiple polymorphic genomic imbalances may contribute to mild MR. In addition, the development of high-throughput molecular

**Table 1.** Losses and Gains in Idiopathic MR Patient in Primary HWG Arrays Screening and Secondary Subarray Verification

| | Primary HWG array screen | | Subarray verification | |
|---|---|---|---|---|
| Sample | loss BAC | gain BAC | loss BAC | gain BAC |
| 26–03 | 20 | 80 | 0 | 15 |
| 27–03 | 25 | 155 | 0 | 32 |
| 35–03 | 9 | 87 | 0 | 4 |
| 37–03 | 119 | 156 | 3 | 2 |
| 14–03 | 193 | 218 | 5 | 16 |
| 18–03 | 112 | 341 | 2 | 16 |
| 19–03 | 25 | 97 | 1 | 0 |
| 17–03 | 21 | 200 | 0 | 81 |

methods for validation of detected gains and losses in the second or even third screen will be necessary to efficiently eliminate false-positive detection. For genomic regions that heavily cross-hybridize, it will be necessary to develop specific arrays with unique DNA fragments to compensate for potential false-negative detection in these regions in the whole-genome BAC array screening.

## ACKNOWLEDGMENTS

We thank Qian Li for technical help in array production, Dr. Wei Yu for providing the 1p contig arrays, Dr. Rue Chen for the updated BAC clone position information, and Dr. Sau Wai Cheung for advice on result validation. This work was supported by a grant from the Mental Retardation Research Center of Baylor College of Medicine.

## REFERENCES

Beheshti B., Braude I., Marrano P., Thorner P., Zielenska M., and Squire J.A. 2003. Chromosomal localization of DNA amplifications in neuroblastoma tumors using cDNA microarray comparative genomic hybridization. *Neoplasia* **5:** 53.

Brewer C., Holloway S., Zawalnyski P., Schinzel A., and FitzPatrick D. 1998. A chromosomal deletion map of human malformations. *Am. J. Hum. Genet.* **63:** 1153.

———. 1999. A chromosomal duplication map of malformations: Regions of suspected haplo- and triplolethality—and tolerance of segmental aneuploidy—in humans. *Am. J. Hum. Genet.* **64:** 1702.

Cai W.W., Mao J.H., Chow C.W., Damani S., Balmain A., and Bradley A. 2002. Genome-wide detection of chromosomal imbalances in tumors using BAC microarrays. *Nat. Biotechnol.* **20:** 393.

Cheung V.G., Dalrymple H.L., Narasimhan S., Watts J., Schuler G., Raap A.K., Morley M., and Bruzel A. 1999. A resource of mapped human bacterial artificial chromosome clones. *Genome Res.* **9:** 989.

Cheung V.G., Nowak N., Jang W., Kirsch I.R., Zhao S., Chen X.-N., Furey T.S., Kim U.-J., Kuo W.-L., Olivier M., Conroy J., Kasprzyk A., Massa H., Yonescu R., Sait S., Thoreen C., Snijders A., Lemyre E., Bailey J.A., Bruzel A., Burrill W.D., Clegg S.M., Collins S., Dhami P., and Friedman C., et al. 2001. Integration of cytogenetic landmarks into the draft sequence of the human genome. *Nature* **409:** 953.

Dean F.B., Hosono S., Fang L., Wu X., Faruqi A.F., Bray-Ward P., Sun Z., Zong Q., Du Y., Du J., Driscoll M., Song W., Kingsmore S.F., Egholm M., and Lasken R.S. 2002. Comprehensive human genome amplification using multiple displacement amplification. *Proc. Natl. Acad. Sci.* **99:** 5261.

Emmert-Buck M.R., Bonner R.F., Smith P.D., Chuaqui R.F., Zhuang Z., Goldstein S.R., Weiss R.A., and Liotta LA. 1996. Laser capture microdissection. *Science* **274:** 998.

Fare T.L., Coffey E.M., Dai H., He Y.D., Kessler D.A., Kilian K.A., Koch J.E., LeProust E., Marton M.J., Meyer M.R., Soughton R.B., Tokiwa G.Y., and Wang Y. 2003. Effects of atmospheric ozone on microarray data quality. *Anal. Chem.* **75:** 4672.

Flint J., Wilkie A.O.M., Buckle V., Winter R.M., Holland A.J., and McDermid H.E. 1995. The detection of subtelomeric chromosomal rearrangements in idiopathic mental retardation. *Nat. Genet.* **9:** 132.

Gardner R.J.M. and Sutherland G.R. 1996. *Chromosome abnormalities and genetic counseling,* 2nd edition. Oxford University Press, Oxford, United Kingdom.

Heiskanen M.A., Bittner M.L., Chen Y., Khan J., Adler K.E., Trent J.M., and Meltzer P.S. 2000. Detection of gene amplification by genomic hybridization to cDNA microarrays. *Cancer Res.* **60:** 799.

Kallioniemi A., Kallioniemi O.P., Sudar D., Rutovitz D., Gray J.W., Waldman F., and Pinkel D. 1992. Comparative genomic hybridization for molecular cytogenetic analysis of solid tumors. *Science* **258:** 818.

Klein C.A., Schmidt-Kittler O., Schardt J.A., Pantel K., Speicher M.R., and Riethmuller G. 1999. Comparative genomic hybridization, loss of heterozygosity, and DNA sequence analysis of single cells. *Proc. Natl. Acad. Sci.* **96:** 4494.

Knight S.J.L., Regan R., Nicod A., Horsley S.W., Kearney L., Homfray T., Winter R.M., Bolton P., and Flint J. 1999. Subtle chromosomal rearrangements in children with unexplained mental retardation. *Lancet* **354:** 1676.

Korenberg J.R., Chen X.N., Sun Z., Shi Z.Y., Ma S., Vataru E., Yimlamai D., Weissenbach J.S., Shizuya H., Simon M.I., Gerety S.S., Nguyen H., Zemsteva I.S., Hui L., Silva J., Wu X., Birren B.W., and Hudson T.J. 1999. Human genome anatomy: BACs integrating the genetic and cytogenetic maps for bridging genome and biomedicine. *Genome Res.* **9:** 994.

Lage J.M., Leamon J.H., Pejovic T., Hamann S., Lacey M., Dillon D., Segraves R., Vossbrinck B., Gonzalez A., Pinkel D., Albertson D.G., Costa J., and Lizardi P.M. 2003. Whole genome analysis of genetic alterations in small DNA samples using hyperbranched strand displacement amplification and array-CGH. *Genome Res.* **13:** 294.

Lander E.S., Linton L.M., Birren B., Nusbaum C., Zody M.C., Baldwin J., Devon K., Dewar K., Doyle M., FitzHugh W., Funke R., Gage D., Harris K., Heaford A., Howland J., Kann L., Lehoczky J., LeVine R., McEwan P., McKernan K., Meldrim J., Mesirov J.P., Miranda C., Morris W., and Naylor J., et al. (International Human Genome Sequencing Consortium). 2001. Initial sequencing and analysis of the human genome. *Nature* **409:** 860.

McPherson J.D., Marra M., Hillier L., Waterston R.H., Chinwalla A., Wallis J., Sekhon M., Wylie K., Mardis E.R., Wilson R.K., Fulton R., Kucaba T.A., Wagner-McPherson C., Barbazuk W.B., Gregory S.G., Humphray S.J., French L., Evans R.S., Bethel G., Whittaker A., Holden J.L., McCann O.T., Dunham A., Soderlund C., and Scott C.E., et al. (International Human Genome Mapping Consortium). 2001. A physical map of the human genome. *Nature* **409:** 934.

Pinkel D., Segraves R., Sudar D., Clark S., Poole I., Kowbel D., Collins C., Kuo W.L., Chen C., Zhai Y., Dairkee S.H., Ljung B.M., Gray J.W., and Albertson D.G. 1998. High resolution analysis of DNA copy number variation using comparative genomic hybridization to microarrays. *Nat. Genet.* **20:** 207.

Pollack J.R., Perou C.M., Alizadeh A.A., Eisen M.B., Pergamenschikov A., Williams C.F., Jeffrey S.S., Botstein D., and Brown P.O. 1999. Genome-wide analysis of DNA copy-number changes using cDNA microarrays. *Nat. Genet.* **23:** 41.

Schena M., Shalon D., Davis R.W., and Brown P.O. 1995. Quantitative monitoring of gene expression patterns with a complementary DNA microarray. *Science* **270:** 467.

Shapira S.K. 1998. An update on chromosome deletion and microdeletion syndromes. *Curr. Opin. Pediatr.* **10:** 622.

Telenius H., Carter N.P., Bebb C.E., Nordenskjold M., Ponder B.A., and Tunnacliffe A. 1992. Degenerate oligonucleotide-primed PCR: General amplification of target DNA by a single degenerate primer. *Genomics.* **13:** 718.

Venter J.C., Adams M.D., Myers E.W., Li P.W., Mural R.J., Sutton G.G., Smith H.O., Yandell M., Evans C.A., Holt R.A., Gocayne J.D., Amanatides P., Ballew R.M., Huson D.H., Wortman J.R., Zhang Q., Kodira C.D., Zheng X.H., Chen L., Skupski M., Subramanian G., Thomas P.D., Zhang J., Gabor Miklos G.L., and Nelson C., et al. 2001. The sequence of the human genome. *Science* **291:** 1304.

Wells D., Sherlock J.K., Handyside A.H., and Delhanty J.D. 1999. Detailed chromosomal and molecular genetic analysis of single cells by whole genome amplification and comparative genomic hybridisation. *Nucleic Acids Res.* **27:** 1214.

Zhang L., Cui X., Schmitt K., Hubert R., Navidi W., and Arnheim N. 1992. Whole genome amplification from a single cell: Implications for genetic analysis. *Proc. Natl. Acad. Sci.* **89:** 5847.

# Annotation of Novel Proteins Utilizing a Functional Genome Shotgun Coupled with High-throughput Protein Interaction Mapping

J.A. MALEK,* J.M. WIERZBOWSKI,* G.A. DASCH,[†] M.E. EREMEVA,[‡]
P.J. MCEWAN,* AND K.J. MCKERNAN*
*Agencourt Bioscience Corporation, Beverly, Massachusetts 01915; [†]Centers for Disease Control and Prevention, Atlanta, Georgia 30332; and [‡]University of Maryland, Baltimore, Maryland 21201

It is quoted frequently that the amount of genomic sequence data is increasing at a tremendous rate while functional methods for studying proteins have not kept pace. Many groups have attempted to address this issue by use of microarrays, yeast two-hybrid screens, and protein complex purification with subsequent identification. The underlying theme of these approaches is their use of "guilt-by-association" (Oliver 2000) methods for annotation of proteins of unknown function. The association of a protein of unknown function with proteins of known function is used to derive a potential function for the protein of unknown function. Although large-scale microarray experiments have increased dramatically and are carried out in numerous large and small laboratories, proteome-wide two-hybrid experiments have only been carried out on yeast (Uetz et al. 2000; Ito et al. 2001) with some large studies in *Caenorhabditis elegans* (Walhout et al. 2000), and *Helicobacter pylori* (Rain et al. 2001), among others. Numerous review papers have been written on these large-scale two-hybrid studies, analyzing the data, testing the data's validity, and using the data to train in silico protein interaction prediction software. The need for further, validated, protein interaction information is clear. Among the challenges in generating proteome-wide interaction data are the lack of fully automated processes, the sheer amount of screening necessary to complete one map for one organism, and an incomplete grasp of what constitutes a true, physiologically important protein interaction. It is our belief that using comparative interaction data will allow deciphering of what interactions are physiologically valid. Validation of interactions has centered around comparisons to databases of individually obtained and presumably more verified interactions, the presence of interactions among proteins with similar expression profiles, and the frequency of interactions among proteins sharing similar biological processes and/or cellular compartments (Deane et al. 2002). Although these methods of verification may add a level of significance to any interaction, their absence should not per se be used to subtract from an interaction's validity. Observing similar expression profiles between two proteins may suggest they are functionally related but does not mean that they physically interact. Observing interactions among proteins of different biological processes may reveal a gap in our knowledge more than an incorrect interaction. Observation of interactions among proteins from different cellular compartments is less meaningful in organelle-free microbes. Physiologically significant interactions with a wide range of strengths have been observed, therefore a significance cutoff based on interaction strength cannot be set at present.

Automated DNA sequencing technology was heavily developed during the Human Genome Project into a robust and cheap process. It would be of benefit to use these developments in advancing proteome-wide interaction data. We have attempted to improve the ease with which such studies can be carried out by adopting a strategy that relies on whole-genome shotgun sequencing and a bacterial two-hybrid system (Dove et al. 1997; Shaywitz et al. 2000). The whole-genome shotgun method generates cloned overlapping fragments of genomic DNA which, if cloned in the proper orientation and frame, can be expressed as a protein. The use of peptide fragments rather than full-length proteins has been shown to reduce false negatives (Ward et al. 2002) while offering the opportunity to localize the domain of a protein responsible for an interaction. Use of the bacterial two-hybrid system allows integration into standard sequencing pipelines. The two vectors used in the system are standard sequencing vectors that are transformed together into an essentially standard cloning strain of *Escherichia coli*. The system, similar to various yeast two-hybrid systems, relies on recruitment of transcriptional machinery to promoters upstream of reporter genes. Briefly, a protein of interest is fused to the λcI protein which binds a λ operator on the reporter construct (Fig. 1a). A second protein of interest is fused to the RNA polymerase α-subunit. An interaction between the proteins of interest stabilizes the transcriptional machinery at a weak promoter upstream of the reporter construct (Fig. 1b). Interactions are observed as a colony able to grow in the presence of an antibiotic and the absence of any carbon source other than lactose. Colonies can enter a standard sequencing pipeline at this point through the automated colony pickers. Sequencing of the bait of prey fragment is conducted with primers specific for either vector.

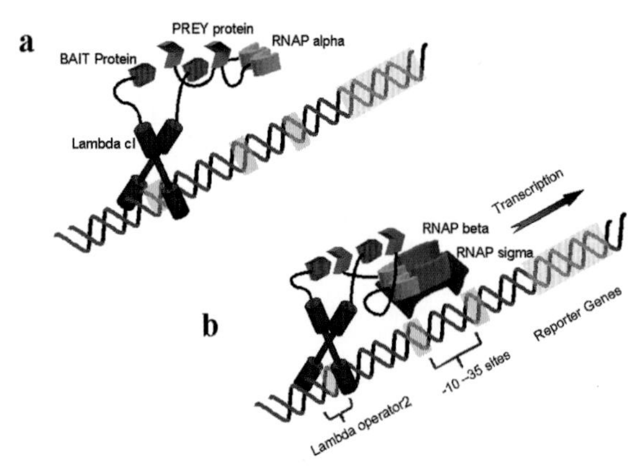


**Figure 1.** The bacterial two-hybrid system. (*a*) A protein of interest is fused to λcI and another protein of interest is fused to RNA polymerase α-subunit. λcI binds the operator upstream of a weak promoter. (*b*) An interaction between the two proteins of interest drives transcription of reporter genes by recruitment of the transcriptional machinery.

## FUNCTIONAL SHOTGUN SEQUENCING AND ANNOTATION

To test the approach, we shotgun-sequenced the genome of *Rickettsia sibirica*, an intracellular human pathogen among the spotted fever group *Rickettsiae*. The general functional shotgun approach involves cloning random fragments from the genome into the pBAIT vector, sequencing to determine clones containing in-frame fragments, and transferring these fragments to the pPREY vector to create a proteome library against which to screen baits of interest (Fig. 2). In this study, random fragments were cloned in the pBAIT, a modified version of the pACλcI vector, using a BstXI linker method. The genome was sequenced to 12.5x sequence coverage from this pBAIT library using a primer designed to allow elucidation of the λcI fusion translation product for each insert. A further 3x coverage was conducted in fosmid paired-end reads to order and orient the contigs. The genome was assembled with the Paracel Genome Assembler™ (Paracel, Pasadena, California), resulting in one scaffold of seven contigs. Assembly revealed a genome of ~1,250,021 bp. Gene regions were determined using standard HMM-based software and assigned function by BLASTP to the GenBank NR database. In this manner, 1,234 protein-coding genes were annotated covering a total of 324,008 amino acids. Translation of the pBAIT inserts revealed that 3,932 contained fragments, translated in-frame with the λcI, matching a portion of an annotated protein-coding gene. These in-frame fragments contained 599,602 amino acids, representing 278,832 unique amino acids, or about 1.85x random proteome coverage. At 1.85x coverage, 85.3% of the proteome should be covered. In-frame fragments were obtained from 986 distinct ORFs with 86% of all amino acids covered at least once. From this set of clones, we were able to select clones spanning regions of interest for screening against either the sheared genomic Prey library or the shuttled ORF fragment Prey library.

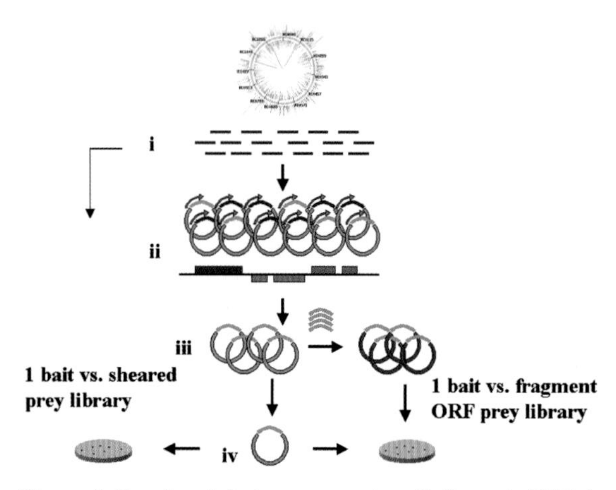


**Figure 2.** Functional shotgun sequencing. (*i*) Genomic DNA is randomly sheared and cloned in the pBAIT vector. (*ii*) Clones are sequenced, the genome assembled and subsequently annotated. (*iii*) Clones containing fragments, translated correctly, from open reading frames, are arrayed and the inserts transferred to the pPREY vector to create a library against which to screen. (*iv*) Clones spanning proteins of interest can be screened against either a random library or the cloned, in-frame ORF library.

## SCREENING OF TYPE IV SECRETION SYSTEM PROTEINS

The Type IV secretion system (T4SS) is a multi-subunit complex frequently involved in virulence effector transport in various gram-negative bacteria. Inter-subunit interactions have been studied using the yeast two-hybrid system for the *Agrobacterium tumefaciens* T4SS (Ward et al. 2002). Prior to screening the entire proteome, we selected proteins from the *R. sibirica* T4SS for screening against the prey random genomic library and prey fragment ORF library. The goal was to observe interactions previously reported among subunits while obtaining novel interactions by screening against the entire proteome. Selected fragments used in the screening covered 6 known T4SS proteins (Fig. 3).

## SCREENING RESULTS

Screening of multiple fragments from the 6 T4SS proteins yielded 285 interactions occurring between the 6 subunits and 155 distinct proteins (Fig. 4). Of the interactions, 162 were observed once, 48 were category-observed more than once, and 74 were observed more than once by different fragments. Of the 162 interactions observed once, 136 (84%) were observed with more than one subunit, adding a level of contextual validation to the in-

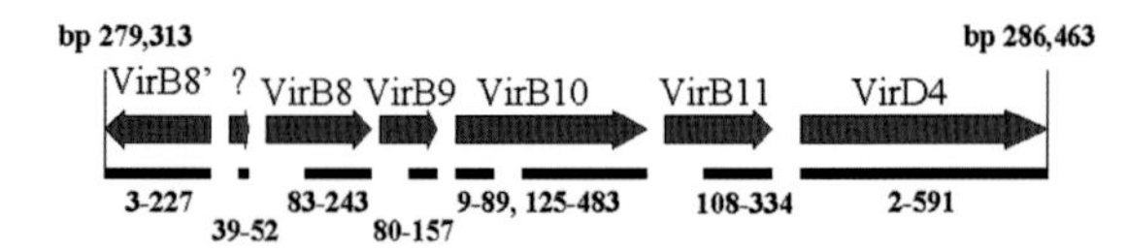


**Figure 3.** Regions of the type 4 secretion system screened. Proteins from the type 4 secretion system in *Rickettsia sibirica* are identified as red arrows. The underlying black lines represent regions for which a clone with a properly cloned protein fragment was available for screening.

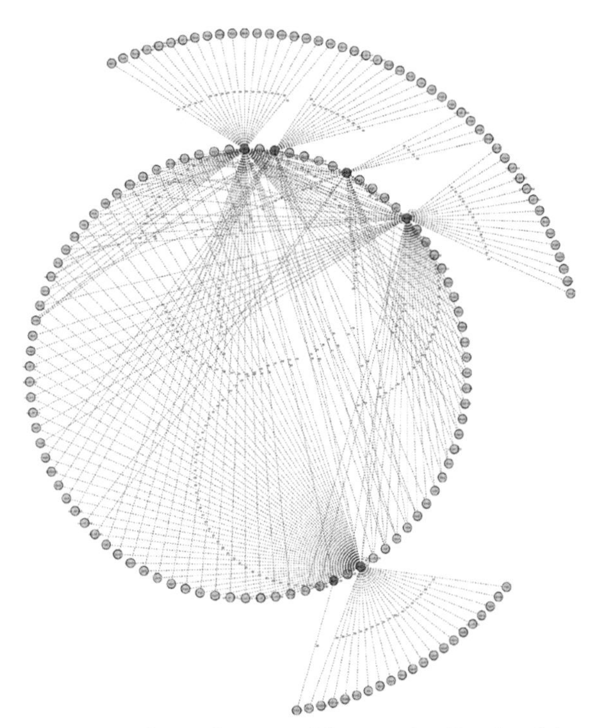

**Figure 4.** An interaction map of the screening. Results of the screening are displayed using Cytoscape (Ideker et al. 2002) with the bait proteins displayed in red. Interactions between two proteins are represented as edges between the two nodes. Interactions occurring with multiple subunits of the T4SS are displayed in the inner circle and subunit-specific interactions are displayed in the outer circle.

teraction. 96% of a randomly selected set of interactions was capable of reconstitution: Purification of plasmid and reintroduction into fresh cells reconstituted the interaction. Approximately 40% of intra-complex interactions reported in a study of *A. tumefaciens* T4SS orthologs using the yeast two-hybrid (Ward et al. 2002) were observed in our study of *R. sibirica* subunits using the bacterial two-hybrid system. This number may improve as we clone and screen regions in the subunits not screened due to the fact that they were missing from our baits. Earlier analysis of the overlap between any two studies of the same interactions using the yeast two-hybrid revealed an ~16–35% overlap (Matthews et al. 2001) showing that our results are toward the upper end of this observation. We observed interactions among subunits not observed in the Ward and colleagues study, and these are being followed up with further investigation. Among the proteins interacting with multiple subunits were proteins involved in cell wall/membrane biogenesis, defense mechanisms, lipid transport and metabolism, and uncharacterized proteins. Strong interactions were also observed with some proteins involved in DNA replication or repair. One potential false positive was observed as a strong interaction between DnaK, a chaperone, and most of the subunits.

## DISCUSSION

Using protein interactions, it is possible to begin assigning functional context to previously unannotated proteins. We have focused our future studies on the proteins interacting with the T4SS in our search for rickettsial host effectors. The use of multiple methods such as microarrays, two-hybrid systems, and large-scale gene knockdowns will be necessary to begin functionally annotating the rapidly increasing number of novel proteins.

On the basis of various studies, the average protein is expected to have ~3 interacting partners. Our study reveals significantly more interactions than this average. A possible explanation is that proteins we investigated are part of a transmembrane complex known to transport various effectors thus increasing the number of proteins they would interact with. This may only be a partial explanation. We believe that previous studies have underestimated the average number of interacting partners per protein. A potential reason for this is the lack of depth of screening, or the stringency used in carrying out interactions, which could have eliminated valid, weaker interactions. If this is true, then the complexity of interactions occurring in any given cell could be immense.

Interactions between apparently unrelated proteins are quite likely meaningless to a protein's direct function, although they may have a critical role in cell physiology and intracellular movement. We believe the approach of capturing as many interactions as possible in a screen is important. In "casting a wide net," the protein interaction space is narrowed without loss of sensitivity. Further validation through other systems such as yeast two-hybrid and protein complex studies would then further narrow what is considered to be of importance. We stress again that comparative protein interaction mapping of organisms will offer a critical method for validating interactions.

The opposite of protein–protein interactions, protein–protein repulsions, may play as important a role in the forming of localized protein complexes in the cell. This is a relatively unexplored possibility, and studying the combination of protein–protein interactions and repulsions may give more insight into cell physiology.

We have reported a method to improve the rate at which necessary proteome-wide interaction studies can be generated. By combining the methods of shotgun sequencing and a bacterial two-hybrid system, we hope that the number of organisms with interaction maps will improve and bring this field into the era of comparative studies.

## REFERENCES

Deane C.M., Salwinski L., Xenarios I., and Eisenberg D. 2002. Protein interactions: Two methods for assessment of the reliability of high throughput observations. *Mol. Cell. Proteomics* **5:** 349.

Dove S.L., Joung J.K., and Hochschild A. 1997. Activation of prokaryotic transcription through arbitrary protein-protein contacts. *Nature* **386:** 627.

Ideker T., Ozier O., Schwikowski B., and Siegel A.F. 2002. Discovering regulatory and signaling circuits in molecular interaction networks. *Bioinformatics* **18**: S233.

Ito T., Chiba T., Ozawa R., Yoshida M., Hattori M., and Sakaki Y. 2001. A comprehensive two-hybrid analysis to explore the yeast protein interactome. *Proc. Natl Acad. Sci.* **98:** 4569.

Matthews L.R., Vaglio P., Reboul J., Ge H., Davis B.P., Garrels

J., Vincent S., and Vidal M. 2001. Identification of potential interaction networks using sequence-based searches for conserved protein-protein interactions or "interologs". *Genome Res.* **11:** 2120.

Oliver S. 2000. Guilt-by-association goes global. *Nature* **403:** 601.

Rain J.C., Selig L., De Reuse H., Battaglia V., Reverdy C., Simon S., Lenzen G., Petel F., Wojcik J., Schachter V., Chemama Y., Labigne A., and Legrain P. 2001. The protein-protein interaction map of *Helicobacter pylori*. *Nature* **409:** 211.

Shaywitz A.J., Dove S.L., Kornhauser J.M., Hochschild A., and Greenberg M.E. 2000. Magnitude of the CREB-dependent transcriptional response is determined by the strength of the interaction between the kinase-inducible domain of CREB and the KIX domain of CREB-binding protein. *Mol. Cell. Biol.* **20:** 9409.

Uetz P., Giot L., Cagney G., Mansfield T.A., Judson R.S., Knight J.R., Lockshon D., Narayan V., Srinivasan M., Pochart P., Qureshi-Emili A., Li Y., Godwin B., Conover D., Kalbfleisch T., Vijayadamodar G., Yang M., Johnston M., Fields S., and Rothberg J.M. 2000. A comprehensive analysis of protein-protein interactions in *Saccharomyces cerevisiae*. *Nature* **403:** 623.

Walhout A.J., Sordella R., Lu X., Hartley J.L., Temple G.F., Brasch M.A., Thierry-Mieg N., and Vidal M. 2000. Protein interaction mapping in *C. elegans* using proteins involved in vulval development. *Science* **287:** 116.

Ward D.V., Draper O., Zupan J.R., and Zambryski P.C. 2002. Peptide linkage mapping of the *Agrobacterium tumefaciens* vir-encoded type IV secretion system reveals protein subassemblies. *Proc. Natl. Acad. Sci.* **99:** 11493.

# Global Predictions and Tests of Erythroid Regulatory Regions

R.C. Hardison,* F. Chiaromonte,* D. Kolbe,* H. Wang,* H. Petrykowska,* L. Elnitski,* S. Yang,* B. Giardine,* Y. Zhang,* C. Riemer,* S. Schwartz,* D. Haussler,[†] K.M. Roskin,[†] R.J. Weber,[†] M. Diekhans,[†] W. J. Kent,[†] M.J. Weiss,[‡] J. Welch,[‡] and W. Miller*

**Departments of Biochemistry and Molecular Biology, Statistics, Health Evaluation Services, and Computer Science and Engineering, and Center for Comparative Genomics and Bioinformatics, The Pennsylvania State University, University Park, Pennsylvania 16802; [†] Center for Biomolecular Science and Engineering and Howard Hughes Medical Institute, University of California at Santa Cruz, Santa Cruz, California 95064; and [‡]Department of Pediatrics, Children's Hospital of Philadelphia and The University of Pennsylvania, Philadelphia, Pennsylvania 19104*

Determinations of the genomic DNA sequences of human, mouse, and other organisms are landmark achievements, but the major changes in biology and medicine anticipated as a result (Lander 1996) require that a function be assigned to all the important segments within those genomes (Collins et al. 2003). It has long been realized that functional sequences change more slowly than nonfunctional (neutral) DNA sequences over evolutionary time (Kimura 1968; Li et al. 1981). Some gene prediction and assessment algorithms incorporate interspecies sequence alignments into their analysis (see, e.g., Korf et al. 2001; Wiehe et al. 2001; Nekrutenko et al. 2002). This slower rate also can be predictive for sequences involved in gene regulation. One of the early approaches for finding critical sequences within bacteriophage promoters used sequence comparison (Pribnow 1975), and highly conserved noncoding DNA sequences are now commonly used as guides for potential gene regulatory elements (for review, see Hardison 2000; Pennacchio and Rubin 2001).

In this paper, we address two complications to the large-scale application of genomic sequence alignments to predicting *cis*-regulatory modules (CRMs), i.e., discrete sequences such as promoters, enhancers, and silencers that control gene expression. The rate at which neutral DNA changes is highly variable within a genome (Wolfe et al. 1989; Hardison et al. 2003), and thus the amount of change observed needs to be corrected for local variation in the neutral rate. Such a corrected score can be used to compute a probability that a sequence is conserved because of purifying selection (Waterston et al. 2002; Chiaromonte et al., this volume). The second complication is that DNA sequences which do not code for protein (noncoding DNA) can be selected for functions other than a role in regulating gene expression. Examples include genes for noncoding RNAs such as tRNAs and microRNAs. Sequences involved in chromosome dynamics may also be under selection. We describe an approach to find patterns characteristic of gene regulatory sequences within the alignments (Elnitski et al. 2003).

We are applying these analyses of whole-genome sequence alignments to predict regulatory elements of genes expressed during late erythroid differentiation. This is a particularly attractive somatic cell model for mammalian differentiation because morphologically distinct cell types are made during the progress of differentiation and maturation, and several abundant red cell proteins, such as hemoglobins and cytoskeletal proteins, are well-characterized markers of later maturation (Migliaccio and Papayannopoulou 2001). Furthermore, cultured cell lines such as murine erythroleukemia (MEL) cells can be chemically induced to undergo a transition similar to that of proerythroblasts to erythroblasts (Friend et al. 1971). More recently, progenitor cell lines missing a particular transcription factor critical for erythroid differentiation, GATA-1, have been isolated and phenotypically rescued using a conditionally active GATA-1 (Weiss et al. 1997). Thus, we can assay globally for genes responding in these two models for erythroid differentiation, and in the latter case, it is highly likely that early-responding genes are direct targets of GATA-1. We report some initial success applying the computational predictions of CRMs in these somatic cell systems.

## ALIGNMENTS OF WHOLE MAMMALIAN GENOMES

The availability of the human (Lander et al. 2001) and mouse (Waterston et al. 2002) genome sequences makes it possible to determine comprehensively which DNA sequences are present in both, which have been inserted or deleted, and which have been altered by nucleotide substitution since primates and rodents diverged. A high-quality assembly of the rat genome sequence is available (International Rat Genome Sequencing Consortium, in prep.), and adding this to the aligned sequences will provide greater resolution on these issues. All the sequences encoding and regulating conserved functions should be found within the sequences common to mouse and human, hence this is the starting point for our search for predicted CRMs.

In our approach to whole-genome alignments, we first find all the meaningful local alignments between the two sequences using the program *blastz*, and then we use *axtBest* to arrange these local alignments into chains that reflect blocks of conserved synteny, which can be many megabases in length (Schwartz et al. 2003). Further layers of chaining reflect duplications, inversions, and other events (Kent et al. 2003). All sequences in one genome are given the opportunity to align with sequences in the other, hence it is an all-versus-all alignment; no prior deductions about blocks of conserved synteny are used. The scoring parameters have been optimized for long mammalian genomic DNA sequences (Chiaromonte et al. 2002) and can be set to achieve high sensitivity with very little noise (Schwartz et al. 2003). Although it is not possible at present to know definitively that all the homologous sequences have been aligned, it is likely that the vast majority have been.

When interpreting the results of interspecies sequence alignments, it is important to distinguish the part of the genome derived from the last common ancestor, which we refer to as the "ancestral" portion, from the part that arose only along one lineage (Fig. 1). Virtually all lineage-specific insertions result from retrotransposition events (Lander et al. 2001); these are not aligned in our procedure (Schwartz et al. 2003). Lineage-specific segmental duplications comprise about 6% of the human genome (Bailey et al. 2002), but all copies can align with the comparison species. The ancestral portion is the non-repetitive DNA plus the repeats that were present in the last common ancestor. Orthologous ancestral repeats are included in alignments that begin in adjacent single-copy regions. The ancestral repeats are relics of transposable elements that were active prior to the primate–rodent divergence but no longer transpose. Although a very small fraction of ancestral repeats have been implicated in regulation of gene expression (Jordan et al. 2003), the vast majority have no identifiable function. Hence, the aligned orthologous copies in human and mouse represent a good model for evolution in neutral DNA (Waterston et al. 2002; Hardison et al. 2003).

To the extent that the alignments are comprehensive,

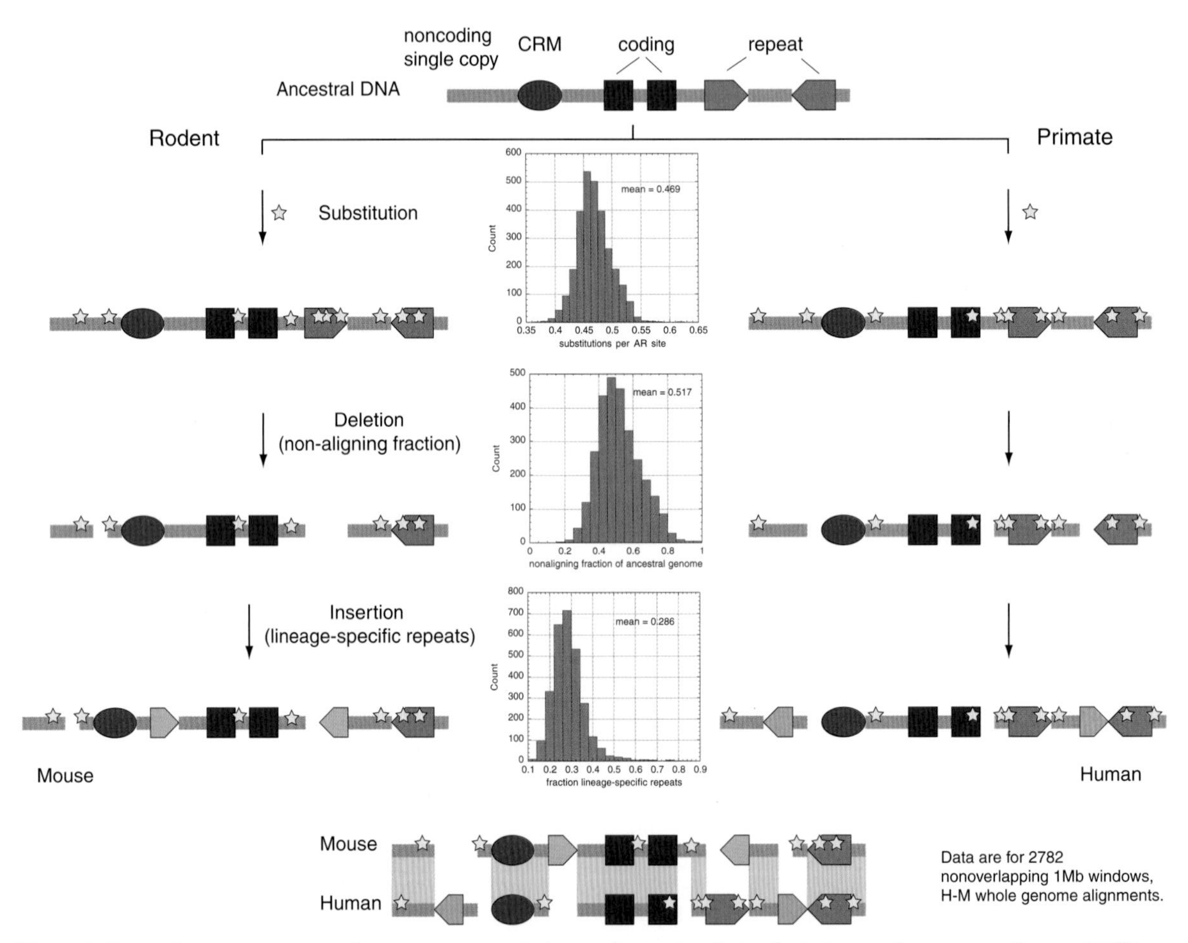


**Figure 1.** Events that cause sequence divergence over evolutionary time and variation in their rate of occurrence. Types of DNA sequences in the ancestor to rodents and primates are diagramed on the top line; these include coding exons, *cis*-regulatory modules (CRMs), and interspersed repeats from transposons (*brown angled boxes*), separated by single-copy DNA. The diagrams along the sides illustrate the accumulation of nucleotide substitutions (*stars*), large deletions (*absence of icons*), and insertions of new classes of transposons (*green* and *purple angled boxes*) in the lineage to mouse (*left*) and human (*right*). Sequences that were in the ancestor and have not been deleted can align between human and mouse (*bottom diagram*). From the human–mouse alignments, one can infer the substitutions per site in the ancestral repeats (AR), the amount deleted (nonaligning fraction of the ancestral genome), and the amount of DNA inserted (fraction lineage-specific repeats) in 1-Mb nonoverlapping windows (2782 in the human genome). The graphs in the center column show the distribution of these values for the three processes of DNA change.

one can draw an informative inference about the nonaligning part of the ancestral portion of a genome—it is not likely to be present in the other genome. Because we do not align lineage-specific insertions, and lineage-specific duplicates can align, the simplest explanation for the sequences not being in the comparison genome is that they were deleted. Other analyses based on a relatively constant genome size in mammals also argue that the nonaligning fraction reveals deletions in the comparison genome (Waterston et al. 2002).

Genome sequences of additional species, such as rat (International Rat Genome Sequencing Consortium, in prep.), are being assembled as large-scale genomic sequence and analysis projects move into more functional and analytical studies (Collins et al. 2003). Pair-wise and multiple alignments of these sequences are regularly updated and made available on the UCSC Genome Browser (Kent et al. 2002) at http://genome.ucsc.edu. Additional mammalian genome sequences substantially improve the power of sequence alignment techniques to resolve functional from nonfunctional DNA sequences (Thomas et al. 2003).

## CONSERVATION AND SELECTION

About 40% of the human genome aligns with sequences in the mouse genome. As expected, almost all (99%) of the genes align between the genomes. These account for at most 2% of the human genome, and they are obviously under selective constraint. The other 38% of the human genome that does not code for protein but still aligns with mouse should include gene regulatory sequences and other functional noncoding sequences. However, these alignments also include much neutral DNA; e.g., about one-fourth of all the ancestral repeats in humans align with orthologs in mouse. All the sequences that align between mouse and human are conserved in the sense that they are present in both species, but the goal is to identify the sequences that are subject to purifying selection. It is the latter sequences that are playing a role in some conserved function.

A major complication to answering this question is that the rate of neutral evolution varies across the genome. The distribution of nucleotide substitutions per site in ancestral repeats (computed on 1-Mb nonoverlapping windows) is quite wide (Fig. 1), reflecting substantial regional variation in the underlying neutral substitution rate. In addition, the amount of DNA inferred to be deleted from mouse and the amount of transposable elements inserted and retained show substantial variation (Fig. 1). Furthermore, the amounts of neutral substitution, deletion, insertion (of LTR repeats), recombination, and single-nucleotide polymorphisms (SNPs) covary dramatically (Fig. 2A). Because the substitutions are measured in neutral DNA, different levels of selection cannot ex-

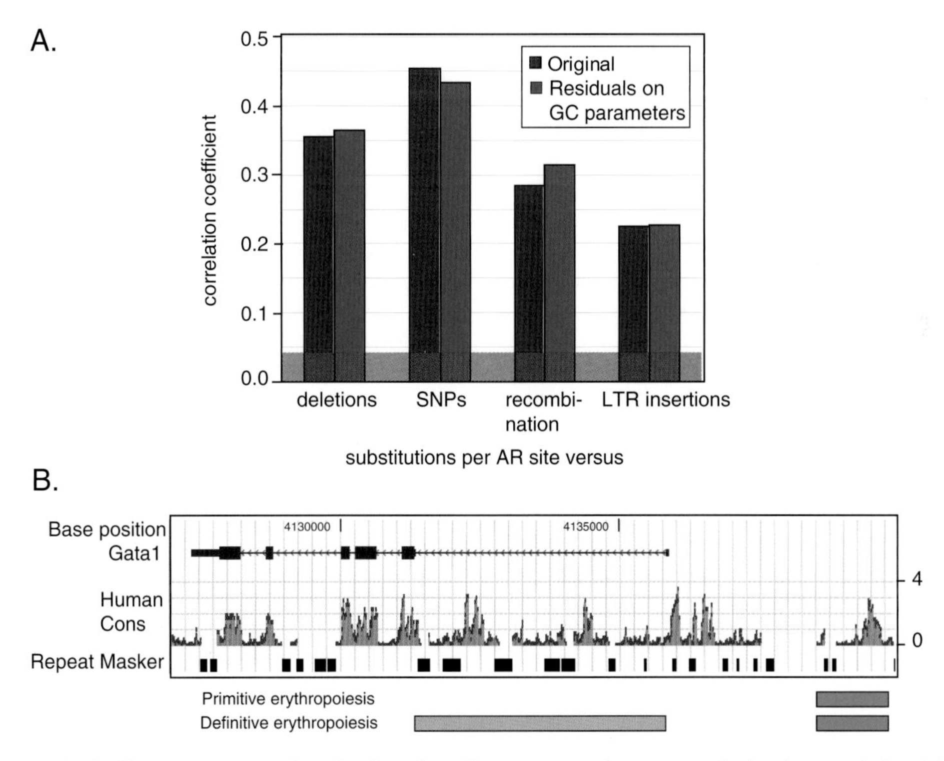


**Figure 2.** Covariation in divergence rates and application of an alignment score that accounts for local rate variation. (*A*) Correlation among the amounts of neutral substitution (in ARs) with large deletions (based on human–mouse alignments), insertions of LTR repeats, recombination, and SNPs in human. Correlations are shown for the original data and for residuals after the quadratic effects of fraction G+C, change in fraction G+C, and CpG island density have been removed (Hardison et al. 2003). The correlations of various divergence processes with insertion and retention of different classes of repeats is the subject of ongoing work. (*B*) The *Human Cons* track plots the *L* score giving the log-likelihood that alignments reflect selection for the mouse *Gata1* gene. The gene is transcribed from right to left. The upstream and intronic noncoding regions with high *L* scores correspond to previously described strong erythroid enhancers (Onodera et al. 1997).

plain the regional differences. Rather, the covariation in the various divergence processes appears to reflect an inherent tendency of large, megabase-sized regions to change at a fast or slow rate (Chiaromonte et al. 2001). The molecular and cellular basis for this inherent tendency to change is unknown, although it is possible that repair of double-stranded breaks could be at least part of the explanation (Lercher and Hurst 2002).

Given the variation in neutral substitution rates, the goal is to find aligning segments whose similarity significantly exceeds that expected from divergence at the local neutral rate. These should be the sequences subject to purifying selection. Indeed, the significance of a particular alignment score will vary substantially depending on the divergence rate of the surrounding DNA (Li and Miller 2002). Thus, the fraction of matching nucleotides for alignments in small (50 bp) windows was adjusted for the local neutral rate, empirically estimated from nearby aligning ancestral repeats. The overall distribution of these adjusted scores is broad; when compared to its neutral component (the distribution for ancestral repeats only) it presents a marked right-skewedness—i.e., increased frequencies on higher score values (Waterston et al. 2002). A statistical decomposition of this skewed overall distribution leads to the conclusion that about 5% of the human genome is under purifying selection (Waterston et al. 2002; Chiaromonte et al., this volume). This is over twice the amount of DNA that codes for protein, showing that the noncoding portion of the genome contributes significantly to the functional DNA. However, it is only about one-eighth of the conserved sequences, so a majority of the aligning sequences do not reflect selection for some function.

To make these scores more useful to biomedical scientists, *L* scores (or *Mouse Cons* and *Human Cons*) have been computed that convert locally adjusted similarity scores into probabilities that alignments in a given 50 bp result from selection. These can be accessed at the UCSC Genome Browser. An example of this track for the mouse *Gata1* gene shows that protein-coding exons, the first intron, the promoter, and a region about 3–4 kb further upstream are not only conserved (align), but are highly likely to be generated by selection (Fig. 2B). The upstream region corresponds to an enhancer that confers erythroid-specific expression during primitive erythropoiesis and collaborates with the intronic enhancer to activate expression during definitive erythropoiesis (Onodera et al. 1997). Thus, these scores, generated from the alignment scores adjusted for local rate variation, can be effective indicators of CRMs.

## DISCRIMINATING CRMS FROM OTHER DNAs

The *Mouse/Human Cons* or *L* scores are measures of alignment quality, where matches are favored more than mismatches, which are favored more than gaps. Noncoding DNA sequences with a high *L* score are more likely to be subject to purifying selection, and this set of sequences should contain CRMs regulating conserved functions. However, it should also contain other functional sequences such as genes encoding structural RNAs and microRNAs.

Therefore, we explored several computational approaches to analyzing interspecies genomic sequence alignments, aiming to develop computational methods to distinguish regulatory regions from neutrally evolving DNA. To do so, we employed statistical models that recognize alignment patterns characteristic of those seen in known CRMs. Alignments rich in these patterns need not be those that score highest in quality (e.g., a similarity score) or a likelihood of being under selection. Known enhancers and other CRMs tend to be clusters of highly conserved binding sites for transcription factors, but sequences between those binding sites are more variable between species. Thus, alignment quality measurements in the CRMs are usually less than those seen in regions under more uniform selection, such as coding exons.

Three training sets were collected from the whole-genome human–mouse alignments: (1) known CRMs, which are a set of 93 experimentally defined mammalian gene regulatory regions (accessible from GALA at http://www.bx.psu.edu/), (2) well-characterized exons (coding sequences, as a positive control), and (3) ancestral interspersed repeats (the major sequence class used for neutrally evolving DNA). Quantitative evaluation of statistical models that potentially could distinguish functional noncoding sequences from neutral DNA showed that discrimination based on frequencies of individual nucleotide pairs or gaps (i.e., of possible alignment columns) is only partially successful. In contrast, scoring procedures that include the alignment context, based on frequencies of short runs of alignment columns, achieve good separation between regulatory and neutral features (Elnitski et al. 2003).

The best-performing scoring function, called regulatory potential (*RP*) score, employs transition probabilities from two Markov models estimated on the training data. In practice, the procedure evaluates short strings of columns in the alignments, giving a higher value to those that occur more frequently in the CRMs training set than in the ancestral repeats set (Fig. 3). In this procedure, alignments are described using a reduced alphabet $A$. In each training set, we compute the frequencies with which short strings of alignment characters are followed by a particular alignment character. As an example, consider alignment columns to consist of two types of matches, those that involve G or C (S) and those that involve A or T (W), plus transitions (I), transversions (V), and gaps (G). A 5-symbol alphabet can thus describe the alignments. For short strings, the number of possible arrangements of these 5 symbols is computationally manageable. Therefore, we estimate the probability that any string of length $T$ is followed by a particular symbol (transition probabilities), where $T$ is the order of the Markov model. For example, the empirical frequencies of a pentamer, say WIISV, followed by a given symbol, say S (or W, I, V, G) are used to estimate the transition probabilities of a fifth-order Markov model.

More generally, for a Markov model of order $T$, we estimate the probability that within a regulatory region an

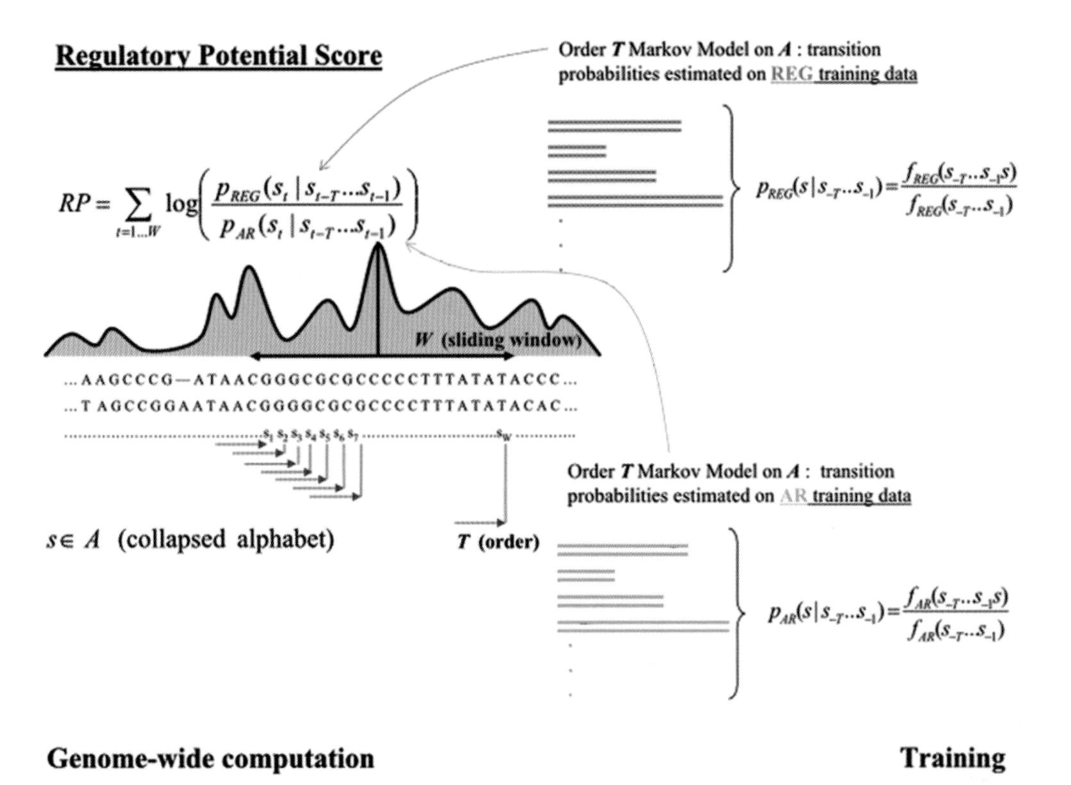


**Figure 3.** Genome-wide computation of regulatory potential (*RP*) scores. Diagrams on the right illustrate the use of two training sets, known regulatory regions (REG) and ancestral repeats (AR), to build Markov models describing the likelihood of a string of *T* alignment characters being followed by a particular alignment character. The alignment characters are from a collapsed alphabet (*A*) that describes mismatches, gaps, and different kinds of matches. The diagram on the left illustrates the application of these Markov models to calculate the log-likelihood that a segment of an alignment (window size *W*) fits with the model for a regulatory region rather than an ancestral repeat. This log-likelihood is the regulatory potential.

alignment character s is preceded by the string of characters $s_{-T}$ to $s_{-1}$ ($p_{REG}[s/s_{-T}...s_{-1}]$ in Fig. 3) as the empirically observed frequency of the string $s_{-T}...s_{-1}s$ divided by that of the string comprising the first *T* positions, $s_{-T}...s_{-1}$, in the CRMs training set. We then repeat the same estimation procedure on the ancestral repeats (*AR*) training set.

The *RP* score is computed for any alignment by dividing the transition probability for regulatory regions, $p_{REG}(s/s_{-T}...s_{-1})$, by that for ancestral repeats, $p_{AR}(s/s_{-T}...s_{-1})$, at each position in the alignment, taking the logarithm, and summing over positions. This log-odds ratio is illustrated in Figure 3 for sliding windows of length *W*. When needed, the score is adjusted for the length of the alignment (Elnitski et al. 2003). The *RP* score has been computed in 50-bp windows (overlapping by 45 bp) for the human–mouse whole-genome alignments, using a 5-symbol collapsed alphabet and a fifth-order Markov model. These scores and plots of them are provided at the UCSC Genome Browser (http://genome.ucsc.edu, Nov. 2002 human assembly), and they are recorded in the database of genomic DNA sequence *al*ignments and *an*notations, GALA (Giardine et al. 2003).

A validation study shows that this approach can separate the reference data set of 93 known regulatory regions from the ancestral repeat segments used in training (Fig. 4). Cross-validation studies also support the discriminatory power of the *RP* score (Elnitski et al. 2003). Of note, the accuracy of our predictive models should become even greater as additional regulatory sequences demonstrated through experimental approaches are added to the training set and as more alignments are added. Moreover, the same computational approach can be applied to discrimination among other functional classes, as training data from them become available.

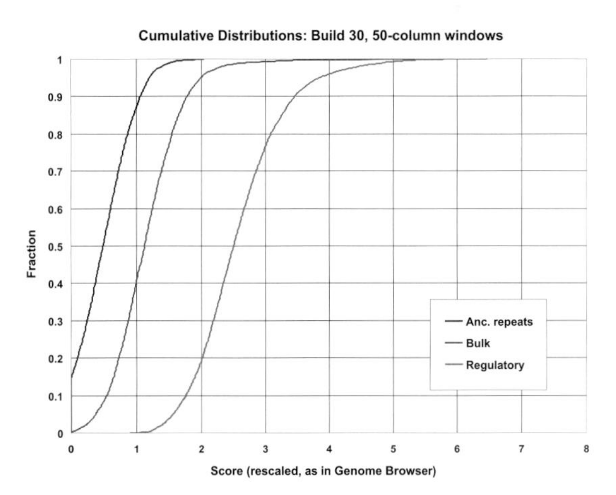


**Figure 4.** Cumulative distribution of *RP* scores of alignments in several classes of DNA, evaluated using fifth-order Markov models and a 5-letter alphabet. Note the complete separation between regulatory regions and neutral DNA (ancestral repeats). The "bulk" alignments are a set picked at random from all alignments.

## CALIBRATION OF THE REGULATORY POTENTIAL SCORE

Realizing that performance on the training set is seldom indicative of performance on new problems not already in the training set, we analyzed the ability of the *RP* score to find known regulatory regions in a well-studied gene complex. The goal is to find an optimal threshold for the *RP* score such that known CRMs are found with high efficiency (high sensitivity) while other noncoding sequences are largely excluded (high specificity). The complex of β-like globin genes (the *HBB* complex) on human Chromosome 11 was chosen for these calibration studies because proximal promoters and upstream regulatory sequences (within a few hundred base pairs of the promoters) have been identified for each active gene, and high-level expression of all the genes is dependent on a distal (as much as 60 kb upstream) enhancer called the locus control region, or LCR (for review, see Forget 2001; Hardison 2001; Stamatoyannopoulos 2001). The LCR is marked by at least four DNase hypersensitive sites (HS1–HS4) that contribute individually and collectively to enhancer function (for review, see Hardison et al. 1997; Li et al. 2002). The five active genes are transcribed right to left in the diagram in Figure 5. A set of eleven intervals was compiled that cover each of the well-characterized CRMs for which experiments show clear, independent effects on regulation. DNA sequences that affect expression levels only in combination with other CRMs were not included. Four of the eleven intervals in the reference set were also in the training set used for the *RP* score. This limits the stringency of this test, but until a larger number of regulatory regions are carefully characterized, some overlap with the training set is difficult to avoid. The reference CRMs are covered by pair-wise and 3-way alignment scores and by the *RP* score, but with different values. Some CRMs, such as the LCR HS3 and the upstream regulatory regions of *HBBG1* and *HBBG2*, have higher *RP* scores than conservation scores.

We used the GALA database (Giardine et al. 2003) to organize and extract the necessary information for the calibration study to find an optimal *RP* threshold. GALA is a relational database with genome-wide information on genes (known and predicted), exons, gene products (including Gene Ontology descriptions; Ashburner et al. 2000), gene expression (including the GNF data using Affymetrix human and mouse gene chips; Su et al. 2002), human–mouse alignments, scores such as *L* and *RP* derived from the alignments, binding sites for transcription factors predicted by matches to *TRANSFAC* weight matrices (Matys et al. 2003), repeats (Smit and Green 1999), and much other information. All data are organized by sequence positions in the human or mouse genome assemblies. GALA allows queries across fields and supports complex queries that combine results from simple queries by conventional set operations (union, intersection, and subtraction) as well as by proximity and by clustering. Thus, it greatly expands the data-mining capacity beyond the conventional one-gene or one-locus view most commonly used at genome browsers. It can be accessed at http://www.bx.psu.edu/.

To determine the *RP* score threshold that works best in identifying the reference set, we queried GALA to find all the ranges of DNA that pass each candidate *RP* score

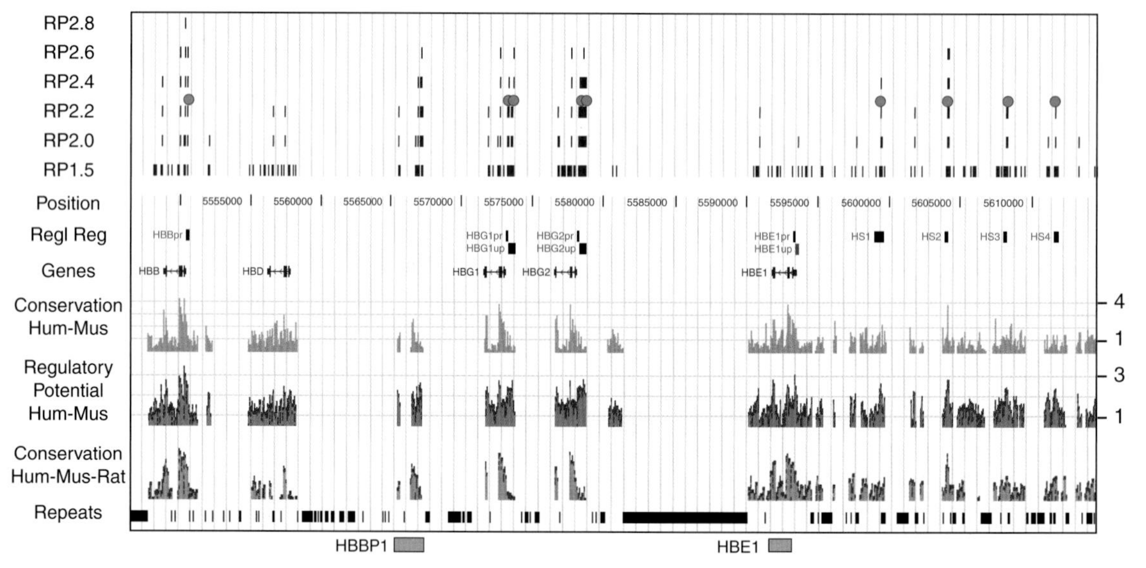


**Figure 5.** Effectiveness of different regulatory potential (*RP*) thresholds in predicting known regulatory elements in the *HBB* complex. Genes are transcribed from right to left. Conservation scores and *RP* scores are plotted, and genes, repeats, and known regulatory regions are shown on the lines below the position track. Above it are the segments whose *RP* scores exceed the designated threshold (after subtracting exons). Small green circles mark the "true positives" for the RP2.2 track. The failure to find the *HBE1* promoter and upstream regulatory region is an artifact of the current genome annotation, and we have installed a work-around in GALA for future analysis. The results are displayed from the UCSC Genome Browser; comparable analyses are available genome-wide. The results for each *RP* threshold were displayed at the Genome Browser, saved as pdfs, and then combined using Adobe Illustrator. The "Conservation Hum-Mus-Rat" track quantifies the level of conservation in human–mouse–rat alignments (M. Blanchette et al., in prep.).

threshold in the 68-kb interval encompassing the *HBB* complex, including the LCR. After subtracting the exons, the set of DNA intervals passing the threshold were diagramed using an automatic connection between GALA and the UCSC Genome Browser (Fig. 5). As expected, higher thresholds returned fewer intervals, and these sets were enriched in the reference CRMs. For *RP* = 2.2, nine of the eleven reference CRMs are returned. The two that are missing (promoter and upstream regulatory region of *HBE1*) are artifacts of the annotation. They were lost because the annotation of this gene uses a minor promoter in the upstream region, thereby including the CRMs in the annotated "first exon." The "false positives," i.e., intervals meeting the filtering thresholds but not annotated as regulatory regions, are a mixture of overprediction, true regulatory regions that have not been tested, and a few artifacts of incomplete annotation, such as the pseudogene *HBBP1*, which is not in the annotation but whose exons pass the filters.

Detailed comparison between the intervals passing the filters and the reference CRMs shows that the specificity reaches a plateau around *RP* = 2.3 whereas sensitivity declines above this threshold (Fig. 6, left). Indeed, *RP* = 2.3 is a minimum in a cost function (Fig. 6, right), and hence we have used it as the threshold in further analysis. Further analysis using the clustering and proximity capabilities in GALA showed that combining this optimal *RP* threshold with a requirement that a DNA segment have a predicted binding site for GATA-1 improved the specificity from about 0.6 to 0.7. The upper limit on specificity is caused partly by incomplete analysis of potential *cis*-regulatory elements even in the *HBB* complex.

## EXPERIMENTAL TESTS OF PREDICTED *cis*-REGULATORY MODULES

The publicly available *RP*, *L*, and other scores can be combined with predictions of binding sites for any relevant transcription factors to predict CRMs genome-wide for a wide variety of mammalian tissues or stages of development. We have begun an extensive set of tests of the predicted CRMs for genes induced during late erythroid differentiation and maturation using the two somatic cell models mentioned in the introduction. Extensive analysis of microarray expression data has revealed a cohort of genes induced along with the β-globin genes (*Hbb-b1* and *Hbb-b2*) in both cell lines. Because the G1E cell line is responding directly to restoration of the activity of the GATA-1 transcription factor, we include predicted binding sites for GATA-1 in our predictions of CRMs. The cohort of coexpressed genes includes some previously known to be induced in erythroid cells, such as *Alas2*, which encodes the enzyme catalyzing the rate-limiting step in heme biosynthesis. Other genes such as *Hipk2* were not well known as erythroid-induced genes.

The gene *Alas2* is an example of our early predictions and tests. Using GALA to search for intervals meeting our criteria (*RP*-score at least 2.3, no exons, and a predicted GATA-1 site within 50 bp), we found four regions in the roughly 25-kb region encompassing human *ALAS2* (Fig. 7, top). (GALA for mouse with mouse–human *RP* scores is now available so that one can perform the analysis entirely from the perspective of the mouse genome.) One predicted CRM is the major promoter, another is in intron 8, which others have shown is an enhancer (Surinya et al. 1998). We focused on a predicted CRM in intron 1 for testing. A more detailed view from our interactive alignment viewer *Laj* (Wilson et al. 2001) shows that the region is strongly conserved and has a predicted GATA-1-binding site in both human and mouse (Fig. 7, bottom).

The strategy for testing the predicted CRMs for enhancer and silencer function is to add them to an expression cassette in which the green fluorescent protein gene is transcribed from a minimal *HBB* promoter, and then we force the test construct to integrate at a marked site in

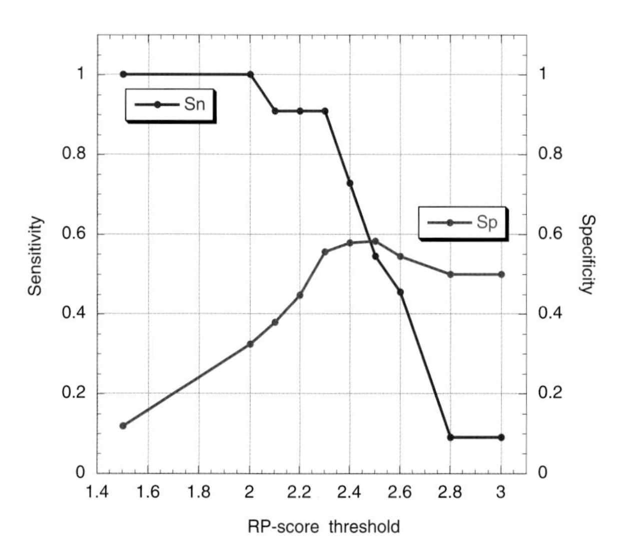


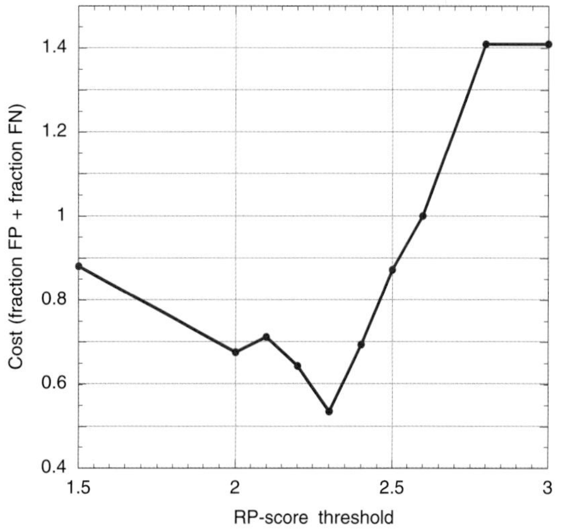


**Figure 6.** Sensitivity and specificity (*left*) and cost (*right*) of *RP*-score thresholds applied to known regulatory elements in the *HBB* complex. The sensitivity (Sn) is the fraction of known elements found above the indicated threshold, and the specificity is the fraction of segments above the indicated threshold that are known regulatory elements. The cost is the fraction of segments above the indicated threshold that are "false positives" plus the fraction of known elements that are false negatives.

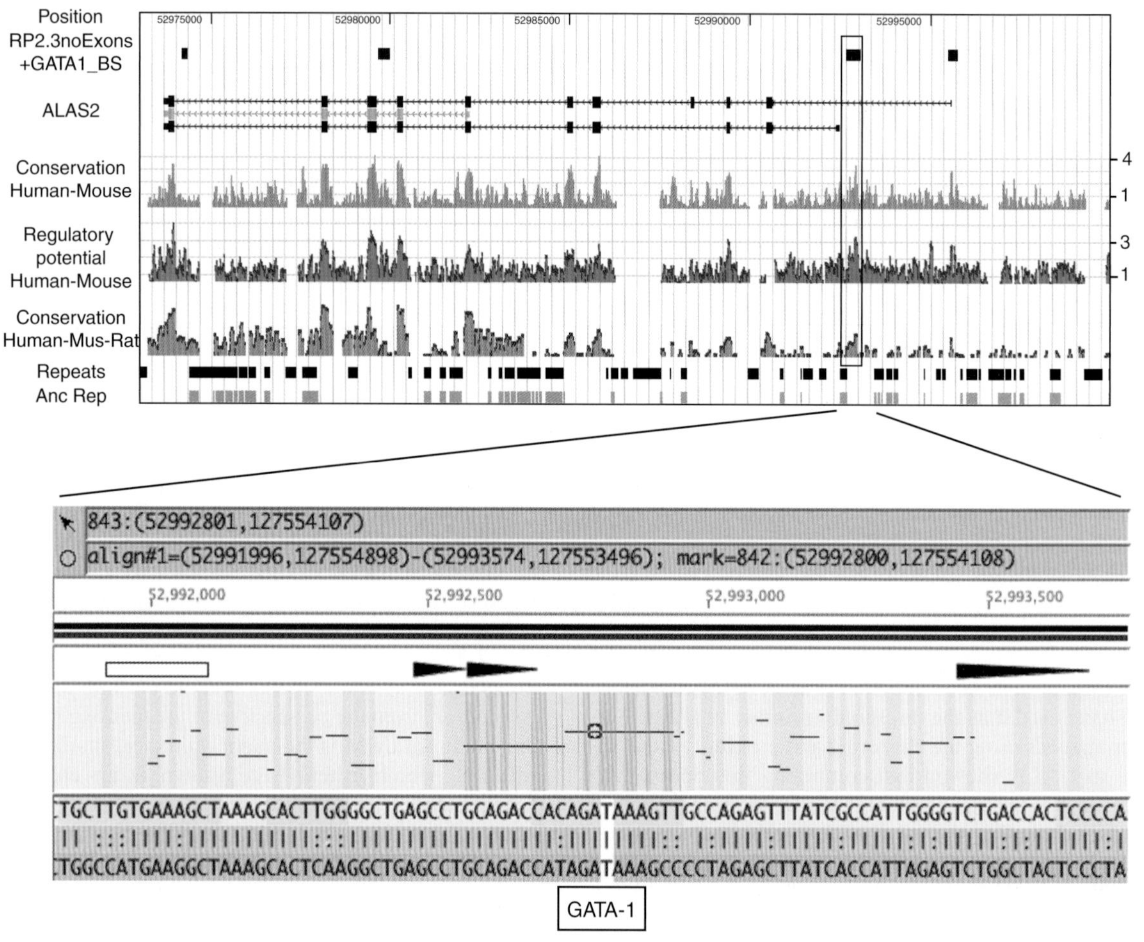


**Figure 7.** Predicted *cis*-regulatory modules for the *ALAS2* gene (first track under the positions) along with the gene structure (transcribed right to left, exons are boxes) and plots of conservation and regulatory potential (*RP*)-scores (*top*). This output was generated by GALA plus the UCSC Genome Browser. The boxed region was selected for experimental tests. The alignment in this region is shown in the lower panel, which is from our interactive alignment viewer *Laj* (Wilson et al. 2001). Note the presence of a conserved predicted GATA-1-binding site. GALA output can be viewed in the UCSC Browser, *Laj*, or other formats.

MEL cells (Bouhassira et al. 1997; Feng et al. 1999; Molete et al. 2001). Thus, we monitor expression levels in a very precise and accurate way, because the test constructs and parental constructs are at the same chromosomal position (Fig. 8A). The predicted CRM in intron 1 of the mouse ortholog, *Alas2*, was amplified by PCR from murine genomic DNA and ligated into the expression cassette. Both the test and parental expression cassettes were forced to integrate at locus RL5 in MEL cells.

The parental cassette with no enhancer expresses GFP, and it induces a small amount when cells are treated with HMBA (Fig. 8B). Addition of the predicted CRM from *Alas2* intron 1 increased the expression both before induction, showing an enhancer function, and after induction, showing that it also affects inducibility. The effect gets stronger with time of induction.

The predicted CRM is thus experimentally verified as an enhancer and as a sequence that confers erythroid inducibility. At this time, we have tested three predicted CRMs from three different genes and have found that two boost expression and one has no effect in this system. As larger numbers of predicted CRMs are tested, we will be able to improve our models for predicting CRMs.

## CONCLUSIONS

Pair-wise and multiple whole-genome alignments of human, mouse, and rat are available and are being continually updated. These alignments, plus scores that provide guidance on the likelihood that a region is under selection or is rich in alignment patterns typical of *cis*-regulatory elements, are available at the Genome Browser and in our GALA database. Users can access these scores alone or in combination with other genomic features of interest (such as predicted binding sites for transcription factors, CpG islands, and exons) to find candidate CRMs throughout the genome. Calibration studies indicate that an *RP* threshold of 2.3 is effective for the *HBB* complex. Our early results on experimental tests based solely on these predictions are encouraging, and we expect continuing improvements in the predictive algorithms. The combination of bioinformatic predictions and

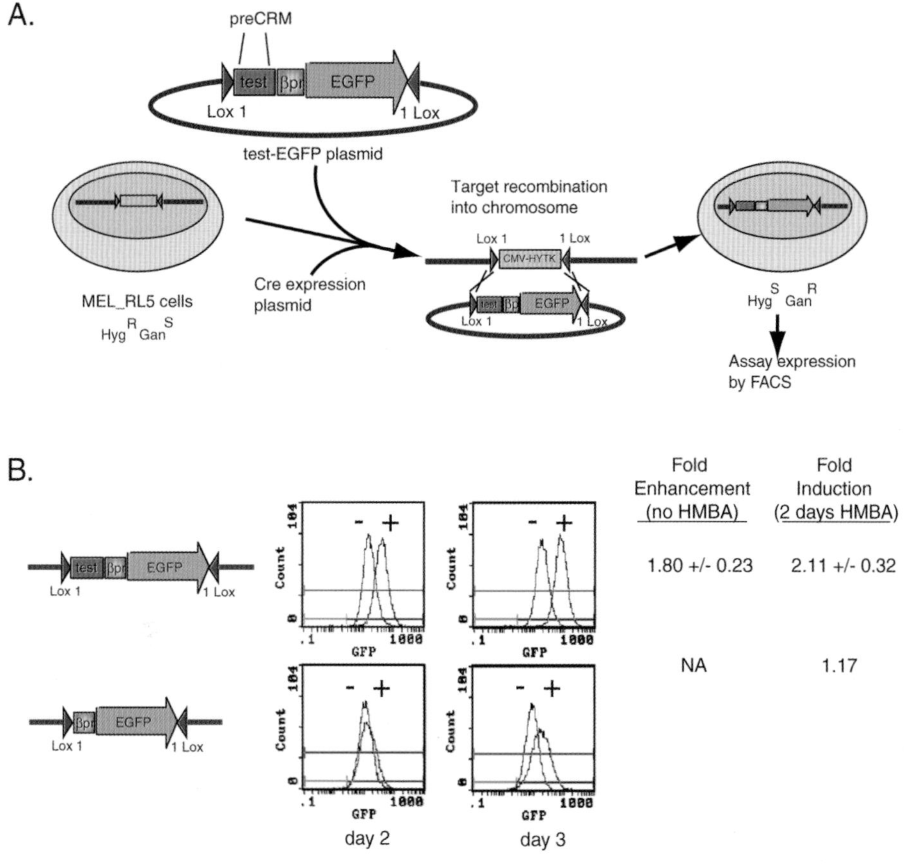


**Figure 8.** The predicted *cis*-regulatory module in intron 1 of *Alas2* enhances expression and increases inducibility. (*A*) The strategy for isolating and testing predicted CRMs using recombinase-mediated cassette exchange at locus RL5 of MEL cells (Feng et al. 1999) is shown. This procedure ensures that expression of all constructs is monitored after integration at the same chromosomal position. (*B*) Maps of the test and parental expression cassettes are on the left, and the FACS profiles of fluorescence from EGFP are plotted for cells uninduced (–) or induced (+) to erythroid maturation by treatment with HMBA. The fold enhancement and induction are shown on the right.

experimental tests in somatic cell developmental models can serve as a paradigm for global analysis of regulation in any tissue.

## ACKNOWLEDGMENTS

We thank the members of all the genome-sequencing consortia for determining the sequences and making them publicly available rapidly. R.C.H., S.Y., F.C., L.E., D.K., S.S., and W.M. were supported by National Human Genome Research Institute grant HG-02238 and the Huck Institute of Life Sciences of Penn State University, with additional support for L.E. from NHGRI grant HG-02325 and for R.C.H. from National Institute of Diabetes and Digestive and Kidney Diseases grant RO1 DK-27635; K.M.R., M.D., and W.J.K by NHGRI grant 1P41HG-02371; D.H. by NHGRI grant 1P41HG-02371 and the Howard Hughes Medical Institute.

## REFERENCES

Ashburner M., Ball C.A., Blake J.A., Botstein D., Butler H., Cherry J.M., Davis A.P., Dolinski K., Dwight S.S., Eppig J.T., Harris M.A., Hill D.P., Issel-Tarver L., Kasarskis A., Lewis S., Matese J.C., Richardson J.E., Ringwald M., Rubin G.M., and Sherlock G. 2000. Gene ontology: Tool for the unification of biology. *Nat. Genet.* **25:** 25.

Bailey J.A., Gu Z., Clark R.A., Reinert K., Samonte R.V., Schwartz S., Adams M.D., Myers E.W., Li P.W., and Eichler E.E. 2002. Recent segmental duplications in the human genome. *Science* **297:** 1003.

Bouhassira E., Westerman K., and Leboulch P. 1997. Transcriptional behavior of LCR enhancer elements integrated at the same chromosomal locus by recombinase-mediated cassette exchange. *Blood* **90:** 3332.

Chiaromonte F., Yap V.B., and Miller W. 2002. Scoring pairwise genomic sequence alignments. *Pac. Symp. Biocomput.* **2002:** 115.

Chiaromonte F., Yang S., Elnitski L., Yap V., Miller W., and Hardison R.C. 2001. Association between divergence and interspersed repeats in mammalian noncoding genomic DNA. *Proc. Natl. Acad. Sci.* **98:** 14503.

Collins F.S., Green E.D., Guttmacher A.E., and Guyer M.S. 2003. A vision for the future of genomics research. *Nature* **422:** 835.

Elnitski L., Hardison R.C., Li J., Yang S., Kolbe D., Eswara P., O'Connor M.J., Schwartz S., Miller W., and Chiaromonte F. 2003. Distinguishing regulatory DNA from neutral sites. *Genome Res.* **13:** 64.

Feng Y.Q., Seibler J., Alami R., Eisen A., Westerman K.A., Leboulch P., Fiering S., and Bouhassira E.E. 1999. Site-specific chromosomal integration in mammalian cells: Highly efficient CRE recombinase-mediated cassette exchange. *J. Mol. Biol.* **292:** 779.

Forget B.G. 2001. Molecular genetics of the human globin genes. In *Disorders of hemoglobin: Genetics, pathophysiology, and clinical management* (ed. M.H. Steinberg et al.), p. 117. Cambridge University Press, Cambridge, United Kingdom.

Friend C., Scher W., Holland J.G., and Sato T. 1971. Hemoglobin synthesis in murine virus-induced leukemic cells in vitro: Stimulation of erythroid differentiation by dimethylsulfoxide. *Proc. Natl. Acad. Sci.* **68:** 378.

Giardine B.M., Elnitski L., Riemer C., Makalowska I., Schwartz S., Miller W., and Hardison R.C. 2003. GALA, a database for genomic sequence alignments and annotations. *Genome Res.* **13:** 732.

Hardison R.C. 2000. Conserved noncoding sequences are reliable guides to regulatory elements. *Trends Genet.* **16:** 369.

———. 2001. Organization, evolution and regulation of the globin genes. In *Disorders of hemoglobin: Genetics, pathophysiology, and clinical management* (ed. M.H. Steinberg et al.), p. 95. Cambridge University Press, Cambridge, United Kingdom.

Hardison R., Slightom J.L., Gumucio D.L., Goodman M., Stojanovic N., and Miller W. 1997. Locus control regions of mammalian β-globin gene clusters: Combining phylogenetic analyses and experimental results to gain functional insights. *Gene* **205:** 73.

Hardison R.C., Roskin K.M., Yang S., Diekhans M., Kent W.J., Weber R., Elnitski L., Li J., O'Connor M., Kolbe D., Schwartz S., Furey T.S., Whelan S., Goldman N., Smit A., Miller W., Chiaromonte F., and Haussler D. 2003. Covariation in frequencies of substitution, deletion, transposition and recombination during eutherian evolution. *Genome Res.* **13:** 13.

Jordan I.K., Rogozin I.B., Glazko G.V., and Koonin E.V. 2003. Origin of a substantial fraction of human regulatory sequences from transposable elements. *Trends Genet.* **19:** 68.

Kent W.J., Baertsch R., Hinrichs A., Miller W., and Haussler D. 2003. Evolution's cauldron: Duplication, deletion, and rearrangement in the mouse and human genomes. *Proc. Natl. Acad. Sci.* **100:** 11484.

Kent W.J., Sugnet C.W., Furey T.S., Roskin K.M., Pringle T.H., Zahler A.M., and Haussler D. 2002. The human genome browser at UCSC. *Genome Res.* **12:** 996.

Kimura M. 1968. Evolutionary rate at the molecular level. *Nature* **217:** 624.

Korf I., Flicek P., Duan D., and Brent M.R. 2001. Integrating genomic homology into gene structure prediction. *Bioinformatics* (suppl. 1) *17:* S140.

Lander E.S. 1996. The new genomics: Global views of biology. *Science* **274:** 536.

Lander E.S., Linton L.M., Birren B., Nusbaum C., Zody M.C., Baldwin J., Devon K., Dewar K., Doyle M., FitzHugh W., Funke R., Gage D., Harris K., Heaford A., Howland J., Kann L., Lehoczky J., LeVine R., McEwan P., McKernan K., Meldrim J., Mesirov J.P., Miranda C., Morris W., and Naylor J., et al. (International Human Genome Sequencing Consortium). 2001. Initial sequencing and analysis of the human genome. *Nature* **409:** 860.

Lercher M.J. and Hurst L.D. 2002. Human SNP variability and mutation rate are higher in regions of high recombination. *Trends Genet.* **18:** 337.

Li J. and Miller W. 2002. Significance of interspecies matches when evolutionary rate varies. *J. Comput. Biol.* **10:** 537.

Li Q., Peterson K., Fang X., and Stamatoyannopoulos G. 2002. Locus control regions. *Blood* **100:** 3077.

Li W.H., Gojobori T., and Nei M. 1981. Pseudogenes as a paradigm of neutral evolution. *Nature* **292:** 237.

Matys V., Fricke E., Geffers R., Gossling E., Haubrock M., Hehl R., Hornischer K., Karas D., Kel A.E., Kel-Margoulis O.V., Kloos D.U., Land S., Lewicki-Potapov B., Michael H., Munch R., Reuter I., Rotert S., Saxel H., Scheer M., Thiele S., and Wingender E. 2003. TRANSFAC: Transcriptional regulation, from patterns to profiles. *Nucleic Acids Res.* **31:** 374.

Migliaccio A.R. and Papayannopoulou T. 2001. Erythropoiesis. In *Disorders of hemoglobin: Genetics, pathophysiology, and clinical management* (ed. M.H. Steinberg et al.), p. 52. Cambridge University Press, Cambridge, United Kingdom.

Molete J.M., Petrykowska H., Bouhassira E.E., Feng Y.Q., MIller W., and Hardison R.C. 2001. Sequences flanking hypersensitive sites of the beta-globin locus control region are required for synergistic enhancement. *Mol. Cell. Biol.* **21:** 2969.

Nekrutenko A., Makova K.D., and Li W.H. 2002. The K(A)/K(S) ratio test for assessing the protein-coding potential of genomic regions: An empirical and simulation study. *Genome Res.* **12:** 198.

Onodera K., Takahashi S., Nishimura S., Ohta J., Motohashi H., Yomogida K., Hayashi N., Engel J., and Yamamoto M. 1997. GATA-1 transcription is controlled by distinct regulatory mechanisms during primitive and definitive erythropoiesis. *Proc. Natl. Acad. Sci.* **94:** 4487.

Pennacchio L.A. and Rubin E.M. 2001. Genomic strategies to identify mammalian regulatory sequences. *Nat. Rev. Genet.* **2:** 100.

Pribnow D. 1975. Nucleotide sequence of an RNA polymerase binding site at an early T7 promoter. *Proc. Natl. Acad. Sci.* **72:** 784.

Schwartz S., Kent W.J., Smit A., Zhang Z., Baertsch R., Hardison R.C., Haussler D., and Miller W. 2003. Human-mouse alignments with *Blastz. Genome Res.* **13:** 103.

Smit A. and Green P. 1999. *RepeatMasker* at: http://ftp.genome.washington.edu/RM/RepeatMasker.html

Stamatoyannopoulos G. 2001. Molecular and cellular basis of hemoglobin switching. *Disorders of hemoglobin: Genetics, pathophysiology, and clinical management* (ed. M.H. Steinberg et al.), p. 131. Cambridge University Press, Cambridge, United Kingdom.

Su A., Cooke M., Ching K., Hakak Y., Walker J., Wiltshire T., Orth A., Vega R., Sapinoso L., Moqrich A., Patapoutian A., Hampton G., Schultz P., and Hogenesch J. 2002. Large-scale analysis of the human and mouse transcriptomes. *Proc. Natl. Acad. Sci.* **99:** 4465.

Surinya K.H., Cox T.C., and May B.K. 1998. Identification and characterization of a conserved erythroid-specific enhancer located in intron 8 of the human 5-aminolevulinate synthase 2 gene. *J. Biol. Chem.* **273:** 16798.

Thomas J.W., Touchman J.W., Blakesley R.W., Bouffard G.G., Beckstrom-Sternberg S.M., Margulies E.H., Blanchette M., Siepel A.C., Thomas P.J., McDowell J.C., Maskeri B., Hansen N.F., Schwartz M.S., Weber R.J., Kent W.J., Karolchik D., Bruen T.C., Bevan R., Cutler D.J., Schwartz S., Elnitski L., Idol J.R., Prasad A.B., Lee-Lin S.Q., and Maduro V.V., et al. 2003. Comparative analyses of multi-species sequences from targeted genomic regions. *Nature* **424:** 788.

Waterston R.H., Lindblad-Toh K., Birney E., Rogers J., Abril J.F., Agarwal P., Agarwala R., Ainscough R., Alexandersson M., An P., Antonarakis S.E., Attwood J., Baertsch R., Bailey J., Barlow K., Beck S., Berry E., Birren B., Bloom T., Bork P., Botcherby M., Bray N., Brent M.R., Brown D.G., and Brown S.D., et al. (Mouse Genome Sequencing Consortium). 2002. Initial sequencing and comparative analysis of the mouse genome. *Nature* **420:** 520.

Weiss M.J., Yu C., and Orkin S.H. 1997. Erythroid-cell-specific properties of transcription factor GATA-1 revealed by phenotypic rescue of a gene-targeted cell line. *Mol. Cell. Biol.* **17:** 1642.

Wiehe T., Gebauer-Jung S., Mitchell-Olds T., and Guigo R. 2001. SGP-1: Prediction and validation of homologous genes based on sequence alignments. *Genome Res.* **11:** 1574.

Wilson M.D., Riemer C., Martindale D.W., Schnupf P., Boright A.P., Cheung T.L., Hardy D.M., Schwartz S., Scherer S.W., Tsui L.C., Miller W., and Koop B.F. 2001. Comparative analysis of the gene-dense ACHE/TFR2 region on human chromosome 7q22 with the orthologous region on mouse chromosome 5. *Nucleic Acids Res.* **29:** 1352.

Wolfe K.H., Sharp P.M., and Li W.H. 1989. Mutation rates differ among regions of the mammalian genome. *Nature* **337:** 283.

# Systems Approaches Applied to the Study of *Saccharomyces cerevisiae* and *Halobacterium sp.*

A.D. Weston,* N.S. Baliga,* R. Bonneau,* and L. Hood
*Institute for Systems Biology, Seattle, Washington 98103-8904*

Integrative systems approaches to studying biological systems have begun to yield striking results. Systems biology has emerged as a powerful new approach over the past 5 years because of (1) the completion of the human and many other genome sequences, which led to the identification and prediction of comprehensive gene lists; (2) the development of high-throughput techniques for genomics, proteomics, metabolomics, and phenomics, leading to the acquisition of global data sets; and (3) the creation of powerful computational methods for storing and assessing different types of global data sets as well as analyzing and integrating them. The integration of different data types is essential because it helps to deal with noise that is inherent in large data sets and it reveals new biological phenomena that are not obvious from the analysis of single data types. Here we summarize the initial applications of systems biology to two systems: one, which has been studied intensively for years (the galactose utilization system in yeast), and a second (the oxygen and light responses in *Halobacterium sp.*) for which little data are available. Systems approaches have allowed us to gain new and fundamental insights into both systems (Ideker et al. 2001; Baliga et al. 2002).

A key aspect of systems biology is viewing biology as an informational science. First, there are two general types of biological information: the digital information of the genome, and environmental information that interacts directly or indirectly with the digital genomic information. Second, the genome has two major types of digital information: the genes that encode protein and RNA molecular machines, and the transcription factor-binding sites in *cis* control regions of genes that create the linkage relationships for gene regulatory networks which control the temporal and spatial parameters of gene expression as well as their amplitude. Proteins may act alone, in loose functional relationships or biomodules (e.g., the enzymes of sugar metabolism), in complex protein machines (e.g., the ribosome), or in larger protein networks. Transcription factors, co-transcription factors, and the factors that mediate changes in chromatin structure all operate via DNA-binding sites in *cis* control regions of genes to regulate gene expression. The protein and gene regulatory networks, although conceptually distinct, obviously are functionally integrated with one another. Third, it should be stressed that biological information is of many different types, starting with the core DNA genomic information and moving out to ecologies (DNA → RNA → protein → protein structures and biomodules → networks of biomodules → cells → organs [tissues] → individual organisms → populations of individual organisms → ecologies), and that environmental signals increasingly modify the basic digital information as one moves outward in this information hierarchy. Therefore, to understand systems, one must have the tools for global measurements of as many information types as possible and the ability to integrate these different types of information. Finally, biological information operates across three distinct time dimensions: evolution, development, and physiology. Accordingly, the patterns of genomic content and organization change across evolution as the patterns of gene expression change across development or throughout a physiological response. Indeed, the informational content of cells or organisms may be viewed as a series of snapshots of changing patterns of information expression.

The systems approach may generally be described as follows:

- A biological system is chosen and all preexisting relevant information is integrated into a model that may be descriptive, graphical, or mathematical.
- A global analysis of the systems elements is carried out. Generally this begins with a genome sequence. Genes (and their corresponding proteins) and transcription factor-binding sites may be cataloged, predicted computationally, or experimentally identified. These data sets are the initial building blocks for constructing protein and gene regulatory networks.
- The system is then perturbed genetically or environmentally, and global data sets are collected from as many different data types as possible. Genetic perturbations include overexpression, underexpression, or knockouts. These data sets are generally collected under steady-state conditions. Environmental perturbations may include, for example, the introduction of substrates that activate metabolic pathways, and hormones that trigger signal transduction pathways. Here, kinetic experiments are required so that the patterns of information change across development or physiological triggering can be characterized.
- The different data types must be integrated (see examples below) and then compared against the initial model. Discrepancies between data and model will emerge. Hypothesis-driven formulations must explain

*These authors contributed equally to this paper.

these differences and then trigger a second round of perturbations generating new global data sets to test the hypothesis. At each round the model will be refined. This process will continue iteratively until the model and experimental data are consistent with one another. The ultimate objective is to move toward an accurate mathematical model. Hence, systems biology is hypothesis-driven, global, quantitative, integrative, and iterative.

## MODEL ORGANISMS FOR SYSTEMS ANALYSIS

We have exploited two organisms for applying systems approaches: the yeast, *Saccharomyces cerevisiae*, and the archaeon, *Halobacterium sp. S. cerevisiae* is an ideal model system in that the organism is single-celled, its genome is fully sequenced, extensive genetic manipulations are possible, it is easy to grow, and an enormous experimental literature is available. Moreover, deletion strains are available for more than 95% of the ~6200 open reading frames (ORFs) in the yeast genome (Giaever et al. 2002). Because of these advantages, genome-wide studies such as microarray analysis, proteomics, genome-wide binding analysis, and large-scale genetic interaction studies became available for yeast much sooner than for most other organisms. Similarly, technologies for characterizing protein–protein interactions are most advanced and easiest to study using yeast. As a result, a vast amount of global information is available, including protein–protein and protein–DNA interactions, changes in mRNA expression, and phenotypic assays.

*Halobacterium sp.* offers a striking contrast to yeast in that far less biological information is available, hence it affords a test of what systems biology can do in a less well developed model organism. It is easily cultured, its genome is sequenced, an array of genetic, biochemical, and genomic tools have been developed, and its robust and interesting physiology make *Halobacterium sp.* ideal for evaluating the response of relevant biological systems to fluctuating environmental factors (DasSarma and Fleischmann 1995; Ng et al. 1998, 2000; Peck et al. 2000; Baliga et al. 2001, 2002). As in other model organisms, certain biological responses such as phototrophy (conversion of light to energy), aerotaxis (movement toward oxygen), and phototaxis have been characterized, one or a few genes or proteins at a time, and serve as benchmarks to help validate the global systems-level interpretations of genome-wide data (Oesterhelt and Stoeckenius 1973; Spudich and Bogomolni 1984; Spudich et al. 1989; Oesterhelt 1995; Oren 1999; Peck et al. 2001). This organism also has unusual protein chemical features that will enormously facilitate global structural studies.

## GALACTOSE UTILIZATION IN YEAST: A BENCHMARK SYSTEM

The galactose system in yeast is one of the best-studied eukaryotic systems of gene regulation, because of its relative simplicity and because the central transcription factor, Gal4, is one of the strongest transcriptional activators described to date. The galactose system refers to a biochemical pathway that enables cells to utilize galactose as a carbon source, and the regulatory network that controls the switching on (in the presence of galactose) and off (in the absence of galactose) of the pathway (for review, see Lohr et al. 1995; Reece 2000). A permease (Gal2) transports galactose into the cell, whereas the enzymatic proteins subsequently convert intracellular galactose to glucose-6-phosphate. These include galactokinase (Gal1), uridylyltransferase (Gal7), epimerase (Gal10), and phosphoglucomutase (Gal5/Pgm2). Three regulatory proteins, Gal3, Gal4, and Gal80, exert transcriptional control over the transporter, the enzymes, and to a certain extent, each other. Gal4 binds to specific sequences upstream of the *GAL* genes (with the exception of the *GAL4* gene itself) to potently activate transcription. In the presence of glucose, this is prevented by the action of Gal80, a co-repressor protein, which binds to and inhibits Gal4. In the most recent model, upon galactose induction, Gal80 is translocated to the cytoplasm where it interacts with Gal3 (Peng and Hopper 2000, 2002). This Gal80–Gal3 interaction is important for relieving the repressive function of Gal80. Gal4, no longer repressed by Gal80, activates the *GAL* genes through a mechanism involving a phosphorylated version of the transcription factor (Mylin et al. 1989, 1990).

Despite years of research on the galactose system, and its well-characterized regulatory network, recent findings indicate that even this relatively simple system is much more complex than previously thought. First, a systems biology study of this system (outlined below) revealed that the cellular response to galactose extends far beyond the activation of the Gal genes (Ideker et al. 2001). Second, the activation and repression of the Gal genes requires a more extensive repertoire of regulatory proteins than simply Gal4, Gal3, and Gal80. A major undertaking in our laboratory has been to characterize the regulatory networks that control the galactose response in yeast, with respect both to the regulation of the Gal genes themselves and to the control of genes indirectly affected by activation of the galactose system.

## STEADY-STATE PERTURBATION OF THE GALACTOSE SYSTEM

As described above, an important task in solving gene regulatory networks for any given pathway is to catalog the changes in gene and protein expression controlled by that pathway. To do this comprehensively requires integrating information both from steady-state perturbation experiments and from kinetic analyses. With respect to the galactose system, the former was carried out through systematic environmental (presence or absence of galactose) and genetic (knockouts) perturbations to the system. Employing nine strains of yeast, each with a different galactose gene knocked out, and the wild type, the mRNA levels of ~6200 yeast genes were monitored, with the system on and off for each genetic perturbation. Nine hundred ninety-seven mRNAs were changed in one or more of these perturbations. In addition, the quantitative changes in protein expression for 300 proteins of the

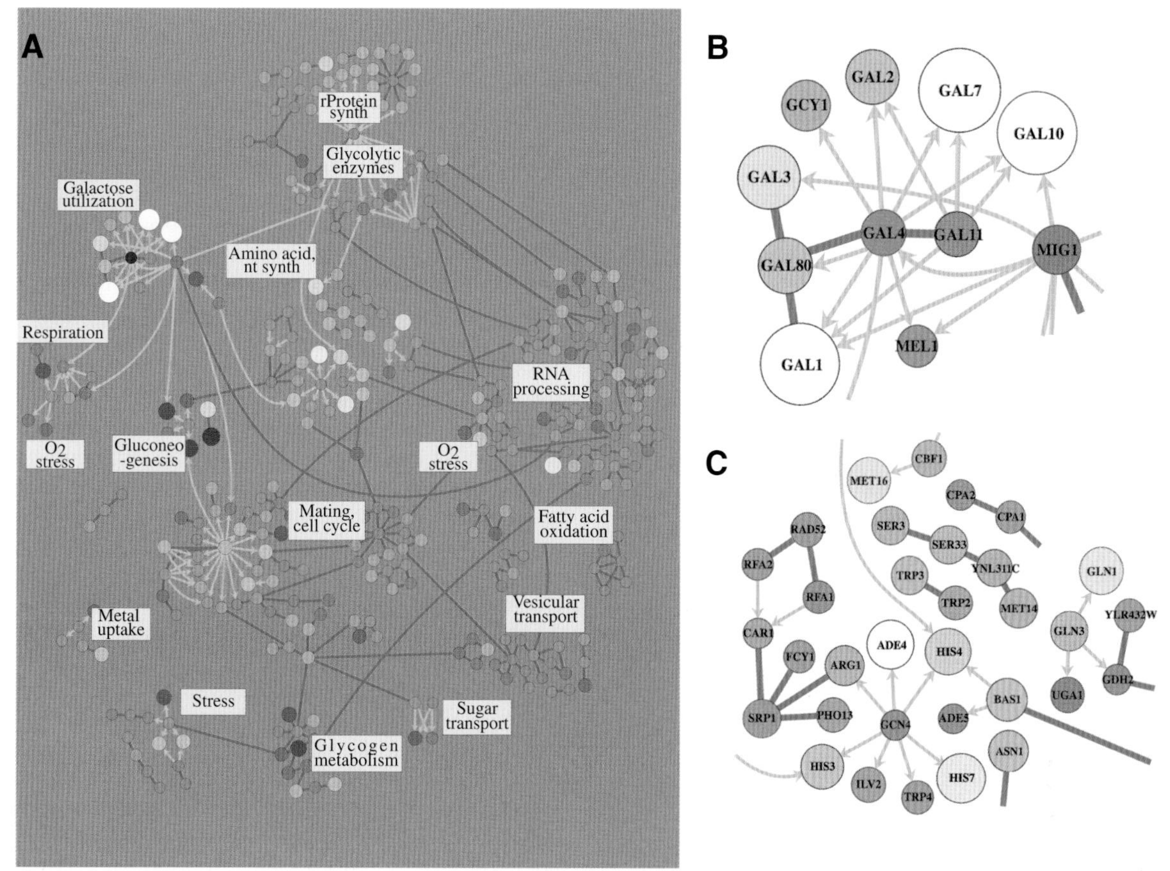


**Figure 1.** Integrated physical interaction network. (*A*) Changes in mRNA expression for the Δ*gal4*+galactose perturbation are superimposed on a network of protein–protein and protein–DNA interactions. Nodes represent genes; a yellow arrow directed from one node to another signifies that the protein encoded by the first gene can influence the transcription of the second by DNA binding (protein–DNA), and a blue line between two nodes signifies that the corresponding proteins can physically interact (protein–protein). The gray-scale intensity of each node represents changes in mRNA levels (with white representing a decrease in expression and black denoting an increase), and Gal4 itself is red. Highly interconnected groups of genes tend to have common biological functions and are labeled accordingly. Regions corresponding to the galactose utilization module are shown in *B*, and those corresponding to the amino acid synthesis biomodule are shown in *C*. Accordingly, this is a Cytoscape graph that integrates three different types of information—mRNA concentrations, protein–protein interactions, and protein–DNA interactions. (Reprinted, with permission, from Ideker et al. 2001 [copyright AAAS].)

wild-type yeast with the system on and off were determined using the isotope-coded affinity tag (ICAT) approach (Gygi et al. 1999). Surprisingly, these studies revealed that, upon manipulation of the galactose pathway, a number of other pathways within the cell are affected at the level of mRNA expression (and some at the level of protein expression), suggesting a network of biomodular connections. This is most dramatically displayed in a network interaction map generated by integrating the information from microarray analysis with known protein–protein and protein–DNA interactions (Fig. 1). This network map highlights the magnitude of the galactose cellular response, and the connections between the galactose pathway and multiple other biomodules.

A number of observations from the large-scale perturbation study had not previously been reported for the galactose system. For instance, manipulating the galactose pathway led to large changes in the expression of genes important for amino acid biosynthesis, many of which are regulated by a common factor, Gcn4 (Fig. 1B). These results suggest a regulatory connection between two metabolic pathways; the galactose pathway and the amino acid synthesis pathway. To understand the nature of this connection will require a more directed study in which the regulation of Gcn4 activity and transcription of the *GCN4* gene are analyzed. A second interesting finding was that in the absence of *GAL7* or *GAL10* (the enzymes that convert galactose-1-phosphate to glucose-1-phosphate), there is a reduction in the mRNA expression for the other Gal enzymes when cells are grown in galactose. The expression of these genes, however, was not reduced in Δ*gal1*/Δ*gal10* cells grown in galactose. Given that, in these cells, there is no production of galactose-1-phosphate, the mRNA changes observed in Δ*gal7* and Δ*gal10* cells are likely due to an accumulation of galactose-1-phosphate or a derivative. These results indicate important regulatory influences of metabolites, adding another layer of complexity to the regulation of the galactose system and highlighting the need to collect data at multiple levels. Finally, a comparison of gene and protein expression patterns across the various perturbations revealed that ~50% of the proteins examined had their levels controlled by posttranscriptional mechanisms. These additional insights, revealed by the systems approach, had never previously been reported despite more than 30 years of research on the galactose utilization pathway.

## GENE REGULATORY NETWORKS UNDERLYING THE GALACTOSE RESPONSE

The large-scale perturbation study was invaluable in that it characterized the global genetic response. A next major challenge is to begin elucidating the gene regulatory networks that underlie this response. This will require two key approaches: (1) a kinetic study in which the global genetic response is dissected according to the timing of mRNA and protein changes and (2) a comprehensive cataloging of the elements that comprise the underlying gene regulatory networks. These networks control which genes are expressed in a cell at any given time and underlie much of the cell's ability to respond to a wide range of environmental cues and cellular signals. Control of gene expression is achieved through a complex interplay between transcription factors, their upstream signals, and the genomic sequences or *cis* elements to which these factors bind. Transcription factors, typically in complexes with other regulatory proteins, bind to *cis* elements upstream of transcriptional start sites and function to modulate the output of the basal transcriptional machinery.

## DEFINING *CIS*-REGULATORY ELEMENTS AND PROTEIN–DNA INTERACTIONS

Our ability to efficiently identify the elements (proteins and *cis* elements) and interactions (protein–protein and protein–DNA) that characterize gene regulatory networks has advanced substantially over the past few years. First, a technology known as genome-wide binding analysis was established to identify the DNA targets of regulatory proteins on a genome-wide scale (Ren et al. 2000). This approach couples chromatin immunoprecipitation with microarray analysis, and was recently used to characterize the binding locations for more than 106 transcriptional regulators in yeast (Lee et al. 2002). Combining the information from this approach with microarray expression data is useful, in that this integration reduces some of the noise generated by each approach and will be important for understanding cause-and-effect relationships not readily apparent from expression data alone. Results from expression array studies and genome-wide binding analyses also provide a useful starting point for identifying, computationally, genes that may share common *cis* elements. For example, a global set of genes may be clustered based on exhibiting common expression profiles across a series of perturbations, suggesting they are regulated by shared elements of a gene regulatory network.

As in all other gene regulatory networks, the response to galactose is largely encrypted within the promoters of the *GAL* genes, namely by *cis*-regulatory elements. Whereas the status of the regulatory proteins (Gal4, Gal80, and Gal3) controls the induction or repression of transcription, the combination and number of *cis* elements within their promoters dictates the response capabilities of the *GAL* genes. This is apparent by comparing the induction of *GAL* genes encoding the structural proteins with that of the *GAL* regulatory genes. As shown in Figure 2, *GAL1, GAL7*, and *GAL10*, all encoding enzymes important for metabolizing galactose, are induced to much higher levels in galactose compared to *GAL3* and *GAL80*, which encode regulatory proteins. This induction is reflective of the distinct numbers of Gal4-binding sites present upstream of each gene. There are four binding sites in the shared promoter of *GAL1* and *GAL10*, two upstream of *GAL7*, and only one within the promoters of *GAL3* and *GAL80* (Johnston 1987). It is interesting to note, however, that within the limits of our resolution, the kinetics of the activation of these genes is similar, indicating coordinated activation by a common mechanism, despite differences in the induction magnitude. There is no Gal4 consensus site upstream of the *GAL4* gene (it does not autoregulate itself), and accordingly, expression of *GAL4* does not noticeably change in response to galactose (Fig. 2).

To determine whether Gal4 binds to genes other than the known targets, Ideker et al. (2001) used search algorithms to look for the well-characterized Gal4-binding site ($UAS_{GAL}$) upstream of the 997 genes whose mRNA levels were significantly affected by the 20 Gal-pathway perturbations. Forty-one genes were found to contain the sequence ($CGGN_{11}GCC$) known to bind Gal4. Three

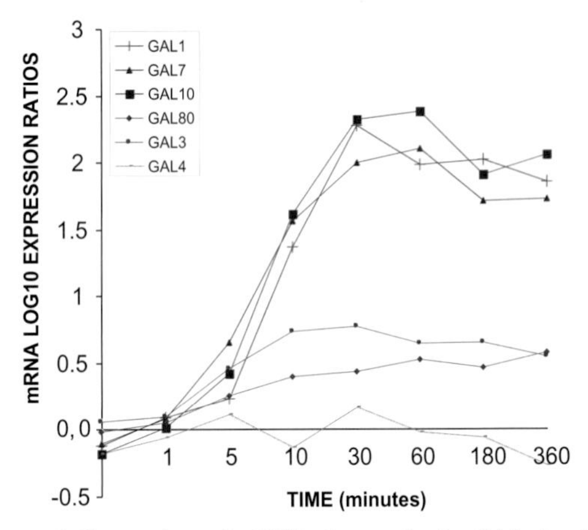


**Figure 2.** Comparison of mRNA changes for the GAL structural genes versus the GAL regulatory genes. *GAL1, GAL7*, and *GAL10*, encoding the enzymes for galactose conversion, are all induced to levels that are comparable to one another, consistent with similar numbers of Gal4-binding sites in their upstream promoters (*GAL7* has two binding sites, and *GAL1* and *GAL10* share four). *GAL3* and *GAL80*, which encode regulatory proteins, each have a single Gal4 site in their promoters and accordingly are not induced to the same extent as *GAL1, GAL7*, and *GAL10*. *GAL4*, which does not contain a binding site for its own encoded protein (it is not autoregulated), does not noticeably change in expression in response to galactose. Despite the differences in induction magnitude between *GAL1, GAL7*, and *GAL10* versus *GAL3* and *GAL80*, the kinetics of activation are similar, suggesting that they are regulated by a common regulatory machinery.

genes on this list (*MTH1, PCL10,* and *FUR4*) not previously known to be targets of Gal4 were confirmed by Ren et al. (2000), using the genome-wide binding assay. *MTH1* encodes a repressor of the hexose transport (*HXT*) genes, *PCL10* encodes a cyclin-dependent protein kinase important for glycogen biosynthesis, and *FUR4* encodes a uracil permease. The identification of these targets of Gal4 therefore revealed potentially new functions for this central transcription factor, and, similar to the results of Ideker et al., indicates that Gal4 coordinates the regulation of multiple pathways in response to galactose. Our analysis of Gal4 targets using the genome-wide binding assay further confirmed Gal4 binding at an additional six (*NAR1, LPE4, OPT2, YEL057c, CDC7*, and *YPS3*) of those genes identified by Ideker et al. (2001). Therefore, the combined use of microarray expression data with location analysis data will be invaluable for identifying the DNA targets of regulators. One wonders how many of the 32 additional Gal4 binding sites identified by Ideker et al. might actually be functional—perhaps only under special environmental conditions. Indeed, filtering binding data with expression data has already proved to be a highly effective process for building a network model of cell cycle regulation (Lee et al. 2002). These combined data sets are also useful in that they enable binning of sequences most likely to contain common transcription factor-binding motifs, providing an ideal start point for the use of computational search motifs.

Our ability to identify functional elements within the genome has advanced even further with the recent release of high-quality draft sequences of three species of yeast that are related to *S. cerevisiae* (*S. paradoxus, S. mikatae,* and *S. bayanus*) (Kellis et al. 2003). Comparative genome analysis of related species is a powerful approach for identifying regulatory elements, as these elements tend to be conserved relative to the remaining genome sequence. We examined the sequence conservation across all four species for the additional six putative targets of Gal4. The predicted Gal4-binding sites upstream of these genes are conserved for four of these targets (*NAR1, YPL066w, YEL057c*, and *YPS3*). Of these four, only *YPS3*, which encodes an aspartic-type endopeptidase activity, has been functionally annotated. It is interesting to note that *NAR1* shares a Gal4-binding site with *GAL6* (*LAP3*), as these two genes are divergently transcribed on the same chromosome. This arrangement is similar to that of *GAL1* and *GAL10*. The lack of conservation for the Gal4 sites upstream of the two other putative targets (*OPT2* and *CDC7*) suggests that the observed changes in the expression of these genes by galactose may be specific to *S. cerevisiae*. To test this experimentally, however, will require expression analysis and genome-wide binding analyses in the other three species. The additional insights into Gal4 targets resulted from the integration of four key approaches: microarray expression analysis, genome-wide binding analysis, the use of search algorithms on a defined list of sequences, and comparative genomics. This integrative approach therefore was useful in selecting four most likely targets of Gal4 from large data sets, which each contain sufficient noise as to preclude a similar identification from any single approach.

## DEFINING THE REGULATORY PROTEINS

Considerable emphasis has been placed on elucidating the regulatory mechanisms controlling transcription of the *GAL* genes. From these studies, it is becoming increasingly apparent that the coordinated activation of the galactose genes is not solely due to the simple induction of Gal4 or the release of the co-repressor protein Gal80. Even this relatively simple system is complemented by a complex regulatory machinery, with multiple proteins acting in a combinatorial fashion to control transcription. To date, the actual switch in this machinery is not known. Clearly, the release of Gal80's repressive effects on Gal4 is important. Evidence also implicates the phosphorylation of Gal4 in the induction of the galactose response (Mylin et al. 1989, 1990). Aside from Gal4 and Gal80, which seem to be specific regulators of the *GAL* genes, general factors known to have a more global role in gene regulation also influence transcriptional activity at the *GAL* promoters. For instance, the *GAL* genes are repressed, in part by the global repressor Mig1 through Mig1-binding sites in their promoters. Similarly, transcriptional activation of the *GAL* genes has been shown to require SAGA, a large chromatin remodeling complex, which acts at several different promoters to initiate transcription (Bhaumik and Green 2001; Larschan and Winston 2001). In this respect, there appear to be two components to regulation of the *GAL* genes. First, there is the specific component, involving factors that are unique to the *GAL* promoters, namely Gal4 and Gal80. Second, there is a nonspecific component that utilizes factors with general repressive or inductive roles at several promoters of very distinct genes. These include the Mig1 repressor and the SAGA complex. In addition, the Cyc8–Tup1 global repressor complex has also been shown to be involved in repression at the *GAL1* promoter and, presumably, at the other *GAL* genes.

The mechanisms whereby specificity is inferred on these global factors such that they affect only those transcriptional programs required under a given circumstance is unclear. For example, it is not yet known how, when cells are grown in galactose, repression is relieved at galactose-inducible genes, while global repressors such as Mig1 and Cyc8–Tup1 retain their repressive function at other gene promoters. Similarly, large chromatin-modifying complexes such as SAGA are brought to promoters of specific genes, depending on the cellular requirement for transcriptional activation of those genes. The basis for this condition-defined specificity, however, remains to be characterized for each subset of genes controlled by these factors. Much of this specificity likely relies on the presence of specific factors such as Gal4, and modifications thereof. For instance, perhaps under gal-inducing conditions, phosphorylation of Gal4 contributes to the recruitment of SAGA. Such a modification occurs only at select promoters under appropriate conditions (the presence of galactose), thereby enabling gene-specific recruitment of SAGA. Although such a hypothesis is likely to be true for certain genes, this seems to be an oversimplification in many instances. Various findings indicate that the availability of cofactors and transcription factor partners figures prominently in the activity of any given repressor or activator. The absence of Gal80 alone, for instance, is sufficient for high-level transcription of the *GAL* genes (Ideker et al. 2001), and this cofactor has been shown to block recruitment of SAGA specifically at Gal4-occupied promoters (Carrozza et al. 2002). In another example, the transcription factor Ste12 binds to distinct program-specific target genes depending on the developmental condition (i.e., filamentation versus mating). This selective binding is dictated by the status of Tec1, a binding partner for Ste12 during filamentation but not under conditions that induce mating (Zeitlinger et al. 2003). Thus, signal transduction pathways, controlling the status of a transcription factor partner, can infer program-specific distribution of Ste12 binding (Zeitlinger et al. 2003). In addition to transcription factor partners, cofactors, which are tethered to DNA indirectly through DNA-binding proteins, often serve as adapters to switch transcription on or off, or to facilitate either process. At the *GAL1* promoter, for example, the global Cyc8–Tup1 co-repressor has been shown to convert from a repressor complex to an activator complex upon binding to the novel protein Cti6 (Papamichos-Chronakis et al. 2002). Cti6 recruits SAGA to the *GAL1* promoter, in turn recruiting the basal transcription machinery. Whether Cti6 affects other genes outside of the galactose pathway is not yet known. If this factor is specific for the *GAL* genes, however, it may infer some level of specificity with respect to repression by the Cyc8–Tup1 global repressor. To understand the role of factors like Cti6 will require genome-wide binding analysis and protein–protein interaction studies.

## FUTURE CHALLENGES FOR CHARACTERIZING GENE REGULATORY NETWORKS

With ongoing improvements in technologies aimed at studying gene regulatory networks, we anticipate enormous progress in the near future toward characterizing these networks on a global scale. Genome-wide analysis has already been employed on a large scale to identify the DNA targets of most transcription factors in yeast. Although these studies, combined with mounting microarray data for gene expression changes, provide an excellent basis for studying regulatory networks, a number of challenges remain to fully understand gene regulation. For instance, characterizing protein–DNA interactions under different conditions will be critical for understanding any given cellular response. This is emphasized by the condition-specific binding distributions for Ste12 (Zeitlinger et al. 2003). Since most transcriptional events take place in response to some environmental change, this is critical. In addition, many transcriptional switches may be of a transient nature, necessitating kinetic studies that can capture transient interactions. Similarly, profiling gene expression changes, both mRNA and protein, over time will be much more useful for extracting cause-and-effect relationships within gene responses. Despite all these future challenges, we have clearly demonstrated the power of systems biology over conventional approaches for characterizing biological systems. After years of intense research on the galactose system, a systems approach revealed new insights and demonstrated a much more global effect to what has traditionally been considered a simple system.

## *HALOBACTERIUM SP.*

*Halobacterium sp.* is an extreme halophile that thrives in saturated brine environments such as those associated with the Dead Sea and solar salterns. It offers a versatile and easily assayed system for an array of well-coordinated physiologies that give it an edge for survival in these harsh environments (DasSarma and Fleischmann 1995; DasSarma and Arora 1999; Oren 1999). It has robust DNA repair systems that can efficiently reverse the damages caused by a variety of mutagens, including UV radiation and cycles of desiccation and rehydration (McCready and Marcello 2003; N.S. Baliga and J. Diruggeiro, unpubl.). *Halobacterium sp.* responds to anaerobic conditions with the synthesis of a purple membrane whose major component, bacteriorhodopsin, facilitates the conversion of light into ATP (energy). The completely sequenced genome of *Halobacterium sp.* has also provided insights into much of its physiological capabilities; however, nearly two-thirds of all genes encoded in the halobacterial genome have no known function (Ng et al. 1998, 2000).

*Halobacterium sp.* can adapt to both aerobic and anaerobic environments in the presence or the absence of light and, therefore, is an ideal model system for deciphering the regulatory circuits that coordinate various metabolic pathways in response to frequent changes in oxygen concentration and light intensity (Fig. 3). In an aerobic environment, *Halobacterium sp.* flourishes as a chemoheterotroph surviving on organic remnants of deceased lesser halophiles incapable of living in extreme 4.5 M salt concentrations. It produces energy predominantly through respiration utilizing oxygen as the terminal electron acceptor (Ng et al. 2000). During the aerobic phase, the regulatory circuits operate through a two-component signal transduction system comprising the retinal-bound sensory rhodopsin II (SRII) and its interacting transducer HtrII. When SRII is activated by blue-green light, it transmits a signal to HtrII exposing methylation sites on the transducer and transiently alters autophosphorylation activity of a bound histidine kinase (CheA). The consequence of the resulting phosphorylation–dephosphorylation cascade is the movement of the organism away from bright light to distance itself from damaging UV radiation (Perazzona and Spudich 1999; Luecke et al. 2001; Spudich and Luecke 2002).

A drop in oxygen tension results in dramatic changes in *Halobacterium sp.* physiology, including repression of the SRII/HtrII complex and induction of an opposing SRI/HtrI complex (Fig. 3). SRI, unlike SRII, translates orange light into an attractant stimulus by communicating with the flagellar motor through the transducer HtrI. The low oxygen tension also induces synthesis of gas vesicles, which are flotation devices that work in conjunction with SRI/HtrI to mediate net upward mobility toward sunlight. SRI, upon absorbing orange light, shifts to a

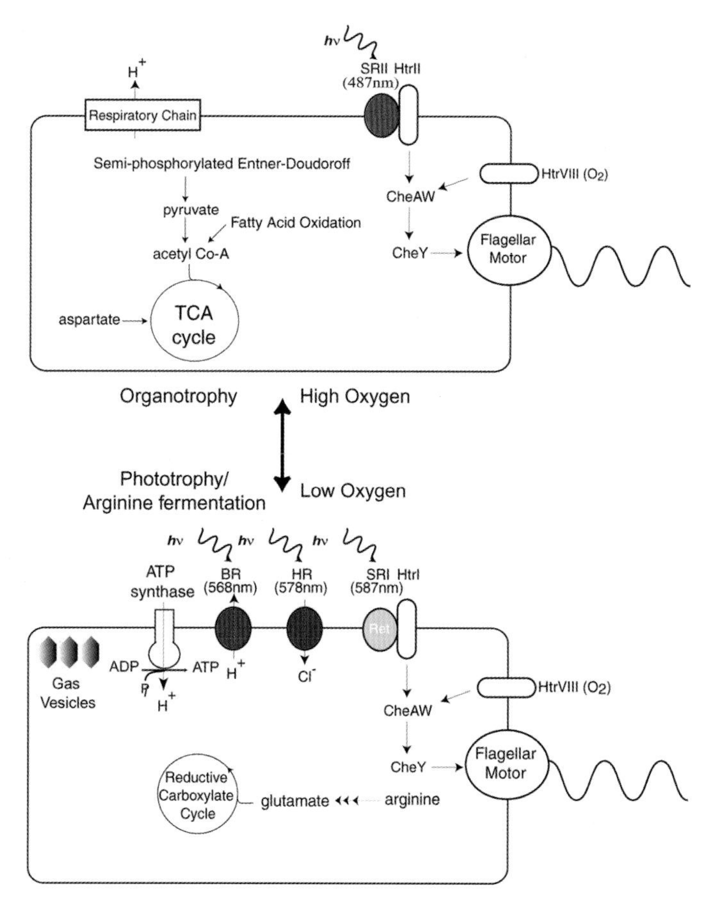


**Figure 3.** Aerobic and anaerobic physiologies in *Halobacterium sp.* related to quality and quantity of light and oxygen in its growth environment (see text).

more stable state called $SR_{373}$ that provides photoprotection by converting near-UV wavelengths into a repellent stimulus (Jung and Spudich 1996; Perazzona et al. 1996; Hoff et al. 1997). Likewise, the transducer HtrVIII mediates aerotaxis (Brooun et al. 1998), and other transducers (at least 17 transducers are encoded in the *Halobacterium sp.* genome) mediate various types of chemotaxis (Storch et al. 1999; Kokoeva and Oesterhelt 2000). Therefore, the net movement of the halobacterial cell is dictated by a combination of the quality and quantity of light, oxygen, and nutrients in the environment.

Other *Halobacterium* biomodules induced by low oxygen tension include bacteriorhodopsin (bR), halorhodopsin, and the arginine fermentation enzymes (Fig. 3) (Betlach et al. 1989; Yang and DasSarma 1990; Ruepp and Soppa 1996) . Synthesis of bacteriorhodopsin is further induced on exposure to higher-intensity light as the cells move upward toward light (Shand and Betlach 1991). Bacteriorhodopsin and halorhodopsin function as light-driven ion pumps and rapidly set up a proton gradient across the membrane. The proton gradient is subsequently utilized for ATP synthesis by a membrane ATP synthase. The phototrophic growth in the suboptimal anaerobic environment is a mechanism for sustenance rather than active growth and doubling. Breakdown of arginine serves as an additional source of energy for *Halobacterium sp.* in the absence of oxygen. *Halobacterium sp.* ferments arginine via the arginine deiminase pathway, which can produce 1 mole of ATP for every mole of arginine fermented (Ruepp and Soppa 1996). Therefore, both the light-driven ion-pumping activity of bR and arginine fermentation operate concurrently in an anaerobic state.

## SYSTEMS BIOLOGY OF *HALOBACTERIUM SP.*: STEADY-STATE ANALYSIS

The data acquired from the oxygen and light response systems of *Halobacterium sp.* described above, studied by conventional approaches, one gene at a time, led to an attractive preliminary model suitable for testing by systems approaches. Integration of the existing knowledge base with new global information has been key to defining the biomodules that participate in this process and their interrelationships.

The phototrophy operon (a set of linked genes controlling phototrophy) has at least four genes, respectively coding a transcription regulator, Bat; a structural protein, Bop (for bacterioopsin); and two key enzymes in the carotenoid biomodule, CrtB1 and Brp. Bat regulates the expression of the other genes in the operon in response to changes in at least two environmental stimuli, light and oxygen (Gropp and Betlach 1994; Baliga and DasSarma 1999, 2000; Baliga et al. 2001). The gene regulatory network including Bat contains at least two feedback loops (positive and negative), coordinating the synthesis of retinal and Bop, which come together in a 1:1 stoichiometry to form the light-driven ion pump bacteriorhodopsin (bR) (Deshpande and Sonar 1999). Our systems analysis employed three genetic perturbations, two in *bat* and one in *bop*. One of the *bat* perturbations increases the expression of bacteriorhodopsin, and the second (a knockout) shuts it down completely. The *bop* perturbation is a knockout (Yang et al. 1996).

We compared the three mutants under steady-state conditions against one another and the wild type by using DNA arrays containing the ~2400 *Halobacterium sp.* genes (Baliga et al. 2002). We also compared, in the two *bat* mutations, the changing concentrations of ~300 proteins by the ICAT method (Baliga et al. 2002). We then integrated the above experimental data with several global data sets and classifications obtained from the literature.

## ELUCIDATING NETWORKS OF PROTEIN INTERACTIONS AND FUNCTIONAL RELATIONSHIPS FOR *HALOBACTERIUM SP.*

Extensive databases for protein–protein interactions are available for several organisms. We have searched these databases for *Halobacterium sp.* orthologs and identified pairs of interacting orthologs (1431 interactions were mapped to *Halobacterium sp.* from the yeast data, and 178 were mapped from the *H. pylori* data). We identified these pairs of interacting orthologs using the database of Clusters of Orthologous Genes (COGs) (http://www.ncbi.nlm.nih.gov/COG1). This phylogenetic classification identifies orthologs existing in three or more major lineages of 43 complete genomes and 30 major lineages, thus identifying ancient conserved genes or domains (Tatusov et al. 2001). Putative protein-binding pairs in *Halobacterium sp.* were inferred in three stages: (1) COG members of the protein–protein interaction pairs were determined; (2) all corresponding "COG partners" of yeast and *H. pylori* proteins were identified in the *Halobacterium sp.* genome; and (3) the interacting pairs were given a confidence level determined by the strength of the match to the COG pair and the confidence of the original protein interaction measurements. A total of 1143 nonredundant interacting pairs were inferred by this method. Finally, SCOP (structural classification of proteins) classification for *Halobacterium sp.* proteins has been identified, when possible, based on homology modeling. The SCOP classification is then used to determine likely interactions based on the prior observation that some pairs of SCOP families are known to physically interact in protein complexes in the PDB with a disproportionately high frequency (Park and Teichmann 2001). The SCOP classification of *Halobacterium sp.* proteins has been used to identify 562 likely interactions. Support by an independent predictive method increases confidence in the putative interactions. Clearly, all of the predicted interactions must be verified experimentally.

- *Chromosomal proximity.* Consistent chromosomal colocalization of pairs of conserved homologs (PCHs) across many different genomes correlates well with their functional coupling (Overbeek et al. 1999). These putative functional couplings include proteins that may (1) participate in the same biochemical pathway, (2) interact with each other, and/or (3) be co-regulated in response to a common stimulus. This approach added 327 interactions to our list.

  To identify these relationships, we use the method of Marcotte et al. (1999), commonly referred to as the phylogenetic profile method. This added 525 interactions to our list.
- *Phylogenetic profile.* Similar profiles of presence or absence of pairs of orthologs in fully sequenced genomes are indicative of their functional coupling.
- *Domain fusion.* The activity of a protein is a function of one or more distinct structural units referred to as domains. For example, the protein Bat in *Halobacterium sp.* is composed of a redox-sensing PAS/PAC domain, a light-sensing GAF domain, and a helix-turn-helix (HTH) DNA-binding motif, which together impose the light- and oxygen-sensing regulatory control of Bat on the phototrophy genes. PAS/PAC domains are also found fused with domains exhibiting other functions, such as the kinase domain in the halobacterial protein KinA1, leading to oxygen concentration-mediated control of a phosphorylation cascade. Enright et al. (1999) have demonstrated that domain fusions are often correlated with functional interactions among the corresponding domains. Domains within *Halobacterium sp.* proteins fused in other genomes have been used as a metric to predict functional interactions between the corresponding proteins. We identified 2460 putative interactions with this approach.

  In all, 5017 interactions were inferred for *Halobacterium sp. NRC-1*.
- *Metabolic networks.* Metabolic networks are the most detailed and well studied of all protein networks. Although a number of metabolic pathway databases are now available, KEGG (www.kegg.org) (Kanehisa 2002) provides a convenient starting point because it already contains the proteins of known function for *Halobacterium sp.*

Researchers at the Institute for Systems Biology, together with other collaborators, have developed a graphical network platform, Cytoscape (www.cytoscape.org), that has the ability to integrate many distinct experimental and computational global data types. Cytoscape can display any biological network as a graph of proteins and their relationships, where the proteins are depicted as nodes and the relationships between proteins as edges. Once this network is displayed, microarray and proteomics data, as well as a wide variety of available annotation types (including the KEGG annotations), can be visually displayed on the network (see Fig. 1 for a Cytoscape display). Figure 4 shows the data-integration/annotation platform, currently implemented here at the Institute for Systems Biology, as applied to *Halobacterium sp.* All data types are stored, curated, and managed using our database, SBEAMS, and displayed and analyzed using Cytoscape. The integration of these different global data sets and inferred relationships in the Cytoscape environment has led to several key conclusions:

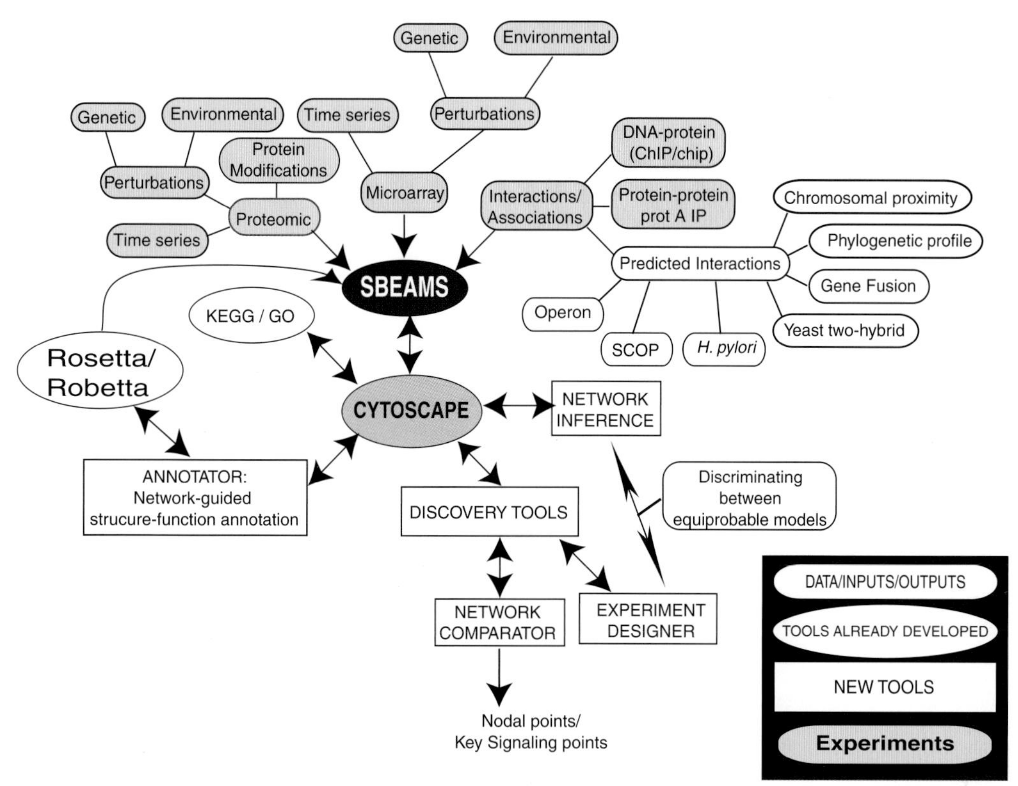


**Figure 4.** Systems biology information flow and data integration. Schematic detailing of the flow of experimental data (*yellow*) and data inferred from computational approaches into Cytoscape and SBEAMS. Predicted protein–protein interactions are calculated as described in the text. Microarray and proteomics data can be mapped onto any biological network using Cytoscape and SBEAMS. Once disparate data sets are integrated in Cytoscape, a number of methods can then be used to design further iterations of experiments (these methods are indicated by "discovery tools" and "experimental designer"; *lower right*). Function information for the proteins encoded in the genome is obtained primarily from the KEGG and GO databases; the Rosetta/Robetta module provides additional structure–function annotations for proteins of unknown function by interfacing with the ANNOTATOR, which simultaneously processes information from many sources; the NETWORK INFERENCE tool facilitates discovery of regulatory networks based on function, steady-state and time-series mRNA and protein levels, and protein–protein and protein–DNA interactions; the DISCOVERY TOOLS facilitate exploration of the network, resolving conflicting interactions by proposing additional experiments and discovery of new biology. Several tools are under development and are indicated as rectangular nodes (NEW TOOLS).

1. Perturbation of transcriptional regulators leads to far more striking alterations of the protein and gene regulatory networks than does the perturbation of structural genes. Bat overexpression and knockout mutants each lead to a statistically significant perturbation of gene expression (compared to wild type) in 7% of the *Halobacterium sp.* genes. In contrast, the *bop* knockout only perturbs about 1% of the genes.
2. Posttranscriptional regulation plays a key role in regulating protein levels. In the two bat mutants, 272 proteins were analyzed quantitatively (by ICAT); 50 changed in a statistically significant manner, but only 17 changed in a manner corresponding to their mRNA level changes. Thus, 33/50 proteins (65%) appear to be regulated by posttranscriptional mechanisms—an observation that could only be made by integrating quantitative mRNA and protein expression data.
3. The integration of diverse data types and inferred relationships illustrates that there are three distinct biomodules that lead to the synthesis of bacteriorhodopsin or purple membrane: the isoprenoid, the carotenoid, and the bacteriorhodopsin biomodules (Fig. 5b). The *Bat* control of the latter two biomodules is clearly depicted in Figure 5b. This is mediated through bat control of the genes in its operon (Fig. 5a). In the face of constant mRNA expression, the changing levels of protein expression suggest that the isoprenoid biomodule may be regulated primarily by posttranscriptional control.
4. The reciprocal relationship between the two modes for ATP synthesis, phototrophy and arginine fermentation, under anaerobic conditions was revealed. Inactivation of *bat* suppressed many of the mRNAs related to phototrophy, whereas *bat* overexpression led to increased mRNA levels for these genes. Arginine fermentation was regulated in an inverse manner, presumably to maintain a relatively constant ATP supply under anaerobic conditions. In contrast, the inactivation of the *bop* structural gene had no effect on arginine fermentation—hence, its regulatory control appears to be correlated with Bat function.
5. Of all the genes whose expression patterns were changed in the *bat* mutants, only five appeared to have *cis*-control elements with the correct sequence motif for bat binding. The remainder of the changes must occur as a consequence of downstream modulations.
6. The arginine fermentation biomodule (Fig. 5c), as well as other genes controlling arginine and glutamate metabolism, appear to be coordinately regulated (Fig.

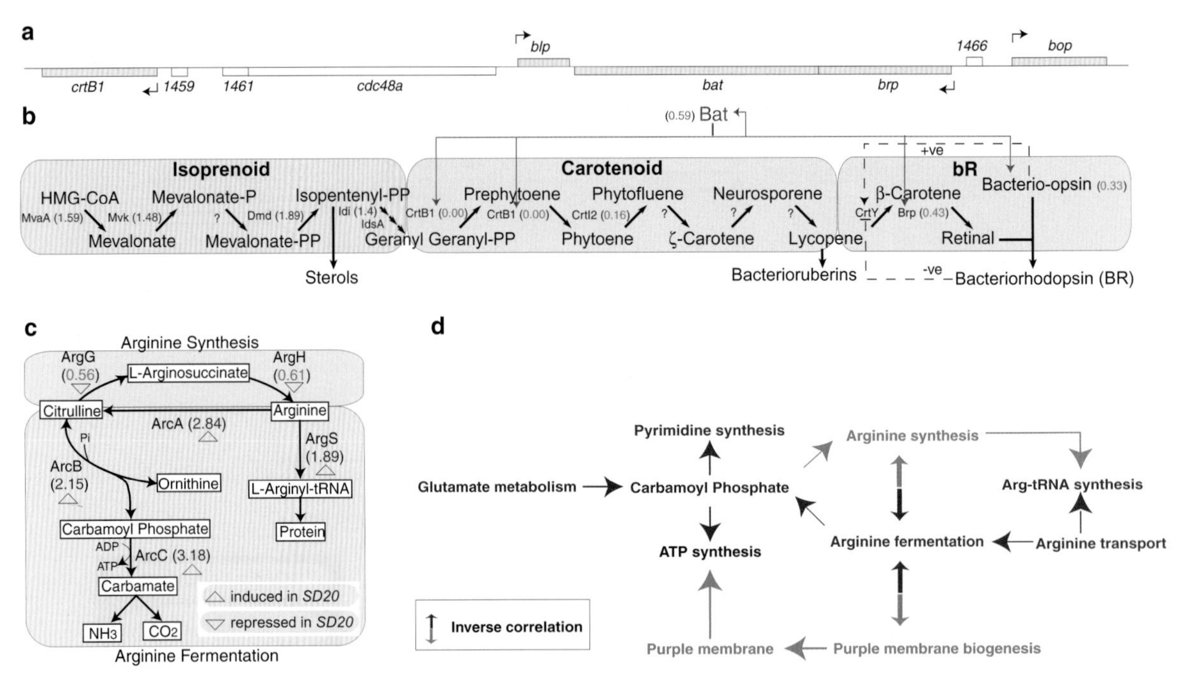

**Figure 5.** Systems-level insights obtained through analyses of three phototrophy mutants in *Halobacterium sp.* (*a*) The *bop* gene cluster; genes shaded in red are under the control of Bat. Direct and indirect effects of Bat depletion on regulation of three biomodules responsible for phototrophy (*b*) and two biomodules involved in arginine metabolism (*c*) in *Halobacterium sp.* Biomodules that are up-regulated are shaded in red and those down-regulated are shaded in green. (*d*) Functional interrelationships among the various biomodules directly or indirectly involved in anaerobic energy transduction that are up (*red*) or down (*green*) regulated as a consequence of Bat inactivation. (Reprinted, with permission, from Baliga et al. 2002 [copyright National Academy of Sciences].)

5d). Two of the three arginine fermentation genes together with three other genes important in glutamate metabolism and other aspects of arginine metabolism share a highly conserved *cis*-regulatory element, presumably the binding site for an unidentified transcription factor that regulates these modules much like *bat* regulates two of the three biomodules of phototrophy. Thus, integration of DNA motif information with mRNA and protein expression profiles provides additional insights into the gene regulatory and protein networks associated with ATP production.

## *HALOBACTERIUM SP.* OFFERS A POWERFUL MODEL FOR DETERMINING STRUCTURE/FUNCTION RELATIONSHIPS

To survive in a hypersaline environment, *Halobacterium sp.* has adapted so that most of its proteins have an extensive negative charge distribution on their external surfaces. For example, the average protein domain (150 residues) in *Halobacterium sp.* has a net charge of –17 in contrast to –3 for *S. cerevisiae*. This negative surface charge has two important consequences. First, many proteins that are insoluble in most other organisms will be soluble in *Halobacterium sp.* This will facilitate their purification, crystallization, and structural determinations either by X-ray crystallography or nuclear magnetic resonance (NMR). Second, the extensive negative surface charge will be useful in ab initio computational predictions of the three-dimensional structure (e.g., higher surface negative charge and fewer surface hydrophobic residues will result in fewer incorrect predicted conformations).

As genome sequences continue to be determined for microbes (and metazoa), often half or more of the predicted genes do not have sequence homologies that permit functional assignments. The unknown genes (proteins) fall into two categories: (1) those sufficiently diverged from orthologs whose functions are known but whose primary sequence homology relationships cannot be identified and (2) those genes with truly unique new functions. We are employing computational approaches to attempt to discern distant homology relationships in genes of the first type. We elucidate these distant functional relationships by using ab initio structure prediction and fold recognition methods to predict the structure of proteins of unknown structure and function. These predicted structures are then used to search for structure matches between the predicted structures and proteins of known structure, resulting in weak structure/function information. We integrate this weak structure/function information with other global data sets (including weak primary sequence homology) to strengthen our preliminary conclusion (see below). For primary sequence homology searches we employ a variety of programs (PSI-BLAST, Altschul et al. 1997; Pfam-hmmer, Bateman et al. 2000). Likewise, we employ a variety of computational tools to predict three-dimensional structures (PCONS [Lundstrom et al. 2001], MODBASE [Pieper et al. 2002], Rosetta [Bonneau et al. 2001, 2002], Robetta [Chivian et al. 2003]). The strategy employed by these programs is outside the boundaries of this paper, apart from a discussion of Rosetta, a tool for ab initio predictions of protein folding—a strategy chosen because Rosetta predicts three-dimensional protein structures in the absence of detectable primary sequence homology. Rosetta employs the following steps. (1) Proteins are subdivided into putative domains. (2) Three- and nine-residue local structure fragments are estimated based on local sequence similarity to corresponding three- and nine-residue substrates from protein of known structure. (3) These precomputed local structure fragments are assembled into global structures by minimizing a scoring function that favors hydrophobic burial/packaging, strand-pairing, compactness, and energetically favorable residue pairings. Since the predictions from Rosetta are lower in resolution than experimentally solved structures, they need to be supported by other data types (microarrays, proteomics, predicted networks, etc.). In the third, fourth, and fifth community-wide *c*ritical *a*ssessments of *s*tructure *p*rediction (CASP3, 4, and 5), Rosetta was one of the most effective methods employed (Bonneau et al. 2001). We are using Rosetta to predict the three-dimensional structures of all of the *Halobacterium sp.* proteins and protein domains of unknown function that are less than 150 residues in length.

When sequence similarities by PSI-BLAST or Pfam-hmmer searches have statistically marginal relationships, structure-based methods such as Rosetta can help confirm or negate the putative relationships. For example, a Pfam search on the *Halobacterium* protein VNG0511H yields a weak hit (homology) (e-value: 0.61) to PF00392, a family of bacterial transcription factors. The Rosetta-predicted structure for this protein, which has a strong match to a CATH family of helix-turn-helix DNA-binding proteins (CATH I.D.: 1.10.10.10), supported this initial tentative predicted function. CATH, like SCOP, is a hierarchical classification of protein domain structures. Thus, the agreement of these two procedures (weak Pfam and Rosetta) imparts a higher degree of confidence to this prediction not obtainable with either individual method. We can use the knowledge that VNG0551H is a DNA-binding protein to tentatively assign the role of a putative mediator of the changes observed in the gene(s) downstream to its position in the inferred gene regulatory network. Likewise, all function annotations that point to regulatory or DNA-binding roles should be examined in the context of one or more gene regulatory networks.

A second example, the protein VNG1302H, has a weak BLAST match (e-value: 0.091) to one of three domains of a protein disulfide isomerase from *Cricetulus griseus*. The Rosetta-predicted structure for the protein VNG1302H had a strong match to 1A8L (the structure for a protein disulfide oxidoreductase from the hyperthermophile *Pyrococcus furiosus*), which has a thioredoxin-

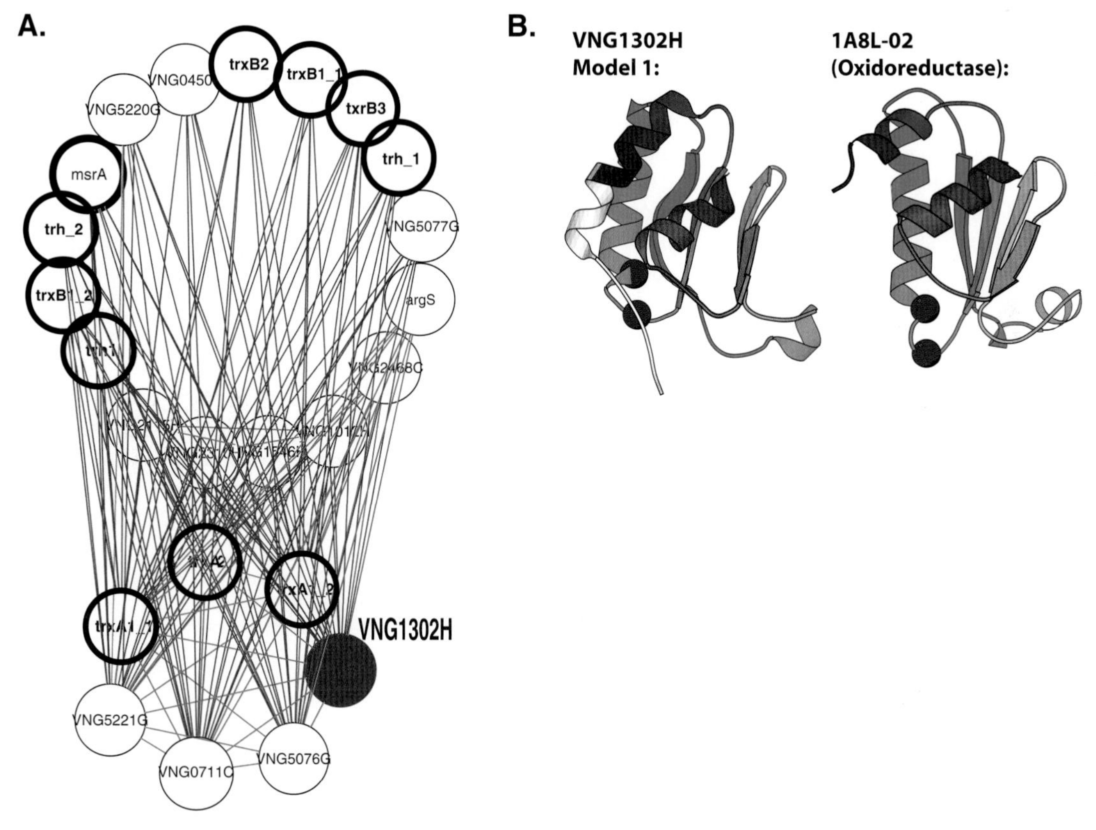


**Figure 6.** Structure/function annotation for VNG1302H. (*A*) The physical interaction network for the *Halobacterium sp.* VNG1302H. Blue edges indicate domain fusion type interactions, green edges are yeast interaction data mapped onto *Halobacterium sp.* via the COG database. VNG1302H is indicated with the sign of the rising sun. All nodes shown have one or more direct connections to VNG1302H. Proteins/nodes in this subnetwork that are known to have redox-related functions are indicated in bold. These functions include thioredoxin, thioredoxin-like, and a peptide methionine sulfoxide reductase (msrA). (*B*) Rosetta structure prediction for VNG1302H and the closest match in the PDB to this model. This fold similarity (note the structural alignment of two key cysteines, indicated by *black spheres*) indicates a thioredoxin (redox) function.

like fold (Fig. 6). Moreover, eleven of all predicted interactions (inferred on the basis of interactions between corresponding pairs in yeast as well as the domain fusion method) for VNG1302H are to proteins with known thioredoxin or thioredoxin-like functions. Thus, the inferred interaction network, the predicted general function, and evidence that thioredoxin-like domains excised from larger proteins are functional (Darby and Creighton 1995) together provide us with a strong function annotation for the VNG1302H protein (Fig. 6).

## SUMMARY: SYSTEMS BIOLOGY USING *HALOBACTERIUM SP.*

Systems approaches offer many striking opportunities. *Halobacterium sp.*, because of its small gene number, offers the opportunity to define the gene regulatory networks employing most of *Halobacterium*'s transcription factors (~160). Moreover, it offers an unusual opportunity for defining many of the key signal transduction pathways that operate in response to changing light, oxygen, and other environmental conditions. The mechanisms behind its UV repair capacity will shed new insights into this important area. Finally, *Halobacterium sp.* offers an exceptional opportunity for structural genomics—that is, defining many (most) of the three-dimensional structures of proteins in this organism.

## CODA

The ultimate objective of systems biology is to develop accurate models for biological systems—ultimately, these will be mathematical. The key to generating accurate models is the integration of many different data types and inferred functional relationships. Our approach to these integrations is summarized in Figure 4. When this is accomplished, we will achieve two important objectives: (1) We will predict the behavior of the system given any perturbation and (2) we will redesign the protein and gene regulatory networks so that completely new systems or emergent behaviors are generated.

## ACKNOWLEDGMENTS

We thank Paul Shannon for his help with Cytoscape, Eric Deutsch for his help on SBEAMS, and Vesteinn Thorsson for helpful discussions. We also acknowledge Alistair Rust for computational assistance. A.D.W. is supported by a fellowship from the National Science and Engineering Research Council (NSERC). This work was supported by National Science Foundation Grants 0220153 to L.H. and N.S.B. and 0223056 to L.H. and A.D.W.

## REFERENCES

Altschul S.F., Madden T.L., Schaffer A.A., Zhang J.H., Zhang Z., Miller W., and Lipman D.J. 1997. Gapped BLAST and PSI-BLAST: A new generation of protein database search programs. *Nucleic Acids Res.* **25:** 3389.

Baliga N.S. and DasSarma S. 1999. Saturation mutagenesis of the TATA box and upstream activator sequence in the haloarchaeal bop gene promoter. *J. Bacteriol.* **181:** 2513.

———. 2000. Saturation mutagenesis of the haloarchaeal bop gene promoter: Identification of DNA supercoiling sensitivity sites and absence of TFB recognition element and UAS enhancer activity. *Mol. Microbiol.* **36:** 1175.

Baliga N.S., Kennedy S.P., Ng W.V., Hood L., and DasSarma S. 2001. Genomic and genetic dissection of an archaeal regulon. *Proc. Natl. Acad. Sci.* **98:** 2521.

Baliga N.S., Pan M., Goo Y.A., Yi E.C., Goodlett D.R., Dimitrov K., Shannon P., Aebersold R., Ng W.V., and Hood L. 2002. Coordinate regulation of energy transduction modules in *Halobacterium sp.* analyzed by a global systems approach. *Proc. Natl. Acad. Sci.* **99:** 14913.

Bateman A., Birney E., Durbin R., Eddy S.R., Howe K.L., and Sonnhammer E.L. 2000. The Pfam protein families database. *Nucleic Acids Res.* **28:** 263.

Betlach M.C., Shand R.F., and Leong D.M. 1989. Regulation of the bacterio-opsin gene of a halophilic archaebacterium. *Can. J. Microbiol.* **35:** 134.

Bhaumik S.R. and Green M.R. 2001. SAGA is an essential in vivo target of the yeast acidic activator Gal4p. *Genes Dev.* **15:** 1935.

Bonneau R., Tsai J., Ruczinski I., Chivian D., Rohl C., Strauss C.E., and Baker D. 2001. Rosetta in CASP4: Progress in ab initio protein structure prediction. *Proteins* (suppl.) **5:** 119.

Bonneau R., Strauss C.E., Rohl C.A., Chivian D., Bradley P., Malmstrom L., Robertson T., and Baker D. 2002. De novo prediction of three-dimensional structures for major protein families. *J. Mol. Biol.* **322:** 65.

Brooun A., Bell J., Freitas T., Larsen R.W., and Alam M. 1998. An archaeal aerotaxis transducer combines subunit I core structures of eukaryotic cytochrome c oxidase and eubacterial methyl-accepting chemotaxis proteins. *J. Bacteriol.* **180:** 1642.

Carrozza M.J., John S., Sil A.K., Hopper J.E., and Workman J.L. 2002. Gal80 confers specificity on HAT complex interactions with activators. *J. Biol. Chem.* **277:** 24648.

Chivian D., Kim D.E., Malmstrom L., Bradley P., Robertson T., Murphy P., Strauss C.E., Bonneau R., Rohl C.A., and Baker D. 2003. Automated prediction of CASP-5 structures using the Robetta server. *Proteins* **6:** 524.

Darby N.J. and Creighton T.E. 1995. Functional-properties of the individual thioredoxin-like domains of protein disulfide-isomerase. *Biochemistry* **34:** 11725.

DasSarma S. and Arora P. 1999. In *Encyclopedia of life sciences*, vol. 8, p. 458. Nature Publishing Group, New York.

DasSarma S. and Fleischmann E.M., Eds. 1995. *Halophiles.* Cold Spring Harbor Laboratory Press, Cold Spring Harbor, New York.

Deshpande A. and Sonar S. 1999. Bacterioopsin-triggered retinal biosynthesis is inhibited by bacteriorhodopsin formation in *Halobacterium salinarium*. *J. Biol. Chem.* **274:** 23535.

Enright A.J., Iliopoulos I., Kyrpides N.C., and Ouzounis C.A. 1999. Protein interaction maps for complete genomes based on gene fusion events. *Nature* **402:** 86.

Giaever G., Chu A.M., Ni L., Connelly C., Riles L., Veronneau S., Dow S., Lucau-Danila A., Anderson K., Andre B., Arkin A.P., Astromoff A., El-Bakkoury M., Bangham R., Benito R., Brachat S., Campanaro S., Curtiss M., Davis K., Deutschbauer A., Entian K.D., Flaherty P., Foury F., Garfinkel D.J., and Gerstein M., et al. 2002. Functional profiling of the *Saccharomyces cerevisiae* genome. *Nature* **418:** 387.

Gropp F. and Betlach M.C. 1994. The bat gene of *Halobacterium halobium* encodes a trans-acting oxygen inducibility factor. *Proc. Natl. Acad. Sci.* **91:** 5475.

Gygi S.P., Rist B., Gerber S.A., Turecek F., Gelb M.H., and Aebersold R. 1999. Quantitative analysis of complex protein mixtures using isotope-coded affinity tags. *Nat. Biotechnol.* **17:** 994.

Hoff W.D., Jung K.H., and Spudich J.L. 1997. Molecular mechanism of photosignaling by archaeal sensory rhodopsins. *Annu. Rev. Biophys. Biomol. Struct.* **26:** 223.

Ideker T., Thorsson V., Ranish J.A., Christmas R., Buhler J., Eng J.K., Bumgarner R., Goodlett D.R., Aebersold R., and

Hood L. 2001. Integrated genomic and proteomic analyses of a systematically perturbed metabolic network. *Science* **292:** 929.

Johnston M. 1987. A model fungal gene regulatory mechanism: The GAL genes of *Saccharomyces cerevisiae. Microbiol. Rev.* **51:** 458.

Jung K.H. and Spudich J.L. 1996. Protonatable residues at the cytoplasmic end of transmembrane helix-2 in the signal transducer HtrI control photochemistry and function of sensory rhodopsin I. *Proc. Natl. Acad. Sci.* **93:** 6557.

Kanehisa M. 2002. The KEGG database. *Novartis Found. Symp.* **247:** 91.

Kellis M., Patterson N., Endrizzi M., Birren B., and Lander E.S. 2003. Sequencing and comparison of yeast species to identify genes and regulatory elements. *Nature* **423:** 241.

Kokoeva M.V. and Oesterhelt D. 2000. BasT, a membrane-bound transducer protein for amino acid detection in *Halobacterium salinarum. Mol. Microbiol.* **35:** 647.

Larschan E. and Winston F. 2001. The *S. cerevisiae* SAGA complex functions in vivo as a coactivator for transcriptional activation by Gal4. *Genes Dev.* **15:** 1946.

Lee T.I., Rinaldi N.J., Robert F., Odom D.T., Bar-Joseph Z., Gerber G.K., Hannett N.M., Harbison C.T., Thompson C.M., Simon I., Zeitlinger J., Jennings E.G., Murray H.L., Gordon D.B., Ren B., Wyrick J.J., Tagne J.B., Volkert T.L., Fraenkel E., Gifford D.K., and Young R.A. 2002. Transcriptional regulatory networks in *Saccharomyces cerevisiae. Science* **298:** 799.

Lohr D., Venkov P., and Zlatanova J. 1995. Transcriptional regulation in the yeast GAL gene family: A complex genetic network. *FASEB J.* **9:** 777.

Luecke H., Schobert B., Lanyi J.K., Spudich E.N., and Spudich J.L. 2001. Crystal structure of sensory rhodopsin II at 2.4 angstroms: Insights into color tuning and transducer interaction. *Science* **293:** 1499.

Lundstrom J., Rychlewski L., Bujnicki J., and Elofsson A. 2001. Pcons: A neural-network-based consensus predictor that improves fold recognition. *Protein Sci.* **10:** 2354.

Marcotte E.M., Pellegrini M., Ng H.L., Rice D.W., Yeates T.O., and Eisenberg D. 1999. Detecting protein function and protein-protein interactions from genome sequences. *Science* **285:** 751.

McCready S. and Marcello L. 2003. Repair of UV damage in *Halobacterium salinarum. Biochem. Soc. Trans.* **31:** 694.

Mylin L.M., Bhat J.P., and Hopper J.E. 1989. Regulated phosphorylation and dephosphorylation of GAL4, a transcriptional activator. *Genes Dev.* **3:** 1157.

Mylin L.M., Johnston M., and Hopper J.E. 1990. Phosphorylated forms of GAL4 are correlated with ability to activate transcription. *Mol. Cell. Biol.* **10:** 4623.

Ng W.V., Ciufo S.A., Smith T.M., Bumgarner R.E., Baskin D., Faust J., Hall B., Loretz C., Seto J., Slagel J., Hood L., and DasSarma S. 1998. Snapshot of a large dynamic replicon in a halophilic archaeon: Megaplasmid or minichromosome? *Genome Res.* **8:** 1131.

Ng W.V., Kennedy S.P., Mahairas G.G., Berquist B., Pan M., Shukla H.D., Lasky S.R., Baliga N.S., Thorsson V., Sbrogna J., Swartzell S., Weir D., Hall J., Dahl T.A., Welti R., Goo Y.A., Leithauser B., Keller K., Cruz R., Danson M.J., Hough D.W., Maddocks D.G., Jablonski P.E., Krebs M.P., Angevine C.M., and Dale H. 2000. From the cover: Genome sequence of *Halobacterium* species NRC-1. *Proc. Natl. Acad. Sci.* **97:** 12176.

Oesterhelt D. 1995. Structure and function of halorhodopsin. *Isr. J. Chem.* **35:** 475.

Oesterhelt D. and Stoeckenius W. 1973. Functions of a new photoreceptor membrane. *Proc. Natl. Acad. Sci.* **70:** 2853.

Oren A. 1999. Bioenergetic aspects of halophilism. *Microbiol. Mol. Biol. Rev.* **63:** 334.

Overbeek R., Fonstein M., D'Souza M., Pusch G.D., and Maltsev N. 1999. The use of gene clusters to infer functional coupling. *Proc. Natl. Acad. Sci.* **96:** 2896.

Papamichos-Chronakis M., Petrakis T., Ktistaki E., Topalidou I., and Tzamarias D. 2002. Cti6, a PHD domain protein, bridges the Cyc8-Tup1 corepressor and the SAGA coactivator to overcome repression at GAL1. *Mol. Cell* **9:** 1297.

Park J. and Teichmann S.A. 2001. Mapping protein family interactions: Intramolecular and intermolecular protein family interaction repertoires in the PDB and yeast. *J. Mol. Biol.* **307:** 929.

Peck R.F., DasSarma S., and Krebs M.P. 2000. Homologous gene knockout in the archaeon *Halobacterium salinarum* with ura3 as a counterselectable marker. *Mol. Microbiol.* **35:** 667.

Peck R.F., Echavarri-Erasun C., Johnson E.A., Ng W.V., Kennedy S.P., Hood L., DasSarma S., and Krebs M.P. 2001. brp and blh are required for synthesis of the retinal cofactor of bacteriorhodopsin in *Halobacterium salinarum. J. Biol. Chem.* **276:** 5739.

Peng G. and Hopper J.E. 2000. Evidence for Gal3p's cytoplasmic location and Gal80p's dual cytoplasmic-nuclear location implicates new mechanisms for controlling Gal4p activity in *Saccharomyces cerevisiae. Mol. Cell. Biol.* **20:** 5140.

———. 2002. Gene activation by interaction of an inhibitor with a cytoplasmic signaling protein. *Proc. Natl. Acad. Sci.* **99:** 8548.

Perazzona B. and Spudich J.L. 1999. Identification of methylation sites and effects of phototaxis stimuli on transducer methylation in *Halobacterium salinarum. J. Bacteriol.* **181:** 5676.

Perazzona B., Spudich E.N., and Spudich J.L. 1996. Deletion mapping of the sites on the HtrI transducer for sensory rhodopsin I interaction. *J. Bacteriol.* **178:** 6475.

Pieper U., Eswar N., Stuart A.C., Ilyin V.A., and Sali A. 2002. MODBASE, a database of annotated comparative protein structure models. *Nucleic Acids Res.* **30:** 255.

Reece R.J. 2000. Molecular basis of nutrient-controlled gene expression in *Saccharomyces cerevisiae. Cell Mol. Life Sci.* **57:** 1161.

Ren B., Robert F., Wyrick J.J., Aparicio O., Jennings E.G., Simon I., Zeitlinger J., Schreiber J., Hannett N., Kanin E., Volkert T.L., Wilson C.J., Bell S.P., and Young R.A. 2000. Genome-wide location and function of DNA binding proteins. *Science* **290:** 2306.

Ruepp A. and Soppa J. 1996. Fermentative arginine degradation in *Halobacterium salinarium* (formerly *Halobacterium halobium*): Genes, gene products, and transcripts of the arcRACB gene cluster. *J. Bacteriol.* **178:** 4942.

Shand R.F. and Betlach M.C. 1991. Expression of the bop gene cluster of *Halobacterium halobium* is induced by low oxygen tension and by light. *J. Bacteriol.* **173:** 4692.

Spudich E.N., Takahashi T., and Spudich J.L. 1989. Sensory rhodopsins I and II modulate a methylation/demethylation system in *Halobacterium halobium* phototaxis. *Proc. Natl. Acad. Sci.* **86:** 7746.

Spudich J.L. and Bogomolni R.A. 1984. Mechanism of colour discrimination by a bacterial sensory rhodopsin. *Nature* **312:** 509.

Spudich J.L. and Luecke H. 2002. Sensory rhodopsin II: Functional insights from structure. *Curr. Opin. Struct. Biol.* **12:** 540.

Storch K.F., Rudolph J., and Oesterhelt D. 1999. Car: A cytoplasmic sensor responsible for arginine chemotaxis in the archaeon *Halobacterium salinarum. EMBO J.* **18:** 1146.

Tatusov R.L., Natale D.A., Garkavtsev I.V., Tatusova T.A., Shankavaram U.T., Rao B.S., Kiryutin B., Galperin M.Y., Fedorova N.D., and Koonin E.V. 2001. The COG database: New developments in phylogenetic classification of proteins from complete genomes. *Nucleic Acids Res.* **29:** 22.

Yang C.F. and DasSarma S. 1990. Transcriptional induction of purple membrane and gas vesicle synthesis in the archaebacterium *Halobacterium halobium* is blocked by a DNA gyrase inhibitor. *J. Bacteriol.* **172:** 4118.

Yang C.F., Kim J.M., Molinari E., and DasSarma S. 1996. Genetic and topological analyses of the bop promoter of *Halobacterium halobium:* Stimulation by DNA supercoiling and non-B-DNA structure. *J. Bacteriol.* **178:** 840.

Zeitlinger J., Simon I., Harbison C.T., Hannett N.M., Volkert T.L., Fink G.R., and Young R.A. 2003. Program-specific distribution of a transcription factor dependent on partner transcription factor and MAPK signaling. *Cell* **113:** 395.

# Implications of Genomics for Public Health: The Role of Genetic Epidemiology

K.R. Merikangas
*National Institute of Mental Health, National Institutes of Health, Department of Health and Human Services, Bethesda, Maryland 20892*

Advances in molecular genetics have generated substantial progress in identifying the genetic basis of Mendelian diseases; however, the pace of discovery of genes for complex disorders has been less rapid (Glazier et al. 2002). The slow progress has generated considerable debate regarding optimal strategies and priorities for genetic studies of complex human disorders.

Linkage studies that led to the exciting discoveries of major genes underlying some of the most devastating human disorders, such as Huntington's disease and cystic fibrosis, have continued to be highly successful in identifying susceptibility genes for most of the Mendelian disorders. However, linkage strategies have been less successful when applied to complex diseases. In a summary of linkage findings for 6 of 32 complex diseases with large samples (i.e., asthma, bipolar affective disorder, psoriasis, schizophrenia, type I diabetes, and type II diabetes), only 2 of 52 studies confirmed linkage (type I diabetes and psoriasis). Increased success in the replication of linkage studies was associated with two study features: an increase in the sample size and ethnically homogeneous samples (Altmuller et al. 2001).

Although association studies have been shown to be far more powerful in detecting susceptibility genes for complex disorders (Risch and Merikangas 1996), few associations have been replicated consistently (Hirschhorn et al. 2002). Reasons for the lack of replicable findings have been widely discussed, but one of the most compelling explanations is the remarkably high false-positive rates induced by the extremely low prior probability of associations (i.e., thousands of genes may be involved in any complex disorder) (Wacholder et al. 2002).

## IMPEDIMENTS TO GENE IDENTIFICATION FOR COMPLEX DISEASES

### Lack of Evidence for Role of Major Genes

Although it is assumed that there is a priori evidence for the role of genes in disease etiology before embarking on large-scale gene-finding efforts, there are a surprising number of complex diseases for which gene discovery studies have been mounted without such evidence. Familial recurrence risks, measured by $\lambda$ (i.e., the ratio of the risk of disease in relatives of affected cases to the population prevalence or to relatives of controls, based on controlled family studies of first-degree relatives) should be well-established before searching for specific susceptibility genes (Risch 1990). In fact, successful identification of genes for several complex disorders has resulted from discrimination of disease subtypes based on disease clustering within families (e.g., type I vs. type II diabetes, early vs. late onset of breast cancer and of Alzheimer's disease).

Although the absolute risk of disease in first-degree relatives of cases may be low, elevated risk with respect to the population prevalence (as measured by $\lambda$) can indicate the importance of genetic susceptibility. For example, although the absolute risk to relatives of multiple sclerosis cases is only 4%, the risk to relatives is 20 times greater than the general population risk. In fact, twin, adoption, and half-sibling studies have shown that the familial clustering of multiple sclerosis is almost completely attributable to shared genes (Sadovnick 2002). Likewise, although the $\lambda$ value for breast cancer is of relatively low magnitude, empirical evidence implicates a genetic basis for the familial aggregation and age of onset of most cancers (Risch 2001).

### Role of Environmental Factors

There is abundant evidence for the significance of environmental contributions to complex diseases. Environmental exposure is prerequisite to numerous complex disorders such as acquired immunodeficiency syndrome (AIDS), alcohol and nicotine dependence, and cervical cancer (Akerblom et al. 2002; ACS 2003). The marked increase in the incidence of several complex disorders during the past few decades and pronounced regional differences in population prevalence (e.g., type II diabetes, AIDS, and alcohol and nicotine dependence) also strongly implicate the role of environmental factors (Kenny et al. 1995; Rewers and Hamman 1995). For example, there is a dramatic elevation (i.e., fivefold) in obesity and type II diabetes in U.S. Pima Indians compared with their Mexican counterparts (Ravussin et al. 1994). Because the genetic background of these two groups appears to be identical (K. Kidd et al., pers. comm.), the differential rates of diabetes must be attributed to environmental variation. Despite evidence that environmental factors play a major role in numerous diseases, the identification of these specific environmental factors has been even more elusive than the identification of genes. This is primarily attributable to the indirect nature of the evidence that environmental factors play a key etiologic role.

On the basis of family and twin studies, it is possible to identify the relative contribution of genetic and environmental factors. Evidence for environmental influences includes: discordant monozygotic twins; greater than a 50% diminution in risk in relatives than expected by degree of genetic relatedness; and greater concordance for monozygotic twins reared together compared to those reared apart. For example, our recent review of a series of complex diseases revealed that multiple sclerosis, type I diabetes, autism, and schizophrenia are more strongly influenced by genetic risk factors; breast cancer, type II diabetes, and Alzheimer's disease are vulnerable to both genetic and environmental susceptibility; and environmental exposures are necessary for the development of alcohol and drug abuse, cervical cancer, and AIDS, but genes may determine the extent to which exposure is associated with disease, infection, or cancer (Merikangas and Risch 2003).

## CONTRIBUTION OF EPIDEMIOLOGY TO THE FUTURE OF GENETICS

The importance of epidemiology to the future of genetics has been described by numerous geneticists and epidemiologists who conclude that the best strategy for gene identification will ultimately involve large epidemiologic studies in diverse populations (Risch and Merikangas 1996; Khoury and Yang 1998; Risch 2000; Thomas 2000; Yang et al. 2000; Merikangas 2002; Merikangas et al. 2002; Khoury et al. 2003). It is likely that population-based association studies will assume increasing importance in translating the products of genomics to public health (Risch and Merikangas 1996). The term "human genome epidemiology" was coined by Khoury (Khoury et al. 2003) to denote the emerging field that employs systematic applications of epidemiologic methods in population-based studies to study the impact of human genetic variation on health and disease.

### Genetic Epidemiology

Applying the tools of genetic epidemiology, particularly when coupled with continued progress in basic biomedical sciences, is likely to be one of the most fruitful approaches to resolving etiologic factors underlying disorders and translating the progress being made in genomics to the public (Merikangas 2002). Figure 1 shows the classic epidemiologic triangle that illustrates the major focus of epidemiologic investigations: the products of the interaction between the host, an infectious or other type of agent, and the environment that promotes the exposure (Gordis 2000). The field of genetic epidemiology focuses on the role of genetic factors that interact with other domains of risk to enhance vulnerability or protection against disease (King et al. 1984; Khoury et al. 1993; Ellsworth and Manolio 1999a,b). It is quite conceivable that several combinations of these risk factors could produce similar phenotypes in susceptible individuals. The test for epidemiology over the next decades will be the extent to which its tools can be refined to capture these situations. The key aspects of the epidemiologic method that discriminate it from traditional genetic study designs are described below.

***Study designs.*** Epidemiologic studies generally proceed from retrospective case-control studies designed to develop specific hypotheses to prospective cohort studies that can test causal associations. The major goal of ana-

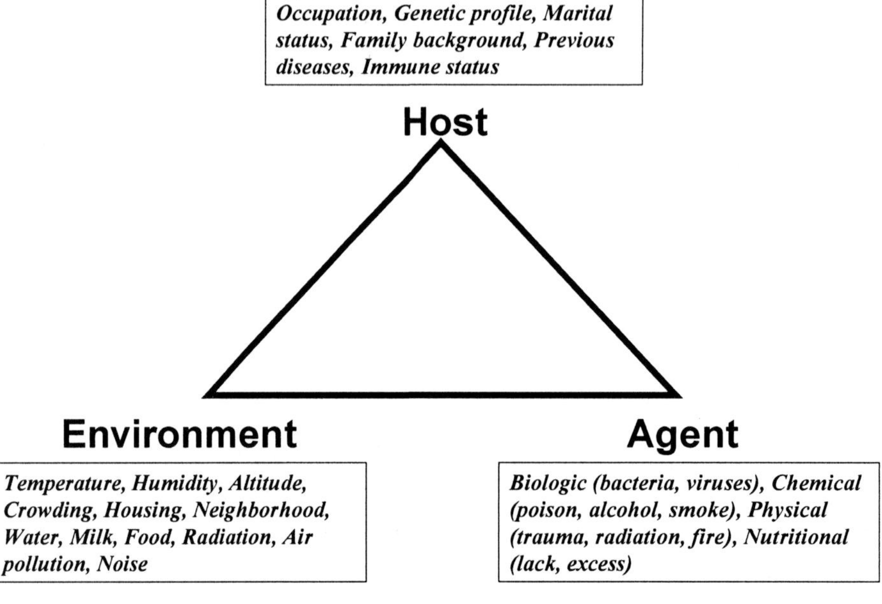


**Figure 1.** The epidemiologic triangle.

lytic epidemiology is to identify risk and protective factors and their causal links to disease, with the ultimate goal of disease prevention. Genetic epidemiology employs traditional epidemiologic study designs to identify explanatory factors for aggregation in groups of relatives ranging from twins to migrant cohorts. In general, study designs in genetic epidemiology either control for genetic background while letting the environment vary (e.g., migrant studies, half siblings, separated twins) or control for the environment while allowing variance in the genetic background (e.g., siblings, twins, adoptees– nonbiologic siblings). Since epidemiology has developed sophisticated designs and analytic methods for identifying disease risk factors, these methods can now be extended to include both genes and environmental factors as gene identification proceeds (Yang and Khoury 1997). The tools of genetic epidemiology will be employed in the era of genomics *to derive estimates of the population distribution of disease genes, to test modes of disease transmission in systematic samples that are representative of the population, and to identify sources of gene–environment interactions for diseases.*

Despite the critical role of the traditional case-control family studies that served as the core of human genetics during the earlier part of this century, systematically ascertained family studies have largely been abandoned in favor of studies designed solely for gene finding. Although the latter approaches do play an important role in studying Mendelian diseases, they are far less informative for complex diseases, since it is impossible to incorporate sources of heterogeneity into the study design, particularly because most of the factors that lead to genetic complexity are still unknown.

Although these sampling approaches do increase the power for detecting genes, they have diminished the generalizability of the study findings, and contribute little else to the knowledge base if genes are not discovered. It may be advisable in the future to collect both families and controls from representative samples of the population in order to enable estimation of population risk parameters, enhance generalizability, refine definitions of complex phenotypes, and examine the specificity of endophenotypic transmission. The *nested case-control study* built on an established cohort is likely to be a key study design in identifying the joint contribution of genetic and environmental risk factors. Prospective cohort studies are also valuable sources of diagnostic stability, causal associations between risk factors and disease, and developmental aspects of complex disorders. Langholz et al. (1999) describe some of the world's prospective cohort studies that may serve as a basis for studies of gene–disease associations or gene–environment interactions. Finally, the half-sibling approach may eventually replace the adoption paradigm to investigate genetic and environmental effects because of the recent trends toward selective adoption and the diminishing frequency of adoptions in the US and in numerous other countries.

***Population-based studies.*** There are several reasons that population-based studies will be critical to the future of genetics. First, the prevalence of newly identified polymorphisms, whether single nucleotide polymorphisms (SNPs) or other variants, especially in particular population subgroups, is not known. Second, current knowledge of genes as risk factors is based nearly exclusively on clinical and nonsystematic samples. The significance of the susceptibility alleles that have been identified for cancer, heart disease, diabetes, and so forth is relatively unknown in the population at large. To provide accurate risk estimates, the next stage of research needs to move beyond samples identified through affected individuals to the general population in order to obtain estimates of the risk of specific polymorphisms for the population as a whole. Third, identification of risk profiles will require very large samples to assess the significance of vulnerability genes with relatively low expected population frequencies. Fourth, similar to the role of epidemiology in quantifying risk associated with traditional disease risk factors, applications of human genome epidemiology can provide information on the specificity, sensitivity, and impact of genetic tests to inform science and the individual (Yang et al. 2000).

Several types of risk estimates are used in public health. The most common is *relative risk*, defined as the magnitude of the association between an exposure and disease. It is independent of the prevalence of the exposure. The *absolute risk* is the overall probability of developing a disease in an individual or in a particular population (Gordis 2000). The *attributable risk* is the difference in the risk of the disease in those exposed to a particular risk factor compared to the background risk of a disease in a population (i.e., in the unexposed). The *population-attributable risk* relates to the risk of a disease in a total population (exposed and unexposed) and indicates the amount the disease can be reduced in a population if an exposure is eliminated. The population-attributable risk depends on the prevalence of the exposure, or in the case of genes, the gene frequency. *Genetic attributable risk* would indicate the proportion of a particular disease that would be eliminated if a particular gene or genes were not involved.

Because genetic polymorphisms involved in complex diseases are likely to be nondeterministic (i.e., the marker predicts neither disease nor non-disease with certainty), traditional epidemiologic risk factor designs can be used to estimate their impact (Ellsworth and Manolio 1999b). As epidemiologists add genes to their risk equations, it is likely that the contradictory findings from studies that have generally employed solely environmental risk factors, such as diet, smoking, and alcohol use, will be resolved. Likewise, the studies that seek solely to identify genes will also continue to be inconsistent without considering the effects of nongenetic biologic parameters as well as environmental factors that contribute to the diseases of interest.

***Identification of environmental factors.*** The identification of gene–environment interactions is one of the most important future goals of genetic epidemiology. Study designs and statistical methods should focus increasingly on gene–environment interaction (Smith and

Day 1984; Ottman 1990; Hwang et al. 1994; Foppa and Spiegelman 1997; Yang and Khoury 1997; Khoury 1998; Garcia-Closas and Lubin 1999). There is accumulating evidence that gene–environment interaction will underlie many of the complex human diseases. Some examples include inborn errors of metabolism, individual variation in response to drugs (Nebert et al. 1999), substance use disorders (Dick et al. 2001; Heath et al. 2001), and the protective influence of a deletion in the CC-chemokine receptor gene on exposure to HIV (Michael 1999; Guttmacher and Collins 2002). Over the next decades, it will be important to identify and evaluate the effects of specific environmental factors on disease outcomes and to refine measurement of environmental exposures to evaluate specificity of effects. Once susceptibility genes have been identified, case-control studies defined by vulnerability genotypes can be employed to identify environmental factors that lead to either the potentiation or suppression of particular genotypes.

Evidence for environmental factors can be gleaned from migration studies, a particularly powerful tool to identify environmental exposures that may modify gene expression. Migration studies have shown that the rates of breast cancer, heart disease, and type II diabetes have increased substantially in several subgroups who migrated to new cultural settings (Shaper and Elford 1991; Hanley et al. 1995; Trevisan et al. 1998).

Intervention studies are another method to test environmental etiology. Systematic trials of weight loss and exercise among high-risk offspring of parents with type II diabetes have shown a significant decrease in incidence among those in the weight loss group (Knoller et al. 2002). The increased incidence of type II diabetes among offspring exposed prenatally to maternal diabetes also suggests that intrauterine exposure should be incorporated into genetic studies of this disorder (Sobngwi et al. 2003).

## ISSUES IN THE TRANSLATION OF GENOMICS TO THE PUBLIC

Scientists, health care providers, ethicists, regulators, patient groups, and the pharmaceutical industry are challenged by the rapid changes in the science and technology of genetic research. Researchers and policymakers are continually updating and ensuring that they keep pace with the rapidly evolving field by instilling ethically sound, workable guidelines for both the research and clinical applications of human genetics. As a result of the increased education, discussion, and understanding of the relevant issues, one area that will be particularly important in genetic epidemiology research is informed consent. It is important to address consent issues in a reasoned, practical, and consistent way, including input from patients and their families, health care providers, ethicists, scientists, regulatory bodies, research sponsors, and the lay community. Responsibility for assessing issues related to family consent for research should remain with local investigators, ethics boards, and study sponsors. A "one-size-fits-all" perspective in the form of new regulations, for example, would likely be a disservice to all (Renegar et al. 2001). Specific issues that need to be considered in applying genomics in the general population include public comprehension of genetic information, confidentiality, and accurate risk estimation for complex disorders.

### Confidentiality

There has been substantial attention to ethical issues as part of the activities involved in mapping the human genome (Human Genome Project Information 2003: http://www.ornl.gov/sci/techresources/Human_Genome/elsi/elsi.shtml). Principles of ethical research, confidentiality, and communication have been well-established. There are several aspects of human subject protection that differ when collecting DNA from samples of individuals recruited from the general population compared with from the disease-based samples for which genetic counseling has been most sought. With respect to complex disorders, and mental disorders in particular, several ethical issues require attention including (1) protection of the privacy of the proband in the ascertainment process; (2) permission to contact other family members; (3) specific problems associated with informed consent such as the ability of an individual in acute stages of illness to provide informed consent; and (4) therapeutic intervention (Alexander 2003). Successful and unsuccessful approaches that have been applied in ongoing population-based studies may ultimately be used to develop standards for the ethical application of genomics in general population samples.

### Comprehension

One major impediment to the application of the tools of genomics to public health is a lack of public understanding of the implications of genetic markers. Although there has been a dramatic increase in awareness of the role of genes in human behavior stemming from the accomplishments in genomics, some evidence suggests that there has not been a concomitant increase in public comprehension of the role of genes in disease etiology. For example, comparison of the results of a telephone survey on basic principles of genetics for some common human diseases administered in 1990 and 2000 showed no difference in accuracy between administration in 1990 and 2000 (Institute for Social Research, University of Michigan) (Weiss 2003). Moreover, the public tends to attribute far more of the causality of human behavior and complex diseases to genes than warranted by empirical evidence. As knowledge regarding specific roles of genes in complex diseases evolves, public education regarding risk estimation will be critical in the translation of genomics to public health.

### Risk Estimation for Complex Diseases

The lack of knowledge regarding genes underlying complex diseases reduces our ability to communicate accurate risk information associated with specific alleles at candidate gene loci. A shift to a risk factor-based epidemiologic approach in evaluating the roles of candidate genes will be necessary to provide accurate information regarding the risk associated with particular gene markers (Slooter and van Duijn 1997). For example, differential pathways to Alzheimer's disease illustrate the importance of understanding differences between *deterministic genes* (β-amyloid precursor, presenilin-1 and -2), and the *susceptibility* genes such as apolipoprotein-E ε-4 (APOE ε4), which increase the risk of Alzheimer's disease in a dose-dependent fashion (Tol et al. 1999). Whereas individuals with mutations in deterministic genes appear to have nearly a 100% chance (i.e., fully penetrant) for the development of Alzheimer's disease, those with the APOE ε4 allele have an increased risk for this disease, but only in combination with other background genes or environmental exposures. As research defines specific environmental exposures, prevention may ultimately be a realistic goal for those individuals who harbor the APOE ε4 allele (Farrer et al. 1997; Munoz and Feldman 2000; Kivipelto et al. 2001, 2002).

### Gap between Health Knowledge and Behavior

One of the greatest impediments to the application of genomics in public health is the gap between health knowledge and behavior. Even though there is widespread knowledge regarding the environmental exposures that confer increased disease risk and even precipitate mortality, it has been extremely difficult to identify effective strategies for inducing changes in negative health behaviors. For example, prevention of exposure to cigarettes would have a dramatic effect on the incidence of respiratory diseases, several forms of cancer, and cardiovascular disease; yet there is still a substantial proportion of individuals who start to smoke despite this knowledge. An important area of future research will involve studies of health behavior cognition, behavioral motivation, and effective methods for inducing changes in human behavior in order to maximize prevention efforts for malleable risk behaviors.

## REFERENCES


ACS. 2003. *Cancer facts and figures 2003*, p. 1. American Cancer Society, Atlanta, Georgia.

Akerblom H.K., Vaarala O., Hyoty H., Ilonen J., and Knip M. 2002. Environmental factors in the etiology of type 1 diabetes. *Am. J. Med. Genet.* **115:** 18.

Alexander D.R. 2003. Uses and abuses of genetic engineering. *Postgrad. Med. J.* **79:** 249.

Altmuller J., Palmer L.J., Fischer G., Scherb H., and Wjst M. 2001. Genomewide scans of complex human diseases: True linkage is hard to find. *Am. J. Hum. Genet.* **69:** 936.

Dick D.M., Rose R.J., Viken R.J., Kaprio J., and Koskenvuo M. 2001. Exploring gene-environment interactions: Socioregional moderation of alcohol use. *J. Abnorm. Psychol.* **110:** 625.

Ellsworth D.L. and Manolio T.A. 1999a. The emerging importance of genetics in epidemiologic research. II. Issues in study design and gene mapping. *Ann. Epidemiol.* **9:** 75.

———. 1999b. The emerging importance of genetics in epidemiologic research. III. Bioinformatics and statistical genetic methods. *Ann. Epidemiol.* **9:** 207.

Farrer L.A., Cupples L.A., Haines J.L., Hyman B., Kukull W.A., Mayeux R., Myers R.H., Pericak-Vance M.A., Risch N., and van Duijn C.M. 1997. Effects of age, sex, and ethnicity on the association between apolipoprotein E genotype and Alzheimer disease. A meta-analysis. APOE and Alzheimer Disease Meta Analysis Consortium. *J. Am. Med. Assoc.* **278:** 1349.

Foppa I. and Spiegelman D. 1997. Power and sample size calculations for case-control studies of gene-environment interactions with a polytomous exposure variable. *Am. J. Epidemiol.* **146:** 596.

Garcia-Closas M. and Lubin J.H. 1999. Power and sample size calculations in case-control studies of gene-environment interactions: Comments on different approaches. *Am. J. Epidemiol.* **149:** 689.

Glazier A.M., Nadeau J.H., and T.J. Aitman T.J. 2002. Finding genes that underlie complex traits. *Science* **298:** 2345.

Gordis L. 2000. *Epidemiology*. W.B. Saunders, Philadelphia, Pennsylvania.

Guttmacher A.E. and Collins F.S. 2002. Genomic medicine - A primer. *N. Engl. J. Med.* **347:** 1512.

Hanley A.J., Choi B.C., and Holowaty E.J. 1995. Cancer mortality among Chinese migrants: A review. *Int. J. Epidemiol.* **24:** 255.

Heath A.C., Whitfield J.B., Madden P.A., Bucholz K.K., Dinwiddie S.H., Slutske W.S., Bierut L.J., Statham D.B., and Martin N.G. 2001. Towards a molecular epidemiology of alcohol dependence: Analysing the interplay of genetic and environmental risk factors. *Br. J. Psychiatry* (suppl.) **40:** s33.

Hirschhorn J.N., Lohmueller K., Byrne E., and Hirschhorn K. 2002. A comprehensive review of genetic association studies. *Genet. Med.* **4:** 45.

Human Genome Project Information. 2003. Ethical, legal, and social issues: http://www.ornl.gov/sci/techresources/Human_Genome/elsi/elsi.shtml

Hwang S.J., Beaty T.H., Liang K.Y., Coresh J., and Khoury M.J. 1994. Minimum sample size estimation to detect gene-environment interaction in case-control designs. *Am. J. Epidemiol.* **140:** 1029.

Kenny S.J., Aubert R.E., and Geiss L.S. 1995. Prevalence and incidence of non-insulin-dependent diabetes. In *Diabetes in America,* 2nd edition, p. 47. National Diabetes Data Group, National Institutes of Health, National Institute of Diabetes and Digestive and Kidney Diseases, NIH Publication No. 95-1468.

Khoury, M.J. 1998. Genetic epidemiology. In *Modern epidemiology*, 2nd edition (ed. K.J. Rothman and S. Greenland), p. 609. Lippincott-Raven, Philadelphia, Pennsylvania.

Khoury M.J. and Yang Q. 1998. The future of genetic studies of complex human disease. An epidemiologic perspective. *Epidemiology* **9:** 350.

Khoury M.J., Beaty T.H., and Cohen B.H. 1993. *Fundamentals of genetic epidemiology*. Oxford University Press, New York.

Khoury M.J., McCabe L.L., and McCabe E.R. 2003. Population screening in the age of genomic medicine. *N. Engl. J. Med.* **348:** 50.

King M.C., Lee G.M., Spinner N.B., Thomson G., and Wrensch M.R. 1984. Genetic epidemiology. *Annu. Rev. Public Health* **5:** 1

Kivipelto M., Helkala E.L., Laakso M.P., Hanninen T., Hallikainen M., Alhainen K., Soininen H., Tuomilehto J., and Nissinen A. 2001. Midlife vascular risk factors and Alzheimer's disease in later life: Longitudinal, population based study. *Br. Med. J.* **322:** 1447.

Kivipelto M., Helkala E.L., Laakso M.P., Hanninen T., Hallikainen M., Alhainen K., Iivonen S., Mannermaa A., Tuomilehto J., Nissinen A., and Soininen H. 2002. Apolipoprotein E ε4 allele, elevated midlife total cholesterol level, and high midlife systolic blood pressure are independent risk factors for late-life Alzheimer disease. *Ann. Intern. Med.* **137:** 149.

Knoller W.C., Barrett-Connor E., Fowler S.E., Hamman R.F., Lachin J.M., Walker E.A., and Nathan D.M. 2002. Diabetes

Prevention Research Group. Reduction in the incidence of type II diabetes with lifestyle intervention or metformin. *N. Engl. J. Med.* **346:** 393.

Langholz B., Rothman N., Wacholder S., and Thomas D.C. 1999. Cohort studies for characterizing measured genes. *J. Natl. Cancer Inst. Monogr.* **26:** 39.

Merikangas K.R. 2002. Genetic epidemiology: Bringing genetics to the population—The NAPE Lecture 2001. *Acta Psychiatr. Scand.* **105:** 3.

Merikangas K.R. and Risch N. 2003. Genomic priorities and public health. *Science* **302:** 599.

Merikangas K.R., Chakravarti A., Moldin S.O., Araj H., Blangero J., Burmeister M., Crabbe J.C.J., DePaulo J.R.J., Foulks E., Freimer N.B., Koretz D.S., Lichtenstein W., Mignot E., Reiss A.L., Risch N.J., and Takahashi J. 2002. Future of genetics of mood disorders research: Workgroup on genetics for NIMH strategic plan for mood disorders. *Biol. Psychiatry* **52:** 457.

Michael N.L. 1999. Host genetic influences on HIV-1 pathogenesis. *Curr. Opin. Immunol.* **11:** 466.

Munoz D.G. and Feldman H. 2000. Causes of Alzheimer's disease. *Can. Med. Assoc. J.* **162:** 65.

Nebert D.W., Ingelman-Sundberg M., and Daly A.K. 1999. Genetic epidemiology of environmental toxicity and cancer susceptibility: Human allelic polymorphisms in drug-metabolizing enzyme genes, their functional importance, and nomenclature issues. *Drug Metab. Rev.* **31:** 467.

Ottman R. 1990. An epidemiologic approach to gene-environment interaction. *Genet. Epidemiol.* **7:** 177.

Ravussin E., Valencia M.E., Esparza J., Bennett P.H., and Schulz L.O. 1994. Effects of a traditional lifestyle on obesity in Pima Indians. *Diabetes Care* **17:** 1067.

Renegar G., Rieser P., and Manasco P. 2001. Family consent and the pursuit of better medicines through genetic research. *J. Contin. Educ. Health Prof.* **21:** 265.

Rewers M.R. and Hamman R.F. 1995. Risk factors for non-insulin-dependent diabetes. In *Diabetes in America,* 2nd edition, p. 179. National Diabetes Data Group, National Institutes of Health, National Institute of Diabetes and Digestive and Kidney Diseases, NIH Publication No. 95-1468.

Risch N. 1990. Linkage strategies for genetically complex traits. I. Multilocus models. *Am. J. Hum. Genet.* **46:** 222.

———. 2000. Searching for genetic determinants in the new millenium. *Nature* **405:** 847.

———. 2001. The genetic epidemiology of cancer: Interpreting family and twin studies and their implications for molecular genetic approaches. *Cancer Epidemiol. Biomark. Prev.* **10:** 733.

Risch N. and Merikangas K.R. 1996. The future of genetic studies of complex human diseases. *Science* **273:** 1516.

Sadovnick A.D. 2002. The genetics of multiple sclerosis. *Clin. Neurol. Neurosurg.* **104:** 199.

Shaper A.G. and Elford J. 1991. Place of birth and adult cardiovascular disease: The British Regional Heart Study. *Acta Paediatr. Scand. Suppl.* **373:** 73.

Slooter A. and van Duijn C. 1997. Genetic epidemiology of Alzheimer disease. *Epidemiol. Rev.* **19:** 107.

Smith P.G. and Day N.E. 1984. The design of case-control studies: The influence of confounding and interaction effects. *Int. J. Epidemiol.* **13:** 356.

Sobngwi E., Boudou P., Mauvais-Jarvis F., Leblanc H., Velho G., Vexiau P., Porcher R., Hadjadj S., Pratley R., Tataranni P.A., Calvo F., and Gautier J.-F. 2003. Effect of a diabetic environment in utero on predisposition to type 2 diabetes. *Lancet* **361:** 1861.

Thomas D.C. 2000. Genetic epidemiology with a capital "E". *Genet. Epidemiol.* **19:** 289.

Tol J., Roks G., Slooter A.J., and van Duijn C.M. 1999. Genetic and environmental factors in Alzheimer's disease. *Rev. Neurol.* (suppl. 4) **155:** S10.

Trevisan R., Vedovato M., and Tiengo A. 1998. The epidemiology of diabetes mellitus. *Nephrol. Dial. Transplant.* (suppl. 8) **13:** 2.

Wacholder S., Chatterjee N., and Hartge P. 2002. Joint effect of genes and environment distorted by selection biases: Implications for hospital-based case-control studies. *Cancer Epidemiol. Biomark. Prev.* **11:** 885.

Weiss R. 2003. Science Notebook, May 19, p. 7. In *The Washington Post*, Washington, D.C.

WHO. 2002. Reducing risks, promoting healthy life. In *World Health Report 2002: Preventing risks and taking action*, chapter 7. World Health Organization.

Yang Q. and Khoury M.J. 1997. Evolving methods in genetic epidemiology III. Gene-environment interaction in epidemiologic research. *Epidemiol. Rev.* **19:** 33.

Yang Q., Khoury M.J., Coughlin S.C., Sun F., and Flanders W.D. 2000. On the use of population-based registries in the clinical validation of genetic tests for disease susceptibility. *Genet. Med.* **2:** 186.

# A Model System for Identifying Genes Underlying Complex Traits

D. DRAYNA,* U.-K. KIM,* H. COON,[†] E. JORGENSON,[‡] N. RISCH,[‡] AND M. LEPPERT[†]
*National Institute on Deafness and Other Communication Disorders, National Institutes of Health, Rockville, Maryland 20850; [†]University of Utah School of Medicine, Salt Lake City, Utah 84112; and [‡]Department of Genetics, Stanford University School of Medicine, Stanford, California 94305

Advances in genomics and positional cloning have allowed us to identify the genes that underlie all of the major Mendelian disorders in humans. However, despite many efforts, these methods have not been extended to complex disorders, which typically show non-Mendelian inheritance patterns and are thought to have nongenetic components in their etiology. Our goal has been to address this problem using a well-characterized model human trait, the inability to taste the substance phenylthiocarbamide (PTC). The inability to taste PTC was documented in a large portion of the population more than 70 years ago (Anonymous 1931) and was quickly shown to be genetic in nature (Blakeslee 1931; Snyder 1931). Despite initial conclusions that this trait demonstrated simple recessive inheritance, clear exceptions to this pattern gradually emerged, and large-scale epidemiologic studies suggested a significant non-Mendelian component in the transmission pattern (Olson et al. 1989; Reddy and Rao 1989). In addition, genetic linkage studies produced conflicting results, and initial reports of linkage to markers on chromosome 7q (Chautard-Freire-Maia 1974; Conneally et al. 1976) were followed by studies that failed to confirm this linkage (Spence et al. 1984). It also became clear that PTC taste ability was not a simple dichotomous trait in the population, but instead displayed a broad and continuous, albeit bimodal, distribution of taste sensitivity in the population (Kalmus 1958). Thus, PTC emerged as a phenotype that displayed many of the characteristics that have stymied traditional linkage and positional cloning studies of complex disorders in humans. Moreover, this phenotype has been exceptionally well characterized in the human population, aided by a reliable quantitative noninvasive measurement. This has facilitated a very large number of studies in populations worldwide, which cumulatively have reported phenotypes in over 130,000 individuals (Guo and Reed 2001). In addition, this phenotype has been shown to be associated with many dietary preferences and thus may have important health implications (Tepper 1998).

## GENETIC LINKAGE STUDIES

We undertook studies of this trait in a series of genetically well-characterized families that were originally ascertained for the construction of the normal human linkage map, known as the Utah CEPH families (Dausset et al. 1990). These medically normal families were chosen because they possess a structure that is especially powerful for linkage studies—large sibships (often 10 or more) and their parents and all four of their grandparents. Members of these families are among the most intensively genotyped individuals in the world. Genotype information at 3,000 to over 10,000 marker loci exists for each individual, and this information has been analyzed in a variety of linkage, association, and population studies (NIH/CEPH Collaborative Mapping Group 1992; Dib et al. 1996). In a systematic phenotypic evaluation of these families, we performed PTC taste testing in a standardized clinical setting, and set out to resolve the conflicts regarding the genetics of this trait.

A total of 27 families, consisting of 269 individuals, were phenotyped. The phenotype measure was a modification of the classical method (Kalmus 1958) in which subjects were required to distinguish a series of PTC solutions from plain water. A subject's taste threshold was defined as the most dilute solution of PTC that he could correctly sort when presented with three cups of PTC at a given concentration and three cups of water. Scores ranged from 14 (the most dilute solution, containing 1 μM PTC) to 1 (the most concentrated solution, containing 8 mM PTC). The distribution of scores in the Utah CEPH families is shown in Figure 1. A clear bimodal distribution was observed, corresponding to the classically defined tasters and non-tasters. However, the distribution is continuous, and a significant number of individuals fall into the range between these two groups and thus display intermediate taste sensitivity. In addition, the variance within the two groups is large. For example, the range within the taster group is about as large as that between the peaks of the tasters and non-tasters.

Using the preexisting genotypic data in these families, we performed linkage analyses with both parametric methods using a dichotomous phenotype classification (taster and non-taster), and non-parametric analyses using the full range of quantitative phenotypes. Initial linkage studies were performed in GENEHUNTER combining all families and scoring phenotype as a simple dichotomous trait, tasters and non-tasters. In this analysis, we observed linkage to distal chromosome 7q with a LOD score of 4.45, strong evidence for linkage. We also performed linkage analyses using the full range of quantitative values for PTC threshold phenotype. These were initially

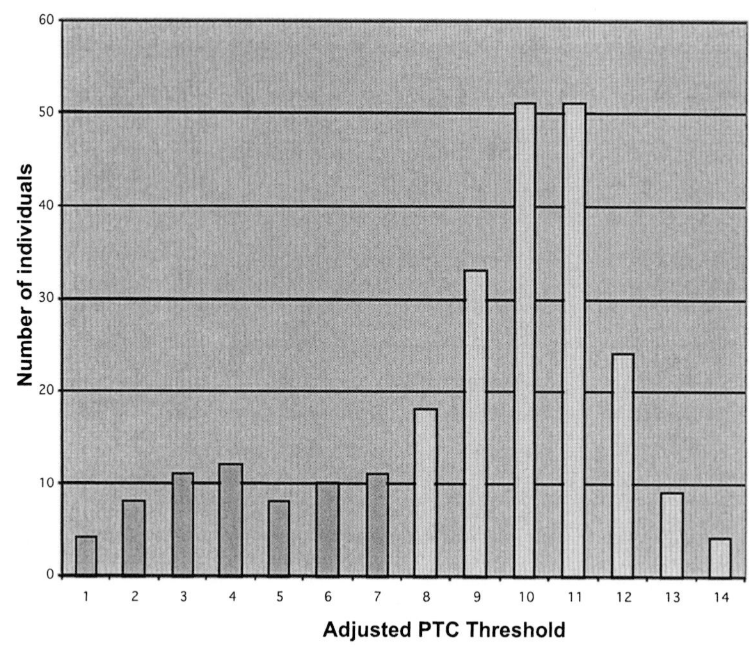


**Figure 1.** Distribution of PTC taste thresholds in the Utah CEPH population. Individuals classified as tasters by the method of Kalmus (1958) are indicated by light gray bars; those classified as non-tasters are indicated by dark gray bars.

performed using GENEHUNTER with variance components, and subsequently using SOLAR, again combining all families. The results of these analyses were highly consistent; linkage to markers on chromosome 7q was observed in both analyses, with a maximum LOD score of 8.85 (Drayna et al. 2003). When families were analyzed one at a time, none of the families was large enough to provide a significant LOD score for linkage individually. However, a large fraction of families (about 50%) showed highest scores genome-wide on chromosome 7q. Another fraction of families displayed highest scores on both 7q and another locus elsewhere in the genome, and a third small fraction displayed negative scores across the length of 7q. A second linkage analysis performed conditioning on the chromosome 7q linkage resulted in highest scores observed on the distal portion of the short arm of chromosome 16.

Linkage analysis in the Utah CEPH families was thus consistent with the view that a single major locus, acting as a largely Mendelian recessive allele and residing on chromosome 7q, is responsible for PTC taste blindness. In addition, a significant fraction of the phenotype is due to additional factors, one of which may reside on chromosome 16. This finding is in contrast to previous conflicting linkage results, which were performed on numerous families that were generally much smaller than the Utah families we have studied, and thus provided much less resolving power for linkage studies. These early studies nevertheless were successful in directing attention to the long arm of chromosome 7, at the same time that they demonstrated aspects of complex inheritance for this trait.

## FINDING THE CHROMOSOME 7 GENE

We initially focused our efforts on identifying the gene residing on chromosome 7q. We performed additional genotyping for high-resolution linkage and haplotype studies in all of the CEPH families that showed linkage only on chromosome 7q and nowhere else in the genome. An example of the results of this genotyping is shown in Figure 2. The assumption for this work was that in these families, PTC taste blindness behaved as a simple Mendelian recessive trait, and recombination breakpoints localized the causative gene to an interval of ~4 cM, bounded by D7S684 on the proximal side and D7S498 on the distal side. This region contains over 5 Mb of genomic DNA, and ~156 known and predicted genes. Among these is the gene encoding the KEL blood group antigen, thus confirming previous linkage results between PTC and the KEL blood group locus (Chautard-Freire-Maia 1974; Conneally et al. 1976). To further refine the location of this gene, we performed additional genotyping with SNPs across this region and further refined the recombination breakpoints in these families to a 2.6-Mb interval, from 139.2 Mb to 141.8 Mb on the chromosome 7 genomic sequence. Despite this reduction in size of the interval, many genes remained as candidates. Of particular interest were a number of candidate genes involved in chemosensory perception. These included 7 members of the TAS2R bitter taste receptor gene family (Adler et al. 2000), and 9 genes annotated as Odorant Receptor (OR)-like, indicating they were candidate odorant receptors (Buck and Axel 1991). Because both of these gene families encode 7 transmem-

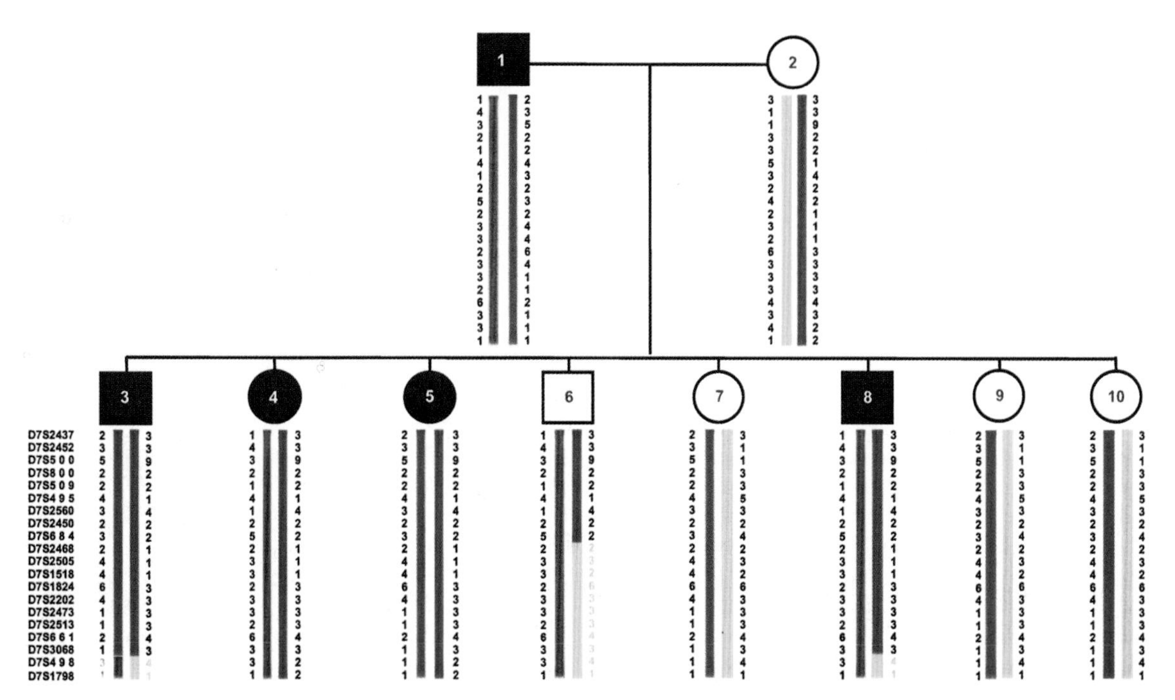


**Figure 2.** Haplotype analysis in Utah CEPH family showing linkage to chromosome 7q. Chromosome 7 marker loci typed are listed at left. The four parental chromosome segments are indicated in different colors. Within this family, the minimal region segregating with the phenotype is bounded by D7S684 on the proximal side and by D7S498 on the distal side.

brane domain G-protein-coupled receptors, it was possible that genes described as OR-like were in fact involved in bitter taste perception. We sequenced these 16 candidate genes exhaustively in a group of 8 individuals, and many sequence differences were observed. One sequence difference, a G versus C residing in a putative member of the TAS2R bitter taste receptor family, segregated absolutely with the phenotype in the chromosome-7-linked CEPH families. However, given the large critical interval defined by recombination in the CEPH families, it was uncertain whether there were other differences in other genes that segregated equally well with the phenotype.

A solution to this problem was suggested when we typed a series of 49 SNPs distributed at an average of 50-kb resolution across the interval. We observed that unrelated non-tasters carried not only the same sequence difference in the TAS2R candidate gene, but they were also homozygous for SNP markers in the immediately surrounding region. This suggested a founder effect, giving rise to identity by descent within and around our candidate TAS2R gene. To better characterize this finding, we enrolled a second independent group of subjects (the NIH sample), consisting of 94 unrelated individuals of all races/ethnicities. Selecting unrelated individuals from both the CEPH and NIH samples, we measured differences in allele frequency between a group of 23 tasters and 22 non-tasters and found a subregion of 150 kb across which SNP markers showed statistically significant difference in allele frequency between tasters and non-tasters. We further analyzed this region in a group of 37 unrelated non-tasters, chosen from both the NIH and Utah CEPH populations. We typed a group of 60 SNPs at an average spacing of 50 kb, and found a subregion in which almost all of these 37 individuals were homozygous for a conserved haplotype, as would be predicted for a recessive allele. The 2 individuals who did not share this haplotype were individuals from Utah CEPH families that showed no linkage of the phenotype to chromosome 7. Among the 35 individuals who shared at least some of the haplotype, the minimal region shared by all of them was 29.6 kb in length. Bioinformatics and gene-finding efforts within this region revealed only one gene, the TAS2R bitter taste receptor containing the original sequence difference we observed in the Utah CEPH families.

## THE PTC TASTE RECEPTOR GENE

The PTC receptor gene consists of a single coding exon 1002 bp long, encoding a 333-amino acid 7-transmembrane domain G-protein-coupled receptor (GPCR). Three variable sites were observed in this gene, summarized in Table 1. In most individuals, these three sites were arrayed in two haplotypes, denoted the AVI and PAV haplotypes. These haplotypes were very strongly associated with phenotype as shown in Figure 3 and Table 2. Figure 3 shows that, consistent with Mendelian recessive inheritance, carrying two copies of the AVI haplotype is largely associated with non-taster status, whereas either one or two copies of the PAV haplotype are strongly associated with the taster phenotype. A modest heterozygote effect is apparent, and PAV/AVI heterozygotes have a mean threshold about 1 unit less than the mean of PAV homozygotes, which is statistically significant. The statistical associations between haplotype and phenotype were highly significant, with chi-squared $p$ values as low as $10^{-17}$. The NIH sample showed a larger effect than the CEPH sample in all cases.

**Table 1.** Summary of Polymorphisms in the PTC Gene

| SNP | | | Amino acid | | Location in |
|---|---|---|---|---|---|
| position (b.p.) | allele | frequency | position (a.a.) | a.a. encoded | predicted protein |
| | C | .36 | | proline | 1st |
| 145 | G | .64 | 49 | alanine | intracellular loop |
| 785 | C | .38 | 262 | alanine | 6th transmembrane |
| | T | .62 | | valine | |
| 886 | G | .38 | 296 | valine | 7th transmembrane |
| | A | .62 | | isoleucine | |

In addition to the common PAV and AVI haplotypes, a third haplotype, designated AAV, was observed at lower frequency. Note this haplotype differs from the PAV haplotype by a single amino acid difference, a proline versus alanine at residue 49. This creates a significant difference in phenotype. In a simple dichotomous phenotypic classification, AVI/AAV heterozygotes are almost evenly divided between tasters and non-tasters. In comparison, we would expect AVI/PAV heterozygotes, which differ from AVI/AAV heterozygotes by one amino acid difference in one copy of the gene, to be all tasters. Further analysis showed that all of the bimodality of the taste threshold distribution was explained by these haplotypes, and that these haplotypes explained from 55% (in the CEPH sample) to 85% (in the NIH sample) of the total variance in taste threshold score.

Overall, these data indicate that this TAS2R gene is the major determinant of PTC taste threshold. However, other factors are clearly involved. For example, Figure 3 shows three individuals who are homozygous for the non-taster haplotype AVI and have taste thresholds of 8 and above, clearly in the taster range. Conversely, among the AVI/PAV heterozygotes is one individual whose taste threshold is under 4, well into the non-taster range. Our linkage studies in the Utah CEPH families support the view that additional genes outside of chromosome 7q contribute to this phenotype, and that one such gene may reside on the short arm of chromosome 16.

### Gene Relationships

The TAS2R bitter taste receptor family is highly divergent. Among the 24 functional TAS2R genes predicted in the human genomic sequence, the average amino acid identity between any two members is ~25%. The major non-taster allele of the human PTC receptor gene is identical to the human TAS2R38 sequence (Gen Bank accession number AF494231), with the exception of nucleotide 557, which is an A (encoding Asn-186) in TAS2R38 and a T (encoding Ile-186) in the PTC receptor. The next closest relative is human TAS2R37, which displays 30% amino acid identity with the PTC receptor.

### Worldwide Distribution

The PTC taste phenotype has been studied in hundreds of populations worldwide. In an effort to measure correlations between phenotypes and genotypes in diverse populations, we studied variation in the PTC receptor gene in seven major populations: European, sub-Saharan African, Pakistani, Chinese, Korean, Japanese, and Southwest American Indians. In this study, we determined haplotypes in unrelated individuals by using PCR to amplify both copies of the PTC receptor gene from genomic DNA and cloning the PCR product en masse into a plasmid vector. This was then transformed into competent cells, and 10 transformants were picked from each PCR. The inserts in these clones, which each represented one allele or the other, were then sequenced. In doing so, we identified two additional haplotypes, designated AAI and PVI. Both of these haplotypes were observed only in individuals of sub-Saharan origin. The frequency of each of these five haplotypes in different populations is shown in Figure 4.

The worldwide distribution of haplotypes confirms the phenotype data gathered over the past 60 years and provides a number of additional insights into the history of these variants. Outside of sub-Saharan Africa, most pop-

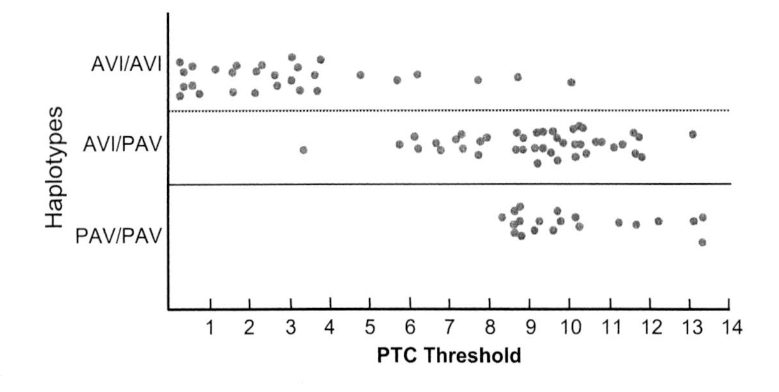


**Figure 3.** Correlation of haplotypes with PTC taste threshold phenotype. PTC taste thresholds of 90 subjects from the NIH population sorted by haplotype. AVI/PAV heterozygotes (mean threshold 8.81) display a significantly reduced PTC sensitivity compared to PAV homozygotes (mean threshold 10.00), indicating a heterozygote effect.

**Table 2.** Haplotype Association with Taste Phenotypes

| Haplotypes | Sample | No. of subjects non-tasters | tasters |
|---|---|---|---|
| AVI/AVI | Utah | 38 | 14 |
| | NIH | 21 | 0 |
| AVI/AAV | Utah | 10 | 7 |
| | NIH | 1 | 3 |
| */PAV | Utah | 3 | 108 |
| | NIH | 1 | 58 |

*Indicates any haplotype found in the sample. No AAV homozygotes were observed in either sample. Dichotomous assignment of subjects into tasters/non-tasters was done using the classic PTC threshold value of 8.0 (Kalmus 1958).

ulations surveyed contained only the PAV and AVI haplotypes, which we refer to as the major taster and major non-taster haplotypes, respectively. A sample of Southwest American Indians were virtually all homozygous for the major taster haplotype, consistent with the very low reported frequency of non-tasters in this population. The highest frequency of non-tasters reported worldwide is in the Asian subcontinent, and this finding is consistent with the very high frequency of the major non-taster allele we observed in the Pakistani population.

In contrast to the rest of the world, sub-Saharan Africa contains much more diversity. Five common haplotypes were observed, the PAV and AVI, plus AAV, AAI, and PVI. This finding is consistent with the view that most of the world's human genetic diversity exists in Africa and is of relatively ancient origin. It also supports the hypothesis that from this ancestral population, a small subpopulation emerged from Africa and spread to populate the remainder of the world.

### Primate Studies

The five haplotypes present in Africa differ from each other by varying degrees. For example, the major taster and major non-taster haplotypes differ at three positions, indicating that multiple steps separate these two haplotypes. In an effort to gain additional information on the origins of the different haplotypes, we sequenced the PTC receptor in a series of primates. This sample consisted of one animal each from chimpanzee, gorilla, orang, crab-eating macaque (an old world monkey), black-handed spider monkey (a new world monkey), and ring-tailed lemur (a prosimian). The human PCR primers failed to generate complete PCR products from prosimian and New World monkey DNAs, probably due to the great sequence divergence in this group of genes. Predicted protein sequences of the PTC gene in human (taster allele), chimp, and gorilla are shown in Figure 5. Although chimp and gorilla show a number of different amino acid differences from humans, both these animals were homozygous for the proline allele at residue 49, the alanine allele at residue 262, and the valine allele at residue 296, equivalent to the major taster haplotype in humans. Orang and crab-eating macaque were similar to these primates, carrying a number of amino acid differences but identical to the human taster allele at residues 49, 262, and 296 (data not shown).

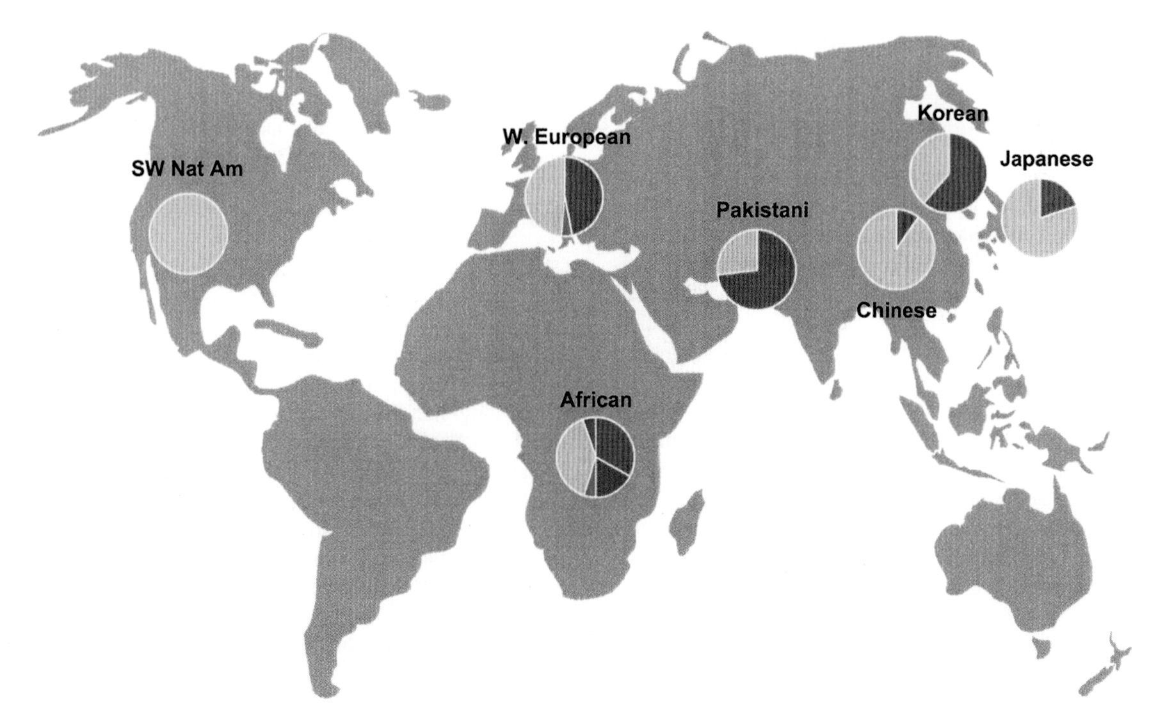


**Figure 4.** Frequency of five haplotypes of the PTC gene in populations worldwide. Frequencies were determined in samples of 12–48 chromosomes. (*Yellow*) PAV, major taster haplotype; (*red*) AVI, major non-taster haplotype; (*green*) AAI haplotype; (*light blue*) AAV haplotype; (*dark blue*) PVI haplotype.

Human

MLTLTRIRTVSYEVRSTFLFISVLEFAVGFLTNAFVFLVNFWDVVKRQPLSNSDCVLLCLSISRLFLHG
LLFLSAIQLTHFQKLSEPLNHSYQAIIMLWMIANQANLWLAACLSLLYCSKLIRFSHTFLICLASWVSR
KISQMLLGIILCSCICTVLCVWCFFSRPHFTVTTVLFMNNNTRLNWQIKDLNLFYSFLFCYLWSVPPFL
LFLVSSGMLTVSLGRHMRTMKVYTRNSRDPSLEAHIKALKSLVSFFCFFVISSCAAFISVPLLILWRDK
IGVMVCVGIMAACPSGHAAVLISGNAKLRRAVMTILLWAQSSLKVRADHKADSRTLC

Chimpanzee

MLTLTRIHTVSYEVRSTFLFISVLEFAVGFLTNAFVFLVNFWDVVKRQPLSNSDCVLLCLSISRLFLHG
LLFLSAIQLTHFQKLSEPLNHSYQAIIMLWMIANQANLWLAACLSLLYCSKLIRFSHTFLICLASWVSR
KISQMLLGIILCSCICTVLCVWCFFSRPHFTVTTVLFMNNNTRLNWQIKDLNLFYSFLFCYLWSVPPFL
LFLVSSGMLTVSLGRHMRTMKVYTRDSRDPSLEAHIKALKSLVSFFCFFVISSCAAFISVPLLILWRDK
IGVMVCVGIMAACPSGHAAVLISGNAKLRRAVTTILLWAQSSLKVRADHKADSRTLC

Gorilla

MLTLTRIRTVSYEVRSTFLFISVLEFAVGFLTNAFVFLVNFWDVVKRQPLSNSDCVLLCLSISRLFLHG
LLFLSAIQLTHFQKLSEPLNHSYQAIIMLWMIANQANLWLAACLSLLYCSKLIRFSHTFLICLASWVSR
KISQMLLGIILCSCICTVLCVWCFFSRPHFTVTTVLFMNNNTRLNWQIKDLNLFYSFLFCYLWSVPPFL
LFLVSSGMLTVSLGRHMRTMKVYIRDSRDPSLEAHIKALKSLVSFFCFFVISSCAAFISVPLLILWRDK
IGVMVCVGIMAACPSGHAAVLISGNAKLRRAVTTILLWAQSSLKVRADHKADSRTPC

**Figure 5.** Predicted protein sequence of the PTC gene in one individual from human (taster allele), chimpanzee, and lowland gorilla. Amino acid residues in the human sequence that differ between the human taster and non-taster alleles are highlighted in green. Chimp and gorilla amino acid residues that differ from the human sequence are highlighted in orange.

### Evolutionary Considerations

The sense of bitter taste is believed to serve as a protection against the ingestion of toxic compounds in plant material, most of which are perceived as bitter. Although PTC has not been observed in nature, it is structurally related to a group of compounds that occur in cruciferous vegetables and are toxic in large quantities, with the thyroid as the primary organ affected. For example, cabbage contains the thyrotoxic compound goitrin, which, like many members of this chemical family, is not strongly perceived as bitter by PTC non-tasters. A question that arises is how the non-taster allele rose to such high frequencies in some populations given the apparent selective disadvantage this imposes. One possibility is suggested by the fact that the alterations present in the non-taster allele are not obviously disabling mutations, such as stop codons or deletions. This gives rise to the possibility that the PTC non-taster allele encodes a receptor that is fully functional for a different but as yet unknown bitter toxic substance. Subsequent investigation of DNA sequence variation may provide evidence for selection of the non-taster allele, as opposed to drift, that can account for its current high frequency. Additional studies would be required to test hypotheses about the nature of such positive selective forces.

### Lessons for Identifying Genes Underlying Complex Traits

Although a large fraction of the variance in PTC taste ability is attributable to this single locus on chromosome 7, the long-standing uncertainties regarding the genetics of PTC non-tasting suggest that this trait can serve as a model for common complex traits in humans, including some diseases. What lessons for complex disease gene discovery can we draw from the PTC example? First, an important advantage in our linkage studies was provided by the large nuclear families present in the Utah CEPH population, which produced strong statistical support for linkage and clarified the uncertainties from previous linkage studies. In addition, the worldwide distribution of PTC gene haplotypes is consistent with the view that the non-taster allele is of ancient origin, having arisen before the emergence of anatomically modern humans from Africa. Taking the non-taster phenotype as a model for a disease, this trait is consistent with the common disease/common variant hypothesis, which states that common medical disorders are the result of relatively ancient variants that are now common in the population. This is in contrast to an alternative hypothesis, which suggests that common diseases are due to the cumulative effect of many different mutations, each of which is uncommon and has arisen more recently in the population. The PTC example implies that many of the world's predisposing disease alleles are present in a fraction of the sub-Saharan African population, and their spread into the rest of the world may be accompanied by local genomic regions of identity by descent that can be used to fine-map and ultimately identify these disease genes.

## ACKNOWLEDGMENTS

We thank the members of the Utah CEPH families and the National Institutes of Health study population for their generous participation. We thank Brith Otterud, Kevin Cromer, and Tami Elsner for assistance in the genetic linkage studies. This work was supported by the Division of Intramural Research, the National Institute on Deafness and Other Communication Disorders, Z01-000046-04, by the National Human Genome Research Institute, T32 HG00044, by the W.M. Keck Foundation, and by the National Institute of General Medical Sciences.

## REFERENCES

Adler E., Hoon M.A., Mueller K.L., Chandrashekar J., Ryba N.J.P., and Zuker C.S. 2000. A novel family of mammalian taste receptors. *Cell* **100:** 693.

Anonymous. 1931. Tasteblindness. *Science* **73:** 14.

Blakeslee A.F. 1931. Genetics of sensory thresholds: Taste for phenyl thio carbamide. *Proc. Acad. Natl. Acad. Sci.* **18:** 120.

Buck L. and Axel R. 1991. A novel multigene family may encode odorant receptors: A molecular basis for odor recognition. *Cell* **65:** 175.

Chautard-Freire-Maia E.A. 1974. Linkage relationships between 22 autosomal markers. *Ann. Hum. Genet.* **38:** 191.

Conneally P.M., Dumont-Driscoll M., Huntzinger R.S., Nance W.E., and Jackson C.E. 1976. Linkage relations of the loci for Kell and phenylthiocarbamide taste sensitivity. *Hum. Hered.* **26:** 267.

Dausset J., Cann H., Cohen D., Lathrop M., Lalouel J.-M., and White R. 1990. Centre d'etude du polymorphisme humain (CEPH): Collaborative mapping of the human genome. *Genomics* **6:** 575.

Dib C., Faure S., Fizames C., Samson D., Drouot N., Vignal A., Millasseau P., Marc S., Hazan J., Seboun E., Lathrop M., Gyapay G., Morissette J., and Weissenbach J.A. 1996. A comprehensive genetic map of the human genome based on 5,264 microsatellites. *Nature* **380:** 152.

Drayna D., Coon H., Kim U.-K., Elsner T., Cromer K., Otterud B., Baird L., Peiffer A., and Leppert M. 2003. Genetic analysis of a complex trait in Utah Genetic Reference Project: A major locus for PTC taste ability on chromosome 7q and a secondary locus on chromosome 16p. *Hum. Genet.* **112:** 567.

Guo S.-W. and Reed D. 2001. The genetics of phenylthiocarbamide perception. *Ann. Hum. Biol.* **28:** 111.

Kalmus H. 1958. Improvements in the classification of the taster genotypes. *Ann. Hum. Genet.* **22:** 222.

NIH/CEPH Collaborative Mapping Group. 1992. A comprehensive genetic linkage map of the human genome. *Science* **258:** 67.

Olson J.M., Boehnke M., Neiswanger K., Roche A.F., and Siervogel R.M. 1989. Alternative genetic models for the inheritance of the phenylthiocarbamide taste deficiency. *Genet. Epidemiol.* **6:** 423.

Reddy B.M. and Rao D.C. 1989. Phenylthiocarbamide taste sensitivity revisited; complete sorting test supports residual family resemblance. *Genet. Epidemiol.* **6:** 413.

Snyder L.H. 1931. Inherited taste deficiency. *Science* **74:** 151.

Spence M.A., Falk C.T., Neiswanger K., Field L.L., Marazita M.L., Allen F.H., Siervogel R.M., Roche A.F., Crandall B.F., and Sparkes R.S. 1984. Estimating the recombination frequency for the PTC-Kell linkage. *Hum. Genet.* **67:** 183.

Tepper B. 1998. 6-n-Propylthiouracil: A genetic marker for taste, with implications for food preference and dietary habits. *Am. J. Hum. Genet.* **63:** 1271.

# Genomic Variation in Multigenic Traits: Hirschsprung Disease

A.S. McCallion,* E.S. Emison,* C.S. Kashuk,* R.T. Bush,* M. Kenton,*
M.M. Carrasquillo,* K.W. Jones,† G.C. Kennedy,† M.E. Portnoy,‡ E.D. Green,‡
and A. Chakravarti*

*McKusick-Nathans Institute of Genetic Medicine, Johns Hopkins University School of Medicine, Baltimore, Maryland 21205; †Affymetrix, Santa Clara, California 95051; ‡Genome Technology Branch, National Human Genome Research Institute, National Institutes of Health, Bethesda, Maryland 20892

Recent advances in genomic technology and the availability of a finished human genome sequence have greatly facilitated the identification of genes underlying human Mendelian disorders. The contemporary challenge lies in the elucidation of complex disorders. Classically, the transmission of a Mendelian disorder is explained by the exact co-segregation of a single mutation with the phenotype. Such mutations are absent in controls and, most frequently, involve conserved coding sequences. These observations are in stark contrast to the complex non-Mendelian diseases, wherein mutations at single genes do occur in unaffecteds, and variants with weak or moderate quantitative effects on the phenotype play a significant role. Consequently, such mutations may exist at relatively high frequency in the general population.

Complex organisms tolerate genetic variation, suppressing the potential expression of deleterious alleles by requiring the combined influence of hypomorphic alleles at multiple loci. Thus, a disease phenotype is revealed only in some multilocus genotypes and physiological and genetic homeostasis is maintained. However, this complicates genetic dissection of complex disease. First, multiple genes contribute to clinical expression, yet they may not contribute equally to the phenotypic variation. One must therefore identify multiple contributing genes to explain the majority of the phenotypic variation. This may be further complicated by locus heterogeneity wherein components of multiple pathways or multiple components within a single pathway may contribute to a phenotype. Second, the presence of disease-causing variants in cases and controls necessitates functional analyses to establish their respective contributions to disease transmission. Notably, the ultimate proof lies in the synthesis of the corresponding complex phenotype in a model organism. Consequently, understanding the genetic basis of complex traits necessitates integration of multiple genetic strategies, including studies of human populations, comparative genomics, and model organism-based approaches. We report a synergistic set of approaches aimed at understanding the genetic mechanisms underlying Hirschsprung disease (HSCR), a complex trait and relatively common birth defect.

HSCR is a congenital malformation with an incidence in the general population of 1 per 5,000 live births, and is characterized by an absence of neural crest (NC)-derived intrinsic ganglia along a variable length of the distal intestinal tract. Patients typically present in the neonatal period with intestinal obstruction and abdominal distension resulting from an inability to propagate peristaltic waves in the distal gut. HSCR also displays several hallmarks of complex genetic disease, including incomplete penetrance and pleiotropic effects of mutant genotypes, a marked sex-difference in clinical expression, and variation in penetrance with extent of aganglionosis. Pathological examination permits classification as long (L)- or short (S)-segment HSCR dependent on whether aganglionosis extends beyond the upper descending colon (L-HSCR) or not (S-HSCR). Significantly, the observed sex bias differs between these classifications, with a fourfold excess of males among S-HSCR patients compared to their twofold excess in L-HSCR patients (Badner et al. 1990). Although it is assumed to be a sex-modified multifactorial trait (Bodian and Carter 1963; Passarge 1967), molecular genetic studies have shown HSCR to be oligogenic, the reduced penetrance arising from the segregation of one or a few genes. Genetic interaction between hypomorphic mutations at known genes explains why single-gene inheritance is rare and non-Mendelian patterns of inheritance are common in HSCR.

To date, mutations in at least 8 genes encoding members of the RET receptor tyrosine kinase (RET) pathway, and the endothelin receptor type B (EDNRB) pathway, and the SOX10 and ZFHX1B transcription factors, have been reported in patients with HSCR (Fig. 1). The RET and EDNRB receptors are expressed in the same NC-derived cell populations during development (McCallion et al. 2003), contrasting with their corresponding ligands that are expressed in the mesenchymal cell populations of the developing gut. Consequently, although HSCR is primarily recognized as a defect in NC-derived enteric neurons, it may not be described as a cell-autonomous defect. Targeted and spontaneous mutations of HSCR genes in mice, embryologic analyses of enteric development, and functional analyses of gene mutations have also been critical in illuminating the molecular processes contributing

**Conflict-of-Interest Disclosure Statement:** Aravinda Chakravarti is a paid member of the Scientific Advisory Board of Affymetrics. The terms of this arrangement are being managed by the Johns Hopkins University in accordance with its conflict-of-interest policies.

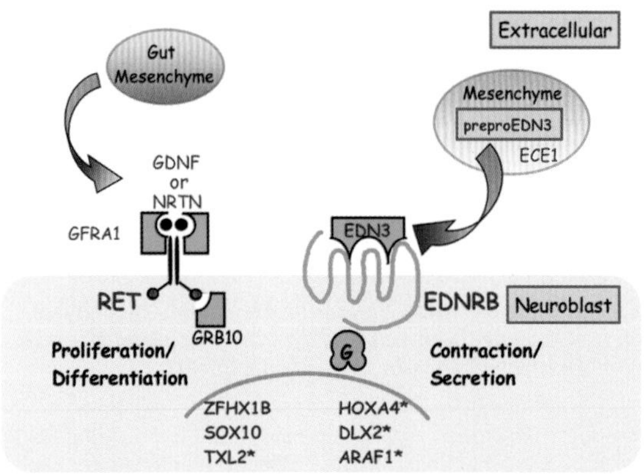


**Figure 1.** Schematic diagram of the RET and EDNRB pathways. Mutations in *RET* and *EDNRB*, as well as components of their respective signaling pathways (*GDNF, NRTN, EDN3, ECE-1*) have been identified in HSCR patients. Additionally, mutations in *SOX10* and *ZFHX1B* have also been demonstrated in HSCR patients. Protein names are colored red if mutations in the corresponding gene result in HSCR or a HSCR-like phenotype in human or mouse. * indicates HSCR-associated phenotype in mouse only. (See Table 1 for further details.)

to HSCR pathology (for review, see Chakravarti et al. 2001). However, unlike the mouse mutations, human mutations in the corresponding genes are never fully penetrant and are sex-dependent (Table 1). The major unresolved question is the mechanism of reduced penetrance. This might be due to stochastic factors, developmental differences between the sexes, diversity of the multiple mutant genes in patients, or a combination of these factors. Our studies favor the hypothesis of multigenic inheritance.

Although it was initially thought to be a heterogeneous disorder primarily with mutations at either *RET* or *EDNRB* in different families (Angrist et al. 1995; Chakravarti 1996), recent studies have clarified that mutations in individual HSCR genes are neither necessary nor sufficient for disease expression. We have previously shown that HSCR patients harbor mutations at multiple genes; most frequently, mutations at *RET* acting in concert with those at other gene(s) (Bolk et al. 2000; Carrasquillo et al. 2002; Gabriel et al. 2002).

One way to simplify genetic dissection of complex inherited disease is to study them within population isolates wherein genetic heterogeneity is decreased. We began our genetic dissection of HSCR in a kindred of inbred Old Order Mennonites with a population incidence of HSCR tenfold higher than the general population, hypothesizing that a single recessive mutation shared identical-by-descent (IBD) would explain disease inheritance (Puffenberger et al. 1994b). Initially, a single mutation in *EDNRB*, W276C, was identified in the Mennonites (Puffenberger et al. 1994a). However even in this isolate, this mutation was insufficient to explain disease transmission in all individuals. Importantly, these early studies also revealed a nonrandom transmission of alleles at *RET* and a role for a locus within the interval 21q21-tel. Although mutations in *RET* and *EDNRB* both led to HSCR, it was unclear whether in combination they might explain transmission in all individuals in whom disease could not be attributed to a single locus. To test this hypothesis, we performed a whole-genome association scan in Mennonites using a combination of microsatellite and single-nucleotide polymorphism (SNP) genetic markers. We confirmed the role of EDNRB in this population, established the existence of a relatively common HSCR-susceptibility haplotype at *RET*, and uncovered a novel contributing locus at 16q23 (Carrasquillo et al. 2002). Despite thorough examination of the entire coding sequence, the putative *RET* minimal promoter, intron–exon boundaries, the sequences within the introns deemed necessary for lariat structure formation, and the 5′ and 3′ UTRs, no putatively functional *RET* variants were identi-

**Table 1.** Comparison of Human and Mouse Mutations Resulting in HSCR or HSCR-like Phenotypes

| Gene (map position, human) | Number of mutations (human) | Phenotype, human | | Phenotype, mouse[c] |
|---|---|---|---|---|
| | | homozygotes | heterozygotes | |
| *RET* (10q11) | >100 | HSCR | HSCR | HSCR[d] |
| *GDNF* (5p13) | 5 | unobserved | HSCR | HSCR[d] |
| *NRTN* (19p13) | 1 | unobserved | HSCR | mild hypoganglionosis |
| *GFRA1* (10q26) | 0 | unobserved | unobserved | HSCR[d] |
| *GRB10* (7p12-p11.2) | 0 | unobserved | unobserved | not examined |
| *EDNRB* (13q22) | 15 | HSCR[a] | HSCR | HSCR[e] |
| *EDN3* (20q13) | 2 | HSCR[a] | HSCR | HSCR[e] |
| *ECE-1* (1p36) | 1 | unobserved | HSCR[b] | HSCR[e,f,g] |
| *SOX10* (22q13) | 6 | unobserved | HSCR[a] | HSCR[e] |
| *ZFHX1B* (2q22) | 3 | unobserved | HSCR | unknown[h] |
| *ARAF1* (Xp11) | unknown | ? | ? | Megacolon[i] |
| *TLX2* (2p13) | unknown | ? | ? | HSCR[f] |
| *HOXA4* (7p15-p14) | unknown | ? | ? | Megacolon[i] |
| *DLX2* (2q32) | unknown | ? | ? | HSCR |

[a]Shah-Waardenburg syndrome (WS4). [b]Associated neurocristopathies. [c]All mutations represented in mouse are knockouts except for the spontaneously occurring *Dom* (SOX10), piebald lethal, $Ednrb^{s-l}$, and lethal spotting, $Edn3^{ls}$ alleles. All mutations demonstrate 100%, sex-independent penetrance. [d]Renal agenesis. [e]Coat spotting. [f]Craniofacial defects. [g]Cardiac defects. [h]Homozygous mutation is embryonic lethal, resulting from neurulation failure at the vagal level. [i]No evidence of aganglionosis.

fied. Furthermore, joint transmission tests of alleles from *EDNRB* and *RET* strongly suggested that mutations at these loci do not act independently. These data prompted us to assay for noncomplementation between these genes using extant mutations in the mouse. We have successfully established mouse strains, simultaneously mutant at both *Ret* and *Ednrb*, which completely recapitulate the clinical and genetic features of HSCR and now provide a unique system with which to test hypotheses about the molecular interactions that result in disease (McCallion et al. 2003). However, in the absence of identifying the novel 16q locus and examining the potential involvement of others, we could not explain disease transmission in all affected individuals.

As predicted, the disease-causing variants identified to date in the Mennonites are common (10–20%) in the population, suggesting that if other common variants exist at additional loci, they may also be detected in association scans of sufficient resolution. Here we report the completion of such a genome-wide screen that implicates three additional loci in disease susceptibility (1p34, 4q31, and 11p15) in this isolated population. However, despite detection of additional loci, our genome scan lacks the resolution necessary to reveal the specific genes underlying disease.

Often the ability to refine a genetic locus and narrow the list of possible candidate genes is dependent on the availability of biological and functional annotation for the corresponding genes in the interval. Unfortunately, at this stage, this information is incomplete. To reduce the list of gene candidates in HSCR susceptibility regions, we hypothesized that the pertinent loci may comprise downstream components of RET and EDNRB signaling and that disruption of RET and/or EDNRB signaling may thus influence the transcript levels of these genes. One powerful approach to assay such changes is through array-based transcript profiling technologies, permitting the simultaneous analysis of thousands of transcripts. We posited that RET- and EDNRB-responsive genes lying within known HSCR susceptibility genomic intervals would comprise ideal disease gene candidates. To test this hypothesis, we examined the transcript profile in the intestinal tracts of our two-locus HSCR mouse strains, mice harboring *Ret* and *Ednrb* mutations independently, and in wild-type mice. Our data demonstrate that a group of 9 genes, whose human orthologs localize to a known HSCR susceptibility locus on 21q22 (Puffenberger et al. 1994b), are differentially expressed in the colons of wild-type and HSCR mutant mice. This report affirms the value of integrating genomic approaches in dissecting the mechanisms underlying complex inherited disease. The ultimate goal, of understanding the molecular nature of the variants contributing to clinical expression, will also be facilitated by the integration of genomic information from multiple organisms.

The absence of a coding sequence change in *RET* in the Mennonite population is not completely surprising. Despite the identification of over 100 *RET* mutations in HSCR and related syndromes, the majority of *RET*-linked families lack a frank coding sequence mutation. We recently completed a genome screen in L-HSCR families, demonstrating linkage at *RET* in 11/12 (92%) families but identifying *RET* mutations in only 6/11 linked families (Bolk et al. 2000). Similarly, the majority (88%) of S-HSCR sib pairs, in a second study, demonstrated allele sharing at *RET*, but mutations were found in only 40% of these families (Gabriel et al. 2002). The lack of identifiable mutations in *RET* in the outbred population suggests that noncoding mutations in *RET* probably underlie most forms of HSCR. This is consistent with the observation of a role for a relatively common *RET* haplotype with no coding sequence mutation in the Mennonites. We have used these observations to posit that the same (or similar) common haplotype may be present at elevated frequency in HSCR patients in the general population. To test this hypothesis, we used the transmission disequilibrium test (TDT) in samples drawn from a general, outbred population of HSCR patients. We demonstrate here, using an 8-marker study, that the HSCR-susceptibility haplotype observed in the Mennonites is also found in association with disease in the general population. Therefore, a noncoding mutation at *RET* likely underlies the genesis of most cases of HSCR.

Perhaps the greatest challenge lies ahead as we seek to narrow the variants in noncoding sequence to determine which may be causative. We posited that sequence conservation between orthologous sequences from distantly related organisms is a reliable predictor of functional constraint (Thomas et al. 2003). Thus, we have aligned human and mouse *RET* genomic sequences, resulting in the identification of both coding and noncoding multispecies conserved sequences (MCSs) within the *RET* locus. By overlapping the peak transmission bias from our TDT data with MCSs in noncoding regions, we have discovered the specific functional element involved in HSCR.

Hence, through an integrated approach to the study of complex disease, using all available genomic resources in mouse and human, we have made significant progress in elucidation of the genetic basis of HSCR. We expect that the causative variants at the remaining linked loci will be identified by similar studies and that our work will serve as a paradigm for the field of complex disease research.

## MATERIALS AND METHODS

***Genome scan.*** We genotyped 43 trios from the Old Order Mennonite community in Lancaster and Berks Counties, Pennsylvania. Ascertainment was conducted under protocols approved by the Institutional Review Board of Johns Hopkins University School of Medicine. A complete description of the study population has been published elsewhere (Carrasquillo et al. 2002). Following the manufacturer's protocol, we genotyped 4,363 SNPs using the WGSA-*Eco*RI-p502 array (Affymetrix). Briefly, following *Eco*RI digest of genomic DNA and ligation with adapters, samples were PCR-amplified as a means of size-selecting fragments up to 1 kb. Details of sample labeling and array hybridization have been published previously (Kennedy et al. 2003).

Also included in the analysis are data from an additional 569 microsatellites and 1,384 SNPs genotyped earlier (Carrasquillo et al. 2002). We used CRIMAP to generate a linkage map based on the June 2002 release of the human genome build as described previously (Carrasquillo et al. 2002). The total map length is 3839 cM and available at http://chakravarti.igm.jhmi.edu.

The genome scan confirmed our previous findings of a disease-associated locus on 16q23. In the most recent genome build (HG#34), the map position of the associated markers is 16q23.1, a slight difference from the previously reported map position of 16q23.3.

***MLD analysis.*** The MLD method was implemented as described previously (Carrasquillo et al. 2002). MLD is a multi-marker test based on association by common descent. Our test of linkage disequilibrium assesses whether marker alleles have a higher frequency on mutant gene-enriched (transmitted, T) versus mutant gene-depauperate (untransmitted, U) chromosomes in the two parents in each family. We designated alleles as T and U based on the proband genotypes. If $y$ and $x$ are the frequencies of the associated allele on T and U chromosomes, respectively, $\theta$ the recombination value between the mutation and a marker locus, $\alpha$ the proportion of mutant chromosomes attributable to a particular mutation, and $g$ the time to origin (in generations) of the mutation, then: $y = x + \alpha(1-x)(1-\theta)^g$ for the associated allele and $y = (1-x)[1-\alpha(1-\theta)^g]$ for all other alleles. By assuming values of $\alpha$, $g$, and $\theta$ (based on putative map location of a susceptibility allele), the likelihood of observed values of T and U alleles is calculated. Since the associated allele is unknown, the posterior likelihood is calculated assuming a prior distribution that the probability of the associated allele is proportional to its allele frequency. This likelihood is combined across all marker loci on a chromosome by taking their products. A LOD score is finally computed by contrasting the $\log_{10}$ likelihood assuming $\theta$ from the putative location on the map versus $\theta = 1/2$. The marker frequencies $x$ are estimated as nuisance parameters. Intermarker distances were taken from the linkage map described above. We analyzed the genotype data under the assumptions (1) $g = 12$ or 48, (2) $\alpha = 0.2$ (0.2) 1.0. We estimated the $p$ value by randomizing T and U haplotype data for each parent but keeping the map constant and estimating the probability of a LOD score greater than or equal to that observed in 1,000 replications.

***RET SNP genotyping.*** We genotyped 8 SNPs across the coding sequence that defined previously identified *RET* haplotypes (Carrasquillo et al. 2002). Genotypes were generated using the fluorogenic 5′ nuclease assay (Taqman, ABI). Detailed primer and probe sequence information is available online at http://chakravarti.igm.jhmi.edu. A TECAN Genesis workstation was used for all liquid-handling steps, thermal cycling was completed on MJ Research Tetrads, and end-point reads were made on an ABI 7900. Genotyping calls were made using SDS 2.1 (ABI) and verified by the instrument operator. 5% of the samples were genotyped in duplicate; no discrepancies were observed among the paired replicates. The TDT test statistic was used to identify significant deviation from expected Mendelian transmission (Spielman et al. 1993).

***Comparative sequence analysis.*** Genomic sequences were obtained from http://genome.ucsc.edu - Human genome build HG32 (chr10:43347672-43401001) and Mouse build (UCSC mm3, chr6:118828273-118871640). Pair-wise sequence alignments of FASTA-formatted human and mouse genomic sequences (100 kb encompassing *RET*, extending 35 kb 5′ and 20 kb 3′) were subjected to several alignment algorithms: MULTIPIP (Schwartz et al. 2000; Thomas et al. 2003), a BLAST-based sequence alignment algorithm (Altschul et al. 1990); mVISTA, an AVID-based algorithm (Bray et al. 2003); and MLAGAN (Brudno et al. 2003), a sequence alignment algorithm that permits use of identified open reading frames to anchor alignments between distantly related species. All these algorithms facilitate alignment of multiple large-sequence contigs simultaneously, uncovering sequence elements conserved at a predetermined level of identity over a predetermined length of sequence. We identified MCSs using the functionally validated criteria of sequences demonstrating ≥70% identity over ≥100 bp in pair-wise comparisons of human versus mouse (Loots et al. 2000).

***Expression microarrays.*** We assayed the RNA profiles of distal colon segments isolated from $Ret^{+/+}$; $Ednrb^{+/+}$, $Ret^{+/-}$; $Ednrb^{+/+}$, $Ret^{+/-}$; $Ednrb^{s\text{-}l/s}$ and $Ret^{+/-}$; $Ednrb^{s\text{-}l/s}$ mice (two males and two females of each genotype; see Fig. 3A) and from microdissected embryonic gut at two developmental stages. We used Affymetrix MG-U74Av oligonucleotide arrays, containing ~6,000 known genes and 6,000 ESTs. To calculate expression indices for every probe set on our arrays, we utilized an established linear model for the computation of expression indices (Li and Wong 2001), accounting for experimental variance and conducting a simple $t$-test of computed values when comparing samples. We have initially limited our inclusion of differentially regulated transcripts to those with a $p$ value of $p \leq 0.01$ in any pair-wise comparison of samples. To identify genes critical for the normal development and function of the intestinal tract, we compared the RNA profiles of the distal intestine of all aganglionic ($Ret^{+/-}$; $Ednrb^{s\text{-}l/s}$) mice with those normally innervated ($Ret^{+/+}$; $Ednrb^{+/+}$), and with phenotypically normal mice harboring heterozygous mutant genotypes at *Ret* and *Ednrb* ($Ret^{+/-}$; $Ednrb^{+/+}$ *and* $Ret^{+/+}$; $Ednrb^{s\text{-}l/s}$).

## RESULTS

### Identification of Six New Loci Linked to HSCR in Old Order Mennonites

We performed a genome scan for disease-marker associations in the Old Order Mennonites, an inbred population with a tenfold increased incidence of HSCR. We selected a chip-based method to genotype over 4,000 SNPs per individual, in a single-tube assay. In particular, the array-based method (whole-genome sampling analysis; Kennedy et al. 2003) used here permits multiplexing such that for each individual only one sample is processed to generate genotypes at over 4,000 loci. This is a significant advance over the chip-based method we used and published only one year earlier, which required 96 tubes per

**Table 2.** Genome-wide Disease Marker Association in Mennonites

| MLD analysis | LOD score | Proportion mutant | *p* |
|---|---|---|---|
| 13q22.3 (*EDNRB*) | 110.96 | 0.6 | 0.000 |
| 10q11.21 (*RET*) | 18.53 | 0.2 | 0.016 |
| 11p15.3 | 6.42 | 0.2 | 0.011 |
| 16q23.1 | 4.15 | 0.4 | 0.046 |
| 1p34.3 | 3.86 | 0.4 | 0.037 |
| 4q31.1 | 3.71 | 0.2 | 0.009 |

Genome-wide LOD scores were calculated under assumptions that locus heterogeneity was 0.2, 0.4, 0.6, 0.8, or 1.0 and that the disease variant arose 12 or 48 generations ago. Calculations for 12 generations are shown here. Calculations under 48 generations did not appreciably alter results. *p* is a bootstrap *p* value based on 1,000 randomizations of transmitted and untransmitted haplotypes.

individual and generated data on ~1,400 SNPs. Thus, the genome screen served as both an evaluation of novel technology and a platform to identify new genomic intervals of interest.

We analyzed a total of 5,747 SNP markers and 569 microsatellites distributed across the genome. MLD analysis of the combined data set detected 6 loci with LOD scores greater than 3.0 (Table 2). Three of these loci (10q11.21, 13q22.3-q31.1, and 16q23.1) have been previously observed in this population (Carrasquillo et al. 2002). We have previously demonstrated that the genes within 10q11.21 and 13q22.3-q31.1 are *RET* and *EDNRB*, respectively (Carrasquillo et al. 2002, Puffenberger et al. 1994a). The gene underlying HSCR in 16q23.1 has not been identified. None of the 3 newly identified loci (1p34.3-p34.2, 4q31.1-q31.21, and 11p15.3) contains genes previously shown to be associated with HSCR in the human (i.e., *ECE1, EDN3, GDNF, NRTN, SOX10*, and *ZFHX1B*). Furthermore, genomic intervals previously identified in linkage studies of S-HSCR (3p21 and 19q12; Bolk et al. 2000) and L-HSCR (9q31; Gabriel et al. 2002) families did not show association within the Mennonites. Thus, in addition to *RET* and *EDNRB*, we demonstrate evidence for additional genes modifying HSCR expression in the Mennonites and proving multigenic inheritance. Replication of these findings is necessary.

### A Common *RET* Haplotype Associated with HSCR in an Outbred Population

We, and others, have previously proposed that common variants may underlie a significant fraction of common disorders (Lander 1996; Risch and Merikangas 1996; Chakravarti 1999). The existence of common variants, within a gene of interest, can be assayed in either a case-control (CS) study or transmission disequilibrium test (TDT) using affected individuals and their parents. Recent CS studies have associated specific haplotypes in *RET* in the genesis of HSCR (Borrego et al. 2003; Sancandi et al. 2003). On the basis of our identification of a single predisposing *RET* haplotype in the Old Order Mennonites, we hypothesized that the same haplotype exists in the general population of HSCR patients.

We tested this hypothesis using TDT analyses of >150 outbred trios and ~43 Mennonite trios, which include individuals affected with all forms of HSCR (e.g., S-HSCR, L-HSCR, syndromic). We have assayed eight SNPs across the *RET* gene haplotypes in the Mennonites (Carrasquillo et al. 2002). Our previously published analysis of *RET* polymorphisms in the Mennonites used the haplotype relative risk test (HRR). Here, we present data from the same individuals, reanalyzed with the TDT statistic, so that the results can be compared directly to the TDT analysis of the outbred population. This analysis eliminates false positives that may arise from population substructure rather than true association (Spielman et al. 1993). The TDT test excludes parents who are homozygous at a given marker, thus designating each allele carried by heterozygous parents as either transmitted or untransmitted. Under strictly random segregation, each allele is expected to be transmitted 50% of the time. However, markers in linkage disequilibrium with and, indeed, the disease-causing variant itself, will appear as overtransmitted from heterozygous parents.

We assayed 8 SNPs within the coding sequence of *RET* using the Taqman assay for allelic discrimination. The results of the TDT analysis are given in Table 3 and highlight two important observations. First, in the outbred population, overtransmitted alleles are identical to those overtransmitted in the Mennonites, suggesting the disease-associated haplotype previously identified in the Mennonites also underlies the genesis of HSCR in the general outbred population. The 8-marker haplotype shared in common between the Mennonites and the outbred population is AGAGGCAT. Second, in both populations, maximal overtransmission is observed with the intron-1 and exon-2 SNPs. These markers correspond to the 5′ portion of the *RET* haplotype in our previous work (Carrasquillo et al. 2002). Additional marker data will be required to determine the 5′ boundary of the association at *RET*. Because our previous sequencing of *RET* in the Mennonites failed to identify disease-causing coding sequence variants, we hypothesize that a noncoding *RET* variant exists on the shared haplotype.

### Comparative Sequence Analysis Reveals Coding and Noncoding MCSs within the *RET* Locus

A major question is how to identify critical noncoding elements which, when mutant, would lead to a distinct phenotype. We posit that such noncoding, potentially regulatory, sequences are subject to selective pressure and would be characterized by a nonneutral pattern of sequence evolution, being more highly conserved across multiple species than their nonfunctional neighbors. The availability of genomic sequence has revolutionized the way in which we might systematically search for these noncoding sequences. Specifically, genomic sequence comparisons of multiple closely and/or distantly related species can identify all evolutionarily conserved regions, termed multispecies conserved sequences (MCSs) (Thomas et al. 2003). Notably, identification of MCSs can uncover all coding exons, including alternatively spliced exons that may not have been known, additional

**Table 3.** HSCR Association at *RET* in Two Populations

| Assay name | dbSNP ID | General outbred population N | X | T | U | τ | $\chi^2$ | *p* value | Old Order Mennonites N | T | U | τ | $\chi^2$ | *p* value |
|---|---|---|---|---|---|---|---|---|---|---|---|---|---|---|
| INT1.1SfcI | rs2435362 | 76 | A | 64 | 12 | 84.2 | 39.06 | 4.1 $e^{-10}$ | 33 | 24 | 9 | 72.7 | 7.07 | 0.0078 |
| INT1.4b | rs2505535 | 74 | A | 59 | 15 | 79.7 | 27.98 | 1.2 $e^{-7}$ | 33 | 24 | 9 | 72.7 | 7.07 | 0.0078 |
| X2EagI | rs1800858 | 77 | A | 62 | 15 | 80.5 | 30.80 | 2.8 $e^{-8}$ | 33 | 24 | 9 | 72.7 | 7.07 | 0.0078 |
| INT8 | rs3026750 | 66 | A | 41 | 25 | 62.1 | 3.92 | 0.048 | 29 | 20 | 9 | 69.0 | 4.28 | 0.0386 |
| X13TaqI | rs1800861 | 62 | G | 39 | 23 | 62.9 | 4.18 | 0.041 | 27 | 19 | 8 | 70.4 | 4.61 | 0.0317 |
| INT18BbvI | rs2742237 | 55 | C | 39 | 16 | 70.9 | 9.92 | 0.0016 | 27 | 19 | 8 | 70.4 | 4.61 | 0.0317 |
| INT18StyI | rs2742239 | 49 | A | 33 | 16 | 67.3 | 6.02 | 0.014 | 21 | 16 | 5 | 76.2 | 6.06 | 0.0138 |
| INT19BsgI | rs2075912 | 51 | C | 37 | 14 | 72.6 | 10.76 | 0.001 | 21 | 15 | 6 | 71.4 | 3.98 | 0.0459 |

(N) Number of heterozygous parents studied; (X) associated allele; (T) number of associated alleles on transmitted chromosomes; (U) number of associated alleles on untransmitted chromosomes; (τ) TDT statistic.

genes in the same genomic locus that might also be disease gene candidates, and conserved noncoding sequences. These comparisons, for a small number of mammals at specific loci, show that MCSs comprise 4.5% of the genome, with coding and noncoding sequences contributing 1.5% and 3%, respectively (Thomas et al. 2003). The potential function of identified noncoding MCSs is not restricted to gene regulation but may also include roles in chromosomal assembly and replication, among others (Pennacchio and Rubin 2001). There is now significant evidence, from a diverse collection of experiments, that the function of regulatory elements is conserved over great evolutionary distances. One recent estimate suggests that analyses over the mammalian radiation, e.g., human versus mouse, alone are capable of capturing >80% of the noncoding MCSs (Margulies et al. 2003). We conducted pair-wise sequence alignments of human and mouse genomic sequences at *RET*, using several alignment algorithms, including MULTIPIP, mVISTA, and MLAGAN. Analysis of 100 kb encompassing the *RET* gene (35 kb 5′ and 20 kb 3′ from *RET*) has uncovered 48 MCSs. BLAST, TBLASTX (Altschul et al. 1990), and GENSCAN (Burge and Karlin 1997) analyses revealed that 20/48 MCSs correspond to *RET* exons (Fig. 2) and that 28/48 MCSs have no known ESTs or predicted open reading frames. These data constitute the starting point for comprehensive mutational and functional analyses of the genomic region encompassing *RET*. To further aid our studies, we are now in the process of generating genomic sequence from 13 vertebrate species, uncovering all known and predicted exons within the 350 kb encompassing the *RET* proto-oncogene.

### Identification of Genes Regulated in Concert with *RET* and *EDNRB*

Complex traits result from the contribution of multiple genes, whether comprising inherited variation or not, presenting a unique set of problems in the dissection of their underlying network of interactions. Classical genetic mapping has power to map, but not refine, implicated regions and permit positional cloning, and candidate gene-based strategies are predicated on a platform of existing biological understanding that is incomplete for most genes. One feasible approach is to assess the impact of mutations, in one or more disease-associated genes, on the genomic signature of the cell (RNA or protein). Array-based transcript analyses aim to identify expression patterns that may be correlated with tissue type, pathological status, treatment, genotype, etc. Deficiencies in pathways compromised in the pathogenesis of disease may influence the abundance of transcripts/proteins corresponding to other genes with roles in the function of that tissue and the genesis of disease. Although array-based profiling assays may uncover hundreds of genes, we suggest that the list of real candidates may be enriched by identifying those harbored within known susceptibility loci.

We have assayed and compared the transcript profiles of postnatal mouse colons derived from wild-type ($Ret^{+/+}$; $Ednrb^{+/+}$) and our aganglionic mice ($Ret^{+/-}$; $Ednrb^{s-l/s}$). These two-locus HSCR mouse strains provide an ideal model system in which to ask questions regarding the genetic mechanisms underlying neuronal losses, smooth muscle defects, dysmotility, and sex bias observed in HSCR patients (Carrasquillo et al. 2002; McCallion et al. 2003). To assess the independent influence of genotypes at *Ret* and *Ednrb* we have also assayed transcript profiles of postnatal distal colon from $Ret^{+/-}$; $Ednrb^{+/+}$ and $Ret^{+/-}$; $Ednrb^{s-l/s}$ mice.

In brief, we compared the RNA profiles of the distal intestine of all aganglionic ($Ret^{+/-}$; $Ednrb^{s-l/s}$) mice with those normally innervated ($Ret^{+/+}$; $Ednrb^{+/+}$), identifying 200 transcripts whose levels differ significantly ($p \leq 0.01$), including genes whose products function in G-protein-coupled receptor (GPCR) and receptor tyrosine kinase (RTK) signaling pathways, apoptosis, cell adhesion, immune response, and protein biosynthesis. Of 200 differentially regulated transcripts, 106 correspond to known genes, many of which are involved in signaling pathways (15/106), transcriptional regulation (11/106), protein biosynthesis (16/106), ion transport (9/106), immune response (8/106), and development/cell cycle/apoptosis (8/106). We further hypothesized that phenotypically normal mice harboring heterozygous mutant genotypes at *Ret* and *Ednrb* ($Ret^{+/-}$; $Ednrb^{+/+}$ and $Ret^{+/+}$; $Ednrb^{s-l/s}$) may yield differences in transcript profile that may be equally important to our understanding of HSCR pathology and may also be less obscured by transcriptional response to the pathological state of the tissue; e.g., colitis, necrosis. Consequently, we also compared the RNA profiles of the distal intestine of wild-type ($Ret^{+/+}$; $Ednrb^{+/+}$) mice with those of normally innervated ($Ret^{+/-}$; $Ednrb^{+/+}$ and $Ret^{+/+}$; $Ednrb^{s-l/s}$) mice, identifying 240 transcripts ($Ret^{+/+}$; $Ednrb^{+/+}$ versus $Ret^{+/-}$; $Ednrb^{+/+}$) and 67 tran-

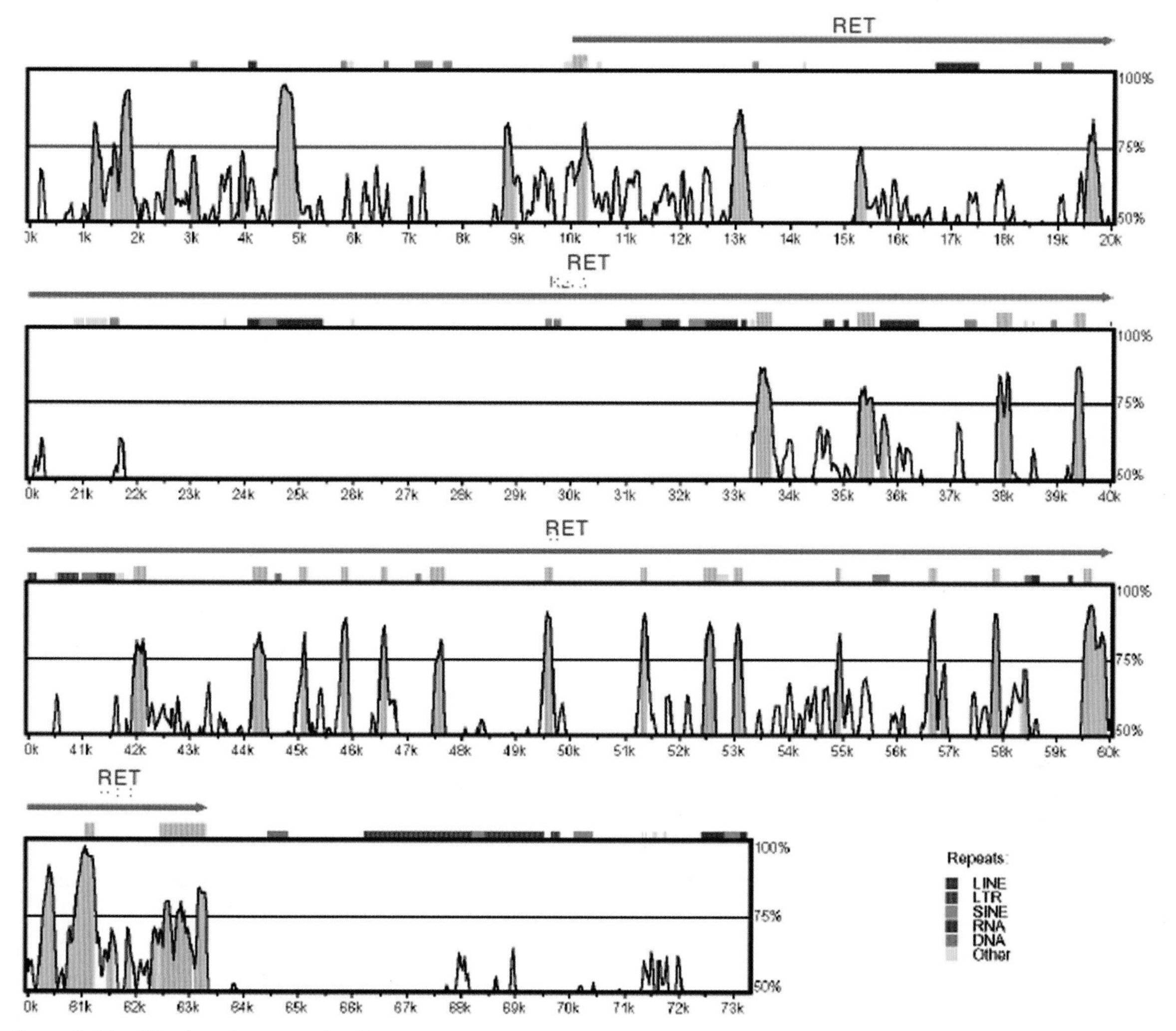


**Figure 2.** Identification of conserved coding and noncoding sequences at the *RET* gene. VISTA pair-wise analysis, aligning 73 kb encompassing human *RET* with orthologous sequence from the mouse. (*Blue*) Exons; (*pink*) noncoding multispecies conserved (MCS; 70% identity over ≥100 bp). Repeats are indicated by colored boxes above the plot with a color key embedded in the lower right corner, indicating their identity. Sequence identity is displayed in the 50–100% range.

scripts (*Ret*$^{+/+}$; *Ednrb*$^{+/+}$ versus *Ret*$^{+/+}$; *Ednrb*$^{s-l/s}$) whose levels differ significantly ($p$ <0.01).

We sought to refine our putative candidate gene list by localizing all genes demonstrating significant alteration in transcript levels in our comparison within the human genome and identifying those lying within known HSCR susceptibility regions. We have localized 88/106 known genes within the human genome (April 2003, HG#33 assembly), 11 of which localize to known HSCR susceptibility regions. We have also attempted to determine the genomic positions of all genes/ESTs whose transcripts demonstrate significant differences in our pair-wise comparisons, identifying a total of 38 genes whose positions fall within HSCR regions. Interestingly, the orthologs corresponding to 24% (9/38) of these localize within human Chromosome 21q22-tel (*SOD1, IL10*β*, IFNGR2B, SON, CBR1, TTC3, TFF3, CSTB, PFKL*). This observation does not represent a bias inherent in the array; the distribution of all genes present on the array for which a position within the human genome could be identified is not significantly different from the distribution of known genes on human Chromosome 21. However, the distribution of genes identified as differentially expressed in all pair-wise comparisons was markedly different from that of all Chromosome-21 genes represented on the array, localizing exclusively within 21q22.

Figure 3 illustrates a subset of the results of these analyses, indicating that 89% (8/9) of the genes whose orthologs localize to human Chromosome 21 are up-regulated in aganglionic *Ret*$^{+/-}$; *Ednrb*$^{s-l/s}$ mice compared to wild-type. Significantly, all 9 genes are detected in the embryonic gut during intestinal colonization by enteric ganglia (Reymond et al. 2002). Of these transcripts, 7 are demonstrably more abundant and/or regionally restricted within the developing gut. Although this is, in and of itself, not evidence of a role in neuronal development in the gut, it is noteworthy that 4/9 genes (*Ifnar2, Son, Ttc3*, and *Pfkl*) are also expressed in the developing CNS in a regionally restricted pattern within the ventricular zone of the forming forebrain/midbrain. These data indicate that the corresponding genes may be expressed in dividing and migrating populations of neuronal precursors. SOD1

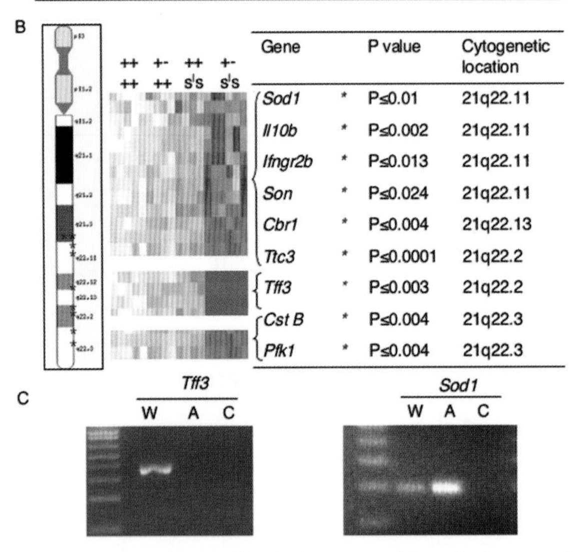


**Figure 3.** Transcript profiling in the postnatal intestinal tracts of mice wild-type and mutant for key HSCR genes. (*A*) Table of genotypes and genders of animals whose colon RNA profiles were interrogated on transcript profiling microarrays. (*B*) Genes whose orthologs localize to human chromosome 21 and are dysregulated in HSCR mutant mice compared to wild type. Also reported are *p* values for these comparisons and cytogenetic positions for each gene. (*C*) Examples of semi-quantitative RT-PCR assays used to confirm the differential expression observed in our array comparisons. W, $Ret^{+/+}$; $Ednrb^{+/+}$; A, $Ret^{+/-}$; $Ednrb^{s\text{-}l/s}$; C, non RT control.

has been implicated in the degeneration of neuronal populations (Rosen et al. 1993; Tu et al. 1996), and overexpression of this key protein has been shown to impede the dimerization of RET receptor tyrosine kinase (Kato et al. 2000). We have used semi-quantitative RT-PCR assays to confirm the differential expression observed in our array comparisons (Fig. 3). In light of the elevated incidence of HSCR among individuals with Down syndrome, we considered this observation to be particularly significant. We hypothesize that increased dosage of genes contained within human Chromosome 21 may compromise ENS integrity and/or increase susceptibility to variants at other HSCR loci.

## DISCUSSION

Observations on phenotypic resemblance between relatives clearly show that most traits are inherited but that the patterns are decidedly non-Mendelian. Thus, Mendelian inheritance of traits is the exception rather than the rule. However, it is not clear why any phenotype should be genetically complex. Is this simply a result of multiple gene segregation? Does the nature and effect of the underlying mutation matter? Which properties of a gene render its mutations likely to contribute to non-Mendelian patterns of inheritance? In other words, the major question in human disease genetics is, Why complex inheritance? Before we can answer this with any generality, we must first decipher the molecular details of a number of complex traits and disorders.

The choice of these traits is critical. There is currently great interest in understanding the genetic basis of many common and chronic diseases of humans. This is crucial, because the disease and economic burdens of these disorders are immense. However, studying a number of model disorders in which genetic dissection is currently feasible is just as important. We have selected HSCR as an example of a complex disease in which one may construct tractable questions. A key component in elucidating the genetic basis of non-Mendelian disorders is an understanding of the nature of genetic variation in outbred populations, such as humans. As in nature, the mutations that jointly contribute to HSCR include variants at multiple genes encoding different components of an integrated pathway(s). Mutations therein may independently lead to disease but most frequently require contributions from (and interaction with) mutations in other genes. These variants run the gamut of simple nucleotide substitutions to larger insertions/deletions and chromosomal rearrangements, indicating that differences in both the quality and quantity of gene product play a critical role in disease expression. Consistent with this observation, non-Mendelian inheritance patterns may also arise from nontraditional methods of gene action, including imprinting mutations and the dynamic expansion of a coding sequence (e.g., triplet repeats). Furthermore, different genetic effects of Watson and Crick sister strands may also lead to profoundly different, non-Mendelian phenotypes (Klar 1996). Thus, non-Mendelian should not necessarily be equated with multiple gene inheritance. These hypotheses may be distinguished using genomic approaches. However, our studies suggest that a comprehensive understanding of a complex disease will not result from a monolithic "genome scan" but will require integrated analyses, scanning the entire genome for substitutional mutations, insertion/deletion mutations, gene dosage changes, and/or gene expression changes at the level of RNA and protein. Importantly, scanning for changes in specific chemical modifications of the genome (e.g., methylation of DNA) will also be required.

An important characteristic of Mendelian disease is the high correlation between the mutation and phenotype, although other aspects of the phenotype (severity, onset, etc.) may not be as predictable based on the mutation type. Conversely, for complex disorders, proving that a set of genes thought to be necessary and sufficient for disease onset is difficult in humans. However, this has to be proven in animal or cellular model systems. Indeed, synthesis of the complex trait from the constituent gene mutations is an absolute necessity in complex disease research. In other words, the genetic dissection of a complex trait should be considered incomplete in the absence of trait synthesis. HSCR has proven this point clearly.

## ACKNOWLEDGMENTS

This work was supported by grants from the National Institute of Child Health and Human Development. We thank the current members of the Chakravarti lab for their helpful discussions and comments on this manuscript.

## REFERENCES

Altschul S.F., Gish W., Miller W., Myers E.W., and Lipman D.J. 1990. Basic local alignment search tool. *J. Mol. Biol.* **215:** 403.

Angrist M., Bolk S., Thiel B., Puffenberger E.G., Hofstra R.M., Buys C.H., Cass D.T. and Chakravarti A. 1995. Mutation analysis of the RET receptor tyrosine kinase in Hirschsprung disease. *Hum. Mol. Genet.* **4:** 821.

Badner J.A., Sieber W.K., Garver K.L., and Chakravarti A. 1990. A genetic study of Hirschsprung disease. *Am. J. Hum. Genet.* **46:** 568.

Bodian M. and Carter C. 1963. A family study of Hirschsprung disease. *Ann. Hum. Genet.* **26:** 261.

Bolk S., Pelet A., Hofstra R.M., Angrist M., Salomon R., Croaker D., Buys C.H., Lyonnet S., and Chakravarti A. 2000. A human model for multigenic inheritance: Phenotypic expression in Hirschsprung disease requires both the *RET* gene and a new 9q31 locus. *Proc. Natl. Acad. Sci* **97:** 268.

Borrego S., Wright F.A., Fernandez R.M., Williams N., Lopez-Alonso M., Davuluri R., Antinolo G., and Eng C. 2003. A founding locus within the *RET* proto-oncogene may account for a large proportion of apparently sporadic Hirschsprung disease and a subset of cases of sporadic medullary thyroid carcinoma. *Am. J. Hum. Genet.* **721:** 88.

Bray N., Dubchak I., and Pachter L. 2003. AVID: A global alignment program. *Genome Res.* **13:** 97.

Brudno M., Do C.B., Cooper G.M., Kim M.F., Davydov E., Green E.D., Sidow A., and Batzoglou S. 2003. LAGAN and Multi-LAGAN: Efficient tools for large-scale multiple alignment of genomic DNA. *Genome Res.* **13:** 721.

Burge C. and Karlin S. 1997. Prediction of complete gene structures in human genomic *DNA J. Mol. Biol.* **268:** 78.

Carrasquillo M.M., McCallion A.S., Puffenberger E.G., Kashuk C.S., Nouri N., and Chakravarti A. 2002. Genome-wide association study and mouse model identify interaction between RET and EDNRB pathways in Hirschsprung disease. *Nat. Genet.* **32:** 237.

Chakravarti A. 1996. Endothelin receptor-mediated signaling in Hirschsprung disease. *Hum. Mol. Genet.* **5:** 303.

———. 1999. Population genetics—Making sense out of sequence. *Nat. Genet.* **21:** 56.

Chakravarti A., McCallion A.S., and Lyonnet S. 2001. Hirschsprung disease. In *The metabolic and molecular bases of inherited disease,* 8th edition (ed. C.R. Scriver et al.), p. 6231 (see update [2003] at http://www.genetics.accessmedicine.com/). McGraw-Hill, New York.

Gabriel S.B., Salomon R., Pelet A., Angrist M., Amiel J., Fornage M., Attie-Bitach T., Olson J.M., Hofstra R., Buys C., Steffann J., Munnich A., Lyonnet S., and Chakravarti A. 2002. Segregation at three loci explains familial and population risk in Hirschsprung disease. *Nat. Genet.* **31:** 89.

Kato M., Iwashita T., Takeda K., Akhand A.A., Liu W., Yoshihara M., Asai N., Suzuki H., Takahashi M., and Nakashima I. 2000. Ultraviolet light induces redox reaction-mediated dimerization and superactivation of oncogenic Ret tyrosine kinases. *Mol. Biol. Cell* **11:** 93.

Kennedy G.C., Matsuzaki H., Dong S., Liu W.M., Huang J., Liu G., Su X., Cao M., Chen W., Zhang J., Liu W., Yang G., Di X., Ryder T., He Z., Surti U., Phillips M.S., Boyce-Jacino M.T., Fodor S.P., and Jones K.W. 2003. Large-scale genotyping of complex DNA. *Nat. Biotechnol.* **21:**1233.

Klar A.J. 1996. A single locus, RGHT, specifies preference for hand utilization in humans. *Cold Spring Harbor Symp. Quant. Biol.* **61:** 59.

Lander E.S. 1996. The new genomics: Global views of biology. *Science* **274:** 536.

Li C. and Wong W.H. 2001. Model-based analysis of oligonucleotide arrays: Expression index computation and outlier detection. *Proc. Natl. Acad. Sci.* **98:** 31.

Loots G.G., Locksley R.M., Blankespoor C.M., Wang Z.E., Miller W., Rubin E.M., and Frazer K.A. 2000. Identification of a coordinate regulator of interleukins 4, 13, and 5 by cross-species sequence comparisons. *Science* **288:** 136.

Margulies E.H., Blanchette M., NISC Comparative Sequencing Program, Haussler D., and Green E.D. 2003. Identification and characterization of multi-species conserved sequences. *Genome Res.* **13:** 2507.

McCallion A.S., Stames E., Conlon R.A., and Chakravarti A. 2003. Phenotype variation in two-locus mouse models of Hirschsprung disease: Tissue-specific interaction between Ret and Ednrb. *Proc. Natl. Acad. Sci.* **100:** 1826.

Passarge E. 1967. The genetics of Hirschsprung's disease. Evidence for heterogeneous etiology and a study of sixty-three families. *N. Engl. J. Med.* **276:** 138.

Pennacchio L.A. and Rubin E.M. 2001. Genomic strategies to identify mammalian regulatory sequences. *Nat. Rev. Genet.* **2:** 100.

Puffenberger E.G., Hosoda K., Washington S.S., Nakao K., deWit D., Yanagisawa M., and Chakravarti A. 1994a. A missense mutation of the endothelin-B receptor gene in multigenic Hirschsprung's disease. *Cell* **79:** 1257.

Puffenberger E.G., Kauffman E.R., Bolk S., Matise T.C., Washington S.S., Angrist M., Weissenbach J., Garver K.L., Mascari M., and Ladda R., et al. 1994b. Identity-by-descent and association mapping of a recessive gene for Hirschsprung disease on human chromosome 13q22. *Hum. Mol. Genet.* **3:** 1217.

Reymond A., Marigo V., Yaylaoglu M.B., Leoni A., Ucla C., Scamuffa N., Caccioppoli C., Dermitzakis E.T., Lyle R., Banfi S., Eichele G., Antonarakis S.E., and Ballabio A. 2002. Human chromosome 21 gene expression atlas in the mouse. *Nature* **420:** 582.

Risch N. and Merikangas K. 1996. The future of genetic studies of complex human diseases. *Science* **273:** 1516.

Rosen D.R., Siddique T., Patterson D., Figlewicz D.A., Sapp P., Hentati A., Donaldson D., Goto J., O'Regan J.P., and Deng H.X., et al. 1993. Mutations in Cu/Zn superoxide dismutase gene are associated with familial amyotrophic lateral sclerosis. *Nature* **362:** 59.

Sancandi M., Griseri P., Pesce B., Patrone G., Puppo F., Lerone M., Martucciello G., Romeo G., Ravazzolo R., Devoto M., and Ceccherini I. 2003. Single nucleotide polymorphic alleles in the 5′ region of the *RET* proto-oncogene define a risk haplotype in Hirschsprung's disease. *J. Med. Genet.* **40:** 714.

Schwartz S., Zhang Z., Frazer K.A., Smit A., Riemer C., Bouck J., Gibbs R., Hardison R., and Miller W. 2000. PipMaker—A web server for aligning two genomic DNA sequences. *Genome Res.* **10:** 577.

Spielman R.S., McGinnis R.E., and Ewens W.J. 1993. Transmission test for linkage disequilibrium: The insulin gene region and insulin-dependent diabetes mellitus (IDDM). *Am. J. Hum. Genet.* **52:** 506.

Thomas J.W., Touchman J.W., Blakesley R.W., Bouffard G.G., Beckstrom-Sternberg S.M., Margulies E.H., Blanchette M., Siepel A.C., Thomas P.J., McDowell J.C., Maskeri B., Hansen N.F., Schwartz M.S., Weber R.J., Kent W.J., Karolchik D., Bruen T.C., Bevan R., Cutler D.J., Schwartz S., Elnitski L., Idol J.R., Prasad A.B., Lee-Lin S.Q., and Maduro V.V., et al. 2003. Comparative analyses of multi-species sequences from targeted genomic regions. *Nature* **424:** 788.

Tu P.H., Raju P., Robinson K.A., Gurney M.E., Trojanowski J.Q., and Lee V.M. 1996. Transgenic mice carrying a human mutant superoxide dismutase transgene develop neuronal cytoskeletal pathology resembling human amyotrophic lateral sclerosis lesions. *Proc. Natl. Acad. Sci.* **93:** 3155.

# Genetics of Schizophrenia and Bipolar Affective Disorder: Strategies to Identify Candidate Genes

D.J. PORTEOUS,* K.L. EVANS,* J.K. MILLAR ,* B.S. PICKARD,* P.A. THOMSON,* R. JAMES,* S. MACGREGOR,† N.R. WRAY,* P.M. VISSCHER,† W.J. MUIR,‡ AND D.H. BLACKWOOD‡

*Medical Genetics Section, University of Edinburgh, Western General Hospital, Edinburgh, Scotland, EH4 2XU; †Institute of Cell, Animal and Population Biology, Ashworth Laboratories, School of Biology, The University of Edinburgh, The King's Buildings, Edinburgh, Scotland, EH9 3JT; ‡Division of Psychiatry, University of Edinburgh, Royal Edinburgh Hospital, Edinburgh, Scotland, EH10 5HF

Schizophrenia (SCZ) and bipolar affective disorder (BPAD) (formerly termed manic-depressive illness) are severe, disabling psychiatric illnesses that feature prominently in the top ten causes of disability worldwide (Lopez and Murray 1998). Each will affect about 1% of the population in their lifetime. The cost of providing treatment for mental illness in the UK National Health Service is estimated at 10% of total expenditure. The total costs (medical, social, economic) are estimated at £32 billion per annum for the population of 50 million in England (Bird 1999; see also www.mentalhealth.org.uk). World Health Organization predictions indicate that major depression will be second only to heart disease in terms of disability-adjusted life years (DALYs) by 2020 (Lopez and Murray 1998). Research into the causes of these devastating disorders and the development of improved interventions is a high scientific, social, individual, and public health priority. Unfortunately, these remain Cinderella disorders compared to cancer and heart disease, both in terms of governmental and societal recognition and national and international research support. Despite their high prevalence and, indeed, decades of neuroscience research, little is known with certainty about their cellular and molecular bases. Consequently, treatments remain largely empirical and palliative. This apparent impasse, however, justifies neither complacency nor despondency, because the one consistent, replicable finding is that family, twin, and adoption studies demonstrate a major genetic component in both SCZ and BPAD (Merikangas and Risch 2003). A decade ago only a handful of genes for monogenic disorders had been positionally cloned through linkage studies. Over the past five years, as the Human Genome Project and associated tools have developed, that number has soared to over 1000. As the Human Genome Project passes from formal completion to full development as a universal biological tool, we can expect accelerated progress in tackling the much more difficult task of genetically dissecting the common complex disorders including major mental illness.

## EVIDENCE FOR A GENETIC COMPONENT TO SCZ AND BPAD

The risk to a first-degree relative of a person affected by SCZ or BPAD is an order of magnitude higher than that of the general population (from ~1% to 10–15% lifetime risk for each disorder) (Kendler and Diehl 1993; Merikangas and Risch 2003). Further evidence for the high heritability of SCZ and of BPAD comes from twin and adoption studies. Concordance rates for monozygotic twins are around 50% for SCZ and 60–80% for BPAD. Importantly, these concordance rates are much higher than for dizygotic twins (10–15%), and the biological risk is unaffected by adoption. Several chromosomal regions that may harbor susceptibility genes for SCZ and for BPAD have been identified by a range of genetic strategies (Owen et al. 2000; Potash and DePaulo 2000; Riley and McGuffin 2000). There is also growing evidence that relatives of sufferers are at higher risk of other psychiatric diagnoses within the schizophrenia–affective disorder spectrum (Kendler et al. 1998; Wildenauer et al. 1999; Berrettini 2000; Valles et al. 2000). This suggests that some genetic risk factors may contribute to a range of psychotic symptoms that cross the traditional diagnostic boundaries of SCZ and affective disorders. In part, this may reflect the fact that diagnosis of psychiatric illness is an imprecise science; psychiatric phenotypes are almost entirely based on profiles of behavioral indices and communication patterns. The use of standardized diagnostic criteria (such as the Diagnostic and Statistical Manual of Mental Disorders, DSM-IV, published by the American Psychiatric Association, and The ICD-10 Classification of Mental Health and Behavioural Disorders, published by the World Health Organization) has ensured good reproducibility of diagnoses between researchers, but there is wide overlap of symptoms between the diagnostic categories of SCZ, BPAD, and recurrent, major (unipolar) depression. In the absence of reliable biological or genetic markers specific for SCZ or BPAD, the validity of existing classification remains uncertain. It would be a major advance if genetic studies yielded molecular diagnostic methods that could not only resolve some of the present uncertainties in psychiatric diagnosis, but also inform treatment choice and disease prognosis. Until such time, we can only speculate that variability in individual diagnoses reflects a combination of genetic (and possibly allelic) heterogeneity, polygenic inheritance, and variation in individual life trajectories/environmental exposures. Nevertheless, post-genome science most certainly offers the best hope for determining the biological basis of these devastating conditions and for developing effective, evidence-based treatments.

## GENETICS OF COMPLEX DISEASE AND THE PARTICULAR CHALLENGES FOR PSYCHIATRIC GENETICS

The most appropriate way forward in dissecting the genetics of common disease is subject to much debate (Terwilliger and Weiss 1998; Evans et al. 2001a; Botstein and Risch 2003). The most pragmatic is to follow both main schools of thought as they each have strengths pertaining to different underlying genetic models. The first focuses on variants, which may be rare in the population, but which have detectable, individual effects. For these, the methods that worked well for single-gene disorders, a combination of molecular cytogenetics, family linkage studies, and candidate gene studies, can be transferred directly. It is, however, worth commenting on how problematic the last of these options is for the psychiatric geneticist. Unlike the common genetic disorders of the other major organs, such as cancer or heart disease, cellular pathology in the brain of SCZ and BPAD patients is poorly defined, and brain biopsies are rarely available. With perhaps a third of all known genes expressed in the brain and a plethora of neurodevelopmental and neurophysiological theoretical constructs to choose between, the number of possible candidates is perhaps greater than for any other class of organic disorder. Add to these considerable problems the more general issue of incomplete penetrance, and the full magnitude of the task will be appreciated. The age at first diagnosis for SCZ is typically in late adolescence/young adulthood, slightly later for BPAD, indicating a biological (and genetically influenced) developmental window for expression or, possibly and related to this, an age-dependent susceptibility to life changes and exposures. The severity and individual response to illness are also highly variable. The history of drug medication will likewise be variable and may result in pathological side effects. The field would benefit greatly from a set of validated biomarkers.

Linkage studies are dependent on the availability of informative families, and application to complex traits can be problematic, because it may be hard to define a genetic model that explains the observed inheritance pattern. Replication of linkage results is needed to provide credibility to initial linkage reports, which are likely to overestimate the size of effect, whereas subsequent studies should reflect the true size of effect (Lander and Kruglyak 1995; Terwilliger and Weiss 1998; Botstein and Risch 2003). Lander and Kruglyak (1995) provided a generally accepted yardstick by which to assess claims for genome-wide linkage. They recommended that a LOD of 3.3 be taken as primary evidence for significant linkage accounting for multiple testing issues. Where that value was matched or exceeded, they argued that additional studies with $p$ values of 0.01 (or LOD = 1.2) can be taken as evidence for replication (as testing is of a prior hypothesis). Failure to replicate does not necessarily imply an initial report is a false positive; it may reflect lack of power in the replication study, population heterogeneity, diagnostic differences, or statistical fluctuation (Lander and Kruglyak, 1995). We explain below the extent to which these yardsticks have been matched.

A second school of thought argues from a theoretical and experimental perspective for a polygenic model to explain the genetic variation in complex traits. By virtue of their frequency in the human population, the argument goes that the origins of common genetic disease in humans most probably lie in common and ancient sequence variants and in multiple variants of additive action. As Risch and Merikangas (1996) pointed out, if the genotype relative risk (GRR) is small (as predicted by the common ancient variant hypothesis), then unachievable numbers of simplex families (sib pairs or trios) would be required to reach genome-wide significance (see also Botstein and Risch 2003). On the other hand, association studies, the formal comparison of allele frequencies in cases and matched controls, do have the power to detect even a very modest GRR with realistic numbers of unrelated samples. The association mapping strategy is powerful and attractive, yet problematic. It relies on the scored variant, typically a single-nucleotide polymorphism (SNP), of which there are some 3 million known examples to choose between in the public domain (http://www.ncbi.nlm.nih.gov/SNP; http://snp.cshl.org), being a functional (disease-causing) variant itself or, much more likely, in linkage disequilibrium with a causative variant. It is becoming apparent that linkage disequilibrium is unevenly and unpredictably distributed across the genome (varying by over two orders of magnitude) and that although some studies have shown that a limited number of common haplotypes occur across ethnically diverse populations (see, e.g., Gabriel et al. 2002), other studies have shown that there are a limited number of haplotypes in any one population, but that there is significant variation in haplotypes between populations (see, e.g., Kauppi et al. 2003). This emphasizes the importance of case-matching control strategies that avoid the confounder of population stratification (Clayton and McKeigue 2001). Of course, it is feasible to simply increase the number and density of SNPs tested, but this raises both cost and the statistical problem of multiple testing. The problems inherent in map-based association studies have prompted others to promote a sequence-based strategy that focuses on variation within coding elements as an empirical way to reduce cost, a rational way to reduce the problem of multiple testing, and a logical way to maximize the probability that causal SNPs will be discovered and typed (Botstein and Risch 2003). We and others would extend the logic to include regulatory elements, identified empirically or highlighted by virtue of comparative genome analysis. There is thus no doubt about the importance of association studies as part of the end game in positional cloning in complex trait genetics, but the value of extended family-based studies should not be underestimated.

As already alluded to, there are several alternative family-based linkage approaches, which typically test segregation in multiplex families (multiple affecteds in three- or more generation families), affected sib pairs or trios (one affected and both parents). To identify susceptibility loci for highly heritable, genetically complex diseases such as SCZ or BPAD, appropriate study design is essential. This topic has been the subject of intense debate as to which of the two main family-based strategies should be employed: large numbers of small families, or smaller numbers of large extended families. The differential efficacy of the two depends on the distribution of genetic risk for the disease of interest. If the risk loci are common in the population and have a small effect on disease risk, the small-families strategy is preferred, largely because the overall sample size can be maximized. On the other hand, if the susceptibility loci are more genetically heterogeneous (multiple susceptibility loci), the best strategy is to identify larger, extended families. Such large families minimize heterogeneity by collecting large numbers of related individuals, most of whom share the same disease risk alleles. Although collecting large numbers of small families may seem a cost-effective strategy, if heterogeneity is expected (as in the case of SCZ and BAPD), the power to detect susceptibility loci will be limited. This point was made clearly by MacGregor et al. (2002) in response to Levinson et al. (2002); it is true that a very large ($n$ = 900) sib pair collection has 80% power to detect linkage with a LOD >3 if 75% of families are segregating mutations at the gene of interest, but realistically, the power drops to 40% even if there is only 50% heterogeneity, and to just 5% power at a still modest level of 33% heterogeneity (MacGregor et al. 2002). Thus, unless selected from a single population isolate, the power to replicate linkage detected in multiplex families is quite limited, even with very large sib pair collections. This theoretical analysis justifies our decision to adopt a twin-pronged approach to the difficult task of identifying susceptibility genes for SCZ and BPAD, through genome-wide surveys for cytogenetic lesions in probands and through genome-wide linkage studies on multiplex families.

The usefulness of this approach is exemplified by the genetics of Alzheimer's disease (AD) (http://www.ncbi.nlm.nih.gov:80/entrez/dispomim.cgi?id=104300). The high incidence of AD in individuals with Down syndrome (DS) implicated a gene or genes on Chromosome 21 and, subsequently, a mutation was found in the Chromosome 21 amyloid precursor protein (APP) gene in an early-onset extended AD family. Linkage analysis in other early-onset families then also implicated the presenilin 1 (PS1) gene. Presenilin 2 (PS2) was then discovered by virtue of its homology with PS1. Linkage analysis identified a locus on 19q13.1-q13.3, and subsequent association studies identified the apolipoprotein E (ApoE) gene from this region. The ApoE4 allele has been shown to be a significant risk factor in both familial and sporadic AD. Downstream of these genetic discoveries, research has linked the presenilins to the γ-secretase activity that regulates APP processing (Hardy and Israel 1999) and has shown that the learning deficit in mice that overexpress human mutant APP can be rescued by immunization with APP antibodies (Chen et al. 2000; Janus et al. 2000). Thus, the genetics of AD illustrates the value of combining cytogenetics, linkage, and association to dissect late-onset, neurological disorders of complex behavioral phenotype and inheritance pattern and connecting these to the characteristic pathology through biological study.

## GENOME-WIDE STRATEGIES FOR MAPPING SUSCEPTIBILITY GENES: MOLECULAR CYTOGENETICS

### Genetic Evidence for the DISC1 Locus as a Risk Factor in Psychosis

We first reported evidence for linkage (LOD = 3.3) between a balanced reciprocal translocation between human Chromosomes 1 and 11 and major mental illness (SCZ, schizoaffective disorder, and recurrent major depression) in a single Scottish family (St Clair et al. 1990) as the Human Genome Project was formally launched. Follow-up investigations of the reciprocal translocation identified the two breakpoints as t(1;11) (1q42;q14) (Muir et al. 1995). Positional cloning of the breakpoint identified two novel genes disrupted by the translocation on Chromosome 1 (DISC1/DISC2, for *d*isrupted *i*n *sc*hizophrenia 1 and 2) (Fig. 1, top) (Millar et al. 2000b). A third gene, TRAX, undergoes intergenic splicing to DISC1 and may affect DISC1 expression (Fig. 1, top) (Millar et al. 2000a). There are no genes near the breakpoint on Chromosome 11. A clinical follow-up on the family (Blackwood et al. 2001), which described additional family members with psychosis, reported a LOD score of 7.1 when subjects with recurrent major depression, BPAD, or SCZ were classed as affected (Fig. 2).

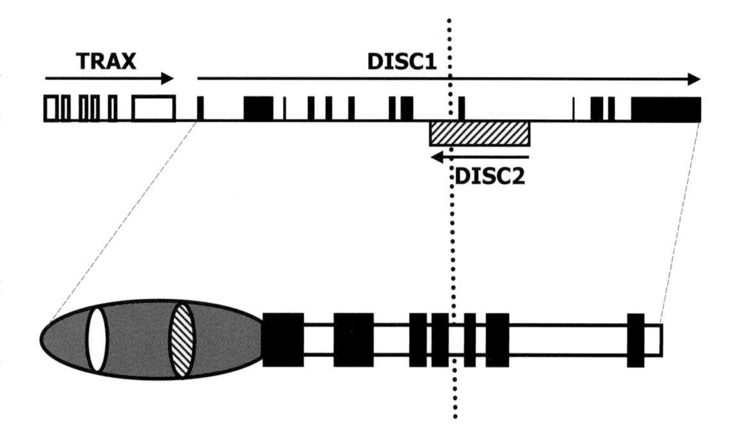


**Figure 1.** The genomic organization of the DISC1 locus and predicted protein structure of DISC1. (*Top*) Schematic of the transcriptional orientation and exon/intron structure of TRAX (*open boxes*) lying immediately centromeric (5′) of DISC1, the exon/intron structure of DISC1 (*solid black boxes*), the antiparallel and antisense position of DISC2 (*hatched box*) and, as a dotted vertical line, the position of the t(1;11) translocation breakpoint. (*Bottom*) Schematic view of the predicted protein structure of DISC1, indicating the amino-terminal globular head domain (*solid gray ellipse*), including putative nuclear localization signals (*open ellipse*), a serine/phenylalanine-rich domain (*hatched ellipse*), and seven domains (*solid black boxes*) with coiled-coil-forming potential within the carboxy-terminal tail domain (*open*).

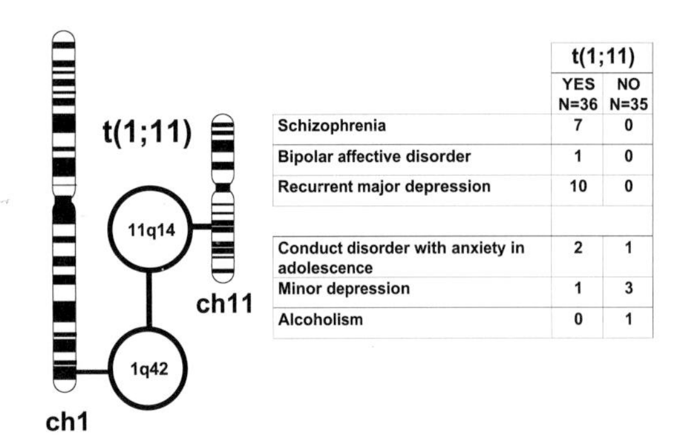


| | t(1;11) | |
|---|---|---|
| | YES N=36 | NO N=35 |
| Schizophrenia | 7 | 0 |
| Bipolar affective disorder | 1 | 0 |
| Recurrent major depression | 10 | 0 |
| | | |
| Conduct disorder with anxiety in adolescence | 2 | 1 |
| Minor depression | 1 | 3 |
| Alcoholism | 0 | 1 |

**Figure 2.** The t(1:11) translocation and segregation with psychosis. Diagram of the t(1;11), which breaks at ch1q42 and ch11q14. The diagnosis of patients (ascertained blinded to karyotype status) is tabulated. The LOD score for SCZ alone is 3.4 and for SCZ plus BPAD and recurrent major depression is 7.1 (Blackwood et al. 2001).

The most parsimonious explanation for the correlation between the t(1;11) and psychosis is as a consequence of DISC1/DISC2 gene disruption (Millar et al. 2003b), but brief mention should be made of an alternative explanation proposed by Klar (2002), who seized on the observation that about half of the t(1:11) subjects developed a major psychotic diagnosis, and about half did not. Klar (2002) proposes an elegant, if elaborate, hypothesis that requires three hypothetical genes, *DOH1, SEG*, and *RGHT*. *DOH1* is required for brain laterality specification and is active on one chromatid, but switched off on the other by imprinting; *SEG* exists on the same chromosome, but is unlinked to DISC1; and a *trans*-acting factor, *RGHT* (*right* hander), utilizes *SEG* to govern nonrandom segregation of sister chromatids at the critical period of development when brain hemisphere asymmetry is determined. It is then hypothesized that the translocation separates *DOH1* from *SEG*. The net consequence of the t(1;11) according to this scheme is that 50% of individuals carrying the translocation will have abnormal brain laterality and develop psychiatric illness, and the other 50% will be normal. The clinical picture is not so simple. This five-generation family has been under clinical observation and care for over 30 years (Blackwood et al. 2001). Overall, 62% of t(1;11) carriers are affected, but if the youngest generation is set aside (many of the individuals not yet having reached the average age of onset at the time of ascertainment), the figure rises to 70% of translocation carriers being affected. When the P300 event-related potential (P300 ERP), a trait marker of risk, is taken into account, the proportion of affected translocation carriers increases still further (Blackwood et al. 2001). The P300 ERP is considered to be a measure of the pace and efficiency of information processing in the brain and is abnormal in translocation carriers, with or without a major psychiatric diagnosis. This indicates the presence of central nervous system abnormalities in *all* translocation carriers, including those who are clinically unaffected, and is thus inconsistent with the strand segregation hypothesis. In contrast, the proportion of clinically affected translocation carriers and the P300 ERP data are consistent with our model of dominant inheritance coupled with reduced penetrance and variable expressivity dependent on the action of modifiers (genetic and/or environmental). Indeed, we propose that disruption of *DISC1* and *DISC2*, plus the actions of genetic and environmental modifying factors, are sufficient to explain the psychiatric illness arising from inheritance of the t(1;11) chromosome (Millar et al. 2003b). In support of this argument, the relative risk of the t(1;11) is equivalent to that of the monozygotic co-twin of an affected individual.

## Replication of the Genetic Evidence for the DISC1 Locus as a Risk Factor

The most convincing independent genetic evidence so far for the involvement of the TRAX/DISC1 region in mental illness has come from the Finnish population. Initial linkage evidence was found in an internal isolate of Finland (LOD = 3.7) and a common haplotype spanning 6.6 cM reported in 3 of the 20 families multiply affected with SCZ and schizoaffective disorder (Hovatta et al. 1999). A later analysis of 134 sib pairs collected from throughout Finland gave additional support for the same 1q32.2-q41 region on Chromosome 1 with a maximum LOD of 2.6 for a diagnostic class that included SCZ, schizophrenia spectrum disorders, and bipolar and unipolar disorders (Ekelund et al. 2000). Fine mapping of both Finnish populations resulted in maximum LOD scores for SCZ and schizoaffective disorder at D1S2709, a microsatellite located in intron 9 of DISC1, in both the combined sample ($Z_{max}$ = 2.71) and outside the internal isolate ($Z_{max}$ = 3.21) (Ekelund et al. 2001).

Other data implicating this region in psychosis have come from linkage analysis in families of diverse ethnic origin. Detera-Wadleigh et al. (1999) reported a maximum LOD of 2.67 in a 30-cM region spanning 1q25-1q42 in bipolar families, some of whose members were affected by SCZ or schizoaffective disorder. Gejman et al. (1993) reported a maximum LOD of 2.39 at D1S103 (1q42.2) in one North American family with bipolar disorder. Most recently, a suggestive nonparametric linkage score (NPL = 2.18), equating approximately to a LOD score of 1.2, has been reported in Taiwanese families with diagnoses of SCZ, schizoaffective disorder, and other non-affective psychotic disorders (Hwu et al. 2003). Our own studies have also found evidence for linkage in a subset of BPAD families in Scotland (S. MacGregor et al., unpubl.).

Finally, as discussed earlier in the context of the t(1;11) clinical follow-up study, the use of endophenotypes in molecular studies may overcome the effect of reduced penetrance on the ability to detect linkage and association and help to resolve the affected status of family members. In this context, it is worth mentioning that Gasperoni et al. (2003) undertook a QTL analysis of spatial working memory, an endophenotype for SCZ (Cannon et al. 2000; Glahn et al. 2003). Gasperoni et al. (2003) provide evidence of linkage and association between D1S283 at ch1q42 and variation in spatial working memory between individuals affected by SCZ and their unaffected co-twins ($p = 0.007$ and $p = 0.003$, respectively). This marker was the most telomeric of the markers tested and lies in 1q42.2 approximately 1 Mb centromeric to DISC1.

In light of the growing evidence of independent linkage of 1q42 to major mental illness, we have reanalyzed data initially published in Devon et al. (2001). In that study, we described 15 SNP variants in the DISC1 gene, of which 4 were carried forward for an association study, but no evidence for statistically significant association with any one SNP, or pair of consecutive SNPs, was found (Devon et al. 2001). There is, however, weak evidence of association ($p = 0.0034$; $p = 0.044$, corrected for multiple testing) of a 3-SNP haplotype and BPAD. None of the other 3 or 4 SNP haplotypes showed significant association with BPAD or SCZ (data not shown). This preliminary finding provides tentative evidence of association in the Scottish population between BPAD and polymorphisms in DISC1/DISC2. Further higher resolution association studies stratified by clinical phenotype may allow us to define more specifically the nature and effect size of DISC1/DISC2 variants.

### Biology and Predicted Function of DISC1

Full-length *DISC1* is composed of 13 exons and covers over 50 kb of genomic sequence and is subject to alternative splicing. The full-length or long (L) transcript utilizes all 13 exons. A commonly spliced variant of full-length *DISC1*, the long variant (Lv) transcript, utilizes a proximal splice site in exon 11, thereby reducing the transcript by 66 nucleotides. Additional *DISC1* transcripts have been identified by RT-PCR and Northern blotting, and protein isoforms consistent with these alternative splice forms have been detected by Western blotting. *DISC1* transcripts are detected in all tissues tested (Millar et al. 2000b), and similarly, DISC1 protein shows a widespread pattern of expression.

In silico analysis was performed on the protein sequence of the human L DISC1 isoform (Millar et al. 2000b; Taylor et al. 2003). DISC1 consists of amino- and carboxy-terminal domains (Fig. 1, bottom). The two termini approximate to exon boundaries, with the first two exons encoding the amino terminus and the remainder of the gene encoding the carboxyl terminus. The two termini can also be distinguished on the basis of secondary structure prediction and levels of conservation between species. The amino terminus is made up of one or more globular domains, with two nuclear localization signals (NLS) (Ma et al. 2002; Taylor et al. 2003). The amino terminus shares no homologies with any known proteins. Conversely, the carboxyl terminus consists of α-helical and looped structures, interspersed with regions of coiled-coil-forming potential. Similarities exist between the carboxyl terminus and structural proteins or proteins involved in transport and motility (Millar et al. 2000b). However, the similarities are within the coiled-coil regions and are unlikely to be functionally relevant. Due to the modular structure of the coiled coils, these regions are anticipated to represent the protein interaction domains of DISC1 (Taylor et al. 2003). In addition, three leucine zippers are present within the carboxyl terminus (Ma et al. 2002) and may also have a role in mediating DISC1 protein interactions (Fig. 1, bottom).

*DISC1* orthologs have been identified in mouse, rat, and fish species (Ma et al. 2002; Ozeki et al. 2003; Taylor et al. 2003). The overall gene structure is maintained, and there is evidence for alternative splicing in mouse, rat, and pufferfish. DISC1 is poorly conserved across species, with the amino terminus being less well conserved than the carboxyl terminus. The amino terminus shows 52% identity and 63% similarity, and the carboxyl terminus 61% identity and 78% similarity, between human and mouse (Taylor et al. 2003). The amino and carboxyl termini are conserved, as are the amino-terminal NLS and carboxy-terminal coiled-coil regions, suggesting that these are functionally important.

Preliminary evidence for protein–protein interactions comes from Ozeki et al. (2003), who reported the results of yeast two-hybrid studies, using full-length and truncated DISC1 protein as bait. They identified NudE-like (NUDEL), a cytoskeletal protein expressed in the cortex, as a strong interactor with DISC1. Our own results support the NUDEL interaction and identify several other potential interactors of known neuronal function (Millar et al. 2003a).

To further investigate the function of DISC1, we have raised antibodies to both the amino and carboxyl termini of the protein. The subcellular distribution of DISC1 is cell-type-specific and most likely reflects the cell shape and concomitant organization of the cytoskeleton (data not shown). DISC1 is predominantly, but not exclusively, expressed in the mitochondrion (Ozeki et al. 2003; James et al. 2004). In differentiated neuroblastoma cells, DISC1 redistributes to the shafts and tips of developing neurites, suggesting potential involvement of DISC1 in neurite outgrowth (Fig. 3).

### Other Cytogenetic Rearrangements Associated with Psychosis

The potential opportunities provided by rare chromosome aberrations, such as those used to pinpoint single-gene disorders in the first phase of the Human Genome Project (Collins 1992) and as described here for SCZ in the t(1;11) family, have led to the widespread use of karyotype screening of psychiatric patients. Many karyotypic rearrangements have been defined at the visual, "Giemsa-band" level, but relatively few as yet at the molecular level (for review, see MacIntyre et al. 2003). We have extended our studies to 12 additional cases and families in whom chromosome rearrangements are associated with diagnoses ranging from SCZ to BPAD as well as "co-morbid" diagnoses where SCZ is coupled with learning disability (US: mental retardation) (B.S. Pickard et al., in prep.). Nearly all of the constituent chromosomal breakpoints from each of these abnormalities have now been mapped and their genomic environments searched for

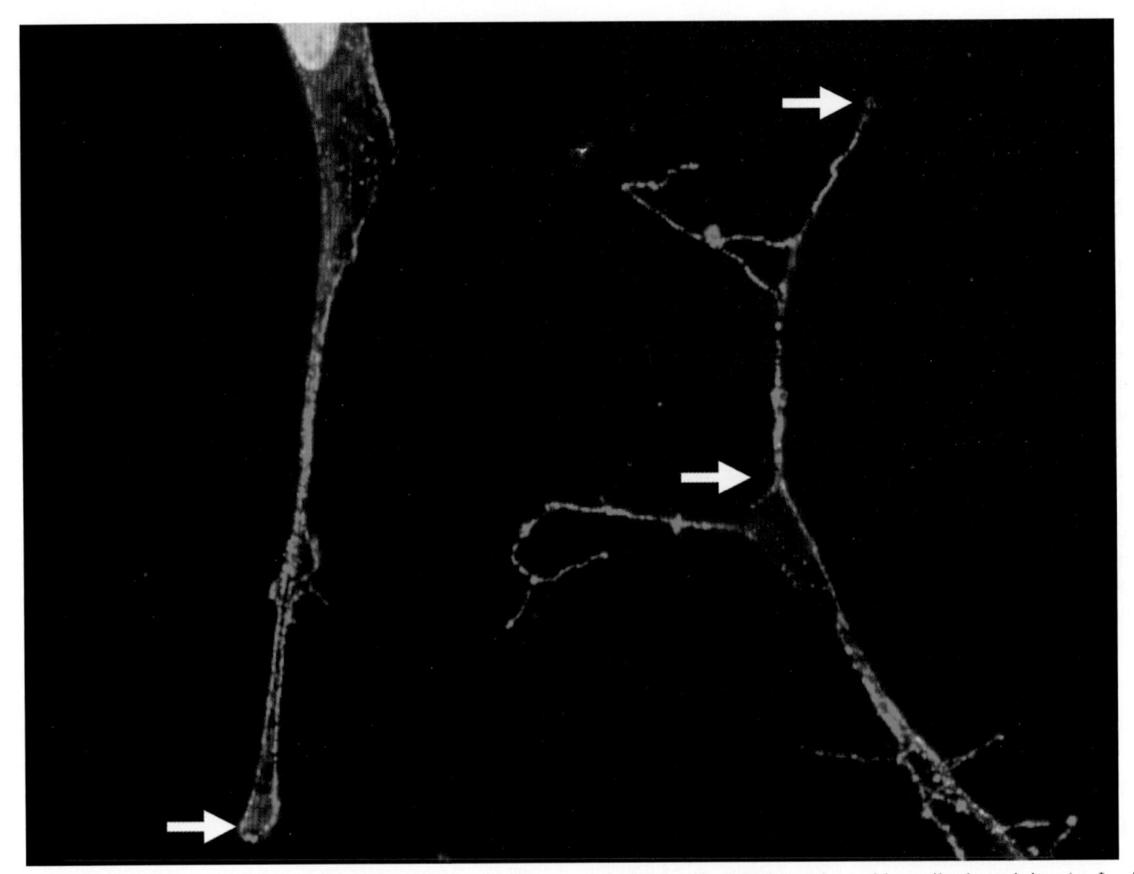

**Figure 3.** DISC1 protein expression in differentiated neuroblastoma-derived cells. DISC1 anti-peptide antibody staining (*red*) with anti-actin staining (*green*) and DAPI-stained nucleus (*blue*) shows that DISC1 protein is concentrated at the branch points (*arrowed*) and growing neurite tips (*arrowed*) of neuroblastoma-derived SH-SY5Y cells induced to differentiate in the presence of retinoic acid.

candidate genes. Emerging data from the Human Genome Project have been instrumental in this process in two ways. First, the physical map of BAC and PAC clones has provided cytogenetically cross-referenced molecular probes for fluorescence in situ hybridization (FISH) to patient chromosomes, positioning the breakpoints to ~200-kb windows (Fig. 4). Second, the collation, positioning, and annotation of expressed sequence tags (ESTs) and defined gene transcripts have allowed the rapid identification of potential candidate genes. Of 11 abnormalities where breakpoints have been experimentally defined, 8 directly disrupt or are predicted to positively influence at least one gene. The remaining 3 have breakpoints positioned where perturbations of nearby genes could be postulated. One such gene, *NPAS3*, encodes a transcription factor expressed in the central nervous system (Kamnasaran et al. 2003; B.S. Pickard et al., in prep.). Intriguingly, its close homolog, *NPAS2*, has been implicated in synaptic long-term potentiation (LTP) and the cellular energy-state-dependent modification of circadian rhythms (Garcia et al. 2000; Reick et al. 2001; Rutter et al. 2001; Dioum et al. 2002). Recently, *DIBD1* (for *d*isrupted *i*n *b*ipolar *d*isorder 1), encoding a component of the pathway responsible for adding carbohydrate moieties to membrane-bound and secreted proteins, has been cloned by a similar strategy from a family where BPAD was prevalent (Baysal et al. 2002). These suc-

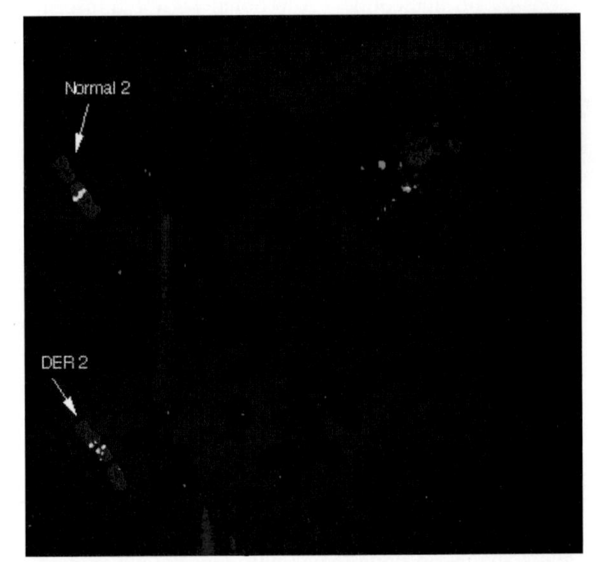


**Figure 4.** FISH of a metaphase chromosome spread from a patient with SCZ and mild learning disability (US: mental retardation). The red signal is obtained from a Chromosome-2 "paint" probe mix. The gap in the long arm of one Chromosome 2 represents the insertion of a portion of the short arm of Chromosome 5. This is highlighted by the yellow yeast artificial chromosome (YAC) probe which spans the point of insertion on Chromosome 2. Placing this YAC onto the Human Genome Project BAC/PAC physical map led to the identification of a Chromosome-2 gene disrupted by the insertion event.

cesses have been mirrored in the study of other complex neurological/psychiatric conditions such as autism (Vincent et al. 2000; Tentler et al. 2001; Sultana et al. 2002) and speech and language development (Fisher et al. 1998; Lai et al. 2001), where chromosome abnormalities have pinpointed individual susceptibility genes. Functional studies of these and other candidate genes resulting from this approach are certain to shed light on the developmental and molecular pathways and processes that go awry in mental illness, as well as suggesting rational choices for future population-based association and mutation detection studies.

## GENOME-WIDE STRATEGIES FOR MAPPING SUSCEPTIBILITY GENES: LINKAGE STUDIES IN MULTIPLEX FAMILIES

Genome-wide linkage studies in multiplex families have been criticized because of inconsistent replications in genome scans of affected sib pairs or large collections of small families. As discussed above, failure to replicate could be due to lack of power of existing sample sets to detect alleles of small effect, and/or to the presence of substantial locus heterogeneity. Thus, it remains to be determined where the balance lies between the models of Mendelian inheritance with genetic heterogeneity and a fully quantitative model. In BPAD, significant linkage has been reported in extended pedigrees on Chromosomes 1q, 4p, 4q, 12q, 18q, and 21q (Potash and DePaulo 2000). The original 4p report came from our genome-wide linkage study in a large Scottish pedigree (F22) that segregates major affective disorder (Blackwood et al. 1996). A whole-genome scan of F22 found significant linkage to Chromosome 4p16 (LOD score = 4.8). Subsequent to that report, a number of other groups have also found evidence for linkage of major psychiatric illness to this region. Detera-Wadleigh et al. (1999) carried out a genome-wide scan of 22 BPAD families, and their largest family (F48) generated a LOD of 3.24. Asherson et al. (1998) found linkage (LOD = 1.96) in schizoaffective family CF50. Ewald et al. (1998) reported linkage in BPAD families (LOD = 2.0). Williams et al. (1999) found increased allele sharing (LOD = 1.73) in SCZ. Polymeropoulos and Schaffer (1996) described LODs between 1 and 2 in a BPAD family. In our follow-up studies of 57 Scottish families, we found evidence for linkage in a second Scottish BPAD family (F59, LOD = 1.15). The LOD score in this family comes very close to meeting the replication criteria proposed by Lander and Kruglyak (1995), but the maximum possible LOD score is limited by the small size of the family. It is not possible to estimate accurately the proportion of BPAD families linked to 4p, but a first approximation is 4%, based on the figures currently available; i.e., 2/58 in a Scottish sample, 1/24 in a Welsh sample (Asherson et al. 1998), and 1/22 in a US sample (Detera-Wadleigh et al. 1999).

The high LOD score of 4.8 generated by the linked markers in F22 indicates that the disease haplotype is very likely to contain a susceptibility locus for psychiatric illness. Further statistical evidence for this conclusion

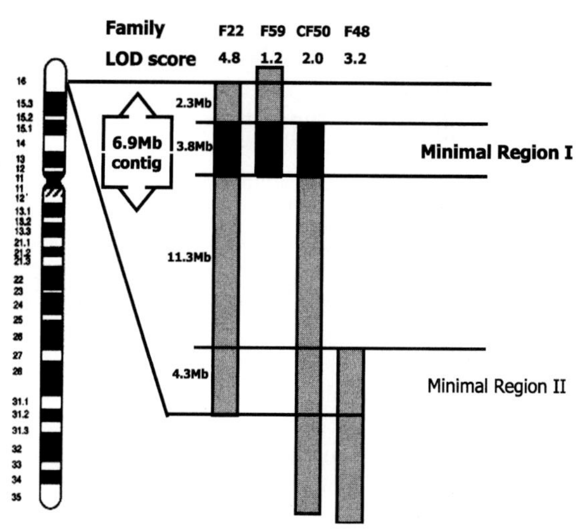


**Figure 5.** Genetic linkage to psychosis to Chromosome 4p16. The figure shows the LOD scores and a representation of the recombination intervals for each of four 4p16 linked families. Their diagnostic features are, respectively: F22 and F59, major affective disorder; F50, SCZ and schizoaffective disorder; F48, SCZ and major affective disorder. A BAC and PAC clone contig of 6.9 Mb with a single gap (estimated size 150–350 kb) covers all of Minimal Region I, plus proximal and distal sequence. A contig for the entire 20-Mbp region is curated in a custom version of AceDB.

comes from variance component analysis of the same data, which found significant evidence (LOD = 3.7) for a BAPD locus in the region (Visscher et al. 1999). Family F48 is also of a size that by itself can generate a significant LOD score (Lander and Kruglyak 1995). The use of smaller families that do not have the ability to generate significant LOD scores is a higher risk strategy, due to the increased likelihood of false positives. However, if such families contain affected individuals with recombination events that are complementary to those of the larger families, this allows the division of the candidate region into subregions with different priority levels, as described for Chromosome 4p in Figure 5. The proximal and distal boundary of the candidate region is defined by the recombination breakpoints in F22 (see Fig. 5). Subregions have been prioritized for candidate gene analysis on the strength of evidence provided by the other linked families. On this basis, Minimal Region I is perhaps the most promising, being common to three out of four families (all of Celtic origin), followed by Minimal Region II (3/4, including F48 of Ashkenazi Jewish origin).

As a platform for genome annotation and association mapping (collaboration with Dr. Mark Ross, Sanger Institute, UK), we combined large-insert genomic library screening with computational analysis to construct and curate in ACeDB (Eeckman and Durbin 1995) and SAM (Soderlund et al. 1997) a 6.9-Mb contig (consisting of 460 overlapping BAC and PAC clones) that encompasses and extends proximal and distal to Minimal Region I (Evans et al. 2001b). Advances in the Human Genome Project mean that these can now be largely constructed in silico. However, it was our experience that successive builds of the human genome (http://genome.ucsc.edu/

cgi-bin/hgGateway) resulted in quite major changes in the public domain view of the region, which were inconsistent with the physical map built "in-house" by detailed STS mapping. A small gap in the contig, estimated at 150–350 kb, still remains to be convincingly filled. This region is repeat-dense and underrepresented in YAC, PAC, and BAC libraries (K.L. Evans et al., unpubl.).

A clone contig of the region facilitates the effective use of genomic sequence to construct a transcript map and to conduct comparative genetic analyses that focus on the search for DNA sequence variants within coding or regulatory regions (Semple et al. 2002). We have set up our ACeDB database to display the automated results of exon and gene prediction programs and BLAST searches of EST databases and genomic sequence from human and other organisms, assisting recognition of both regulatory and coding regions, as well as prediction of biological function. The assembled evidence is being used to direct laboratory work to confirm these annotations and generate reagents for biological analyses, including accessing multiple, full-length cDNA libraries (in collaboration with Dr. Kate Rice and colleagues, Sanger Institute, UK). The physical and transcript maps form the basis for SNP discovery and association studies. The rest of the shared haplotype region has been built in silico from the Golden Path (http://genome.ucsc.edu/cgi-bin/hgGateway) and by in-house annotation and experimentation.

The most parsimonious explanation for the convergence of linkage evidence on Minimal Region I in the families of Celtic origin is a founder mutation. If there is a founder mutation common to more than one family, this will be flanked by a region of haplotype sharing that may vary in extent between families, but will include a functional variant that is common to and specific to all cases. With this in mind, we have looked for evidence of allele sharing between families. We have preliminary evidence for allele sharing in Minimal Region I, but definitive evidence awaits a much higher density of marker analysis. We have drawn upon the public domain SNP databases (dbSNP at NCBI, http://www.ncbi.nlm.nih.gov/SNP and the SNP Consortium, http://snp.cshl.org), but in practice, a significant proportion are uninformative in our samples. We are therefore undertaking SNP discovery, focusing on coding and putative regulatory regions in affected and control chromosomes from the linked families, and testing these SNPs for allele sharing between families prior to large-scale association analysis (Le Hellard et al. 2002). To enhance our capacity for SNP discovery and the power to detect associations with rare SNP variants (Thomas et al. 2004), we have adapted a somatic cell hybrid approach (Douglas et al. 2001) to derive haploid reagents for key probands and random cases.

A strength of our overall strategy is that the case and control samples, like F22 and F59, are drawn from the Scottish population, which has a low attrition rate (<2% per annum) and which, from HLA analyses, is thought to be one of the most genetically homogeneous populations in Europe (Cavalli-Sforza and Piazza 1993). This minimizes the possibility of admixture and increases the possibility of detecting a true association. The association study is being carried out in two phases. The first phase involves screening a subset of the sample in the form of parent offspring trios to allow selection of SNPs that represent haplotype blocks. The construction of this preliminary LD map also allows an examination of the extent to which all haplotype blocks in the Scottish population have been covered. Within haplotype blocks, SNP selection for phase 2 favors those that show preliminary evidence for association. In the second phase, association analysis is carried out on a much larger set of unrelated cases and controls, allowing replication of the first data set with high power to detect significant association. This two-stage strategy has a number of advantages: Replication is important in ruling out false-positive association results; the need for correction for multiple testing is reduced, as only a subset of markers are genotyped on the second set, and savings of both DNA and money are made. Our ACeDB database has been extended to facilitate the management and analysis of the data generated by association studies.

Once an associated region has been identified, candidate variants are prioritized by their predicted effect on protein function and the extent to which they are shared by linked families. We have invested in establishing lymphoblastoid cell lines for a large number of patients so that we have the capacity to undertake proteomic studies. If the region identified contains large numbers of variants, it may be necessary also to look for evidence for association with illness in other populations, particularly in populations that display lower levels of linkage disequilibrium, thus decreasing the number of associated variants. Prioritization of the variants will be followed by (1) assessment of individual and population attributable risk, (2) identification and analysis of gene(s) function, (3) study of how function might be affected by the variant(s), and (4) identification of biochemical pathways, which should also identify targets for the development of novel treatments and additional candidates for further genetic study.

## CONCLUSIONS

There have been several false dawns in the early history of psychiatric genetics, but there is growing confidence, backed by persuasive statistical findings and biological associations, that there is at last cause for cautious optimism (Botstein and Risch 2003; Merikangas and Risch 2003). The recent evidence for linkage and association to the neuregulin (NRG1) gene by Stefansson and colleagues (2002, 2003) at deCODE Genetics both supports our position on the value of genome-wide studies of multiplex families and emphasizes the need for functional evaluation of gene discoveries.

Our own results provide clear statistical evidence for a BPAD locus of major effect in family F22 and persuasive evidence for replication in other families. The task of defining the locus in molecular terms is still challenging. It remains to be seen whether this locus represents a rare or common variant in the Scottish population and, indeed, other populations. If there is a founder effect, then alleles sharing between (distantly related) families narrow the locus. On the other hand, allelic heterogeneity, as seen often in Mendelian disorders, may be valuable for confirming a candidate gene but poses additional, statistical challenges for discovery (Pritchard 2001). Nevertheless, within the overall framework of the Human Genome Project (Lander et al. 2001) and the steady impact of emergent genomic technologies, we can be confident that this is now a tractable problem.

The cytogenetic approach, which was so successful in the early stages of the Human Genome Project in pinpointing single-gene disorders, has, in our opinion, received insufficient attention as a strategy for dissecting complex genetic disorders. This is all the more surprising, given the fact that somatic rearrangement has been the single most productive approach to gene discovery in cancer and that phenotypically complex disorders of the nervous system such as Down syndrome, neurofibromatosis, AD, autism, and specific language disorder have all benefited from a cytogenetic approach. The explanation may reside in part in a passionate debate over the underlying genetic model for SCZ and for BPAD. In the absence of empirical evidence, however, it is unwise to settle on a preferred model or oblige the data to fit that model. Our empirical evidence from molecular cloning of a balanced t(1;11) breakpoint in a multigeneration family with a high loading of psychosis points to DISC1 as a very promising candidate gene. This finding, made through molecular cytogenetics, has been replicated by conventional family- and population-based studies in both Finland and Scotland. Bioinformatic analysis of DISC1 identified potential structural and interaction domains, but few clues as to function (Taylor et al. 2003). Protein–protein interaction studies (Millar et al. 2003a; Ozeki et al. 2003) will be important in establishing the biological pathway(s) in which DISC1 acts. Immunocytochemical studies point to a possible role in neurite outgrowth (Fig. 4) (James et al. 2004). The emergence of a growing list of equally interesting genes identified by molecular cytogenetic studies in other families further vindicates the general approach. Combining the availability of a BAC contig with high-resolution fluorescence in situ hybridization techniques to analyze metaphase chromosomes is a low-tech solution to genome-wide screens that, on our evidence, are likely to bring valuable benefits to other researchers and, we conjecture, are entirely applicable to other common, complex genetic disorders.

Figure 6 summarizes our broad hopes for where this research will eventually lead. We aim through family- and population-based studies to be able to nominate candidate genes for biological study and to feed a rational drug-development pipeline. The gene variants so discovered will also be valuable in the near term as molecular diagnostics, for disease surveillance, for monitoring response to treatment, and for treatment choice. Low compliance and adverse drug reaction are major problems in psychiatry. This is one specific and important area where the successful application of pharmacogenetic principles (Roses 2000) would be of immense value. If the gene-to-drug pathway is indeed successful, this opens the possibility for tailored drug prescription of rational medicines. The ethical, legal, and social issues that are fundamental to all applications of the new genetics are nowhere more sensitive than in relation to behavioral disorders. There is, in our view, a realistic possibility that genetic research will lead in time to completely novel insights into these perplexing and distressing disorders and that may in turn lead to completely novel therapeutic strategies which take account of both genetic *and* environmental risk factors. Scientifically sound and ethical application of this promised knowledge to the benefit of the many who suffer from mental illness is a major challenge in itself, but is the ultimate goal that drives the research.

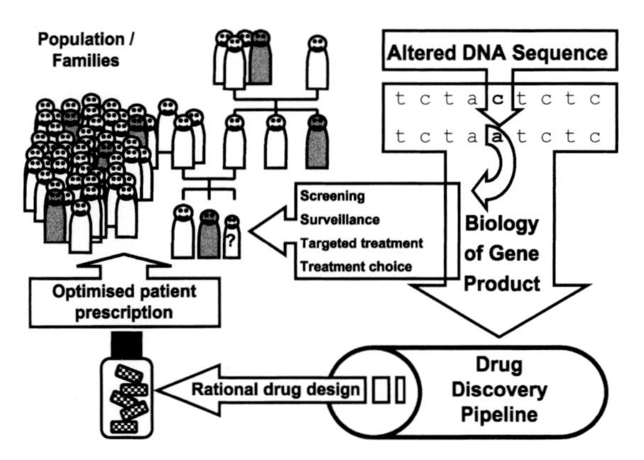


**Figure 6.** Putting it all together. Summary of the overall research objectives. Family- and population-based studies are used to nominate candidate genes for biological study and rational drug development. Gene variants can be used as molecular diagnostics, for disease surveillance, for monitoring response to treatment, and for treatment choice. If the gene-to-drug pathway is successful, this opens the possibility of tailored drug prescription of evidence-based medicines.

## ACKNOWLEDGMENTS

This work was supported in part by the UK Medical Research Council, the Caledonian Research Foundation, the Chief Scientists Office in Scotland, the Scottish Hospitals Research Endowment Trust, the Stanley Medical Research Institute, the Cunningham Trust, the PPP Foundation, Organon NV, and Merck Sharpe Dohme. We acknowledge Professor Mike Owen for access to F50 and Dr. Sevilla Detera-Wadleigh for access to F48. We acknowledge Dr. Ole Mors and Dr. Henrik Ewald, Aarhus, Denmark, for access to materials from patients with suspected cytogenetic rearrangements. We thank the Genetics Core at the Wellcome Trust Clinical Research Facility, Edinburgh (www.wtcrf.med.ed.ac.uk), for genotyping support and Drs. David Bentley, Mark Ross, and Kate Rice, Sanger Institute, UK, for help with establishing ACeDB, SAM, and access to genomic and cDNA libraries. We thank colleagues at the Medical Genetics Section and Division of Psychiatry, Edinburgh University, for support and critical comment and the many families and individuals who have helped this research.

## REFERENCES


Asherson P., Mant R., Williams N., Cardno A., Jones L., Murphy K., Collier D.A., Nanko S., Craddock N., Morris S., Muir W., Blackwood B., McGuffin P., and Owen M.J. 1998. A study of chromosome 4p markers and dopamine D5 receptor gene in schizophrenia and bipolar disorder. *Mol. Psychiatry* **3:** 310.

Baysal B.E., Willett-Brozick J.E., Badner J.A., Corona W., Ferrell R.E., Nimgaonkar V.L., and Detera-Wadleigh S.D. 2002. A mannosyltransferase gene at 11q23 is disrupted by a translocation breakpoint that co-segregates with bipolar affective disorder in a small family. *Neurogenetics* **4:** 43.

Berrettini W.H. 2000. Susceptibility loci for bipolar disorder: Overlap with inherited vulnerability to schizophrenia. *Biol. Psychiatry* **47:** 245.

Bird I. 1999. *The fundamental facts: All the latest facts and figures on mental illness*. The Mental Health Foundation, London, United Kingdom.

Blackwood D.H., Fordyce A., Walker M.T., St Clair D.M., Porteous D.J., and Muir W.J. 2001. Schizophrenia and affective disorders—Cosegregation with a translocation at chromosome 1q42 that directly disrupts brain-expressed genes: Clinical and P300 findings in a family. *Am. J. Hum. Genet.* **69:** 428.

Blackwood D.H., He L., Morris S.W., McLean A., Whitton C., Thomson M., Walker M.T., Woodburn K., Sharp C.M., Wright A.F., Shibasaki Y., St Clair D.M., Porteous D.J., and Muir W.J. 1996. A locus for bipolar affective disorder on chromosome 4p. *Nat. Genet.* **12:** 427.

Botstein D. and Risch N. 2003. Discovering genotypes underlying human phenotypes: Past successes for Mendelian disease, future approaches for complex disease. *Nat. Genet.* **33:** 228.

Cannon T.D., Huttunen M.O., Lonnqvist J., Tuulio-Henriksson A., Pirkola T., Glahn D., Finkelstein J., Hietanen M., Kaprio J., and Koskenvuo M. 2000. The inheritance of neuropsychological dysfunction in twins discordant for schizophrenia. *Am. J. Hum. Genet.* **67:** 369.

Cavalli-Sforza L.L. and Piazza A. 1993. Human genomic diversity in Europe: A summary of recent research and prospects for the future. *Eur. J. Hum. Genet.* **1:** 3.

Chen G., Chen K.S., Knox J., Inglis J., Bernard A., Martin S.J., Justice A., McConlogue L., Games D., Freedman S.B., and Morris R.G. 2000. A learning deficit related to age and beta-amyloid plaques in a mouse model of Alzheimer's disease. *Nature* **408:** 975.

Clayton D. and McKeigue P.M. 2001. Epidemiological methods for studying genes and environmental factors in complex diseases. *Lancet* **358:** 1356.

Collins F.S. 1992. Positional cloning: Let's not call it reverse anymore. *Nat. Genet.* **1:** 3.

Detera-Wadleigh S.D., Badner J.A., Berrettini W.H., Yoshikawa T., Goldin L.R., Turner G., Rollins D.Y., Moses T., Sanders A.R., Karkera J.D., Esterling L.E., Zeng J., Ferraro T.N., Guroff J.J., Kazuba D., Maxwell M.E., Nurnberger J.I., Jr., and Gershon E.S. 1999. A high-density genome scan detects evidence for a bipolar-disorder susceptibility locus on 13q32 and other potential loci on 1q32 and 18p11.2. *Proc. Natl. Acad. Sci.* **96:** 5604.

Devon R.S., Anderson S., Teague P.W., Burgess P., Kipari T.M., Semple C.A., Millar J.K., Muir W.J., Murray V., Pelosi A.J., Blackwood D.H., and Porteous D.J. 2001. Identification of polymorphisms within Disrupted in Schizophrenia 1 and Disrupted in Schizophrenia 2, and an investigation of their association with schizophrenia and bipolar affective disorder. *Psychiatr. Genet.* **11:** 71.

Dioum E.M., Rutter J., Tuckerman J.R., Gonzalez G., Gilles-Gonzalez M.A., and McKnight S.L. 2002. NPAS2: A gas-responsive transcription factor. *Science* **298:** 2385.

Douglas J.A., Boehnke M., Gillanders E., Trent J.M., and Gruber S.B. 2001. Experimentally-derived haplotypes substantially increase the efficiency of linkage disequilibrium studies. *Nat. Genet.* **28:** 361.

Eeckman F.H. and Durbin R. 1995. ACeDB and macace. *Methods Cell Biol.* **48:** 583.

Ekelund J., Lichtermann D., Hovatta I., Ellonen P., Suvisaari J., Terwilliger J.D., Juvonen H., Varilo T., Arajarvi R., Kokko-Sahin M.L., Lonnqvist J., and Peltonen L. 2000. Genome-wide scan for schizophrenia in the Finnish population: Evidence for a locus on chromosome 7q22. *Hum. Mol. Genet.* **9:** 1049.

Ekelund J., Hovatta I., Parker A., Paunio T., Varilo T., Martin R., Suhonen J., Ellonen P., Chan G., Sinsheimer J.S., Sobel E., Juvonen H., Arajarvi R., Partonen T., Suvisaari J., Lonnqvist J., Meyer J., and Peltonen L. 2001. Chromosome 1 loci in Finnish schizophrenia families. *Hum. Mol. Genet.* **10:** 1611.

Evans K.L., Muir W.J., Blackwood D.H.R., and Porteous D.J. 2001a. Nuts and bolts of psychiatric genetics: Building on the Human Genome Project. *Trends Genet.* **17:** 35.

Evans K.L., Le Hellard S., Morris S.W., Lawson D., Whitton C., Semple C.A., Fantes J.A., Torrance H.S., Malloy M.P., Maule J.C., Humphray S.J., Ross M.T., Bentley D.R., Muir W.J., Blackwood D.H., and Porteous D.J. 2001b. A 6.9-Mb high-resolution BAC/PAC contig of human 4p15.3-p16.1, a candidate region for bipolar affective disorder. *Genomics* **71:** 315.

Ewald H., Degn B., Mors O., and Kruse T.A. 1998. Support for the possible locus on chromosome 4p16 for bipolar affective disorder. *Mol. Psychiatry* **3:** 442.

Fisher S.E., Vargha-Khadem F., Watkins K.E., Monaco A.P., and Pembrey M.E. 1998. Localisation of a gene implicated in a severe speech and language disorder. *Nat. Genet.* **18:** 168.

Gabriel S.B., Schaffner S.F., Nguyen H., Moore J.M., Roy J., Blumenstiel B., Higgins J., DeFelice M., Lochner A., Faggart M., Liu-Cordero S.N., Rotimi C., Adeyemo A., Cooper R., Ward R., Lander E.S., Daly M.J., and Altshuler D. 2002. The structure of haplotype blocks in the human genome. *Science* **296:** 2225.

Garcia J.A., Zhang D., Estill S.J., Michnoff C., Rutter J., Reick M., Scott K., Diaz-Arrastia R., and McKnight S.L. 2000. Impaired cued and contextual memory in *Npas2*-deficient mice. *Science* **288:** 2226.

Gasperoni T.L., Ekelund J., Huttunen M., Palmer C.G., Tuulio-Henriksson A., Lonnqvist J., Kaprio J., Peltonen L., and Cannon T.D. 2003. Genetic linkage and association between chromosome 1q and working memory function in schizophrenia. *Am. J. Med. Genet.* **116B:** 8.

Gejman P.V., Martinez M., Cao Q., Friedman E., Berrettini W.H., Goldin L.R., Koroulakis P., Ames C., Lerman M.A., and Gershon E.S. 1993. Linkage analysis of fifty-seven microsatellite loci to bipolar disorder. *Neuropsychopharmacology* **9:** 31.

Glahn D.C., Therman S., Manninen M., Huttunen M., Kaprio J., Lonnqvist J., and Cannon T.D. 2003. Spatial working memory as an endophenotype for schizophrenia. *Biol. Psychiatry* **53:** 624.

Hardy I. and Israel A. 1999. Alzheimer's disease: In search of gamma-secretase. *Nature* **398:** 466.

Hovatta I., Varilo T., Suvisaari J., Terwilliger J.D., Ollikainen V., Arajarvi R., Juvonen H., Kokko-Sahin M.L., Vaisanen L., Mannila H., Lonnqvist J., and Peltonen L. 1999. A genomewide screen for schizophrenia genes in an isolated Finnish subpopulation, suggesting multiple susceptibility loci. *Am. J. Hum. Genet.* **65:** 1114.

Hwu H.G., Liu C.M., Fann C.S., Ou-Yang W.C., and Lee S.F. 2003. Linkage of schizophrenia with chromosome 1q loci in Taiwanese families. *Mol. Psychiatry* **8:** 445.

James R., Adams R.R., Christie S., Buchanan D., Porteous D.J., and Millar J.K. 2004. DISC1 is a multicompartmentalized protein that predominantly localises to mitochondia. *Mol. Cell. Neurosci.* (in press).

Janus C., Pearson J., McLaurin J., Mathews P.M., Jiang Y., Schmidt S.D., Chishti M.A., Horne P., Heslin D., French J., Mount H.T., Nixon R.A., Mercken M., Bergeron C., Fraser P.E., St George-Hyslop P., and Westaway D. 2000. A beta peptide immunization reduces behavioural impairment and plaques in a model of Alzheimer's disease. *Nature* **408:** 979.

Kamnasaran D., Muir W.J., Ferguson-Smith M.A., and Cox D.W. 2003 Disruption of the neuronal PAS3 gene in a family affected with schizophrenia. *J. Med. Genet.* **40:** 325.

Kauppi L., Sajantila A., and Jeffreys A.J. 2003. Recombination hotspots rather than population history dominate linkage disequilibrium in the MHC class II region. *Hum. Mol. Genet.* **12:** 33.

Kendler K.S. and Diehl S.R. 1993. The genetics of schizophrenia: A current, genetic-epidemiologic perspective. *Schizophr. Bull.* **19:** 261.

Kendler K.S., Karkowski L.M., and Walsh D. 1998. The structure of psychosis: Latent class analysis of probands from the Roscommon Family Study. *Arch. Gen. Psychiatry* **55:** 492.

Klar A.J. 2002. The chromosome 1;11 translocation provides the best evidence supporting genetic etiology for schizophrenia and bipolar affective disorders. *Genetics* **160:** 1745.

Lai C.S., Fisher S.E., Hurst J.A., Vargha-Khadem F., and Monaco A.P. 2001. A forkhead-domain gene is mutated in a severe speech and language disorder. *Nature* **413:** 519.

Lander E. and Kruglyak L. 1995. Genetic dissection of complex traits: Guidelines for interpreting and reporting linkage results. *Nat. Genet.* **11:** 241.

Lander E.S., Linton L.M., Birren B., Nusbaum C., Zody M.C., Baldwin J., Devon K., Dewar K., Doyle M., FitzHugh W., Funke R., Gage D., Harris K., Heaford A., Howland J., Kann L., Lehoczky J., LeVine R., McEwan P., McKernan K., Meldrim J., Mesirov J.P., Miranda C., Morris W., and Naylor J., et al. (International Human Genome Sequencing Consortium). 2001. Initial sequencing and analysis of the human genome. *Nature* **409:** 860.

Le Hellard S., Ballereau S.J., Visscher P.M., Torrance H.S., Pinson J., Morris S.W., Thomson M.T., Semple C.A.M., Muir W.J., Blackwood D.H.R., Porteous D.J., and Evans K.L. 2002. SNP genotyping on pooled DNAs: Comparison of genotyping technologies and a semi automated method for data storage. *Nucleic Acids Res.* **30:** e74.

Levinson D.F., Holmans P.A., Laurent C., Riley B., Pulver A.E., Gejman P.V., Schwab S.G., Williams N.M., Owen M.J., Wildenauer D.B., Sanders A.R., Nestadt G., Mowry B.J., Wormley B., Bauche S., Soubigou S., Ribble R., Nertney D.A., Liang K.Y., Martinolich L., Maier W., Norton N., Williams H., Albus M., and Carpenter E.B., et al. 2002. No major schizophrenia locus detected on chromosome 1q in a large multicenter sample. *Science* **296:** 739.

Lopez A.D. and Murray C.C. 1998. The global burden of disease. *Nat. Med.* **4:** 1241.

Ma L., Liu Y., Ky B., Shughrue P.J., Austin C.P., and Morris J.A. 2002. Cloning and characterization of *Disc1*, the mouse ortholog of DISC1 (Disrupted-in-Schizophrenia 1). *Genomics* **80:** 662.

MacGregor S., Visscher P.M., Knott S., Porteous D., Muir W., Millar K., and Blackwood D. 2002. Is schizophrenia linked to chromosome 1q? *Science* **298:** 2277.

MacIntyre D.J., Blackwood D.H., Porteous D.J., Pickard B.S., and Muir W.J. 2003. Chromosomal abnormalities and mental illness. *Mol. Psychiatry* **8:** 275.

Merikangas K.R. and Risch N. 2003. Will the genomics revolution revolutionize psychiatry? *Am. J. Psychiatry* **160:** 625.

Millar J.K., Christie S., and Porteous D.J. 2003a. Yeast two-hybrid screens implicate DISC1 in brain development and function. *Biochem. Biophs. Res. Commun.* **311:** 1019.

Millar J.K., Christie S., Semple C.A., and Porteous D.J. 2000a. Chromosomal location and genomic structure of the human translin-associated factor X gene (TRAX; TSNAX) revealed by intergenic splicing to DISC1, a gene disrupted by a translocation segregating with schizophrenia. *Genomics* **67:** 69.

Millar J.K., Thomson P.A., Wray N.R., Muir W.J., Blackwood D.H., and Porteous D.J. 2003b. Response to Amar J. Klar: The chromosome 1;11 translocation provides the best evidence supporting genetic etiology for schizophrenia and bipolar affective disorders. *Genetics* **163**: 833.

Millar J.K., Wilson-Annan J.C., Anderson S., Christie S., Taylor M.S., Semple C.A., Devon R.S., Clair D.M., Muir W.J., Blackwood D.H., and Porteous D.J. 2000b. Disruption of two novel genes by a translocation co-segregating with schizophrenia. *Hum. Mol. Genet.* **9:** 1415.

Muir W.J., Gosden C.M., Brookes A.J., Fantes J., Evans K.L., Maguire S.M., Stevenson B., Boyle S., Blackwood D.H., St Clair D.M., Porteous D.J., and Weith A. 1995. Direct microdissection and microcloning of a translocation breakpoint region, t(1;11)(q42.2;q21), associated with schizophrenia. *Cytogenet. Cell Genet.* **70:** 35.

Owen M.J., Cardno A.G., and O'Donovan M.C. 2000. Psychiatric genetics: Back to the future. *Mol. Psychiatry* **5:** 22.

Ozeki Y., Tomoda T., Kleiderlein J., Kamiya A., Bord L., Fujii K., Okawa M., Yamada N., Hatten M.E., Snyder S.H., Ross C.A., and Sawa A. 2003. Disrupted-in-Schizophrenia-1 (DISC-1): Mutant truncation prevents binding to NudE-like (NUDEL) and inhibits neurite outgrowth. *Proc. Natl. Acad. Sci.* **100:** 289.

Polymeropoulos M.H. and Schaffer A.A. 1996. Scanning the genome with 1772 microsatellite markers in search of a bipolar disorder susceptibility gene. *Mol. Psychiatry* **1:** 404.

Potash J.B. and DePaulo J.R., Jr. 2000. Searching high and low: A review of the genetics of bipolar disorder. *Bipolar Disord.* **2:** 8.

Pritchard J.K. 2001. Are rare variants responsible for susceptibility to complex diseases? *Am. J. Hum. Genet.* **69:** 124.

Reick M., Garcia J.A., Dudley C., and McKnight S.L. 2001. NPAS2: An analog of clock operative in the mammalian forebrain. *Science* **293:** 506.

Riley B.P. and McGuffin P. 2000. Linkage and associated studies of schizophrenia. *Am. J. Med. Genet.* **97:** 23.

Roses A. 2000. Pharmacogenetics and the practice of medicine. *Nature* **405:** 857.

Rutter J., Reick M., Wu L.C., and McKnight S.L. 2001. Regulation of clock and NPAS2 DNA binding by the redox state of NAD cofactors. *Science* **293:** 510.

Semple C.A.M., Morris S.W., Porteous D.J., and Evans K.L. 2002. Computational comparison of human genomic sequence assemblies for a region of chromosome 4. *Genome Res.* **12:** 424.

Soderlund C., Longden I., and Mott R. 1997. FPC: A system for building contigs from restriction fingerprinted clones. *CABIOS* **13:** 523.

St Clair D., Blackwood D.H., Muir W.J., Carothers A., Walker M., Spowart G., Gosden C., and Evans H.J. 1990. Association within a family of a balanced autosomal translocation with major mental illness. *Lancet* **336:** 13.

Stefansson H., Sarginson J., Kong A., Yates P., Steinthorsdottir V., Gudfinnsson E., Gunnarsdottir S., Walker N., Petursson H., Crombie C., Ingason A., Gulcher J.R., Stefansson K., and St Clair D. 2003. Association of neuregulin 1 with schizophrenia confirmed in a Scottish population. *Am. J. Hum. Genet.* **72:** 83.

Stefansson H., Sigurdsson E., Steinthorsdottir V., Bjornsdottir S., Sigmundsson T., Ghosh S., Brynjolfsson J., Gunnarsdottir S., Ivarsson O., Chou T.T., Hjaltason O., Birgisdottir B., Jonsson H., Gudnadottir V.G., Gudmundsdottir E., Bjornsson A., Ingvarsson B., Ingason A., Sigfusson S., Hardardottir H., Harvey R.P., Lai D., Zhou M., Brunner D., and Mutel V., et al. 2002. Neuregulin 1 and susceptibility to schizophrenia. *Am. J. Hum. Genet.* **71:** 877.

Sultana R., Yu C.E., Yu J., Munson J., Chen D., Hua W., Estes A., Cortes F., de la Barra F., Yu D., Haider S.T., Trask B.J., Green E.D., Raskind W.H., Disteche C.M., Wijsman E., Dawson G., Storm D.R., Schellenberg G.D., and Villacres E.C. 2002. Identification of a novel gene on chromosome 7q11.2 interrupted by a translocation breakpoint in a pair of autistic twins. *Genomics* **80:** 129.

Taylor M.S., Devon R.S., Millar J.K., and Porteous D.J. 2003. Evolutionary constraints on the Disrupted in Schizophrenia locus. *Genomics* **81:** 67.

Tentler D., Brandberg G., Betancur C., Gillberg C., Anneren G., Orsmark C., Green E.D., Carlsson B., and Dahl N. 2001. A balanced reciprocal translocation t(5;7)(q14;q32) associated with autistic disorder: Molecular analysis of the chromosome 7 breakpoint. *Am. J. Med. Genet.* **105:** 729.

Terwilliger J.D. and Weiss K.M. 1998. Linkage disequilibrium mapping of complex disease: Fantasy or reality? *Curr. Opin. Biotechnol.* **9:** 578.

Thomas S., Porteous D.J., and Visscher P.M. 2004. Power of direct vs. indirect haplotyping in association studies. *Genet. Epidemiol.* **26:** 116.

Valles V., Van Os J., Guillamat R., Gutierrez B., Campillo M., Gento P., and Fananas L. 2000. Increased morbid risk for schizophrenia in families of in-patients with bipolar illness. *Schizophr. Res.* **42:** 83.

Vincent J.B., Herbrick J.A., Gurling H.M., Bolton P.F., Roberts W., and Scherer S.W. 2000. Identification of a novel gene on chromosome 7q31 that is interrupted by a translocation breakpoint in an autistic individual. *Am. J. Hum. Genet.* **67:** 510.

Visscher P.M., Haley C.S., Heath S.C., Muir W.J., and Blackwood D.H. 1999. Detecting QTLs for uni- and bipolar disorder using a variance component method. *Psychiatr. Genet.* **9:** 75.

Wildenauer D.B., Schwab S.G., Maier W., and Detera-Wadleigh S.D. 1999. Do schizophrenia and affective disorder share susceptibility genes? *Schizophr. Res.* **39:** 107.

Williams N.M., Rees M.I., Holmans P., Norton N., Cardno A.G., Jones L.A., Murphy K.C., Sanders R.D., McCarthy G., Gray M.Y., Fenton I., McGuffin P., and Owen M.J. 1999. A two-stage genome scan for schizophrenia susceptibility genes in 196 affected sibling pairs. *Hum. Mol. Genet.* **9:** 1729.

## WEB SITES

www.mentalhealth.org.uk
http://www.ncbi.nlm.nih.gov:80/entrez/dispomim.cgi?id=104300
http://genome.ucsc.edu/cgi-bin/hgGateway
http://www.ncbi.nlm.nih.gov/SNP
http://snp.cshl.org
www.wtcrf.med.ed.ac.uk

# The Genetics of Common Diseases: 10 Million Times as Hard

D.B. Goldstein, G.L. Cavalleri, and K.R. Ahmadi
*Department of Biology (Galton Lab), University College London, London, United Kingdom*

The transition taking place in human genetics could not be more pronounced. Over the past 20 odd years, human geneticists have identified more than 1200 genes causing Mendelian diseases, and with few exceptions, the identifications have been unambiguous and uncontroversial. On the other hand, we know next to nothing about the genetics of common diseases that are influenced by multiple genes in a complicated interaction with the environment. There are only about 20 different polymorphisms that are generally accepted as risk factors for these so-called "complex" diseases (Glazier et al. 2002; Hirschhorn et al. 2002; Botstein and Risch 2003). It is not only in the disparity between the remarkable successes in the Mendelian cases, and the striking failure in the case of common diseases, that the contrast is apparent. It is also in the cultures of the two research enterprises. When a new Mendelian mutation is identified, the evidence usually requires little qualification: The mutation is found in multiple affected members of a pedigree, it is never observed in unaffected individuals, and it knocks out or alters the properties of an important and relevant protein, and so on. There is no follow-up flood of publications debating whether or not it is really the disease gene. When results on common diseases are presented, it is not uncommon for researchers to say that "this association looks biologically plausible so I tend to believe it." A literature debate invariably ensues.

Common disease genetics is clearly a different sort of research enterprise. The community of human geneticists, however, has been molded by its successes in studying Mendelian diseases. It is not clear that the prevailing perspectives and expectations resulting from this experience are relevant and useful in the study of complex traits such as common disease and variable drug responses. Here we review genetic association studies, expected to be one of the main tools for elucidating the genetics of complex traits. Along the way, we comment on changes in perspective that we think will help to address the new challenges associated with studying common diseases as opposed to the rare and genetically simpler Mendelian diseases.

## WHAT WE ARE LOOKING FOR

The study of Mendelian diseases led to widespread usage of terms such as "disease-causing mutation" or even the "gene for" a given disease. Some have noted that it is not appropriate to speak of a gene for a disease, since the normal function of the gene often has, at best, an indirect relationship to the disease phenotype. But to the extent that certain mutations in a given gene (sometimes only that gene) result invariably in a disease, speaking of disease-causing mutations or even a disease gene does not seem inappropriate.

This terminology is much harder to justify in the study of common diseases, yet geneticists often describe their work as a search for "susceptibility genes" for a given complex condition. The usefulness and accuracy of this terminology, however, as a description of how genetic differences among people moderate their common-disease risks is far from clear.

What little we do know about the genetics of common diseases, and of responses to their treatment, is that derived variants at a given gene can lead to increased or decreased susceptibility or result in good or bad responses to a given medicine (e.g., in the ABCB1 multi-drug resistance gene, Siddiqui et al. [2003] showed that the derived variant results in a better response to anti-epileptic drugs while, e.g., the CCR5 deletion mutation confers resistance to HIV [Dean et al. 1996]). More fundamentally, it seems likely that there is a continuum of effect sizes for variants that influence common diseases, ranging from moderate effects such as that of ApoE4 for Alzheimer's, to apparently more marginal effects such as that of PPARγ and Calpain-10 on Type II diabetes (odds ratios estimated at 3.3, 1.23, and 1.19, respectively) (Weedon et al 2003; McCarthy 2004; Zondervan and Cardon 2004). Looked at this way, it is hard to see where one would draw the line to declare a particular gene a "susceptibility" gene for the condition. ApoE4 is certainly a risk factor for Alzheimer's, although interestingly, there are suggestions that it is deleterious only in certain environments (Pastor et al. 2003). There is no reason to believe there is a small set of polymorphisms with clear and strong effects, like apoE4, and that the remainder of the polymorphisms have no effect. Rather, it would appear more likely that there is a full continuum of effect sizes (and degrees of environmental dependency) and that whether a gene registers as a susceptibility gene will depend on the power of detection. Consistent with this view, estimated odds ratios do not appear clustered above any threshold (see Fig. 1). For current sample sizes and study designs, it would seem that few genes carry polymorphisms that are consistently identified as risk factors (Lohmueller et al. 2003). For larger sample sizes, or designs which consider appropriate environmental interactions, it is probable that a much more sizable minority of the genes in the genome will have polymorphisms in or near them that have some effect on risk in some environments (Zondervan and Cardon 2004). Currently, there is no way to guess how large

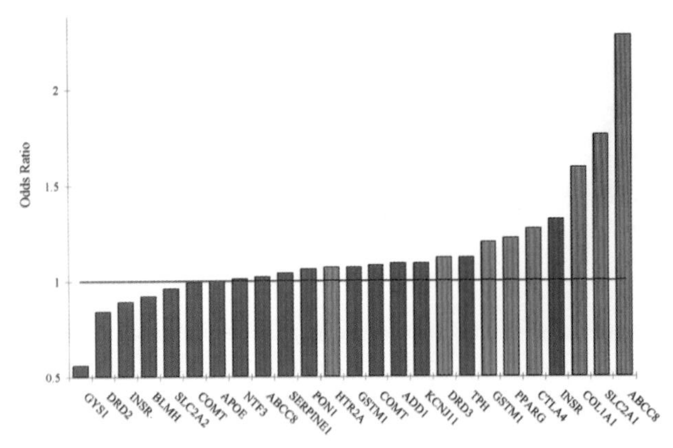


**Figure 1.** Pooled odds ratios for 25 polymorphisms reported as associated with common diseases for variants that are (*dark blue*) and are not (*red*) replicated by meta-analysis (data from Lohmueller et al. 2003). Note that the values indicated are for the associations studied in Lohmueller et al. (2003), and not necessarily for the condition with which the given polymorphism is most strongly associated. For example, the odds ratio estimate for APOE relates to schizophrenia, not Alzheimer's disease.

this proportion might be. It thus seems premature to cast the search as one for a small number of susceptibility genes for each condition.

We therefore think that study of the genetics of common diseases is better cast not as a search for "disease" genes, but rather as an assessment of how polymorphisms in the human genome influence disease risk and drug responses in specific environments and genetic backgrounds. This may sound like a subtle or pedantic distinction, but there are implications both in the way research is motivated and interpreted, and in how it is perceived by the public. For example, current interpretations of association studies tend to be typological in assessing whether a variant really is or is not a risk factor. We believe that this is too driven by the Mendelian experience, and that the real point is to understand the effects of the polymorphism, as opposed to reach conclusive agreement that a polymorphism has *some* effect. Understanding whether and how polymorphisms have their effects in given environments and genetic backgrounds is a long-term process that does not have any clear stopping point. Similarly, the Mendelian experience leads the public to the view that there are good and bad variants, whereas what little we know about complex disease causation suggests that few variants are uniformly bad or uniformly good.

Looked at this way, the research program ahead of us is open-ended, and we should be clear about how hard it will be. There are more than 10 million polymorphic sites out of the approximately 3 billion bases in our genome in typical human populations (where polymorphic is defined as having a minor allele frequency greater than 1%; cf. Kruglyak and Nickerson 2001). The job is not to find the few disease polymorphisms out of these 10 million for any given condition. Instead, the challenge is to understand how these 10 million polymorphisms collectively moderate disease risk, and how they influence response to treatment, in their appropriate genetic and environmental backgrounds.

A focus on even this large set of sites assumes that the most important genetic contributions are due to sites that carry classic polymorphisms (i.e., minor allele frequency greater than 1%). Clearly, these polymorphic sites contribute primarily to genetic differences among people on average, but there is considerable debate about their importance relative to less common variation in the genetics of common disease predisposition (Wright and Hastie 2001). When the problem is cast this way, it is apparent that very large scale and high-quality genetic association studies, with all their problems (Zondervan and Cardon 2004), are indispensable in the effort to understand the genetics of common diseases and variable drug reactions.

## AND HOW TO FIND IT

Most genes causing Mendelian diseases have been identified using linkage analyses that assess the co-inheritance of disease and genetic markers in pedigrees. Linkage analyses, however, are known to have limited power to identify variants that have only modest effects on disease risks. Risch and Merikangas (1996) showed that for variants of only modest effects, association studies, which assess correlations between diseases and gene variations in random population samples, have considerably greater power than linkage-based methods (Risch and Merikangas 1996). Two basic strategies have been advocated. Botstein and Risch have recently argued for a *Sequence-based* or *direct* approach, which focuses on the identification of all variants in and near exons, and in core promoter regions of genes (Botstein and Risch 2003). They argue that the experience of Mendelian disease suggests that this is where most of the important variants will reside. All identified variants in these putatively functional genomic regions would then be genotyped in individuals of known phenotype, and correlations assessed in a case-control or related design.

The alternative, *Map-based* approach, refrains from making any assumptions about the precise genomic location of causal variants. Instead, the aim is to identify a map of markers that is designed to be sufficient to be associated with all other variants in a gene or gene region of interest. Both approaches have strengths and weaknesses. The principal advantages of the map-based approach are economy and the fact that it does not require strong assumptions about where the important variants are. The principal weakness is that indirect approaches will have difficulty in accurately representing variants with low minor allele frequency, as discussed below. For these reasons, we see the two approaches as more complementary than competing. Here, we describe the development of map-based approaches.

## LINKAGE DISEQUILIBRIUM GENE MAPPING

Linkage disequilibrium (LD) gene mapping was first used successfully to fine-localize Mendelian mutations within genomic regions implicated through linkage analyses, as for example, in the cloning of the gene responsible for cystic fibrosis (Riordan et al. 1989).

Two factors greatly increased interest in the application of LD mapping to common diseases. One was the demonstration of Risch and Merikangas (1996) of the greater power of association mapping over linkage studies. They showed that for modest effect sizes, affected sib pair analyses, for example, would require unrealistically large sample sizes. The other was the rapidly expanding coverage of the human genome with polymorphic markers.

The potential application of LD mapping for common disease generated intense interest in patterns of linkage disequilibrium in human populations. Until 2001, this work was predicated in many ways on expectations of gradual, though noisy, decay of LD with physical distance in the genome. It had long been appreciated that certain genes, such as the β-globin gene cluster (Chakravati et al. 1984; Jeffries et al. 2000), harbored hot spots of recombination—regions of intense homologous recombination—and also that LD decay was often poorly correlated with distance, either because of uneven recombination in some regions or because of the stochastic effects.

Nevertheless, many empirical studies looked at the average decay in LD with distance over different genomic regions, effectively ignoring the precise pattern of decay within a given region. In addition, models used to generate expectation of the pattern of LD assumed uniform recombination. In his influential report, Leonid Kruglyak established a sort of null model for the expected extent of LD by assuming an idealized human demographic history and assuming homogeneous recombination rates. Under these assumptions, Kruglyak (1999) showed that usable levels of LD (at that time set at $d^2>0.1$, which corresponds to a $r^2<0.2$; see below) would not extend farther than about 3 kb in the human genome, implying that a genome-wide map would require more than 1 million markers.

This conclusion was at variance, however, with empirical descriptions of LD; for example, Reich et al. (2001), looking at multiple genomic regions, published conclusive evidence of much higher levels of LD extending over relatively long sequence tracts. The reasons for the discrepancy between early modelling and empirical patterns depend on both the ways in which human demographics differ from the models assumptions, and also the fact that recombination in the human genome is not uniform. The relative importance of these two factors in shaping patterns of human LD remains unclear.

## BLOCKS OF LINKAGE DISEQUILIBRIUM

The tendency to view LD decay with distance as gradual (although noisy) was dramatically changed by a set of papers published in the October 2001 issue of *Nature Genetics*. The challenge to the "gradual-decay" view had two components. Analyzing 103 common SNPs (MAF>5%) across 500 kb of Chromosome 5q31, Daly et al. (2001) argued that the pattern was better viewed as discrete, with stretches of sequence showing little or no LD breakdown interspersed by regions of sharper LD breakdown. The stretches of limited haplotype diversity were called blocks, and within them up to 95% of the observed chromosomes were accounted for by 3 or 4 haplotypes (Daly et al. 2001). Similar patterns were reported by Johnson et al., who observed a similar block-like pattern of LD among 122 SNPs across 135 kb of 9 genes in European populations. Haplotype diversity within blocks was low, with a maximum of 6 common haplotypes (frequency >5%) observed within any one block (Johnson et al. 2001).

These publications led directly to the idea that blocks of LD, within which haplotype diversity is limited, are a prevailing characteristic of the genome. This in turn led to the concept that a set of SNPs could represent, or tag, each of the common haplotypes in a given region. These SNPs were first referred to as haplotype-tagging SNPs (htSNPs). In an accompanying supplement to the Johnson et al. study, David Clayton introduced an approach for selecting htSNPs that focused on the proportion of the haplotype diversity that could be explained by a set of tags (Johnson et al. 2001). This definition of htSNPs was not explicitly tied to the idea of a block, but given its focus on tagging a small number of haplotypes, it has often been perceived in the community as applying to the selection of htSNPs within blocks (see below).

In the same issue, Jeffries et al. published data related to a possible cause of uneven decay of LD. Through single sperm typing in the class II major histocompatibility complex (MHC) region, they showed recombination events to be clustered in narrow, discrete regions (Jeffreys et al. 2001).

This shows that at least in one genomic region the location of recombination hot spots corresponds with, and drives discontinuities in, the pattern of LD. There is currently a great deal of debate about the extent to which LD discontinuities coincide with recombination rate hot spots. However, this should not distract from the central contribution of the Johnson et al., Daly et al., and Jeffries et al. papers. In breaking with the strong tradition of viewing LD decay as relatively smooth and homogeneous, this work represented a striking and important conceptual advance. After these papers, it was immediately unacceptable to focus on descriptions of LD that ignored the precise pattern of decay within a region, and similarly unacceptable to make predictions based on models that assumed uniform recombination rates.

Whatever the causes of sharp LD discontinuities, they have important implications for the design and interpretation of association studies. The first large-scale study explicitly incorporating the idea of very uneven patterns of LD decay, or blocks of LD, was reported less than a year later by Gabriel et al. (2002), who analyzed patterns of haplotype diversity from over 3700 SNPs in 51 regions spanning 13 megabases of the genome in four population groups (European, African-American, Nigerian, and Chinese). The same apparent block-like pattern of LD with accompanying lack of within-block haplotype diversity as reported by Daly et al. (2001) was observed across all populations. Interestingly, the size of blocks varied across populations reflecting differing population histories, with European and Chinese populations having, on average, larger block sizes than African-American and Yoruban samples (maximum block size 173 kb vs. 94 kb) (Gabriel et al. 2002).

## TAGGING METHODOLOGY AND RELATED ISSUES

As noted, the first use of the term "tagging" was by Johnson and colleagues, who suggested the term htSNPs. In the case of all 9 genes examined in the Johnson paper, 2–5 htSNPs per gene were sufficient to tag the common haplotypes. That is, instead of typing the full set of 122 SNPS, close to the same haplotypic variation could be captured by typing a subset of 34 htSNPs (Johnson et al. 2001).

Tagging common haplotypes is only one of many possible ways to select a subset of SNPs that retain as much information as possible about the other SNPs. Broadly speaking, the approaches that have been evaluated can be divided into two groups (Weale et al. 2003): those based on maximizing the haplotype diversity present in the tagging set compared to the tagged set (*diversity based*) and those based on establishing as high an association as possible between the "tagging" and "tagged" set (*association based*). To avoid the close identification with haplotype diversity in the selection of tags, some have suggested that tags be referred to as tSNPs rather than htSNPs (see, e.g., Weale et al. 2003).

The primary motivation for tSNP selection is their application in LD-based gene mapping. For this reason, a tSNP selection criterion focused on the $r^2$ measure of LD seems the most directly relevant because it allows quantification of the loss of power in typing the tSNPs instead of all the SNPs. Pritchard and Prezeworski showed that for two biallelic loci, power scales with $r^2$, such that typing the associated marker with $n/r^2$ individuals would have approximately the same power as $n$ individuals in which the causative variant itself was typed, where $r^2$ is the association between the two variants (Pritchard and Przeworski 2001). This finding has been extended by Chapman et al. (2003) to include generalized $r^2$, including haplotype $r^2$ (see below).

## MULTIMARKER VERSUS PAIR-WISE APPROACHES

There still remains the question of how to define the $r^2$ value. Relying on pair-wise measures is straightforward, but may be inefficient. This is because pairs of SNPs will only have high pair-wise association when their minor allele frequencies are very closely matched, thus meaning that SNPs which exhibit a full range of frequencies will need to be selected as tags. This can be overcome if combinations of the tSNPs are used to predict the other SNPs. One approach to this is to use the haplotype $r^2$ value (Chapman et al. 2003; Goldstein et al. 2003; Weale et al. 2003; and the D. Clayton Web site: http://www-gene.cimr.cam. ac.uk/clayton/software).

Haplotype $r^2$ is defined as the proportion of variance in a "tagged" SNP of interest that is explained by an analysis of variance based on the $G$ haplotypes formed by the set of tSNPs.

$$Yi = x_{i1}b_1 + x_{i2}b_2 + .... + x_{iG}b_G$$

Where $Yi$ is the predicted state of the tagged SNP of interest on the $i$th chromosome, $x_{i1}...x_{iG}$ are indicator variables for the $G$ haplotypes, and $b_1...b_G$ are coefficients estimated by standard least squares from the observed data.

This approach is more efficient in the sense of requiring fewer tags because it relies on combinations of haplotypes generated by tagging SNPs to predict the state of tagged SNPs. These combinations are identified by selecting the appropriate coefficients in a linear regression. The haplotype $r^2$ criterion therefore appears an appropriate measure if one of the aims is to reduce the number of tags that must be typed in phenotyped material (e.g., cases and controls).

The haplotype $r^2$ approach focuses on the prediction of haploid allelic state (0 or 1) on the basis of the tSNP haplotype observed on a given chromosome. As such, it does not explicitly address the issue of haplotype inference in phenotyped individuals. One approach that simultaneously considers both aspects is from Stram et al. (2003), who define a coefficient of determination for predicting the haplotypes observed in an individual based on the tSNP configuration. At present, it is hard to predict which approaches to tSNP selection will prove the most useful in practice.

## BLOCK-BASED AND BLOCK-FREE SELECTION OF TAGS

Although the discovery of the block-like nature of LD and its effect on haplotype distribution inspired the idea of tags for given haplotypes (Johnson et al. 2001), the use of tSNPs in no way depends on blocks of LD. Indeed, even if there is such a block structure, it is not apparent that tag selection should make reference to blocks. As noted above, the early suggestions for the definition of htSNPs did not address this issue directly. More recently, however, we have argued that block-based identification of tags will always be less efficient (sometimes considerably) than methods that select across block boundaries. This is because tagging within blocks limits the effective range of a set of tags, and means that cross-block associations cannot be exploited (Goldstein et al. 2003). For this reason, we advocate the selection of tSNPs across large contiguous sequence stretches, independently of any underlying block structure in the region. Even so, some questions remain in this approach. For example, computational issues make it difficult to select across very large regions without some sort of subdivision. In addition, selecting across large regions may result in a set of tSNPs that are not optimized for specific subregions, as for example, a candidate gene (Goldstein et al. 2003). These are just some of the issues that will need to be addressed in order to develop appropriate, efficient strategies for genome-wide tSNP selection.

## SELECTING AND EVALUATING TAGGING SNPS

A typical way to currently apply tSNPs is to define them on the basis of incomplete genotype data (perhaps 1 SNP every 5 kb or so) available in a relatively small population sample (up to about 60 unrelated individuals, ethnically matched to case/control cohort), and then to apply them in usually much larger phenotyped populations. Increasing the number of tags increases performance, but also increases expense. A consensus seems to be emerging that the coefficient of determination should be at least 0.80, which means that the increased sample size is $n/0.85$, in comparison with exhaustive typing. It is a striking demonstration of how fast the field has advanced to note that only 4 years ago a coefficient of determination of 0.2 was seen as a reasonable goal (Kruglyak 1999). It is necessary, however, to test whether the selected tSNPs will (1) represent other SNPs not yet known and (2) tag as efficiently in a new sample of individuals from the same population.

To address the first point, we introduced a SNP dropping procedure (Goldstein et al. 2003; Weale et al. 2003). The basic approach is to take the set of known SNPs and for each SNP $i$ drop it from the analysis in turn. For each reduced set of $N$–1 SNPs, new tags are selected, and their ability to represent (predict) the dropped SNP $i$ is assessed. In this way, a statistical estimate is obtained of how well the tSNPs can represent SNPs that are not observed (for example, SNPs that are not yet discovered) in the region.

Goldstein et al. (2003) carried out analysis along these lines on the Gabriel et al. data set with SNP densities of up to 4 SNPs per 10 kb using SNPs with a minor allele frequency above 8%. Averaged over the regions considered, we found that performance increased up to a SNP density of 1.5 SNPs per 10 kb, after which no further improvement was noted (Goldstein et al. 2003). We estimated using this preliminary evidence that an average marker density of somewhere between 1 and 2 SNPs per 10 kb would be sufficient to identify tSNPs that capture most of the common allelic variation in the human genome (Goldstein et al. 2003).

The regions Goldstein et al. (2003) studied from the original Gabriel et al. (2002) paper, however, did not cover a sufficiently broad range of densities, and the average SNP density was 1 SNP approximately every 7 kb. The question still remains, therefore, of how tSNP performance changes as density of the original genotype data set increases to densities higher than 1 every 7 kb. A direct way to address this is by assessing the performance of tSNPs in data sets with manually adjusted densities, similar to the way in which Ke et al. (2004) have approached the question of block boundary identifications. If the performance of tSNPs improves (for example, assessed by SNP dropping and testing in independent samples, as described below) only modestly when densities are adjusted from 1 SNP every 5 kb to higher densities, as is implied by the asymptotic performance for the less-dense regions studied in Goldstein et al. (2003), this would imply that higher densities may not be justified for most genomic regions. These experiments, however, have not yet been reported, although the HapMap project will shortly make available data sets ideal for this purpose: that is, data sets in which all SNPs have been genotyped (International HapMap Consortium 2003).

One shortcoming of our original implementation of the SNP dropping procedure was that it ignored LD which may have been generated by the sampling procedure itself. More recently, we have carried out a similar experiment but extended it by selecting the tSNPs in one population sample, in which SNP $i$ has been dropped, but then evaluating the performance of the tSNP set in predicting the state of SNP $i$ in a second, independent sampled population of the same size (K.R. Ahmadi et al., unpubl.). This approach more closely mimics the real situation. We found that the results of this experiment and the earlier version of the SNP dropping procedure are similar for SNPs with MAF greater than about 6% and a sample size of 32 individuals. For SNPs with lower MAF, however, the performance of the tSNPs in one sample is not a good guide to their performance in a new sample (K.R. Ahmadi et al., unpubl.).

Some critics of haplotype mapping have argued that rare variants may be the main genetic factor influencing disease, and that these will be difficult to document using haplotype mapping. This raises the question of how well tags can capture rare variation. Most analyses to date have simply discarded SNPs with low MAF. Although rare alleles are often young, and thus resident on relatively long haplotypes, this does not mean that a set of tSNPs provides high power of detection when used in the ordinary way. Although we consider it well established that tSNPs can be used to adequately represent common variation, we share the concerns of some critics that it is unclear how well rare variants can be represented, although adjustments in tSNP selection and use may help. Considerably more work will be required to resolve the issues of how well SNPs with low MAF can be represented by tSNPs.

Finally, although tagging rare variants may be possible in one population, the tSNP sets that tag such variants are expected to behave largely as private rather than cosmopolitan, and therefore, tagging rare variation in multiple populations of close ancestry may not be possible, with each population requiring a unique set of tSNPs (see below).

## GENOME-WIDE TAG

A map-based approach incorporating tags can be applied either at a local (tagging variation across a gene/region) or genomic (tagging variation across the entire genome) level. Various estimates of the number of tags required to cover the genome have been made (Judson et al. 2002). More recently, using the same data set, Gabriel et al. (2002) and Goldstein et al. (2003) reached different conclusions regarding the number of tSNPs required to tag the human genome; Gabriel et al. predicted that approximately 300,000 are needed and Goldstein et al. predicted that only about 170,000 tags are sufficient for tagging a European population sample.

At least part of the discrepancy between the results is due to the way in which blocks were used in the identification of the tSNPs. Gabriel et al. estimated the total number of common haplotypes residing in blocks of high LD using their 51 regions and extrapolated the results across the genome, then calculated the total number of tags needed to tag these block-derived haplotypes using the diversity-based haplotype-tagging method (D. Clayton Web site: http://www-gene.cimr.cam.ac.uk/clayton/). Goldstein et al., however, chose to ignore the underlying block structure of LD and selected tSNPs across each of the entire regions they studied, thus capitalizing on long-range associations that would be ignored in the block-based identification of tags. They also used the haplotype $r^2$ criterion (Weale et al. 2003), which appears to increase tagging efficiency (see above). It would seem likely that much of the difference between the two estimates has to do with whether tSNPs are selected within blocks, but more detailed comparisons of different methods of selecting tSNPs are still required.

## COSMOPOLITAN TAGS

A major goal of the HapMap project is to provide the means by which a minimum set of tSNPs can be ascertained and used in future association studies. However, the question remains as to how well tSNPs from one population can represent variation in others. It is already clear that selecting tags in a population from one geographic region (e.g., Europe) and applying them to another (e.g., East Asia) can result in a substantial decrease in performance (Weale et al. 2003). It is possible, however, that tSNPs can be selected which are specifically designed to be cosmopolitan without requiring nearly as many tSNPs as the sum of those required for each constituent population.

For example, a recent study of 105 genes identified the haplotype-tagging SNPs (htSNPs) necessary to represent all the observed haplotypes in population samples from Africa and Europe (Sebastiani et al. 2003). They found that the htSNPs identified in Europe were largely a subset of those identified in Africa—a total of 538 and 379 htSNPs were found to represent all the observed haplotypes in the African and European samples, respectively, and 360 of the European htSNPs (95%) were also observed in the Africans. Recent analyses from our lab also show that only a 10–15% increase in the number of tSNPs from a northern European population (CEPH) is sufficient to tag the common genetic variation in the Japanese (K.R. Ahmadi et al., unpubl.). Overall, these results indicate that although the patterns and magnitude of LD can be markedly different across diverse populations, largely due to population history, it may be possible to identify a set of tags that work adequately in multiple human population groups without excessive increases in the number of markers that must be typed.

## CONCLUSIONS

Map-based studies incorporating tagging offer an economical, yet powerful, tool to identify common variation contributing to complex disease. With the inevitable publication of dense SNP maps covering large genomic regions, typed in large cohorts from different global populations, tagging SNP design will become more refined as the issues raised here and elsewhere are addressed with empirical data. Although considerable work remains to be done to optimize the selection and use of tSNPs, we expect relatively efficient and standardized approaches to become available soon. We are therefore already entering an era in which the constraints for genetic association studies relate to the phenotype. The complexity of the genetic control of common disease and responses to their treatment means that genetic association studies will need to be carried out in very large population cohorts with detailed information not only about disease outcome, but also about intermediate phenotypes.

## REFERENCES

Botstein D. and Risch N. 2003. Discovering genotypes underlying human phenotypes: Past successes for Mendelian disease, future approaches for complex disease. *Nat. Genet.* (suppl.) **33:** 228.

Chakravarti A., Buetow K.H., Antonarakis S.E., Waber P.G., Boehm C.D., and Kazazian H.H. 1984. Nonuniform recombination within the human β-globin gene cluster. *Am. J. Hum. Genet.* **36:** 1239.

Chapman J.M., Cooper J.D., Todd J.A., and Clayton D.G. 2003. Detecting disease associations due to linkage disequilibrium: A class of tests and the determinants of statistical power. *Hum. Hered.* **56:** 18.

Daly M.J., Rioux J.D., Schaffner S.F., Hudson T.J., and Lander E.S. 2001. High-resolution haplotype structure in the human genome. *Nat. Genet.* **29:** 229.

Dean M., Carrington M., Winkler C., Huttley G.A., Smith M.W., Allikmets R., Goedert J.J., Buchbinder S.P., Vittinghoff E., Gompert E., Donfield S., Vlahov D., Kaslow R., Saah A, Rinaldo C., Detels R., and O'Brien S.J. 1996. Genetic restriction of HIV-1 infection and progresssion to AIDS by a deletion allele of the CKR5 structural gene Hemophilia Growth and Development Study, Multicenter AIDS Cohort Study, Multicenter Hemophila Cohort Study, San Francisco City Cohort, ALIVE Study (erraturm in *Science* [1996] **274:** 1069). *Science* **273:** 1856.

Gabriel S.B., Schaffner S.F., Nguyen H., Moore J.M., Roy J., Blumenstiel B., Higgins J., DeFelice M., Lochner A., Faggart M., Liu-Cordero S.N., Rotimi C., Adeyemo A., Cooper R., Ward R., Lander E.S., Daly M.J., and Altshuler D. 2002. The structure of haplotype blocks in the human genome. *Science* **296:** 2225.

Glazier A.M., Nadeau J.H., and Aitman T.J. 2002. Finding genes that underlie complex traits. *Science* **298:** 2345.

Goldstein D.B., Ahmadi K.R., Weale M.E., and Wood N.W. 2003. Genome scans and candidate gene approaches in the study of common diseases and variable drug responses. *Trends Genet.* **19:** 615.

Hirschhorn J.N., Lohmueller K., Byrne E., and Hirschhorn K. 2002. A comprehensive review of genetic association studies. *Genet. Med.* **4:** 45.

International HapMap Consortium. 2003. The International HapMap Project. *Nature* **426:** 789.

Jeffreys A.J., Kauppi L., and Neumann R. 2001. Intensely punctate meiotic recombination in the class II region of the major histocompatibility complex. *Nat. Genet.* **29:** 217.

Jeffreys A.J., Ritchie A., and Neumann R. 2000. High resolution analysis of haplotype diversity and meiotic crossover in the human TAP2 recombination hotspot. *Hum. Mol. Genet.* **9:** 725.

Johnson G.C., Esposito L., Barratt B.J., Smith A.N., Heward J., Di Genova G., Ueda H., Cordell H.J., Eaves I.A., Dudbridge

F., Twells R.C., Payne F., Hughes W., Nutland S., Stevens H., Carr P., Tuomilehto-Wolf E., Tuomilehto J., Gough S.C., Clayton D.G., and Todd J.A. 2001. Haplotype tagging for the identification of common disease genes. *Nat. Genet.* **29:** 233.

Judson R., Salisbury B., Schneider J., Windemuth A., and Stephens J.C. 2002. How many SNPs does genome-wide haplotype map require? *Pharmacogenomics* **3:** 379.

Ke X., Hunt S., Tapper W., Lawrence R., Stavrides G., Ghori J., Whittaker P., Collins A., Morris A.P., Bentley D., Cardon L.R., and Deloukas P. 2004. The impact of SNP density on fine-scale patterns of linkage disequilibrium. *Hum. Mol. Genet.* (in press).

Kruglyak L. 1999. Prospects for whole-genome linkage disequilibrium mapping of common disease genes. *Nat. Genet.* **22:** 139.

Kruglyak L. and Nickerson D.A. 2001. Variation is the spice of life. *Nat. Genet.* **27:** 234.

Lohmueller K.E., Pearce C.L., Pike M., Lander E.S., and Hirschhorn J.N. 2003. Meta-analysis of genetic association studies supports a contribution of common variants to susceptibility to common disease. *Nat. Genet.* **33:** 177.

McCarthy M.I. 2004. Progress in defining the molecular basis of type 2 diabetes mellitus through susceptibility-gene identification. *Hum. Mol. Genet.* (in press).

Pastor P., Roe C.M., Villegas A., Bedoya G., Chakraverty S., Garcia G., Tirado V., Norton J., Rios S., Martinez M., Kosik K.S., Lopera F., and Goate A.M. 2003. Apolipoprotein Eepsilon4 modifies Alzheimer's disease onset in an E280A PS1 kindred. *Ann. Neurol.* **54:** 163.

Pritchard J.K. and Przeworski M. 2001. Linkage disequilibrium in humans: Models and data. *Am. J. Hum. Genet.* **69:** 1.

Reich D.E., Cargill M., Bolk S., Ireland J., Sabeti P.C., Richter D.J., Lavery T., Kouyoumjian R., Farhadian S.F., Ward R., and Lander E.S. 2001. Linkage disequilibrium in the human genome. *Nature* **411:** 199.

Riordan J.R., Rommens J.M., Kerem B., Alon N., Rozmahel R., Grzelczak Z., Zielenski J., Lok S., Plavsic N., Chou J.L., et al. 1989. Identification of the cystic fibrosis gene: Cloning and characterization of complementary DNA. *Science* **245:** 1066.

Risch N. and Merikangas K. 1996. The future of genetic studies of complex human diseases. *Science* **273:** 1516.

Sebastiani P., Lazarus R., Weiss S.T., Kunkel L.M., Kohane I.S., and Ramoni M.F. 2003. Minimal haplotype tagging. *Proc. Natl. Acad. Sci.* **100:** 9900.

Siddiqui A., Kerb R., Weale M.E., Brinkmann U., Smith A., Goldstein D.B., Wood N.W., and Sisodiya S.M. 2003. Association of multidrug resistance in epilepsy with a polymorphism in the drug-transporter gene ABCB1. *N. Engl. J. Med.* **348:** 1442.

Stram D.O., Haiman C.A., Hirschhorn J.N., Altshuler D., Kolonel L.N., Henderson B.E., and Pike M.C. 2003. Choosing haplotype-tagging SNPS based on unphased genotype data using a preliminary sample of unrelated subjects with an example from the Multiethnic Cohort Study. *Hum. Hered.* **55:** 27.

Weale M.E., Depondt C., Macdonald S.J., Smith A., Lai P.S., Shorvon S.D., Wood N.W., and Goldstein D.B. 2003. Selection and evaluation of tagging SNPs in the neuronal-sodium-channel gene SCN1A: Implications for linkage-disequilibrium gene mapping. *Am. J. Hum. Genet.* **73:** 551.

Weedon M.N., Schwarz P.E., Horikawa Y., Iwasaki N., Illig T., Holle R., Rathmann W., Selisko T., Schulze J., Owen K.R., Evans J., Del Bosque-Plata L., Hitman G., Walker M., Levy J.C., Sampson M., Bell G.I., McCarthy M.I., Hattersley A.T., and Frayling T.M. 2003. Meta-analysis and a large association study confirm a role for calpain-10 variation in type 2 diabetes susceptibility. *Am. J. Hum. Genet.* **73:** 1208.

Wright A.F. and Hastie N.D. 2001. Complex genetic diseases: Controversy over the Croesus code. *Genome Biol.* **2:** COMMENT2007.

Zondervan K.T. and Cardon L.R. 2004. The complex interplay among factors that influence allelic association. *Nat. Rev. Genet.* **5:** 89.

# Genetics of Quantitative Variation in Human Gene Expression

V.G. Cheung,[*†] K.-Y. Jen,[*] T. Weber,[*] M. Morley,[*] J.L. Devlin,[†]
K.G. Ewens,[†] and R.S. Spielman[†]
*Departments of *Pediatrics and †Genetics, University of Pennsylvania,
Philadelphia, Pennsylvania 19104*

The extent of variation among individuals at the DNA sequence level has been well characterized. The goal of many genetic studies is to determine the consequences of these sequence variants, for both normal and disease phenotypes. We have extended the study of genome variation from the sequence to mRNA transcript level, with the goal of understanding natural variation in gene expression in humans. We began by measuring the quantitative differences in expression levels of genes among normal individuals and determining whether there is an inherited component to this variation. We found a set of genes whose expression levels are highly variable in lymphoblastoid cells prepared from white blood cells of normal individuals. For these genes, we observed that genetically related individuals tend to have more similar transcript levels than unrelated individuals. This suggests that there is a genetic component in gene expression phenotype. Next, we are identifying the sequence differences that control variation in gene expression phenotype in a *cis*- or *trans*-acting manner. Like other quantitative traits, baseline variation in gene expression levels is likely to be regulated by a variety of genetic determinants, as well as environmental effects.

## GENOME VARIATION

The study of genetic polymorphisms in the human genome has evolved from analysis of variation in proteins to DNA sequence and now mRNA. Early characterization of the extent of variation was performed on blood proteins using electrophoretic techniques. During the 1960s, polymorphic variants of many proteins were discovered (Harris 1966, 1969). Subsequently, as methods for analysis of DNA became available, the frequency of DNA sequence variants was estimated for specific regions of the genome and then extrapolated to genome-wide estimates (Jeffreys 1979; Ewens et al. 1981). These studies provided estimates of the extent of natural variation in DNA sequence and in proteins in humans. Information about the frequency of DNA sequence variants (about 1 per 1000 bp) has been important for understanding population structure and also for disease gene mapping.

Recent advances in microarray technology have allowed extension of the study of variation at the mRNA level to a genomic scale. The extent of intra- and inter-species variation in gene expression has been assessed in primates (Enard et al. 2002), and various approaches have shown that there is appreciable variation in gene expression in other species, including mice, fish, and yeast (Cowles et al. 2002; Enard et al. 2002; Oleksiak et al. 2002; Steinmetz et al. 2002; Townsend et al. 2003). The genetic *control* of variation in gene expression has been explored in various organisms from yeast to man (Cowles et al. 2002; Enard et al. 2002; Yan et al. 2002; Cheung et al. 2003; Schadt et al. 2003; Yvert et al. 2003), mainly by focusing on preferential expression of one allele in heterozygous individuals (Cowles et al. 2002; Yan et al. 2002; Lo et al. 2003). The systematic study of natural variation in human gene expression is still in its infancy.

## HUMAN GENE EXPRESSION AS A COMPLEX TRAIT

There are several reasons for focusing our interest on *natural* variation in gene expression in humans. First, with development of high-throughput tools such as microarrays and serial analysis of gene expression, there have been many studies that compared expression profiles of normal versus diseased cells. However, few studies have analyzed natural variation in unaffected control individuals. This baseline information is important for assessing the significance of the gene expression in disease. Second, expression level of genes is a phenotype that can be measured quite precisely in a large number of unrelated and related individuals. Therefore, the "expression phenotype" can be analyzed genetically as a quantitative trait in order to identify the determinants of the variation in gene expression. The mechanisms that control transcription remain largely unknown. Globally, some control mechanisms are known, such as regulation (1) at the synthesis step, by modulating transcription initiation and elongation and (2) at the decay step, by changing the stability of transcripts. However, for most individual genes, the specific regulatory mechanisms are unknown. Linkage-based methods allow us to map genetically variable transcriptional control elements in the genome without having to know in advance whether regulation occurs via a *cis*- or *trans*-acting mechanism.

Finally, expression levels of genes are intermediate phenotypes that will be useful in developing better methods for analyzing quantitative traits. Studies of the genetic basis of monogenic (qualitative) conditions have been very successful. However, the genetic basis of most common complex traits remains poorly understood, in part because of the difficulties they pose for statistical genetic analysis. Expression phenotypes are good models for developing molecular and statistical tools for analyzing quantitative traits more generally. In comparison to other human phenotypes, it is relatively easy to measure a large number of expression phenotypes at once in the same person. There are various mechanisms that might regulate gene expression in humans. In some cases, the expression level of a gene is regulated only by a *cis*-acting element. In others, the expression level is regulated by several genes that act in *trans* or a combination of regulatory elements. These levels of complexity make expression phenotypes good models for the many components underlying complex traits and diseases in humans.

## ANALYSIS OF NATURAL VARIATION IN HUMAN GENE EXPRESSION

Our goals are (1) to define the extent of variation in gene expression in normal individuals and (2) to determine the genetic basis of that variation. We analyzed the expression levels of genes in lymphoblastoid cells from 50 unrelated individuals from the Utah pedigrees of the Centre d'Etude du Polymorphisme Humain (CEPH) using microarrays (Affymetrix Human Genome Focus Chip) that contain about 8500 human genes. Expression profiling for each sample was performed in duplicate. As a measure of variation in gene expression, for each gene, we calculated the variance ratio (variance among individuals divided by mean variance between microarray replicates on same individual). This allows us to characterize variation among individuals relative to measurement noise (Cheung et al. 2003). We ranked the genes by variance ratio and focused on those that are the most variable (i.e., with the highest variance ratio) since they are likely to be more amenable to genetic dissection. Then, we measured the expression levels of some of these "variable" genes in individuals in large families and also in sets of monozygotic twin pairs.

## LYMPHOBLASTOID CELLS AS RNA SOURCE FOR GENE EXPRESSION ANALYSIS

We have chosen to use lymphoblastoid cells in our study for several reasons. First, we need an RNA source that can be obtained from a large number of normal individuals in large pedigrees. Immortalized lymphocytes (transformed by EBV) are available from all the members of the CEPH pedigrees. These are exceptionally large three-generation families that have been studied extensively. Genotypes for genetic mapping are available for many of these families, which will facilitate our effort to map the genetic determinants of variation in gene expression. Second, lymphoblastoid cells from our study subjects can be grown under near-identical conditions. It has been shown that large differences are found in expression levels of genes that are studied on different occasions in fresh blood samples from the same individual (Whitney et al. 2003). With the lymphoblastoid cells, we can control the growth conditions in order to minimize environmental variation.

We were concerned about how transformation may affect gene expression. To examine this issue, we compared the differences in expression levels of 15 highly variable genes among unrelated individuals to those between monozygotic (MZ) twins. We found that in these transformed cells, the variance within MZ twin pairs, as a fraction of variance among unrelated individuals, ranged from 0.002 to 0.57 (mean 0.19, median 0.19). These findings indicated that the expression levels of genes are highly correlated among monozygotic twins compared to unrelated individuals. Differences in expression levels of genes in lymphoblastoid cells reflect germ-line genetic differences, despite the transformation process.

## VARIABLE GENES

We analyzed data for ~3800 (45%) of the 8500 genes, restricting attention to those that were expressed in at least 10 out of the 50 unrelated individuals who were studied. Among these genes, the variance ratios ranged from 0.2 to 48.9 (mean 2.6, median 1.6). For most of the genes, the variance of expression levels between individuals is higher than the variance between microarray replicates (Fig. 1). This shows the reproducibility of the microarray data. In our data, some genes have variance ratios that are approximately 1; for these genes, there is no evidence for meaningful variation among individuals. Therefore, in order to maximize the chance to detect genetic differences that account for variation in gene expression, we focus on the genes with variance ratio appreciably greater than 1. We expect that these are the genes where variation is most likely to be biologically meaningful, and therefore potentially due to genetic differences. The points in red in Figure 1 represent the genes (top 5%) with the highest variance ratios.

We examined the genomic location of these "variable" genes (Fig. 2). We found that they were not clustered in the genome; instead, they mapped to many sites across the genome. Figure 2 shows the chromosomal locations of the 50 most variable genes. For these genes, the range

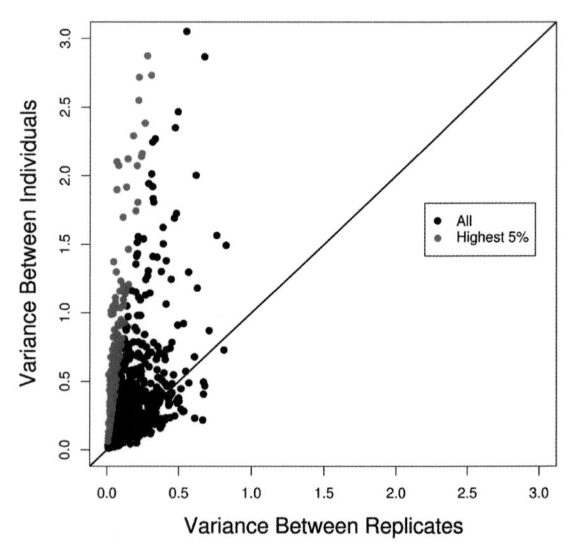


**Figure 1.** Scatter plot of variance in expression levels between individuals and between microarray replicates for ~3800 genes. The genes with the highest variance ratios (top 5%) are highlighted in red. The solid line indicates a variance ratio of 1.0.

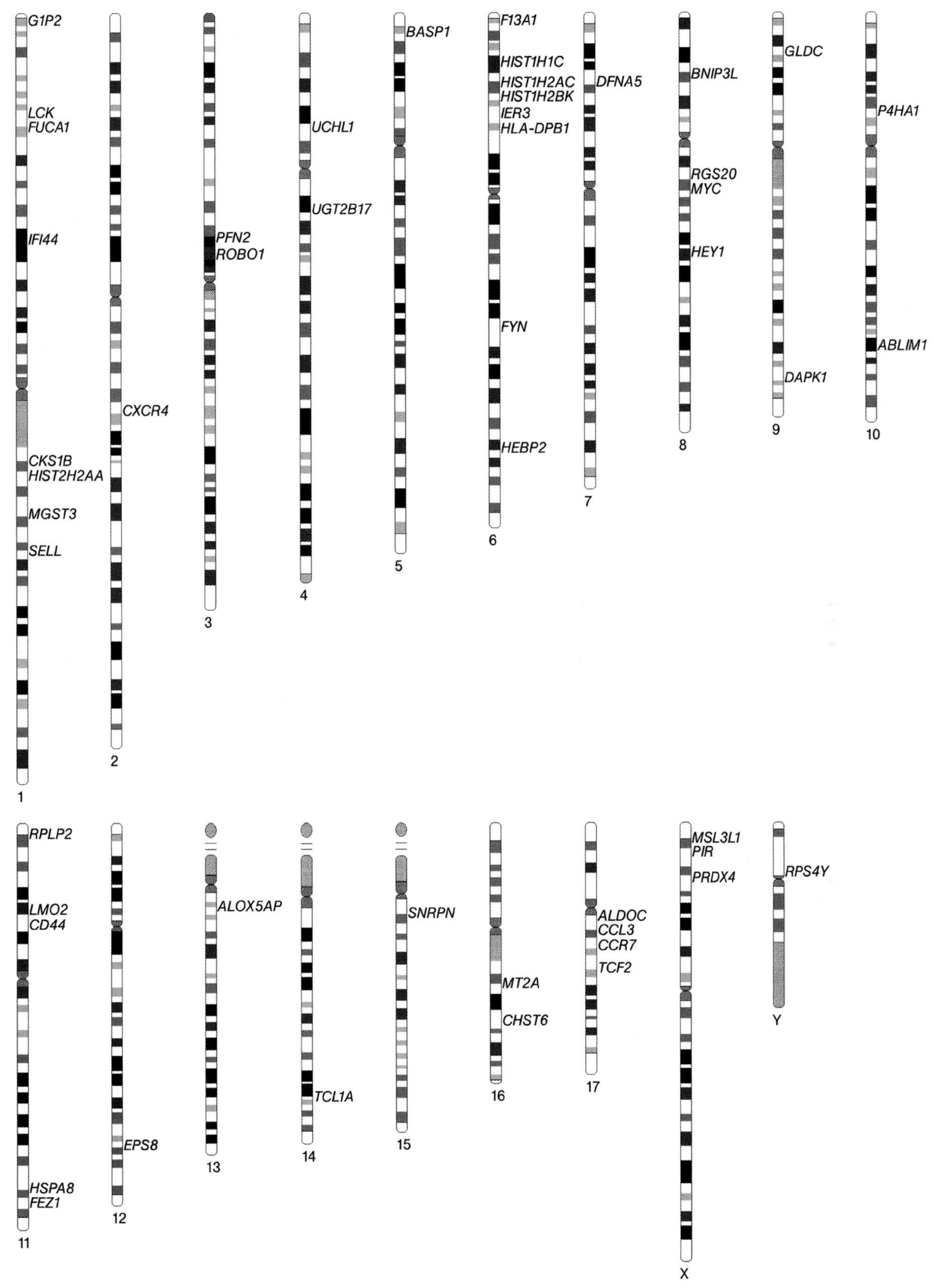


**Figure 2.** Genomic locations of 50 most variable genes. Chromosomes that do not contain any of these genes are not included in the figure.

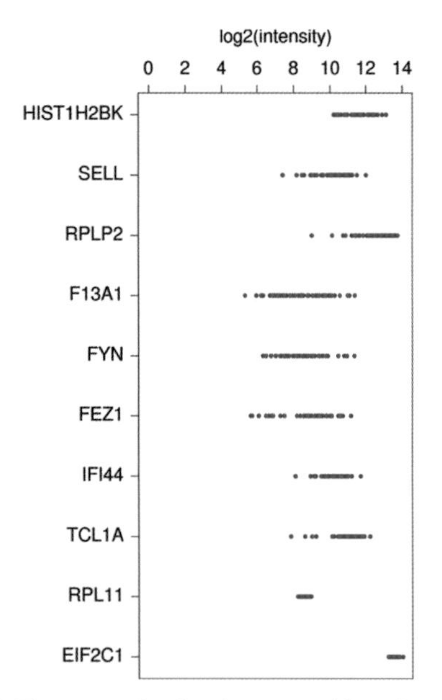


**Figure 3.** The expression levels measured by microarrays for eight highly variable and two relatively nonvariable genes in 50 individuals. Each point represents the expression level for an individual.

of expression levels among the individuals studied was from 14- to 48-fold. The expression levels of 8 of the highly variable genes and 2 relatively nonvariable genes for the 50 unrelated subjects are shown in Figure 3.

## EVIDENCE FOR HERITABILITY

We assessed the heritability of expression variation in two ways. In the first method, we used data from sets of individuals with different degrees of relatedness, and in the second, we used the resemblance between offspring and parent. Figure 4 shows the results from the first approach and compares the variance in expression level among three groups of subjects: 50 unrelated members of CEPH families (parents of the large CEPH sibships), 10 sets of siblings (also CEPH sibships), and 10 pairs of MZ twins. We found that the variance is largest among the unrelated individuals (a sample from the European population) and smallest between members of the MZ twin pairs. For all 5 genes shown in Figure 4 (and for 10 others that we have tested), we found the following pattern: as the degree of relatedness increases, the variance in expression levels decreases, suggesting that genetic (germ-line) differences contribute to variation in gene expression.

Although this kind of comparison of variances provides a valid initial assessment of evidence for heritability in some sense, other, more standard methods have the advantage of providing numerical estimates of the conventional measure, the so-called "narrow-sense" heritability. In addition to data for the CEPH parents, we have obtained gene expression data for their parents (CEPH grandparents). It is known that the regression of offspring on mid-parent value (in standard linear regression) provides an estimate of the desired heritability (Falconer and MacKay 1996), so we used the expression data for the unrelated CEPH parents and *their* parents to estimate this regression coefficient. A striking example is shown for glutathione *S*-transferase M2, *GSTM2*, in Figure 5. We have estimated the regression of parent on mid-grandparent this way for all ~3800 genes on the microarray that were expressed in the lymphoblastoid cells. Approximately 50% of these values are negative, and thus give no evidence for a heritable component. Viewing the collection as a whole, if no genes showed evidence for heritability, we would expect that the distribution of regression estimates would have as many negative as positive values. Instead, we find an excess of large positive values. The regression coefficient ($b$) is greater than 0.5 for 94 genes, but $b < -0.5$ for only 48. Similarly, we find $b > 0.75$ for 9 genes, but $b < -0.75$ for only 4 genes.

Of course, we consider these estimates at best crude indicators of heritability, especially since we have looked at so many genes. However, our interest in the estimate of heritability is not mainly as an end in itself; we use it in conjunction with the variance ratio to select genes for further study. In the follow-up studies, usually by RT-PCR, we will carry out group comparisons (sibships, twins, etc.) like those shown in Figure 4, tests of closely linked

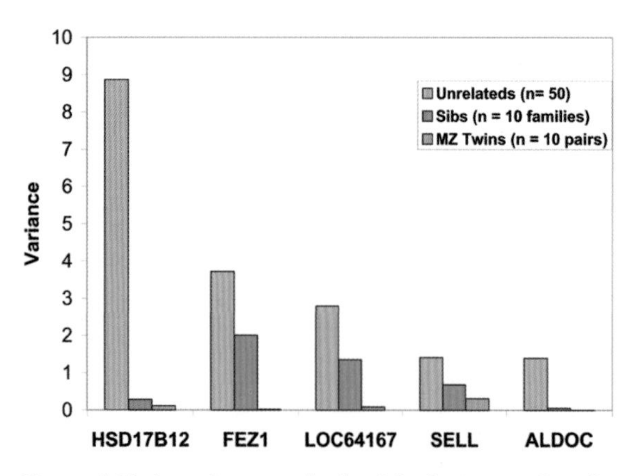


**Figure 4.** Variance in expression level for five genes. Quantitative RT-PCR data for 50 unrelated individuals, 89 offspring in 10 CEPH sibships, and 10 sets of monozygotic twins.

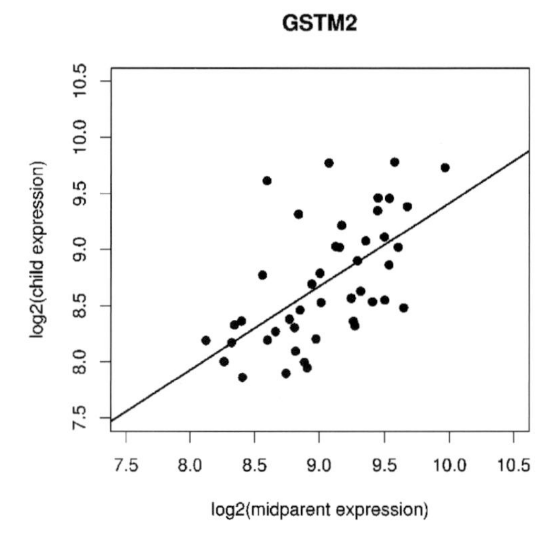


**Figure 5.** Regression plot of expression level of *GSTM2* for 50 offspring–midparent pairs. The slope (represented by *solid line*) for the regression is 0.75 with a standard error of 0.45.

SNPs for association with expression level, and genome scans to identify unlinked genetic determinants that influence expression.

## CONCLUSIONS

In this study, we have assessed the extent of natural variation in gene expression in humans. Our results suggest that there is a genetic component to this variation. Next, we will identify the genetic determinants for this variation. Identification and characterization of these determinants, called "expression control elements (ECEs)" by Cheung et al. (2003), will lead to a better understanding of transcriptional control. Less than 10% of the genome represents coding regions. Comparative genomic studies have shown that a substantial portion of the noncoding sequence is conserved between species, so it is believed that these conserved noncoding regions play a role in regulating gene expression and function. By using genetic approaches such as genome scans and methods of quantitative trait locus (QTL) analysis to map the determinants of gene expression, we expect to identify new elements that regulate gene expression. Some of these ECEs will regulate transcription in a *cis*-acting manner, whereas others will act via *trans*-acting mechanisms. We expect that a combination of association studies and genome scans will allow us to discover both types of determinants.

So far, we have considered the expression level of each gene as a separate phenotype. However, it is also possible to consider the coordinated expression of correlated genes as a single complex phenotype. It is likely that *trans*-acting regulators influence expression of several to many genes, directly or indirectly. Thus, it is reasonable to expect to find these determinants by clustering the genes by their expression phenotypes, and using these clusters as "super-phenotypes" in QTL mapping. Clustering of correlated genes has been done in many studies to find coregulated genes (Eisen et al. 1998; Golub et al. 1999; Yvert et al. 2003), in most cases, the correlations between genes were useful for identification of common pathways that were defective in diseased cells, or in the case of tumor samples, for classification purposes. In our study, the coordinated expression should minimize the number of genome scans that need to be performed, since the genes can be analyzed as groups rather than as singletons.

Observations in experimental organisms including plants, yeast, fish, and mice (Cavalieri et al. 2000; Jansen and Nap 2001; Brem et al. 2002; Oleksiak et al. 2002; Yvert et al. 2003) reveal that levels of gene expression, like the genes themselves, show abundant natural variation. This variation is viewed here as an expression phenotype which is itself under genetic control. By combining the power of microarray and classical genetics, we expect to identify the determinants for natural variation in human gene expression.

## ACKNOWLEDGMENTS

We are grateful to Warren J. Ewens for advice throughout this project. This work was supported by grants from the National Institutes of Health (HG-02386, HG-01880) and from the W.W. Smith Endowed Chair (to V.G.C.).

## REFERENCES

Brem R.B., Yvert G., Clinton R., and Kruglyak L. 2002. Genetic dissection of transcriptional regulation in budding yeast. *Science* **296:** 752.

Cavalieri D., Townsend J.P., and Hartl D.L. 2000. Manifold anomalies in gene expression in a vineyard isolate of *Saccharomyces cerevisiae* revealed by DNA microarray analysis. *Proc. Natl. Acad. Sci.* **97:** 12369.

Cheung V.G., Conlin L.K., Weber T.M., Arcaro M., Jen K.-Y., Morley M., and Spielman R.S. 2003. Natural variation in human gene expression assessed in lymphoblastoid cells. *Nat. Genet.* **33:** 422.

Cowles C.R., Hirschhorn J.N., Altshuler D., and Lander E.S. 2002. Detection of regulatory variation in mouse genes. *Nat. Genet.* **32:** 432.

Eisen M.B., Spellman P.T., Brown P.O., and Botstein D. 1998. Cluster analysis and display of genome-wide expression patterns. *Proc. Natl. Acad. Sci.* **95:** 14863.

Enard W., Khaitovich P., Klose J., Zollner S., Heissig F., Giavalisco P., Nieselt-Struwe K., Muchmore E., Varki A., Ravid R., Doxiadis G.M., Bontrop R.E., and Paabo S. 2002. Intra- and interspecific variation in primate gene expression patterns. *Science* **296:** 340.

Ewens W.J., Spielman R.S., and Harris H. 1981. Estimation of genetic variation at the DNA level from restriction endonuclease data. *Proc. Natl. Acad. Sci.* **78:** 3748.

Falconer D.S. and MacKay T.F.C. 1996. *Introduction to quantitative genetics*, 4th edition, Longman, London.

Golub T.R., Slonim D.K., Tamayo P., Huard C., Gaasenbeek M., Mesirov J.P., Coller H., Loh M.L., Downing J.R., Caligiuri M.A., Bloomfield C.D., and Lander E.S. 1999. Molecular classification of cancer: Class discovery and class prediction by gene expression monitoring. *Science* **286:** 531.

Harris H. 1966. Enzyme polymorphisms in man. *Proc. R. Soc. Lond. B Biol. Sci.* **164:** 298.

———. 1969. Enzyme and protein polymorphism in human populations. *Br. Med. Bull.* **25:** 5.

Jansen R.C. and Nap J.P. 2001. Genetical genomics: The added value from segregation. *Trends Genet.* **17:** 388.

Jeffreys A.J. 1979. DNA sequence variants in the G gamma-, A gamma-, delta- and beta-globin genes of man. *Cell* **18:** 1.

Lo H.S., Wang Z., Hu Y., Yang H.H., Gere S., Buetow K.H., and Lee M.P. 2003. Allelic variation in gene expression is common in the human genome. *Genome Res.* **13:** 1855.

Oleksiak M.F., Churchill G.A., and Crawford D.L. 2002. Variation in gene expression within and among natural populations. *Nat. Genet.* **32:** 261.

Schadt E.E., Monks S.A., Drake T.A., Lusis A.J., Che N., Colinayo V., Ruff T.G., Milligan S.B., Lamb J.R., Cavet G., Linsley P.S., Mao M., Stoughton R.B., and Friend S.H. 2003. Genetics of gene expression surveyed in maize, mouse and man. *Nature* **422:** 297.

Steinmetz L.M., Sinha H., Richards D.R., Spiegelman J.I., Oefner P.J., McCusker J.H., and Davis R.W. 2002. Dissecting the architecture of a quantitative trait locus in yeast. *Nature* **416:** 326.

Townsend J.P., Cavalieri D., and Hartl D.L. 2003. Population genetic variation in genome-wide gene expression. *Mol. Biol. Evol.* **20:** 955.

Whitney A.R., Diehn M., Popper S.J., Alizadeh A.A., Boldrick J.C., Relman D.A., and Brown P.O. 2003. Individuality and variation in gene expression patterns in human blood. *Proc. Natl. Acad. Sci.* **100:** 1896.

Yan H., Yuan W., Velculescu V.E., Vogelstein B., and Kinzler K.W. 2002. Allelic variation in human gene expression. *Science* **297:** 1143.

Yvert G., Brem R.B., Whittle J., Akey J.M., Foss E., Smith E.N., Mackelprang R., and Kruglyak L. 2003. Trans-acting regulatory variation in *Saccharomyces cerevisiae* and the role of transcription factors. *Nat. Genet.* **3:** 3.

# Regulation of α-Synuclein Expression: Implications for Parkinson's Disease

O. Chiba-Falek and R.L. Nussbaum
*Genetic Disease Research Branch, National Human Genome Research Institute, National Institutes of Health, Bethesda, Maryland 20892-4472*

Parkinson's disease (PD) is the second most common neurodegenerative disorder, after Alzheimer's disease. The disease has a prevalence of ~0.5–1% among individuals 65–69 years of age, rising to 1–3% among individuals 80 years of age and older (Tanner and Goldman 1996). Clinically, it is characterized by parkinsonism: resting tremor, bradykinesia, rigidity, and postural instability (Hoehn and Yahr 1967). Neuropathologically, it is characterized by loss of neurons in the substantia nigra and elsewhere, accompanied by the presence of intracytoplasmic inclusion bodies termed Lewy bodies in many brain regions (Pollanen et al. 1993).

Although the etiology of PD remains largely unknown, several genes involved in PD have been mapped and/or identified in the last 6 years (Table 1). Seven different chromosomal loci (PARK 1–8) have been implicated in PD inherited as a Mendelian trait in a few families (Gasser et al. 1998; Valente et al. 2001; van Duijn et al. 2001; Funayama et al. 2002). Genes for familial PD (α-synuclein, parkin, UCHL 1, and DJ-1) were identified for four of the loci (PARK 1, 2, 5, and 7, respectively) (Polymeropoulos et al. 1997; Kitada et al. 1998; Kruger et al. 1998; Leroy et al. 1998; Bonifati et al. 2003). In addition, mutations in NR4A2 (Le et al. 2003) and synphilin-1 (Marx et al. 2003) have been associated with familial and sporadic PD, respectively. In this paper, we review the involvement of α-synuclein in PD from a genetic point of view and, in particular, the regulation of α-synuclein expression and the potential role of variation of α-synuclein expression in predisposing to the disease.

## THE *SNCA* GENE AND ITS PROTEIN α-SYNUCLEIN

The first gene that was found to be involved in PD was the *SNCA* gene, encoding the protein α-synuclein. α-Synuclein, a small (14 kD) presynaptic nerve terminal protein, was originally identified as the precursor protein for the *n*on-β-*a*myloid *c*omponent (NAC) of Alzheimer's disease amyloid plaques (Ueda et al. 1993); its function is still a puzzle. The first clue linking α-synuclein to PD came from the observation that missense mutations in the *SNCA* gene caused an early-onset, Lewy body-positive, autosomal dominant Parkinson's disease in a few rare families of Mediterranean and German origin (Ala53Thr and Ala30Pro, respectively) (Polymeropoulos et al. 1997; Kruger et al. 1998). Although efforts to identify α-synuclein mutations in sporadic PD have failed, the importance of α-synuclein in PD was reinforced by the identification of the protein as a major component of Lewy bodies and Lewy neurites, the pathological hallmark of PD, in sporadic PD patients with no mutation in the *SNCA* gene (Spillantini et al. 1997). Thus, one of the most intriguing questions in the field of the molecular basis of PD is the role of α-synuclein in sporadic PD. Although the normal function of α-synuclein remains to be discovered, there is accumulating evidence in both in vitro and in vivo systems suggesting that expression levels of this protein are critical to the development of the disease.

A very recent report supports this notion. Singleton et al. (2003) reported a genomic triplication of the region containing α-synuclein in PD-affected individuals of the Iowan kindred, in which PD presented as an autosomal dominant disease. This triplication resulted in four fully functional copies of α-synuclein and twofold overexpression of α-synuclein mRNA and protein. This finding emphasizes the importance of α-synuclein levels on PD, showing that elevated levels of wild-type α-synuclein are sufficient to cause familial early-onset PD.

Many in vitro and cell-culture models have been used to examine the role of α-synuclein in neurodegeneration and toxicity. In vitro studies have shown that increasing

**Table 1.** Genetic Factors in PD

| Locus | Gene | Location | Mode of inheritance | Population |
|---|---|---|---|---|
| PARK1 | α-synuclein | 4q21 | autosomal dominant | Greek, Italian, German |
| PARK2 | Parkin | 6q25-27 | autosomal recessive; may be also autosomal dominant | vary |
| PARK3 | unknown | 2p13 | autosomal dominant | German |
| PARK5 | ubiquitin C-terminal hydrolase | 4p14 | may be autosomal dominant | German |
| PARK6 | unknown | 1p35 | autosomal recessive | Italian, Spanish |
| PARK7 | DJ-1 | 1p36 | autosomal recessive | Netherlands |
| PARK8 | unknown | 12p11.2-q13.1 | autosomal dominant | Japanese |
| | NR4A2 (Nurr1) | 2q22-23 | may be autosomal dominant | European |
| | Synphilin | 5q | sporadic | German |

concentrations of recombinant α-synuclein form fibers that aggregate (Hashimoto et al. 1998). α-Synuclein aggregates induce apoptotic cell death in human neuroblastoma cells (El-Agnaf et al. 1998). Another study showed that α-synuclein overexpression in a hypothalamic neuronal cell line resulted in formation of α-synuclein-immunopositive inclusion-like structures and mitochondrial alterations accompanied by oxidative stress, which eventually can lead to cell death (Hsu et al. 2000). Transgenic mice overexpressing wild-type human α-synuclein develop motor impairments, α-synuclein-containing intraneuronal inclusions, and loss of dopaminergic terminals in the striatum (Masliah et al. 2000). In a *Drosophila* model, wild-type overexpression driven by a cell-type-specific promoter led to aggregate formation, relative selective dopaminergic cell loss, and locomotor dysfunction (Feany and Bender 2000). More recently, it was shown that targeted overexpression of α-synuclein in the nigrostriatal dopamine neurons of adult rat induced Parkinson-like neurodegeneration (Kirik et al. 2002). These studies, among others, imply that overexpression of wild-type α-synuclein can be toxic to cells (Ostrerova et al. 1999) and that dopaminergic neurons might be particularly vulnerable to toxicity from α-synuclein overexpression. However, the molecular basis for this specific sensitivity is still an enigma. One commonly proposed hypothesis is that overexpression of α-synuclein leads to mitochondrial dysfunction resulting in the production of free radicals, which is higher in dopaminergic neurons because of the presence of both dopamine and iron, leading to neurodegeneration. Thus, the regulation of α-synuclein expression might play an important role in the susceptibility to develop PD.

## ENVIRONMENTAL AND *TRANS* FACTORS THAT REGULATE α-SYNUCLEIN EXPRESSION LEVELS

Environmental and cellular *trans*-acting factors can contribute to the modulation of α-synuclein expression. Several studies demonstrate that exposure to various stress conditions, chemicals, or biological agents can affect α-synuclein expression. Cell stress, such as that induced by serum deprivation, caused increased expression of α-synuclein in 293 HEK cells (Ostrerova et al. 1999) and human astrocytic glioma cells (Tanji et al. 2001). Up-regulation of α-synuclein was also observed in the substantia nigra of a rat model of developmental injury to the striatum (Kholodilov et al. 1999). Furthermore, α-synuclein levels increased in brains of paraquat-exposed mice (Manning-Bog et al. 2002). MPTP (1-methyl-4-phenyl-1, 2, 3, 6-tetrahydropyridine) is a neurotoxin that induces dopaminergic neurodegeneration; exposure to this chemical, or its active metabolite MPP+, replicates most of the biochemical and pathological features seen in PD. The expression of α-synuclein increased in human neuroblastoma cells exposed to MPP+ (Gomez-Santos et al. 2002) and in the substantia nigra of rodents (Vila et al. 2000) and nonhuman primates treated with MPTP (Kowall et al. 2000). Finally, a variety of growth factors can also alter α-synuclein expression. The rat ortholog of human α-synuclein is highly up-regulated in rat pheochromocytoma PC12 cells upon treatment with nerve growth factor (Stefanis et al. 2001). A subsequent study showed that bFGF promotes α-synuclein expression in cultured ventral midbrain dopaminergic neurons from rat (Rideout et al. 2003). We have observed that treatment with human β-NGF resulted in increased luciferase expression driven by the α-synuclein promoter/enhancer (O. Chiba-Falek, unpubl.), suggesting that some growth factor effects on α-synuclein expression may be mediated through *SNCA* transcription. Last, it was also reported that α-synuclein levels increased following stimulation with lipopolysaccharide and interleukin-1β (Tanji et al. 2001, 2002). These results suggest that a variety of biological agents, including neurotrophic factors, may affect α-synuclein levels in the nervous system.

## GENOMIC VARIATIONS UPSTREAM OF THE α-SYNUCLEIN TRANSCRIPTION START SITE

*cis*-Acting factors, i.e., sequence variation, in the putative promoter/enhancer region of the *SNCA* gene might also contribute to different expression levels of α-synuclein. Nine polymorphic variant sites (Table 2) have been identified within the *SNCA* promoter/enhancer region: A complex microsatellite repeat termed NACP-Rep1 (Xia et al. 1996), six SNPs, and two insertions (Farrer et al. 2001). NACP-Rep1 is a complex polymorphic microsatellite repeat located ~10 kb upstream of the translational start of *SNCA* (Xia et al. 1996; Touchman et al. 2001). The human NACP-Rep1 alleles are composed of the following dinucleotides: $(TC)_x(TT)_1(TC)_y(TA)_z(CA)_w$ with variable numbers of TC, TA, and CA dinucleotide repeats (Accession numbers AC015529, AC022357, AP001947). Five alleles were identified with a size difference of two nucleotides as determined by gel electrophoresis of the PCR products containing the repeats (Xia et al. 1996; Hellman et al. 1998). The basis for the size difference among the different NACP-Rep1 alleles is mainly due to the length of the CA portion of the repeat, with 10, 11, 12, and 13 repeats present in alleles 0, 1, 2, and 3, respectively (Table 3) (Chiba-Falek and Nussbaum 2001). However, there are sequence differences within same-sized NACP-Rep1 alleles, and thus, the number of alleles at the NACP-Rep1 site is greater than the five defined only by size (Farrer et al. 2001; Tan et al. 2003;

**Table 2.** Polymorphism Upstream of the α-Synuclein Transcriptional Start Site

| Polymorphism | Type | Positive association studies/ Total association studies |
|---|---|---|
| –109G>T | SNP | NA |
| –116C>G | SNP | 0/2 |
| –165insA | insertion | NA |
| –668T>C | SNP | 0/1 |
| –770C>A | SNP | 0/1 |
| –1608T>C | SNP | NA |
| –1912(A)13-20 | insertion | NA |
| –1942T>G | SNP | NA |
| NACP-Rep1 | microsatellite repeat | 3/8 |

(NA) Not available.

**Table 3.** Sequence of the Different Alleles at the NACP-Rep1 Locus in Human

| Allele size | Different variants | | | | |
|---|---|---|---|---|---|
| | TC | TT | TC | TA | CA |
| –1 (76 bp) | | | NA | | |
| 0 (78 bp) | 10 | 1 | 10 | 8 | 10 |
| | 10 | 1 | 11 | 7 | 10 |
| | 10 | 1 | 9 | 9 | 10 |
| 1 (80 bp) | 10 | 1 | 10 | 8 | 11 |
| | 10 | 1 | 11 | 7 | 11 |
| | 11 | 1 | 10 | 8 | 10 |
| | 11 | 1 | 11 | 7 | 10 |
| | 11 | 1 | 9 | 9 | 10 |
| 2 (82 bp) | 10 | 1 | 10 | 8 | 12 |
| | 10 | 1 | 11 | 7 | 12 |
| 3 (84 bp) | 10 | 1 | 10 | 8 | 13 |

Indicated are only the sequences of the variant alleles that were found more than once to date, based on sequencing of independent cloned PCR products at the NACP-Rep1 region (Chiba-Falek and Nussbaum 2001; Chiba-Falek et al. 2003). (NA) Not available.

Chiba-Falek et al. 2003). Extended sequence analysis indicates that sequence variation also occurs at positions other than within the CA repeat itself and that alleles of the same size can differ in their sequence combination. The intra-size sequence difference is contributed mainly by variation in the $(TC)_y(TA)_z$ component of the repeat. Where the number y + z is constant (=18) for a given allele, the most frequent combination is y = 10 and z = 8, followed by 11 and 7, with the least common being 9 and 9 (Table 3) (Chiba-Falek et al. 2003). It is interesting to note that allele size 1, which is the most frequent size among the Caucasian population (Farrer et al. 2001), has the largest number of intra-sequence variations. Some of the allele size 1 variants demonstrate an even higher degree of variation with n = 11 in the first $(TC)_n$ repeat and w = 10 in the last $(CA)_w$ dinucleotide vs. n = 10 and w = 11 (Table 3).

## SNPs AT THE α-SYNUCLEIN PROMOTER REGION FAIL TO BE ASSOCIATED WITH SPORADIC PD

Three SNPs in the promoter region of α-synuclein were tested for association with sporadic PD (Farrer et al. 2001; Holzmann et al. 2003). Farrer et al. (2001) found no significant association of –116C>G and –770C>A with sporadic PD ($p = 0.317$ and 0.763, respectively) in Caucasian residents of the US. Recently, the –116C>G SNP was studied again in the Caucasian German population as well as the –668T>C polymorphism. In this report, alleles and genotype frequencies of both loci also did not reveal significant differences between the groups of sporadic PD patients and controls (Holzmann et al. 2003).

## ASSOCIATION STUDIES BETWEEN NACP-Rep1 AND SPORADIC PD

Several association studies have shown that certain alleles of the NACP-Rep1 locus may confer an increased risk of sporadic PD in Caucasian (Kruger et al. 1999; Tan et al. 2000; Farrer et al. 2001) and Asian (Mizuta et al. 2002; Tan et al. 2003) populations, but other studies failed to replicate these findings (Parsian et al. 1998; Izumi et al. 2001; Khan et al. 2001; Tan et al. 2003). Unfortunately, inconsistencies in the designation and naming of the NACP-Rep1 alleles in the literature make an attempt at a meta-analysis a great challenge. Thus, we decided to perform our analysis separately for Caucasian and Asian populations (Table 4). Because differences in allele sizes reported in the different studies made cross-study comparisons of particular alleles difficult, we attempted to assign a uniform designation of allele 0, 1, 2, and 3 within each study based on the frequency of the allele relative to the frequency of the other alleles within that same study. Overall, the distribution of alleles 0, 1, and 2+3 was significantly different among Caucasians with PD and controls ($p = 0.002$, $\chi^2 = 12.05$, 2 df); because allele 3 is so rare, we combined 2 and 3 together for chi-square analysis. There was a significantly lower frequency of allele size 0 in sporadic Caucasian PD patients as compared to controls ($p = 0.0007$, $\chi^2 = 11.25$, 1 df), resulting in a reduced relative risk for PD for allele 0 carriers of 0.89 (95% C.I. 0.81–0.95). There is also a higher frequency of allele 1 ($p = 0.02$, $\chi^2 = 5.14$, 1 df) compared to the other alleles, although the relative risk was only 1.08 (95% C.I. 1.01–1.16). These results suggest a negative association of allele 0 and a positive association with allele 1 with PD in the studies from Caucasian populations.

In the case of reports based on studies among populations in Asia, once again, the distribution of alleles 0, 1 and 2, and 3 was significantly different between PD and controls ($p = 0.0158$, $\chi^2 = 10.33$, 3 df); however, the particular alleles and their effect on PD susceptibility are different and appear to contradict the results from Caucasian populations. No significant reduced risk was seen for allele 0 carriers versus noncarriers in Asian populations ($p = 0.3$, $\chi^2 = 1.02$, 1 df), giving a relative risk of 1.044 (95% C.I. 0.96–1.14). There also appeared to be a *reduced* risk for carriers of allele 1 versus other alleles ($p = 0.0018$, $\chi^2 = 9.64$, 1 df), resulting in a *lower* relative risk for PD for allele 1 carriers of 0.86 (95% C.I. 0.78–0.95). However, it should be noted that the allele frequencies of the four alleles *among the controls* are vastly different between the Caucasian and Asian populations. Restricting our analysis to control populations only, the allele frequencies for alleles 0, 1, 2, and 3 among Caucasians were reportedly 28%, 65%, 7%, and <<1%, respectively, whereas allele frequencies among Asian controls were 41%, 32%, 26% and 1%. This difference is highly significant ($p < 10^{-10}$, $\chi^2 = 330$, 3 df). It is unknown whether this difference in allele frequencies is a real population genetic phenomenon or is the result of misclassification of alleles because of difficulties comparing PCR fragment sizes across platforms or lack of uniformity in the particular PCR primers used for genotyping. Thus, it remains an open question as to whether the apparent contradiction between studies performed with Caucasian and Asian populations is real or an artifact of allelic misclassification.

**Table 4.** Association Studies of NACP-Rep1 with Sporadic PD

| Study | $\chi^2$ | Population | Allele[a] | –1[a] | 0[a] | 1[a] | 2[a] | 3[a] |
|---|---|---|---|---|---|---|---|---|
| Farrer et al. (2001) | $p = 0.005$ | Caucasian | PD | 1(0.2) | 126(20.6) | 430(70.3) | 54(8.8) | 1(0.2) |
| | | (European) | control | 3(0.8) | 110(29.1) | 236(62.4) | 28(7.4) | 1(0.3) |
| Kruger et al. (1999) | $p = 0.16$ | Caucasian | PD | | 33(20.2) | 113(69.1) | 17(10.7) | 0 |
| | | German | control | | 44(22.8) | 139(72.0) | 10(5.2) | 0 |
| Tan et al. (2000) | $p = 0.009$ | Caucasian | PD | | 42(21) | 140(70) | 18(9) | 0 |
| | | | control | | 56(28) | 139(69.5) | 5(2.5) | 0 |
| Parsian et al. (1998) | $p = 0.19$ | Caucasian | PD | | 118(27.3) | 275(63.7) | 39(9) | 0 |
| | | | control | | 114(32.8) | 203(58.3) | 31(8.9) | 0 |
| Khan et al. (2001) | $p = 0.64$ | UK | PD | | 77(25.1) | 212(69.4) | 17(5.5) | 0 |
| | | | control | | 86(26.1) | 216(65.5) | 28(8.4) | 0 |
| TOTAL | $p = 0.0007$ | Caucasian | PD | | 396(23.1) | 1170(68.3 | 145(8.5) | 1 |
| | | | control | | 410(28.3) | 933(64.5) | 102(7) | 1 |
| Izumi et al. (2001) | $p = 0.14$ | Japanese | PD | 1(0.25) | 163(40.75) | 99(24.75) | 132(33) | 5(1.25) |
| | | | control | 1(0.2) | 195(39) | 160(32) | 136(27.2) | 6(1.2) |
| Mizuta et al. (2002) | $p = 0.017$ | Japanese | PD | | 171(51.8) | 84(25.5) | 66(20) | 6(1.8) |
| | | | control | | 126(40.6) | 110(35.5) | 65(21) | 5(1.6) |
| Tan et al. (2003) | $p = 0.20$ | Singaporean | PD | | 148(36) | 115(28) | 140(34) | |
| | | | control | | 169(41) | 119(29) | 115(28) | |
| TOTAL | $p = 0.0158$ | Asian | PD | | 482(42.7) | 298(26.4) | 338(30) | 11(0.9) |
| | | | control | | 490(40.6) | 389(32.2) | 316(26.2) | 11(0.9) |

[a]Alleles as defined by size using Xia et al. (1996) original designation. The results are presented in each row as follows: number of chromosomes (% of chromosomes).

## NACP-Rep1 AS A TRANSCRIPTIONAL FUNCTIONAL ELEMENT

To determine whether the genetic association studies correlated with biological function, we examined the activity of the entire 10.7-kb upstream region of *SNCA* in a luciferase reporter assay in 293T and SH-SY5Y cells to determine the role of the NACP-Rep1 repeat and its various alleles on *SNCA* promoter function (Chiba-Falek and Nussbaum 2001). Luciferase expression levels varied very significantly among the different alleles over a 3-fold range in the human neuroblastoma SH-SY5Y cells but showed little or no significant variation in the 293T cells (Fig. 1). In SH-SY5Y cells, the highest promoter activity was seen with the construct containing the repeat with 11 CA (allele size 1), leading to a 3-fold increase in the expression relative to the allele carrying 10 CA (allele size 0) (Fig. 1) (Chiba-Falek and Nussbaum 2001). It should be noted that allele 1 and allele 0, respectively, confer the most significant increased and decreased relative risk for PD in Caucasian populations. The construct containing the largest repeat (13 CA- size 3) resulted in a 2.5-fold increase in activity over that seen with the construct containing 10 CA (size 0) repeats. The promoter activity was suppressed as the length of the repeat decreased to 12 CA (size 2), resulting in only a 1.5-fold increase in activity as compared to the shortest repeat.

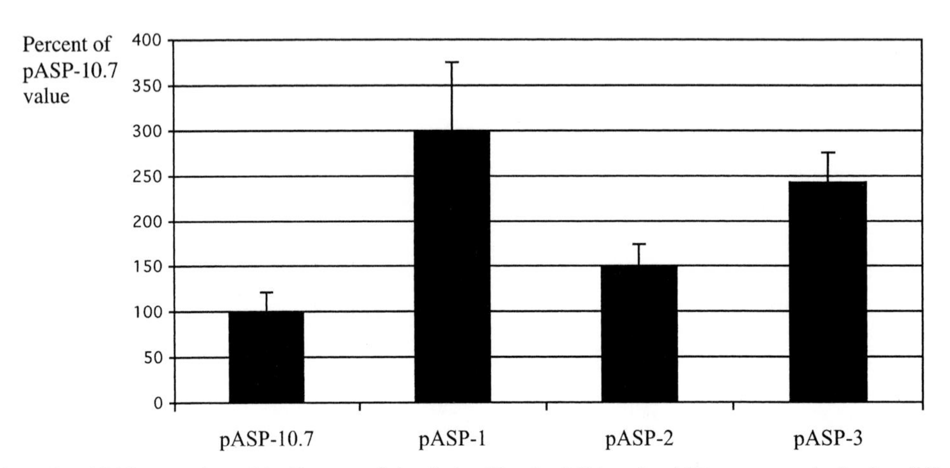


**Figure 1.** The ratio of fold expression of luciferase activity derived by the full-length pASP constructs harboring different NACP-Rep1 alleles to pASP-10.7 in SH-SY5Ycells. The fold expression for each pASP was determined by dividing the average relative activity of each construct to that of the average obtained with the empty pGL 3-Basic. Then, the ratio in percentage of the fold expression for each of the pASP-1, 2, and 3 relative to pASP-10.7 was determined. The average of the ratios of four to five independent experiments performed on separate days was calculated. The fold expression of pASP-10.7 is arbitrarily assigned 100%. The data represented here are the average ratios in percentage ± 1 S.E.M. for the pASP-1, 2, and 3 constructs relative to pASP-10.7. Student's t-test comparing the "fold expression" of each of the pASP-1, 2, and 3 constructs to pASP-10.7 revealed $p = 5 \times 10^{-6}$, 0.002, and $5 \times 10^{-9}$, respectively, in SH-SY5Y cells. (Adapted, with permission, from Chiba-Falek and Nussbaum 2001.)

Thus, the effect of different alleles on promoter strength in the luciferase assay was not linear with respect to repeat length. Deletion analysis of the NACP-Rep1 locus and surrounding DNA suggested that two domains flanking the repeat interact to enhance expression while the repeat itself modulates this interaction to a greater or lesser extent depending on which allele is present at the NACP-Rep1 locus (Chiba-Falek and Nussbaum 2001).

We then examined the effect on *SNCA* promoter activity of sequence variation within a single size "1" allele. Two intra-allelic variants, defined by y = 10 z = 8 versus y = 11 and z = 7 in the segment $(TC)_y(TA)z$ within the NACP = Rep1 repeat were studied. We found only a very small, insignificant difference in luciferase expression levels (Chiba-Falek et al. 2003). These findings obtained using the luciferase reporter system in human neuroblastoma cells imply that the overall length of the NACP-Rep1 allele plays the main role in the transcriptional regulation by the NACP-Rep1 element, whereas the intra-size variation has only a minor contribution to the function of the NACP-Rep1 element as a transcriptional regulator.

## THE MOUSE NACP-Rep1 REGION

Examination of the mouse sequence reveals a complex repeat similar to the human NACP-Rep1, located ~6.1 kb upstream of the transcriptional start site of *Snca* (Touchman et al. 2001). The human and mouse repeats are only 40% identical but contain similar dinucleotide elements, although the human element contains a CA dinucleotide not present in the mouse element. DNAs from 22 inbred mouse strains derived from *Mus musculus musculus*, two from *M. musculus* subspecies (*CAST/Ei* and *MOLG/Dn*), and one from the species *Mus spretus* (*SPRET/Ei*) were examined by PCR for polymorphisms at this complex dinucleotide repeat (Touchman et al. 2001). The *M. musculus*-derived strains were not polymorphic; all can be denoted as $(CT)_8N_2(AT)_9N_5(GT)_4N_8(GT)_3$. The *CAST/Ei, MOLG/Dn*, and *SPRET/Ei*, however, do differ in size. The *CAST/Ei* and *MOLG/Dn* sequences differ from the other inbred musculus strains only in the size of the AT repeat, while showing the identical sequence and spacing for the other dinucleotide elements of the complex repeat. The *CAST/Ei* product contains $(AT)_{29}$ instead of $(AT)_9$, whereas that from *MOLG/Dn* has $(AT)_{22}$. *SPRET/Ei* shows a more complex polymorphism that can be denoted as $(CT)_{13}(AT)_{35}N_9(GT)_5N_8(GT)_3$ (Touchman et al. 2001). It will be of great interest to study the levels of α-synuclein mRNA in the different mouse strains that were shown to carry variant alleles at the NACP-Rep1 site and to correlate the α-synuclein expression level to the composition of the mouse NACP-Rep1 allele.

## OTHER GENES HAVING A FUNCTIONAL POLYMORPHIC MICROSATELLITE IN THEIR PROMOTER REGIONS

Several studies in the past have implicated different-length dinucleotide repeats at promoter regions in regulating variable transcription activity. Using reporter gene expression systems, the expression of several genes was shown to be regulated by polymorphic dinucleotide repeats in their 5′-flanking region, and some alleles of these polymorphic repeat sequences were shown to enhance their expression. Some of these studies are listed below.

1. The dinucleotide repeat $(CA)_lN(CG)_m(CA)_n$ upstream of the human *COL1A2* has an enhancing activity on gene transcription, and that variation in the number of repetitions may be responsible for the difference in the transcription activity of the gene (Akai et al. 1999).
2. Variation in the length of a (CA) repeat element in the promoter of the human *MMP-9* gene was shown to modulate its promoter activity (Peters et al. 1999; Shimajiri et al. 1999).
3. Variants of the *PAX-6* polymorphic dinucleotide repeat $(AC)_m(AG)_n$ were shown to have different transcriptional efficiencies and to drive variable mRNA expression levels in human brain (Okladnova et al. 1998).
4. Four alleles in the promoter of the human *NRAMP1* gene consist of $T(GT)_xAC(GT)_yAC(GT)_zG$ and differ in their ability to drive gene expression (Searle and Blackwell 1999).
5. A $(GT)_n$ dinucleotide repeat upstream of the human *HO-1* gene shows length polymorphism, which modulates the level of transcription (Yamada et al. 2000).
6. Certain alleles at the $(CA)_n$ polymorphic site located 2.1 kb upstream of the human *AR2* gene lead to significantly higher expression level (Ikegishi et al. 1999).
7. A functional polymorphic dinucleotide repeat $(TCTCT(TC)_n)$ upstream of the translational start codon of human *HMGA2* was found to regulate strongly the human *HMGA2* promoter with an activation pattern that correlates with its TC-repeat length (Borrmann et al. 2003).

The mechanisms by which dinucleotide repeats might affect gene expression in *cis* are largely unknown. In one case, however, a dinucleotide-binding protein, angiogenin, has been implicated in regulation of expression of the human rRNA gene. Angiogenin binds $(CT)_n$ repeats specifically, in a length-dependent manner, and the affinity of its binding increases for longer (CT) repeats. Moreover, the CT repeats exhibit angiogenin-dependent promoter activity in a luciferase reporter system, and the level of the promoter activity depends on the numbers of the CTs. This study suggested that angiogenin may play a role in regulating expression of genes containing CT repeats in their 5′-flanking regions, which are common in the eukaryotic genome (Xu et al. 2003).

It is important to note that in most of the genes studied, the combined dinucleotide repeat was at a distance of ≤2 kb upstream of the transcriptional start site. The complex dinucleotide repeat, NACP-Rep1, is unusual in that it can influence gene expression from a much longer distance of ~10 kb, such as might be seen with a long-range enhancer or a locus control region (LCR).

## CONCLUSIONS

We propose the following hypotheses:

1. The levels of α-synuclein protein may be critical to the development of PD; even small changes in α-synu-

clein in neurons may, over many decades, predispose to the disease.

2. The regulation of *SNCA* transcription may be one important factor in regulating the expression of the α-synuclein protein.
3. Alleles at the NACP-Rep1 microsatellite, through their differential effects on *SNCA* expression, may confer differential susceptibility to PD, thereby providing an explanation for an allelic association between certain NACP-Rep1 alleles and sporadic disease.

Further genetic and cell biological studies need to be done to determine whether a direct connection exists in vivo between different NACP-Rep1 alleles and α-synuclein levels in neurons and whether such differences are significant enough to cause damage and neurodegeneration. It is also imperative that a robust, inexpensive, and universally accepted method for accurate genotyping of NACP-Rep1 alleles be developed and adopted so that correct comparisons can be made between different studies within and between population groups.

## REFERENCES


Akai J., Kimura A., and Hata R.I. 1999. Transcriptional regulation of the human type I collagen alpha2 (COL1A2) gene by the combination of two dinucleotide repeats. *Gene* **239:** 65.

Bonifati V., Rizzu P., van Baren M.J., Schaap O., Breedveld G.J., Krieger E., Dekker M.C., Squitieri F., Ibanez P., Joosse M., van Dongen J.W., Vanacore N., van Swieten J.C., Brice A., Meco G., van Duijn C.M., Oostra B.A., and Heutink P. 2003. Mutations in the DJ-1 gene associated with autosomal recessive early-onset Parkinsonism. *Science* **299:** 256.

Borrmann L., Seebeck B., Rogalla P., and Bullerdiek J. 2003. Human HMGA2 promoter is coregulated by a polymorphic dinucleotide (TC)-repeat. *Oncogene* **22:** 756.

Chiba-Falek O. and Nussbaum R.L. 2001. Effect of allelic variation at the NACP-Rep1 repeat upstream of the alpha-synuclein gene (SNCA) on transcription in a cell culture luciferase reporter system. *Hum. Mol. Genet.* **10:** 3101.

Chiba-Falek O., Touchman J.W., and Nussbaum R.L. 2003. Functional analysis of intra-allelic variation of NACP-Rep1 in the alpha-synuclein gene. *Hum. Genet.* **113:** 426.

El-Agnaf O.M., Jakes R., Curran M.D., Middleton D., Ingenito R., Bianchi E., Pessi A., Neill D., and Wallace A. 1998. Aggregates from mutant and wild-type alpha-synuclein proteins and NAC peptide induce apoptotic cell death in human neuroblastoma cells by formation of beta-sheet and amyloid-like filaments. *FEBS Lett.* **440:** 71.

Farrer M., Maraganore D.M., Lockhart P., Singleton A., Lesnick T.G., de Andrade M., West A., de Silva R., Hardy J., and Hernandez D. 2001. α-Synuclein gene haplotypes are associated with Parkinson's disease. *Hum. Mol. Genet.* **10:** 1847.

Feany M.B. and Bender W.W. 2000. A *Drosophila* model of Parkinson's disease. *Nature* **404:** 394.

Funayama M., Hasegawa K., Kowa H., Saito M., Tsuji S., and Obata F. 2002. A new locus for Parkinson's disease (PARK8) maps to chromosome 12p11.2-q13.1. *Ann. Neurol.* **51:** 296.

Gasser T., Muller-Myhsok B., Wszolek Z.K., Oehlmann R., Calne D.B., Bonifati V., Bereznai B., Fabrizio E., Vieregge P., and Horstmann R.D. 1998. A susceptibility locus for Parkinson's disease maps to chromosome 2p13. *Nat. Genet.* **18:** 262.

Gomez-Santos C., Ferrer I., Reiriz J., Vinals F., Barrachina M., and Ambrosio S. 2002. MPP+ increases alpha-synuclein expression and ERK/MAP-kinase phosphorylation in human neuroblastoma SH-SY5Y cells. *Brain Res.* **935:** 32.

Hashimoto M., Hsu L.J., Sisk A., Xia Y., Takeda A., Sundsmo M., and Masliah E. 1998. Human recombinant NACP/alpha-synuclein is aggregated and fibrillated in vitro: Relevance for Lewy body disease. *Brain Res.* **799:** 301.

Hellman N.E., Grant E.A., and Goate A.M. 1998. Failure to replicate a protective effect of allele 2 of NACP/alpha-synuclein polymorphism in Alzheimer's disease: An association study. *Ann. Neurol.* **44:** 278.

Hoehn M.M. and Yahr M.D. 1967. Parkinsonism: Onset, progression and mortality. *Neurology* **17:** 427.

Holzmann C., Kruger R., Saecker A.M., Schmitt I., Schols L., Berger K., and Riess O. 2003. Polymorphisms of the alpha-synuclein promoter: Expression analyses and association studies in Parkinson's disease. *J. Neural Transm.* **110:** 67.

Hsu L.J., Sagara Y., Arroyo A., Rockenstein E., Sisk A., Mallory M., Wong J., Takenouchi T., Hashimoto M., and Masliah E. 2000. α-Synuclein promotes mitochondrial deficit and oxidative stress. *Am. J. Pathol.* **157:** 401.

Ikegishi Y., Tawata M., Aida K., and Onaya T. 1999. Z-4 allele upstream of the aldose reductase gene is associated with proliferative retinopathy in Japanese patients with NIDDM, and elevated luciferase gene transcription in vitro. *Life Sci.* **65:** 2061.

Izumi Y., Morino H., Oda M., Maruyama H., Udaka F., Kameyama M., Nakamura S., and Kawakami H. 2001. Genetic studies in Parkinson's disease with an alpha-synuclein/NACP gene polymorphism in Japan. *Neurosci. Lett.* **300:** 125.

Khan N., Graham E., Dixon P., Morris C., Mander A., Clayton D., Vaughan J., Quinn N., Lees A., Daniel S., Wood N., and de Silva R. 2001. Parkinson's disease is not associated with the combined alpha-synuclein/apolipoprotein E susceptibility genotype. *Ann. Neurol.* **49:** 665.

Kholodilov N.G., Neystat M., Oo T.F., Lo S.E., Larsen K.E., Sulzer D., and Burke R.E. 1999. Increased expression of rat synuclein in the substantia nigra pars compacta identified by mRNA differential display in a model of developmental target injury. *J. Neurochem.* **73:** 2586.

Kirik D., Rosenblad C., Burger C., Lundberg C., Johansen T.E., Muzyczka N., Mandel R.J., and Bjorklund A. 2002. Parkinson-like neurodegeneration induced by targeted overexpression of alpha-synuclein in the nigrostriatal system. *J. Neurosci.* **22:** 2780.

Kitada T., Asakawa S., Hattori N., Matsumine H., Yamamura Y., Minoshima S., Yokochi M., Mizuno Y., and Shimizu N. 1998. Mutations in the parkin gene cause autosomal recessive juvenile Parkinsonism. *Nature* **392:** 605.

Kowall N.W., Hantraye P., Brouillet E., Beal M.F., McKee A.C., and Ferrante R.J. 2000. MPTP induces alpha-synuclein aggregation in the substantia nigra of baboons. *Neuroreport* **11:** 211.

Kruger R., Kuhn W., Muller T., Woitalla D., Graeber M., Kosel S., Przuntek H., Epplen J.T., Schols L., and Riess O. 1998. Ala30Pro mutation in the gene encoding alpha-synuclein in Parkinson's disease. *Nat. Genet.* **18:** 106.

Kruger R., Vieira-Saecker A.M., Kuhn W., Berg D., Muller T., Kuhnl N., Fuchs G.A., Storch A., Hungs M., Woitalla D., Przuntek H., Epplen J.T., Schols L., and Riess O. 1999. Increased susceptibility to sporadic Parkinson's disease by a certain combined alpha-synuclein/apolipoprotein E genotype. *Ann. Neurol.* **45:** 611.

Le W.D., Xu P., Jankovic J., Jiang H., Appel S.H., Smith R.G., and Vassilatis D.K. 2003. Mutations in NR4A2 associated with familial Parkinson disease. *Nat. Genet.* **33:** 85.

Leroy E., Boyer R., Auburger G., Leube B., Ulm G., Mezey E., Harta G., Brownstein M.J., Jonnalagada S., Chernova T., Dehejia A., Lavedan C., Gasser T., Steinbach P.J., Wilkinson K.D., and Polymeropoulos M.H. 1998. The ubiquitin pathway in Parkinson's disease. *Nature* **395:** 451.

Manning-Bog A.B., McCormack A.L., Li J., Uversky V.N., Fink A.L., and Di Monte D.A. 2002. The herbicide paraquat causes up-regulation and aggregation of alpha-synuclein in mice: Paraquat and alpha-synuclein. *J. Biol. Chem.* **277:** 1641.

Marx F.P., Holzmann C., Strauss K.M., Li L., Eberhardt O., Gerhardt E., Cookson M.R., Hernandez D., Farrer M.J., Kachergus J., Engelender S., Ross C.A., Berger K., Schols L., Schulz J.B., Riess O., and Kruger R. 2003. Identification and functional characterization of a novel R621C mutation in the synphilin-1 gene in Parkinson's disease. *Hum. Mol. Genet.* **12:** 1223.

Masliah E., Rockenstein E., Veinbergs I., Mallory M., Hashimoto M., Takeda A., Sagara Y., Sisk A., and Mucke L. 2000. Dopaminergic loss and inclusion body formation in alpha-synuclein mice: Implications for neurodegenerative disorders. *Science* **287:** 1265.

Mizuta I., Nishimura M., Mizuta E., Yamasaki S., Ohta M., and Kuno S. 2002. Meta-analysis of alpha synuclein/ NACP polymorphism in Parkinson's disease in Japan. *J. Neurol. Neurosurg. Psychiatry* **73:** 350.

Okladnova O., Syagailo Y.V., Tranitz M., Stober G., Riederer P., Mossner R., and Lesch K.P. 1998. A promoter-associated polymorphic repeat modulates PAX-6 expression in human brain. *Biochem. Biophys. Res. Commun.* **248:** 402.

Ostrerova N., Petrucelli L., Farrer M., Mehta N., Choi P., Hardy J., and Wolozin B. 1999. α-Synuclein shares physical and functional homology with 14-3-3 proteins. *J. Neurosci.* **19:** 5782.

Parsian A., Racette B., Zhang Z.H., Chakraverty S., Rundle M., Goate A., and Perlmutter J.S. 1998. Mutation, sequence analysis, and association studies of alpha-synuclein in Parkinson's disease. *Neurology* **51:** 1757.

Peters D.G., Kassam A., St. Jean P.L., Yonas H., and Ferrell R.E. 1999. Functional polymorphism in the matrix metalloproteinase-9 promoter as a potential risk factor for intracranial aneurysm. *Stroke* **30:** 2612.

Pollanen M.S., Dickson D.W., and Bergeron C. 1993. Pathology and biology of the Lewy body. *J. Neuropathol. Exp. Neurol.* **52:** 183.

Polymeropoulos M.H., Lavedan C., Leroy E., Ide S.E., Dehejia A., Dutra A., Pike B., Root H., Rubenstein J., Boyer R., Stenroos E.S., Chandrasekharappa S., Athanassiadou A., Papapetropoulos T., Johnson W.G., Lazzarini A.M., Duvoisin R.C., Di Iorio G., Golbe L.I., and Nussbaum R.L. 1997. Mutation in the alpha-synuclein gene identified in families with Parkinson's disease. *Science* **276:** 2045.

Rideout H.J., Dietrich P., Savalle M., Dauer W.T., and Stefanis L. 2003. Regulation of alpha-synuclein by bFGF in cultured ventral midbrain dopaminergic neurons. *J. Neurochem.* **84:** 803.

Searle S. and Blackwell J.M. 1999. Evidence for a functional repeat polymorphism in the promoter of the human NRAMP1 gene that correlates with autoimmune versus infectious disease susceptibility. *J. Med. Genet.* **36:** 295.

Shimajiri S., Arima N., Tanimoto A., Murata Y., Hamada T., Wang K.Y., and Sasaguri Y. 1999. Shortened microsatellite d(CA)21 sequence down-regulates promoter activity of matrix metalloproteinase 9 gene. *FEBS Lett.* **455:** 70.

Singleton A.B., Farrer M., Johnson J., Singleton A., Hauge S., Kachergus J., Hulihan M., Peuralinna T., Dutra A., Nussbaum R., Lincoln S., Crawley A., Hanson M., Maraganore D., Adler C., Cookson M.R., Muenter M., Baptista M., Miller D., Blancato J., Hardy J., and Gwinn-Hardy K. 2003. alpha-Synuclein locus triplication causes Parkinson's disease. *Science* **302:** 841.

Spillantini M.G., Schmidt M.L., Lee V.M., Trojanowski J.Q., Jakes R., and Goedert M. 1997. Alpha-synuclein in Lewy bodies. *Nature* **388:** 839.

Stefanis L., Kholodilov N., Rideout H.J., Burke R.E., and Greene L.A. 2001. Synuclein-1 is selectively up-regulated in response to nerve growth factor treatment in PC12 cells. *J. Neurochem.* **76:** 1165.

Tan E.K., Matsuura T., Nagamitsu S., Khajavi M., Jankovic J., and Ashizawa T. 2000. Polymorphism of NACP-Rep1 in Parkinson's disease: An etiologic link with essential tremor? *Neurology* **54:** 1195.

Tan E.K., Tan C., Shen H., Chai A., Lum S.Y., Teoh M.L., Yih Y., Wong M.C., and Zhao Y. 2003. Alpha synuclein promoter and risk of Parkinson's disease: Microsatellite and allelic size variability. *Neurosci. Lett.* **336:** 70.

Tanji K., Imaizumi T., Yoshida H., Mori F., Yoshimoto M., Satoh K., and Wakabayashi K. 2001. Expression of alpha-synuclein in a human glioma cell line and its up-regulation by interleukin-1beta. *Neuroreport* **12:** 1909.

Tanji K., Mori F., Imaizumi T., Yoshida H., Matsumiya T., Tamo W., Yoshimoto M., Odagiri H., Sasaki M., Takahashi H., Satoh K., and Wakabayashi K. 2002. Upregulation of alpha-synuclein by lipopolysaccharide and interleukin-1 in human macrophages. *Pathol. Int.* **52:** 572.

Tanner C.M. and Goldman S.M. 1996. Epidemiology of Parkinson's disease. *Neurol. Clin.* **14:** 317.

Touchman J.W., Dehejia A., Chiba-Falek O., Cabin D.E., Schwartz J.R., Orrison B.M., Polymeropoulos M.H., and Nussbaum R.L. 2001. Human and mouse alpha-synuclein genes: Comparative genomic sequence analysis and identification of a novel gene regulatory element. *Genome Res.* **11:** 78.

Ueda K., Fukushima H., Masliah E., Xia Y., Iwai A., Yoshimoto M., Otero D.A., Kondo J., Ihara Y., and Saitoh T. 1993. Molecular cloning of cDNA encoding an unrecognized component of amyloid in Alzheimer disease. *Proc. Natl. Acad. Sci.* **90:** 11282.

Valente E.M., Bentivoglio A.R., Dixon P.H., Ferraris A., Ialongo T., Frontali M., Albanese A., and Wood N.W. 2001. Localization of a novel locus for autosomal recessive early-onset Parkinsonism, PARK6, on human chromosome 1p35-p36. *Am. J. Hum. Genet.* **68:** 895.

van Duijn C.M., Dekker M.C., Bonifati V., Galjaard R.J., Houwing-Duistermaat J.J., Snijders P.J., Testers L., Breedveld G.J., Horstink M., Sandkuijl L.A., van Swieten J.C., Oostra B.A., and Heutink P. 2001. Park7, a novel locus for autosomal recessive early-onset Parkinsonism, on chromosome 1p36. *Am. J. Hum. Genet.* **69:** 629.

Vila M., Vukosavic S., Jackson-Lewis V., Neystat M., Jakowec M., and Przedborski S. 2000. Alpha-synuclein up-regulation in substantia nigra dopaminergic neurons following administration of the Parkinsonian toxin MPTP. *J. Neurochem.* **74:** 721.

Xia Y., Rohan de Silva H.A., Rosi B.L., Yamaoka L.H., Rimmler J.B., Pericak-Vance M.A., Roses A.D., Chen X., Masliah E., DeTeresa R., Iwai A., Sundsmo M., Thomas R.G., Hofstetter C.R., Gregory E., Hansen L.A., Katzman R., Thal L.J., and Saitoh T. 1996. Genetic studies in Alzheimer's disease with an NACP/alpha-synuclein polymorphism. *Ann. Neurol.* **40:** 207.

Xu Z.P., Tsuji T., Riordan J.F., and Hu G.F. 2003. Identification and characterization of an angiogenin-binding DNA sequence that stimulates luciferase reporter gene expression. *Biochemistry* **42:** 121.

Yamada N., Yamaya M., Okinaga S., Nakayama K., Sekizawa K., Shibahara S., and Sasaki H. 2000. Microsatellite polymorphism in the heme oxygenase-1 gene promoter is associated with susceptibility to emphysema. *Am. J. Hum. Genet.* **66:** 187.

# Genomic Perspective and Cancer

D. BOTSTEIN
*Lewis-Sigler Institute, Princeton University, Princeton, New Jersey 08544*

The discovery of the double-helical structure of DNA, and the concomitant realization that DNA molecules carry genetic information digitally encoded in their nucleotide sequences (Watson and Crick 1953a,b,c), neatly divide the history of biology in the 20th century. The first half featured the rise of classical genetics: analysis of the inheritance of traits obtained at first from natural variation and later by induced or selected mutations. Much was learned from breeding studies in plants and simple "model" organisms such as *Drosophila* and *Neurospora,* including quite detailed genetic linkage maps. The second half saw the rise of molecular biology: the elucidation, in considerable detail, of the information pathway that begins with the nucleotide sequence in DNA and ends with the specification of the phenotypes of cells and organisms.

The determination of the complete nucleotide sequences of the human, mouse, and many bacterial and eukaryotic model organism genomes is the logical culmination of both molecular biology and classical genetics. Just as the DNA structure transformed classical genetics into a molecular science, the genomic sequences are transforming molecular biology into an information science, marking the beginning of a third era in the history of biology. In each case, the newer science is of necessity firmly based in its predecessor, but thinking and research become transformed both in style and substance; each era adds a new perspective to our understanding of biology.

## INTELLECTUAL ORIGINS OF GENOMICS

The ideas underlying the science of genomics can be traced back to the early years of molecular biology. The first truly "genomic" paper was presented at the Cold Spring Harbor Symposium in 1963; it summarized the results of a deliberate program to identify all the genes of bacteriophage T4 (Epstein et al. 1964). The paper described, in a general way, what each of the T4 genes does for the organism. This program was based on the idea that one might be able to obtain mutations in all the essential T4 genes by isolating conditional-lethal mutations. Two kinds of conditional-lethal mutations (chain-terminating and temperature-sensitive) had recently been described; strong arguments were made for the idea that either or both of these kinds of mutations could be found in any essential phage gene if one looked hard enough for them. The T4 genes themselves were then defined and enumerated genetically: The mutations were classified into genes by complementation and recombination mapping using their common conditional-lethal phenotypes. This was clearly a lot of work, even for an organism expected to have no more than about 100 genes. It thus seems worth noting that, like modern genomics papers, Epstein et al. (1964) was the result of an international collaboration among several laboratories and had 10 authors (remarkably many, in 1963).

Figure 1 is a composite, consisting of an electron micrograph of phage T4 on the left, and a diagram from Epstein et al. (1964). The genes are shown in the order they appeared on the circular T4 genetic linkage map, along with an abbreviation or ideogram that describes the outcomes of infections with mutant phages under nonpermissive circumstances: D0 for no DNA synthesis, DA for DNA synthesis arrest, MD for maturation defective, and ideograms for the presence of heads, tails, unassembled heads and tails, and various kinds of assembled, but defective, phage particles. Updates of this figure served, for many years, as the genome database for bacteriophage T4. A similar genomics program was successfully carried out with several other bacterial viruses as well, notably bacteriophages λ, P22, and ϕX174, and a few animal viruses (e.g., polyoma, adenovirus, and herpesvirus). In each of these cases, most (if not quite all) essential viral genes were identified in advance of any sequencing, and their biological roles were defined in at least a general way.

Shortly after the appearance of Epstein et al. (1964), two substantial efforts were undertaken to identify all the genes of two free-living organisms (the yeast *Saccharomyces cerevisiae* and the nematode worm *Caenorhabditis elegans*), despite the expectation that their genes would number in the many thousands. Once again, gene enumeration was to be via conditional-lethal mutations and classical genetic methods (complementation and recombination mapping). These efforts, led by Leland Hartwell and Sydney Brenner, respectively, also had substantial success well before the DNA sequence era (Hartwell 1970, 1974, 1978; Brenner 1974). It was this success that attracted a large and productive research community of molecular biologists to the study of these organisms. These active research communities and the progress they made in biology made it logical, even inevitable, that these would be the model organisms whose genomes would be sequenced first.

As molecular sequences accumulated, it became clear that the sequences and functions of most genes and proteins are strongly conserved in evolution. Today, the findings of the bacterial, yeast, worm, and other model organism research communities about individual genes and

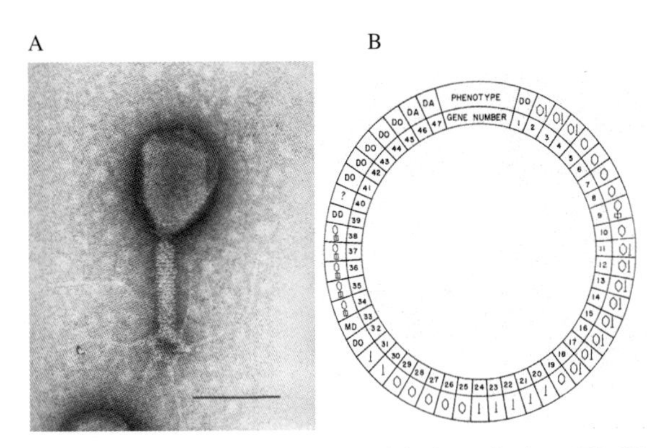


**Figure 1.** (*A*) Electron micrograph of the bacteriophage T4. (*B*) Summary of the T4 genes in linkage map order with ideograms and abbreviations indicating mutant phenotypes of conditional-lethal alleles. (*A*, Reprinted, with permission, from Büchen-Osmond 2003.)

proteins are the basis for most of what is known about the roles they play in the biology of all organisms, including the human. This "grand unification" of biology is part of the genomic perspective.

## EXTRACTING BIOLOGICAL INFORMATION FROM GENOME SEQUENCE

Even before the Human Genome Project had been organized, the need for suitable archives for the onrushing flood of genomic data became obvious, as did the need for ways to compare and display data in a manner useful to biologists. Computation quickly became indispensable; thanks to the rapid pace of advance in the productivity of computers, computation per se has rarely if ever been limiting in genomics. It was the provision of suitable biological context for sequences and computed results about sequences that became the challenge. At first, most effort went into "primary annotation," which includes finding the open reading frames, splice junctions, homology and synteny with other organisms, etc. Most of this annotation today is being done essentially automatically by an increasingly sophisticated and powerful set of computer programs.

It soon became clear that if biologists were to have useful access to the fruits of genomic sequencing, another level of "biological" annotation would be necessary. Databases were organized to meet this larger challenge, ranging from the very basic and important archival sequence databases (NCBI, EBI, SwissProt, etc.) to the more specialized organism-specific databases (SGD, MGD, FlyBase, WormBase, etc.). Today, these have grown into a veritable armamentarium, including many more focused databases that catalog such things as sequence motifs or mutations in particular gene families. These databases have already become indispensable to working biologists of every kind.

A considerable part of the challenge facing biological annotators concerns nomenclature and language. The classical methods for naming and describing the functions of genes, proteins, protein assemblies, and even biological processes themselves remain different for each species and for each sub-field of biology, producing something of a Tower of Babel. The genomic database organizations recognized the need for a common language describing the biology associated with genes and proteins, and banded together to produce what is now called the "Gene Ontology" (GO; this is not, in a strict sense, an ontology, but the name has caught on nevertheless; Ashburner et al. 2000; Harris et al. 2004). GO, which is described elsewhere in this volume (Ashburner et al.), emerged as a limited vocabulary organized in a set of directed acyclic graphs that represent the "biological processes," "molecular activities," and "subcellular locations" associated with genes and proteins of an organism. GO has rapidly become popular with genome biologists, as it facilitates biological annotation in a way that allows, among other things, computational connections among the functional annotations of orthologs and makes it possible to begin to assess quantitatively the significance, in the context of biological function, of the coexpression of two genes (see, e.g., Raychaudhuri et al. 2003; Troyanskaya et al. 2003).

## ASSESSING GENE EXPRESSION GENOME-WIDE

Complete genomic sequences have provided biologists with a finite universe of genes and proteins for each organism. For the first time, it has become possible to design experiments that interrogate every gene for its activity in a biological process. It is this kind of comprehensive experiment that provides a global perspective and that we think of as "genomic." The genomic technology that has advanced the most rapidly in recent years is DNA microarray hybridization. Many variants of this technology have come into use. All have in common the intent to measure, by hybridization, the relative amounts of nucleic acid in a sample corresponding to each gene. As with any method, there are limitations in practice, some of which apply to all the technologies, and others that affect some methods more than others; we do not discuss further the technology per se; instead, the reader is directed to a collection of recent reviews (Brown and Botstein 1999; *Nature Genetics* [supplement] 2002). Despite these limitations, DNA microarray technology has provided a wide-ranging and comprehensive view of gene expression patterns both in experimental model systems and in normal and diseased human tissues. DNA microarrays have also been used to study, genome-wide, changes in DNA copy number, once again in both model systems and human tissues.

As with DNA sequences themselves, the value of DNA microarray analysis depends on the ability to connect results with biology. The large numbers of measurements represented in a single array (typically tens of thousands) require considerable computation not only to recover and organize the data, but also to present them in a form that is simultaneously comprehensive and intuitive. In 1998, Eisen et al. described a system for analysis and display of microarray data, many features of which have come into common use. The most important and general feature is

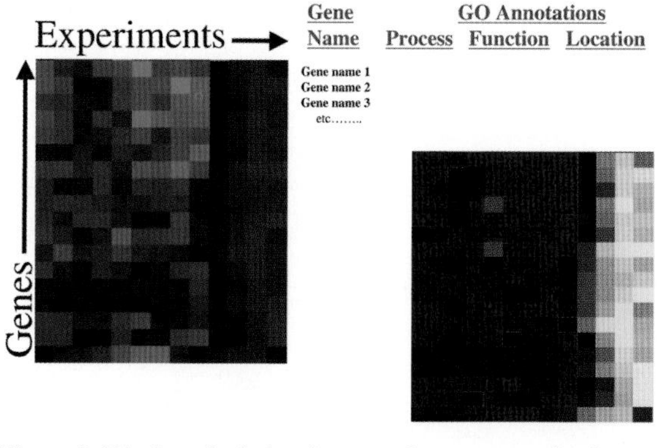


**Figure 2.** Display of relative degrees of gene expression in a set of DNA microarray experiments. A table of $\log_2$ of ratios of gene expression between an experimental sample and (usually) a common reference is colored according to the relationship of each cell in a row to the median (or mean) for that row. Increasing intensity of red indicates higher ratios, and increasing intensity of green indicates lower ratios; often yellow and blue are substituted for red and green, respectively (*inset*). The display is connected to biology by the text annotations of gene name and abbreviations of appropriate GO terms. For more detail, see Eisen et al. (1998) and Ashburner et al. (2000).

the method of display (adapted from Weinstein et al. 1997): Tables of suitably analyzed gene expression values are presented with cells colored according to the magnitude of the difference between the value in the cell and the mean or median for that gene in the group of arrays being compared. Generally, each row of the table represents a single gene, and each column a single array. Much useful analysis can be performed, without removal of any data, just by manipulating the order of the rows and columns according to an analysis scheme (usually some form of clustering), after which patterns of gene expression become manifest as patches of color. Viewing the entire colored table, one can see an overview of patterns consisting of literally millions of individual gene expression values. One can also zoom in on portions of the pattern. As shown in Figure 2, the gene names are listed next to each row along with summary descriptions of what is known about the genes (e.g., GO annotations). At this level, biologists can often not only see relationships among the genes in their experiments, but also begin to make inferences based on what is held in common by the annotations for the genes clustered together by the analysis. A Windows implementation (TreeView) and an enhanced platform-independent version (JavaTreeView) of this display system are freely available from genome-www.stanford.edu.

Several points are worth emphasizing about the practical advantages of this style of analysis and display. First, and probably most important, the analyis preserves the comprehensive nature of experiments intended to interrogate the entire genome. In this way, it provides and maintains a genomic perspective. The experimenter gets an overview, through the patterns of color, of all the data as analysis proceeds. Second, because data are not removed, it facilitates the unsupervised discovery of relationships of the patterns of gene expression between uncharacterized genes and those that have been well-studied. Inclusion of an uncharacterized gene in a cluster of coexpressed genes has become one of the most common leads to characterization of such a gene's role in the cell. Third, it allows the comparison (and, under the right circumstances, even the amalgamation) of data from many different kinds of experiments. For example, the Eisen et al. analysis and display system facilitated the discovery that the so-called "proliferation cluster" observed in a number of studies of tumors consists of genes periodically expressed in synchronized HeLa cells (Whitfield et al. 2002). The ability to usefully compare diverse data, collected by different groups under different conditions, is an important property that microarray data share with molecular sequence data. The data have cumulative value, which makes it important that all data, not just the subsets used to make a point in a paper, should be made freely available at the time of publication.

## MOLECULAR PORTRAITS OF CELLS, TISSUES, AND TUMORS

DNA microarrays that contain many thousands of different human cDNA sequences can be used to assess patterns of gene expression, producing a highly detailed and nuanced map of gene expression across the genome. Each individual microarray shows the relative abundance of transcripts of each of the genes represented on the array, and thereby gives a characteristic and nuanced picture of the biological state of the cells or tissues from which the mRNA was extracted. After application of clustering algorithms, the patterns of a number of microarrays can be assessed together, not only visually, but also quantitatively, using a variety of statistical methods and computer algorithms that relate gene expression patterns to each other and to external information, including the identities of the cells or tissues, their environment, their response to previously applied stimuli, or disease state. An example of such a map is shown in Figure 3, in which the patterns of gene expression of about 6000 different human genes in 440 different cell lines and tissues are shown together, after the data had been clustered in both the gene and array dimension. It is easy to discern visually that similar cell types and tissues, collected under similar conditions, display similar patterns of gene expression. Likewise, despite the extreme diversity of cell and tissue types and environmental conditions, it is easy to discern groups of genes that appear regularly to be expressed similarly over the entire gamut of cell type and condition.

What can be learned from clustering of gene expression patterns of large numbers of cell and tissue samples? First, as pointed out above, one can obtain, for relatively uncharacterized genes, quite specific suggestions regarding their role in the biology of the tissue or organism. Second, clustering of arrays according to the patterns of gene expression allows inferences to be made about the biology of the cells from which the RNA was drawn. A good example of this was the demonstration of substantial and reproducible biological differences among apparently similar cell types (e.g., fibroblasts or endothelial cells), depending on their anatomical site of origin (Chang et al. 2002; Chi

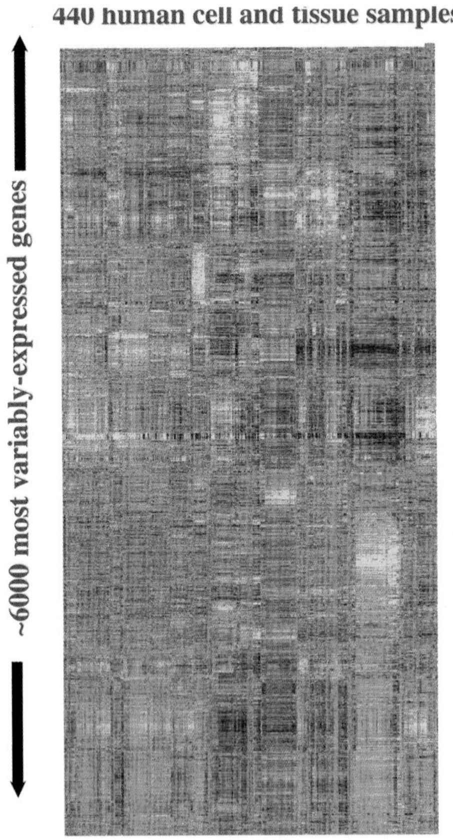

440 human cell and tissue samples

~6000 most variably-expressed genes

Figure 3. Cluster diagram as described in Fig. 2 that includes gene expression data from more than 400 cell and tissue samples whose gene expression was measured relative to a common reference; about 6000 of the most variably expressed genes are represented. (This figure was made by Pat Brown and Mike Eisen and includes data collected by Max Diehn, Xin Chen, Jon Pollack, Chuck Perou, Therese Sorlie, Mitch Garber, Marci Schaner, Matt van de Rijn, Gavin Sherlock, and Mike Fero.)

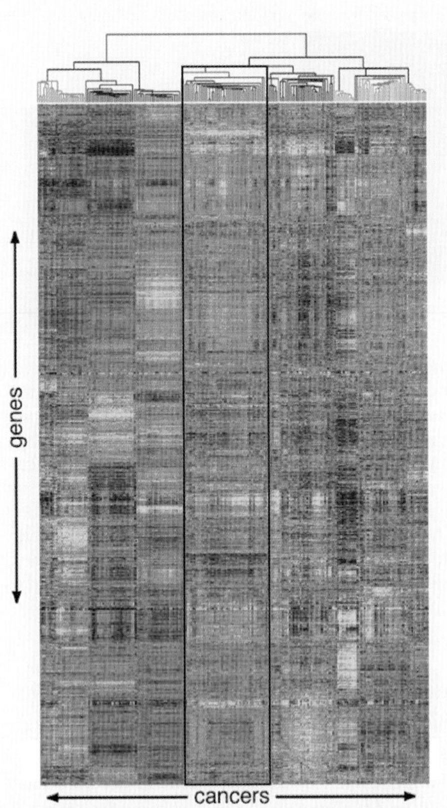


Figure 4. Cluster diagram as described in Fig. 2 that includes samples of about 500 diverse tumors relative to a common reference; about 6000 of the most variably expressed genes are represented. Source of the data is the same as in Fig. 3.

et al. 2003). This conclusion emerged when gene expression patterns of fibroblast or endothelial cell cultures from several individuals were compared by cluster analysis. The patterns for cells derived from similar anatomical sites but from different individuals clustered tightly together, indicating very little interindividual variation compared to the variation found between similar cell cultures derived from different parts of the human body.

Similarly strong biological inferences could be drawn from the analysis of gene expression profiles of human tumors. A large set of arrays representing the patterns of expression of about 6000 genes in a variety of diverse, crudely dissected tumors is shown in Figure 4. Once again, the clustering was done in both the gene and array dimension. Inspection of the figure shows that tumors of similar tissue of origin, but from many different patients, cluster together, indicating that tumors of each type (e.g., breast) are more similar in pattern of gene expression to each other than any of them is to another tumor type (e.g., ovarian or liver). This is despite the fact that these tumors consist of many different cell types, each of which contributes characteristic patterns of gene expression to the overall portrait of the tumor (for a fuller discussion of this point, see Perou et al. 1999, 2000).

Gene expression patterns thus appear to reflect accurately, as might have been expected, the biological differences among cell types and tumor tissues. Considering the many thousands of genes whose expression varies among the various cell and tissue tumor types, and the relatively small variation in interindividual gene expression for each different cell type and tissue, these molecular portraits may represent the best and most nuanced distinctions that can today be made among human cells and tissues. Molecular profiles thus provide a genomic perspective, faithfully representing, in considerable detail, the genomic contribution to cell and tissue phenotype, identity, and developmental history.

## TUMOR SUBTYPES BASED ON MOLECULAR PORTRAITS

Gene expression patterns also appear to reflect the genomic contribution to the development of tumors, consistent with everything that is known about the genetic events that underlie tumor initiation, progression, and metastasis. It therefore seemed particularly significant that among the portraits of tumors of similar origin and pathological diagnosis (e.g., breast tumors), there appeared, based on the clustering patterns, clear indications of distinguishable subtypes.

Figure 5 (Sørlie et al. 2003) shows the molecular portraits of breast tumors derived from 115 different patients. From the dendrogram (Fig. 5B), one can see that breast tumors are very diverse, especially when compared with the patterns of three typical normal breast samples (shown in black). Similarly wide diversity in tumor profiles and relatively minimal variation in normal tissue profiles have been found not only for breast cancers (Perou et al. 2000; Sørlie et al. 2001, 2003), but also for lung (Garber et al. 2001), liver (Chen et al. 2002), and gastric (Leung et al. 2002) cancers. Another common feature (not shown) is that tumor samples from the same breast cancer patient, either by repeated surgical sampling or from lymph node metastases, tend to have profiles very similar to each other (Perou et al. 2000; Sørlie et al. 2001, 2003); similar results were obtained in our studies of lung and liver tumors (Garber et al. 2001; Chen et al. 2002). This property is useful for defining subsets of genes whose expression patterns contain the most information for distinguishing subtypes, as was done for Figure 5.

The simplest interpretation of the dendrogram in Figure 5 is to suppose that there are five subtypes corresponding to the top-level nodes of the dendrogram. The samples whose patterns are best correlated in each sub-

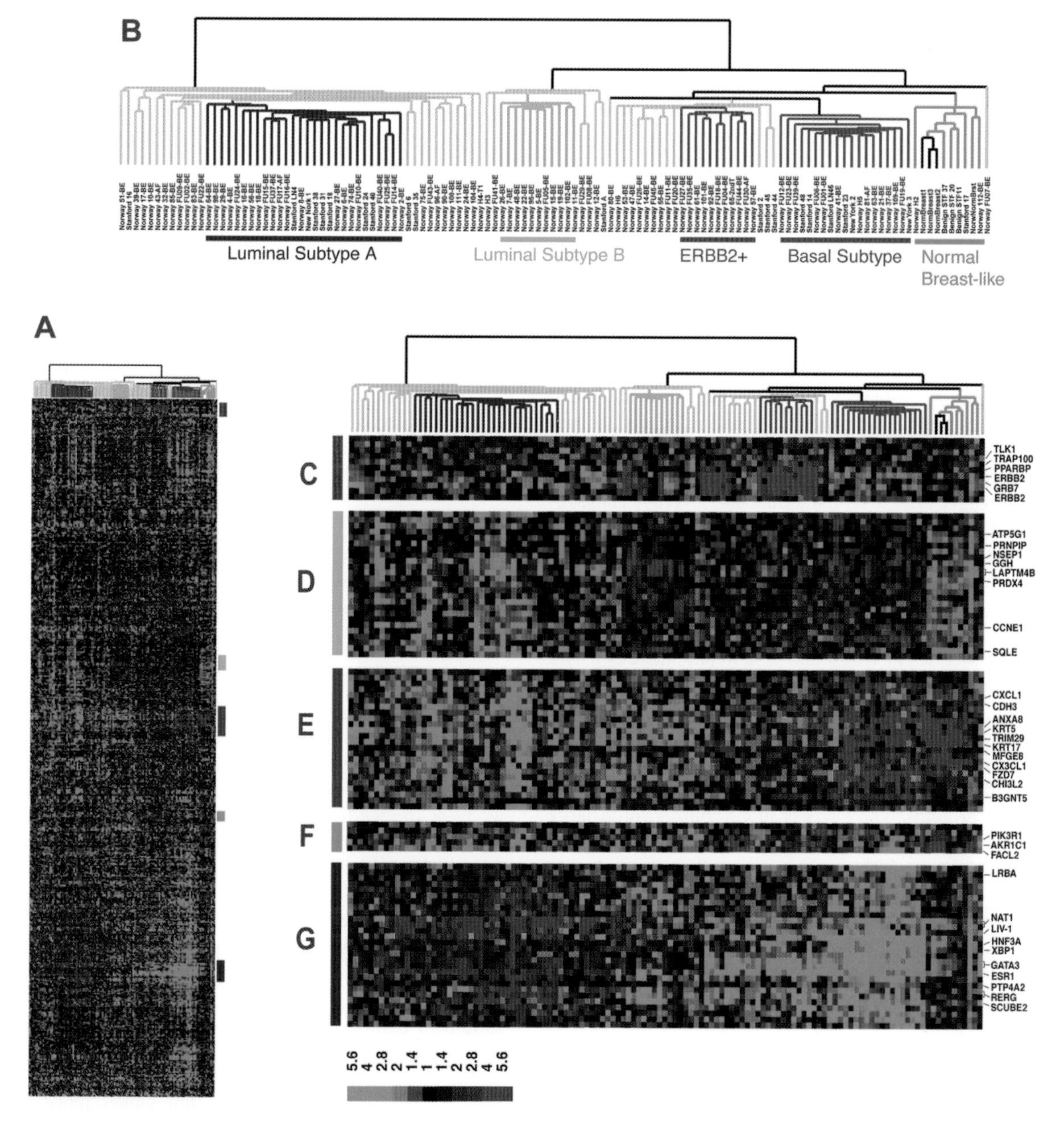


**Figure 5.** Cluster analysis of 115 breast tumors. (*A*) Representation of the entire data set clustered according to the expression of the 534 genes that vary least in repeated samples from the same individual and most across all the samples. (*B*) Dendrogram showing the clustering of the tumor samples into five groups, color coded as indicated. Black indicates the three normal breast samples. *C,D,E,F,* and *G* show the clusters of genes whose expression is characteristic of the ERBB2+, luminal B, basal, normal-like, and luminal A subtypes, respectively. Scale bar shows the fold difference relative to the median for each gene. (Reprinted, with permission, from Sørlie et al. 2003.)

group are color-coded. Each subtype has been named according to previous practice (Perou et al. 2000, Sørlie et al. 2001). The "luminal" tumor subtypes express genes (e.g., cytokeratins 5 and 17) normally expressed in the epithelial cells that normally line the lumen of breast, whereas the "basal" tumor subtypes express genes (e.g., cytokeratins 8 and 18) normally expressed by the basal epithelial cells that normally are located one or more cell diameters away from the lumen (Perou et al. 2000; van de Rijn et al. 2002). This suggests that the origins of luminal and basal tumors are somehow related to the differences between the development of normal basal and luminal epithelial cell types in the breast.

Three additional lines of evidence support the biological significance of at least some of the distinctions among the subtypes. First, the several subtypes are associated with different disease severity. Second, some of the subtype distinctions, and their different clinical consequences, are reproducible in completely separate cohorts of patients (Sørlie et al. 2003). Third, the tumors of patients genetically predisposed to breast cancer appear always to be of the basal subtype, suggesting this subtype is biologically distinct from the others.

Figure 6 shows disease outcomes for women with different breast tumor subtypes. Data for two separate patient cohorts, differing in age at onset and methods of treatment, are shown. In both cohorts the relationship of subtype to clinical course is similar. The most prominent features of the Kaplan-Meier curves are that women with tumors of the luminal A subtype have markedly less severe outcomes than do those with the basal subtype. In both cohorts, women with luminal B subtype tumors appear to have disease with intermediate severity. These results are in considerable agreement with previous studies relating expression of particular individual genes or proteins (e.g., estrogen receptor or Her2/neu) to disease outcome (cf. Henson et al. 1995; Allred et al. 1998).

Our ability to discern a relatively small number of biologically coherent breast tumor subtypes suggests a systematic explanation of such results: The different subtypes have many correlated differences in gene expression, and the differences in outcome are related to the difference in subtype, and not generally the expression of individual genes or the presence or absence of particular proteins. The success in correlating, on a large scale (more than 600 patients), the presence of cytokeratin 17 (by immunohistochemistry) with outcome underscores this point (van de Rijn et al. 2002).

As indicated above, many different tumor types have been studied using genome-wide gene expression profiling. In our experience, subtypes are more often discernible in such studies than not: We have found evidence for subtypes in diffuse large-cell lymphomas (Alizadeh et al. 2000); lung (Garber et al. (2001), liver (Chen et al. 2002), gastric (Chen et al. 2003), soft tissue (Nielsen et al. 2002), ovarian (Schaner et al. 2003), and follicular lymphoma (Bohen et al. 2003). Interestingly, in the case of follicular lymphoma, the subtypes appeared to have dichotomous responses to treatment with rituximab, providing a somewhat different line of evidence for the biological and clinical significance of the distinction between the subtypes of follicular lymphoma.

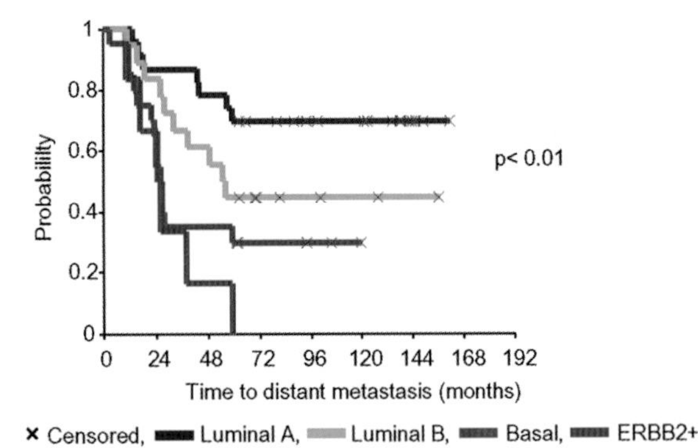


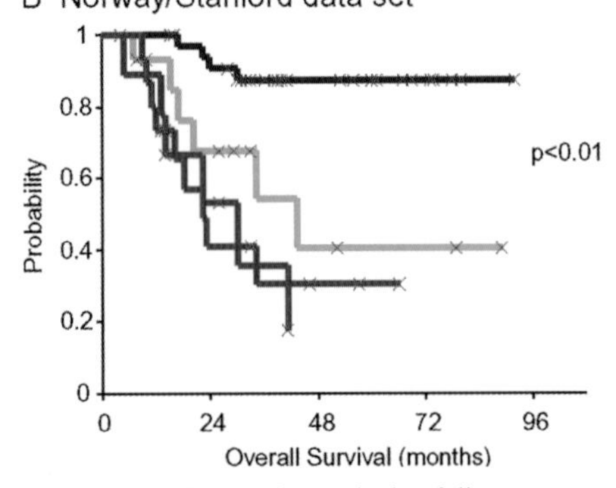


**Figure 6.** Kaplan-Meier analysis of disease outcome in two patient cohorts (from Sørlie et al. 2003). Color codes are the same as in Fig. 5. (*A*) Data from van't Veer et al. (2002) showing time to metastasis for 97 sporadic cases. (*B*) Data from Sørlie et al. (2003) showing overall survival for 72 patients with locally advanced breast cancer. The normal-like class was omitted from this analysis.

## GENOMICS AND BIOLOGICAL PERSPECTIVE

The availability of genomic sequences has changed the way in which we think about biology, even as it has changed the way in which we do research. Where once we were limited to studies applicable directly only to a limited set of organisms, we now can make inferences that apply, with high likelihood, to most organisms; where once we were limited to a view, in an experiment, of only a few genes and/or gene products, we are beginning to be able to see our experimental results in the context of all the genes and gene products.

The genome sequences have resulted in a "grand unification" of biology based on molecular sequence conservation. Molecular sequence comparisons have all but ended the intellectual fragmentation along taxonomic lines that has been a feature of the biological sciences for centuries. It is now routine to make detailed studies of common ancestry at the whole-genome level, at the level of individual gene and protein sequences, and even at the level of oligonucleotide-length sequence motifs. Results from such studies have facilitated and stimulated the development of limited vocabularies and a common language about biological functions and relationships (e.g., GO) that allows information in experimentally tractable systems to be used effectively in understanding, and devising experimental tests of that understanding, under less tractable circumstances, including humans and human disease.

The genome sequences themselves have made possible comprehensive, genome-wide experimentation where previously only a few genes and proteins could be studied simultaneously. The most advanced of these technologies are the genome-wide gene expression techniques, but others, such as the production of comprehensive sets of deletion (or "knockout") mutations in all genes (Winzeler et al. 1999; Giaever et al. 2002), comprehensive sets of fluorescently labeled proteins in vivo (Ghaemmaghami et al. 2003), comprehensive two-hybrid protein interaction screens (Uetz et al. 2000), and genome-wide synthetic lethality tests (Tong et al. 2001) are coming into use. Such methods are qualitatively different because they provide relatively complete information in context. This context is causing a new appreciation of the global consequences of phenomena where only the behavior of a few genes had been examined before. Taking a few very basic examples just from our own experience, this approach expanded severalfold the number of known cell-cycle-regulated genes in yeast (Spellman et al. 1998) and animal cells (Whitfield et al. 2002), as well as the stress response genes in yeast (Gasch et al. 2000, 2001).

The context provided by genomics has stimulated great interest in understanding globally interactions among genes and proteins. It is, for example, routine in gene expression studies to find genes and proteins that interact or participate in a process or pathway simply because they show characteristic coexpression with the other genes and proteins involved under many different conditions. New fields (called integrative genomics or system biology) are emerging whose explicit goal is to understand biological function and regulation in context, capitalizing on the new perspective and technology provided by the genome sequences.

The study of molecular portraits of tumors provides a good illustration of the change in perspective provided by the genomic view. It was difficult to distinguish what turn out to be reproducible and robust subtypes of tumors on the basis of expression of one or a few genes or proteins. Only when it became possible to study in parallel the expression of thousands of genes was it possible to see these subtypes. Instead of thinking of each new molecular marker as a central actor in tumorigenesis, progression, or metastasis, one can now see that there may be hundreds of genes with the same expression patterns. Similarly, only by following many genes at once could one distinguish differences in apparently normal fibroblasts or endothelial cells based on their anatomical origin. It is the perspective provided by the still novel ability to study and appreciate biological phenomena in a global context that will characterize biological thinking and research for years to come.

## ACKNOWLEDGMENTS

I am indebted, first of all, to P.O. Brown for a decade-long collaboration at Stanford that produced many of the ideas and results summarized above. The illustrations contain the work of virtually all of the members of our joint laboratory and our many collaborators. I am also indebted to Mike Cherry and the staff of the Saccharomyces Genome Database. Research was supported by grants from the National Cancer Institute, the National Human Genome Research Institute, and the National Institute of General Medical Sciences.

## REFERENCES

Alizadeh A.A., Eisen M.B., Davis R.E., Ma C., Lossos I.S., Rosenwald A., Boldrick J.C., Sabet H., Tran T., Yu X., Powell J.I., Yang L., Marti G.E., Moore T., Hudson J., Jr., Lu L., Lewis D.B., Tibshirani R., Sherlock G., Chan W.C., Greiner T.C., Weisenburger D.D., Armitage J.O., Warnke R., Levy R., Wilson W., Grever M.R., Byrd J.C., Botstein D., Brown P.O., and Staudt L.M. 2000. Distinct types of diffuse large B-cell lymphoma identified by gene expression profiling. *Nature* **403:** 503.

Allred D.C., Harvey J.M., Berardo M., and Clark G.M. 1998. Prognostic and predictive factors in breast cancer by immunohistochemical analysis. *Mod. Pathol.* **11:** 155.

Ashburner M., Ball C.A., Blake J.A., Botstein D., Butler H., Cherry J.M., Davis A.P., Dolinski K., Dwight S.S., Eppig J.T., Harris M.A., Hill D.P., Issel-Tarver L., Kasarskis A., Lewis S., Matese J.C., Richardson J.E., Ringwald M., Rubin G.M., and Sherlock G. 2000. Gene ontology: Tool for the unification of biology. The Gene Ontology Consortium. *Nat. Genet.* **25:** 25.

Bohen S.P., Troyanskaya O.G., Alter O., Warnke R., Botstein D., Brown P.O., and Levy R. 2003. Variation in gene expression patterns in follicular lymphoma and the response to rituximab. *Proc. Natl. Acad. Sci.* **100:** 1926.

Brenner S. 1974. The genetics of *Caenorhabditis elegans*. *Genetics* **77:** 71.

Brown P.O. and Botstein D. 1999. Exploring the new world of the genome with DNA microarrays. *Nat. Genet.* (suppl. 1) **21:** 33.

Büchen-Osmond C., Ed. 2003. Myoviridae. In *ICTVdB—The Universal Virus Database*, version 3. ICTVdB Management, The Earth Institute, Biosphere 2 Center, Columbia University, Oracle, Arizona.

Chang H.Y., Chi J.T., Dudoit S., Bondre C., van de Rijn M., Botstein D., and Brown P.O. 2002. Diversity, topographic differentiation, and positional memory in human fibroblasts. *Proc. Natl. Acad. Sci.* **99:** 12877.

Chen X., Cheung S.T., So S., Fan S.T., Barry C., Higgins J., Lai K.M., Ji J., Dudoit S., Ng I.O., Van De Rijn M., Botstein D., and Brown P.O. 2002. Gene expression patterns in human liver cancers. *Mol. Biol. Cell* **13:** 1929.

Chen X., Leung S.Y., Yuen S.T., Chu K.M., Ji J., Li R., Chan A.S., Law S., Troyanskaya O.G., Wong J., So S., Botstein D., and Brown P.O. 2003. Variation in gene expression patterns in human gastric cancers. *Mol. Biol. Cell* **14:** 3208.

Chi J.T., Chang H.Y., Haraldsen G., Jahnsen F.L., Troyanskaya O.G., Chang D.S., Wang Z., Rockson S.G., van de Rijn M., Botstein D., and Brown P.O. 2003. Endothelial cell diversity revealed by global expression profiling. *Proc. Natl. Acad. Sci.* **100:** 10623.

Eisen M.B., Spellman P.T., Brown P.O., and Botstein D. 1998. Cluster analysis and display of genome-wide expression patterns. *Proc. Natl. Acad. Sci.* **95:** 14863.

Epstein R.H., Bolle A., Steinberg C.M., Kellenberger E., Boy de la Tour E., Chevalley R., Edgar R.S., Susman M., Denhardt G.H., and Lielausis A. 1964. Physiological studies of conditional lethal mutations of bacteriophage T4D. *Cold Spring Harbor Symp. Quant. Biol.* **28:** 375.

Garber M.E., Troyanskaya O.G., Schluens K., Petersen S., Thaesler Z., Pacyna-Gengelbach M., van de Rijn M., Rosen G.D., Perou C.M., Whyte R.I., Altman R.B., Brown P.O., Botstein D., and Petersen I. 2001. Diversity of gene expression in adenocarcinoma of the lung. *Proc. Natl. Acad. Sci.* **98:** 13784.

Gasch A.P., Huang M., Metzner S., Botstein D., Elledge S.J., and Brown P.O. 2001. Genomic expression responses to DNA-damaging agents and the regulatory role of the yeast ATR homolog Mec1p. *Mol. Biol. Cell* **12:** 2987.

Gasch A.P., Spellman P.T., Kao C.M., Carmel-Harel O., Eisen M.B., Storz G., Botstein D., and Brown P.O. 2000. Genomic expression programs in the response of yeast cells to environmental changes. *Mol. Biol. Cell* **11:** 4241.

Ghaemmaghami S., Huh W.K., Bower K., Howson R.W., Belle A., Dephoure N., O'Shea E.K., and Weissman J.S. 2003. Global analysis of protein expression in yeast. *Nature* **425:** 737.

Giaever G., Chu A.M., Ni L., Connelly C., Riles L., Veronneau S., Dow S., Lucau-Danila A., Anderson K., Andre B., Arkin A.P., Astromoff A., El-Bakkoury M., Bangham R., Benito R., Brachat S., Campanaro S., Curtiss M., Davis K., Deutschbauer A., Entian K.D., Flaherty P., Foury F., Garfinkel D.J., and Gerstein M., et al. 2002. Functional profiling of the *Saccharomyces cerevisiae* genome. *Nature* **418:** 387.

Harris M.A., Clark J., Ireland A., Lomax J., Ashburner M., Foulger R., Eilbeck K., Lewis S., Marshall B., Mungall C., Richter J., Rubin G.M., Blake J.A., Bult C., Dolan M., Drabkin H., Eppig J.T., Hill D.P., Ni L., Ringwald M., Balakrishnan R., Cherry J.M., Christie K.R., Costanzo M.C., and Dwight S.S., et al. (Gene Ontology Consortium). 2004. The Gene Ontology (GO) database and informatics resource. *Nucleic Acids Res.* **32:** D258.

Hartwell L.H. 1970. Biochemical genetics of yeast. *Annu. Rev. Genet.* **4:** 373.

———. 1974. *Saccharomyces cerevisiae* cell cycle. *Bacteriol. Rev.* **38:** 164.

———. 1978. Cell division from a genetic perspective. *J. Cell Biol.* **77:** 627.

Henson D.E., Fielding L.P., Grignon D.J., Page D.L., Hammond M.E., Nash G., Pettigrew N.M., Gorstein F., and Hutter R.V. 1995. College of American Pathologists Conference XXVI on clinical relevance of prognostic markers in solid tumors. Summary. Members of the Cancer Committee. *Arch. Pathol. Lab. Med.* **119:** 1109.

Leung S.Y., Chen X., Chu K.M., Yuen S.T. Mathy J., Ji J., Chan A.S., Li R., Law S., Troyanskaya O.G., Tu I.P., Wong J., So S., Botstein D., and Brown P.O. 2002. Phospholipase A2 group IIA expression in gastric adenocarcinoma is associated with prolonged survival and less frequent metastasis. *Proc. Natl. Acad. Sci.* **99:** 16203.

*Nature Genetics*. 2002. Supplement, volume 32, pp. 461–552. Nature Publishing Group, Nature America, New York.

Nielsen T.O., West R.B., Linn S.C., Alter O., Knowling M.A., O'Connell J.X., Zhu S., Fero M., Sherlock G., Pollack J.R., Brown P.O., Botstein D., and van de Rijn M. 2002. Molecular characterisation of soft tissue tumours: A gene expression study. *Lancet* **359:** 1301.

Perou C.M., Jeffrey S.S., van de Rijn M., Rees C.A., Eisen M.B., Ross D.T., Pergamenschikov A., Williams C.F., Zhu S.X., Lee J.C., Lashkari D., Shalon D., Brown P.O., and Botstein D. 1999. Distinctive gene expression patterns in human mammary epithelial cells and breast cancers. *Proc. Natl. Acad. Sci.* **96:** 9212.

Perou C.M., Sørlie T., Eisen M.B., van de Rijn M., Jeffrey S.S., Rees C.A., Pollack J.R., Ross D.T., Johnsen H., Akslen L.A., Fluge O., Pergamenschikov A., Williams C., Zhu S.X., Lonning P.E., Børresen-Dale A.L., Brown P.O., and Botstein D. 2000. Molecular portraits of human breast tumours. *Nature* **406:** 747.

Raychaudhuri S., Chang J.T., Imam F., and Altman R.B. 2003. The computational analysis of scientific literature to define and recognize gene expression clusters. *Nucleic Acids Res.* **31:** 4553.

Schaner M.E., Ross D.T., Ciaravino G., Sørlie T., Troyanskaya O., Diehn M., Wang Y.C., Duran G.E., Sikic T.L., Caldeira S., Skomedal H., Tu I.P., Hernandez-Boussard T., Johnson S.W., O'Dwyer P.J., Fero M.J., Kristensen G.B., Børresen-Dale A.L., Hastie T., Tibshirani R., van de Rijn M., Teng N.N., Longacre T.A., Botstein D., Brown P.O., and Sikic B.I. 2003. Gene expression patterns in ovarian carcinomas. *Mol. Biol. Cell* **14:** 4376.

Sørlie T., Tibshirani R., Parker J., Hastie T., Marron J.S., Nobel A., Deng S., Johnsen H., Pesich R., Geisler S., Demeter J., Perou C.M., Lønning P.E., Brown P.O., Børresen-Dale A.L., and Botstein D. 2003. Repeated observation of breast tumor subtypes in independent gene expression data sets. *Proc. Natl. Acad. Sci.* **100:** 8418.

Sørlie T., Perou C.M., Tibshirani R., Aas T., Geisler S., Johnsen H., Hastie T., Eisen M.B., van de Rijn M., Jeffrey S.S., Thorsen T., Quist H., Matese J.C., Brown P.O., Botstein D., Eystein-Lønning P., and Børresen-Dale A.L. 2001. Gene expression patterns of breast carcinomas distinguish tumor subclasses with clinical implications. *Proc. Natl. Acad. Sci.* **98:** 10869.

Spellman P.T., Sherlock G., Zhang M.Q., Iyer V.R., Anders K., Eisen M.B., Brown P.O., Botstein D., and Futcher B. 1998. Comprehensive identification of cell cycle-regulated genes of the yeast *Saccharomyces cerevisiae* by microarray hybridization. *Mol. Biol. Cell* **9:** 3273.

Tong A.H., Evangelista M., Parsons A.B., Xu H., Bader G.D., Page N., Robinson M., Raghibizadeh S., Hogue C.W., Bussey H., Andrews B., Tyers M., and Boone C. 2001. Systematic genetic analysis with ordered arrays of yeast deletion mutants. *Science* **294:** 2364.

Troyanskaya O.G., Dolinski K., Owen A.B., Altman R.B., and Botstein D. 2003. A Bayesian framework for combining heterogeneous data sources for gene function prediction (in *Saccharomyces cerevisiae*). *Proc. Natl. Acad. Sci.* **100:** 8348.

Uetz P., Giot L., Cagney G., Mansfield T.A., Judson R.S., Knight J.R., Lockshon D., Narayan V., Srinivasan M., Pochart P., Qureshi-Emili A., Li Y., Godwin B., Conover D., Kalbfleisch T., Vijayadamodar G., Yang M., Johnston M., Fields S., and Rothberg J.M. 2000. A comprehensive analysis of protein-protein interactions in *Saccharomyces cerevisiae. Nature* **403:** 623.

van de Rijn M., Perou C.M., Tibshirani R., Haas P., Kallioniemi O., Kononen J., Torhorst J., Sauter G., Zuber M., Kochli O.R., Mross F., Dieterich H., Seitz R., Ross D., Botstein D., and Brown P. 2002. Expression of cytokeratins 17 and 5 identifies a group of breast carcinomas with poor clinical outcome. *Am. J. Pathol.* **161:** 1991.

van'T Veer L.J., Dai H., van de Vijver M.J., He Y.D., Hart A.A., Mao M., Peterse H.L., van der Kooy K., Marton M.J., Witteveen A.T., Schreiber G.J., Kerkhoven R.M., Roberts C., Linsley P.S., Bernards R., and Friend S.H. 2002. Gene expression profiling predicts clinical outcome of breast cancer. *Nature* **415:** 530.

Watson J.D. and Crick F.H.C. 1953a. A structure for deoxyribose nucleic acid. *Nature* **171:** 737.

———. 1953b. Genetic implications of the structure of deoxyribonucleic acid. *Nature* **171:** 964.

———. 1953c. The structure of DNA. *Cold Spring Harbor Symp. Quant. Biol.* **18:** 123.

Weinstein J.N., Myers T.G., O'Connor P.M., Friend S.H., Fornace A.J., Jr., Kohn K.W., Fojo T., Bates S.E., Rubinstein L.V., Anderson N.L., Buolamwini J.K., van Osdol W.W., Monks A.P., Scudiero D.A., Sausville E.A., Zaharevitz D.W., Bunow B., Viswanadhan V.N., Johnson G.S., Wittes R.E., and Paull K.D. 1997. An information-intensive approach to the molecular pharmacology of cancer. *Science* **275:** 343.

Whitfield M.L., Sherlock G., Saldanha A.J., Murray J.I., Ball C.A., Alexander K.E., Matese J.C., Perou C.M., Hurt M.M., Brown P.O., and Botstein D. 2002. Identification of genes periodically expressed in the human cell cycle and their expression in tumors. *Mol. Biol. Cell* **13:** 1977.

Winzeler E.A., Shoemaker D.D., Astromoff A., Liang H., Anderson K., Andre B., Bangham R., Benito R., Boeke J.D., Bussey H., Chu A.M., Connelly C., Davis K., Dietrich F., Dow S.W., El Bakkoury M., Foury F., Friend S.H., Gentalen E., Giaever G., Hegemann J.H., Jones T., Laub M., Liao H., and Davis R.W. 1999. Functional characterization of the *S. cerevisiae* genome by gene deletion and parallel analysis. *Science* **285:** 901.

# Chromosome 21 and Down Syndrome: The Post-sequence Era

S.E. Antonarakis, A. Reymond, R. Lyle, S. Deutsch, and E.T. Dermitzakis
*Division of Medical Genetics, NCCR Frontiers in Genetics, University of Geneva Medical School and University Hospitals, Geneva, Switzerland*

Trisomy 21, the etiology of Down syndrome (DS), is the most common cause of genetic mental retardation, and a model for the numerous aneuploidy syndromes. DS was first described by John Langdon Down in 1866 (Down 1866), and the trisomy for a group G acrocentric chromosome was recognized by Lejeune et al. in 1959 (Lejeune et al. 1959). Many recent studies have used human chromosome 21 as a model for genomic studies such as determination of haplotype block (Patil et al. 2001), identification of transcribed sequences (Kapranov et al. 2002), and comparative genomics (Frazer et al. 2003).

DS can be viewed as a collection of various phenotypes that are directly or indirectly related to the supernumerary copy of genes or other functional DNA elements on human chromosome 21 (Hsa21). The entire phenotypic expression of the syndrome is therefore a polygenic disease in which all primarily of the contributing genes map to Hsa21. One striking clinical observation is that for the majority of the DS phenotypes the penetrance is variable; i.e., not all DS patients manifest all the phenotypes. For example, the characteristic atrioventricular septal heart defect (AVSD) is only present in 16% of patients with DS. In addition, the degree of severity of the phenotypes varies among patients; for example, the IQs of affected individuals vary from the low 20s to the 70s.

The working hypotheses to explain the phenotypic variability include the following:

1. There are two categories of genes on Hsa21; those that are dosage-sensitive and contribute to the phenotypes of DS, and those that are not dosage-sensitive and therefore do not contribute to any of the phenotypes.
2. The effect of the dosage-sensitive genes could either be allele-specific or allele-nonspecific. In other words, the combination of certain but not all alleles could be contributory to the phenotype. This could be either qualitative (alleles with amino acid variation) or quantitative (alleles with variation in gene expression/protein levels). In the latter case, a threshold effect of total transcript output or amount of protein could be envisaged: A phenotype is only present if the total transcript/protein level of the three alleles reaches a critical amount.
3. The effect of the dosage-sensitive genes could either have a direct or indirect effect on the phenotype. The indirect effect may be due to the interaction of Hsa21 genes or gene products with non-Hsa21 genes or gene products. This interaction could again be allele-specific; maybe only certain combinations of non-Hsa21 alleles contribute to susceptibility of specific phenotypes. Therefore, the global dysregulation of the individual transcriptome or proteome could contribute to the DS phenotypes.
4. Triplication of certain conserved non-genic sequences (CNGs, see below) on Hsa21 may contribute to the DS phenotypes. This contribution could also be allele-specific.

The completion of the sequence of chromosome 21 (Hsa21) and the comparison of these sequences with those of other mammalian genomes, such as mouse, provide an unprecedented opportunity to identify not only the protein-coding sequences, but also other functional segments of Hsa21. In addition, the study of population variation of Hsa21 will reveal functional variation (quantitative or qualitative) that may contribute to the various phenotypes of DS. Finally, the development of an expression atlas of all mouse orthologs of the Hsa21 genes has provided another useful tool for prioritizing the candidate genes that may be involved in the various phenotypes of trisomy 21. These studies will serve as model for the other partial or full aneuploidies. In this paper, we briefly discuss three different aspects of Hsa21/DS research in our laboratory.

## CONSERVED NON-GENIC SEQUENCES

The completion of the sequencing of Hsa21 (Hattori et al. 2000) and the mouse genome (Mural et al. 2002; Waterston et al. 2002) provided for the first time the opportunity to compare the two and to recognize functionally conserved genomic elements. We compared the 33.5-Mb genome of Hsa21q with the syntenic mouse genomic regions on Mmu16, Mmu17, and Mmu10; the initial goal was to complete the genic annotation of Hsa21 by recognizing novel genes (coding or noncoding RNAs), and/or to update and correct the description of known genes. This is of importance because the full gene catalog is necessary for the understanding of the molecular pathophysiology of DS. We used the program PipMaker (Schwartz et al. 2000) and focused on sequences ≥100 nucleotides in length and ≥70% identity without gaps. A total of 3491 such sequences were identified, and only 1229 of those corresponded to exons of previously known genes. The mapping position of the remaining 2262 conserved se-

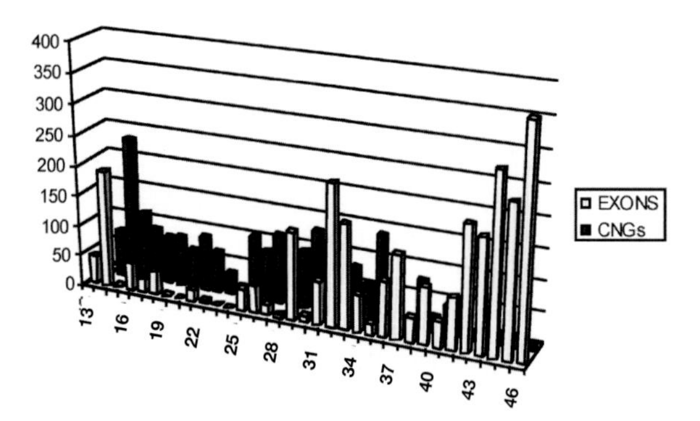


**Figure 1.** Distribution of exons and CNGs along human chromosome 21. X-axis indicates the position on the chromosome in megabases; Y-axis indicates the number of sequences.

quences was remarkable. They were preferentially placed in the gene-poor regions of the chromosome. About 80% of these 2262 sequences are located in intergenic regions and the remaining 20% in introns (see Fig. 1 for distribution of exons and CNGs along the chromosome). Extensive attempts to show that these conserved sequences were transcribed were unsuccessful. In an experiment to test 123 putative gene models of the 2262 conserved sequences by RT-PCR in 20 human tissues, only 2 gave a positive signal of a transcribed product. In addition, several characteristics clearly separated them from coding sequences. The distribution of Dn/Ds (rate of replacement/rate of silent substitutions) was clearly much lower in the known exons than the remaining conserved sequences (Fig. 2). Furthermore, the distribution of the frequency of distances between substitutions showed the expected periodicity of 3 in the 1229 known exons (due to the high occurrence of silent substitutions in the third codon position), whereas this pattern was not present in the remaining 2262 conserved sequences. Figure 3 shows schematically the transcription potential of the 2262 conserved sequences, based on six criteria that could classify these sequences as transcribed. The majority of those (63%) have a very poor potential for transcription (none of the six criteria satisfied). We therefore named these sequences Conserved Non-Genic (CNG). In a second experiment, we could retrieve more than 50% of them in rabbit by PCR using conserved primers. This highly conservative estimate shows that most of the CNGs are probably shared by multiple species. The conservation between human and mouse over the 70 My of evolution from the common ancestor strongly indicates that the majority of CNGs, which account for ~1% of the Hsa21 sequence, are functional.

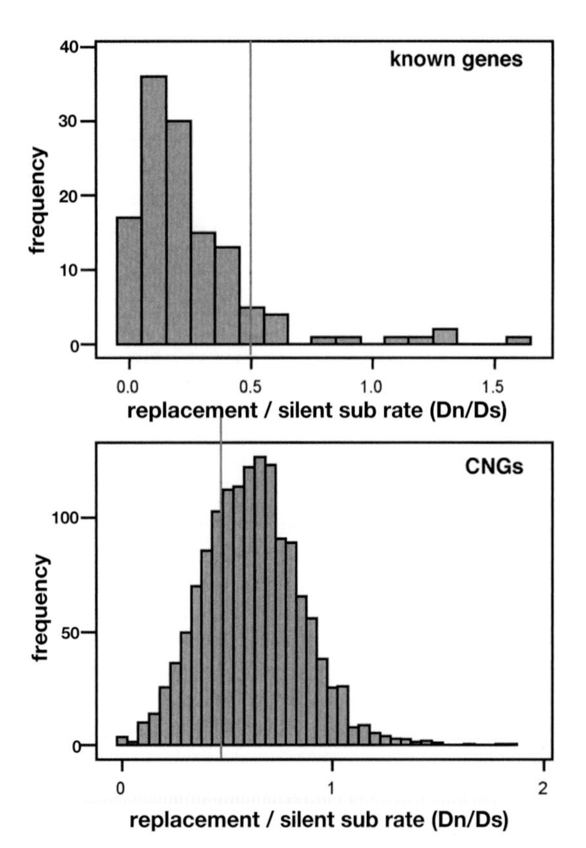


**Figure 2.** Distribution of Dn/Ds in known Hsa21 genes (*top*) and distribution of the minimum Dn/Ds values from each of the six frames of CNGs on Hsa21.

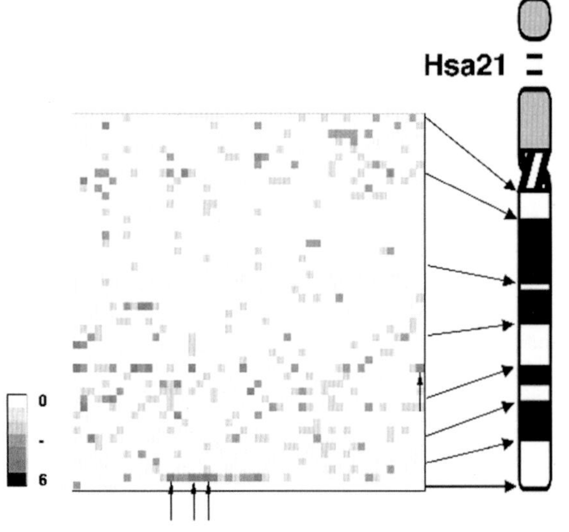


**Figure 3.** Transcription potential of 2262 CNGs on Hsa21 as determined by six criteria (Grail; Affymetrix, Genomescan, Human ESTs, Mouse ESTs, QRNA). See Dermitzakis et al. (2002) for explanation of criteria. Arrows indicate subsequently confirmed genes. (Adapted from Dermitzakis et al. 2002.)

The function of CNGs is presently unknown. Their hypothetical functions include (1) regulatory elements that exert their function either in *cis* (in nearby genes) or in *trans* (in genes at a considerable distance from them or even on other chromosomes); (2) structural elements; and (3) totally unknown or unpredictable function. The levels of conservation and position of CNGs in the genome suggest that if they are regulatory they do not represent traditional *cis* elements, since known regulatory regions are not as highly conserved as CNGs between human and mouse (Dermitzakis et al. 2002).

The extensive and systematic functional and evolutionary study of CNGs will enhance our understanding of these important genomic elements. Further studies will address the contribution of CNGs to health and disease, both Mendelian and complex phenotypes; in addition, they may reveal their potential involvement in the different phenotypes of trisomy 21.

## EXPRESSION ATLAS OF MOUSE ORTHOLOGS OF Hsa21 GENES

The completion of the human genome, and the continued effort to identify all the human genes, present now the formidable task of the elucidation of the function of all these genes. The next logical level of gene annotation is to determine the expression pattern of each individual gene in terms of tissue, cellular, and temporal specificity. High-resolution methods, such as RNA in situ hybridization (ISH) provide an accurate description of the spatiotemporal distribution of transcripts as well as a three-dimensional "in vivo" gene expression overview.

We set out to analyze systematically the expression pattern of genes from the entire Hsa21. We chose to develop this gene expression atlas using the mouse orthologs of the Hsa21 genes; these genes map in the syntenic region of the mouse genome; i.e., segments of Mmu16, Mmu17, and Mmu10. A total of 161 murine genes out of the 178 confirmed human genes were studied. The expression atlas, which was a collaborative effort with the laboratories of A. Ballabio, TIGEM, Naples, and G. Eichele, Max Planck Institute, Hannover, consists of (1) whole-mount ISH of mouse embryos at E9.5 and E10.5; (2) ISH on serial sagittal sections of E14.5; and (3)RT-PCR in four developmental stages (E8.5, E9.5, E12.5, and E19) and 12 adult tissues (brain, heart, kidney, thymus, liver, stomach, muscle, lung, testis, ovary, skin, and eyes). All the data (gene list, original ISH images, annotation tables, details on probes and oligonucleotide primers) are publicly available through Web sites http://www.tigem.it/ch21exp/ or http://www.nature.com/nature (Reymond et al. 2002).

By whole-mount ISH at E10.5, patterned (regional) gene expression was observed in 28%, and ubiquitous expression in 24%, of genes; furthermore, weak ubiquitous and strong regional expression was observed in 47% of genes. The ISH in tissue sections at E14.5 revealed patterned (regional) gene expression in 42%, ubiquitous expression in 13%, and combined weak ubiquitous and strong regional expression in 9% of genes. The highest numbers of genes with a restricted expression pattern were observed in the brain, the eye, and the gut at all stages (Fig. 4a). The RT-PCR analysis resulted in an average of eight adult tissues per gene (Fig. 4b). The transcriptomes of the brain and kidney showed the highest complexity, each tissue expressing 85% of the 161 genes examined. Muscle, as expected, had the lowest transcriptome complexity, since only 21% of genes examined were expressed there.

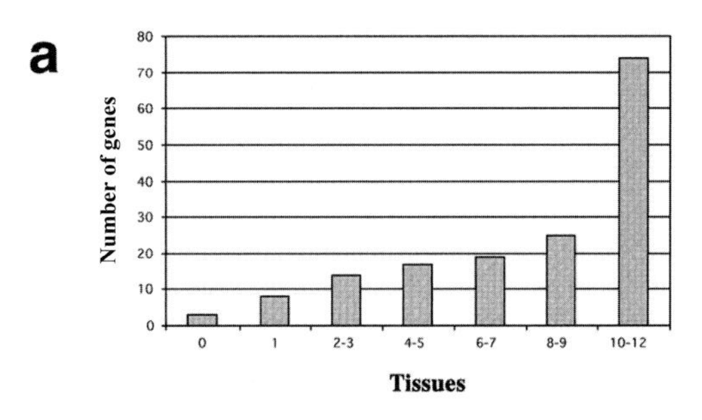


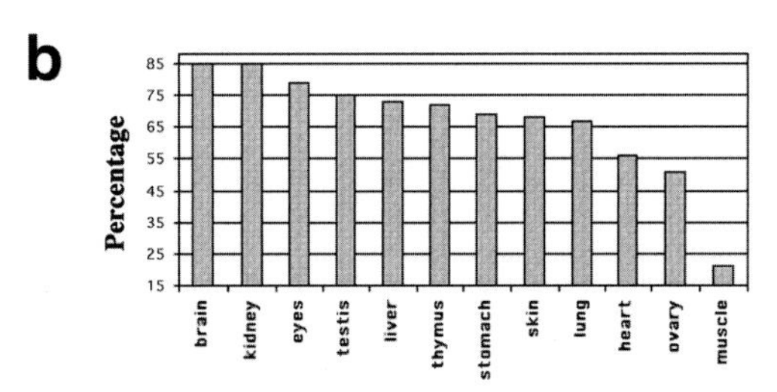


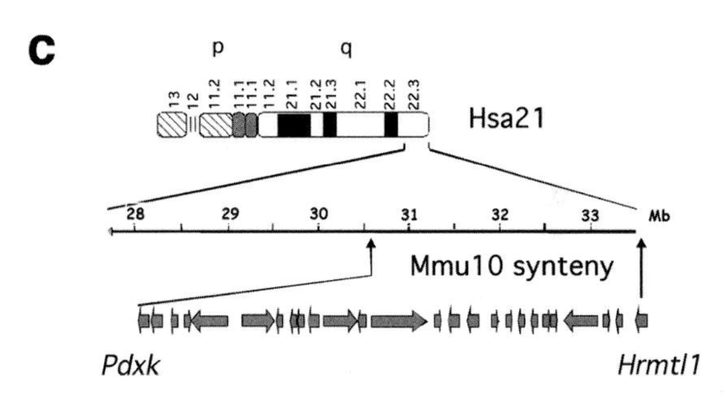


**Figure 4.** (*a*) Number of the Hsa21 mouse orthologous genes analyzed by RT-PCR expressed in 0, 1, 2–3, 4–5, 6–7, 8–9, and 10–12 adult tissues. (*b*) Percentage of the Hsa21 mouse orthologous genes analyzed by RT-PCR identified in each murine adult tissue. (*c*) Schematic representation of a cluster of Hsa21 orthologous genes (*Pdxk* to *Hrmtl1*) mapping to Mmu10 that show a significant absence of expression in muscle ($p = 0.02$). Mb distances were published in Hattori et al. (2000). Genes are represented by gray arrows. (Adapted from Reymond et al. 2002.)

Unexpectedly, we found genomic regions containing clusters of genes with similar expression patterns. Regions containing either co-silenced or co-expressed genes were observed. For example, RT-PCR analysis identified a 3.9-Mb (from genes *B3galt5* to *Hsf2bp*) and a 4.3-Mb region (from genes *Pdxk* to *Hrmtl1*, Fig. 4c) in which all genes were not expressed in heart ($p<0.001$) and in muscle ($p = 0.028$), respectively. The molecular mechanism accounting for this observation is unknown.

The combination of gene mapping with expression analysis is important for the identification of positional candidate genes for human diseases. The expression atlas of Hsa21 provides a rich resource for candidate genes for both monogenic and multifactorial diseases mapping to chromosome 21. The atlas will also have an impact in assessing the contribution of specific candidate genes to Down syndrome traits and phenotypes.

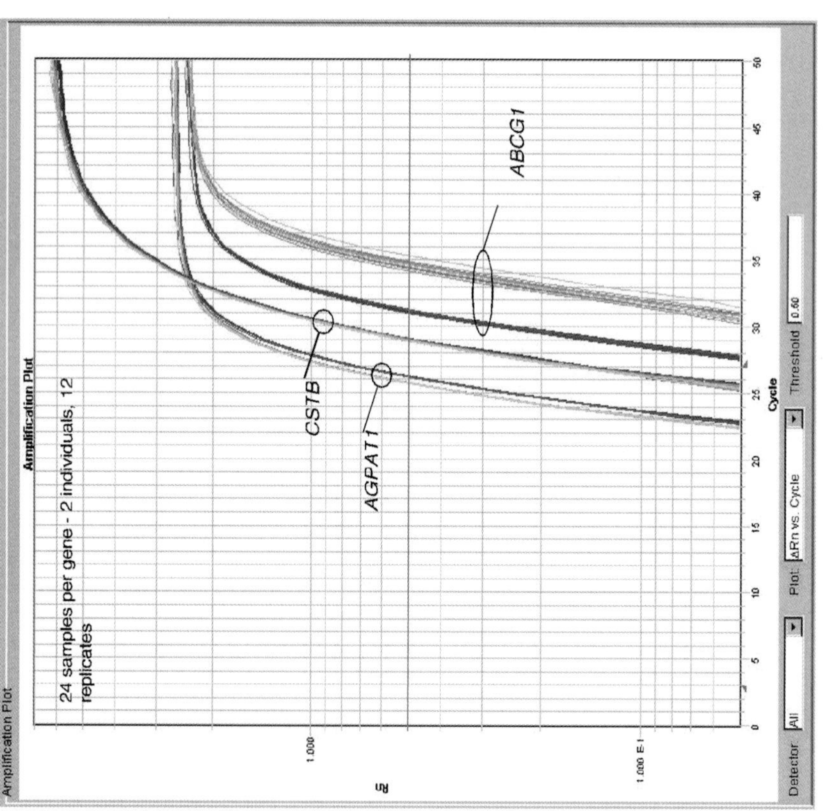


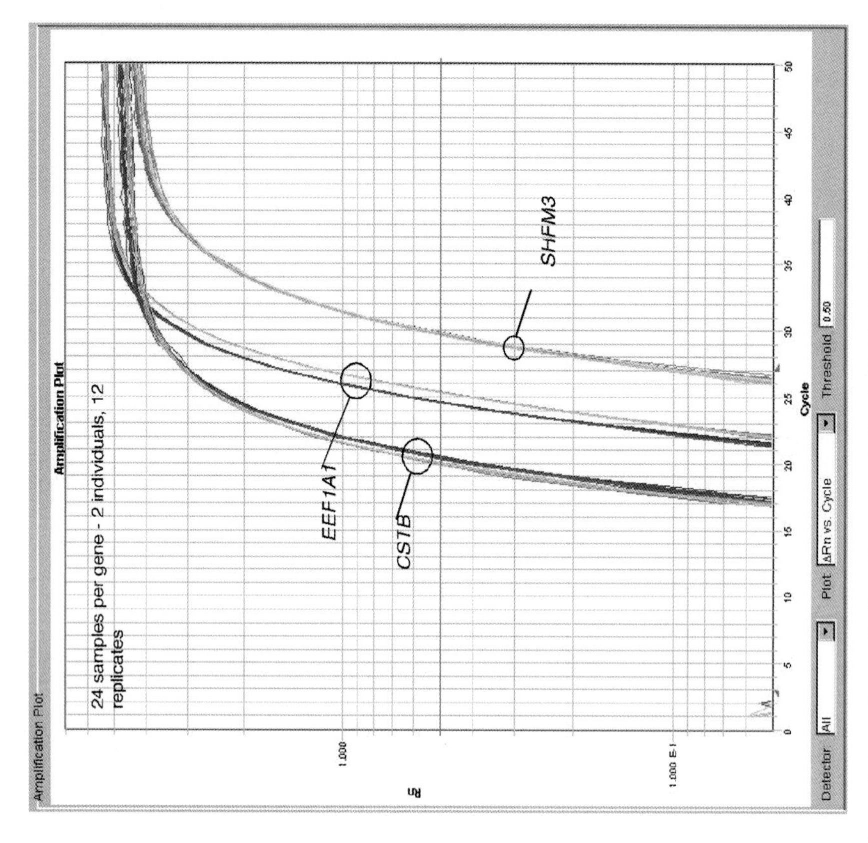


**Figure 5.** Sample amplification plots from human pilot study. Each panel shows the results from three genes in two individuals, with 12 replicates per gene. For each gene, the different shadings represent two individuals. Note, for example, the large expression difference in the two individuals for *ABCG1* relative to *CSTB* and *AGPAT1*.

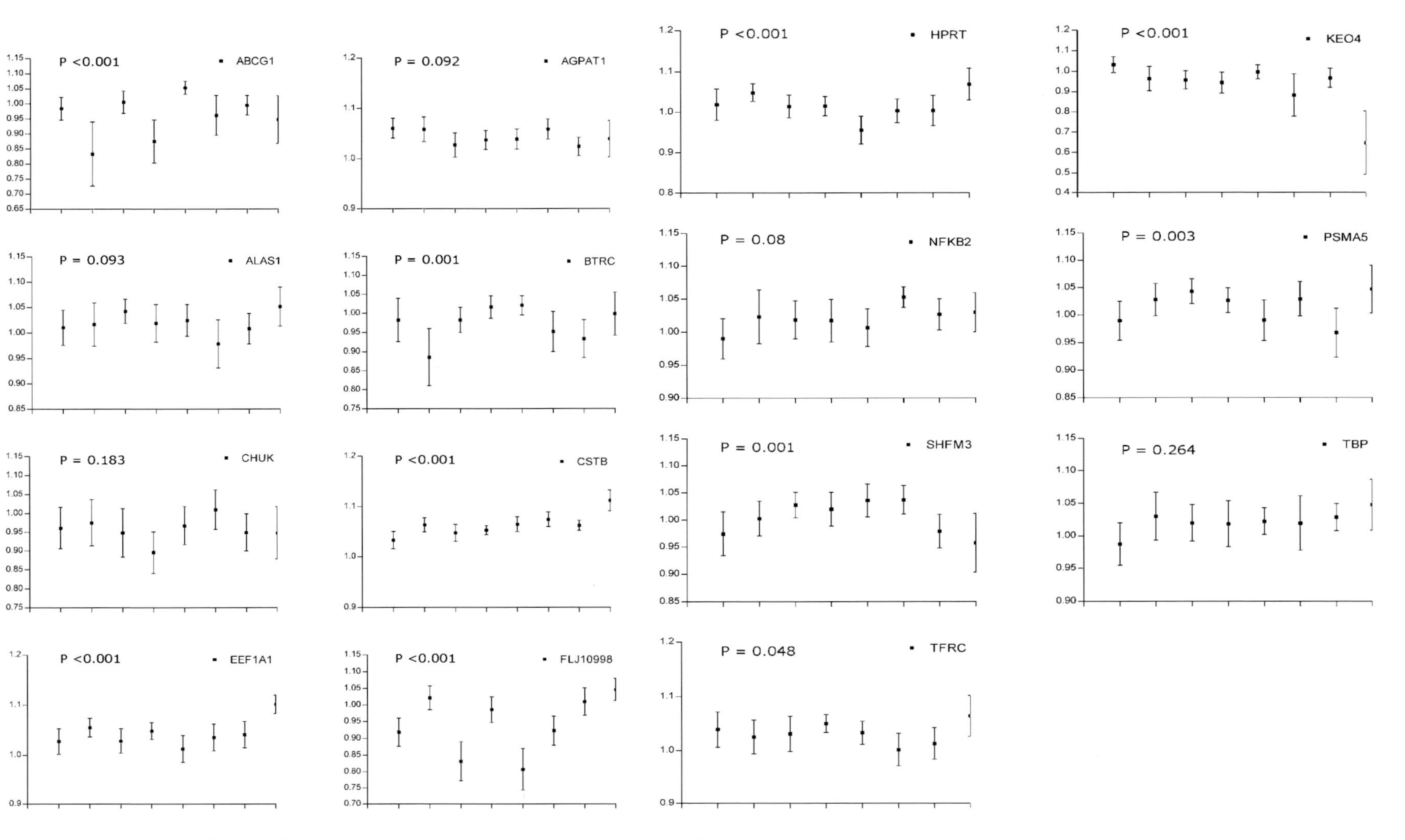


**Figure 6.** Plots of relative expression levels of 15 genes in eight individuals. Significance of variation between individuals was determined with one-way ANOVA. Error bars represent the 95% CI. The $x$ axis represents relative normalized expression levels between individuals. The eight individuals are plotted in the same order for each gene on the $y$ axis.

## POPULATION VARIATION IN GENE EXPRESSION

The allele-specific gene-dosage effect mainly relies on the hypothesis that there are allelic differences in normal gene expression. Some initial studies confirmed that (1) there are differences in the allelic expression of genes in different mouse strains and humans (Schadt et al. 2003), and (2) there are differences in allelic gene expression in lymphoblastoid cell lines from individuals of the CEPH families (Cheung et al. 2003). Both studies further showed that some of this variation is genetically determined. However, these studies used oligonucleotide microarray analysis, which is not ideal to detect small differences from genes producing less than ~5 RNA copies per cell. Real-time quantitative PCR (RT-qPCR) is more sensitive than microarray hybridization, and the dynamic range of gene expression detection is much larger. Therefore, RT-qPCR can reveal subtle differences in gene expression that cannot be detected with microarrays, but could still be causative for phenotypic variation. In a preliminary analysis, we used RT-qPCR to determine whether there is natural variation in gene expression. We tested 15 genes in lymphoblastoid cell lines from 8 different normal individuals. Each measurement was done in 12 replicates. Figure 5 shows an example of the RT-qPCR confirming the accuracy and reproducibility of the assay. Figure 6 shows the normalized relative expression levels of all the genes tested in all individuals. There is indeed wide variation of transcript amount among these individuals; this variation was statistically significant in 10 (66%) of these genes (Table 1).

These results may serve as a basis for association and linkage studies between genomic variation *cis* or *trans* and gene expression variation. Thus, the genomic variation that controls transcript variation could be identified. This in turn may lead to a better understanding of genomic variation that confirms susceptibility to common disorders, to the identification of a plethora of regulatory elements, and to the characterization of genes and alleles that contribute to the different phenotypes of Down syndrome.

**Table 1.** Variation of Transcript Amounts among Genes

| Gene | *p* value[a] |
|---|---|
| *ABCG1* | <0.001 |
| *AGPAT1* | 0.092 |
| *ALAS1* | 0.093 |
| *BTRC* | 0.001 |
| *CHUK* | 0.183 |
| *CSTB* | <0.001 |
| *EEFIA1* | <0.001 |
| *FLJ10998* | <0.001 |
| *HPRT* | <0.001 |
| *KEO4* | <0.001 |
| *NFKB2* | 0.08 |
| *PSMA5* | 0.003 |
| *SHFM3* | 0.001 |
| *TBP* | 0.264 |
| TFRC | 0.048 |

[a]15 genes, 8 individuals (lymphoblastoid cell lines). *p* values are for one-way ANOVA: 9/15 genes show significant variation.

## ACKNOWLEDGMENTS

We thank C. Ucla, N. Scamuffa, C. Rossier, A. Ballabio, G. Eichele, V. Marigo, M. Yaylaoglu, S. Banfi, V. Jongeneel, B. Stevenson, V. Flegel, P. Bucher, and C. Gehrig for their invaluable contributions to this body of work. We also thank the Swiss National Science Foundation, the European Union, the Lejeune and Childcare Foundations for their support.

## REFERENCES

Cheung V.G., Conlin L.K., Weber T.M., Arcaro M., Jen K.Y., Morley M., and Spielman R.S. 2003. Natural variation in human gene expression assessed in lymphoblastoid cells. *Nat. Genet.* **33:** 422.

Dermitzakis E.T., Reymond A., Lyle R., Scamuffa N., Ucla C., Deutsch S., Stevenson B.J., Flegel V., Bucher P., Jongeneel C.V., and Antonarakis S.E. 2002. Numerous potentially functional but non-genic conserved sequences on human chromosome 21. *Nature* **420:** 578.

Down J.L.H. 1866. Observations on an ethnic classification of idiots. *Lond. Hosp. Clin. Lect. Rep.* **3:** 259.

Frazer K.A., Chen X., Hinds D.A., Pant P.V., Patil N., and Cox D.R. 2003. Genomic DNA insertions and deletions occur frequently between humans and nonhuman primates. *Genome Res.* **13:** 341.

Hattori M., Fujiyama A., Taylor T.D., Watanabe H., Yada T., Park H.S., Toyoda A., Ishii K., Totoki Y., Choi D.K., Soeda E., Ohki M., Takagi T., Sakaki Y., Taudien S., Blechschmidt K., Polley A., Menzel U., Delabar J., Kumpf K., Lehmann R., Patterson D., Reichwald K., Rump A., and Schillhabel M., et al. 2000. The DNA sequence of human chromosome 21. The chromosome 21 mapping and sequencing consortium. *Nature* **405:** 311.

Kapranov P., Cawley S.E., Drenkow J., Bekiranov S., Strausberg R.L., Fodor S.P., and Gingeras T.R. 2002. Large-scale transcriptional activity in chromosomes 21 and 22. *Science* **296:** 916.

Lejeune J., Gautier M., and Turpin R. 1959. Etudes des chromosomes somatiques de neufs enfants mongoliens. *C.R. Acad. Sci.* **248:** 1721.

Mural R.J., Adams M.D., Myers E.W., Smith H.O., Miklos G.L., Wides R., Halpern A., Li P.W., Sutton G.G., Nadeau J., Salzberg S.L., Holt R.A., Kodira C.D., Lu F., Chen L., Deng Z., Evangelista C.C., Gan W., Heiman T.J., Li J., Li Z., Merkulov G.V., Milshina N.V., Naik A.K., and Qi R., et al. 2002. A comparison of whole-genome shotgun-derived mouse chromosome 16 and the human genome. *Science* **296:** 1661.

Patil N., Berno A.J., Hinds D.A., Barrett W.A., Doshi J.M., Hacker C.R., Kautzer C.R., Lee D.H., Marjoribanks C., McDonough D.P., Nguyen B.T., Norris M.C., Sheehan J.B., Shen N., Stern D., Stokowski R.P., Thomas D.J., Trulson M.O., Vyas K.R., Frazer K.A., Fodor S.P., and Cox D.R. 2001. Blocks of limited haplotype diversity revealed by high-resolution scanning of human chromosome 21. *Science* **294:** 1719.

Reymond A., Marigo V., Yaylaoglu M.B., Leoni A., Ucla C., Scamuffa N., Caccioppoli C., Dermitzakis E.T., Lyle R., Banfi S., Eichele G., Antonarakis S.E., and Ballabio A. 2002. Human chromosome 21 gene expression atlas in the mouse. *Nature* **420:** 582.

Schadt E.E., Monks S.A., Drake T.A., Lusis A.J., Che N., Colinayo V., Ruff T.G., Milligan S.B., Lamb J.R., Cavet G., Linsley P.S., Mao M., Stoughton R.B., and Friend S.H. 2003. Genetics of gene expression surveyed in maize, mouse and man. *Nature* **422:** 297.

Schwartz S., Zhang Z., Frazer K.A., Smit A., Riemer C., Bouck J., Gibbs R., Hardison R., and Miller W. 2000. PipMaker - A web server for aligning two genomic DNA sequences. *Genome Res.* **10:** 577.

Waterston R.H., Lindblad-Toh K., Birney E., Rogers J., Abril J.F., Agarwal P., Agarwala R., Ainscough R., Alexandersson M., An P., Antonarakis S.E., Attwood J., Baertsch R., Bailey J., Barlow K., Beck S., Berry E., Birren B., Bloom T., Bork P., Botcherby M., Bray N., Brent M.R., Brown D.G., and Brown S.D., et al. 2002. Initial sequencing and comparative analysis of the mouse genome. *Nature* **420:** 520.

# Harvesting the Genome's Bounty: Integrative Genomics

P. JORGENSEN,*† B.-J. BREITKREUTZ,* K. BREITKREUTZ,* C. STARK,* G. LIU,* M. COOK,*†
J. SHAROM,*† J.L. NISHIKAWA,*† T. KETELA,¶ D. BELLOWS,* A. BREITKREUTZ,† I. RUPES,*
L. BOUCHER,*† D. DEWAR,* M. VO,* M. ANGELI,* T. REGULY,* A. TONG,†¶
B. ANDREWS,† C. BOONE,†¶ AND M. TYERS*†

*Samuel Lunenfeld Research Institute, Mount Sinai Hospital, Toronto, Ontario, Canada M5G 1X5; †Department of Medical Genetics and Microbiology, University of Toronto, Toronto, Ontario, Canada, M4G 1A8; and ¶Banting and Best Department of Medical Research, University of Toronto, Toronto, Ontario, Canada M5G 1L6*

In this post-genomic era, we face the daunting challenge of assigning function to tens of thousands of uncharacterized open reading frames. Furthermore, the physical and functional connections between gene products must be elucidated if we are to understand how the linear DNA sequence encodes dynamic cellular function. An unexpected result of the Human Genome Project is the low number of predicted human genes, roughly fivefold more than in single-celled yeasts and approximately double that in flies or worms (Lander et al. 2001; Venter et al. 2001). Increased connectivity between this relatively constant number of genes may underlie the massive increase in system-level complexity that distinguishes yeast from humans. Genome-wide approaches to discovery of gene function now include systematic analysis of genetic interactions, protein interactions, genome-wide expression profiles, and mutant phenotypes. More than any single approach, each of which is subject to caveats in interpretation and reproducibility, the intersection of orthogonal genome-scale data sets provides a robust means to interrogate gene function. Here, we summarize advanced genome-scale methods for biological discovery in the budding yeast *Saccharomyces cerevisiae*, many of which will be transportable to interrogation of human gene function.

## OVERVIEW OF FUNCTIONAL GENOMICS TOOLS IN YEAST

The *S. cerevisiae* genome sequence, completed in 1996, was the first reported for an autonomous life form. Since then, the powerful molecular genetics of the budding yeast system has afforded the first detailed analysis of a genome and its encoded proteins. Even though few yeast genes have introns, reliable open reading frame (ORF) prediction has proven non-trivial, due to both sequencing errors and a preponderance of short ORFs. Recently, the gene annotation problem has been effectively solved by comparative genome sequence analysis of *S. cerevisiae* and three closely related yeast species, *S. paradoxus, S. mikatae*, and *S. bayanus* (Kellis et al. 2003). Through this approach, 5,726 conserved ORFs have been annotated as bona fide genes, including 148 new ORFs of less than 100 residues. Notably, 500 of the originally predicted ORFs fail to meet the criterion of conservation and are likely spurious. Comparative genome analysis has also pinpointed conserved sequence elements in promoter regions, which are presumed to correspond to transcription factor-binding sites (Cliften et al. 2003; Kellis et al. 2003).

The yeast genome sequence has spawned a host of systematic approaches, many of which have since been applied to other model systems. For example, the genome sequence allowed the first genome-wide transcriptional profiles to be recorded by DNA microarray technology (DeRisi et al. 1997). Public databases now contain expression data for over 1,000 different experimental conditions (Sherlock et al. 2001). This sizable compendium of genome-wide expression data allows many gene functions and small-molecule targets to be assigned simply by clustering of gene expression profiles (Hughes et al. 2000). Insight into transcriptional regulatory circuits has been further enhanced by the development of array-based chromatin-immunoprecipitation methods that in principle allow identification of all DNA sequence elements bound by transcription factors and other chromatin-associated proteins (Ren et al. 2000; Iyer et al. 2001).

The ease with which genes can be deleted or otherwise manipulated by homologous recombination is a key attribute of budding yeast. This feature was exploited by an international consortium of laboratories to systematically delete ~95% of all predicted yeast genes (Giaever et al. 2002). The resulting collection of ~6,000 gene deletion strains has allowed numerous high-throughput approaches for characterization of gene function. Phenotypic analysis using the deletion set is augmented by an imaginative design feature, namely the inclusion of unique 20-mer oligonucleotide tags that flank each PCR deletion cassette, aptly termed barcodes (Winzeler et al. 1999). The relative abundance of all gene deletion strains in a population pool can then be determined by PCR amplification of all barcodes from genomic DNA using common primers, followed by hybridization of labeled PCR products to a microarray of all ~12,000 barcode sequences. Competitive growth of the deletion pool under any given selection thus identifies gene deletions that confer sensitivity by reduction in the cognate barcode signals. This method has been used to quantify strain growth defects and strain sensitivity to a variety of stress conditions, including osmotic stress, DNA damage, and bioactive small molecules (Giaever et al. 1999, 2002). Below, we describe application of the barcode technique to the problem of cell-size homeostasis.

A plethora of other reagent sets for yeast functional genomics have now been developed. A complete set of activation domain–ORF fusions has been constructed for systematic two-hybrid screens in an array format (Uetz et al. 2000). Fusions to different epitope tags, including a *lacZ* reporter, generated by random transposon insertions into yeast genomic DNA, have allowed a survey of the localization and expression of most predicted ORFs (Kumar et al. 2002). Recently, green fluorescent protein (GFP) and tandem affinity purification (TAP) fusions have been constructed for all yeast ORFs at endogenous chromosomal loci, thereby allowing systematic assessment of protein localization and abundance (Ghaemmaghami et al. 2003; Huh et al. 2003). Comprehensive plasmid sets have been built for conditional high-level expression of each ORF as a GST fusion in yeast (Martzen et al. 1999; Zhu et al. 2000). These fusions are readily purified from yeast and can be spotted into a microarray format, which can then be probed with proteins or small molecules or used in enzymatic assays such as phosphorylation reactions (Zhu and Snyder 2003). The cloning of yeast ORFs into flexible recombinational vectors also allows a variety of protein expression formats in bacterial, yeast, and insect cells (Ho et al. 2002). Finally, high-density arrays of small molecules open up an entirely parallel chemical genetic approach for dissection of protein function (Kuruvilla et al. 2002). In total, when combined with the molecular genetic power of the yeast system, these genome-scale reagent sets have greatly accelerated discovery of gene function, genetic networks, and biochemical pathways (Bader et al. 2003b).

## GENETIC INTERACTIONS

Genetic interactions identify components that perform related functions by enhancement or suppression of an original mutant phenotype. Synthetic lethality is an extreme case of enhancement in which two single mutations that cause minimal phenotypes individually are lethal in combination (Bender and Pringle 1991). Synthetic lethal interactions may reflect one of several situations at the physical level (Fig. 1A). Synthetic lethality is typically interpreted to arise from lesions in separate pathways that converge on the same essential function. However, in other instances, the interacting genes may encode nonessential components of the same essential complex or reflect an essential damage-response relationship. The observation that only ~20% of the genes in budding yeast are essential for viability (Giaever et al. 2002) provides a striking illustration of how genetic networks buffer the cellular phenotype against environmental and mutational perturbation (Hartman et al. 2001). The resilience of the yeast genetic network to mutation derives from both redundant biochemical function and gene duplication (Wagner 2000; Gu et al. 2003). The manifold consequences of complex genetic interactions likely form the basis for phenotypic variation (Hartman et al. 2001). The

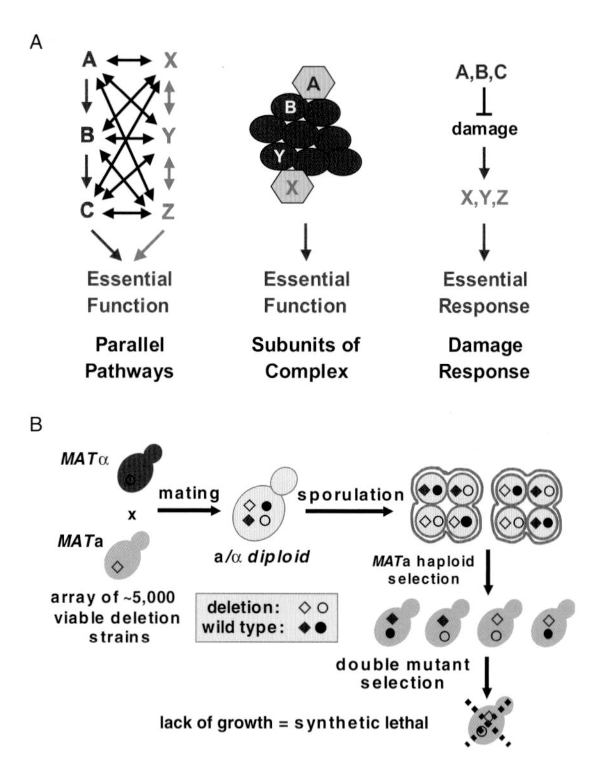


**Figure 1.** Principles of synthetic lethal genetic screens. (*A*) Underlying mechanisms of synthetic lethal interactions. (*B*) Synthetic genetic array (SGA) selection scheme. All steps may be carried out in high-density arrays by manual or robotic pinning.

phenotypic divergence of laboratory yeast strains, such as the commonly used W303 and S288C backgrounds, from wild-type *S. cerevisiae* isolates, such as Σ1278, provides an opportunity for investigating the genetic basis of variation. For example, over 600 genes are expressed at more than twofold different levels in W303 versus Σ1278 strains (A. Breitkreutz et al. 2003). Deconstruction of strain backgrounds by meiotic segregation of global gene expression patterns has identified a number of genetic alterations in transcriptional regulatory cascades within various *S. cerevisiae* strains (Yvert et al. 2003).

## GENETIC INTERACTION NETWORKS

To interrogate the global genetic interaction network of budding yeast, we have recently developed a high-throughput (HTP) method for the systematic identification of genetic interactions, termed the synthetic genetic array (SGA) method (Fig. 1B). In the SGA scheme, a strain bearing a marked query allele is individually mated to each of the ~5,000 haploid deletion mutants, followed by selection for heterozygous double-mutant diploids and sporulation (Tong et al. 2001). Haploid progeny of the **a**-cell mating type are then selected through expression of an **a**-cell-specific reporter, along with selection for markers linked to each deletion allele and the query mutation. This procedure isolates double-mutant meiotic progeny en masse by simple replica-pinning of colonies at each step, a task accomplished either manually or robotically with high-density pinning devices. Failure to recover a double-mutant combination in SGA identifies a candidate synthetic lethal interaction, which must then be confirmed by direct tetrad analysis to rule out false-positive and -negative interactions. The SGA approach is readily applied to mapping of dominant mutations, dosage lethality screens, and chemical genetic screens (Tong et al. 2001; Jorgensen et al. 2002b). Recently, an alternative systematic method termed synthetic lethal analysis by microarray (SLAM) has been developed (Ooi et al. 2003). SLAM employs high-efficiency transformation of the pool of all viable deletion strains with a marked query allele, followed by identification of nonviable combinations by barcode microarray analysis of the selected population. The SGA and SLAM approaches thus enable comprehensive HTP mapping of the yeast genetic interaction network.

An initial implementation of the SGA method for eight genes implicated in cytoskeletal control and the DNA damage response recovered 291 bona fide synthetic lethal interactions (Tong et al. 2001). Recent application of SGA to >100 query mutations in a variety of cellular processes has identified nearly 4,000 synthetic lethal interactions, fourfold more than has previously been reported in the entire biomedical literature (Tong et al. 2004). On average, each SGA screen uncovers >30 synthetic lethal interactions, considerably more than might have been anticipated from conventional forward screens (Hartwell et al. 1997; Tong et al. 2001, 2004). If the general principles of genetic networks are conserved, the unexpected density of genetic interactions revealed in yeast will greatly complicate the detection of human polygenic disorders by single-nucleotide polymorphism (SNP) mapping (Hartman et al. 2001; Tong et al. 2004). Genetic interaction maps reveal functional connections between and within pathways, and thereby assign roles to uncharacterized genes by their position in the network. As data sets increase in size, such connections are readily revealed by clustering algorithms (Tong et al. 2004).

## PROTEIN INTERACTIONS

Most genetic interactions reflect either direct or indirect functional connections between the encoded protein products. Protein interactions span a vast range of affinities, from the stable complexes that form cellular machines such as the ribosome, RNA polymerases, or the proteasome, to the ephemeral regulatory interactions such as those that control cell division, signaling, transcription, ubiquitin-dependent proteolysis, directed transport, and vesicle fusion. Transient protein interactions are inevitably dictated by reversible posttranslation modifications such as phosphorylation, methylation, acetylation, and ubiquitination. Modified protein sequences are usually bound in a reversible manner by dedicated recognition domains on interaction partners, which thereby bring pathway dynamics under posttranslational control (Pawson and Nash 2003). Mapping of protein interactions by various biochemical means has thus been a major engine for pathway discovery.

Methods for detection of protein interactions have become increasingly sophisticated and sensitive, such that it is usually no longer necessary to spend arduous months in the cold room fractionating complex cell lysates for desired activities. Two primary approaches are amenable to HTP detection of interactions; namely, the two-hybrid system and mass spectrometric analysis of protein complexes. The two-hybrid system senses a binary interaction between bait and prey fusion proteins by virtue of transcriptional activation of reporter genes or restoration of other forms of reporter protein activity (Fields and Song 1989; Uetz 2002). The cost-effectiveness of the two-hybrid system and its ability to delineate interaction regions through partial-length clones have made it an attractive system for HTP interaction studies. Mass spectrometric methods have recently been applied to HTP characterization of protein complexes. In this approach, an epitope-tagged protein complex is affinity-purified from cell lysate, resolved into its constituents by SDS-PAGE, and each species is identified by mass spectrometry (Fig. 2). Despite the conceptual simplicity of this scheme, HTP implementation of protein tagging, complex isolation, mass spectrometry, and informatics is far from trivial.

## TWO-HYBRID PROTEIN INTERACTION MAPS

The first systematic studies for mapping protein interaction networks were based on the two-hybrid method. Large-scale two-hybrid screens may be carried out either in a library pool format, whereby interacting clones are selected from mass transformation of a cDNA or genomic library, or in an array-based format, whereby each bait is directly tested against all possible prey fusions via a robotic mating procedure. Library-based methods are rapid and convenient, whereas array-based formats allow systematic coverage and are less prone to spurious interactions. The yeast proteome served as a test bed for comprehensive library pool and array-based two-hybrid screens. In the pioneering study by Uetz et al. (2000), 957 interactions were recovered from screens with 1,004 baits, whereas extensive library pool screens by Ito et al. (2001) yielded a set of 4,549 interactions. Together, these two sets of two-hybrid data contain >5,506 potential protein interactions, providing both an important resource of interaction information for biological studies and a benchmark for subsequent HTP studies. Because of relative ease and low cost, systematic two-hybrid analysis has been carried for other unicellular organisms, including the gut pathogen *Helicobacter pylori* (Rain et al. 2001). Recently, the first large-scale two-hybrid protein interaction map for a metazoan species was reported for *Drosophila melanogaster* (Giot et al. 2003). Using a pooled library approach, 12,000 baits cloned in a recombinational vector system were used to identify ~20,000 interactions, some 5,000 of which were considered to be highly probable based on their local connectivity. This study elaborated known and novel pathways in many areas of metazoan cell biology.

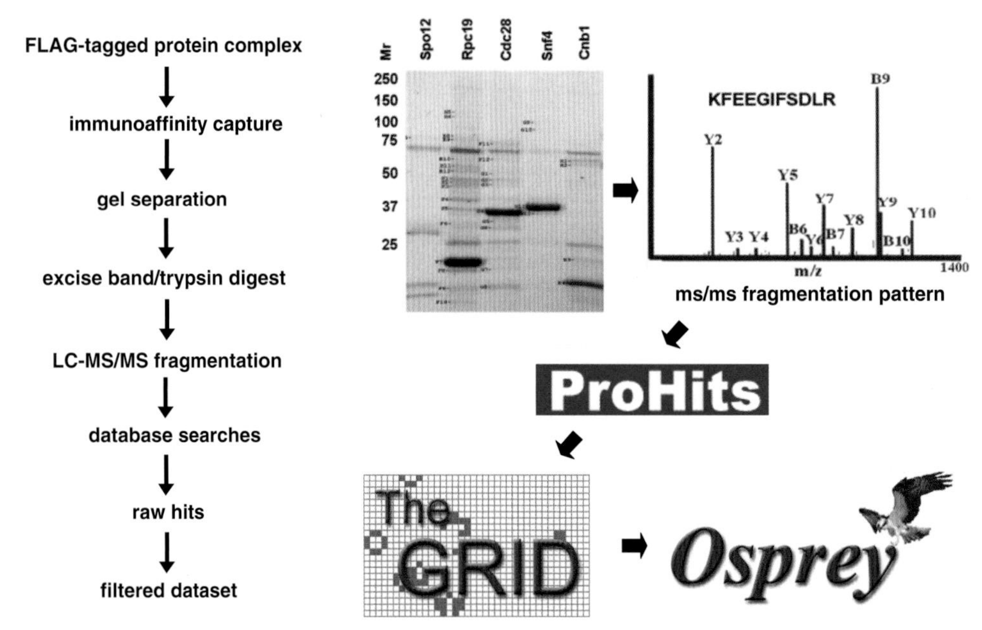


**Figure 2.** High-throughput identification of protein interactions by mass spectrometric analysis of proteins captured by FLAG-tagged bait proteins. For details, see Ho et al. (2002). Software is available at http://www.mshri.on.ca/tyers/.

## MASS SPECTROMETRIC ANALYSIS OF PROTEIN COMPLEXES

To date, only two HTP mass spectrometric studies have been reported, each of which employed complementary methods to interrogate the yeast proteome. A study conducted by CellZome AG (Heidelberg) used integration of a TAP epitope tag into each chromosomal locus that encodes a bait protein (Gavin et al. 2002). The advantage of integration is that tagged proteins are expressed at natural abundance levels under the culture conditions employed. A reciprocal disadvantage, however, is that weakly expressed or nonexpressed proteins fail as baits. The TAP purification method is a two-step strategy that yields highly purified complexes through initial capture of the protein A component of the tag onto IgG resin, followed by TEV protease release of the complex and recapture of a second tag component, calmodulin-binding domain (CBD), onto a CBD-binding peptide resin (Rigaut et al. 1999). Although the TAP method yields an impressively low background of nonspecific proteins, its drawbacks include loss of weak interactions in the dilution step upon release from the IgG and the relatively low initial capture efficiency of TAP protein fusions. The inherent low background of the TAP method allows facile identification of the single protein species usually present in excised gel slices by peptide mass fingerprinting. By beginning with 1,739 TAP-tagged fusion proteins, 1,440 proteins were identified in 459 successfully purified complexes, representing 25% coverage of the predicted proteome.

A study from our group in conjunction with MDS Proteomics (Toronto) used recombinational cloning to generate ~700 FLAG epitope-tagged ORFs under control of the inducible *GAL1* promoter (Ho et al. 2002). Many of these baits were selected to correspond to interesting regulatory categories, such as kinases, phosphatases, and proteins implicated in the DNA damage response (DDR). Inducible expression obviates concerns of bait toxicity and may help capture weakly associated proteins through production of excess bait. The FLAG tag is well suited for complex isolation because of its small size, its high capture efficiency, and the ability to elute the captured complex from antibody beads with either FLAG peptide or acidic conditions. The one-step FLAG purification procedure also facilitates HTP sample processing. A further advantage of a single-step affinity-capture method is that weak interactions tend to be retained by virtue of local concentration effects on the bait–antibody resin surface. A drawback of single-step purification methods is that a number of proteins adhere nonspecifically to resin and tube surfaces. Due in part to background, even upon mixture resolution by SDS-PAGE, many excised gel bands contain multiple proteins, a feature that confounds identification by peptide mass fingerprinting alone. To unambiguously identify proteins in such moderately complex mixtures, it is necessary to sequence individual peptide species (Ho et al. 2002). Our study was enabled by implementation of a HTP LC-MS/MS platform, through which 940,000 MS/MS individual spectra were acquired, corresponding to 35,000 protein identifications, at an average of 3.6 protein identifications per excised gel slice. We have developed software tools to cope with this glut of data, including a primary curation database for mass spectrometry called ProHits, a database for management of large-scale data sets called the GRID, and an associated visualization tool called Osprey (Fig. 2; see below). In total, 493 of 600 baits that were successfully expressed contained 3,617 specific interaction partners, as defined by frequency of occurrence in the entire data set (Ho et al. 2002). This data set also connected 25% of the predicted yeast proteome. We have continued to elaborate this net-

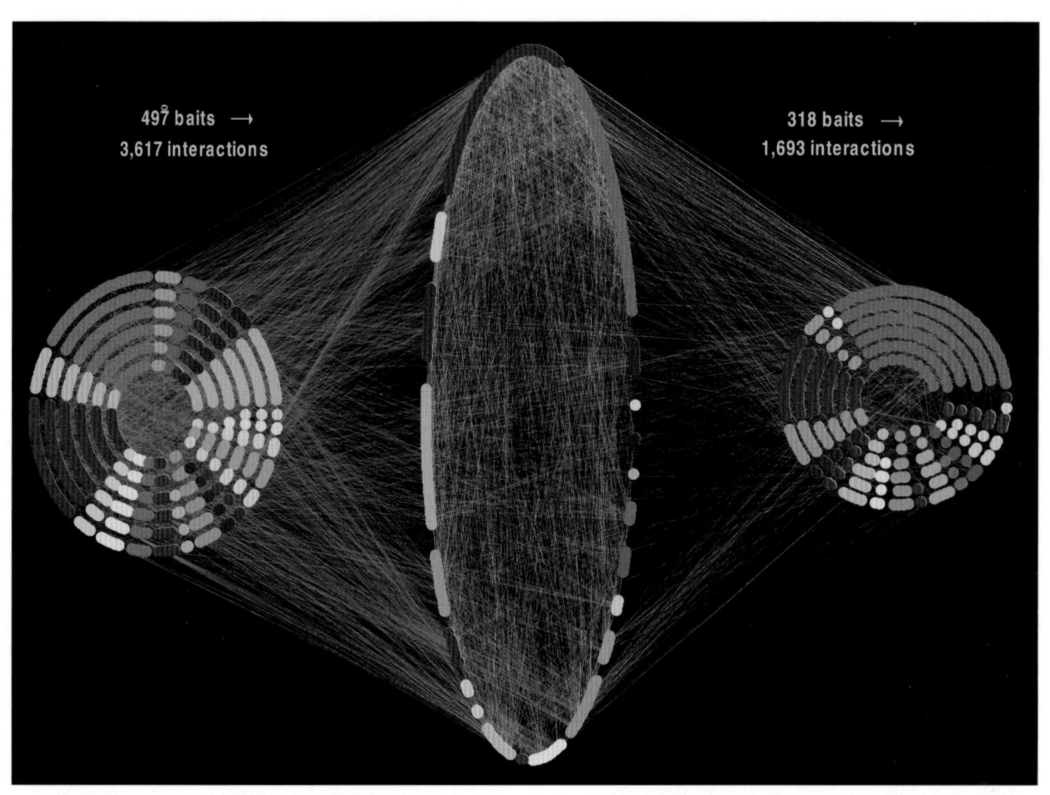


**Figure 3.** Current yeast protein interaction data set derived from mass spectrometric analysis of FLAG-tagged protein complexes, composed of 3,617 filtered interactions from Ho et al. (2002) and 1,693 filtered interactions from an unpublished data set. Colors indicate GO annotation (nodes) and data sources (edges).

work: At present, 318 productive baits and 1,693 additional interactions have been added, many of which link back to the original network (Fig. 3).

## VISUALIZATION AND ANNOTATION OF LARGE-SCALE DATA SETS

Representation of the highly interconnected genetic and protein interaction networks revealed by HTP approaches is a challenge in itself. Not only must many thousands of binary interactions be displayed in a readily navigable format, but these interactions should be linked to phenotypic information, expression profiles, localization, and post-translational modifications, all of which may have one or more quantititative attributes. In addition to dedicated HTP interaction studies, a vast body of data embedded in the biomedical literature remains to be incorporated into the growing databases. For example, the initial installment of the Human Protein Reference Database houses curated interactions from over 300,000 publications, all of which were individually read by curators (Peri et al. 2003). Although various methods enable machine-based interpretation of biomedical abstracts (Yeh et al. 2003), with simple Web-based interfaces, an expert curator can rapidly scour published interactions and provide a much sounder assessment of data reliability. Our initial efforts to curate interactions from the current ~60,000 *S. cerevisiae* publications suggest that the task will require only 1,000 hours of effort. High-quality interaction data sets derived from the literature, such as the known interactions reported in MIPS (Mewes et al. 2002), provide an important benchmark for gauging HTP approaches (Bader and Hogue 2002; Ho et al. 2002; von Mering et al. 2002).

To display and manipulate large-scale interaction data, we have developed a flexible database format called the General Repository for Interaction Datasets (GRID), designed to house all publicly available large-scale interaction data sets from any organism (see http://biodata.mshri.on.ca/grid). The GRID is built on a MySQL platform with a Java interface for Web-based transactions (B.J. Breitkreutz et al. 2003a). A key feature of the GRID is a comprehensive look-up table for gene nomenclature and gene ontology (GO) annotation, derived for *S. cerevisiae* from the Saccharomyces Genome Database (Cherry et al. 1998) and for *D. melanogaster* from FlyBase (FlyBase Consortium 2003). Each entry in a GRID table contains the ORF designation, associated gene names, a description of gene function, GO terms, experimental source of the interaction, and links to PubMed. Yeast GRID currently holds 14,114 physical and 4,925 genetic interactions, and Fly GRID holds 20,509 physical and 6,087 genetic interactions. *Schizosaccharomyces pombe*, *Caenorhabditis elegans*, mouse, and human versions of the GRID are in progress. Private versions of the GRID may also be implemented in academic labs for manipulation of unpublished data.

To explore interaction data housed in the GRID in a two-dimensional graphical format, we created a flexible visualization tool called Osprey, which affords a bird's-eye view of interaction networks and the ability to hunt down data (B.J. Breitkreutz et al. 2003b). Interaction pairs are displayed in conventional node (gene or protein) and edge format (genetic or physical interaction), which are colored according to GO annotation and experimental system. Gene descriptions and additional interactions can be viewed in pop-up tables linked to each node. Customized networks are rapidly constructed by addition of user-defined interactions from tables and are fully searchable by gene name. Networks can be automatically rendered in a variety of graphical layouts and saved in an Osprey format that contains all interaction-associated data for archiving and file exchange. Networks can also be saved in JPG, PNG, or SVG formats for construction of publication-quality figures. All networks represented in this paper were constructed from the GRID database with Osprey.

Other databases, such as Cytoscape, Intact, and BIND, also allow two-dimensional depictions of interaction data sets, some of which can be elaborated to contain coexpression, localization, and modification data (Ideker et al. 2002; Bader et al. 2003a; Hermjakob et al. 2004). Cytoscape has been engineered on a versatile plug-in platform such that custom tools and viewers can be built and linked to the core database (Ideker et al. 2002). The implementation of a standardized systems biology mark-up language (SBML) should facilitate the development of new database and visualization platforms (Hucka et al. 2003). The next generation of visualization tools will undoubtedly incorporate dynamic attributes into network representations, including transient interactions and regulated posttranslational modifications, as well as tissue, developmental, and evolutionary contexts.

## PROTEIN INTERACTION NETWORKS

Effective databases and visualization tools enable the general properties of interaction networks to be readily discerned. The current combined HTP mass spectrometric interaction data sets map some 6,894 interactions to 1,002 bait proteins, thereby nominally connecting ~40% of the yeast proteome. Similarly, combined HTP two-hybrid data sets map 6,382 interactions to 2,126 baits, linking ~60% of the proteome. In contrast, literature-curated interactions contained in the MIPS database connect only ~15% of the proteome (Mewes et al. 2002). The average connection density per bait for two-hybrid and mass spectrometric approaches are 4.5 and 9.4, respectively. The extent of overlap between these data sets has been analyzed in detail elsewhere (Bader and Hogue 2002; Ho et al. 2002; von Mering et al. 2002).

Several interesting features are evident in the protein interaction network. Many of the interactions detected occur between proteins conserved from yeast to human (Fig. 4A). Model organism networks thus provide a crucial framework for predicting and building analogous networks in human cells, as well as charting the evolution of biological circuitry. Although protein interactions are undoubtedly powerful predictors of protein function, in the absence of other data, local interactions may not be informative. For example, the density of connections for uncharacterized proteins is lower than for characterized proteins, 0.9 versus 2.7 interactions per bait, and unknowns are more likely to connect to other unknowns. Given that protein interactions are often mediated by discrete modules, such as SH3 domains (Pawson and Nash 2003), it is somewhat surprising that there is no evident enrichment for proteins with such domains in the network, either in an overall sense, or in terms of density of interactions per bait (Fig. 4B). In addition to interaction information, protein data sets directly validate gene prediction. For example, our study provided the first direct evidence for expression of nearly 500 ORFs for which no prior data at the protein level existed (Ho et al. 2002). Interestingly, mass spectrometric analysis strongly confirms a recent re-annotation of the *S. cerevisiae* genome by comparative genome sequence analysis (Kellis et al. 2003). Of the ~500 originally predicted *S. cerevisiae* ORFs that have been designated as non-protein-coding because of lack of sequence conservation, only a single such ORF was identified by mass spectrometric methods.

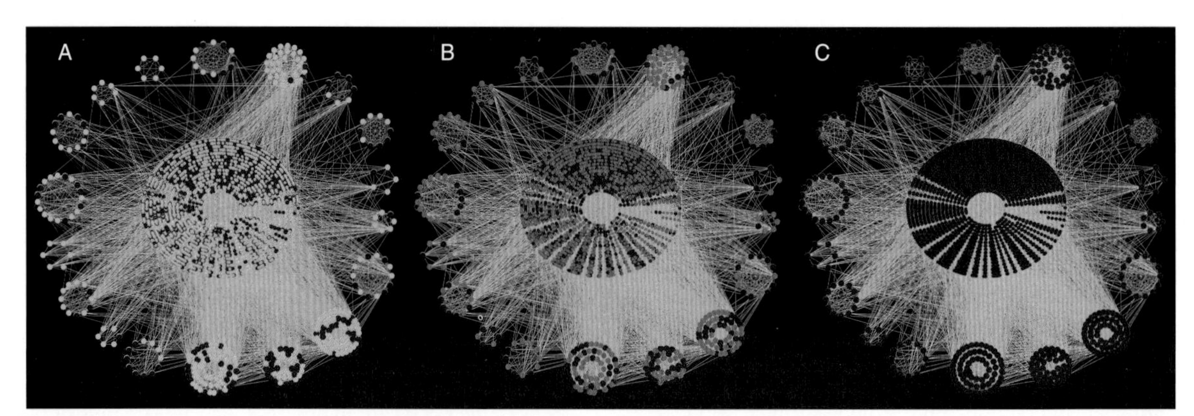


**Figure 4.** Properties of current combined yeast two-hybrid and mass spectrometric protein interaction data sets housed in the GRID database. The full yeast interaction data set was rendered into 18 highly connected complexes, shown around the perimeter of the rest of the data set. (*A*) Human orthologs of yeast proteins are shown in cyan (>50% sequence identity; average connectivity of 7.4; 1,974 of 4,301 proteins in the network). (*B*) Interaction domain-containing proteins are shown in green (<–10 RPS blast score; average connectivity of 6.3; 2,275 of 4,301 proteins in the network). (*C*) Occurrence of spurious ORFs is shown in blue (1 from mass spectrometric data set; 240 from two-hybrid data set). See Fig. 5 for annotation of most highly connected complexes.

In contrast, 240 of the dubious ORFs were reported in two-hybrid studies; these must now be viewed as spurious interactions (Fig. 4C).

A vast store of biological information is obviously embedded in large-scale interaction data sets. Local sub-networks that reflect protein complexes may be identified visually or in an automated fashion by algorithms such as MCODE (Bader and Hogue 2003). The most dramatic example to emerge de novo from nondirected HTP protein interaction analysis was a previously undetected large network of nucleolar proteins implicated in ribosome biogenesis and cell-size control (Jorgensen et al. 2002a). Comprehensive selection of baits across the known DDR network allowed elaboration of connections within the network. In particular, many previously undiscovered substrates for DDR kinases such as Dun1 were discovered (Ho et al. 2002). In addition to discrete local modules, a supramodular regulatory connection between many kinases and phosphatases can be traced through the interaction network, thereby buttressing the notion that cellular behavior is controlled in large part by phosphorylation cascades (Bader et al. 2003b).

## BIOLOGICAL NETWORK STRUCTURE

Although large-scale interaction data sets contain significant false-positive and false-negative interactions (Bader and Hogue 2002; von Mering et al. 2002), they nevertheless allow a glimpse of the topology and structure of the global protein interaction network. These networks have scale-free properties, in which the distribution of connections per node obeys a power law, similar to communication networks like the Internet (Jeong et al. 2001). Scale-free networks contain a small number of highly connected hubs that interact with a large number of partners with low connectivity, and therefore display "small-world" behavior; that is, a relatively short path length between any two nodes and neighborhood clustering (Watts and Strogatz 1998). Although this class of network is tolerant to the removal of random nodes, it is highly sensitive to the failure of the most highly connected hubs (Albert et al. 2000). Many aspects of biological systems exhibit scale-free properties, perhaps as an inescapable consequence of their evolutionary growth by incremental attachment (Luscombe et al. 2002; Wolf et al. 2002). Indeed, networks that tend to build new connections around existing well-connected nodes, such as the Internet, will spontaneously self-organize into scale-free structures (Barabasi and Albert 1999). Not surprisingly, highly connected hubs are statistically enriched for essential proteins (Jeong et al. 2001). Although essential hubs tend not to physically interact (Maslov and Sneppen 2002), they do tend to engage in genetic interactions with other hubs (Ozier et al. 2003). Physically interacting proteins evolve at similar rates, whereas proteins with many interacting partners tend to evolve more slowly because of lower tolerance for mutational perturbation of interaction surfaces (Fraser et al. 2002; Valencia and Pazos 2003). Finally, networks can be partitioned into regulatory modules, qualitatively defined as a discrete unit of function that can be separated from the remainder of the system (Hartwell et al. 1999; Rives and Galitski 2003). Such modular organization is also evident in metabolic networks (Ravasz et al. 2002) and transcriptional regulatory networks (Lee et al. 2002; Milo et al. 2002).

## INTEGRATION OF NETWORK DATA

The combined use of multiple orthogonal data sets increases confidence in the validity of any given interaction and allows useful predictions to be made, an approach now embodied by systems biology (Selinger et al. 2003). For example, a combination of high-throughput two-hybrid interaction analysis, genome-wide expression profiles, and systematic RNAi-based phenotypes has yielded a high confidence functional network specific for the *C. elegans* germ line (Walhout et al. 2002). Predictive methods that assess multiple unlinked attributes may be used to detect false-positive and false-negative interactions in large-scale data sets. For example, a Bayesian network approach that incorporates interaction, expression, localization, and phenotypic data has been successful in discerning bona fide protein interactions embedded in noisy data sets (Jansen et al. 2003). In a less-biased approach, analysis of connectivity across different data sets independent of other parameters has proven a useful criterion to discern true interactions (Bader 2003; Przulj et al. 2004). As data sets grow, reiterative application of these methods should refine network quality.

The yeast genetic and physical interaction networks exhibit a striking degree of non-overlap. Of the 14,114 physical interactions and 4,925 genetic interactions currently in yeast GRID, only 135 are shared (Fig. 5A–C). Because much of the genetic network represents interactions between null alleles of nonessential genes discovered by SGA, in reality there may be more overlap. Indeed, almost all current overlap is due to genetic interactions between conditional alleles of essential components of the same complex (Mewes et al. 2002). Comparison of the 20,509 two-hybrid interactions recently reported for *D. melanogaster* (Giot et al. 2003) to 6,087 genetic interactions stored in FlyBase (FlyBase Consortium 2003) similarly yields a paltry overlap of 23 interactions (Fig. 5D–F). This minimal overlap may derive from the facts that most genes are nonessential and that genetic interactions among nonessential genes typically arise between pathways, whereas physical interactions occur within pathways. Nevertheless, because genes within the same pathway often show similar patterns of genetic interactions, nodes that share many genetic interactions (typically >30 in yeast) often do show a physical interaction (Tong et al. 2004). In summary, genetic interaction networks establish crucial regulatory connections that are simply not evident in physical interaction networks.

Comparisons of physical, genetic, and phenotypic interaction networks between different species is a powerful predictive method. For example, analysis of several thousand genome-wide expression profiles from a variety of model organisms has uncovered a large number of previously unrecognized metagene sets associated with biological processes (Stuart et al. 2003). Sequence conservation strongly implies that physical interaction networks will be

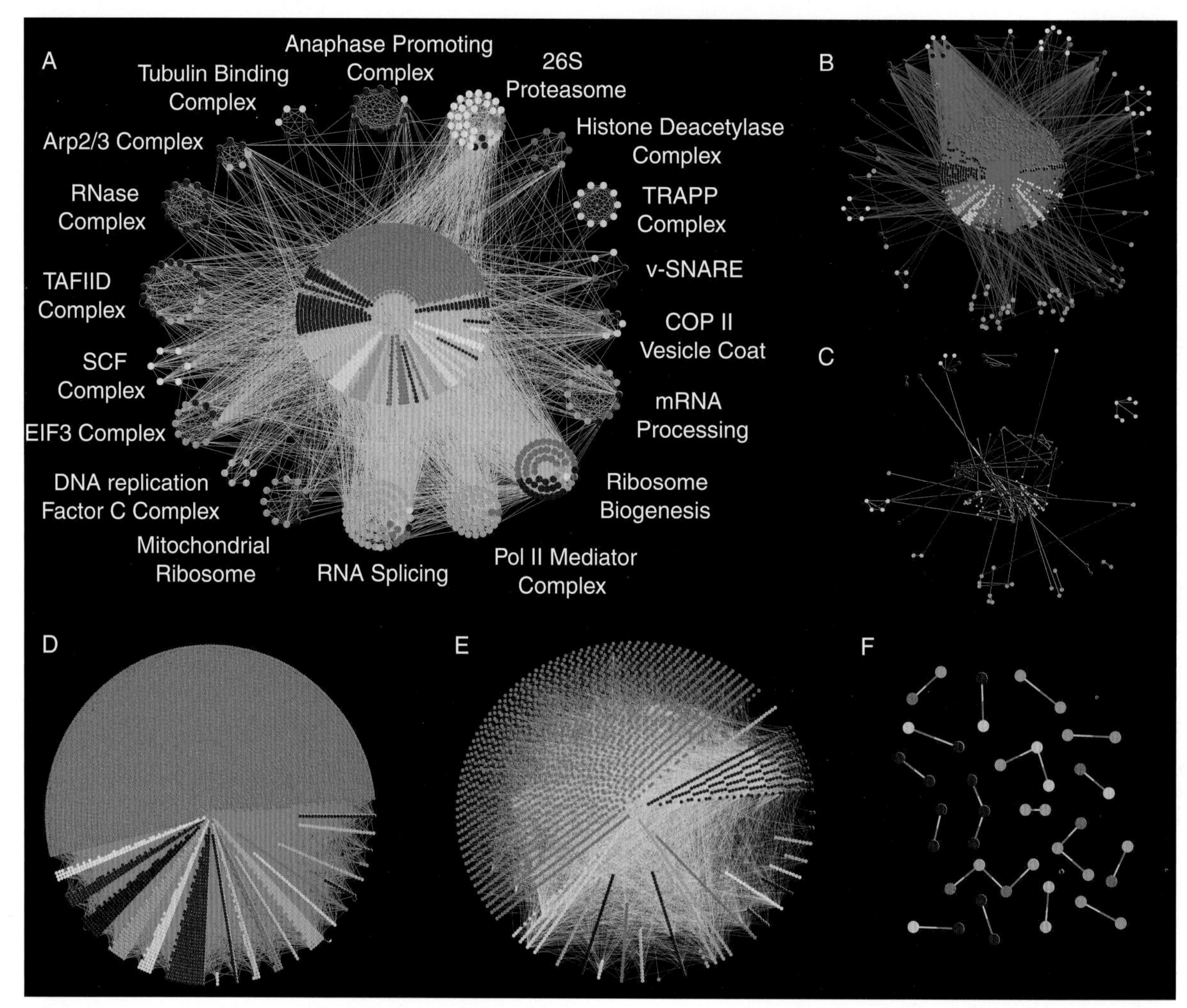


**Figure 5.** Minimal overlap between protein and genetic interaction networks. (*A–C*) Yeast networks. (*A*) Complete protein interaction network (12,379 interactions). (*B*) Genetic interaction network (3,404 interactions). (*C*) Overlap between protein and genetic networks (135 interactions). (*D–F*) Fly networks. (*D*) Protein interaction network derived from two-hybrid interaction data (Giot et al. 2003). (*E*) Genetic interaction network (curated from FlyBase). (*F*) Overlap between protein and genetic networks (23 interactions).

partially superimposable across distantly related species. To illustrate, the *S. cerevisiae* and *D. melanogaster* protein interaction networks exhibit substantial overlap at a high stringency of >50% sequence identity (Fig. 6). Indeed, sequence-based comparison methods across the evolutionary spectrum also have a surprising degree of predictive power based on coevolution of interaction partners (Valencia and Pazos 2003). All told, model organism interaction networks provide an informative scaffold for assembly of the corresponding human networks, which in turn will facilitate the dissection of human polygenic traits and disorders (Hartwell et al. 1997).

## BIOLOGICAL DISCOVERY IN INTERACTION NETWORKS: CRITICAL CELL SIZE

We have used the power of integrated functional genomic approaches to uncover myriad new pathways that influence the timing of cell cycle commitment, an event called Start in yeast and the Restriction Point in mammalian cells (Jorgensen et al. 2002a). Because cell cycle commitment in budding yeast requires that cells achieve a minimum critical cell size in late $G_1$ phase to pass Start, growth and division are coupled at this point (Johnston et al. 1977). The coupling between growth and division is poorly understood in metazoan species, but there is now a general sentiment that oncogenic factors such as cyclin D, c-Myc, and the AKT/TOR signaling network primarily dictate cell growth, which in turn drives proliferation (Saucedo and Edgar 2002; Ruggero and Pandolfi 2003). In yeast, mutations that perturb the timing of Start have little or no growth defect, but instead affect cell size (Rupes 2002). That is, a delay at Start increases cell size, and acceleration of Start reduces cell size. Because there is no easily selectable growth phenotype for cell size mutants, and because such mutants arise spontaneously with very high frequency, we and other investigators have found that this problem is largely refractory to conventional genetics. Thus, only a handful of genes that regulate Start have been identified over the past two decades. Start is currently viewed as a largely transcriptional event, limited by the activity of two key transcription factor com-

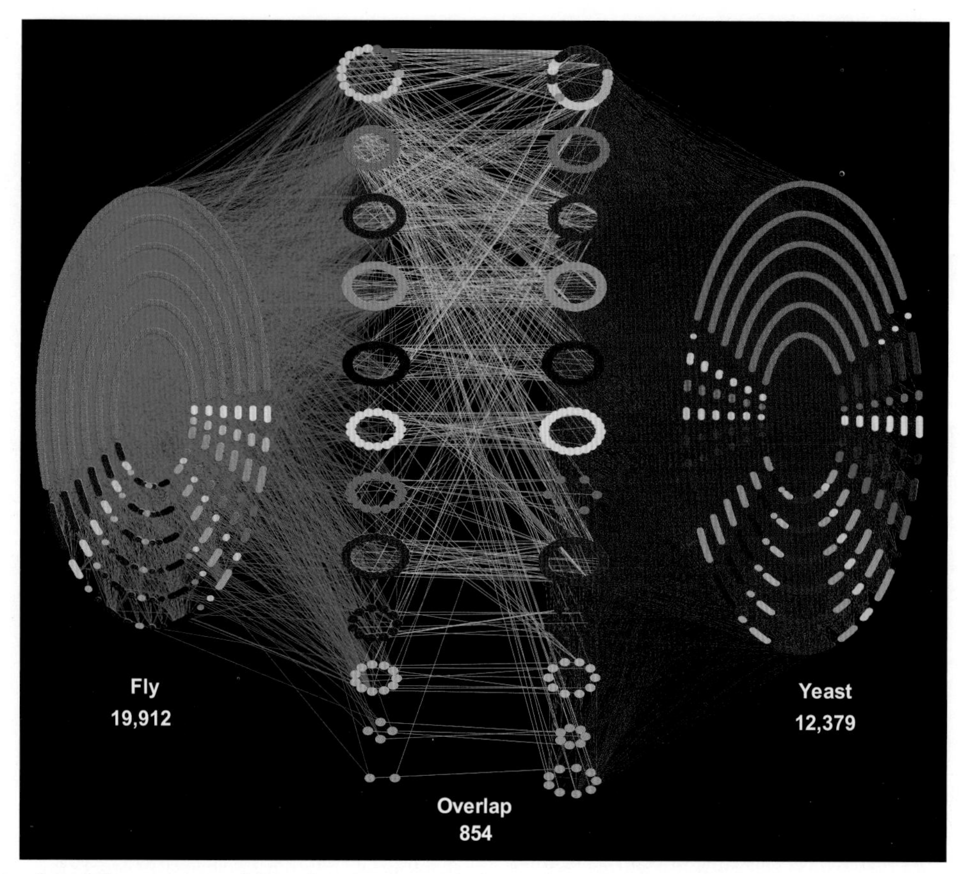


**Figure 6.** Overlap of *S. cerevisiae* and *D. melanogaster* protein interaction networks. 854 proteins of sequence identity >50% are shown between the fly network (19,912 interactions derived from Giot et al. [2003]) and the yeast networks (12,379 interactions in the GRID).

plexes called SBF and MBF, which control the expression of ~120 genes at Start (Fig. 7A). Although it is clear that the cyclin-dependent kinase (CDK) complex, Cln3-Cdc28, and another factor of unknown function, Bck2, somehow activate Start, despite much effort, the mechanism of SBF/MBF activation has remained a mystery.

The advent of the complete set of yeast gene deletion strains afforded a direct means to identify cell size regulators by simply determining the cell size distribution of each individual deletion strain. Through this rather brute-force approach, we and other workers uncovered many hundreds of genes that influence cell size (Jorgensen et al. 2002a; Zhang et al. 2002). Hierarchical clustering of size profiles immediately reveals small (Whi) or large (Lge) mutants (Fig. 7B). We have recently repeated this cell size screen by using barcode technology to compare the strain composition of a full deletion set pool before and after physical separation of small cell-sized mutants by centrifugal elutriation. In contrast to the many months needed to size individual cultures, within a single week the highly parallel barcode approach recapitulated the full set of Whi and Lge mutants (Fig.7B).

To identify the cell size mutants that act directly at Start, we employed pair-wise genetic tests using the SGA method, microarray expression profiles, and protein interaction studies. Two quite remarkable findings came from this analysis. First, a formal genetic equivalent of the retinoblastoma tumor suppressor (*Rb*), called *WHI5*, was genetically placed upstream of SBF (Jorgensen et al. 2002a). We are currently testing the model that Whi5 inhibits SBF, and that Cln3-Cdc28 relieves this inhibition, analogous to the cyclin D-Rb-E2F axis that is a primary target for deregulation in cancer (Sherr and McCormick 2002). If this model holds, it will substantially unify our perception of the yeast and metazoan cell cycle machineries. A second insight is a potential direct connection between the ribosome biogenesis and cell cycle machineries (Jorgensen et al. 2002a). We identified 15 different ribosome biogenesis factors as Whi mutants, which are partially uncoupled for growth and division compared to control strains. Furthermore, the two strongest Whi mutants corresponded to deletions in *SFP1*, which encodes a split zinc finger transcription factor, and *SCH9*, which encodes the closest yeast homolog of the protein kinase AKT/PKB, previously implicated in cell survival and growth control in metazoans. In an unexpected convergence, genome-wide expression profiles revealed that Sfp1 and Sch9 control the expression of many components of the ribosome biogenesis machinery. This suite of over 200 genes, which we termed the Ribi regulon, comprises the largest regulon in yeast and is tightly coregulated over virtually all nutrient and stress conditions tested to date (Fig. 8A). Despite extensive informatics analysis of the Ribi regulon, including identifi-

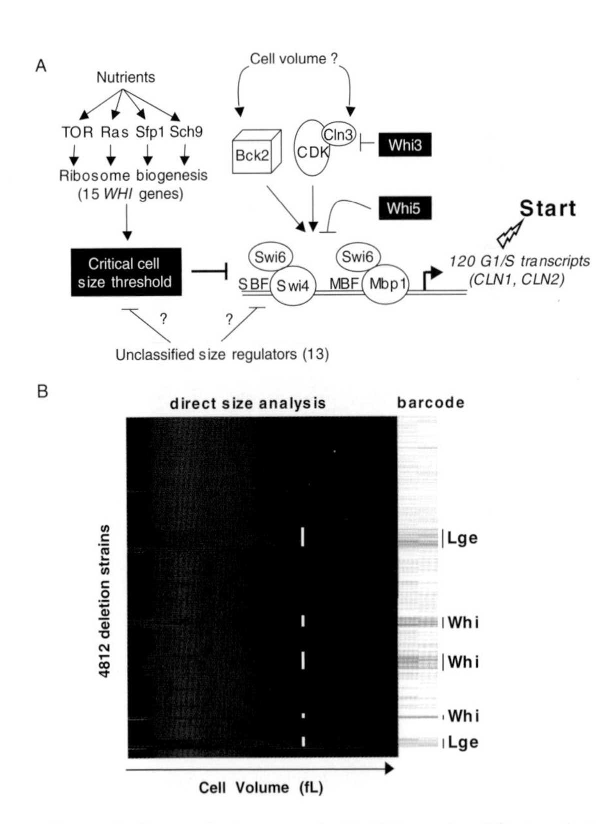


**Figure 7.** Start and size control. (*A*) Schematic of factors that control Start. (*B*) Hierarchical clustering of cell size distributions for the complete set of *S. cerevisiae* deletion strains determined by direct size analysis (*left*) revealed deletion strains with large (Lge) or small (Whi) cell size. Barcode microarray analysis of pooled deletion strain populations before and after small cell size selection by centrifugal elutriation. Red lines indicate >3-fold enrichment in the small size fraction, green lines indicate <3-fold depletion in the small size fraction (*right*).

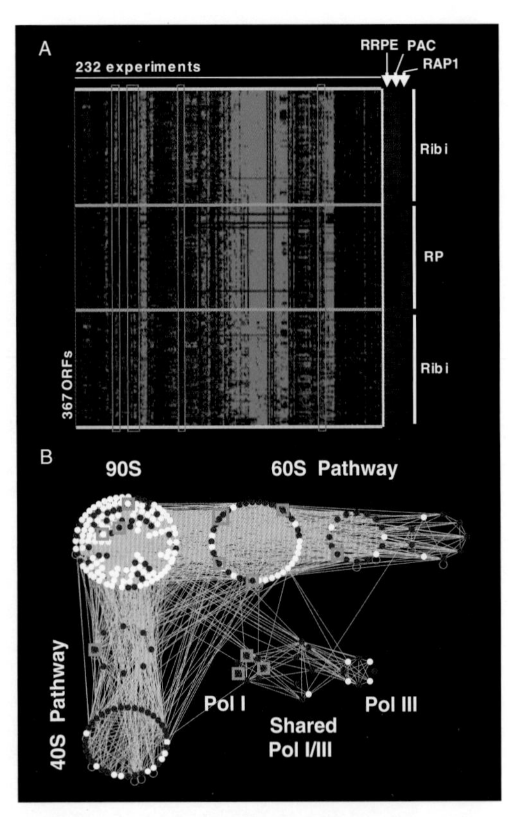


**Figure 8.** The ribosome biogenesis (Ribi) regulon and Start. (*A*) Hierarchical clustering of genome-wide expression data reveals that Ribi and ribosomal protein (RP) genes are coregulated over many experimental conditions. (*B*) Interaction network of ribosome biogensis factors as assembled from large-scale protein interaction data sets. Network connectivity is partially segregated into known 40S and 60S biogenesis pathways, as well as RNA Pol I and RNA Pol III subcomplexes. Sfp1/Sch9 regulated genes are shown in red and genes encoding potential Start regulators are marked with blue squares.

cation of two highly correlated promoter elements, termed PAC and RRPE, prior to the identification of Sfp1 in our phenotypic screen the cognate transcription factors were not known. Because expression of the Ribi regulon is obviously essential for viability, there must be additional transcriptional regulators, which may be revealed in synthetic lethal SGA screens with *sfp1Δ* or *sch9Δ* strains. In yet another unexpected connection, many of the encoded ribosome biogenesis factors in the Ribi regulon turned out to be physically connected within the large-scale protein interaction data sets described above (Fig. 8B). The Ribi network assembled from HTP data has since been discovered to correspond to distinct protein complexes in the nucleolus that catalyze various steps in the 40S and 60S ribosome biogenesis pathways (Fatica and Tollervey 2002). Taken together, these observations suggest an active regulatory connection between ribosome biogenesis and the cell cycle machinery, in contrast to previous models in which the timing of Start was simply viewed as a passive readout of protein synthetic rate. Overall, this systematic study identified over 30 new genes that appear to regulate Start (Fig. 7A), demonstrating the power of the integrated functional genomics approach to dissect complex biological processes.

## CURRENT LIMITATIONS: DYNAMICS AND EMERGENT BEHAVIOR

Despite the impressive quantity of data generated by HTP interaction studies, these data sets provide only an imperfect snapshot of the global protein interaction network. In particular, HTP approaches fail to provide detailed insight into the mechanisms that lie at the heart of cellular regulation, such as the higher-order properties of biochemical modules and circuits (Ferrell 2002). In addition, HTP analysis must be carried out on cell populations, and thus inevitably obscures effects at the single-cell level (McAdams and Arkin 1997; Levsky et al. 2002). As ever more sensitive mass spectrometric methods are developed (Aebersold and Mann 2003), the acquisition of comprehensive data on the temporal and spatial dynamics of protein interactions and posttranslational modifications will certainly lend insight. Even then, however, there is little doubt that the main contribution of such data sets will be to provide a point of departure for biological discovery. As with genome sequencing, the impact of HTP functional genomics approaches will be to empower rather than overwhelm small-scale investigative biology.

## IMPLICATIONS FOR DRUG DISCOVERY: CHEMICAL GENETIC NETWORKS

An understanding of cellular networks provides an essential foundation for rational drug discovery. The field of chemical genetics rests on the premise that inactivation of a protein by a specific small-molecule ligand is equivalent to mutational inactivation of the corresponding gene (Mitchison 1994). The concept of synthetic lethality is readily extended to small molecule–gene interactions, which may be used to selectively target genetic disorders, in particular, cancer (Hartwell et al. 1997). Indeed, proof-of-concept synthetic lethal screens have identified compounds that selectively kill engineered cell lines with defined mutant genotypes (Simons et al. 2001; Torrance et al. 2001). Moreover, most successful therapeutics marketed to date have been discovered by screening for the desired phenotypic effects rather than specific biochemical activities, such that off-target effects may often make critical contributions to drug action through synthetic interactions (Huang 2001). Mass-spectrometry-based chemiproteomics approaches that directly identify protein targets are beginning to reveal the scope of such effects (Graves et al. 2002). With the advent of genome- and proteome-wide data sets, it seems propitious to apply cellular logic to the drug discovery process, using genetic and physical interaction networks as a guide (Sharom et al. 2004).

## CONCLUSION: PROSPECTS FOR HUMAN FUNCTIONAL GENOMICS

Many of the functional genomics first developed in yeast are readily extended to other organisms. The discovery and application of RNA interference (RNAi) will enable synthetic genetic interactions to be systematically investigated in *C. elegans, D. melanogaster*, and mammalian tissue-culture cells (Grishok and Mello 2002; Kamath et al. 2003). Notably, if the density of digenic interactions observed for deletion alleles in yeast holds for natural alleles in metazoan populations, including humans, the sheer number of genetic interactions underlying a specific phenotype may compromise efforts to map complex disease traits by SNP-based haplotype maps (Hartman et al. 2001; Tong et al. 2004). The use of physical and biochemical pathway interaction information to focus on relevant polymorphisms predicted to underlie genetic interactions may become an essential component of efforts to map polygenic traits. A compelling case can be made for systematic elucidation of genetic and protein interaction networks in both model organisms and human cells.

## ACKNOWLEDGMENTS

We thank Don Gilbert and Rachel Drysdale for invaluable assistance in curating all genetic interactions from FlyBase. B.A., C.B., and M.T are supported by grants from the Canadian Institutes of Health Research, the National Cancer Institute of Canada, and Genome Canada.

## REFERENCES

Aebersold R. and Mann M. 2003. Mass spectrometry-based proteomics. *Nature* **422:** 198.

Albert R., Jeong H., and Barabasi A.L. 2000. Error and attack tolerance of complex networks. *Nature* **406:** 378.

Bader G.D. and Hogue C.W. 2002. Analyzing yeast protein-protein interaction data obtained from different sources. *Nat. Biotechnol.* **20:** 991.

———. 2003. An automated method for finding molecular complexes in large protein interaction networks. *BMC Bioinformatics* **4:** 2.

Bader G.D., Betel D., and Hogue C.W. 2003a. BIND: The Biomolecular Interaction Network Database. *Nucleic Acids Res.* **31:** 248.

Bader G.D., Heilbut A., Andrews B., Tyers M., Hughes T., and Boone C. 2003b. Functional genomics and proteomics: Charting a multidimensional map of the yeast cell. *Trends Cell Biol.* **13:** 344.

Bader J.S. 2003. Greedily building protein networks with confidence. *Bioinformatics* **19:** 1869.

Barabasi A.L. and Albert R. 1999. Emergence of scaling in random networks. *Science* **286:** 509.

Bender A. and Pringle J.R. 1991. Use of a screen for synthetic lethal and multicopy suppressee mutants to identify two new genes involved in morphogenesis in *Saccharomyces cerevisiae. Mol. Cell. Biol.* **11:** 1295.

Breitkreutz A., Boucher L., Breitkreutz B.-J., Sultan M., Jurisica I., and Tyers M. 2003. Phenotypic and transcriptional plasticity directed by a yeast mitogen-activated protein kinase network. *Genetics* **165:** 997.

Breitkreutz B.J., Stark C., and Tyers M. 2003a. The GRID: The General Repository for Interaction Datasets. *Genome Biol.* **4:** R23.

———. 2003b. Osprey: A network visualization system. *Genome Biol.* **4:** R22.

Cherry J.M., Adler C., Ball C., Chervitz S.A., Dwight S.S., Hester E.T., Jia Y., Juvik G., Roe T., Schroeder M., Weng S., and Botstein D. 1998. SGD: *Saccharomyces* Genome Database. *Nucleic Acids Res.* **26:** 73.

Cliften P., Sudarsanam P., Desikan A., Fulton L., Fulton B., Majors J., Waterston R., Cohen B.A., and Johnston M. 2003. Finding functional features in *Saccharomyces* genomes by phylogenetic footprinting. *Science* **301:** 71.

DeRisi J.L., Iyer V.R., and Brown P.O. 1997. Exploring the metabolic and genetic control of gene expression on a genomic scale. *Science* **278:** 680.

Fatica A. and Tollervey D. 2002. Making ribosomes. *Curr. Opin. Cell Biol.* **14:** 313.

Ferrell J.E., Jr. 2002. Self-perpetuating states in signal transduction: Positive feedback, double-negative feedback and bistability. *Curr. Opin. Cell Biol.* **14:** 140.

Fields S. and Song O. 1989. A novel genetic system to detect protein-protein interactions. *Nature* **340:** 245.

FlyBase Consortium. 2003. The FlyBase database of the *Drosophila* genome projects and community literature. *Nucleic Acids Res.* **31:** 172.

Fraser H.B., Hirsh A.E., Steinmetz L.M., Scharfe C., and Feldman M.W. 2002. Evolutionary rate in the protein interaction network. *Science* **296:** 750.

Gavin A.C., Bosche M., Krause R., Grandi P., Marzioch M., Bauer A., Schultz J., Rick J.M., Michon A.M., Cruciat C.M., Remor M., Hofert C., Schelder M., Brajenovic M., Ruffner H., Merino A., Klein K., Hudak M., Dickson D., Rudi T., Gnau V., Bauch A., Bastuck S., Huhse B., and Leutwein C., et al. 2002. Functional organization of the yeast proteome by systematic analysis of protein complexes. *Nature* **415:** 141.

Ghaemmaghami S., Huh W.K., Bower K., Howson R.W., Belle A., Dephoure N., O'Shea E.K., and Weissman J.S. 2003. Global analysis of protein expression in yeast. *Nature* **425:** 737.

Giaever G., Shoemaker D.D., Jones T.W., Liang H., Winzeler E.A., Astromoff A., and Davis R.W. 1999. Genomic profiling of drug sensitivities via induced haploinsufficiency. *Nat. Genet.* **21:** 278.

Giaever G., Chu A.M., Ni L., Connelly C., Riles L., Veronneau S., Dow S., Lucau-Danila A., Anderson K., Andre B., Arkin A.P., Astromoff A., El-Bakkoury M., Bangham R., Benito R., Brachat S., Campanaro S., Curtiss M., Davis K., Deutschbauer A., Entian K.D., Flaherty P., Foury F., Garfinkel D.J., and Gerstein M., et al. 2002. Functional profiling of the *Saccharomyces cerevisiae* genome. *Nature* **418:** 387.

Giot L., Bader J.S., Brouwer C., Chaudhuri A., Kuang B., Li Y., Hao Y.L., Ooi C.E., Godwin B., Vitols E., Vijayadamodar G., Pochart P., Machineni H., Welsh M., Kong Y., Zerhusen B., Malcolm R., Varrone Z., Collis A., Minto M., Burgess S., McDaniel L., Stimpson E., Spriggs F., and Williams J., et al. 2003. A protein interaction map of *Drosophila melanogaster. Science* **302:** 1727.

Graves P.R., Kwiek J.J., Fadden P., Ray R., Hardeman K., Coley A.M., Foley M., and Haystead T.A. 2002. Discovery of novel targets of quinoline drugs in the human purine binding proteome. *Mol. Pharmacol.* **62:** 1364.

Grishok A. and Mello C.C. 2002. RNAi (Nematodes: *Caenorhabditis elegans*). *Adv. Genet.* **46:** 339.

Gu Z., Steinmetz L.M., Gu X., Scharfe C., Davis R.W., and Li W.H. 2003. Role of duplicate genes in genetic robustness against null mutations. *Nature* **421:** 63.

Hartman J.L., Garvik B., and Hartwell L. 2001. Principles for the buffering of genetic variation. *Science* **291:** 1001.

Hartwell L.H., Hopfield J.J., Leibler S., and Murray A.W. 1999. From molecular to modular cell biology. *Nature* **402:** C47.

Hartwell L.H., Szankasi P., Roberts C.J., Murray A.W., and Friend S.H. 1997. Integrating genetic approaches into the discovery of anticancer drugs. *Science* **278:** 1064.

Hermjakob H., Montecchi-Palazzi L., Lewington C., Mudali S., Kerrien S., Orchard S., Vingron M., Roechert B., Roepstorff P., Valencia A., Margalit H., Armstrong J., Bairoch A., Cesareni G., Sherman D., and Apweiler R. 2004. IntAct: An open source molecular interaction database. *Nucleic Acids. Res.* **32:** D452.

Ho Y., Gruhler A., Heilbut A., Bader G.D., Moore L., Adams S.L., Millar A., Taylor P., Bennett K., Boutilier K., Yang L., Wolting C., Donaldson I., Schandorff S., Shewnarane J., Vo M., Taggart J., Goudreault M., Muskat B., Alfarano C., Dewar D., Lin Z., Michalickova K., Willems A.R., and Sassi H., et al. 2002. Systematic identification of protein complexes in *Saccharomyces cerevisiae* by mass spectrometry. *Nature* **415:** 180.

Huang S. 2001. Genomics, complexity and drug discovery: Insights from Boolean network models of cellular regulation. *Pharmacogenomics* **2:** 203.

Hucka M., Finney A., Sauro H.M., Bolouri H., Doyle J.C., Kitano H., Arkin A.P., Bornstein B.J., Bray D., Cornish-Bowden A., Cuellar A.A., Dronov S., Gilles E.D., Ginkel M., Gor V., Goryanin I.I., Hedley W.J., Hodgman T.C., Hofmeyr J.H., Hunter P.J., Juty N.S., Kasberger J.L., Kremling A., Kummer U., and Le Novere N., et al. 2003. The systems biology markup language (SBML): A medium for representation and exchange of biochemical network models. *Bioinformatics* **19:** 524.

Hughes T.R., Marton M.J., Jones A.R., Roberts C.J., Stoughton R., Armour C.D., Bennett H.A., Coffey E., Dai H., He Y.D., Kidd M.J., King A.M., Meyer M.R., Slade D., Lum P.Y., Stepaniants S.B., Shoemaker D.D., Gachotte D., Chakraburtty K., Simon J., Bard M., and Friend S.H. 2000. Functional discovery via a compendium of expression profiles. *Cell* **102:** 109.

Huh W.K., Falvo J.V., Gerke L.C., Carroll A.S., Howson R.W., Weissman J.S., and O'Shea E.K. 2003. Global analysis of protein localization in budding yeast. *Nature* **425:** 686.

Ideker T., Ozier O., Schwikowski B., and Siegel A.F. 2002. Discovering regulatory and signalling circuits in molecular interaction networks. *Bioinformatics* (suppl. 1) **18:** S233.

Ito T., Chiba T., Ozawa R., Yoshida M., Hattori M., and Sakaki Y. 2001. A comprehensive two-hybrid analysis to explore the yeast protein interactome. *Proc. Natl. Acad. Sci.* **98:** 4569.

Iyer V.R., Horak C.E., Scafe C.S., Botstein D., Snyder M., and Brown P.O. 2001. Genomic binding sites of the yeast cell-cycle transcription factors SBF and MBF. *Nature* **409:** 533.

Jansen R., Yu H., Greenbaum D., Kluger Y., Krogan N.J., Chung S., Emili A., Snyder M., Greenblatt J.F., and Gerstein M. 2003. A Bayesian networks approach for predicting protein-protein interactions from genomic data. *Science* **302:** 449.

Jeong H., Mason S.P., Barabasi A.L., and Oltvai Z.N. 2001. Lethality and centrality in protein networks. *Nature* **411:** 41.

Johnston G.C., Pringle J.R., and Hartwell L.H. 1977. Coordination of growth with cell division in the yeast *Saccharomyces cerevisiae. Exp. Cell Res.* **105:** 79.

Jorgensen P., Nishikawa J.L., Breitkreutz B.J., and Tyers M. 2002a. Systematic identification of pathways that couple cell growth and division in yeast. *Science* **297:** 395.

Jorgensen P., Nelson B., Robinson M.D., Chen Y., Andrews B., Tyers M., and Boone C. 2002b. High-resolution genetic mapping with ordered arrays of *Saccharomyces cerevisiae* deletion mutants. *Genetics* **162:** 1091.

Kamath R.S., Fraser A.G., Dong Y., Poulin G., Durbin R., Gotta M., Kanapin A., Le Bot N., Moreno S., Sohrmann M., Welchman D.P., Zipperlen P., and Ahringer J. 2003. Systematic functional analysis of the *Caenorhabditis elegans* genome using RNAi. *Nature* **421:** 231.

Kellis M., Patterson N., Endrizzi M., Birren B., and Lander E.S. 2003. Sequencing and comparison of yeast species to identify genes and regulatory elements. *Nature* **423:** 241.

Kumar A., Cheung K.H., Tosches N., Masiar P., Liu Y., Miller P., and Snyder M. 2002. The TRIPLES database: A community resource for yeast molecular biology. *Nucleic Acids Res.* **30:** 73.

Kuruvilla F.G., Shamji A.F., Sternson S.M., Hergenrother P.J., and Schreiber S.L. 2002. Dissecting glucose signalling with diversity-oriented synthesis and small-molecule microarrays. *Nature* **416:** 653.

Lander E.S., Linton L.M., Birren B., Nusbaum C., Zody M.C., Baldwin J., Devon K., Dewar K., Doyle M., FitzHugh W., Funke R., Gage D., Harris K., Heaford A., Howland J., Kann L., Lehoczky J., LeVine R., McEwan P., McKernan K., Meldrim J., Mesirov J.P., Miranda C., Morris W., and Naylor J., et al. (International Human Genome Sequencing Consortium). 2001. Initial sequencing and analysis of the human genome. *Nature* **409:** 860.

Lee T.I., Rinaldi N.J., Robert F., Odom D.T., Bar-Joseph Z., Gerber G.K., Hannett N.M., Harbison C.T., Thompson C.M., Simon I., Zeitlinger J., Jennings E.G., Murray H.L., Gordon D.B., Ren B., Wyrick J.J., Tagne J.B., Volkert T.L., Fraenkel E., Gifford D.K., and Young R.A. 2002. Transcriptional regulatory networks in *Saccharomyces cerevisiae. Science* **298:** 799.

Levsky J.M., Shenoy S.M., Pezo R.C., and Singer R.H. 2002. Single-cell gene expression profiling. *Science* **297:** 836.

Luscombe N.M., Qian J., Zhang Z., Johnson T., and Gerstein M. 2002. The dominance of the population by a selected few: Power-law behaviour applies to a wide variety of genomic properties. *Genome Biol.* **3:** RESEARCH0040.

Martzen M.R., McCraith S.M., Spinelli S.L., Torres F.M., Fields S., Grayhack E.J., and Phizicky E.M. 1999. A biochemical genomics approach for identifying genes by the activity of their products. *Science* **286:** 1153.

Maslov S. and Sneppen K. 2002. Specificity and stability in topology of protein networks. *Science* **296:** 910.

McAdams H.H. and Arkin A. 1997. Stochastic mechanisms in gene expression. *Proc. Natl. Acad. Sci.* **94:** 814.

Mewes H.W., Frishman D., Guldener U., Mannhaupt G., Mayer K., Mokrejs M., Morgenstern B., Munsterkotter M., Rudd S., and Weil B. 2002. MIPS: A database for genomes and protein sequences. *Nucleic Acids Res.* **30:** 31.

Milo R., Shen-Orr S., Itzkovitz S., Kashtan N., Chklovskii D., and Alon U. 2002. Network motifs: Simple building blocks of complex networks. *Science* **298:** 824.

Mitchison T.J. 1994. Towards a pharmacological genetics. *Chem. Biol.* **1:** 3.

Ooi S.L., Shoemaker D.D., and Boeke J.D. 2003. DNA helicase

gene interaction network defined using synthetic lethality analyzed by microarray. *Nat. Genet.* **35:** 277.

Ozier O., Amin N., and Ideker T. 2003. Global architecture of genetic interactions on the protein network. *Nat. Biotechnol.* **21:** 490.

Pawson T. and Nash P. 2003. Assembly of cell regulatory systems through protein interaction domains. *Science* **300:** 445.

Peri S., Navarro J.D., Amanchy R., Kristiansen T.Z., Jonnalagadda C.K., Surendranath V., Niranjan V., Muthusamy B., Gandhi T.K., Gronborg M., Ibarrola N., Deshpande N., Shanker K., Shivashankar H.N., Rashmi B.P., Ramya M.A., Zhao Z., Chandrika K.N., Padma N., Harsha H.C., Yatish A.J., Kavitha M.P., Menezes M., Choudhury D.R., and Suresh S., et al. 2003. Development of human protein reference database as an initial platform for approaching systems biology in humans. *Genome Res.* **13:** 2363.

Przulj N., Wigle D., and Jurisica I. 2004. Functional topology in a network of protein interactions. *Bioinformatics* **20:** 340.

Rain J.C., Selig L., De Reuse H., Battaglia V., Reverdy C., Simon S., Lenzen G., Petel F., Wojcik J., Schachter V., Chemama Y., Labigne A., and Legrain P. 2001. The protein-protein interaction map of *Helicobacter pylori. Nature* **409:** 211.

Ravasz E., Somera A.L., Mongru D.A., Oltvai Z.N., and Barabasi A.L. 2002. Hierarchical organization of modularity in metabolic networks. *Science* **297:** 1551.

Ren B., Robert F., Wyrick J.J., Aparicio O., Jennings E.G., Simon I., Zeitlinger J., Schreiber J., Hannett N., Kanin E., Volkert T.L., Wilson C.J., Bell S.P., and Young R.A. 2000. Genome-wide location and function of DNA binding proteins. *Science* **290:** 2306.

Rigaut G., Shevchenko A., Rutz B., Wilm M., Mann M., and Seraphin B. 1999. A generic protein purification method for protein complex characterization and proteome exploration. *Nat. Biotechnol.* **17:** 1030.

Rives A.W. and Galitski T. 2003. Modular organization of cellular networks. *Proc. Natl. Acad. Sci.* **100:** 1128.

Ruggero D. and Pandolfi P.P. 2003. Does the ribosome translate cancer? *Nat. Rev. Cancer* **3:** 179.

Rupes I. 2002. Checking cell size in yeast. *Trends Genet.* **18:** 479.

Saucedo L.J. and Edgar B.A. 2002. Why size matters: Altering cell size. *Curr. Opin. Genet. Dev.* **12:** 565.

Selinger D.W., Wright M.A., and Church G.M. 2003. On the complete determination of biological systems. *Trends Biotechnol.* **21:** 251.

Sharom J., Bellows D., and Tyers M. 2004. From large networks to small molecules. *Curr. Opin. Chem. Biol.* **8:** 81.

Sherlock G., Hernandez-Boussard T., Kasarskis A., Binkley G., Matese J.C., Dwight S.S., Kaloper M., Weng S., Jin H., Ball C.A., Eisen M.B., Spellman P.T., Brown P.O., Botstein D., and Cherry J.M. 2001. The Stanford Microarray Database. *Nucleic Acids Res.* **29:** 152.

Sherr C.J. and McCormick F. 2002. The RB and p53 pathways in cancer. *Cancer Cell* **2:** 103.

Simons A., Dafni N., Dotan I., Oron Y., and Canaani D. 2001. Establishment of a chemical synthetic lethality screen in cultured human cells. *Genome Res.* **11:** 266.

Stuart J.M., Segal E., Koller D., and Kim S.K. 2003. A gene-coexpression network for global discovery of conserved genetic modules. *Science* **302:** 249.

Tong A.H., Evangelista M., Parsons A.B., Xu H., Bader G.D., Page N., Robinson M., Raghibizadeh S., Hogue C.W., Bussey H., Andrews B., Tyers M., and Boone C. 2001. Systematic genetic analysis with ordered arrays of yeast deletion mutants. *Science* **294:** 2364.

Tong A., Lesage G., Bader G., Ding H., Xu H., Xin X., Young J., Berriz G., Brost R., Chang M., Chen Y., Cheng X., Chua G., Friesen H., Goldberg D., Haynes J., Humphries C., He G., Hussein S., Ke L., Krogan N., Li Z., Levinson J., Lu H., and Manard P., et al. 2004. Global mapping of the Yeast Genetic Interaction Network. *Science* **303:** 808.

Torrance C.J., Agrawal V., Vogelstein B., and Kinzler K.W. 2001. Use of isogenic human cancer cells for high-throughput screening and drug discovery. *Nat. Biotechnol.* **19:** 940.

Uetz P. 2002. Two-hybrid arrays. *Curr. Opin. Chem. Biol.* **6:** 57.

Uetz P., Giot L., Cagney G., Mansfield T.A., Judson R.S., Knight J.R., Lockshon D., Narayan V., Srinivasan M., Pochart P., Qureshi-Emili A., Li Y., Godwin B., Conover D., Kalbfleisch T., Vijayadamodar G., Yang M., Johnston M., Fields S., and Rothberg J.M. 2000. A comprehensive analysis of protein-protein interactions in *Saccharomyces cerevisiae. Nature* **403:** 623.

Valencia A. and Pazos F. 2003. Prediction of protein-protein interactions from evolutionary information. *Methods Biochem. Anal.* **44:** 411.

Venter J.C., Adams M.D., Myers E.W., Li P.W., Mural R.J., Sutton G.G., Smith H.O., Yandell M., Evans C.A., Holt R.A., Gocayne J.D., Amanatides P., Ballew R.M., Huson D.H., Wortman J.R., Zhang Q., Kodira C.D., Zheng X.H., Chen L., Skupski M., Subramanian G., Thomas P.D., Zhang J., Gabor Miklos G.L., and Nelson C., et al. 2001. The sequence of the human genome. *Science* **291:** 1304.

von Mering C., Krause R., Snel B., Cornell M., Oliver S.G., Fields S., and Bork P. 2002. Comparative assessment of large-scale data sets of protein-protein interactions. *Nature* **417:** 399.

Wagner A. 2000. Robustness against mutations in genetic networks of yeast. *Nat. Genet.* **24:** 355.

Walhout A.J., Reboul J., Shtanko O., Bertin N., Vaglio P., Ge H., Lee H., Doucette-Stamm L., Gunsalus K.C., Schetter A.J., Morton D.G., Kemphues K.J., Reinke V., Kim S.K., Piano F., and Vidal M. 2002. Integrating interactome, phenome, and transcriptome mapping data for the *C. elegans* germline. *Curr. Biol.* **12:** 1952.

Watts D.J. and Strogatz S.H. 1998. Collective dynamics of 'small-world' networks. *Nature* **393:** 440.

Winzeler E.A., Shoemaker D.D., Astromoff A., Liang H., Anderson K., Andre B., Bangham R., Benito R., Boeke J.D., Bussey H., Chu A.M., Connelly C., Davis K., Dietrich F., Dow S.W., El Bakkoury M., Foury F., Friend S.H., Gentalen E., Giaever G., Hegemann J.H., Jones T., Laub M., Liao H., and R.W. Davis, et al. 1999. Functional characterization of the *S. cerevisiae* genome by gene deletion and parallel analysis. *Science* **285:** 901.

Wolf Y.I., Karev G., and Koonin E.V. 2002. Scale-free networks in biology: New insights into the fundamentals of evolution? *Bioessays* **24:** 105.

Yeh A.S., Hirschman L., and Morgan A.A. 2003. Evaluation of text data mining for database curation: Lessons learned from the KDD Challenge Cup. *Bioinformatics* (suppl. 1) **19:** I331.

Yvert G., Brem R.B., Whittle J., Akey J.M., Foss E., Smith E.N., Mackelprang R., and Kruglyak L. 2003. Trans-acting regulatory variation in *Saccharomyces cerevisiae* and the role of transcription factors. *Nat. Genet.* **35:** 57.

Zhang J., Schneider C., Ottmers L., Rodriguez R., Day A., Markwardt J., and Schneider B.L. 2002. Genomic scale mutant hunt identifies cell size homeostasis genes in *S. cerevisiae. Curr. Biol.* **12:** 1992.

Zhu H. and Snyder M. 2003. Protein chip technology. *Curr. Opin. Chem. Biol.* **7:** 55.

Zhu H., Klemic J.F., Chang S., Bertone P., Casamayor A., Klemic K.G., Smith D., Gerstein M., Reed M.A., and Snyder M. 2000. Analysis of yeast protein kinases using protein chips. *Nat. Genet.* **26:** 283.

# Genomic Disorders: Genome Architecture Results in Susceptibility to DNA Rearrangements Causing Common Human Traits

P. Stankiewicz,* K. Inoue,* W. Bi,* K. Walz,* S.-S. Park,* N. Kurotaki,* C.J. Shaw,* P. Fonseca,* J. Yan,* J.A. Lee,* M. Khajavi,* and J.R. Lupski*†‡

*Departments of *Molecular and Human Genetics and †Pediatrics, Baylor College of Medicine and ‡Texas Children's Hospital, Houston, Texas 77030*

The beginnings of molecular medicine can perhaps be traced to Pauling's work that recognized sickle cell anemia as a molecular disease, or to Sanger's demonstration of a specific amino acid sequence for insulin. During the four to five decades that followed these discoveries, and that of the chemical basis of the gene, molecular medicine investigations have focused on genes (genocentric)—how mutations specifically alter DNA and how these changes affect the structure, function, and expression of encoded proteins. Recently, however, advances in the Human Genome Project and completion of the genomes for several model organisms have enabled investigators to view genetic information in the context of the entire genome. As a result, we recognize that the mechanisms for some genetic diseases are best understood at the genomic level. Human diseases recognized to result from DNA rearrangements involving unstable genomic regions have been termed genomic disorders (Lupski 1998). These rearrangements are not random events but rather reflect genome architecture consisting of region-specific low-copy repeats (LCRs), which contribute to the susceptibility to DNA rearrangements. LCRs constitute at least 5–10% of the sequenced genome, perhaps >20% of some autosomal regions, and comprise 30–45% of the Y chromosome. These LCRs, also termed segmental duplications or duplicons, usually span ~10–400 kb of genomic DNA and share >95–97% sequence identity (Bailey et al. 2001; Cheung et al. 2001; Eichler 2001; Lander et al. 2001; Hurles and Jobling 2003). By stimulating and mediating nonallelic homologous recombination (NAHR), LCRs are responsible for genome instability and can lead to DNA rearrangements associated with several genomic disorders.

In the last few years, the number of diseases recognized as resulting from constitutional genomic rearrangements due to an LCR/NAHR-mediated (or unequal crossing-over) mechanism has significantly increased (Table 1) (Emanuel and Shaikh 2001; Stankiewicz and Lupski 2002a). This mechanism predicts a reciprocal duplication counterpart for each deletion syndrome described. However, probably because of ascertainment bias and milder phenotypes, reciprocal duplication events anticipated to occur at the same frequency have been identified in only a few cases to date.

The vast majority of rearrangements responsible for genomic disorders are 30 kb–4 Mb in size and can only be analyzed by methods that enable the resolution of changes in the human genome of these magnitudes. Among the most powerful methods capable of resolving genomic changes of such sizes have been pulsed-field gel electrophoresis (PFGE) and fluorescence in situ hybridization (FISH) techniques (Fig. 1). Breakpoint analysis of the rearrangements and the identification of junction fragments that result from the products of recombination in patients with genomic disorders have provided insights into the functional consequences of human genome architecture.

Genomic disorders are frequent diseases (~1 per 1000 births) and often sporadic resulting from de novo rearrangements (Shaffer and Lupski 2000). The clinical phenotype is a consequence of abnormal dosage of a gene(s) located within the rearranged genomic fragment(s). The mutation event occurs via recombination because the architecture of the genome provides ample substrates for NAHR; it does not result from errors in DNA replication/repair that may lead to population-specific alleles. Thus, no frequency differences among world populations were initially anticipated (Shaffer and Lupski 2000). However, recently, there have been reports of significantly different frequencies of a common deletion (flanked by LCRs) in patients with Sotos syndrome among Japanese, British, and French populations (Kurotaki et al. 2002; Douglas et al. 2003; Rio et al. 2003), suggesting potential ongoing evolution of segmental duplications.

## CONSTITUTIONAL RECURRENT GENOMIC DISORDERS IN PROXIMAL 17p

The unstable and LCR- and gene-rich human genomic region 17p11.2-p12 is associated with a wide variety of structural chromosome aberrations. We have documented LCRs and unique genome architecture at the breakpoints for rearrangements in proximal 17p responsible for at least four genomic disorders: Charcot-Marie-Tooth disease type 1A (CMT1A), hereditary neuropathy with liability to pressure palsies (HNPP), Smith-Magenis syndrome (SMS), and dup(17)(p11.2p11.2) syndrome, and have obtained evidence for the involvement of genome

**Table 1.** Selected Common Traits Resulting from Genomic Rearrangements

| Trait | Gene | Chromosome location | Rearrangement type | Rearrangement size (kb) | LCR size (kb) |
|---|---|---|---|---|---|
| Neurobehavioral | | | | | |
| WBS | *ELN, GTF2I, ?* | 7q11.23 | del | 1600 | 320 |
| PWS | ? | 15q12pat | del | 3500 | 500 |
| AS | *UBE3A* | 15q12mat | del | 3500 | 500 |
| dup(15)(q11.2q13) | *GABRB3?* | 15q12mat | dup | 3500 | 500 |
| SMS | *RAI1* | 17p11.2 | del | 4000 | 200 |
| dup(17)(p11.2p11.2) | *RAI1?* | 17p11.2 | dup | 4000 | 200 |
| DGS/VCFS | *TBX1* | 22q11.2 | del | 3000/1500 | 225–400 |
| dup(22)(q11.2q11.2) | *TBX1?* | 22q11.2 | dup | 3000 | 225–400 |
| Peripheral neuropathy | | | | | |
| CMT1A | *PMP22* | 17p12 | dup | 1400 | 24 |
| HNPP | *PMP22* | 17p12 | del | 1400 | 24 |
| Hypertension | *CYP11B1/2* | 8q21 | dup | 45 | 10 |
| Color blindness | *RCP & GCP* | Xq28 | del | 0 | 39 |
| Male infertility | | | | | |
| *AZF*a | *DBY, USP9Y* | Yq11.2 | del | 800 | 10 |
| *AZF*c | *RBMY, DAZ?* | Yq11.2 | del | 3500 | 229 |
| Overgrowth | | | | | |
| Sotos syndrome | *NSD1* | 5q35.3 | del | 1300[a] | 560–600[a] |

(del) Deletion; (dup) duplication; (inv) inversion; (WBS) Williams-Beuren syndrome; (PWS) Prader-Willi syndrome; (AS) Angelman syndrome; (DGS/VCFS) DiGeorge syndrome/Velocardiofacial syndrome.
[a]N. Kurotaki and J.R. Lupski, unpubl.

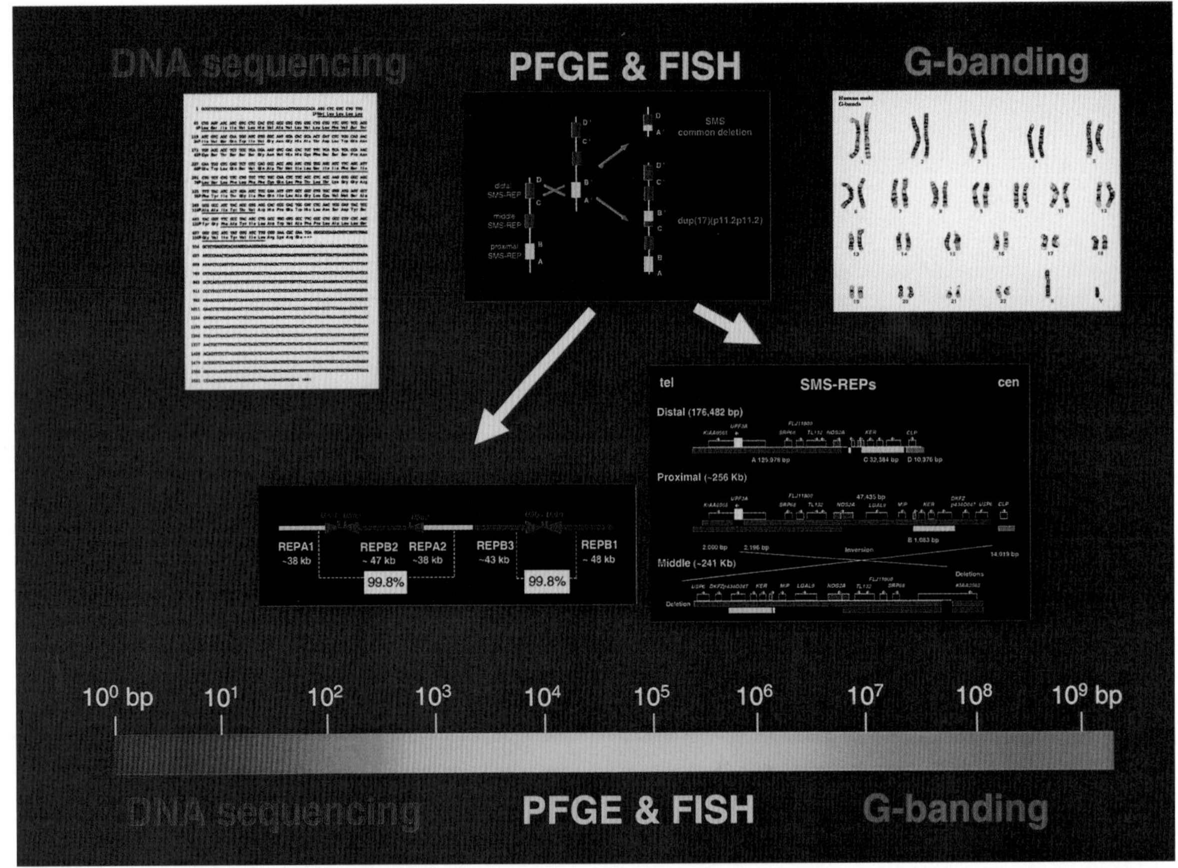


**Figure 1.** Genome architecture and methods to resolve structure of varying DNA sizes. Above are shown three levels of genome architecture, from viewing the entire human genome, resolved by conventional G-banding, PFGE, FISH, and direct DNA sequencing. Below is a scale of the human genome from 1 bp ($10^0$ bp) to ~$3 \times 10^9$ bp and the size ranges (color coded) in which the different methods can physically resolve differences. Note that the genome architecture and rearrangements responsible for genomic disorders occur in the size range of ~30 kb–4 Mb that cannot be resolved either by DNA sequencing and agarose gel electrophoresis or by conventional G-banding, but has been successfully identified by PFGE and/or FISH.

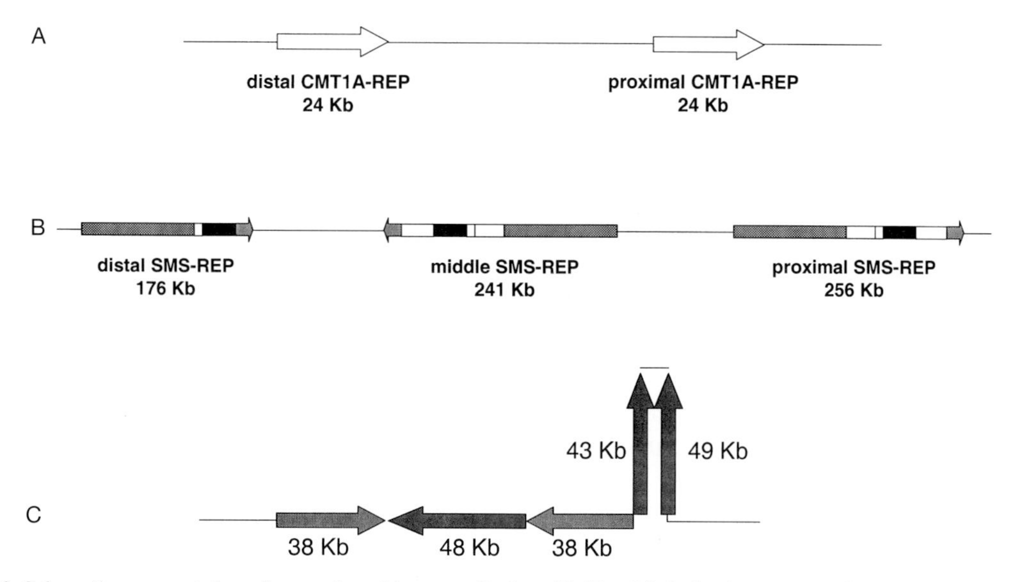


**Figure 2.** Schematic representation of genomic architecture of selected LCRs. (*A*) A simple structure of two directly oriented 24-kb LCRs, the CMT1A-REP copies. Unequal crossing-over between them results in reciprocal duplication (CMT1A) or deletion (HNPP). (*B*) A complex structure of three large modular LCRs—the 176–256-kb SMS-REPs. Note the distal SMS-REP is deleted for two interstitial segments. The middle SMS-REP is inverted with reference to proximal and distal SMS-REPs. The recombination between proximal and distal SMS-REPs leads to reciprocal deletion (SMS) or duplication dup(17)(p11.2p11.2). (*C*) The breakpoints of idic(17)(p11.2) frequently found in a number of tumors (e.g., leukemias, medulloblastoma) cluster within an LCR consisting of five subunits. Two of them form a large (>80 kb) palindromic structure of 99.96% sequence identity, which likely stimulates and mediates the somatic recombination between sister chromatids, resulting in isodicentric chromosomes found in tumors.

architecture in the genomic evolution of this region during primate speciation.

About a decade ago, we recognized that the mechanisms for some genetic diseases are best understood at a genomic level when we identified an ~1.4-Mb duplicated DNA fragment within chromosome 17p12 associated with the common inherited peripheral neuropathy CMT1A (Lupski et al. 1991). The same genomic segment was found to be deleted in patients with HNPP (Chance et al. 1994; Reiter et al. 1996), and is flanked by two ~24-kb, 98.7% identical LCRs, termed proximal and distal CMT1A-REPs (Fig. 2A) (Pentao et al. 1992; Reiter et al. 1997). The peripheral nerve dysfunction in CMT1A and HNPP results from segmental trisomy or monosomy, respectively, of the dosage-sensitive peripheral myelin protein 22, *PMP22* (Lupski and Garcia 2001). Of the 21 genes located within the 1.4-Mb duplicated/deleted segment, *PMP22* appears to be the only dosage-sensitive gene (Inoue et al. 2001).

The same reciprocal deletion/duplication LCR/NAHR mechanism is responsible for two other genomic disorders in proximal chromosome 17p: SMS and dup(17)(p11.2p11.2) syndrome. SMS is characterized by mental retardation, neurobehavioral abnormalities such as aggressive and self-injurious behaviors, sleep disturbances, delayed speech and motor development, multiple congenital anomalies (midface hypoplasia, short stature, and brachydactyly), and is associated with an interstitial deletion of chromosome 17p11.2 (Smith et al. 1986; Stratton et al. 1986; Greenberg et al. 1991, 1996; Chen et al. 1997). The majority (>80–90%) of patients with SMS carry a common ~4-Mb deletion, as defined by a unique de novo junction fragment identified by PFGE (Chen et al. 1997; Bi et al. 2002). The same ~4-Mb genomic interval was found to be duplicated in patients with dup(17)(p11.2p11.2) syndrome, who have a milder, predominantly neurobehavioral phenotype (Potocki et al. 2000). Genetic marker studies and the analysis of the crossovers resulting in recombinant SMS-REPs demonstrate the reciprocal nature of the recombination (Shaw et al. 2002; Bi et al. 2003). This genomic segment is flanked by ~98% homologous LCRs, termed proximal (~256 kb) and distal (~176 kb) SMS-REPs (Chen et al. 1997; Potocki et al. 2000; Park et al. 2002). A third LCR copy, middle SMS-REP (~241 kb) maps between them and is inverted in orientation (Fig. 2B) (Park et al. 2002). Recently, mutations in the retinoic acid-induced-1 gene *RAI1* have been found in SMS-like patients without deletion (Slager et al. 2003), suggesting that *RAI1* may be a dosage-sensitive gene, manifesting with SMS or dup(17)(p11.2p11.2) syndrome when deleted or duplicated, respectively.

The relatively high frequency of these genomic disorders in proximal chromosome 17p is further substantiated by the identification of a patient with two such DNA rearrangements. This patient, with mild delay and a familial history of autosomal dominant carpal tunnel syndrome, had both a de novo dup(17)(p11.2p11.2) and inherited the HNPP deletion on the other homolog (Potocki et al. 1999).

To model SMS and dup(17)(p11.2p11.2) syndromes and to study dosage effects of the genes in the region, we engineered (Ramirez-Solis et al. 1995) chromosomes carrying the deletion/deficiency [*Df(11)17*] or duplication [*Dp(11)17*] of the mouse chromosome 11, syntenic to the ~1-Mb SMS critical region including the *RAI1* gene (Bi et al. 2002; Walz et al. 2003). *Df(11)17/+* mice exhibit craniofacial abnormalities, seizures, marked obesity, male-specific reduced fertility (Walz et al. 2003), and a shorter period length when compared to wild-type littermates (Walz et al. 2004). *Dp(11)17/+* animals are underweight and do not present seizures, craniofacial abnormalities, or reduced fertility (Walz et al. 2003), but do show hyperactivity and learning disabilities (Walz et al. 2004). *Df(11)17/Dp(11)17* animals are normal, indicating that most of the observed phenotypes result from gene dosage effects and not position effects. To refine regions responsible for different SMS phenotypic features, we have generated mice with smaller-sized chromosome deletions. BAC transgenic mice have been constructed also to allow complementation analysis and to study the consequences of gene overexpression. Our murine models represent a powerful tool to analyze the consequences of gene dosage imbalance in this genomic interval and to investigate the molecular genetic bases of both SMS and dup(17)(p11.2p11.2) syndrome.

## EVOLUTION OF PROXIMAL 17p DURING PRIMATE SPECIATION

From the analysis of different genomes, it has been known for over three decades that whole-genome duplications occurred during early evolution, whereas single-gene duplications have enabled the creation of novel protein functions. In contrast to other mammalian species such as mice, rats, and possibly lower nonhuman primates, the human genome appears to be particularly rich in LCRs.

The comparison among karyotypes of humans and great apes reveals a high degree of similarity. Human chromosomes differ from chimpanzee chromosomes by only 1 translocation and 9 pericentric inversions, and from gorilla chromosomes by 2 translocations, 2 paracentric inversions, and 10 pericentric inversions (Yunis and Prakash 1982). At the nucleotide level, human and chimpanzee functionally important nonsynonymous sequence reveals as much as 99.4% identity (Wildman et al. 2003). However, it has been proposed that genomic rearrangements, rather than single-nucleotide mutations/polymorphisms, through creation of novel fusion/fission genes, may have been the driving force for primate chromosome evolution (Fig. 3) (Inoue et al. 2001; Inoue and Lupski 2002; Samonte and Eichler 2002; Stankiewicz and Lupski 2002b). Genome sequencing studies have estimated the divergence between human and chimpanzee may be as high as 5%, with the majority of differences resulting not from single base-pair changes, but from indel events (Britten 2002; Britten et al. 2003), as has been shown for operon structure in bacterial species (Versalovic et al. 1993).

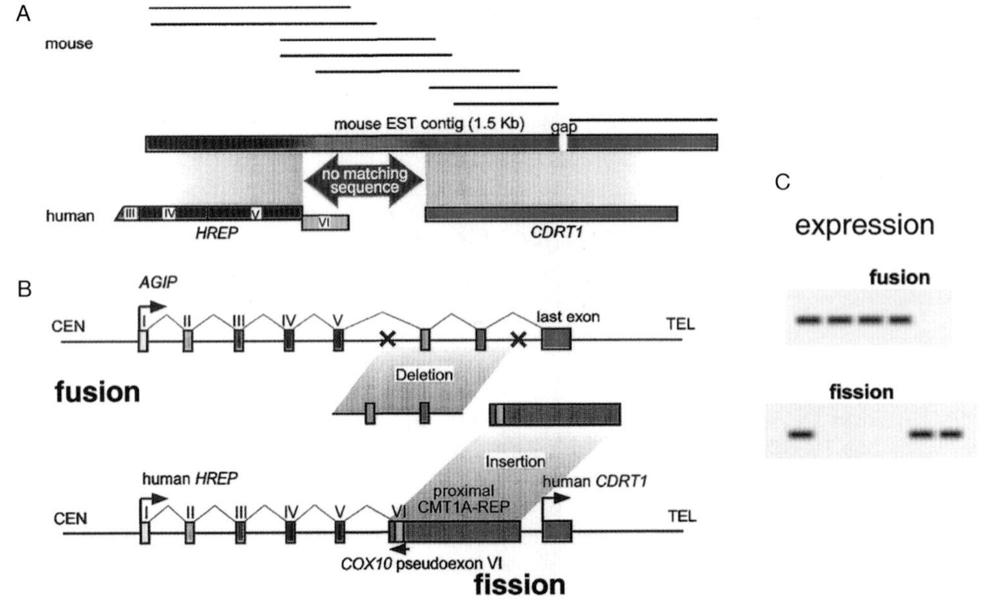


**Figure 3.** Gene evolution surrounding the proximal CMT1A-REP region. (*A*) A comparison of EST contigs between mouse and human. Eight mouse ESTs were constructed into a 1.5-kb contig with a 24-bp gap (*horizontal rectangle with gradient colors*) that aligns with two human genes, *HREP* (the numbers represent each exon of *HREP*; exon VI does not align with the mouse EST contig) and *CDRT1*. Between the alignment of these two genes, there is a 269-bp region in the mouse clone that does not match any human sequence (*purple arrow*). The conceptual translation of this region does not identify a known functional protein motif. (*B*) A model for the evolution of new genes and the genomic structure surrounding the proximal CMT1A-REP region. (*Top*) Figure represents the genomic structure of a hypothetical ancient gene *AGIP* (*A*ncestral *G*ene before the *I*ntegration of *P*roximal CMT1A-REP) modeled in mice. One or more exons originally contained in *AGIP* are predicted to be lost by the integration of the proximal CMT1A-REP. (*Bottom*) Figure shows human genomic structure in which *HREP* and *CDRT1* are separated by the inserted proximal CMT1A-REP (*dark rectangle*). The pseudoexon of *COX10* is utilized as the last exon of *HREP* from the opposite direction (*green box*). A model showing gene fusion and fission by transposition of LCR. A part of the ancestral gene may be fused with the potential exons within the LCR to generate a new gene. Other portions of the ancestral gene may become a new gene by itself or by fusion with potential exons within the LCR. (*C*) Schematic representation of the different expression patterns between the newly created fusion and fission genes, shown as a multi-tissue northern blot. Note different expression pattern (Inoue et al. 2001; Inoue and Lupski 2002).

Evolutionary studies have revealed that LCRs arose recently, apparently during primate speciation (Stankiewicz and Lupski 2002b). Recent data suggest that LCR-associated genome architecture does not represent simple segmental duplications, but rather has evolved through consecutive complex rearrangements, potentially representing multiple events. These serial segmental duplications can result in a complex shuffling of genomic sequences (Lupski 2003).

FISH studies using primate cell lines in conjunction with molecular clock analysis enabled us to construct a working model that most parsimoniously explains how higher-order genomic architecture in proximal 17p has evolved through a series of segmental duplications during primate speciation (Lupski 2003). About 50 million years ago (Mya), LCR17pA was duplicated into LCR17pC and LCR17pD copies that subsequently were split by the insertion of proximal SMS-REP which, in turn, resulted in middle and distal SMS-REPs (Park et al. 2002). After the divergence of orangutan and gorilla 7–12 Mya, the LCR17pB copy arose, and it was inverted together with the middle SMS-REP, followed by the gorilla evolutionary translocation t(4;19) (Stankiewicz et al. 2001a). As the most recent event after the divergence of chimpanzee and gorilla 3–7 Mya, the segmental duplication of distal CMT1A-REP resulted in proximal CMT1A-REP, splitting the LCR17pA copy (Kiyosawa and Chance 1996; Reiter et al. 1997; Boerkoel et al. 1999; Inoue et al. 2001; Lupski 2003).

To confirm the presence of two CMT1A-REPs in chromosome 17p12 among different world populations, we performed Southern analysis with the CEPH DNA panel (48 different world populations) (Cann et al. 2002). Interestingly, in all analyzed populations we identified two CMT1A-REPs; however, in the samples from Central Africa we observed the presence of different-sized bands. In addition, we found similar bands in 6 out of 93 African-Americans analyzed. Using the same approach on DNA samples from 12 chimpanzees (obtained from Dr. Stephen Warren), we found different-sized CMT1A-REP bands (M. Khajavi and J.R. Lupski, unpubl.). These findings suggest ongoing evolution of CMT1A-REP LCRs in human and nonhuman primate species. Moreover, the analysis of chromosome breakpoints involving LCRs in the gorilla evolutionary translocation t(4;19) identified genes that in humans are expressed in brain (P. Stankiewicz and J.R. Lupski, unpubl.). Interestingly, the common features found in patients with genomic disorders are neurobehavioral anomalies, further suggesting that genome architecture involving LCRs is responsible for continuous human genome evolution. In addition, the dosage-sensitive genes mapping within the rearranged genomic interval may be excellent candidates for studying common behavior traits (Inoue and Lupski 2003).

## SOMATIC RECURRENT REARRANGEMENTS

Although a great deal of information has accumulated regarding the mechanisms underlying constitutional DNA rearrangements associated with inherited disorders, very little is known about the molecular processes involved in acquired neoplasia-associated chromosomal rearrangements (Stankiewicz and Lupski 2002b). Recently, Saglio et al. (2002) identified two large duplicons or LCRs in the proximity of the breakpoints in t(9;22)(q34;q11) associated with the Philadelphia chromosome translocation seen in chronic myelogenous leukemia.

Isochromosome 17q, i(17q), is one of the most common neoplasia-associated structural abnormalities and has been described as both a primary and a secondary chromosomal abnormality, indicating that it plays an important pathogenetic role in tumorigenesis as well as in tumor progression. A breakpoint cluster region for i(17q) formation (acute myelogenous leukemia, chronic myelogenous leukemia, and myelodysplastic syndrome) has been previously mapped within the SMS common deletion region in 17p11.2, and it has been hypothesized that genomic architectural features could be responsible for this clustering (Fioretos et al. 1999). Subsequently, using FISH with several BAC and PAC clones, the i(17q) breakpoints in 10 out of 11 hematologic malignancies have been mapped precisely within one BAC clone, RP11-160E2. DNA sequence analysis revealed a >215-kb complex genomic architecture in the i(17q) breakpoint cluster region characterized by five large (~38–49 kb), highly homologous (>99.8%) palindromic LCRs. These findings implicate genomic structure in somatic rearrangements; like constitutional rearrangements, somatic rearrangements are likely not random events (Fig. 2C) (Barbouti et al. 2004).

## NONRECURRENT CHROMOSOME ABERRATIONS

In addition to recurrent constitutional and somatic genomic disorders, we have obtained evidence that genome architecture may also be involved in nonrecurrent genomic rearrangements (Inoue et al. 2002; Stankiewicz et al. 2003b).

### Deletions/Duplications

Pelizaeus-Merzbacher disease (PMD) is an X-linked recessive disorder characterized by arrest of oligodendrocyte differentiation and failure to produce myelin in the CNS (Hudson 2001). In the majority (60–70%) of patients with PMD, duplication of the dosage-sensitive proteolipid protein gene *PLP1* on chromosome Xq22.2 is responsible for the abnormal phenotype (Ellis and Malcolm 1994; Inoue et al. 1996, 1999). In contrast to other genomic disorders, *PLP1* duplications reveal variation in size (Inoue et al. 1999). Recently, rare families with *PLP1* deletions have been described (Inoue et al. 2002). Interestingly, the deletion breakpoint mapping enabled the identification of two novel inverted LCRs, ~45-kb LCR-PMDA and ~32-kb LCR-PMDB, that were associated with the genomic recombination resulting in deletions. However, these LCRs did not serve as substrates for NAHR, as they are not flanking the deleted genomic segments. Rather, they may have been associated with increased susceptibility to initiate DNA rearrangements, perhaps by stimulating double-strand breaks (DSB). In addition, the recombination of nonhomologous sequences from the proximal/distal flanking regions, and the insertion of short sequence stretches at the recombination breakpoints, are both consistent with the mechanism of nonhomologous end joining (NHEJ) (Inoue et. al. 2002). Of interest, preliminary data indicate that at least some of the *PLP1* duplication breakpoints are also located within the intervals containing LCR-PMDA and LCR-PMDB.

Results supporting the involvement of genome architecture in nonrecurrent chromosome aberrations were obtained when analyzing the breakpoints of 18 unusual sized SMS deletions in 17p11.2-p12. We identified seven novel LCRs, which we termed LCR17ps. LCR17pA-G are highly homologous (~98%), and their sizes vary between 23 and 383 kb (Stankiewicz et al. 2003b). Remarkably, LCRs including CMT1A-REPs, SMS-REPs, LCR17ps, and RNU3-REPs (Gao et al. 1997) constitute >23% of the genome sequence in proximal 17p—an experimental observation two- to fourfold higher than predictions based on virtual analysis of the genome. Interestingly, 64% of analyzed breakpoints within 17p11.2 occur in these LCRs (Stankiewicz et al. 2003b).

For chromosome deletions with both breakpoints mapping within nonhomologous copies of LCRs, or those with one breakpoint mapping within an LCR and the other in LCR-free unique DNA sequence, we proposed that the deletion is stimulated, but not mediated, by the LCR(s) and may occur via either NAHR utilizing small repeat segments or by NHEJ. Deletions in which breakpoints did not involve LCRs may occur through NHEJ between repeat-free DNA fragments (Inoue et al. 2002; Stankiewicz et al. 2003b). NAHR between palindromic LCRs has been proposed also as a mechanism responsible for frequent deletions involving subtelomeric regions of chromosome 1p36 (Ballif 2003).

### Translocations

To date, only a few constitutional reciprocal chromosome translocation breakpoints have been shown to be associated with LCRs (Kehrer-Sawatzki et al. 1997; Kurahashi et al. 2000; Edelmann et al. 2001; Giglio et al. 2002). Recently, we demonstrated that 7/8 analyzed breakpoints of chromosome translocation involving chromosome 17p11.2-p12 were associated with higher-order genomic architectural features such as LCRs and/or (peri)centromeric heterochromatin (Stankiewicz et al. 2003b). Similarly, Spiteri et al. (2003) reported that out of 14 different reciprocal translocations involving chromosome 22q11, 5 breakpoints mapped within LCR22-3 and an additional 4 occurred within the vicinity of other LCR22s. Interestingly, 18 partner chromosome breakpoints mapped within the most telomeric bands (Stankiewicz et al. 2003b; Spiteri et al. 2003). These data further indicate that higher-order genomic architecture also plays a significant role in the origin of nonrecurrent chromosome rearrangements.

### Marker Chromosomes and Jumping Translocations

LCR-related genome architecture appears to be involved in the origin of other nonrecurrent chromosome aberrations, such as marker chromosomes (Stankiewicz et al. 2001b) and jumping translocations (Stankiewicz et al. 2003a).

## MOLECULAR MECHANISMS OF GENOMIC DISORDERS

Depending on the location and orientation of nonallelic LCR copies, the stimulated NAHR may involve the same chromatid, different chromatids on the same chromosome, different chromosome homologs, or even different chromosomes. As a result, a wide variety of chromosome aberrations may arise (for review, see Stankiewicz and Lupski 2002a). Most often, unequal exchange between directly (tandemly) oriented LCRs is predicted to result in deletion and duplication, whereas the recombination using inverted copies as recombination substrates would lead to the inversion of the DNA segment flanked by the repeats (Lupski 1998).

### Unequal Crossing-over

NAHR between paralogous LCR copies (unequal crossing-over) within one chromosome (intrachromosomal) or between different chromosomes (interchromosomal) has been proven to be a molecular mechanism resulting in genomic deletions and duplications in several genomic disorders (Pentao et al. 1992; Reiter et al. 1996; Edelmann et al. 1999; Amos-Landgraf et al. 1999; Dorschner et al. 2000; Shaw et al. 2002; Bayés et al. 2003).

### Double-strand DNA Breaks Initiate Recombination Events

DNA sequence analysis and comparison between paralogous LCRs in regions of strand exchange revealed that crossovers occur within selected stretches of perfect sequence identity (Lagerstedt et al. 1997; Reiter et al. 1998; Lopes et al. 1999). The exact molecular mechanism and requirements for strand exchange remain unknown. It is thought that the minimal efficient processing segment (MEPS), or stretch of sequence identity required by the cellular recombination machinery, is probably ~200–300 bp in mitosis (Waldman and Liskay 1988) and potentially ~300–500 bp in meiosis (Reiter et al. 1998). Evidence for gene conversion, consisting of an admixture of sequence derived from two recombined LCRs, has been observed in several genomic disorders. This finding suggests that DSBs initiate NAHR because the DSB-repair process utilizes allelic homologs (or nonallelic paralogs in the case of NAHR) as the template to reconstruct the sequence that was lost in the DSB and subsequent exonuclease excision (Inoue and Lupski 2002). Double exchange within LCRs may also lead to their homogenization, thus in-

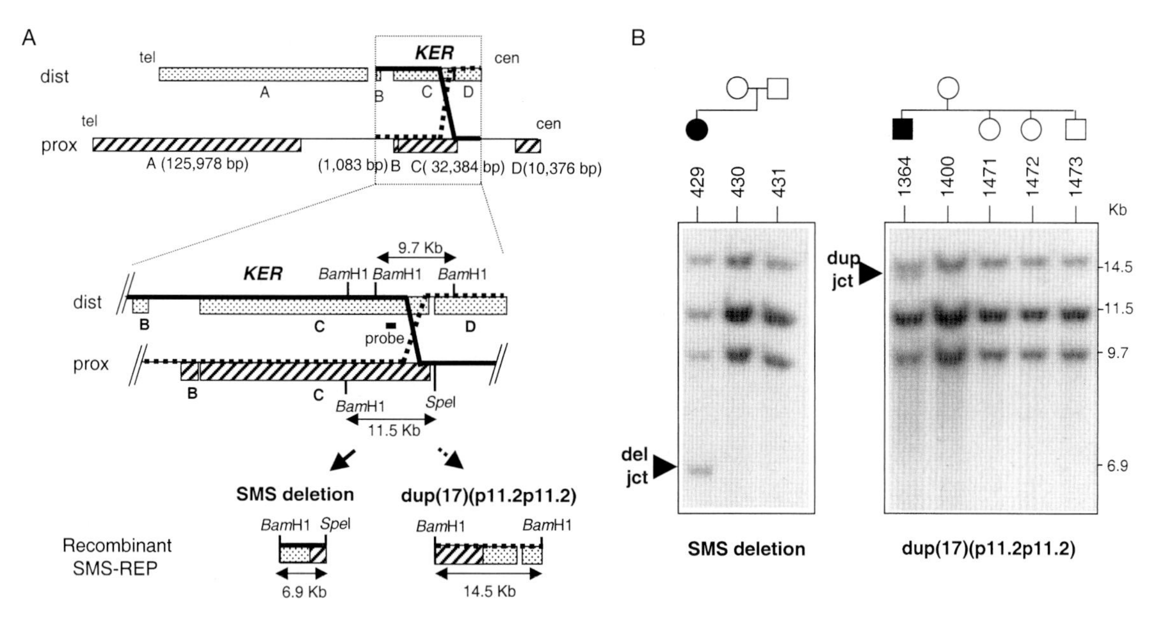


**Figure 4.** Novel junction fragments from reciprocal recombination were identified in SMS and dup(17)(p11.2p11.2) patients. (*A*) Sequence prediction of novel junction fragments from recombinant SMS-REPs. A portion of the *KER* gene cluster of distal (dist) and proximal (prox) SMS-REPs are shown with the horizontal filled rectangles depicting homology segments. A 9.7-kb fragment from the distal SMS-REP and an 11.5-kb fragment from the proximal copy will be detected when genomic DNA is double-digested with *Bam*HI and *Spe*I and hybridized with a 906-bp probe (a short horizontal bar under the distal SMS-REP). In addition, a 6.9-kb junction fragment from patients with SMS deletion and a 14.5-kb junction fragment in its reciprocal duplication from dup(17)(p11.2p11.2) patients are predicted when the strand exchange events occur in block C and centromeric to the distal specific *Bam*HI site. (*B*) The predicted novel junction fragments were detected in SMS patients and dup(17)(p11.2p11.2) patients. A 6.9-kb junction fragment was observed in a patient (429) with the SMS common deletion, but not in her unaffected parents. A 14.5-kb junction fragment was observed in a dup(17)(p11.2p11.2) patient (1364) but not in his unaffected mother and siblings. The novel junction fragments are indicated by the arrowheads. (Modified from Bi et al. 2003.)

creasing their similarity and the potential rate of subsequent NAHR events between them (Blanco et al. 2000; Saunier et al. 2000; Hurles 2001).

### Positional Preference for Strand Exchange

Clustering of breakpoints has been observed for the few LCR/NAHR-mediated events studied at the nucleotide sequence level. HNPP and CMT1A recombination product characterization has revealed, respectively, 557-bp and 2-kb regions of strand exchange at recombination hot spots (Reiter et al. 1998; Lopes et al. 1999). Similar positional recombination hot spots have been identified also within LCRs associated with *AZF*a (azoospermia factor a) microdeletion (Kamp et al. 2000), neurofibromatosis type 1 (Jenne et al. 2001; López-Correa et al. 2001), and Williams-Beuren syndrome (Bayés et al. 2003). Recently, we have identified similar strand exchange clustering in SMS-REPs, in which ~50% of breakpoints clustered in a 12-kb region of the ~170-kb homology intervals within SMS-REPs. As anticipated, the same 12-kb interval is the preferred region for strand exchange in the reciprocal crossover resulting in dup(17)(p11.2p11.2) (Fig. 4). Interestingly, we identified inverted repeats of ~2 kb in size directly flanking this hot spot. We hypothesize that these repeats, or the large loop secondary DNA structures missing in the distal SMS-REP, may make the hot spot region more sensitive to DSBs, and therefore stimulate homologous recombination (Bi et al. 2003).

## CONCLUSIONS

Genomic disorder-associated DNA rearrangements in proximal chromosome 17p are an excellent model to investigate the role of genome architecture in constitutional, evolutionary, and somatic rearrangements (Fig. 5). Molecular studies of the CMT1A duplication, the SMS deletion, their reciprocal recombination products, and other disease-causing chromosomal rearrangements have revealed common mechanisms for genomic disorders. The rearrangements are not random events, but rather reflect genome architecture. This genome architecture consists of region-specific LCRs that act as substrates for NAHR (unequal crossing-over) and thus contribute to the susceptibility to DNA rearrangements and genomic instability. The LCRs appear to have arisen recently during primate speciation through serial segmental duplications and likely play a role in genome evolution. The human genome has evolved an architecture that may make us as a species more susceptible to rearrangements causing genomic disorders. Because LCRs comprise a significant portion of the human genome, these recombination-based disorders contribute substantially to genetic disease burden, and the number of conditions that are recognized as genomic disorders continues to grow.

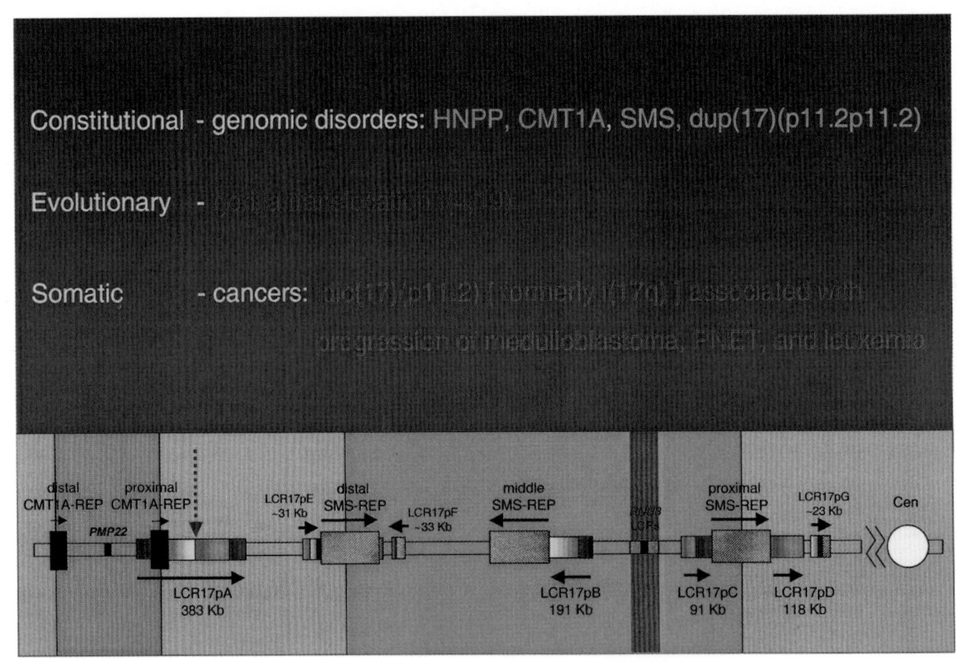


**Figure 5.** Schematic representation of summarized genomic architecture in proximal 17p. The LCR-rich genomic architecture in chromosome 17p11.2p12 results in susceptibility to recurrent and nonrecurrent constitutional [CMT1A, HNPP, SMS, dup(17)(p11.2p11.2)], evolutionary (gorilla translocation, LCR17ps), and somatic [idic(17)(p11.2)] genomic rearrangements.


## ACKNOWLEDGMENTS

This work has been generously supported by The National Institute of Child Health and Development (PO1 HD-39420), the Baylor College of Medicine Mental Retardation Research Center (HD-2406407), the Texas Children's Hospital General Clinical Research Center (MO1RR-00188), the National Institute for Neurological Disorders and Strokes (RO1 NS-27042), and the Muscular Dystrophy Association.


## REFERENCES


Amos-Landgraf J.M., Ji Y., Gottlieb W., Depinet T., Wandstrat A.E., Cassidy S.B., Driscoll D.J., Rogan P.K., Schwartz S., and Nicholls R.D. 1999. Chromosome breakage in the Prader-Willi and Angelman syndromes involves recombination between large, transcribed repeats at proximal and distal breakpoints. *Am. J. Hum. Genet.* **65:** 370.

Bailey J.A., Yavor A.M., Massa H.F., Trask B.J., and Eichler E.E. 2001. Segmental duplications: Organization and impact within the current human genome project assembly. *Genome Res.* **11:** 1005.

Ballif B.C. 2003. "Monosomy 1p36 as a model for the molecular basis of terminal deletions." Ph.D. thesis, Baylor College of Medicine, Houston, Texas.

Barbouti A., Stankiewicz P., Nusbaum C., Cuomo C., Cook A., Hoglund M., Johansson B., Hagemeijer A., Park S.-S., Mitelman F., Lupski J.R., and Fioretos T. 2004. The breakpoint region of the most common isochromosome, i(17q), in human neoplasia is characterized by a complex genomic architecture with large, palindromic, low-copy repeats. *Am. J. Hum. Genet.* **74:** 1.

Bayés M., Magano L.F., Rivera N., Flores R., and Pérez Jurado L.A. 2003. Mutational mechanisms of Williams-Beuren syndrome deletions. *Am. J. Hum. Genet.* **73:** 131.

Bi W., Park S.-S., Shaw C.J., Withers M.A., Patel P.I., and Lupski J.R. 2003. Reciprocal crossovers and a positional preference for strand exchange in recombination events resulting in deletion or duplication of chromosome 17p11.2. *Am. J. Hum. Genet.* **73:** 1302.

Bi W., Yan J., Stankiewicz P., Park S.-S., Walz K., Boerkoel C.F., Potocki L., Shaffer L.G., Devriendt K., Nowaczyk M.J.M., Inoue K., and Lupski J.R. 2002. Genes in a refined Smith-Magenis syndrome critical deletion interval on chromosome 17p11.2 and the syntenic region of mouse. *Genome Res.* **12:** 713.

Blanco P., Shlumukova M., Sargent C.A., Jobling M.A., Affara N., and Hurles M.E. 2000. Divergent outcomes of intrachromosomal recombination on the human Y chromosome: male infertility and recurrent polymorphism. *J. Med. Genet.* **37:** 752.

Boerkoel C.F., Inoue K., Reiter L.T., Warner L.E., and Lupski J.R. 1999. Molecular mechanisms for CMT1A duplication and HNPP deletion. *Ann. N.Y. Acad. Sci.* **883:** 22.

Britten R.J. 2002. Divergence between samples of chimpanzee and human DNA sequences is 5%, counting indels. *Proc. Natl. Acad. Sci.* **99:** 13633.

Britten R.J., Rowen L., Williams J., and Cameron R.A. 2003. Majority of divergence between closely related DNA samples is due to indels. *Proc. Natl. Acad. Sci.* **100:** 4661.

Cann H.M., de Toma C., Cazes L., Legrand M.-F., Morel V., Piouffre L., Bodmer J., Bodmer W.F., Bonne-Tamir B., Cambon-Thomsen A., Chen Z., Chu J., Carcassi C., Contu L., Du R., Excoffier L., Ferrara G.B., Friedlaender J.S., Groot H., Gurwitz D., Jenkins T., Herrera R.J., Huang X., Kidd J., and Kidd K.K., et al. 2002. A human genome diversity cell line panel. *Science* **296:** 261.

Chance P.F., Abbas N., Lensch M.W., Pentao L., Roa B.B., Patel P.I., and Lupski J.R. 1994. Two autosomal dominant neuropathies result from reciprocal DNA duplication/deletion of a region on chromosome 17. *Hum. Mol. Genet.* **3:** 223.

Chen K.-S., Manian P., Koeuth T., Potocki L., Zhao Q., Chinault A.C., Lee C.C., and Lupski J.R. 1997. Homologous recombination of a flanking repeat gene cluster is a mechanism for a common contiguous gene deletion syndrome. *Nat. Genet.* **17:** 154.

Cheung V.G., Nowak N., Jang W., Kirsch I.R., Zhao S., Chen X.-N., Furey T.S., Kim U.-J., Kuo W.-L., Olivier M., Conroy J., Kasprzyk A., Massa H., Yonescu R., Sait S., Thoreen C., Snijders A., Lemyre E., Bailey J.A., Bruzel A., Burrill W.D., Clegg S.M., Collins S., Dhami P., and Friedman C., et al.

2001. Integration of cytogenetic landmarks into the draft sequence of the human genome. *Nature* **409:** 953.

Dorschner M.O., Sybert V.P., Weaver M., Pletcher B.A., and Stephens K. 2000. *NF1* microdeletion breakpoints are clustered at flanking repetitive sequences. *Hum. Mol. Genet.* **9:** 35.

Douglas J., Hanks S., Temple I.K., Davies S., Murray A., Upadhyaya M., Tomkins S., Hughes H.E., Cole T.R.P., and Rahman N. 2003. *NSD1* mutations are the major cause of Sotos syndrome and occur in some cases of Weaver syndrome but are rare in other overgrowth phenotypes. *Am. J. Hum. Genet.* **72:** 132.

Edelmann L., Spiteri E., Koren K., Pulijaal V., Bialer M.G., Shanske A., Goldberg R., and Morrow B.E. 2001. AT-rich palindromes mediate the constitutional t(11;22) translocation. *Am. J. Hum. Genet.* **68:** 1.

Edelmann L., Pandita R.K., Spiteri E., Funke B., Goldberg R., Palanisamy N., Chaganti R.S.K., Magenis E., Shprintzen R.J., and Morrow B.E. 1999. A common molecular basis for rearrangement disorders on chromosome 22q11. *Hum. Mol. Genet.* **8:** 1157.

Eichler E.E. 2001. Recent duplication, domain accretion and the dynamic mutation of the human genome. *Trends Genet.* **17:** 661.

Ellis D. and Malcolm S. 1994. Proteolipid protein gene dosage effect in Pelizaeus-Merzbacher disease. *Nat. Genet.* **6:** 333.

Emanuel B.S. and Shaikh T.H. 2001. Segmental duplications: An 'expanding' role in genomic rearrangements. *Nat. Rev. Genet.* **2:** 791.

Fioretos T., Strömbeck B., Sandberg T., Johansson B., Billström R., Borg Å., Nilsson P.-G., Van Den Berghe H., Hagemeijer A., Mitelman F., and Höglund M. 1999. Isochromosome 17q in blast crisis of chronic myeloid leukemia and in other hematologic malignancies is the result of clustered breakpoints in 17p11 and is not associated with coding *TP53* mutations. *Blood* **4:** 225.

Gao L., Frey M.R., and Matera A.G. 1997. Human genes encoding U3 snRNA associate with coiled bodies in interphase cells and are clustered on chromosome 17p11.2 in a complex inverted repeat structure. *Nucleic Acids Res.* **25:** 4740.

Giglio S., Calvari V., Gregato G., Gimelli G., Camanini S., Giorda R., Ragusa A., Guerneri S., Selicorni A., Stumm M., Tonnies H., Ventura M., Zollino M., Neri G., Barber J., Wieczorek D., Rocchi M., and Zuffardi O. 2002. Heterozygous submicroscopic inversions involving olfactory receptor-gene clusters mediate the recurrent t(4;8)(p16;p23) translocation. *Am. J. Hum. Genet.* **71:** 276.

Greenberg F., Guzzetta V., Montes de Oca-Luna R., Magenis R.E., Smith A.C.M., Richter S.F., Kondo I., Dobyns W.B., Patel P.I., and Lupski J.R. 1991. Molecular analysis of the Smith-Magenis syndrome: A possible contiguous-gene syndrome associated with del(17)(p11.2). *Am. J. Hum. Genet.* **49:** 1207.

Greenberg F., Lewis R.A., Potocki L., Glaze D., Parke J., Killian J., Murphy M.A., Williamson D., Brown F., and Dutton R. 1996. Multi-disciplinary clinical study of Smith-Magenis syndrome (deletion 17p11.2). *Am. J. Med. Genet.* **62:** 247.

Hudson L.D. 2001. Pelizaeus-Merzbacher disease and the allelic disorder X-linked spastic paraplegia type 2. In *The metabolic and molecular basis of inherited diseases* (ed. C.R. Scriver et al.), p. 5789. McGraw-Hill, New York.

Hurles M.E. 2001. Gene conversion homogenizes the CMT1A paralogous repeats. *BMC Genomics* **2:** 11.

Hurles M.E. and Jobling M.A. 2003. A singular chromosome. *Nat. Genet.* **34:** 246.

Inoue K. and Lupski J.R. 2002. Molecular mechanisms for genomic disorders. *Annu. Rev. Genomics Hum. Genet.* **3:** 199.

———. 2003. Genetics and genomics of behavioral and psychiatric disorders. *Curr. Opin. Genet. Dev.* **13:** 303.

Inoue K., Dewar K., Katsanis N., Reiter L.T., Lander E.S., Devon K.L., Wyman D.W., Lupski J.R., and Birren B. 2001. The 1.4-Mb CMT1A duplication/HNPP deletion genomic region reveals unique genome architectural features and provides insights into the recent evolution of new genes. *Genome Res.* **11:** 1018.

Inoue K., Osaka H., Sugiyama N., Kawanishi C., Onishi H., Nezu A., Kimura K., Kimura S., Yamada Y., and Kosaka K. 1996. A duplicated *PLP* gene causing Pelizaeus-Merzbacher disease detected by comparative multiplex PCR. *Am. J. Hum. Genet.* **59:** 32.

Inoue K., Osaka H., Thurston V.C., Clarke J.T.R., Yoneyama A., Rosenbarker L., Bird T.D., Hodes M.E., Shaffer L.G., and Lupski J.R. 2002. Genomic rearrangements resulting in *PLP1* deletion occur by nonhomologous end joining and cause different dysmyelinating phenotypes in males and females. *Am. J. Hum. Genet.* **71:** 838.

Inoue K., Osaka H., Imaizumi K., Nezu A., Takanashi J., Arii J., Murayama K., Ono J., Kikawa Y., Mito T., Shaffer L.G., and Lupski J.R. 1999. Proteolipid protein gene duplications causing Pelizaeus-Merzbacher disease: Molecular mechanism and phenotypic manifestations. *Ann. Neurol.* **45:** 624.

Jenne D.E., Tinschert S., Reimann H., Lasinger W., Thiel G., Hameister H., and Kehrer-Sawatzki H. 2001. Molecular characterization and gene content of breakpoint boundaries in patients with neurofibromatosis type 1 with 17q11.2 microdeletions. *Am. J. Hum. Genet.* **69:** 516.

Kamp C., Hirschmann P., Voss H., Huellen K., and Vogt P.H. 2000. Two long homologous retroviral sequence blocks in proximal Yq11 cause AZFa microdeletions as a result of intrachromosomal recombination events. *Hum. Mol. Genet.* **9:** 2563.

Kehrer-Sawatzki H., Häussler J., Krone W., Bode H., Jenne D.E., Mehnert K.U., Tümmers U., and Assum G. 1997. The second case of a t(17;22) in a family with neurofibromatosis type 1: Sequence analysis of the breakpoint regions. *Hum. Genet.* **99:** 237.

Kiyosawa H. and Chance P.F. 1996. Primate origin of the CMT1A-REP repeat and analysis of a putative transposon-associated recombinational hotspot. *Hum. Mol. Genet.* **5:** 745.

Kurahashi H., Shaikh T.H., Hu P., Roe B.A., Emanuel B.S., and Budarf M.L. 2000. Regions of genomic instability on 22q11 and 11q23 as the etiology for the recurrent constitutional t(11;22). *Hum. Mol. Genet.* **9:** 1665.

Kurotaki N., Imaizumi K., Harada N., Masuno M., Kondoh T., Nagai T., Ohashi H., Naritomi K., Tsukahara M., Makita Y., Sugimoto T., Sonoda T., Hasegawa T., Chinen Y., Tomita Ha H.A., Kinoshita A., Mizuguchi T., Yoshiura K., Ohta T., Kishino T., Fukushima Y., Niikawa N., and Matsumoto N. 2002. Haploinsufficiency of *NSD1* causes Sotos syndrome. *Nat. Genet.* **30:** 365.

Kurotaki N., Harada N., Shimokawa O., Miyake N., Kawame H., Uetake K., Makita Y., Kondoh T., Ogata T., Hasegawa T., Nagai T., Ozaki T., Touyama M., Shenhav R., Ohashi H., Medne L., Shiihara T., Ohtsu S., Kato Z., Okamoto N., Nishimoto J., Lev D., Miyoshi Y., Ishikiriyama S., and Sonoda T., et al. 2003. Fifty microdeletions among 112 cases of Sotos syndrome: A new genomic disorder mediated by low copy repeats? *Hum. Mutat.* **22:** 378.

Lagerstedt K., Karsten S.L., Carlberg B.-M., Kleijer W.J., Tönnesen T., Pettersson U., and Bondeson M.-L. 1997. Double-strand breaks may initiate the inversion mutation causing the Hunter syndrome. *Hum. Mol. Genet.* **6:** 627.

Lander E.S., Linton L.M., Birren B., Nusbaum C., Zody M.C., Baldwin J., Devon K., Dewar K., Doyle M., FitzHugh W., Funke R., Gage D., Harris K., Heaford A., Howland J., Kann L., Lehoczky J., LeVine R., McEawan P., McKernan K., Meldrim J., Mesirov J.P., Miranda C., Morris W., and Naylor J., et al. (International Human Genome Sequencing Consortium). 2001. Initial sequencing and analysis of the human genome. *Nature* **409:** 860.

Lopes J., Tardieu S., Silander K., Blair I., Vandenberghe A., Palau F., Ruberg M., Brice A., and LeGuern E. 1999. Homologous DNA exchanges in humans can be explained by the yeast double-strand break repair model: A study of 17p11.2 rearrangements associated with CMT1A and HNPP. *Hum. Mol. Genet.* **8:** 2285.

López-Correa C., Dorschner M., Brems H., Lázaro C., Clementi M., Upadhyaya M., Dooijes D., Moog U., Kehrer-Sawatzki H., Rutkowski J.L., Fryns J.-P., Marynen P., Stephens K., and

Legius E. 2001. Recombination hotspot in *NF1* microdeletion patients. *Hum. Mol. Genet.* **10:** 1387.

Lupski J.R. 1998. Genomic disorders: Structural features of the genome can lead to DNA rearrangements and human disease traits. *Trends Genet.* **14:** 417.

———. 2003. Genomic disorders: Recombination-based disease resulting from genome architecture. *Am. J. Hum. Genet.* **72:** 246.

Lupski J.R. and Garcia C.A. 2001. Charcot-Marie-Tooth peripheral neuropathies and related disorders. In *The metabolic and molecular bases of inherited diseases* (ed. C.R. Scriver et al.), p. 5759. McGraw-Hill, New York.

Lupski J.R., de Oca-Luna R.M., Slaugenhaupt S., Pentao L., Guzzetta V., Trask B.J., Saucedo-Cardenas O., Barker D.F., Killian J.M., Garcia C.A., Chakravarti A., and Patel P.I. 1991. DNA duplication associated with Charcot-Marie-Tooth disease type 1A. *Cell* **66:** 219.

Park S.-S., Stankiewicz P., Bi W., Shaw C., Lehoczky J., Dewar K., Birren B., and Lupski J.R. 2002. Structure and evolution of the Smith-Magenis syndrome repeat gene clusters, SMS-REPs. *Genome Res.* **12:** 729.

Pentao L., Wise C.A., Chinault A.C., Patel P.I., and Lupski J.R. 1992. Charcot-Marie-Tooth type 1A duplication appears to arise from recombination at repeat sequences flanking the 1.5 Mb monomer unit. *Nat. Genet.* **2:** 292.

Potocki L., Chen K.-S., Koeuth T., Killian J., Iannaccone S.T., Shapira S.K., Kashork C.D., Spikes A.S., Shaffer L.G., and Lupski J.R. 1999. DNA rearrangements on both homologues of chromosome 17 in a mildly delayed individual with a family history of autosomal dominant carpal tunnel syndrome. *Am. J. Hum. Genet.* **64:** 471.

Potocki L., Chen K.-S., Park S.-S., Osterholm D.E., Withers M.A., Kimonis, V., Summers, A.M., Meschino W.S., Anyane-Yeboa K., Kashork C.D., Shaffer L.G., and Lupski J.R. 2000. Molecular mechanism for duplication 17p11.2—The homologous recombination reciprocal of the Smith-Magenis microdeletion. *Nat. Genet.* **24:** 84.

Ramirez-Solis R., Liu P., and Bradley A. 1995. Chromosome engineering in mice. *Nature* **378:** 720.

Reiter L.T., Murakami T., Koeuth T., Gibbs R.A., and Lupski, J.R. 1997. The human *COX10* gene is disrupted during homologous recombination between the 24 kb proximal and distal CMT1A-REPs. *Hum. Mol. Genet.* **6:** 1595.

Reiter L.T., Hastings P.J., Nelis E., De Jonghe P., Van Broeckhoven C., and Lupski J.R. 1998. Human meiotic recombination products revealed by sequencing a hotspot for homologous strand exchange in multiple HNPP deletion patients. *Am. J. Hum. Genet.* **62:** 1023.

Reiter L.T., Murakami T., Koeuth T., Pentao L., Muzny D., Gibbs R.A., and Lupski J.R. 1996. A recombination hotspot responsible for two inherited peripheral neuropathies is located near a *mariner* transposon-like element. *Nat. Genet.* **12:** 288. (erratum in *Nat. Genet.* [1998] **19:** 303).

Rio M., Clech L., Amiel J., Faivre L., Lyonnet S., Le Merrer M., Odent S., Lacombe D., Edery P., Brauner R., Raoul O., Gosset P., Prieur M., Vekemans M., Munnich A., Colleaux L., and Cormier-Daire V. 2003. Spectrum of *NSD1* mutations in Sotos and Weaver syndromes. *J. Med. Genet.* **40:** 436.

Saglio G., Storlazzi C.T., Giugliano E., Surace C., Anelli L., Rege-Cambrin G., Zagaria A., Jimenez Velasco A., Heiniger A., Scaravaglio P., Torres Gomez A., Roman Gomez J., Archidiacono N., Banfi S., and Rocchi M.A. 2002. A 76-kb duplicon maps close to the BCR gene on chromosome 22 and the ABL gene on chromosome 9: Possible involvement in the genesis of the Philadelphia chromosome translocation. *Proc. Natl. Acad. Sci.* **99:** 9882.

Samonte R.V. and Eichler E.E. 2002. Segmental duplications and the evolution of the primate genome. *Nat. Rev. Genet.* **3:** 65.

Saunier S., Calado J., Benessy F., Silbermann F., Heilig R., Weissenbach J., and Antignac C. 2000. Characterization of the *NPHP1* locus: Mutational mechanism involved in deletions in familial juvenile nephronophthisis. *Am. J. Hum. Genet.* **66:** 778.

Shaffer L.G. and Lupski J.R. 2000. Molecular mechanisms for constitutional chromosomal rearrangements in humans. *Annu. Rev. Genet.* **34:** 297.

Shaw C.J., Bi W., and Lupski J.R. 2002. Genetic proof of unequal meiotic crossovers in reciprocal deletion and duplication of 17p11.2. *Am. J. Hum. Genet.* **71:** 1072.

Slager R.E., Newton T.L., Vlangos C.N., Finucane B., and Elsea S.H. 2003. Mutations in *RAI1* associated with Smith-Magenis syndrome. *Nat. Genet.* **33:** 466.

Smith A.C., McGavran L., Robinson J., Waldstein G., Macfarlane J., Zonona J., Reiss J., Lahr M., Allen L., and Magenis E. 1986. Interstitial deletion of (17)(p11.2p11.2) in nine patients. *Am. J. Med. Genet.* **24:** 393.

Spiteri E., Babcock M., Kashork C.D., Wakui K., Gogineni S., Lewis D.A., Williams K.M., Minoshima S., Sasaki T., Shimizu N., Potocki L., Pulijaal V., Shanske A., Shaffer L.G., and Morrow B.E. 2003. Frequent translocations occur between low copy repeats on chromosome 22q11.2 (LCR22s) and telomeric bands of partner chromosomes. *Hum. Mol. Genet.* **12:** 1823.

Stankiewicz P. and Lupski J.R. 2002a. Genome architecture, rearrangements and genomic disorders. *Trends Genet.* **18:** 74.

———. 2002b Molecular-evolutionary mechanisms for genomic disorders. *Curr. Opin. Genet. Dev.* **12:** 312.

Stankiewicz P., Park S.-S., Inoue K., and Lupski J.R. 2001a. The evolutionary chromosome translocation 4;19 in *Gorilla gorilla* is associated with microduplication of the chromosome fragment syntenic to sequences surrounding the human proximal CMT1A-REP. *Genome Res.* **11:** 1205.

Stankiewicz P., Cheung S.W., Shaw C.J., Saleki R., Szigeti K., and Lupski J.R. 2003a. The donor chromosome breakpoint for a jumping translocation is associated with large low-copy repeats in 21q21.3. *Cytogenet. Genome Res.* **101:** 118.

Stankiewicz P., Park S.-S., Holder S.E., Waters C.S., Palmer R.W., Berend S.A., Shaffer L.G., Potocki L., and Lupski J.R. 2001b. Trisomy 17p10-p12 resulting from a supernumerary marker chromosome derived from chromosome 17: Molecular analysis and delineation of the phenotype. *Clin. Genet.* **60:** 336.

Stankiewicz P., Shaw C.J., Dapper J.D., Wakui K., Shaffer L.G., Withers M., Elizondo L., Park S.-S., and Lupski J.R. 2003b. Genome architecture catalyzes nonrecurrent chromosomal rearrangements. *Am. J. Hum. Genet.* **72:** 1101.

Stratton R.F., Dobyns W.B., Greenberg F., DeSana J.B., Moore C., Fidone G., Runge G.H., Feldman P., Sekhon G.S., Pauli R.M., and Ledbetter D.H. 1986. Interstitial deletion of (17)(p11.2p11.2): Report of six additional patients with a new chromosome deletion syndrome. *Am. J. Med. Genet.* **24:** 421.

Versalovic J., Koeuth T., Britton R., Geszvain K., and Lupski J.R. 1993. Conservation and evolution of the *rpsU-dnaG-rpoD* macromolecular synthesis operon in bacteria. *Mol. Microbiol.* **8:** 343.

Waldman A.S. and Liskay R.M. 1988. Dependence of intrachromosomal recombination in mammalian cells on uninterrupted homology. *Mol. Cell. Biol.* **8:** 5350.

Walz K., Spencer C., Kaasik K., Lee C.C., Lupski J.R., and Paylor R. 2004. Behavorial characterization of mouse models for Smith-magenis syndrome and dup17(p11.2p11.2). *Hum. Mol. Genet.* (in press).

Walz K., Caratini-Rivera S., Bi W., Fonseca P., Mansouri D.L., Lynch J., Vogel H., Noebels J.L., Bradley A., and Lupski J.R. 2003. Modeling del(17)(p11.2p11.2) and dup(17)(p11.2p11.2) contiguous gene syndromes by chromosome engineering in mice: Phenotypic consequences of gene dosage imbalance. *Mol. Cell Biol.* **23:** 3646.

Wildman D.E., Uddin M., Liu G., Grossman L.I., and Goodman M. 2003. Implications of natural selection in shaping 99.4% nonsynonymous DNA identity between humans and chimpanzees: Enlarging genus Homo. *Proc. Natl. Acad. Sci.* **100:** 7181.

Yunis J.J. and Prakash O. 1982. The origin of man: A chromosomal pictorial legacy. *Science* **215:** 1525.

# Human Versus Chimpanzee Chromosome-wide Sequence Comparison and Its Evolutionary Implication

Y. Sakaki,* H. Watanabe,* T. Taylor,* M. Hattori,* A. Fujiyama,* A. Toyoda,* Y. Kuroki,* T. Itoh,* N. Saitou,† S. Oota,† C.-G. Kim,† T. Kitano,† H. Lehrach,‡ M.-L. Yaspo,‡ R. Sudbrak,‡ A. Kahla,‡ R. Reinhardt,‡ M. Kube,‡‡ M. Platzer,¶¶ S. Taenzer,¶¶ P. Galgoczy,¶¶ A. Kel,§ H. Blöecker,** M. Scharfe,** G. Nordsiek,** I. Hellmann,†† P. Khaitovich,†† S. Pääbo,†† Z. Chen,‡‡ S.-Y. Wang,‡‡ S.-X. Ren,‡‡ X.-L. Zhang,‡‡ H.-J. Zheng,‡‡ G.-F. Zhu,‡‡ B.-F. Wang,‡‡ G.-P. Zhao,‡‡ S.-F. Tsai,¶¶ K. Wu,¶¶ T.-T. Liu,§§ K.-J. Hsiao,*** H.-S. Park,††† Y.-S. Lee,††† J.-E. Cheong,††† and S.-H. Choi††† (The Chimpanzee Chromosome 22 Sequencing Consortium)

*RIKEN, Genomic Sciences Center, Yokohama 230-0045, Japan; †National Institute of Genetics, Japan; ‡Max-Planck-Institut for Molekulare Genetik, Germany; ¶Institute of Molecular Biotechnology, Germany; §BIOBASE GmbH, Germany; German Research Centre for Biotechnology, Germany; ††Max-Planck-Institut of Evolutionary Anthropology, Germany; ‡‡Chinese National Human Genome Center at Shanghai, China; ¶¶National Health Research Institutes, Taiwan; §§Veterans General Hospital-Taipei, Taiwan; ***National Yang-Ming University, Taiwan; and †††Genome Research Center, KRIBB, Korea*

*Homo sapiens* is a unique organism characterized by its highly developed brain, use of complex languages, bipedal locomotion, and so on. These unique features have been acquired by a series of mutation and selection events during evolution in the human lineage and are mainly determined by genetic factors encoded in the human genome. It is of great interest and also of great importance from biological and medical viewpoints to understand what kind of genetic factors are involved in these complex human features and how they have been established during human evolution (Carrol 2003). Recent completion of the human genome sequence provided a solid platform for addressing these issues. However, the information obtained from the human genome alone is insufficient to discover genetic changes specific to human. The genomes of several experimental organisms such as mouse, fly, and nematode have successfully been used to characterize the human genome, but they are evolutionarily too distant to zoom in on the human-specific changes. Therefore, we definitely need the genome sequence of the closest organism to human. Detailed molecular anthropological studies have now established that the chimpanzee (and bonobo) is the closest organism to human, followed by the gorilla (see, e.g., Sibley and Ahlquist 1984; Saitou 1991; Chen and Li 2001; Wildman et al. 2003). The chimpanzee genome is thus the best for comparison with the human genome to elucidate the genetic changes that have occurred on the human lineage in the past 5–6 million years, providing us with important clues to address the above issues.

A number of pilot studies have already been done comparing human and chimpanzee genomes (see Olson and Varki 2003). For example, we previously showed through chimpanzee BAC end sequencing that the genomic difference between human and chimpanzee in terms of nucleotide substitution is 1.23% (Fujiyama et al. 2002). Frequent insertions and deletions were detected (Frazer et al. 2003). When insertions and deletions (indels) were compared, about 5% of the genome was shown to differ between human and chimpanzee (Britten 2002). Duplication, translocation, and transposition events have been also reported (Bailey et al. 2002). However, these data were obtained from various parts of the genomes by using several different technologies including sequence-based comparison, chip technology, and cytogenetic analysis, so that it is difficult to draw an integrated picture of dynamic changes of the genome to evaluate the overall consequence of these genetic changes to human evolution. For these reasons, we conducted a human–chimpanzee whole-chromosome comparison at the nucleotide sequence level. We chose human chromosome 21 and its genomic ortholog in chimpanzee, namely chromosome 22, because human chromosome 21 is one of the most well-characterized human chromosomes (Hattori et al 2000) and contains regions and units representing characteristic features of the human genome such as GC-rich/gene-rich regions and AT-rich/gene-poor regions, many repeated structures, duplications, housekeeping genes and tissue-specific genes, genes with a variety of functions such as transcriptional factors and receptors, members of large gene families and singleton genes.

## RESULTS AND DISCUSSION

### Mapping and Sequencing

At first, a BAC clone map of chimpanzee chromosome 22 was constructed based on the sequence similarity of BAC end sequences to human chromosome 21. Some gaps were then filled by screening chromosome-22-specific libraries or by PCR amplification. Finally, only two clone gaps remained at positions corresponding to gaps in human chromosome 21 (Hattori et al. 2000). A nucleotide sequence totaling 32.7 Mb of the mapped clones was then determined. We paid special attention to the data quality, because high-quality data are essential for comparison between human and chimpanzee genomic sequences in order to avoid false-positive differences and to identify even subtle differences within the sequences that may represent functionally and evolutionarily significant changes between the species. For quality assessment of the finished sequence, we first counted actual errors in the overlapping regions between sequenced chimpanzee clones. Overlaps ranged in size from 6 bp to 129 kb. In 249 overlaps, of which the total length was 9,890,299 bases (both clones) representing about 15% of the entire non-redundant sequence, errors totaling 165 bp including substitutions and indels were found and corrected. From this result, the accuracy of the overlapping sequence was estimated to be 99.9983%. These high-quality data enabled us for the first time to conduct a chromosome-wide sequence comparison between human and chimpanzee.

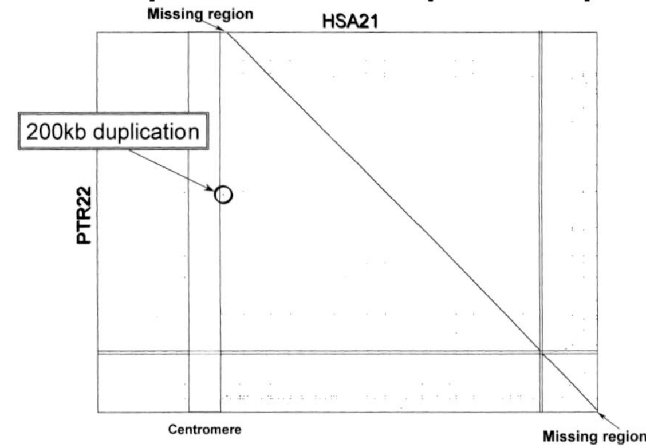


**Figure 1.** Harrplot between HSA21q and PTR 22q. Human chromosome 21 sequence (Hattori et al. 2000) was used for comparison by BLAST.

## Overall Comparison

The high-quality chimp sequence was compared to the human chromosome 21 sequence essentially by using the local alignment algorithm BLAST. Since there are many tandem duplications in the chromosome, global alignment algorithms are not suitable for this study. Figure 1 shows a so-called Harrplot analysis for overall comparison, demonstrating that there are no large rearrangements between human chr. 21 and chimp chr. 22, except for a 200-kb duplication that was found at the pericentromeric region in human chr. 21 but is missing in chimp chr. 22. To identify genetic changes of smaller sizes, we conducted more regional analyses as summarized below:

1. Base substitutions: The rate of overall base substitutions in comparable regions (except the centromeric and telomeric regions) is about 1.69%, which is slightly higher than the previously reported average substitution rate (1.23%) with the human genome (Fujiyama et al 2002). Interestingly, a significant bias was observed between A+T/L1-rich regions (1.62%) and G+C/Alu-rich regions (1.77%). This A+T vs. G+C bias may be partly explained by different substitution rates of some sequence units such as L1 (1.53%), Alu (2.69%), and CpG islands (3.83%). Centromeric and telomeric regions seem more unstable because of their repeated structures and show higher substitution rates. These regional differences in base substitution rate may reflect some structural and also some physiological differences of each unit, element, and region, but this remains to be resolved.
2. Insertion/deletions: Detailed comparisons also revealed the presence of many insertion/deletions (indels), ranging from one base to more than 5 kb. PCR amplification verification using an out group as a reference can distinguish whether these indels are insertions or deletions that occurred in the human or chimpanzee lineages. As shown in Figure 2, there are human-specific insertions and deletions as well as chimpanzee-specific insertions and deletions. In some cases, more complex combinations of insertions and deletions were found. Most indels seem to be derived from the length differences of simple repeats, but transposable elements also comprise a considerable portion of indels. Interestingly, some subfamilies of transposable elements were preferentially found in one lineage as insertions. For example, L1Hs (11 vs. 2), MER83B (11 vs. 0), AluYa5 (23 vs. 3), and AluYb8 (37 vs. 2) are preferentially found as insertions in the human lineage (Table 1). On the other hand, LTR/ERV1 and LTR/MaLR are found preferentially in chimpanzee. In addition to transposable elements, some large human-specific (or chimp-specific) "insertion" sequences were observed. For instance, a large (nearly 18 kb) insertion in a GRIk1 gene first intron was found in human, potentially affecting the expression profile of this gene (data not shown). The origin of this sequence is obscure as no homologous sequence was found in all the known sequences in the public database.

## Gene Structure

The above-described study demonstrated that there are a large number and variety of genetic changes between human and chimpanzee. The average rate of those changes suggested that almost every gene (including pro-

**Table 1.** Significantly Deviated Distribution of Repeats

| Family | Subfamily | HS21 | PTR22 |
|---|---|---|---|
| LINE/L1 | L1HS | 11 | 2 |
| LTR/ERV1 | HERVIP10FH | 14 | 5 |
| | MER41A-int | 10 | 2 |
| | MER4A1-int | 5 | 0 |
| | MER83B-int | 11 | 0 |
| | MER87 | 32 | 12 |
| SINE/Alu | AluYa5 | 23 | 3 |
| | AluYb8 | 37 | 2 |
| | AluYb9 | 7 | 1 |
| DNA/MER2 | Tigger3 | 42 | 67 |
| LTR/ERV1 | LTR49-int | 11 | 23 |
| LTR/MaLR | MLT1E-int | 0 | 5 |

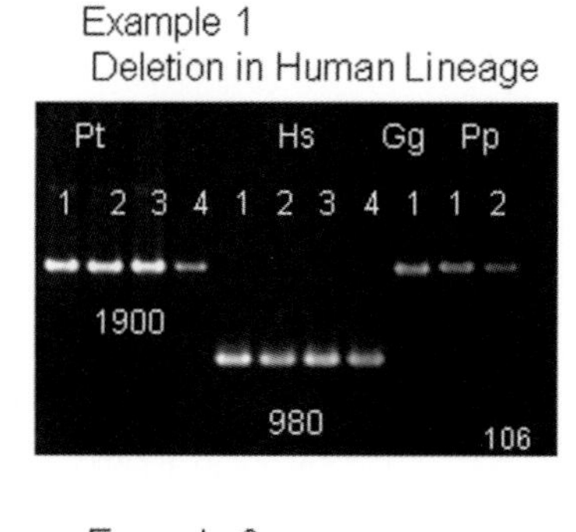


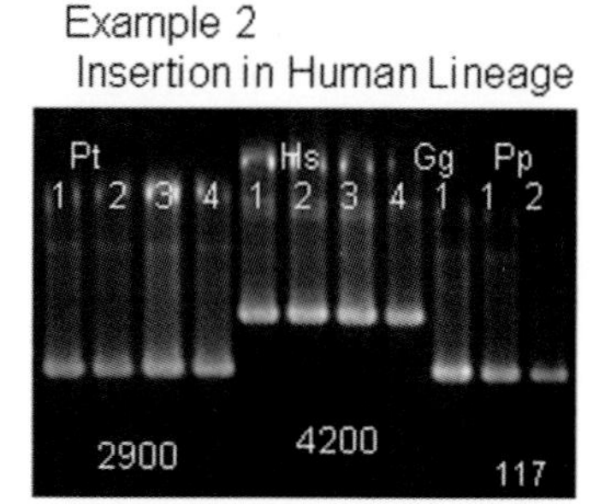


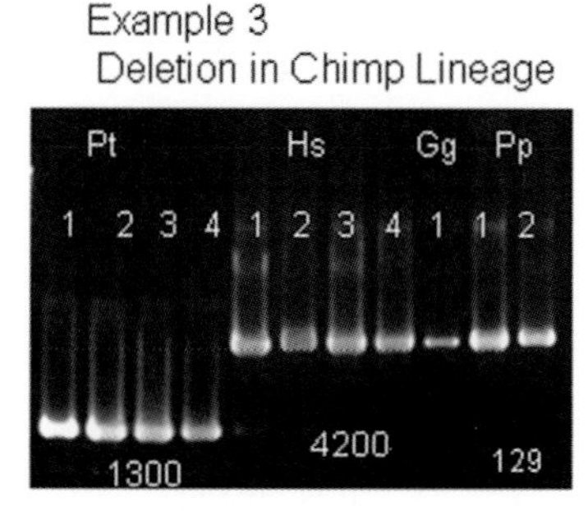

**Figure 2.** PCR-amplification test of "insertion" and "deletion." PCR primers were designed according to human sequence, and genomic DNAs of each species were amplified by PCR and analyzed by agarose gel electrophoresis. (Pt) Chimpanzee, (Hs) human, (Gg) gorilla, (Pp) orangutan.

moter regions) has some type of genetic change between human and chimpanzee. An obvious and important question is, "What are the biological and evolutionary consequences caused by these genetic changes?" To address this question, we checked all the changes in gene regions. Genetic changes that may affect biological functions were found in more than 30 genes in chimpanzee as compared with their human counterparts (Table 2). These changes included premature stop codons, missing start sites, and amino acid insertions or deletions. The biological consequences caused by these changes remain to be further studied by experimental approaches.

Among so many changes, it is of great interest to ask what types of changes are positively or actively involved in human (and chimp) evolution. It is known that the ratio of base substitutions that cause amino acid (nonsynonymous) changes (Ka) against base substitutions that cause synonymous changes (Ks) is correlated to the strength of evolutionary constraint. A low value of Ka/Ks means that the gene evolved under strong constraint. We calculated Ka/Ks values of all the genes according to Nei and Gojobori (1986), and found that more than 20 genes showed high Ka/Ks values of 1.0 or more, suggesting the possibility that these genes are positively selected during evolution (Table 3). Interestingly, Table 3 lists three keratin-associated protein genes. It is also of interest to note that 5 out of 10 genes listed in Table 3 are so-called predicted genes. Those genes may be more free from constraint or poorly predicted. The total number of genes examined is too small to be statistically significant, but it is likely that some specific types of genes are more free from constraint and positively involved in establishing species-specific phenotypes.

**Promoter and Gene Expression**

Genetic changes may cause two types of effects on biological processes: One is qualitative changes of proteins and the other is quantitative changes of gene expression (protein production). It has been reported that the gene expression profiles in human and chimpanzee are significantly different in brain but not so in other tissues examined (Enard et al. 2002). The difference in gene expression profiles may be caused by both the difference of regulatory *cis*-elements, and the difference of regulatory

**Table 3.** Genes Showing High Ka/Ks Value (top 10)

| Minimum Ka/Ks value | Gene name (Hs) | Locus link ID (Hs) * | Description |
|---|---|---|---|
| 3.37 | KRTAP23-1 | 337963 | keratin-associated protein 23-1 |
| 2.78 | C21ohr87 | 257357 | chromosome 21 open reading frame 87 |
| 1.98 | C21ohr81 | 114035 | chromosome 21 open reading frame 81 |
| 1.79 | FIJ33471 | 150147 | hypothetical protein FLJ33471 |
| 1.76 | C21ohr11 19 | | not found |
| 1.73 | RPS5L | 54022 | ribosomal protein S5-like |
| 1.71 | PRED62 | | not found |
| 1.67 | KRTAP15-1 | 254950 | keratin-associated protein 15-1 |
| 1.57 | KRTAP21-1 | 337977 | keratin-associated protein 21-1 |
| 1.47 | ABCC13 | 150000 | "ATP-binding cassette, sub-family C (CFTR/MRP). Member 13" |

**Table 2.** Base Substitutions Affecting the Protein Structures

| Chimp clone | Gene Symbol | Specific cDNA reference | Position |
|---|---|---|---|
| Missing Methionine: | | | |
| PTB-047A24 | C21orf18 | | substitution (51593<>51806) |
| PTB-084L06 | C21orf15 | | substitution (101445<>102078) |
| PTB-091H17 | C21orf9 | | substitution (67625<>67626) |
| PTB-111H18 | D21S2056E | AY033999 | substitution (31975<>32024) |
| Missing Stop codon: | | | |
| PTB-005B20 | PRED5 | | deletion (65024<>66013) |
| PTB-034G05 | C21orf62 | | 1 nt insertion (46310<>46359) |
| PTB-034G05 | C21orf49 | | 1 nt insertion (49441<>49490) |
| PTB-061A04 | C21orf30 | | substitution (65282<>65331) |
| PTB-292C11 | TMPRSS3 | AB038160 | deletion (79823<>79871) |
| RP43-117B17 | C21orf71 | | substitution (59715<>59764) (GCA/TCA) |
| Premature Stop codon: | | | |
| CH251-388O03 | PCNT2 | PCNT2 | lots of differences (48484<>59924) |
| PTB-003F15 | PRED75 | | insertion (120053<>120102) |
| PTB-013J17 | PRED48 | XM-071328 | 1 nt insertion (86107<>87630) |
| PTB-028I09 | PRED78 | | lots of differences |
| PTB-045O21 | C21orf11 | HSA409094 | big deletion (73480<>73519) |
| PTB-047A24 | C21orf27 | | substitution (55008<>57947) |
| PTB-047A24 | C21orf19 | | 2 nt insertion (124956<>125005) |
| PTB-051O03 | KIAA0653 | KIAA0653 | large deletion (135145<>135178) |
| PTB-058L13 | PRED48 | XM-071328 | 1 nt insertion (43697<>45220) |
| PTB-099D17 | C21orf104 | | substitution (202361<>202410) |
| PTB-120M01 | LSS | AK092334 | lots of differences (49204<>50027) |
| PTB-148P14 | PRED61 | P704-101D08-model17 | deletion (38514<>38562) |
| PTB-155J22 | C21orf79 | | 1 nt insertion (97084<>97133) |
| PTB-187O16 | C21orf104 | | 1 nt insertion (64532<>64581) |
| PTB-190I13 | PRED48 | XM-071328 | 1 nt insertion (135221<>136744) |
| RP43-001G01 | DSCR6 | DSCR6 | substitution (98219<>98268) |
| RP43-006A20 | PRED42 | | 2 nt insertion (90951<>91000) |
| RP43-006A20 | PRED78 | | lots of differences |
| RP43-015P20 | PDE9A | AF067226 | substitution (99938<>136172) |
| Additional amino acids: | | | |
| PTB-063M23 | C21orf96 | | 48 nt deletion (64157<>64208) |
| PTB-073G05 | IFNAR1 | IFNAR1 | 3 nt insertion (48299<>48348) |
| PTB-083K24 | USP16 | | 9 nt insertion (28318<>28367) |
| PTB-126B09 | PRED77 | | 6 nt insertion (191328<>191398) |
| PTB-153H02 | PRED58 | AA872876 P704-1023B21-model18-mpi | 6 nt insertion (73500<>73549 |
| PTB-196A08 | IFNAR1 | IFNAR1 | 3 nt insertion (87798<>87847) |
| RP43-012J05 | SYNJ1 | SYNJ1 | 6 nt insertion (41371<>41420) |

*trans*-acting factors. We examined whether the genetic differences found in the present work have any correlation to gene expression. By using Gene-chip technology, the gene expression profiles in brain and liver were compared between human and chimp, and it was found that several among 189 genes examined showed significant differences in brain and/or liver (data not shown). For example, IFNAR2, IFNGR2, and ETS2 showed enhanced expression in chimpanzee brain. On the other hand, C21orf97 was expressed more preferentially in human brain. Comparative analysis suggested the presence of some significant differences in the promoter regions that may affect gene expression. Figure 3 shows differences in the promoter region of the INFAR2 gene, which is expressed at significantly higher level in chimpanzee brain (and liver). The conclusive view must wait for further experimental verification, but it is likely that the genetic changes in intergenic regions also have made considerable contribution to human evolution through their quantitative effects on gene expression.

## CONCLUSIONS

The present study showed for the first time the chromosome-wide sequence comparison between human and nonhuman primates. The study revealed several interesting and important features of the human genome from evolutionary viewpoints. The overall study showed that a significant bias in base substitution exists from region to region and from element to element. These biases may have been generated from the regional difference of frequency of base substitution and the regional difference of the strength of constraint. Careful analysis of the base substitution rate may reveal the strength of evolutionary constraint on each element and region in the genome. For example, some CpG islands might be under more strict constraint than others. Overall comparison also showed transposable elements seem to have been distributed in species-specific manners, suggesting that expansion of transposable elements occurred discontinuously during evolution in each species. The most important finding of

**Alignment of upstream regions of IFNAR2 (Interferon alpha receptor 2) gene**

```
I
  IFNAR2_human   AGGGGCTGCTTATAACTATATTTTTTGGTTTACACTTCTTTTCTTGTGTATTCATTTAAT   3767
               2            <-----------V$NKX62_Q2(0.964,0.077)
  IFNAR2_chimp   AGGGGCTGCTTATAATTATATTTTTTGGTTTACACTTCTTTTCTTGTGTATTCATTTAAT   3453
                 *************** ********************************************

II
  IFNAR2_human   TTCAAATTTGTTAGCATGAGTTTTATGGCCCAGAATGTGGTCTATTTT------------   4055
               4                                            < ---------V$CEBP_Q3(0.975,0.116)
               7                                                       < -----...
  IFNAR2_chimp   TTCAAATTTGTTAGCATGAGTTTTATGGCCCAGAATGTGGTCTATTTTGGAAACTATTGT   3753
                 ************************************************

  IFNAR2_human   ------------------------------------------------------AATTAA   4061
               1                          <---------V$CEBP_Q3(0.990,0.131)
               4                     -----------›V$NKX62_Q2(0.969,0.082)
               7----V$CEBP_Q3(0.988,0.129)         <---------V$CEBP_Q3(0.981,0.122)
              10        <---------V$CEBP_Q3(0.981,0.122)
  IFNAR2_chimp   GAAATAGTTTCTTGTGAAACTATTAATTGTGAAATAGTTTCTTGTGAAACTATTAATTAA   3813
                                                                       * *****

III
  IFNAR2_human   TATTTTATATGATTCCATTCTCCCCTCACTTAGCACATTAATTGTATTAGTCAGGATTCT   5368
               1                             < -------------V$OCT1_02(0.916,0.203)
  IFNAR2_chimp   TATTTTATATGATTCCATTCTCCCCTCACTTAGCATATTAATTGTATTAGTCAGGATTCT   5132
                 *********************************** ************************

IV
               2    <-----------V$EGR1_01(0.856,0.159)
               4          <-----------V$EGR1_01(0.854,0.157)
  IFNAR2_human   GGCTCCGCCCCCGCCCCCGCGCCGGCGGCGGCGCGGCGCCCGCGCTTCCGTATCGCTCCT   9975
  IFNAR2_chimp   GGCTCCGCCCCCGACCCCGCGCCGGCGGCGGCGCGGCGCCCGCGCTTCCGCAGCGCTCCT   9749
                 ************* ************************************ * ******
```

**Figure 3.** Sequence alignment of upstream regions of IFNAR2 gene. Identical bases are shown by *. Nonidentical bases are shown by open gothic, and deleted region is shown by underline. Arrows show possible regulatory elements.

this study came from the Ka/Ks study showing that more than 5% of genes are relatively free from constraint, and some of them have been positively selected. The number of genes examined by the present study is too small to draw general conclusions, but there is the suggestion that some tissue-specific genes may play a positive role in establishing species-specific phenotypes.

In summary, the present study revealed a complex set of genetic differences between human and chimpanzee. Further experimental examination is required to say exactly what biological differences are caused by these genetic differences, but these data suggest that the biological consequences derived from the genetic differences may be more complicated than previously speculated.

## ACKNOWLEDGMENTS

We are grateful to T. Ito, T. Kawagoe, T. Kojima, X. Son, A. Beck, K. Borzym, S. Gelling, V. Gimmel, K. Heitmann, A. Kel, S. Klages, N. Lang, I. Mueller, M. Sontag, R. Yildirimman, Jean Wickings, Cornelia Baumgart, Oliver Mueller, T.T. Liao, H. Tsai, Y. Huang, Y. Liu, and all the technical staffs of the contributing genome centers. This work was supported in part by a Special grant from RIKEN Genomic Sciences Center and Grant-in-Aid for Scientific Research on Priority Areas "Genome Science" from the Ministry of Education, Culture, Sports, Science and Technology, Japan; The Ministry of Education and Research, Germany; The Chinese International Science and Technology Cooperation Project, Ministry of Science and Technology, China; The Chinese High-Tech Research and Development Program, Shanghai Commission for Science and Technology; and The National Research Program for Genomic Medicine of National Science Council, Taiwan.

## REFERENCES

Bailey J.A., Yavor A.M., Viggiano L., Misceo D., Horvath J.E., Archidiacono N., Schwartz S., Rocchi M., and Eichler E.E. 2002. Human-specific duplication and mosaic transcripts: The recent paralogous structure of chromosome 22. *Am. J. Hum. Genet.* **70:** 83.

Britten R.J. 2002. Divergence between samples of chimpanzee and human DNA sequences is 5%, counting indels. *Proc. Natl. Acad. Sci.* **99:** 13633.

Carrol S.B. 2003. Genetics and making of *Homo sapiens*. *Nature* **422:** 849.

Chen F.-C. and Li W.-H. 2001. Genomic divergence between humans and other hominoids and the effective population size of the common ancestor of humans and chimpanzees. *Am. J. Hum. Genet.* **68:** 444.

Enard W., Khaitovich P., Klose J., Zöllner S., Heissig F., Giavalisco P., Nieselt-Struwe K., Muchmore E., Varki A., Ravid R., Doxiadis G.M., Bontrop R.E., and Pääbo S. 2002. Intra- and interspecific variation in primate gene expression patterns. *Science* **296:** 340.

Frazer K.A., Chen X., Hinds D.A., Pant K., Patil N., and Cox

D.R. 2003. Genomic DNA insertions and deletions occur frequently between humans and nonhuman primates. *Genome Res.* **13:** 341.

Fujiyama A., Watanabe H., Toyoda A., Taylor T.D. Itoh T., Tsai S.-F., Park H.-S., Yaspo M.-L., Lehrach H., Chen Z., Fu G., Saitou N., Osoegawa K., de Jong P., Suto Y., Hattori M., and Sakaki Y. 2002. Construction and analysis of a human-chimpanzee comparative clone map. *Science* **295:** 131.

Hattori M., Fujiyama A., Taylor T.D., Watanabe H., Yada T., Park H.S., Toyoda A., Ishii K., Totoki Y., Choi D.K., Soeda E., Ohki M., Takagi T., Sakaki Y., Taudien S., Blechschmidt K., Polley A., Menzel U., Delabar J., Kumpf K., Lehmann R., Patterson D., Reichwald K., Rump A., and Schillhabel M., et al. 2000. The DNA sequence of human chromosome 21. The chromosome 21 mapping and sequencing consortium. *Nature* **405:** 311.

Nei M., and Gojobori T. 1986. Simple methods for estimating the numbers of synonymous and nonsynonymous nucleotide substitutions. *Mol. Biol. Evol.* **5:** 418.

Olson M. and Varki A. 2003. Sequencing the chimpanzee genome: Insights into human evolution and diseases. *Nat. Rev. Genet.* **4:** 20.

Saitou N. 1991. Reconstruction of molecular phylogeny of extant hominoids from DNA sequence data. *Am. J. Phys. Anthropol.* **84:** 75.

Sibley C.G. and Ahlquist J.E. 1984. The phylogeny of the hominoid primates, as indicated by DNA-DNA hybridization. *J. Mol. Evol.* **20:** 2.

Wildman D.E., Uddin M., Liu G., I. Grossman L.I., and Goodman M. 2003. The role of natural selection in shaping 99.4% identity between humans and chimpanzees at non-synonymous DNA sites: Implications for enlarging the genus *Homo*. *Proc. Natl. Acad. Sci.* **100:** 7181.

# Novel Transcriptional Units and Unconventional Gene Pairs in the Human Genome: Toward a Sequence-level Basis for Primate-specific Phenotypes?

L. LIPOVICH AND M.-C. KING
*Department of Genome Sciences, University of Washington, Seattle, Washington 98195-7730*

Despite the availability of a highly accurate human genome sequence (Lander et al. 2001) that has been comprehensively annotated (Hubbard et al. 2002; Kent et al. 2002), the functional definition of a mammalian gene remains in flux (Okazaki and Hume 2003). Furthermore, mammalian transcriptomes contain single-copy RNA species other than mRNAs of protein-coding genes. These RNAs have been shown to function in imprinting, X-inactivation, snoRNA hosting, and *trans*- and *cis*-antisense posttranscriptional regulation of other transcripts (Numata et al. 2003). Sequencing of random clones from normalized and subtracted cDNA libraries continues to uncover previously uncharacterized transcripts at a nearly linear rate, and a substantial number of the newly sequenced transcripts are unlikely to represent mRNAs of conserved protein-coding genes (Carninci et al. 2003). Since the loci giving rise to such transcripts may not fit conventional definitions of a gene, the term transcriptional unit (TU) is used to refer to them (Carninci et al. 2003). However, cDNA sequences are not the only line of evidence supporting the hypothesis that more of the genome is transcribed as exons than is accounted for by reference protein-coding gene sets. Multiple independent experiments in which bulk cDNA was hybridized to tiled-oligonucleotide microarrays representing nonrepetitive portions of the human genomic sequence also suggest that transcription is more widespread than is implied by conservative estimates based on full-length, protein-coding transcripts (Shoemaker et al. 2001; Kapranov et al. 2002; Rinn et al. 2003).

As increasing numbers of nonredundant cDNA sequences are produced by large-scale cDNA discovery projects and mapped onto progressively higher-quality drafts of the corresponding genomes, an intriguing feature of genome organization is emerging: frequent incidence of unconventional gene pairs (UGPs) in which two expressed features (which may be two protein-coding genes, a gene and a TU, or two TUs) overlap or are in close proximity to one another in the genomic locus which they share, in a manner suggesting that one of the features has the potential to regulate the expression of the other. Two classes of UGPs of particular interest are naturally occurring antisense pairs and putative bidirectional promoters. Gene regulation by antisense transcripts has been demonstrated in *Escherichia coli* (Delihas and Forst 2001), *Dictyostelium discoideum* (Hildebrandt and Nellen 1992), *Caenorhabditis elegans* (Lee and Ambros 2001), and *Drosophila melanogaster*, leading to the conclusion that antisense-related regulatory mechanisms are more prevalent than previously thought (Misra et al. 2002). Translational down-regulation of a sense transcript by antisense RNA induction has been observed (Stuart et al. 2000), consistent with the expectation that hybridization of two RNAs *cis*-antisense to one another results in translation blockage via steric hindrance and/or RNase-mediated degradation of the duplex (Vanhee-Brossollet and Vaquero 1998). In mammals, naturally occurring *cis*-antisense transcripts influence aspects of genome dynamics as diverse as imprinting (Sleutels et al. 2002), pathogenesis of human-specific neurodegenerative disorders (Andres et al. 2003), and gene vestigialization (Millar et al. 1999). *cis*-Antisense pairs have been recently recognized as a major feature of mammalian genomic architecture (Kiyosawa et al. 2003; Yelin et al. 2003). Bidirectional promoters have been experimentally verified in genomes ranging from viral (Moriyama et al. 2003) to human (Adachi and Lieber 2002). They have been shown to contain *cis*-elements utilized simultaneously by the genomically paired genes in the course of expression coregulation (R. Myers, unpubl.).

An interesting attribute of TUs and UGPs in mammalian genomes is that subsets of both classes of phenomena are lineage-specific. Although currently accepted explanations for the drastic differences in phenotypes between closely related species such as chimpanzees and humans invoke regulatory element differences responsible for distinct gene expression profiles in organisms with largely identical proteomes (King and Wilson 1975) or lineage-specific phenotypes related to the loss of function of particular genes during evolution (Olson and Varki 2003), it is conceivable that some phenotypic differences might be due to a gain of function related to the presence in one lineage of a gene absent in the other, or to a novel regulatory modality resulting, for instance, from a *cis*-antisense overlap in one lineage of two genes, the orthologs of which in other lineages lack the potential for transcript overlap. In fact, recent comparative genomic analysis of mammalian UGPs has confirmed the existence of a locus at which one member of a bidirectionally promoted pair is conserved while the other member is lineage-specific (Lipovich et al. 2002) and has implicated sequence-level mechanisms, such as substitutions that create or destroy polyadenylation signals in the genomic DNA sequence (Dan et al. 2002), in the generation of lineage-specific *cis*-antisense UGPs. Against a background of evidence for substantial numbers of lineage-specific genes in completely sequenced prokaryotic (Jordan et al. 2001) and eukaryotic (Lespinet et al. 2002) genomes, human–mouse comparisons have shown that even in two species sharing a common ancestor only ~ 70 mya, lineage-specific members of gene families related to transcription regulation, olfaction, behavior, and immunity have appeared (Young et al. 2002; Emes et al. 2003; Shannon et al. 2003), spurring an entire field of "genome zoology" focused on the genomic basis for interspecific differences (Emes et al. 2003).

The present study began more than 4 years ago as a simple attempt to construct and refine a comprehensive in silico physical and transcript map of a 5.5-Mb human genomic region at 5q31. At the time, the genome draft was highly fragmented, the cDNA databases rudimentary, and published analyses of TUs and UGPs beyond isolated single-locus examples practically nonexistent. Nevertheless, our extensive manual annotation uncovered substantial evidence for the existence of numerous TUs and UGPs in the region. As our in silico models of these TUs and UGPs continued to be supported by progressively larger amounts of increasingly higher-quality genomic and cDNA sequences over time, even while gap-filling by the Human Genome Project permitted completion of the data set with additional genes mapping to our 5q31 region, we set out to determine whether the patterns of incidence and genomic distribution of TUs and UGPs observed at 5q31 would be also detectable over larger intervals elsewhere in the genome. Our development of an automated TU and UGP discovery pipeline and subsequent validation of the 5q31 observations over the entire 35-Mb euchromatic sequence of human chromosome 22 constitute the balance of the study.

## OPERATIONAL DEFINITIONS

1. *Transcriptional unit (TU).* A TU is a transcribed feature in the genome other than a known gene. It is predicted in silico from analyzing EST-to-genomic DNA alignments in which the ESTs do not correspond to known or undocumented exons of known genes. ESTs comprising a TU must be canonically spliced (GT-AG introns) and/or canonically polyadenylated (AATAAA or ATTAAA polyadenylation signal within 40 bp of the submitter-indicated 3′ end). In defining TUs, we excluded ESTs from the ORESTES data set (Strausberg et al. 2002) and the RAGE data set (Harrington et al. 2001) because the former contains large numbers of unspliced, singleton, and chimeric ESTs, and the latter is derived from cell lines with artificial promoter insertions and therefore is not representative of naturally occurring transcription.

   Since our definition of a TU was developed prior to and independently from that of Carninci et al. (2003), it is not identical to that of Carninci et al. Due to the absence of an experimental component, our analysis cannot distinguish functionally important TUs from nonfunctional, stochastically transcribed TUs.
2. *Unconventional gene pair (UGP) type 1:* cis-*antisense.* The term "*cis*-antisense" means that both members of the pair are encoded within the same genomic locus. *cis*-antisense overlaps included in this analysis must be exon-to-exon, meaning that the predicted mature RNAs must overlap (intronic intercalation alone is insufficient). Two transcribed features *cis*-antisense to one another, residing on the opposite strands of the same locus, may be two genes, two TUs, or one of each. If only the first exons of the two features overlap, then the overlap is categorized as head-to-head. If only the last exons overlap, then the overlap is tail-to-tail. The "other" category is for all remaining possibilities.
3. *Unconventional gene pair (UGP) type 2: putative promoter-sharing.* This is a pair of divergently transcribed features whose transcription start sites are separated by <1 kb of genomic sequence. For multiple transcription start sites, the in silico cDNA- and EST-inferred transcript models are adjusted to reflect the minimum possible distance between the 5′ ends of the features.

## MANUAL ANNOTATION OF THE 5.5-Mb DAMS-DIAPH1 REGION OF HUMAN 5q31

Upon completion of the physical map, every genomic clone in the interval was preannotated using SeqHelp (Lee et al. 1998). In the course of the annotation, a large number of ESTs with BLAST (Altschul et al. 1990) hits to nonrepetitive genomic sequence outside the exons of known genes were found. Incorporation of the remaining ESTs into the transcript map led to our definition of a TU as given above. The entire annotation process was conducted without consulting precomputed annotation portals, which were accessed at the end of the project solely in order to compare our results with publicly available analyses of the same region.

### High Complexity and Primate Specificity of UGPs at 5q31: Case Studies

Our annotation demonstrated that novel TUs at 5q31 frequently participate in highly complex UGPs, and that some of this complexity is primate-specific. Two examples illustrate this point.

Example 1 (Fig. 1A) concerns a 415-kb region of 5q31 that contains six known genes. For clarity, their orientation is shown, but their splicing is not. Our annotation of this region uncovered a single novel alternatively spliced TU, 38I10.TU1, represented by 13 EST clones, the structure of which is shown. The TU did not contain any ORFs >100 amino acids in length, and most of its ORFs were either unique or located inside expressed repetitive elements, supporting the notion that if this TU is functional, its function is not to encode a protein. The TU is *cis*-antisense to an internal, translated exon of the TTID gene, which may be unusual because most *cis*-antisense in hu-

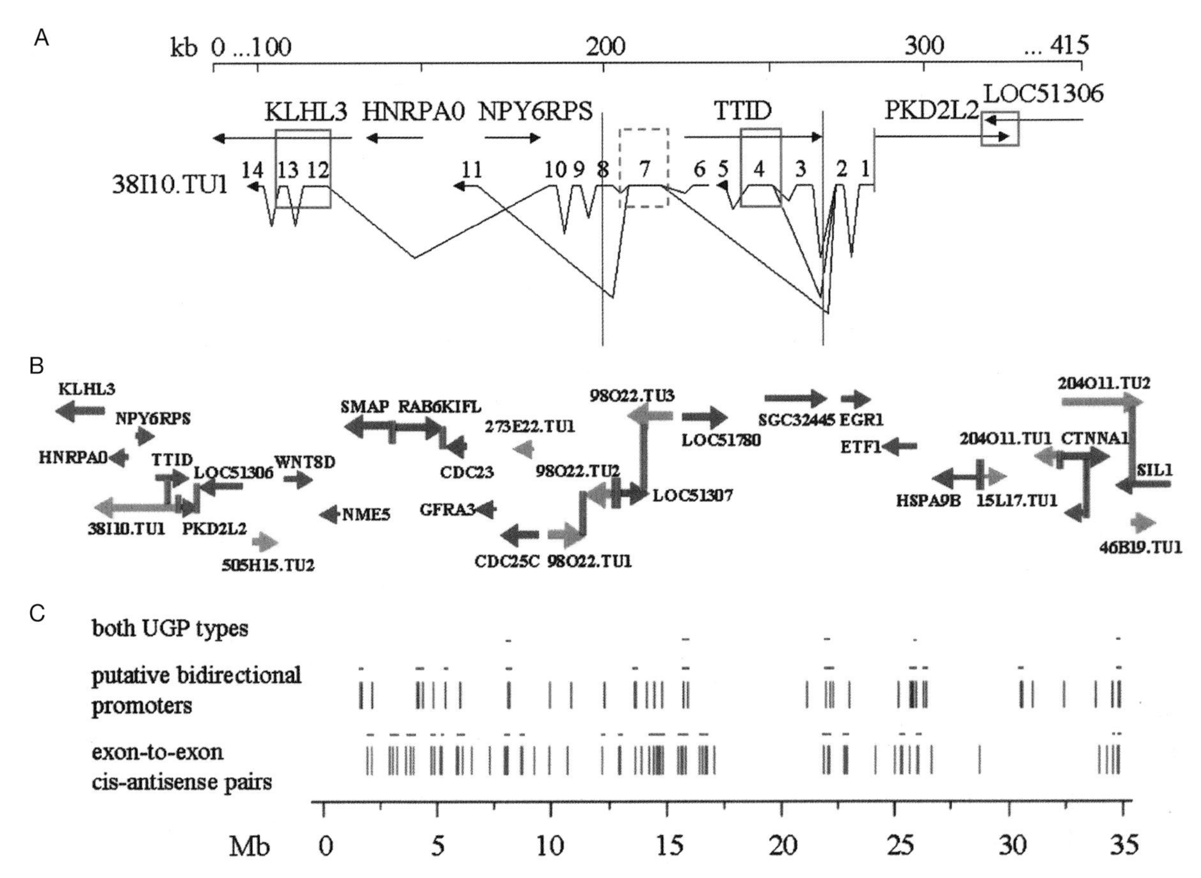


**Figure 1.** (*A*) Genomic complexity at human 5q31: the case of 38I10.TU1. The solid horizontal line at top represents the genomic sequence. 5cen is on the left. Arrows indicate the directions of transcription. The complete exon–intron structure is shown only for the TU, not for the known genes. Solid red boxes indicate antisense overlaps. The green vertical line indicates a putative bidirectional promoter. The magenta box indicates same-strand exon sharing. The region between the blue lines is the homology to the 10q21 myopalladin gene within the 5q31 genomic sequence. The dotted red box highlights an exon of 38I10.TU1 which is *cis*-antisense to an upstream ORF homologous to the myopalladin amino terminus and absent from the known TTID sequence. The figure was derived from our SeqHelp-aided manual annotation of genomic clones AC004021, AC073206, AC004820, and AC006084. The ESTs originally used to define the TU are AL707026, BI755671, AI822067, AI822069, BF238994, BI462983, BF217923, BF977167, AA459981, N66667, N98951, AA805667, BX097833, and BG704696. To allow maximum detail, the drawing is not to scale, although relative coordinates of feature boundaries along the 415-kb genomic interval are indicated. (*B*) A 1,580-kb interval of 5q31 enriched in unconventional gene pairs. 5cen is to the left. Arrows indicate the direction of transcription. Known genes are in blue, EST-supported TUs are in orange, *cis*-antisense overlaps are in green, and putative bidirectional promoters are in red. To allow maximum detail, the drawing is not to scale. (*C*) The distribution of unconventional gene pairs along human chr. 22. The horizontal line at the bottom represents the chromosome. Each short colored horizontal line represents an island (a region enriched in a particular type of UGP, as defined by the presence of at least two UGPs and a distance of <250 kb between any two consecutive UGPs). (*Blue*) Exon-to-exon *cis*-antisense. (*Red*) Putative bidirectional promoters. (*Magenta*) Regions enriched in both.

mans and mice seems to occur in terminal regions of genes (Shendure and Church 2002). This TU simultaneously participates in a second UGP, because it shares a 146-bp putative bidirectional promoter, associated with a CpG island, with the PKD2L2 gene. Manual curation of all EST-to-genomic alignments comprising the TU revealed that all of its introns except one were GT-AG, supporting the biological validity of the transcript. However, eight of the splice sites resided inside six expressed repetitive elements, five of which belonged to the primate-specific repeat families Alu (Greally 2002) and Mer1 (Kawashima et al. 1992).

In addition to a putative regulatory role suggested by its placement in UGPs, 38I10.TU1 may be relevant to modification of gene structure during evolution. Genomic sequence corresponding to exon 7 of the TU was subjected to TBLASTX (Gish and States 1993) in both orientations. In the sense orientation, which is opposite to the transcriptional orientation of the TU, the sequence appeared homologous to amino-terminal sequences of myopalladin, a human paralog of TTID mapping to 10q21. TTID is amino-terminally shorter than myopalladin due to the lack of sense-strand transcription of this sequence. This suggests that the robust transcription of 38I10.TU1 *cis*-antisense to this unused, potentially amino-terminus-encoding genomic sequence upstream of TTID is correlated with, and may even have caused, the elimination of an amino-terminal exon of TTID during evolution.

Two additional instances of genomic complexity were exposed by our annotation. The PKD2L2 gene was tail-to-tail antisense to the LOC51306 gene. The KLHL3 gene shared two exons in the sense orientation with 38I10.TU1 (Fig. 1A). Although the multiple UGPs associated with 38I10.TU1 and other genes in the locus were evident from manual curation of EST-to-genomic alignments, no mention of these UGPs, or of the existence of 38I10.TU1, is seen in published analyses of the genomic structures of genes which we show to be adjacent to or overlapped by the TU (Godley et al. 1999; Guo et al. 2000; Lai et al. 2001).

Comparative annotation of the syntenic region in the mouse genome was undertaken to determine whether 38I10.TU1 has a direct homolog or a positional equivalent in the mouse. The mouse Ttid and Pkd2l2 genes were found to be 10 Mb from each other due to a local inversion within a larger syntenic stretch. This sharply contrasts with the 2-kb distance separating the human orthologs of these genes and precludes the existence of a mouse TU that would connect or span the two genes. Along with the major contribution of expressed primate-specific repeats to the definition of splice junctions and hence exons of 38I10.TU1, these results indicate that the TU, along with its potential impact on expression regulation and structural modification of its multiple overlapping and adjacent coding genes, is likely primate-specific. If true, this would not be the only case of a lineage-specific TU simultaneously involved in multiple UGPs (Ohinata et al. 2003).

Example 2 was discovered during our annotation of the 5q31 protocadherin (PCDH) gene cluster. PCDHs are members of the cadherin superfamily, known to represent major structural and functional components of synapses. Cell-specific combinations of PCDHs are expressed at synaptic junctions. PCDHs likely account for some of the neuronal combinatorial complexity in both developing and adult brain, affecting brain development and possibly memory formation (Noonan et al. 2003). Open questions regarding the mechanism of cell specificity and splicing regulation of PCDH expression remain (Wang et al. 2002). Our annotation demonstrated EST support for 8 TUs *cis*-antisense to human PCDH exons (Table 1). In all cases, ESTs comprising the TUs were erroneously grouped by UniGene together with the sense-strand PCDH ESTs from the appropriate locus, even though their antisense nature is clear from a distinct pattern of EST-to-genomic alignments coupled with unambiguous assignment of GT-AG splice sites and/or canonical polyadenylation signals to the strand opposite that from which the PCDH exons are transcribed.

Two of the eight anti-PCDH TUs overlap newly pseudogenic PCDH exons inactivated during mammalian evolution, in a fashion reminiscent of that exhibited by 38I10.TU1 in its overlap of TTID's apparently silenced amino-terminal-coding exon. Specifically, the human PCDHβΨ5 pseudogene, which has an antisense TU, and the human PCDHβ15 gene are the putative products of a lineage-specific duplication of the ancestral PCDHβV gene, which is single-copy in mouse (Vanhalst et al. 2001). Similarly, the human PCDHγ variable exon Ψ3, which also has an antisense TU, is pseudogenic, but its mouse ortholog, Pcdhγ variable exon b8, has an intact ORF (Wu et al. 2001). This again suggests a role for *cis*-antisense in gene structure evolution.

Lineage specificity of such gene-structure evolution is accentuated by the absence of equivalents of any human anti-PCDH TUs in the mouse, revealed by manual annotation of all mouse ESTs representing true orthologs or nearest homologs of human PCDH exons. It is highly intriguing that not only are PCDHs directly relevant to neuronal and behavioral complexity (which is more extensive in mammals, and especially primates, than in any other animals), but they may also be subject to an antisense-mediated regulatory mechanism that arose after the primate–rodent divergence. Such a mechanism would be in agreement with suggestions of species-specific evolutionary pressures on PCDH genes (Vanhalst et al. 2001).

## Major Properties of UGPs and TUs at 5q31

The results of our 5q31 annotation can be summarized in four major trends. First, the number of EST-supported novel TUs approximately equals that of known genes. Second, certain genes and TUs are simultaneously involved in multiple UGPs. Third, UGPs occur more frequently in specific genomic intervals and may be nonran-

**Table 1.** Putative Novel Transcriptional Units (TUs) *cis*-Antisense to Protocadherin (PCDH) Genes at 5q31

| Human PCDH gene or exon | Characteristics | Mouse ortholog | Expression human | Expression mouse | Antisense ESTs human | Antisense ESTs mouse | Exons in human antisense transcript |
|---|---|---|---|---|---|---|---|
| PCDHα9 | variable exon | Pcdhα7 | + | + | 10 | 0 | 1 |
| PCDHα11 | variable exon | Pcdhα11 | + | + | 1 | 0 | 1 |
| PCDHα12 | variable exon | none | + | na | 2 | na | 1 |
| PCDHαc1 | variable exon | Pcdhαc1 | + | + | 1 | 0 | 1 |
| PCDHβ3 | single-exon gene | Pcdhβ3 | + | + | 6 | 0 | 2 |
| PCDHβ16 | single-exon gene | Pcdhβ16 | + | + | 4 | 0 | 1 |
| PCDHΨ5 | single-exon unprocessed β-class pseudogene | none | – | na | 16 | na | 3 |
| PCDHΨ3 | unprocessed γ-class pseudogenic variable exon | Pcdhγb8 | – | + | 2 | 0 | 1 |

Table was derived from our SeqHelp-aided manual annotation of genomic clones AC005609, AC010223, AC025436, AC008727, AC005618, and AC005366.

(na) Not applicable.

**Table 2.** Clustering of Genes and TUs Involved in Unconventional Gene Pairs at 5q31

| Interval | [DAMS-TRP7] | (TRP7-KLHL3) | [KLHL3-SIL1] | (SIL1-25P15.TU1) | [25P15.TU1-HARSL] |
|---|---|---|---|---|---|
| Size (kb) | 225 | 1260 | 1580 | 1200 | 300 |
| Transcript models | 6 | 6 | 32 | 42 | 16 |
| Transcript models per 100 kb | 2.7 | 0.5 | 2.0 | 3.5 | 5.0 |
| Transcript models involved in UGPs | 4 | 0 | 18 | 4 | 11 |
| % of transcript models in UGPs | 67% | 0% | 56% | 10% | 69% |

The term "transcript models" refers to both genes and TUs.

domly distributed along the genomic sequence in a way that is independent from gene density. Fourth, relative to genes, TUs are enriched in expressed repetitive elements, including primate-specific Alu and Mer1 repeats.

An analysis of UGP distribution in the 5q31 region exclusive of the PCDH clusters demonstrated that UGPs cluster in well-defined genomic intervals (Table 2). The intervals differ in the proportion of expressed features participating in UGPs, which represent the majority of features in some intervals and a very small minority in others. The UGP-enriched genomic intervals contain multiple types of genomic complexity. For example, the interval containing 38I10.TU1 also contained four consecutive features, each of which was oriented opposite to its neighbors and which formed three antisense pairs (Fig. 1B), a rare arrangement analogous to, but even more complex than, that seen in the human Surfeit locus (Duhig et al. 1998).

Table 3 indicates differences between known genes and novel TUs in ~4.5 Mb of 5q31. The biological reality of TUs is suggested by canonical transcript processing and the presence of multiple ESTs. Nonetheless, TUs represent a radically different fraction of the transcriptome. For example, BLASTN and TBLASTX analysis of all TUs mapping to this 5q31 region, performed against the NT, EST, GSS, and HTGS databases, found nonhuman homology for most of the genes but for less than half of the TUs.

## LARGE-SCALE VERIFICATION OF TRENDS FROM CHR. 5q31 ON CHR. 22

### Perl-based High-throughput TU Discovery and UGP Analysis Pipeline

We hypothesized that these four trends are global and not region-specific. Therefore, we utilized 5q31 as a training set for chromosome-scale analysis of TUs and UGPs. We codified criteria developed for the 5q31 annotation into a three-stage Perl-based high-throughput automated annotation pipeline (Fig. 2). We modified the bioperl.org open-source BLAST parsers (Stajich et al. 2002) to make them aware of the transcriptional orientation of cDNAs and ESTs matching genomic sequences. The first stage utilized these parsers to analyze all cDNA and EST BLAST matches against the query genomic sequence and determined which nongenic EST matches were TU-worthy (i.e., completely and precisely satisfied our operational definition of a TU). Only primary EST and cDNA evidence was used. We did not use third-party annotations or any curated reference transcripts bearing NM and XM designations. In the second stage, all cDNA and TU-worthy EST matches were subjected to BLASTN against the entire human genomic sequence in the NR and HTGS databases. Matches with homologies to genomic regions other than the region in which they were originally identified, and with equal or higher BLAST scores associated with homologies to those other regions, were automatically eliminated due to their putatively segmentally duplicated or pseudogenic nature. The third stage automatically compiled complete exon–intron structures for every gene and TU, quoting exact coordinates of every element of the structure on the genomic sequence, accession numbers of cDNAs or ESTs supporting each exon of each gene and TU, and the extent of apparent involvement of the gene or TU in UGPs in a table suitable for manual curation. We selected human chr. 22 for chromosome-scale validation of 5q31 trends because chr. 22 is small and thoroughly annotated, facilitating comparisons with other algorithms (Collins et al. 2003) and representative of other chromosomes in terms of segmental duplications (Bailey et al. 2002), low-copy repeats (McDermid and Morrow 2002), and gene family expansions (Coggan et al. 1998; Jarmuz et al. 2002).

**Table 3.** Differences between Known Genes and Novel TUs at 5q31

| | Known genes | Novel TUs | P-value for difference of genes and TUs |
|---|---|---|---|
| N | 54 | 47 | |
| With homology to nonhuman DNA | 52 | 17 | 0.0001 |
| Length (amino acid) of longest sense-strand ORF | 496 ± 47.3 | 64 ± 4.4 | 0.0001 |
| % of reference transcript in expressed repeats | 5.3 ± 1.6 | 21.4 ± 3.8 | 0.0011 |
| ESTs per gene or TU | 201 ± 26.7 | 6 ± 1.0 | 0.0001 |

Standard errors (calculated by SPSS v10, GLM parameter estimates module) are given after ±.

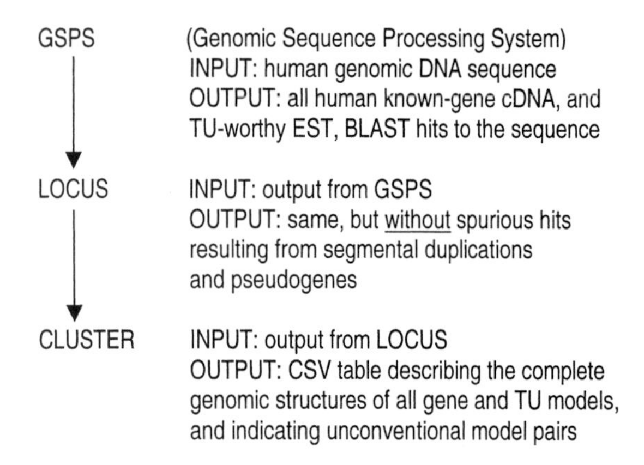


**Figure 2.** Perl-based, high-throughput gene verification, TU discovery, and UGP identification pipeline. Following stage 3 (CLUSTER) output, manual curation was performed.

### Automated Identification of Known Genes and Novel TUs on Chr. 22

Our algorithm identified 1012 nonredundant transcript models on chr. 22. Of these, 495 (49%) represented known genes and 517 (51%) represented novel TUs supported solely by ESTs. This result was consistent with the relative proportions of genes and TUs at 5q31, as well as with recent findings indicating that, due to large numbers of TUs, the total number of expressed features comprising a mammalian transcriptome is likely to be more than twice the number of coding genes (Carninci et al. 2003). We automatically excluded transcripts homologous to immunoglobulin λ gene segments, because existing annotations generally group them into a special category separate from the rest of expressed features on chr. 22 (Collins et al. 2003).

The sensitivity and specificity of our approach were assessed using the most current Sanger Centre chr. 22 annotation (Collins et al. 2003). Of the 577 genes identified by the Sanger annotation, 469 were found by our algorithm, a sensitivity of 81%. The discrepancy was due in part to the fact that our definition of a known gene was based solely on the presence of a full-length cDNA in Genbank, and did not include genes based solely on ORFs or ab initio exon prediction. Of the 108 Sanger genes missed by our algorithm, the majority were either transcriptionally silent, segmentally duplicated paralogous copies of genes we identified, or completely devoid of sense-strand cDNA and EST support. Of the 234 features identified by the Sanger analysis as pseudogenes, 206 were missing from the gene and TU sets created by our algorithm, a specificity of 88%. The remaining 28 were transcribed and had full-length cDNA or EST-only support, therefore fitting our definition of genes and TUs, respectively.

### Characterization of UGPs on Chr. 22

Of the 1012 transcript models on chr. 22, 209 (21%) participated in UGPs. 77 *cis*-antisense pairs and 42 putative bidirectional promoters were found.

Of the 77 antisense pairs, 23 were tail-to-tail and 13 head-to-head, roughly consistent with the proportion at 5q31 outside of the PCDH clusters (7 and 4, respectively) and with published evidence that in mammals tail-to-tail gene overlaps are more common than head-to-head overlaps (Edgar 2003). Surprisingly, the remaining 41 pairs did not fit either category (this was the case for only 2 pairs in the non-PCDH part of our 5.5-Mb 5q31 region), which argues for a substantial diversity and complexity of gene and TU structures participating in antisense overlaps. Of the 77 pairs, 36 were gene–gene, 38 were gene–TU, and 3 were TU–TU. Hence, a gene-only approach to chr. 22 annotation would miss more than half of the *cis*-antisense pairs. The 77 pairs accounted for only 145 transcript models rather than the expected 154, because 8 models participated in *cis*-antisense overlaps with multiple other models.

We addressed whether any of the 77 pairs had potential for hybridization of sense and antisense transcripts in vivo due to the expression of both members of the pair in the same tissue or cell type. Complete lists of cDNAs and TU-worthy ESTs for every pair were obtained by BLAST and manual curation, and were examined for commonalities in expression profiles. ESTs from pooled libraries or total fetus were eliminated, since their precise origin was unknown. Normal tissues were considered as different from corresponding tumors; e.g., a gene–TU pair in which the gene was expressed only in normal brain but the TU was expressed only in brain tumors would be characterized as lacking any overlap in expression profiles of the two. For 35 of the 77 antisense pairs (45%), EST evidence suggested expression of both members of the pair in the same tissue or cell type. In 19 of these 35, genomic organization of the locus was conserved between human and mouse. However, in the other 16, one or both members of each human pair lacked orthologs and positional equivalents in mouse (Table 4). Examples include the acrosin precursor gene, whose head-to-head antisense TU in humans has no mouse equivalent, and the CHK2/BC000004 head-to-head antisense pair, the orthologs of whose members in the mouse are in a head-to-head orientation but do not overlap. Of the 16 cases of human–mouse differences in antisense-containing loci, 5 were characterized by the expression of both members of the antisense pairs in human brain, raising the intriguing possibility that some *cis*-regulatory effects on gene expression in human brain are lineage-specific and are not universally conserved in mammals.

Of the 42 putative bidirectional promoters, 34 (79%) occurred at CpG islands, confirming existing reports of divergent transcription initiation at mammalian CpG islands (Adachi and Lieber 2002). Of the 42 bidirectionally promoted pairs, 21 were gene–gene, 18 were gene–TU, and 3 were TU–TU. Therefore, similar to the case with *cis*-antisense, a gene-only annotation would miss approximately half of the putative bidirectionally promoted transcript model pairs.

Chromosomewide, 20 transcript models participated in both *cis*-antisense and putative promoter-sharing pairs. This is significantly more than expected under the null hypothesis that involvement in the two types of UGPs is

**Table 4.** Human Chr. 22 Antisense Pairs in Which Both Members of the Pair Are Expressed in the Same Human Tissue and Genomic Organization Differs in the Mouse

| Type of genomic organization | | Pairs | Genes and TUs in antisense pairs | | Expression of both members of pair |
|---|---|---|---|---|---|
| human → ← | mouse → | 9 | FLJ32500 | BID | kidney; skin |
| | | | HIRA | tu_BF589683 | skin (cancer) |
| | | | MIF | BC036909 | brain |
| | | | ADORA2A | tu_AI198582 | testis |
| | | | PISD | tu_BI915399 | brain |
| | | | TIMP3 | tu_BM990787 | trabecular bone; placenta; uterus |
| | | | UNC84B | tu_AW504307 | B-cells |
| | | | UNC84B | tu_BG057310 | colon (tumor) |
| | | | ACR | BC050343 | testis |
| human → ← | mouse *neither* | 4 | RFPL1 | RFPLL1ANT | brain |
| | | | RFPL3 | RFPL3ANT | brain |
| | | | FLJ30933 | tu_AA844700 | testis |
| | | | tu_AA383102 | tu_BU585030 | testis |
| human ← → | mouse ← → | 2 | FLJ30119 | SNRPD3 | skin (cancer) |
| | | | SF3AI | BC018040 | brain |
| Exon skip in mouse relative to human | | 1 | AL365514 | HSC3 | testis |

a chance combination of two independent events ($p$ = 0.01 by $\chi^2$). Therefore, for a given expressed feature, participation in one UGP type increases the probability that the feature is also involved in the other.

It is notable that more than half of our antisense data set does not overlap the antisense pairs identified in the independent and parallel study by Yelin et al. (2003), which was performed while the present work was in preparation, on 5q31 and chr. 22. At 5q31, 9 of the 13 pairs identified by us outside the PCDH clusters, along with 1 of the 8 pairs within the PCDH clusters, were also identified by Yelin et al. (48%). On chr. 22, 34 of our 77 pairs were also identified by Yelin et al. (44%). This suggests that our manual and automated approaches, respectively, are capable of identifying UGPs missed by independently designed algorithms.

### UGP Distribution along Chr. 22

On the basis of the 5q31 results, we hypothesized that UGPs on chr. 22 would cluster, rather than be randomly distributed along the chromosome or distributed proportional to gene density. We operationally defined a "UGP island" as a region with at least two UGPs of the same type, and in which any two consecutive UGPs are <250 kb from one another. The choice of this somewhat arbitrary island-size criterion was guided by the observation that at 5q31 most instances of proximity of multiple UGPs along the genomic sequence occurred at the sub-250-kb scale, even within larger UGP-enriched domains.

The distribution of both types of UGPs on chr. 22 is illustrated in Figure 1C. The close clustering of most UGPs in well-defined small intervals is visually apparent. The sum of the genomic regions enriched in UGPs (the islands, indicated by horizontal bars in the figure), represents a minority of the total chr. 22 sequence: 26 of the 42 putative bidirectional promoters (62%) reside in 1.5 Mb (4%) of the sequence, and 62 of the 77 *cis*-antisense pairs (81%) reside in 3.4 Mb (10%) of the sequence. Five specific regions were simultaneously enriched in both types of UGPs.

Islands of putative bidirectional promoters were weakly correlated with locally high CpG island density. There was no evidence for correlation of islands of either UGP type with GC content, SINE or LINE density, recombination hot spots, human–mouse synteny breakpoints, or recent segmental duplications.

We tested the hypothesis that UGP incidence and clustering are proportional to gene density by a nonparametric assessment of the probability that the observed incidence and clustering of UGPs were due to chance. We partitioned chr. 22 into 20 intervals containing approximately equal total numbers of transcript models per interval, but differing in size due to variations in transcript model density along the chromosome. Information on the transcriptional orientation (toward vs. away from the centromere), genomic size, and complete exon–intron structure was obtained for all reference transcripts representing the 1012 transcript models. This information was left intact in the subsequent simulations of transcript model distribution, in which a random number generator was used to assign positions of the simulated equivalents of the actual observed transcripts within the interval. Interval boundary spanning, sense-strand intercalation including exon sharing, and transcript models with >2 independent antisense partners were not allowed. For each interval, 10,000 simulations were performed. The simulations were implemented with a Perl script we designed specifically for this purpose. The actual biological complexity of an interval was defined as the number of UGPs of both types, and the number of islands of UGPs of either type, within the interval. The proportion of total simulations per interval in which the actual biological complexity of that interval was met or exceeded was tabulated.

Simulation results suggest that nonoverlapping intervals of chr. 22 containing ~50 transcript models per interval can be assigned to one of two types. In intervals of the first type, which contain most UGP islands, the actual biological complexity is far greater than that expected by chance, as it is matched or exceeded in <0.1% of simulations. In intervals of the second type, which are less common, the actual biological complexity is less than that expected by chance, and is matched or exceeded in a majority of the simulations. In these intervals, most genes and TUs reside singly and are neither co-promoted with, nor overlap, other genes and TUs. This implies that UGP-poor genomic domains, just like UGP-rich ones, should be unlikely under a model stipulating that UGPs occur by chance and proportionally to gene density. Therefore, the observed mosaic of UGP-rich and UGP-poor segments along chr. 22 may be due to functional constraints that act over contiguous regions hundreds of kilobases or greater in size and either favor or restrict the existence of UGPs in the regions, rather than to chance.

### Expressed Primate-specific Repetitive Elements on Chr. 22

From our 5q31 results, we expected that a subset of primate-specific Alu and Mer1 repeats on chr. 22 would be transcribed as exons, and that evidence for transcriptional recruitment of these sequences in the course of primate evolution would be more abundant in TUs than in genes. Our results indicated that 71,702 bp of chr. 22 genomic sequence (0.21% of the chromosome) consisted of expressed primate-specific repeats localized to exons in in silico models of genes and TUs. This number is a lower bound on the extent of sequence recruitment into exons during primate evolution, because alternatively spliced and alternatively polyadenylated isoforms of all genes and TUs and nonrepetitive primate-specific exonic sequences were not included. As on 5q31, on chr. 22 novel TUs were significantly enriched in expressed primate-specific repeats relative to known genes: 3.6% of the reference transcript for an average known gene consisted of repeats of this type (which in almost all coding-gene cases were in UTRs) vs. 9.4% of the reference transcript for an average TU ($p = 0.002$).

## CONCLUSIONS

Chr. 22 analysis qualitatively validated all four trends discovered during manual annotation of chr. 5q31. Novel TUs appear as abundant as known genes; a significant number of genes and TUs participate simultaneously in both types of UGPs; the UGPs are not randomly distributed along the genomic sequence; and TUs are enriched in expressed primate-specific sequences, relative to genes. These trends would not be apparent from manual examination of precomputed annotations at Ensembl or UCSC, because those annotation portals do not resolve pseudogene- and duplication-related mapping ambiguities and do not emphasize UGP discovery.

This work raises at least three questions. First, which TUs and UGPs are functionally important? TUs that participate in UGPs would seem better candidates for functional, vs. stochastic, transcription, especially if the UGP is an antisense pair in which EST evidence indicates that both members are expressed in the same tissue. Of course, even in cases of *cis*-antisense corresponding to spatiotemporally exclusive expression profiles of the pair members, it is possible that one member's expression profile was altered in the course of evolution because of the appearance of an antisense counterpart, leading to lineage-specific modification of function even in the absence of any in vivo hybridization between the UGP-encoded transcripts. Only experimental analysis can test TU and UGP functionality, with TU validation and cloning by RT-PCR, RACE, and other approaches. Potential experimental directions include evaluating the effects of anti-TU RNAi in primate cell lines and assessing expression of primate-specific TUs in regions of the brain responsible for primate behavioral complexity using custom-designed microarrays with probes derived from nonrepetitive portions of the TUs.

Second, which TUs are evolutionarily young genes? Although mouse sequences are useful in evaluating whether a TU arose before or after the mammalian radiation, answering this question requires more nonhuman primate genomic and cDNA sequences than are presently available. Putative primate orthologs would be useful for establishing that TUs are in fact young genes rather than human-specific transcripts potentially due to transcription initiation inefficiency or stochastic initiation from weak, perhaps repeat-supplied, promoters.

Third, the genomic structures of certain TUs and UGPs strongly suggest that the existence of those TUs and UGPs is made possible by primate-specific sequences such as Alu repeats, primate-specific positional proximity of genes that may be distant from one another in non-primate mammals because of synteny breakpoints, or both. It is therefore tempting to speculate that certain primate-specific TUs and UGPs comprise an essential part of the genomic basis of primate, including human, phenotypic uniqueness.

## ACKNOWLEDGMENTS

We thank Phil Green and Debbie Nickerson for guidance and help, and Ming K. Lee for assistance with SeqHelp and advice on Perl programming. L.L. was supported by National Institutes of Health training grants HG-00035 and CA-09437. This work was also supported in part by National Institutes of Health grant R01 CA-27632 to M.C.K.

## REFERENCES

Adachi N. and Lieber M.R. 2002. Bidirectional gene organization: A common architectural feature of the human genome. *Cell* **109:** 807.

Altschul S.F., Gish W., Miller W., Myers E.W., and Lipman D.J. 1990. Basic local alignment search tool. *J. Mol. Biol.* **215:** 403.

Andres A.M., Soldevila M., Saitou N., Volpini V., Calafell F., and Bertranpetit J. 2003. Understanding the dynamics of spinocerebellar ataxia 8 (SCA8) locus through a comparative genetic approach in humans and apes. *Neurosci. Lett.* **336:** 143.

Bailey J.A., Yavor A.M., Viggiano L., Misceo D., Horvath J.E., Archidiacono N., Schwartz S., Rocchi M., and Eichler E.E. 2002. Human-specific duplication and mosaic transcripts: The recent paralogous structure of chromosome 22. *Am. J. Hum. Genet.* **70:** 83.

Carninci P., Waki K., Shiraki T., Konno H., Shibata K., Itoh M., Aizawa K., Arakawa T., Ishii Y., Sasaki D., Bono H., Kondo S., Sugahara Y., Saito R., Osato N., Fukuda S., Sato K., Watahiki A., Hirozane-Kishikawa T., Nakamura M., Shibata Y., Yasunishi A., Kikuchi N., Yoshiki A., Kusakabe M., Gustincich S., Beisel K., Pavan W., Aidinis V., Nakagawara A., Held W.A., Iwata H., Kono T., Nakauchi H., Lyons P., Wells C., Hume D.A., Fagiolini M., Hensch T.K., Brinkmeier M., Camper S., Hirota J., Mombaerts P., Muramatsu M., Okazaki Y., Kawai J., and Hayashizaki Y. 2003. Targeting a complex transcriptome: The construction of the mouse full-length cDNA encyclopedia. *Genome Res.* **13:** 1273.

Coggan M., Whitbread L., Whittington A., and Board P. 1998. Structure and organization of the human theta-class glutathione S-transferase and D-dopachrome tautomerase gene complex. *Biochem. J.* **334:** 617.

Collins J.E., Goward M.E., Cole C.G., Smink L.J., Huckle E.J., Knowles S., Bye J.M., Beare D.M., and Dunham I. 2003. Reevaluating human gene annotation: A second-generation analysis of chromosome 22. *Genome Res.* **13:** 27.

Dan I., Watanabe N.M., Kajikawa E., Ishida T., Pandey A., and Kusumi A. 2002. Overlapping of MINK and CHRNE gene loci in the course of mammalian evolution. *Nucleic Acids Res.* **30:** 2906.

Delihas N. and Forst S. 2001. MicF: An antisense RNA gene involved in response of *Escherichia coli* to global stress factors. *J. Mol. Biol.* **313:** 1.

Duhig T., Ruhrberg C., Mor O., and Fried M. 1998. The human Surfeit locus. *Genomics* **52:** 72.

Edgar A.J. 2003. The gene structure and expression of human ABHD1: Overlapping polyadenylation signal sequence with Sec12. *BMC Genomics* **4:** 18.

Emes R.D., Goodstadt L., Winter E.E., and Ponting C.P. 2003. Comparison of the genomes of human and mouse lays the foundation of genome zoology. *Hum. Mol. Genet.* **12:** 701.

Gish W. and States D.J. 1993. Identification of protein coding regions by database similarity search. *Nat. Genet.* **3:** 266.

Godley L.A., Lai F., Liu J., Zhao N., and Le Beau M.M. 1999. TTID: A novel gene at 5q31 encoding a protein with titin-like features. *Genomics* **60:** 226.

Greally J.M. 2002. Short interspersed transposable elements (SINEs) are excluded from imprinted regions in the human genome. *Proc. Natl. Acad. Sci.* **99:** 327.

Guo L., Schreiber T.H., Weremowicz S., Morton C.C., Lee C., and Zhou J. 2000. Identification and characterization of a novel polycystin family member, polycystin-L2, in mouse and human: Sequence, expression, alternative splicing, and chromosomal localization. *Genomics* **64:** 241.

Harrington J.J., Sherf B., Rundlett S., Jackson P.D., Perry R., Cain S., Leventhal C., Thornton M., Ramachandran R., Whittington J., Lerner L., Costanzo D., McElligott K., Boozer S., Mays R., Smith E., Veloso N., Klika A., Hess J., Cothren K., Lo K., Offenbacher J., Danzig J., and Ducar M. 2001. Creation of genome-wide protein expression libraries using random activation of gene expression. *Nat. Biotechnol.* **19:** 440.

Hildebrandt M. and Nellen W. 1992. Differential antisense transcription from the *Dictyostelium* EB4 gene locus: Implications on antisense-mediated regulation of mRNA stability. *Cell* **69:** 197.

Hubbard T., Barker D., Birney E., Cameron G., Chen Y., Clark L., Cox T., Cuff J., Curwen V., Down T., Durbin R., Eyras E., Gilbert J., Hammond M., Huminiecki L., Kasprzyk A., Lehvaslaiho H., Lijnzaad P., Melsopp C., Mongin E., Pettett R., Pocock M., Potter S., Rust A., Schmidt E., Searle S., Slater G., Smith J., Spooner W., Stabenau A., Stalker J., Stupka E., Ureta-Vidal A., Vastrik I., and Clamp M. 2002. The Ensembl genome database project. *Nucleic Acids Res.* **30:** 38.

Jarmuz A., Chester A., Bayliss J., Gisbourne J., Dunham I., Scott J., and Navaratnam N. 2002. An anthropoid-specific locus of orphan C to U RNA-editing enzymes on chromosome 22. *Genomics* **79:** 285.

Jordan I.K., Makarova K.S., Spouge J.L., Wolf Y.I., and Koonin E.V. 2001. Lineage-specific gene expansions in bacterial and archaeal genomes. *Genome Res.* **11:** 555.

Kapranov P., Cawley S.E., Drenkow J., Bekiranov S., Strausberg R.L., Fodor S.P., and Gingeras T.R. 2002. Large-scale transcriptional activity in chromosomes 21 and 22. *Science* **296:** 916.

Kawashima I., Mita-Honjo K,, and Takiguchi Y. 1992. Characterization of the primate-specific repetitive DNA element MER1. *DNA Seq.* **2:** 313.

Kent W.J., Sugnet C.W., Furey T.S., Roskin K.M., Pringle T.H., Zahler A.M., and Haussler D. 2002. The human genome browser at UCSC. *Genome Res.* **12:** 996.

King M.C. and Wilson A.C. 1975. Evolution at two levels in humans and chimpanzees. *Science* **188:** 107.

Kiyosawa H., Yamanaka I., Osato N., Kondo S., and Hayashizaki Y. 2003. Antisense transcripts with FANTOM2 clone set and their implications for gene regulation. *Genome Res.* **13:** 1324.

Lai F., Godley L.A., Joslin J., Fernald A.A., Liu J., Espinosa R., III, Zhao N., Pamintuan L., Till B.G., Larson R.A., Qian Z., and Le Beau M.M. 2001. Transcript map and comparative analysis of the 1.5-Mb commonly deleted segment of human 5q31 in malignant myeloid diseases with a del(5q). *Genomics* **71:** 235.

Lander E.S. et al. (members of the International Human Genome Sequencing Consortium). 2001. Initial sequencing and analysis of the human genome. *Nature* **409:** 860.

Lee M.K., Lynch E.D., and King M.C. 1998. SeqHelp: A program to analyze molecular sequences utilizing common computational resources. *Genome Res.* **8:** 306.

Lee R.C. and Ambros V. 2001. An extensive class of small RNAs in *Caenorhabditis elegans*. *Science* **294:** 862.

Lespinet O., Wolf Y.I., Koonin E.V., and Aravind L. 2002. The role of lineage-specific gene family expansion in the evolution of eukaryotes. *Genome Res.* **12:** 1048.

Lipovich L., Hughes A.L., King M.C., Abkowitz J.L., and Quigley J.G. 2002. Genomic structure and evolutionary context of the human feline leukemia virus subgroup C receptor (hFLVCR) gene: Evidence for block duplications and de novo gene formation within duplicons of the hFLVCR locus. *Gene* **286:** 203.

McDermid H.E. and Morrow B.E. 2002. Genomic disorders on 22q11. *Am. J. Hum. Genet.* **70:** 1077.

Millar R., Conklin D., Lofton-Day C., Hutchinson E., Troskie B., Illing N., Sealfon S.C., and Hapgood J. 1999. A novel human GnRH receptor homolog gene: Abundant and wide tissue distribution of the antisense transcript. *J. Endocrinol.* **162:** 117.

Misra S., Crosby M.A., Mungall C.J., Matthews B.B., Campbell K.S., Hradecky P., Huang Y., Kaminker J.S., Millburn G.H., Prochnik S.E., Smith C.D., Tupy J.L., Whitfied E.J., Bayraktaroglu L., Berman B.P., Bettencourt B.R., Celniker S.E., de Grey A.D., Drysdale R.A., Harris N.L., Richter J., Russo S., Schroeder A.J., Shu S.Q., Stapleton M., Yamada C., Ashburner M., Gelbart W.M., Rubin G.M., and Lewis S.E. 2002. Annotation of the *Drosophila melanogaster* euchromatic genome: A systematic review. *Genome Biol.* **3:** RESEARCH0083.

Moriyama K., Hayashida K., Shimada M., Nakano S., Nakashima Y., and Fukumaki Y. 2003. Antisense RNAs transcribed from the upstream region of the precore/core promoter of hepatitis B virus. *J. Gen. Virol.* **84:** 1907.

Noonan J.P., Li J., Nguyen L., Caoile C., Dickson M., Grimwood J., Schmutz J., Feldman M.W., and Myers R.M. 2003. Extensive linkage disequilibrium, a common 16.7-kilobase deletion, and evidence of balancing selection in the human protocadherin alpha cluster. *Am. J. Hum. Genet.* **72:** 621.

Numata K., Kanai A., Saito R., Kondo S., Adachi J., Wilming L.G., Hume D.A., Hayashizaki Y., and Tomita M. 2003. Identification of putative noncoding RNAs among the RIKEN mouse full-length cDNA collection. *Genome Res.* **13:** 1301.

Ohinata Y., Sutou S., and Mitsui Y. 2003. Peas-Mea1-Ppp2r5d overlapping gene complex: A transposon mediated-gene formation in mammals. *DNA Res.* **10:** 79.

Okazaki Y. and Hume D.A. 2003. A guide to the mammalian genome. *Genome Res.* **13:** 1267.

Olson M.V. and Varki A. 2003. Sequencing the chimpanzee genome: Insights into human evolution and disease. *Nat. Rev. Genet.* **4:** 20.

Rinn J.L., Euskirchen G., Bertone P., Martone R., Luscombe N.M., Hartman S., Harrison P.M., Nelson F.K., Miller P., Gerstein M., Weissman S., and Snyder M. 2003. The transcriptional activity of human chromosome 22. *Genes Dev.* **17:** 529.

Shannon M., Hamilton A.T., Gordon L., Branscomb E., and Stubbs L. 2003. Differential expansion of zinc-finger transcription factor loci in homologous human and mouse gene clusters. *Genome Res.* **13:** 1097.

Shendure J. and Church G.M. 2002. Computational discovery of sense-antisense transcription in the human and mouse genomes. *Genome Biol.* **3:** RESEARCH0044.

Shoemaker D.D., Schadt E.E., Armour C.D., He Y.D., Garrett-Engele P., McDonagh P.D., Loerch P.M., Leonardson A., Lum P.Y., Cavet G., Wu L.F., Altschuler S.J., Edwards S., King J., Tsang J.S., Schimmack G., Schelter J.M., Koch J., Ziman M., Marton M.J., Li B., Cundiff P., Ward T., Castle J., Krolewski M., Meyer M.R., Mao M., Burchard J., Kidd M.J., Dai H., Phillips J.W., Linsley P.S., Stoughton R., Scherer S., and Boguski M.S. 2001. Experimental annotation of the human genome using microarray technology. *Nature* **409:** 922.

Sleutels F., Zwart R., and Barlow D.P. 2002. The non-coding Air RNA is required for silencing autosomal imprinted genes. *Nature* **415:** 810.

Stajich J.E., Block D., Boulez K., Brenner S.E., Chervitz S.A., Dagdigian C., Fuellen G., Gilbert J.G., Korf I., Lapp H., Lehvaslaiho H, Matsalla C., Mungall C.J., Osborne B.I., Pocock M.R., Schattner P., Senger M., Stein L.D., Stupka E., Wilkinson M.D., and Birney E. 2002. The Bioperl toolkit: Perl modules for the life sciences. *Genome Res.* **12:** 1611.

Strausberg R.L., Camargo A.A., Riggins G.J., Schaefer C.F., de Souza S.J., Grouse L.H., Lal A., Buetow K.H., Boon K., Greenhut S.F., and Simpson A.J. 2002. An international database and integrated analysis tools for the study of cancer gene expression. *Pharmacogenomics J.* **2:** 156.

Stuart J.J., Egry L.A., Wong G.H., and Kaspar R.L. 2000. The 3′ UTR of human MnSOD mRNA hybridizes to a small cytoplasmic RNA and inhibits gene expression. *Biochem. Biophys. Res. Commun.* **274:** 641.

Vanhalst K., Kools P., Vanden Eynde E., and van Roy F. 2001. The human and murine protocadherin-beta one-exon gene families show high evolutionary conservation, despite the difference in gene number. *FEBS Lett.* **495:** 120.

Vanhee-Brossollet C. and Vaquero C. 1998. Do natural antisense transcripts make sense in eukaryotes? *Gene* **211:** 1.

Wang X., Su H., and Bradley A. 2002. Molecular mechanisms governing Pcdh-gamma gene expression: Evidence for a multiple promoter and cis-alternative splicing model. *Genes Dev.* **16:** 1890.

Wu Q., Zhang T., Cheng J.F., Kim Y., Grimwood J., Schmutz J., Dickson M., Noonan J.P., Zhang M.Q., Myers R.M., and Maniatis T. 2001. Comparative DNA sequence analysis of mouse and human protocadherin gene clusters. *Genome Res.* **11:** 389.

Yelin R., Dahary D., Sorek R., Levanon E.Y., Goldstein O., Shoshan A., Diber A., Biton S., Tamir Y., Khosravi R., Nemzer S., Pinner E., Walach S., Bernstein J., Savitsky K., and Rotman G. 2003. Widespread occurrence of antisense transcription in the human genome. *Nat. Biotechnol.* **21:** 379.

Young J.M., Friedman C., Williams E.M., Ross J.A., Tonnes-Priddy L., and Trask B.J. 2002. Different evolutionary processes shaped the mouse and human olfactory receptor gene families. *Hum. Mol. Genet.* **11:** 535.

# Positive Selection in the Human Genome Inferred from Human–Chimp–Mouse Orthologous Gene Alignments

A.G. CLARK,* S. GLANOWSKI,† R. NIELSEN,‡ P. THOMAS,¶ A. KEJARIWAL,¶ M.J. TODD,‡ D.M. TANENBAUM,§ D. CIVELLO,** F. LU,§ B. MURPHY,† S. FERRIERA,† G. WANG,† X. ZHENG,¶ T.J. WHITE,** J.J. SNINSKY,** M.D. ADAMS,§,†† AND M. CARGILL**,††

*Molecular Biology & Genetics, Cornell University, Ithaca, New York 14853; †Applied Biosystems, Rockville, Maryland 20850; ‡Biological Statistics & Computational Biology, Cornell University, Ithaca, New York 14853; ¶Protein Informatics, Celera Genomics, Foster City, California 94404; §Celera Genomics, Rockville, Maryland 20850; **Celera Diagnostics, Alameda, California 94502.

The availability of genomic sequence from diverse organisms allows the opportunity to identify genes that have undergone evolutionary divergence from our most recent common ancestors. By fitting aligned DNA sequences from multiple species to models of sequence divergence it is possible to distinguish divergence due to random drift from that caused by nonneutral processes such as natural selection. The key to this problem is to realize that nucleotide sites can be partitioned a priori according to whether substitutions at these sites change the encoded amino acid or are silent. Under neutrality, these two types of substitutions are expected to be distributed at random, and a variety of tests have been devised to test this null hypothesis. The identification of genes that have undergone positive Darwinian evolution (inferred from an excess of amino acid-changing substitutions) might lead to hypotheses of physiological mechanisms that underlie the specialization of species and their reproductive isolation. Furthermore, discovery of genes that appear to show adaptive evolution in humans may lead to the identification of genes important in human disease.

Although humans and chimpanzees differ by only 1.2% in their coding regions (Chen and Li 2001; Ebersberger et al. 2002), this small level of sequence divergence can still provide clues to adaptive evolution. However, the addition of a third, out-group species provides significantly more information. By adding the orthologous mouse sequence to each gene alignment, it is possible to identify the most likely ancestral allele and therefore information about the direction of DNA substitutions. If human and chimpanzee differ at a nucleotide position, and, say, the chimpanzee and mouse nucleotides are identical at this site, then we can infer that the change occurred most likely on the human lineage after the divergence from our common ancestor with chimpanzee. This logic is formalized and made statistically rigorous by maximum likelihood procedures that are widely available. In this paper, we analyze alignments of 7,645 chimpanzee gene sequences to their unambiguous human and mouse orthologs and identify genes that appear to have undergone adaptive evolution in the human genome.

††Present address: Department of Genetics, Case Western Reserve University, 10900 Euclid Avenue, Cleveland, Ohio 44106.

## IDENTIFICATION OF HUMAN–MOUSE ORTHOLOGS

An accurate and unambiguous identification of human–mouse orthologous gene pairs is necessary for the evolutionary analysis. Incorrectly paired orthologs, paralogous proteins, and inaccurate annotation can all corrupt the evolutionary analysis described here. Our initial set of orthologs was taken from Mural et al. (2002), who had identified 32,598 transcript and 21,638 gene pairs from an analysis of Celera's human and mouse genome assemblies and annotations (Venter et al. 2001; Mural et al. 2002). Each gene pair was scored for four lines of evidence of orthology: sequence identity from tblastx, syntenic anchor or syntenic block evidence, or shared PANTHER protein family classification (Thomas et al. 2003b). This ortholog set provided unambiguous transcript pairs, but there were ambiguities when the transcripts were collapsed into gene pairs. To avoid duplicate evolutionary analysis of the same base, we derived a subset of unambiguous gene orthologs by selecting the gene pair with the most lines of evidence or the pair with the highest tblastx identity. Furthermore, we required that each ortholog be located in a human–mouse syntenic block (Mural et al. 2002), reducing the entire set to 14,104 gene/transcript pairs (85% with tblastx evidence, 90% with syntenic anchor evidence, and 75% with shared PANTHER family evidence; 9,017 pairs had PANTHER assignments). The set of 14,104 mouse–human orthologs was further reduced to 7,645 for methodological reasons (Table 1). The final breakdown of evidence classes was 87% with tblastx evidence, 90% with syntenic anchor evidence, and 80% with shared PANTHER family evidence (5,136 pairs had PANTHER assignments).

We compared the set of 7,645 human–mouse orthologs analyzed in this paper with orthologs from other sources (Table 2). Specifically, April 2003 downloads of mouse–human orthologs from HomoloGene (ftp.ncbi.nih.gov/pub/HomoloGene/hmlg.ftp), Homologous sequence pairs (ftp.ncbi.hih.gov/refseq/LocusLink/homol_seq_pairs.gz), Homology Map (ftp.ncbi.nlm.nih.gov/Homolgy), and the Mouse Genome Database (ftp.informatics.jax.org/ pub/reports.html) were used for validation. We compared the concordance rate of mouse–hu-

**Table 1.** Derivation of 7645 Human–Mouse Orthologs

| Number of genes | Number of transcripts | Justification for removal |
|---|---|---|
| 21,638 | 32,598 | starting set of human–mouse orthologous pairs |
| –7,534 | –18,494 | removal of ambiguous gene pairs |
| –992 | –992 | removal of genes with an absence of human PCR sequence data |
| –424 | –424 | removal of genes with an absence of chimp PCR sequence data |
| –426 | –426 | removal of genes that failed chimp alignment QC |
| –3,821 | –3,821 | removal of genes that failed mouse alignment QC |
| –796 | –796 | removal of genes with less than 50 amino acids between human, chimp, and mouse |
| = 7,645 | = 7,645 | final set of human–mouse orthologs analyzed |

man orthologs between the different sets and found that the data sets were 97% identical (Table 2). Manual examination of the discordant pairs indicated that the discordances were due to methodological issues rather than evidence supporting the ortholog.

## CONSTRUCTION OF HUMAN–MOUSE–CHIMP ALIGNMENTS

The chimpanzee sequence was produced using 201,805 primer pairs designed to 23,363 human coding sequences based on expertly annotated genes in Celera's human genome assembly (Venter et al. 2001). The primer pairs covered 27.6 Mb of human coding sequence, with the coding sequences of most gene families covered at 92% of the genes. Primer pairs were amplified in one male common chimpanzee (*Pan troglodytes*) (4X0033, Southwest National Primate Research Center) and 39 human females (19 African-Americans and 20 Caucasians, Coriell Cell Repositories), and were sequenced using standard sequence chemistry. Approximately 85% of the human amplicons and 75% of the chimp amplicons resulted in good-quality sequence data that could be analyzed further.

Quality-trimmed chimp traces for a human gene were blasted against human exon sequence (Venter et al. 2001). Matches were ordered by decreasing gap count and placed in exon order to create virtual chimp transcripts. Care was taken to maintain the reading frame by internal trimming of low-quality bases and insertion of unambiguous placeholders for unmatched human exon segments. After assembly, 73% of the human coding sequence was covered with chimp sequence. The gaps in this coverage tended to be spread across genes fairly uniformly. In particular, every gene recovered at least some chimpanzee sequence.

The mouse sequences were originally derived from Celera's mouse genome assembly (Mural et al. 2002). Since the gene annotations were not expertly reviewed, we attempted to improve the automated mouse annotation by generating a mouse transcript based on human–mouse alignments. For human exons that appeared to be "missing" from the orthologous mouse transcript (identified by blasting the human exon set for a gene to the orthologous mouse transcript), we blasted the human exon sequence to surrounding mouse genome sequence. Reconstruction of each mouse transcript based on the human–mouse alignments yielded a "humanized" mouse transcript. Compared to human proteins, humanized mouse transcripts had the same level of protein identity (average 80%) as computationally predicted mouse tran-

**Table 2.** Comparison of Different Mouse–Human Ortholog Sets

| | Number of orthologs | Number of orthologs with NM accessions and unambiguous gene pairs | Number of comparisons possible (either human or mouse gene is present in both data sets) | Percent of data set that can be compared | Percent of orthologs that are the same |
|---|---|---|---|---|---|
| JAX[a] | 8326 | 4116 | 1341 | 33 | 97 |
| HomoloGene[b] | 2692 | 2691 | 1337 | 50 | 97 |
| JAX[a] | 8326 | 4116 | 939 | 23 | 99 |
| Homol seq pairs[c] | 3460 | 1977 | 939 | 47 | 99 |
| HomoloGene[b] | 2692 | 2691 | 715 | 27 | 99 |
| Homol seq pairs[c] | 3460 | 1977 | 715 | 36 | 99 |
| JAX[a] | 8326 | 4116 | 3041 | 74 | 99 |
| Homology Map[d] | 9136 | 5434 | 3041 | 56 | 99 |
| JAX[a] | 8326 | 4116 | 2212 | 54 | 97 |
| This paper | 7645 | 5045 | 2216 | 44 | 97 |
| HomoloGene[b] | 2692 | 2691 | 1527 | 57 | 97 |
| This paper | 7645 | 5045 | 1534 | 30 | 97 |

[a]Mouse Genome Informatics Web Site, The Jackson Laboratory, Bar Harbor, Maine. (ftp.informatics.jax.org/pub/reports.html, 4/2003); at least two lines of evidence.

[b]NCBI Homologene (ftp.ncbi.nih.gov/pub/HomoloGene/hmlg.ftp; 4/2003).

[c]NCBI Homol_seq_pairs (ftp.ncbi.nih/gov/refseq/LocusLink/homol_seq_pairs.gz, 4/2003).

[d]NCBI Homology Map (www.ncbi.nlm.nih.gov/Homology, 4/2003).

scripts, but generated twice as many human–mouse alignments that passed quality control.

The multiple alignment program ClustalW, run with default parameters from the ClustalX package v1.83 α (Thompson et al. 1997), was used to align human, chimp, and mouse coding sequences. Mouse–human and chimp–human alignments were independently examined for the introduction of alignment gaps that would result in a frame-shifted human protein. Although these alignment gaps could represent real differences between two species, other causes (incorrect base calls, annotation, or ortholog inference) are more likely, especially between human vs. mouse alignments. Only alignments that produced either zero insertion/deletions, or those that had gaps whose length was a multiple of three bases were analyzed further. The alignment failure rate for chimp–human and mouse–human pairs was 3.3% and 31%, respectively (Table 1). The alignments passing quality control were converted into Phylip format (Felsenstein 1981) and are available at http://panther.celera.com/appleraHCM_alignments/index.jsp for download.

## EVOLUTIONARY MODELS

A commonly used measure to identify genes undergoing adaptive protein evolution involves comparing the ratio of nonsynonymous to synonymous substitution rates for each gene ($d_N/d_S$). If selection pressures favor proteins with altered amino acid sequence, then nonsynonymous changes are favored at the nucleotide level, and $d_N/d_S$ will be greater than 1. However, since humans and chimps have a relatively recent common ancestor, the overall sequence divergence is only about 1.2% (Chen and Li 2001), and coding regions show less than half this level of divergence (Shi et al. 2003). This results in wide variation from gene to gene in the absolute count of synonymous nucleotide changes, and low values of $d_S$ in the denominator result in large variance in $d_N/d_S$ ratios. Requiring that the predicted protein sequence must have diverged at more than one residue from the most parsimonious ancestral sequence can partly control this variance. There were 363 human genes (4.7% of the total) that had more than one amino acid difference and a $d_N/d_S > 1$. Formal statistical models to test for significant departure from a neutral evolutionary model were also applied to go beyond this ad hoc description of the genes.

A model specifying sequence divergence with parameters fitted by maximum likelihood (Felsenstein 1981) can be applied to multispecies sequence alignments and an evolutionary tree that describes the ancestral history of those sequences. This approach can be extended by allowing separate parameters specifying rates of synonymous and nonsynonymous substitution (Goldman and Yang 1994; Muse and Gaut 1994), variability among amino acid residues in their degree of constraint, and lineage-specific differences in the divergence rates (Yang and Nielsen 2002).

In the first of two evolutionary models applied to the set of 7,645 alignments, we applied a classical test of the null hypothesis of $d_N/d_S = 1$ in the human lineage (Nielsen and Yang 1998; Yang 2002). This test may be rejected if $d_N/d_S > 1$, showing evidence of positive selection, or $d_N/d_S < 1$, showing strong conservation of the gene in the human lineage. The neutral null hypothesis of model 1 was rejected by 72 genes (0.94%) at $p < 0.001$, 414 (5.4%) at $p < 0.01$, and 1216 genes (15.9%) at $p < 0.05$. There were 6 genes (0.08%) with $p < 0.05$ and $d_N/d_S > 1$.

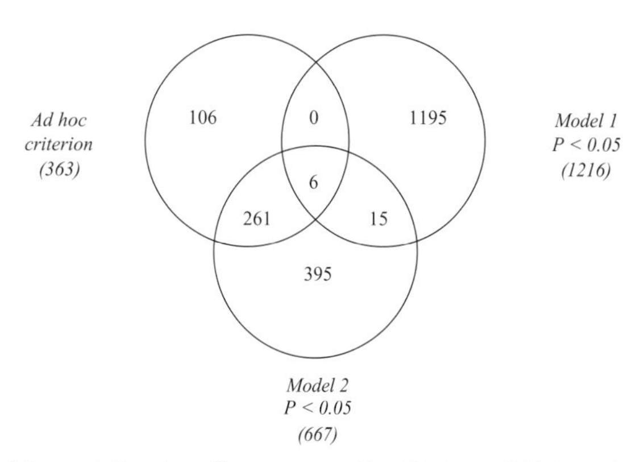


**Figure 1.** Overlap of human genes identified as exhibiting a signature of positive selection by the ad hoc criterion, a model testing for departure of $d_N/d_S$ from 1 (Model 1) and a model testing for excess nonsynonymous substitution within a domain of the protein in the human lineage only (Model 2).

The second formal model applied to test for positive selection is modified from Yang and Nielsen (2002) and allows variation in the $d_N/d_S$ ratio among lineages and among sites at the same time (see also Yang and Swanson 2002). In this method (Model 2), a likelihood ratio test of the hypothesis of neutrality is performed by comparing the likelihood values for two hypotheses. Under the null hypothesis, it is assumed that all sites are either neutral ($d_N/d_S = 1$) or evolve under negative selection ($d_N/d_S < 1$). Under the alternative hypothesis, some of the sites are allowed to evolve by positive selection in the human lineage only. The neutral null hypothesis of Model 2 was rejected by 28 genes (0.37%) at $p < 0.001$, 178 genes (2.3%) at $p < 0.01$, and 667 genes (8.7%) at $p < 0.05$.

The overlap between these three sets of genes is high (Fig. 1), but differences reflect the different attributes of the data that the tests consider. For example, small genes or genes with few substitutions may be flagged by the ad hoc criterion, but not attain statistical significance by the evolutionary models. Importantly, Model 2 can detect cases where a portion of the protein (perhaps a protein domain) is undergoing positive selection, but the overall $d_N/d_S$ may not be elevated, resulting in those genes being missed by the ad hoc criterion and by Model 1. For this reason, the remainder of the analysis considers only Model 2 test results.

## THE IMPACT OF LOCAL SEQUENCE COMPOSITION

Genome sequence composition such as GC content, gene density, repeat density, and local recombination rate can influence patterns and rates of sequence divergence (Hellmann et al. 2003; Webster et al. 2003). Before we go

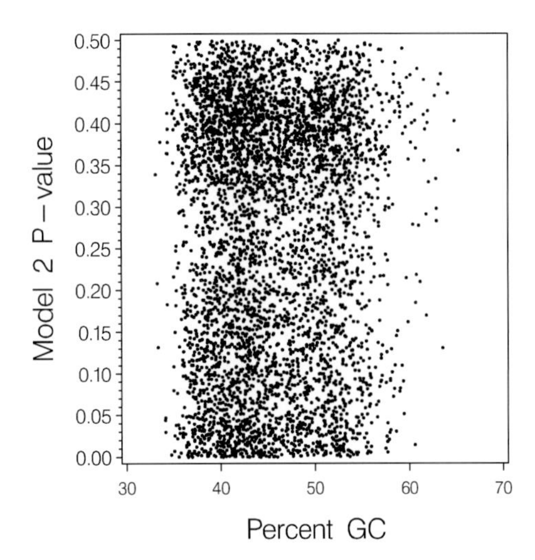


**Figure 2.** The relationship between local GC content for each of the 7,645 genes in this study and the Model 2 $p$-value, expressing the probability that the gene displays a signature of accelerated protein evolution in humans. The lack of correlation is one indication that base composition variation is not spuriously driving the test results.

into the details of which genes appear to exhibit unusual evolutionary patterns in the human lineage, it is important to test whether our model is particularly sensitive to local base composition changes. Although the maximum likelihood estimation should, in principle, take into account variation in base composition, it is easy to directly examine whether any residual correlation exists between GC content and the test results. For the 7,645 set the synonymous substitution rate was significantly correlated with: GC content (0.164, $p < 0.0001$), local recombination rate (Kong et al. 2002) (0.100, $p < 0.001$), and LINE element density (–0.091, $p < 0.0001$). None of these factors is significantly correlated with the nonsynonymous substitution rate or with Model 2 $p$-values (Fig. 2).

Segmental duplications (Bailey et al. 2002) do not appear to cause distortions in our analysis, since genes with close duplicates were underrepresented in our set due to the requirement of strict human–mouse orthology. This is confirmed by the observation that 10% of the genes in our set have at least one coding base pair in a segmental duplication (http://humanparalogy.gene.cwru.edu/SDD, UCSC Aug 2001 release) compared to 13% of genes in the entire human genome (Bailey et al. 2002). There is not an enrichment of segmental duplicated genes in the tail of the Model 2 $p$-value distribution.

## CLASSES OF GENES WITH ATYPICAL SEQUENCE DIVERGENCE

Given the large number of genes that exhibit a signature of natural selection along the human lineage, it becomes important to identify common features of these genes. A powerful approach to organizing such genetic information is to classify genes based on the inferred biological process in which they function or based on the molecular attributes of the gene product. In this analysis, we employ the assignment of the 7,645 genes into classes

**Table 3.** Biological Processes Showing Significant ($p < 0.01$) Positive Selection in the Human Lineage

| Biological Process | Number of genes[a] | P value[a] |
|---|---|---|
| Olfaction | 48 | 0 |
| Sensory perception | 146 (98) | 0 (0.026) |
| Cell surface receptor-mediated signal transduction | 505 (464) | 0 (0.0386) |
| Chemosensory perception | 54 (6) | 0 (0.1157) |
| Nuclear transport | 26 | 0.0003 |
| G-protein mediated signaling | 252 (211) | 0.0003 (0.1205) |
| Signal transduction | 1030 (989) | 0.0004 (0.0255) |
| Amino acid catabolism | 16 | 0.0041 |

[a]Excluding olfactory receptor genes.

based on biological processes and molecular functions using the PANTHER classification system (Thomas et al. 2003b; http://panther.celera.com), which is similar to categories in the Gene Ontology classification (http://www.geneontology.org). Functional classifications of genes were used only if the human protein sequence had a significant score to a PANTHER Hidden Markov Model (NLL-NULL score < –0.50). The accuracy of gene-function associations is shown to be comparable for well-curated model organism databases (Thomas et al. 2003a). For each functional category, a cumulative distribution of Model 2 $p$-values was compared to the cumulative distribution of all genes using the Mann-Whitney U test. Categories of biological processes and molecular functions with $p < 0.01$ under this Mann-Whitney U test were considered significant.

In the human lineage, genes involved in two different biological process, olfaction and amino acid catabolism, have a significant tendency to show human-specific accelerations of protein evolution (Table 3, Fig. 3). The olfaction biological process contains mostly olfactory receptors (OR), and it is reasonable to hypothesize that the different lifestyles of human, mouse, and chimp might lead to selective pressure on these genes. Since there has been a rapid rate of loss of function of OR genes in hu-

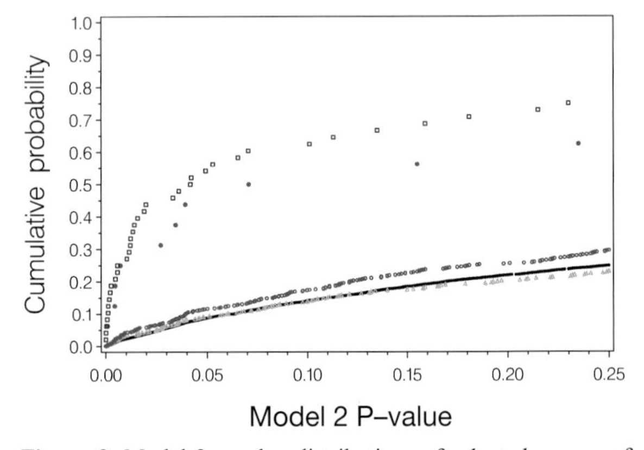


**Figure 3.** Model 2 $p$-value distributions of selected groups of genes. The plot gives the cumulative fraction of selected biological processes showing the excess of cases of significant positive selection in genes for olfaction (*open squares*), amino acid catabolism (*closed circles*), and Mendelian disease genes (*open circles*) relative to the overall distribution of genes (*dots*, fused into one line). The distribution of developmental genes (*open triangles*) that do not show significant excess is shown for comparison.

mans (Gilad et al. 2003), which would be expected to show increased nonsynonymous substitution, we verified that most of the OR genes in our set are bona fide genes (http://bioinformatics.weizmann.ac.il/HORDE). Our results, suggesting that many of the still active OR genes display a signature of positive selection, are supported by the observation that there is a discordance between levels of human polymorphisms in many OR genes from interspecific divergence (Gilad et al. 2003).

Genes involved in amino acid catabolism also show evidence of adaptive evolution. It is possible that the radical change in diet between human and chimps might be partly responsible for this pattern of divergence. Of the eight protein catabolism genes (GSTZ1, HGD, PAH, BCKDHA, PCCB, HAL, ALDH6A1, AMT) with the lowest Model 2 $p$-values (http://panther.celera.com/appleraHCM_alignments/index.jsp), all have been implicated in human metabolic disorders. In fact, there is a significant tendency ($p < 0.001$, Kolmogorov-Smirnov test) for genes implicated in human Mendelian disorders (http://www.ncbi.nlm.nih.gov/htbin-post/Omim/getmorbid) to exhibit significant ($p < 0.01$) Model 2 $p$-values.

## GENES WITH ATYPICAL SEQUENCE DIVERGENCE

Of the 667 genes showing evidence of adaptive evolution under Model 2 ($p < 0.05$), there are several categories of genes, such as developmental, hearing, and reproductive, that are particularly interesting considering the physiological differences between humans and chimps (Table 4). Considering the importance and general conservation across evolution of these traits, it is not surprising that the classes as a whole do not show adaptive evolution (Fig. 3). However, given the important functions of genes in these classes, each one may account disproportionately for specific phenotypic differences.

Most of the human developmental genes that appear to

**Table 4.** Selected Genes Showing Evidence of Adaptive Evolution in the Human Lineage

| Biological process | Gene symbol | Gene name | Model 2 $p$-value |
|---|---|---|---|
| Development | ALPL | alkaline phosphatase, liver/bone/kidney | 3.69E-03 |
| | BMP4 | bone morphogenetic protein 4 | 2.50E-02 |
| | CDX4 | caudal type homeobox transcription factor 4 | 1.57E-03 |
| | DIAPH1 | diaphanous homolog 1 (*Drosophila*) | 2.01E-02 |
| | EPHB6 | EphB6 | 3.53E-02 |
| | EYA1 | eyes absent homolog 1 (*Drosophila*) | 6.49E-03 |
| | EYA4 | eyes absent homolog 4 (*Drosophila*) | 3.48E-03 |
| | FOXI1 | forkhead box I1 | 2.87E-03 |
| | FOXP2 | forkhead box P2 | 2.67E-03 |
| | HOXA5 | homeobox A5 | 2.16E-02 |
| | HOXD4 | homeobox D4 | 4.55E-03 |
| | MEOX2 | mesenchyme homeobox 2 | 3.45E-02 |
| | MGP | matrix Gla protein | 3.94E-02 |
| | MIXL1 | Mix1 homeobox-like 1 (*Xenopus laevis*) | 4.61E-02 |
| | MMP20 | matrix metalloproteinase 20 (enamelysin) | 3.18E-02 |
| | NEUROG1 | neurogenin 1 | 3.57E-02 |
| | NLGN3 | neuroligin 3 | 2.80E-02 |
| | NTF3 | neurotrophin 3 | 8.92E-03 |
| | OTOR | otoraplin | 3.46E-02 |
| | PHTF | putative homeodomain transcription factor 1 | 5.24E-02 |
| | PLXNC1 | plexin C1 | 5.93E-03 |
| | POU2F3 | POU domain, class 2, transcription factor 3 | 3.34E-02 |
| Development | SEMA3B | Semaphorin 3B | 3.57E-03 |
| | SIM2 | single-minded homolog 2 (*Drosophila*) | 4.57E-02 |
| | SNAI1 | snail homolog 1 (*Drosophila*) | 6.74E-03 |
| | TECTA | tectorin alpha | 1.57E-05 |
| | TLL2 | tolloid-like 2 | 3.28E-03 |
| | TRAF5 | TNF receptor-associated factor 5 | 6.35E-04 |
| | WHN | winged-helix nude | 5.57E-04 |
| | WIF1 | WNT inhibitory factor 1 | 2.71E-02 |
| | WNT2 | wingless-type MMTV integration site family member 2 | 2.99E-02 |
| Amino acid catabolism | ALDH6A1 | aldehyde dehydrogenase 6 family, member A1 | 2.72E-02 |
| | AMT | aminomethyltransferase (glycine cleavage system protein T) | 3.45E-02 |
| | BCKDHA | branched chain keto acid dehydrogenase E1$\alpha$ polypeptide | 4.64E-03 |
| | GSTZ1 | glutathione transferase $\zeta$ 1 (maleylacetoacetate isomerase) | 9.04E-04 |
| | HAL | histidine ammonia-lyase | 7.18E-03 |
| | HGD | homogentisate 1,2-dioxygenase (homogentisate oxidase) | 4.39E-03 |
| | PAH | phenylalanine hydroxylase | 7.08E-02 |
| | PCCB | propionyl Coenzyme A carboxylase, beta polypeptide | 3.93E-02 |
| Reproduction | GNRHR | gonadotropin-releasing hormone receptor | 8.48E-03 |
| | MTNR1A | melatonin receptor 1A | 3.97E-02 |
| | PAPPA | pregnancy-associated plasma protein A | 8.18E-04 |
| | PGR | progesterone receptor | 1.05E-03 |

be under adaptive evolution fall into four main categories: skeletal development, neurogenesis, reproduction, and homeotic transcription factor genes (Table 4). For example, the homeotic transcript factor genes CDX4, HOXA5, HOXD4, MEOX2, POU2F3, MIXL1, and PHTF play key roles in early development and have Model 2 $p$-values less than 0.05. TRAF5 plays a key role in osteoclast proliferation and may be implicated in accelerated growth of the long bones in the leg (Kanazawa et al. 2003). TRAF5 shows adaptive evolution along with six other skeletal developmental genes. At least 10 genes involved in neurogenesis processes, including axonal guidance and synapse remodeling, have low Model 2 $p$-values. For example, the SIM2 transcription factor has been implicated in human Down syndrome and memory defects in mice (Chrast et al. 2000). FOXN1, or winged helix nude, encodes a transcription factor involved in keratin gene expression. Mutations in this gene cause athymia, resulting in a severely compromised immune system. Developmental defects in *Drosophila* and *Caenorhabditis elegans* are also observed when this gene is mutated. A plausible hypothesis is that the relative hairlessness of humans compared to chimps is in part determined by FOXNI. Hypotheses like this are generated in abundance by studies such as this, and an exciting aspect of the work is that such hypotheses are amenable to future testing.

The anatomy and physiology of reproduction are strikingly different between humans and chimpanzees. Several genes involved in pregnancy appear to exhibit nonneutral evolution (Table 4). For example, the progesterone receptor (PGR) is involved with maintenance of the uterus and may be involved in the acrosome reaction (Gadkar et al. 2003). The reproductive hormone receptors GNRHR and MTNR1A also have significant Model 2 $p$-values.

Several genes associated with the development of hearing appear to have undergone adaptive evolution (Table 4). α-Tectorin, which shows the most significant Model 2 $p$-value, plays an important role in the tectorial membrane of the inner ear. When it is mutated, humans show high-frequency hearing loss (Mustapha et al. 1999) and mouse knockout mice are deaf. Other genes under human-specific selection, DIAPH1, FOXI1, and EYA4, cause hearing loss in humans when mutated.

## CONCLUSIONS

There has been considerable interest in obtaining the genome sequence of the chimpanzee, our closest relative, because of the notion that, by comparing our two genomes, it might be possible to infer which genetic differences are responsible for the morphological, physiological, and behavioral factors that differentiate us. At 1% sequence divergence, however, we expect there to be roughly 3 million base pairs of sequence difference, and the discrimination between substitutions that are totally unimportant and substitutions that are causal to our biological differences appears to be a steep challenge. Fortunately, the phylogenetic approach offers a promise to make progress on this problem. With multiple related species arranged on a phylogenetic tree, models of molecular evolution can place the mutations on particular lineages of the tree. This information can be used to infer what DNA sequence changes have occurred specifically along the lineage subsequent to the node representing our common ancestry with chimpanzee, and reflecting changes that occurred in our line of descent since that time. The challenge that remains is that many of these changes will have arisen purely because our population size is finite, and because random mutations, provided they are not too deleterious, may go to fixation by random drift. It is humbling to consider that potentially a large portion of the genomic differences between humans and chimps have arisen by such a purely neutral process. If one asks, "What are the genes that make us human?", these random changes may surely be an important class of genes that carry this label.

In this paper, we apply methods that have been used by many others to infer which genes have been undergoing positive or adaptive evolution. The idea is based on the relative rates of substitution at silent (synonymous) and at replacement (nonsynonymous) nucleotide positions in the gene. Strictly neutral genes are expected to have equal rates of substitution for these two classes of sites, while most genes have some selective constraint and show considerable deficit of replacement changes. Formal statistical approaches allow us to test the null hypothesis that the changes are compatible with neutrality, and to make quite incisive tests of alternative hypotheses about the way that selection has acted (Yang and Nielsen 2002).

The application of this inferential approach identified a long list of genes for which it is all too easy to tell an evolutionary story about how these genes are important for human–chimp differences. However, it should be emphasized that these approaches are strictly exploratory and that they really only highlight hypotheses to be tested by additional data collection at several levels. The finding of positive selection in genes such as α-tectorin suggests that there may be differences in the hearing acuity of humans and chimpanzees, and the data available on the subject are too sparse to properly address the issue. This motivates specific hearing tests of chimpanzees, with the idea that aspects of vocal speech may place additional requirements on hearing not faced by speechless chimpanzees. For every gene cited in this paper, there are additional experiments that must be done to solidify the evidence that these genes may be involved in human–chimpanzee differences. In the case of significant differences in biological processes, the case based only on DNA sequences becomes more compelling, mostly because each test involves many genes showing aberrant (nonneutral) behavior. That amino acid catabolism should be a biological process showing rapid adaptive evolution suggests further research into the physiology of digestion of low- versus high-protein diets, and consideration of the differences among primates in diet. Dietary changes are not the only thing that might be driving this difference. Demands on protein synthesis during brain development might be the driver of this signal of past natural selection. Despite all these uncertainties, and the worry that this approach only raises possibilities rather than proving anything, it seems nearly a certainty that comparative genomic methods like this will serve as a powerful generator of hypotheses that will admit further analysis of the differences in all levels of biological function between humans and our near relatives.

## ACKNOWLEDGMENTS

We thank the employees of the Celera Genomics sequencing center for their excellent technical participation; J. Duff, C. Gire, M.A. Rydland, C. Forbes, and B. Small for development and maintenance of software systems, laboratory information management systems, and analysis programs. We also thank S. Hannenhalli and S. Levy for helpful discussions.

## REFERENCES

Bailey J.A., Gu Z., Clark R.A., Reinert K., Samonte R.V., Schwartz S., Adams M.D., Myers E.W., Li P.W., and Eichler E.E. 2002. Recent segmental duplications in the human genome. *Science* **297:** 1003.

Chen F.C. and Li W.-H. 2001. Genomic divergences between humans and other hominoids and the effective population size of the common ancestor of humans and chimpanzees. *Am. J. Hum. Genet.* **68:** 444.

Chrast R., Scott H.S., Madani R., Huber L., Wolfer D.P., Prinz M., Aguzzi A., Lipp H.P., and Antonarakis S.E. 2000. Mice trisomic for a bacterial artificial chromosome with the single-minded 2 gene (Sim2) show phenotypes similar to some of those present in the partial trisomy 16 mouse models of Down syndrome. *Hum. Mol. Genet.* **9:** 1853.

Ebersberger I., Metzler D., Schwarz C., and Pääbo S. 2002. Genomewide comparison of DNA sequences between humans and chimpanzees. *Am. J. Hum. Genet.* **70:** 1490.

Felsenstein J. 1981. Evolutionary trees from DNA sequences: A maximum likelihood approach. *J. Mol. Evol.* **17:** 368.

Frazer K.A., Chen X., Hinds D.A., Pant P.V., Patil N., and Cox D.R. 2003. Genomic DNA insertions and deletions occur frequently between humans and nonhuman primates. *Genome Res.* **13:** 341.

Gadkar S., Shah C.A., Sachdeva G., Samant U., and Puri C.P. 2003. Progesterone receptor as an indicator of sperm function. *Biol. Reprod.* **67:** 1327.

Gilad Y., Man O., Pääbo S., and Lancet D. 2003. Human specific loss of olfactory receptor genes. *Proc. Natl. Acad. Sci.* **100:** 3324.

Goldman N. and Yang Z. 1994. A codon-based model of nucleotide substitution for protein-coding DNA sequences. *Mol. Biol. Evol.* **11:** 725.

Hellmann I., Zöllner S., Enard W., Ebersberger I., Nickel B., and Pääbo S. 2003. Selection on human genes as revealed by comparisons to chimpanzee cDNA. *Genome Res.* **13:** 831.

Kanazawa K., Azuma Y., Nakano H., and Kudo A. 2003. TRAF5 functions in both RANKL- and TNFalpha-induced osteoclastogenesis. *J. Bone Miner. Res.* **18:** 443.

King M.C. and Wilson A.C. 1975. Evolution at two levels in humans and chimpanzees. *Science* **188:** 107.

Kong A., Gudbjartsson D.F., Sainz J., Jonsdottir G.M., Gudjonsson S.A., Richardsson, B., Sigurdardottir S., Barnard J., Hallbeck B., Masson G., Shlien A., Palsson S.T., Frigge M.L., Thorgeirsson T.E., Gulcher J.R., and Stefansson K. 2002. A high-resolution recombination map of the human genome. *Nat. Genet.* **31:** 241.

Mural R.J., Adams M.D., Myers E.W., Smith H.O., Miklos G.L., Wides R., Halpern A., Li P.W., Sutton G.G., Nadeau J., Salzberg S.L., Holt R.A., Kodira C.D., Lu F., Chen L., Deng Z., Evangelista C.C., Gan W., Heiman T.J., Li J., Li Z., Merkulov G.V., Milshina N.V., Naik A.K., and Qi R., et al. 2002. A comparison of whole-genome shotgun-derived mouse chromosome 16 and the human genome. *Science* **296:** 1661.

Muse S.V. and Gaut B.S. 1994. A likelihood approach for comparing synonymous and nonsynonymous nucleotide substitution rates, with application to the chloroplast genome. *Mol. Biol. Evol.* **11:** 715.

Mustapha M., Weil D., Chardenoux S., Elias S., El-Zir E., Beckmann J.S., Loiselet J., and Petit C. 1999. An alpha-tectorin gene defect causes a newly identified autosomal recessive form of sensorineural pre-lingual non-syndromic deafness, DFNB21. *Hum. Mol. Genet.* **8:** 409.

Nielsen R. and Yang Z. 1998. Likelihood models for detecting positively selected amino acid sites and applications to the HIV-1 envelope gene. *Genetics* **148:** 929.

Shi J., Xi H., Wang Y., Zhang C., Jiang Z., Zhang K., Shen Y., Jin L., Zhang K., Yuan W., Wang Y., Lin J., Hua Q., Wang F., Xu S., Ren S., Xu S., Zhao G., Chen Z., Jin L., and Huang W. 2003. Divergence of the genes on human chromosome 21 between human and other hominoids and variation of substitution rates among transcription units. *Proc. Natl. Acad. Sci.* **100:** 8331.

Thomas P.D., Campbell M.J., Kejariwal A., Mi H., Karlak B., Daverman R., Diemer K., Muruganujan A., and Narechania A. 2003a. PANTHER: A library of protein families and subfamilies indexed by function. *Genome Res.* **13:** 2129.

Thomas P.D., Kejariwal A., Campbell M.J., Mi H., Diemer K., Guo N., Ladunga I., Ulitsky-Lazareva B., Muruganujan A., Rabkin S., Vandergriff J.A., and Doremieux O. 2003b. PANTHER: A browsable database of gene products organized by biological function, using curated protein family and subfamily classification. *Nucleic Acids Res.* **31:** 334.

Thompson J.D., Gibson T.J., Plewniak F., Jeanmougin F., and Higgins D.G. 1997. The ClustalX windows interface: Flexible strategies for multiple sequence alignment aided by quality analysis tools. *Nucleic Acids Res.* **24:** 4876.

Venter J.C., Adams M.D., Myers E.W., Li P.W., Mural R.J., Sutton G.G., Smith H.O., Yandell M., Evans C.A., Holt R.A., Gocayne J.D., Amanatides P., Ballew R.M., Huson D.H., Wortman J.R., Zhang Q., Kodira C.D., Zheng X.H., Chen L., Skupski M., Subramanian G., Thomas P.D., Zhang J., Gabor Miklos G.L., and Nelson C., et al. 2001. The sequence of the human genome. *Science* **291:** 1304.

Webster M.T., Smith N.G., and Ellegren H. 2003. Compositional evolution of noncoding DNA in the human and chimpanzee genomes. *Mol. Biol. Evol.* **20:** 278.

Yang Z. 2002. Inference of selection from multiple species alignments. *Curr. Opin. Genet. Dev.* **12:** 688.

Yang Z. and Nielsen R. 2002. Codon-substitution models for detecting molecular adaptation at individual sites along specific lineages. *Mol. Biol. Evol.* **19:** 908.

Yang Z. and Swanson W.J. 2002. Codon-substitution models to detect adaptive evolution that account for heterogeneous selective pressures among site classes. *Mol. Biol. Evol.* **19:** 49.

# mtDNA Variation, Climatic Adaptation, Degenerative Diseases, and Longevity

D.C. WALLACE, E. RUIZ-PESINI, AND D. MISHMAR
*Center for Molecular and Mitochondrial Medicine and Genetics, University of California, Irvine, California 92697-3940*

The mtDNA encodes 13 polypeptides that are critical for mitochondrial energy metabolism. These include 7 of the approximately 46 polypeptides of respiratory complex I (ND1, 2, 3, 4L, 4, 5, 6), 1 of the 11 polypeptides of complex III (cytb), 3 of the 13 polypeptides of complex IV (COI, II, III), and 2 of the 17 polypeptides of complex V (ATP6, 8). In addition, the mtDNA encodes the small and large rRNAs and the 22 tRNAs required for mitochondrial protein synthesis (Fig. 1).

The mitochondrial energy-generating pathway, oxidative phosphorylation (OXPHOS), oxidizes the carbohydrates and fats of our diet with the oxygen that we breathe to generate energy in the form of ATP and heat to maintain our body temperatures. Electrons from dietary calories pass sequentially through complex I, coenzyme Q (CoQ), complex III, cytochrome c, complex IV, and then atomic oxygen (1/2 $O_2$) to produce water ($H_2O$). The energy that is released from this electron transport chain (ETC) is used to pump protons across the mitochondrial inner membrane through complexes I, III, and IV to generate an electrochemical gradient, $\Delta P = \Delta\Psi + \Delta pH$. The potential energy stored in $\Delta P$ is then used to generate ATP by passage through the proton channel in the ATP synthase (complex V). The efficiency with which complexes I, III, and IV pump protons out of the mitochondrial inner membrane and with which the passage of protons back through complex V is converted into ATP is called the coupling efficiency. Highly coupled mitochondria produce the maximum amount of ATP and minimum heat, whereas loosely coupled mitochondria generate less ATP and more heat.

As a toxic by-product of OXPHOS, the mitochondria generate most of the endogenous reactive oxygen species (ROS) of the cell (Kagawa et al. 1999; Wallace and Lott 1999; Kadenbach 2003). The rate of ROS production is tied to the degree to which complex I, coenzyme Q, and complex III are reduced and thus have excess electrons that can be transferred directly to molecular oxygen ($O_2$) to generate superoxide anion ($O_2^-$), the first of the ROS. Superoxide anion is converted to $H_2O_2$ by manganese superoxide dismutase (MnSOD), and $H_2O_2$ is converted to •OH or $H_2O$. The mitochondrial ROS can react with the DNA, proteins, and lipids of the mitochondrion and the nucleus-cytosol. Ultimately, the mitochondria become sufficiently impaired that the cell malfunctions. This leads to the activation of the mitochondrial permeability transition pore (mtPTP), which initiates apoptosis, killing the cell and its defective mitochondria. When sufficient cells in a tissue are lost, organ dysfunction occurs, resulting in degenerative diseases and aging (Wallace 1999, 2001b; Friberg and Wieloch 2002).

The maternally inherited mtDNA has a very high mutation rate, perhaps because of the high oxidative damage that it sustains. As a consequence, pathogenic mtDNA mutations are common (Jacobs 2003). Examples of pathogenic mutations in mtDNA protein synthesis genes include a 12S rRNA mutation at nucleotide position (np) 1555G associated with sensory neural hearing loss, a $tRNA^{Leu(UUR)}$ mutation at np 3243G associated with cardiomyopathy and stroke-like episodes at high percentage of mutant mtDNAs (heteroplasmy), or adult-onset diabetes at a low percentage of heteroplasmy; a $tRNA^{Gln}$ mutation at np 4336C associated with late-onset Alzheimer disease (AD) and Parkinson disease (PD); and a $tRNA^{Lys}$ mutation at np 8344G associated with epilepsy and progressive muscle weakness. Examples of pathogenic protein missense mutations include several complex I mutations (ND1 np 3460A, ND4L np 10663C, ND4 np 11778A, and ND6 np 14484C) that cause Leber hereditary optic neuropathy (LHON), an ND6 np 14459A mutation that causes generalized dystonia and LHON, and the ATP6 np 8993G mutation that can cause retinitis pigmentosa, macular degeneration, mental retardation, and lethal childhood Leigh syndrome, depending on the percentage heteroplasmy (Fig. 1) (Tatuch et al. 1992; Wallace 2001a).

Since nucleotide substitutions occur at random, a wide variety of mutations accumulate over time. Most of these will be deleterious and eliminated by purifying selection, resulting in genetic diseases. Others could be neutral, such as the third codon position changes, and these could accumulate by chance through genetic drift. Occasionally, a mutation could be advantageous in a new environment, becoming established through adaptive selection.

Both neutral and adaptive mtDNA mutations have become established in different populations. For example, an African variant at np 3594T defines a group of related mtDNA haplotypes (haplogroup) specific for Africa (Fig. 1). This variant is common to African haplogroups L0, L1, and L2, which encompass 2/3 of all sub-Saharan African mtDNAs, and are thus designated macro-haplogroup L (Fig. 2). Other variants include a polymorphism at np 7028T in COI that marks European haplogroup H; variants at np 6392 in COI and at np 10310 in

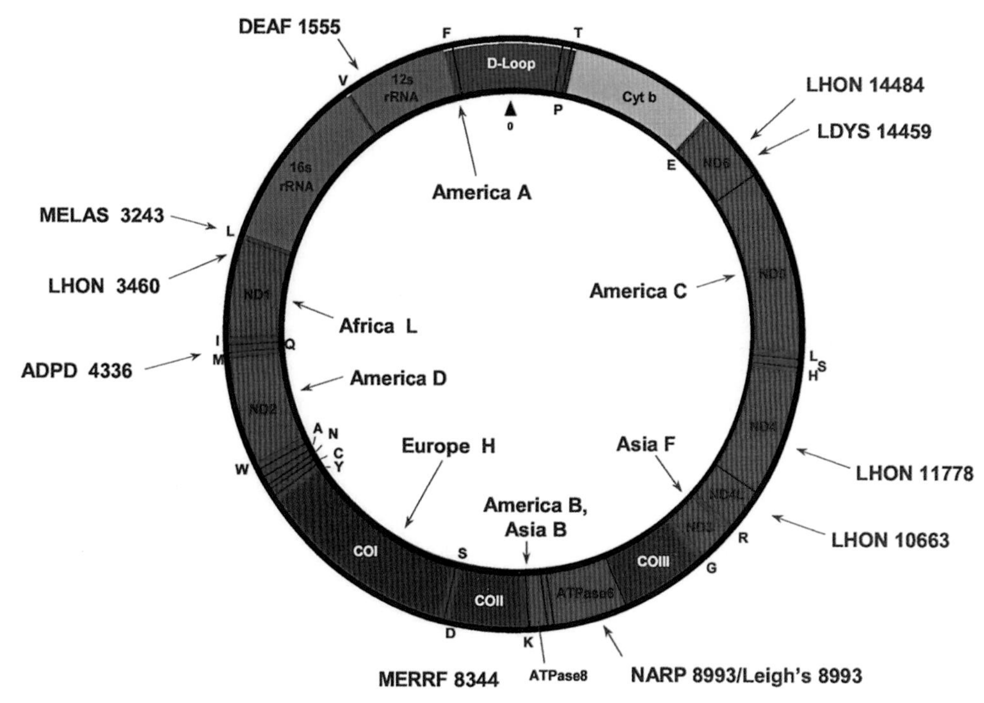


**Figure 1.** The mitochondrial genome—pathological mutations and region-specific polymorphisms. (Adapted from www. mitomap.org.)

ND3 which define the Asian-specific haplogroup F, and variants in the 12S rRNA at np 663G, in the intergenic region between COII and $tRNA^{Lys}$ (9-bp deletion), in ND5 at np 13263G, and in ND2 at np 5178A, that define haplogroups A, B, C, and D, respectively (Figs. 1 and 2). These later haplogroups arose in Asia and crossed into America to found the Native Americans.

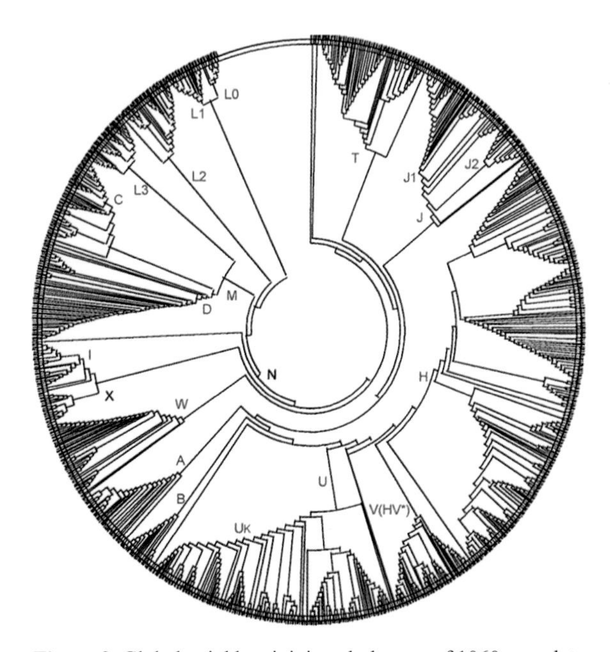


**Figure 2.** Global neighbor-joining phylogeny of 1060 complete human mitochondrial DNA sequences. (Adapted from Ruiz-Pesini et al. 2004.)

A phylogenetic analysis of 1125 mtDNA coding region sequences (Fig. 2) provides an overall outline of the radiation of mtDNAs around the world (Ruiz-Pesini et al. 2004.). Humans arose in Africa between 150,000 and 200,000 years before present (YBP) and radiated in that continent to generate haplogroups L0, L1, L2, and L3. Macro-haplogroup L was established as the first human mtDNA lineage by analyzing the sequences of nuclear DNA (nDNA)-encoded mtDNA pseudogenes. These are fragments of the mtDNA that have been transferred to the nDNA at various times in the past where they become "molecular fossils" preserving their ancient mtDNA sequence motifs. Analysis of two recently integrated mtDNA pseudogenes revealed the characteristic markers of macro-haplogroup L. Further analysis of one of these pseudogenes revealed its presence in all global populations. Therefore, the proto-human mtDNA was the progenitor of African macro-haplogroup L, making Africa the origin of our species (Mishmar et al. 2003a,b).

From macro-halogroup L, two lineages arose in northeastern Africa about 70,000 YBP, giving rise to macrohaplogroups M and N, the only two mtDNA lineages that left Africa to colonize Eurasia (Fig. 3) (Quintana-Murci et al. 1999). Macro-haplogroup N radiated into the middle East and Europe to generate the European-specific haplogroups H, I, J, T, U, Uk, V, W, and X. Macro-haplogroups N and M radiated into Asia to give a plethora of mtDNA lineages including A, B, and F from N and C, D, and G from M. Haplogroup F is at high frequency in southeastern Asia, and haplogroup B occurs in central Asia, along the Asian coast, and out into the Pacific Islands. Haplogroups A, C, D, and G represent only 14% of the mtDNAs in central Asia, but increase in frequency

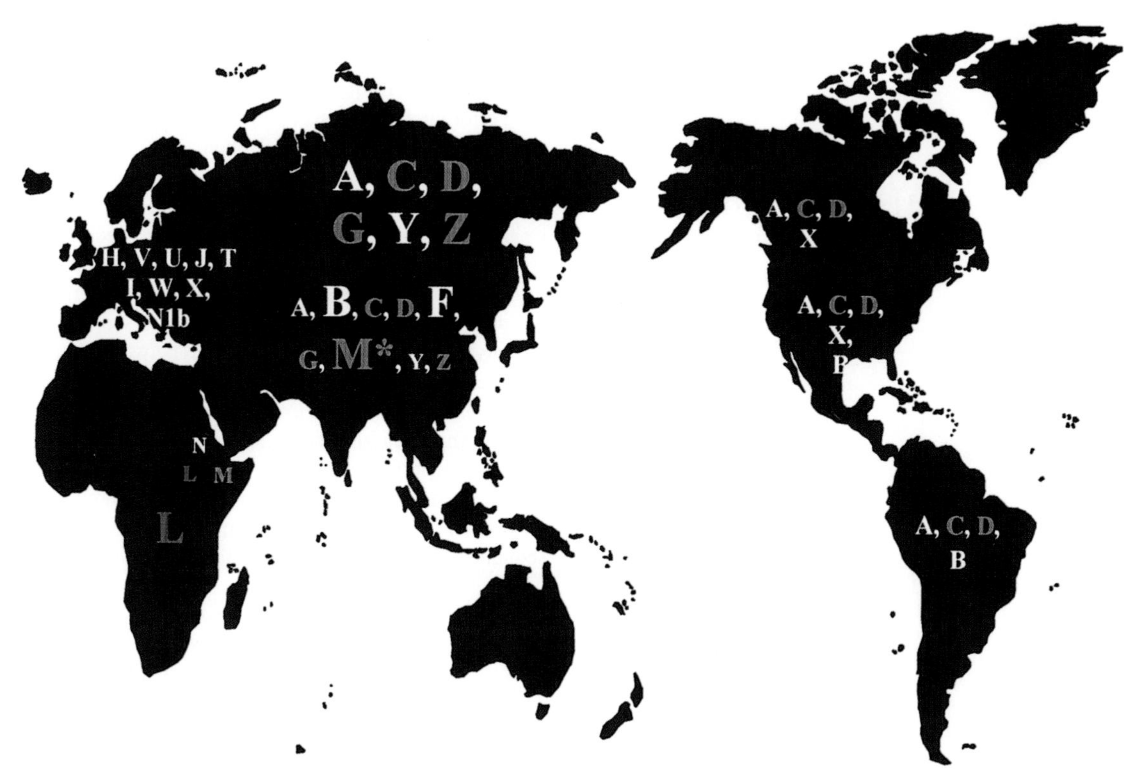


**Figure 3.** Regional distribution of the mtDNA haplogroups throughout the world. Font size indicates contribution of the corresponding haplogroup in Asia. (*Red*) Macro-haplogroup L; (*white*) macro-haplogroup N and derived haplogroups; (*green*) macro-haplogroup M and derived haplogroups.

fivefold to about 75% of the mtDNAs in northeastern Siberia (Schurr and Wallace 2002). When the Bering land bridge became exposed, only people carrying haplogroups A, C, and D were in a position to cross into the Americas to found the Paleo-Indian populations. Haplogroup B joined A, C, and D later, but this haplogroup is absent from northeastern Asia, and may have come to the Americas by a more southern coastal route. Finally, haplogroup X, which is more prevalent in Europe than in Asia, migrated to the New World and came to reside in the Great Lakes region (Fig. 3) (Brown et al. 1998; Stone and Stoneking 1998; Wallace et al. 1999; Malhi and Smith 2002).

Every major human transition from one latitude to another has thus been accompanied by a striking change in the mtDNA haplogroups: macro-haplogroup L confined to Africa, macro-haplogroups M and N radiating in temperate Eurasia, and haplogroups A, C, and D coming to occupy the arctic zone (Mishmar et al. 2003b). Since changes in latitude are also correlated with decreases in the ambient temperature, we hypothesized that mtDNA mutations, which reduced the mitochondrial coupling, permitted adaptation to increasingly cold climates.

## EVIDENCE OF mtDNA CLIMATIC ADAPTATION THROUGH mtDNA SEQUENCE VARIATION

If mtDNA variation were associated with climatic adaptation, then variants in specific mtDNA proteins would be expected to be associated with different climatic zones. This proved to be the case when we analyzed the protein variation in 104 complete mtDNA sequences representing the major mtDNA haplogroups from around the world (Mishmar et al. 2003b). The African haplogroups L0, L1, L2 of the macro-haplogroup L were taken as representative of the tropical and subtropical regions, European haplogroups (H, I, J, T, U, Uk, V, W, X) of the temperate zone, and Siberia and the trans-Beringia haplogroups (A, C, D, X) of the arctic and subarctic. All mtDNA haplotypes from each climatic zone were compared pair-wise for the nonsynonymous (NS) versus synonymous (S) mutation rates (Ka/Ks), and the Ka/Ks values plotted for each climatic zone (Fig. 4). This revealed a striking result. The ATP6 gene showed little amino acid sequence variation in the tropical or temperate zone populations, but enormous variation in the arctic populations. The cytb gene was conserved in the tropical and arctic zones, but was highly variable in the temperate zone; and the COI gene was constant in the temperate and arctic zones, but was more variable in the tropical zone. Similar region-specific variation was found in COIII, ND2, ND5, and ND6 (Fig. 4) (Mishmar et al. 2003b). Thus, regional differences do exist in the mtDNA protein gene sequences, and these correlate strikingly with climatic zones.

To better understand the nature of these regional mtDNA polypeptide variants, we assembled the complete mtDNA coding region sequences from 1125 mtDNAs from around the world and generated a global, neighbor-

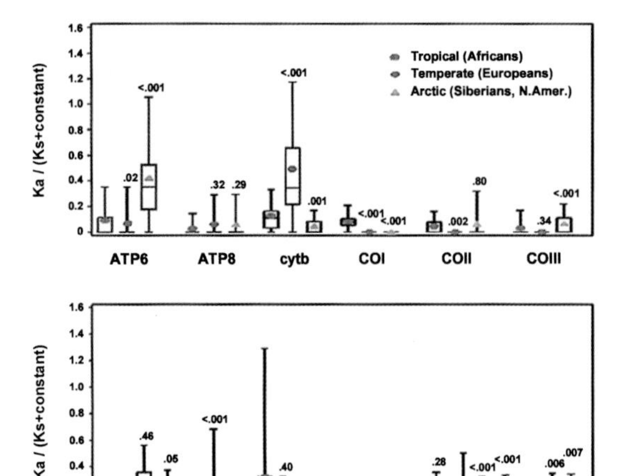


**Figure 4.** Geographic variation in mtDNA polypeptides. The haplogroups included in the "Tropical" category are L0–L3; "Temperate" are H, V, U, J, T, X, N1b, and W; and "arctic" are A, C, D, G, X, Y, and Z. The distribution of the Ka/(Ks + constant) values is given by the vertical line. The colored dot is the mean; the square encompasses the 25–75% range. The statistical difference of distribution of Ka/Ks values of the temperate and arctic zones, relative to the tropical zone, are presented above the distributions. (Adapted from Mishmar et al. 2003b.)

joining, phylogeny (Fig. 2). This phylogeny was then used to position all nucleotide substitutions in the tree. Mutations that were shared by a number of related mtDNA haplotypes were placed at the internal branch of their common ancestor. Mutations that were found in only one mtDNA were placed at the ends of the branches. Therefore, the position of each mutation within the tree is indicative of its relative age.

Three different categories of nucleotide variants would be expected. The first is neutral mutations. These mutations could be either synonymous or nonsynonymous, yet have minimal phenotypic effect. Neutral variants would accumulate in the mtDNA phylogeny by chance at a relatively constant rate and be uniformly distributed throughout the tree. Hence, the number of synonymous mutations in a segment of the tree would be proportional to that segment's age. The second class of mutations would be deleterious replacement substations. These mutations would be rapidly eliminated by purifying selection, experienced at the individual level as mitochondrial disease. The third class of mutations would be advantageous replacements. These would become enriched in a population by adaptive selection. Such adaptive mutations would be rare, but they would tend to found branches of the tree that were successful in new environments. Thus, adaptive mutations would be common in internal branches but rare at the terminal branches (Templeton 1996).

By this logic, we can distinguish between deleterious and adaptive mutations by their distribution in the mtDNA tree. If only purifying selection has been acting on a region-specific branch of the tree, the frequency of replacement mutations will be low at the internal branches (Nodes). Consequently, the ratio of the replacement mutation frequencies (NS/S) of the Nodes divided by Terminal Branches (Tips) will be low. However, if adaptive selection has acted, then the frequency of replacement mutations at the Nodes will be increased and the Nodes/Tips replacement mutation frequency will also be increased. Purifying selection should predominate in stable environments, whereas adaptive selection should become more important in changing environments.

Since neutral mutations would be dispersed throughout the phylogeny, we had to distinguish between the neutral and the adaptive replacement mutations. This was accomplished by analyzing the inter-specific conservation of the mutant amino acid, since functionally important amino acids would tend to be conserved through organismal evolution. To determine whether a mutation altered a highly conserved amino acid, we examined the mtDNA protein sequences from 39 vertebrates listed in our MITOMAP database (www.mitomap.org). The number of these species that have the ancestral human mtDNA sequence is defined as that amino acid's Conservation Index.

We then calculated the mean Conservation Index of the Tip and Node replacement mutations and found that the Conservation Index of the Tips was greater than that of the Nodes (Fig. 5). This is probably due to the fact that deleterious mutations can alter any amino acid, but adaptive mutations can only modify a function, not eliminate it. Hence, adaptive mutations are more constrained.

To determine the level of Conservation Index that would result in a deleterious phenotype, we turned to the MITOMAP collection of all known human pathogenic missense mutations. The Conservation Indices of all 22 reported pathogenic replacement mutations were then calculated. The average of these values proved to be $36.4 \pm 5.2$ ($93 \pm 13\%$), the maximum value being 39.

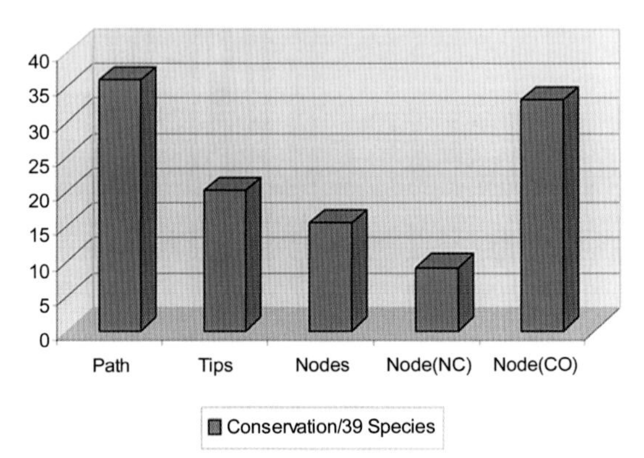


**Figure 5.** Global amino acid Conservation Indices for pathogenic, nodal, and terminal replacement mutations. "Path" equals the mean Conservation Index for 22 known pathogenic mutations, "Tips" and "Nodes" equal Conservation Indices for amino acid substitutions at internal branches versus terminal branches, "Node (NC)" and "Node (CO)" represent the average conservation of the nodal replacement mutations that are outside and within two standard deviations from the pathogenic mutations mean.

The average replacement mutation Conservation Index of the Global Phylogeny's Nodes and Tips is considerably lower than the average of the pathogenic mutations (Fig. 5).

The lower average Global Phylogeny Node and Tip Conservation Indices, relative to the pathogenic mutations, reflect in part the presence of the neutral mutations scattered across the Global Phylogeny. These depress the mean Conservation Index of the functionally important replacement mutations. To distinguish between the adaptive versus the neutral mutations in the Nodes, we used two standard deviations from the mean (the 95% confidence limit) of the pathogenic mutations as an indicator of replacement mutations that might have phenotypic consequences. Mutations below this limit were presumed to be relatively neutral (Fig. 5). Using these criteria, we found that 26% of the nodal replacement mutations were highly conserved with an average amino acid conservation index of 33.2 ± 3.6 (85 ± 9%), slightly below the average of the pathogenic mutations (Ruiz-Pesini et al. 2004). The remaining 74% of the nodal replacement mutations had a much lower average Conservation Index of 9.1 ± 5.8 (23 ± 15%). The Conservation Index of the Tips was comparable, 34.4 (CI = 88%) (42%) versus 9.9 (CI = 25%) (58%).

The high Conservation Index of amino acid replacements occurring at the Tips represents primarily deleterious mutations in the process of being eliminated by purifying selection (Templeton 1996). The highly conserved nodal replacement mutations have a Conservation Index comparable to that of pathogenic mutations, yet they have been retained in the human population for tens of thousands of years in the face of purifying selection. Consequently, these mutations could not be deleterious. Therefore, they must have been advantageous in the new environments confronted by our ancestors.

## ANALYSIS OF ADAPTIVE SELECTION FOR REGIONAL mtDNA LINEAGES

Direct evidence for the regional accumulation of adaptive mutations was obtained by the analysis of amino acid replacement mutation frequencies at the Nodes and the Tips of the mtDNA macro-haplogroups. The replacement mutation frequency of the Nodes of the regional macro-haplogroups increased progressively from its lowest level in African L to intermediate levels for the Eurasian M and N(R) to their highest level for N(nonR). N(nonR) encompasses the arctic haplogroups A and X (Fig. 6). This south-to-north increase in Nodal replacement mutations is also reflected in the Node/Tip ratio, which was lowest in African L (0.7), intermediate in Eurasia M and N(R) (0.8), and highest in Northern N(nonR) (0.94) (Fig. 6). Thus, more northern latitudes are associated with more Nodal replacement mutations.

The excess of Nodal mutations in the higher, colder latitudes was further confirmed by analysis of the haplogroups that reside in the arctic and subarctic (Fig. 7). Using the African macro-haplogroup L nodal replacement mutation frequency (0.31) as the reference, the northeastern Siberian–northern Native American haplogroups A, C, D, and X all had higher nodal values rang-

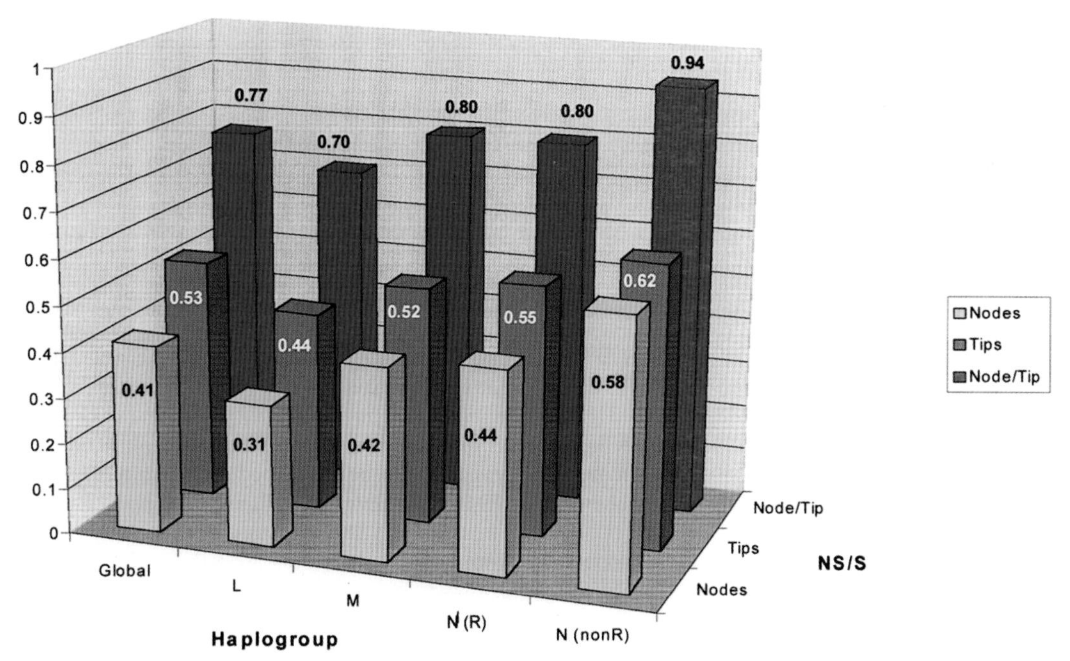


**Figure 6.** Global and regional replacement mutation frequencies at Nodes and Tips of the mitochondrial phylogeny. Global encompasses the entire mtDNA tree. L, M, N(R), and N(nonR) are individual region-specific macro-haplogroups. Replacement mutation frequencies, listed on the bars in the Nodes (*first row*) and Tips (*second row*) are calculated by the ratio of nonsynonymous (NS) to synonymous (S) nucleotide substitutions. The ratio of the Node and Tip NS/S ratios is shown above the back row of bars.

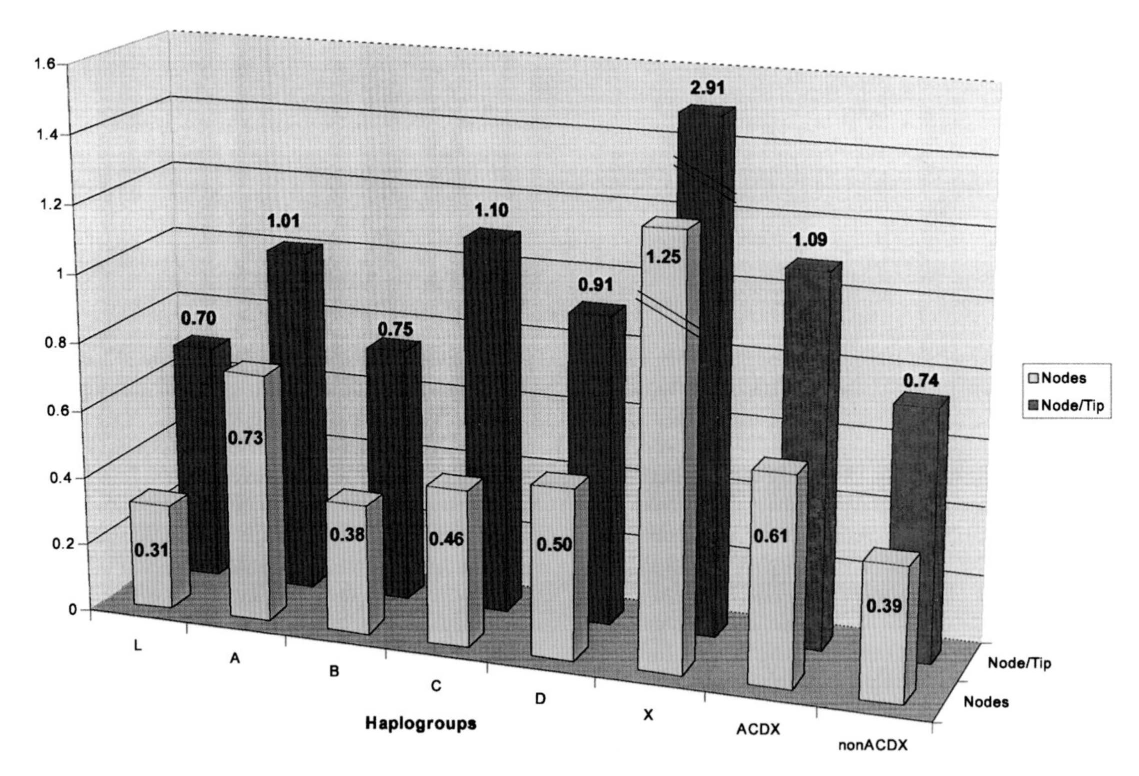


**Figure 7.** Arctic zone replacement mutation frequencies. L equals African-specific macro-haplogroup L. A, B, C, D, and X represent Asian or Eurasia mtDNA haplogroups that crossed from Eurasia into the Americas. ACDX equals the average of these haplogroups. nonACDX equals the average of all mtDNA haplogroups but A, C, D, and X. (*First row*) Nodal NS/S ratios. (*Second row*) Node/Tip ratios.

ing from 0.46 to 1.25. In contrast, haplogroup B, which reached America without going through Siberia, had a nodal replacement frequency similar to that of Africa (0.38). These same trends are seen in the Node/Tip replacement frequency ratios. For African L, the ratio is 0.70. For haplogroups A, C, D, and X the range of ratios is 0.91–2.91. Haplogroup B is 0.75. Combining the data from haplogroups A, C, D, and X (ACDX) gives an average nodal value of 0.61 and a Node/Tip ratio of 1.09, whereas all nonACDX haplogroups give an average nodal ratio of 0.39 and a Node/Tip ratio of 0.74.

This striking increase in the replacement mutation frequency in the arctic and subarctic haplogroups is also reflected in the average Conservation Index of the nodal replacement mutations. The nodal Conservation Indices of haplogroups A, C, and D were 20.8 (53%), 28.3 (73%), and 16.3 (42%), respectively, whereas those for African L and haplogroup B were 14.1 (36%) and 12.0 (31%). Thus, the nodal replacement mutations of the arctic haplogroups have both a higher frequency of replacement mutations and a higher Conservation Index of the altered amino acid.

A comparable result was obtained for the European (temperate zone) haplogroups (Fig. 8). Relative to the Node/Tip ratio of African L (0.70), the ratios of the European haplogroups H, V(HV*), J, T, and IWX were markedly higher (0.79–2.89). Likewise, compared to the African 14.1 (36%), the nodal Conservation Indices of haplogroups H, J, T, and IWX were increased to the range of 16.5–20.3 (42–52%).

Clearly, adaptive replacement mutations have accumulated in human populations of both the arctic and temperate zones. Therefore, these mutations must have been selected in the human mtDNAs as people moved out of Africa into colder climates.

## THE NATURE OF ADAPTIVE mtDNA MUTATIONS

From this analysis we can conclude that highly conserved nodal replacement mutations from the temperate and arctic haplogroups were important in adaptation to higher, colder latitudes. The specific adaptive mutations would be expected to occur close to the base of the region-specific mtDNA haplogroup and to change a highly conserved amino acid. Using this logic, arctic haplogroup A was found to encompass two adaptive replacement mutations: ND2 np 4824G (T119A) and ATP6 np 8794T (H90Y). Similarly, the Siberian haplogroup C lineage was found to be associated with two adaptive variants: ND4 np 11969A (A404T) and cytb np 15204C (I153T). The cytb np 15204C mutation is two amino acids away from a pathogenic mutation S151P that alters the outer CoQ-binding site (Qo). Haplogroup D has only one likely adaptive variant: ND2 np 5178A (L237M), and a major subbranch of haplogroup X harbored another likely adaptive mutation: ND5 np 13708A (A458T), one of the same variants that occurs at the base of European haplogroup J.

Similar functional mutations were found at the base of European haplogroups. For example, the sister hap-

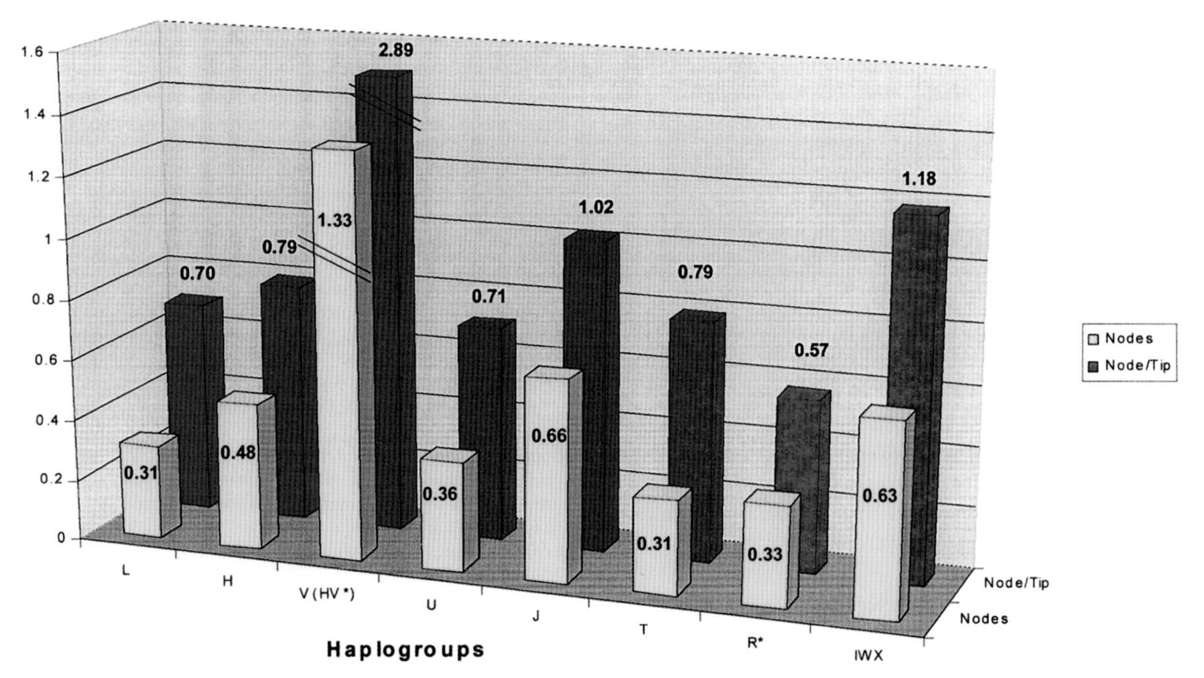


**Figure 8.** Temperate zone replacement mutation frequencies. L equals African macro-haplogroup L. H, V (HV*), U, J, T, R, IWX equal European haplogroups. Other conventions are the same as in Fig. 7.

logroups J and T shared a common root involving two amino acid substitutions: ND1 np 4216C (Y304H) and cytb np 15452A (L236I). J and T then diverge.

The root of haplogroup T encompasses 9 base substitutions, one of which is clearly adaptive: ND2 np 4917G (N150D). This is the most conserved ND2 polymorphism that exists in the global human phylogeny, having a Conservation Index of 35 (90%) (Ruiz-Pesini et al. 2004).

The root of haplogroup J has two replacement mutations: ND3 np 10398G (T114A) and ND5 np 13708A (A458T), the latter being the same missense mutation found in a haplogroup X. Haplogroup J then splits into sub-haplogroups, J1 and J2, each defined by a major cytb mutation. The J2 cytb mutation is at np 15257A (D171N) and the J1 cytb mutation is at np 14798C (F18L). The np 14798C mutation is also found at the root of sub-haplogroup Uk. The np 15257 and np 14798 variants alter well-conserved amino acids with Conservation Indices of 37 (95%) and 30 (77%), respectively. The 15257 variant occurs at Qo and contacts the Rieske iron–sulfur protein. The np 14798 mutation occurs at the inner CoQ-binding site (Qi) of cytb, interacts directly with CoQ, and is the site of amino acid changes that cause resistance of diuron (Ruiz-Pesini et al. 2004). Since the Qo and Qi binding sites are fundamental to proton pumping by the Q-cycle of complex III, these variants are likely to affect the coupling efficiency of mitochondrial OXPHOS.

## CLINICAL CORRELATES WITH ADAPTIVE mtDNA VARIANTS

Adaptive replacement mutations that change the functional state of the mitochondrion would be expected to have phenotypic consequences. Since these variants are adaptive for different environments, the variant must be beneficial in one context but deleterious in others.

The deleterious components of some of the adaptive mutations have been detected in studies of certain clinical diseases. Haplogroup T has been shown to be associated with moderate asthenozoospermia and reduced complex I and IV activities (Ruiz-Pesini et al. 2000). Haplogroup J has been shown to increase the penetrance of the milder LHON primary complex I mutations: ND4 np 11778A, ND6 np 14484C, and ND4L np 10663C (Brown et al. 1997, 2002; Hofmann et al. 1997). Haplogroups J and Uk have been linked to susceptibility to multiple sclerosis (Kalman et al. 1996), and haplogroup T has been associated with Wolfram syndrome (Hofmann et al. 1997).

In contrast, these same haplogroups (J, Uk, and T) have been found to be protective for the ravages of aging, AD, and PD (Chagnon et al. 1999; van der Walt et al. 2003). Moreover, sub-haplogroup Uk is protective of the adverse effects of the apoE-ε4 allele in AD (Carrieri et al. 2001). Finally, haplogroup J and sub-haplogroup Uk, which share the np 14798 cytb mutation, have been associated with increased longevity in several European studies (Ivanova et al. 1998; De Benedictis et al. 1999; Ross et al. 2001; Coskun et al. 2003; Niemi et al. 2003).

How could the same mutation be both deleterious and advantageous? The answer follows logically from the multifunctional nature of the mitochondrion. A mtDNA mutation that reduces the coupling efficiency of OXPHOS would diminish maximal ATP production. This would increase the susceptibility of an individual to pathogenic mutations that cause disease by ATP deficiency, perhaps like LHON. On the other hand, uncoupled mutations would keep the mitochondrial ETC more oxidized, thus reducing ROS production. This, in turn, would be protective for diseases caused by oxidative stress. Since oxidative stress has been hypothesized to be a causative factor in AD, PD, and aging, the uncoupled mutation would be protective for these problems.

Thus, ancient adaptive mutations accumulated in the mtDNA of our ancestors that permitted them to survive and prosper in the colder climates of northern Europe and northeastern Asia. Today, however, modern technology has changed these environmental constraints with real consequences on our susceptibility to modern diseases.

## ACKNOWLEDGMENTS

This work was supported by the Israeli Academy of Sciences Bikura fellowship awarded to D.M. and by National Institutes of Health grants AG-13154, NS-21328, and NA-37167, as well as an Ellison Medical Foundation Senior Investigator Award awarded to D.C.W.

## REFERENCES

Brown M.D., Sun F., and Wallace D.C. 1997. Clustering of Caucasian Leber hereditary optic neuropathy patients containing the 11778 or 14484 mutations on an mtDNA lineage. *Am. J. Hum. Genet.* **60:** 381.

Brown M.D., Starikovskaya E., Derbeneva O., Hosseini S., Allen J.C., Mikhailovskaya I.E., Sukernik R.I., and Wallace D.C. 2002. The role of mtDNA background in disease expression: A new primary LHON mutation associated with Western Eurasian haplogroup J. *Hum. Genet.* **110:** 130.

Brown M.D., Hosseini S.H., Torroni A., Bandelt H.J., Allen J.C., Schurr T.G., Scozzari R., Cruciani F., and Wallace D.C. 1998. mtDNA haplogroup X: An ancient link between Europe/Western Asia and North America? *Am. J. Hum. Genet.***63:** 1852.

Carrieri G., Bonafe M., De Luca M., Rose G., Varcasia O., Bruni A., Maletta R., Nacmias B., Sorbi S., Corsonello F., Feraco E., Andreev K.F., Yashin A.I., Franceschi C., and De Benedictis G. 2001. Mitochondrial DNA haplogroups and APOE4 allele are non-independent variables in sporadic Alzheimer's disease. *Hum. Genet.* **108:** 194.

Chagnon P., Gee M., Filion M., Robitaille Y., Belouchi M., and Gauvreau D. 1999. Phylogenetic analysis of the mitochondrial genome indicates significant differences between patients with Alzheimer disease and controls in a French-Canadian founder population. *Am. J. Med. Genet.* **85:** 20.

Coskun P.E., Ruiz-Pesini E., and Wallace D.C. 2003. Control region mtDNA variants: Longevity, climatic adaptation, and a forensic conundrum. *Proc. Natl. Acad. Sci.* **100:** 2174.

De Benedictis G., Rose G., Carrieri G., De Luca M., Falcone E., Passarino G., Bonafe M., Monti D., Baggio G., Bertolini S., Mari D., Mattace R., and Franceschi C. 1999. Mitochondrial DNA inherited variants are associated with successful aging and longevity in humans. *FASEB J.* **13:** 1532.

Friberg H. and Wieloch T. 2002. Mitochondrial permeability transition in acute neurodegeneration. *Biochimie* **84:** 241.

Hofmann S., Bezold R., Jaksch M., Obermaier-Kusser B., Mertens S., Kaufhold P., Rabl W., Hecker W., and Gerbitz K.D. 1997. Wolfram (DIDMOAD) syndrome and Leber hereditary optic neuropathy (LHON) are associated with distinct mitochondrial DNA haplotypes. *Genomics* **39:** 8.

Ivanova R., Lepage V., Charron D., and Schachter F. 1998. Mitochondrial genotype associated with French Caucasian centenarians. *Gerontology* **44:** 349.

Jacobs H.T. 2003. The mitochondrial theory of aging: Dead or alive? *Aging Cell* **2:** 11.

Kadenbach B. 2003. Intrinsic and extrinsic uncoupling of oxidative phosphorylation. *Biochim. Biophys. Acta* **1604:** 77.

Kagawa Y., Cha S.H., Hasegawa K., Hamamoto T., and Endo H. 1999. Regulation of energy metabolism in human cells in aging and diabetes: FoF(1), mtDNA, UCP, and ROS. *Biochem. Biophys. Res. Commun.* **266:** 662.

Kalman B., Lublin F.D., and Alder H. 1996. Characterization of the mitochondrial DNA in patients with multiple sclerosis. *J. Neurol. Sci.* **140:** 75.

Malhi R.S. and Smith D.G. 2002. Brief communication: Haplogroup X confirmed in prehistoric North America. *Am. J. Phys. Anthropol.* **119:** 84.

Mishmar D., Ruiz-Pesini E., Brandon M., and Wallace D.C. 2003a. Mitochondrial DNA-like sequences in the nucleus (NUMTS): Insights into our African origins and the mechanism of foreign DNA integration. *Hum. Mutat.* (in press).

Mishmar D., Ruiz-Pesini E., Golik P., Macaulay V., Clark A.G., Hosseini S., Brandon M., Easley K., Chen E., Brown M.D., Sukernik R.I., Olckers A., and Wallace D.C. 2003b. Natural selection shaped regional mtDNA variation in humans. *Proc. Natl. Acad. Sci.* **100:** 171.

Niemi A.K., Hervonen A., Hurme M., Karhunen P.J., Jylha M., and Majamaa K. 2003. Mitochondrial DNA polymorphisms associated with longevity in a Finnish population. *Hum. Genet.* **112:** 29.

Quintana-Murci L., Semino O., Bandelt H.J., Passarino G., McElreavey K., and Santachiara-Benerecetti A.S. 1999. Genetic evidence of an early exit of *Homo sapiens sapiens* from Africa through eastern Africa. *Nat. Genet.* **23:** 437.

Ross O.A., McCormack R., Curran M.D., Duguid R.A., Barnett Y.A., Rea I.M., and Middleton D. 2001. Mitochondrial DNA polymorphism: Its role in longevity of the Irish population. *Exp. Gerontol.* **36:** 1161.

Ruiz-Pesini E., Mishmar D., Brandon M., Procaccio V., and Wallace D.C. 2004. Effects of purifying and adoptive selection on regional variation in human mtDNA. *Science* (in press).

Ruiz-Pesini E., Lapena A.C., Diez-Sanchez C., Perez-Martos A., Montoya J., Alvarez E., Diaz M., Urries A., Montoro L., Lopez-Perez M.J., and Enriquez J.A. 2000. Human mtDNA haplogroups associated with high or reduced spermatozoa motility. *Am. J. Hum. Genet.* **67:** 682.

Schurr T.G. and Wallace D.C. 2002. Mitochondrial DNA diversity in Southeast Asian populations. *Hum. Biol.* **74:** 431.

Stone A.C. and Stoneking M. 1998. mtDNA analysis of a prehistoric Oneota population: Implications for the peopling of the New World. *Am. J. Hum. Genet.* **62:** 1153.

Tatuch Y., Christodoulou J., Feigenbaum A., Clarke J.T.R., Wherret J., Smith C., Rudd N., Petrova-Benedict R., and Robinson B.H. 1992. Heteroplasmic mtDNA mutation (T-G) at 8993 can cause Leigh disease when the percentage of abnormal mtDNA is high. *Am. J. Hum. Genet.* **50:** 852.

Templeton A.R. 1996. Contingency tests of neutrality using intra/interspecific gene trees: The rejection of neutrality for the evolution of the mitochondrial cytochrome oxidase II gene in the hominoid primates. *Genetics* **144:** 1263.

van der Walt J.M., Nicodemus K.K., Martin E.R., Scott W.K., Nance M.A., Watts R.L., Hubble J.P., Haines J.L., Koller W.C., Lyons K., Pahwa R., Stern M.B., Colcher A., Hiner B.C., Jankovic J., Ondo W.G., Allen F.H., Jr., Goetz C.G., Small G.W., Mastaglia F., Stajich J.M., McLaurin A.C., Middleton L.T., Scott B.L., Schmechel D.E., Pericak-Vance M.A., and Vance J.M. 2003. Mitochondrial polymorphisms significantly reduce the risk of Parkinson disease. *Am. J. Hum. Genet.* **72:** 804.

Wallace D.C. 1999. Mitochondrial diseases in man and mouse. *Science* **283:** 1482.

———. 2001a. Mitochondrial defects in neurodegenerative disease. *Ment. Retard. Dev. Disabil. Res. Rev.* **7:** 158.

———. 2001b. A mitochondrial paradigm for degenerative diseases and ageing (also see discussion). *Novartis Found. Symp.* **235:** 247.

Wallace D.C. and Lott M.T. 1999. Mitochondrial bioenergetics and reactive oxygen species in degenerative diseases and aging. In *Molecular biology of aging* (ed. V.A. Bohr et al.). *Alfred Benzon Symp.* **44:** 125. Munksgaard, Copenhagen.

Wallace D.C., Brown M.D., and Lott M.T. 1999. Mitochondrial DNA variation in human evolution and disease. *Gene* **238:** 211.

# Inferring Human History: Clues from Y-Chromosome Haplotypes

P.A. UNDERHILL
*Department of Genetics, Stanford University School of Medicine, Stanford, California 94305-5120*

DNA molecules are organic elements of information storage imbued with imperfect copying processes. Thus, they are fundamental repositories of an organism's evolutionary history. The field of human molecular evolution is predicated on the concept that patterns of DNA sequence variation in living populations encode aspects of human heritage shaped by a constellation of evolutionary influences. The framework of genetic variability in the genome reflects both evolutionary adaptive processes that are locus-specific and population-level forces that affect all the components of the genome equally. Genetic research often focuses on distinguishing inconsistencies in patterns of variation between genomic regions to help bridge the gap between particular genes and traits, including matters of function and malfunction. Alternatively, genomic DNA is also an archive of those aspects of human evolutionary processes reflective of population-level forces like drift, subdivision, size fluctuation, and migration. By studying the degree of genetic molecular variation, one can, in principle, reconstruct past events such as expansions and settlements from which origins of specific populations can be predicted (Cavalli-Sforza et al. 1994). However, since the bulk of common variation in the genome occurs between individuals, the difference between populations is low, making it more challenging to investigate ambiguities concerning affinities and origins of populations. It is the component of between-population variance that best provides insights into the evolution of the spectrum of extant populations (Cavalli-Sforza and Feldman 2003). In addition, determining the migratory patterns of our ancestors and their timing, as well as the amount of population admixture due to such migrations, has been experimentally less tractable. Consequently, such issues are often simply ignored when reconstructing population phylogenies where little or no migration is implicitly assumed.

Progress in understanding the spectrum of human DNA sequence variation and its causes, especially when integrated with other knowledge from historians, archaeologists, anthropologists, and linguists, can help recapitulate human population histories (Owens and King 1999; Cann 2001). In particular, informative haplotypes, immune from the scrambling effects of recurrent mutation and recombination, provide a promising way forward. In the early genetic studies, both protein and matrilineal inherited mitochondrial DNA (mtDNA) loci have provided much of the initial evidence. However, considerable recent progress in elucidating Y-chromosome sequence variation from the nonrecombining region (NRY) has made it possible to more fully investigate the parallel paternal heritage that underlies the central theme of this paper. Although extrapolating variation associated with a single gene to population history must be done cautiously, the phylogeographic reconstruction of haplotypes offers one such interpretation that is amenable to testing by further studies from a number of disciplines. Although nongenetic evidence provides constraints that improve confidence in phylogeographic-based interpretations, the development of statistical methods to objectively evaluate such geographic and phylogenetic patterns remains incomplete (Knowles and Maddison 2002). The growing accumulation of Y-chromosome haplotype data in geographic context will provide a substantial test bed for future statistical modeling.

## Y-CHROMOSOME HAPLOTYPES CONVEY SPATIAL AND TEMPORAL INFORMATION

Although explorations into prehistory have been traditionally archeological, additional perspectives have been provided by linguistic and genetic studies. Although the records of these nonrecombining sex-specific loci may diverge because of natural selection or differences between male and female behaviors, the accumulation of sequence variation during the lineal life spans of these haplotypic systems provides a powerful way to recover genetic prehistory. Specifically, the particular distinctive clinal patterns of NRY haplotypes, together with patterns of associated genetic diversification with geography, mark trajectories of gene flow (and by inference, the movement of populations). The lower effective population size of Y chromosomes relative to other components of the human genome make the Y chromosome particularly sensitive to the influences of drift and founder effect. Whatever the causes of this property (e.g., localized natural selection, gender-based differential reproductive success, and/or migratory behavior), it is particularly useful since it explains the characteristic high stratification of NRY diversity with geography relative to other genes including mtDNA. Although nearly all Y chromosomes have shared common ancestry, the succession of accumulating genetic markers reveals a cascade of differentiation that randomly coincides with various population origin episodes, each with specific temporal and geographical context (Underhill et al. 2001b).

The Y chromosome is now the most informative haplotyping system known, making this locus an attractive reservoir of gene lineages permeated with nonrandom geographic structure. The determination of evolutionary stable haplotypes with geographic appellation provides clues to geographic ancestry. Considerable potential for additional informative haplotype resolution remains, since only a small fraction of the Y chromosome has been surveyed for informative DNA sequence variants and only a fraction of populations have been surveyed. Consequently, Y-chromosome haplotypes are ideal for assessing the origins of contemporary population affinities.

## GEOGRAPHIC PATERNAL ANCESTRY

Binary DNA sequence variants like single nucleotide substitutions (SNPs) and small insertions or deletions (DIPs) associated with the nonrecombining portion of the haploid Y chromosome provide a unique metric into population affinity and substructure. These evolutionarily stable mutations accrue throughout the lineal life span of the molecule such that their sequential accumulation across the generations can be deduced. Since most of the Y chromosome escapes recombination, one can construct an unequivocal genealogy and observe the geographic relationships of various lineages (Fig. 1). The tree is rooted using great ape sequences to deduce ancestral allele status. The lower effective population size of the Y chromosome in relation to other components in the gene pool translates into increased levels of subdivision creating the strongest geographic and greatest diversification signal among populations.

What emerges is a molecular narrative of population demography, relatedness, and genetic substrata reflecting possible signatures of initial colonists, novel micro-evolutionary differentiations during isolation, and subsequent dispersals overlaying previous ranges. These patterns of Y-chromosome lineal origins, affinity, and substructure provide a portrait of human paternal history.

The aim here is to present examples of Y-chromosome affinity and diversification. These interpretations of Y-chromosome heritage provide independent perspectives to theories of prehistorical events and resemblances based on material culture, linguistic, and other genetic knowledge. The inference of putative origins of lineages should be possible, as well as deducing their subsequent dispersal routes, by localizing regions of highest associated diversity. When high-resolution binary lineages are coupled to more rapidly mutating microsatellites or short tandem repeat loci, the combination of linked polymorphic systems provides a powerful system for understanding diversity across both intermediate and recent time frames (de Knijff 2000). These geographic patterns of genetic affinity and diversification provide intriguing insights into human history, especially the population dynamics associated with migration, population subdivision, fluctuations in population size, and subsequent gene flow episodes.

An overall picture, reflective of modern human origins, affinity, differentiation, and demographic history, has been reconstructed. Components of the Y-chromosome binary haplotype phylogeny are described according to a recently formulated standard nomenclature (Y Chromosome Consortium 2002).

## AFRICAN HERITAGE

Over 400 binary polymorphisms currently describe the Y-chromosome tree. Several mutually reinforcing binary mutations divide the Y-chromosome haplotype phylogeny into two distinctive components, haplogroup A and the remainder of all other haplogroups, specifically B through R. The ancestral (i.e., nonhuman primate) alleles associated with these ancient polymorphisms are localized exclusively to a minority of both extant north African and sub-Saharan populations, whereas the majority of other Africans, and all non-Africans, carry only the derived mutant alleles. This mode indicates that almost all modern Y chromosomes trace their ancestry to a common primogenitor, as expected in a stable genealogy. These Y-chromosome data contradict the possibility that early hominids contributed significantly, if at all, to the gene pool of anatomically modern humans of the region (Capelli et al. 2001; Ke et al. 2001). This is evidence that all modern extant human Y chromosomes trace their ancestry to Africa and that the descendants of the derived lineage left Africa and eventually completely replaced previous archaic human Y-chromosome lineages.

A second distinctive monophyletic haplogroup called B, defined by several binary polymorphisms, is also restricted to African populations. Both A and B lineages are diverse and suggest a deeper genealogical heritage than other haplotypes. Representatives of these lineages are distributed across Africa, but generally at low frequencies. Populations represented in A and B clades include some Khoisan and Bantu speakers from South Africa, pygmies from central Africa, and lineages in Sudan, Ethiopia, and Mali (Underhill et al. 2000; Semino et al. 2002). One group B-associated lineage is shared by click speaking San of South Africa and Hadzabe of Tanzania. The genetic distance between these lineal representatives indicates that both populations have been separated for a considerable time but share ancient genetic and possibly linguistic heritage (Knight et al. 2003). The phylogenetic position of A and B lineages nearest the root of the Y tree, their survivorship in isolated populations, and accumulated variation are suggestive of an early diversification and dispersal of human populations within Africa, and an early widespread distribution of human populations in that continent. The discovery of *Homo sapiens* fossils in Ethiopia dating to 160,000 years ago is consistent with an African origin of our species (White et al. 2003).

## OUT OF AFRICA

At least three mutations lie at the root of all the remaining Y-chromosome haplotypes that compose the majority of African and non-African lineages, namely

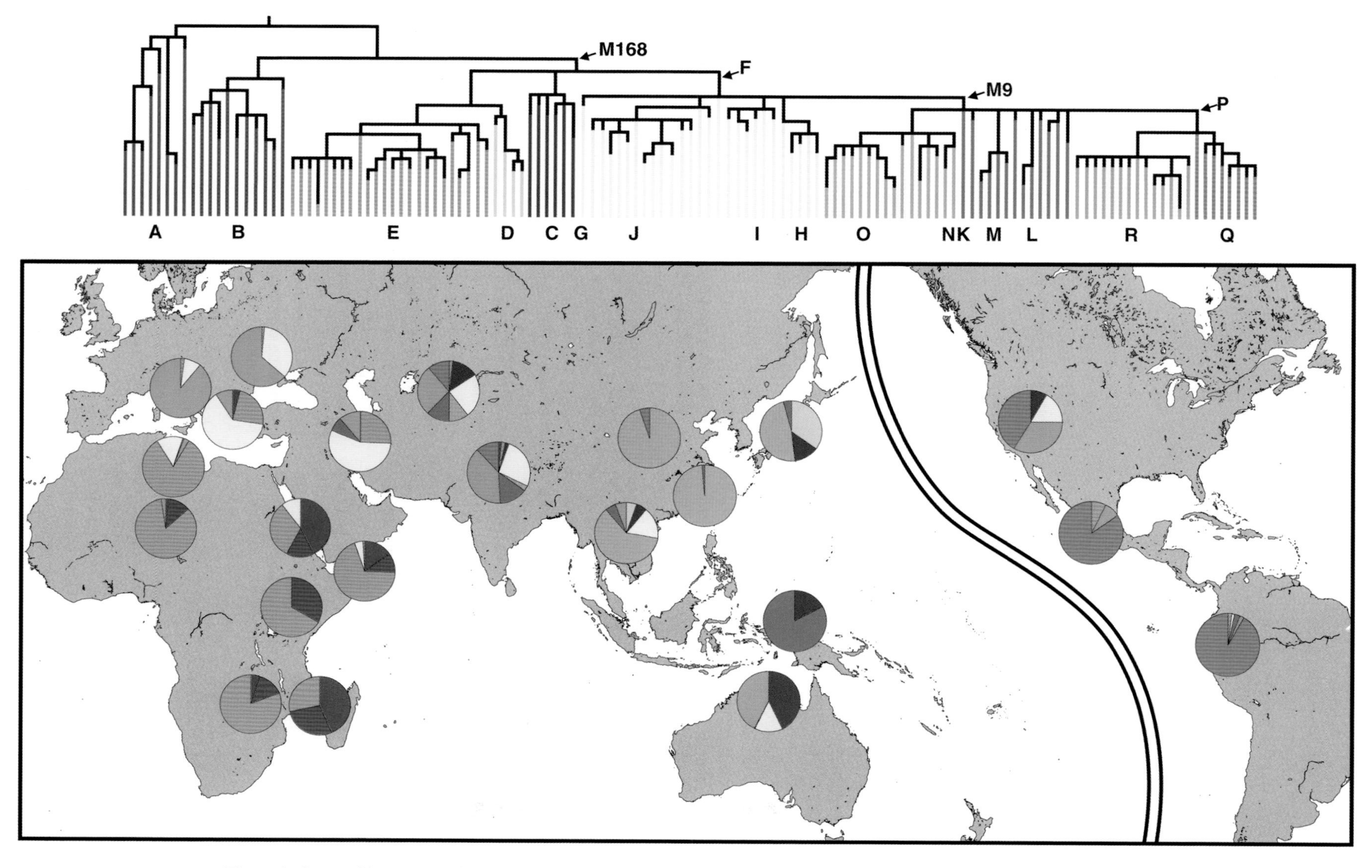


**Figure 1.** Geographic patterns of Y-chromosome haplogroups according to recently standardized nomenclature in 22 global geographic regions. (Adapted, with permission, from Underhill et al. 2001b [copyright Blackwell Scientific].)

haplogroups C through R (Underhill et al. 2001b). The mutations that define this node deep within the interior of the Y tree reflect descendants of males who successfully left Africa and formed the scaffold on which all other Y-chromosome diversification with geography has accumulated. The geographical distribution of this diversification allows us to try to understand some of the major movements that occurred after anatomically modern humans left Africa. The original founders diversified into important lineages that display an irregular geographic distribution. The majority of Y lineages in the world are composed of a tripartite assemblage consisting of (1) haplogroup C, (2) haplogroups D and E, and (3) overarching haplogroup F that defines the internal node of all remaining haplogroups G through R. These geographic patterns of genetic affinity and diversification provide insights into the population dynamics associated with migration, population subdivision, fluctuations in population size, and more recent gene flow episodes. A synopsis of the relevant features of these haplogroups is discussed below.

## ASIAN HAPLOGROUP C

Since the mutations that define haplogroup C have not been observed in any African populations, it has been postulated that this haplogroup likely arose somewhere in Asia on an M168 lineage sometime after an early departure event prior to the arrival of modern humans to Sahul in Southeast Asia (Capelli et al. 2001; Underhill et al. 2001b). Haplogroup C comprises a collection of sublineages that display irregular geographic patterning. Most notable are C lineages that carry the M217 transversion mutation which are common in eastern Asia and Siberia with representatives in north America (Bergen et al. 1999; Karafet et al. 2001; Lell et al. 2002). Interestingly, these M217-derived lineages are absent in haplogroup C lineages seen in Indonesia (Underhill et al. 2001a), Oceania (Kayser et al. 2000, 2001), and Yunnan, China, where numerous minority populations reside (Karafet et al. 2001). Recently, such related C lineages have been observed in India at 4.6% (Kivisild et al. 2003). Their persistence in India is consistent with the model of an early coastal migration route via southwest Asia to insular southeast Asia and Oceania (Stringer 2000). The phylogeography suggests that early male colonizers to Australia were haplogroup C descendants, and the presence of the M217 sublineage in Siberia (Karafet et al. 2001, 2002; Lell et al. 2002) is consistent with diversification and northward migration since the last ice age.

## ASIAN AND AFRICAN HAPLOGROUPS D AND E

Although each monophyletic haplogroup displays continental separate distributions, both share three phylogenetically equivalent binary markers indicative of unequivocal shared heritage. The ancestors who accumulated these three mutations could have just as well arisen in Africa as in Asia. Despite the apparent absence of any intermediate haplotypes based on these three binary polymorphisms, Africa remains the most plausible geographic origin of these three relatively old polymorphisms (Underhill and Roseman 2001). It appears that some descendants with these three mutations remained and some left Africa to become part of the gene pool of the early successful colonizers in Asia. Following geographic separation, subsequent continent-specific mutations arose creating the two monophyletic D and E clades in Asia and Africa, respectively. Haplogroup E lineages are the most frequent in Africa and display subsequent binary and microsatellite diversification. Conversely, Asian haplogroup D generally occurs at low frequencies throughout eastern Asia, except in peripheral locations like Tibet, Japan, and the Andaman Islands, where significant frequencies have been observed, most likely because of founder effects (Underhill et al. 2001b; Thangaraj et al. 2003). The Ainu of Japan are composed of both C and most likely D representatives (Tajima et al. 2002). The phylogeography suggests that Asian D lineages are likely the descendants of early Asian colonizers who arrived from Africa. To a large degree, they have been subsequently displaced to geographic margins by pressures from demic expansions by ensuing peoples.

## HAPLOGROUP F ACROSS THE ENTIRE WORLD EXCEPT SUB-SAHARAN AFRICA

The third major and most peripatetic subcluster of M168 lineages is characterized by at least three mutations (one of which is M89) that define the root of haplogroup F from which all other haplogroups (G through R) deploy. This F subcluster (Fig. 2a) is suggested to have evolved outside Africa early in the diversification and migration of modern humans (Kivisild et al. 2003). Early Upper Paleolithic peoples throughout Eurasia provide sources from which later populations derive. The differentiation of haplogroup F (Fig. 2a–h) within Eurasia helps to begin understanding this complex period of the peopling of the world (Underhill et al. 2001b). The Middle East has major representatives of haplogroups G, J, and R. India also has J, which may have arrived with agriculturalists. Interestingly, it has some F and H lineages seldom observed elsewhere. An expansion of F lineages toward central Asia or the Caucasus would have given rise to a population that acquired the M9 mutation which defines a major bifurcation in the phylogeny (Fig. 2b). Haplogroup K, L, and M lineages all descend from an M9 ancestor and are widespread with some distinctive K lineages being observed in India, the Middle East, and Europe. Haplogroup L has greatest frequency in southwest Asia, and distinctive K and M lineages are restricted to Oceania (Fig. 2c). The M9 mutation also lies at the root of haplogroups O through R, all of which are Eurasian and American in distribution. The population carrying the M9 mutation must have expanded widely (Fig. 2b), with one in north Asia characterized by the haplogroup P, which encompasses distinctive eastward expanding Q (Siberian and American) and westward expanding R (Eurasian) lineages (Fig. 2d), and another one in eastern Asia characterized by monophyletic haplogroups N and

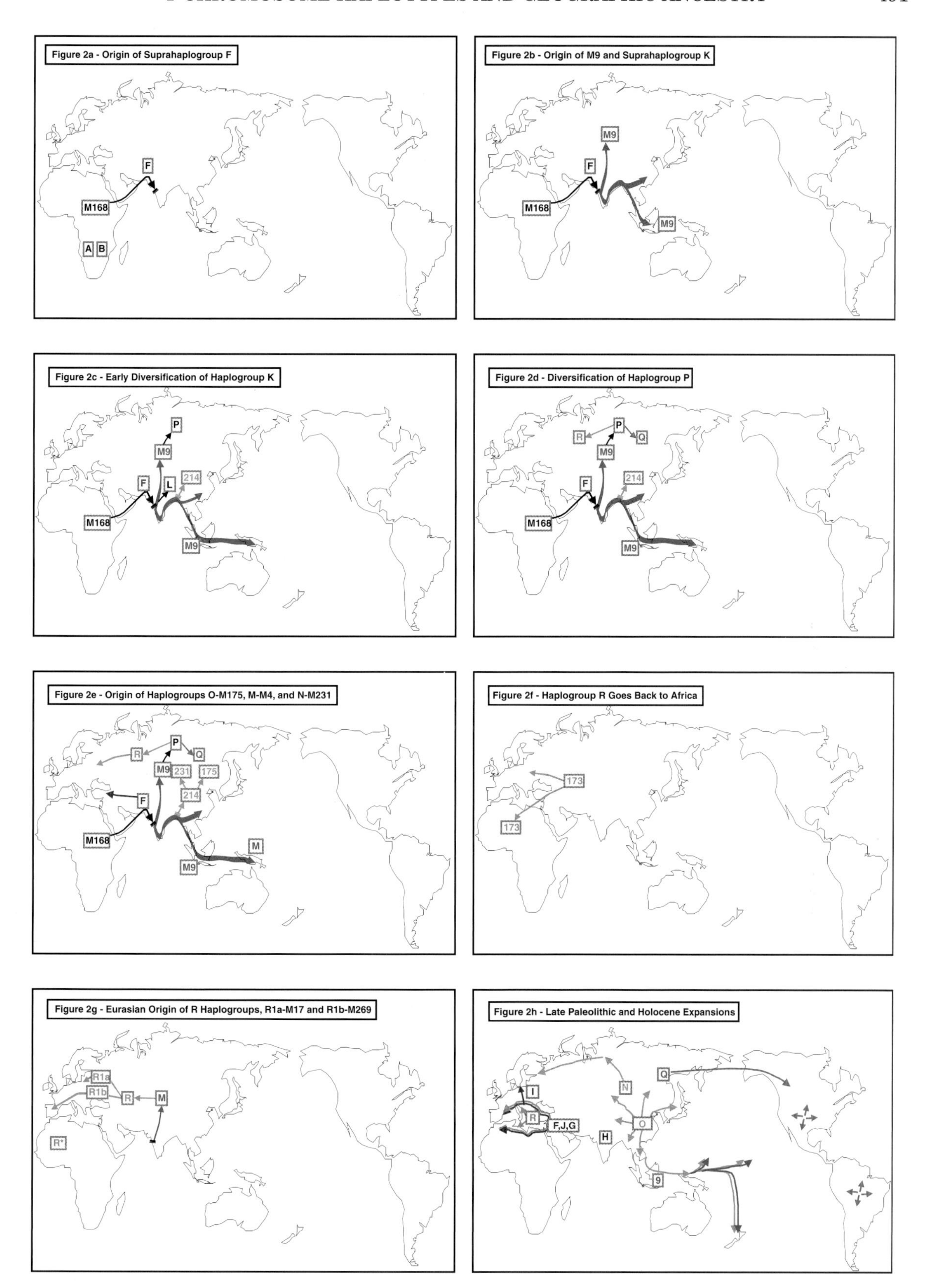


**Figure 2.** Phylogeographic inferences of the origin of suprahaplogroup F and its subsequent diversification across the world. (Adapted, with permission, from Underhill et al. 2001b [copyright Blackwell Scientific].)

O (Fig. 2e) that share a common unifying mutation. In summary, the early diversification of a haplogroup F population in Eurasia between 40,000 and 30,000 years ago would have given rise to at least six Y-chromosome populations (Underhill et al. 2001b). Thus, there were multiple independent formations and fragmentations of populations carrying F-related lineages throughout most of Asia, displacing the earlier haplogroup C and D lineages toward the margins.

More recent expansion events following population contraction associated with the last Ice Age 18,000–16,000 years ago are detectable in the Y-chromosome phylogeny (Fig. 2h). These include two main R lineages, R1a and R1b, with distinctive European geography (Semino et al. 2000), and Neolithic farmers from the Near East (haplogroups E and J). Interestingly, the diversification and phylogeographic patterns of Y chromosomes indicate a rather old back-to-Africa migration of Euroasian R lineages (Fig. 2f) prior to the widespread dispersion of high-frequency R1a and R1b lineages (Fig. 2g) that are not observed in Africa (Cruciani et al. 2002). Haplogroup O and N lineages are common in eastern Asia and may reflect the impact of millet and rice (Cavalli-Sforza et al. 1994). The apparent success of O and N lineages in eastern Asia appear analogous to certain widespread African E lineages that are common in Bantu-speaking agriculturalists (Diamond and Bellwood 2003).

## CONCLUSIONS

The haploid Y chromosome is unusual in that it is depauperate in genes relative to other nuclear chromosomes. However, the other unusual innate properties of being largely nonrecombining as well as having a low effective population size relative to other loci, combine both to preserve haplotypes over evolutionary time scales and to record numerous episodes of population divergence, even on micro-geographic scales, making it perhaps the single most insightful haplotype system known to characterize population affinity, substructure, and history. Some Y-chromosome polymorphisms could become part of a genome-wide inventory of genomic control markers useful in assessing the influences of population stratification. Both the Y chromosome and autosomes can be evaluated as SNPSTR systems with the empirical determination of phase providing an index of haplotype deterioration (Mountain et al. 2002). The Y chromosome provides a comparative model for evaluating haplotypes from other regions of the genome. The recovery of complex scenarios can be best advanced via an integrative approach, since the totality of the evidence should be reflective of an overall history and some correlation should be expected. When the story lines from multiple genes reinforce one another, overall population histories are revealed. Conversely, when different genes yield different haplotype patterns, locus-specific forces are in play. The recent and ongoing progress in deciphering the Y-chromosome structure in contemporary populations provides new opportunities to formulate specific testable hypotheses involving human evolutionary population genetics. Although the genetic legacy of *Homo sapiens* remains incomplete, the recent ability to unearth new levels of shared Y-chromosome haplotypic heritage and subsequent diversification provide not only an index of contemporary population structure, but also a preamble to human prehistory and substantial foundation for comparisons with other genomic regions.

## ACKNOWLEDGMENTS

Although many colleagues have helped to accelerate progress in understanding Y-chromosome variation in human populations, the contributions of Drs. P.J. Oefner and L.L. Cavalli-Sforza deserve particular recognition.

## REFERENCES

Bergen A.W., Wang C.-Y., Tsai J., Jefferson K., Dey C., Smith K.D., Park S.-C., Tsai S.-J., and Goldman D. 1999. An Asian-Native American paternal lineage identified by RPS4Y resequencing and by microsatellite haplotyping. *Ann. Hum. Genet.* **63:** 63.

Cann R.L. 2001. Genetic clues to dispersal in human population: Retracing the past from the present. *Science* **291:** 1742.

Capelli C., Wilson J.F., Richards M., Stumpf M.P.H, Gratrix F., Oppenheimer S., Underhill P.A., Pascali V.L., Ko T.-M., and Goldstein D.B. 2001. A predominantly indigenous paternal heritage for the Austronesian speaking peoples of insular South East Asia and Oceania. *Am. J. Hum. Genet.* **68:** 432.

Cavalli-Sforza L.L. and Feldman M.W. 2003. The application of molecular genetic approaches to the study of human diversity. *Nat. Genet.* **33:** 267.

Cavalli-Sforza L.L., Menozzi P., and Piazza A. 1994. *The history and geography of human genes.* Princeton University Press, Princeton, New Jersey.

Cruciani F., Santolamazza P., Shen P., Macaulay V., Moral P., Olckers A., Modiano D., Destro-Bisol G., Holmes S., Coia V., Wallace D.C., Oefner P.J., Torroni A., Cavalli-Sforza L.L., Scozzari R., and Underhill P.A. 2002. An Asia to Sub-Saharan Africa back migration is supported by high-resolution analysis of human Y chromosome haplotypes. *Am. J. Hum. Genet.* **70:** 1197.

de Knijff P. 2000. Messages through bottlenecks: On the combined use of slow and fast evolving polymorphic markers on the human Y chromosome. *Am. J. Hum. Genet.* **67:** 1055.

Diamond J. and Bellwood P. 2003. Farmers and their languages: The first expansions. *Science* **300:** 597.

Karafet T.M., Osipova L.P., Gubina M.A., Posukh O.L., Zegura S.L., and Hammer M.F. 2002. High levels of Y-chromosome differentiation among native Siberian populations and the genetic signature of a boreal hunter-gatherer way of life. *Hum. Biol.* **74:** 761.

Karafet T., Xu L., Du R., Wang W., Feng S., Wells R.S., Redd A.J., Zegura S.L., and Hammer M.F. 2001. Paternal population history of East Asia: Sources, patterns, and microevolutionary processes. *Am. J. Hum. Genet.* **69:** 615.

Kayser M., Brauer S., Weiss G., Schiefenhövel W., Underhill P.A., and Stoneking M. 2001. Independent histories of human Y chromosomes from Melanesia and Australia. *Am. J. Hum. Genet.* **68:** 173.

Kayser M., Brauer S., Weiss G., Underhill P.A., Roewer L., Schiefenhövel W., and Stoneking M. 2000. Melanesian origin of Polynesian Y chromosomes. *Curr. Biol.* **10:** 1237.

Ke Y., Su B., Son X., Lu D., Chen L., Li H., Qi C., Marzuki S., Deka R., Underhill P., Xiao C., Shriver M., Lell J., Wallace D., Wells R.S., Seielstad M., Oefner P., Zhu D., Jin J., Huang W., Chakraborty R., Chen Z., and Jin L. 2001. African origin of modern humans in East Asia: A tale of 12,000 Y chromosomes. *Science* **292:** 1151.

Kivisild T., Rootsi S., Metspalu M., Mastana S., Kaldma K., Parik J., Metspalu E., Adojaan M., Tolk H.-V., Stepanov V., Gölge M., Usanga E., Papiha S.S., Cinnioglu C., King R., Cavalli-Sforza L., Underhill P.A., and Villems R. 2003. The genetic heritage of earliest settlers persist in both the Indian tribal and caste populations. *Am. J. Hum. Genet.* **72:** 313.

Knight A., Underhill P.A., Zhivotovsky L.A., Mortensen H.M., Ruhlen M., and Mountain J.L. 2003. African Y chromosome and mtDNA diversity and the antiquity of click languages. *Curr. Biol.* **13:** 464.

Knowles L.L. and Maddison W.P. 2002. Statistical phylogeography. *Mol. Ecol.* **11:** 2623.

Lell J.T., Sukernik R.I., Starikovskaya Y.B., Su B., Jin L., Schurr T.G., Underhill P.A., and Wallace D.C. 2002. The duel origin and Siberian affinities of Native American Y chromosomes. *Am. J. Hum. Genet.* **70:** 192.

Mountain J.L., Knight A., Jobin M., Gignoux C., Miller A., Lin A.A., and Underhill P.A. 2002. SNPSTRs: Empirically derived, rapidly typed, autosomal haplotypes for inference of population history and mutational processes. *Genome Res.* **12:** 1766.

Owens K. and King M.-C. 1999. Genomic views of human history. *Science* **286:** 451.

Semino O., Santachiara-Benerecetti A.S., Falaschi F., Cavalli-Sforza L.L., and Underhill P.A. 2002. Ethiopians and Khoisan share the deepest clades of the human Y-chromosome phylogeny. *Am. J. Hum. Genet.* **70:** 265.

Semino O., Passarino G., Oefner P.J., Lin A.A., Arbuzova S., Beckman L.E., De Benedictis G., Francalacci P., Kouvatsi A., Limborska S., Marcikiae M., Mika A., Mika B., Primorac D., Santachiara-Benerecetti A., Cavalli-Sforza L.L., and Underhill P.A. 2000. The genetic legacy of Palaeolithic *Homo sapiens sapiens* in extant Europeans: A Y-chromosome perspective. *Science* **290:** 1155.

Stringer C. 2000. Coasting out of Africa. *Nature* **405:** 24.

Tajima A., Pan I.-H., Fucharoen G., Fucharoen S., Matsuo M., Tokunaga K., Juji T., Hayami M., Omoto K., and Horai S. 2002. Three major lineages of Asian Y chromosomes: Implications for the peopling of east and southeast Asia. *Hum. Genet.* **110:** 80.

Thangaraj K., Singh L., Reddy A.G., Rao V.R., Underhill P.A., Pierson M., Frame I.G, and Hagelberg E. 2003. Genetic affinities of the Andaman Islanders, a vanishing human population. *Curr. Biol.* **13:** 86.

Underhill P.A. and Roseman C.C. 2001. The case for an African rather than an Asian origin of the human Y-chromosome YAP insertion. In *Recent advances in human biology,* vol. 8: *Genetic, linguistic and archaeological perspectives on human diversity in Southeast Asia* (ed. L. Jin et al.), p. 43. World Scientific, River Edge, New Jersey.

Underhill P.A., Passarino G., Lin A.A., Marzuki S., Cavalli-Sforza L.L., and Chambers G. 2001a. Maori origins, Y chromosome haplotypes and implications for human history in the Pacific. *Hum. Mutat.* **17:** 271.

Underhill P.A., Passarino G., Lin A.A., Shen P., Foley R.A., Mirazón Lahr M., Oefner P.J., and Cavalli-Sforza L.L. 2001b. The phylogeography of Y chromosome binary haplotypes and the origins of modern human populations. *Ann. Hum. Genet.* **65:** 43.

Underhill P.A., Shen P., Lin A.A., Jin L., Passarino G., Yang W.H., Kauffman E., Bonné-Tamir B., Bertranpetit J., Francalacci P., Ibrahim M., Jenkins T., Kidd J.R., Mehdi S.Q., Seielstad M.T., Wells R.S., Piazza A., Davis R.W., Feldman M.W., Cavalli-Sforza L.L., and Oefner P.J. 2000. Y chromosome sequence variation and the history of human populations. *Nat. Genet.* **26:** 358.

White T.D., Asfaw B., DeGusta D., Gilbert H., Richards G.D., Suwa G., and Howell F.C. 2003. Pleistocene *Homo sapiens* from Middle Awash, Ethiopia. *Nature* **423:** 742.

Y Chromosome Consortium. 2002. A nomenclature system for the tree of human Y-chromosomal binary haplogroups. *Genome Res.* **12:** 339.

# The New Quantitative Biology

M.V. Olson
*Departments of Medicine and Genome Sciences, University of Washington, Seattle, Washington 98195*

Late on Monday afternoon, June 2, 2003, Bruce Stillman stepped to the podium in Grace Auditorium to introduce my talk summarizing the 68th Cold Spring Harbor Symposium on Quantitative Biology. In the midst of his remarks, the power failed. It quickly became evident that the problem was not local—the whole laboratory and surrounding areas had gone dark. Hence, there was little to do but carry on. With only emergency runner lights along the outside corridors, I stared out into the dark auditorium, abandoned my PowerPoint slides, raised my unamplified voice and began. The symbolism was inauspicious for a meeting that took a first, exhilarating look at ways in which the human-genome sequence was casting new light on biology.

A sense of history enveloped the occasion. Artificial as it may have been to coordinate the announcement—or, more precisely, the latest in a series of announcements—that the human genome had been sequenced with the 50th anniversary of the Watson-Crick *Nature* paper, there was a clear sense at the meeting that we were witnessing a major transition in the history of biology. For 50 years, the focus of molecular biology had been on understanding the flow of information within cells, a process that ultimately led to the sequencing of whole genomes. Now, the gears of this vast enterprise were shifting before our eyes. At the end of his Nobel Lecture in 1955, Hugo Theorell referred to the "yawning gulf" between biochemistry and morphology (Theorell 1964). At the end of CSHSQB LXVIII, in the darkness of Grace Auditorium, we were looking out at the still larger gulf between the genome and organismal biology.

The terms of encounter between biologists and living systems had changed. As Jim Watson said to me at the meeting:

> It's like 1953. Once we had the double helix, everything had changed. Now we have the human-genome sequence and everything has changed again.

In her more down-to-earth style, Jane Rogers described the meeting as having been "a pretty good start to the human genomic era." Surely the main interest in these volumes, both now and in the future, will be as a record of that start. Perhaps it is true that "everything has changed" in biology. If so, the Symposium was a prime opportunity to see how biologists were seizing the moment.

A taxonomy of the meeting's talks is presented in Table 1. The conceptual scheme is arbitrary, and I have made debatable judgments about where to assign particular papers. Nonetheless, themes emerge from this rough view of the topics emphasized by the 79 speakers who preceded me to the podium. For example, genome-analysis technology was only lightly represented. This omission was all the more striking since this year's Symposium subsumed the annual Cold Spring Harbor meeting on genome sequencing and biology, a traditional forum for talks on genomic technology.

All authors cited here without dates refer to papers in this volume.

## GENOME SEQUENCING

Although there were five talks on genome sequencing, three were retrospective views of the Human Genome Project. Each concluded that the April, 2003, release of the human-genome sequence was a high-quality product. Rogers (Rogers) reported that nucleotides randomly sampled from the 2.85 Gbp of euchromatic DNA in the April, 2003, release reside on contigs of average size 27 Mbp. Fewer than 400 gaps, unspanned by analyzed clones, remain. This level of completeness resulted from the largely unsung efforts of the expert "finishers" associated with ~20 genome centers. This army of finishers hand-curated the final sequences of 26,000 BAC clones. Schmutz (Schmutz et al.) reported that a quality-control program, based on resequencing randomly selected BACs, detected errors at a rate $<10^{-5}$. T. Furey followed with a computationally based analysis that found few inconsistencies between the April, 2003, release and other sources of data: The genome sequence has a "near perfect" alignment with the human genetic map and also correlates well with the STS-based physical map. However, Furey did note a significant number of discrepancies with available full-length cDNA sequences; experimental follow-up will be needed to sort out which of these discrepancies are due to errors in the cDNA data, errors in the genome sequence, or biological causes.

## COMPARATIVE GENOMICS

Hence, the consensus was that the researchers at this Symposium on "The Genome of *Homo sapiens*" have quite a good genome sequence to work with. A high priority now is to increase the functional annotation of the genome. It was apparent at the meeting that comparative genomics will play a central role in this endeavor. The link between comparative genomics and function is rooted in the discovery, discussed by Haussler (Chiaromonte et al.), that ~5% of the human genome has been "conserved" since the last common ancestor between the human and mouse (~75 Mya). This estimate is

**Table 1.** Talks Grouped by Primary Emphasis

| Primary emphasis | Number | Speakers* |
|---|---|---|
| Molecular evolution | 23 | Antonarakis (Antonarakis et al.); Brent (Wang et al.); Clark (Clark et al.); Eichler (Bailey and Eichler); C. Fraser; K. Frazer; E. Green (Margulies et al.); P. Green; Hardison (Hardison et al.); Haussler (Chiaromonte et al.); Kent (Chiaromonte et al.); Koonin (Rogozin et al.); S. Pääbo; D. Page; Parkhill (Parkhill and Thomson); C. Ponting; Riethman (Riethman et al.); E. Rubin (Pennacchio et al.); Sakaki (Sakaki et al.); Stubbs (Hamilton et al.); Waterman (Zhang and Waterman); Waterston (Waterston et al.); Weissenbach (Jaillon et al.) |
| Genotype–phenotype correlations | 15 | Chakravarti (McCallion et al.); Cheung (Cheung et al.); Chiba-Falek (Chiba-Falek and Nussbaum); D. Cox; Drayna (Drayna et al.); Georges (Georges and Andersson); Little (Cotsapas et al.); Loots (Ovcharenko and Loots); Lupski (Stankiewicz et al.); Merikangas (Merikangas); Ostrander (Guyon et al.); Porteous (Porteous et al.); J. Singer; Wallace (Wallace et al.); Windemuth (Windemuth et al.) |
| Functional genomics | 14 | Baillie (McKay et al.); Botstein (Botstein); Cai (Li et al.); Friddle (Friddle et al.); G. Hannon; Hayashizaki (Hayashizaki); Hood (Weston et al.); E. Lander; Malek (Malek et al.); Roe (Roe et al.); Snyder (Lian et al.); Tyers (Jorgensen et al.); M. Wigler; R. Young |
| Genetic variation | 12 | D. Altshuler; Bentley (Bentley); Bertranpetit (Bertranpetit et al.); Chee (Fan et al.); M. Daly; P. Donnelly; Goldstein (Goldstein et al.); Kwok (Kwok and Xiao); M. Olson; N. Risch; Underhill (Underhill); Wilson (Wilson et al.) |
| Genome annotation | 6 | Ashburner (Ashburner et al.); Birney (Birney et al.); Lipovich (Lipovich and King); G. Rubin; Stein (Joshi-Tope et al.); Zhang (Zhang) |
| Genome sequencing | 5 | T. Furey; Gibbs (Gibbs and Weinstock); E. Myers; Rogers (Rogers); Schmutz (Schmutz et al.) |
| Ecological genomics | 2 | D. Rokhsar; J.C. Venter |
| Genome function | 1 | Willard (Rudd et al.) |
| Overview | 1 | Collins (Collins) |

*Author names in parentheses indicate chapters included in this volume.

based on an analysis of the distribution of sequence divergence across alignable regions of the human and mouse genomes. Alignable regions include not only exons and other conserved sequences, but also all neutrally evolving sequences that were present in the last common ancestor between human and mouse and are still present in both contemporary genomes. Because of a high rate of genomic turnover, due to constant insertions and deletions, only ~40% of the two genomes are alignable; however, this portion of both genomes is presumed to contain nearly all the functionally important sequences, since such sequences are expected to be protected from loss by purifying selection. The distribution of sequence divergence within the alignable regions has a low-divergence tail that is too large to explain without invoking selection. The 5% estimate for the portion of the genomes under purifying selection comes from an analysis of the size of this tail. Haussler acknowledged that the estimate was "still a little squishy," and put the bounds at 4–7%.

Obviously the real question is not whether the number is 4% or 7%, but which sequences it includes. This question received so much attention at the meeting, much of it based on analyses of data from Eric Green's Comparative Vertebrate Sequencing Project, that Bob Waterston jokingly introduced Green with the comment, "Now Eric gets to talk about his data after everyone else has already talked about it!" The data set was based on sequencing a sampling of genomic regions, each comprising a million or more base pairs, from a wide variety of vertebrates. The immediate challenge has been to use these data to identify what Green (Margulies et al.) referred to as *m*ulti-species-*c*onserved *s*equences (MCSs); Weissenbach (Jaillon et al.) offered the alternate name *e*volutionarily *co*nserved *re*gions (ECOREs). Green reported, on the basis of his group's statistical model, that 4% of the sequences in the region of the human CFTR gene lie in MCSs of average size 58 bp. Of these MCSs, 27% code for protein, 5% are in untranslated portions of mRNAs, 2% are in ancient repeats (a presumed indication of the false-positive rate), and 66% have unknown functional significance. In a detailed report of evolutionarily conserved sequences on human Chromosome 21, Antonarakis (Antonarakis et al.) also emphasized the large proportion of these sequences whose functions are unknown. A joke at the meeting was that the genome is full of "dark matter." Beyond the difficulty of assigning functions to the mysterious majority of MCSs, there is the sobering reality that existing methods of analyzing comparative data miss many functionally important sites. To emphasize this point, Green commented that 20% of CF-causing mutations in the CFTR gene lie outside of currently detectable MCSs.

One possible explanation for false negatives when scanning for MCSs is that some aspects of genome function evolve too rapidly to be detectable in comparisons between humans and distantly related mammals. K. Frazer presented experimental support for this view from a chip-based resequencing study that involved comparisons of one human locus with horse, cow, pig, dog, and cat: Only a quarter of the evolutionarily conserved regions were detectable via human–mouse comparisons alone. There was much discussion at the Symposium about which set of organisms would provide the ideal inputs for comparative analysis. However, given the diversity of questions that can be addressed through evolutionary comparisons, consensus on this point seems unlikely. As Eddy Rubin put it, "There is no perfect distance for evolutionary comparisons." At one extreme, Weissenbach (Jaillon et al.) emphasized the utility of distant comparisons for tracking down the stray exons that continue to elude standard annotation methods. For this purpose, long divergence times decrease the "noise" associated with sequences that evolve more rapidly than typical exons. On the basis of a sophisticated statistical analysis of exon-detection methods, Brent (Wang et al.) concluded that the mouse–chicken divergence was nearly ideal: At this evolutionary distance, intron alignments have just been eliminated, and exon alignments are still quite good. For detection of functional elements that evolve rapidly, Rubin (Pennacchio et al.) advocated comparative analysis of many primates, or even many individuals of a single species. This approach has the advantage of constraining comparisons to biologically similar organisms but requires far more data than a reliance on distant comparisons. In a similar vein, Birney (Birney et al.) advocated analysis of single-nucleotide differences (SNDs) between closely related species to detect the effects of selection on wild populations such as *Anopheles* mosquitoes. Brent (Wang et al.) predicted that it will ultimately be possible to use multiple, simultaneous comparisons of genome sequences to infer the rate of evolution at every nucleotide position. However, most would agree with his assessment that "We are not there yet."

The two main analyses of closely related metazoan genomes both involved human–chimpanzee comparisons. Clark (Clark et al.) compared rates of nonsynonymous versus synonymous substitutions in three-way comparisons between human, chimpanzee, and mouse. Nearly 10% of mammalian genes show statistically significant evidence of positive selection that has accelerated on the human lineage. However, this figure is reminiscent of Haussler's (Chiaromonte et al.) estimate that 4–7% of mammalian genomes are under purifying selection. Both estimates are based on decomposing two heavily overlapping distributions; hence, neither produces anything like a clean list of which elements do or do not meet the specified criteria. S. Pääbo presented evidence for positive selection at the *FOXP2* gene, which has been implicated in language development through studies of rare families with *FOXP2* mutations. He also reported on loss of function in many olfactory-receptor genes in both the human and chimpanzee lineages, with the most pronounced loss having occurred on the human lineage.

In contrast to the many talks that attempted to infer the functional importance of genomic elements through sequence conservation, Phil Green presented data showing that transcribed sequences are often discernable on the basis of an entirely different phenomenon. Promiscuous gene transcription in the germ line appears to affect nucleotide-substitution patterns during evolution, presumably via operation of a transcription-coupled repair system. This mechanism leads to a statistically significant excess of G's and T's in the coding strand of ~70% of human genes. These data are perhaps the most convincing evidence available that most genes are transcribed, at least at minimal levels, in the germ line. Comparative-sequence analysis is often treated as a signal-to-noise problem in which evolutionarily conserved sequences are the "signal" and the much larger number of neutrally evolving sequences are the "noise." Green pointed out that his findings indicate that the noise actually carries a signal about the locations of functional units in genomes.

The basic data structure for all comparative genomics is the sequence alignment. Indeed, behind every sequence-based phylogenetic tree is a multi-sequence alignment, often constructed by less rigorous methods than those used to build the actual tree. Waterman reported on a new approach to sequence alignment based on Eulerian paths. This method appears to be less easily confused by the ubiquitous repeated sequences that pose the major challenge to sequence-alignment algorithms. The new method also has better computational performance than competing approaches and, therefore, may offer a practical path toward multiple alignment of thousands of sequences. Waterman jokingly disparaged talks that include "shameless used-car selling" and then proceeded to promote his method as outperforming the widely used ClustalW algorithm. Kent (Chiaromonte et al.) discussed another aspect of the alignment problem, long-range alignment of relatively diverged sequences whose alignable segments are interrupted by long insertions and deletions. A rich source of examples is provided by human–mouse alignments, which fail altogether in "synteny-breakpoint hot spots" that can span hundreds of thousands of base pairs.

The impressive focus on comparative genomics at this meeting, by experimentalists and bioinformaticians alike, foreshadows an era in which analyses based on selected regions will be displaced by simultaneous, comparative analyses of multiple whole genomes. The potential power of such methods for higher organisms was foreshadowed at the meeting by reports on their application to bacteria. C. Fraser gave a broad overview of the many ways in which whole-genome comparisons are transforming microbiology. With 105 complete genomes available—including an increasing number of clusters of sequences from closely related bacteria—it is becoming possible to address the interrelated questions "How did these genomes evolve?" and "How do they work?" Parkhill (Parkhill and Thomson) gave several examples of ways in which a better understanding of genome evolution illuminates genome function and vice versa. A dramatic finding is that such feared human pathogens as *Salmonella typhi* and *Bordetella pertussis* show strong signatures of sudden, extreme specialization. By acquiring some new genes and losing function in many old ones, these pathogens have lost the ability of their ancestral forms to survive in general environments, to interact with diverse hosts, and to maintain relatively benign interactions with their hosts. Much of this evolutionary activity appears to be recent, possibly indicating that these bacterial lineages prospered by piggybacking on the huge expansion of the human population that followed the development of agriculture.

Although available data are much sparser for metazoans, analogous evolutionary processes appear to have occurred throughout the tree of life. Evolution appears to involve repeated cycles of specialization in which individual species and whole evolutionary lineages periodically shed genes that were present in more generalist ancestors. Koonin reported on large-scale comparisons of the gene content of available metazoan genomes. These comparisons reveal a pattern of many, independent gene-loss events on different lineages. This finding, which applies on a timescale of hundreds of millions of years, is reminiscent of S. Pääbo's report of recent large-scale loss of olfactory receptor genes on the chimpanzee and human lineages. In all likelihood, the recent loss of these genes is simply an instance of the general process described by Koonin (Rogozin et al.).

Given the desirability of basing the comparative genomics of all evolutionary groups on comparisons of multiple whole-genome sequences, the meeting was oddly lacking in serious assessments of the rate at which such data will become available and the extent to which they will be of adequate quality to meet future needs. This void appears to have reflected lack of activity rather than the tastes of the organizers. E. Myers did report on incremental progress with his whole-genome assembler. He also expressed the view that reasonable whole-genome assemblies were only attainable when the sequence coverage was at least 8x, a sampling depth higher than that reached in many current whole-genome-shotgun projects. Gibbs (Gibbs and Weinstock) was a lone voice who actually addressed the technical challenges associated with adding value to the raw output of whole-genome-shotgun projects. His main message was that, although it may be impractical to keep the army of finishers who produced the current human-genome sequence in the field, there are numerous lower-cost options for improving on raw whole-genome assemblies.

## GENOME ANNOTATION AND RESOURCE DEVELOPMENT

This Symposium was dominated by the consumers, not the producers, of genomic sequence. These consumers want, first and foremost, better annotation. Gerald Rubin emphasized challenges and successes in annotating the *Drosophila* genome. Challenges include the presence of hundreds of genes in heterochromatic regions that remain difficult to sequence reliably, much less to annotate. There are also many instances of genes embedded within genes, transcribed either from the same or the opposite strand, a theme also developed by Lipovich (Lipovich and King) for the human. Rubin's emphasis on these complex gene structures stimulated a lively exchange with Brent about the prospects of detecting and disentangling overlapping and embedded genes computationally. Whereas Brent expressed optimism about the pace at which gene-detection programs are improving, Rubin countered with the observation that "sequencing is getting cheaper." The implication of cheaper sequencing is that the comparative genomics of *Drosophila* species may provide the best path toward comprehensive gene identification for *D. melanogaster*. Rubin forcefully advocated the view that annotation and resource creation (e.g., full-length cDNA clones, comprehensive collections of transposon mutations) should be taken to the same level of completion as the sequence itself.

Many talks emphasized the breadth of commitment to this goal, at least for intensively studied organisms. Hayashizaki described impressive progress on development of a full-length-cDNA resource for the mouse. Indeed, he indicated that the resource is now sufficiently rich that more attention is needed to distribution mechanisms. Hayashizaki (Hayashizaki) illustrated the increasing disparity between the ease of distributing data and the difficulty of distributing biological resources by describing his experiences sending out the FANTOM2 set of 60,770 mouse full-length-cDNA clones packaged with 100 kg of dry ice. To solve this problem, the Riken group has developed DNA-printing technology that may allow routine distribution of DNA samples, designed for easy PCR amplification, on printed pages. Perhaps this application will provide a long-term future for hard-copy editions of biological journals!

Other examples of large-scale biological resources included Friddle's (Friddle et al.) report on high-throughput mouse-knockout technology. His talk described an insertion-mutagenesis protocol based on retroviral constructs that disrupt mouse transcription units. This method has yielded over 200,000 ES cell lines containing unique insertion events, which disrupt the function of ~60% of mouse genes. The positions of the insertions are determined by sequencing the exon downstream from the insertion site following PCR amplification of reverse transcripts of mRNAs transcribed from a promoter within the inserted element. RNAi offers a potential alternative to knockout technology for functional annotation of genomes. Hannon reported on major progress in understanding the mechanism of RNAi, a step that should facilitate increased reliance on this relatively cheap alternative to traditional gene-inactivation methods.

High-throughput in situ hybridization to developing embryos provides still another experimental means of functional annotation. Roe (Roe et al.) described a project to gather such data for the zebrafish, with an initial focus on the orthologs of genes present on human Chromosome 22. This talk was of technical interest in that Roe, a genome-sequencing veteran, retooled his sequencing pipeline for this purpose. Malek (Malek et al.) described a way of coupling sequencing technology even more directly to functional annotation. In his system, which has been pioneered on the bacterium *Rickettsia sibirica*, the subclone libraries that provide sequencing templates for whole-genome-shotgun sequencing are constructed in a vector that supports their use as "bait" and "prey" sequences in a bacterial implementation of the two-hybrid method for detecting protein–protein interactions. Hence, simply as a by-product of genome sequencing, one obtains a saturating collection of sequenced clones for use in protein–protein-interaction mapping.

There was ample evidence at the Symposium that microarrays will also continue to play a major role in the functional annotation of genomes. Snyder (Lian et al.) described initial results of hybridizing placental poly(A) RNA to a human Chromosome 22 array manufactured by spotting unique genomic sequences on glass. Fully half of the hybridizing spots do not correspond to known genes. These data raise the obvious question of whether or not the current annotation of the human genome, even in terms of gene content, is grossly incomplete. Lively discussions of this issue ensued after a presentation by Birney (Birney et al.) that was equally memorable for its scientific content and entertainment value.

## HUMAN GENE NUMBER

Birney had the unenviable charge of declaring a winner of a three-year lottery on the number of human-protein-coding genes. The lottery had been started in 2000, when a $1 bet bought full participation. It continued in 2001 and 2002 with the betting price rising first to $5, then $20. The idea was that knowledge of the gene number would be increasingly reliable during this period so the odds on an accurate guess should be adjusted accordingly. David Stewart, the Director of Meetings and Courses at Cold Spring Harbor, had served as lottery director and promoter. From the beginning, he had informed participants that "The winner will be declared in 2003, by which time the scientific community is expected to reach a consensus." At the Symposium, it was apparent that Stewart's expectation had been overly optimistic. Birney asked for another year or two, but Stewart held firm. Hence, after a careful assessment of the many uncertainties in counting genes, Birney shocked the audience by supporting the startlingly low estimate of 24,500. Even this number, he indicated, could include as many as 3,000 miscalled pseudogenes. Nonetheless, for purposes of the lottery, further downward adjustments would be irrelevant: The overall winner was Lee Rowen, whose 2001 bet on 25,947 genes was the lowest in the entire pool. Snyder, still impressed with the many physical transcripts that do not correspond to annotated genes, expressed the opinion that "It is still too early to be giving out disposable prizes for estimates of the human gene number." Nonetheless, the days are past when the complexity of human biology can be explained by postulating that our species has an unusually large number of genes. Botstein provided one humorous indication of how rapidly our knowledge of the genome of *Homo sapiens* has advanced by referring to a 1971 *Scientific American* article that estimated the human gene number as 10 million! He attributed this extravagance to neurobiologists, some of whom apparently felt in 1971 that we must need "lots of genes for thinking."

## BIOINFORMATICS

Residual uncertainties in the human gene count are just one of many indications that the computational and experimental annotation of genomes will continue to be a growth industry for the foreseeable future. Consequently, there was intense interest in how all the resultant data would be integrated. Botstein (Botstein) declared that "Genomics has made biology into an information science." He emphasized that one consequence of this makeover was the need to back away from extreme specialization both in education and research. Botstein advocated a major emphasis on building skills and systems that will allow generalists to compute on all available data when testing high-level hypotheses. Ashburner (Ashburner et al.) opined that "we are in deep trouble" building the needed infrastructure sufficiently rapidly. Genomic data are accumulating at a rate faster than Moore's Law, a description of the rate at which transistor densities on chips have increased since the invention of integrated circuits. In Ashburner's view, many past efforts to address the biological information explosion have failed because technical solutions have been implemented before conceptual issues were adequately addressed. The Gene Ontology Project, which attempts to describe gene products in a biologically meaningful, controlled vocabulary, is an effort to shore up the conceptual foundations of large biological databases. Stein (Joshi-Tope et al.) extended this theme with his discussion of the Genome Knowledgebase Project, which employs gene ontologies while adding further value through expert curation of existing biological knowledge.

Tyers (Jorgensen et al.) and E. Lander described the challenges and opportunities confronting potential users of integrated information resources. Tyers's project involves efforts to bring diverse sources of data to bear on cell-size homeostasis in budding yeast. Through integrated analyses of knockout phenotypes, patterns of synthetic lethality, protein–protein-interaction networks, and gene-expression data, his laboratory has identified over 40 new genes that regulate cell size. Lander described applications of similarly diverse resources to a variety of human phenotypes, including type II diabetes.

## SYSTEMS BIOLOGY

The meeting saw little disagreement about the pressing need to improve the ease with which large, comprehensive, and diverse sources of data can be mobilized during the everyday conduct of biological research. More controversial was the question of whether or not we are on the cusp of a true "systems biology." The first probings in this direction involve transcriptional regulatory networks. Hardison (Hardison et al.) described an integrated bioinformatic and experimental approach to identifying *cis*-acting sequences that mediate erythroid development. R. Young reported on "chip-chip" experiments that reveal, both in yeast and mammalian cells, the sites at which specific regulatory proteins interact with genomic DNA. Snyder (Lian et al.) described similar experiments and offered the optimistic prediction that "Transcriptional networks are complex but still figurable-outable, if that is a word." Hood (Weston et al.) presented the only talk in which the "figuring out" phase was formulated explicitly as a problem in systems biology. He presented elaborate schematics of regulatory networks that control sea urchin development and claimed that the models behind these schematics already allow predictions of how the system will respond to perturbations. This talk generated a vigorous discussion of where we really are on the path from the types of heuristic models that have dominated molecular biology since the early days of the *lac* operon to true predictive modeling. During this discussion, Peter Little (Cotsapas et al.) advocated the stringent standard that predictions must be non-obvious, quantitative, and falsifiable; speaking for many skeptics, he indicated that he had yet to see any results from systems biology that meet this test.

This debate captured one clear axis of tension at the meeting. Certainly there was optimism that the reference data, computational and experimental tools, and global perspective of genome research will successfully shift the center of gravity of biology toward the analysis of increasingly complex aspects of life. However, balancing the sense of new possibilities was a clear recognition that the challenges ahead are immense. Hence, one question that hovered over CSHSQB LXVIII was "If total victory in developing a molecular understanding of life remains a distant dream, what are the more proximate goals of the genomic era?" One answer is that genome researchers may increasingly pursue practical goals. In his overview of future directions in genome research, Collins (Collins) advocated an increased focus on achieving medical benefits, and many other speakers described efforts in this direction.

## HUMAN GENETICS

Most medically related genome research continues to be directed toward the elucidation of pathogenic mechanisms. Although population-based and case-control studies received most of the attention, Porteous (Porteous et al.) described family-based approaches to schizophrenia and bipolar affective disorder. These diseases, and psychiatric disorders in general, have proven particularly frustrating targets for "positional cloning." By analyzing one extended family with multiple cases of bipolar affective disorder, Porteous's group has defined a candidate region on Chromosome 4. Even when supporting evidence from other families is also used, the candidate region spans several megabase pairs. Interestingly, the audience voted for immediate sequencing of affected chromosomes across the entire region as the preferred next step in the search for causal mutations.

D. Cox reported on one actual effort to carry out a whole-genome-association scan. His project sought to identify genetic variants contributing to the occurrence of cirrhosis in heavy drinkers. Cox expressed the view that the technology still did not exist to allow such scans to be carried out by genotyping individual samples at an adequate density of SNPs. However, by pooling case and control samples, Cox's group identified a modest number of SNPs that defined potential "disease-enhancing" loci. All the SNPs were common in both case and control populations: Hence, if they prove to be true positives, the model for enhancement of disease susceptibility must involve the combinatoric effects of many loci.

Most talks on whole-genome-association scans focused on technology and infrastructure rather than applications. Chee (Fan et al.) and Kwok (Kwok and Xiao) both described advances in genotyping methods that address one of the major obstacles to such studies. Other talks addressed the question of how SNPs should be chosen for association studies. Goldstein (Goldstein et al.) expressed optimism that as few as $10^5$ SNPs might be sufficient for whole-genome scans, at least in some populations. D. Altshuler reported on haplotype mapping in multiple populations. On the whole, his studies suggest that SNPs which adequately capture diversity in African, Asian, and European populations also work well in more narrowly defined groups. Linkage disequilibrium extends over the longest distances in Native Americans and Pacific Islanders, and these populations also have the least haplotype diversity of all groups examined. Bertranpetit (Bertranpetit et al.) expressed skepticism about the adequacy of SNPs acquired in a few major populations to capture worldwide diversity. In a study of linkage disequilibrium on Chromosome 22, his group found substantial heterogeneity in allele frequencies for many SNPs and the extent of linkage disequilibrium between SNPs in different populations.

## ETHICAL AND SOCIAL ISSUES

Discussions of social concerns about current approaches to human genetics were interspersed with those of technical issues. Jan Witkowski moderated an interactive panel that sought to explore the sensitive issues surrounding such work. The panel largely explored familiar topics: potential conflicts of interest for scientists who straddle the divide between academic and commercial research, the risk that geneticists will reinforce racial stereotypes as they characterize human-population structure, and the need for more realism in discussing the health relevance of advances in genome research. Nonetheless, I perceived an unspoken tension at the meeting that transcended these traditional ethical, legal, and social issues. A brief exchange between Ewan Birney and David Altshuler, following Altshuler's talk, was exemplary. Birney asked whether, in designing the haplotype-mapping project, it would not have made sense to collect some phenotypic data about the individuals whose genomes were being mapped. In this way, assessment of the number and choice of SNPs required for whole-genome-association scans could be informed by simultaneous efforts to associate these SNPs with phenotypes. Altshuler responded that this approach was impossible because of Institutional Review Board restrictions on the haplotype-mapping project.

In my summary talk, I revisited this point. While acknowledging that Altshuler's reply was factually correct, I expressed the view that the tight restrictions under which most current research on human genotype–phenotype correlations is conducted are largely self-imposed by geneticists. From a scientific standpoint, Birney's point is incontrovertible. However, there is a comfort level, to which human geneticists have perhaps too readily acceded, in blaming Institutional Review Boards for the difficulty of applying genomic tools effectively to studies of the genetic basis of human phenotypes. In reality, the risks of most such studies to individual human subjects are quite theoretical and, in comparison to traditional standards of human-subjects protection, insignificant. I expressed the view that the real source of reluctance to use more effective research strategies is a largely unspoken fear by geneticists of what they might learn if they adopted scientifically optimum approaches to the analysis of diverse human phenotypes. There is perhaps a hope that more cautious approaches will allow geneticists to skim off the "medically relevant" information about genotype–phenotype correlations without inflaming the nature–nurture debate. To the extent that this dynamic is actually in play, I suggested, it is a cause for concern. The idea that geneticists can prejudge what is medically relevant, while remaining profoundly ignorant of the genes that influence most aspects of human individuality, draws little support from the history of science.

## THE NEW QUANTITATIVE BIOLOGY

CSHSQB LXVIII captured a transitional phase of biology. Unlike many of its predecessors, this meeting will not be remembered for the "discoveries" announced. I suspect that most attendees would be hard pressed to identify a single, significant example of new biological knowledge that they acquired during the 5-day span of the meeting. Yet there was a palpable sense of enthusiasm about the opportunities before us. An observation about this Symposium that is particularly appropriate to the venerable series of maroon books in which its proceedings appear is that we are suddenly practicing a new form of "quantitative biology." This new quantitative biology differs entirely from the vision articulated by the founders of this series in 1933. Their dream was of a quantitative physiology that has yet to emerge for any broad range of biological processes. Indeed, if systems biology has its day, it will be the ultimate realization of that early dream. By the middle of the last century, structural biology had gradually displaced quantitative physiology as the most important application of quantitative theory to biology. Of course, the influence of structural biology on the Cold Spring Harbor Symposia was immense, if largely indirect. By providing access to macromolecular structures, most notably that of DNA, structural biology created the whole field of molecular biology, in whose history these Symposia have played such a prominent role. What we now have, rather suddenly, is a new quantitative biology rooted in genome sequences. Gene-finding, sequence alignment, and phylogenetic-tree construction are its defining tools. Statistical genetics and computational methods for representing and retrieving diverse sources of biological information are its natural adjuncts. Molecular evolution and population genetics are its unifying conceptual principles.

Real integration of this new quantitative biology with biological function—phenotype in the case of genetic approaches and systems biology in the case of molecular physiology—is largely a task for the future. However, genome sequences are now firmly installed as the central data structures of biology. Phil Green commented to me at the meeting that it was stimulating to see the merging of disciplines reflected by the Symposium's program. Terms such as "parsimony," "coalescence," "Markov chain," and "maximum likelihood" were freely intermixed with "transcription factor," "operon," "regulatory network" and "mutant."

## CONCLUSION

This meeting was a tribute to the enormous sea change that genome sequences have brought to biology. Genome sequences are not just one more tool in the biologist's armamentarium. Our understanding of genomes—and, now, our detailed access to the digital messages they contain—provide constraints on biological complexity that we are just learning to exploit. Suddenly, biology is finite. Indeed, it is the finite bounds that genomes place on biological complexity that are the central legacy of the double helix, in this year of its Golden Jubilee.

Toward the end of my summary lecture, I highlighted this message of the genomic revolution by telling a personal anecdote. The year was 1972, and I was an Assistant Professor of Chemistry at Dartmouth College. One day, when I should have been doing something else, I picked up a copy of the Second Edition of *The Molecular Biology of the Gene* (Watson 1970). Chapter 3, entitled "A Chemist's View of the Bacterial Cell" read as though it had been written directly for me. Indeed, in an early section of the chapter entitled "Even Small Cells are Complex," Watson echoed my nascent view of biology exactly: "At first sight, the problem of soon, if ever, understanding the essential features of *E. coli* should seem insuperable to an honest chemist." He followed this cautionary note with a survey of the types of molecules present in cells and an introduction to biochemical pathways and metabolic charts. Impressive as this biochemical knowledge of the 1970s may have been, Watson acknowledged that it would not dispel an "honest chemist's" skepticism. How do we know, he asked, that cells do not have 50–100 pathways for metabolizing glucose or 20 different ways to make histidine? Why should a "sophisticated pure chemist" believe that any tidy description of cellular processes would ever be possible? Perhaps existing knowledge only captures the most biochemically tractable tip of a huge iceberg.

Then, on the chapter's last page, the tone changed. With a back-of-the-envelope calculation of the number of proteins encoded in the *E. coli* genome—one that would still suffice for the purpose—Watson concluded that the progress that had been made in understanding the chemistry of life was real, not illusory:

> ... even a cautious chemist, when properly informed, need not look at a bacterial cell as a hopelessly complex object. Instead he might easily adopt an almost joyous enthusiasm, for it is clear that he, unlike his nineteenth-century equivalent, at last possesses the tools to describe completely the essential features of life.

In short, much remained to be done, but the task was finite. For me, this insight hit with career-altering force.

Just as I reached the climax of this story, my dark-adapted eyes were blinded by the sudden restoration of power in Grace Auditorium. The audience was as startled as I was, and there was palpable interest as to how I would react. I blinked away the glare, reconnected visually with the congregation, most notably with Watson himself in a front-row seat, and pronounced the benediction: "And then there was light..."

## REFERENCES

Theorell H. 1964. *Nobel lectures: Physiology or medicine, 1942–1962*, p. 495. Elsevier, Amsterdam.

Watson J.D. 1970. *Molecular biology of the gene*, 2nd edition. W.A. Benjamin, New York, New York.

# Author Index

# Subject Index

## Q



## R



## S

**T**